MW00514223

Nonuniform Sampling

Theory and Practice

Information Technology: Transmission, Processing, and Storage

---

Multi-Carrier Digital Communications: Theory and Applications of OFDM
Ahmad R. S. Bahai and Burton R. Saltzberg

Nonuniform Sampling: Theory and Practice
Edited by Farokh Marvasti

Principles of Digital Transmission: With Wireless Applications
Sergio Benedetto and Ezio Biglieri

Simulation of Communication Systems, Second Edition: Methodology, Modeling, and Techniques
Michel C. Jeruchim, Philip Balaban, and K. Sam Shanmugan

Nonuniform Sampling

Theory and Practice

Edited by

Farokh Marvasti

King's College
London, United Kingdom
and
Sharif College of Technology
Tehran, Iran

Kluwer Academic / Plenum Publishers
New York, Boston, Dordrecht, London, Moscow

Library of Congress Cataloging-in-Publication Data

Marvasti, Farokh A.
Nonuniform sampling: theory and practice/Farokh A. Marvasti.
p. cm. — (Information technology: transmission, processing, and storage)
Includes bibliographical references and index.
ISBN 0-306-46445-4
1. Signal theory (Telecommunication)—Mathematics. 2. Sampling. 3. Time-series analysis. I. Title. II. Series.

TK5102.5. .M2953 2000
621.382′23—dc21

00-062213

ISBN: 0-306-46445-4

233 Spring Street, New York, New York 10013

http://www.wkap.nl

10 9 8 7 6 5 4 3 2 1

A C.I.P. record for this book is available from the Library of Congress

Printed in the United States of America

Dedicated to:

The late Prof. J. L. Yen who introduced interpolation formulae from special nonuniform samples[1]

The late Claude Shannon, father of information theory, who popularized the field of sampling theory in the engineering community[2]

The Portuguese government, University of Aveiro, and the Portuguese people who hosted me while I was writing several chapters of this book

My family, Maryam, Salman, Laleh, Ali, and Narges for their understanding and help

My mother who instilled motivation and curiosity in me

My late father who cared so much about education

---

[1]See his biography and the photo in Chapter 1.
[2]See more comments and photos in Chapters 1 and 2.

Contributors

Andreas Austeng • University of Oslo, Oslo, Norway

Sonali Bagchi • Mobilian Corporation, Hillsboro, OR, USA

John Thomas Barnett • Space and Naval Warfare Systems Center, San Diego, CA, USA

John Benedetto • University of Maryland, College Park, MD, USA

M. Bourgeois • Université LYON I-CPE, Villeurbanne, France

P. L. Butzer • RWTH Aachen, Aachen, Germany

Willem L. de Koning • Delft University of Technology, Delft, The Netherlands

A. J. W. Duijndam • Philips Medical Systems, The Netherlands

Paulo J. S. G. Ferreira • Universadade de Aveiro, Aveiro, Portugal

D. Graveron-Demilly • Université LYON I-CPE, Villeurbanne, France

Karlheinz Gröchenig • University of Connecticut, Storrs, CT, USA

Mohammed Hasan • King's College London, London, UK

C. O. H. Hindriks • Delft University of Technology, Delft, The Netherlands

Sverre Holm • University of Oslo, Oslo, Norway

Jon-Fredrik Hopperstad • University of Oslo, Oslo, Norway

Kamran Iranpour • University of Oslo, Oslo, Norway

Timo I. Laakso • Helsinki University of Technology, Espoo, Finland

B. Lacaze • Lab TéSA, Cedex, France

Farokh Marvasti • King's College London, London, UK & Sharif University of Technology, Tehran, Iran

Sanjit K. Mitra • University of California at Santa Barbara, Santa Barbara, CA, USA

Dale H. Mugler • University of Akron, Akron, OH, USA

M. Sandler • King's College London, London, UK

G. Schmeisser • University of Erlangen-Nurenberg, Erlangen, Germany

M. A. Schonewille • PGS Onshore, Houston, TX, USA

Sherry Scott • University of Maryland, College Park, MD, USA

Atif Sharaf • King's College London, London, UK

R. L. Stens • RWTH Aachen, Aachen, Germany

G. H. L. A. Stijnman • Delft University of Technology, Delft, The Netherlands

Thomas Strohmer • University of California at Davis, Davis, CA, USA

Vesa Välimäki • Helsinki University of Technology, Espoo, Finland

D. van Ormont • Delft University of Technology, Delft, The Netherlands

L. Gerard van Willigenburg • Wageningen University, Wageningen, The Netherlands

F. T. A. W. Wajer • Delft University of Technology, Delft, The Netherlands

Yan Wu • University of Akron, Akron, OH, USA

Ahmed I. Zayed • University of Central Florida, Orlando, FL, USA

Foreword

I was very pleased when Farokh Marvasti asked me to write a Foreword to this volume. I can certainly recommend it to the scientific community in the confidence that its readers will get a real sense of the flavour, style and innovative character of nonuniform sampling and its applications.

The importance of sampling as a scientific principle, both in theory and practice, can hardly be in doubt. The reader who cares to glance through the Table of Contents cannot fail to be convinced as to the ubiquity of the subject and its broad scope; and if we allow that the subject finds its root in the finite interpolation problems that were already being studied in the middle part of the seventeenth century (and there are cogent reasons for taking this view), nonuniformity of the distribution of sample points has been present in sampling from its very beginnings.

A basic idea, one that lies at the foundations of sampling, is that it is very convenient to consider a signal, or function, as consisting merely of a collection of discrete samples, that is, values taken by the function at some countable set of sample points. When this can be done one is saying effectively that the information contained in the function's samples is equivalent to, or at least approximately equivalent to, that present in the whole function. In order to reach such a desirable position one needs to ask whether the set of sample points is a set of uniqueness for some class of functions, and if so, how a member of the class could be reconstructed from data which would often consist of samples of the function, or perhaps of pre-processed versions of it. Going beyond the finite case to that of infinitely many sample points, the simplest, indeed the "classical", case involves uniformly distributed sample points along the real line and is embodied in the famous sampling theorem associated with the names of E. T. Whittaker, V. A. Kotel'nikov and C. E. Shannon (and others); a theorem that can be found in many places throughout this book.

It is from such basic ideas that sampling developed, ever more rapidly, throughout the last century and into the present. Indeed, sampling is now a multi-disciplinary activity, and is often found where some or all of: harmonic analysis including classical Fourier analysis, function theory, approximation theory, prediction theory, stochastic processes, information theory and practical considerations of signal processing are seen to overlap (and that is just a short list!).

Movements of scientific thought develop continuously, and are indeed intelligible to us as continuity. The past informs the present, which in turn will inevitably inform the future. However, we live in an era when these developments are, more often than not, accompanied by changes that are huge and rapid. We are forced, therefore, to ask the question: how do we cope with this change; how are we to deal with the new? I believe that books such as the present one have an important role to play here (and it will surely be a sad day for the human race when there is no role for books to play any more). We can pause with them; we can gain breathing space, a time to ponder, a time to sift foreground from background before the tidal waves of change press us forward once again.

On a more day-to-day level, the usefulness of a book such as this will depend in part on its selection from the masses of material available in the literature and in the minds of its contributors; in part on how it improves access to the subject, and in part on whether its contributions contain careful analysis, informative bibliographies and a high quality of elucidation. I confidently predict that this book will not be found wanting in any of these attributes.

It seems that there is no shortage of new and interesting problems in non-uniform sampling, and I am sure that the present volume will help to maintain that interest.

Happy sampling!
J. R. Higgins, Cambridge, UK

Preface

Shannon never claimed any credit for originating The Sampling Theorem, he simply realized as part of his original development of a mathematical Theory of Communication that a band-limited signal could (in theory) be reconstructed from uniformly spaced values provided the samples were closer than a critical amount. In my first encounters with The Sampling Theorem, it was called the "Shannon-Whittaker-Kotel'nikov Sampling Theorem", with honors shared by Shannon with the radar pioneer and originator of Signal Detection Theory. Regardless of the origins of the result—which are considered in this book—these three, and Shannon in particular, had a profound effect on the rapid spread of both understanding and applications of sampling to communications and signal processing theory and practice.

On the theoretical side, sampling provided the means of converting continuous time signals to discrete-time signals without loss of information. This permitted tools for discrete time signals such as linear algebra and time-series analysis to be applied to the evaluation of channel capacities and source rates of continuous time processes. When sampling, the discretization of time, was combined with quantization, the discretization of amplitude, the result was analog-to-digital conversion and the true beginnings of the modern digital revolution, as embodied in the first digital communication technique for continuous waveforms—pulse coded modulation (PCM), as popularized by Oliver, Pierce, and Shannon. The descendants of this technique are ubiquitous in the Internet and modern wireless communication.

Sampling has grown in many directions, especially nonuniform sampling and generalization incorporating transforms such as Fourier, Karhunen-Loeve, wavelets, and filter-banks. For all of these extensions, however, the basic issue for an engineer remains the discretization of time or space, whether the goal be to prove a theorem or transmit speech or video. Along with quantization, sampling resides at the border between the analog world of nature, and the digital world of communication, signal processing and computing. This book provides a thorough and varied tour of that border.

Prof. Robert M. Gray
Department of Electrical and Computer Engineering
Stanford University
Stanford, California

Contents

Chapter 1. Introduction
F. Marvasti

Chapter 2. An Introduction to Sampling Analysis
P. L. Butzer, G. Schmeisser, and R. L. Stens

Chapter 7. The Nonuniform Discrete Fourier Transform

S. Bagchi and S. K. Mitra

Chapter 8. Reconstruction of Stationary Processes Sampled at Random Times
B. Lacaze

Chapter 9. Zero-Crossings of Random Processes with Application to Estimation and Detection
J. T. Barnett

Chapter 10. Magnetic Resonance Image Reconstruction from Nonuniformly Sampled *k*-Space Data

F. T. A. W. Wajer, G. H. L. A. Stijnman,
M. Bourgeois, D. Graveron-Demilly, D. van Ormondt

Chapter 11. Irregular and Sparse Sampling in Exploration Seismology

A. J. W. Duijndam, M. A. Schonewille and C. O. H. Hindriks

Chapter 12. Randomized Digital Optimal Control

W. L. De Koning and L. G. van Willigenburg

Chapter 13. Prediction of Band-Limited Signals from Past Samples and Applications to Speech Coding

D. H. Mugler and Y. Wu

Chapter 14. Frames, Irregular Sampling, and a Wavelet Auditory Model

J. J. Benedetto and S. Scott

Chapter 16. Applications of Nonuniform Sampling to Nonlinear Modulation, A/D and D/A Techniques

F. Marvasti and M. Sandler

Chapter 17. Applications to Error Correction Codes

F. Marvasti

Chapter 18. Application of Nonuniform Sampling to Error Concealment

M. Hasan and F. Marvasti

Chapter 19. Sparse Sampling in Array Processing

S. Holm, A. Austeng, K. Iranpour and J.-F. Hopperstad

Chapter 20. Fractional Delay Filters—Design and Applications

V. Välimäki and T. I. Laakso

1

Introduction

F. Marvasti

So many signals around *Fortunate those who decode*
Plenty points to be found *Where does the secret abode?*[1]

1.1. Preliminaries

After nearly 14 years I could not have come up with a better introduction than the one I wrote in my previous book on zero-crossings and nonuniform sampling [1]:

> *Man's understanding of nature is often through nonuniform observations in space or time. In space, one normally observes the important features of an object (such as edges); the less important features are interpolated... History is a collection of important events that are nonuniformly spaced in time. Historians infer between events (interpolation) and politicians, similar to stock market analysts, forecast the future from past and present events (extrapolation). This fact can be generalized to other realms of human endeavor. Indeed most readers of this book may read the introduction, the conclusion, and possibly a few sections of a few chapters of this book and then try to figure out the rest by interpolation!*

Coming more down to earth, I wanted to write another book which is more practical with more relevant theories as opposed to [1] which was more comprehensive with cursory treatment of each subject. I realized the theory

[1] Adapted from Hafez, one of the greatest Persian poets (14th century). Goethe, the great German poet, was inspired by Hafez to write the 12 books of West-Eastern Divan (compiled during 1814–1820).

F. Marvasti • Multimedia Laboratory, King's College, London, United Kingdom.
E-mail: farokh.marvasti@kcl.ac.uk

Nonuniform Sampling: Theory and Practice, edited by Marvasti
Kluwer Academic/Plenum Publishers, New York, 2001.

and application in the field of nonuniform sampling is so vast that it is impossible to master diversified topics such as seismology, magnetic resonance imaging, array signal processing, filter design, modulation and coding, sampled data control, and multimedia applications [31]; see the accompanying CD-ROM file: 'Chapter1\Sampta95.txt. Hence I had to ask for help from the experts.

This 21st century book is the expansion, the continuation and the update of the contents of [1] and [3] with a tutorial outlook for practicing engineers in industry and advanced students in science, engineering and mathematics. Specifically, it is intended to be used as a reference book for scientists and engineers in Medical Imaging, Geophysics, Astronomy, Biomedical Engineering, Computer Graphics, Digital Filter Design, Speech and Video Processing, and Phased Array Radars. This book consists of 20 chapters where I have been the author or the co-author of 6 of them (Chapters 1, 4, 15–18). All the other chapters have been written by experts in the field, something I could not have possibly done by myself. Most of the contributors of this book are either the sole authors or chapter contributors of other related books [1], [3–5] and [7]. The C codes, Matlab and Mathcad programs for the algorithms are also provided in an attached CD-ROM for the first time.

Since I wrote a monograph on nonuniform sampling and zero-crossings in 1987 [1], there have been several books on sampling theory [2–6, 10] and applications on wavelets and multirate signal processing [7]–[8], and image recovery from partial information [9]. Reference [1] deals with zero-crossings, random and nonuniform sampling of single and two-dimensional signals and systems. Reference [2] is an undergraduate/MS level textbook on uniform sampling theory. Reference [3] is a collection of contributed chapters by experts in the field on advanced research topics in uniform and nonuniform sampling theory with some applications. Reference [4] is on advanced mathematical topics in sampling theory. Reference [5] is on random sampling for digital signal processing and instrumentation. Reference [6] is an advanced mathematical treatment on sampling theory. Reference [7] is a mathematical treatment on wavelets with several chapters on nonuniform sampling while Reference [8] is on multirate digital signal processing with one section on practical implementation of signal recovery from periodic nonuniform samples. Reference [9] is on recovery methods using the technique known as Projection Onto Convex Sets (POCS). Reference [10] is a recent book on nonuniform discrete Fourier transforms based on the PhD thesis of Dr. Bagchi. The basic elements of that book are summarized in Chapter 7 of the present book.

The idea of this book was conceived during the summer of 1996, and I then contacted some publishers and individuals for help. At the meeting of SampTA'97 in Aveiro, Portugal, agreements between most of the authors of the present book and myself were made. At the same time Profs. Benedetto and Ferreira planned to publish a polished version of the Conference Proceedings, which later turned into a separate book on modern sampling theory [11]. While there might be some overlap between that book and the present book, the emphasis of this book is on nonuniform sampling and a tutorial exposition with practical algorithms including C codes, Matlab and Mathcad programs available on an accompanying CD- ROM.

I have done my best to make all the chapters uniform in format, notations, and readability despite the nonuniform nature of the authors and the topics! I have read all the chapters several times and have given my comments to the corresponding authors. Most of the authors have accepted my comments and have made a very good job of revising their chapters; for that I am very thankful to all the authors.

I take pride in the international nature of the authors from various universities who have contributed to this book and I think it has enhanced the technical content as well as given an international flavor to the book; indeed this book is another attempt for an East-West Divan, to use Goethe's phrase. A photo of some of the authors who attended SampTA'99 in Norway is shown below:

Figure 1. Photo of some of the contributors of this book on an excursion trip from SampTA'99 in Norway, August 99. From left to right: D. Mugler, F. Marvasti, A. Zayed, R. Stens, P. Butzer, J. Benedetto, and G. Schmeisser.

1.2. Historical Perspective

About 50 years ago, among the engineers, the knowledge on nonuniform sampling was limited to that of Cauchy [19] as quoted by Black in his book [20]. There is also some evidence that Shannon was aware of Cauchy's work from his paper [21]. Black quoted Cauchy as saying:

> *If a signal is a magnitude-time function, and if time is divided into equal parts forming subintervals, such that each subdivision comprises an interval T seconds long where T is less than half the period of the highest significant frequency component of the signal; and if one sample is taken from each subinterval in any manner; then a knowledge of the instantaneous magnitude of each sample plus a knowledge of the instant within each subinterval at which the sample is taken contains all of the information of the original signal (when the chosen instant of sampling falls at one extremity of a subinterval, that instant is excluded from the adjacent subinterval in order to insure 1/T distinct instants of sampling as well as one sample from each subinterval).*

Black's motivation in quoting this 'folklore Theorem' was related to natural Pulse Position Modulation PPM [20, p. 42] where nonuniform sampling is used (see Chapter 16 Fig. 1). He also discussed irregular sampling in the context of bunch sampling [20, p. 50] and random sampling for nonsynchronous multiplexing [20, p. 338].

This 'folklore Theorem', which is not always true, was later narrated by Yen in [22] and subsequently quoted by many authors. Later Higgins [23] did some historical research and was not able to find the above quotation in any of Cauchy's work. However, recent research (Chapter 2 Section 10 and the introduction of Chapter 8) shows that Cauchy did indeed have some ideas (most likely indirectly) related to sampling.

Some of the earlier theoretical work on Lagrange interpolation from nonuniform samples (see Chapter 3) was done in 1934 by Paley and Wiener [24]. Also, the theoretical background of many nonuniform sampling theorems was laid by Levinson in 1940 [25] and Duffin and Schaeffer in 1952 [26]. Levinson laid the foundations of completeness of the sampling set and the zeros of a band-limited signal (see Chapter 9), and Duffin and Schaeffer laid the foundations of the theory of frames (see Chapter 14), stability of the nonuniform sampling set (see Section 4.4.1.), and past sampling (see Section 4.4. and Chapter 13).

From the engineering point of view, modern nonuniform sampling interpolation started with Yen's seminal papers in 1956 [22] and 1957 [27]. In his classical and highly quoted work, he derived the interpolation formulas for special cases of irregular sampling such as migration of a finite number of uniform samples, a single gap in an otherwise uniform samples, and periodic nonuniform sampling.

To my dismay, I found out that Prof. J. L. Yen had passed away a few years ago. I would have liked him to have contributed a section about the origins of his work to this book. His papers were my initial exposure to the area of nonuniform sampling when I was a graduate student at Rensselaer Polytechnic Institute in the early seventies. Yen's work has been very influential in the area of electrical engineering as evidenced in some of the early textbooks [28] and this is the reason I am dedicating this book to him; a brief biography along with his photo are given below[2].

Professor Jui Lin (Allen) Yen was born in Canton, China in 1925, and graduated from Chiao Tung University, Shanghai, in 1948. His long association with the University of Toronto began in 1949 as a graduate student; he received the M.A.Sc. and Ph.D. degrees in the now vanished Department of Applied Physics in 1950 and 1953, respectively. He immediately joined the Department of Electrical Engineering as an instructor, becoming Professor in 1966 and University Professor in 1980. Beginning in the late 1950's he played a pivotal

Figure 2. Photo of Prof. J. L. Yen, the pioneering researcher in engineering, who wrote two seminal papers on nonuniform sampling [22], [27] back in 1956–1957.

[2] With thanks to Prof. P. Pasupathy from the University of Toronto for providing the biography and the photo.

role in establishing radio astronomy research at the University of Toronto, which has continued to the present. From 1968 he was cross-appointed to the Department of Astronomy. He reached retirement age in 1991.

Prof. Yen was a leading figure both in microwave engineering and its applications to satellite communications and in radio astronomy, where his pioneering work in long baseline interferometry was recognized by his being awarded, as a co-recipient, the Rumford Award of the American Academy of Arts and Sciences in 1971. In recent years Professor Yen concentrated on the development of a very high density recording system based on VCR technology for use in radio astronomy. These recording systems are currently deployed throughout the world. Prof. Yen was a Fellow of the Royal Society of Canada, and was much sought after as an adviser to international scientific bodies, to the National Research Council and to industry. He died suddenly on May 30 1993.

It would have been remiss of me not to mention and acknowledge the contribution of Claude Shannon, one of the greatest scientists of our time, in sampling theory as a tool for information theory. He certainly popularized the uniform sampling theorem in the engineering community. I had always wondered about the relation between information theory and sampling theory until I discovered that the fundamental channel theorem of information theory is indirectly related to the nonuniform sampling theorem (see Section 17.1.).

Figure 3. Photo of Claude Shannon, the man who popularized sampling theory in the engineering community.

I recently wrote to him and asked him for a photo and a brief description of how he came across the idea of sampling. Unfortunately, his wife Mary Shannon replied with a photo and a letter saying that Claude Shannon "is an Alzheimer's patient and suffering serious loss of memory". In a consequent telephone call, she said that he had developed the Alzheimer's disease about 10 years ago and now at the age of 83 he barely recognized her, and could not recognize his children*. I am sure he and his work remain in the memory of thousands of his admirers; he has always been humble, private and disliked public speeches—see a copy of his letter to me in 1987 in the accompanying CD-ROM, directory: 'Chapter1\Shannon-letter.zip. Figure 2 is the photo that he used for publication and was kindly given to me by his wife. A more recent picture of Claude Shannon is given in Chapter 2.

1.3. Summary of the Chapters

The general format of the book is as follows: Chapters 2–9 cover theoretical issues such as uniform/nonuniform sampling theory, missing samples, random sampling, zero-crossing problem, and nonuniform discrete Fourier transforms. Chapters 10 to 20 are applications to various areas of science and engineering. Having said this, I should emphasize that some of the chapters such as Chapter 7 include applications to digital filter design that fit closely to Chapter 20 on fractional delay filter design, and some other chapters in the application part such as Chapters 10–14 contain some theory. In fact Chapters 13–14 contain mathematical theories of prediction and wavelets as applied to speech compression.

An informal and brief description of each chapter with some of the backgrounds of the authors will now be given.

Chapter 2 is mainly review material on the history of sampling theory, the basics of sampling theory, error analysis and the fundamentals of Fourier transform as applied to linear time invariant systems and sampling theory. Professors Paul Butzer and Rudolf Stens have done an excellent job on this topic. They are also leading researchers in the area of sampling theory at Aachen Technical University in Germany. When the chapter was written, we found out at SampTA'99 in Loen, Norway, that Prof. Schmeisser from the University of Erlangen had a very deep knowledge of the history of sampling theory. Prof. Butzer and I kindly requested that he should contribute a section to Chapter 2; thus, Chapter 2 is a German contribution to the basics of uniform sampling theory.

Chapter 3 covers the basics of Lagrange interpolation written by Prof. Ahmed Zayed from the University of Central Florida and Prof. Paul Butzer, the first author of Chapter 2. This chapter discusses the Lagrange interpolation for

*While reading the proofs, I heard the sad news that he died on February 25, 2001 at the age of 84.

finite polynomials and band-limited signals for nonuniform samples; nonuniform samples with derivative sampling is also discussed in this chapter. The relationship between Lagrange interpolation and Sturm Liouville boundary-value problems is established. Prof. Ahmed Zayed has written a book on sampling theory [4] and has contributed significantly to the field of sampling theory. I thus label this chapter an Egyptian/German contribution with an American flavor!

Chapter 4 entitled "Random Topics in Nonuniform Sampling" is the extension of my previous book [1] and of a book-chapter contribution [3] on nonuniform sampling and zero-crossings. As the name of the chapter implies the topics are not related except that they are related to nonuniform sampling. The first six sections are similar to the chapter contribution in [3] except that it covers 2-D signals. Section 7 is an original discussion on Wavelets and Quadrature Mirror Filters (QMF); the closest discussion to this topic is given in [8]. Section 8 and 9 are related to theorems and recovery methods related to nonuniform sampling at half the Nyquist rate for low-pass and band-pass signals. Section 10 is a totally new topic on Extremum sampling for finite polynomials. The derivations of discrete Lagrange interpolation and other interpolation and recovery techniques and simulations methods are all new and have not been published yet. If you are curious to know where I come from, I am an Iranian, with an American education, and have teaching and work experience in Iran, the USA and the UK.

Chapter 5 is the contribution of Prof. Paulo Ferreira from the University of Aveiro, Portugal on analysis, stability and recovery methods for the problem of 1-D signals with missing samples. Prof. Ferreira was the technical chair of SampTA'97 in Aveiro and is one of the editors of a recent book on sampling theory [11]. He has widely published in the area of stability, recovery of finite trigonometric polynomials with missing samples using iterative and non-iterative techniques. Prof. Ferreira was instrumental in getting me a grant to go to Portugal and my chapter contributions (Chapters 4, 15–18) are essentially due to the hospitality of Prof. Ferreira, his former student Jose Vieira, University of Aveiro, and the Portuguese government. I thus welcome this Portuguese contribution.

Chapter 6 is the continuation of Chapter 5 for 2-D signals. This chapter is, however, more general and covers nonuniform sampling for 2-D images including the special case of missing samples. I originally asked Prof. Hans Feichtinger from the University of Vienna to contribute a chapter but due to his commitments, he referred this task to his former student Dr. Thomas Strohmer—who is now a Professor at the University of California at Davis—and Dr. Karlheinz Gröchenig, a former student of the University of Vienna at Austria who is now a Professor at the University of Connecticut at Storrs in the USA. Prof. Gröchenig together with Prof. Feichtinger have written a book chapter on the 1-D aspects of nonuniform sampling in Benedetto and Frazier's book on Wavelets [7]. Recently, he has also written another book on sampling theory [29]. Prof. Gröchenig has respectable theoretical publications in the area of nonuniform sampling. Prof. Strohmer did his MS and his PhD degree at the University of Vienna under Prof. Feichtinger in the area of nonuniform sampling. He has several impressive papers in the area of efficient and numerical recovery of 1-D and 2-D signals. This chapter is essentially an Austrian contribution from the USA.

Chapter 7 is on Nonuniform Discrete Fourier transforms with applications to filter design both for the 1-D and the 2-D filters. In this chapter the assumption is made that the signal in the time or space domain is uniform but the z-transform (or the frequency domain) is nonuniform. The authors are Dr. Sonali Bagchi from Lucent Technology and Prof. Sanjit Mitra from the University of California, Santa Barbara. This chapter is a condensed version of the book published by the same authors [10] based on the PhD thesis of Dr. Bagchi while she was working at Santa Barbara under the supervision of Prof. Mitra. This chapter is the result of hard negotiations with both authors due to their other commitments. I credit this to my insistence and my negotiating skills with our Indian neighbors from USA!

Chapter 8 is on random sampling of stationary random signals and recovery of signals from finite and infinite random samples when the position of random samples may or may not be observed. Initially I asked Prof. Beutler from the University of Michigan to contribute a chapter on random sampling related to some of the original works that he and his associate Dr. Leneman had written in 1960's and 1970's, e.g., [17]–[18]. He refused on the basis that he had been retired for several years and his research was no longer current since his interests had shifted to other areas. Fortunately, I found a very capable person Prof. Bernard Lacaze from INSAT (Institut National des Sciences Appliquées de Télécommunication), a department of TéSA (Special Telecom and Aeronautics) in Toulouse, France. This chapter is a theoretical treatment of the subject by him. He is regarded as an expert in the field and has many publications in the area of random sampling of stochastic signals. He certainly lives up to the French reputation of being very good in mathematics.

Chapter 9 is on signal recovery and spectral estimation of deterministic and random signals from its zero-crossings and higher order crossings. This chapter is contributed by Dr. John Barnett from the Naval Research Center in San Diego, California. He was referred to me by his former PhD advisor, Prof. Ben Kedem from the Mathematics Department of the University of Maryland. Dr. Barnett's PhD thesis is on this topic and he has also summarized some of the topics of Prof. Kedem's book [12] in this chapter. Dr. Barnett managed to do a good job on the topic despite his extreme engagements with other activities; I thus welcome this American contribution.

Chapter 10 is on applications of nonuniform sampling to Magnetic Resonance Imaging (MRI). Computer Tomography is also briefly described. My main contact was with Prof. Dirk van Ormondt from the Faculty of Applied Physics, the Delft University of Technology, The Netherlands. I am delighted that I had the pleasure to meet Dirk at SampTA'97 in Aveiro, Portugal. I was very much impressed with the PhD thesis of one of the students of Prof. Ormondt on this topic, along with many other publications, and thus requested him to contribute a chapter on this topic. He wavered at the beginning and then he declined due to his other commitments but finally I managed to convince him by delaying the deadline. Prof. Ormondt consequently convinced several other authors, Wajer and Stijnman from his department, and Bourgeois and Graveron-Demilly from Université Lyon, France, to coauthor with him. I am happy with this collaborative Dutch/French contribution.

Chapter 11 is the application of nonuniform sampling to seismic data by Prof. Adri Duijndam and his students M. Schonewille and Hindriks from the Faculty of Geology, the Delft University of Technology, The Netherlands. My original contact was with Mr. Schonewille at SampTA'97 in Aveiro. He then contacted Prof. Duijndam. Prof. Duijndam has many publications in this area, and in this chapter he discusses the nature of seismic data and how the concept of Discrete Fourier Transform for nonuniform data can be used for interpolation. A similar concept is also discussed in Chapter 10 for MRI images. Needless to say I am impressed with the quality of research at the Delft University of Technology.

Chapter 12 is another contribution by the Delft University of Technology by Prof. Willem de Koning from the Department of Information Technology and Systems on random sampled data control systems. This chapter was originally scheduled to be written by Prof. Graham Goodwin from the New Castle University of Australia based on the recommendation of my good friend Dr. H. Javaherian. However, unforseen commitments forced Prof. Goodwin to decline. I was fortunate to find Prof. De Koning at the last minute through my colleagues Prof. Lakmal Senevratne and James Whidborne from the Mechatronics Department of King's College London. I briefly discussed this topic in my previous book [1, pp. 102–104]. In control systems, random samples are used to either model a continuous system by a nonuniform discrete system, to estimate the state of a system from the nonuniform observation of the output, or to sample adaptively a sampled data control system. Prof. De Koning has published extensively in the area of randomized optimal digital control, and I am very grateful for his contribution.

Chapters 13 to 18 are related applications of nonuniform sampling to multimedia data compression, modulation, coding and error concealment techniques; while Chapters 19–20 are on signal processing aspects. Below I briefly discuss each chapter.

Chapter 13 is on speech prediction for compression purposes by Prof. Dale Mugler from the Mathematical Sciences Department of the University of Akron, Ohio. Prof. Mugler in this chapter introduces his technique of prediction that depends only on the bandwidths of a signal such as speech. He then compares his method to the traditional techniques of Linear Predictive Coding based on the autocorrelation of the speech samples. This chapter is also coauthored by one of his students Yan Wu. Prof. Mugler was the first author to submit his first and final drafts, thus confirming American efficiency and productivity.

Chapter 14 is on frames, irregular sampling, and a wavelet auditory model by Prof. John Benedetto and his student Sherry Scott. Prof. Benedetto is one of the editors of the book on wavelets [7], which contains several interesting chapters on nonuniform sampling theory. He is also one of the editors of the recently published book on sampling [11]. In addition, he has widely published in the areas of frames, nonuniform sampling and wavelets. He is one of the few mathematicians with a patent in the area of a speech coder using wavelets and irregular samples. This chapter also discusses this invention. I am also very glad to convince him to write this interesting chapter despite his initial and later reluctance; it was not that hard to twist my American friend's arm at the age of 60!

Chapter 15 is an application of nonuniform sampling to video compression using geometrically transformed motion compensation. The first author is my former PhD student, Dr. Atif Sharaf, a Palestinian/Egyptian, who is now working in the Aviation Agency of Saudi Arabia. His work is based on my original publications on nonlinear (time varying) and iterative techniques for the recovery of images, and he has also published this result in [13]. This chapter is essentially similar to [13], which is coming from a chapter of Dr. Atif's thesis. Dr. Hasan, my other former Palestinian/Jordanian student, has kindly reproduced this work.

Chapter 16 is the application of nonuniform sampling to demodulation of Frequency Modulated (FM), Pulse Position Modulation (PPM), Pulse Width Modulation (PWM), and Delta Modulated (DM) signals. These types of modulation are encountered in telecommunication systems and novel demodulation methods are developed based on nonuniform sampling theory (especially the discrete case for finite trigonometric polynomials). The author are Prof. Mark Sandler and myself. I had done some original work for the analog case back in the early 1970's during my PhD thesis; some of it, in bits and pieces, was published in journal and conference papers as referenced in [1]. The novelty of this chapter is the development of interpolation formulas and simulations based on discrete sampling of finite polynomials. Since PWM and a variation of DM called Sigma DM are also widely used in Digital-to-Analog Conversion (DAC) and Analog-to-Digital Conversion (ADC), I have also requested my colleague Prof. Sandler from the Department of Electronics of King's College London to add a section on these topics at the end. Prof. Sandler has published widely in this area since his PhD thesis. I am glad that he has kindly accepted my request. Take this as an Iranian/British contribution.

Chapter 17, written by myself, is on error correction codes using real or complex numbers. While I was at Illinois Institute of Technology in Chicago, I discovered the close relationship between coding theory and nonuniform sampling theory [14]–[15]. The main difference is that coding theory has been traditionally developed for finite Galois Fields while sampling theory is for real and complex functions. The two areas can mutually help each other. I am also glad that one of the former students of Prof. Ferreira, Jose Veiera, who was at SampTA'95 in Riga, Latvia was persuaded to work in this area for his PhD thesis. I had asked both Prof. Ferreira and now Dr. Veiera to contribute a section but unfortunately they could not find time. Applications of nonuniform sampling to error concealment for speech is briefly discussed in Chapter 17. Also Section 17.2. succinctly describes the difference between error correction and error concealment concepts.

Chapter 18 is on error concealment for image signals as opposed to speech signals. It is written by Dr. Mohammad Hasan and myself. Dr Hasan was my former PhD student (also one of the authors of Chapter 15) who did his PhD in this area and is now working in the UK. In this chapter, we discuss the concealment of lost blocks (or even slices) from still images for compressed or non-compressed signals. Various methods giving impressive results are discussed. Some of the topics in this chapter have already been published by the authors and some are unpublished results.

Chapter 19 is on sparse sampling in array signal processing contributed by Prof. Sverre Holm from the Department of Informatics, University of Oslo, Norway. The other contributors are Prof. Holm's students. Sparse or random arrays have been suggested in several fields such as radar, sonar, ultrasound imaging, and seismology. Prof. Holm has contributed many papers and book chapters in this area. While attending SampTA'99 in Loen, Norway, I was very much impressed by the quality of Norwegian academics as well as the natural beauty of the country. I am very much pleased with this Norwegian contribution. He has certainly written an authoritative tutorial on this subject.

Chapter 20 is on fractional delay filters written by Profs. Vesa Välimäki and Timo Laakso from Helsinki University of Technology. Fractional delay filters interpolate the samples at non-integer positions with wide-range applications in music, speech, antennas, and time delay estimation. Originally, I had asked Prof. Laakso to contribute a chapter on filter design using nonuniform samples in the frequency domain but he suggested that he would write a tutorial with his colleague Prof. Välimäki on fractional delay filters. They also have a section on nonuniform sampling reconstruction using polynomial filtering and fractional delay filters. The two authors have widely published in this area including a tutorial on the subject [16]. This chapter has been initiated and contributed by my Finnish colleagues.

Topics not discussed here but that need to be addressed in the revisions of this book are the important areas of computer graphics, biomedical engineering [1, p. 114], optics [1, p. 113] and [30, pp. 37–83], phased locked loops [1, p. 106], multirate signal processing [1, pp. 109–111] and [8], and antenna and radar systems [1, pp. 105–106], statistical multiplexers [1, p. 104], and astronomy [101].

1.4. Acknowledgment

Inherent in my informal introduction of the authors of the above 20 chapters are my deepest appreciation of their knowledge, effort and interests they have shown to make this book a reality. Again I would like to thank the Portuguese Government for their support; specially Prof. Ferreira and his former student Dr. Vieira and the whole staff of INESCA at the University of Aveiro for their invaluable help while this book was under preparation.

I would also like to thank my former and present PhD students in the Multimedia Lab of King's College London for their help; especially Siamak Talebi for proof-reading and providing some figures and simulations to my chapters, Dr. Mohammed Hasan for contributing to Chapters 15 and 18, Dr. Atif Sharaf for Chapter 15, Mr. Theo. Nikolaidis for providing his paper in the accompanying CD-ROM, and Dr. Mohammad Reza Nakhai for freeing my time by taking responsibility of my teaching tutorials and providing his paper and thesis to the CD-ROM. Also, my thanks to my former MSc students: Mr. Panos Livanos, Alireza Majdnia, Said Said, and Dimitris Meleas for providing their Master theses and the codes in the accompanying CD-ROM. I would also like to thank my PhD student Khoo Swee and Research Assistant Dr. Alireza Ahmadian for their general help and for redrawing one of the figures of Chapter 4.

I would like to thank Prof. Hans Feichtinger and his student Tobias Werther for providing the Matlab files and the corresponding paper in the accompanying CD-ROM, directory: Chapter4\Nuhag. The same appreciation goes to Profs. Redinbo[3], Wolberg[4] and Schmeisser/Voss[5] for providing the relevant papers and PhD thesis, respectively.

Also, I would like to thank my colleague Dr. William Chambers for his comments about the introduction, and Prof. Garth Swanson, the Head of the Electronics Department of King's College for allowing me to take a sabbatical leave to finish this book; I will always remember him for his honesty and integrity.

Finally, my deepest gratitude to my family who left me alone so that I could concentrate on writing and editing this book. I missed my family during my long trips to Portugal. In fact my wife Maryam played the role of the author's widow by staying abroad with my 2 smaller children Ali and Narges. My older children Salman and Laleh went to residence halls and flats in London while I was writing and editing this book. Also, my thanks to Salman for redrawing some of the figures of Chapter 4; my appreciation of Laleh for translating the Persian poem of Hafez into English given at the beginning of this introduction.

[3] See the acknowledgment in Chapter 17.

[4] See the accompanying CD-ROM, directory: 'Chapter6\IEEE_pdf.gz.'

[5] See the accompanying CD-ROM, directory: 'Chapter2\PhD-Thesis-Voss.zip.

Also, my special thanks go to my friend Dr. Soroush for bringing the Hafez's verse to my attention; his mental data base was more useful than my computer data base search.

References

[1] F. Marvasti. *A Unified Approach to Zero-Crossings and Nonuniform Sampling of Single and Multidimensional Signals and Systems*. NONUNIFORM Publication, 5 Bayberry Drive, Princeton, NJ 08540, USA.

[2] Robert J. Marks II. *Shannon Sampling and Interpolation Theory I: Foundations*. Springer Verlag, 1990.

[3] F. Marvasti. In R. Marks II, Ed., *Advanced Topics in Shannon Sampling and Interpolation Theory*, R. Marks II, Ed., Springer-Verlag, New York, 1993.

[4] A. Zayed. *Advances in Shannon Sampling Theory*. CRC Press Inc., Boca Raton, FL, 1993.

[5] I. Bilinskis and A. Mikelsons. *Randomized Signal Processing*. Prentice-Hall International, Hemel Hempstead, Herts., 1992.

[6] J. R. Higgins. *Sampling Theory in Fourier and Signal Analysis: Foundations*. Clarendon Press, Oxford, 1996.

[7] J. Benedetto and M. Frazier, Eds. *Wavelets: Mathematics and Applications*. CRC Press, Inc., Boca Raton, 1994.

[8] P. P. Vaidyanathan. *Multirate Systems and Filter Banks*. Prentice-Hall, New Jersey, 1993.

[9] H. Stark and Y. Yang. *Vector Space Projections: A Numerical Approach to Signal and Image Processing, Neural Nets, and Optics*. John Wiley & Sons, New York, 1998.

[10] S. Bagchi and S. Mitra. *The Nonuniform Discrete Fourier Transform and Its Applications in Signal Processing*. Kluwer Academic Publishers, Norwell, MA, 1999.

[11] J. J. Benedetto and P. J. S. G. Ferreira, Eds. *Modern Sampling Theory: Mathematics and Applications*. Birkhauser, Boston, 2000.

[12] B. Kedem. *Time Series Analysis by Higher Order Crossings*. IEEE Press, NJ, 1993.

[13] A. Sharaf and F. Marvasti. Motion Compensation using Spatial Transformations with Forward Mapping. *Signal Processing: Image Communication*, 14:209–227, 1999.

[14] F. Marvasti and L. Chuande. Equivalence of the Sampling Theorem and the Fundamental Theorem of Information Theory. In *Proc. International Communication Conference*, ICC'92, Chicago, IL, June 1992.

[15] F. Marvasti and M. Nafie. Sampling Theorem: A Unified Outlook on Information Theory, Block and Convolutional Codes. Special Issue on Information Theory and its Applications. *IEICE Trans. of Japan on Fundamentals of Electronics, Commun. and Computer Sciences*, Section E, September 1993.

[16] T. I. Laakso, V. Välimäki *et al.* Splitting the Unit Delay Tools for Fractional Delay Filter Design. *IEEE SP Magazine*, 13(1):30–60, January 1996.

[17] F. J. Beutler and O. Leneman. The Theory of Stationary Point Processes. *Acta Math*, 116:159–197, September 1966.

[18] F. J. Beutler. Alias Free Randomly Timed Sampling of Stochastic Processes. *IEEE Trans. Information Theory*, IT-16, (2):147–152, March 1970.

[19] A.-L. Cauchy. Mémoir sur Diverses Formulaes dé Analyse. Comptes Rendus, Paris, France, vol. 12, pp. 283–298, 1841.

[20] H. S. Black. *Modulation Theory*. Van Nostrand, Princeton, N.J., 1953.

[21] C. E. Shannon. Communication in the Presence of Noise. *Proc. of IRE*, 37(1):10–21, 1949.

[22] J. L. Yen. On Nonuniform Sampling of Bandwidth-limited Signals. *IRE Trans. on Circuit Theory*, vol. CT-3, December 1956, pp. 251–257.

[23] J. R. Higgins. Five Short Stories about Cardinal Series. *Bulletin of Am. Math. Soc.*, 12(1) 1985.

[24] R. E. A. Paley and N. Wiener. *Fourier Transform in Complex Domain*. Colloquium Publications, 19, *Am. Math. Soc.*, New York, 1934.

[25] N. Levinson. *Gap and Density Theorems*. Colloquium Publication 26, *Am. Math. Soc.*, New York, 1940.

[26] R. J. Duffin and A. C. Schaeffer. A Class of Nonharmonic Fourier Series. *Trans. Am. Math. Soc.*, 1952, pp. 341–366.

[27] J. L. Yen. On the Synthesis of Line Sources and Infinite Strip Sources. *IRE Trans. Antennas and Propagation*, January 1957, pp. 40–46.

[28] Herbert Freeman. *Discrete Time Systems*. John Wiley & Sons Inc., 1965.

[29] K. Gröchenig. *Foundations of Time Frequency Analysis*. To be published by Birkhauser.

[30] Franco Gori. Sampling in Optics. In R. Marks II, Ed., *Advanced Topics in Shannon Sampling and Interpolation Theory*, Springer-Verlag, New York, 1993.

[31] F. Marvasti. Applications of Nonuniform Sampling. In *Proc. of SampTA'95*, Riga, Latvia, September 1995.

2

An Introduction to Sampling Analysis

P. L. Butzer, G. Schmeisser,[1] and R. L. Stens

2.1. Introduction and Some History

2.1.1. Prelude

A basic property of algebraic polynomials $P_n(t) := \sum_{k=0}^{n} a_k t^k$ of degree $\leq n$ is that any such polynomial is fully and uniquely determined by its values $P_n(t_\nu)$ at an arbitrary given set of $n+1$ distinct points $t_0, t_1, \ldots, t_\nu, \ldots, t_n$. The converse question, whether it is always possible to find a suitable polynomial $P_n(x)$ which takes on the function (polynomial) values $y_0, y_1, \ldots, y_n$ associated with any $n+1$ distinct *abscissas* (interpolation or nodal points) $t_0, t_1, \ldots, t_n$, is answered by the *Lagrangian interpolation formula:*

$$P_n(t) := \sum_{k=0}^{n} y_k l_k(t) = \sum_{k=0}^{n} P_n(t_k) l_k(t) \tag{1}$$

where, for $k = 0, 1, \ldots, n$,

$$l_k(t) := l_{k,n}(t) = \frac{\omega_{n+1}(t)}{(t-t_k)\omega'_{n+1}(t_k)}, \quad \omega_{n+1}(t) := (t-t_0)\ldots(t-t_n). \tag{2}$$

[1]After the first and third named authors had completed Sections 1 to 9 of this chapter, they invited Gerhard Schmeisser, Erlangen, during the workshop "SampTA '99" (Loen, Norway, 1999) to contribute some material on sampling in connection with partial fraction decomposition and quadrature formulae. He supplied the last two sections in record time.

P. L. Butzer, R. L. Stens • Lehrstuhl A für Mathematik, RWTH Aachen, Templergraben 55, D-52056 Aachen, Germany. E-mail: butzer@rwth-aachen.de, stens@matha.rwth-aachen.de
G. Schmeisser • Mathematisches Institut, Universität Erlangen-Nürnberg, Bismarckstraße 1 1/2, D-91054 Erlangen, Germany. Email: schmeisser@mi.uni-erlangen.de

Nonuniform Sampling: Theory and Practice, edited by Marvasti
Kluwer Academic/Plenum Publishers, New York, 2001.

Indeed, $l_k(t)$ is the only algebraic polynomial of degree n which possesses the property

$$l_k(t_\nu) = \delta_{k,\nu} = \begin{cases} 0 & \text{for } \nu \neq k \\ 1 & \text{for } \nu = k. \end{cases} \tag{3}$$

Hence the polynomial $P_n(t)$ of degree n, thus determined, is unique.

Formula (1), (2), presented by J.-L. Lagrange in his lectures at the École Normale in Paris in 1795, was discovered by E. Waring in 1779 [228] in order to give the polynomial interpolant directly in terms of the samples (thus without having any recourse to finding successive divided differences—the approach of Gregory and Newton of 1668–1676).

Concerning the corresponding *trigonometric interpolation* problem, the polynomial

$$T_n(t) = \frac{a_0}{2} + \sum_{k=1}^{n}(a_k \cos kt + b_k \sin kt) = \sum_{k=-n}^{n} c_k e^{jkt}$$

of degree n has $2n+1$ coefficients, so that $2n+1$ constraints suffice in principle to determine $T_n(t)$. Given $2n+1$ interpolation points $x_0, x_1, \ldots, x_{2n}$ distinct modulo 2π, and real (complex) numbers $y_0, y_1, \ldots, y_{2n}$, there always exists a unique trigonometric polynomial $T_n(t)$ such that $T_n(t_k) = y_k$ for $k = 0, 1, \ldots, 2n$. It is given by

$$T_n(t) = \sum_{k=0}^{2n} y_k \tau_k(t), \tag{4}$$

where the so-called *fundamental* polynomials $\tau_k(t)$ of degree n have the property

$$\tau_k(t_\nu) = \delta_{k,\nu},$$

which corresponds to (3); see also Section 2.6.2 below.

Interpolation of given data was also E. T. Whittaker's (see Photo 1) point of view in his famous paper of 1915 [230]. He posed the problem of finding a function which passes through the points (t_k, f_k) where $t_k := a + kw$, $f_k := f(t_k)$, a, w complex, $k \in \mathbb{Z}$ (= set of all positive and negative integers). He called the set of all such functions the *cotabular set* associated with the sequence $\{f_k\}$, and picked out a special member, namely

$$C(t) = \sum_{k=-\infty}^{\infty} f(a+kw) \frac{\sin \frac{\pi}{w}(t-a-kw)}{\frac{\pi}{w}(t-a-kw)}, \tag{5}$$

Photo 1. Edmund T. Whittaker (1873–1956) in the year 1932, with his signature. By courtesy of Prof. J.R. Higgins, Cambridge.

which he called the *cardinal function* of this cotabular set. In fact, one obviously has $C(t_k) = f_k$ for all k, which is again due to the interpolatory property

$$\frac{\sin \dfrac{\pi}{w}(a + \nu w - a - kw)}{\dfrac{\pi}{w}(a + \nu w - a - kw)} = \delta_{k,\nu}. \tag{6}$$

Whittaker's new idea was that $C(t)$ is special because it has the properties of being entire in the sense of complex function theory and that it is what we call band-limited nowadays. (For the fact that the cotabular set has infinitely many members see Section 2 of Chapter 3).

However, nowhere in his paper did Whittaker state explicitly that $C(t)$ is equal to $f(t)$ or give any conditions yielding this equality. It seems that it was

Photo 2. Claude E. Shannon (1916–2001) in the year 1991 when he received the Eduard Rhein Award (of DM 200,000) for his "fundamental research on information theory." Prof. H.D. Lüke, Aachen, a member of the Board of Curators, supplied this information to the authors. The photo was kindly provided by Dr. Rolf Gartz, Managing Chairman of the Foundation and nephew of Eduard Rudolph Rhein (1900–1993), the founder of the award.

K. Ogura (1920) [162] who first noted that $C(t) = f(t)$ provided f is band-limited or, as he called it, if *f is a cardinal function*. He was the first mathematician ever to have stated and indicated a rigorous proof of the theorem which is known today as the *Whittaker-Kotel'nikov-Shannon (or WSK) sampling theorem*, namely:

Every signal function $f(t)$ defined on $\mathbb{R}$ that is band-limited to an interval $[-\Omega, \Omega]$ for some $\Omega > 0$, can be completely reconstructed for all $t \in \mathbb{R}$ from its sampled values $f(k\pi/\Omega)$, taken at the nodal points $k\pi/\Omega$, $k \in \mathbb{Z}$, equally spaced apart on the real axis $\mathbb{R}$, in terms of

$$f(t) = \sum_{k=-\infty}^{\infty} f\left(\frac{k\pi}{\Omega}\right) \frac{\sin(\Omega t - k\pi)}{\Omega t - k\pi} = \sin(\Omega t) \sum_{k=-\infty}^{\infty} f\left(\frac{k\pi}{\Omega}\right) \frac{(-1)^k}{\Omega t - k\pi}. \quad (7)$$

Photo 3. Vladimir Aleksandrovich Kotel'nikov (b. 1908) in the year 1999. He received the Eduard Rhein Award 1999 (DM 150,000), in the presence of the Bavarian Prime Minister, Dr. E. Stoiber, in the Hall of Fame of the Deutsche Museum in Munich on October 16, 1999. He was honoured "for the first theoretically exact formulation of the sampling theorem." Prof. Lüke kindly supplied the photo.

If we choose $a = 0$ and $w = \pi/\Omega$ in (5), then (7) is exactly $C(t) = f(t)$. Here *band-limited* means that the signal f contains no frequencies higher than Ω or, in mathematical terminology, that f is continuous and square integrable on $\mathbb{R}$ (i.e., finite energy $E := \int_{\mathbb{R}} |f(u)|^2 du < \infty$), and its Fourier transform (15) vanishes outside $[-\Omega, \Omega]$.

Ogura also pointed out inaccuracies in Whittaker's paper and his lack of restrictions on the growth of the entire function f (see Zayed [237, p. 20]).

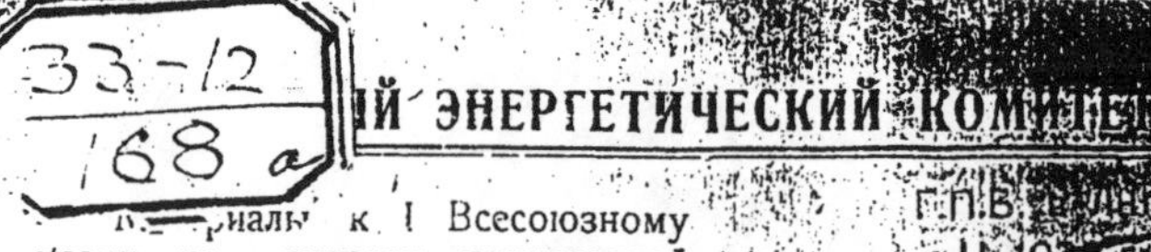

...риалы к I Всесоюзному с'езду по ...опросам технической ...онстру... дела связи и разви- ... промышленности.

ПО РАДИОСЕКЦИИ

Инж. ... ОТЕЛЬНИКОВ.

О пропускной способности „эфира“ и проволоки в электросвязи.

Как в радио, так и в проволочной технике для каждой передачи требуется не одна какая-либо частота, а целый диапазон частот. Это ведет к тому, что одновременно может работать лишь ограниченное количество радиостанций (передающих разные программы). По одной паре проводов также нельзя передавать сразу больше определенного количества передач, так как нельзя, чтобы полоса частот одной передачи перекрывала полосу частот другой,—такое перекрытие привело бы к взаимным помехам.

Чтобы увеличить пропускную способность „эфира“ и проволоки (а это имело бы колоссальное практическое значение в особенности в связи с бурным развитием радиотехники и таких передач, как телевидение), нужно как-то сократить диапазон частот, требуемый для данной передачи, не вредя ее качеству, или изобрести способ разделения передач с налезающими друг на друга частотами, а может быть применить даже способ разделения передач не по частотному признаку, как это делалось до сих пор, а по какому-нибудь другому [1].

По настоящее время никакие ухищрения в этих направлениях не позволяли, даже теоретически, увеличить пропускную способность „эфира“ и проволоки в большей степени, чем это позволяет сделать передача „на одной боковой полосе“.

Поэтому возникает вопрос: возможно ли вообще это сделать? Или же все попытки в этом направлении будут равносильны попыткам построить perpetuum mobile?

Этот вопрос в настоящее время имеет актуальное значение в радиотехнике ввиду с каждым годом все увеличивающейся „тесноты в эфире“. И сейчас особенно важно в нем разобраться в связи с планированием научно-исследовательских работ, так как при планировании важно знать, что возможно и что совершенно невозможно сделать, чтобы направить силы в нужном направлении.

В настоящей работе разбирается этот вопрос и доказывается, что для телевидения и передачи изображений со всеми полутонами, а также для телефонной передачи, существует вполне определенная минимально необходимая полоса частот, которую, не вредя качеству передачи и скорости, нельзя никакими средствами уменьшить, а также доказывается, что для этих передач нельзя увеличить пропускную

[1] Это, правда, можно сделать иногда направленными антеннами, но мы сейчас будем лишь рассматривать случай, когда антеннами это сделать почему-либо нельзя.

Photo 4. Facsimile of the first page of Kotel'nikov's paper of 1933.

THE ALL-UNION ENERGY COMMITTEE.

Proceedings of the first All-Union congress on the technological reconstruction of the communication trade and the development of low-current industry.

RADIO SECTION

Engineer V.A.KOTEL'NIKOV

On the channel capacity of the "ether" and of the wire in the electrical communications.

To broadcast a program by radio or by wire one needs not a single frequency but a whole range of frequencies. Because of this only a limited number of radio stations (which broadcast different programs) may work simultaneously. It is also possible to transmit simultaneously no more than a limited number of programs by one pair of wires because it is impossible for the frequency band of a program to overlap with the frequency band of another program. Such overlapping may cause mutual interferences.

To increase the channel capacity of the "ether" and of the wire (this would be of the utmost importance, especially in connection with the rapid growth of the radio engineering and of the telecasting) it is necessary either to decrease somehow the frequency range which is required for the transmission of a program (but in such a way as not to deteriorate the quality of the program) or to invent a method of the separation of programs with the overlapping frequency bands. May be one needs even to use a method of separation of programs not on the basis of frequencies but on another basis [1].

Until the present time no contrivances in these directions allow (even theoretically) to increase the channel capacity of the ether and of the wire more than it may be done using the single side band transmission. Therefore the question arises: is it possible to realize this at all. May be, all attempts in this direction are equivalent to attempts to build the perpetuum mobile?

At the present time this is a topical question in radio engineering because the "crowding of the ether" is growing year after year. Now, in connection with planning of scientific researches, it is extremaly important to investigate this problem, because to direct efforts in the right directions we need to know what is possible and what is impossible at all.

In the present paper we investigate this problem and prove that for the television program and image transmission (with all semi-tones) as well as for telephone program transmission there exists certain completely determined minimal frequency band. This frequency band can not be decreased by any means without deterioration of the transmission quality and lowering of the speed of transmission. We also prove that for such programs it is impossible to increase the channel capacity neither of the ether nor of the wire by means of non-frequency selections of various kind, as well as by means of any other methods (except of course the directional selection by means of the directional antennas). The maximal possible channel capacity for

[1]Sometimes it is possible to do this using directional antennas, but now we consider only the case where it is impossible to do this for some reasons.

Photo 5. The beginning of Kotel'nikov's paper, translated by Prof. Victor Katsnelson, Jerusalem

Photo 6. Herbert Raabe (b. 1909) taken in the last years. Courtesy of Prof. Lüke.

The sampling theorem was introduced into the engineering literature on communication theory by C. E. Shannon (see Photo 2) in 1948/9 [197, 198, 200]. Some years earlier V. K. Kotel'nikov (1933) (see Photo 3) published in Russia the theorem [128] (see Photos 4, 5 for the first page of this manuscript, which is most difficult to locate in libraries, together with its translation); it became known by his name there. He observed in this respect that the set of information which is contained in all of the values of a band-limited function on an interval $[-T, T]$ is completely equivalent to the set of information contained in the function values at just $2WT$ discrete points, provided that T is large (cf. [126, 127]).

It should also be mentioned that work of Raabe (1939) [177] (see Photo 6) in this matter also predated that of Shannon (see Lüke [140, 142, 144] for firm-set historical accounts). There exist other independent introductions to the sampling theorem, for example that due to Someya (1949) [201].

Let us now compare the sampling theorem with the Lagrangian interpolation formula, (1), (2) now put into a more convenient form:

A polynomial $L_{2n}f(t)$ of degree $2n$ which takes on the same values as a given signal $f(t)$ for the $2n+1$ distinct nodal points $t_k = k\pi/\Omega$ for $k = 0, \pm 1, \dots, \pm n$, is given by

$$L_{2n}f(t) = \sum_{k=-n}^{n} f(t_k) \frac{G_n(t)}{(t-t_k)G_n'(t_k)},$$

where $G_n(t)$ is the canonical product of the points, namely,

$$G_n(t) := t\prod_{\mu=1}^{n}\left(1-\left(\frac{t}{t_\mu}\right)^2\right) = t\prod_{\mu=1}^{n}\left(1-\left(\frac{\Omega t}{\mu\pi}\right)^2\right)$$
$$= \omega_{n+1}(t)\left(\frac{\Omega}{\pi}\right)^{2n}\frac{(-1)^n}{(n!)^2}.$$

Concerning the sampling series, it can be regarded as a Lagrange interpolation series with *infinitely many* knots. In fact, the right side of (7) can be obtained as the limiting case of Lagrange's formula as the number of knots tends to infinity, thus as $\lim_{n\to\infty} L_{2n}f(t)$. Indeed, take $G(t) := (\sin \Omega t)/\Omega$ and $t_k := k\pi/\Omega$; then $G(t_k) = 0$, and $G'(t) = \cos \Omega t$ with $G'(t_k) = (-1)^k$ for all $k \in \mathbb{Z}$. So (7) can be rewritten as

$$f(t) = \sum_{k=-\infty}^{\infty} f(t_k)\frac{G(t)}{(t-t_k)G'(t_k)},$$

noting that

$$\frac{G(t)}{(t-t_k)G'(t_k)} = \frac{(-1)^k \sin \Omega t}{\Omega(t-t_k)} = \frac{\sin \Omega(t-t_k)}{\Omega(t-t_k)}.$$

Furthermore, in view of the infinite product representation of $\sin \Omega t$,

$$\frac{G(t)}{(t-t_k)G'(t_k)} = \frac{(-1)^k t}{(t-t_k)}\prod_{\mu=1}^{\infty}\left(1-\left(\frac{\Omega t}{\mu\pi}\right)^2\right) = \lim_{n\to\infty}\frac{G_n(t)}{(t-t_k)G_n'(t_k)}. \tag{8}$$

The analogy between the sampling theorem and the Lagrange interpolation formula has been observed already by E. T. Whittaker [230] and W. L. Ferrar [80]; the latter showed in particular (8). For the details see Hinsen-Klösters [114], Higgins [109, Chapter 5] or [238]. A more detailed presentation of the connections between sampling and Lagrange interpolation can be found in Section 2 of Chapter 3 in this volume; there also is treated the more intricate case of non-uniformly spaced samples.

2.1.2. Some Further History

Concerning the further history, taking $\Omega = \pi$ for simplicity, series of the form

$$\sum_{k=-\infty}^{\infty} c_k \frac{\sin \pi(t-k)}{\pi(t-k)} = \frac{\sin \pi t}{\pi} \sum_{k=-\infty}^{\infty} c_k \frac{(-1)^k}{t-k} \tag{9}$$

are called a *cardinal series*, a name used by J. M. Whittaker (E. T. Whittaker's second son) in the 1920s [231, 232]. This series is known to converge absolutely, hence unconditionally, for every $t \in \mathbb{C}$ iff $\sum_{k=-\infty, k\neq 0}^{\infty} |c_k/k| < \infty$. In this form the series was studied by E. Borel already in 1899 (even a more general form of the series with the factor $(z/k)^p$ in the summand, to improve convergence) as well as by J. Hadamard in 1901. In 1897 Borel had already shown that if f is representable as

$$f(t) = \int_{-\pi}^{\pi} g(u) e^{jut} du \tag{10}$$

with $g(u)$ satisfying Dirichlet's condition for convergence of a Fourier series (see e.g. [43, p. 52], [241, p. 52]) and if the set $\{f(k)\}_{k\in\mathbb{Z}}$ is known, then the function g is known, so that the function f is fully determined by its values at the integers $k \in \mathbb{Z}$.

As J. R. Higgins has pointed out in his excellent account of the history of the sampling theorem [104], this result of Borel can be regarded as the first half (part (A) of Higgins) of the sampling theorem for band-limited functions. However, Borel does not seem to have linked his statement with his own results on cardinal series, which would have established the second half (part (B) of Higgins, see below).

To see this, expanding g into a Fourier series

$$g(u) = \sum_{k=-\infty}^{\infty} \left\{ \frac{1}{2\pi} \int_{-\pi}^{\pi} g(x) e^{-jkx} dx \right\} e^{jku},$$

one notes that the Fourier coefficients (curly brackets, see Definition (26) below) of g are the "sampled" values $\{1/2\pi f(-k)\}_{k\in\mathbb{Z}}$ in view of (10). Thus g is determined by $\{f(k)\}_{k\in\mathbb{Z}}$, and so f. This gives Borel's statement, or part (A).

Part (B) can be stated as follows: A signal f band-limited in the form (10) is the sum of its cardinal series, i.e.,

$$f(t) = \sum_{k=-\infty}^{\infty} f(k)\frac{\sin \pi(t-k)}{\pi(t-k)}. \tag{11}$$

Now to the missing link between parts (A) and (B), namely that (A) implies (B). If a function f of the form (10) is such that the sum of the cardinal series on the right of (11) is band-limited to $[-\pi, \pi]$, then this sum coincides with f at the points $t = k \in \mathbb{Z}$ by the interpolatory property (6). Hence this sum is actually $f(t)$ itself by part (A). In fact, it follows easily from the property (33) of band-limited functions that the sum of the series in (11) is band-limited to $[-\pi, \pi]$ as is each of its terms separately (cf. (34)).

This gives at the same time (at least) a formal proof of the sampling theorem for band-limited functions. Thus had Borel linked his part (A) with the cardinal series (9), he would have been the true founder of the famous theorem.

Perhaps the next basic contribution to the field is the Paley-Wiener theorem (1934, [166, p. 12]):

The class of functions which are entire, of exponential type Ω (see the Definition of the Bernstein spaces B_Ω^p in Section 2.2.1), and whose restrictions to the real axis belong to $L^2(\mathbb{R})$, is identical to the class of functions f representable as $f(z) = \int_{-\Omega}^{\Omega} g(u)e^{jzu}du$ for some $g \in L^2(-\Omega, \Omega)$.

Then G. H. Hardy (1941) called members of this class *Paley-Wiener functions*, showed that they form a Hilbert subspace of $L^2(\mathbb{R})$, usually denoted by PW_Ω, as well as that the functions set $\{\sin \pi(t-k)/\pi(t-k)\}_{k\in\mathbb{Z}}$ forms a complete orthonormal set in PW_π. Hardy also showed that PW_π is a *reproducing kernel Hilbert space* with *reproducing kernel* $\sin \pi(t-u)/\pi(t-u)$. This means that the inner product of $f \in PW_\pi$ with this kernel reproduces f, i.e., (cf. (66) and [109, Chapter 3], [98, 103, 107, 157–158, 190]),

$$\left\langle f(u), \frac{\sin \pi(t-u)}{\pi(t-u)} \right\rangle := \int_{\mathbb{R}} f(u)\frac{\sin \pi(t-u)}{\pi(t-u)}du = f(t). \tag{12}$$

The similarity between the sampling series (11), actually a *convolution sum*, and the integral (12) is striking. Indeed, the integral can be regarded as a continuous analogue of the series, a convolution integral.

2.1.3. Some Literature, Aims and Contents

J. Neveu (1965) [160] states "... le théorème d'échantillonage dit de Shannon mais dû en fait a Poisson". But he gives no reference. However it is true that the Poisson summation formula of Fourier analysis can be exploited to derive the sampling theorem, a fact apparently first carried out by R. P. Boas (1972) [20] (see also [122]). As to Poisson's formula, considered in Section 2.2.1 below, it was given in a general form without proof by C. F. Gauss on the cover of a book titled "Opuscula mathematica, 1799–1813". Gauss indicates its use in proving the Fourier inversion formula, needed by him in probability theory; see [91, p. 88], also [26, p. 1338]. C. G. J. Jacobi in his work on the transformation formula for elliptic functions [21, p. 260] attributes it to Poisson. According to H. Burkhardt [26, p. 1341], Poisson used it (in a paper of 1923) for the transformation formula of the theta-functions. Further remarks concerning the history of the Poisson summation formula are to be found in [217, pp. 36 f.], [109, p. 19].

Since band-limited signals cannot be simultaneously time-limited (i.e., of finite duration), and since both limitations are rather natural for real physical signals, it is important to consider sampling series for signals which are just approximately band-limited or just time-limited. In regard to the sampling theorem the former will mean that the series in (7) will not be equal to $f(t)$, but that it will tend to $f(t)$ for $\Omega \to \infty$, namely

$$f(t) = \lim_{\Omega\to\infty} \sum_{k=-\infty}^{\infty} f\left(\frac{k\pi}{\Omega}\right) \operatorname{sinc}\left(\frac{\Omega t}{\pi} - k\right). \tag{13}$$

The problem then is to choose the bandwidth parameter Ω so large that the error

$$R_\Omega f(t) := f(t) - \sum_{k=-\infty}^{\infty} f\left(\frac{k\pi}{\Omega}\right) \operatorname{sinc}\left(\frac{\Omega t}{\pi} - k\right) \tag{14}$$

becomes sufficiently small. Details will follow in Sections 2.6, 2.7.

De la Vallée Poussin was the first (1908) [224] to deal with the representations of type (13) and error (14). In fact he employed sampling sums for the interpolation and approximation of *time-limited* signals. His first major result reads:

If f is of bounded variation on some finite interval (a, b), zero outside, and continuous at $t_0 \in (a, b)$, then (13) holds for $t = t_0$. As to the error of convergence, it is of order $\mathcal{O}(\Omega^{-1})$ for suitable f.

This work was continued especially by M. Theis (1919) [218] and J. M. Whittaker (1927) [232]. The former replaced sinc(t) by $\operatorname{sinc}^2(t)$, the latter gave infinite interval versions of the original de la Vallée Poussin results.

As to the mathematicians, electrical and communication engineers working in the broad area of sampling analysis as sketched above and to be handled in the first two chapters of this volume, a division into two periods is suggested by the first major survey paper on the subject, that by A. J. Jerri of 1977 [120]. Shannon's two papers of 1948/9 [197–198] marked the beginnings of wide-scale studies of sampling analysis, at least in the West. In fact, between 1950 and c. 1975 at least 250 articles appeared in the many engineering journals dealing with various aspects of sampling analysis, written by some 176 different authors.

Among the mathematicians of all fields spear-heading research in the broad area of sampling and associated interpolation theory from the beginnings until circa 1975 there were, as partly indicated above, J. Gregory (1668/70), I. Newton (c. 1675/76), L. Euler (1732), C. Maclaurin (1742), E. Waring (1779), J. L. Lagrange (1795), C. F. Gauss (c. 1805), J. B. J. Fourier (1807/20), S. D. Poisson (1820), A. Cauchy (1827), P. Cazzaniga (1892), M. Guichard (1884), E. Borel (1897–99), J. Hadamard (1901), Ch. de la Vallée Poussin (1908), F. J. W. Whipple (1910), E. T. Whittaker (1915), T. A. Brown (1915/16), M. Theis (1919), J. F. Steffensen (1914), K. Ogura (1920), N.E. Nörlund (1924), G. Valiron (1925), W. L. Ferrar (1925/26), J. M. Whittaker (1929/35), L. Tschakaloff (1933), R. E. A. C. Paley - N. Wiener (1934), M. Cartwright (1936), A. J. MacIntyre (1938), B. Spain (1940/58), G. H. Hardy (1941), J. Korevaar (1949), J. D. Weston (1949), A. O. Gelfond (1952), R. P. Boas (1954/72), A. N. Kolmogorov (1956), H. P. Kramer (1957), A. V. Balakrishnan (1957), Kolmogorov – V. M. Tichomirov (1959–), H. S. Shapiro – R. A. Silverman (1960), H. J. Landau – H. O. Pollack (1962), R. P. Gosselin (1963/72), L. L. Campbell (1964/68), M. I. Kadec (1964), I. Kluvánek (1965), H. J. Landau (1967–), J. L. Brown, Jr. (1967–), J. B. Kioustelidis (1969), I. J. Schoenberg (1969/72), R. Kress (1970), J. McNamee – F. Stenger – E. L. Whitney (1971), H. Pollard – O. Shisha (1972), J. R. Higgins (1972–), Boas – Pollard (1974), L. B. Sofman (1974), V. I. Buslaw – A. G. Vituškin (1974), Vituškin (1976), Stenger (1976–), D. H. Mugler (1976–).

According to the literature list in R. J. Marks II two-volume work [147–148] by 1991/3 almost a thousand papers written by some 800 authors of all fields had appeared, the greater part of which is devoted to sampling. Thus since the appearance of Jerri's survey paper, in a period of 25 years, almost 750 further papers appeared, and it is the period when not only engineers but also mathematicians in ever increasing numbers turned to the field and made it so popular.

There exist many monographs and textbooks intended for an engineering audience which cover sampling theory. These include, at least A. N. Akansu – R. A. Haddad (1992) [1], A. V. Balakrishnan [6], M. Bellanger (1987,1989) [12, 13], I. Bilinskis – A. Mikelsons (1992) [17], R. N. Bracewell (1978) [23], J. I. Churkin – C. P. Jakowlew – G. Wunsch (1966) [64], L. W. Couch II (1990) [65], R. E. Crochière – L. R. Rabiner (1983) [66], D. E. Dudgeon – R. M. Mersereau

(1984) [74], A. Fettweis (1996) [85], L. Franks (1969) [88], H. D. Lüke (1995) [143], S. L. Marple and M. Marietta (1987) [149], F. A. Marvasti (1987) [151], A. V. Oppenheim – R. W. Schafer (1975) [164], A. Papoulis (1977) [169], A. Peled and B. Liu (1976), [172], B. Picinbono (1988) [173], J. G. Proakis – C. M. Rader – F. Ling – C. L. Nikias (1992) [176], L. R. Rabiner – B. Gold (1975) [178], H. W. Schüßler (1994) [195], G. Wunsch (1971) [234], R. E. Ziemer – W. H. Tranter (1995) [240].

Books intended for a mathematical audience treating the sampling theorem include, at least, H. Babovsky – T. Beth – H. Neunzert – M. Schulz-Reese (1987) [5], J. J. Benedetto – M. W. Frazier (1994) [14], R. P. Boas (1954) [19], J. R. Higgins (1977) [102], J. R. Higgins (1996) [109], J. R. Higgins – R. L. Stens (1999) [111], R. F. Hoskins and J. Sousa Pinto (1994) [115], W. Krabs (1995) [129], R. Lasser (1996) [134], F. Natterer (1986) [159], J. R. Partington (1997) [171], A. Schönhage (1971) [193], A. Terras (1985) [217], H. Triebel (1977) [219], L. A. Wainstein – V. D. Zubakov (1962) [227], R. W. Young (1980) [236], A. I. Zayed (1993) [237].

There also exist several proceedings of workshops or conferences that were devoted to signal processing, sampling and related areas. Many of them concentrated on the interplay between the mathematical principles and the approaches by electrical and communication engineers in the broad area. These include the proceedings edited by H. D. Lüke (1981) [141], P. L. Butzer (1984) [29], T. S. Durrani *et al.* (1987) [75], D. Meyer-Ebrecht (1987) [155], W. Ameling (1990) [2], P. Vary (1994) [225], F. A. Marvasti (1995) [152], J. M. Blackledge (1997) [18], M. E. H. Ismail *et al.* (1995) [116], P. J. S. G. Ferreira (1997) [83], Y. Lyubarskii (1999) [146], A.I. Zayed (2001) [242].

Let us finally mention that there also exist a number of survey or overview papers in the general area. We have already mentioned that by A. J. Jerri [120]. There are further those by J. McNamee – F. Stenger – E. L. Whitney (1971) [154], F. Stenger (1981) [209], P. L. Butzer (1983) [27, 28], P. L. Butzer – R. L. Stens – S. Ries (1983) [44], J. R. Higgins (1985) [104], P. L. Butzer – W. Splettstößer – R.L. Stens (1988) [49], P. L. Butzer – R. L. Stens (1992) [54], J. L. Brown, Jr. (1993) [25], P. L. Butzer – G. Nasri-Roudsari (1997) [42], P. L. Butzer – J. R. Higgins – R. L. Stens (2000) [37].

The present introduction to sampling analysis is not really a survey or overview paper of the field intended for a mathematical (research) audience, but rather more a tutorial style paper primarily written for an audience of electrical and communication engineers. The authors, in particular the senior author, have worked together with electrical engineers since 1972, especially closely in the period 1977 to 1993. As the "Index of Papers on Signal Analysis 1972–1994" [50] reveals, which lists 148 research papers, master, doctoral and "Habilitation" theses written by members of the chair "Lehrstuhl A für Mathematik" during the given period, many of the members worked together with engineers in joint national and international research projects. In fact, many of the problems solved in these papers answer questions raised by engineers, seismologists and medical doctors in the many meetings held. These papers set in with the work of W. Splettstößer and P. L. Butzer (1977) [47], Splettstößer (1978) [202], R. L. Stens (1980) [210, 211].

Based upon our long experience in this respect as well as in our capacity as members of the Programme Committees of the international workshops in Jurmala, Latvia, (1995) [152], Aveiro, Portugal, (1997) [83], and Loen, Norway, (1999) [146], Sections 2.2 and 2.3 aim to treat the various topics presented, also their choice, in the spirit of the audience in mind. Thus Section 2.2 contains the basic facts on Fourier analysis and distributions needed, usually written in engineering terminology, as well as material on many elementary functions needed in the applications, they being pictured in figures. Section 2.3 is devoted to half a dozen different proofs of the sampling theorem, two of them being the standard engineering textbook proofs. Included is a geometrical demonstration of the sampling theorem by means of seven figures of signals together with their transforms; they also deal with over-sampling and under-sampling. The authors feel that also engineers should be made aware of the fact whether a proof is just a sketch, formal or complete and fully mathematically sound. Many results can in fact be only well understood if they are connected with at least a sketch of the proof that reveals the core of the result in question. Section 2.4 focuses on sampling representations of derivatives as well as Hilbert transforms, so important in the applications (e.g. Bracewell [23, pp. 267 ff.] in regard to causality and a generalization of the phasor idea beyond pure alternating current). Section 2.5 contains some results on multi-channel sampling, and Section 2.6 deals with sampling theory for signals which are not necessarily band-limited. These include duration-limited signals (which cannot be simultaneously band-limited), as well as signals which need only be "approximately" band-limited; it is a much weaker hypothesis. Section 2.7 is devoted to the various errors which occur naturally in applications, such as the aliasing errors, truncation errors, amplitude and quantization errors, time jitter errors. The results presented are based on a recent novel approach.

Section 2.8 is concerned with generalized sampling series, which are discrete analogues of singular Fourier convolution integrals on $\mathbb{R}$, for which the sinc function is replaced by other kernel functions; these have better properties and so are more suitable for fast and efficient implementations. The applications deal with examples of special band-limited kernels which do not seem to have been considered so far, as well as with time limited kernels, namely linear combinations of B-splines. There is also a short section on the important area of prediction of signals. Section 2.9 contains information concerning connections of the sampling theorem with other basic theorems of mathematics and its applications. The chapter concludes with a reference list of approximately 180 research papers and books.

Section 2.10 is devoted to sampling theory in connection with partial fraction decompositions. Basic here is a theorem of Cauchy of 1827 which can be used to establish the sampling theorem and even some of its variants. The most general theorem in this respect is the partial fraction decomposition due to Mittag-Leffler (1877). Section 2.11 is concerned with quadrature formulae, in particular with generalized Gaussian quadrature formulae. Special emphasis is placed on quadrature of non-band-limited signals over the real line $\mathbb{R}$ as well as over the semi-axis $\mathbb{R}_+$. The remainders occurring in this case are estimated in detail.

Chapter 3 of this volume [238] includes an alternative introduction to sampling theory not via Fourier analysis as in the present Chapter 2 but via Lagrange polynomial interpolation theory and Sturm-Liouville differential equations. It is almost self-contained and can be read independently of Chapter 2.

Let us end this introduction with the very appropriate "Conclusion" of Lüke's paper *The Origins of the Sampling Theorem* [144]:

> *The sampling theorem for low-pass functions plays an important role in communication engineering as a connecting link between continuous-time and discrete-time signals. The numerous different names to which the sampling theorem is attributed in the literature—Shannon, Nyquist, Kotelnikov, Whittaker, to Someya—gave rise to the above discussion of its origins. However, this history also reveals a process which is often apparent in theoretical problems in technology or physics: first the practicians put forward a rule of thumb, then the theoreticians develop the general solution, and finally someone discovers that the mathematicians have long since solved the mathematical problem which it contains, but in "splendid isolation".*

2.2. Foundations in Fourier Analysis

2.2.1. Fourier Transforms, Fourier Series

As usual, let $\mathbb{N}$, $\mathbb{Z}$, $\mathbb{R}$, $\mathbb{C}$ be the sets of all naturals, integers, real and complex numbers, respectively. $L^p(\mathbb{R})$, $1 \le p < \infty$, is the space of all functions $f : \mathbb{R} \to \mathbb{C}$, which are absolutely (Lebesgue) integrable to the pth power, endowed with the norm

$$\|f\|_{L^p(\mathbb{R})} := \left\{ \int_{-\infty}^{\infty} |f(u)|^p du \right\}^{1/p};$$

here $L^\infty(\mathbb{R})$ is the space of all measurable, essentially bounded functions (bounded almost everywhere) with norm

$$||f||_{L^\infty(\mathbb{R})} := \operatorname*{ess\,sup}_{u\in\mathbb{R}} |f(u)|.$$

Further, $C(\mathbb{R})$ is the space of all uniformly continuous and bounded functions f on $\mathbb{R}$ with the usual supremum norm $\|f\|_{C(\mathbb{R})} := \sup_{u\in\mathbb{R}} |f(u)|$, and $C^{(r)}(\mathbb{R}) := \{f \in C(\mathbb{R}); f^{(r)} \in C(\mathbb{R})\}$ for some $r \in \mathbb{N}$. See e.g., [43, pp. 1, 2] or [216, p. 59] for the definitions of L^p-spaces.

The *Fourier transform* (or *integral*) of $f \in L^1(\mathbb{R})$, also called the *spectrum*[1] of the signal $f(t)$, will be defined by

$$F(v) \equiv f^\wedge(v) \equiv \mathcal{F}[f](v) := \int_{-\infty}^{\infty} f(u)e^{-jvu}du \quad (v \in \mathbb{R}). \tag{15}$$

The same notation will be used for the Fourier transform of $f \in L^p(\mathbb{R})$, $1 < p \leq 2$, defined by

$$\lim_{R\to\infty} \left\| F(v) - \int_{-R}^{R} f(u)e^{-jvu}du \right\|_{L^q(\mathbb{R})} = 0, \tag{16}$$

where $1/p + 1/q = 1$. If $f \in L^p(\mathbb{R}) \cap C(\mathbb{R})$ is such that $F \in L^1(\mathbb{R})$, then there holds the *inversion formula*, permitting the representation of $f(t)$ in terms of $F(v)$, namely

$$f(t) = \frac{1}{2\pi}\int_{-\infty}^{\infty} F(v)e^{jvt}dv, \quad (t \in \mathbb{R}). \tag{17}$$

In view of this equation $f(t)$ will be also called the *inverse Fourier transform* of $F(v)$. For the basic properties of the Fourier transform, one may consult any textbook on Fourier analysis, e.g., [43, Chapter 5], [76, Chapter 2], [167, Chapter 2], [169, Chapter 3], [189, Chapter 7], [241, II, Chapter 16].

This definition of the Fourier transform is restricted to $f \in L^p(\mathbb{R})$ for $1 \leq p \leq 2$. In order to have the Fourier transform available for $p > 2$, too, one has to turn to the theory of distributions. It is assumed that the reader is familiar with the basic properties of the spaces $\mathcal{S}$ of rapidly decreasing testing functions, and $\mathcal{S}'$ of tempered distributions; see [115, Chapters 1–3], [119, Chapters 9, 10], [129, Chapter 4], [215, Chapter 2], [239, Chapters 4, 5]. If $T \in \mathcal{S}'$, we designate

[1] The definition of the Fourier transform is not uniform in the literature. Many authors define it by $f^\wedge(v) := (1/\sqrt{2\pi})\int_{-\infty}^{\infty} f(u)e^{-jvu}du$ and its inverse by $f(t) := (1/\sqrt{2\pi})\int_{-\infty}^{\infty} F(v)e^{jvt}dv$ (as e.g., in [43], [241, II, p. 247]), others by $f^\wedge(v) := \int_{-\infty}^{\infty} f(u)e^{-j2\pi vu}du$, the inverse by $f(t) := \int_{-\infty}^{\infty} F(v)e^{j2\pi vt}dv$ (e.g. [23, 143]). Here $j = \sqrt{-1}$ is the complex unit.

the complex number that T assigns to a particular testing function ϕ by $\langle T, \phi \rangle$ or $\langle T(u), \phi(u) \rangle$. We sometimes use the symbolic notation

$$\langle T, \phi \rangle = \int_{-\infty}^{\infty} T(u)\phi(u)du,$$

although the integral in general has no meaning as an ordinary integral. This notation, however, indicates that the functional T has many properties in common with an integral.

For $T \in \mathcal{S}'$ the Fourier transform $T^\wedge \in \mathcal{S}'$ is defined via

$$\langle T^\wedge(v), \phi(v) \rangle := \langle T(t), \phi^\wedge(t) \rangle \quad (\phi \in \mathcal{S}), \tag{18}$$

where $\phi^\wedge$ is to be understood in the sense of (15). The Fourier transform maps $\mathcal{S}'$ one to one onto itself with

$$T^{\wedge\wedge}(t) = 2\pi T(-t). \tag{19}$$

Furthermore, since the $L^p(\mathbb{R})$-spaces, $1 \le p \le \infty$, are contained in $\mathcal{S}'$, (18) extends the definitions (15) and (16) to $L^p(\mathbb{R})$ for $p > 2$. For the Fourier transform on $\mathcal{S}'$ see e.g. [239, Chapter 7], [189, Chapter 7], [119, Chapter 11], [129, Chapter 7].

The notation $f(t) \longleftrightarrow F(v) \equiv f^\wedge(v)$ will often be used to indicate that the functions or distributions $f(t)$ and $F(v)$ are related by (15), (16) and (18), respectively. The variables t and v will often be referred to as *time* and *frequency*. By using this notation, (19) yields

$$T(t) \longleftrightarrow S(v) \quad \text{if and only if} \quad S(t) \longleftrightarrow 2\pi T(-v). \tag{20}$$

Note that relations (19) and (20) hold in L^p-spaces as well, provided the functions are interpreted as distributions if necessary. In particular, one has for $f \in L^2(\mathbb{R})$ that in the sense of ordinary functions

$$f^{\wedge\wedge}(t) = 2\pi f(-t),$$

$$f(t) \longleftrightarrow F(v) \quad \text{if and only if} \quad F(t) \longleftrightarrow 2\pi f(-v).$$

Next to the inversion formula (17), the convolution theorem is the most powerful tool in Fourier analysis. Given two time functions $f_1, f_2 : \mathbb{R} \to \mathbb{C}$, with $f_1(t) \longleftrightarrow F_1(v), f_2(t) \longleftrightarrow F_2(v)$, then their *convolution product* is defined by

$$g(t) \equiv f_1(t) * f_2(t) \equiv (f_1 * f_2)(t) := \int_{-\infty}^{\infty} f_1(t-u) f_2(u) du \tag{21}$$

whenever the integral exists, and the *time convolution theorem* states that

$$G(v) \equiv [f_1 * f_2]^\wedge(v) = f_1^\wedge(v) \cdot f_2^\wedge(v) = F_1(v) F_2(v), \tag{22}$$

or

$$\int_{-\infty}^{\infty} f_1(t-u) f_2(u) du \longleftrightarrow F_1(v) F_2(v).$$

Thus convolution of two time functions means multiplication of their transforms. Observe that if $f_1 \in L^1(\mathbb{R})$ and $f_2 \in L^p(\mathbb{R})$, $1 \le p \le \infty$, or $f_2 \in C(\mathbb{R})$, then $f_1 * f_2$ belongs to $L^p(\mathbb{R})$ or $C(\mathbb{R})$, respectively. If $f_1 \in L^1(\mathbb{R})$ and $f_2 \in L^p(\mathbb{R})$, $1 \le p \le 2$, then the convolution theorem is valid almost everywhere (a.e.).

From (22) and the *symmetry property* $F(t) \longleftrightarrow 2\pi f(-v)$ of the pair $f(t) \longleftrightarrow F(v)$ (cf. (19)), it follows that the transform of the product $f_1(t)f_2(t)$ equals the convolution product $(1/2\pi)(F_1 * F_2)(v)$. Thus

$$f_1(t) f_2(t) \longleftrightarrow \frac{1}{2\pi} \int_{-\infty}^{\infty} F_1(v-y) F_2(y) dy \tag{23}$$

or

$$[f_1 \cdot f_2]^\wedge(v) = \frac{1}{2\pi} (F_1 * F_2)(v).$$

which holds if, e.g., f_1 and f_2 belong to $L^2(\mathbb{R})$. This is referred to as the *frequency convolution theorem* or *multiplication theorem.*

Denoting the complex conjugate of f by $\overline{f}$ (i.e., $\overline{f}(t) = f_1(t) - jf_2(t)$ for $f(t) = f_1(t) + jf_2(t)$), then the *power* or *energy theorem* (or *generalized Parseval's theorem*[2]) reads for $f_1,\ f_2 \in L^2(\mathbb{R})$,

$$\int_{-\infty}^{\infty} f_1(t) \overline{f_2(t)} dt = \frac{1}{2\pi} \int_{-\infty}^{\infty} F_1(v) \overline{F_2(v)} dv. \tag{24}$$

[2] As to Marc-Antoine Parseval (1755–1836), his paper [170], dealing with the Fourier series version of (23), is hardly ever cited.

An important special case is *Rayleigh's theorem* of 1889 or *Plancherel's theorem* (1910) in mathematical circles (cf. [23, p. 112]):

$$\int_{-\infty}^{\infty} |f(t)|^2 dt = \frac{1}{2\pi}\int_{-\infty}^{\infty} |F(v)|^2 dv.$$

We also need the concept of a *Fourier series* associated with a periodic function $f(t)$ on $\mathbb{R}$, with period $2T, f(t+2T) = f(t)$, $t \in \mathbb{R}$. To this end, let L^p_{2T}, $1 \le p < \infty$, be the space of all $2T$-periodic functions $f : \mathbb{R} \to \mathbb{C}$ with the norm

$$||f||_{L^p_{2T}} \equiv ||f||_p := \left\{\int_{-T}^{T} |f(u)|^p du\right\}^{1/p}.$$

The spaces L^∞_{2T} and C_{2T} are defined appropriately. The Fourier series of $f \in L^1_{2T}$ is given by

$$f(t) \sim \sum_{k=-\infty}^{\infty} c_k e^{jk\pi t/T}, \tag{25}$$

where the *Fourier coefficients* (or *finite Fourier transform*) of f are defined by

$$c_k = [f]^\wedge_T(k) = \frac{1}{2T}\int_{-T}^{T} f(u)e^{-jk\pi u/T} du \quad (k \in \mathbb{Z}). \tag{26}$$

There exists a formula linking a signal f with its transform F in a fundamental way; it also enables one to replace the infinite limits in the integral of $F(v)$ by finite limits of an associated integral. It is the famous *Poisson summation formula:*

If $g \in L^1(\mathbb{R})$ and

$$g^*(t) := 2T \sum_{k=-\infty}^{\infty} g(t + 2kT), \tag{27}$$

then g^* belongs to L^1_{2T}, and

$$\int_{-\infty}^{\infty} f(t-u)g(u)du = \frac{1}{2T}\int_{-T}^{T} f(t-u)g^*(u)du \tag{28}$$

for every $f \in C_{2T}$. In particular, for $f(t) = \exp(-jk\pi t/T)$,

$$[g^*]_T^{\wedge}(k) = G\left(\frac{k\pi}{T}\right) \quad (k \in \mathbb{Z}).$$

The actual summation formula reads

$$g^*(t) \sim \sum_{k=-\infty}^{\infty} G\left(\frac{k\pi}{T}\right) e^{jk\pi t/T} \tag{29}$$

in the sense that the series on the right is the Fourier series of g^*. Equality holds in (29) provided, e.g., $g \in L^1(\mathbb{R})$ is absolutely continuous and $g' \in L^1(\mathbb{R})$, i.e., in this case one even has:

$$2T \sum_{k=-\infty}^{\infty} g(t + 2kT) = \sum_{k=-\infty}^{\infty} G\left(\frac{k\pi}{T}\right) e^{jk\pi t/T} \quad (t \in \mathbb{R}). \tag{30}$$

To establish (28), substitute (27) into the right-hand side of (28) and interchange summation and integration,

$$\frac{1}{2T}\int_{-T}^{T} f(t-u)g^*(u)du = \sum_{k=-\infty}^{\infty} \int_{-T}^{T} f(t-u)g(u+2kT)du$$
$$= \int_{-\infty}^{\infty} f(t-u)g(u)du.$$

For (30) see, e.g., [43, p. 202]. Another, more formal proof of (30) will be given at the end of Section 2.2.2 below.

This summation formula, named after S. D. Poisson, but already known to C. F. Gauss (recall Section 2.1.3) enables one to discretize certain integrals and, in particular, to write a convolution *integral* as a discrete convolution *sum*.

In this respect we introduce for $\Omega \geq 0$ and $1 \leq p \leq \infty$ the Bernstein spaces B_Ω^p which consist of those functions $f \in L^p(\mathbb{R})$ having an extension to the complex plane as an entire function of exponential type Ω. This means that f is an entire function in the sense of complex analysis (i.e., f is differentiable (analytic) in the whole complex plane) and satisfies the inequality

$$|f(z)| \leq ||f||_{C(\mathbb{R})} e^{\Omega|y|} \quad (z = x + jy \in \mathbb{C}).$$

As usual we do not distinguish between f and its extension. For $p = 2$ the space B_Ω^2 is also known as Paley-Wiener space. It follows immediately from the definition that

$$f(t) \in B_\Omega^p \Rightarrow f(\alpha t) \in B_{|\alpha|\Omega}^p \quad (\alpha \in \mathbb{R}), \tag{31}$$

$$f \in B_{\Omega_1}^p, g \in B_{\Omega_2}^p \Rightarrow f \cdot g \in B_{\Omega_1+\Omega_2}^p,$$

and one has

$$B_\Omega^1 \subset B_\Omega^p \subset B_\Omega^{p'} \subset B_\Omega^\infty \quad (1 \le p \le p' \le \infty).$$

More important is the Bernstein-type inequality stating that $f \in B_\Omega^p$ implies $f' \in B_\Omega^p$ and

$$\|f'\|_{L^p(\mathbb{R})} \le \Omega \|f\|_{L^p(\mathbb{R})} \quad (f \in B_\Omega^p; 1 \le p \le \infty). \tag{32}$$

Moreover, this inequality can be shown to imply for any $h > 0$ and $1 \le p < \infty$ the estimates

$$\|f\|_{L^p(\mathbb{R})} \le \sup_{u \in \mathbb{R}} \left\{ h \sum_{k=-\infty}^{\infty} |f(u - hk)|^p \right\}^{1/p} \le (1 + h\Omega) \|f\|_{L^p(\mathbb{R})} \quad (f \in B_\Omega^p), \tag{33}$$

which means that the expression in the middle defines a norm on B_Ω^p, which is equivalent to the usual L^p-norm (see [161, Chapter 3]).

Let us consider some examples. In view of the estimate $|e^{jz}| = e^{-y} \le e^{|y|}$ for $z = x + jy \in \mathbb{C}$ and (31), the functions

$$e^{j\alpha z}, \quad \cos \alpha z, \quad \sin \alpha z$$

with $\alpha \in \mathbb{R}$ obviously belong to $B_{|\alpha|}^\infty$. Similarly, the function $f(z) := (\sin \alpha z)/(\alpha z)$ for $z \in \mathbb{C} \setminus \{0\}$ and $f(0) := 1$ lies in $B_{|\alpha|}^p$ for all $1 < p \le \infty$, $\alpha \in \mathbb{R} \setminus \{0\}$. The particular case $\alpha = \pi$ seems to be one of the most important functions in sampling theory. Nowadays it is usually denoted by sinc as an abbreviation for the Latin *sinus cardinalis* (cf. [233, p. 29], [109, p. 4]), thus

$$\operatorname{sinc}(z) := \begin{cases} \dfrac{\sin \pi z}{\pi z}, & z \in \mathbb{C} \setminus \{0\} \\ 1, & z = 0. \end{cases} \tag{34}$$

The attribute *cardinal* in connection with sinc goes back to E. T. Whittaker [230], who is considered by many as one of the fathers of sampling theory. Whittaker called the series (64) below the cardinal function associated to the function *f*. See also Sections 2.1.1 and 2.2.2(d).

As an example of a integrable function *f* one may take

$$f(z) = \begin{cases} \left(\dfrac{\sin \alpha z}{\alpha z}\right)^n, & z \in \mathbb{C} \setminus \{0\} \\ 1, & z = 0, \end{cases} \tag{35}$$

where $\alpha \in \mathbb{R} \setminus \{0\}$ and $n \in \mathbb{N} \setminus \{1\}$. It follows that $f \in B^1_{n|\alpha|}$ and $n|\alpha|$ cannot be replaced by any smaller number.

In our definition of "exponential type" we have implicitly assumed that the functions in question are bounded along the real axis. A more general approach allowing one also to consider unbounded functions is described in the chapter by A. I. Zayed and P. L. Butzer of this volume [238]; see also [19, Chapter 2], [161, Chapter 3], [236, Chapter 2], [156].

In the language of electrical engineers a signal function $f : \mathbb{R} \to \mathbb{C}$ is called *band-limited*, if it belongs to $L^2(\mathbb{R})$, i.e., it has finite energy, and its Fourier transform F vanishes outside a certain interval, say $F(v) = 0$ for $|v| > \Omega$. The connection between the Bernstein spaces introduced above and band-limited functions is given by the famous Paley-Wiener Theorem: *A function f belongs to* B^p_Ω *if and only if its Fourier transform F vanishes a.e. outside the interval* $[-\Omega, \Omega]$. Using the definition of the Fourier transform given above in (15) and (16), respectively, one has to assume $1 \le p \le 2$, but the result remains true for $p > 2$ if the Fourier transform is understood in the distributional sense (cf. [76, Section 3.3], [115, Section 3.3.2], [188, Chapter 19], [189, Chapter 7]. Thus the B^p_Ω-spaces generalize the concept of band-limitation from $L^2(\mathbb{R})$ to $L^p(\mathbb{R})$, $1 \le p \le \infty$.

Now to the discretization of integrals announced above:

If $g \in B^1_{2\Omega}$ *for some* $\Omega > 0$, *then*

$$\int_{-\infty}^{\infty} g(u)du = \frac{\pi}{\Omega} \sum_{k=-\infty}^{\infty} g\left(\frac{k\pi}{\Omega}\right) \tag{36}$$

the series being absolutely convergent.

Indeed, applying the summation formula (29) to g, replacing T by $\pi/2\Omega$ and setting $t = 0$ yields (36), since the infinite series on the right-hand side of (29) reduces to the term for $k = 0$ in view of $G(v) = 0$ for all $|v| > 2\Omega$, and hence $G(2k\Omega) = 0$ for $k \neq 0$. Note that the assumptions sufficient for (30) to hold are satisfied in view of (32). The absolute convergence of the series in (36) follows from (33).

As a particular case of (36) one obtains: *If* $f_1 \in B^p_\Omega, f_2 \in B^q_\Omega$ *for some* $\Omega > 0$, $1 \le p \le \infty$ *and* $1/p + 1/q = 1$, *then*

$$\int_{-\infty}^{\infty} f_1(u) f_2(t-u)\,du = \frac{\pi}{\Omega} \sum_{k=-\infty}^{\infty} f_1\left(\frac{k\pi}{\Omega}\right) f_2\left(t - \frac{k\pi}{\Omega}\right) \quad (t \in \mathbb{R}), \tag{37}$$

the series converging absolutely and uniformly for all $t \in \mathbb{R}$. *In particular,*

$$\sum_{k=-\infty}^{\infty} f_1\left(\frac{k\pi}{\Omega}\right) f_2\left(t - \frac{k\pi}{\Omega}\right) = \sum_{k=-\infty}^{\infty} f_1\left(t - \frac{k\pi}{\Omega}\right) f_2\left(\frac{k\pi}{\Omega}\right) \quad (t \in \mathbb{R}). \tag{38}$$

Formula (37) is just (36), applied to $f_1(\cdot)f_2(t-\cdot) \in B^1_{2\Omega}$, and the commutativity (38) follows from the commutativity of the convolution integral $f_1 * f_2$ of (21). The absolute and uniform convergence of all series in (37) and (38) is again a consequence of (33); indeed,

$$\begin{aligned} \sum_{|k|>N} \left| f_1\left(\frac{k\pi}{\Omega}\right) f_2\left(t - \frac{k\pi}{\Omega}\right) \right| &\le \left(\sum_{|k|>N} \left| f_1\left(\frac{k\pi}{\Omega}\right) \right|^p \right)^{1/p} \left(\sum_{k=-\infty}^{\infty} \left| f_2\left(t - \frac{k\pi}{\Omega}\right) \right|^q \right)^{1/q} \\ &\le \left(\sum_{|k|>N} \left| f_1\left(\frac{k\pi}{\Omega}\right) \right|^p \right)^{1/p} (1+\pi)\|f_2\|_{L^q} \end{aligned} \tag{39}$$

and the right-hand side can be made arbitrarily small by (33).

In connection with sampling of non-band-limited functions in Sections 2.6 and 2.8, as well as with the various errors occurring in practice which will be considered in Section 2.7, we will make use of the so-called *modulus of continuity* of a function $f \in L^p(\mathbb{R})$ or $f \in C\mathbb{R}$. It is a measure of smoothness of the signal function f. If X is one of the spaces $L^p(\mathbb{R})$ or $C(\mathbb{R})$ with norm $\|f\|_X$, then the modulus of continuity of $f \in X$ and the associated *Lipschitz class* of order $\alpha > 0$ are defined by

$$\omega(f, X, \tau) \equiv \omega(f, \tau) := \sup_{|h| \le \tau} \| f(t+h) - f(t) \|_X \quad (\tau > 0), \tag{40}$$

$$\mathrm{Lip}(\alpha; X) := \{ f \in X; \omega(f, X, \tau) = \mathcal{O}(\tau^\alpha), \tau \to 0+ \}. \tag{41}$$

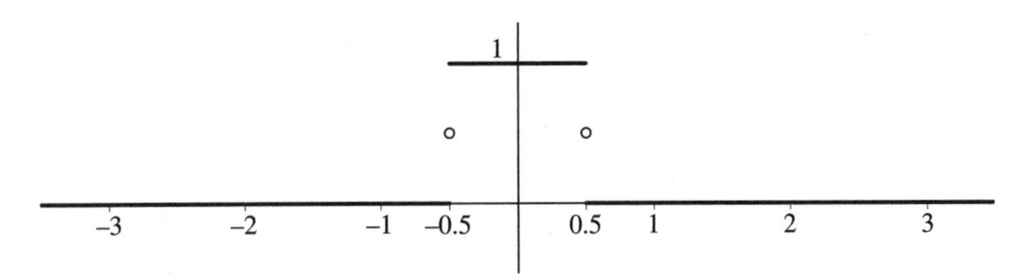

Figure 1. Rectangle function $\Pi(t)$.

Moreover, we will need the Lipschitz class of order $\alpha > 0$ with Lipschitz constant $L > 0$, namely,

$$\mathrm{Lip}_L(\alpha; X) := \{f \in X; \omega(f, X, \tau) \leq L\tau^{\alpha}, \tau > 0\}. \tag{42}$$

2.2.2. Elementary Functions and Distributions

In this section we consider a number of elementary functions and distributions as well as their Fourier transforms. They occur throughout sampling analysis.

(a) The *rectangle function (Fig. 1) (pulse)* of unit height and base,

$$\Pi(t) := \begin{cases} 1, & |t| < 1/2 \\ 1/2, & |t| = 1/2 \\ 0, & |t| > 1/2. \end{cases}$$

Its Fourier transform (Fig. 2) is given by

$$\Pi^{\wedge}(v) = \int_{-1/2}^{1/2} e^{-jvu} du = \frac{\sin v/2}{v/2}. \tag{43}$$

Observe that convolving $f(t)$ with the pulse $\Pi(t)$ results in the *smoothing* of $f(t)$, since

$$(f * \Pi)(t) = \int_{t-1/2}^{t+1/2} f(u) du.$$

(b) The *triangle function (Fig. 3) (pulse)* of unit height and area

$$\Lambda(t) := \left\{ \begin{array}{cc} 1 - |t|, & |t| < 1 \\ 0, & |t| > 1 \end{array} \right\} = (1 - |t|)\Pi\left(\frac{t}{2}\right). \tag{44}$$

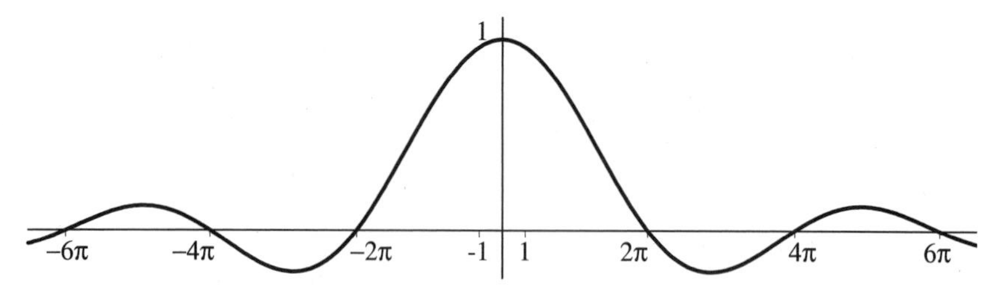

Figure 2. Fourier transform $\Pi^\wedge(v) = \frac{\sin v/2}{v/2}$.

Noting that

$$\Lambda(t) = (\Pi * \Pi)(t),$$

the Fourier transform (Fig. 4) can be obtained from the time convolution theorem (22), namely,

$$\Lambda(t) \longleftrightarrow \left(\frac{\sin v/2}{v/2}\right)^2. \tag{45}$$

(c) The *signum function (Fig. 5)*

$$\operatorname{sgn}(t) := \begin{cases} -1, & t < 0 \\ 0, & t = 0\,, \\ 1, & t > 0 \end{cases}$$

and *Heaviside's unit step function (Fig. 6)*

$$U(t) := \begin{cases} 0, & t < 0 \\ 1/2, & t = 0\,, \\ 1, & t > 0 \end{cases}$$

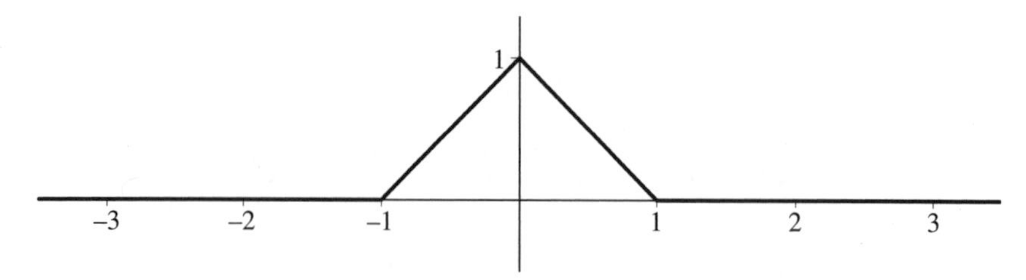

Figure 3. Triangle function $\Lambda(t)$.

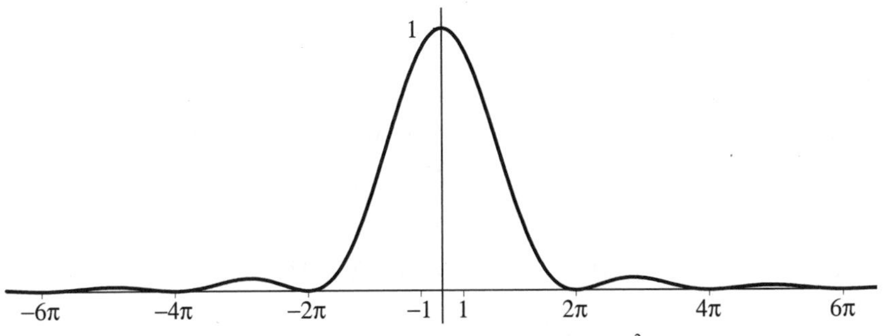

Figure 4. Fourier transform $\Lambda^{\wedge}(v) = \left(\frac{\sin v/2}{v/2}\right)^2$.

which are related in terms of $\text{sgn}(t) = 2U(t) - 1$, have transforms to be given below, in (54) and (55), respectively.

Note that the primitive of a signal f can be written as a convolution $\int_{-\infty}^{t} f(u)du = (f * U)(t)$.

(d) The sinc function (Fig. 7) introduced in (34) for which $\text{sinc}(k) = \delta_{k,0}$. As to its Fourier transform, it follows again from (43) and (19) that

$$\text{sinc}^{\wedge}(v) = \Pi\left(\frac{v}{2\pi}\right), \tag{46}$$

so that the Fourier transform (Fig. 8) equals 1 for any v with $|v| < \pi$, but zero for any $|v| > \pi$. In particular, one has $\int_{-\infty}^{\infty} \text{sinc}(u)du = 1$. This curious behaviour is

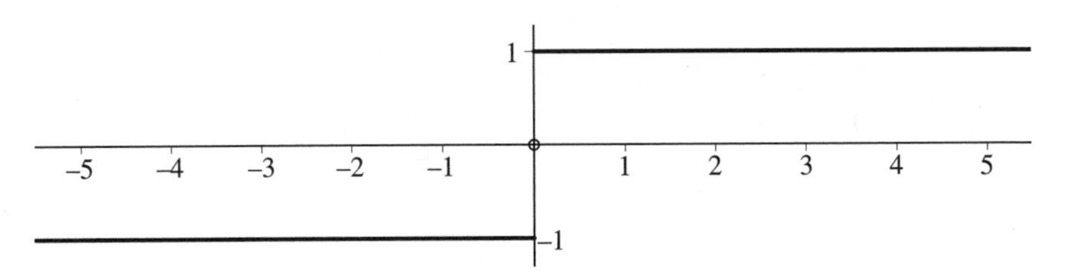

Figure 5. Signum function $\text{sgn}(t)$.

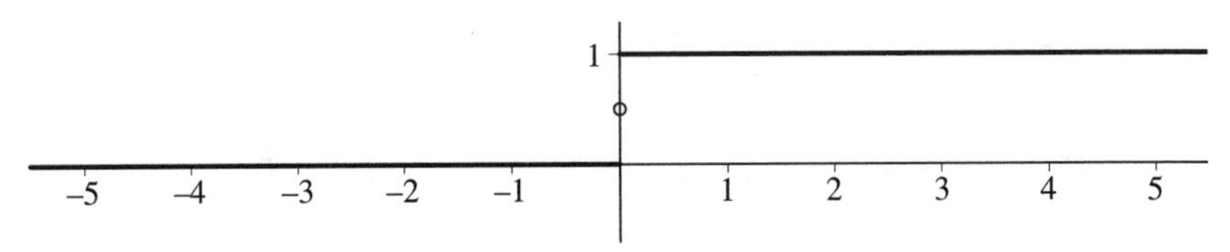

Figure 6. Heaviside function $U(t)$.

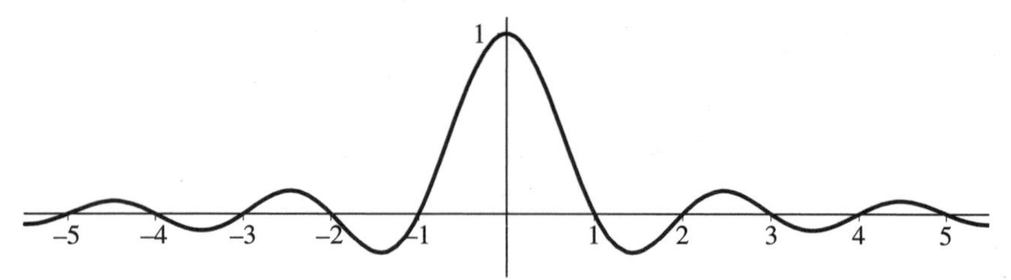

Figure 7. Sinc function $\operatorname{sinc}(t) = \frac{\sin \pi t}{\pi t}$.

typical in Fourier analysis, connecting ordinary continuous and smooth functions with awkward, piecewise abrupt functions. A slight generalization of (43) and (46) is the transform pair,

$$T \operatorname{sinc}(Tt - h) \longleftrightarrow \Pi\left(\frac{v}{2\pi T}\right) e^{-jhv/T} \quad (T > 0; h \in \mathbb{R}), \tag{47}$$

We have already mentioned the importance of the sinc function in signal theory, but it plays a fundamental role in Fourier analysis, too. Using Parseval's theorem (24) and the transform pair (47), we can rewrite the Fourier inversion integral (17) as

$$\begin{aligned} f(t) &= \lim_{R\to\infty} \frac{1}{2\pi} \int_{-R}^{R} F(v) e^{jvt} dv \\ &= \lim_{R\to\infty} \frac{R}{\pi} \int_{-\infty}^{\infty} f(u) \operatorname{sinc}\left(\frac{R}{\pi}(t-u)\right) du. \end{aligned} \tag{48}$$

In this connection sinc is better known as *Dirichlet's kernel* and the right-hand integral in (48) as *singular convolution integral of Dirichlet.*

Concerning the square of the sinc function, also called *Fejèr's kernel*, one has as above in view of formulae (45) and (19) that

$$(\operatorname{sinc}^2)^\wedge(u) = \Lambda\left(\frac{v}{2\pi}\right).$$

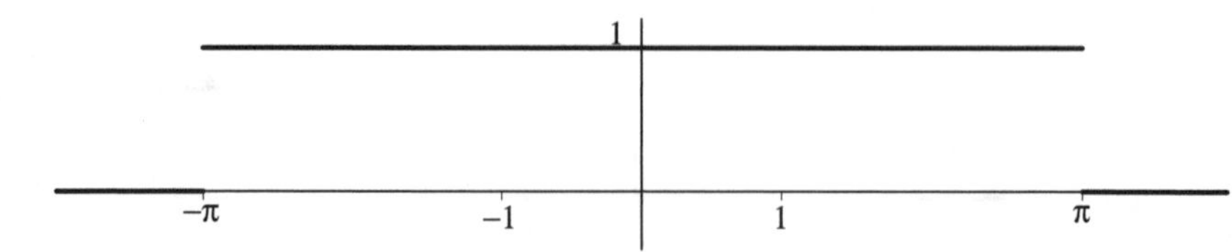

Figure 8. Fourier transform of the sinc function $\operatorname{sinc}^\wedge(v) = \Pi\left(\frac{v}{2\pi}\right)$.

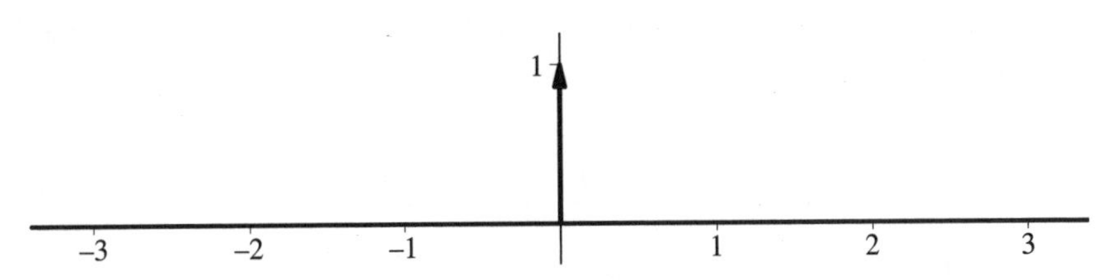

Figure 9. Dirac delta function.

(e) The *Dirac delta function* (Fig. 9) or better *delta distribution* $\delta(t)$, also called the unit area impulse, is defined by

$$\langle \delta, \phi \rangle = \int_{-\infty}^{\infty} \delta(u)\phi(u)du := \phi(0) \quad (\phi \in \mathcal{S}). \tag{49}$$

As said above, the integral has a symbolic meaning only, but here it can be understood as the limit

$$\int_{-\infty}^{\infty} \delta(u)\phi(u)du = \lim_{\varepsilon \to 0+} \int_{-\infty}^{\infty} g_\varepsilon(u)\phi(u)du,$$

where $(g_\varepsilon)_{\varepsilon>0}$ is a so-called *approximate identity*, i.e.,

$$\int_{-\infty}^{\infty} g_\varepsilon(u)du = 1, \quad \lim_{\varepsilon \to 0+} \int_{|u|\geq\eta} |g_\varepsilon(u)|du = 0$$

for any $\eta > 0$. Thus δ can be viewed as the limit of an approximate identity $(g_\varepsilon(t))_{\varepsilon>0}$ for $\varepsilon \to 0+$. Particular approximate identities are

$$\frac{1}{\varepsilon}\Pi\left(\frac{t}{\varepsilon}\right), \quad \frac{1}{\varepsilon}\operatorname{sinc}^2\left(\frac{t}{\varepsilon}\right), \quad \frac{1}{\varepsilon\sqrt{\pi}}\exp\left\{-\left(\frac{t}{\varepsilon}\right)^2\right\}.$$

The following properties of δ are well know (cf. [239, Chapters 1, 2], [119, Chapter 6], [167, pp. 269–282]). There is the *shifting property*

$$\langle \delta(u-t), \phi(u) \rangle = \int_{-\infty}^{\infty} \delta(t-u)\phi(u)du = \phi(t), \tag{50}$$

which can be rewritten in terms of a convolution as

$$(\delta * f)(t) = f(t), \tag{51}$$

so that δ is the unit element with respect to the operation of convolution. Further, there are the *scaling property*, and the connection of $U(t)$ with $\delta(t)$,

$$\delta(at) = \frac{1}{|a|}\delta(t), \quad \frac{d}{dt}U(t) = \delta(t),$$

where the derivative has to be interpreted in the distributional sense. Also there is the property

$$g(t)\delta(t-a) = g(a)\delta(t-a) \tag{52}$$

if $g(t)$ is a continuous and bounded function.

The Fourier transform (Fig. 10) of $\delta(t)$ follows from (49) and (18), namely

$$\langle \delta^\wedge, \phi \rangle = \langle \delta, \phi^\wedge \rangle = \phi^\wedge(0) = \int_{-\infty}^{\infty} \phi(u)du = \langle 1, \phi \rangle,$$

so that $\delta(t) \longleftrightarrow 1$ or, more generally,

$$\delta(t-T) \longleftrightarrow e^{-jvT}. \tag{53}$$

As to the Fourier transform pair for sgn(t), it is given by

$$\operatorname{sgn}(t) \longleftrightarrow -2j\,\mathrm{PV}\frac{1}{v}, \tag{54}$$

where PV $1/t$ denotes the distribution generated by Cauchy's principal value of the corresponding integral. Indeed, the Fourier transform of PV $1/t$ can be evaluated, at least formally, to be (see [43, p. 310], [119, pp. 224, 343, 344]),

$$\begin{aligned} \mathrm{PV}\int_{-\infty}^{\infty} \frac{e^{-jvt}}{t}dt &= \lim_{\varepsilon\to 0+}\lim_{R\to\infty}\int_{\varepsilon\le|t|\le R}\frac{e^{-jvt}}{t}dt \\ &= \lim_{\varepsilon\to 0+}\lim_{R\to\infty}\left\{-2j\operatorname{sgn}(v)\int_{\varepsilon|v|}^{R|v|}\frac{\sin y}{y}dy\right\} = -\pi j\operatorname{sgn}(v), \end{aligned}$$

and (54) now follows from (19). As to the Fourier transform of $U(t)$

$$U(t) \longleftrightarrow -j\,\mathrm{PV}\frac{1}{v} + \pi\delta(v), \tag{55}$$

recalling $U(t) = (1/2)(\operatorname{sgn}(t) + 1)$ and (53), (54), (19).

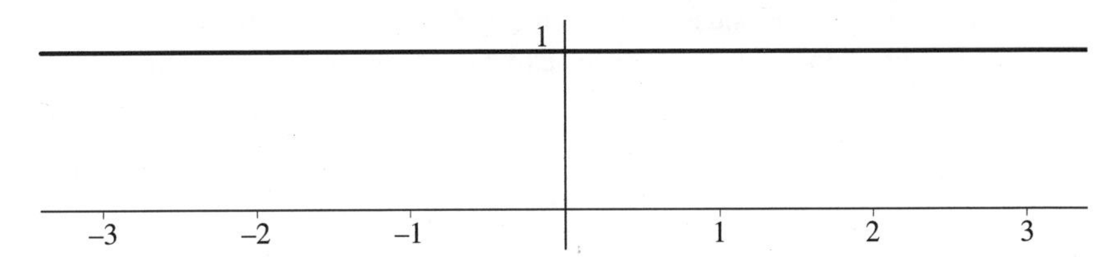

Figure 10. Fourier transform of the delta function.

(f) A pulse train (Fig. 11) which is fundamental to the study of sampling, consisting of an infinite sequence of equidistant pulses $\delta(t+k)$, sometimes called the *comb function* or *Shah function* (functional), often also denoted by III(t) (cf. [23, p. 77 ff.]), is

$$\operatorname{comb}(t) := \sum_{k=-\infty}^{\infty} \delta(t-k),$$

or, with a scaling factor $T > 0$,

$$\operatorname{comb}\left(\frac{t}{T}\right) = T \sum_{k=-\infty}^{\infty} \delta(t-kT), \tag{56}$$

which means that for $\phi \in \mathcal{S}$,

$$\left\langle \operatorname{comb}\left(\frac{t}{T}\right), \phi(t) \right\rangle = T \sum_{k=-\infty}^{\infty} \phi(kT). \tag{57}$$

On the other hand, one has

$$\left\langle \sum_{k=-\infty}^{\infty} e^{jk2\pi t/T}, \phi(t) \right\rangle = \sum_{k=-\infty}^{\infty} \phi^{\wedge}\left(\frac{2k\pi}{T}\right), \tag{58}$$

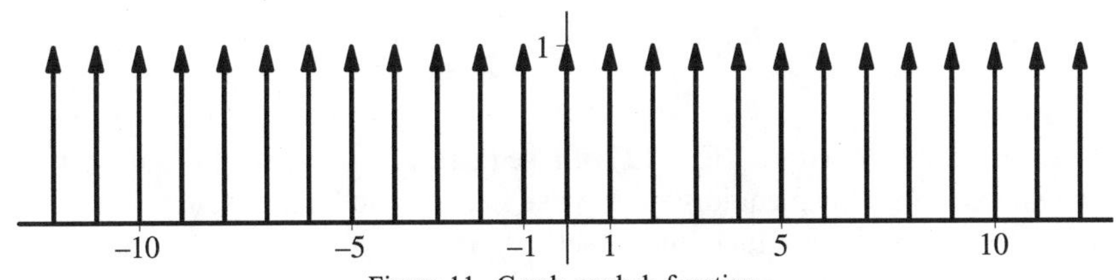

Figure 11. Comb or shah function.

and since the right-hand sides of (57) and (58) coincide by Poisson's formula (30), we end up with the Fourier series expansion for $\mathrm{comb}(t/T)$, namely,

$$\mathrm{comb}\left(\frac{t}{T}\right) = \sum_{k=-\infty}^{\infty} e^{jk2\pi t/T}. \tag{59}$$

Whereas multiplication of a signal $f(t)$ by $\mathrm{comb}(t/T)$ effectively samples it at intervals of length T,

$$f(t)\frac{1}{T}\mathrm{comb}\left(\frac{t}{T}\right) = \sum_{k=-\infty}^{\infty} f(kT)\delta(t-kT), \tag{60}$$

noting (52), the convolution of f with comb results in the *replicating* property

$$T\,\mathrm{comb}(tT) * f(t) = \sum_{k=-\infty}^{\infty} f\left(t-\frac{k}{T}\right), \tag{61}$$

recalling (51). The convolution in (61) is well-defined as an element of $\mathcal{S}'$, e.g., if f is locally integrable with compact support.

The Fourier transform of $\mathrm{comb}(t)$, which follows from (53), is given by

$$\mathrm{comb}\left(\frac{t}{T}\right) \longleftrightarrow T\sum_{k=-\infty}^{\infty} e^{-jvkT} \quad (T>0),$$

and, by using representation (59), one obtains immediately that (cf. [239, pp. 51, 189])

$$\mathrm{comb}\left(\frac{t}{T}\right) \longleftrightarrow T\,\mathrm{comb}\left(\frac{vT}{2\pi}\right), \tag{62}$$

or, in terms of the delta distribution (cf. (56)),

$$T\sum_{k=-\infty}^{\infty}\delta(t-kT) \longleftrightarrow 2\pi\sum_{k=-\infty}^{\infty}\delta\left(v-\frac{2\pi k}{T}\right).$$

Thus the Fourier transform (Fig. 12) of a sequence of equidistant δ-pulses in the time domain is again an equidistant δ-pulse sequence in the frequency domain. In other words, the comb function is selfreciprocal with respect to the Fourier transform, except for some normalization factors.

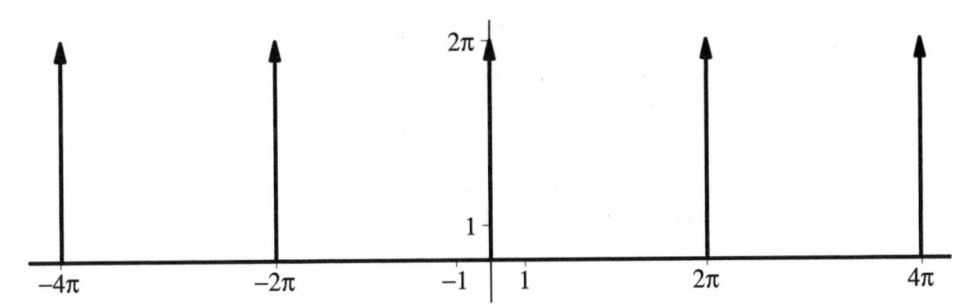

Figure 12. Fourier transform of the comb^$(v) = \mathrm{comb}(\frac{v}{2\pi})$.

Let us close this section with the engineering textbook proof of the Poisson summation formula. Observing (61) and (59),

$$2T\sum_{k=-\infty}^{\infty} g(t+2kT) = g(t) * \mathrm{comb}\left(\frac{t}{2T}\right) = g(t) * \sum_{k=-\infty}^{\infty} e^{jk\pi t/T},$$

and noting that $g(t) * e^{jk\pi t/T} = G(k\pi/T)e^{jk\pi t/T}$, one obtains (30). For the matter from a mathematical point of view see e.g., [43, pp. 201, 202], [134, p. 253].

For the foregoing subsection the reader may also consult [147, pp. 13–21, 23, pp. 51–84], [167, pp. 18–25], [143, pp. 32–38].

2.3. The Sampling Theorem Itself

2.3.1. Two Simple Proofs and an Engineering Demonstration

The sampling theorem, as it is attributed to E. T. Whittaker [230], V. A. Kotel'nikov [128], C. E. Shannon [198], and many others, states:

Each signal function f that is band-limited to $[-\Omega, \Omega]$ for some $\Omega > 0$, i.e., f is square integrable (finite energy) and contains no frequencies higher than Ω, thus

$$f(t) = \frac{1}{2\pi}\int_{-\Omega}^{\Omega} F(u)e^{jut}du \quad (t \in \mathbb{R}), \tag{63}$$

can be completely reconstructed from its sampled values $f(k\pi/\Omega)$ taken at the instants $k\pi/\Omega$, $k \in \mathbb{Z}$, equally spaced apart on $\mathbb{R}$, in terms of

$$f(t) = S_\Omega f(t) := \sum_{k=-\infty}^{\infty} f\left(\frac{k\pi}{\Omega}\right)\mathrm{sinc}\left(\frac{\Omega t}{\pi} - k\right) \quad (t \in \mathbb{R}). \tag{64}$$

The series $S_\Omega f$ converges absolutely and uniformly for $t \in \mathbb{R}$.

First we mention that (64) is valid not only for $f \in B^2_\Omega$ but even for $f \in B^p_\Omega$, $1 \le p < \infty$, as is shown by the following proof, which is based upon the commutativity (38) of the semi-discrete convolution product, which in turn is a consequence of the particular case (36) of Poisson's summation formula (29), (30). Indeed, since $\operatorname{sinc}(\Omega t/\pi) \in B^q_\Omega$, $1 < q \le \infty$, one can apply (38) with $f_1(t) = f(t)$ and $f_2(t) = \operatorname{sinc}(\Omega t/\pi)$ to deduce

$$S_\Omega f(t) = \sum_{k=-\infty}^{\infty} f\left(\frac{k\pi}{\Omega}\right) \operatorname{sinc}\left(\frac{\Omega}{\pi}\left(t - \frac{k\pi}{\Omega}\right)\right)$$
$$= \sum_{k=-\infty}^{\infty} f\left(t - \frac{k\pi}{\Omega}\right) \operatorname{sinc}\left(\frac{\Omega}{\pi} \cdot \frac{k\pi}{\Omega}\right) = f(t). \tag{65}$$

The latter equality is valid in view of the interpolation property (6), i.e., $\operatorname{sinc}(k) = \delta_{k,0}$, and the absolute and uniform convergence of the series follows as in (39) from (33).

Considered formally, equation (65) is probably the simplest "demonstration" of the sampling theorem. The interpolation property (6), which is basic in the proof given above, also shows that the sampling series (64) itself interpolates f at the nodes $t = k\pi/\Omega$, regardless of the conditions satisfied by f.

The infinite series in (64) can be regarded as an infinite Riemann sum of the Fourier inversion integral (singular integral of Dirichlet) (48), which can be rewritten in case of band-limited functions without the limit, namely,

$$f(t) = \int_{-\Omega}^{\Omega} F(v) e^{jvt} dv = \frac{\Omega}{\pi} \int_{-\infty}^{\infty} f(u) \operatorname{sinc}\left(\frac{\Omega}{\pi}(t-u)\right) du. \tag{66}$$

In the following we will present seven further (different) proofs of (64), most of them applying to a restricted range of p only. Let us begin with the standard engineering textbook proof, namely the "delta method"; see e.g., [23, Chapter 10], [109, pp. 13, 14], [143, Section 3.1], [147, Section 3.2.1], [125, 157]. Sampling consists of reconstructing a signal function $f(t)$ just by knowing its values $f(kT)$ taken at whole multiples of the time interval T, called the *sampling period*.

For this purpose consider the T-periodic function

$$s(t) := \sum_{k=-\infty}^{\infty} f(kT)\delta(t - kT) = f(t)\frac{1}{T}\operatorname{comb}\left(\frac{t}{T}\right) \tag{67}$$

noting (60). Here the information about $f(t)$ is conserved only at the sampling points $t = kT$. To recover the intervening values, which do not appear in $s(t)$, this suggests that f itself can be represented as

$$f(t) = \sum_{k=-\infty}^{\infty} f(kT)g(t - kT) \tag{68}$$

where the reconstruction function g (independent of f) as well as the sampling period T have to be determined, provided f is band-limited to $[-\Omega, \Omega]$, say. Since, by (50),

$$f(kT)g(t - kT) = \int_{-\infty}^{\infty} f(kT)\delta(u - kT)g(t - u)du,$$

(68) can be rewritten as,

$$f(t) = (s * g)(t). \tag{69}$$

Let us now examine what happens in the frequency domain when we sample. Here one has in view of (67), (23), (62) and (61),

$$S(v) = F(v) * \frac{1}{2\pi}\text{comb}\left(\frac{Tv}{2\pi}\right) = \frac{1}{T}\sum_{k=-\infty}^{\infty} F\left(v - \frac{2k\pi}{T}\right). \tag{70}$$

Now, since $S(v)$ is periodic with period $2\pi/T$, the right-hand side of (70) is a so-called spectrum repetition of f. If we choose $T \leq \pi/\Omega$, these repetitions do not overlap and we have a sequence of copies of the spectrum of f along the real axis. In order to single out just one copy we apply a low-pass filter with transfer function $T\Pi(Tv/2\pi)$ or, in mathematical language, we multiply both sides of (70) by $T\Pi(Tv/2\pi)$. Hence we obtain

$$S(v) \cdot T\Pi\left(\frac{Tv}{2\pi}\right) = F(v). \tag{71}$$

Together with (22) and (47) this yields

$$f(t) = s(t) * \text{sinc}\left(\frac{t}{T}\right). \tag{72}$$

By comparing (69) with (72) the function g is given by $g(t) = \mathrm{sinc}(t/T)$. Hence it follows in view of (67) and (51) that

$$f(t) = s(t) * \mathrm{sinc}\left(\frac{t}{T}\right) = \sum_{k=-\infty}^{\infty} f(kT)\delta(t - kT) * \mathrm{sinc}\left(\frac{t}{T}\right)$$
$$= \sum_{k=-\infty}^{\infty} f(kT)\mathrm{sinc}\left(\frac{1}{T}(t - kT)\right). \tag{73}$$

As indicated above, the largest possible sampling period T to avoid overlapping is $T = T_N := \pi/\Omega$. In the limiting case $T = T_N$ the sampling frequency or sampling rate $1/T_N = \Omega/\pi$ is called the Nyquist rate. Then $g(t) = (\sin \Omega t)/(\Omega t) = \mathrm{sinc}(\Omega t/\pi)$, and (73) is the desired (64), completing the proof.

Figure 13 demonstrates the essential steps of this proof in a geometrical fashion. The pictures on the right-hand side show the Fourier transforms of the functions or distributions on the left. Sampling is first performed in time space by multiplying the function $f(t)$ of line (a) by $(1/T)\,\mathrm{comb}(t/T)$ of line (b) to yield the function $s(t)$ of line (c), recall (67). The corresponding transform $S(v)$ is the above mentioned spectrum repetition, cf. (70). As long as we sample above the Nyquist rate, i.e. $T \leq T_N := \pi/\Omega$, these repetitions do not overlap, but there may be 'gaps' in between, as shown in the right-hand picture of line (c). The function $f(t)$ is regained by multiplying $S(v)$ of line (c) by the transform of $\mathrm{sinc}(t/T)$, namely by $T\Pi(Tv/2\pi)$ of line (d), in frequency space. This singles out exactly one copy of the spectrum $F(v)$, cf. (71), and we obtain $f(t)$ of line (a) by the inverse Fourier transform. The situation of this stage in case T is strictly less than T_N is also referred to as *over-sampling*.

If we sample at the Nyquist rate $1/T = 1/T_N = \Omega/\pi$, then the 'gaps' between the copies of the spectrum of $f(t)$ in line (c) vanish, and lines (c) and (d) of Fig. 13 change to the corresponding lines (c)′ and (d)′ in Fig. 14. Again, multiplying lines (c)′ and (d)′ in frequency domain singles out one copy of $F(v)$, and $f(t)$ can be regained as above.

Figure 15 demonstrates sampling below the Nyquist rate, also known *under-sampling*. Here the copies of $F(v)$ overlap as indicated by the dashed line in (c)″. This prevents an exact recovering of $f(t)$, and leads to the so-called *aliasing error*, which arises if the sampling time T is incompatible with the band-region of the function $f(t)$, in particular, if $f(t)$ is non-band-limited (cf. Section 2.7.2).

Let us give another proof, one based on a connection between the Fourier transform of a (complex) valued function f, defined by (15), or (16), and the *Fourier coefficients* (or finite *Fourier transform*) of a 2Ω-periodic function f on $\mathbb{R}$, given by (26), the associated Fourier series of this f being

$$f(t) \sim \sum_{k=-\infty}^{\infty} [f]_{\Omega}^{\wedge}(k)e^{jk\pi t/\Omega}. \tag{74}$$

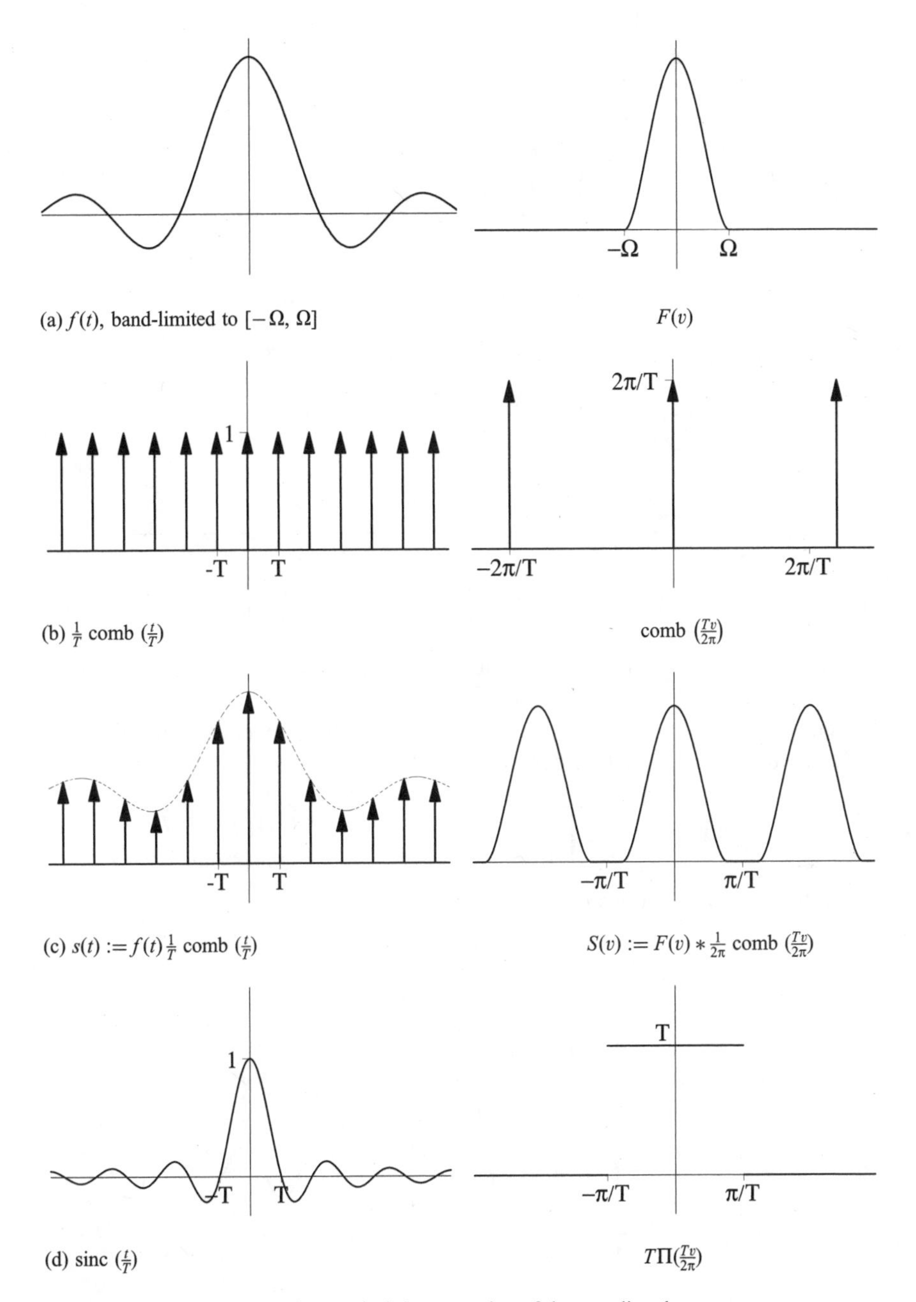

(a) $f(t)$, band-limited to $[-\Omega, \Omega]$ $F(v)$

(b) $\frac{1}{T}$ comb $(\frac{t}{T})$ comb $(\frac{Tv}{2\pi})$

(c) $s(t) := f(t)\frac{1}{T}$ comb $(\frac{t}{T})$ $S(v) := F(v) * \frac{1}{2\pi}$ comb $(\frac{Tv}{2\pi})$

(d) sinc $(\frac{t}{T})$ $T\Pi(\frac{Tv}{2\pi})$

Figure 13. Geometrical demonstration of the sampling theorem.

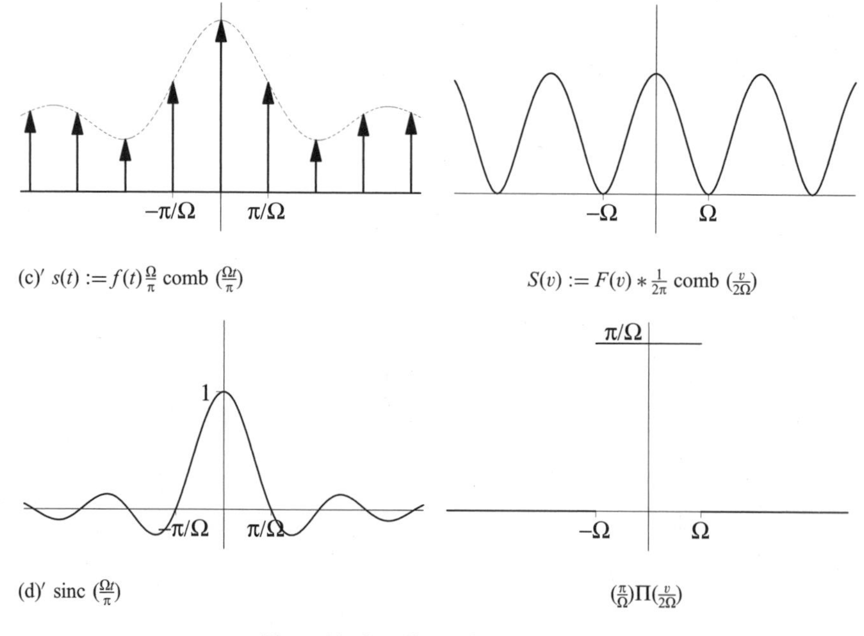

Figure 14. Sampling at the Nyquist rate.

Indeed, consider the Fourier series of the 2Ω-periodic extension of e^{jvt} from $(-\Omega, \Omega]$ to the whole axis $\mathbb{R}$, denoted by $g(v)$. Then

$$[g]^{\wedge}_{\Omega}(k) = \frac{1}{2\Omega}\int_{-\Omega}^{\Omega} e^{jvt}e^{-jk\pi v/\Omega}dv = \operatorname{sinc}\left(\frac{\Omega t}{\pi} - k\right), \tag{75}$$

so that g has Fourier series expansion

$$g(v) = \sum_{k=-\infty}^{\infty} \operatorname{sinc}\left(\frac{\Omega t}{\pi} - k\right)e^{jk\pi v/\Omega} \quad a.e. \tag{76}$$

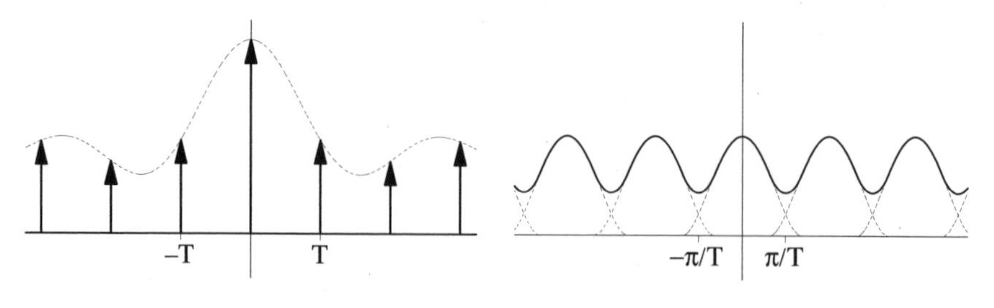

(c)″ $s(t) := f(t)\frac{1}{T}\operatorname{comb}\left(\frac{t}{T}\right)$ $S(v) := F(v) * \frac{1}{2\pi}\operatorname{comb}\left(\frac{Tv}{2\pi}\right)$

Figure 15. Sampling below the Nyquist rate — Under-sampling.

Here $\sim$ of (25) can be replaced be equality a.e., since g is of bounded variation. Moreover, the partial sums of the infinite series are uniformly bounded (cf. [241, pp. 57, 90]). Hence the series expansion can be multiplied on both sides by $F(v)$ and integrated term by term over $(-\Omega, \Omega)$ so that

$$\int_{-\Omega}^{\Omega} g(v)F(v)dv = \int_{-\Omega}^{\Omega} e^{jvt}F(v)dv$$
$$= \sum_{k=-\infty}^{\infty} \operatorname{sinc}\left(\frac{\Omega t}{\pi} - k\right) \int_{-\Omega}^{\Omega} e^{jk\pi v/\Omega}F(v)dv. \tag{77}$$

But according to the Fourier inversion theorem (17), noting that f is band-limited to $[-\Omega, \Omega]$, f has the representation (63). So the integrals on the left-hand side of (77) have the value $2\pi f(t)$ and those on the right-hand side equal $2\pi f(k\pi/\Omega)$. This proves (64) again.

2.3.2. Four Further Proofs

In textbooks on communication theory one often establishes the sampling theorem by expanding the Fourier transform F of a band-limited signal f into its 2Ω-periodic Fourier series, noting (74) and (63),

$$F(v) \sim \sum_{k=-\infty}^{\infty} \left\{\frac{1}{2\Omega}\int_{-\Omega}^{\Omega} F(u)e^{-jk\pi u/\Omega}du\right\} e^{jk\pi v/\Omega} = \sum_{k=-\infty}^{\infty} \frac{\pi}{\Omega} f\left(\frac{-k\pi}{\Omega}\right) e^{jk\pi v/\Omega}.$$

Then one multiplies both sides by $(1/2\pi)e^{jvt}$, integrates from $-\Omega$ to Ω and uses (63) to deduce

$$f(t) = \frac{1}{2\pi}\int_{-\Omega}^{\Omega} e^{jvt}F(v)dv = \sum_{k=-\infty}^{\infty} \frac{1}{2\Omega} f\left(\frac{-k\pi}{\Omega}\right) \int_{-\Omega}^{\Omega} e^{jv(t+k\pi/\Omega)}dv$$
$$= \sum_{k=-\infty}^{\infty} f\left(\frac{-k\pi}{\Omega}\right) \operatorname{sinc}\left(\frac{\Omega t}{\pi} + k\right). \tag{78}$$

The equality sign in (78) is easily justified, see e.g., [241, p. 160].

Another method of establishing the sampling theorem is by means of the periodic version of the *generalized Parseval formula* (cf. [43, p. 175]), which is related to (24) (see e.g. [35]): *If* $f_1, f_2 \in L^2_{2\Omega}$, *then*

$$\frac{1}{2\Omega}\int_{-\Omega}^{\Omega} f_1(u)\overline{f_2(u)}du = \sum_{k=-\infty}^{\infty} [f_1]^{\wedge}_{\Omega}(k)\ \overline{[f_2]^{\wedge}_{\Omega}(k)}.$$

Applying it to the functions $f_1 = F, f_2 = g$ as in formula (75), one has

$$\frac{1}{2\Omega}\int_{-\Omega}^{\Omega} F(v)e^{jvt}dv = \sum_{k=-\infty}^{\infty}\left\{\frac{1}{2\Omega}\int_{-\Omega}^{\Omega} F(v)e^{-jk\pi v/\Omega}dv\right\}\text{sinc}\left(\frac{\Omega t}{\pi} - k\right).$$

This yields again (64), noting (63).

In the foregoing proof the theory of Fourier series with respect to the 2Ω-periodical orthogonal system $\{e^{-jk\pi v/\Omega}\}_{k\in\mathbb{Z}}$ was applied to $F = f^\wedge$. But it is also possible to regard the sampling series as a Fourier series with respect to the orthogonal system $\{\sqrt{\Omega/\pi}\,\text{sinc}(\Omega t/\pi - k)\}_{k\in\mathbb{Z}}$. Indeed, noting that $\{(2\Omega)^{-1/2}\Pi(v/2\Omega)e^{-jk\pi v/\Omega}\}_{k\in\mathbb{Z}}$ is a complete orthonormal system in $L^2(\Omega,\Omega)$, it follows from Parseval's formula (24) and (47) that $\{\sqrt{\Omega/\pi}\,\text{sinc}(\Omega t/\pi - k)\}_{k\in\mathbb{Z}}$ is a complete orthonormal system on B^2_Ω (cf. e.g. [102], [216, pp. 76 ff.], [236, pp. 6 ff.]). Furthermore, the associated Fourier coefficients of $f \in B^2_\Omega$ are given by, using (24) and (47) again,

$$\sqrt{\frac{\Omega}{\pi}}\int_{-\infty}^{\infty} f(u)\text{sinc}\left(\frac{\Omega u}{\pi} - k\right)du = \sqrt{\frac{\pi}{\Omega}}\frac{1}{2\pi}\int_{-\Omega}^{\Omega} F(v)e^{jk\pi v/\Omega}dv = \sqrt{\frac{\pi}{\Omega}}f\left(\frac{k\pi}{\Omega}\right).$$

So the general theory of orthogonal expansions in Hilbert spaces (e.g. [216, pp. 86 ff.]) can be employed, B^2_Ω being such a space, to give

$$\lim_{N\to\infty}\left\| f(t) - \sum_{k=-N}^{N} f\left(\frac{k\pi}{\Omega}\right)\text{sinc}\left(\frac{\Omega t}{\pi} - k\right)\right\|_{L^2(\mathbb{R})} = 0. \tag{79}$$

So far we have shown that the sampling series converges in the norm of $L^2(\mathbb{R})$ towards f. The fact that it converges also uniformly follows from (33) or from the theory of Hilbert spaces with reproducing kernels (cf. [109, p. 29, 236, pp. 15 ff.]). It further follows that the series in (79) give the best approximation to f among all "polynomials" $\sum_{k=-N}^{N}\gamma_k\,\text{sinc}(\Omega t/\pi - k)$ in the norm of $L^2(\mathbb{R})$. So the sampling series results too from a certain minimum problem, the functions $\text{sinc}(\Omega t/\pi - k)$ being given and the coefficients γ_k having to be determined.

We will finally derive the sampling theorem from an associated minimum problem. Define a truncation error by

$$T_N f(t) := \left| f(t) - \sum_{k=-N}^{N} f\left(\frac{k\pi}{\Omega}\right)s_k(t,\Omega)\right|$$

where s_k stands for a "sampling functions" which are to be determined. Using the Fourier inversion formula (63) at the points t and $k\pi/\Omega$, respectively, then by the Cauchy-Schwarz inequality,

$$T_N f(t) := \left| \frac{1}{2\pi} \int_{-\Omega}^{\Omega} F(v) \left[e^{jvt} - \sum_{k=-N}^{N} e^{jk\pi v/\Omega} s_k(t, \Omega) \right] dv \right|$$

$$\leq \frac{1}{2\pi} \left\{ \int_{-\Omega}^{\Omega} |F(v)|^2 dv \right\}^{1/2} \cdot \left\{ \int_{-\Omega}^{\Omega} \left| e^{jvt} - \sum_{k=-N}^{N} e^{jk\pi v/\Omega} s_k(t, \Omega) \right|^2 dv \right\}^{1/2}.$$

Now since $\{e^{jk\pi v/\Omega}\}_{k \in \mathbb{Z}}$ forms a complete orthogonal system in $L^2_{2\Omega}$, $T_N f(t)$ will be minimized and tends to zero for $N \to \infty$ provided the $s_k(t, \Omega)$ are chosen to be the Fourier coefficients of the 2Ω-periodic extension of e^{jvt}. But these coefficients are given by (75), and we end up with (64) again.

In the above we have proved that the sampling series $S_\Omega f$ of $f \in B^2_\Omega$ converges towards f uniformly as well as in $L^2(\mathbb{R})$-norm. An analogous result is true for $f \in B^p_\Omega$, $1 < p < \infty$, i.e.,

$$\lim_{N \to \infty} \left\| f(t) - \sum_{k=-N}^{N} f\left(\frac{k\pi}{\Omega}\right) \operatorname{sinc}\left(\frac{\Omega t}{\pi} - k\right) \right\|_{L^p(\mathbb{R})} = 0.$$

For a proof which is based on the boundedness of a semi-discrete Hilbert transform see [78].

Entirely different proofs of the sampling theorem (64) make use of complex function theory methods, in particular, residue theory. One of these based on a theorem of Cauchy (1827) is presented in Section 2.10; for others see e.g. [45, 106, 118], [234, pp. 114 ff.]). This approach even allows one to establish (64) for $f \in B^\infty_\sigma$ provided $\sigma < \Omega$.

Observe that the sampling expansion may also be written in the form

$$f(t) = \sum_{k=-\infty}^{\infty} f\left(\frac{k\pi}{\Omega} + \tau\right) \operatorname{sinc}\left(\frac{\Omega}{\pi}(t - \tau) - k\right) \tag{80}$$

where τ is an arbitrary real constant. Indeed, (80) follows by an application of (64) to $f(t + \tau)$, which is also a band-limited function since $[f(\cdot + \tau)] \wedge (v) = e^{j\tau v} F(v)$, and then replacing t by $t - \tau$. This means that the samples are not taken at the instants $k\pi/\Omega$, but at points shifted by τ.

There exists another representation for band-limited functions in term of samples. It is the so-called *Valiron interpolation series* of 1925 also called

Tschakaloff's series (cf. [104], [109, pp. 60 ff.], [193, p. 137, 206], [237, p. 31]), namely

$$f(t) = \{tf'(0) + f(0)\}\mathrm{sinc}\left(\frac{\Omega t}{\pi}\right) + \sum_{k=-\infty}^{\infty}{}' f\left(\frac{k\pi}{\Omega}\right)\frac{\Omega t}{k}\mathrm{sinc}\left(\frac{\Omega t}{\pi} - k\right) \quad (t \in \mathbb{R}), \tag{81}$$

the prime indicating that the term $k = 0$ is missing. This formula, which can be deduced by applying (64) to $g(t) := \{f(t) - f(0)\}/t$, is valid under weaker assumptions, namely that the function f, band-limited to $[-\Omega, \Omega]$, is merely bounded on $\mathbb{R}$ (whence $g(t)$ is square integrable over $\mathbb{R}$). The advantage here is that the additional factor $\Omega t/k$ increases the speed of convergence of the partial sums of this infinite series. However, one item about derivative information, $f'(0)$, is also required for the reconstruction of f. This representation can also be deduced by the above mentioned theorem of Cauchy; see Section 2.10.

For $\Omega = \pi$ one may rewrite (81) in the form (cf. [241, II, p. 275, [222])

$$\begin{aligned} f(t) &= f(0) + \frac{\sin \pi t}{\pi}\left\{f'(0) + \sum_{k=-\infty}^{\infty}{}'(-1)^k t\frac{f(k) - f(0)}{k(t-k)}\right\} \\ &= \frac{\sin \pi t}{\pi}\left\{f'(0) + \frac{f(0)}{t} + \sum_{k=-\infty}^{\infty}{}'(-1)^k f(k)\left(\frac{1}{t-k} + \frac{1}{k}\right)\right\}. \end{aligned}$$

The latter equality follows from the well-known *partial fraction* expansion

$$\frac{\pi}{\sin \pi z} = \frac{1}{z} + \sum_{k=-\infty}^{\infty}{}'(-1)^k\left(\frac{1}{z-k} + \frac{1}{k}\right) = \frac{1}{z} + z\sum_{k=-\infty}^{\infty}{}'\frac{(-1)^k}{k(z-k)}.$$

In the above it was always assumed that the signal function f is band-limited to some interval $[-\Omega, \Omega]$, say. This can be generalized to f being band-limited to some set $A \subset \mathbb{R}$, i.e., the Fourier transform of f vanishes outside A. The set A has to be measurable and to satisfy the *disjoint translates* condition

$$A \cap (A + 2k\Omega) = \emptyset \qquad (k \in \mathbb{Z}\backslash\{0\})$$

for some $\Omega > 0$. In this case the sampling series expansion of f is given by

$$f(t) = \frac{\pi}{\Omega}\sum_{k=-\infty}^{\infty} f\left(\frac{k\pi}{\Omega}\right)\chi_A^{\vee}\left(t - \frac{k\pi}{\Omega}\right) \qquad (t \in \mathbb{R}), \tag{82}$$

where χ_A is the characteristic function of the set A, i.e., $\chi_A(t) := 1$ for $t \in A$ and $\chi_A(t) := 0$ otherwise, and $^{\vee}$ denotes the inverse Fourier transform (17); see [9, 69,

100, 138, 214]. For $A = [-\Omega, \Omega]$ expansion (82) reduces to (64). The disjoint translates condition is sufficient for sampling with band-region A, and necessary for sampling points with arithmetic progression, but it is not always necessary if one allows more general sets of sampling points; see [109, Chapter 13].

The type of sampling considered above was carried out in the time domain. There also exists so-called *frequency sampling*:

If $f(t) = 0$ for $|t| > T$, then its Fourier transform $F(v)$ can be uniquely determined from its values $F(k\pi/T)$ at the sequence of equidistant points $k\pi/T$ in terms of

$$F(v) = \sum_{k=-\infty}^{\infty} F\left(\frac{k\pi}{T}\right) \operatorname{sinc}\left(\frac{tT}{\pi} - k\right).$$

For the proof observe that F has the representation $F(v) = \int_{-T}^{T} f(-u) e^{jvu} du$ for $v \in \mathbb{R}$, i.e., F is band-limited to $[-T, T]$ (cf. (63)). Now apply the sampling theorem (64) to F; see e.g. [143, Section 3.2], [167, p. 52].

2.4. Sampling Representations for Derivatives and Hilbert Transforms

In this section we present sampling series expansions of derivatives $f^{(r)}$ and the Hilbert transform $\tilde{f}$ in terms of samples of f only. Such representations for $f^{(r)}$ can be obtained from (64) by termwise differentiation, indeed: *Under the hypotheses of the sampling theorem one has, uniformly in $t \in \mathbb{R}$,*

$$f^{(r)}(t) = \sum_{k=-\infty}^{\infty} f\left(\frac{k\pi}{\Omega}\right) \left(\frac{d}{dt}\right)^r \operatorname{sinc}\left(\frac{\Omega t}{\pi} - k\right). \tag{83}$$

The interchange of summation and differentiation can be justified by observing that the series in (83) is uniformly convergent in view of (32) and (33). For the first derivative f' $(r = 1)$ this leads to the sampling representation

$$f'(t) = \Omega \sum_{k=-\infty}^{\infty} f\left(\frac{k\pi}{\Omega}\right) \left\{ \frac{\cos(\Omega t - k\pi)}{\Omega t - k\pi} - \frac{\sin(\Omega t - k\pi)}{(\Omega t - k\pi)^2} \right\} \tag{84}$$

valid uniformly in $t \in \mathbb{R}$.

A different proof of (84) works with the Fourier inversion formula for the derivative, namely,

$$f'(t) = \frac{1}{2\pi}\int_{-\Omega}^{\Omega}(jv)e^{jvt}F(v)dv, \tag{85}$$

and proceeds as above by expanding the function $(jv)\exp(jvt)$ on $(-\Omega, \Omega)$ into a Fourier series.

R. P. Boas in his book [19, p. 211] established a different series representation for f'. He rewrote $\exp(jvt)$ in (85) as the product $\exp(jv(t+\pi)/(2\Omega))\exp(-jv\pi/2\Omega)$, and expanded $(jv)\exp(-jv\pi/2\Omega)$ into a Fourier series. This led him to the identity

$$f'(t) = \frac{4\Omega}{\pi^2}\sum_{k=-\infty}^{\infty} f\left(t + \frac{(2k+1)\pi}{2\Omega}\right)\frac{(-1)^k}{(2k+1)^2}. \tag{86}$$

Note that (86) could also be deduced by applying (84) with $t = -\pi/2\Omega$ to $f(x+\cdot)$ and replacing x by $t + \pi/2\Omega$.

A similar representation can be obtained by starting with (85) and using the commutativity of the discrete convolution (cf. (38)); this yields (cf. [145])

$$f'(t) = \frac{\Omega}{\pi}\sum_{k=-\infty}^{\infty}{}' f\left(t + \frac{k\pi}{\Omega}\right)\frac{(-1)^{k+1}}{k}. \tag{87}$$

Observe that neither (86) nor (87) are true sampling series of type (64) since the sampling nodes depend on t.

Also the Hilbert transform on the line $\mathbb{R}$, which is defined by

$$\widetilde{f}(t) = \lim_{\eta\to 0+}\frac{1}{\pi}\int_{|u|\geq\eta}\frac{f(t-u)}{u}du \tag{88}$$

and which plays an important role in electrical engineering (see e.g. [23, pp. 267 ff.]) can be sampled in terms of f. Indeed, using the representation (cf. (66) and [43, p. 318])

$$\widetilde{f}(t) = \frac{1}{2\pi}\int_{-\Omega}^{\Omega}F(v)(-j\,\mathrm{sgn}\,v)e^{jvt}dv = f(t) * \frac{\Omega}{\pi}\mathrm{sinc}^{\sim}\left(\frac{\Omega t}{\pi}\right) \tag{89}$$

in the band-limited case, and noting that

$$\operatorname{sinc}^{\sim}\left(\frac{\Omega t}{\pi}\right) = \frac{1-\cos\Omega t}{\Omega t} = \frac{\sin^2 \Omega t/2}{\Omega t/2},$$

one can apply (37) to deduce the expansions

$$\tilde{f}(t) = \sum_{k=-\infty}^{\infty} f\left(\frac{k\pi}{\Omega}\right) \frac{1-\cos(\Omega t - k\pi)}{\Omega t - k\pi} = \sum_{k=-\infty}^{\infty} f\left(\frac{k\pi}{\Omega}\right) \frac{\sin^2\left(\frac{1}{2}(\Omega t - k\pi)\right)}{\frac{1}{2}(\Omega t - k\pi)}. \quad (90)$$

The same result can be achieved by expanding $(-j \operatorname{sgn} v)\exp(jvt)$ in the integral in (89) on $(-\Omega, \Omega)$ into its Fourier series expansion, whereas the expansion of $\operatorname{sgn} v$ alone gives

$$\tilde{f}(t) = \sum_{k=-\infty}^{\infty} f\left(t + \frac{(2k+1)\pi}{\Omega}\right) \frac{-2}{(2k+1)\pi}. \quad (91)$$

Note that not only the sampling series (90) but also the Boas-type series (91) can be interpreted as a Riemann sum of an appropriate integral. Thus the representation (90) is the discrete form of the convolution integral (cf. (89))

$$\tilde{f}(t) = f(t) * \frac{\Omega}{\pi} \operatorname{sinc}^{\sim}\left(\frac{\Omega t}{\pi}\right) = \frac{\Omega}{\pi} \int_{-\infty}^{\infty} f(u) \frac{\sin^2\left(\frac{\Omega}{2}(t-u)\right)}{\frac{\Omega}{2}(t-u)} du,$$

and (91) is the discrete form of (88). More details and proofs can be found in [49, 145, 205, 212].

2.5. Multi-Channel Sampling

In most applications it is usually of interest to keep the sampling rate as low as possible. This can be achieved, for example, by taking a sampling series which involves sampled values of the function and its first derivative. In this case it suffices to take samples at every second node $2k/\Omega$:

Under the hypotheses of the sampling theorem one has, uniformly in $t \in \mathbb{R}$,

$$f(t) = \sum_{k=-\infty}^{\infty} \left\{ f\left(\frac{2k\pi}{\Omega}\right) + \left(t - \frac{2k\pi}{\Omega}\right) f'\left(\frac{2k\pi}{\Omega}\right) \right\} \left[\operatorname{sinc} \frac{1}{2}\left(\frac{\Omega t}{\pi} - 2k\right)\right]^2. \quad (92)$$

For the proof one can use the Fourier expansion method used above already several times. One decomposes $\exp(jvt)$ on $(-\Omega, \Omega)$ into (cf. [48, 202])

$$e^{jvt} = \varepsilon_1(v) + jv\varepsilon_2(v) \quad (v \in (-\Omega, \Omega)) \tag{93}$$

whereby

$$\varepsilon_1(v) = \left\{1 - \frac{|v|}{\Omega}\left(1 - e^{-j\Omega t \operatorname{sgn} v}\right)\right\}e^{jvt}$$

$$\varepsilon_2(v) = \frac{j \operatorname{sgn} v}{\Omega}\left(e^{-j\Omega t \operatorname{sgn} v} - 1\right)e^{jvt}$$

for which $\varepsilon_\mu(v) = \varepsilon_\mu(v - \Omega)$ for $v \in (0, \Omega)$, $\mu = 1, 2$. Hence both functions may be expanded in their Ω-periodic boundedly converging Fourier series on $(-\Omega, \Omega)\backslash\{0\}$, namely

$$\varepsilon_1(v) = \sum_{k=-\infty}^{\infty}\left[\operatorname{sinc}\frac{1}{2}\left(\frac{\Omega t}{\pi} - 2k\right)\right]^2 e^{j2k\pi v/\Omega}$$

$$\varepsilon_1(v) = \sum_{k=-\infty}^{\infty}\left[\operatorname{sinc}\frac{1}{2}\left(\frac{\Omega t}{\pi} - 2k\right)\right]^2\left(t - \frac{2k\pi}{\Omega}\right)e^{j2k\pi v/\Omega}.$$

After substituting these expansions into (93) and (63), one now continues along lines similar to those mentioned.

Whatever the nature of f, the series (92) interpolates f at the nodes $2k\pi/\Omega$, and the termwise differentiated series interpolates f' there. This means that (92) gives an interpolation formula of Hermite type.

This type of reconstructing a function f from samples is called *two-channel* or, more generally, *multi-channel* sampling, which should indicate that f has been passed through various processing channels before sampling. In case of (64) one may speak of *single channel* sampling. The idea of sampling f together with its derivative can already be found in Shannon's paper of 1949 [198]. For sampling formulae involving samples not only of f and f' but samples of f and all its derivative up to some order $r \in \mathbb{N}$ see Section 2.10 below.

One can also deduce an expansion of f using samples of f and its Hilbert transform $\tilde{f}$. It is given by

$$f(t) = \sum_{k=-\infty}^{\infty}\left\{f\left(\frac{2k\pi}{\Omega}\right)\cos\left(\frac{1}{2}(\Omega t - 2\pi k)\right)\right.$$
$$\left.-\tilde{f}\left(\frac{2k\pi}{\Omega}\right)\sin\left(\frac{1}{2}(\Omega t - 2\pi k)\right)\right\}\operatorname{sinc}\left(\frac{1}{2}\left(\frac{\Omega t}{\pi} - 2k\right)\right).$$

For a general approach to multi-channel sampling and the related multi-band sampling see, e.g., [109, Chapter 12], and for particular results see [7, 8, 16, 105, 108, 121, 125, 199, 233].

2.6. Sampling Theory for Non-Band-Limited Functions

2.6.1. The General Situation

So far we have exclusively dealt with band-limited functions f. The assumption that a signal f is band-limited is, however, a mathematical idealization. As we have seen above, band-limitation implies that f is an entire function and such functions cannot vanish outside a finite interval unless they are zero everywhere. So there do not exist band-limited signals of finite duration, although both limitations are rather natural for real physical signals. In this respect it seems to be convenient to consider signals, which are approximately band-limited, thus for which the Fourier inversion formula does not hold in the form (63) but only in the form (48) with the limit for $R \to \infty$. If we replace R by the bandwidth Ω this can be interpreted as f being the limit of a sequence of band-limited signals.

Comparing the representations (63) and (64) for band-limited signals with the representation (48) for approximately band-limited signals, one may expect that there holds

$$f(t) = \lim_{\Omega\to\infty} \sum_{k=-\infty}^{\infty} f\left(\frac{k\pi}{\Omega}\right) \operatorname{sinc}\left(\frac{\Omega t}{\pi} - k\right) \quad (t \in \mathbb{R}). \tag{94}$$

In fact, (94) can be proved under various different assumptions upon the signal f. We state here only one result due to P. Weiss [229] and J. L. Brown [24]:

If $f \in C(\mathbb{R}) \cap L^1(\mathbb{R})$ with $F \in L^1(\mathbb{R})$, then

$$f(t) = \sum_{k=-\infty}^{\infty} f\left(\frac{k\pi}{\Omega}\right) \operatorname{sinc}\left(\frac{\Omega t}{\pi} - k\right) + R_\Omega f(t) \tag{95}$$

where the so-called aliasing error $R_\Omega f(t)$ is given by

$$R_\Omega f(t) := \frac{1}{2\pi} \int_{|v|>\Omega} F(v) \left\{ e^{jvt} - \sum_{k=-\infty}^{\infty} \operatorname{sinc}\left(\frac{\Omega t}{\pi} - k\right) e^{jk\pi v/\Omega} \right\} dv \tag{96}$$

$$= \frac{1}{2\pi} \sum_{\mu=-\infty}^{\infty} (1 - e^{-j2\mu t\Omega}) \int_{(2\mu-1)\Omega}^{(2\mu+1)\Omega} F(v) e^{jvt} dv \tag{97}$$

$$\|R_\Omega f\|_C \leq \frac{1}{\pi} \int_{|v|>\Omega} |F(v) dv. \tag{98}$$

In particular, (94) *holds uniformly for* $t \in \mathbb{R}$.

For the proof we use again the Fourier expansion method. We first divide the Fourier inversion integral (17) into two parts, substitute the Fourier series expansion of e^{jvt}, namely (76), into the first part to obtain

$$\begin{aligned} f(t) &= \frac{1}{2\pi}\int_{-\Omega}^{\Omega} F(v)e^{jvt}dv + \frac{1}{2\pi}\int_{|v|>\Omega} F(v)e^{jvt}dv \\ &= \frac{1}{2\pi}\int_{-\Omega}^{\Omega} F(v)\left\{\sum_{k=-\infty}^{\infty} \operatorname{sinc}\left(\frac{\Omega t}{\pi} - k\right)e^{jk\pi v/\Omega}\right\}dv + \frac{1}{2\pi}\int_{|v|>\Omega} F(v)e^{jvt}dv. \end{aligned} \tag{99}$$

Now one has for the first integral on the right,

$$\sum_{k=-\infty}^{\infty}\left\{\frac{1}{2\pi}\int_{-\infty}^{\infty} F(v)e^{jk\pi v/\Omega}dv\right\}\operatorname{sinc}\left(\frac{\Omega t}{\pi} - k\right)$$

$$- \frac{1}{2\pi}\int_{|v|>\Omega} F(v)\left\{\sum_{k=-\infty}^{\infty} \operatorname{sinc}\left(\frac{\Omega t}{\pi} - k\right)e^{jk\pi v/\Omega}\right\}dv \tag{100}$$

$$= \sum_{k=-\infty}^{\infty} f\left(\frac{k\pi}{\Omega}\right)\operatorname{sinc}\left(\frac{\Omega t}{\pi} - k\right)$$

$$- \frac{1}{2\pi}\int_{|v|>\Omega} F(v)\left\{\sum_{k=-\infty}^{\infty} \operatorname{sinc}\left(\frac{\Omega t}{\pi} - k\right)e^{jk\pi v/\Omega}\right\}dv, \tag{101}$$

the last equality being valid in view of the Fourier inversion formula (17) taken at $t = k\pi/\Omega$. Inserting this into (99) yields (96).

As to (97), we split the integral in (101) into a series of integrals over $((2\mu - 1)\Omega, (2\mu + 1)\Omega)$, $\mu \in \mathbb{Z}\backslash\{0\}$, and recall that the infinite series in (101) defines a 2Ω-periodic function which equals $\exp(j(v - 2\mu\Omega)t)$ on each of these intervals. Hence,

$$\int_{|v|>\Omega} F(v)\left\{\sum_{k=-\infty}^{\infty} \operatorname{sinc}\left(\frac{\Omega t}{\pi} - k\right)e^{jk\pi v/\Omega}\right\}dv = \sum_{\mu=-\infty}^{\infty}{}' \int_{(2\mu-1)\Omega}^{(2\mu+1)\Omega} F(v)e^{jvt}dv\ e^{-j2\Omega t},$$

and (97) now follows from (96). Inequality (98) is obvious.

Another proof of this result due to R. P Boas [20] (see also [49] makes use of Poisson's summation formula in order to justify the interchange of summation and integration in (100)–(101). Note that the constant π^{-1} in (98) is best possible, see e.g. [10].

As to the rate of convergence in (94) it follows by an elementary estimate of the integral in (95) (cf. [56]):

If the transform $F(v)$ has a certain rate of decay at infinity, i.e., $F(v) = \mathcal{O}(|v|^{-r-1})$ for $|v| \to \infty$, some $r \in \mathbb{N}$, then f is approximately band-limited in the sense

$$\left| f(t) - \frac{1}{2\pi} \int_{-\Omega}^{\Omega} F(v) e^{jvt} dt \right| = \mathcal{O}(\Omega^{-r}) \quad (t \in \mathbb{R}) \tag{102}$$

for $\Omega \to \infty$, and f has the same rate of approximation by $S_\Omega f(t)$, namely

$$|R_\Omega f(t)| \equiv |f(t) - S_\Omega f(t)| = \mathcal{O}(\Omega^{-r}) \quad (t \in \mathbb{R}). \tag{103}$$

A sufficient condition for the hypothesis of (102) and (103) to hold, namely that $F(v) = \mathcal{O}(|v|^{-r-1})$ for $r \geq 1$ and $v \to \infty$, is that the rth derivative $f^{(r)}$ belongs to the Lipschitz class $\mathrm{Lip}(1; L^1(\mathbb{R})$; recall (41).

The behaviour of the series in (94) for $\Omega \to \infty$ has much in common with the behaviour of the Fourier inversion integral (48) for $R \to \infty$. So it is not surprising that (94) is valid essentially under the same assumptions as for (48). The first result of type (94) seems to be due to Ch.-J. de la Vallée Poussin [224] (see also [54, 57]), who established (94) for functions of bounded variation f (for the definition of bounded variation see [43, pp. 10 ff.], [188, pp. 160 ff.]). This corresponds to Jordan's theorem for the convergence of the Fourier inversion integral.

2.6.2. Duration-Limited Signals

Let us now consider time-limited functions in more detail. Since these functions cannot be band-limited simultaneously, the sampling theorem does not hold in the form (64) but in the form (94) at most.

Let $C_T(\mathbb{R})$ be the space of continuous functions $f: \mathbb{R} \to \mathbb{C}$ vanishing outside the compact interval $[-T, T]$ endowed with the sup-norm. In this case, the sampling series $S_\Omega f$ of (64) reduces to a finite one, namely,

$$S_\Omega f(t) := \sum_{|k| < T\Omega/\pi} f\left(\frac{k\pi}{\Omega}\right) \operatorname{sinc}\left(\frac{\Omega t}{\pi} - k\right) \quad (t \in \mathbb{R}). \tag{104}$$

The operators $S_\Omega: C_T(\mathbb{R}) \to C(\mathbb{R})$ defined in this way are bounded linear operators with operator norm

$$\|S_\Omega\|_{[C_T(\mathbb{R}), C(\mathbb{R})]} = \sup_{t \in \mathbb{R}} \sum_{|k| < T\Omega/\pi} \left| \operatorname{sinc}\left(\frac{\Omega t}{\pi} - k\right) \right| \cong \log \Omega \quad (\Omega \to \infty). \tag{105}$$

This means that the operator norms go to infinity for $n \to \infty$, and the Banach-Steinhaus theorem (see e.g., [43, p. 19]) implies that there exist functions f in $C_T(\mathbb{R})$ for which (94) does not hold in the sense of uniform convergence. So one needs additional assumptions upon f in order to ensure (94). As seen above, $f^\wedge \in L^1(\mathbb{R})$ or f being of bounded variation are sufficient.

In case of time-limited functions it follows easily from (105) and the Banach-Steinhaus theorem with rates (cf. [30, 68]) that (94) holds provided f satisfies a *Dini-Lipschitz condition*, i.e., (cf. (40)),

$$\lim_{\tau \to 0+} \omega(f, C(\mathbb{R}), \tau) \log \tau = 0.$$

The same holds true, if $f(t)$ is not time-limited but has a certain rate of decay at infinity. For those and further results in this respect, e.g., a localization principle, sampling of functions which are analytic in a strip, and pointwise convergence of sampling series, see [28, 47–48, 185, 187, 203–204, 206, 208, 210, 212–213].

Now we rewrite (104) so that it is comparable with well-known results on interpolation of 2π-periodic functions. If we assume that f vanishes outside $[0, 2\pi]$ rather than $[-T, T]$, and choose $\Omega = (2n+1)/2$, then (104) takes on the form

$$S_n f(t) := \sum_{k=0}^{2n} f\left(\frac{2\pi k}{2n+1}\right) \frac{\sin\left(\frac{2n+1}{2}\left(t - \frac{2\pi k}{2n+1}\right)\right)}{\left(t - \frac{2\pi k}{2n+1}\right)} \quad (t \in \mathbb{R}). \tag{106}$$

In comparison, let us recall the trigonometric Lagrange interpolating polynomial of a continuous 2π-periodic function f considered in Section 2.1.1. In general, the fundamental polynomials $\tau_k(t)$ in (4) are given by (cf. [193, Section 5.3], [241, Section X.1], [79])

$$\tau_k(t) = \frac{\Delta_{2n+1}(t)}{2\sin\frac{1}{2}(t - t_k)\Delta'_{2n+1}(t_k)}, \quad \Delta_{2n+1}(t) := \prod_{k=0}^{n} 2\sin\frac{1}{2}(t - t_k). \tag{107}$$

But in the particular case of equidistant nodal points $t_k = 2\pi k/(2n+1)$ for $k = 0, 1, \ldots, 2n$, the $\tau_k(t)$ reduce to $(2n+1)^{-1} D_n(t - t_k)$, where $D_n(t)$ is the trigonometric *Dirichlet kernel*

$$D_n(t) = 1 + 2\sum_{k=1}^{n} \cos kt = \frac{\sin\frac{(n+1)t}{2}}{\sin\frac{t}{2}}. \tag{108}$$

Hence the trigonometric polynomial $T_n f$ of degree n interpolating the function f at the nodal points t_k can be written as

$$T_n f(t) := \frac{1}{2n+1} \sum_{k=0}^{2n} f\left(\frac{2\pi k}{2n+1}\right) \frac{\sin\left(\frac{2n+1}{2}\left(t - \frac{2\pi k}{2n+1}\right)\right)}{\sin\frac{1}{2}\left(t - \frac{2\pi k}{2n+1}\right)} \quad (t \in \mathbb{R}). \tag{109}$$

Note that the $T_n f$ are trigonometric polynomials, thus in particular periodic, whereas the $S_n f$ are not. However, the norm of the operators T_n behaves like $\log n$ for $n \to \infty$ as does the norm of the S_n. Consequently, the $T_n f$ fail to converge to f for $n \to \infty$ unless f has certain smoothness properties; see [28], [193, Section 5.4]. For different connections between sampling and interpolation see M. H. Annaby [3].

2.7. Error Analysis in Sampling

There are several types of errors which might influence the accuracy of the reconstruction of a signal from its sampled values and so impair the practical implementation of sampling theorems. Apart from the *aliasing error* $R_\Omega f$ (cf. (96)) treated in Section 2.6, which arises if f is not exactly band-limited or if its bandwidth is larger than $[-\Omega, \Omega]$ (the situation of "under-sampling"), there are in addition

(i) the *truncation error* $T_N f$, arising if only a finite number N of samples are taken into account instead of the infinitely many of the sampling series. More general is the *information loss error*, occurring if some of the samples are missing;

(ii) the *amplitude error* $Q_\varepsilon f$, arising if the exact sampled values $f(k\pi/\Omega)$ are not at one's disposal but only approximate values, say $\bar{f}(k\pi/\Omega)$, differing from the exact (or correct) values by not more than ε; this falsification may be due to *quantization* (see below), round off error or noise;

(iii) the *time-jitter error* $J_\delta f$, arising if the Nyquist sampling instants $t_k = k\pi/\Omega$ are not met correctly but are perturbed, thus $t_k + \delta_k$, which differ by $|\delta_k| \le \delta$.

All of these errors can, of course, occur in combination.

2.7.1. Errors for Band-Limited Signals

Let us first consider the situation for band-limited signals. Truncation errors, which occur naturally in applications, have been studied rather intensively in the engineering literature. It is given by

$$T_N f(t) = f(t) - \sum_{k=-N}^{N} f\left(\frac{k\pi}{\Omega}\right) \operatorname{sinc}\left(\frac{\Omega t}{\pi} - k\right) = \sum_{|k|>N} f\left(\frac{k\pi}{\Omega}\right) \operatorname{sinc}\left(\frac{\Omega t}{\pi} - k\right), \tag{110}$$

and can be controlled by imposing some extra conditions on f besides being band-limited. A rather common estimate of (110) seems to be that of D. Jagerman [117], given by:

If $f \in B^2_\Omega$ satisfies the additional condition $t^r f(t) \in L^2(\mathbb{R})$, then one has the pointwise estimate $|(T_N f)(t)| = \mathcal{O}(N^{-r-1/2})$ for $N \to \infty$.

If one renounces Hilbert space methods then P. L. Butzer *et al.* [32] showed that: *If $f \in B^1_\Omega$ with $F^{(r)} \in Lip(\alpha; C(\mathbb{R}))$, $0 < \alpha \le 1$, then $|(T_N f)(t)| = \mathcal{O}(N^{-r-\alpha})$ for $N \to \infty$.* A recent improvement of this approach, with uniform bounds, due to Xin Min Li [137] states:

If $f \in B^2_\Omega$ such that $|f(t)| = \mathcal{O}(|t|^{-\gamma})$ for $|t| \to \infty$, some $\gamma > 0$, then $\|(T_N f)(t)\|_C = \mathcal{O}(N^{-\gamma-1}\log N)$ for $N \to \infty$.

For a good treatment of truncation errors see the book by A. Zayed [237, pp. 86–92], also [169, p. 142], [15, 59, 96, 97, 99, 168, 174, 175, 235]. For the recovery of signals if some samples are missing and the corresponding errors see, [81, 82, 84].

In order to study amplitude, quantization, jitter errors and related ones in a unified setting, let us follow the approach by P. L. Butzer and J. Lei [40–41]. Thus let us view the coefficients in a sampling series as a result of measurements taken from the signal. To mathematically describe measured sampled values, we use linear functionals acting on the undergoing signal. Let $\lambda = (\lambda_k)_{k\in\mathbb{Z}}$ be a sequence of continuous linear functionals mapping $C_0(\mathbb{R})$ into $\mathbb{C}$ ($C_0(\mathbb{R})$ = set of functions $g \in C(\mathbb{R})$ for which $g(t) \to 0$ as $|t| \to \infty$). For a function $f \in C_0(\mathbb{R})$, and for $k \in \mathbb{Z}$, $\Omega > 0$, we apply λ_k to $f(\cdot + k\pi/\Omega)$ to obtain measured sampled values to approximate $f(k\pi/\Omega)$, namely $\lambda_k f(\cdot + k\pi/\Omega) \approx f(k\pi/\Omega)$. The quality of the measured sampled values, for each given $\Omega > 0$, may be measured by the error

$$E(f,\lambda) = E_\Omega(f,\lambda) = \sup_{k\in\mathbb{Z}} \left| \lambda_k f\left(\cdot + \frac{k\pi}{\Omega}\right) - f\left(\frac{k\pi}{\Omega}\right) \right|. \tag{111}$$

It is to be expected that under reasonable assumptions upon f and λ the sampling series with the measured sampled values

$$S_{\Omega,\lambda} f(t) := \sum_{k=-\infty}^{\infty} \lambda_k f\left(\cdot + \frac{k\pi}{\Omega}\right) \operatorname{sinc}\left(\frac{\Omega t}{\pi} - k\right) \tag{112}$$

will be a good approximation to $S_\Omega f = f$.

According to the Riesz representation theorem (cf. [43, p. 22], [188, p. 131] [216, pp. 146 ff.]) there exist functions $\mu_k : \mathbb{R} \to \mathbb{C}$ of bounded variation such that $\lambda_k g = \int_\mathbb{R} g(t) d\mu_k(t)$ for $g \in C_0(\mathbb{R})$. Below $\lambda = (\lambda_k)_{k\in\mathbb{Z}}$ is said to be *finitely supported* if $\operatorname{supp}(\lambda) := \cup_{k\in\mathbb{Z}} \operatorname{supp}(\lambda_k) = \cup_{k\in\mathbb{Z}} \operatorname{supp}(\mu_k)$ is compact. The general result here is:

Let $\lambda = (\lambda_k)_{k\in\mathbb{Z}}$ be a sequence of continuous linear functionals having a sufficiently small support. If $f \in C_0(\mathbb{R})$ satisfies the decay condition

$$|f(t)| \leq M_f|t|^{-\gamma} \quad (t \in \mathbb{R} \setminus \{0\}) \tag{113}$$

for a constant M_f (depending on f) and some $\gamma > 0$, then

$$\|f - S_{\Omega,\lambda}f\|_{C(\mathbb{R})} \leq KE(f,\lambda)\log\frac{1}{E(f,\lambda)} \tag{114}$$

provided that $E(f,\lambda) \leq \min\{e^{-1/2}, \pi/\Omega\}$, $\Omega \geq \pi$, and that $E(g,\lambda) < \infty$ for every $g \in C_0(\mathbb{R})$, where K is a constant independent of $E(f,\lambda)$.

For the proof, which is intricate and quite long, the interested reader may consult [41]. The constant K in (114) can be computed in concrete instances as shown in the following examples.

As a **first example**, consider the measured sampled values given by averages of the signal, thus let the sampled values f_k^1 of f be given by the rule

$$f_k^1 = \frac{1}{2\sigma_k}\int_{-\sigma_k}^{\sigma_k} f\left(t + \frac{k\pi}{\Omega}\right)dt, \tag{115}$$

where σ_k are numbers satisfying $0 < \sigma_k \leq \sigma$ for all $k \in \mathbb{Z}$ and some $\sigma > 0$. Then the error $E(f,\lambda)$ can be given in terms of the modulus of continuity $\omega(f, C(\mathbb{R}), \tau) \equiv \omega(f,\tau)$ of f (cf. (40)).

Let the signal $f \in B_\Omega^\infty$ satisfy the decay condition (113), *and let its sampled values be f_k^1 of* (115). *If $\sigma < 1/2$ and $\omega(f,\sigma) \leq \min\{e^{-1}, \pi/\Omega\}$, $\Omega \geq \pi$, then*

$$\left\| f(t) - \sum_{k=-\infty}^{\infty} f_k^1 \operatorname{sinc}\left(\frac{\Omega t}{\pi} - k\right)\right\|_{C(\mathbb{R})} \leq K_1\omega(f,\sigma)\log\frac{1}{\omega(f,\sigma)}, \tag{116}$$

where $K_1 = (66 + 31M_f)/\gamma$, and M_f and γ are constants as in (113).

Above, the λ_k are the averages (115), i.e., $\lambda_k f(\cdot + k\pi/\Omega) = f_k^1$, and $|f(t + k\pi/\Omega) - f(k\pi/\Omega)| \leq \omega(f, \sigma_k) \leq \omega(f,\sigma)$ for $t \in [-\sigma_k, \sigma_k]$, so that $E(f,\lambda) \leq \omega(f,\sigma)$.

The **second example** is concerned with jitter errors. *Let $f \in B_\Omega^\infty$ satisfy the decay condition* (113), *and suppose its sampled values f_k^2 are obtained as*

$$f_k^2 = f\left(\frac{k\pi}{\Omega} + \sigma_k\right) \quad (k \in \mathbb{Z}),$$

where $\sigma_k \in \mathbb{R}$ *satisfy* $0 < \sigma_k \le \sigma$ *for all* $k \in \mathbb{Z}$ *and some* $0 < \sigma < 1/2$. *If* $\omega(f, \sigma) \le \min\{e^{-1/2}, \pi/\Omega\}$, $\Omega \ge \pi$, *then*

$$\left\| f(t) - \sum_{k=-\infty}^{\infty} f_k^2 \operatorname{sinc}\left(\frac{\Omega t}{\pi} - k\right) \right\|_{C(\mathbb{R})} \le K_2 \omega(f, \sigma) \log \frac{1}{\omega(f, \sigma)},$$

where again $K_2 = (66 + 31M_f)/\gamma$.

Here λ is the sequence of functionals given by $\lambda_k f(\cdot + k\pi/\Omega) = f(k\pi/\Omega + \sigma_k) = f_k^2$ for each $k \in \mathbb{Z}$. Thus $E(f, \lambda) \le \omega(f, \sigma)$.

Jitter error is usually considered as a random distortion, stochastic methods being used to estimate it; the σ_k's are regarded as a weak sense stationary discrete-parameter random process having finite variance (see [49] and the literature cited there). The deterministic approach used above was apparently first employed in [28] (see also [37, 49] and the literature cited therein). It is based on the sole basic assumption that the local deviations σ_k have an upper bound $\sigma > 0$, i.e., $|\sigma_k| \le \sigma$ for $k \in \mathbb{Z}$.

Our **third example** to the above concerns amplitude errors.

Let $f \in B_\Omega^\infty$ *satisfy* (113), *and suppose its sampled values are the numbers* f_k^3 *satisfying*

$$\left| f_k^3 - f\left(\frac{k\pi}{\Omega}\right) \right| \le \min\left\{ \varepsilon_0, \left| f\left(\frac{k\pi}{\Omega}\right) \right| \right\} \tag{117}$$

for all $k \in \mathbb{Z}$, $\varepsilon_0 > 0$ *being a constant. If* $\varepsilon_0 \le \min\{e^{-1/2}, \pi/\Omega\}$, *then*

$$\left\| f(t) - \sum_{k=-\infty}^{\infty} f_k^3 \operatorname{sinc}\left(\frac{\Omega t}{\pi} - k\right) \right\|_{C(\mathbb{R})} \le K_3 \varepsilon_0 \log \frac{1}{\varepsilon_0},$$

where $K_3 = (66 + 52M_f)/\gamma$.

As to the proof, define $\lambda_k = \delta f_k^3 / f(k\pi/\Omega)$, δ being the δ-distribution, whence $\lambda = (\lambda_k)_{k\in\mathbb{Z}}$ are linear functionals on $C_0(\mathbb{R})$, with $\lambda_k f(\cdot + k\pi/\Omega) = f_k^3$, and $|\lambda_k f(\cdot + k\pi/\Omega) - f(k\pi/\Omega)| \le \varepsilon$ for all $k \in \mathbb{Z}$. Thus $E(f, \lambda) \le \varepsilon_0$.

The amplitude error is also often dealt with by stochastic methods, particularly when interpreted as some sort of "noise". However, if this error results from round off or quantization noise, i.e., the sampled values are replaced by the nearest discrete values, then it is especially preferable to employ deterministic methods. See [27, 37, 49], [109, Chapter 11] and the literature cited there for the engineering background of the imposed condition (117).

2.7.2. Errors for Not-Necessarily Band-Limited Signals

In the previous subsection we considered error analysis for the sampling theorem in the band-limited case. The aim now is to extend the matter not only to non-band-limited functions but also to take into account *truncation errors*, the usual instance in practice. Again we present a new, unified approach to the field which now covers four types of errors, including the aliasing error. The proofs, not to be carried out here, are fully different to those for the band-limited case and depend upon deep results in classical approximation theory and Fourier analysis, specifically upon the approximation behaviour of so-called de la Vallée Poussin (delayed) means of Fourier convolution integrals (see [49, 206]).

We consider the sampling series (112) as well as the truncated sampling series with measured sampled values,

$$S_{\Omega,\lambda,N}f(t) := \sum_{k=-N}^{N} \lambda_k f\left(\cdot + \frac{k\pi}{\Omega}\right) \operatorname{sinc}\left(\frac{\Omega t}{\pi} - k\right) \quad (t \in \mathbb{R}).$$

Let $\lfloor x \rfloor$ stand for the smallest integer that is greater than or equal to an $x \in \mathbb{R}$, and let $E_\Omega(f, \lambda)$ again be given by (111). The general result covering the total matter is:

Let $f \in \mathrm{Lip}_L(1; C(\mathbb{R}))$ (recall (42)*) satisfy the decay condition* (113) *for some $0 < \gamma \leq 1$, and let again $\lambda = (\lambda_k)_{k\in\mathbb{Z}}$ be a sequence of continuous linear functionals on $C_0(\mathbb{R})$. Then for each $\Omega \geq \pi e^2$ there follows the truncation error*

$$\| f(t) - S_{\Omega,\lambda,N}f(t) \|_{C(\mathbb{R})} \leq K' \frac{\log \Omega}{\Omega}, \tag{118}$$

provided $N := \lfloor \frac{1}{2}(\Omega/\pi)^{1+1/\gamma} - \frac{1}{2} \rfloor$ and $E_\Omega(f, \lambda) \leq c_0\pi/\Omega$ for some constant $c_0 > 0$. Here $K' = K'(f, \lambda, c_0)$ is the constant

$$K' = 14L/\pi + e(1 + 1/\gamma)(7L/\pi + c_0 + 2^\gamma 6(M_f + \|f\|_{C(\mathbb{R})})). \tag{119}$$

The **first application** again deals with measured sampled values given by the average f_k^1 of (115).

Let $f \in \mathrm{Lip}_L(1; C(\mathbb{R}))$ satisfy (113) *for some $0 < \gamma \leq 1$, and suppose the sampled values are obtained by the rule* (115)*, where σ_k are numbers satisfying $0 < \sigma_k \leq \sigma < 1/2$ for all $k \in \mathbb{Z}$. If $\omega(f, \sigma) \leq \min\{e^{-1}, c_0\pi/\Omega\}$ for some constant c_0, then*

$$\left\| f(t) - \sum_{k=-N}^{N} f_k^1 \operatorname{sinc}\left(\frac{\Omega t}{\pi} - k\right) \right\|_{C(\mathbb{R})} \leq K' \omega(f, \sigma) \log \frac{1}{\omega(f, \sigma)},$$

K' again being the constant given by (119).

As to the proof, just as that of (116), again $\lambda_k f(\cdot + k\pi/\Omega) = f_k^1$ with $E(f, \sigma) \le \omega(f, \sigma)$.

The **second application** example covers simultaneously the aliasing, truncation, amplitude and time-jitter errors. As to the amplitude error which results from quantization, meaning that the function value $f(t)$ of a signal f at time t is replaced by the nearest discrete value or machine number $\bar{f}(t)$, the quantization size is known beforehand or can be chosen arbitrarily. We may assume that the local error at any time t is bounded by a constant $\varepsilon > 0$, i.e., $|\bar{f}(t) - f(t)| \le \varepsilon$; also assume that $f(t) = 0$ implies $\bar{f}(t) = 0$.

We therefore consider the combined error

$$f(t) - \sum_{k=-N}^{N} \bar{f}\left(\frac{k\pi}{\Omega} + \sigma_k\right) \operatorname{sinc}\left(\frac{\Omega t}{\pi} - k\right),$$

where again $\sigma_k \in \mathbb{R}$ satisfy $0 < \sigma_k \le \sigma$ for all k and some constant $0 < \sigma < 1/2$.

Let again $f \in \mathrm{Lip}_L(1; C(\mathbb{R}))$ *satisfy* (113) *for some* $0 < \gamma \le 1$. *If* $\omega(f, \sigma) \le c_1\pi/\Omega$ *and* $|f(t) - \bar{f}(t)| \le c_2\pi/\Omega$ *for all* $t \in \mathbb{R}$ *with* $\Omega \ge \pi e^2$ *for some constants* $c_1, c_2 > 0$, *then*

$$\left\| f(t) - \sum_{k=-N}^{N} \bar{f}(k\pi/\Omega + \sigma_k) \operatorname{sinc}\left(\frac{\Omega t}{\pi} - k\right)\right\|_{C(\mathbb{R})} \le K' \frac{\log \Omega}{\Omega},$$

where $N := \lfloor \frac{1}{2}(\Omega/\pi)^{1+1/\gamma} - \frac{1}{2} \rfloor$ *and* K' *is the constant of* (119) *with* $c_0 := c_1 + c_2$.

As to the proof, define $\lambda_k := \delta(\cdot - \sigma_k)\bar{f}(k\pi/\Omega + \sigma_k)/f(k\pi/\Omega + \sigma_k)$.

2.8. Generalized Sampling Series

The present section is devoted to generalizations of the Shannon sampling series; the generalized series will turn out to be discrete analogs of singular Fourier convolution integrals on $\mathbb{R}$. Recall that the classical Shannon series is a discrete-time version ("infinite Riemann sum") of the singular convolution integral of Dirichlet (cf. (48) and the remark above formula (66)). There are several reasons which motivate the following generalizations.

Firstly, they are especially useful when dealing with non-band-limited functions. On account of the non-conformity of time and band-limitation, the Shannon reconstruction is not very appropriate. For this reason we built up the theory of Section (2.6) for not-necessarily band-limited functions for which the sampling series (64) is not equal to $f(t)$ for all t but only tends to $f(t)$ as $\Omega \to \infty$. In addition, this approximation of f by the Shannon series only holds under more or less restrictive conditions upon f; continuity of f alone does not suffice. To this end, we introduced in Section (2.6) the concept of approximately band-limited functions; if f is approximately band-limited with a certain rate, then f has the same rate of approximation by the sampling series (cf. (102, 103)).

Whereas the foregoing result is of practical importance, there is a *second* drawback associated with the Shannon series: the reconstruction of a signal by this series or its convergence to f as $\Omega \to \infty$ is strongly influenced by the various errors that may occur in practice, as observed in Section (2.7). For example, the reconstruction of a signal when only a finite number N of sampled values is used (instead of an infinite number in the Shannon series, an unpractical case) involves a truncation error which decreases rather slowly for $N \to \infty$ since the sinc function involved behaves only like $\mathcal{O}(|t|^{-1})$ for $|t| \to \infty$.

A suitable modification of the sinc function will improve the convergence behaviour for $\Omega \to \infty$ as well as for $N \to \infty$.

Thirdly, the Shannon series $S_\Omega f$ of (64) provides an interpolation formula for recovering the time signal f from its samples in the sense that $S_\Omega f(t)$ interpolates $f(t)$ at $t = k\pi/\Omega$ for all $k \in \mathbb{Z}$. This interpolation is trivial in case of band-limited signals since $S_\Omega f(t)$ "interpolates" $f(t)$ for *all* t. However, the approximate sampling theorem for non-band-limited functions given by (95), is a "true" interpolation theorem in the sense that the Shannon series interpolates $f(t)$ in the form just stated but only converges to f as $\Omega \to \infty$. Thus (95) together with (98) is simultaneously an interpolation as well as an approximation theorem (without an exact reconstruction) provided the continuous function f satisfies additional conditions, e.g., $f \in L^1(\mathbb{R})$ with $F \in L^1(\mathbb{R})$ in case of (95). The question arises whether this exceptional simultaneous behaviour can be preserved provided the signal is just continuous. It will indeed be the case if the sinc function in the Shannon series is replaced by a suitable function φ. However, for "non-suitable" φ the interpolation property may be violated, whereas the associated approximation behaviour may be (much) better.

Fourthly, the sinc function itself is not very suitable for fast and efficient implementations; in fact, an ideal low-pass is not realizable in practical applications.

More details of the material presented in the following sections can be found in [5, 31, 49, 55, 135, 211, 214]. For a multivariate setting see [33–34, 87, 214].

2.8.1. General Convergence Theorems

The foregoing four difficulties will be overcome by replacing the sinc function by so-called kernel functions φ. For this purpose we study *(generalized) sampling series* of the form

$$S_\Omega^\varphi f(t) := \sum_{k=-\infty}^{\infty} f\left(\frac{k\pi}{\Omega}\right)\varphi\left(\frac{\Omega t}{\pi} - k\right) \quad (t \in \mathbb{R}), \tag{120}$$

which should have the property that $S_\Omega^\varphi f(t)$ exists for all $f \in C(\mathbb{R})$, $t \in \mathbb{R}$, $\Omega > 0$, and that

$$\lim_{\Omega \to \infty} S_\Omega^\varphi f(t) = f(t) \tag{121}$$

uniformly in $t \in \mathbb{R}$, provided $f \in C(\mathbb{R})$ (recall the definition of $C(\mathbb{R})$ in Section 2.2.1).

This series can be regarded as a discrete counterpart of the singular convolution integral

$$I_\Omega^\varphi f(t) = \frac{\Omega}{\pi} \int_{-\infty}^{\infty} f(u) \varphi\left(\frac{\Omega}{\pi}(t-u)\right) du. \tag{122}$$

In fact,

$$\lim_{\Omega \to \infty} I_\Omega^\varphi f(t) = f(t) \quad (t \in \mathbb{R}) \tag{123}$$

holds for all $f \in C(\mathbb{R})$ in the sense of uniform convergence on $\mathbb{R}$ if and only if the function φ is a *kernel*, thus satisfies the conditions $\varphi \in L^1(\mathbb{R})$ and $\varphi^\wedge(0) = \int_{-\infty}^{\infty} \varphi(u)du = 1$ (see [43, pp. 120 f.]).

Concerning to the original situation of the generalized convolution sum (120) with the desired property (121), the conditions $\varphi \in L^1(\mathbb{R})$ with $\varphi^\wedge(0) = 1$ are not sufficient for (121) to be valid. Take e.g. $\varphi(t) = (2 - 4|t|)$ for $|t| \le 1/2$, $= 0$ for $|t| > 1/2$, and $f(t) \equiv 1$. Although $\varphi^\wedge(0) = 1$, (121) fails for e.g., $t = 0$.

We will assume that φ is continuous and

$$\sum_{k=-\infty}^{\infty} |\varphi(t-k)| < \infty \tag{124}$$

uniformly for $t \in [0, 1]$. Further, we need the αth *absolute (sum) moment* of φ, namely

$$m_\alpha(\varphi) := \sup_{t \in \mathbb{R}} \sum_{k=-\infty}^{\infty} |t-k|^\alpha |\varphi(t-k)| \quad (\alpha \ge 0; t \in \mathbb{R}). \tag{125}$$

One can readily show that if φ satisfies (124), then $\varphi \in L^1(\mathbb{R}) \cap C(\mathbb{R})$, the series in (124) converges absolutely for all $t \in \mathbb{R}$, and uniformly on each compact subinterval of $\mathbb{R}$, with $m_0(\varphi) < \infty$; further, one has uniformly for all $t \in \mathbb{R}$,

$$\lim_{R \to \infty} \sum_{|k-t| > R} |\varphi(t-k)| = 0. \tag{126}$$

Hence the operators S_Ω^φ, defined by (120) for $\Omega > 0$, are bounded linear operators mapping $C(\mathbb{R})$ into itself, having operator norm

$$\|S_\Omega^\varphi\|_{[C(\mathbb{R}),C(\mathbb{R})]} = m_0(\varphi) \quad (\Omega > 0), \tag{127}$$

which corresponds to the norm of the operators I_Ω^φ of (122), namely,

$$\|I_\Omega^\varphi\|_{[C(\mathbb{R}),C(\mathbb{R})]} = \|\varphi\|_{L^1(\mathbb{R})} \quad (\Omega > 0). \tag{128}$$

As to condition (125), $m_\alpha(\varphi) < \infty$ implies $m_\beta(\varphi) < \infty$ for any $0 \le \beta < \alpha$ together with $\int_{-\infty}^{\infty} |u|^\beta |\varphi(u)| du < \infty$. As a consequence, the Fourier transform $\varphi^\wedge$ has continuous derivatives of any order $s \in \mathbb{N}$ with $s \le \alpha$ (cf. [43, p. 197]). A sufficient condition for $m_\alpha(\varphi) < \infty$ is $|\varphi(t)| = \mathcal{O}(|t|^{-\alpha-\gamma})$ for $|t| \to \infty$ and for some $\gamma > 1$. The first basic result as to the generalized sampling operator S_Ω^φ reads:

Let φ be a continuous and bounded function on $\mathbb{R}$, and let $\sum_{k=-\infty}^{\infty} |\varphi(t-k)|$ converge uniformly on $[0,1]$. The following four conditions are equivalent:

(i) $\sum_{k=-\infty}^{\infty} \varphi(t-k) = 1 \quad (t \in \mathbb{R})$,

(ii) $\lim_{\Omega\to\infty} S_\Omega^\varphi f(t) = f(t)$
uniformly on $\mathbb{R}$ for all $f \in C(\mathbb{R})$, (129)

(iii) $\lim\limits_{\Omega\to\infty} S_\Omega^\varphi f(t_0) = f(t_0)$
for all bounded signals f that are continuous at $t = t_0$,

(iv) $\varphi^\wedge(2k\pi) = \delta_{k,0} \quad (k \in \mathbb{Z})$.

Let us first show that (i) and (ii) are equivalent. Indeed, (i) yields for $f \in C(\mathbb{R})$ and any $\eta > 0$,

$$\begin{aligned}|S_\Omega^\varphi f(t) - f(t)| &\le \left\{ \sum_{|\frac{k\pi}{\Omega}-t|<\eta} + \sum_{|\frac{k\pi}{\Omega}-t|\ge\eta} \right\} \left| f\left(\frac{k\pi}{\Omega}\right) - f(t) \right| \left| \varphi\left(\frac{\Omega t}{\pi} - k\right) \right| \\ &\le m_0(\varphi) \sup_{|h|\le\eta} \|f(t+h) - f(t)\|_{C(\mathbb{R})} \\ &\quad + 2\|f\|_{C(\mathbb{R})} \sum_{|k\pi - t\Omega|\ge\eta\Omega} \left| \varphi\left(\frac{\Omega t}{\pi} - k\right) \right|.\end{aligned}$$

By choosing $\eta > 0$ appropriately, the first term can be made arbitrarily small since f is uniformly continuous. The second tends uniformly to zero for $\Omega \to \infty$ by (126). Conversely, if (ii) holds, then one has for $f(t) = 1$ and $t = 1$ the limit relation $\lim_{\Omega\to\infty}\sum_{k=-\infty}^{\infty}\varphi(\Omega/\pi-k)=1$. Since the series defines a periodic function, it must be identically equal to 1, which is (i).

The equivalence of (i) and (iii) follows along the same lines. Now to the equivalence of (i) and (iv). The series in (i) defines a continuous function of period one, so that by Poisson's summation formula (29) its Fourier series is given by

$$\sum_{k=-\infty}^{\infty}\varphi(t-k)\sim\sum_{k=-\infty}^{\infty}\varphi^{\wedge}(2k\pi)e^{j2k\pi t}. \tag{130}$$

If (iv) holds, the series on the right side of (130) reduces to the term $k=0$, which is equal to 1, and so represents the function on the left. Conversely, if (i) holds, then the Fourier series of the sum function on the left has only one nonzero term, namely that for $k=0$ which is equal to 1; comparing it with (130), one obtains (iv).

A function φ satisfying (124) uniformly for $t\in[0,1]$, and condition (i) or (iv), will be called an *approximate identity.*

2.8.2. Rate of Convergence

In this section we investigate the rate of convergence in (121). It turns out that it is convenient to distinguish between band-limited and non-band-limited kernels φ. In the former case one can transfer the desired results easily from corresponding results on convolution integrals (122), whereas in the latter case one uses a Taylor expansion approach which is well known in approximation theory. At first to band-limited φ.

Let $\varphi\in B_{\pi}^{1}$ *with* $\varphi^{\wedge}(0)=1$. *There exist constants* $c_1,c_2>0$, *depending on* φ *only, such that for all* $f\in C(\mathbb{R})$ *and* $\Omega>0$,

$$c_1\|I_{\Omega}^{\varphi}f-f\|_{C(\mathbb{R})}\le\|S_{\Omega}^{\varphi}f-f\|_{C(\mathbb{R})}\le c_2\|I_{\Omega}^{\varphi}f-f\|_{C(\mathbb{R})}. \tag{131}$$

First we note that φ is a kernel in view of (33) and (iv) above. Moreover, $I_{\Omega}^{\varphi}f$ as well as $S_{\Omega}^{\varphi}f$ both belong to B_{Ω}^{∞} (see [161, pp. 127, 136]). Hence one obtains by (37) that

$$S_{\Omega}^{\varphi}I_{\Omega}^{\varphi}f=I_{\Omega}^{\varphi}I_{\Omega}^{\varphi}f,\quad I_{\Omega}^{\varphi}S_{\Omega}^{\varphi}f=S_{\Omega}^{\varphi}S_{\Omega}^{\varphi}f,$$

and further

$$\begin{aligned}\|S_{\Omega}^{\varphi}f-f\|_{C(\mathbb{R})}&\le\|S_{\Omega}^{\varphi}f-S_{\Omega}^{\varphi}I_{\Omega}^{\varphi}f\|_{C(\mathbb{R})}+\|I_{\Omega}^{\varphi}I_{\Omega}^{\varphi}f-I_{\Omega}^{\varphi}f\|_{C(\mathbb{R})}+\|I_{\Omega}^{\varphi}f-f\|_{C(\mathbb{R})}\\&\le\{\|S_{\Omega}^{\varphi}\|_{[C(\mathbb{R}),C(\mathbb{R})]}+\|I_{\Omega}^{\varphi}\|_{[C(\mathbb{R}),C(\mathbb{R})]}+1\}\|I_{\Omega}^{\varphi}f-f\|_{C(\mathbb{R})}.\end{aligned}$$

Since the operators S_Ω^φ and I_Ω^φ are uniformly bounded by (127) and (128), the right-hand inequality is established, and the other one follows along the same lines.

This result enables one to transfer all results regarding the approximation by singular convolution integrals to that by generalized sampling series. For examples see Section 2.8.3 below. If the kernel φ belongs to B_σ^1 for some $\sigma \in \mathbb{R}$ with $\pi < \sigma < 2\pi$ rather than to B_π^1, then one can obtain somewhat weaker results, see [52].

For the study of rates of approximation in case of non-band-limited kernels the moments (125) are basic. As a first result we have (cf. (40, 41)):

Let φ be an approximate identity.

(*a*) *If $m_\alpha(\varphi) < \infty$ for some $0 < \alpha \leq 1$, then*

$$f \in \mathrm{Lip}(\alpha; C(\mathbb{R})) \Rightarrow \|S_\Omega^\varphi f - f\|_{C(\mathbb{R})} = \mathcal{O}(\Omega^{-\alpha}) \quad (\Omega \to \infty). \tag{132}$$

(*b*) *If $m_1(\varphi) < \infty$, then*

$$\|S_\Omega^\varphi f - f\|_{C(\mathbb{R})} \leq M\omega(f, C(\mathbb{R}), \Omega^{-1}) \quad (\Omega > 0),$$

the constant M depending only on φ. Furthermore, there holds (132) *for* $0 < \alpha \leq 1$.

As to the proof, if $f \in \mathrm{Lip}(\alpha; C(\mathbb{R}))$ there exists a constant $L > 0$ such that $|f(k\pi/\Omega) - f(t)| \leq L|k\pi/\Omega - t|^\alpha$. Thus

$$\begin{aligned} |S_\Omega^\varphi f(t) - f(t)| &\leq \sum_{k=-\infty}^{\infty} \left| f\left(\frac{k\pi}{\Omega}\right) - f(t) \right| \left| \varphi\left(\frac{\Omega t}{\pi} - k\right) \right| \\ &\leq L \sum_{k=-\infty}^{\infty} \left| \frac{k\pi}{\Omega} - t \right|^\alpha \left| \varphi\left(\frac{\Omega t}{\pi} - k\right) \right| \leq L' m_\alpha(\varphi) \Omega^{-\alpha}. \end{aligned}$$

Concerning part (b), it suffices to establish the *Jackson-type inequality*

$$\|S_\Omega^\varphi f - f\|_{C(\mathbb{R})} \leq \frac{M_1}{\Omega} \|f'\|_{C(\mathbb{R})} \quad (f \in C^{(1)}(\mathbb{R}); \Omega > 0)$$

which can be deduced from the estimate $|f(k\pi/\Omega) - t| \leq |k\pi/\Omega - t| \|f'\|_{C(\mathbb{R})}$ as in the proof of part (a). The assertion of (b) then follows by standard arguments involving the K-functional (see e.g. [68]).

Thus the best possible order which can be achieved by this result is $\mathcal{O}(\Omega^{-1})$ unless $f =$ *constant*. For better orders the high order absolute moments of (125) are needed.

Let φ be an approximate identity with $m_r(\varphi) < \infty$ for some $r \in \mathbb{N}\backslash\{1\}$. The following assertions are equivalent:

$$
\begin{aligned}
&\text{(i)} \quad \sum_{k=-\infty}^{\infty} (t-k)^s \varphi(t-k) = 0 \quad (s = 1, 2, \ldots, r-1; t \in \mathbb{R}), \\
&\text{(ii)} \quad (\varphi^\wedge)^{(s)}(2k\pi) = 0 \quad (s = 1, 2, \ldots, r-1; k \in \mathbb{Z}), \\
&\text{(iii)} \quad \|S_\Omega^\varphi f - f\|_{C(\mathbb{R})} \leq M\|f^{(r)}\|_{C(\mathbb{R})}\Omega^{-r} \quad (f \in C^r(\mathbb{R}); \Omega > 0).
\end{aligned}
\tag{133}
$$

If φ satisfies additionally one of the three conditions above, then there holds for $s = 0, 1, \ldots, r-1$ and $0 < \alpha \leq 1$:

$$
\|S_\Omega^\varphi f - f\|_{C(\mathbb{R})} \leq M\omega(f^{(s)}, C(\mathbb{R}), \Omega^{-1}) \quad (\Omega > 0), \tag{134}
$$

$$
f^{(s)} \in \operatorname{Lip}(\alpha; C(\mathbb{R})) \Longrightarrow \|S_\Omega^\varphi f - f\|_{C(\mathbb{R})} = \mathcal{O}(\Omega^{-s-\alpha}) \quad (\Omega \to \infty). \tag{135}
$$

To prove that (i) implies (iii) apply the operator S_Ω^φ to the Taylor expansion

$$
f(u) - f(t) = \sum_{s=1}^{r-1} \frac{f^{(s)}(t)}{s!}(u-t)^s + \frac{1}{r!} f^{(r)}(\xi)(u-t)^r
$$

to yield

$$
\begin{aligned}
|S_\Omega^\varphi f(t) - f(t)| = &\sum_{s=1}^{r-1} \frac{f^{(s)}(t)}{s!} \sum_{k=-\infty}^{\infty} \left(\frac{k\pi}{\Omega} - t\right)^s \varphi\left(\frac{\Omega t}{\pi} - k\right) \\
&+ \frac{1}{r!} \sum_{k=-\infty}^{\infty} f^{(r)}(\xi) \left(\frac{k\pi}{\Omega} - t\right)^r \varphi\left(\frac{\Omega t}{\pi} - k\right),
\end{aligned}
$$

where ξ depends on k, Ω and t. By (i) the double series vanishes, so that

$$
\|S_\Omega^\varphi f - f\|_{C(\mathbb{R})} \leq \frac{1}{r!} \|f^{(r)}\|_C m_r(\varphi)\Omega^{-r}.
$$

This is (iii). For the proof of the converse direction see [186]. Concerning the equivalence of (i) and (ii) note that $[t^s\varphi(t)]^\wedge(v) = (-j)^s[\varphi^\wedge]^{(s)}(v)$ (cf. [43, p. 197]), and apply Poisson's summation formula (29) to the function $t^s\varphi(t)$, namely,

$$
\sum_{k=-\infty}^{\infty} (t-k)^s \varphi(t-k) \sim (-j)^s \sum_{k=-\infty}^{\infty} [\varphi^\wedge]^{(s)}(2k\pi)e^{j2k\pi t}.
$$

The proof is now analogous to that of the equivalence (i) $\Longleftrightarrow$ (iv) of (129).

As to the theoretical part, the rate of approximation in (135) cannot be improved, at least for $0 < \alpha < 1$. In fact:

If φ is given as above, then for each $s = 0, 1, \ldots, r-1$ and each $0 < \alpha < 1$ there exists a function f_0 with $f_0^{(s)} \in \mathrm{Lip}(\alpha; C(\mathbb{R}))$ such that (see [186])

$$\lim_{\Omega\to\infty} (\Omega^{s+\alpha}) \|S_\Omega^\varphi f_0 - f_0\|_{C(\mathbb{R})} \neq 0.$$

2.8.3. Applications

2.8.3.1. Band-Limited Kernels

Let us consider the kernel

$$\vartheta_{a,r}(t) = c_{a,r}\left(\frac{\sin at}{at}\right)^{2r}, \quad c_{a,r} = \left\{\int_{-\infty}^{\infty}\left(\frac{\sin au}{au}\right)^{2r} du\right\}^{-1} \quad (t \in \mathbb{R}) \qquad (136)$$

with $r \in \mathbb{N}$ and $0 < a \leq \pi/r$. If $r = 1$ it is the famous Fejér kernel for $a = \pi/2$, and that considered by M. Theis [218] for $a = \pi$. If $r = 2$ it is known as kernel of Jackson and de la Vallée Poussin (see [43, p. 130] and the literature cited there).

As we have seen in example (35) the kernel $\vartheta_{a,r}$ belongs to $B^1_{2ar} \subset B^1_{2\pi}$. As to its moments, one has for $0 \leq \beta < 2r - 1$ and $n \leq t < n + 1$,

$$\begin{aligned}
&c_{a,r}^{-1} \sum_{k=-\infty}^{\infty} |t-k|^\beta |\vartheta_{a,r}(t-k)|^{2r} \\
&\leq |t-n|^\beta + |t-(n+1)|^\beta + \sum_{\substack{k=-\infty \\ k\neq n,\, k\neq n+1}}^{\infty} |t-k|^\beta (a|t-k|)^{-2r} \\
&< 2 + a^{-2r} 2 \sum_{k=1}^{\infty} k^{-2r+\beta} < \infty.
\end{aligned}$$

Further, it follows from (36) with $\Omega = \pi$ and the definition of $c_{a,r}$ that

$$\sum_{k=-\infty}^{\infty} \vartheta_{a,r}(t-k) = \int_{-\infty}^{\infty} \vartheta_{a,r}(t-u)du = 1.$$

Noting that $\vartheta_{a,r} \in B^1_{2\pi}$ we also have that $\vartheta^\wedge_{a,r}(2k\pi) = 0$ for $k \in \mathbb{Z}\backslash\{0\}$ so that condition (iv) of (129) is satisfied. Hence we obtain, uniformly on $\mathbb{R}$,

$$\lim_{\Omega\to\infty} S_\Omega^{\vartheta_{a,r}} f(t) = f(t).$$

As to the rate of convergence, first consider $\vartheta_{a,r}$ for $r = 1$, $0 < a \leq \pi$. Then the moments of order $\beta \in (0, 1)$ exist, and one has for $0 < \beta < 1$ that (cf. (132))

$$f \in \mathrm{Lip}(\beta; C(\mathbb{R})) \Longrightarrow \|S_{\Omega}^{\vartheta_{a,1}} f - f\|_{C(\mathbb{R})} = \mathcal{O}(\Omega^{-\beta}) \quad (\Omega \to \infty). \tag{137}$$

This estimate is valid for all $0 < a \leq \pi$, in particular for both $a = \pi/2$ (Fejér's kernel) and $a = \pi$ (Theis's kernel). Since Fejér's kernel $\vartheta_{\pi/2,1}$ belongs to B_{π}^{1}, all results concerning this kernel can also be deduced from (131) and well-known results on the corresponding singular integral. Using this approach one can also treat the limiting case $\beta = 1$ in (137), which is not covered by (132), since the first moment of Fejér's kernel is infinite.

Observe that in case $a = \pi$ the kernel $\vartheta_{\pi,1}(t)$ equals $\mathrm{sinc}^2(t)$ and the sampling series interpolates the function at the nodes $n\pi/\Omega$ for $n = 0, \pm 1, \pm 2, \ldots,$ thus

$$S_{\Omega}^{\vartheta_{\pi,1}} f\left(\frac{n\pi}{\Omega}\right) = \sum_{k=-\infty}^{\infty} f\left(\frac{k\pi}{\Omega}\right) \mathrm{sinc}^2(n-k) = f\left(\frac{n\pi}{\Omega}\right).$$

This is not the case for $\vartheta_{\pi/2,1}$.

Let us now consider the case $r = 2$. The kernel $\vartheta_{a,2}$ for $0 < a \leq \pi/2$ possesses a finite moment of order 2 and, since $\vartheta_{a,2}$ is even, $\int_{-\infty}^{\infty} t\vartheta_{a,2}(t)dt = 0$. Thus we may now apply (134) and (135) to yield:

There holds for $s = 0, 1$ and $0 < \alpha \leq 1$,

$$\|S_{\Omega}^{\vartheta_{a,2}} f - f\|_{C(\mathbb{R})} \leq M\omega(f^{(s)}, C(\mathbb{R}), \Omega^{-1}) \quad (\Omega > 0), \tag{138}$$

$$f^{(s)} \in \mathrm{Lip}(\alpha; C(\mathbb{R})) \Longrightarrow \|S_{\Omega}^{\vartheta_{a,2}} f - f\|_{C(\mathbb{R})} = \mathcal{O}(\Omega^{-s-\alpha}) \quad (\Omega \to \infty). \tag{139}$$

Although the third moment $m_3(\vartheta_{a,2})$ is finite, too, the kernel $\vartheta_{a,2}$ does not provide a better rate of convergence, since $\int_{-\infty}^{\infty} t^2\vartheta_{a,2}(t)dt \neq 0$.

Using the same arguments as above one can prove the estimates (138) and (139) for the kernels $\vartheta_{a,r}$ in case $r > 2$. Again s is restricted to 1 or 2 since $\int_{-\infty}^{\infty} t^2\vartheta_{a,r}(t)dt \neq 0$. So the best possible order which can be achieved in (135) for $\varphi = \vartheta_{a,r}$ is $\mathcal{O}(\Omega^{-2})$ for $\Omega \to \infty$. On the other hand, these kernels have a bigger rate of decay for $t \to \pm\infty$ which gives a smaller truncation error when replacing the infinite series by a finite one. For a band-limited kernel having a faster rate of decay than any power $|t|^{-r}$, $r \in \mathbb{N}$ for $t \to \pm\infty$ see [93].

Note that formula (36) can be used to compute the constants $c_{a,r}$. For example, for $r = 1$ one may take $a = \pi$ to deduce

$$c_{\pi,1}^{-1} = \int_{-\infty}^{\infty} \left(\frac{\sin \pi t}{\pi t}\right)^2 dt = \sum_{k=-\infty}^{\infty} \left(\frac{\sin k\pi}{k\pi}\right)^2 = 1,$$

and by a linear substitution it follows that

$$c_{a,1}^{-1} = \int_{-\infty}^{\infty} \left(\frac{\sin at}{at}\right)^2 dt = \frac{\pi}{a}.$$

For $r = 2$ one has for $a = \pi/2$ that

$$c_{\pi/2,2}^{-1} = \sum_{k=-\infty}^{\infty} \left(\frac{\sin k\pi/2}{k\pi/2}\right)^4 = 1 + 2\sum_{\substack{k=1\\ k \text{ odd}}}^{\infty} \left(\frac{2}{\pi}\right)^4 \frac{1}{k^4} = 1 + 2\left(\frac{2}{\pi}\right)^4 \frac{\pi^4}{96} = \frac{4}{3}.$$

Hence

$$c_{a,2}^{-1} = \int_{-\infty}^{\infty} \left(\frac{\sin at}{at}\right)^4 dt = \frac{2\pi}{3a}.$$

For applications to a Rogosinski-type kernel as well as other kernels of approximation theory see [123].

2.8.3.2. Non-Band-Limited Kernels

The band-limited kernels φ discussed above provided a smaller truncation error than the sinc kernel since they are of order $\mathcal{O}(t^{-2r})$ for $t \to \pm\infty$, but in any case the corresponding series $S_\Omega^\varphi f$ are in general infinite ones.

This can be overcome by choosing φ to be *time-limited*. The *B*-splines of order $n \geq 2$ are very suitable choices. They are defined by

$$M_n(t) := \begin{cases} \displaystyle\sum_{\nu=0}^{[\frac{n}{2}-|t|]} \frac{(-1)^\nu n\left(\frac{n}{2} - |t| - \nu\right)^{n-1}}{\nu!(n-\nu)!}, & |t| \leq \dfrac{n}{2} \\ 0, & |t| > \dfrac{n}{2}, \end{cases} \tag{140}$$

their Fourier transform being given by

$$M_n^\wedge(v) = (\text{sinc}(v/2\pi))^n. \tag{141}$$

The M_n are piecewise polynomials of degree $n-1$ vanishing outside $(-n/2, n/2)$. Hence the associated sampling series take the form

$$(S_\Omega^{M_n} f)(t) = \sum_{|\Omega t/\pi - k| < n/2} f\left(\frac{k\pi}{\Omega}\right) M_n\left(\frac{\Omega t}{\pi} - k\right),$$

so that in this case one has only n nonzero terms at most. If $n = 2$, then M_2 is the triangle function $M_2(t) = \Lambda(t)$ (recall (44), (45)), and the sampling series is just the linear spline interpolation of f at the knots $k\pi/\Omega$. The B-spline kernels have the additional advantage that they are fast to compute in contrast to the transcendental kernels in the band-limited case above.

Concerning error estimates, it follows from the Fourier transform (141) that the M_n are approximate identities, which satisfy (ii) of (133) for $r = 2$, so that by (134) and (135) for $s = 1, 2$ and $0 < \alpha \le 1$:

$$\|S_\Omega^{M_n} f - f\|_{C(\mathbb{R})} \le M\omega(f^{(s)}, C(\mathbb{R}), \Omega^{-1}) \quad (\Omega > 0),$$

$$f^{(s)} \in \mathrm{Lip}(\alpha; C(\mathbb{R})) \Longrightarrow \|S_\Omega^{M_n} f - f\|_{C(\mathbb{R})} = \mathcal{O}(\Omega^{-s-\alpha}) \quad (\Omega \to \infty).$$

In order to obtain higher order estimates one has to take linear combinations of B-splines (see Fig. 16), e.g.,

$$\varphi_1(t) := \frac{5}{4} M_3(t) - \frac{1}{8}\{M_3(t+1) + M_3(t-1)\},$$
$$\varphi_2(t) := 4M_3(t) - 3M_4(t).$$

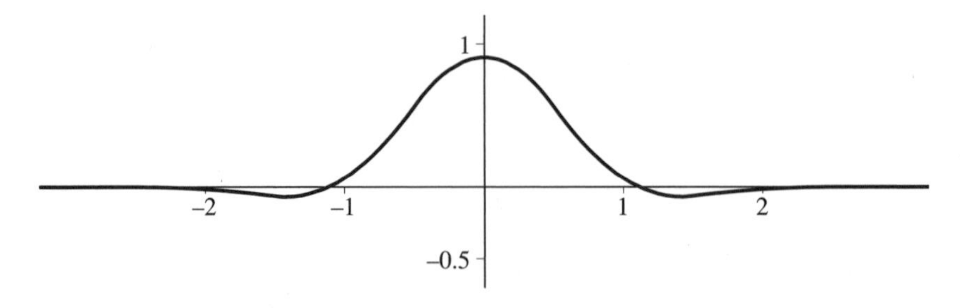

Figure 16. Kernel $\varphi_1(t)$, a linear combination of B-splines of order 3.

These are again approximate identities (Figs. 16 and 17), but satisfying now (ii) of (133) for $r = 3$. Hence one has for $s = 0, 1, 2$ and $0 < \alpha \leq 1$ that

$$\|S_\Omega^{\varphi_1} f - f\|_{C(\mathbb{R})} \leq M\omega(f^{(s)}, C(\mathbb{R}), \Omega^{-1}) \quad (\Omega > 0),$$

$$f^{(s)} \in \mathrm{Lip}(\alpha; C(\mathbb{R})) \Longrightarrow \|S_\Omega^{\varphi_1} f - f\|_{C(\mathbb{R})} = \mathcal{O}(\Omega^{-s-\alpha}) \quad (\Omega \to \infty);$$

the same results hold true for φ_2.

For a general approach to B-spline kernels see [49, 55], and for a multivariate setting [33, 34, 87, 214]. For sampling series based upon wavelets see [86] and the literature cited there. Furthermore, one can also approximate discontinuous functions by generalized sampling series cf. [46, 214].

2.8.3.3. Prediction of Signals

Kernels having compact support (Fig. 18) or having at least support bounded from below can also be used for prediction of signals, i.e., for evaluating $f(t)$ using samples from the past only. Indeed, if the kernel φ in (120) vanishes for $t \leq 0$, then $\varphi(t - k/W)$ vanishes for $k/W \leq t$, and only samples $f(k/W)$ for $k/W \leq t$ are needed. Choosing, e.g.,

$$\varphi_3(t) := 3M_2(t - h - 1) - 2M_2(t - h - 2)$$

for some $h \geq 0$, one has $\varphi(t) = 0$ for $t \notin (h, h+3)$, and the associated sampling series reads

$$(S_W^{\varphi_3} f)(t) = \sum_{\frac{\Omega t}{\pi} - h - 3 < k < \frac{\Omega t}{\pi} - h} f\left(\frac{k\pi}{\Omega}\right) \varphi\left(\frac{\Omega t}{\pi} - k\right). \tag{142}$$

This shows once more that one needs exclusively samples for $k\pi/\Omega < t - h\pi/\Omega \leq t$, thus samples from the past only.

Error estimates follow again from (134) and (135) with $r = 2$; hence there holds for $s = 0, 1$ and $0 < \alpha \leq 1$,

$$\|S_\Omega^{\varphi_3} f - f\|_{C(\mathbb{R})} \leq M\omega(f^{(s)}, C(\mathbb{R}), \Omega^{-1}) \quad (\Omega > 0), \tag{143}$$

$$f^{(s)} \in \mathrm{Lip}(\alpha; C(\mathbb{R})) \Longrightarrow \|S_\Omega^{\varphi_3} f - f\|_{C(\mathbb{R})} = \mathcal{O}(\Omega^{-s-\alpha}) \quad (\Omega \to \infty). \tag{144}$$

Since the parameter $h \geq 0$ is arbitrary, one can predict the signal f *arbitrarily far ahead*, at least theoretically. In practice, however, this is restricted by the fact that the constant M in (143) and the $\mathcal{O}$-constant in (144) depend on the kernel φ_3

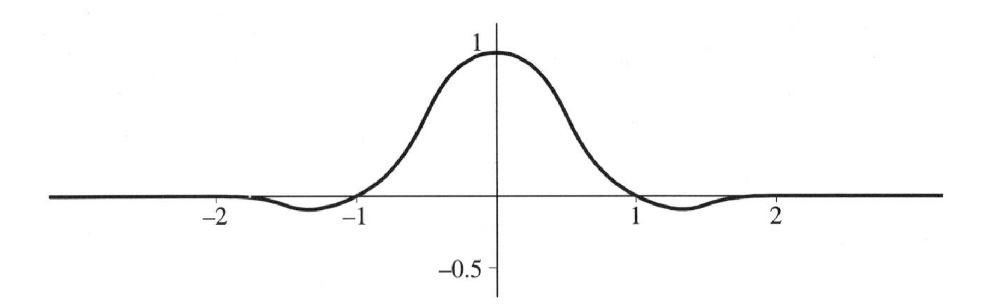

Figure 17. Kernel $\varphi_2(t)$, a linear combination of B-splines of orders 3 and 4.

and hence on h. These constants contain the moment $m_2(\varphi_3)$ as a factor which increases with h, and so may become quite large. Of course, this behaviour is to be expected. The farther you predict ahead the bigger is the error you make.

On the other hand, one can decrease the error by taking more samples into account. Whereas the kernel φ_3 needs just 3 samples in order to evaluate the series (142), the following kernel (Fig. 19)

$$\varphi_4(t) := \tfrac{1}{8}\{47M_3(t-h-1) - 62M_3(t-h-2) + 23M_3(t-h-3)\}$$

needs 5. It provides the same error estimates (143) and (144) as the kernel φ_3 but for $s = 0, 1, 2$. This yields smaller errors, at least for smoother functions. For the details see [53, 55].

2.9. Connections of the Sampling Theorem with other Basic Theorems

The aim of this section is to observe that five fundamental theorems in four fields of mathematics are equivalent to the sampling theorem in the sense that each can be deduced from any of the others by straightforward methods. These include Poisson's summation formula of Fourier analysis, Cauchy's integral formula of complex function theory, as well as the Euler-Maclaurin and Abel-Plana summation formulae of numerical analysis, and the functional equation of the Riemann zeta function of number theory.

The results will be presented at first in the case of band-limited functions—when certain results are equivalent to the classical sampling theorem, then for not-necessarily band-limited functions—when these are equivalent to the approximate sampling theorem. Some of the sampling formulae will be stated in a more general form than given above, namely for sampling not on the real axis $\mathbb{R}$, but on $\mathbb{C}$.

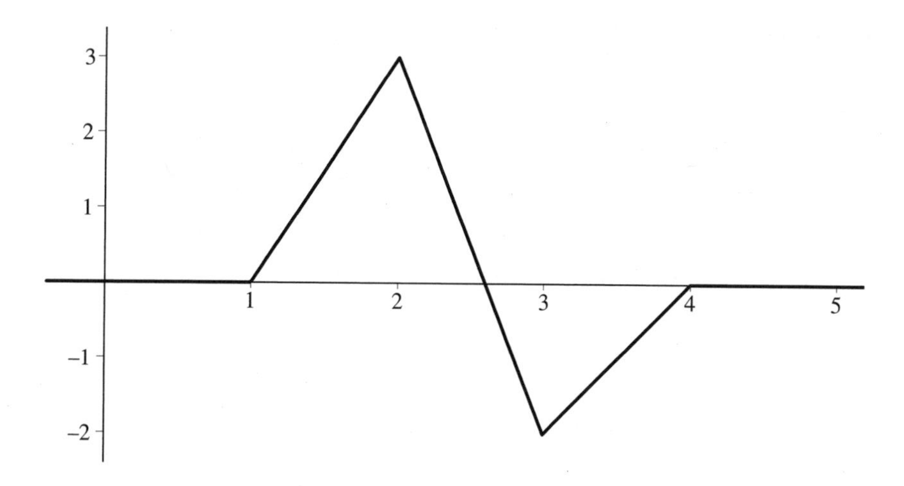

Figure 18. Kernel $\varphi_3(t)$ for $h = 1$, suitable for prediction.

2.9.1. Band-Limited Situation

The following four formulae are equivalent to another for functions $f : \mathbb{C} \to \mathbb{C}$. For the proofs we refer in particular to the recent paper by Higgins *et al.* [110], which also fills some gaps in the original proofs of the various equivalences first presented in several papers by the authors together with M. Hauss and S. Ries [36, 45, 51]; see also [109, Chapter 9].

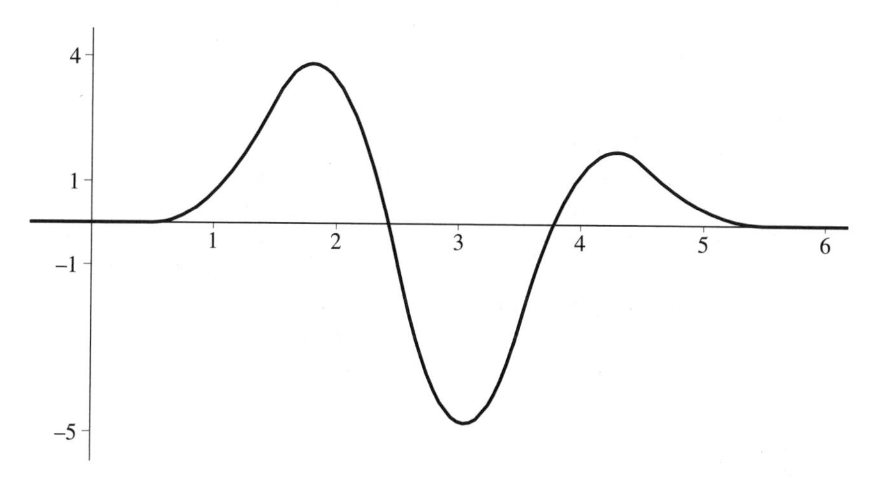

Figure 19. Kernel $\varphi_4(t)$ for $h = 1$, another one for prediction.

Whittaker-Kotel'nikov-Shannon sampling formula (cf. (64)): *Let* $f \in B_\Omega^p$ *for some* $1 \leq p < \infty$, *then*

$$f(z) = \sum_{k=-\infty}^{\infty} f\left(\frac{k\pi}{\Omega}\right) \operatorname{sinc}\left(\frac{\Omega z}{\pi} - k\right) \quad (z \in \mathbb{C}),$$

the convergence being absolute and uniform on compact subsets of $\mathbb{C}$.

Derivative sampling formula (cf. (92)): *Let* $f \in B_\Omega^p$ *for some* $1 \leq p < \infty$, *then*

$$f(z) = \sum_{k=-\infty}^{\infty} f\left\{\left(\frac{2k\pi}{\Omega}\right) + \left(z - \frac{2k\pi}{\Omega}\right) f'\left(\frac{2k\pi}{\Omega}\right)\right\} \left[\operatorname{sinc}\frac{1}{2}\left(\frac{\Omega z}{\pi} - 2k\right)\right]^2 \quad (z \in \mathbb{C}),$$

the convergence being absolute and uniform on compact subsets of $\mathbb{C}$.

Cauchy's integral formula: *Let* $f \in B_\Omega^\infty$, C *be a simple, closed, rectifiable (finite length), positively oriented curve in* $\mathbb{C}$. *Then*

$$\frac{1}{2\pi j} \int_C \frac{f(u)}{(u-z)} du = \begin{cases} f(z), & z \in \operatorname{int} C \\ 0, & z \in \operatorname{ext} C. \end{cases}$$

Poisson's summation formula—band-limited case (cf. (30)): *Let* $f \in B_\Omega^1$, *then*

$$\sum_{k=-\infty}^{\infty} f\left(\frac{2k\pi}{\Omega}\right) = \frac{\Omega}{2\pi} f^\wedge(0) = \frac{\Omega}{2\pi} \int_{-\infty}^{\infty} f(u)du,$$

the series being absolutely convergent.

Note that in Higgins et al. [110] one may also find discussions of the mutual equivalence of the classical Whittaker-Kotel'nikov-Shannon sampling formula with the Gaussian integral $\int_{-\infty}^{\infty} \exp(-u - jx)^2 du = \sqrt{\pi}$, $x \in \mathbb{R}$ and with certain properties of the weighted Hermite polynomials. These authors also treat the equivalence of Shannon's sampling formula with the Phragmèn-Lindelöf principle, and with the maximum principle of function theory.

2.9.2. General Case

Finally to the various formulae which are equivalent to the approximate sampling formula (cf. [42, 45]). Let us take $t \in \mathbb{R}$ as at the beginning.

Approximate sampling formula (cf. (95)): *If* $f \in C(\mathbb{R}) \cap L^1(\mathbb{R})$ *with* $F \in L^1(\mathbb{R})$, *then*

$$f(t) = \sum_{k=-\infty}^{\infty} f\left(\frac{k\pi}{\Omega}\right) \operatorname{sinc}\left(\frac{\Omega t}{\pi} - k\right) + R_\Omega f(t)$$

$$R_\Omega f(t) := \frac{1}{2\pi} \sum_{\mu=-\infty}^{\infty} (1 - e^{-j2\mu t\Omega}) \int_{(2\mu-1)\Omega}^{(2\mu+1)\Omega} F(v) e^{jvt} dv.$$

General Poisson summation formula (cf. (30)): *If* $f \in L^1(\mathbb{R})$ *is absolutely continuous and* $f' \in L^1(\mathbb{R})$, *then*

$$\sum_{k=-\infty}^{\infty} f\left(t + \frac{2k\pi}{\Omega}\right) = \frac{\Omega}{2\pi} \sum_{k=-\infty}^{\infty} F(k\Omega) e^{jkt\Omega}.$$

Euler-Maclaurin summation formula: *If* $f \in C^{2r}[n, m]$, *then*

$$\sum_{k=n}^{m} f(k) = \int_n^m f(u)\, du + \frac{1}{2}[f(n) + f(m)]$$

$$+ \sum_{k=1}^{r} \frac{B_{2k}}{(2k)!} \left\{ f^{(2k-1)}(m) - f^{(2k-1)}(n) \right\} + R_r,$$

$$R_r := \frac{(-1)^r}{(2\pi)^r} \sum_{k=-\infty}^{\infty}{}' \frac{1}{k^{2r}} \int_n^m f^{(2r)}(u) e^{-2jk\pi u} du,$$

the B_{2k} *being the Bernoulli numbers, see* [*51, 125*].

Abel-Plana summation formula ([184]): *Let* f *be analytic in* $\{z \in \mathbb{C};\ \mathrm{Re}(z) \geq 0\}$. *Then*

$$\sum_{k=0}^{\infty} f(k) - \int_0^\infty f(u)\, du = \frac{1}{2} f(0) + j \int_0^\infty \frac{f(jy) - f(-jy)}{e^{2\pi y} - 1} dy$$

if either the series or integral converges, and $\lim_{y \to \infty} |f(x \pm jy)| e^{-2\pi y} = 0$ *uniformly in* x *on each finite interval, and* $\int_0^\infty |f(x \pm jy)| e^{-2\pi y} dy$ *exists for every* $x > 0$ *and tends to zero for* $x \to \infty$.

Functional equation of the Riemann zeta function:

$$\pi^{z/2} \Gamma\left(\frac{z}{2}\right) \zeta(z) = \pi^{-(1-z)/2} \Gamma\left(\frac{1-z}{2}\right) \zeta(1-z)$$

or, setting $\xi(z) := (1/2)z(z-1)\pi^{-z/2}\Gamma(z/2)\zeta(z)$,

$$\xi(z) = \xi(1-z) \qquad (z \in \mathbb{C}).$$

There is no appropriate form of Cauchy's theorem in this context, but cf. [45] for the equivalence of a modified approximate sampling formula with a particular Cauchy theorem.

2.10. Sampling and Partial Fraction Decomposition

2.10.1. Cauchy's Partial Fraction Decomposition

Occasionally the question has been raised whether Cauchy knew about the sampling theorem. A discussion in [104, §1.2] ends with doubts. That topic has been taken up again in a recent article of Lacaze [133], where the author highlights two papers from Cauchy's "Exercices de Mathématiques" of 1827 (see [62] and [63]) since they include all the necessary material for deriving the sampling theorem. Here we favour another paper from the "Exercices de Mathématiques" of 1827 (see [61]) where Cauchy had a theorem on partial fraction decomposition of meromorphic functions. That result contains the sampling theorem and some of its variants as special cases. However, Cauchy did not explicitly point out that his formula allows a reconstruction of functions from *samples*. This may still prevent us from seeing Cauchy as one of the fathers of the sampling theorem.

Let us denote by res(f, a) the residue of a meromorphic function f at a (for these notions see [112] or any other textbook on complex analysis). In a more modern terminology, Cauchy's result [61, Théorème II] may be stated as follows:

Let ϕ be a meromorphic function. Suppose that there is a sequence $\{R_n\}_{n\in\mathbb{N}}$ of positive numbers approaching infinity such that $|\phi(R_n e^{j\theta})|$ remains bounded for all $\theta \in [0, 2\pi]$ and tends to zero as $n \to \infty$ except for certain special angles θ. Then

$$\phi(z) = \lim_{n\to\infty} \sum_{|\lambda_k|<R_n} \operatorname{res}\left(\frac{\phi}{z-\cdot}, \lambda_k\right), \tag{145}$$

where the numbers λ_k denote the poles of ϕ.

It was known to Cauchy [60, pp. 128–129] that the residues in (145) may be expressed as

$$\operatorname{res}\left(\frac{\phi}{z-\cdot}, \lambda_k\right) = \frac{A_{k,1}}{z-\lambda_k} + \frac{A_{k,2}}{(z-\lambda_k)^2} + \cdots + \frac{A_{k,\mu}}{(z-\lambda_k)^{\mu}} \tag{146}$$

with some positive integer μ and complex numbers $A_{k,1}, \ldots, A_{k,\mu}$. Therefore, (145) gives a partial fraction decomposition of ϕ.

Cauchy's proof is based on the formula

$$\phi(z) - \sum_{|\lambda_k|<R_n} \operatorname{res}\left(\frac{\phi}{z-\cdot}, \lambda_k\right) = \frac{1}{2\pi}\int_0^{2\pi} \frac{R_n e^{j\theta}\phi(R_n e^{j\theta})}{R_n e^{j\theta} - z}\, d\theta \qquad (|z| < R_n), \quad (147)$$

which originates from the residue theorem. The integral on the right-hand side is derived from the contour integral

$$\frac{1}{2\pi j}\oint_{|\zeta|=R_n} \frac{\phi(\zeta)}{\zeta - z}\, d\zeta. \qquad (148)$$

Therefore, we may say that Cauchy used the method of contour integration (see Section 2.3.2).

If in Cauchy's hypothesis on $|\phi(R_n e^{j\theta})|$ the "special angles" constitute a set of measure zero, then Lebesgue's dominated convergence theorem (see, e.g., [43, p. 3, Proposition 0.1.3]) guarantees that the integral on the right-hand side of (147) approaches zero as $n \to \infty$; thus (145) follows. However, Lebesgue's theorem is from 1910, and so it was definitely not available to Cauchy[3]. A suitable hypothesis for Cauchy to succeed without Lebesgue's theorem would be *that* $|\phi(R_n e^{j\theta})|$ *remains bounded for* $\theta \in [0, 2\pi]$ *and that for every* $\varepsilon \in (0, 2\pi)$, *there exists a subset of* $[0, 2\pi]$, *of measure* $2\pi - \varepsilon$, *on which* $|\phi(R_n e^{j\theta})|$ *tends to zero as* $n \to \infty$, *uniformly in* θ.

Either of the above two hypotheses on $|\phi(R_n e^{j\theta})|$ restricts the applicability of (145) to functions ϕ which are quotients of entire functions of exponential type. Cauchy did consider $\phi = f/g$ as a quotient of entire functions f and g, and in all of his examples f and g were of exponential type, but the significance of the exponential type, which could have led him to band-limited signals, was perhaps not recognized. Moreover, that the constants $A_{k,1}, \ldots, A_{k,\mu}$ in (146) are given by samples of f and its derivatives, was not emphasized.

We shall now show that Cauchy's theorem contains the sampling theorem and various variants and generalizations. This method of proof naturally extends the real variable results of Sections 2.3–2.5 to a complex variable. As far as uniform sampling is concerned, we may restrict ourselves to sampling on the integers since the results can be generalized by scaling.

[3] Preliminary forms of Lebesgue's theorem, which would suffice for Cauchy's proof, are due to Arzelà (1885) [4] and to Osgood (1897) [165], but they were not available to Cauchy either.

Sampling in B_Ω^∞. Let f belong to the Bernstein space B_Ω^∞ with $0 \le \Omega < \pi$. Then

$$|f(x+jy)| \le Ce^{\Omega|y|} \quad \text{with} \quad C = \|f\|_{C(\mathbb{R})}.$$

Setting $\phi(z) = f(z)/\sin \pi z$, and noting that

$$|\sin \pi(x+jy)| \ge \frac{1-e^{-2\pi|y|}}{2} e^{\pi|y|} \quad (x, y \in \mathbb{R}),$$

we find by a short reflection that the hypotheses of Cauchy's theorem are satisfied for

$$R_n := \frac{2n+1}{2} \quad (n \in \mathbb{N}). \tag{149}$$

Since ϕ has its poles at the integers, and

$$\operatorname{res}\left(\frac{\phi}{z-\cdot}, k\right) = \frac{(-1)^k f(k)}{\pi(z-k)} \quad (k \in \mathbb{Z}),$$

we obtain (cf. (64))

$$f(z) = \sin \pi z \lim_{n\to\infty} \sum_{k=-n}^{n} \frac{(-1)^k f(k)}{\pi(z-k)} = \lim_{n\to\infty} \sum_{k=-n}^{n} f(k)\operatorname{sinc}(z-k).$$

Here $\lim_{n\to\infty} \sum_{k=-n}^{n}$ cannot be replaced by $\sum_{k=-\infty}^{\infty}$, in general. Although $f \in B_\Omega^\infty$ with $\Omega < \pi$, the sampling rate here is 1, i.e., strictly larger than the Nyquist rate Ω/π. Indeed, the result fails if one samples at the Nyquist rate, as can be seen from the example $\sin \Omega z \in B_\Omega^\infty$, since

$$\sin \Omega z \ne \lim_{n\to\infty} \sum_{k=-n}^{n} \sin\left(\Omega \cdot \frac{k\pi}{\Omega}\right) \operatorname{sinc}(z-k) = 0.$$

In other words, if $f \in B_\Omega^\infty$ but $f \notin B_\Omega^p$ for any $1 \le p < \infty$, then one has to oversample in order to recover f from its samples.

The classical sampling theorem. Let $f \in B_\pi^2$. Proceeding as before, we find that $|\phi(R_n e^{j\theta})|$ remains bounded, but unfortunately it may not converge to zero as $n \to \infty$. However, by the basic properties of the space B_π^2,

$$\sin \pi z \sum_{k=-\infty}^{\infty} \frac{(-1)^k f(k)}{\pi(z-k)}$$

converges uniformly on compact subsets of $\mathbb{C}$ and represents an entire function which belongs to L^2 on the real line. In the expression

$$F(z) := \frac{f(z)}{\sin \pi z} - \sum_{k=-\infty}^{\infty} \frac{(-1)^k f(k)}{\pi(z-k)}$$

all the poles have been removed. Thus (147) shows that F is a bounded entire function and hence, by a theorem of Liouville, it is a constant c. Consequently,

$$f(z) - \sin \pi z \sum_{k=-\infty}^{\infty} \frac{(-1)^k f(k)}{\pi(z-k)} = c \sin \pi z.$$

As a function of z, the left-hand side belongs to $L^2(\mathbb{R})$. Therefore the same must be true for the right-hand side. This implies that $c = 0$. Thus the classical sampling theorem (cf. (64)) is derived.

It should be mentioned that Cauchy [61, Théorème 1] did consider the case that the right-hand side of (147) does not converge to zero as $n \to \infty$ but to some constant $c \neq 0$.

Valiron—Tschakaloff interpolation formula. Let $f \in B_\pi^\infty$, then Cauchy's theorem applies to

$$\phi(z) := \frac{f(z)}{z \sin \pi z}$$

with R_n given by (149). Since

$$\operatorname{res}\left(\frac{\phi}{z - \cdot}, k\right) = \begin{cases} (-1)^k \dfrac{f(k)}{k(z-k)} & \text{if } k \in \mathbb{Z}\setminus\{0\}, \\ \dfrac{1}{\pi}\left(\dfrac{f'(0)}{z} + \dfrac{f(0)}{z^2}\right) & \text{if } k = 0, \end{cases}$$

we obtain the interpolation formula of Valiron [223] and Tschakaloff [220] (cf. (81))

$$\begin{aligned} f(z) &= \frac{\sin \pi z}{\pi}\left(f'(0) + \frac{f(0)}{z} + \lim_{n\to\infty} \sum_{0<|k|\le n} (-1)^k \frac{\pi z f(k)}{k(z-k)}\right) \\ &= \{zf'(0) + f(0)\}\operatorname{sinc}(z) + \lim_{n\to\infty} \sum_{0<|k|\le n} f(k) \frac{\pi z}{k} \operatorname{sinc}(z-k). \end{aligned} \tag{150}$$

Here $\lim_{n\to\infty} \sum_{0<|k|\le n}$ may be replaced by $\sum_{k=-\infty}^{\prime\infty}$, which means that the summation extends over all integers different from zero.

Sampling with a multiplier. Let f be an entire function of exponential type Ω with $0 \le \Omega < \pi$. Now f need not be bounded on the real line, but let us suppose that $f(x) = \mathcal{O}(|x|^{\ell})$ as $x \to \pm\infty$ with some positive integer ℓ. For an integer m at least as large as ℓ, we choose a positive ε less than $(\pi - \Omega)/(\pi m)$ and set

$$\varphi(z) := \operatorname{sinc}^m(\varepsilon z).$$

Then Cauchy's theorem applies to

$$\phi(z) := \frac{\varphi(\zeta - z)f(z)}{\sin \pi z}$$

with R_n given by (149), where ζ is an arbitrary complex number. Thus we find that

$$\varphi(\zeta - z)f(z) = \sin \pi z \lim_{n\to\infty} \sum_{k=-n}^{n} \frac{(-1)^k \varphi(\zeta - k)f(k)}{\pi(z-k)}.$$

On substituting $\zeta = z$, we obtain

$$f(z) = \lim_{n\to\infty} \sum_{k=-n}^{n} f(k)\varphi(z-k)\operatorname{sinc}(z-k). \tag{151}$$

The summation on the right-hand side may be replaced by $\sum_{k=-\infty}^{\infty}$ if $m > \ell$.

There exist multipliers φ which allow one to sample functions that grow on the real line faster than any polynomial (see [93]). However, the multiplier technique is also of interest if f is bounded on $\mathbb{R}$ since it improves the properties of the sampling series considerably. For the various advantages of sampling with a rapidly decaying band-limited multiplier φ, see the results in [179].

Derivative sampling. Let $f \in B_{\Omega}^{\infty}$ with $0 \le \Omega < 2\pi$, then Cauchy's theorem applies to

$$\phi(z) = \frac{f(z)}{\sin^2 \pi z}$$

with R_n given by (149). A calculation yields that

$$\operatorname{res}\left(\frac{\phi}{z - \cdot}, k\right) = \frac{1}{\pi^2}\left(\frac{f(k)}{(z-k)^2} + \frac{f'(k)}{z-k}\right).$$

Hence there follows (cf. (92))

$$
\begin{aligned}
f(z) &= \frac{\sin^2 \pi z}{\pi^2} \lim_{n\to\infty} \sum_{k=-n}^{n} \left(\frac{f(k)}{(z-k)^2} + \frac{f'(k)}{z-k} \right) \\
&= \lim_{n\to\infty} \sum_{k=-n}^{n} \{f(k) + (z-k)f'(k)\} \operatorname{sinc}^2(z-k).
\end{aligned}
\tag{152}
$$

An analogous consideration for

$$
\phi(z) = \frac{f(z)}{\sin^m \pi z} \qquad (m \in \mathbb{N})
$$

leads to a sampling formula involving samples of f and its derivatives f' up to $f^{(m-1)}$. This formula holds for all $f \in B_\Omega^\infty$ with $0 \le \Omega < m\pi$.

In all of the preceding examples, we have applied Cauchy's result to the fraction

$$
\frac{f(z)}{\sin \pi z}
$$

or to a modification with an additional multiplier in the numerator or in the denominator. However, the multipliers in the denominator were quite special; in particular, they kept all the poles at the integers. It seems that Cauchy's theorem is tailored for these situations. Now we want to replace the sine function by a more general denominator G. Then it may become difficult to find a sequence of radii R_n such that the hypotheses on $|\phi(R_n e^{j\theta})|$ are satisfied. But a modification of Cauchy's approach will help. In fact, it will suffice to replace the circles of radii R_n in (148) by more general curves γ_n whose shape allows some flexibility near the zeros of G.

Irregular sampling. According to Levin (e.g., [136]), an entire function G of exponential type is called a *sine-type function* if its zeros are simple and separated and there exist constants C_1, C_2, and η such that

$$
C_1 e^{\pi|y|} \le |G(x+jy)| \le C_2 e^{\pi|y|} \quad \text{for } x, y \in \mathbb{R},\ |y| \ge \eta.
$$

Now, let $f \in B_\Omega^\infty$ with $0 \le \Omega < \pi$, let G be a sine-type function with zeros $\lambda_k (k \in \mathbb{Z})$, and set $\phi := f/G$. Then the above mentioned modification of Cauchy's approach yields

$$
\frac{f(z)}{G(z)} = \lim_{\rho\to\infty} \sum_{|\lambda_k|<\rho} \operatorname{res}\left(\frac{f}{(z-\cdot)G}, \lambda_k \right).
$$

Noting that

$$\operatorname{res}\left(\frac{f}{(z-\cdot)G}, \lambda_k\right) = \frac{f(\lambda_k)}{(z-\lambda_k)G'(\lambda_k)},$$

we obtain the Lagrange type formula

$$f(z) = G(z) \lim_{n\to\infty} \sum_{k=-n}^{n} \frac{f(\lambda_k)}{(z-\lambda_k)G'(\lambda_k)}.$$

For extensions to sequences $\{\lambda_k\}_{k\in\mathbb{Z}}$ which are not necessarily the set of zeros of a sine-type function and combinations with the multiplier technique as well as extensions to derivative sampling, see Hinsen [113] and Voss [226, Theorem 2.2.1, Corollary 2.2.3, Corollary 2.3.1 and Corollary 2.3.3].

2.10.2. An Improvement of Cauchy's Method

Some textbooks on complex analysis (e.g., Behnke & Sommer [11, pp. 241–244], Kneser [124, § 3.05] present an improvement of Cauchy's method and attribute it also to Cauchy without giving a precise reference.

Let ϕ be a meromorphic function, and let λ_k be a pole of ϕ. Denote by $h_k(z)$ the principal part (or singular part) of the Laurent expansion of ϕ about λ_k. Then $h_k(z)$ is equal to the right-hand side of (146). Hence (147) may be rewritten as

$$\phi(z) - \sum_{|\lambda_k|<R_n} h_k(z) = \frac{1}{2\pi j}\oint_{|\zeta|=R_n} \frac{\phi(\zeta)}{\zeta - z}\, d\zeta. \tag{153}$$

Equation (153) remains true if we expand both sides into a Laurent series about $z = 0$ and cancel the constant terms and the terms containing z, z^2 up to z^{m-1}. We find it convenient to have a short notation for that operation. For an arbitrary meromorphic function ψ with a Laurent series

$$\psi(z) = \frac{a_{-\mu}}{z^\mu} + \cdots + \frac{a_{-1}}{z} + \sum_{k=0}^{\infty} a_k z^k,$$

we define

$$\psi(z)_{(m)} := \frac{a_{-\mu}}{z^\mu} + \cdots + \frac{a_{-1}}{z} + \sum_{k=m}^{\infty} a_k z^k.$$

Observing that

$$\frac{1}{\zeta - z} = \sum_{k=0}^{\infty} \frac{z^k}{\zeta^{k+1}} \quad \text{for } |z| < |\zeta|,$$

we find that

$$\left(\frac{1}{\zeta - z}\right)_{(m)} = \frac{z^m}{\zeta^m} \cdot \frac{1}{\zeta - z},$$

and so an application of the (m) operation to (153) yields that

$$\phi(z)_{(m)} - \sum_{|\lambda_k| < R_n} h_k(z)_{(m)} = \frac{z^m}{2\pi j} \oint_{|\zeta| = R_n} \frac{\phi(\zeta)}{\zeta^m (\zeta - z)} d\zeta.$$

By integrating along more general curves than circles, we may state the following theorem (see [124, p. 170]).

Let ϕ be a meromorphic function with poles λ_k and associated principal parts $h_k(z)$. Let $\{\mathcal{G}_n\}_{n \in \mathbb{N}}$ be a sequence of bounded regions whose boundaries γ_n are piecewise smooth Jordan curves not containing any of the poles of ϕ, and suppose that all the points of γ_n approach infinity as $n \to \infty$. If

$$\lim_{n \to \infty} \int_{\gamma_n} \frac{\phi(\zeta)}{\zeta^m (\zeta - z)} d\zeta = 0 \tag{154}$$

for some $m \in \mathbb{N}$, then

$$\phi(z)_{(m)} = \lim_{n \to \infty} \sum_{\lambda_k \in \mathcal{G}_n} h_k(z)_{(m)}. \tag{155}$$

An application. Let f be an entire function of exponential type π. Suppose that $f(x) = \mathcal{O}(|x|^\ell)$ as $x \to \pm\infty$ with some positive integer ℓ. Set

$$\phi(z) := \frac{f(z)}{\sin \pi z}.$$

Then (154) holds for every integer $m > \ell$ with γ_n being a circle of radius $n + 1/2$ centred at the origin. The principal part of $\phi(z)$ at the pole k is

$$h_k(z) := \frac{(-1)^k f(k)}{\pi(z - k)}.$$

From this we deduce that

$$h_0(z)_{(m)} = \frac{f(0)}{\pi z}$$

and

$$h_k(z)_{(m)} = \frac{(-1)^k z^m f(k)}{\pi k^m (z - k)} \quad \text{for} \quad k \neq 0.$$

Defining

$$a_\mu := \frac{1}{(\mu+1)!}\left(\frac{d^{\mu+1}}{d\zeta^{\mu+1}}\frac{\zeta f(\zeta)}{\sin \pi\zeta}\right)_{\zeta=0},$$

we obtain from (155)

$$f(z) = \frac{\sin \pi z}{\pi}\left(\frac{f(0)}{z} + \lim_{n\to\infty}\sum_{0<|k|\le n}\frac{(-1)^k z^m f(k)}{k^m(z-k)}\right) + \sin \pi z \sum_{\mu=0}^{m-1} a_\mu z^\mu. \tag{156}$$

This is Valiron's interpolation formula of rank m (see [223, p. 204, formula (9)]). For $\ell = 0$ and $m = 1$, we obtain the Valiron–Tschakaloff interpolation formula (150). Comparing with the multiplier technique presented in Section 2.10.1, we note that in (156) the exponential type is allowed to be π, that is, over-sampling is not required. Moreover, the basis functions in the interpolation formula (156) are of exponential type π, while in (151) they are of exponential type $\pi(1+m\varepsilon)$.

On replacing $\sin \pi z$ by a sine-type function G, we can deduce from (155) an interpolation formula analogous to (156) but with nonuniform nodes. In fact, let λ_k ($k \in \mathbb{Z}$) be the zeros of G, and suppose that $\lambda_0 = 0$. Then, setting

$$a_\mu := \frac{1}{(\mu+1)!}\left(\frac{d^{\mu+1}}{d\zeta^{\mu+1}}\frac{\zeta f(\zeta)}{G(\zeta)}\right)_{\zeta=0},$$

we obtain

$$f(z) = G(z)\left(\frac{f(0)}{zG'(0)} + \lim_{n\to\infty}\sum_{0<|k|\le n}\frac{z^m f(\lambda_k)}{\lambda_k^m G'(\lambda_k)(z-\lambda_k)}\right) + G(z)\sum_{\mu=0}^{m-1} a_\mu z^\mu.$$

2.10.3. The Mittag-Leffler Theorem

The most general theorem on partial fraction decomposition is due to Mittag-Leffler. It does not need any growth conditions on the meromorphic function ϕ. Using the notations of Section 2.10.2, we may state it as follows (e.g., [112, p. 219, Theorem 8.5.2]):

Let ϕ be a meromorphic function with poles λ_k and associated principal parts $h_k(z)$. Then there exists a sequence of positive integers m_k and an entire function H such that

$$\phi(z) = \sum_k h_k(z)_{(m_k)} + H(z),$$

where the series converges absolutely and uniformly on compact subsets of $\mathbb{C}$.

This formula may be applied as follows. Let f be an arbitrary entire function and let $\{\lambda_k\}_{k\in\mathbb{Z}}$ be an arbitrary sequence of distinct complex numbers without an accumulation point. By a theorem of Weierstrass there exists an entire function G whose zeros are the numbers λ_k ($k \in \mathbb{Z}$). Denote by $h_k(z)$ the principal part of f/G with respect to the pole λ_k. Then there exists a sequence of positive integers $\{m_k\}_{k\in\mathbb{Z}}$ and an entire function H such that

$$f(z) = G(z) \sum_{k=-\infty}^{\infty} h_k(z)_{(m_k)} + G(z)H(z). \tag{157}$$

Obviously, this formula generalizes Valiron's interpolation formula (156). Moreover, every interpolation formula flowing from (155) has the form (157) with H being always a polynomial and the numbers m_k being all the same.

In (157), we may regard

$$S[f](z) := G(z) \sum_{k=-\infty}^{\infty} h_k(z)_{(m_k)} \tag{158}$$

as a sampling series of f with respect to the nodes λ_k ($k \in \mathbb{Z}$), and then $G(z)H(z)$ will be the remainder (more precisely, the *aliasing error*) in the representation of f by that sampling series.

Alternatively, we may regard H as the "integer part" in the division f/G and then $S[f](z)$ takes the role of a remainder. The equation

$$f(z) = H(z)G(z) + S[f](z)$$

describes a division transformation just as $19/7$ leads to

$$19 = 2 \times 7 + 5.$$

Hence f is representable by a sampling series of the form (158) with the zeros of G as nodes if and only if the division f/G has no integer part, i.e., $H = 0$. The determination of H is often a serious problem which limits the practical use of the Mittag-Leffler theorem. When in textbooks on complex analysis the Mittag-Leffler theorem is applied to concrete functions such as the cotangent function or the reciprocal of the sine function, special efforts are needed for showing that $H = 0$. Some of the common techniques amount to an application of Cauchy's method.

2.11. Sampling and Quadrature

2.11.1. Quadrature of Band-Limited Signals over the Real Line

A common method of constructing quadrature formulae consists in integrating both sides of an interpolation formula. The resulting quadrature formulae are said to be *interpolatory.*

If we take the sampling formula for band-limited signals f, namely

$$f(t) = \sum_{k=-\infty}^{\infty} f\left(\frac{k\pi}{\Omega}\right) \operatorname{sinc}\left(\frac{\Omega t}{\pi} - k\right) \qquad (t \in \mathbb{R}), \tag{159}$$

and integrate both sides formally over $\mathbb{R}$, interchanging integration and summation on the right-hand side, we obtain

$$\int_{-\infty}^{\infty} f(t)dt = \frac{\pi}{\Omega} \sum_{k=-\infty}^{\infty} f\left(\frac{k\pi}{\Omega}\right). \tag{160}$$

This is nothing but the trapezoidal rule on $\mathbb{R}$. Hence the trapezoidal rule on $\mathbb{R}$ is interpolatory for certain classes of band-limited signals.

Although (160) was formally derived from (159), the two formulae have somewhat different ranges of validity. We know that (159) holds for functions belonging to the Bernstein space B_Ω^p with $1 \leq p < \infty$. But if $p > 1$, then $\int_{-\infty}^{\infty} f(t)dt$ may not exist. On the other hand, we shall see that for the validity of (160) it is not necessary that f be band-limited to $[-\Omega, \Omega]$.

For exact statements we recall that an integral $\int_{-\infty}^{\infty} f(t)\, dt$ is said to exist as a *Cauchy principal value* if

$$\lim_{T\to\infty} \int_{-T}^{T} f(t)dt$$

exists. Analogously, a series $\sum_{k=-\infty}^{\infty} a_n$ is said to exist as a *Cauchy principal value* if

$$\lim_{N\to\infty} \sum_{k=-N}^{N} a_k$$

exists.

Concerning the validity of (160), the following statements hold:

If f belongs to the Bernstein space $B_{2\Omega}^1$, then (160) *holds with the integral existing in the sense of Lebesgue and the series converges absolutely* (see [20]).

If f is an entire function of exponential type less than 2Ω and the integral in (160) *exists as an improper Riemann integral, then* (160) *holds with the series converging conditionally* (see [90, Theorem 1], [182, Corollary 3]).
If f is an entire function of exponential type less than 2Ω and the integral in (160) *exists as a Cauchy principal value, then* (160) *holds in the sense of Cauchy principal values* (see [182, Corollary 4]).

In these statements, the hypotheses cannot be relaxed. However, the assumption concerning the existence of the integral in (160) can be replaced by a corresponding assumption on the convergence of the series along with an asymptotic condition on f (see [182]).

The preceding results can be generalized. Instead of (159), we may consider a corresponding formula for derivative sampling with the same nodes, or equivalently, a *Hermite interpolation formula*. If the formula involves derivatives up to order $2m$, then integration over $\mathbb{R}$ leads to the quadrature formula

$$\int_{-\infty}^{\infty} f(t)dt = \sum_{k=-\infty}^{\infty} \sum_{\mu=0}^{m} c_{m,2\mu}\left(\frac{\pi}{\Omega}\right)^{2\mu+1} f^{(2\mu)}\left(\frac{k\pi}{\Omega}\right), \tag{161}$$

where the constants $c_{m,2\mu}$ $(\mu = 0, \ldots, m)$ are given by (see [131])

$$\prod_{\ell=1}^{m}\left(1 + \left(\frac{t}{2\pi\ell}\right)^2\right) =: \sum_{\mu=0}^{m} c_{m,2\mu} t^{2\mu}.$$

Concerning the validity of (161), the statements for (160) hold analogously with 2Ω replaced by $2\Omega(m+1)$; see [163] and [182]. For $m = 0$ all the results for (161) reduce to those for (160).

2.11.2. Generalized Gaussian Quadrature Formulae

In the previous section, we have seen that the quadrature formula (160) holds for signals with a bandwidth which is twice as large as that allowed in the underlying interpolation formula (159). A similar phenomenon occurs in connection with Gaussian quadrature. While a polynomial of degree $n-1$ is uniquely determined by n samples, the classical Gaussian quadrature formula with n nodes is interpolatory and is exact for polynomials up to degree $2n-1$. Therefore (160) may be seen as an analogue of the Gaussian quadrature formula.

Generally, we may proceed as follows. Let $\{V_\Omega\}$ be a family of function spaces such that $V_\Omega \subset V_{\Omega'}$ for $\Omega < \Omega'$, where Ω and Ω' are positive real numbers. Suppose that we have an interpolation formula of the form

$$f(t) = \sum_k f(t_{\Omega,k}) L_{\Omega,k}(t) \tag{162}$$

which holds for $f \in V_\Omega$, whereas for $\Omega' > \Omega$, there exists a $g \in V_{\Omega'}$ for which (162) fails. Then we say that

$$\int_I f(t)dt = \sum_k f(t_{\Omega,k}) \int_I L_{\Omega,k}(t)dt \tag{163}$$

is a *generalized Gaussian quadrature formula* associated with V_Ω if (163) holds for all $f \in V_{2\Omega}$. Roughly speaking, a Gaussian formula is an interpolatory quadrature formula whose degree of exactness is at least twice the degree of exactness of the underlying interpolation formula.

Of course, only an appropriate choice of the nodes $\{t_{\Omega,k}\}$ in (162) will lead to a Gaussian formula. Those nodes are called *Gaussian nodes*. It is a fortunate incidence that, for band-limited signals, uniform nodes are Gaussian.

In the classical polynomial case, the Gaussian nodes are complicated algebraic numbers which have been characterized in various ways (see [77, Sections 2.3–2.5, 7.1, 7.3]). We mention two of them. The first includes an explanation for the Gaussian phenomenon of doubled degree of exactness.

(1) *Given distinct points $t_1, \ldots, t_n$, there exists a unique interpolation formula of the form*

$$f(t) = \sum_{k=1}^{n} f(t_k)G_k(t) + f'(t_k)H_k(t) \tag{164}$$

which holds for polynomials up to degree $2n-1$. Then the nodes $t_1, \ldots, t_n$ are Gaussian for quadrature over an interval I if and only if

$$\int_I H_k(t)dt = 0 \quad \text{for} \quad k = 1, \ldots, n. \tag{165}$$

(2) *Among all the points $t_1, \ldots, t_n$, the Gaussian nodes for quadrature over I are the only ones that minimize the integral*

$$\Phi(t_1, \ldots, t_n) := \int_I \prod_{k=1}^{n} |t - t_k|^2 dt. \tag{166}$$

Analogous characterizations hold for the generalized Gaussian quadrature formula (160). For example, Kress [131] derived (160) from the formula for derivative sampling (152), which corresponds to (164). The question whether for quadrature over $\mathbb{R}$ the uniform nodes are the only Gaussian nodes and whether there is a characterization by an external problem analogous to (166), is answered in [180]. For a presentation of these results, we need some terminology.

By US(Ω) we denote the class of all increasing sequences $\boldsymbol{t} = \{t_k\}_{k\in\mathbb{Z}}$ of real numbers with the following properties:

(a) $t_0 = 0$ and $t_k \to \pm\infty$ as $k \to \pm\infty$.
(b) If an entire function f of exponential type less than Ω vanishes on the sequence $\boldsymbol{t}$, then $f \equiv 0$.
(c) There exists an entire function $\phi_{\boldsymbol{t}}$ of order 1 and type Ω such that $\phi_{\boldsymbol{t}}(\mathbb{R}) \subset \mathbb{R}$, $\phi'_{\boldsymbol{t}}(0) = 1$, and $\phi_{\boldsymbol{t}}(t_k) = 0$ for $k \in \mathbb{Z}$.

We call US (Ω) the class of *uniqueness sequences* of type Ω.

Within the class US(Ω), the Gaussian nodes for band-limited signals can be characterized as follows (see [180, Theorem 2]):

For sequences $\boldsymbol{t} = \{t_k\}_{k\in\mathbb{Z}} \in \mathrm{US}(\Omega)$, *the following statements are equi-valent:*

(i) $t_k = k\pi/\Omega \quad (k \in \mathbb{Z})$.
(ii) *There exists a quadrature formula*

$$\int_{-\infty}^{\infty} f(t)dt = \sum_{k=-\infty}^{\infty} A_k f(t_k)$$

which holds for all entire functions of exponential type 2Ω *belonging to* $L^1(\mathbb{R})$.
(iii) *Every function* $f := \phi_{\boldsymbol{t}}$ *with* $\phi_{\boldsymbol{t}}$ *as specified in* (c) *minimizes the integral*

$$\int_{-\infty}^{\infty} t^{-2}|f(t)|^2 dt$$

over all entire functions f of exponential type Ω *satisfying* $f(0) = 0$ *and* $f'(0) = 1$.

If (ii) holds, then $A_k = \pi/\Omega$ for all $k \in \mathbb{Z}$. It should be observed that the class US (Ω) contains the sequences which are commonly used in irregular sampling with real nodes in case of functions band-limited to $[-\Omega, \Omega]$; see [101, 113, 196].

In the definition of the class US(Ω), the condition that $t_0 = 0$ is a convenient normalization for the formulation of property (c) and statement (iii). Two sequences $\{s_k\}_{k\in\mathbb{Z}}$ and $\{t_k\}_{k\in\mathbb{Z}}$ may be considered as equivalent if there is a $c \in \mathbb{R}$ such that $s_k = t_k - c(k \in \mathbb{Z})$. Of course, an obvious modification of the statements (i)–(iii) holds if the sequence $\boldsymbol{t}$ is only equivalent to one in US(Ω).

The concept of Gaussian quadrature formulae was generalized by Turán [221] in another way. He considered quadrature formulae of the form

$$\int_a^b f(t)dt = \sum_{k=1}^{n} \sum_{\mu=0}^{m-1} \lambda_{k,\mu} f^{(\mu)}(t_k) \tag{167}$$

and asked for nodes $t_1, \ldots, t_n$ and coefficients $\lambda_{k,\mu}$ ($k = 1, \ldots, n$; $\mu = 0, \ldots, m-1$) so that (167) holds for polynomials up to a degree N as large as possible. He showed that for even $m \in \mathbb{N}$ there is no system of real nodes so that (167) holds for all polynomials of degree $N \leq mn$. For odd $m \in \mathbb{N}$ there is unique system of nodes such that (167) holds for all polynomials of degree $N \leq (m+1)n - 1$. These nodes minimize the integral

$$\Phi_m(t_1, \ldots, t_n) := \int_a^b \prod_{k=1}^{n} |t - t_k|^{m+1} dt.$$

The associated quadrature formula is called *Turán's formula*. The case $m = 1$ reduces to the classical Gaussian formula.

Analogues of Turán's results were established for entire functions of exponential type or band-limited signals (see [181, Theorem 1]). It turned out that in this situation, formula (161) is the uniquely determined Turán formula. Dryanov [70] explored a class of generalizations of (161) and showed in [71] that, within this class, formula (161) is optimal in some sense.

2.11.3. Quadrature of Non-Band-Limited Signals over the Real Line

If f is not band-limited to $[-2\Omega, 2\Omega]$, then the right-hand side of (160) needs an additional term $R_\Omega[f]$, defined by

$$R_\Omega[f] := \int_{-\infty}^{\infty} f(t)\, dt - \frac{\pi}{\Omega} \sum_{k=-\infty}^{\infty} f\left(\frac{k\pi}{\Omega}\right), \tag{168}$$

which is called the *remainder*.

Since the interval of integration is unbounded, the classical representation of the remainder of the trapezoidal rule (e.g., [67, formula (2.1.11)] is not applicable. Fortunately, an extremely efficient representation for $R_\Omega[f]$ arises from the generalized Poisson summation formula (see (30) and Section 2.9.2). In fact:

If $f \in L^1(\mathbb{R})$ is absolutely continuous and $f' \in L^1(\mathbb{R})$, then

$$R_\Omega[f] := -2 \sum_{k=1}^{\infty} F_c(2k\Omega), \tag{169}$$

where

$$F_c(v) := \int_{-\infty}^{\infty} f(t) \cos(tv) dt$$

is the cosine transform of f.

For this result and slight variants, see [20, 51] and [52]. The hypotheses on f can be relaxed by applying a summability method to the series in (169). Two typical results are as follows:

Let $f \in C(\mathbb{R}) \cap L^1(\mathbb{R})$ be such that

$$\sum_{k=-\infty}^{\infty} \left| f\left(\frac{k\pi}{\Omega} - u\right) \right|$$

is uniformly convergent in $u \in$ for each fixed $\Omega > 0$. Then

$$R_\Omega[f] = -2 \lim_{N \to \infty} \sum_{k=1}^{N} \left(1 - \frac{|k|}{N}\right) F_c(2k\Omega) \tag{170}$$

and

$$R_\Omega[f] = -2 \lim_{\delta \to 0} \sum_{k=1}^{\infty} e^{-2k\Omega\delta} F_c(2k\Omega). \tag{171}$$

For (170), see [51, Theorem 2]; representation (171) was obtained in [73, Lemmma 4.3] under a weaker condition on f. Both results are covered by a general theorem of Borgen [22] on the summability of Poisson's formula.

The advantage of the representations (170) and (171) is that they hold under relatively weak conditions on f. If f is analytic in a strip

$$S_d := \{z \in \mathbb{C} : |\text{Im } z| < d\} \quad (d > 0),$$

then alternative representations for $R_\Omega[f]$, based on methods of complex analysis, have been established (see [130, 132, 150, 153, 207, 208], and [209]). Two typical error estimates deduced from these representations or those of (169)–(171) are as follows:

Let $f \in C(\mathbb{R}) \cap L^1(\mathbb{R})$ be such that $f^{(m)} \in L^1(\mathbb{R})$ for some $m \in \mathbb{N}$. *If the modulus of continuity of $f^{(m)}$ satisfies*

$$\omega(\delta, f^{(m)}; L^1(\mathbb{R})) \leq L\delta^\alpha$$

for some $\alpha \in (0, 1]$, then

$$|R_\Omega[f]| \leq \frac{M}{\Omega^{m+\alpha}} \quad (\Omega > 0) \tag{172}$$

with a constant M depending only on L, m, and α; see [51, Corollary 3].

Let $\mathcal{B}_d^1$ denote the set of all functions f analytic in the strip S_d such that

$$\int_{-d}^{d} |f(t+jy)|dy \longrightarrow 0 \quad \text{as } t \to \pm\infty$$

and

$$N(f, S_d) := \limsup_{y \to d-} \int_{-\infty}^{\infty} (|f(t+jy)| + |f(t-jy)|)dt < \infty.$$

If $f \in \mathcal{B}_d^1$, then

$$R_\Omega[f] = \mathcal{O}\left(e^{-2d\Omega}\right) \quad \text{as } \Omega \to \infty; \tag{173}$$

see [208, 209] and [52].

Towards converse results, one may ask whether in these statements the hypotheses on f are *necessary.* The answer is "no" since for an odd functions f, i.e., $f(t) \equiv -f(-t)$, we have $R_\Omega[f] = 0$ by symmetry, and so (172) and (173) hold no matter whether f is very regular or rather erratic. However, regularity properties of a function f are inherited by its translates $f_h := f(\cdot + h)$, but the same is not true for symmetries such as f being an odd function. Moreover, the integral $\int_{-\infty}^{\infty} f(t)dt$ is invariant under translation. For these reasons, it seems appropriate to consider the rate of convergence of $R_\Omega[f_h]$ as $\Omega \to 0$, with h varying in some interval, in order to establish converse results.

On the basis of representation (171), the following characterizations hold.

(i) *The remainders satisfy*

$$R_\Omega[f_h] = 0$$

for all $\Omega \geq \Omega_0/2$ and $h \in [-\pi/2\Omega, \pi/2\Omega]$, where $\Omega_0 > 0$, if and only if f is band-limited to $[-\Omega_0, \Omega_0]$.

(ii) *The remainders satisfy*

$$|R_\Omega[f_h]| \leq \frac{g(\Omega)}{\Omega^m}$$

for all $\Omega \geq \Omega_0$ and $h \in [-\pi/2\Omega, \pi/2\Omega]$ with an integer $m \in \mathbb{N}$, a real number $\Omega_0 > 0$, and a function $g \in L^2[\Omega_0, +\infty)$ if and only if f belongs to the Sobolev space $W^m_{L^2(\mathbb{R})}$.

Furthermore, there exists a normed linear space $\widetilde{\mathcal{B}}_d$ of functions analytic in S_d with $\mathcal{B}_d^1 \subset \widetilde{\mathcal{B}}_d$, which allows the following characterization:

(iii) *The remainders satisfy*

$$|R_\Omega[f_h]| \le M e^{-2d\Omega}$$

for all $\Omega \ge \Omega_0$ *and* $h \in [-\pi/2\Omega, \pi/2\Omega]$ *with positive constants M, d, and* Ω_0 *if and only if f is the restriction to* $\mathbb{R}$ *of a function belonging to* $\widetilde{\mathcal{B}}_d$.

For the details we refer to [73, Theorems 3.1–3.3]. In [72], the characterizations (I)–(III) were extended to the Turán type formula (161).

Another way of excluding odd functions consists in considering the even part

$$f_e(t) := \frac{f(t) + f(-t)}{2}$$

of f only. It is easily seen that $R_\Omega[f] = R_\Omega[f_e]$. Moreover, denoting by F_e the Fourier transform of f_e and, as before, by F_c the cosine transform of f, we have

$$F_e(v) = \int_{-\infty}^{\infty} f_e(t) e^{-jtv}\, dt = \int_{-\infty}^{\infty} f(t) \cos(tv)\, dt = F_c(v).$$

Hence (169) may be rewritten as

$$R_\Omega[f] = -2 \sum_{k=1}^{\infty} F_e(2k\Omega). \tag{174}$$

This formula can be inverted with the help of the Möbius inversion formula. Under appropriate assumptions on f (see [139, Thoerem 2]), we obtain

$$F_e(2\Omega) = -\frac{1}{2} \sum_{k=1}^{\infty} \mu(k) R_{k\Omega}[f], \tag{175}$$

where μ is the Möbius function defined by

$$\mu(k) := \begin{cases} 1 & \text{if } k = 1, \\ (-1)^m & \text{if } k \text{ is a product of } m \text{ distinct primes}, \\ 0 & \text{if } k \text{ is divisible by a square of a prime.} \end{cases}$$

Formulae (174) and (175) show that there is a one-to-one correspondence between the speed of convergence of $R_\Omega[f]$ and the decay of the Fourier transform F_e. On the other hand, the decay of a Fourier transform G can be characterized by regularity properties of g; e.g., [43, Theorem 5.2.21] and [73,

Theorem 2.1]. This allows us to characterize certain rates of convergence of $R_\Omega[f]$ by function spaces containing the even part of f. No translates of f are needed. Results analogous to the above statements (I)–(III) can be established.

Let $f \in C(\mathbb{R}) \cap L^1(\mathbb{R})$ be such that for each $h > 0$ the series

$$\varphi(h, u) := h \sum_{k=-\infty}^{\infty} f(h(k+u))$$

converges uniformly in $u \in [-1/2, 1/2]$ to a piecewise continuously differentiable function $\varphi(h, \cdot)$. Then the following statements hold.

(I′) *The remainders satisfy*

$$R_\Omega[f] = 0$$

for all $\Omega \geq \Omega_0/2$, where $\Omega_0 > 0$, if and only if, the even part of f is band-limited to $[-\Omega_0, \Omega_0]$.

(II′) *The remainders satisfy*

$$|R_\Omega[f]| \leq \frac{g(\Omega)}{\Omega^m}$$

for all $\Omega \geq \Omega_0$ with an integer $m \in \mathbb{N}$, a real number $\Omega_0 > 0$, and a function $g \in L^2[\Omega_0, +\infty)$, if and only if the even part of f belongs to the Sobolev space $W^m_{L^2(\mathbb{R})}$.

(III′) *The remainders satisfy*

$$|R_\Omega[f]| \leq Me^{-2d\Omega}$$

for all $\Omega \geq \Omega_0$ with positive constants M, d, and Ω_0, if and only if the even part of f is the restriction to $\mathbb{R}$ of a function belonging to $\widetilde{B}_d$.

The general hypotheses on f can be varied. It suffices to guarantee that representation (174) holds for all $\Omega > 0$ with the series converging absolutely.

2.11.4. Weighted quadrature over the real line

Of course, the interpolation formula (159) may be multiplied by a weight $w(t)$, and then formal integration produces the weighted quadrature formula

$$\int_{-\infty}^{\infty} w(t) f(t) dt = \sum_{k=0}^{\infty} A_k f\left(\frac{k\pi}{\Omega}\right) \tag{176}$$

with

$$A_k = \int_{-\infty}^{\infty} w(t)\, \operatorname{sinc}\left(\frac{\Omega t}{\pi} - k\right) dt \quad (k \in \mathbb{Z}),$$

provided that the integrals exist. Formulae of this type with some interesting properties were established in [92] and [179, Theorem 7]. However, due to the appearance of a weight, the Gaussian property of (160) gets lost, i.e., (176) will not hold for signals of twice the band-width allowed in (159).

It is an open question whether for each reasonable weight, there exist a sequence of nodes $\{t_k\}_{k\in\mathbb{Z}}$ equivalent to a uniqueness sequence of type Ω (see the definition in Section 2.11.2) and a sequence of coefficients $\{A_k\}_{k\in\mathbb{Z}}$ such that the quadrature formula

$$\int_{-\infty}^{\infty} w(t) f(t)\, dt = \sum_{k=-\infty}^{\infty} A_k f(t_k)$$

holds for a non-trivial class of signals band-limited to $[-2\Omega, 2\Omega]$. However, for a special set of weights, namely

$$w(t) = |t|^{2\alpha+1} \quad (\alpha > -1),$$

an affirmative answer has been given.

In fact, let J_α denote the Bessel function of the first kind of order α. It is known that

$$\frac{J_\alpha(z)}{z^\alpha} = \sum_{k=0}^{\infty} (-1)^k \frac{z^{2k}}{2^{\alpha+2k} k!\, \Gamma(k+\alpha+1)}$$

is an even entire function of exponential type 1. Its zeros $j_k = j_k(\alpha)$ $(k = \pm 1, \pm 2, \ldots)$ may be ordered such that $j_{-k} = -j_k$ and $0|j_1| \leq |j_2| \leq \cdots$. It can be shown that for $\alpha > -1$ the sequence $\{j_k/\Omega\}_{k\in\mathbb{Z}}$ is equivalent to a uniqueness sequence of type Ω. Frappier and Olivier [89] discovered that for Re $\alpha > -1$, there exists a quadrature formula with nodes $j_k/\Omega (k \in \mathbb{Z})$ which is exact for functions of exponential type up to 2Ω. A variant of their result, due to Grozev and Rahman [95, Theorem 1], may be stated as follows:

Let $\alpha > -1$ and let f be an entire function of exponential type 2Ω such that $|\cdot|^{2\alpha+1} f \in L^1(\mathbb{R})$. Then

$$\int_{-\infty}^{\infty} |t|^{2\alpha+1} f(t) dt = \frac{\pi}{\Omega^{2\alpha+2}} \sum_{k\in\mathbb{Z}\setminus\{0\}} C_k f\left(\frac{j_k}{\Omega}\right), \tag{177}$$

where

$$C_k := \frac{2|j_k|^{2\alpha}}{\pi(J'_\alpha(j_k))^2} \quad (k \in Z \setminus \{0\}).$$

Furthermore, the series on the right-hand side of (177) *converges absolutely.*

Formula (177) has all the properties of a generalized (weighted) Gaussian formula (see Section 2.11.2). The Gaussian nodes j_k/Ω ($k \in \mathbb{Z} \setminus \{0\}$) can be characterized by the property that for every $m \in \mathbb{N}$ the function

$$\phi(t) := \frac{j_m^\alpha}{J'_\alpha(j_m)} \cdot \frac{J_\alpha(\Omega t)}{t^\alpha(\Omega t - j_m)}$$

minimizes the integral

$$\int_{-\infty}^{\infty} |t|^{2\alpha+1}(\phi(t))^2 dt$$

over a certain class of entire functions ϕ of exponential type Ω; see see [89, Theorem 2] and [95, Theorem 3].

Under a relaxed hypothesis on f, formula (177) holds in the sense of Cauchy principle values (see [95, Theorem 2] and [94]). Moreover, (177) has been extended to all $\alpha \in \mathbb{C} \setminus \{-1, -2, \ldots\}$; see [94]. In [191, Theorem 4.2] a representation for the remainder of (177) in case of non-band-limited functions has been established. In all these results, the case $\alpha = -1/2$ leads to statements for formula (160), as presented in Sections 2.11.1–2.11.3.

2.11.5. Quadrature over a Semi-Infinite Interval

When we look for a quadrature formula over a semi-infinite interval such as $\mathbb{R}_+ := [0, +\infty)$, we may consider again the trapezoidal rule, which takes the form

$$\int_0^\infty f(t)\,dt = \frac{\pi}{2\Omega} f(0) + \frac{\pi}{\Omega}\sum_{k=1}^{\infty} f\left(\frac{k\pi}{\Omega}\right) + R_\Omega[f]. \tag{178}$$

Two classical summation formulae provide a representation for the remainder $R_\Omega[f]$.

Under appropriate assumptions on f, the Abel-Plana formula (see Section 2.9.2) yields that

$$R_\Omega[f] = -j \int_0^\infty \frac{f(jy) - f(-jy)}{e^{2\Omega y} - 1}\, dy.$$

Another representation for $R_\Omega[f]$ can be deduced from the Euler-Maclaurin summation formula (see Section 2.9.2).

If f is an entire function of exponential type τ less than 2Ω and the integral in (178) *exits as an improper Riemann integral, then*

$$R_\Omega[f] = \sum_{k=1}^{\infty} \left(\frac{\pi}{\Omega}\right)^{2k} \frac{f^{(2k-1)}(0)}{\Omega^{2k}} \frac{B_{2k}}{(2k)!}, \tag{179}$$

where B_{2k} $(k \in N)$ are the Bernoulli numbers. The series converges absolutely; see [183, Theorem 1].

Unfortunately, the remainder does not vanish for band-limited signals. On the contrary, the initial terms of the series in (179) may be relatively large. In numerical applications, a partial sum

$$\sum_{k=1}^{N-1} \left(\frac{\pi}{\Omega}\right)^{2k} \frac{f^{(2k-1)}(0)}{\Omega^{2k}} \frac{B_{2k}}{(2k)!}$$

should therefore be added to the approximation by the trapezoidal sum. If, in addition to the previous hypotheses, f is bounded by M on the real line, then

$$\left| \sum_{k=N}^{\infty} \left(\frac{\pi}{\Omega}\right)^{2k} \frac{f^{(2k-1)}(0)}{\Omega^{2k}} \frac{B_{2k}}{(2k)!} \right| \leq \frac{M\zeta(2N)}{\pi(1-(\tau/2\Omega)^2)} \left(\frac{\tau}{2\Omega}\right)^{2N-1},$$

where $\zeta(t) := \sum_{k=1}^{\infty} k^{-t}$ denotes the Riemann zeta function (see [183, Theorem 2]). By choosing both Ω and N relatively large, good approximations to $\int_0^\infty f(t)dt$ can be obtained. However, the trapezoidal rule over $\mathbb{R}_+$ is far from being Gaussian.

The search for a generalized Gaussian formula for quadrature over $\mathbb{R}_+$ leads to the theory of the *Mellin transform*. For $c \in \mathbb{R}$, we introduce the space

$$X_c := \{f : \mathbb{R}_+ \to \mathbb{C} : f(x)x^{c-1} \in L^1(\mathbb{R}_+)\}.$$

Then, for each $f \in X_c$, the Mellin transform

$$\mathcal{M}[f](c+jt) := \int_0^\infty f(u)u^{c+jt-1}du$$

exists for all $t \in \mathbb{R}$. For $T \in \mathbb{R}_+$, the space

$$\mathcal{B}_c^T := \{f \in X_c \cap C(\mathbb{R}_+) : \mathcal{M}[f](c+jt) = 0 \quad \text{for } |t| > T\}$$

is the class of functions which are *c-Mellin band-limited* to the interval $[-T, T]$. Finally, we denote by $\mathcal{K}_c(\omega)$ the class of all functions $f \in X_c \cap C(\mathbb{R}_+)$ such that

$$\sum_{k=-\infty}^{\infty} |f(e^{k\pi/\Omega})e^{k\pi c/\Omega} < \infty$$

and

$$\sum_{k=-\infty}^{\infty} |\mathcal{M}[f](c + j2k\Omega)| < \infty$$

for all $\Omega \geq \omega$.

Butzer and Jansche [39, Theorem 7.3] established a Mellin analogue of the Poisson summation formula, which allows to deduce the following result just as (168)–(169) were deduced from Poisson's formula.

Let $f \in \mathcal{K}_c(\omega)$. Then, for every $\Omega \geq \omega$,

$$\int_0^{\infty} f(t)t^{c-1}dt = \frac{\pi}{\Omega}\sum_{k=-\infty}^{\infty} f(e^{k\pi/\Omega})e^{k\pi c/\Omega} + R_{c,\Omega}[f], \tag{180}$$

with

$$R_{c,\Omega}[f] = -\sum_{k=1}^{\infty} (\mathcal{M}[f](c + j2k\Omega) + \mathcal{M}[f](c - j2k\Omega)). \tag{181}$$

The series in (180) *and* (181) *converge absolutely.*

Formula (180) is *interpolatory* since it can be obtained by formal integration of the exponential sampling theorem [38, Theorem 6.3] and it has the *Gaussian property* since it is exact for functions having twice the Mellin band-width of those admissible in the underlying exponential sampling theorem. Hence (180) is a weighted generalized Gaussian formula associated with the class $\mathcal{B}_c^{\Omega}$ of c-Mellin band-limited functions.

For $c = 1$ and functions f which are analytic in a strip, formula (180) was already obtained by Stenger (1973) [207, Corollary 1.4]. Representation (181) allows one to establish characterizations of the speed of convergence analogous to those stated in Section 2.11.3. For details see [192].

Acknowledgment

The authors would like to thank Ms Angelika Seves for her great patience in typing various versions of the manuscript, Mr Thorsten Heck for his kind help in preparing the many figures, and especially Professor Dieter Lüke for his willing help and encouragement throughout the years. The conferences which Professor Maurice Dodson organized at York University, UK, and his interest in our work were a steady inspiration to the authors from Aachen. These authors also extend special thanks to Professor J. R. Higgins, Cambridge, UK, who began working in sampling theory already in 1972, for his profound advice on all questions they posed in the broad area.

References

[1] A. N. Akansu and R. A. Haddad. *Multiresolution Signal Decomposition; Transforms, Subbands, and Wavelets*. Academic Press, New York, 1992.

[2] W. Ameling Ed. 7. *Aachener Symposium für Signaltheorie, Modellgestützte Signalverarbeitung, Proc. Conf., Aachen, Germany, 1990*, Informatik-Fachberichte, Vol. 253. Springer-Verlag, Berlin, 1990.

[3] M. H. Annaby. Sampling Expansions for Discrete Transforms and their Relationship with Interpolation Series. *Analysis* (Munich), 18(1):55–64, 1998.

[4] C. Arzelà. *Sulla Integrazione per Serie. Rend. Accad. Lincei Roma.*, 1:532–537, 566–569, 1885.

[5] H. Babovsky, T. Beth, H. Neunzert, and M. Schulz-Reese. Mathematische Methoden in der Systemtheorie: Fourieranalysis. *Mathematische Methoden in der Technik*, Vol. 5. Teubner, Stuttgart, 1987.

[6] A. V. Balakrishnan. *Communication Theory*. Inter-University Electronics Series, Vol. 6. McGraw-Hill, New York, 1968.

[7] M. G. Beaty. Multichannel Sampling for Multiband Signals. *Signal Process*, 36(1):133–138, 1994.

[8] M. G. Beaty and M. M. Dodson. Derivative Sampling for Multiband Signals. *Numer. Funct. Anal. Optim.*, 10(9):875–898, 1989.

[9] M. G. Beaty, M. M. Dodson, and J. R. Higgins. Approximating Paley-Wiener Functions by Smoothed Step Functions. *J. Approx. Theory*, 78(3):433–445, 1994.

[10] M. G. Beaty and J. R. Higgins, Aliasing and Poisson Summation in the Sampling Theory of Paley-Wiener Spaces. *J. Fourier Anal. Appl.*, 1(1):67–85, 1994.

[11] H. Behnke and F. Sommer. *Theorie der analytischen Funktionen einer Komplexen Veränderlichen*. Third edition, Springer-Verlag, Berlin, 1976.

[12] M. Bellanger. *Adaptive Filters and Signal Analysis*. Marcel Dekker, Inc., New York, 1987.

[13] M. Bellanger. *Digital Processing of Signals: Theory and Practice*. (2nd edn), John Wiley & Sons, New York and Teubner, Stuttgart, 1989.

[14] J. J. Benedetto and M. W. Frazier, Eds. *Wavelets: Mathematics and Applications*. Studies in Advanced Mathematics, CRC Publishers, Boca Raton, FL, 1994.

[15] F. J. Beutler. On the Truncation Error of the Cardinal Sampling Expansion. *IEEE Trans. Inform. Theory*, IT-22:568–573, 1976.

[16] L. Bezuglaya and V. E. Katsnelson. The Sampling Theorem for Functions with Limited Multi-Band Spectrum. I. *Z. Anal. Anwendungen*, 12(3):511–534, 1993.

[17] I. Bilinskis and A. Mikelsons. *Randomized Signal Processing*. Prentice Hall International Series in Acoustics, Speech and Signal Processing. Prentice-Hall, Hemel Hempstead, Herts, 1992.

[18] J. M. Blackledge, Ed. Image Processing: Mathematical Methods and Applications, *Proc. Conf., Cranfield, G. B., 1994*. The Institute of Mathematics and Its Applications Conference Series, New Series, Vol. 61. Clarendon Press, Oxford, 1997.

[19] R. P. Boas, Jr. *Entire Functions*. Academic Press, New York, 1954.

[20] R. P. Boas, Jr. Summation Formulas and Band-Limited Signals. *Tôhoku Math. J.* 24(2):121–125, 1972.
[21] C. W. Borchardt, Ed. *C. G. J. Jacobi, Gesammelte Werke*, Vol. 1, G. Reimer, Berlin, 1881.
[22] S. Borgen. Note on the Summability of Poisson's Formula. *J. London Math. Soc.*, 19:100–105, 1944.
[23] R. N. Bracewell. *The Fourier Transform and Its Applications*. (2nd edn), McGraw-Hill, New York, 1978.
[24] J. L. Brown, Jr. On the Error in Reconstructing a Non-Bandlimited Function by Means of the Bandpass Sampling Theorem. *J. Math. Anal. Appl.*, 18(1):75–84, 1967; Erratum. Ibid. 21(4):699, 1968.
[25] J. L. Brown, Jr. Sampling of Bandlimited Signals: Fundamental Results and Some Extensions. In N. K. Bose and C. R. Rao, Eds. *Signal Processing and its Applications*. Handbook of Statistics, North Holland, Amsterdam, 1993, Vol. 10, pp. 59–101.
[26] H. Burkhardt. Trigonometrische Reihen und Integrale. In H. Burkhardt, W. Wirtinger, and R. Fricke, Eds. *Encyklopädie der mathematischen Wissenschaften mit Einschluß ihrer Anwendungen*, Vol. 2, Analysis, erster Teil, zweite Hälfte (II A), Teubner, Leipzig, 1904–1916, pp. 819–1354.
[27] P. L. Butzer. The Shannon Sampling Theorem and some of its Generalizations. An Overview. In B. Sendov, B. Boyanov, D. Vačov, R. Maleev, S. Markov, and T. Boyanov, Eds., *Constructive Function Theory '81, Proc. Conf., Varna, Bulgaria, 1981*, Publishing House of the Bulgarian Academy of Sciences, Sofia, 1983, pp. 258–274.
[28] P. L. Butzer. A Survey of the Whittaker-Shannon Sampling Theorem and some of its Extensions. *J. Math. Res. Exposition*, 3(1):185–212, 1983.
[29] P. L. Butzer Ed., *5. Aachener Kolloquium, Mathematische Methoden in der Signalverarbeitung, Proc., Conf., Aachen, 1984*. Lehrstuhl A für Mathematik. RWTH Aachen, Aachen, 1984.
[30] P. L. Butzer. Some Recent Applications of Functional Analysis to Approximation Theory. In E. Knobloch, I. S. Louhivaara, and J. Winkler, Eds., *Zum Werk Leonhard Eulers. Vorträge des Euler-Kolloquiums im Mai 1983 in Berlin*, Birkhäuser Verlag, Basel, 1984, pp. 133–155.
[31] P. L. Butzer, W. Engels, S. Ries, and R. L. Stens. The Shannon Sampling Series and the Reconstruction of Signals in Terms of Linear, Quadratic and Cubic Splines. *SIAM J. Appl. Math.*, 46(2):299–323, 1986.
[32] P. L. Butzer, W. Engels, and U. Scheben. Magnitude of the Truncation Error in Sampling Expansions of Band-Limited Signals. *IEEE Trans. Acoust. Speech Signal Process*, ASSP-30(6):906–912, 1982.
[33] P. L. Butzer, A. Fischer, and R. L. Stens. Generalized Sampling Approximation of Multivariate Signals; Theory and some Applications. *Note Mat.*, 10 Suppl. No. 1:173–191, 1990.
[34] P. L. Butzer, A. Fischer, and R.L. Stens. Generalized Sampling Approximation of Multivariate Signals; General Theory. *Atti Sem. Mat. Fis. Univ. Modena*, 41(1):17–37, 1993.
[35] P. L. Butzer and A. Gessinger. The Approximate Sampling Theorem, Poisson's Sum Formula, a Decomposition Theorem for Parseval's Equation and their Interconnections. *Ann. Numer. Math.*, 4(1–4):143–160, 1997.
[36] P. L. Butzer, M. Hauss, and R. L. Stens. The Sampling Theorem and its Unique Role in Various Branches of Mathematics. *Mitt. Math. Ges. Hamburg* 12(3):523–547, 1991.
[37] P. L. Butzer, J. R. Higgins, and R. L. Stens. Sampling Theory of Signal Analysis. In J.-P. Pier, Ed., *Development of Mathematics*. 1950–2000. Birkhäuser Verlag, Basel, 2000.
[38] P. L. Butzer and S. Jansche. The Exponential Sampling Theorem of Signal Analysis. *Atti Sem. Mat. Fis. Univ. Modena*, Suppl. 46:99–122, 1998.
[39] P. L. Butzer and S. Jansche. The Finite Mellin Transform, Mellin-Fourier Series, and the Mellin-Poisson Summation Formula. *Rend. Circ. Mat. Palermo* (2), Suppl. 52:55–81, 1998.

[40] P. L. Butzer and J. Lei. Errors in Truncated Sampling Series with Measured Samples for Not-Necessarily Bandlimited Signals. *Funct. Approx. Comment. Math.*, 26:18–32, 1998.

[41] P. L. Butzer and J. Lei. Approximation of Signals Using Measured Sampled Values and Error Analysis. *Commun. Appl. Anal.*, 4:245–255, 2000.

[42] P. L. Butzer and G. Nasri-Roudsari. *Kramer's Sampling Theorem in Signal Analysis and its Role in Mathematics*. In Blackledge [18] Ed., pp. 49–95.

[43] P. L. Butzer and R. J. Nessel. *Fourier Analysis and Approximation*. Vol. I. Birkhäuser Verlag, Basel, and Academic Press, New York, 1971.

[44] P. L. Butzer, S. Ries, and R. L. Stens. The Whittaker-Shannon Sampling Theorem, Related Theorems and Extensions; a Survey. *Proc. IEEE Conf.*, pp. 50–56. Amman, Jordan, 1983.

[45] P. L. Butzer, S. Ries, and R. L. Stens. *Shannon's Sampling Theorem, Cauchy's Integral Formula, and Related Results*. In Butzer *et al.* [58] Ed., pp. 363–377.

[46] P. L. Butzer, S. Ries, and R. L. Stens. Approximation of Continuous and Discontinuous Functions by Generalized Sampling Series. *J. Approx. Theory*, 50(1):25–39, 1987.

[47] P. L. Butzer and W. Splettstößer. A Sampling Theorem for Duration-Limited Functions with Error Estimates. *Information and Control*, 34(1):55–65, 1977.

[48] P. L. Butzer and W. Splettstößer. *Approximation und Interpolation durch Verallgemeinerte Abtastsummen*. Forschungsberichte des Landes Nordrhein-Westfalen, Vol. 2708. Westdeutscher Verlag, Opladen, 1977.

[49] P. L. Butzer, W. Splettstößer, and R. L. Stens. The Sampling Theorem and Linear Prediction in Signal Analysis. *Jahresber. Deutsch. Math.-Verein*, 90:1–70, 1988.

[50] P. L. Butzer, W. Splettstößer, and R. L. Stens. *Index of Papers on Signal Analysis 1972–1994*. Lehrstuhl A für Mathematik. RWTH Aachen, Aachen, 1994.

[51] P. L. Butzer and R. L. Stens. The Euler-MacLaurin Summation Formula, the Sampling Theorem, and Approximate Integration over the Real Axis. *Linear Algebra and Appl.*, 52/53:141–155, 1983.

[52] P. L. Butzer and R. L. Stens. The Poisson Summation Formula, Whittaker's Cardinal Series and Approximate Integration. In Z. Ditzian, A. Meir, S. Riemenschneider, and A. Sharma, Eds., *Second Edmonton Conference on Approximation Theory, Proc. Conf., Edmonton, Canada, 1983*, Canad. Math. Soc. Conf. Proc., American Mathematical Society, Providence, RI, 1983, Vol. 3, pp. 19–36.

[53] P. L. Butzer and R. L. Stens. Prediction from Past Samples in Terms of Splines of Low Degree. *Math. Nachr.*, 132:115–130, 1987.

[54] P. L. Butzer and R. L. Stens. Sampling Theory for not Necessarily Band-Limited Functions; a Historical Overview. *SIAM Rev.*, 34(1):40–53, 1992.

[55] P. L. Butzer and R. L. Stens. In Marks II [148] Ed., *Linear Prediction by Samples from the Past*, pp. 157–183.

[56] P. L. Butzer and R. L. Stens. An Extension of Kramer's Sampling Theorem for not Necessarily "Bandlimited" Signals – the Aliasing Error. *Acta Sci. Math. (Szeged)*, 60(1–2):59–69, 1995.

[57] P. L. Butzer and R. L. Stens. De la Vallèe Poussin's Paper of 1908 on Interpolation and Sampling Theory, and its Influence. In P. L. Butzer, J. Mawhin, and P. Vetro, Eds., Charles Baron de la Vallèe Poussin, Collected Works *Rend. Circ. Mat.* Palermo (2) Suppl., (to appear).

[58] P. L. Butzer, R. L. Stens, and B. Sz-Nagy (editors). *Anniversary Volume on Approximation Theory and Functional Analysis, Proc. Conf., Oberwolfach, Germany, 1983*. ISNM, Vol. 65, Birkhäuser Verlag, Basel, 1984.

[59] S. Cambanis and E. Masry. Truncation Error Bounds for the Cardinal Sampling Expansion of Band-Limited signals. *IEEE Trans. Inform. Theory*, IT-28:605–612, 1982.

[60] A. Cauchy. Sur Diverses Relations qui Existent Entre les Résidus des Fonctions et les Intégrales Définies. In *Œuvres de Cauchy (2)*, Vol. 6, pp. 124–145. Gauthier-Villars, Paris, 1887.

[61] A. Cauchy. Sur le développement des Fonctions d'Une Seule Variable en Fractions Rationnelles. In *Œuvres de Cauchy (2)*, Gauthier-Villars, Paris, 1889, Vol. 7, pp. 324–344.

[62] A. Cauchy. Usage du Calcul des Résidus pour la Sommation ou la Transformation des Séries dont le Terme Général est une Fonction Paire du Nombre qui Représente le Rang de ce Terme. In *Œuvres de Cauchy (2)*, Gauthier-Villars, Paris, 1889, Vol. 7, pp. 346–362.
[63] A. Cauchy. Méthode pour Développer des Fonctions d'une ou de Plusieurs Variables en Séries Composées de Fonctions de Même Espéce. In *Œuvres de Cauchy (2)*, Gauthier-Villars, Paris, 1889, Vol. 7, p. 366–392.
[64] J. I. Churkin, C. P. Jakowlew, and G. Wunsch. *Theorie und Anwendung der Signalabtastung*. Theoretische Grundlagen der Technischen Kybernetik, Verlag Technik, Berlin, 1966.
[65] L. W. Couch, II. *Digital and Analog Communication Systems*. (5th edn), Prentice-Hall, Englewood Cliffs, N.J., 1997.
[66] R. E. Crochiére and L. R. Rabiner. *Multirate Digital Signal Processing*. Prentice-Hall, Englewood Cliffs, N.J., 1983.
[67] P. J. Davis and P. Rabinowitz. *Methods of Numerical Integration*. Academic Press, Orlando, FL, 1984.
[68] W. Dickmeis and R. J. Nessel. On Uniform Boundedness Principles and Banach-Steinhaus Theorems with Rates. *Numer. Funct. Anal. Optim.*, 3(1):19–52, 1981.
[69] M. M. Dodson and A. M. Silva. Fourier Analysis and the Sampling Theorem. *Proc. Roy. Irish Acad. Sect*, A, 85(1):81–108, 1985.
[70] D. P. Dryanov. Quadrature Formulae for Entire Functions of Exponential Type. *J. Math. Anal. Appl.*, 152(2):488–495, 1990.
[71] D. P. Dryanov. Optimal Quadrature Formulae on the Real Line. *J. Math. Anal. Appl.*, 165(2):556–564, 1992.
[72] D. P. Dryanov, Q. I. Rahman, and G. Schmeisser. Equivalence Theorems for Quadrature on the Real Line. In H. Braß and G. Hämmerlin, Eds., *Numerical Integration III, Proc. Conf., Oberwolfach, Germany, 1987*, ISNM, Birkhäuser Verlag, Basel, 1988, Vol. 85, pp. 202–215.
[73] D. P. Dryanov, Q. I. Rahman, and G. Schmeisser. Converse Theorems in the Theory of Approximate Integration. *Constr. Approx.*, 6(3):321–334, 1990.
[74] D. E. Dudgeon and R. M. Mersereau. *Multidimensional Digital Signal Processing*. Prentice-Hall Signal Processing Series, Prentice-Hall, Englewood Cliffs, N.J., 1984.
[75] T. S. Durrani, J. B. Abbiss, J. E. Hudson, R. N. Madan, J. G. McWhirter, and T. A. Moore (editors). *Mathematics in Signal Processing, Proc. Conf., Bath, U.K., 1985*. The Institute of Mathematics and its Applications Conference Series, New Series, Vol. 12, Clarendon Press, Oxford, 1987.
[76] H. Dym and H. P. McKean. *Fourier Series and Integrals*. Probability and Mathematical Statistics, Vol. 14. Academic Press, New York, 1972.
[77] H. Engels. *Numerical Quadrature and Cubature*. Academic Press, London, 1980.
[78] G. Fang. Whittaker-Kotelnikov-Shannon Sampling Theorem and Aliasing Error. *J. Approx. Theory*, 85(2):115–131, 1996.
[79] E. Feldheim. Théorie de la Convergence des Procédés d'Interpolation et de Quadrature Mécanique. *Mémor. Sci. Math.*, 95:1–90, 1939.
[80] W. L. Ferrar. On the Cardinal Function of Interpolation Theory. *Proc. Roy. Soc. Edinburgh Sect. A*, 45:269–282, 1925.
[81] P. J. S. G. Ferreira. The Stability of a Procedure for the Recovery of Lost Samples in Band-Limited Signals. *Signal Process*, 40(2–3):195–205, 1994.
[82] P. J. S. G. Ferreira. *Sampling Series with an Infinite Number of Unknown Samples*. In Marvasti [152] Ed., pp. 268–271.
[83] P. J. S. G. Ferreira, Ed. *SAMPTA '97: 1997 International Workshop on Sampling Theory and Applications, Proc. Conf., Aveiro, Portugal*. Departamento de Electrónica e Telecommunicações, Universidade de Aveiro, Aveiro, 1997.

[84] P. J. S. G. Ferreira. *Superresolution, the Recovery of Missing Samples, and Vandermonde Matrices on the Unit Circle.* In Lyubarskii [146] Ed., pp. 216–220.

[85] A. Fettweis. *Elemente nachrichtentechnischer Systeme.* (2nd edn), Teubner-Studienbücher: Elektrotechnik. Teubner, Stuttgart, Leipzig, 1996.

[86] A. Fischer. *Sampling Theory and Wavelets.* In Higgins and Stens [111] Eds., pp. 158–186.

[87] A. Fischer and R. L. Stens. Generalized Sampling Appoximation of Multivariate Signals; Inverse Approximation Theorems. In J. Szabados and K. Tandori, Eds., *Approximation Theory, Proc. Conf., Kecskemét, Hungary, 1990,* Colloquia Mathematica Societatis János Bolyai, North-Holland Publishing Company, Amsterdam, and János Bolyai Mathematical Society, Budapest, 1991, Vol. 58, pp. 275–286.

[88] L. E. Franks. *Signal Theory.* Prentice-Hall, Englewood Cliffs, N.J., 1969.

[89] C. Frappier and P. Olivier. A Quadrature Formula Involving the Zeros of Bessel Functions. *Math. Comput.*, 60(201):303–316, 1993.

[90] C. Frappier and Q. I. Rahman. Une Formule de Quadrature Pour les Fonctions Entiéres de Type Exponentiel. *Ann. Sci. Math. Québec*, 10:17–26; 1986.

[91] C. F. Gauss. *Werke.* Vol. VIII, Königliche Gesellschaft der Wissenschaften zu Göttingen. Teubner, Leipzig, 1900.

[92] R. Gervais, Q. I. Rahman, and G. Schmeisser. A Quadrature Formula of Infinite Order. In G. Hämmerlin, Ed., *Numerical Integration II, Proc. Conf., Oberwolfach, Germany, 1981*, ISNM, Birkhäuser Verlag, Basel, 1982, Vol. 57, pp. 107–118.

[93] R. Gervais, Q. I. Rahman, and G. Schmeisser. *A Bandlimited Function Simulating a Duration-Limited One.* In Butzer *et al.* [58] Ed. pp. 355–362.

[94] R. Ben Ghanem. Quadrature Formulae using Zeros of Bessel Functions as Nodes. *Math. Comput.*, 67(221):323–336, 1998.

[95] G. R. Grozev and Q. I. Rahman. A Quadrature Formula with Zeros of Bessel Functions as Nodes. *Math. Comput.*, 64(210):715–725, 1995.

[96] P. Găvruţă. Bounds of Truncation Error in Sampling Theorem of Band-Limited Signals. *Lucr. Semin. Mat. Fiz.*, 1988:27–30, 1988.

[97] P. Găvruţă. On the Sampling Theorem. *Bull. Appl. Math.*, 850:205–212, 1993.

[98] P. Găvruţă. Sampling Theorems and Frames. *Bull. Appl. Math.*, 980:203–210, 1994.

[99] H. D. Helms and J. B. Thomas. Truncation Error of Sampling-Theorem Expansions. *Proc. IRE,* 50:179–184, 1962.

[100] U. Hettich and R. L. Stens. Approximating a Bandlimited Function in Terms of its Samples. *Comput. Math. Appl.* 40:107–116, 2000.

[101] J. R. Higgins. A Sampling Theorem for Irregularly Spaced Sample Points. *IEEE Trans. Inform. Theory,* IT-22:621–622, 1976.

[102] J. R. Higgins. *Completeness and Basis Properties of Sets of Special Functions.* Cambridge University Press, Cambridge, 1977.

[103] J. R. Higgins. Bases for the Hilbert Space of Paley-Wiener Functions. In Butzer [29] Ed., pp. 274–278.

[104] J. R. Higgins. Five Short Stories about the Cardinal Series. *Bull. Amer. Math. Soc. (N.S.),* 12(1):45–89, 1985.

[105] J. R. Higgins. A Fresh Approach to the Derivative Sampling Theorem. In Durrani *et al.* [75] Ed., pp. 25–31.

[106] J. R. Higgins. Sampling Theorems and the Contour Integral Method. *Appl. Anal.*, 41(1–4):155–169, 1991.

[107] J. R. Higgins. Sampling Theory for Paley-Wiener Spaces in the Riesz Basis Setting. *Proc. Roy. Irish Acad. Sect.* A, 94(2):219–236, 1994.

[108] J. R. Higgins. Sampling for Multi-Band Functions. In Ismail *et al.* [116] Ed., pp. 165–170.

[109] J. R. Higgins. *Sampling Theory in Fourier and Signal Analysis: Foundations*. Oxford Science Publications, Clarendon Press, Oxford, 1996.

[110] J .R. Higgins, G. Schmeisser, and J. J. Voss. The Sampling Theorem and Several Equivalent Results in Analysis. *J. Comput. Anal. Appl.*, 2:333–371, 2000.

[111] J. R. Higgins and R. L. Stens (editors). *Sampling Theory in Fourier and Signal Analysis: Advanced Topics*. Oxford Science Publications, Oxford University Press, Oxford, 1999.

[112] E. Hille. *Analytic Function Theory.* (2nd edn), Vol.1, Chelsea Publishing Company, New York, 1973.

[113] G. Hinsen. Irregular Sampling of Bandlimited l^p-Functions. *J. Approx. Theory*, 72(3):346–364, 1993.

[114] G. Hinsen and D. Klösters. The Sampling Series as a Limiting Case of Lagrange Interpolation, *Appl. Anal.*, 49(1–2):49–60, 1993.

[115] R. F. Hoskins and J. Sousa Pinto. *Distributions, Ultradistributions and Other Generalised Functions*. Ellis Horwood Series in Mathematics and its Applications. Ellis Horwood, New York, 1994.

[116] M. E. H. Ismail, M. Z. Nashed, A. I. Zayed, and A. F. Ghaleb, Eds. *Mathematical Analysis, Wavelets, and Signal Processing, Proc. Conf., Cairo, Egypt, 1994*. Contemporary Mathematics, Vol. 190. American Mathematical Society, Providence, RI, 1995.

[117] D. Jagerman. Bounds for Truncation Error of the Sampling Expansion. *SIAM J. Appl. Math.*, 14(4):714–723, 1966.

[118] D. Jagerman and L. J. Fogel. Some General Aspects of the Sampling Theorem. *IRE Trans. Inform. Theory*, IT-2:139–156, 1956.

[119] L. Jantscher. *Distributionen*. Walter de Gruyter, Berlin, 1971.

[120] A. J. Jerri. The Shannon Sampling Theorem – its Various Extension and Applications: A Tutorial Review. *Proc. IEEE*, 65(11):1565–1596, 1977.

[121] V. E. Katsnelson. Sampling and Interpolation for Functions with Multi-Band Spectrum: The Mean-Periodic Continuation Method. In W. Eisenberg *et al.*, Ed., *Wiener-Symposium zum100. Geburtstag von Norbert Wiener, Proc. Conf., Grossbothen, Germany, 1994*, Verlag im Wissenschaftszentrum Leipzig, Grossbothen, 1996, pp. 91–132.

[122] J. B. Kioustelidis. Fehlerbetrachtungen beim Abtasttheorem. *Arch. Elektron.* Uebertrag. Tech., 23:629–630, 1969.

[123] A. Kivinukk and G. Tamberg. *Subordination in Generalized Sampling Series by Rogosinski-Type Sampling Series*. In Ferreira [83] Ed., pp. 397–402.

[124] H. Kneser. *Funktionentheorie*. Studia Mathematica/Mathematische Lehrbücher, Vol. 13. Vandenhoek & Ruprecht, Göttingen, 1958.

[125] A. Kohlenberg. Exact Interpolation of Band-Limited Functions. *J. Appl. Phys.*, 24:1432–1436, 1953.

[126] A. N. Kolmogorov and V. M. Tikhomirov. ε-Entropy and ε-Capacity of Sets of Functions (Russian). *Uspekhi Mat. Nauk*, 14:3–86, 1959.

[127] A. N. Kolmogorov and V. M. Tikhomirov. *ε-Entropie und ε-Kapazität von Mengen in Funktionenräumen*. Arbeiten zur Informationstheorie, Vol. III, VEB Deutscher Verlag der Wissenschaften, Berlin, 1960, German translation of [126].

[128] V. A. Kotel'nikov. *On the Carrying Capacity of the "Ether" and Wire in Telecommunications*. Material for the first All-Union Conference on Questions of Communications (Russian), Izd. Red. Upr. Svyazi RKKA, Moscow, 1933.

[129] W. Krabs. *Mathematical Foundations of Signal Theory.* Sigma Series in Applied Mathematics, Vol. 6. Heldermann Verlag, Berlin, 1995.

[130] R. Kress. Interpolation auf einem unendlichen Intervall. *Computing*, 6:274–288, 1970.

[131] R. Kress. On the General Hermite Cardinal Interpolation. *Math. Comput.*, 26:925–933, 1972.

[132] R. Kress. Zur Quadratur uneigentlicher Integrale bei Analytischen Funktionen. *Computing*, 13:267–277, 1974.

[133] B. Lacaze. La Formule d'Échantillonnage et A. L. Cauchy. *Traitement du Signal*, 15(4):289–295, 1998.

[134] R. Lasser. *Introduction to Fourier Series*. Marcel Dekker, New York, 1996.

[135] A. J. Lee. Approximate Interpolation and the Sampling Theorem. *SIAM J. Appl. Math.*, 32(4):731–744, 1977.

[136] B.Y. Levin. *Lectures on Entire Functions*. Translations of Mathematical Monographs, Vol. 150. American Mathematical Society, Providence, RI, 1996.

[137] Xin Min Li. Uniform Bounds for Sampling Expansions. *J. Approx. Theory*, 93(1):100–113, 1998.

[138] S. P. Lloyd. A Sampling Theorem for Stationary (Wide Sense) Stochastic Processes. *Trans. Amer. Math., Soc.* 92:1–12, 1959.

[139] J. H. Loxton and J. W. Sanders. On an Inversion Theorem of Möbius. *J. Austral. Math. Soc. Ser. A*, 30:15–32, 1980.

[140] H. D. Lüke. Zur Entstehung des Abtasttheorems, *ntz*. 31:271–274, 1978.

[141] H. D. Lüke, Ed. *4. Aachener Kolloquium, Theorie und Anwendung der Signalverarbeitung, Proc. Conf. Aachen, 1881*. IENT, Aachen, 1981.

[142] H. D. Lüke. Herbert P. Raabe, der "Vater" des Abtasttheorms. *ntz*. 43:46, 1990.

[143] H. D. Lüke. *Signalübertragung*. (6th edn), Springer-Verlag, Berlin, 1995.

[144] H. D. Lüke. The Origins of the Sampling Theorem. *IEEE Comm. Mag.*, 37:106–108, 1999.

[145] L. Lundin and F. Stenger. Cardinal-Type Approximations of a Function and its Derivatives. *SIAM J. Math. Anal.*, 10(1):139–160, 1979.

[146] Y. Lyubarskii, Ed. *SAMPTA '99, 1999 International Workshop on Sampling Theory and Applications*. Proc. Conf., Loen, Norway, Norwegian University of Science and Technology, Trondheim, 1999.

[147] R. J. Marks II. *Introduction to Shannon Sampling and Interpolation Theory*. Springer-Verlag, New York, 1991.

[148] R. J. Marks II, Ed. *Advanced Topics in Shannon Sampling and Interpolation Theory*. Springer-Verlag, New York, 1993.

[149] S. L. Marple and M. Marietta. *Digital Spectral Analysis with Applications*. Prentice-Hall Signal Processing Series, Prentice-Hall, Englewood Cliffs, N.J., 1987.

[150] E. Martensen. Zur numerischen Auswertung uneigentlicher Integrale. *Z. Angew. Math. Mech.*, 48:T83–T85, 1968.

[151] F. Marvasti. *A Unified Approach to Zero-Crossing and Nonuniform Sampling of Single and Multidimesional Signals and Systems*. Nonuniform Publications, Oak Park, IL, 1987.

[152] F. A. Marvasti (editor). *SAMPTA '95, 1995 Workshop on Sampling Theory & Applications, Proc. Conf., Jurmala, Latvia*, Institute of Electronics and Computer Science, Riga, 1995.

[153] J. McNamee. Error-Bounds for the Evaluation of Integrals by the Euler-Maclaurin Formula and by Gauss-Type Formulae. *Math. Comput.*, 18:368–381, 1964.

[154] J. McNamee, F. Stenger, and E. L. Whitney. Whittaker's Cardinal Function in Retrospect. *Math. Comput.*, 25:141–154, 1971.

[155] D. Meyer-Ebrecht, Ed. *6. Aachener Symposium für Signaltheorie, Mehrdimensionale Signale und Bildverarbeitung, Proc. Conf., Aachen, Germany, 1987*. Informatik-Fachberichte, Vol. 153. Springer-Verlag, Berlin, 1987.

[156] D. H. Mugler. Convolution, Differential Equations, and Entire Functions of Exponential Type. *Trans. Amer. Math. Soc.*, 216:145–187, 1976.

[157] M. Z. Nashed and G. Walter. General Sampling Theorems for Functions in Reproducing Kernel Hilbert Spaces. *Math. Control Signals Systems*, 4(4):363–390, 1991.

[158] M. Z. Nashed and G. G. Walter. *Reproducing Kernel Hilbert Spaces from Sampling Expansions*, In Ismail *et al.* [116] Ed., pp. 221–226.

[159] F. Natterer. *The Mathematics of Computerized Tomography.* Teubner, Stuttgart, and John Wiley & Sons, New York, 1986.

[160] J. Neveu. Le Problème de l'Échantillonnage et de l'Interpolation d'un Signal. *C. R. Acad. Sci. Paris*, 260:49–51, 1965.

[161] S. M. Nikol'skiĭ, *Approximation of Functions and Imbedding Theorems*. Springer-Verlag, Berlin, 1975.

[162] K. Ogura. On a Certain Transcendental Integral Function in the Theory of Interpolation. *Tôhoku Math. J.*, 17(2):64–72, 1920.

[163] P. Olivier and Q. I. Rahman. Sur une Formule de Quadrature pour des Fonctions Entières. *RAIRO Modél. Math. Anal. Numér.*, 20:517–537, 1986.

[164] A. V. Oppenheim and R. W. Schafer, *Discrete Time Signal Processing*. Prentice Hall Signal Processing Series, Prentice-Hall, Englewood Cliffs, N.J., 1989.

[165] W. F. Osgood. Non-Uniform Convergence and the Integration of Series Term by Term. *Amer. J. Math.*, 19:155–190, 1897.

[166] R. E. A. C. Paley and N. Wiener. *Fourier Transforms in the Complex Domain*. Amer. Math. Soc. Colloq. Publ., Vol. 19. American Mathematical Society, New York, 1934.

[167] A. Papoulis, *The Fourier Integral and its Applications*. McGraw-Hill, New York, 1962.

[168] A. Papoulis. Error Analysis in Sampling Theory. *Proc. IEEE*, 54:947–955, 1966.

[169] A. Papoulis. *Signal Analysis*. McGraw-Hill, New York, 1977.

[170] M. -A. Parseval des Chênes. Mémoire sur les séries et sur l'Intégration Complète d'une Équation aux Différences Partielles Linéaires du Second Ordre, à Coefficiens Constans, Mémoires Présent'es à l'Institut des Sciences, Lettres et Arts, par Divers Savants, et lus dans ses Assemblées. *Sciences math. et phys.* (savans étrangers) 1:638–648, 1806.

[171] J. R. Partington. *Interpolation, Identification, and Sampling*. London Mathematical Society Monographs. New Series, Vol. 17. Clarendon Press, Oxford, 1997.

[172] A. Peled and B. Liu. *Digital Signal Processing: Theory, Design and Implementation*. John Wiley & Sons, New York, 1976.

[173] B. Picinbono. *Principles of Signals and Systems: Deterministic Signals*. Artech House, London, 1988.

[174] H. S. Piper. Jr. Bounds for Truncation Error in Sampling Expansions of Finite Energy Band-Limited Signals. *IEEE Trans. Inform. Theory*, IT-21:482–484, 1975.

[175] H. S. Piper. Jr. Best Asymptotic Bounds for Truncation Error in Sampling Expansions of Band-Limited Signals. *IEEE Trans. Inform. Theory*, IT-21:687–690, 1975.

[176] J. G. Proakis, C. M. Rader, F. Ling, and C. L. Nikias. *Advanced Digital Signal Processing*. Macmillan, New York, 1992.

[177] H. Raabe. Untersuchungen an der wechselseitigen Mehrfachübertragung (Multiplexübertragung). *Elektrische Nachrichtentechnik*, 16:213–228, 1939.

[178] L. R. Rabiner and B. Gold. *Theory and Application of Digital Signal Processing*. Prentice-Hall, Englewood Cliffs, N.J., 1975.

[179] Q. I. Rahman and G. Schmeisser. Reconstruction and Approximation of Functions from Samples. In G. Meinardus and G. Nürnberger, Eds., *Delay Equations, Approximation and Application, Proc. Conf., Mannheim, Germany, 1984*, ISNM, Vol. 74. Birkhäuser Verlag, Basel, 1985, pp. 213–233.

[180] Q. I. Rahman and G. Schmeisser, On a Gaussian Quadrature Formula for Entire Functions of Exponential Type. In L. Collatz, G. Meinardus, and G. Nürnberger, Eds., *Numerical Methods of Approximation Theory, Vol. 8, Proc. Conf., Oberwolfach, Germany, 1986*, ISNM, Vol. 83, Birkhäuser Verlag, Basel, 1987, pp. 155–168.

[181] Q. I. Rahman and G. Schmeisser. An Analogue of Turán's Quadrature Formula. In Bl. Sendov, P. Petrushev, K. Ivanov, and R. Maleev, Eds., *Constructive Theory of Functions, Proc. Conf.,*

Varna, Bulgaria, 1987, Publishing House of the Bulgarian Academy of Sciences, Sofia, 1988, pp. 405–412.

[182] Q. I. Rahman and G. Schmeisser. Quadrature Formulae and Functions of Exponential Type. *Math. Comput.*, 54(189):245–270, 1990.

[183] Q. I. Rahman and G. Schmeisser. A Quadrature Formula for Entire Functions of Exponential Type. *Math. Comput.*, 63(207):215–227, 1994.

[184] Q. I. Rahman and G. Schmeisser. The Summation Formulae of Poisson, Plana, Euler-Maclaurin and their Relationship. *J. Math. Sci. (Calcutta)*, 28:151–171, 1994.

[185] S. Ries and R. L. Stens. *Pointwise Convergence of Sampling Series*, In Schüßler [194] Ed., pp. 5–7.

[186] S. Ries and R. L. Stens. Approximation by Generalized Sampling Series. In Bl. Sendov, P. Petrushev, R. Maleev, and S. Tashev, Eds., *Constructive Theory of Functions, Proc. Conf., Varna, Bulgaria. 1984*, Publishing House of the Bulgarian Academy of Sciences, Sofia, 1984, pp. 746–756.

[187] S. Ries and R. L. Stens. A Localization Principle for the Approximation by Sampling Series. In N.P. Korneĭchuk, S. B. Stečkin, and S. A. Telyakovskiĭ, Eds., *Theory of the Approximation of Functions (Russian), Proc. Conf., Kiev, 1983*, pp. 507–510. "Nauka", Moscow, 1987.

[188] W. Rudin. *Real and Complex Analysis*. McGraw-Hill, New York, 1966.

[189] W. Rudin. *Functional Analysis*. McGraw-Hill, New York, 1973.

[190] S. Saitoh. *Theory of Reproducing Kernels and its Applications*. Pitman Research Notes in Mathematics Series, Vol. 189. Longman Scientific and Technical, Harlow, Essex, and John Wiley & Sons, New York, 1988.

[191] G. Schmeisser. Sampling, Gaussian Quadrature, and Poisson Summation Formula, In Ferreira [83] Ed., pp. 327–332.

[192] G. Schmeisser. Quadrature over a Semi-Infinite Interval and Mellin Transform, In Lyubarskii [146] Ed., pp. 203–208.

[193] A. Schönhage. *Approximationstheorie*. Walter de Gruyter, Berlin, 1971.

[194] H. W. Schüßler Ed., *Signal Processing II: Theories and Applications; Proc. EUSIPCO-83, Erlangen, Germany, 1983*. North-Holland Publishing Company, Amsterdam, 1983.

[195] H. W. Schüßler. *Digitale Signalverarbeitung*, Vol. 1. Springer-Verlag, Berlin, 1994.

[196] K. Seip. An Irregular Sampling Theorem for Functions Bandlimited in a Generalized Sense. *SIAM J. Appl. Math.*, 47:1112–1116, 1987.

[197] C. E. Shannon. A Mathematical Theory of Communication. *Bell System Tech. J.*, 27:379–423, 623–656, 1948.

[198] C. E. Shannon. Communication in the Presence of Noise. *Proc. IRE*, 37:10–21, 1949.

[199] B. D. Sharma and F. C. Mehta. Generalized Band-Pass Sampling Theorem. *Math. Balkanica (N.S.)*, 6:204–217, 1976.

[200] N. J. A. Sloane and A. D. Wyner, Eds. *Claude Elwood Shannon: Collected Papers*. IEEE Press, Piscataway, N.J., 1993.

[201] I. Someya. *Waveform Transmission (Japanese)*. Shyukyoo Ltd., Tokyo, 1949.

[202] W. Splettstößer. Some Extensions of the Sampling Theorem. In P. L. Butzer and B. Sz.-Nagy, Eds., *Linear Spaces and Approximation, Proc. Conf., Oberwolfach, Germany, 1977*, ISNM, Vol. 40, Birkhäuser Verlag, Basel, 1978, pp. 615–628.

[203] W. Splettstößer. Error Estimates for Sampling Approximation of Non-Bandlimited Functions. *Math. Methods Appl. Sci.*, 1(2):127–137, 1979.

[204] W. Splettstößer. 75 Years Aliasing Error in the Sampling Theorem. In Schüßler [194] Ed., pp. 1–4.

[205] W. Splettstößer. Some Aspects on the Reconstruction of Sampled Signal Functions. In H. Neunzert, Ed. *The Road-Vehicle-System and Related Mathematics, Proc. Conf., Lambrecht, Germany, 1985*, Teubner, Stuttgart, 1985, pp. 126–142.

[206] W. Splettstößer, R.L. Stens, and G. Wilmes. On Approximation by the Interpolating Series of G. Valiron. *Funct. Approx. Comment. Math.*, 11:39–56, 1981.

[207] F. Stenger. Integration Formulae Based on the Trapezoidal Formula. *J. Inst. Math. Appl.*, 12:103–114, 1973.

[208] F. Stenger. Approximations via Whittaker's Cardinal Function. *J. Approx. Theory*, 17(3):222–240, 1976.

[209] F. Stenger. Numerical Methods Based on Whittaker Cardinal, or Sinc Functions. *SIAM Rev.*, 23(2):165–224, 1981.

[210] R. L. Stens. Approximation to Duration-Limited Functions by Sampling Sums. *Signal Process*, 2(2):173–176, 1980.

[211] R. L. Stens. Error Estimates for Sampling Sums Based on Convolution Integrals. *Inform. and Control*, 45(1):37–47, 1980.

[212] R. L. Stens. A Unified Approach to Sampling Theorems for Derivatives and Hilbert Transforms. *Signal Process*, 5(2):139–151, 1983.

[213] R. L. Stens. Approximation of Functions by Whittaker's Cardinal Series. In W. Walter, Ed., *General Inequalities 4, Proc. Conf., Oberwolfach, Germany, 1983*, ISNM, Vol. 71. Birkhäuser Verlag, Basel, 1984, pp. 137–149.

[214] R. L. Stens. Sampling with Generalized Kernels. In Higgins and Stens [111] Eds., pp. 130–157.

[215] Z. Szmydt. *Fourier Transformation and Linear Differential Equations*. D. Reidel, Dordrecht, Boston, and PWN-Polish Scientific Publishers, Warszawa, 1977.

[216] A. E. Taylor and D. C. Lay. *Introduction to Functional Analysis*. (2nd edn), John Wiley & Sons, New York, 1980.

[217] A. Terras. *Harmonic Analysis on Symmetric Spaces and Applications I*. Springer, New York, 1985.

[218] M. Theis. Über eine Approximationsformel von de la Vallée-Poussin. *Math. Z.*, 3:93–113, 1919.

[219] H. Triebel. *Fourier Analysis and Function Spaces*. Teubner, Leipzig, 1977.

[220] L. Tschakaloff. Zweite Lösung der Aufgabe 105. *Jahresber. Deutsch. Math.-Verein*, 43:11–12, 1933.

[221] P. Turán. On the Theory of Mechanical Quadrature. *Acta Sci. Math. (Szeged)*, 12:30–37, 1950.

[222] J. D. Vaaler. Some Extremal Functions in Fourier Analysis. *Bull. Amer. Math. Soc. (N.S.)*, 12(2):183–216, 1985.

[223] G. Valiron. Sur la Formule d'Interpolation de Lagrange. *Bull. Sci. Math.*, 49(2):203–224, 1925.

[224] Ch. -J. de la Vallée Poussin. Sur la Convergence des Formules d'Interpolation entre Ordonnées Équidistantes. *Acad. Roy. Belg. Bull. Cl. Sci.*, 4:319–410, 1908.

[225] P. Vary, Ed. *8. Aachener Kolloquium für Signaltheorie, Mobile Kommunikationssysteme, Proc. Conf., Aachen, Germany, 1994*. vde Verlag, Berlin, 1994.

[226] J. J. Voss. Irregular Sampling: Error Analysis, Applications, and Extensions. *Mitt. Math. Sem. Giessen*, 238:1–86, 1999.

[227] L. A. Wainstein and V. D. Zubakov. *Extraction of Signals from Noise. Internat. Ser. Appl. Math.*, Prentice-Hall, Englewood Cliffs, N.J., 1962.

[228] E. Waring. Problems Concerning Interpolations. *Philos. Trans. Roy. Soc. London Ser. A*, 69:59–67, 1779.

[229] P. Weiss. An Estimate of the Error Arising From Misapplication of the Sampling Theorem. *Notices Amer. Math. Soc.*, 10:351, 1963.

[230] E. T. Whittaker. On the Functions which are Represented by the Expansion of the Interpolation Theory. *Proc. Roy. Soc. Edinburgh Sect. A*, 35:181–194, 1915.

[231] J. M. Whittaker. On the Cardinal Function of Interpolation Theory. *Proc. Edinburgh Math. Soc.*, 1(2):41–46, 1927–1929.

[232] J. M. Whittaker. The "Fourier" Theory of the Cardinal Function. *Proc. Edinburgh Math. Soc.*, 1(2):169–176, 1927–1929.

[233] P. M. Woodward. *Probability and Information Theory, with Applications to Radar.* (2nd edn), International Series of Monographs on Electronics and Instrumentation, Vol. 3. Pergamon, New York, 1964.

[234] G. Wunsch. *Systemtheorie der Informationstechnik*, Bücherei der Hochfrequenztechnik, Vol. 20. Akademische Verlagsgesellschaft Geest & Portig, Leipzig, 1971.

[235] K. Yao and J. B. Thomas. On Truncation Error Bounds for Sampling Representations of Bandlimited Signals. *IEEE Trans. Aerospace Electron. Systems*. AES 2:640–647, 1966.

[236] R. M. Young. *An Introduction to Nonharmonic Fourier Series*, Pure and Applied Mathematics, Vol. 93. Academic Press, New York, 1980.

[237] A. I. Zayed. *Advances in Shannon's Sampling Theory.* CRC Publishers, Boca Raton, FL, 1993.

[238] A. I. Zayed and P. L. Butzer. Lagrange Interpolation and Sampling Theorems. In F. Marvasti, Ed., *Nonuniform Sampling: Theory and Practice*, Plenum Publishers, New York, 2001, pp. 123–168.

[239] A. H. Zemanian. *Distribution Theory and Transform Analysis*. McGraw-Hill, New York, 1965.

[240] R. E. Ziemer and W. H. Tranter. *Principles of Communications: Systems, Modulation, and Noise*. (4th edn), John Wiley & Sons, New York, 1995.

[241] A. Zygmund. *Trigonometric Series*, Reprint of the 2nd edn, Vol. I and II. Cambridge University Press, Cambridge, 1968.

[242] A. Zayed, Ed. SAMPTA '2001: International Workshop on Sampling Theory and Applications, Proc. Conf., Orlando, Florida. Department of Mathematics, University of Central Florida, Orlando, 2001.

3

Lagrange Interpolation and Sampling Theorems

A. I. Zayed and P. L. Butzer

3.1. Introduction

The aim of this chapter is to discuss the relationship between Lagrange interpolation and sampling theorems. The first three sections can be regarded as an alternative introduction to sampling theory, avoiding the Fourier analysis of Chapter 2, at least to begin with. Because Lagrange interpolation is the central theme of the chapter, it is natural to start off with an introduction to the Lagrange interpolation method and then proceed to a more general form of it, which we call Lagrange-type interpolation. Having done that, we can then investigate the relationship between Lagrange-type interpolation and sampling theorems, in particular, the Whittaker-Shannon-Kotel'nikov (WSK) sampling theorem.

Although this chapter is written as a tutorial, it contains some new results that have not been published before. It also includes some open questions that may spur the reader's interest and open new directions for research on sampling theory.

The rest of the chapter is divided into six sections. In Section 2, we discuss the Lagrange interpolation method and show how one can construct a polynomial of degree n that coincides with a given function at $n + 1$ points. In Section 3, we extend this method from polynomial interpolation to entire function interpolation. We call this extension a Lagrange-type interpolation. In the first part of Section 3,

A. I. Zayed • Department of Mathematics and Department of Electrical and Computer Engineering, University of Central Florida, Orlando, Florida 32816, U.S.A.
E-mail: Zayed@pegasus.cc.ucf.edu
P. L. Butzer • Lehrstuhl A für Mathematik, RWTH, D-52056 Aachen, Germany.
E-mail: Butzer@rwth-aachen.de

Nonuniform Sampling: Theory and Practice, edited by Marvasti
Kluwer Academic/Plenum Publishers, New York, 2001.

Subsection 3.1, we discuss uniform sampling and introduce the Whittaker-Shannon-Kotel'nikov (WSK) sampling theorem. We reserve Subsection 3.2, for nonuniform sampling, where we introduce the Paley-Wiener-Levinson sampling theorem, which is a generalization of the Whittaker-Shannon-Kotel'nikov sampling theorem.

In Section 4 we introduce sampling expansions involving samples of the signal and its derivatives taken at uniformly spaced points, as well as at non-uniformly spaced points. It will be shown that a bandlimited signal can be reconstructed from its samples and the samples of its $m-1$ derivatives taken at $1/m$ times the Nyquist rate. Section 5 is devoted to another generalization of the WSK sampling theorem, known as Kramer's sampling theorem. This section contains some new results and examples that have not been published before, such as a generalization of Kramer's theorem that will enable us to derive sampling expansions associated with boundary-value problems having not necessarily simple eigenvalues. We shall focus more on the relationship between Kramer's theorem and Lagrange-type interpolation.

Since Kramer's theorem is closely related to boundary-value problems, we devote Section 6 for investigating that relationship, as well as the connection between boundary-value problems and Lagrange-type interpolation. A summary of the major results is presented in Subsection 6.1 leaving Subsection 6.2 entirely for examples. Finally, we conclude the chapter with Section 7, which is concerned with the question as to what happens to the WSK and Kramer's theorems when the condition of band-limitedness is weakened.

Throughout this chapter we shall adopt the following standard notation. The set of complex numbers is denoted by $\mathbf{C}$, the set of real number by $\mathbb{R}$, and the set of integers by $\mathbb{Z}$. The set of all polynomials of degree at most n will be denoted by $\mathcal{P}_n$, the set of all continuous functions on a set E will be denoted by $C(E)$, and the set of all functions having k ($k \geq 0$) continuous derivatives on E will be denoted by $C^k(E)$. The space l^p, $1 \leq p < \infty$, consists of all sequences $\{a_n\}_{n\in\mathbb{Z}}$ of complex numbers satisfying $(\sum_{n\in\mathbb{Z}} |a_n|^p) < \infty$, and the space $L^p(I)$, where I is an interval, consists of all measurable functions f on I satisfying $(\int_I |f(x)|^p dx) < \infty$.

3.2. Lagrange (Polynomial) Interpolation

In this section, we discuss the Lagrange interpolation method and show how one can construct a polynomial of degree n that coincides with a given function at $n+1$ points.

Theorem 1 (Lagrange interpolation theorem) *Given $n+1$ distinct (real or complex) points, $z_0, z_1, \ldots, z_n$ and $n+1$ (real or complex) values,*

$w_0, w_1, \ldots, w_n$, *there exists a unique polynomial* $p_n(z) \in \mathcal{P}_n$ *for which*

$$p_n(z_i) = w_i, \qquad i = 0, 1, \ldots, n. \tag{1}$$

Proof: Let $p_n(z) = a_0 + a_1 z + \cdots + a_n z^n$, where the a_n's are coefficients to be determined. The conditions (1) lead to the system of $n+1$ linear equations in the a_i's:

$$a_0 + a_1 z_i + \cdots + a_n z_i^n = w_i, \qquad i = 0, \ldots, n. \tag{2}$$

The determinant of the system is the Vandermonde determinant:

$$U(z_0, z_1, \ldots, z_n) = \begin{vmatrix} 1 & z_0 & z_0^2 & \cdots & z_0^n \\ 1 & z_1 & z_1^2 & \cdots & z_1^n \\ \cdot & \cdot & \cdot & & \cdot \\ \cdot & \cdot & \cdot & & \cdot \\ 1 & z_n & z_n^2 & \cdots & z_n^n \end{vmatrix}. \tag{3}$$

To evaluate U, consider the function

$$U(z) = U(z_0, z_1, \ldots, z_{n-1}, z) = \begin{vmatrix} 1 & z_0 & z_0^2 & \cdots & z_0^n \\ \cdot & & & & \cdot \\ \cdot & & & & \cdot \\ \cdot & & & & \cdot \\ 1 & z_{n-1} & z_{n-1}^2 & \cdots & z_{n-1}^n \\ 1 & z & z^2 & \cdots & z^n \end{vmatrix}, \tag{4}$$

which is obviously in $\mathcal{P}_n$. Furthermore, it vanishes at $z_0, z_1, \ldots, z_{n-1}$, since substituting z_i ($i = 0, 1, 2, \ldots, n-1$) for z yields two identical rows in the determinant. Thus,

$$U(z_0, z_1, \ldots, z_{n-1}, z) = C(z - z_0)(z - z_1)\ldots(z - z_{n-1}), \tag{5}$$

where C is a constant that depends only on $z_0, z_1, \ldots, z_{n-1}$.
By comparing the coefficient of z^n in (4) and (5), we conclude that $C = U(z_0, z_1, \ldots, z_{n-1})$. Thus, we have

$$U(z_0, z_1, \ldots, z_{n-1}, z) = U(z_0, z_1, \ldots, z_{n-1})(z - z_0)(z - z_1)\ldots(z - z_{n-1}) \tag{6}$$

and hence we have the recursion formula

$$U(z_0, z_1, \ldots, z_{n-1}, z_n) = U(z_0, \ldots, z_{n-1})(z_n - z_0)(z_n - z_1)\ldots(z_n - z_{n-1}). \tag{7}$$

But since

$$U(z_0, z_1, z_2) = U(z_0, z_1)(z_2 - z_0)(z_2 - z_1),$$
$$U(z_0, z_1) = z_1 - z_0,$$

we have by repeated applications of (7),

$$U(z_0, z_1, \dots, z_n) = \prod_{i>j}^{n} (z_i - z_j). \tag{8}$$

By assumption, the points $z_0, z_1, \dots, z_n$ are distinct. Therefore $U \neq 0$ and consequently the system (2) has a unique solution, i.e., there is a unique polynomial $p_n(z)$ satisfying (1). ■

The form of the polynomial $p_n(z)$ obtained above is not very convenient. A better representation of it can be obtained as follows: Let $z_0, z_1, \dots, z_n$ be distinct and introduce the following polynomials of degree n:

$$l_k(z) = \frac{(z - z_0)(z - z_1) \cdots (z - z_{k-1})(z - z_{k+1}) \cdots (z - z_n)}{(z_k - z_0)(z_k - z_1) \cdots (z_k - z_{k-1})(z_k - z_{k+1}) \cdots (z_k - z_n)}, \tag{9}$$

where $k = 0, 1, 2, \dots, n$. It is clear that

$$l_k(z_j) = \delta_{k,j} = \begin{cases} 0 & \text{if } k \neq j \\ 1 & \text{if } k = j. \end{cases} \tag{10}$$

For given values, $w_0, w_1, \dots, w_n$, the polynomial

$$p_n(z) = \sum_{k=0}^{n} w_k \, l_k(z) \tag{11}$$

is in $\mathcal{P}_n$ and takes on these values at the points z_i:

$$p_n(z_j) = w_j, \quad j = 0, 1, \dots, n. \tag{12}$$

Since the interpolation problem (1) has a unique solution, all other representations of the solution must, upon rearrangement of terms, coincide with (11).

Another alternative form of p_n that is more useful may be introduced as follows: Let

$$G_n(z) = (z - z_0)(z - z_1) \cdots (z - z_n). \tag{13}$$

Then,

$$G'_n(z_k) = (z_k - z_0)(z_k - z_1) \cdots (z_k - z_{k-1})(z_k - z_{k+1}) \cdots (z_k - z_n) \tag{14}$$

and hence it follows from (9) that

$$l_k(z) = \frac{G_n(z)}{(z - z_k)G'_n(z_k)}. \tag{15}$$

The formula (11) becomes

$$p_n(z) = \sum_{k=0}^{n} w_k \frac{G_n(z)}{(z - z_k)G'_n(z_k)}. \tag{16}$$

Formula (16) is known as the *Lagrange Interpolation Formula* and the polynomials $l_k(z)$ are called the *interpolating* or *sampling*, or *fundamental* polynomials of the Lagrange interpolation.

The most important property of the interpolating polynomials l_k is the relation (10), which enables us to write the simple explicit solution (16) of the interpolation problem (1).

When the numbers, w_i, are the values of some function $f(z)$ at the points z_i, i.e., $w_i = f(z_i)$, the polynomial $p_n(z)$ given by (16) coincides with the function $f(z)$ at the points $z_0, z_1, \ldots, z_n$. That is, if

$$p_n(z) = \sum_{k=0}^{n} f(z_k) l_k(z) = \sum_{k=0}^{n} f(z_k) \frac{G_n(z)}{(z - z_k)G'_n(z_k)}, \tag{17}$$

then

$$p_n(z_j) = f(z_j), \qquad j = 0, 1, \ldots, n. \tag{18}$$

If the function $f(z)$ itself is a polynomial of degree n, i.e., $f \in \mathcal{P}_n$, then, in view of the fact that polynomials of degree n are completely determined by their values at $n + 1$ points, we immediately have the trivial relation

$$f(z) = \sum_{k=0}^{n} f(z_k) \frac{G_n(z)}{(z - z_k)G'_n(z_k)}. \tag{19}$$

In fact, $p_n(z) = f(z)$ if and only if f is a polynomial of degree at most n. As a special case, if we set $f(z) = (z - t)^j$, $j = 0, 1, 2, \ldots, n$, for some fixed t, we obtain from (19)

$$(z - t)^j = \sum_{k=0}^{n} (z_k - t)^j l_k(z).$$

By selecting $t = z$, we obtain for $j = 0$

$$\sum_{k=0}^{n} l_k(z) = 1 \tag{20}$$

and for $j = 1, 2, \ldots, n$

$$\sum_{k=0}^{n} (z_k - z)^j l_k(z) = 0. \tag{21}$$

The identities (20) and (21) are called the *Cauchy relations* for the fundamental polynomials $l_k(z)$.

3.3. Lagrange-Type Interpolation

3.3.1. Uniform Sampling

3.3.1.1. Foundations

In many applications a function $f(z)$ needs to be reconstructed from its values at a discrete set of points $f(z_n)$. Formula (19) is almost a perfect candidate for that, except for the fact that the equality in (19) holds only if f is a polynomial of degree at most n. If f is not a polynomial, a generalization of (19) is needed. This necessitates that the interpolating polynomials l_k be replaced by more general functions.

A natural generalization of polynomials is entire functions. As is known, a polynomial is an expression of the form

$$p(z) = a_0 + a_1 z + a_2 z^2 + \cdots + a_n z^n,$$

which is clearly defined for every complex number z. The degree (order) of a polynomial is the largest power of z appearing in it. That is to say the degree of

p is n, provided that $a_n \neq 0$. Likewise, an entire function is a function given by a Taylor series of the form

$$f(z) = \sum_{k=0}^{\infty} a_k z^k,$$

that converges for every complex number z. The coefficients a_k are easily seen to be

$$a_k = f^{(k)}(0)/k! \quad \text{with} \lim_{k\to\infty} \sqrt[k]{|a_k|} = 0.$$

If we set

$$M_f(r) = \max_{|z|=r} |f(z)|,$$

then the *maximum modulus principle* states that $M_f(r)$ is an increasing function of r. Because there is no largest power of z in the Taylor series of f, unless it degenerates to a polynomial, the order of an entire function is defined differently. We say that an entire function $f(z)$ is of *finite order* if there exists a positive number k such that

$$M_f(r) < e^{r^k},$$

for sufficiently large r. The greatest lower bound of such numbers k is called the *order of the entire function f*. If the order of f is denoted by ρ, it can be shown [3, p. 8] that

$$\rho = \limsup_{r\to\infty} \frac{\ln \ln M_f(r)}{\ln r}.$$

Entire functions of order 1 are called *entire functions of exponential type.*

Sometimes the order of entire functions is not sufficient to classify them. Another quantity, called the *type of entire functions*, is needed. If the order of an entire function is ρ, we define its type as the greatest lower bound of all positive numbers c such that

$$M_f(r) < e^{cr^{\rho}},$$

for sufficiently large r. Similarly, it can be shown [3, p. 8] that if the type of an entire function, f, is σ, then

$$\sigma = \limsup_{r\to\infty} \frac{\ln M_f(r)}{r^\rho}.$$

If $\sigma = 0$, we say that f is of minimal type and if $\sigma = \infty$, we say that f is of maximal type. Polynomials are of exponential type zero and $f(z) = \exp(\sigma z^m)$, where m is an integer, is of order m and type σ. In particular, $f(z) = \sin(\sigma z)$ is of exponential type with type equals σ.

For $\sigma > 0$ and $1 \le p \le \infty$, we denote by B_σ^p the set of all entire functions f of exponential type with type at most σ that belong to $L^p(\mathbb{R})$ when restricted to the real line. That is, $f \in B_\sigma^p$ if and only if f is an entire function satisfying

$$|f(z)| \le \sup_{x\in\mathbb{R}} |f(x)| \exp(\sigma|y|), \quad z = x + jy, \quad j = \sqrt{-1},$$

and

$$\int_{-\infty}^{\infty} |f(x)|^p \, dx < \infty, \quad \text{if } 1 \le p < \infty$$

and

$$\text{ess.}\sup_{x\in\mathbb{R}} |f(x)| < \infty, \quad \text{if } p = \infty,$$

where ess. sup means the essential supremum. The space B_σ^p was called the Bernstein space in Chapter 2.

We say that a function f is *band-limited* to $[-\sigma, \sigma]$ if and only if $f \in B_\sigma^2$. Although the class of band-limited functions has been extended to include B_σ^p for any $1 \le p \le \infty$, throughout the rest of this chapter we shall be mainly dealing with the space B_σ^2. A nice characterization of the space B_σ^2 is given by a theorem of R. Paley and N. Wiener [33], known as the Paley-Wiener theorem. It may be stated as follows:

Theorem 2 (Paley-Wiener) *A function f belongs to B_σ^2 if and only if it is representable in the form*

$$f(t) = \int_{-\sigma}^{\sigma} e^{jxt} g(x) \, dx \qquad (t \in \mathbb{R})$$

for some function $g \in L^2(-\sigma, \sigma)$.

As is known, a polynomial $p(z)$ of exact degree n has exactly n zeros and can be factored in the form

$$p(z) = c \prod_{k=1}^{n} (z - z_k), \tag{22}$$

where z_k, $k = 1, 2, \ldots, n$, are the zeros of $p(z)$. But unlike polynomials, an entire function may not have any zeros at all, unless its order is not an integer, and in such a case it must have infinitely many zeros [3, p. 24]. An example of an entire function with no zeros is $f(z) = e^z$.

If an entire function has zeros, it can be factored as in (22) by using Hadamard's Factorization Theorem [27, p. 24], which can be stated as follows:

Theorem 3 (Hadamard) *Let $f(z)$ be an entire function of order ρ, and $\{z_n\}_{n=1}^{\infty}$ be its non-zero zeros, i.e., $z_n \neq 0$ for all n. Let p be the smallest integer for which the series*

$$\sum_{n=1}^{\infty} \frac{1}{|z_n|^{p+1}}$$

converges. Set

$$G(u; p) = (1 - u) \exp\left(u + \frac{u^2}{2} + \cdots + \frac{u^p}{p}\right),$$

and $G(u; 0) = (1 - u)$. Then

$$f(z) = z^m e^{P(z)} \prod_{n=1}^{\infty} G\left(\frac{z}{z_n}; p\right),$$

where m is the multiplicity of the zero at the origin and $P(z)$ is a polynomial of degree not exceeding ρ.

We shall call the product

$$\prod_{n=1}^{\infty} G\left(\frac{z}{z_n}; p\right)$$

the *canonical product* of the $\{z_n\}_{n=1}^{\infty}$. As a special case, we have

$$\sin \pi z = \pi z \prod_{n=1}^{\infty} \left(1 - \frac{z^2}{n^2}\right).$$

3.3.1.2. Whittaker-Shannon-Kotel'nikov Sampling Theorem

Having seen many similarities between polynomials and entire functions, we now ask whether it would be possible to extend the Lagrange interpolation formula to entire functions. More precisely, given the values of some function $f(z)$ at a set of points $\mathcal{A} = \{z_n\}_{n=1}^{\infty}$, is it possible to find an entire function $g(z)$ that coincides with $f(z)$ at these points, i.e., $f(z_n) = g(z_n)$, for all n?

To the best of our knowledge, E.T. Whittaker [40] was the first to address that question. He first pointed out that $f(z)$ itself should be analytic, and then proceeded to show that there might be infinitely many entire functions $g(z)$, which he called *cotabular functions*, that would coincide with f at those points. For example, if $f(z_n) = g(z_n)$ for all n, then by choosing an entire function $h(z)$ such that $h(z_n) = 0$ for all n, we have another entire function, namely, $H(z) = g(z) + h(z)$, coinciding with f at the points $\{z_n\}_{n=1}^{\infty}$.

The existence of h depends on the set $\mathcal{A}$. For example, if $\mathcal{A}$ has a finite accumulation point, then no such h exists, unless it is identically zero. On the other hand, if $\mathcal{A} = \{z_n = a + nW\}_{n=-\infty}^{\infty}$, where a and W are constants and $W \neq 0$, we can take $h(z) = \sin[\pi(z - a)/W]$.

Whittaker confined his attention to the case in which $z_n = a + nW$, $n = 0, \pm 1, \pm 2, \ldots$, and constructed the entire function

$$\begin{aligned} C_f(z) &= \sum_{n=-\infty}^{\infty} f(a + nW) \frac{\sin[\pi(z - a - nW)/W]}{[\pi(z - a - nW)/W]} \\ &= \sum_{n=-\infty}^{\infty} f(a + nW) \operatorname{sinc}[(z - a - nW)/W], \end{aligned} \tag{23}$$

which coincides with f at the points $z_n = a + nW$, $n = 0, \pm 1, \pm 2, \ldots$, where

$$\operatorname{sinc}(z) = \begin{cases} \sin \pi z/(\pi z) & z \neq 0 \\ 1, & z = 0. \end{cases}$$

He called $C_f(z)$ the *cardinal function* of the cotabular set of f. He also said that among all functions that are cotabular with f, $C_f(z)$ is the "*simplest*". However, he did not elaborate more on that. But it should be noted that nowhere in his paper did Whittaker discuss the conditions under which $C_f(z) = f(z)$.

It is interesting to observe that the series in (23) can be easily put in a form similar to (16) as

$$C_f(z) = \sum_{n=-\infty}^{\infty} f(z_n)S_n(z) = \sum_{n=-\infty}^{\infty} f(z_n)\frac{G(z)}{G'(z_n)(z-z_n)}, \tag{24}$$

where $G(z) = \sin[\pi(z-a)/W]$. The sampling functions

$$S_n(z) = \frac{G(z)}{G'(z_n)(z-z_n)},$$

are readily seen to be band-limited. Indeed,

$$S_n(z) = \int_{-\sigma}^{\sigma} s_n(u)e^{jzu}\,du,$$

where

$$s_n(u) = \frac{1}{2\sigma}e^{-jz_n u}, \qquad \sigma = \pi/W, \qquad z_n = a + nW.$$

The sampling functions $S_n(z)$ play the role of the sampling (fundamental) polynomials $l_n(z)$ of the Lagrange interpolation; see (9) and (15). And because of the similarity between (16) and (24), we have the following definition.

Definition 1 *We shall call any series of the form*

$$\sum_{n=-\infty}^{\infty} f(z_n)\frac{G(z)}{G'(z_n)(z-z_n)} \tag{25}$$

whether $G(z) = \sin[\pi(z-a)/W]$ *or not, a Lagrange-type interpolation series.*

Under what conditions does the relation $C_f(z) = f(z)$ hold? An answer to this question came five years later when Ogura [32] published a paper in which he refined Whittaker's paper and pointed out some inaccuracies in it.

Ogura considered the case in which $a = 0$ and $W = 1$, and showed that if f is an entire function of exponential type with type at most π, then $C_f(z) = f(z)$, i.e.,

$$f(z) = \sum_{n=-\infty}^{\infty} f(n)\frac{\sin\pi(z-n)}{\pi(z-n)} = \sum_{n=-\infty}^{\infty} f(n)\frac{G(z)}{G'(n)(z-n)}, \tag{26}$$

where $G(z) = \sin \pi z$.

This is almost the same as the modern formulation of the Whittaker-Shannon-Kotel'nikov (WSK) sampling theorem which may be stated as follows.

Theorem 4 (Whittaker-Shannon-Kotel'nikov) *If a function f is band-limited to $[-\sigma, \sigma]$, i.e., it is representable as*

$$f(t) = \int_{-\sigma}^{\sigma} e^{jxt} g(x)\, dx \qquad (t \in \mathbb{R}) \tag{27}$$

for some function $g \in L^2(-\sigma, \sigma)$, then f can be reconstructed from its samples, $f(k\pi/\sigma)$, that are taken at the equally spaced nodes $k\pi/\sigma$ on the time axis $\mathbb{R}$. The construction formula is

$$f(t) = \sum_{k=-\infty}^{\infty} f\left(\frac{k\pi}{\sigma}\right) \frac{\sin(\sigma t - k\pi)}{(\sigma t - k\pi)} \qquad (t \in \mathbb{R}), \tag{28}$$

the series being absolutely and uniformly convergent on $\mathbb{R}$. See, e.g., [10], [45, p.16].

For different proofs of this theorem, see Chapter 2. For $\sigma = \pi$, Equation (28) can be put in the form

$$f(t) = \sum_{k=-\infty}^{\infty} f(k) \frac{G(t)}{(t-k)G'(k)}, \tag{29}$$

where

$$G(t) = \frac{\sin \pi t}{\pi} = t \prod_{k=1}^{\infty} \left(1 - \frac{t^2}{k^2}\right) = \lim_{n \to \infty} \tilde{G}_n(t),$$

and

$$\tilde{G}_n(t) = t \prod_{k=1}^{n} \left(1 - \frac{t^2}{k^2}\right). \tag{30}$$

For a given f, let us denote the series on the right-hand side of (28) by $S_\sigma[f](t)$, and to simplify the notation, when $\sigma = \pi$, we set $S[f](t) = S_\pi[f](t)$. Therefore, from the WSK theorem we may say that if f is band-limited to $[-\sigma, \sigma]$, then $f(t) = S_\sigma[f](t)$.

Let us now return to Formula (17) and assume that the values of f are known at the points $z_k = k = 0, \pm 1, \pm 2, \ldots, \pm n$. Let us also set

$$G_n(t) = \prod_{k=-n}^{n} (t-k) = t \prod_{k=1}^{n} (t^2 - k^2),$$

so that

$$p_n(t) = L_n[f](t) = \sum_{k=-n}^{n} f(k) \frac{G_n(t)}{(t-k)G_n'(k)}, \tag{31}$$

where L_n is the nth Lagrange interpolation operator. But it is easy to see that $G_n(t) = (n!)^2(-1)^n \tilde{G}_n(t)$, where $\tilde{G}_n(t)$ is given by (30); therefore, we have

$$L_n[f](t) = \sum_{k=-n}^{n} f(k) \frac{\tilde{G}_n(t)}{(t-k)\tilde{G}_n'(k)}, \tag{32}$$

from which, by taking the limit as $n \to \infty$, we formally conclude (29), i.e.,

$$\lim_{n \to \infty} L_n[f](t) = f(t). \tag{33}$$

Although formula (33) seems to hold formally, showing that it actually holds in some rigorous sense was proved rather recently by G. Hinsen and D. Klösters [22]; see also [19, pp. 42–47]. They showed that (33) holds uniformly on each compact interval $[a, b]$, and that the result is best possible in the sense that it does not hold on unbounded intervals.

We state their result as a theorem:

Theorem 5 *Let $[a,b]$ be a finite closed interval, $1 \le p \le \infty$, and let f be a function defined on $\mathbb{Z}$ such that $\{f(n)\}_{n\in\mathbb{Z}} \in l^p$. Furthermore, let $S[f](t)$ and $L_n[f](t)$ be defined as in* (29) *and* (32). *Then*

$$\lim_{n \to \infty} \|L_n[f] - S[f]\|_{C[a,b]} = 0,$$

where

$$\|g(x)\|_{C[a,b]} = \sup_{a \le x \le b} |g(x)|.$$

In particular, if $f \in B_\pi^p$ *for some* $p \in [1, \infty)$, *then for each finite interval* $[a, b]$, *we have*

$$\lim_{n\to\infty} L_n[f](t) = S[f](t) = f(t),$$

uniformly on $[a,b]$.

The result is best possible in the sense that it cannot, in general, hold on any unbounded region. For, by Hölder's inequality for $1 < p < \infty$

$$|S[f](t)| \le \left(\sum_{n=-\infty}^{\infty} |f(n)|^p\right)^{1/p} \left(\sum_{n=-\infty}^{\infty} \left|\frac{\sin \pi(t-n)}{\pi(t-n)}\right|^q\right)^{1/q}$$
$$\le Cp < \infty, \quad (t \in \mathbb{R})$$

where $\frac{1}{p} + \frac{1}{q} = 1$, and for $p = 1$ we have

$$|S[f](t)| \le \left(\sum_{n=-\infty}^{\infty} |f(n)|\right) \sup_{t\in\mathbb{R}} \left|\frac{\sin \pi(t-n)}{\pi(t-n)}\right|$$
$$\le C < \infty, \quad (t \in \mathbb{R})$$

but for any n, $\lim_{|t|\to\infty} \|L_n[f](t)\| = \infty$, whenever $L_n[f](t)$ is not a constant. Here we have used the fact that

$$\left(\sum_{n=-\infty}^{\infty} \left|\frac{\sin \pi(t-n)}{\pi(t-n)}\right|^q\right)^{1/q} \le p \qquad (1 < p < \infty),$$

for a proof of this relation see [10].

Closely related to entire functions are *meromorphic functions*. A function is said to be *meromorphic* if it is single-valued and has no singularities other than poles. Such a function has no more than a finite number of poles in any bounded region. Every meromorphic function is the quotient of two entire functions having no zeros in common. A fundamental theorem for meromorphic functions is the Mittag-Leffler theorem [35, p. 291].

Theorem 6 (Mittag-Leffler) *Let* $\{z_n\}$ *be a given set of points, finite or infinite, but having no finite limit point. Associate with every point* z_n *a finite sequence of complex numbers* $\{a_{n,i}\}_{i=1}^{k_n}$ *and a rational function* $h_n(z)$, *called the principle part associated with* z_n, *defined by*

$$h_n(z) = \frac{a_{n,1}}{(z-z_n)} + \frac{a_{n,2}}{(z-z_n)^2} + \cdots + \frac{a_{n,k_n}}{(z-z_n)^{k_n}},$$

where k_n is a positive integer. Then one can construct a meromorphic function that has poles exactly at the points $\{z_n\}$ with the prescribed principle parts. If $f(z)$ is such a function, then the most general meromorphic function having the same poles with their corresponding principle parts as $f(z)$ is a function of the form $f(z) + g(z)$, where $g(z)$ is an entire function.

Let us return momentarily to the Cauchy relations (20) and (21). There is an analogue of the Cauchy relation (20) for the Lagrange-type interpolation (29), namely,

$$\sum_{n=-\infty}^{\infty} S_n(t) = \sum_{n=-\infty}^{\infty} \frac{G(t)}{(t-n)G'(n)} = 1.$$

This is an immediate consequence of the Mittag-Leffler expansion of the meromorphic function $\csc t$,

$$\csc \pi t = \frac{1}{\sin \pi t} = \sum_{n=-\infty}^{\infty} \frac{(-1)^n}{\pi(t-n)}.$$

The other Cauchy relations (21) do not seem to have an obvious extension to (29), except in the limit.

3.3.2. Nonuniform Sampling

Theorem 5 has been extended to the case of nonuniform sampling by the same authors. But first, let us introduce the following generalization of the WSK sampling theorem which allows us to reconstruct a band-limited signal from its samples at nonuniformly spaced points.

Theorem 7 (Paley-Wiener-Levinson) *Let $\{t_k\}_{k\in\mathbb{Z}}$ be a sequence of real numbers such that*

$$D = \sup_{k\in\mathbb{Z}} |t_k - k| < \frac{1}{4},$$

and let

$$G(t) = (t - t_0)\prod_{k=1}^{\infty}\left(1 - \frac{t}{t_k}\right)\left(1 - \frac{t}{t_{-k}}\right).$$

Then for any $f \in B^2_\pi$, we have

$$f(t) = \sum_{k=-\infty}^{\infty} f(t_k) \frac{G(t)}{(t - t_k)G'(t_k)} \qquad (t \in \mathbb{R}), \tag{34}$$

and the series being uniformly convergent on compact sets ([28], [45, p. 24]).

When $t_n = n$, $G(t)$ reduces to $\sin \pi t/\pi$ and we obtain the WSK theorem. The following definition is needed.

Definition 2 *Let T_L denote the set of all real sequences $\{t_n\}_{n=-\infty}^{\infty}$ that satisfy the following conditions:*

$$t_0 = 0, \quad t_1 < t_2 < t_3 < \cdots, \quad t_{-n} = -t_n, \quad n = 1, 2, \ldots,$$

and

$$|t_n - n| \le L, \qquad n = 0, \pm 1, \pm 2, \ldots.$$

If $L < 1/2$, we associate with any such sequence the entire function

$$G(t) = t \prod_{k=1}^{\infty} \left(1 - \frac{t^2}{t_k^2}\right),$$

which is the canonical product of the $\{t_n\}_{n=-\infty}^{\infty}$.

Now we can state the analogue of Theorem 5 for the case of nonuniform sampling.

Theorem 8 *Let $[a, b]$ be a finite interval, $1 \le p < \infty$, $\{t_n\}_{n=-\infty}^{\infty} \in T_L$ for some L with*

$$L \begin{cases} < 1/(4p) & \text{if } 1 < p < \infty \\ \le 1/4 & \text{if } p = 1. \end{cases}$$

Let f be any function defined on $\{t_n\}_{n=-\infty}^{\infty}$ such that $\{f(t_n)\}_{n=-\infty}^{\infty} \in l^p$. Moreover, let L_n denote the n-th Lagrange interpolation operator with respect to the sequence $\{t_n\}_{n=-\infty}^{\infty}$ i.e.,

$$L_n[f](t) = \sum_{k=-n}^{n} f(t_k) \frac{G_n(t)}{(t - t_k)G_n'(t_k)}, \tag{35}$$

where

$$G_n(t) = t\prod_{k=1}^{n}(t^2 - t_k^2),$$

and let S denote the sampling operator with respect to $\{t_n\}_{n=-\infty}^{\infty}$*; that is*

$$S[f](t) = \sum_{k=-\infty}^{\infty} f(t_k)\frac{G(t)}{(t-t_k)G_n'(t_k)}.$$

Then

$$\lim_{n\to\infty} \|L_n[f] - S[f]\|_{C[a,b]} = 0.$$

As a special case, we obtain the following corollary:

Corollary 1 *Let* $1 \le p < \infty$ *and* $\{t_n\}_{n=-\infty}^{\infty} \in T_L$ *where*

$$L < \begin{cases} 1/4 & \text{if } 1 \le p \le 2 \\ 1/(2p) & \text{if } 2 \le p < \infty. \end{cases}$$

Then for any $f \in B_\pi^p$,

$$f(t) = \sum_{k=-\infty}^{\infty} f(t_k)\frac{G(t)}{(t-t_k)G'(t_k)}.$$

In addition, for $1 \le p < \infty$, *and* $\{t_n\}_{n=-\infty}^{\infty} \in T_L$, *with* $L < 1/(4p)$, *we have for each* $f \in B_\pi^p$,

$$\lim_{n\to\infty} L_n[f](t) = S[f](t) = f(t)$$

uniformly on any finite interval $[a,b]$.

There is also an analogue of Cauchy's relation (20) for the Lagrange-type interpolation in the case of nonuniform sampling, namely,

$$\sum_{n=-\infty}^{\infty} S_n(t) = \sum_{n=-\infty}^{\infty} \frac{G(t)}{(t-t_n)G'(t_n)} = 1, \tag{36}$$

provided that the series converges. This also follows from the Mittag-Leffler theorem for meromorphic functions. For, since $G(t)$ is an entire function with simple zeros at $\{t_n\}_{n=-\infty}^{\infty}$, the function $1/G(t)$ is meromorphic with simple poles at $\{t_n\}_{n=-\infty}^{\infty}$. The residue r_n at t_n is easily calculated as follows: Since G is differentiable

$$G'(t_n) = \lim_{t\to t_n} \frac{G(t) - G(t_n)}{t - t_n} = \lim_{t\to t_n} \frac{G(t)}{t - t_n},$$

but on the other hand

$$r_n = \lim_{t\to t_n} \frac{(t - t_n)}{G(t)} = \frac{1}{G'(t_n)}.$$

Therefore, by the Mittag-Leffler expansion of the meromorphic function, $1/G(t)$, we have

$$\frac{1}{G(t)} = \sum_{n=-\infty}^{\infty} \frac{r_n}{(t - t_n)} = \sum_{n=-\infty}^{\infty} \frac{1}{(t - t_n)G'(t_n)},$$

which is (36), provided that the series converges.

3.4. Sampling Expansions Involving Derivatives

Let us now return to Section 2 and try to generalize Theorem 1 in a different direction. Let $\{z_i\}_{i=1}^{n}$ be n distinct points and f be a differentiable function. Let us construct a polynomial, $P_{2n-1}(z)$, of degree $2n - 1$ that coincides with f and f' at the points $\{z_i\}_{i=1}^{n}$, that is

$$P_{2n-1}(z_i) = f(z_i) \quad \text{and} \quad P'_{2n-1}(z_i) = f'(z_i); \quad i = 1, 2, \ldots, n.$$

To this end, let

$$G(z) = (z - z_1)(z - z_2)\cdots(z - z_n), \tag{37}$$

and

$$l_k(z) = \frac{G(z)}{(z - z_k)G'(z_k)}, \quad k = 1, 2, \ldots, n, \tag{38}$$

so that

$$l_k(z_i) = \delta_{k,i}; \quad k, i = 1, 2, \ldots, n. \tag{39}$$

With some easy calculations, one can verify that

$$l_k(z_k) = \frac{G''(z_k)}{2G'(z_k)}, \quad k = 1, 2, \ldots, n. \tag{40}$$

Therefore, by using (37)–(40), we have

$$\begin{aligned} P_{2n-1}(z) &= \sum_{k=1}^{n} f(z_k)\left[1 - \frac{G''(z_k)}{G'(z_k)}(z - z_k)\right]l_k^2(z) + \sum_{k=1}^{n} f'(z_k)(z - z_k)l_k^2(z) \\ &= \sum_{k=1}^{n}\left\{f(z_k) + \left[f'(z_k) - f(z_k)\frac{G''(z_k)}{G'(z_k)}\right](z - z_k)\right\}l_k^2(z). \end{aligned} \tag{41}$$

That $P_{2n-1}(z)$ is a polynomial of exact degree $2n - 1$ follows from the observation that

$$\left[1 - \frac{G''(z_k)}{G'(z_k)}(z - z_k)\right]l_k^2(z) \quad \text{and} \quad (z - z_k)l_k^2(z)$$

are polynomials of exact degree $2n - 1$.

More generally, it is possible to construct a polynomial $P_N(z)$ that coincides with f and its first $q_1 - 1$ derivatives at z_1, and f and its first $q_2 - 1$ derivatives at $z_2, \ldots,$ and f and its first $q_n - 1$ derivatives at z_n, where $q_1, q_2, \ldots, q_n$ are positive integers and $N = q_1 + q_2 + \cdots + q_n - 1$. That is

$$\begin{array}{llll} P_N(z_1) = f(z_1), & P_N^{(1)}(z_1) = f^{(1)}(z_1), & \cdots & P_N^{(q_1-1)}(z_1) = f^{(q_1-1)}(z_1) \\ P_N(z_2) = f(z_2), & P_N^{(1)}(z_2) = f^{(1)}(z_2), & \cdots & P_N^{(q_2-1)}(z_2) = f^{(q_2-1)}(z_2) \\ \vdots & \vdots & \cdots & \vdots \\ P_N(z_n) = f(z_n), & P_N^{(1)}(z_n) = f^{(1)}(z_n), & \cdots & P_N^{(q_n-1)}(z_n) = f^{(q_n-1)}(z_n). \end{array}$$

This problem is known as the *Generalized Hermite Interpolation Problem*. It can be shown [13, p. 37] that its solution is given by

$$P_N(z) = \sum_{i=1}^{n} f(z_i)l_{i,0}(z) + \sum_{i=1}^{n} f^{(1)}(z_i)l_{i,1}(z) + \cdots + \sum_{i=1}^{n} f^{(q_i-1)}(z_i)l_{i,q_i-1}(z),$$

where

$$l_{i,k}(z) = G(z)\frac{(z - z_i)^{k-q_i}}{k!}\frac{d^{(q_i-k-1)}}{dz^{(q_i-k-1)}}\left[\frac{(z - z_i)^{q_i}}{G(z)}\right]_{z=z_i},$$

and

$$G(z) = (z - z_1)^{q_1} \dots (z - z_n)^{q_n}.$$

Constructing a polynomial $P_N(z)$ that coincides with f and its first q derivatives at the points $\{z_i\}_{i=1}^n$, i.e.,

$$P_N(z_i) = f(z_i), \quad P_N^{(1)}(z_i) = f^{(1)}(z_i), \dots, P_N^{(q)}(z_i) = f^{(q)}(z_i), \quad i = 1, \dots, n,$$

is clearly a special case of the generalized Hermite interpolation problem.

Passing from polynomial to entire function interpolation, we surprisingly find that Formula (41) takes on a simpler form for band-limited functions; thanks to the sinc function and the property that

$$\frac{d}{dt}(\operatorname{sinc} t)^2|_{t=n} = 0 \quad \text{for} \quad n = 0, \pm 1, \pm 2, \dots.$$

Before we state the sampling expansion for band-limited functions that uses samples of the function and its derivative, let us recall that if a function is given on the time interval $[0, T]$ and the samples are spaced $1/2W$ seconds apart, there will be a total of $2TW$ samples in the interval. A function is said to be limited to the time interval $[0, T]$ if and only if all the samples outside this interval are exactly zero. Then any function limited to the bandwidth W and the time interval $[0, T]$ can be specified by giving $2TW$ numbers.

In his celebrated paper [37], Shannon indicated that a band-limited signal can be reconstructed from the values of the signal at non-uniformly spaced points and also from the values of the signal and its derivative. In fact, he said: "The $2TW$ numbers used to specify the function need not be equally spaced samples." He then proceeded to say: "One can further show that the value of the function and its derivative at every other sample point are sufficient (to specify the function). The value and first and second derivatives at every third sample point give a still different set of parameters which uniquely determine the function. Generally speaking, any set of $2TW$ independent numbers associated with the function can be used to describe it."

Shannon, however, did not give any proof of his claim. Several years later, Fogel [16], Jagerman and Fogel [23] gave a proof of one of Shannon's claims by deriving a sampling expansion to reconstruct a band-limited signal using the value of the signal and its first derivative at every other sampling point. Their result can be stated as follows:

Theorem 9 *Let $f(t)$ be a band-limited function with maximum frequency W, i.e., its Fourier transform has support in $[-2\pi W, 2\pi W]$, then*

$$f(t) = \sum_{n=-\infty}^{\infty} [f(t_n) + (t - t_n)f'(t_n)] \left[\frac{\sin[\pi(t - t_n)/h]}{\pi(t - t_n)/h} \right]^2 \tag{42}$$

where $0 < h \leq 1/W$ and $t_n = nh$.

By a change of variable, we may write (42) in the form

$$f(t) = \sum [f(t_n) + (t - t_n)f'(t_n)] \left[\frac{\sin \frac{\sigma}{2}(t - t_n)}{\frac{\sigma}{2}(t - t_n)} \right]^2$$

where

$$t_n = \frac{2n\pi}{\sigma}, \quad \sigma = 2\pi W, \quad h = \frac{2\pi}{\sigma},$$

and f is band-limited to $[-\sigma, \sigma]$.

Formula (42) has several applications; one of them is in the field of air traffic control wherein the aircraft estimated velocity as well as position is used to obtain a continuous course plot of the air-path with half the sampling rate.

A proof of Shannon's general claim on sampling with higher derivatives and also a generalization of (42) was obtained by Linden [29] and Linden and Abramson [30]. They proved the following result:

Theorem 10 *Let $f(t)$ be band-limited to $[-\sigma, \sigma]$, i.e., $f \in B_\sigma^2$ and R be an integer ≥ 1.*

Then

$$f(t) = \sum_{k=-\infty}^{\infty} \left[f_0(t_n) + (t - t_n)f_1(t_n) + \cdots + \frac{(t - t_n)^{R-1}}{(R - 1)!} f_{R-1}(t_n) \right] \left(\frac{\sin[\sigma(t - t_n)/R]}{\sigma(t - t_n)/R} \right)^R, \tag{43}$$

where $t_n = nR\pi/\sigma$ and $f_k(t_n)$ are linear combinations of $f(t_n)$, $f^{(1)}(t_n), \ldots, f^{(k)}(t_n)$. In fact, $f_k(t_n)$ can be expressed explicitly in terms of $f(t_n), \ldots, f^{(k)}(t_n)$ as

$$f_k(t_n) = \sum_{i=0}^{k} \binom{k}{i} \left(\frac{\sigma}{R} \right)^{k-i} (\Gamma_R^{k-i}) f^{(i)}(t_n),$$

where

$$\Gamma_R^m = \frac{d^m}{dt^m}\left[\frac{t}{\sin t}\right]^R\Bigg|_{t=0}.$$

Formula (43) reduces to (42) for $R = 2$.

Using the idea of a Riesz basis in a Hilbert space, Rawn [34] extended (43) to nonuniform sampling by obtaining the following theorem.

Theorem 11 *Let $\{t_n\}_{n=-\infty}^{\infty}$ be a sequence of real numbers such that*

$$|t_n - nR| \le d < \frac{1}{4R} \quad \text{for all} \quad n, \tag{44}$$

and let $f \in B_\pi^2$. Then

$$f(t) = \sum_{n=-\infty}^{\infty} [f(t_n) + (t - t_n)f_1(t_n) + \cdots + (t - t_n)^{R-1} f_{R-1}(t_n)]\Psi_n(t), \tag{45}$$

where

$$f_k(t_n) = \frac{1}{k!}\frac{d^k}{dt^k}\left[\frac{f(t)}{\Psi_n(t)}\right]_{t=t_n}, \quad 0 \le k \le R - 1,$$

$$F(t) = \prod_{k=1}^{\infty}\left(1 - \frac{t}{t_k}\right)\left(1 - \frac{t}{t_{-k}}\right), \quad G(t) = tF(t),$$

and

$$\Psi_n(t) = \left[\frac{G(t)}{(t - t_n)G'(t_n)}\right]^R.$$

The convergence of the series (45) *is both uniform on* $(-\infty, \infty)$ *and in the sense of the L^2-norm. Moreover, the expansion is L^2-stable with respect to the class S of all sequences satisfying* (44) *and for each $f \in B_\pi^2$ in the sense that for any sequence $T = \{t_n\} \in S$ and $f \in B_\pi^2$, we have*

$$\int_{-\infty}^{\infty} |f(t)|^2 dt \le A \sum_{n=-\infty}^{\infty}\sum_{i=0}^{R-1} |f^{(i)}(t_n)|^2.$$

J. Voss in his doctoral dissertation (see also the CD-ROM, directory: 'Chapter 3', file: PhD-Thesis-Voss.zip) [39] established the following theorem for Hermite type sampling in the instance of nonuniform, *complex* sampling points $\{z_n\}_{n=-\infty}^{\infty}$ for functions belong to a *weighted* Paley-Wiener space. It includes almost all known irregular sampling theorems, together with derivative sampling.

Theorem 12 *Let $\{z_n\}_{n=-\infty}^{\infty}$ be a sequence of complex numbers satisfying the following conditions:*

$$\sup_{n\in\mathbb{Z}} |\Re e(z_n) - n| \le L, \tag{46}$$

$$|z_m - z_n| \ge \delta \quad \textit{for all} \quad m,\, n \in \mathbb{Z} \textit{ with } m \ne n, \tag{47}$$

$$|\Im m(z_n)| \le M \quad \textit{for all} \quad n \in \mathbb{Z}, \tag{48}$$

where L is a positive real number with

$$L < \min\left\{\frac{1+\alpha}{4(r+1)}, \frac{1+\alpha p}{2p(r+1)}\right\},$$

α and r being non-negative integers, and p and δ positive reals such that $1 \le p < \infty$. Then for every entire function of exponential type $\pi(r+1)$ satisfying $f(\cdot)(1+|\cdot|)^{\alpha} \in L^p(\mathbb{R})$ there holds

$$f(z) = \lim_{K_1,K_2\to\infty} \sum_{\substack{k \\ R_{-K_1} < \Re e(z_k) < R_{K_2}}} \sum_{j=0}^{r} f^{(j)}(z_k) \sum_{l=0}^{r-j} \frac{Q_{r,k,j,l}}{j!(r-j-l)!} \frac{G^{r+1}(z,Z)}{(z-z_k)^{l+1}} \qquad (z \in \mathbb{C})$$

for all strictly increasing sequences $\{R_K\}_{K=-\infty}^{\infty}$ with $\lim_{K\to\pm\infty} R_k = \pm\infty$, where the coefficients $Q_{r,k,j,l}$ and the canonical product $G(z, Z)$ with respect to the nodes $Z = \{z_n\}_{n=-\infty}^{\infty}$ are defined by

$$Q_{r,k,j,l} = \lim_{\zeta\to z_k} \left(\frac{d}{d\zeta}\right)^{r-j-l} \left(\frac{\zeta - z_k}{G(\zeta, Z)}\right)^{r+1},$$

$$G(z, Z) = z^m \lim_{N\to\infty} \prod_{\substack{|n|\le N \\ z_n\ne 0}} \left(1 - \frac{z}{z_k}\right) \quad (m \in \{0, 1\}).$$

The convergence of the sampling sum is uniform on all compact subsets of **C**.

In the particular case that $r = 0$, noting that $Q_{0,k,0,0} = (G'(z_k, Z))^{-1}$ for $k \in \mathbb{Z}$, we immediately obtain the following simplification.

Corollary 2 *Under the hypotheses of Theorem 12 in the case $r = 0$ there holds for every entire function of exponential type π satisfying*

$f(\cdot)(1+|\cdot|)^{\alpha} \in L^p(\mathbb{R})$,

$$f(z) = \lim_{K_1,K_2\to\infty} \sum_{\substack{k \\ R_{-K_1} < \Re e(z_k) < R_{K_2}}} f(z_k) \frac{G(z,Z)}{(z-z_k)G'(z_k,Z)} \qquad (z \in \mathbf{C},)$$

for all strictly increasing sequences $\{R_K\}_{K=-\infty}^{\infty}$ with $\lim_{K\to\pm\infty} R_K = \pm\infty$. The convergence is uniform on all compact subsets of $\mathbf{C}$.

Observe that conditions (46) to (48) can be interpreted geometrically: Inequality (46) restricts the deviation of the real parts of the nodes from the integers; inequality (47) implies that the distance between any two of the nodes is at least δ, called *separated* provided $\delta > 0$; and (48) means that all nodes lie in a strip of width $2M$ parallel to the real line.

Irregular sampling was investigated in particular by M. I. Kadec [24], J. R. Higgins [18] in case $p = 2$, $z_n \in \mathbb{R}$ and $\sup_{n\in\mathbb{Z}} |z_n - n| < 1/4$, Seip [36], and especially thoroughly by G. Hinsen [20–21] using extensive methods of complex analysis including contour integration. The foregoing theorem, the proof of which covers some 30 pages, generalizes Hinsen's results in two ways: firstly, complex nodes are admitted; secondly, the number L which controls the deviation of the real parts of the nodes from the integers may be arbitrarily large provided that the sampled function has a suitable decay on the real axis.

Let us finally mention that condition (46) may be weakened in the sense that $|\Re e(z_n) - n| \le L$ for all $n \in \mathbb{Z}$ with $|n| \ge N$ holds only in an averaged sense. In that case one speaks of *Avdonin* [2] sequences. Hinsen generalized condition (46) in another respect.

3.5. Kramer's Sampling Theorem and Lagrange Interpolation

3.5.1. Kramer's Theorem

In this section we discuss the relationship between the Lagrange-type interpolation and Kramer's sampling theorem, which is a generalization of the WSK sampling theorem. First let us recall that a family of functions $\{\phi_n(x)\}$ in $L^2(I)$, for some measurable set I, is said to be *orthogonal* if

$$\langle \phi_n, \phi_m \rangle = \int_I \phi_n(x)\overline{\phi_m}(x)\,dx = 0, \qquad \text{for } m \neq n.$$

An orthogonal family is said to be *orthonormal* if $\|\phi_n\|^2 = \langle \phi_n, \phi_n \rangle = 1$ for all n. A family of functions $\{\phi_n(x)\}$ in $L^2(I)$ is said to be *complete* if whenever $f \in L^2(I)$ and $\langle f, \phi_n \rangle = 0$ for all n, then $f = 0$. A complete orthogonal family is said to be an *orthogonal basis*. If $\{\phi_n(x)\}$ is an orthogonal basis of $L^2(I)$, then for any $f \in L^2(I)$ we have

$$f(x) = \sum_n \frac{\langle f, \phi_n \rangle}{\|\phi_n\|^2} \phi_n(x).$$

Moreover, if $g \in L^2(I)$, then

$$\langle f, g \rangle = \sum_n \frac{\langle f, \phi_n \rangle \overline{\langle g, \phi_n \rangle}}{\|\phi_n\|^2},$$

which is known as *Parseval's relation*.

Now we can state Kramer's sampling theorem. Although the proof of the theorem is simple and well known, we shall include it because parts of it will be needed later.

Theorem 13 (Kramer) *Let there exist a function $K(x, t)$ continuous in t such that $K(x, t) \in L^2(I)$ for every real number t. Assume that there exists a sequence of real numbers $\{t_n\}_{n\in\mathbb{Z}}$ such that $\{K(x, t_n)\}_{n\in\mathbb{Z}}$ is a complete orthogonal family in $L^2(I)$ for some finite interval $I = [a, b]$. Then for any function of the form*

$$f(t) = \int_a^b F(x)\overline{K}(x, t)\, dx = \langle F, K \rangle, \tag{49}$$

with $F \in L^2(I)$, we have

$$f(t) = \sum_{n=-\infty}^{\infty} f(t_n) S_n^*(t), \tag{50}$$

where

$$S_n^*(t) = \frac{\int_a^b \overline{K}(x, t) K(x, t_n) dx}{\int_a^b |K(x, t_n)|^2 dx}. \tag{51}$$

Proof: Since $F \in L^2[a, b]$ and $\{K(x, t_n)\}_{n\in\mathbb{Z}}$ is an orthogonal basis for $L^2[a, b]$, we have

$$F(x) = \sum_{n=-\infty}^{\infty} \frac{\langle F, \phi_n \rangle}{\|\phi_n\|^2} \phi_n(x),$$

where $\phi_n(x) = K(x, t_n)$ and

$$\langle F, \phi_n \rangle = \int_a^b F(x)\overline{\phi}_n(x)\, dx.$$

By Parseval's relation, we have

$$f(t) = \langle F(x),\ K(x, t) \rangle_x = \sum_{n=-\infty}^{\infty} \frac{\langle F, \phi_n \rangle}{\|\phi_n\|^2} \int_a^b \overline{K}(x, t)\phi_n(x)\, dx,$$

where $\langle ., . \rangle_x$ means that the inner product is taken as an integral over the variable x. But

$$f(t_n) = \int_a^b F(x)\overline{\phi}_n(x)\, dx = \langle F, \phi_n \rangle;$$

thus,

$$f(t) = \sum_{n=-\infty}^{\infty} f(t_n) S_n^*(t). \qquad \blacksquare$$

Note that since $\{K(x, t_n)\}_{n \in \mathbb{Z}}$ is orthogonal, $S_n^*(t)$ has the sampling property (10), i.e., $S_n^*(t_m) = \delta_{n,m}$.

In the case where $I = [-\sigma, \sigma]$, $K(x, t) = e^{jxt}$, $t_n = n\pi/\sigma$, it is easy to see that

$$S_n^*(t) = \frac{\sin \sigma(t - t_n)}{\sigma(t - t_n)} = \operatorname{sinc}(\sigma(t - t_n)/\pi),$$

and Equations (49) and (50) reduce to (27) and (28).

3.5.2. Connections With Boundary-Value Problems

Kramer [25] noted that the function $K(x, t)$ and the sampling points $\{t_n\}_{n \in \mathbb{Z}}$ may be found from certain boundary-value problems. To explain that more precisely, let us first define the differential operator L by

$$L = p_0(x)\frac{d^n}{dx^n} + \cdots + p_{n-1}(x)\frac{d}{dx} + p_n(x), \quad x \in I,$$

where $p_k(x)$ is a complex-valued function with $n-k$ continuous derivatives, $k = 0, 1, \ldots, n$ for any $x \in I = (a, b)$, and $p_0(x) \neq 0$ for any $x \in (a, b)$, with $-\infty \leq a < b \leq \infty$. The adjoint operator L^* is defined as

$$L^* g = (-1)^n \frac{d^n}{dx^n}(\overline{p}_0 g) + (-1)^{n-1} \frac{d^{n-1}}{dx^{n-1}}(\overline{p}_1 g) + \cdots + \overline{p}_n g.$$

The operator L is said to be formally *self-adjoint* if $L = L^*$. If the coefficient functions p_k $(k = 0, 1, \ldots, n)$ are real-valued, then it is easy to see that if L is self-adjoint, then n is even.

The endpoint a is said to be *regular* if it is finite and $p_0(a) \neq 0$, otherwise it is said to be *singular.* A singular endpoint a is classified as *limit-circle* if all solutions of the differential equation $Ly = 0$ are in $L^2(a, c)$ for some $a < c < b$, otherwise it is classified as *limit-point.* The same can be said about b. The operator L is said to be *regular* if both endpoints, a and b, are regular.

Let $U_j(y) = 0, j = 1, \ldots, n$, be linearly independent homogeneous boundary conditions of the form

$$U_j(y) = \sum_{k=1}^{n} \alpha_{j,k}\, y^{(k-1)}(a) + \beta_{j,k}\, y^{(k-1)}(b), \quad j = 1, 2, \ldots, n.$$

To any such system of boundary conditions, there exists an associated system of boundary conditions, known as the *adjoint system*, and if the two systems are equivalent, we say that they are self-adjoint [12, p. 287].

The boundary-value problem

$$Ly = -ty, \quad x \in I, \tag{52}$$

$$U_j(y) = 0, \quad j = 1, \ldots, n, \tag{53}$$

is said to be *self-adjoint* if the differential operator and the boundary conditions are self-adjoint and is said to be *regular* if L is regular.

Kramer noted that if the regular, self-adjoint boundary-value problem (52) and (53) possesses a function $\varphi(x, t)$ that generates the eigenfunctions of the problem $\{\varphi_n(x)\}$ when the eigenvalue parameter t is replaced by the eigenvalues $\{t_n\}$, i.e., $\varphi(x, t_n) = \varphi_n(x)$, then one can take the sampling points to be $\{t_n\}$ and the function $K(x, t)$ to be $\varphi(x, t)$.

It is customary in the theory of boundary-value problems to denote the eigenvalue parameter by λ and the eigenvalues by λ_n; therefore, from now on we shall denote the sampling points by λ_n whenever the sampling expansion is associated with a boundary-value problem.

Definition 3 *We say that the boundary-value problem* (52)–(53) *has the Kramer property if it possesses a function* $\varphi(x, \lambda)$, *entire in* λ, *that generates the eigenfunctions of the problem* $\{\varphi_n(x)\}$ *when the parameter* λ *is replaced by the eigenvalues* $\{\lambda_n\}$, *i.e.*, $\varphi(x, \lambda_n) = \varphi_n(x)$.

Under what conditions does the boundary-value problem (52)–(53) have the Kramer property is still an open question. Some partial but important answers have been obtained in recent years and on which we shall elaborate a little more later.

3.5.3. Connections With Lagrange-Type Interpolation

Although the connection between sampling theorems and boundary-value problems has been the focus of many research papers in the last few years, in this section we shall focus more on that connection as far as the Lagrange-type interpolation is concerned.

The series in (50) does not resemble, and in general is not a Lagrange-type interpolation series since it can not always be put in the form (25). Nevertheless, if the sampling points and functions are obtained from the boundary-value problem (52)–(53), then (50) can be brought closer to the Lagrange-type interpolation series. For, if $\phi(x, \lambda)$ is a function that generates the eigenfunctions of the problem, then

$$L\phi(x, \lambda) = -\lambda\phi(x, \lambda), \quad \text{and} \quad L\phi_n(x) = -\lambda_n\phi_n(x).$$

Therefore, by Lagrange's identity [12, p. 80] for differential operators

$$\int_a^b [\phi(x, \lambda)L\phi_n(x) - \phi_n(x)L\phi(x, \lambda)]\, dx = (\lambda - \lambda_n)\int_a^b \phi(x, \lambda)\phi_n(x)\, dx.$$

But on the other hand, for any two functions $u(x)$ and $v(x)$ in $C^n[a, b]$, we have

$$v(x)Lu(x) - u(x)Lv(x) = \frac{d}{dx}[u(x),\, v(x)],$$

where

$$[u,\, v](x) = \sum_{m=1}^{n} \sum_{j+k=m+1} (-1)^j u^{(k-1)}(x)\big(\overline{p}_{n-m}(x)v(x)\big)^{(j-1)}.$$

Therefore,

$$G_n(\lambda) = (\lambda - \lambda_n)\int_a^b \phi(x, \lambda)\phi_n(x)\, dx = [\phi_n,\, \phi](b) - [\phi_n,\, \phi](a). \tag{54}$$

Since $\phi(x, \lambda)$ is an entire function in λ, so is $G_n(\lambda)$. Clearly, $G_n(\lambda_m) = 0$ if $n \neq m$ by the orthogonality of $\{\phi_n(x)\}$. Differentiating $G_n(\lambda)$ leads to

$$G'_n(\lambda) = (\lambda - \lambda_n) \int_a^b \frac{\partial\phi(x, \lambda)}{\partial\lambda} \phi_n(x)\, dx + \int_a^b \phi(x, \lambda)\phi_n(x)\, dx,$$

and by setting $\lambda = \lambda_n$ we obtain

$$G'_n(\lambda_n) = \|\phi_n\|^2. \tag{55}$$

Hence, by combining (54), (55) and (51), we obtain

$$S_n^*(\lambda) = \frac{G_n(\lambda)}{(\lambda - \lambda_n)G'_n(\lambda_n)},$$

which, upon its substitution in (50), leads to

$$f(\lambda) = \sum_n f(\lambda_n) \frac{G_n(\lambda)}{(\lambda - \lambda_n)G'_n(\lambda_n)}, \tag{56}$$

which is similar, but not exactly the same as (25).

However, if we further assume that

$$G_n(\lambda) = a_n G(\lambda), \tag{57}$$

where a_n is a constant independent of λ, then $G'_n(\lambda_n) = a_n G'(\lambda_n)$ and the series (56) becomes a Lagrange-type interpolation series

$$f(\lambda) = \sum_n f(\lambda_n) \frac{G(\lambda)}{(\lambda - \lambda_n)G'(\lambda_n)}. \tag{58}$$

To illustrate the idea, we begin with the following simple example.

Example 1 *Consider the Sturm-Liouville problem*

$$y'' = -\lambda y, \quad 0 \leq x \leq \pi,$$
$$y'(0) = 0 = y'(\pi).$$

It is easy to see that the eigenvalues are $\lambda_n = n^2$ *and the eigenfunctions are* $\phi_n(x) = \cos nx$. *The function* $\phi(x, \lambda)$ *that generates the eigenfunctions when the parameter* λ *is replaced by the eigenvalues is readily seen to be* $\phi(x, \lambda) = \cos\sqrt{\lambda}x$. *The function* $G(\lambda)$ *is given by*

$$G(\lambda) = \lambda \prod_{n=1}^{\infty}\left(1 - \frac{\lambda}{n^2}\right) = \frac{\sqrt{\lambda}\sin\sqrt{\lambda}\pi}{\pi}.$$

And the sampling theorem takes the form: if

$$f(\lambda) = \int_0^{\pi} F(x)\cos\sqrt{\lambda}x\,dx,$$

for some $F \in L^2(0, \pi)$, *then*

$$f(\lambda) = f(0)\frac{\sin\sqrt{\lambda}\pi}{\sqrt{\lambda}\pi} + 2\sum_{n=1}^{\infty} f(n^2)\frac{\sqrt{\lambda}\sin\pi(\sqrt{\lambda}-n)}{\pi(\lambda - n^2)}.$$

The following definitions are needed.

Definition 4 *We say that the boundary-value problem* (52)–(53) *possesses the Lagrange-type interpolation property if it has the Kramer property and its associated sampling series is a Lagrange-type interpolation series as in* (58).

Definition 5 *The boundary-value problem* (52)–(53) *is said to be simple if all its eigenvalues, except possibly finitely many, are simple, and is said to be strictly simple if all its eigenvalues are simple.*

Recall that an eigenvalue is simple if it has exactly one linearly independent eigenfunction. Let us denote the set of all self-adjoint boundary-value problems (52)–(53) by **B**, the set of all those that have the Kramer property by **K**, and the set of all those that have the Lagrange-type interpolation property by **L**. We denote the set of all self-adjoint boundary-value problems that are simple by **S** and the set of all those that are strictly simple by $\tilde{\mathbf{S}}$.

From the definitions, it follows that $\mathbf{L} \subset \mathbf{K}$ and $\tilde{\mathbf{S}} \subset \mathbf{S}$. Now we show that $\mathbf{K} \subset \tilde{\mathbf{S}}$. For, let $\phi(x, \lambda)$ be a function that generates the eigenfunctions when λ is replaced by the eigenvalues $\{\lambda_n\}$ an assume that the eigenvalue λ_k is not simple. Let us denote its corresponding eigenfunctions by $\phi_{1,k}(x), \ldots, \phi_{p,k}(x)$, where $p > 1$. Since $\phi(x, \lambda)$ is an analytic function in λ, $\lim_{\lambda\to\tilde{\lambda}} \phi(x, \lambda)$ exists for all $\tilde{\lambda}$. But $\lim_{\lambda\to\lambda_k} \phi(x, \lambda)$ does not exist because $\phi_{1,k}(x), \ldots, \phi_{p,k}(x)$, are linearly independent, which is a contradiction. Therefore, we have the following inclusions:

$$\mathbf{L} \subset \mathbf{K} \subset \tilde{\mathbf{S}} \subset \mathbf{S} \subset \mathbf{B}.$$

S is a proper subset of **B** since the boundary-value problem:

$$y'' = -\lambda y, \quad 0 < x < \pi$$

with periodic boundary conditions

$$y(0) = y(\pi), \quad y'(0) = y'(\pi)$$

is self-adjoint, but all its eigenvalues are double. In fact, the eigenvalues are $(2n)^2$ and the corresponding eigenfunctions are $\cos 2nx$ and $\sin 2nx$.

The following more interesting example shows that $\tilde{\mathbf{S}}$ is a proper subset of **S**.

Example 2 *Consider the boundary-value problem*

$$y'' = -\lambda y, \qquad 0 < x < \pi,$$

with

$$y'(0) = y'(\pi) = \frac{y(\pi) - y(0)}{\pi}.$$

It is easy to see that this problem is self-adjoint and the eigenvalues are the solutions of the equation

$$2(1 - \cos s\pi) = s\pi \sin s\pi,$$

or equivalently, they are the solutions of the equations

$$\sin\left(\frac{s\pi}{2}\right) = 0, \quad \text{and} \quad \tan\left(\frac{s\pi}{2}\right) = \frac{s\pi}{2}.$$

There are two sequences of eigenvalues

$$\lambda_{2n} = (2n)^2, \quad \text{and} \quad \lambda_{2n+1} = s_n^2, \; n = 0, 1, 2, \ldots,$$

where s_n is a solution of the equation $\tan(s\pi/2) = s\pi/2$. *All the eigenvalues are simple, except for $\lambda = 0$, which is double. The eigenfunctions corresponding to the eigenvalue $\lambda = 0$ are 1 and x. The remaining eigenfunctions form two sequences*

$$\phi_{2n}(x) = \cos 2nx\,, \quad \text{and} \quad \phi_{2n+1}(x) = 2\sin s_n x - s_n \pi \cos s_n x.$$

Clearly, the boundary-value problem of Example 2 does not belong to the class **K**. *Or in other words, Kramer's sampling theorem does not hold for Example 2.*

3.5.4. Generalization of Kramer's Theorem

Now we introduce the following generalization of Kramer's theorem that will allow us to obtain a sampling theorem for boundary-value problems like that of Example 2.

But first let us recall the following definitions. If M is a subset of $L^2(I)$, then its *closure*, $\overline{M}$, is the smallest closed set containing it, and its *linear span*, $(M) = \text{sp}(M)$, is the set of all finite linear combinations of elements of M. It is easy to see that $\text{sp}(M)$ is a vector space. In particular, if M is finite, then the closure of its linear span is equal to its linear span, i.e., $\overline{\text{sp}(M)} = \text{sp}(M)$ since every finite dimension vector space is closed.

Theorem 14 *Let there exist a function $K(x, t)$ continuous in t such that $K(x, t) \in L^2(I)$ for every real number t. Assume that there exists a sequence of real numbers $\{t_n\}_{n\in\mathbb{Z}}$ such that $\{K(x, t_n)\}_{n\in\mathbb{Z}}$ is an orthogonal family (but not necessarily complete) in $L^2(I)$ for some finite interval $I = [a, b]$. Let $\mathcal{H}$ denote the closure of the linear span of $\{K(x, t_n)\}_{n\in\mathbb{Z}}$. Then for any function of the form*

$$f(t) = \int_a^b F(x)\overline{K}(x, t)dx,$$

with $F \in \mathcal{H}$, we have

$$f(t) = \sum_{n=-\infty}^{\infty} f(t_n)S_n^*(t),$$

where

$$S_n^*(t) = \frac{\displaystyle\int_a^b \overline{K}(x, t)K(x, t_n)dx}{\displaystyle\int_a^b |K(x, t_n)|^2 dx}.$$

In particular, if the boundary-value problem (52)–(53) *possesses a function* $\phi(x, \lambda)$ *that generates all the eigenfunctions of the problem, except for finitely many, say* $\phi_1, \ldots, \phi_p$*, then for any* f *of the form*

$$f(\lambda) = \int_a^b F(x)\overline{\phi}(x, \lambda)dx,$$

where F *is in* $\mathcal{H}$*, the closure of the linear span of the eigenfunctions, and such that* F *is orthogonal to the eigenfunctions* $\phi_1, \ldots, \phi_p$*, i.e.,*

$$\int_a^b F(x)\overline{\phi}_i(x, \lambda)dx = 0, \quad i = 1, 2, \ldots, p,$$

we have

$$f(\lambda) = \sum_{n=p+1}^{\infty} f(\lambda_n)S_n^*(\lambda), \tag{59}$$

where

$$S_n^*(\lambda) = \frac{\displaystyle\int_a^b \overline{\phi}(x, \lambda)\phi(x, \lambda_n)dx}{\displaystyle\int_a^b |\phi(x, \lambda_n)|^2 dx}.$$

Proof: The proof is the same as in Kramer's theorem, except that $F \in \mathcal{H} \subset L^2(I)$ and any function in $\mathcal{H}$ can be expanded in a generalized Fourier series in terms of $\{K(x, t_n)\}_{n\in\mathbb{Z}}$. For the second part of the theorem, let $\{\phi_n(x)\}_{n=1}^{\infty}$ be the eigenfunctions of the problem, and let H be the closure of the linear span of the eigenfunctions $\{\phi_n(x)\}_{n=p+1}^{\infty}$ and $H^{\perp} = sp\{\phi_1(x), \ldots, \phi_p(x)\}$. Hence, $\mathcal{H}$ is the direct sum of H and $H^{\perp}$ and any function in H is orthogonal to $H^{\perp}$. ∎

This new version of Kramer's theorem is particularly useful if the boundary-value problem (52)–(53) *almost* has the Kramer property. For example, let all the eigenvalues of the problem be simple, except for finitely many, say $\lambda_1, \ldots, \lambda_p$, and assume that there is a function $\phi(x, \lambda)$ that generates all the eigenfunctions belonging to the simple eigenvalues. In this case a sampling expansion can be obtained for any function of the form

$$f(\lambda) = \int_a^b F(x)\overline{\phi}(x, \lambda)dx, \tag{60}$$

where F is orthogonal to the linear span of the eigenfunctions corresponding to $\lambda_1, \ldots, \lambda_p$.

Since such is the case for the boundary-value problem of Example 2, we can now apply Theorem 14 to obtain the following example.

Example 3 *Consider the boundary-value problem of Example 2. Let $F \in L^2[0, \pi]$ be orthogonal to 1 and x, i.e.,*

$$\int_0^\pi F(x)\,dx = 0 = \int_0^\pi xF(x)dx,$$

then a sampling expansion of the form (59) *can be obtained for any f of the form* (60)*, where*

$$\phi(x, \lambda) = \left(\sin\left(\frac{s\pi}{2}\right) - \frac{s\pi}{2}\cos\frac{s\pi}{2} + \frac{s\pi}{2}\sin s\pi\right)\cos sx - \sin s\pi \sin sx.$$

The sampling points are the eigenvalues of Example 2, namely,

$$\lambda_{2n} = (2n)^2, \quad \text{and} \quad \lambda_{2n+1} = s_n^2, \quad n = 0, 1, 2, \ldots,$$

where s_n is a solution of the equation $\tan(s\pi/2) = s\pi/2$ *and* $\lambda = s^2$.

Now we investigate the relationship between the classes **K** and **L**. First, we report on cases in which a boundary-value problem in **K** is known to belong to **L**, or in other words, cases in which the series in (50) is a Lagrange-type interpolation series. The first major result in this direction is due to the authors of this article and G. Hinsen [49], where it was shown, among other things, that for regular Sturm-Liouville boundary-value problems, Kramer's sampling series (50) and (51) is nothing more than a Lagrange-type interpolation series of the form (25). This result was refined and extended to singular Sturm-Liouville boundary-value problems in [48]; see also [47].

In [46] Zayed introduced the method of Green's function to derive sampling theorems associated with more general types of boundary-value problems, including non-self-adjoint ones. These results have recently been improved by M. Annaby and A. Zayed [1]. It has also been shown that sampling series associated with boundary-value problems involving one-dimensional Dirac operators are Lagrange-type interpolation series, as in the Sturm-Liouville case [42]. As a special case of the latter, a sampling series for the Hartley transform of bandlimited functions was obtained. The Hartley transform of a function $f(t)$ is defined as

$$F(\omega) = \int_0^\infty [\cos\omega t + \sin\omega t]f(t)dt.$$

It was introduced in [17] by Ralph Hartley, an electrical engineer, as a way to overcome what he considered a drawback of the Fourier transform,

$$F(\omega) = \int_{-\infty}^{\infty} [\cos \omega t + i \sin \omega t] f(t) dt,$$

namely, representing a real-valued signal $f(t)$ by a complex-valued one, $F(\omega)$; for more details, see [44, ch. 13]. The sampling series for a band-limited Hartley transform of the form

$$f(t) = \int_{0}^{\pi} [\cos \omega t + \sin \omega t] F(\omega)\, d\omega$$

is given by

$$f(t) = \sum_{n=-\infty}^{\infty} f(n) \operatorname{sinc}(t-n),$$

which is closely related to the original sampling theorem of Kotel'nikov [26]. Notice that the integral defining the Hartley transform extends from 0 to π rather than from $-\pi$ to π. For sampling expansions associated with other related integral transforms, e.g., the fractional Fourier transform, see [41, 43].

The Green's function method not only yields sampling expansions that resemble the Lagrange-type interpolation series, but it also works for a general class of boundary-value problems; nevertheless it has a drawback. It associates with every boundary-value problem infinitely many Lagrange-type interpolation series. The reason is the Green's function, $G(x, y, \lambda)$ of the boundary-value problem (52)–(53) is a function of three variables. Thus, for each fixed y, a sampling expansion is obtained.

This is in contrast to other direct methods used for Sturm-Liouville problems and some boundary-value problems associated with fourth order self-adjoint differential operators [8, 15, 48, 49], which yield essentially a unique sampling expansion associated with any such boundary-value problem.

In the following section we shall give sufficient conditions for a certain class of boundary-value problems to have a unique Lagrange-type interpolation series associated with it. These conditions also guarantee for a boundary-value problem in the class **K** to be in the class **L**.

3.6. Boundary-Value Problems and Lagrange-Type Interpolation

3.6.1. The Main Result

In this section we discuss the relationship between boundary-value problems and Lagrange-type interpolation. In Subsection 1, we introduce without a proof one of the main results in this direction and in Subsection 2, we give a number of examples.

It is known that a regular self-adjoint boundary-value problem has infinitely (countably) many eigenvalues having no finite limit point, as well as a complete orthogonal set of eigenfunctions and if the boundary conditions are separated, then the eigenvalues are simple. If the boundary conditions are not separated, the eigenvalues are not necessarily simple; see [12, 31].

The boundary conditions are said to be *separated* if m of them involve the values of $y^{(j-1)}(a)$, $j = 1, 2, \ldots, n$ only and the remaining $n - m$ conditions involve the values of $y^{(j-1)}(b)$, $j = 1, 2, \ldots, n$ only. If n is even and $m = n/2$, the boundary conditions are said to be of *Sturm-Liouville* type.

One may wonder if regular self-adjoint boundary-value problems with simple eigenvalues have the Kramer property. This is indeed the case as shown in the next theorem which was proved in [8]. In fact, more is shown in the theorem: any such a problem has not only the Kramer property, but also the Lagrange-type interpolation property.

Theorem 15 *Consider the regular self-adjoint boundary-value problem* (52), (53) *and assume that all its eigenvalues are simple. Then there is a function* $\phi(x, \lambda)$ *having the property that if* f *is representable in the form*

$$f(\lambda) = \int_a^b F(x)\phi(x, \lambda)\, dx$$

for some $F \in L^2[a, b]$, *then* f *is an entire function and admits the sampling representation*

$$f(\lambda) = \sum_n f(\lambda_n) \frac{G(\lambda)}{G'(\lambda_n)(\lambda - \lambda_n)},$$

where $\{\lambda_n\}$ *are the eigenvalues of the problem and*

$$G(\lambda) = \begin{cases} \prod_n \left(1 - \dfrac{\lambda}{\lambda_n}\right) & \text{if zero is not an eigenvalue} \\ \lambda \prod_n \left(1 - \dfrac{\lambda}{\lambda_n}\right) & \text{if zero is an eigenvalue}. \end{cases}$$

The sampling series is uniformly convergent on each compact subset of the real line.

The theorem does not only assert the existence of the function $\phi(x, \lambda)$ that generates the eigenfunctions, but it also shows how to construct it. Let us denote the fundamental set of solutions of the differential equation (52) by $\{y_j(x, \lambda)\}_{j=1}^n$. That is, $y_j(x, \lambda)$, $j = 1, 2, \ldots, n$, are solutions of the differential equation (52) satisfying the initial conditions

$$y_j^{(k-1)}(a, \lambda) = \delta_{j,k}, \quad j, k = 1, 2, \ldots, n.$$

The functions $y_j(x, \lambda)$ are linearly independent, continuous in (x, λ) for $x \in [a, b]$ and all λ, and for fixed x are entire functions of λ of order $1/(2n)$; see [31, p. 13]. Hence, $U_k(y_j)$, $k = 1, \ldots, n$; $j = 1, 2, \ldots, n$, are entire functions in λ. Any function $\tilde{\phi}(x, \lambda)$ that satisfies the assumptions of the theorem can be written in the form

$$\tilde{\phi}(x, \lambda) = \sum_{j=1}^{n} c_j(\lambda) y_j(x, \lambda),$$

where the $c_j(\lambda)$'s are meromorphic functions in λ involving $U_k(y_j)$, $k = 1, \ldots, n$; $j = 1, 2, \ldots, n$. By multiplying $\tilde{\phi}(x, \lambda)$ by an appropriate entire function, we obtain a function $\phi(x, \lambda)$ that is entire in λ and satisfies the assumptions of the theorem. More precisely, let

$$\phi(x, \lambda) = \begin{vmatrix} y_1(x, \lambda) & \ldots & y_n(x, \lambda) \\ U_1(y_1) & \ldots & U_1(y_n) \\ \cdot & \ldots & \cdot \\ \cdot & \ldots & \cdot \\ \cdot & \ldots & \cdot \\ U_{n-1}(y_1) & \ldots & U_{n-1}(y_n) \end{vmatrix}.$$

Then $\phi(x, \lambda)$ is an entire function in λ. It is easily seen that ϕ satisfies the following conditions:

1. $U_j(\phi(x, \lambda)) = 0$ for all $j = 1, 2, \ldots, n - 1$,
2. $U_n(\phi(x, \lambda)) = 0$ if and only if λ is an eigenvalue,
3. $\phi(x, \lambda_n) = \phi_n(x)$, where $\phi_n(x)$ is the eigenfunction corresponding to the eigenvalue λ_n.

Theorem 15 shows that regular self-adjoint boundary-value problems with strictly simple eigenvalues are in the class **L**. The construction of $\phi(x, \lambda)$ given above fails to work if even just one eigenvalue ceases to be simple. For example, let us reconsider Example 2, in which we may rewrite the boundary conditions in the form

$$U_1(y) = y'(0) - y'(\pi), \qquad U_2(y) = y'(0) - \frac{[y(\pi) - y(0)]}{\pi}.$$

The fundamental set of solutions are

$$y_1(x, \lambda) = \cos sx, \qquad y_2(x, \lambda) = \frac{\sin sx}{s}, \quad \text{where } \sqrt{\lambda} = s.$$

Therefore,

$$\phi(x, \lambda) = \begin{vmatrix} \cos sx & \sin sx/s \\ s \sin s\pi & 1 - \cos s\pi \end{vmatrix} = (1 - \cos s\pi) \cos sx - \sin s\pi \sin sx,$$

but when $s_n = \sqrt{\lambda_n} = 2n$, $\phi(x, \lambda)$ fails to reproduce the corresponding eigenfunctions. In fact $\phi(x, \lambda_n)$ is identically zero.

It should be noted that some singular boundary-value problems have both the Kramer and the Lagrange interpolation properties as well. Examples of singular boundary-value problems having the Lagrange-type interpolation property have been derived in both the limit point and the limit circle cases [48], e.g., boundary-value problems associated with the Bessel, Jacobi, Laguerre, and Hermite differential equations.

3.6.2. Examples of Boundary-Value Problems Having the Lagrange Interpolation Property

Now we give examples of boundary-value problems having the Lagrange interpolation property. In each example we shall give the eigenvalues, eigenfunctions, the appropriate function $\phi(x, \lambda)$, and the function $G(\lambda)$.

Example 4 *Let us consider the regular Sturm-Liouville boundary-value problem:*

$$y'' - q(x)y = \lambda y, \quad a \le x \le b, \tag{61}$$

$$\cos \alpha y(a) + \sin \alpha y'(a) = 0, \tag{62}$$

$$\cos \beta y(b) + \sin \beta y'(b) = 0, \tag{63}$$

where $q(x)$ is continuous on $[a, b]$.

Let $\phi(x, \lambda)$ be the solution of (61) that satisfies

$$\phi(a, \lambda) = \sin\alpha, \quad \phi'(a, \lambda) = -\cos\alpha.$$

Then the eigenvalues are the roots of the equation

$$\cos\beta\, \phi(b, \lambda) + \sin\beta\, \phi'(b, \lambda) = 0.$$

Equation (54) takes the form

$$(\lambda - \lambda_n) \int_a^b \phi(x, \lambda)\phi_n(x)\, dx = \phi(b, \lambda)\phi_n'(b) - \phi'(b, \lambda)\phi_n(b), \tag{64}$$

where $\phi(x, \lambda_n) = \phi_n(x)$ are the eigenfunctions. But since the eigenfunctions also satisfy (63), we have

$$\cos\beta\phi_n(b) + \sin\beta\phi_n'(b) = 0,$$

which upon solving for $\phi_n'(b)$, provided that $\sin\beta \neq 0$, and substituting it into (64), we obtain

$$(\lambda - \lambda_n) \int_a^b \phi(x, \lambda)\phi_n(x)\, dx = a_n G(\lambda),$$

where

$$G(\lambda) = \cos\beta\phi(b, \lambda) + \sin\beta\phi'(b, \lambda),$$

and

$$a_n = -\frac{\phi_n(b)}{\sin\beta}.$$

If $\sin\beta = 0$, then $\phi_n(b) = 0$ and $G(\lambda) = a_n\phi(b, \lambda)$, where $a_n = \phi_n'(b)$.

Example 5 *Consider the boundary-value problem:*

$$\begin{aligned} y'' &= -\lambda y, \quad a \le x \le b \\ \cos\alpha y(a) &+ \sin\alpha y'(a) = 0, \\ \cos\beta y(b) &+ \sin\beta y'(b) = 0, \end{aligned}$$

which is a special case of Example 4 with $q(x) = 0$. In this case it is easy to see that a solution of the differential equation and the first boundary condition is

$$\phi(x, \lambda) = \sin\alpha \cos s(x-a) - \frac{1}{s}\cos\alpha \sin s(x-a),$$

where $s = \sqrt{\lambda}$ and that

$$G(\lambda) = \sin(\alpha - \beta)\cos s(b-a) - \frac{\sin s(b-a)}{s}\left[\cos\alpha\cos\beta + s^2 \sin\alpha \sin\beta\right].$$

The eigenvalues are the zeros of $G(\lambda)$ and the eigenfunctions are $\phi(x, \lambda_n)$.

Example 6 *Consider the boundary-value problem:*

$$y'' - \frac{(\nu^2 - 1/4)}{x^2} y = -\lambda y, \quad 0 < a \le x \le b,$$

with boundary conditions

$$y(a) = 0 = y(b).$$

It is easy to see that the solution of the differential equation that satisfies the boundary condition at a is

$$\phi(x, \lambda) = \frac{\pi}{2}\sqrt{ax}\left[J_\nu(sx)Y_\nu(as) - J_\nu(as)Y_\nu(sx)\right],$$

where J_ν and Y_ν are the Bessel functions of the first and second kinds, respectively. Thus,

$$G(\lambda) = \phi(b, \lambda) = \frac{\pi}{2}\sqrt{ab}\left[J_\nu(sb)Y_\nu(sa) - J_\nu(sa)Y_\nu(sb)\right],$$

where $s = \sqrt{\lambda}$, and the eigenvalues $\lambda_n = s_n^2$ are the zeros of $G(\lambda)$. One can verify that

$$G'(\lambda_n) = \frac{\sqrt{ab}}{2s_n^2}\,\frac{J_\nu^2(bs_n) - J_\nu^2(as_n)}{J_\nu(as_n)J_\nu(bs_n)}.$$

Although the following example was obtained in [8] using Theorem 15, it can be obtained directly without it.

Example 7 *Consider the problem:*

$$y^{(4)}(x) = -\lambda y, \quad 0 \le x \le \pi$$
$$y^{(1)}(0) = 0 = y^{(3)}(0)$$
$$y^{(1)}(\pi) = 0 = y^{(3)}(\pi).$$

The eigenvalues are $\lambda_n = n^4$ and the eigenfunctions are $\phi_n(x) = \cos nx$. Here it should be emphasized that although it is tempting to take $\phi(x, \lambda) = \cos sx$, careful inspection shows that this choice of ϕ neither satisfies any three of the boundary conditions, nor is it an entire function in λ, although it is an entire function in s, where $\lambda = s^4$.

So let us take $\phi(x, \lambda)$ to be the solution of the differential equation that satisfies the boundary conditions

$$y^{(1)}(0) = 0 = y^{(3)}(0) \quad \text{and} \quad y^{(1)}(\pi) = 0.$$

One can verify that

$$\phi(x, \lambda) = \frac{\cos sx \sinh s\pi + \sin s\pi \cosh sx}{2s}, \qquad s = \sqrt[4]{\lambda},$$

and it is indeed an entire function in λ. Therefore,

$$G(\lambda) = a_n \phi^{(3)}(\pi, \lambda) = a_n s^2 \sin \pi s \sinh s\pi,$$

where $a_n = -\phi_n(\pi) = (-1)^{n+1}$.

Example 8 *Consider the problem:*

$$y^{(4)}(x) = -\lambda y, \quad 0 \le x \le \pi$$
$$y(0) = 0 = y^{(2)}(0)$$
$$y^{(1)}(\pi) = 0 = y^{(3)}(\pi).$$

The eigenvalues are $\lambda_n = (n + 1/2)^4$ and the eigenfunctions are $\phi_n(x) = \sin(n + 1/2)x$. The solution of the differential equation that satisfies the boundary conditions, $y(0) = y^{(2)}(0) = y^{(1)}(\pi) = 0$ is

$$\phi(x, \lambda) = \frac{\sin sx \cosh s\pi - \sinh sx \cos s\pi}{2s^3},$$

which is easily seen to be an entire function in $\lambda = s^4$, while the function $\phi(x, \lambda) = \sin sx$ is not.

Hence,

$$G(\lambda) = \phi_n(\pi)\phi^{(3)}(\pi, \lambda) = \sin[(n + 1/2)\pi] \cos s\pi \cosh s\pi.$$

Example 9 *Consider the problem:*

$$\begin{aligned} y^{(4)}(x) &= -\lambda y, \quad 0 \le x \le \pi \\ y(0) &= 0 = y^{(2)}(0) \\ y^{(2)}(\pi) &= 0 = y^{(3)}(\pi). \end{aligned}$$

One can verify that the solution of the differential equation that satisfies the boundary conditions, $y(0) = y^{(2)}(0) = y^{(2)}(\pi) = 0$, *is*

$$\phi(x, \lambda) = \frac{\sin sx \sinh s\pi + \sin s\pi \sinh sx}{2s^2},$$

and

$$G(\lambda) = a_n \phi^{(3)}(\pi, \lambda) = \frac{s}{2}[\sin s\pi \cosh s\pi - \cos s\pi \sinh s\pi];$$

hence, the eigenvalues are $\lambda_n = (s_n)^4$, *where* s_n *are the solutions of the equation*

$$\tan s\pi = \tanh s\pi.$$

The eigenfunctions are $\phi_0(x) = x$ *and*

$$\phi_n(x) = \sinh s_n\pi \sin s_n x + \sin s_n\pi \sinh s_n x, \quad n = 1, 2, \ldots.$$

3.7. Extension of Kramer's Theorem to Approximately Band-Limited Signals

This section is concerned with the question as to what happens to both the WSK and Kramer's theorems when the condition that f is "band-limited" is weakened, meaning that conditions (27) and (49) are only "approximately" valid. In regard to Shannon's theorem, this is equivalent to the question of whether $f(t) = \lim_{\sigma\to\infty} \int_{-\sigma}^{\sigma} e^{ixt} g(x)\, dx$ implies $f(t) = \lim_{\sigma\to\infty} S_\sigma f(t)$.

In this respect, J. L. Brown [4] and P. L. Butzer-W. Splettstößer [9] (see also [10], and [5, 11] for a historical overview) established the following result, which is concerned with the *aliasing error* $A_\sigma(t) := |f(t) - S_\sigma f(t)|$; see Chapter 2.

Theorem 16 *If $f \in L^1(\mathbb{R}) \cap C(\mathbb{R})$ is such that the transform $f^\wedge \in L^1(\mathbb{R})$, implying $f(t) = \lim_{\sigma\to\infty}(1/\sqrt{2\pi}) \int_{-\sigma}^{\sigma} e^{ivt} f^\wedge(v)\,dv$, then for all $t \in \mathbb{R}$ and $\sigma > 0$,*

$$|f(t) - S_\sigma f(t)| = \left| \frac{1}{\sqrt{2\pi}} \sum_{k=-\infty}^{\infty} \left(1 - e^{-i2k\sigma t}\right) \int_{(2k-1)\sigma}^{(2k+1)\sigma} e^{ivt} f^\wedge(v)\,dv \right|$$
$$\leq \sqrt{\frac{2}{\pi}} \int_{|v|>\sigma} |f^\wedge(v)|\,dv,$$

which tends to zero uniformly as $\sigma \to \infty$. In particular, if f is actually bandlimited to $[-\sigma, \sigma]$, then $f(t) = S_\sigma f(t)$, all $t \in \mathbb{R}$.

This section is especially concerned with the counterpart of Theorem 16 in the instance of Kramer's setting. We consider kernel functions $K: \mathbb{R} \times \mathbb{R} \to \mathbf{C}$ having the following properties for each $t \in \mathbb{R}$ and some fixed $\sigma > 0$:

(i) $K(\cdot, t)$ is measurable on $\mathbb{R}$ and belongs to $L^2(-\sigma, \sigma)$;

(ii) There exists a sequence of reals $\{t_k\}_{k\in\mathbb{Z}} \subset \mathbb{R}$ and a function $K^*(x, t)$ such that

$$\sum_{k=-\infty}^{\infty} s_k(t) K(x, t_k) = K^*(x, t)$$

holds for each $t \in \mathbb{R}$ and almost all $x \in \mathbb{R}$, where $s_k(t)$ is defined by (51) and $K^*(x, t) = K(x, t)$ a.e. on $(-\sigma, \sigma)$;

(iii) There exists a measurable function h_t and a constant $M_t > 0$, both may depend on $t \in \mathbb{R}$, such that $|h_t(x)/K(x, t)| \leq M_t$ a.e., and

$$\left| \sum_{k=-N}^{N} s_k(t) K(x, t_k) \right| \leq h_t(x) \qquad (x \in \mathbb{R};\ N = 1, 2, \ldots.)$$

Theorem 17 *Let the signal $f: \mathbb{R} \to \mathbf{C}$ be representable as*

$$f(t) = \lim_{\tau\to\infty} \int_{-\tau}^{\tau} K(x, t) g(x)\,dx \equiv \int_{-\infty}^{\infty} K(x, t) g(x)\,dx \qquad (t \in \mathbb{R}), \tag{65}$$

where K is a kernel as described above and g a function such that $g(\cdot)K(\cdot, t) \in L^1(\mathbb{R})$ for all $t \in \mathbb{R}$. Then

a) For all $t \in \mathbb{R}$,

$$\left| f(t) - \sum_{k=-\infty}^{\infty} f(t_k) s_k(t) \right| \leq \int_{|x|>\sigma} |K(x, t) - K^*(x, t)||g(x)|\,dx. \tag{66}$$

b) If f is transform limited, i.e., (49) *is satisfied with* $a = -\sigma$, $b = \sigma$, *then the expansion* (50) *holds.*

Proof: In view of (65) one has for fixed $t \in \mathbb{R}$,

$$\sum_{k=-N}^{N} f(t_k)s_k(t) = \int_{-\infty}^{\infty} \left[\sum_{k=-N}^{N} K(x, t_k)s_k(t) \right] g(x)\, dx,$$

so that for $N \to \infty$,

$$\sum_{k=-\infty}^{\infty} f(t_k)s_k(t) = \int_{-\infty}^{\infty} K^*(x, t)g(x)\, dx, \tag{67}$$

term-by-term integration being justified by Lebesgue's dominated convergence theorem [35, p. 27], observing that by the boundedness condition (iii) above,

$$\left| \sum_{k=-N}^{N} K(x, t_k)s_k(t)g(x) \right| \leq M_t |K(x, t)g(x)| \in L^1(\mathbb{R}) \qquad N = 1, 2, \ldots.$$

By rewriting f in the form $f(t) = \int_{-\infty}^{\infty} K(x, t)g(x)\, dx$, it follows from (67) that

$$f(t) - \sum_{k=-\infty}^{\infty} f(t_k)s_k(t) = \int_{-\infty}^{\infty} [K(x, t) - K^*(x, t)]\, g(x)\, dx,$$

and one obtains the estimate (66) by noting that $K(x, t) = K^*(x, t)$ a.e. on $(-\sigma, \sigma)$.

For a kernel $K(x, t)$ satisfying the properties (i), (ii) and (iii) above for *all* $\sigma > 0$, one may consider the limit $\sigma \to \infty$ in (66). In this case the nodes t_k, the Fourier coefficients $s_k(t)$, and the limit of the Fourier expansion $K^*(x, t)$ in (ii), all depend on σ, i.e., $t_k = t_{k,\sigma}$, $s_k(t) = s_{k,\sigma}(t)$ and $K^*(x, t) = K^*_\sigma(x, t)$. If one however assumes that $|K^*_\sigma(x, t)/K(x, t)| \leq M^*_t$ with M^*_t being independent of σ, then the integral in (66) tends to zero for $\sigma \to \infty$ since $K(\cdot, t)g(\cdot) \in L^1(\mathbb{R})$ and $|K^*(\cdot, t)g(\cdot)| \leq M^*_t |K(\cdot, t)g(\cdot)| \in L^1(\mathbb{R})$. Thus we have also established

Corollary 3 *If the signal* $f\colon \mathbb{R} \longmapsto \mathbf{C}$ *has the representation* (65), *the kernel* $K(x, t)$ *satisfying* (i), (ii) *and* (iii) *for all* $\sigma > 0$, *with* $|K^*(\cdot, t)g(\cdot)| \leq M^*_t$, ($M^*_t$ *being independent of* σ), *then*

$$f(t) = \lim_{\sigma \to \infty} \sum_{k=-\infty}^{\infty} f(t_{k,\sigma})s_{k,\sigma}(t) \qquad (t \in \mathbb{R}).$$

Results corresponding to Theorem 16 and Corollary 3 can be established for kernels $K(x, t)$ which are defined on subsets of $\mathbb{R}^2$ only, such as $\mathbb{R}^+ \times \mathbb{R}^+$, with $\mathbb{R}^+ = [0, \infty)$. In this case the sequence $\{t_k\}_{k=0}^{\infty}$ takes the role of $\{t_k\}_{k \in \mathbb{Z}}$. For more details, see [7].

References

[1] M. H. Annaby and A. I. Zayed. On the Use of Green's Function in Sampling Theory. *J. Integ. Eqns. Appls.*, 10:117–139, 1998.

[2] S. A. Avdonin. On the Question of Riesz Bases of Exponential Functions in L^2. *Vestnik Leningrad Univ. Math.*, 7:203–211, 1979 (translation of Russian version of 1974).

[3] R. Boas. *Entire Functions*. Academic Press, New York, 1954.

[4] J. L. Brown, Jr. On the Error in Reconstructing a Non-Band-Limited Function by Means of the Bandpass Sampling Theorem. *J. Math. Anal. Appl.*, 18:75–84, 1967.

[5] P. L. Butzer, G. Schmeisser, and R. L. Stens. An Introduction to Sampling Theory. In F. Marvasti, Ed., *Nonuniform Sampling: Theory and Practice*, Plenum Press, New York, 2001, pp. 17–121.

[6] P. L. Butzer and G. Nasri-Roudsari. Kramer's Sampling Theorem in Signal Analysis and its Role in Mathematics. In J. M. Blackledge, Ed., *Image Processing: Mathematical Methods and Applications*, Clarendon Press, Oxford, 1997, pp. 49–95.

[7] P. L. Butzer and R. L. Stens. An Extension of Kramer's Sampling Theorem for Not Necessarily Band-Limited Signals—the Aliasing Error. *Acta Sci. Math.* (Szeged), 60:59–69, 1995.

[8] P. L. Butzer and G. Schöttler. Sampling Theorems Associated with Fourth- and Higher-Order Self-Adjoint Eigenvalue Problems. *J. Comput. Appl. Math.*, 51:159–177, 1994.

[9] P. L. Butzer and W. Splettstößer. Approximation Theorems for Duration-Limited Functions with Error Estimates. *Inform. and Control*, 34:55–65, 1977.

[10] P. L. Butzer, W. Splettstößer, and R. L. Stens. The Sampling Theorem and Linear Prediction in Signal Analysis. *Jahresber. Deutsch. Math.-Verein.*, 90:1–70, 1988.

[11] P. L. Butzer and R. L. Stens. Sampling Theory for Not Necessarily Band-Limited Functions: A Historical Overview. *SIAM Rev.*, 34:40–53, 1992.

[12] E. A. Coddington and N. Levinson. *Theory of Ordinary Differential Equations*. McGraw-Hill, New York, 1955.

[13] P. Davis. *Interpolation and Approximation*. Dover Publication, New York, 1975.

[14] W. N. Everitt and G. Nasri-Roudsari. Interpolation and Sampling Theories, and Linear Ordinary Boundary-Value Problems. In J. R. Higgins and R. L. Stens, Eds., *Sampling Theory in Fourier and Signal Analysis: Advanced Topics*, Oxford University Press, Oxford, 1999, pp. 96–129.

[15] W. N. Everitt, G. Schöttler, and P. L. Butzer. Sturm-Liouville Boundary Value Problems and Lagrange Interpolation Series. *J. Rend. Math. Appl.*, 14:87–126, 1994.

[16] L. J. Fogel. A Note on the Sampling Theorem. *IRE Trans. Inform. Theory*, 1:47–48, 1955.

[17] R. V. Hartley. A More Symmetrical Fourier Analysis Applied to Transmission Problems. *Proc. Inst. Radio Eng.*, 30:144–150, 1942.

[18] J. R. Higgins. A Sampling Theorem for Irregularly Spaced Sample Points. *IEEE Trans. Inform. Theory*, IT-22:621–622, 1976.

[19] J. R. Higgins. *Sampling Theory in Fourier and Signal Analysis: Foundations*. Oxford University Press, Oxford, 1996.

[20] G. Hinsen. Abtastsätze mit Unregelmässigen Stützstellen: Rekonstruktionsformeln, Konvergenzaussagen und Fehlerbetrachtungen, Doctoral Dissertation, RWTH Aachen, 1994.

[21] G. Hinsen. Irregular Sampling of Bandlimited L^p-Functions. *J. Approx. Theory*, 72:346–364, 1993.

[22] G. Hinsen and D. Klösters. The Sampling Series as a Limiting Case of Lagrange Interpolation. *Appl. Anal.*, 49:49–60, 1993.

[23] D. L. Jagerman and L. J. Fogel. Some General Aspects of the Sampling Theorem. *IEEE Trans. Inform. Theory*, IT-2:139–156, 1956.

[24] M. I. Kadec. The Exact Value of the Paley-Wiener Constant. *Sov. Math. Dokl.*, 5:559–561, 1964.

[25] H. P. Kramer. A Generalized Sampling Theorem. *J. Math. Phys.*, 38:68–72, 1959.

[26] V. Kotel'nikov. On the carrying capacity of the ether and wire in telecommunications, material for the first All-Union Congress on Questions of Communications. *Izd. Red. Upr. Svyazi RKKA*, Moscow, Russia, 1933.

[27] B. Ja. Levin. Distribution of Zeros of Entire Functions. *Amer. Math. Soc. Transl. Math. Monographs*, Vol. 5, Providence, Rhode Island 1964.

[28] N. Levinson. Gap and Density Theorems. *Amer. Math. Soc. Colloq. Publs. Ser.*, 26, 1940.

[29] D. A. Linden. A Discussion of Sampling Theorems. *Proc. IRE*, 47:1219–1226, 1959.

[30] D. A. Linden and N. M. Abramson. A Generalization of the Sampling Theorem. *Inform. Control* 3:26–31, 1960; Errata, Ibid., 4:95, 1961.

[31] M. Naimark. *Linear Differential Operators*, Vol. 1, Elementary Theory of Linear Differential Operators, George Harrap & Co, London, 1967.

[32] K. Ogura. On Certain Transcendental Integral Function in the Theory of Interpolation. *Tohoku Math. J.*, 17:64–72, 1920.

[33] R. Paley and N. Wiener. The Fourier Transforms in the Complex Domain. *Amer. Math. Soc. Colloq. Publ. Ser.*, 19: Providence, Rhode Island, 1934.

[34] M. Rawn. A Stable Nonuniform Sampling Expansion Involving Derivatives. *IEEE Trans. Inform. Theory*, IT-35:1223–1227, 1989.

[35] W. Rudin. *Real and Complex Analysis*. McGraw-Hill, New York, 1973.

[36] K. Seip. An Irregular Sampling Theorem for Functions Bandlimited in a Generalized Sense. *SIAM J. Appl. Math.*, 47:1112–1116, 1987.

[37] C. E. Shannon. Communication in the Presence of Noise. *Proc. IRE*, 137:10–21, 1949.

[38] J. Voss. Irregular Sampling: Error Analysis, Applications and Extensions. *Milt. Math. Sem. Giessen*, 238:1–86, 1999.

[39] J. Voss. Irreguläres Abtasten: Fehleranalyse, Anwendungen and Erweiterungen. Doctoral Dissertation. University of Erlangen, Nürnberg, 1999.

[40] E. T. Whittaker. On the Functions which are Represented by the Expansion of the Interpolation Theory. *Proc. Roy. Soc. Edinburgh, Sec. A*, 35:181–194, 1915.

[41] A. I. Zayed and A. Garcia. New Sampling Formulae for the Fractional Fourier Transform. *Signal Processing*, 77:111–114, 1999.

[42] A. I. Zayed and A. Garcia. Sampling Theorems Associated with Dirac Operator and the Hartley Transform. *J. Math. Anal. Appls.*, 214:587–598, 1997.

[43] A. I. Zayed. On the Relationship between the Fourier and Fractional Fourier Transforms. *IEEE Signal Processing Letters*, 3:310–311, 1996.

[44] A. I. Zayed. *Function and Generalized Function Transformations*. CRC Press, Boca Raton, 1996.

[45] A. I. Zayed. *Advances in Shannon's Sampling Theory*. CRC Press, Boca Raton, 1993.

[46] A. I. Zayed. A New Role of Green's Function in Interpolation and Sampling Theory. *J. Math. Anal. Appls.*, 175:222–238, 1993.

[47] A. I. Zayed. Kramer's Sampling Theorem for Multidimensional sSgnals and its Relationship with Lagrange-Type Interpolation. *J. Multidim. Systems, Signal Processing*, 3:323–340, 1992.

[48] A. I. Zayed. On Kramer's Sampling Theorem Associated with General Sturm-Liouville Boundary-Value Problems and Lagrange Interpolation. *SIAM J. Appl. Math.*, 51:575–604, 1991.

[49] A. I. Zayed, G. Hinsen and P. L. Butzer. On Lagrange Interpolation and Kramer-Type Sampling Theorems Associated with Sturm-Liouville Problems. *SIAM J. Appl. Math.*, 50:893–909, 1990.

4

Random Topics in Nonuniform Sampling

F. Marvasti

So many secrets in the universe to be discovered
Ample samples, yet the Signal remains to be recovered[1]

Summary

In this chapter, topics that are important but are not fully covered in other chapters are discussed. We shall consider the theorems related to the uniqueness of nonuniform sampling set for 1-D and 2-D signals. The emphasis will be on 2-D signals since the 1-D signals were extensively treated by the author in [1]–[2]. Section 4.2 will be on the Lagrange interpolation of 2-D signals; the Lagrange interpolation for the 1-D signals was discussed in Chapter 3. Section 4.3 will cover the nonuniform sampling for non-band-limited signals and jittered sampling where exact recovery for a certain class of non-band-limited signals is possible. Section 4.4 is on the stability of the nonuniform sampling set. Section 4.5 discusses Parseval relation and spectral analysis of both 1-D and 2-D signals. This spectral analysis leads to some practical methods of recovery such as the time varying and the iterative methods—Section 4.6; variation of these methods will be used in Chapters 5–6 and 16–18. Section 4.7 discusses the relation among the Papoulis generalized sampling theorem, wavelets, sub-band coding, and

[1] Adapted from Hafez—see footnote 1 in Chapter 1.

F. Marvasti • Multimedia Laboratory, King's College London. E-mail:farokh.marvasti@kcl.ac.uk

Nonuniform Sampling: Theory and Practice, edited by Marvasti
Kluwer Academic/Plenum Publishers, New York, 2001.

Quadrature Mirror Filters (QMF). Under certain conditions, it is possible to sample a low-pass or a band-pass signal at lower than the Nyquist rate and be able to recover the signal; these topics are discussed in Sections 7.8–7.9. A new and interesting topic on sampling at the maxima and minima (extrema) points is covered in Section 4.10. This section is based on the Extremum sampling of finite trigonometric polynomials. Discrete and modified Lagrange along with some other novel techniques are discussed in this section. Extremum sampling combined with zero-crossings is another topic covered in this section. The last section (Section 4.11) is a discussion on recovery of deterministic or random signals from random samples, a topic that will be more extensively discussed in Chapter 8.

4.1. General Nonuniform Sampling Theorems

In this section, we would like to address questions like: Under what conditions do the nonuniform samples represent a signal uniquely? For a comprehensive treatment of this subject, see the monograph and the chapter written by the author [1]–[2].

Lemma for 1-D Signals *If the nonuniform samples* $\{t_n\}$ *satisfy the Nyquist rate on the average, it can uniquely represent a band-limited signal (deterministic or random) if the samples are not the zero-crossings of a band-limited signal of the same bandwidth. The set* $\{t_n\}$ *is then called a* sampling set [3].

Proof: Suppose there is one solution to a set of nonuniform samples at instances $\{t_n\}$, and assume that it is possible to interpolate a band-limited function of the same bandwidth at the zero-crossings $\{t_n\}$. Now, if we add this interpolated function to the first solution, we get another band-limited function (of the same bandwidth) having the same nonuniform samples; i.e., the solution is not unique.

Various authors have tried to find conditions on $\{t_n\}$ [4]–[5]. *Lemma 1* implies the following important *Corollary*:

Corollary 1 *If the average sampling rate is higher than the Nyquist rate, irrespective of the set* $\{t_n\}$*, there is always a unique solution.*

Proof: The average density of zero-crossings (real zeros) of a signal band-limited to W is always less than or at most equal to the Nyquist rate ($2W$) for deterministic and random signals [7]. The sampling positions, therefore, cannot be the zero-crossings of a signal band-limited to W. From *Lemma 1*, we conclude that the samples are a *sampling set*. This corollary has also been proved by Beutler [8] by a different method for both deterministic and random signals. □

Another interesting observation is that even if the average sampling rate is less than the Nyquist rate, if $\{t_n\}$ is a sampling set, the reconstruction is unique. That is, for the case of under-sampling, no band-limited signal with a bandwidth smaller than or equal to W can be found where $\{t_n\}$ is a subset of its zeros. This fact has no parallel in the uniform sampling theory [8].

4.1.1. Two-Dimensional Case

In this section, we extend these ideas to 2-D signals based on the papers by the author [9]–[10]. This extension is not straightforward since the zero-crossings of 2-D signals are closed contours (in general) as opposed to isolated points. Also, in the 1-D case, real zeros alone (without the complex zeros) are not sufficient to uniquely determine a band-limited signal to within a constant; this is not the case for 2-D signals. The zero contours (which are all real) of a band-limited 2-D signal determine the signal to within a constant under certain conditions. Some of these conditions are summarized in [9]–[11].

We shall show that, in general, a band-limited 2-D signal can be reconstructed from a set of nonuniform samples if the set is not part of the zero contours of another signal of the same bandwidth. Formally, we state the following proposition[2]:

Proposition 1 *Let F be the space of all 2-D functions band-limited to a region B and let Z be the space of all zero-crossing contours z_i of functions $f_i(x, y)$ that are a member of the set F. The necessary and sufficient condition for a nonuniform sampling set N to uniquely determine a function $h(x, y) \in F$ (i.e., a set of uniqueness) is that N should not be a member of $z_i \in Z$ for any $f_i(x, y) \in F$. (Throughout this section F, Z, z_i, N, $h(x, y)$, $f_i(x, y)$ are defined as in the above.)*

Proof: To prove the sufficient condition, assume the reconstruction is not unique when N is not a member of $z_i \in Z$, i.e., there are at least two functions $h(x, y)$ and $g(x, y) \in F$ that have identical values for the sampling set N. Therefore, $h(x, y) - g(x, y) \in F$ has zero-crossing contours z_{h-g} such that $N \subset z_{h-g}$. But this contradicts our assumption that N is not a member of $z_i \in Z$.

To prove the necessary condition, if $N \in z_i \in Z$, then reconstruction is not unique for functions on F because if $g(x, y) \in F$ is one solution, then $g(x, y) + f_i(x, y) \in F$ is another solution since $f_i(x, y) = 0$ for sampling set $N \in z_i$.

[2] Portions of this section is a modified reprint, with permission, from [9]; © 1989 JOSA.

Note that whether $h(x, y) \in F$ is reducible or not, a nonuniform sampling set N, not being a member of $z_i \in Z$, uniquely represents $h(x, y)$; however, $z_i \in Z$ may be the zero-crossing contours of one irreducible function as well as of many reducible functions on F. □

To check whether the sampling set N is not a member of z_i is a difficult task. However, there are many sufficient conditions where we can easily prove that N is not a member of z_i's. The following corollaries are some of these sufficient conditions.

Corollary 1 *Let F' be the space of all 2-D functions band-limited to a region B', where $B' > B$, and let Z' be the space of all zero-crossing contours z_i' of functions $f_i'(x, y)$ that are a member of the set F'. We assume that $f_i(x, y) \in F$ and $f_i'(x, y) \in F'$ are either irreducible or can be factored into a finite number of real irreducible factors. A sufficient condition for the nonuniform sampling set N not to be a member of z_i is that $N \subset z_i' \in Z'$. Also, not all the points of N should lie on the intersection of Z and Z'.*

Proof: From Theorems A and C in [9], $z_i \neq z_i'$, otherwise, $f_i(x, y) = cf_i'(x, y)$ which contradicts the fact that $f'(x, y)$ is not a member of F. Therefore, if $N \subset z_i'$, N is not a member of z_i. □

Some examples that satisfy sufficient conditions are given below:

Example 1 (Infinite samples in a finite region) An example of a sampling set that satisfies Corollary 1 is the case that an infinite number of samples is chosen from one of the zero-crossing contours of Z'. According to [11], if $f(x, y)$ and $f'(x, y)$ are irreducible, an infinite number of points on Z' in a finite region $D \subset R^2$ uniquely determines Z'. Therefore, the sampling set cannot lie on the zero-crossing contours of Z and hence the infinite set of samples in a finite region is a set of uniqueness.

Example 2 (Counter example) An example of a sampling set that does not satisfy Corollary 1.1 is the case that infinite number of samples is chosen from the zeros of a polynomial of any degree. This example (which has applications in parallel X-ray and diffraction tomography) is ill-posed because it is shown that there is an infinite number of band-limited multidimensional signals that have the same zeros as those of a polynomial of any degree [11].

There are some well known 2-D interpolation functions for special sets of nonuniform samples. Naturally, these sets are sets of uniqueness and satisfy $N \subset z_i' \in Z'$. Some of these examples will be given in the section on the interpolating functions.

Example 3 (Over-sampling) To be a set of uniqueness, one practical condition is that the nonuniform samples are chosen inside the region D while only uniform samples are chosen outside the region D. The sampling rate of the uniform samples should be higher than the Nyquist rate.

The reason is that an infinite number of uniform samples (at greater than or equal to the Nyquist rate) determine a 2-D signal uniquely. If a finite number of uniform samples is replaced by another set of nonuniform samples, the uniqueness does not change—as it can be deduced from the interpolating functions of 2-D signals from uniform samples.

Example 4 (Small deviation) A sufficient condition for the sampling set to be a set of uniqueness is that the nonuniform samples do not deviate from uniform samples by more than a certain amount. The deviation amount follows the constraint

$$|x_{nm} - nT| + |y_{nm} - mT| \leq (\log 2)/\pi, \tag{1}$$

where x_{nm} and y_{nm} are the horizontal and the vertical coordinates of the samples. The proof is given in [12].

This condition is analogous to the 1-D case where a sufficient condition is that $|t_n - nT| < T/4$; where t_n's are nonuniform samples, T is the Nyquist sampling interval, and n is any integer.

4.2. Lagrange Interpolation

The Lagrange interpolation for 1-D signals was discussed in Chapter 3 of this book as well by the author in [2]. In [2], we show that almost all the known interpolation formula for the nonuniform sampling reconstruction can be derived from the Lagrange interpolation; among them are the uniform Shannon sampling reconstruction, the Kramer's interpolation, Bessel interpolation, and Yen's interpolation formulas [13]–[14] such as migration of a finite number of uniform samples, sampling with a single gap in an otherwise uniform distribution, and periodic nonuniform sampling. The periodic nonuniform sampling will be again discussed as a special case of the Papoulis Generalized Sampling Theorem in Section 4.3.

In the next section, we plan to develop interpolating functions for a set of nonuniform samples of a 2-D signal. The first question is whether Lagrange interpolation, which is the root of most interpolating functions for 1-D signals, can be generalized to 2-D signals. Some partial generalizations are now known and will be discussed in detail in the following section.

4.2.1. Lagrange Interpolation for 2-D Signals

Let us take a set of nonuniform samples on parallel lines as shown in Fig. 1. The parallel lines are denoted by $\{x_n\}$, and the samples on the lines are denoted by $\{y_{nm}\}$, where n and m are integers. If x_n is a sampling set satisfying the 1-D Lagrange interpolation (for example, the Kadec condition $|t_n - nT| < T/4$) and if y_{nm} is also a sampling set for each specific n, then the set of 2-D nonuniform samples uniquely determine a signal $f(x, y)$ band-limited to a square $[-W, W] \times [-W, W]$, or more generally, to a parallelogram centered at the origin. The interpolating function is

$$f(x, y) = \sum_{n=-\infty}^{\infty} \sum_{m=-\infty}^{\infty} f(x_n, y_{nm}) \psi_{nm}(x, y), \tag{2}$$

where

$$\psi_{nm}(x, y) = \psi_n(x)\psi_{nm}(y) = \frac{H(x)}{H'(x_n)(x - x_n)} \frac{H_n(y)}{H'_n(y_{nm})(y - y_{nm})}, \tag{3}$$

where

$$H(x) = (x - x_0) \prod_{\substack{k=-\infty \\ k \neq 0}}^{\infty} \left(1 - \frac{x}{x_k}\right), \quad H_n(y) = (y - y_{n0}) \prod_{\substack{k=-\infty \\ k \neq 0}}^{\infty} \left(1 - \frac{y}{y_{nk}}\right). \tag{4}$$

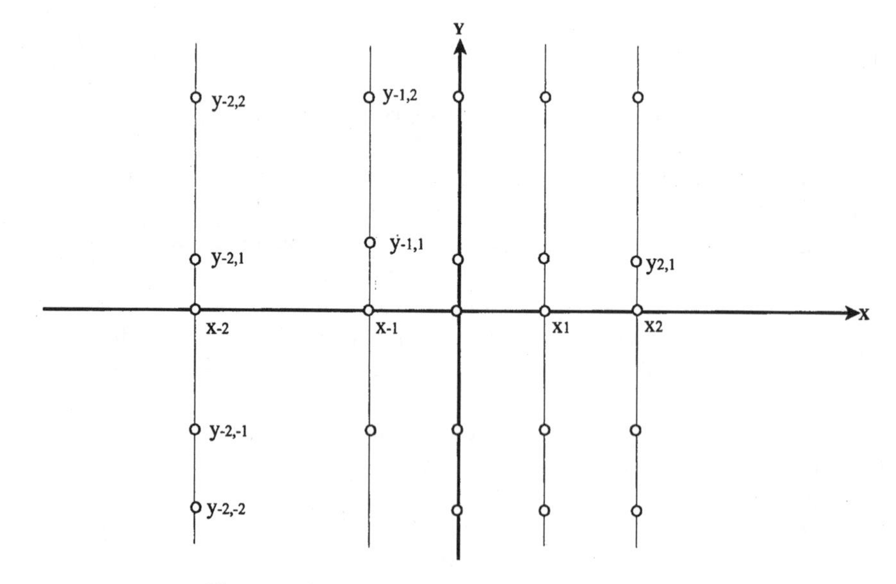

Figure 1. The set of nonuniform samples on vertical lines.

The above series converges uniformly on each compact subset of R^2.

Proof: For each real number $y \in R$, $f(x, y)$ is a 1-D band-limited function $(-W, W)$ in terms of x. Since the set $\{x_n\}$ satisfies the sufficiency condition for Lagrange interpolation, from (2) we get

$$f(x, y) = \sum_{n=-\infty}^{\infty} f(x_n, y)\psi_n(x). \tag{5}$$

Now, for each n, $f(x_n, y)$ is a band-limited 1-D signal in terms of y. Likewise, the 1-D Lagrange interpolation is

$$f(x_n, y) = \sum_{m=-\infty}^{\infty} f(x_n, y_{nm})\psi_{nm}(y). \tag{6}$$

Substituting (6) into (5), we derive (2). It is needless to say that by coordinate transformation, one can get horizontal parallel lines or different orientations as shown in Fig. 2. One example of the above relation is given below.

Example 1 Double-periodic sampling

Let us take $\{x_n\}$ and $\{y_m\}$ to be two sets of independent periodic samples. Using Yen's interpolation [1]–[2], and [14] we get

$$\psi_n(x) = \frac{N \prod_{k=0}^{N-1} \sin \frac{2\pi W}{N}(x - \tau_k)}{2\pi W(x - x_n) \prod_{k \neq n} \sin 2\pi W(x_n - \tau_k)}, \tag{7}$$

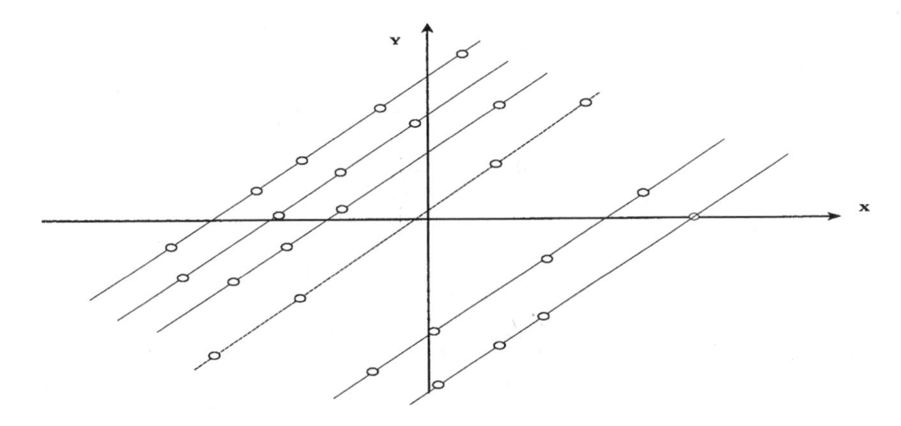

Figure 2. The set of nonuniform samples on slanted lines.

where τ_k's are the horizontal distance of vertical lines in the first periodic interval, and

$$\psi_{nm}(y) = \phi_m = \frac{N' \prod_{k=0}^{N'-1} \sin \frac{2\pi W}{N'}(y - \tau_k)}{2\pi W(y - y_m) \prod_{k \neq m} \sin 2\pi W(y_m - \tau_k)}, \tag{8}$$

where τ_k's are the nonuniform samples on the vertical lines $\{x_n\}$. Substituting (7) and (8) in (2), we get

$$f(x, y) = NN' \prod_{k=0}^{N-1} \sin \frac{2\pi W}{N}(x - \tau_k) \prod_{k=0}^{N'-1} \sin \frac{2\pi W}{N'}(y - \tau_k)$$
$$\times \sum_n \sum_m \frac{f(x_n, y_m)}{2\pi^2 W^2 (x - x_n)(y - y_m) \prod_{k \neq n} \sin 2\pi W(x_n - \tau_k) \prod_{k \neq m} \sin 2\pi W(y_m - \tau_k)}. \tag{9}$$

4.2.2. Lagrange Interpolation in Polar Coordinates

There are two interpolation formulae in polar coordinates; we shall describe both [15]:

Theorem 1 *Assume that a 2-D signal represented by $f(r, \theta)$ in polar coordinates is band-limited to a circular disc of radius A. Also, assume that the periodic signal along the circles $\{f(r_n, \theta),\ r = r_n,\ n \in Z\}$ with respect to θ (Fig. 3) is band-limited (represented by N harmonics). If $\{r_n, n \in Z\}$ satisfies $|r_n - n/2A| < 1/8A$, the set of $2N + 1$ nonuniform samples on each circle uniquely represent $f(r, \theta)$, and the Lagrange interpolation is given by*

$$f(r, \theta) = \sum_{n=-\infty}^{\infty} \sum_{m=0}^{2N} f(r_n, \theta) \frac{G_n(e^{j\theta})}{G_n'(e^{j\theta_{nm}})(e^{j\theta} - e^{j\theta_{nm}})e^{jN\theta}} \frac{H(r)}{H'(r_n)(r - r_n)}, \tag{10}$$

where

$$G_n(e^{j\theta}) = \prod_{m=0}^{2N} (e^{j\theta} - e^{j\theta_{nm}}), \text{ and } H(r) = \prod_{n=-\infty}^{\infty} (r - r_n).$$

Proof: Since $f(r, \theta)$ is band-limited to a disc A, a one-dimensional slice of $f(r, \theta)$ for a fixed θ is band-limited to A. Since $|r_n - n/2A| < 1/8A$, we have the following 1-D Lagrange interpolation:

$$f(r, \theta) = \sum_{n=-\infty}^{\infty} f(r_n, \theta) \frac{H(r)}{H'(r_n)(r - r_n)}, \tag{11}$$

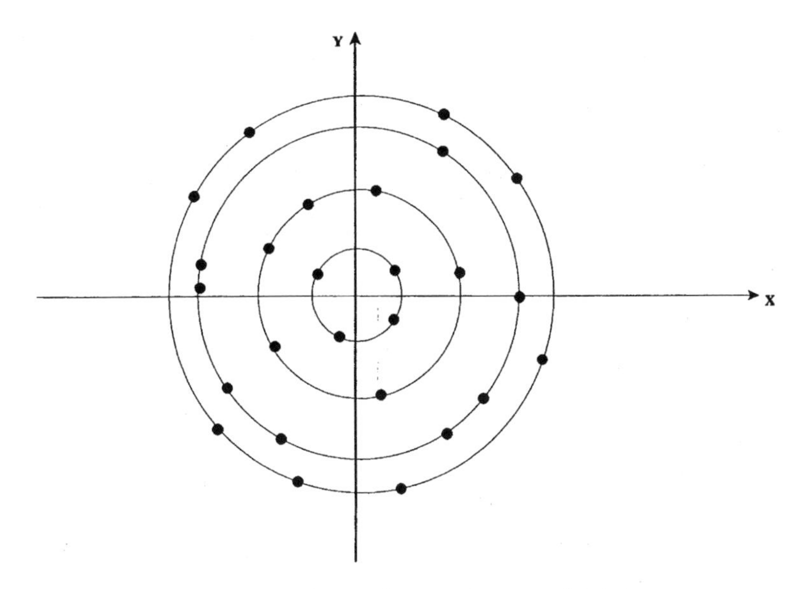

Figure 3. The set of nonuniform samples on parallel circles.

where $H(r)$ is defined in (10). $f(r_n, \theta)$ in (11) is a periodic function and is assumed to be band-limited with N harmonics. Therefore, the Fourier series representation $f(r_n, \theta)$ is a polynomial of degree $2N$ and can be represented by $2N+1$ nonuniform samples. The Lagrange interpolation for this finite polynomial is

$$f(r_n, \theta) = \sum_{m=0}^{2N} f(r_n, \theta_{nm}) \frac{G_n(e^{j\theta})}{G_n'(e^{j\theta_{nm}})(e^{j\theta} - e^{j\theta_{nm}})e^{jN\theta}}. \quad (12)$$

Combining (11) with (12), we get (10). □

Theorem 2 *Assume $f(r, \theta)$ is the same as the one in Theorem 1. If there are $2N+1$ radial lines (Fig. 4) with nonuniform rotational angles and if $\{r_n, n \in Z\}$ is a set of uniqueness satisfying the condition in Theorem 1, the sampling set shown in Fig. 4 uniquely represents $f(r, \theta)$, and the Lagrange interpolation is*

$$f(r, \theta) = \sum_{n=0}^{2N} \sum_{m=-\infty}^{\infty} f(r_{nm}, \theta_n) \frac{H_n(r)}{H_n'(r_{nm})(r - r_{nm})} \frac{G(e^{j\theta})}{G'(e^{j\theta_n})(e^{j\theta} - e^{j\theta_n})e^{jN\theta}}, \quad (13)$$

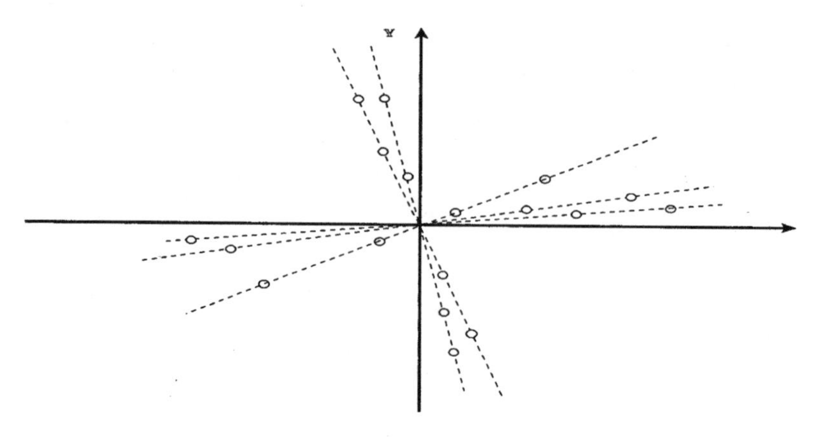

Figure 4. The set of nonuniform samples on radial lines.

where

$$H_n(r) = \prod_{m=\infty}^{\infty} (r - r_{nm}), \text{ and } G(e^{j\theta}) = \prod_{n=0}^{2N} (e^{j\theta} - e^{j\theta_n}).$$

Proof: Analogous to the proof of Theorem 1, the Lagrange polynomial interpolation is

$$f(r, \theta) = \sum_{n=0}^{2N} f(r, \theta_n) \frac{G(e^{j\theta})}{G'(e^{j\theta_n})(e^{j\theta} - e^{-j\theta_n})e^{jN\theta}}, \tag{14}$$

where

$$f(r, \theta_n) = \sum_{n=-\infty}^{\infty} f(r_{nm}, \theta_n) \frac{H_n(r)}{H_n(r_{nm})(r - r_{nm})}. \tag{15}$$

Equations (14) and (15) yield equation (13).

Comments *If the condition* $|r_n - n/2A| < 1/8A$ *is not satisfied but* $|r - n/2A| < L < \infty$, *the Lagrange interpolation functions (10) and (13) are still valid but H(r) in (10) is modified, as follows:*

$$H(r) = e^{ar}(r - r_0) \prod_{n \neq 0} \left(1 - \frac{r}{r_n}\right) e^{r/r_n},$$

where $a = \sum_{n \neq 0} \frac{1}{r_n}$.

Equation (13) reduces to Stark's polar interpolation [16] when nonuniform samples along radial lines are zeros of a Bessel function and θ_k is uniformly spaced.

4.3. Nonuniform Sampling for Non-Band-Limited Signals

It can be shown that certain classes of non-band-limited signals can be represented by a set of uniform samples violating the Nyquist rate, e.g., refer to [17]. By the same token, there is a class of non-band-limited signals that can be represented uniquely by a set of nonuniform samples. We will treat some special cases first and at the end give a general theorem. Assume that a band-limited signal goes through a monotonic nonlinear distortion, $y(t) = f[x(t)]$. Although $y(t)$ is a non-band-limited signal, it can be represented by nonuniform samples if $\{t_n\}$ is a *sampling set* for the band-limited signal $x(t)$. For the necessary and sufficient conditions that a non-band-limited signal can be reduced to a band-limited signal; see [17].

Another example is a set of non-band-limited signals generated by a time varying filter when the input is a band-limited signal. If the system has an inverse, then the samples of the output (the non-band-limited signal) are sufficient to reconstruct the signal. This non-band-limited signal is essentially a time-warped version of the band-limited signal. A nonlinear time varying example is a frequency modulated signal (although non-band-limited) that can be reconstructed from a set of samples that satisfy the Nyquist rate for the modulating band-limited signal, e.g., the zero-crossings of the FM signal.

Uniqueness of a non-band-limited finite energy signal (or in general an entire function) by a set of nonuniform samples is similar to our discussion for band-limited signals. Basically, if the *sampling set* $\{t_n\}$ cannot be the zero-crossings of a non-band-limited signal of finite energy (or an entire function of the same class), then the *sampling set* $\{t_n\}$ uniquely represents the non-band-limited signal. This uniqueness theorem does not give a constructive method of interpolation of the non-band-limited signal from its nonuniform samples.

An exact interpolation for a certain class of non-band-limited signals is possible and will be given in the next section.

4.3.1. Jittered Sampling

By jittered samples, we mean nonuniform samples that are clustered around uniform samples either deterministically or randomly with a given probability distribution. This random jitter is due to uncertainty of sampling at the transmitter end. We will not consider the jitter of uniform samples at the receiver end due to channel delay distortion.

For the deterministic jitter, Papoulis has proposed an interesting method for the recovery [18]. The problem is how to recover $x(t)$ from the jittered samples, $x(nT - \mu_n)$, where μ_n is a known deviation from nT. The main idea is to transform $x(t)$ into another function $g(\tau)$ such that the nonuniform samples at $t_n = nT - \mu_n$ are mapped into uniform samples $\tau = nT$. Consequently, $g(\tau)$ can be reconstructed from $g(nT)$ if $g(\tau)$ is band-limited ($W \leq 1/2T$). Now, $x(t)$ can be found from $g(\tau)$ if the transformation is one to one.

Let us take the one to one transformation as $t = \tau - \theta(\tau)$, where $\theta(\tau)$ is a band-limited function defined as

$$\theta(\tau) = \sum_{n=-\infty}^{\infty} \mu_n \operatorname{sinc} 2\pi W_2(\tau - nT), \tag{16}$$

where $W_2 \leq 1/2T$ is the bandwidth of $\theta(\tau)$. Since we assume the transformation is one to one, the inverse exists and is defined by $\tau = \gamma(t)$. Note that $\theta(nT) = \mu_n$ and $t_n = nT - \theta(nT)$. Let us assume

$$g(\tau) = x[\tau - \theta(\tau)] \Rightarrow g(nT) = x[nT - \theta(nT)] = x(t_n). \tag{17}$$

However, $g(\tau)$ is not band-limited in general. But if μ_n and as a result $\theta(\tau)$ are assumed to be small, $g(\tau)$ is approximately a band-limited function [18]. Using the uniform sampling interpolation for $g(\tau)$, we get

$$g(\tau) = \sum_{-\infty}^{\infty} g(nT) \operatorname{sinc} \frac{\pi}{T}[\tau - nT]. \tag{18}$$

Using the substitution $\tau = \gamma(t)$, we get

$$x(t) = g[\gamma(t)] \approx \sum_{n=-\infty}^{\infty} x(t_n) \operatorname{sinc} \frac{\pi}{T}[\gamma(t) - nT]. \tag{19}$$

Clark et al [19] have suggested that Papoulis transformation for jitter (17) and (19) can also be extended to a certain class of non-band-limited signals. If $x(t)$ is band-limited, $g(\tau)$ in (17) cannot be band-limited in general and, therefore, (19) is only an approximation. But if $g(\tau)$ is band-limited, then $x(t)$ cannot be band-limited; hence (19) is an exact representation for this class of non-band-limited signals, i.e.,

$$x(t) = \sum_{n=-\infty}^{\infty} x(t_n) \operatorname{sinc} \frac{\pi}{T}[\gamma(t) - nT]. \tag{20}$$

4.3.1.1 Jittered Sampling for 2-D Signals

The analysis of jittered samples can be carried over to 2-D signals. The uniform sampling theorem for 2-D signals is given by

$$f(\vec{x}) = \sum_{\{\vec{x}_s\}} f(\vec{x}_s) \tag{21}$$

where the sampling set $\vec{x}_s$ is a hexagonal sampling lattice given by

$$\vec{x}_s = \left\{\vec{x} = n\vec{v}_1 + m\vec{v}_2,\ n, m = 0, \pm 1, \pm 2, \ldots, \vec{v}_1 \neq k\vec{v}_2\right\}. \tag{22}$$

Let $h(\vec{x})$ be the image of $f(\vec{x})$ under the coordinate transformation

$$\vec{\xi} = \gamma(\vec{x}),\ \text{i.e.}, f(\vec{x}) = h(\vec{\xi}) = h[\gamma(\vec{x})]. \tag{23}$$

Let $\gamma(\vec{x})$ be such that the set of nonuniformly spaced samples $\{\vec{x}_s\}$ is transformed into a uniform set of samples $\{\vec{\xi}_s\}$. If $h(\xi)$ is suitably band-limited, then $h(\xi)$ can be interpolated from (21). The signal $f(\vec{x})$ can be obtained by reversing the coordinate transformation (23). The result is

$$f(\vec{x}) = \sum_{\{\vec{x}_s\}} f(\vec{x}_s) \frac{J_1[2\pi B(\gamma(\vec{x}) - (\vec{x}_s)]}{2\sqrt{3}B\,|\,\gamma(\vec{x}) - (\vec{x}_s)\,|} \tag{24}$$

The parameters J_1 and B in the above equation are, respectively, the first-order Bessel function of the first kind and the equivalent radial bandwidth of 2-D Fourier transform. In general, $f(\vec{x})$ is non-band-limited if $g(\vec{\xi})$ is to be band-limited. But an approximate interpolation is possible when $f(\vec{x})$ is band-limited and $g(\vec{\xi})$ is assumed to be approximately band-limited. Reference [20] has verified the feasibility of this method for their specific applications.

4.4. Past Sampling

From the Lemma in Section 4.1, we conclude that when the average sampling rate is higher than that of the Nyquist, then the nonuniform samples represent uniquely the band-limited signal. As a special case, these nonuniform samples could be the past samples. The past samples at a rate higher than the Nyquist is a *sampling set* because no signal of bandwidth W can be found that has zero-crossings of density greater than the Nyquist in an infinite interval. In fact, these past samples could be uniform. That is, past uniform samples of a band-limited signal that are slightly higher than the Nyquist rate, uniquely represent the signal [21]. Prediction of future samples from past samples and their practical implications are given in Chapter 13.

Infinite samples in a finite interval also form a *sampling set* for a band-limited signal. For example, the nonuniform samples at $t_n = 1/n,\ n = 1,\ 2,\ \ldots$ form a sampling set but have no practical value since they have the same problem as the extrapolation of a signal from the knowledge of a portion of the signal in an interval. We can find in the next section that none of these kinds of *sampling sets* are stable.

4.4.1. Stability of Nonuniform Sampling Interpolation

By stability, we mean that a slight perturbation of nonuniform sample amplitudes due to noise does not lead to a large error in the interpolation [22]. The necessary and sufficient condition for nonuniform samples (of a finite energy signal) to be stable is [23]

$$E_x = \int_{-\infty}^{\infty} |x(t)|^2 dt \leq C \sum_{n=-\infty}^{\infty} |x(t_n)|^2, \tag{25}$$

where $x(t)$ is any signal band-limited to W, and E and C are, respectively, the energy of $x(t)$ and a positive constant. In order to see why Equation (25) conforms to our definition of stability, let us take the error in the samples as $e(t_n)$. The energy of the error signal, $e(t)$, is small if the errors in the samples are small. This is because $e(t)$ is a band-limited function derived from a linear interpolation such as Lagrange. Therefore, $e(t)$ satisfies (25), viz.,

$$E_e = \int_{-\infty}^{\infty} |e(t)|^2 dt \leq C \sum_{n=-\infty}^{\infty} |e(t_n)|^2.$$

Equation (25) is derived from inequalities discussed in [24], which shows that for a following condition

$$\begin{aligned} &|t_n - nT| \leq L < \infty,\ n = 0, \pm 1, \pm 2, \ldots, \\ &|t_n - t_m| > \delta > 0,\ \ n \neq m, \end{aligned} \tag{26}$$

the following inequalities are valid:

$$A \leq \frac{\sum_{n=-\infty}^{\infty} |x(t_n)|^2}{\int_{-\infty}^{\infty} |x(t)|^2 dt} \leq B, \tag{27}$$

where A and B are positive constants which depend exclusively on $\{t_n\}$ and the bandwidth of $x(t)$. Equation (27) implies that under the condition (26), $\sum_{n=-\infty}^{\infty} |x(t_n)|^2$ cannot be zero or infinite. The exact value can be calculated from Parseval relationship as will be shown later when we get to spectral analysis of nonuniform samples.

Reference [23] has shown that whenever Lagrange interpolation is possible, the nonuniform samples are stable; but for cases such as past sampling, or infinite number of nonuniform samples in a finite interval, the *sampling set* $\{t_n\}$ is not stable.

Extension of (25) to 2-D signals is straightforward and is discussed in Chapter 6 of this book.

4.5. Spectral Analysis of Nonuniform Samples

In this section we shall discuss Poisson summation formula, Parseval relationship, and frequency spectrum of nonuniform samples of 1-D and 2-D signals. Based on these analyses practical recovery techniques are discussed in the following section.

4.5.1. Extension of Parseval Relationship to Nonuniform Samples

The Poisson Sum Formula for uniform samples is given by

$$\sum_{n=-\infty}^{\infty} |x(nT)|^2 = \frac{1}{T} \sum_{i=-\infty}^{\infty} \left| X\left(\frac{i}{T}\right) \right|^2,$$

where $x(nT)$'s are samples of a real signal and $X(i/T)$'s are samples of the Fourier transform $X(f)$. The above equation reduces to the following when $1/T \geq 2W$, where W is the bandwidth of $x(t)$,

$$\sum_{n=-\infty}^{\infty} |x(nT)|^2 = \frac{1}{T} \int_{-\infty}^{\infty} |x(t)|^2 dt = \frac{1}{T} \int_{-W}^{W} |X(f)|^2 df.$$

[3] Portions of this section are a modified reprint, with permission, from [53]; © 1990 IEEE.

The above equation is the same as Parseval relation for discrete signals when $e^{(j\omega)}$ is used instead of *f*.

It can been shown that the Parseval relationship for a general stable *sampling set* $\{t_n\}$ that satisfies condition (26) is given below, see [53]–[55]

$$\sum_{n=-\infty}^{\infty} |x(t_n)|^2 = \frac{1}{T}\int_{-\infty}^{\infty} x(t)x_{lp}(t)dt = \frac{1}{T}\int_{-\infty}^{\infty} X(f)X_{lp}^{*}(f)df, \qquad (28)$$

where $x_{lp}(t)$ is the low-pass filtered version of the nonuniform samples, $T = 1/2W$, and $X_{lp}(f)$ is the corresponding Fourier transform. For further discussion on Parseval and Poisson Formula, please see [57] and the paper in the CD-ROM, file: 'Chapter4\Butzer.zip'.

The extension of Parseval relationship to 2-D signals is similar to the 1-D case and we shall prove this relationship for the 2-D case. We would like to prove that

$$\begin{aligned}\sum_{m=-\infty}^{\infty}\sum_{n=-\infty}^{\infty} |f(x_{nm}, y_{nm})|^2 &= \frac{1}{T_1T_2}\int_{-\infty}^{\infty}\int_{-\infty}^{\infty} F(u,v)F_{lp}^{*}(u,v)dudv \\ &= \frac{1}{T_1T_2}\int_{-\infty}^{\infty}\int_{-\infty}^{\infty} f(x,y)f_{lp}(x,y)dxdy, \qquad (29)\end{aligned}$$

where $T_1 = 1/2W_1$, $T_2 = 1/2W_2$, W_1 and W_2 are the bandwidths along *u* and *v* coordinates, respectively, and $f_{lp}(x,y)$ is the lowpass filtered version of nonuniform samples given below

$$f_{lp}(x,y) = \sum_{-\infty}^{\infty}\sum_{-\infty}^{\infty} f(x_{nm}, y_{nm})\,\mathrm{sinc}\,[W(x - x_{nm})]\,\mathrm{sinc}\,[W(y - y_{nm})]. \qquad (30)$$

The Fourier transform of (30) is

$$\begin{aligned}F_{lp}(u,v) &= T_1T_2\left[\sum_{m=-\infty}^{\infty}\sum_{n=-\infty}^{\infty} f(x_{nm}, y_{nm})e^{-2j\pi u x_{nm}}e^{-2j\pi v y_{nm}}\right]\prod(uT_1)\prod(vT_2) \\ &= T_1T_2F_s(u,v)\prod(uT_1)\prod(vT_2), \qquad (31)\end{aligned}$$

where $F_s(u,v)$ is the Fourier transform of ideal impulsive nonuniform samples and $\prod(\cdot)$ is an ideal low-pass filter. If the samples are square integrable, the Fourier transform of the samples has finite energy over a finite interval [25].

Substituting (31) into (29), we get

$$\frac{1}{T_1T_2}\int_{-\infty}^{\infty}\int_{-\infty}^{\infty}F(u,v)F_{lp}^*(u,v)dudv = \int_{-\infty}^{\infty}\int_{-\infty}^{\infty}F(u,v)F^*{}_s(u,v)$$

$$\times \prod(uT_1)\prod(vT_2)dudv$$

$$= \int_{-\infty}^{\infty}\int_{-\infty}^{\infty}F(u,v)F_s^*(u,v)dudv. \tag{32}$$

By invoking Parseval theorem for $L^2(R^2)$ signal, (32) becomes

$$\int_{-\infty}^{\infty}\int_{-\infty}^{\infty}F(u,v)F_s^*(u,v)dudv = \int_{-\infty}^{\infty}\int_{-\infty}^{\infty}f(x,y)\sum_{m=-\infty}^{\infty}\sum_{n=-\infty}^{\infty}f(x_{nm},y_{nm})$$

$$\times\, \delta(x-x_{nm}, y-y_{nm})dxdy$$

$$= \sum_{m=-\infty}^{\infty}\sum_{n=-\infty}^{\infty}|f(x_{nm},y_{nm})|^2. \tag{33}$$

That is for any form of uniform or nonuniform samples which uniquely determine a band-limited signal, we have

$$\sum_{n=-\infty}^{\infty}\sum_{m=-\infty}^{\infty}|f(x_{nm},y_{nm})|^2 = \int_{-\infty}^{\infty}\int_{-\infty}^{\infty}F(u,v)F_s^*(u,v)dudv$$

$$= \frac{1}{T_1T_2}\int_{-\infty}^{\infty}\int_{-\infty}^{\infty}F(u,v)F_{lp}^*(u,v)dudv = \frac{1}{T_1T_2}\int_{-\infty}^{\infty}\int_{-\infty}^{\infty}f(x,y)f_{lp}(x,y)dxdy. \tag{34}$$

4.5.2. Spectral Analysis of Nonuniform Samples for 2-D Signals[4]

In this section, we would like to analyze the spectrum of nonuniform samples that deviate from uniform positions by a finite amount. The derivation of the spectrum of nonuniform samples leads to a practical method to recover 1-D and 2-D signals from the nonuniform samples. Since the analysis of 2-D signals is more complex than the 1-D signal, we shall first analyze 2-D signals and then show that 1-D signal is a special case that can be inferred from the 2-D case.

We model the set of 2-D nonuniform samples by [26]–[27]

$$f_s(x,y) = \sum_m\sum_n f(x_{nm},y_{nm})\delta(x-x_{nm}, y-y_{nm}), \tag{35}$$

[4] Portions of this section and Section 4.6 are a modified reprint, with permission, from [27]; © 1994 Signal Processing.

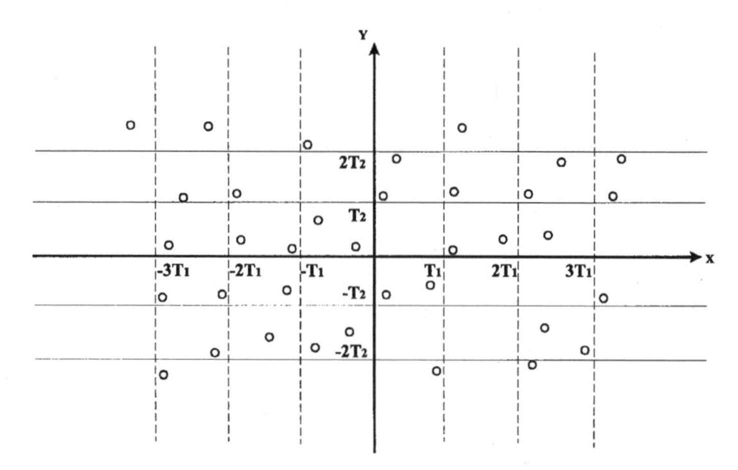

Figure 5. The set of nonuniform samples migrated from a uniform grid.

where $f(x, y)$ is a band-limited 2-D signal. The irregular set $\{x_{nm}, y_{nm}\}$ is assumed to be a deviation from the uniform set $\{nT_1, mT_2\}$ as seen in Figs. 1 and 5–6. Figure 5 is the most general case. Figure 1 represents a non-uniform sampling set that is represented by $\{x_n, y_{nm}\}$. Figure 6 is a special case of a nonuniform sampling set denoted by $\{x_n, y_n\}$.

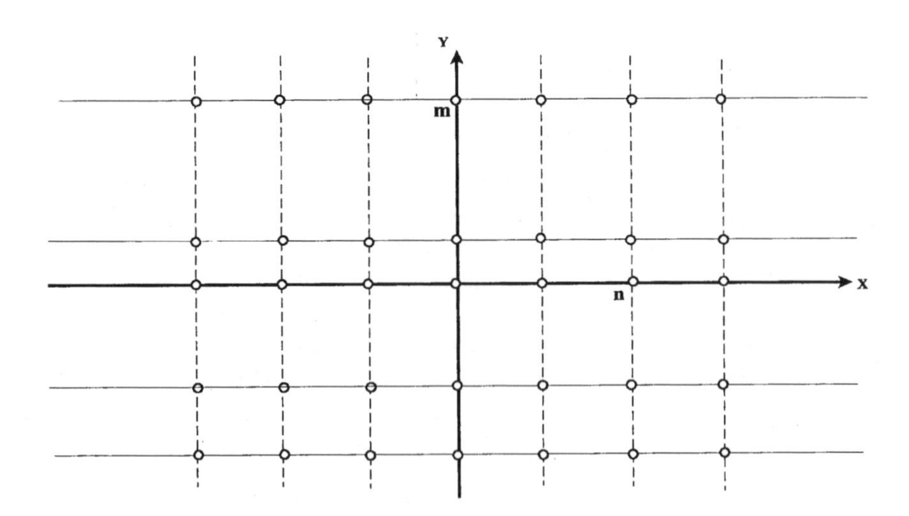

Figure 6. Nonuniform samples on irregular rectangular grids—a special case of Fig. 5.

To analyze the spectrum of (35), we first define the *comb* function as

$$f_p(x, y) = \sum_m \sum_n \delta(x - x_{nm}, y - y_{nm}). \tag{36}$$

Now, define the following functions

$$\begin{cases} g_1(x, y) = x - nT_1 - \theta_1(x, y) \\ g_2(x, y) = y - mT_2 - \theta_2(x, y), \end{cases} \tag{37}$$

such that $g_1(x_{nm}, y_{nm}) = g_2(x_{nm}, y_{nm}) = 0$ for any n and m. Now, we have the following identity:

$$\delta(x - x_{nm}, y - y_{nm}) = \delta[g_1(x, y), g_2(x, y)] \cdot |J|, \tag{38}$$

where J is the Jacobean defined by

$$J = \begin{vmatrix} \frac{\partial g_1}{\partial x} & \frac{\partial g_2}{\partial x} \\ \frac{\partial g_1}{\partial y} & \frac{\partial g_2}{\partial y} \end{vmatrix} = \frac{\partial g_1}{\partial x}\frac{\partial g_2}{\partial y} - \frac{\partial g_1}{\partial y}\frac{\partial g_2}{\partial x}. \tag{39}$$

(It is assumed that $J(x_{nm}, y_{nm}) \neq 0$.) Equation (38) is true because of the definition of distribution functions, i.e.,

$$\begin{aligned} \iint \delta(x - x_{nm}, y - y_{nm}) dx dy &= \iint \delta(g_1, g_2) dg_1 dg_2 \\ &= \iint \delta[x - nT_1 - \theta_1(x, y), y - mT_2 \\ &\quad - \theta_2(x, y)] \, |J| dx dy. \end{aligned} \tag{40}$$

Substituting (37) in (39), we get

$$|J| = \left| \left(1 - \frac{\partial \theta_1}{\partial x}\right)\left(1 - \frac{\partial \theta_2}{\partial y}\right) - \left(\frac{\partial \theta_1}{\partial y}\right)\left(\frac{\partial \theta_2}{\partial x}\right) \right|. \tag{41}$$

Now by using (38) and (39), the *comb* function (36) can be written as

$$f_p(x, y) = \sum_m \sum_n |J| \delta(x - nT_1 - \theta_1(x, y), y - mT_2 - \theta_2(x, y)). \tag{42}$$

Using the notation $x_1 = x - \theta_1(x, y),\ y_1 = y - \theta_2(x, y)$, and using Fourier series expansion, we get

$$f_p(x, y) = \frac{|J|}{T_1 T_2} \sum_i \sum_l e^{j[(i2\pi/T_1)x_1 + (l2\pi/T_2)y_1]}, \tag{43}$$

or,

$$f_p(x, y) = \frac{|J|}{T_1 T_2} \left\{1 + 2\sum_{i=1}^{\infty} \cos\frac{2\pi i}{T_1}[x - \theta_1(x, y)]\right\}\left\{1 + 2\sum_{l=1}^{\infty} \cos\frac{2\pi l}{T_2}[y - \theta_2(x, y)]\right\}. \tag{44}$$

Equation (44) is equivalent to the sum of phase modulated signals in two-dimensions. Therefore, the Fourier spectrum of $f_p(x, y)$ consists of a low-pass component $|J_1|/T_1 T_2$ and bandpass or high-pass components around carrier frequencies $(2\pi n/T_1,\ 2\pi m/T_2)$ as shown in Fig. 7.

The nonuniform samples (35) are derived by multiplying (43) by $f(x, y)$, i.e.,

$$f_s(x, y) = f(x, y) f_p(x, y) = f(x, y) \frac{|J|}{T_1 T_2} \left\{1 + 2\sum_{i=1}^{\infty} \cos\frac{2\pi i}{T_1}[x - \theta_1(x, y)]\right\} \times \left\{1 + 2\sum_{l=1}^{\infty} \cos\frac{2\pi l}{T_2}[y - \theta_2(x, y)]\right\}. \tag{45}$$

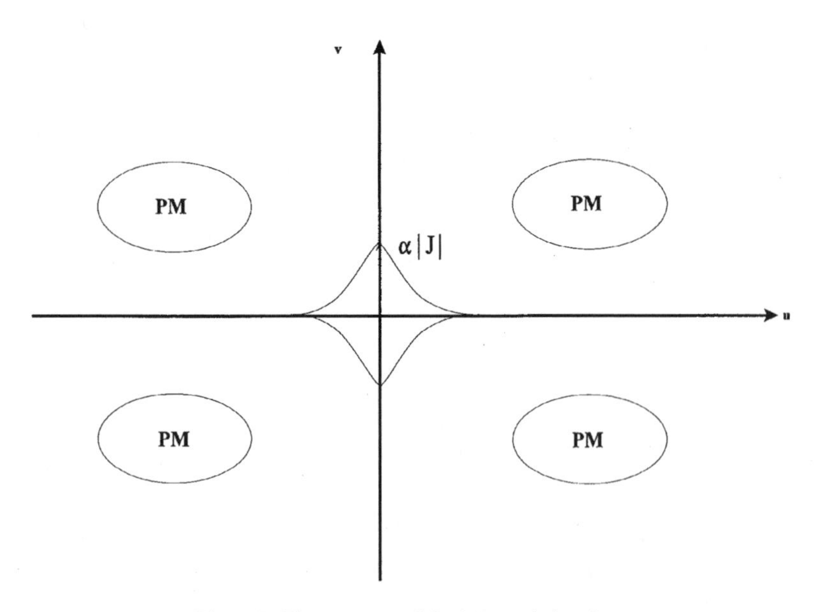

Figure 7. The spectrum of the 2-D comb function.

For the special case shown in Fig. 1, the Jacobean given in (41) can further be reduced to

$$|J| = \left|\left(1 - \frac{d\theta_1(x)}{dx}\right)\left(1 - \frac{\partial\theta_2(x,y)}{\partial y}\right)\right|, \tag{46a}$$

where $\theta_1(x,y) = \theta_1(x)$ and $\theta_2(x,y)$ remains the same. For the special case represented in Fig. 6, the Jacobean reduces to

$$|J| = |[1 - \dot{\theta}(x)][1 - \dot{\phi}(y)]|, \tag{46b}$$

where $\theta_1(x,y) = \theta(x)$ and $\theta_2(x,y) = \phi(y)$.

Notice that for case of uniform samples, (46.b) reduces to the following familiar result

$$\begin{aligned}\phi(y) &= \theta(x) = 0,\ J = 1,\ f_p(x,y)\\ &= \frac{1}{T_1 T_2}\left\{1 + 2\sum_{m=1}^{\infty}\cos\frac{2\pi m}{T_1}x\right\}\left\{1 + 2\sum_{m=1}^{\infty}\cos\frac{2\pi m}{T_2}y\right\}.\end{aligned}$$

The spectrum of the above *comb* function reduces to a periodic function of impulses, and the spectrum of $f_s(x,y)$ becomes a periodic function of $(1/T_1 T_2)F(u,v)$.

4.5.3. The Spectrum of Nonuniform Samples for 1-D signals

For 1-D signals, (35) and (36) become

$$x_s(t) = x(t)x_p(t), \tag{47}$$

where

$$x_p(t) = \sum_k \delta(t - t_k). \tag{48}$$

The Fourier series expansion of nonuniform samples can be derived from (44) and (45) by eliminating the parameters of y, θ_2 and by substituting T for T_1 and T_2 (for exact derivation see [2]). The result is

$$x_p(t) = \frac{|1 - \dot{\theta}(t)|}{T}\left[1 + 2\sum_{k=1}^{\infty}\cos\left(\frac{2\pi kt}{T} - \frac{2\pi k\theta(t)}{T}\right)\right]. \tag{49}$$

and

$$x_s(t) = x(t)x_p(t) = x(t)\frac{|1-\dot{\theta}(t)|}{T}\left[1 + 2\sum_{k=1}^{\infty}\cos\left(\frac{2\pi kt}{T} - \frac{2\pi k}{T}\theta(t)\right)\right]. \quad (50)$$

Equation (48) reveals that $x_p(t)$ has a DC component $|(1-\dot{\theta})/T|$ plus harmonics that resemble phase modulated (PM) signals. The index of modulation is $2\pi k/T$; the bandwidth is proportional to the index of modulation $(2\pi k/T)$, the bandwidth of $\theta(t) \leq \frac{1}{2}T$ and the maximum amplitude of $\theta(t)$, which in our case is related to $\theta(t_k) = t_k - kT$. The spectrum of $x_p(t)$ is sketched in Fig. 8 for the case $1 > \dot{\theta}$. The spectrum of (50) is sketched in Fig. 9 for the case $1 > \dot{\theta}$.

Now if we assume the bandwidth of $\theta(t)$ is less than $1/2T$, the phase modulated signal at the carrier frequency $1/T$ is a narrow band PM and has a bandwidth of approximately twice the bandwidth of $\theta(t)$, i.e., $\leq 1/T$. Therefore, low-pass filtering $x_p(t)$ (see Fig. 8), yields

$$x_{p_{lp}}(t) = \frac{1-\dot{\theta}}{T},$$

where we have assumed that $1 > \dot{\theta}$.

If the bandwidth of $\theta(t)$ is taken to be W_θ, as long as $1/T - W_\theta - W > W + W_\theta$, there is no overlap between the narrow band PM signal and $X(f) * (-\dot{\theta}/T)$, (see Fig. 9). Thus, low-pass filtering the nonuniform samples, $x_s(t)$, (with a bandwidth of $W + W_\theta$), we get

$$x_{s_{lp}}(t) = x(t)\frac{(1-\dot{\theta})}{T} \quad (51)$$

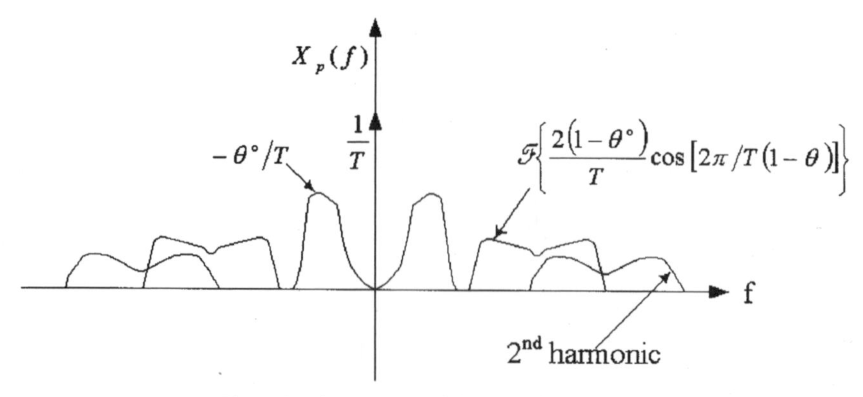

Figure 8. The spectrum of the comb function—x_p.

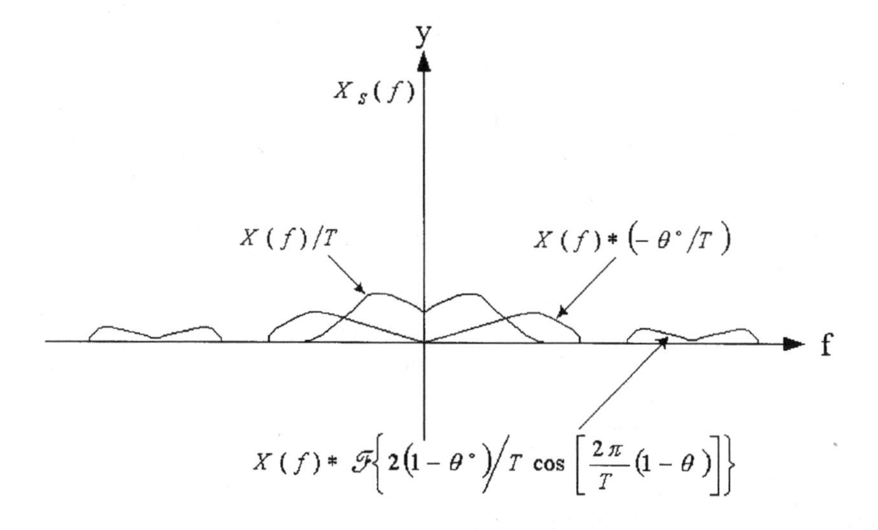

Figure 9. The spectrum of nonuniform samples—x_s.

The above spectral analysis leads to a practical reconstruction method, which is our next topic of discussion.

4.6. Practical Reconstruction Techniques

In this section, we shall discuss two practical reconstruction methods for 1-D (such as speech) and 2-D (such as image) recovery from a set of nonuniform samples, namely, a time varying method, and an iterative technique. Since the recovery techniques for 1-D and 2-D signals are very similar, we shall discuss the recovery methods for the 2-D signal mainly.

4.6.1

#A Time Varying (also called Nonlinear) Method

Spectral analysis reveals that the information about the original signal lies in the low-pass component of the ideal nonuniform samples (45), i.e., lowpass filtering the nonuniform samples (45), we get

$$LPF[f_s(x,y)] \approx f(x,y)\frac{|J|}{T_1 T_2}. \tag{52}$$

The Jacobean in (52) can be evaluated by lowpass filtering the *comb* function (44)—as shown in Fig. 7. The result is

$$LPF[f_p(x, y)] \approx \frac{|J|}{T_1 T_2}. \tag{53}$$

The approximation in (52) and (53) is due to the spectral overlap of phase modulated components as shown in Fig. 7. A division of (52) by (53) yields a good approximation of the original signal $f(x, y)$.

The division is possible if the Jacobean function is not equal to zero. A sufficient condition for this to be true is that the nonuniform samples do not deviate by more than $T/2\pi$ (for the case (41)) and T/π (for the case (46.b)) from their uniform positions in both the x and y directions [27]. Since the denominator in the division process does not depend on the input signal and is only a function of the space (x, y) for 2-D or time t for 1-D signals, the whole process is linear but time varying. However, in the literature it is known as the nonlinear technique [1]–[2], [27]–[28], and [34]. For the applications to video coding and error concealment, see Section 19.3.3 and Chapter 18, respectively.

Now, we can discuss the quality and the error of recovery using this method.

4.6.1.1 Simulation Results Using the Time Varying Technique

As a special case of nonuniform sampling, we have considered a set of uniform samples where some of the samples are lost on a random basis. Simulating this special set of nonuniform samples for 1-D signals (speech), we have got impressive results [28].

For 2-D signals, the above technique have been tested for a 512 × 512 bandlimited image as shown in Fig. 10; the samples are twice the Nyquist rate [27]. To generate nonuniform samples of the image, 50% of the pixels are eliminated on a random basis. To achieve this, a random 512 × 512 array is generated with about 50% 1's and 50% 0's. This array determines the nonuniform positions and therefore, by low-pass filtering this array, one gets the Jacobean function (53). After multiplying this array by the image, we get the nonuniform samples of the image. Low-pass filtering the nonuniform samples, we get a distorted image—Fig. 11. The division of this distorted image by the Jacobean yields the image shown in Fig. 12. As Fig. 12 reveals, the improvement is quite impressive—SNR is 21.22 dB. The results become much better when the sampling rate is increased or, alternatively, the percentage of pixel loss is decreased to, let us say, 30%—Fig. 13.

The Mathcad(TM) simulation for this technique is given in the accompanying CD-ROM under the file: 'Chapter4\Mathcad\Time-Varying-Iterative.mcd'.

Figure 10. The original over-sampled signal.

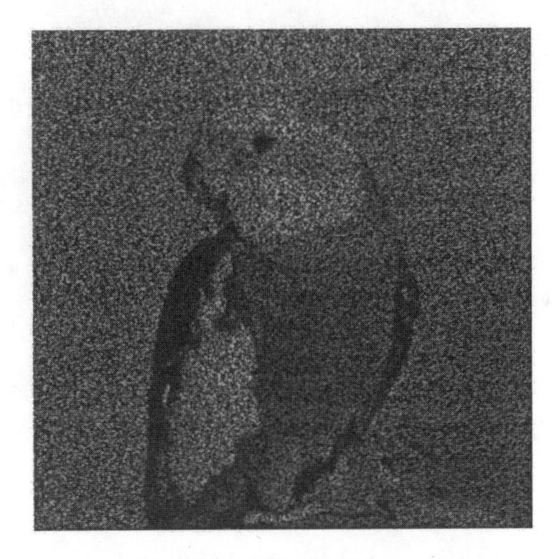

Figure 11. 50% pixel loss.

4.6.2. An Iterative Recovery Method

An iterative procedure has been proposed by Wiley [29] which can recover a signal from a set of non-uniform pulses without any distortion after infinite iterations. Each iteration improves the signal to distortion ratio. The iteration is based on the theorems of Duffin et al [24] and Sandberg [30]. The extensions of this method to ideal impulsive nonuniform sampling, sample-and-hold, flat top

Figure 12. The time varying recovery for 50% loss.

Figure 13. The time varying recovery for 30% loss.

sampling, natural sampling, and random sampling are done by the author [31]–[32].

$$x_{n+1}(t) = \frac{k_1}{k_2}(PSx - PSx_n) + x_n, \text{ and } x_0 = 0. \tag{54}$$

According to Sandberg's theorem, we have

$$x(t) = \lim_{n\to\infty} x_n(t),$$

Let us first discuss an intuitive approach to the iterative method. The iterative technique consists of modules of nonuniform samplers and lowpass filters as depicted in Fig. 14. The result of low-pass filtering the nonuniform samples of 1-D or 2-D signals (51)–(52) is after proper scaling

$$z(t) = x(t) + e(t), \tag{55}$$

where $e(t)$ is the error term. Now, if the error *norm* is less than the signal *norm*[5], the error norm can be reduced with additional iterations. The signal-to-noise ratio after i iterations is

$$\frac{S_i}{N_i} = i \cdot \frac{S_1}{N_1}\,\mathrm{dB}. \tag{56}$$

Equation (56) shows that improvement is possible if $S_1/N_1 > 0\,\mathrm{dB}$.

The 2-D version of the above iterative method is

$$f(x, y) = \lim_{n\to\infty} f_n(x, y),$$

where

$$f_{n+1}(x, y) = \frac{k_1}{k_2}[PSf(x, y) - PSf_n(x, y)] + f_n(x, y), \tag{57}$$

where P and S are, respectively, the low-pass filtering and the nonuniform sampling process for a 2-D signal. For the discrete version of the iterative techniques for the 1-D and the 2-D case and further results see Chapters 5 and 6, respectively; see also [33]. The proof of the convergence is given in [31] for the 1-D case and [27] for the 2-D case. For the applications to video coding and error concealment, see Section 19.3.4 and Chapter 18, respectively.

4.6.2.1 Simulation Results Using the Iterative Technique

For simulation results of 1-D signals, the reader is referred to [31], [34] and [2]. Some of the results are shown in Tables 1–3. The simulation is based on a low pass signal band-limited to $W = 100$ Hz. The nonuniform samples are initially taken at the Nyquist rate. The instances are chosen randomly such that $|t_k - kT| < T/4$; this is a sufficient condition to ensure a stable *sampling* set. The Mean-Square Error (MSE) for the first 10 iterations are shown in Table 1

[5] For deterministic energy signals, $E_e = \| e \|^2 = \int_{-\infty}^{\infty} | e |^2(t)dt$ and for random power signals $S_e = \| e \|^2 = E[e^2(t)]$, where $E[.]$ is the expected value.

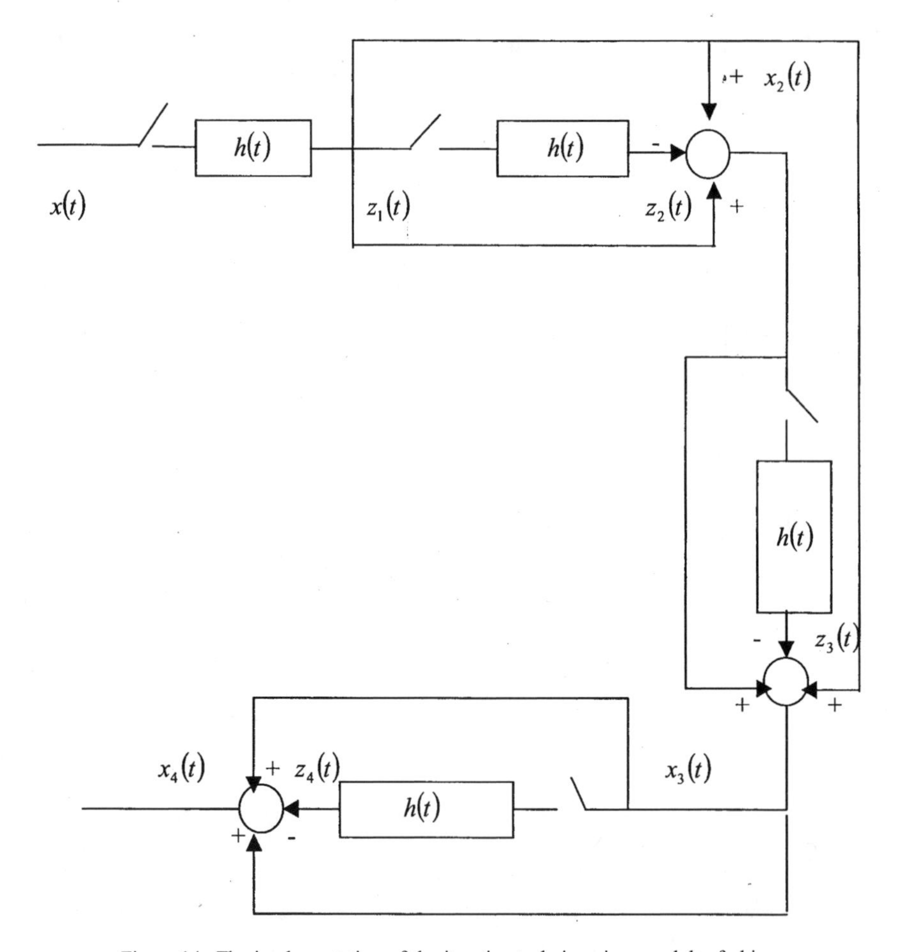

Figure 14. The implementation of the iterative technique in a modular fashion.

Table 1. Mean square error at the Nyquist rate; nonuniform samples are to within $\frac{1}{4}$ of the uniform positions.

Iteration No.	Variable S&H	Constant S&H	Ideal	Natural
10	3.36e-3	4.49	1.38e-3	4.1e-3

Table 2. Mean square error for ideal samples at the Nyquist rate; nonuniform samples are to within $\frac{1}{4}$ of the uniform positions.

Iteration No.	Nyquist Rate	Twice the Nyquist	Thrice the Nyquist
10	1.38e-3	4.35e-4	2.57e-4

Table 3. Mean square error for random sampling at different sampling rates.

Iteration No.	Nyquist Rate	Twice the Nyquist	Thrice the Nyquist
10	6.81e-1	4.24e-2	4.53e-3

under the column "Ideal". If we relax the sufficient condition $|t_k - kT| < T/4$, there is no guarantee that the *sampling set* converges at the Nyquist rate. For a specific *sampling set* that is to within $|t_k - kT| < T/2$, the iterative technique slowly converges. When the samples are totally random, we observe an even slower convergence as shown in Table 3.

At the rates lower than the Nyquist rate, the iterative method diverges. Obviously, if the average sampling rate is higher than the Nyquist rate, the convergence is guaranteed and is faster. For instance, at twice or three times the Nyquist rate, from Table 2 we can deduce that if the nonuniform samples are restricted to within $T/4$, where T is the Nyquist interval, we have a faster convergence after 10 iterations compared to the Nyquist sampling. This conclusion is also true when we sample randomly at twice or three times the Nyquist rate—Table 3. For simulation results for the special case of missing samples for 1-D signals such as speech see Chapters 5 and 17 of this book. We also refer you to [35] for the DSP implementation of speech reconstruction with missing samples.

The reconstructed signal from the nonuniform samples with sample-and-hold (S&H—constant and variable width) and the reconstructed signal from natural samples are listed in Table 1. This table shows that for the cases of ideal samples, the iteration converges slightly faster than other sampling schemes. For good signal recovery, 5 to 30 iterations are sufficient in most cases, depending on the sampling rate.

Simulation results using iterative techniques for 2-D signals are given in [27] and [34]. Further results and improvements for various applications are given in Chapters 6, 17 and 18. Some of the simulation results of [27] are given below; the iterative method is used to recover an image from lost pixels; this is a special type of nonuniform samples. The original image is as shown in Fig. 10. For a pixel loss of 50%, the results are shown in Fig. 15—SNR=20.37 dB. The same

Figure 15. Recovery after 15 iterations for 50% pixel loss-SNR= 20.37.

experiment is repeated for a pixel loss of 30% shown in Fig. 16—SNR = 36.29 dB. The simulation results show that the time varying technique is better than 7–8 iterations for the given sampling losses. The results are also better than linear interpolation. A comparison of the linear interpolation, the time varying method, and the iterative technique is given in Table 4.

Figure 16. Recovery after 15 iterations for 30% pixel loss-SNR= 36.29.

Table 4. SNR (dB) of different image recovery techniques.

Percentage of Lost Samples	Low-Pass Filter	Linear Interpolation	Time Varying Method	After 10 Iterations
10%	19.0	35.5	44.8	51.1
20%	14.2	32.6	39.5	40.5
30%	11.0	30.4	32.6	30.3
40%	8.6	27.8	28.2	22.6
50%	6.7	23.9	24.4	16.5

The Mathcad(TM) simulation for this technique is given in the accompanying CD-ROM under the file: 'Chapter4\Mathcad\Time-Varying-Iterative.mcd'.

4.7. The Generalized Sampling Theorem, Filter Banks, QMF, Wavelets and Sub-band Coding

The relationship among wavelets, multi-resolution analysis and Quadrature Mirror Filters (QMF) as used in sub-band coding in speech compression is known. It is our intention to show the equivalence of the generalized sampling theorem and QMF as used in sub-band coding in speech compression. Besides this relationship, some of the results developed in QMF and filter bank can be used in the sampling theorem and vice versa.

4.7.1. Introduction

QMF was introduced in 1970's as a scheme for sub-band coding of speech signals [36]–[37]. QMF results in perfect reconstruction of transmitted signals [38]. A much later development was on multi-resolution analysis and orthonormal wavelets [39]. It turns out that there is a close relationship between orthonormal wavelets and QMF.

Generalized Sampling Theorem (GST) was developed by Papoulis [40]; for discussions and extensions to 2-D signals see [41]–[42]. The GST of Papoulis consists of N Linear Time Invariant Systems (filters) and samplers at $1/N$ of the Nyquist rate. The problem posed by Papoulis is as follows: Suppose a band-limited (W) input signal $x(t)$ goes through N parallel filters $\{H_k(f),\ k = 1, 2, \ldots, N\}$. The outputs are called $\{g_k(t),\ k = 1, 2, \ldots, N\}$. Now, if each $g_k(t)$ is sampled at $1/N$ the Nyquist rate T, then $x(t)$ can be recovered from $\{g_k(t),\ k = 1, 2, \ldots, N\}$ under certain conditions discussed below. The interpolation formula is [1]

$$x(t) = \sum_{k=1}^{N} \sum_{n=-\infty}^{\infty} g_n(nNT) y_k(t - nNT). \tag{58}$$

The interpolation functions are found by

$$y_k(t) = \int_{-W}^{-W+2W/N} Y_k(f, t) e^{j2\pi ft} df, \ k = 1, 2, \ldots, N, \tag{59a}$$

where $Y_k(f, t)$'s are solutions of the set of equations:

$$\begin{bmatrix} H_1(f) & \cdot\ \cdot & H_N(f) \\ \cdot & \cdot\ \cdot & \cdot \\ \cdot & \cdot\ \cdot & \cdot \\ H_1[f + (N-1)2W/N] & \cdot\ \cdot & H_N[f + (N-1)2W/N] \end{bmatrix} \begin{bmatrix} Y_1(f, t) \\ \cdot \\ \cdot \\ Y_N(f, t) \end{bmatrix} = \begin{bmatrix} 1 \\ \cdot \\ \cdot \\ e^{j(N-1)2\pi(t/NT)} \end{bmatrix} \tag{59b}$$

If the determinant of matrix H is zero, then there is no solution. The interpolation function suggests a reconstruction diagram similar to the one of filter bank configuration of Fig. 17.

The main difference between Papoulis' GST and perfect reconstruction for filter banks is that the first one is based on general analog filters and the second one on discrete bandpass filters, where the inverse of the matrix in (59.b) is guaranteed. QMF (and also orthonormal wavelets) is a special type of filter banks

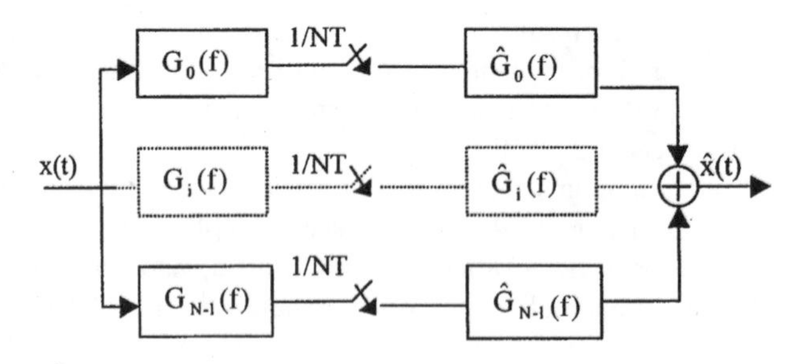

Figure 17. Analysis and synthesis filter banks.

where the parallel filter banks[6] are all related to each other by a scale factor and each filter has the following property (for $N = 2$) [43]:

$$| G_i(z) |^2 + | G_i(-z) |^2 = 2, \tag{60}$$

where $i = 0, 1$ and $G(z)$ is the z transform, and

$$G_0(z)G_1^*(z) + G_0(-z)G_1^*(-z) = 0, \tag{61}$$

where * is the complex conjugate. The above equations are called the orthonormality conditions and imply that the following matrix is unitary

$$A(z) = \begin{bmatrix} G_0(z) & G_1(z) \\ G_0(-z) & G_1(-z) \end{bmatrix}. \tag{62}$$

A unitary matrix has an inverse matrix and is equal to

$$A^{-1}(z) = cA^T(z), \tag{63}$$

where T is the conjugate transpose, and c is any constant. Equations (60)–(61) and hence (63) imply that the synthesis filters $\hat{G}_i(z)$'s are complex conjugate of the analysis filters, i.e.,

$$\hat{G}_i(z) = G_i^*(z), \ i = 0, \text{ and } 1 \tag{64}$$

In the next section, we shall show that some of the results of the filter bank theory can be used in the GST and vice versa.

4.7.2. Interpolation of Various Sampling Schemes

Instead of the impractical interpolation functions given in (59), we can use the orthonormality conditions (61) and (62) developed for filter banks in the Papoulis GST as a sufficient condition and derive much simpler interpolation functions as suggested by (59). From Fig. 17, we would like to find synthesis

[6] We assume N filter banks, in general. However, for QMF and orthonormal wavelets, it is more convenient and computationally more efficient to consider 2 parallel filter banks and iterated versions of them.

filters $\hat{G}_i(f)$, $i = 0, 1, \ldots, N-1$ such that the original signal is recovered without any distortion. The output of Fig. 17 is given by

$$\hat{X}(f) = \sum_{i=0}^{N-1} \sum_{n} X(f - n/NT) G_i(f - n/NT) \hat{G}_i(f) \tag{65}$$

For $N = 2$, (65) reduces to

$$\begin{aligned}\hat{X}(f) = {} & X(f)[G_0(f)\hat{G}_0(f) + G_1(f)\hat{G}_1(f)] + X(f + 1/2T)[G_0(f + 1/2T)\hat{G}_0(f) \\ & + G_1(f + 1/2T)\hat{G}_1(f)] + X(f - 1/2T)[G_0(f - 1/2T)\hat{G}_0(f) \\ & + G_1(f - 1/2T)\hat{G}_1(f)]\end{aligned} \tag{66}$$

Now instead of using (59), we use the filter bank derivations, viz,

$$G_0(f)\hat{G}_0(f) + G_1(f)\hat{G}_1(f) = 2T \tag{67}$$

$$G_0(f + 1/2T)\hat{G}_0(f) + G_1(f + 1/2T)\hat{G}_1(f) = 0 \tag{68}$$

$$G_0(f - 1/2T)\hat{G}_0(f) + G_1(f - 1/2T)\hat{G}_1(f) = 0 \tag{69}$$

Since $G_i(f)$ and $G_i(f)$ are band-limited, from (67)–(69), we have two distinct cases:

$$A\vec{S} = \begin{pmatrix} 2T \\ 0 \end{pmatrix}, \tag{70}$$

where

$$A = \begin{pmatrix} G_0(f) & G_1(f) \\ G_0(f + 1/2T) & G_1(f + 1/2T) \end{pmatrix}, \quad \text{for } -1/2T \leq f \leq 0, \tag{71}$$

or

$$A = \begin{pmatrix} G_0(f) & G_1(f) \\ G_0(f - 1/2T) & G_1(f - 1/2T) \end{pmatrix}, \quad \text{for } 0 \leq f \leq 1/2T, \tag{72}$$

and

$$S = \begin{bmatrix} \hat{G}_0(f) \\ \hat{G}_1(f) \end{bmatrix}. \tag{73}$$

assuming A has an inverse, S can be found from (70). The results are

$$S = 2/D \begin{pmatrix} G_0(f \pm 1/2T) \\ G_1(f \pm 1/2T) \end{pmatrix}, \tag{74}$$

where D is the determinant of A in (71) or (72). The following examples should clarify the above interpretation:

4.7.2.1 Hilbert Transform Sampling

Suppose $G_0 = 1$ and $G_1 = -j\,\mathrm{sgn}(f)$, the Hilbert transform, i.e., we have the samples of the signal and its Hilbert transform at half the Nyquist rate and we would like to find a practical interpolation filter for the recovery of the original signal.

Solution *The matrix (62) in the Fourier domain is unitary (see (63)), hence from (64) the interpolating (synthesis) filters are the conjugate of G_i. Alternatively, we can solve (74) and get*

$$S = \begin{bmatrix} 1 \\ j\,\mathrm{sgn}(f) \end{bmatrix}, \tag{75}$$

which is the conjugate of the analysis filters. The interpolation in time domain from (75) is as follows:

$$\begin{aligned} x(t) &= \sum_k x(2kT)\,\mathrm{sinc}(t/T - 2k) \\ &\quad - \sum_k \hat{x}(2kT)\{\pi/2T(t - 2kT)\,\mathrm{sinc}^2[1/2T(t - 2kT)]\}, \end{aligned} \tag{76}$$

where the term in the second summation is the Hilbert transform of $x(t)$ and the sinc function, respectively.

4.7.3. Periodic Nonuniform Sampling

If the analysis filters are taken as exponential functions, we can derive periodic nonuniform sampling as it was originally derived by Papoulis using non-real interpolation functions (59). For the special 2 filter bank case, if $G_0 = 0$ and $G_1 = e^{-j\omega\tau}$, from (74), we have

$$\begin{aligned} \hat{G}_0 &= \frac{2Te^{j\omega\tau}}{1 - e^{j\tau\pi/T}} \quad \text{for } 0 \le f \le 1/2T, \\ \hat{G}_0 &= \frac{2Te^{j\omega\tau}}{1 - e^{-j\tau\pi/T}} \quad \text{for } -1/2T \le f \le 0 \end{aligned} \tag{77}$$

and

$$\hat{G}_1 = \frac{-2Te^{j\pi\tau/T}}{1 - e^{j\tau\pi/T}} \quad \text{for } 0 \leq f \leq 1/2T,$$

$$\hat{G}_1 = \frac{-2Te^{-j\pi\tau/T}}{1 - e^{-j\tau\pi/T}} \quad \text{for } -1/2T \leq f \leq 0$$

In order to make (77) causal, we add a delay in series at both branches of the synthesis filters. The interpolation formula in time domain is derived in [1] and is given below

$$x(t) = \frac{\cos 2\pi W\tau - \cos 2\pi Wt}{2\pi W \sin 2\pi W\tau} \sum_{n=-\infty}^{\infty} \left[\frac{x(2nT + \tau)}{t - 2nT - \tau} - \frac{x(2nT - \tau)}{t - 2nT + \tau} \right]. \tag{79}$$

We could derive derivative sampling by setting $G_k(f) = (j\omega)^k$. However, the interpolation formula is straightforward from (74).

Using Papoulis' method, one can generalize the periodic nonuniform sampling to a combination of periodic nonuniform samples of a signal at instances $\{t_k\}$ and samples of $N - 1$ derivatives at the same instances taken at $1/N$ times the Nyquist rate. It has been shown that such kinds of sampling under the umbrella of Papoulis generalized sampling theorem might be sensitive to noise (ill-posed) at the Nyquist rate [42]; this reference shows that, in general, by over-sampling, the sensitivity to noise goes away; for more explanation and simulation results using C and DSP implementation, see the MSc thesis in the CD-ROM, directory: 'Chapter4\MSc-Said'. Also, for the extension of GST to multi-input, multi-output systems, please refer to CD_ROM, directory: 'Chapter4\Periodic-Nonuniform'.

4.7.4. Conclusion

Papoulis has to be credited for the original GST. By the use of filter bank theory, we can interpolate various sampling theorems using real time filters. We have shown that GST is equivalent to filter banks, QMF, and wavelets. They all have application in sub-band coding and data compression of signals.

4.8. Interpolation of Lowpass Signals at Half the Nyquist Rate[7]

In this section, we shall describe interpolation of low-pass signals from a class of stable sampling sets at half the Nyquist rate. Practical reconstruction algorithms are also suggested [44].

[7] Portions of this section are a modified reprint, with permission, from [44]; © 1996 IEEE.

4.8.1. Introduction

Although it was known that nonuniform samples at rates below the Nyquist rate is possible [1], issues such as stability, interpolation, and practical reconstruction techniques have not been addressed. In this section, we shall discuss a class of stable nonuniform samples at rates close to half the Nyquist rate. We will develop formulas based on Lagrange interpolation and suggest practical methods for recovery.

4.8.2. Choice of Sampling Set

The analytic signal for a finite energy low-pass signal of bandwidth W is defined as

$$x_a(t) = x(t) + j\hat{x}(t) = A(t)\exp[j\theta(t)] \tag{80}$$

where $\hat{x}$ is the Hilbert transform of $x(t)$, $A(t)$ is the magnitude (envelope) and

$$\tan\theta(t) = \frac{\hat{x}(t)}{x(t)} \tag{81}$$

The nonuniform points are chosen such that

$$x(t_k)\sin\omega_0 t_k = \hat{x}(t_k)\cos\omega_0 t_k, \quad k \in Z \tag{82}$$

where ω_0 is an arbitrary frequency, which determines the sampling rate. In case the sampling rate is higher than half the Nyquist rate, a subset of samples are chosen such that we get W samples on the average, i.e., half the Nyquist rate. For example, if $2f_0/W = N$, where N is an integer, then every N^{th} crossings is chosen to satisfy half the Nyquist rate. Equation (82) is equivalent to the zero-crossings of the single side band modulation of $x(t)$; hence the density of the zero-crossings increases with the carrier frequency ω_0. From (81) and (82), we have

$$\theta(t_k) = \omega_0 t_k - k\pi, \quad k \in z \quad \text{for } -\pi/2 \leq \theta(t) \leq \pi/2. \tag{83}$$

The above equation implies that $|t_k - kT_0| < T_0/2$, where $T_0 = 1/(2\omega_0)$. Deviation of nonuniform points from uniform positions ($t_k - kT$, where T is the sampling interval) decreases if ω_0 is taken to be very large and every $[2f_0/W]$ (rounded up) sample is chosen. For example, if $f_0 = N \cdot W$, where N is a positive integer, every N sample is chosen and we have $|t_k - kT| < T/2N$. This fact implies that the nonuniform samples are well behaved and can be stable. By stability, we mean that a slight perturbation of nonuniform sampling amplitudes due to noise does not lead to a large error in the interpolation. As long as the deviation of nonuniform points is finite, stability is guaranteed [1].

If the analytic signal in (80) is shifted to the left by $W/2$ in the frequency domain, we derive the low pass equivalent signal:

$$x_{lp}(t) = x_a(t)e^{-j\pi Wt} = x_i(t) + jx_q(t) = A(t)e^{j\phi(t)}, \tag{84}$$

where x_i and x_q are real in-phase and quadrature signals of bandwidth $W/2$, and $\phi(t)$ from (80) and (84) is

$$\phi(t) = \theta(t) - \pi Wt \tag{85}$$

From (83) and (85), we derive $\phi(t_k)$ from t_k, i.e.,

$$\phi(t_k) = \omega_0 t_k + k\pi - \pi W t_k \tag{86}$$

From (84) we have

$$x_i(t_k) = A(t_k)\cos\phi_k = (-1)^k A(t_k)\cos(\omega_0 - \pi W)t_k, \tag{87a}$$
$$x_q(t_k) = A(t_k)\sin\phi_k = (-1)^k A(t_k)\sin(\omega_0 - \pi W)t_k. \tag{87b}$$

If the sampling rate satisfies half the Nyquist rate, from (87), $x_i(t)$ and $x_q(t)$ can be recovered from $A(t_k)$ since their corresponding bandwidths are $W/2$. As a result $x_{lp}(t)$ can be recovered from (84). $x(t)$ can then be recovered from $x_{lp}(t)$, viz.,

$$x(t) = \mathrm{Re}[x_{lp}(t)e^{j\pi Wt}]. \tag{88}$$

Thus signal recovery from half the Nyquist rate is possible in principle.

4.8.3. Lagrange Interpolation

The Lagrange interpolation for nonuniform sampling was discussed in Chapter 3. The above discussion can be formulated in terms of Lagrange interpolation, i.e.,

$$x_{lp}(t) = \sum_k x_i(t_k)\psi_k(t) + j\sum_k x_q(t_k)\psi_k(t) \tag{89}$$

In (89), the sampling rate (r) is assumed to be greater than half the Nyquist rate® $\geq W$, and $\psi_k(t)$ is the Lagrange interpolation given by

$$\psi_k(t) = \frac{H(t)}{\dot{H}(t_k)(t - t_k)}, \quad H(t) = \prod_k (t - t_k) \tag{90}$$

A sufficient condition for the convergence of Lagrange interpolation is $t_k - kT < T/4$, [1]. Hence, the convergence of (89) is guaranteed from the discussion on stability after (4). From (84), (86), (87) and (89) we have

$$x_{lp} = \sum_k (-1)^k A(t_k) e^{jt_k(\omega_0 - \pi W)} \psi_k(t) \tag{91}$$

From (88) and (91) we have

$$x(t) = \sum_k (-1)^k A(t_k) \cos[\pi W(t - t_k) + \omega_0 t_k] \psi_k(t) \tag{92}$$

The above formula is the implicit interpolation from nonuniform samples. To derive an explicit interpolation formula, we use (80), (83) and (92) to get an interpolation in terms of $x(t_k)$ and $\hat{x}(t_k)$, i.e.,

$$x(t) = \sum_k x(t_k) \cos[\pi W(t - t_k)] \psi_k(t) - \sum_i \hat{x}(t_k) \sin[\pi W(t - t_k)] \psi_k(t) \tag{93}$$

The implicit formula can then be derived from (81), (83) and (93), i.e.,

$$\hat{x}(t_k) = x(t_k) \tan(\omega_0 t_k),$$

and

$$x(t) = \sum_k x(t_k) \frac{\cos[\pi W(t - t_k) + \omega_0 t_k]}{\cos[\omega_0 t_k]} \psi_k(t). \tag{94}$$

In case $\cos(\omega_0 t_k) = 0$, from (82), we know that $x(t_k) = 0$, and

$$\frac{x(t_k)}{\cos(\omega_0 t_k)} = \frac{\hat{x}(t_k)}{\sin(\omega_0 t_k)}$$

The above equation implies that (94) does not blow up and is well behaved.

In practice, we can recover $x_i(t)$ and $x_q(t)$ from nonuniform samples using iterative techniques; see Section 4.6.2 of this book as well as Chapters 5 and 6. $x(t)$ can thus be recovered from (88). Iterative techniques are guaranteed to converge if the sampling set is stable and the sampling set $\{t_k\}$ is operating at rates above W samples/sec (half the Nyquist rate for $x(t)$).

4.8.4. Conclusion

We developed a class of nonuniform samples at rates close to half the Nyquist rate. Iterative techniques can be used to recover a low pass signal from the sampling set.

4.9. Nonuniform Sampling Theorems for Bandpass Signals[8]

In this section the problem of recovery of bandpass signals from a set of nonuniform samples is discussed. Interpolation formulas for particular sets of irregular samples are given. Issues such as the Nyquist rate are studied, and practical methods for sampling and recovery are suggested.

4.9.1. Introduction

In this section, we would like to study the nonuniform sampling theory for bandpass signals. Extension of uniform sampling to bandpass signals is well known and is given by [41] and [45], periodic nonuniform sampling has been applied to bandpass signals by [46], [47], and [48]. Extension of general nonuniform sampling to bandpass signals is, however, not too well understood [49].

The following section discusses issues related to conditions for the recovery of $x_{bp}(t)$ from nonuniform samples $x_{bp}(t_k)$.

4.9.2. Signal Recovery from the Nonuniform Samples of a Bandpass Signal

We first start with the most general case with a sampling density at or above the Nyquist rate and then discuss the special cases.

Theorem 3 *If the nonuniform sampling set* $\{t_k\}$ *satisfies the following conditions:*

$$1.\quad |t_k - t_i| \geq \alpha \quad \text{for } k \neq i \text{ and } \alpha > 0, \tag{95}$$

$$2.\quad \left| t_k - \delta \frac{k}{B} \right| \leq L < \infty, \tag{96}$$

[8] Portions of this section are a modified reprint, with permission, from [49]; © 1996 IEEE.

where $k \in Z$, B is the bandwidth of the bandpass signal $x_{bp}(t)$, and $0 < \delta \leq 1$, then the nonuniform samples of the complex extension of a bandpass signal[9], $x^c(t) = A(t)e^{j[\omega_c t + \phi(t)]}$, *where $A(t)$ is the envelope and $\omega_c = 2\pi f_c$ is the central frequency of $x_{lp}(t)$ uniquely determines the bandpass signal. The Lagrange interpolation formula is*

$$x^c(t) = \sum_k x^c(t_k)e^{j\omega_c(t-t_k)}\psi_k(t), \tag{97}$$

where $\psi_k(t)$ is the Lagrange interpolating function discussed in Chapter 3.

The bandpass signal $x_{bp}(t)$ is the real part of the above interpolation function. Equivalently, $x_{bp}(t)$ can be interpolated from the sampling sets $\{x_{bp}(t_k)\}$ and $\{\hat{x}_{bp}(t_k)\}$ from the interpolation formula

$$x_{bp}(t) = \sum_k x_{bp}(t_k)\cos\omega_c(t - t_k)\psi_k(t) - \sum_k \hat{x}_{bp}(t_k)\sin\omega_c(t - t_k)\psi_k(t) \tag{98}$$

The proof is straightforward from the Lagrange interpolation of low-pass signals and is omitted here.

Comment 1 *The sampling density is defined as*

$$r = \lim_{\tau\to\infty}\frac{n_\tau}{\tau}, \tag{99}$$

and from (96) we have

$$r = \frac{B}{\delta} \geq B, \tag{100}$$

where n_τ is the number of samples in the interval τ. Equation (100) implies that the density of the complex samples $x^c(t)$ is greater than or equal to B. Since the samples are complex, the effective average sampling density is greater than or equal to $2B$, i.e., the Nyquist rate. Notice that in the above definition, the positions of the samples are not taken into account. For communication systems, the position of the samples is derived from the time of arrival; PPM is an example. For storage purposes, an additional overhead is needed to store the position of the samples in addition to the sample amplitudes. If the position of the samples is included, the effective sampling rate for (97) and (98) is $3/2$ times the Nyquist rate.

Comment 2 *Uniform samples at or greater than the Nyquist rate that are not related to the carrier frequency are a special case of this theorem. For this case, $\psi_k(t)$ in (97) and (98) can be shown to converge to a sinc function.*

[9] Or, equivalently the nonuniform samples of the band-pass signal and its Hilbert transform.

Next, we intend to find a particular sampling set $\{t_k\}$ close to half the Nyquist density (the effective sampling rate is equal to the Nyquist rate if the positions are taken into account) that uniquely represents the bandpass signal $x_{bp}(t)$.

Theorem 4 *Let the $\{t_k\}$ be defined by $\omega_c t_k + \phi(t_k) = 2\pi Nk$ for $N \geq 1$. If $\phi(t)$ is bounded and the density of $\{t_k\}$ is $1/B - 1/f_c \leq 1/r \leq 1/B$, i.e., close to half the Nyquist density, then the following interpolation formula is valid*

$$x_{bp}(t) = \sum_k x_{bp}(t_k) \cos[(\omega_c(t - t_k)]\psi_k(t), \tag{101}$$

where $\psi_k(t)$ is the Lagrange interpolation given in (97). For the proof see [44].

4.9.3. Practical Methods to Find the Sampling Set Close to Half the Nyquist Rate

In this section, we suggest some practical methods to sample a bandpass signal close to half the Nyquist density according to Theorem 4. This implies that the sampling instances are at the maxima points of

$$f(t) = \cos[\omega_c t + \phi(t)] \tag{102}$$

We suggest three methods for the reconstruction of $f(t)$ from $x_{bp}(t)$: The first method is that $x_{bp}(t)$ is first hard limited and then processed by a bandpass filter. The output is approximately equal to $f(t)$ if the band-pass signal is assumed to be narrowband.

An alternative method is to use an envelope detector to detect the envelope of the bandpass signal $A(t)$. A division of $x_{bp}(t)$ by $A(t)$ yields $f(t)$ whenever $A(t) \neq 0$. Once $f(t)$ is found, the sampling set $\{t_k\}$ is determined as a subset of the maxima points of $f(t)$.

The third method bypasses the derivation of $f(t)$ and finds the sampling set $\{t_k\}$ directly. We first derive the Hilbert transform of the bandpass signal, i.e.,

$$x_{bp}(t) = A(t)\cos[\omega_c t + \phi(t)] = A(t)f(t) \tag{103}$$

$$\hat{x}_{bp}(t) = A(t)\sin[\omega_c t + \phi(t)] \tag{104}$$

The alternate zero-crossings of the Hilbert transform $\hat{x}_{bp}(t)$ (excluding the common zeros of the band-pass signal and its Hilbert transform, which is related to the zeros of $A(t)$) are the peaks of $f(t)$ as given in (104). Once the peaks are determined, we choose the sampling set $\{t_k\}$ from the maxima points to satisfy half the Nyquist density; i.e., we choose every N^{th} sample of the maxima points. In our simulations, we used hard-limiting and band-pass filtering technique of the first suggested method to generate $f(t)$; the nonuniform samples are at peaks of $f(t)$. This technique is not as accurate as Hilbert transform technique discussed before.

In [49], we presented simulation results for finding an appropriate non-uniform sampling set and discussed recovery methods for band-pass signals from nonuniform samples using iterative methods—see the Mathcad(TM) simulation in the CD-ROM, file: 'Chapter4\Mathcad\Bandpass.mcd' for finite polynomial at the Nyquist rate.

4.9.4. Conclusion

We have developed a general nonuniform sampling interpolation formula for bandpass signals at the Nyquist density—Theorem 3. Notice that in the definition of the Nyquist density, the positions of the samples are not taken into account. For storage purposes, an additional overhead is needed to store the position of samples in addition to the sample amplitudes. If the position of samples are included, the effective sampling rate for Theorem 3 is $3/2$ times the Nyquist rate. Using an intelligent method of sampling, we have also developed a class of nonuniform samples at half the Nyquist density (the effective sampling rate is equal to the Nyquist rate if the positions are taken into account) that would uniquely determine the original bandpass signal. The interpolation formula for this type of sampling is given in Theorem 4. Practical methods of sampling and recovery techniques at the Nyquist and half the Nyquist density have been suggested and simulated.

4.10. Extremum Sampling

4.10.1. Introduction

Zero-crossings of signals will be covered in Chapter 9. In this chapter, we shall discuss nonuniform samples at the signal maximum and minimum points, i.e., at the zero-crossings of the derivative of the signal; we shall call this sampling process "Extremum Sampling". Our approach would be mainly related to $\mathbb{C}^N$ signals that are band-limited in the Discrete Fourier Transform (DFT) sense. We first derive some new relationships and then go over simulation results.

4.10.2. Description of Extremum Sampling

Extremum sampling consists of the samples of the maximum and minimum of a real signal. Since the derivative of the samples are zero at the Extremum points, if the density of the samples are at least half the Nyquist rate, the recovery can be exact and stable; see Section 4.8 of this chapter. Since the position of Extremum points can be determined from the time of arrival (with an accuracy dependent on the channel bandwidth), the sampling rate is half the Nyquist rate. If we send the amplitude and time position of extremum points as uniform samples, the overall sampling rate would become comparable to the uniform sampling case. In the following, we will consider a signal that is either band-limited in the DFT or the analog Fourier transform sense; i.e., in $\mathbb{C}^N$ or in L_2, respectively. For the application of the Extremum sampling see [56]; the paper and the thesis are in the files: 'Chapter4\RezaThesis.zip in the CD-ROM'.

4.10.3. Discrete Finite Dimensional $\mathbb{C}^N$ Signals

Let us assume that, in a *frame* size of T seconds, we have a set of N real samples $\{\boldsymbol{x} : x_m,\ m = 0, \ldots, N-1\}$ that is a low-pass signal in the DFT sense. This implies that the first K samples and the last $K-1$ samples in the DFT domain are non-zero and the other coefficients are zero. Now, we will show that this set of N samples is derived from an exponential polynomial of order $2K-2$. This implies that $2K-1$ samples of the set $\boldsymbol{x}$ would be sufficient to reconstruct all the N samples of $\boldsymbol{x}$. The $2K-1$ samples could be uniform or nonuniform. If the nonuniform samples happen to be at the Extremum points, only K Extremum samples would be sufficient for the recovery of the set $\boldsymbol{x}$. These interpolation formulae will be derived below.

4.10.3.1 Polynomial Representation

Since $\boldsymbol{x}$ is real, its DFT $\boldsymbol{X}$ has Hermitian symmetry, i.e., $X_i = X^*_{N-i}$ for $i = 1, \ldots, N-1$. Since $\boldsymbol{x}$ is a low-pass signal, we can write

$$x_m = \sum_{i=0}^{K-1} X_i w_m^i + \sum_{i=1}^{K-1} X_{N-i} w_m^{N-i}, \text{ where } w_m = \exp\left(\frac{2\pi mj}{N}\right), \ldots, m = 0, \ldots, N-1. \tag{105}$$

Because of the Hermitian symmetry, we can simplify the above equation and get

$$x_m = \sum_{i=0}^{K-1} X_i w_m^i + \sum_{i=1}^{K-1} X_i^* w_m^{N-i}$$

or

$$x_m = X_0 + \sum_{i=1}^{K-1} (X_i w_m^i + X_i^* w_m^{-i}), \tag{106a}$$

$$w_m^{K-1} x_m = X_0 w_m^{K-1} + \sum_{i=1}^{K-1} (X_i w_m^{K-1+i} + X_i^* w_m^{K-1-i}). \tag{106b}$$

Equation (106) shows that modified samples $w_m^{K-1} x_m$ are derived from a polynomial with respect to w_m of degree $2K-2$; hence $2K-1$ samples should be sufficient to determine $\boldsymbol{x}$; this set of samples is denoted as $\boldsymbol{s}$. The sampling set $\boldsymbol{s}$ could be uniform or nonuniform. If the nonuniform samples happen to be at the Extremum points, then K samples would be sufficient to recover $\boldsymbol{x}$. In general, the band-limited finite dimensional $\mathbb{C}^N$ function $\boldsymbol{x}$ can be reconstructed from the set $\boldsymbol{s}$ by using various methods such as Matrix algebra, Lagrange interpolation, and Iterative methods. Below we shall describe some of these methods.

4.10.4. Lagrange Interpolation (LI)

The Lagrange interpolation for the exponential polynomial represented in (106) is

$$x_k = \frac{1}{w_k^{K-1}} \sum_{i=0}^{2K-1} x_{s_i} (w_{s_i})^{K-1} \psi_{i,k}, \tag{107a}$$

where

$$\psi_{i,k} = \frac{\prod_j (w_k - w_{s_j})}{\prod_{j \neq i} (w_{s_i} - w_{s_j})(w_k - w_{s_i})}, \tag{107b}$$

and w_k's are given in (105) and s_i's are the nonuniform sampling positions.

The Lagrange interpolation for special sets $\boldsymbol{s}$ are depicted in Figs. 18–20. $\boldsymbol{s}$ could be a set of uniform samples—Fig. 18, random samples—Fig. 19, and Extremum samples—Fig. 20. Figures 18–20 represent frames of $N = 32$ samples of a low-pass random source $\boldsymbol{x}$. The bandwidth for this 32 point frame is $K = 4$; thus the resolution is about 4 times the Nyquist rate (see Section 4.10.2 for definitions). Figures 18(a), 19(a), and 20(a) represent the Lagrange interpolation when the samples $\boldsymbol{s}$ are at the Nyquist rate ($2K - 1 = 7$ samples per frame) while Figs. 18(b), 19(b) and 20(b) represent the case when the rate is slightly more than

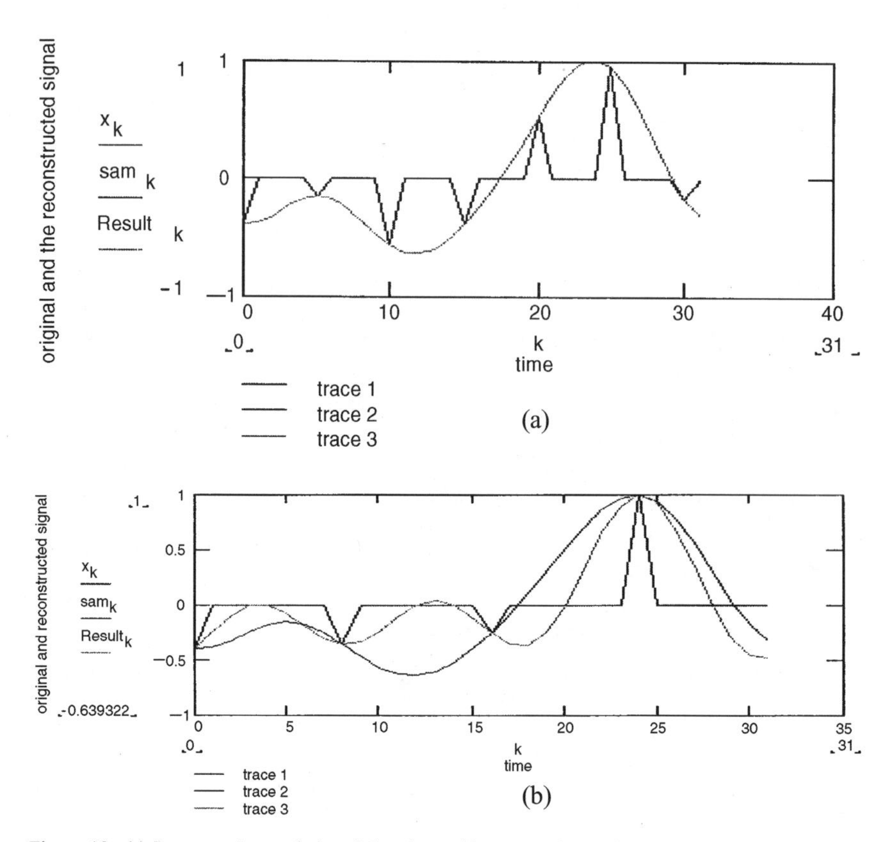

Figure 18. (a) Lagrange Interpolation (LI) using uniform samples at the Nyquist rate. (b) Lagrange Interpolation (LI) using uniform samples at about half the Nyquist rate.

half the Nyquist rate (4 samples per frame). Figures 18(a)–20(a) show that Lagrange interpolation at the Nyquist rate yields perfect recovery (correlation[10] is 1) but at almost half the Nyquist rate (4/7), recovery is not exact. The best performance, as expected, is the reconstruction from Extremum samples shown in Fig. 20(a). The Mathcad(TM) simulation for this technique is given in the accompanying CD-ROM under the file: 'Chapter4\Mathcad\Lagrange.mcd'.

In the next two sections we will show how the extra information of Extremum samples (their derivatives are zero) can be used in improving the reconstruction.

[10] Correlation is defined as: $\dfrac{\sum_i (x_i - mean(x))(y_i - mean(y))}{\sqrt{\sum_i (x_i - mean(x))^2 \sum_i (y_i - mean(y))^2}}$

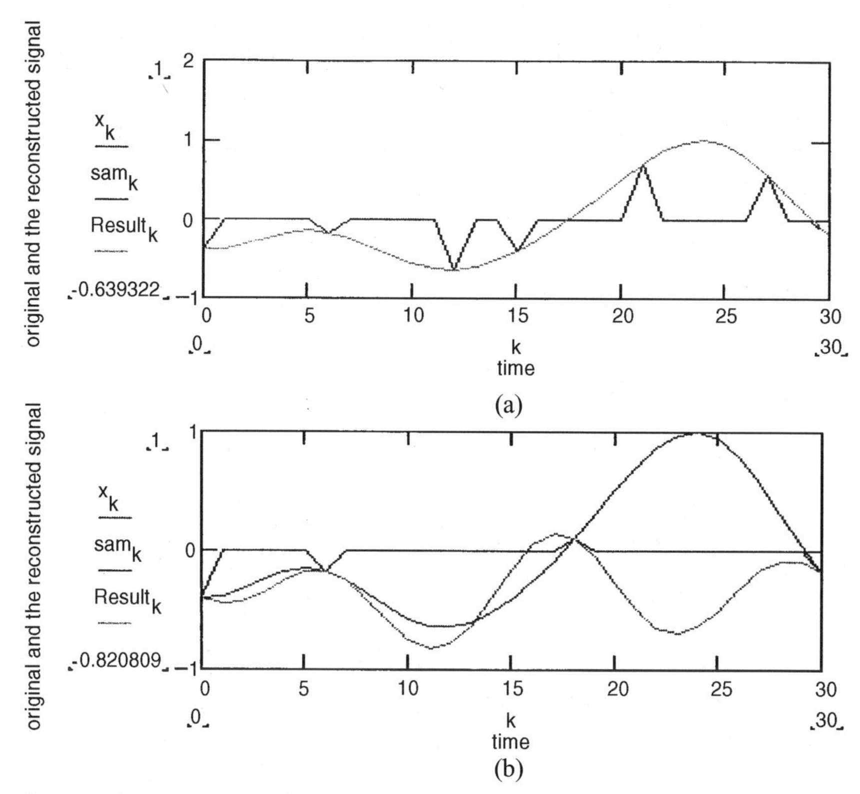

Figure 19. (a) Lagrange Interpolation (LI) using random samples at the Nyquist rate. (b) Lagrange Interpolation (LI) using random samples at about half the Nyquist rate.

4.10.5. Matrix Approximation for the Extremum Reconstruction (MAER)

The interpolation in (107) does not take into consideration that the derivative of the Extremum points are zero. Since the bandwidth of $\boldsymbol{x}$ is K and due to the Hermitian symmetry, there are only K unknown coefficients in the DFT domain: one real X_0, and $K-1$ complex ones $X_1, X_2, \ldots, X_{K-1}$. Therefore, we need $2K-1$ equations to determine these K coefficients. A set of K Extremum points yields $2K$ equations and hence the set is more than sufficient to solve for the K coefficients. From (106a), the K Extremum points can be written as

$$x_k = X_0 + \sum_{i=1}^{K-1} (X_i w_k^i + X_i^* w_k^{-i}), \ x_k \in s. \tag{108a}$$

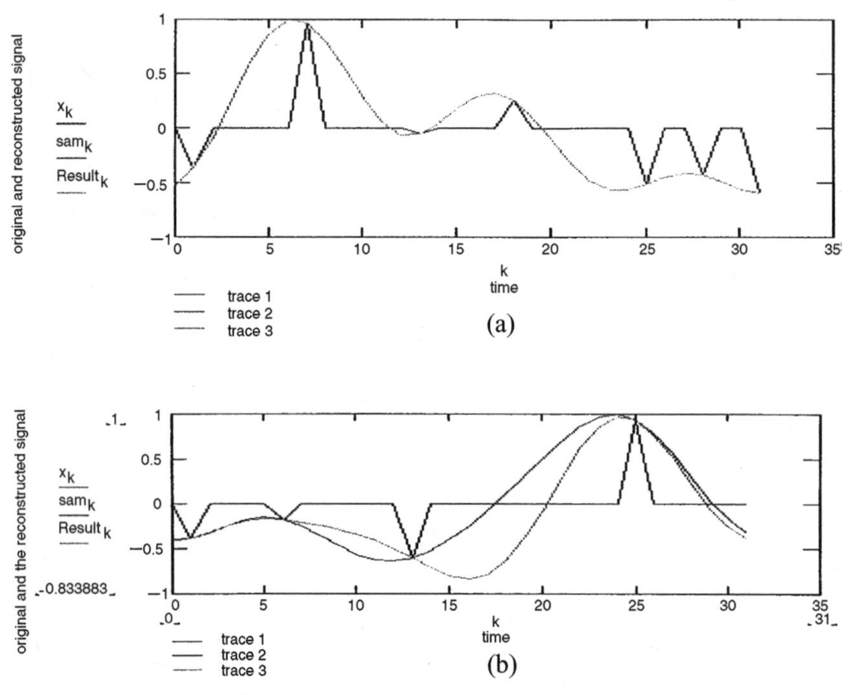

Figure 20. (a) Lagrange Interpolation (LI) using Extremum samples at the Nyquist rate. (b) Lagrange Interpolation (LI) using Extremum samples at about half the Nyquist rate.

The Extremum points imply that

$$x_k - x_{k-1} \approx 0, \ x_k \in s. \tag{108b}$$

The above relations (108) yield $2K$ equations; the DFT coefficients are then determined from $2K - 1$ of these equations. The result for $K = 4$ is shown in Fig. 21. The correlation in this case is 0.993, which compared to that of Fig. 20(b) is definitely an improvement.

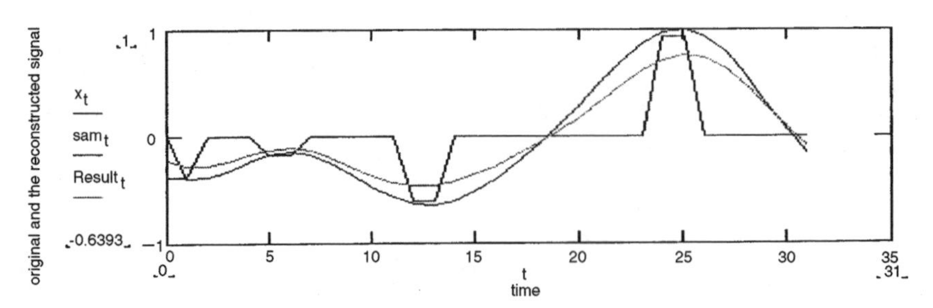

Figure 21. Reconstruction from Extremum samples using Matrix Approximation—MAER.

The Mathcad(TM) simulation for this technique is given in the accompanying CD-ROM under the file: 'Chapter4\Mathcad\MAER.mcd'.

4.10.6. Modified Lagrange with Derivative Sampling (MLDS)

The modification is based on the fact that the Extremum samples convey twice the amount of information than uniform or random samples. The extra information lies on the fact that the derivatives at the Extremum points are zero. Hence we can use the derivative sampling as derived by Rawn [50]–[51] and discussed in Chapter 3 of this book; see also (115)–(120) of this chapter. The modified interpolation formula when we have both the Extremum samples and their derivatives can be shown to be

$$Result_k = w_k^{1-N} \sum_i x_{s_i} w_{s_i}^{N-1} \psi_i^2(k)\theta_i, \text{ where}$$

$$\theta_i = \left[1 + 2(w_{s_i} - w_k)\frac{d}{dk}\psi_i(s_i) - (N-1)(w_{s_i} - w_k)w_{s_i}^{-1}\right], \tag{109}$$

where the parameters in (109) are as defined in (105) and (107). The derivation of the above formula is straightforward from the analog version for L_2 signals developed in Chapter 3.

The Mathcad simulation of (109) is given in Fig. 22. Fig. 22(a) is the usual Lagrange interpolation (107). The correlation for the particular signal as shown in the figure is 0.974. Figure 22(b) is the simulation of (109); the correlation is 0.996. In two other experiments with $K = 4$ Extremum points but two different random signals, the LI yields a correlation of 0.911 and 0.943 but the MLDS technique yields a correlation of 0.993 and 0.999, respectively.

Theoretically, the MLDS method should give perfect results since 4 extremum points are equivalent to 8 random or uniform samples which is above the Nyquist rate ($2K - 1 = 7$ samples per frame). Indeed if we increase the time resolution from 128 points to 512 points per frame (equivalent to increasing the resolution from 16 times up to 64 times the Nyquist rate), the correlation of the MLDS always becomes one. In the next section we shall discuss the discrete version of the MLDS, which even yields better results.

The Mathcad(TM) simulation for this technique is given in the accompanying CD-ROM under the file: 'Chapter4\Mathcad\MLDS.mcd'.

4.10.7. Discrete Lagrange Interpolation (DLI)

The discrete version of differential equations is difference equations. Indeed, the computer simulation for Extremum sampling is performed by finding the difference among the adjacent samples of the signal and then sampling the

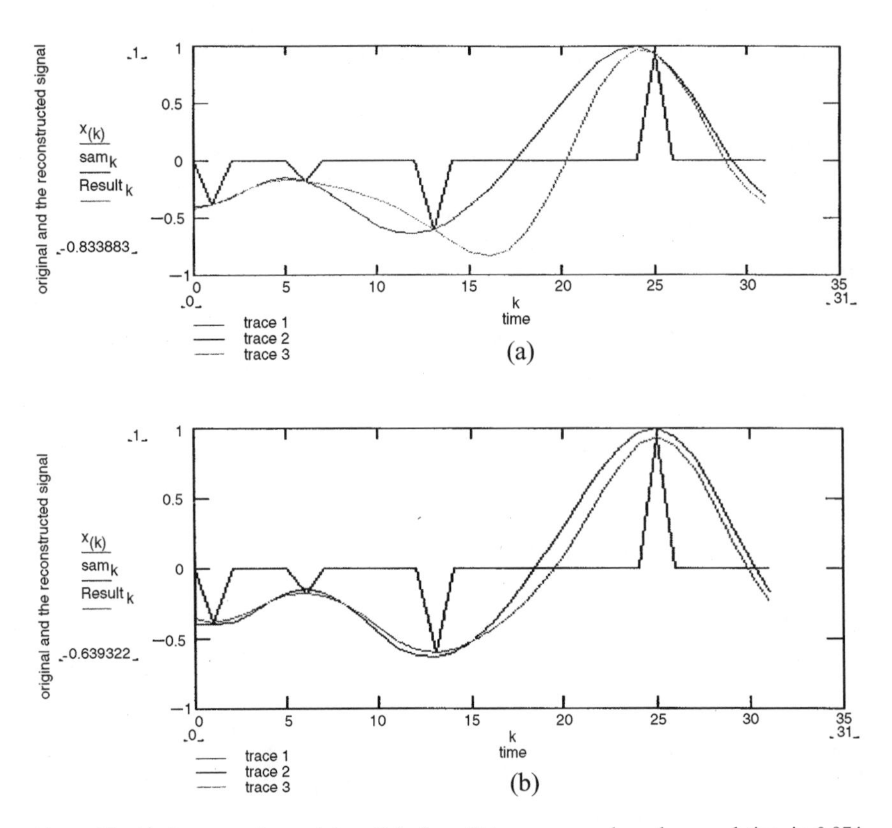

Figure 22. (a) Lagrange Interpolation (LI) from Extremum samples—the correlation is 0.974. (b) Reconstruction based on MLDS from Extremum samples—correlation is 0.996.

original signal when the difference signal is almost zero (when the difference signal goes through a sign change). If we assume that the difference signal is almost zero—which is true if the sampling interval is very close to each other—when there is an Extremum point, the two adjacent samples at the Extremum point are almost the same. Therefore, the knowledge of Extremum points imply that the previous samples right before the Extremum points have almost the same amplitude as the Extremum points.

For the example of our simulation, the knowledge of $K = 4$ Extremum points determines $2K = 8$ samples and hence our frame can be recovered using Discrete Lagrange Interpolation (DLI). Equation (107) is therefore modified to incorporate the position of the adjacent samples with the same amplitude as the Extremum points. The results as expected are excellent. In fact, better than any other method we have discussed so far. The simulation results are shown in Fig. 23. Figure 23(a) shows the result using the DLI method. For the given resolution

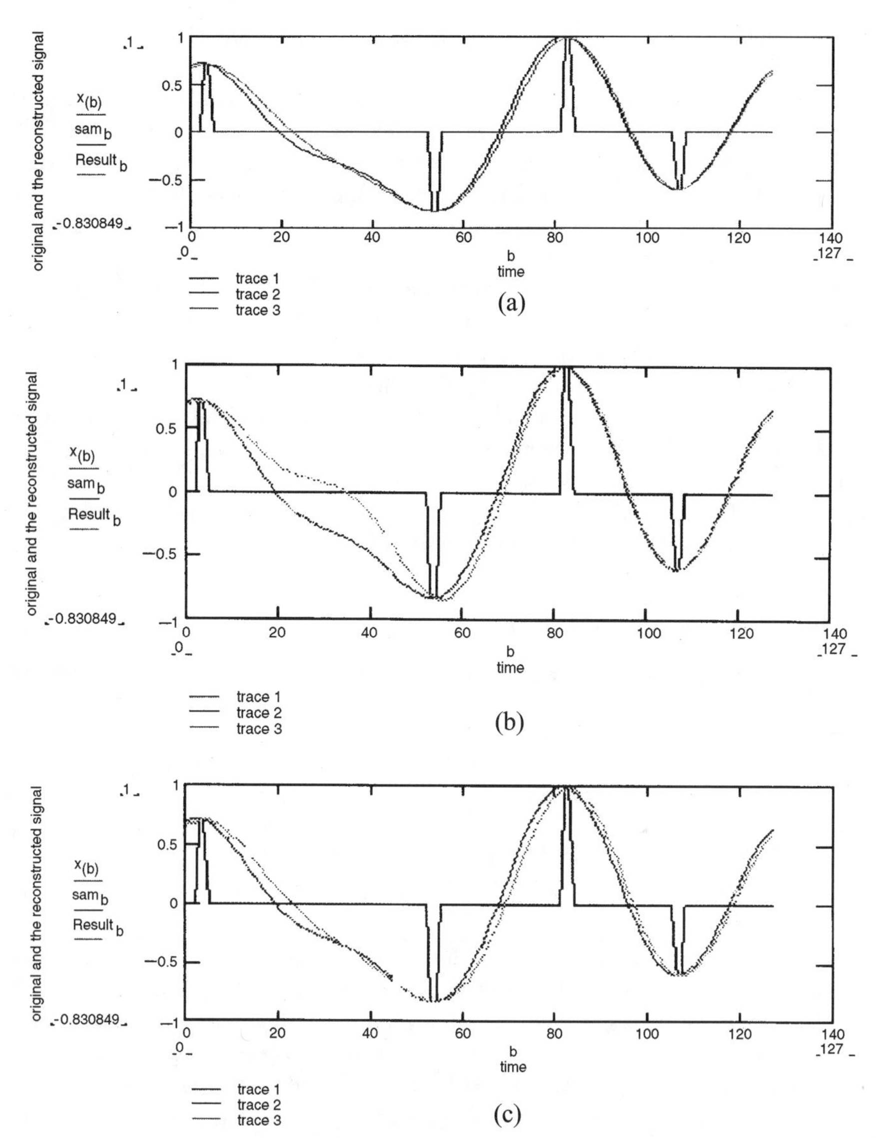

Figure 23. (a) Discrete Lagrange Interpolation (DLI) for Extremum sampling—the correlation is 0.994. (b) Simple Lagrange Interpolation (LI) for Extremum sampling—the correlation is 0.964. (c) MLDS for Extremum sampling—the correlation is 0.988.

(128 samples per frame equivalent to about 16 times the Nyquist rate), the correlation is 0.994.

Clearly, this method outperforms the LI, Fig. 23(b), and MLDS, Fig. 23(c), with correlations 0.954 and 0.988, respectively. We can further improve the DLI method by taking the correlation of the adjacent samples into account. This implies that instead of keeping the previous samples the same as the Extremum points, we divide the Extremum points by the correlation factor determined by the autocorrelation function.

If the resolution of the frame increases, the correlation of adjacent samples will approach one and there is no need to take the correlation factor into consideration. In fact, at high resolutions (greater than 32 times the Nyquist rate), the DLI and MLDS methods yield the same high quality performance. However, in terms of computation, the computational complexity of DLI is much less than that of the MLDS method. In the next section we shall discuss iterative methods for signal recovery for Extremum sampling.

The Mathcad(TM) simulation for this technique is given in the accompanying CD-ROM under the file: 'Chapter4\Mathcad\DLI.mcd'.

4.10.8. Iterative Methods (IM)

We can use iterative methods for the recovery of signals from nonuniform samples as discussed before in Section 4.6.2 (see also Chapter 5 and [31]–[34]). The IM is summarized in the formula below.

$$\begin{aligned} iteration^{\langle 0\rangle} &= LPSAMPLE \\ iteration^{\langle m\rangle} &= iteration^{\langle m-1\rangle} + \lambda \cdot iteration^{\langle 0\rangle} \\ &\quad - \lambda \cdot IFFT(FFT(iteration^{\langle m-1\rangle} \cdot delta) \cdot LP) \end{aligned} \tag{110}$$

The above iterative equation that is the discrete version of the iterative method that is sometimes called Marvasti/Wiley or the frame method (Chapter 6 and [33]) consists of updating the previous iteration—the first term in the right hand side of (110)—with the first guess—the second term in (110)—(low-pass filtering the extremum samples—LPSAMPLE) and with the low-pass filter of the Extremum samples of the previous iteration—the last term in (110). In the above equation *delta* is the position of extremum points, LP is a low-pass filtering operation, and λ is the relaxation parameter that determines the speed of convergence. In the following experimental results, we assume λ is equal to 1.

Instead of LI in Fig. 20(b) with a correlation of 0.916, we can use DFT low-pass filtering, the correlation is 0.93. After 20 iterations, the correlation becomes 0.985. Clearly, low-pass filtering and the iterative technique performs better than the Lagrange interpolation when the extremum is at about half the Nyquist rate.

However, MAER (Section 4.10.5) method yields a correlation of 0.993, and the MLDS (Section 4.10.6) and DLI (Section 4.10.7) yield a correlation of 1. On the other hand, at the Nyquist rate such as the example of Fig. 20(a), the LI yields a correlation of 1 while the correlation of DFT low-pass filtering for the example of Fig. 20(a) is 0.903 and after 20 iterations it becomes 0.991.

The Mathcad™ simulation for this technique is given in the accompanying CD-ROM under the file: 'Chapter4\Mathcad\Iteration.mcd'; its comparison to Lagrange is in the file: 'Chapter4\Mathcad\Lagrange&Iteration.mcd'.

4.10.9. Spline Interpolation (lspline)

In this section we wish to compare classical techniques such as splines and other polynomial interpolations to the previous techniques such as Lagrange (LI), modified Lagrange (MLDS), Discrete Lagrange (DLI), and Iterative methods (IM). For spline interpolations, we will use the Signal Processing Tool of Mathcad. This package has spline curves which are either cubic, parabolic, or linear at the endpoints. The spline with linear end points, *lspline*, yields the best result. If the Extremum points plus the first and last points per frame of 256 points are used for comparison, we get Fig. 24. The *lspline* algorithm performs poorly if the end points are not fixed. Figure 24(a) is the comparison of LI with *lspline*; while Fig. 24(b) is the comparison of *lspline* with LI and MLDS. In general, the previous techniques such as LI, DLI and MLDS yield better results than the lspline method.

The Mathcad™ simulation for this technique is given in the accompanying CD-ROM under the file: 'Chapter4\Mathcad\Spline.mcd'.

4.10.10. Approximations with Other Interpolating Kernels and Polynomials

The most obvious interpolation is low-pass filtering in the DFT domain, which is equivalent to a $\sin(\pi k)/\sin(\pi k/N)$ interpolation, where $k = 0,\ 1,\ N-1$ and N is the block size. In comparison to Fig. 25, the resultant correlation is 0.947, which as expected is not as impressive as the other methods. Another interpolation is a $\text{sinc}(\cdot)$ interpolation; the best sinc yields a correlation of 0.937, which is slightly worse than the DFT low-pass filtering.

Just for the sake of comparison, if low-pass filtering after Sample-and-Hold (S&H–Fig. 25(a)) is performed, a correlation of 0.828 is achieved if proper delay is used to get the best correlation. To get better than S&H, a Balanced S&H (B-S&H) can be used as shown in Fig. 25(b); the correlation in this case is 0.855 without any delay and 0.888 with 2 sample delays. Linear Interpolation (LIN) is shown in Fig. 25(c); the correlation, in this case is 0.954. Note that for S&H and LIN, the beginning and the end points of the frame are not accurate and if these points are excluded, better correlation is achieved. A summary of the results are shown in Table 5. If we assume the adjacent samples of Extremum points are the

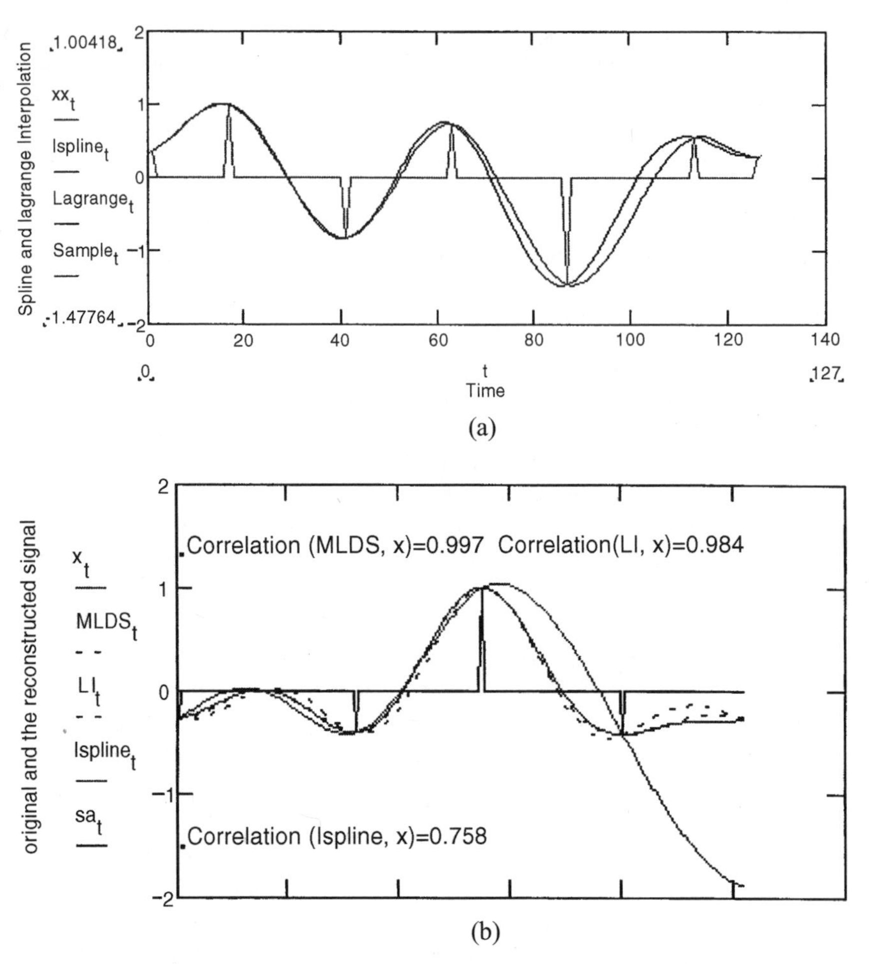

Figure 24. (a) Comparison of spline (correlation is 0.964) and Lagrange (correlation is 1) interpolation for Extremum samples. (b) Comparison of different methods for Extremum reconstruction.

Table 5. Comparison of Correlations of Various Ad-hoc Methods for Recovery from Extremum Samples.

S&H	B-S&H	SINC Interpol.	Lpsample=DFT Filtering	LIN
0.828	0.855–0.888	0.937	0.947	0.954

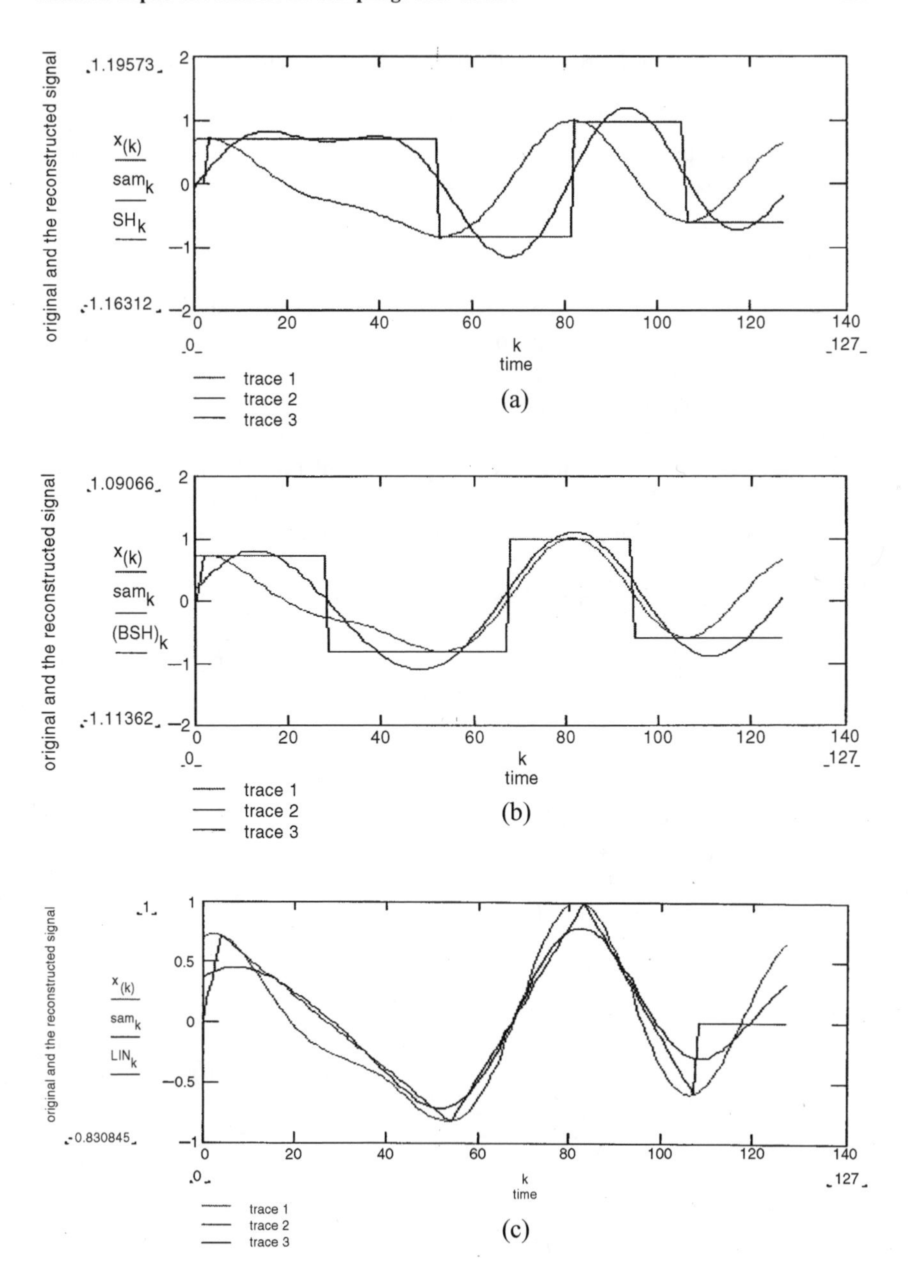

Figure 25. (a) S&H reconstruction; correlation is 0.828 with 14 sample delays. (b) B-S&H reconstruction; correlation is 0.855 with no delays but 0.888 with 2 sample delays. (c) LIN reconstruction; correlation is 0.954 with no delays.

same as the Extremum points, similar to the assumption of DLI, the correlation results in the Table 5 will be improved.

The Mathcad® simulation for this technique is given in the accompanying CD-ROM under the file: 'Chapter4\Mathcad\S&H.mcd'.

4.10.11. Bipolar Sampling

Extremum sampling is very flexible in the sense that the position of the samples convey the zero-crossings of the derivative. The analysis, similar to that of PPM and FM discussed in Chapter 16, shows that it is possible to transmit the pulse position at the Extremum points and get a fairly good replica of the original signal. If we represent the maxima as a unit positive pulse and minima as a unit negative pulse (Bipolar Sampling), we can get a correlation of 0.701 if we use the DFT low-pass filter for the bipolar samples. The results are shown in Fig. 26(a). This correlation is not impressive compared to the sinc interpolation of Table 5.

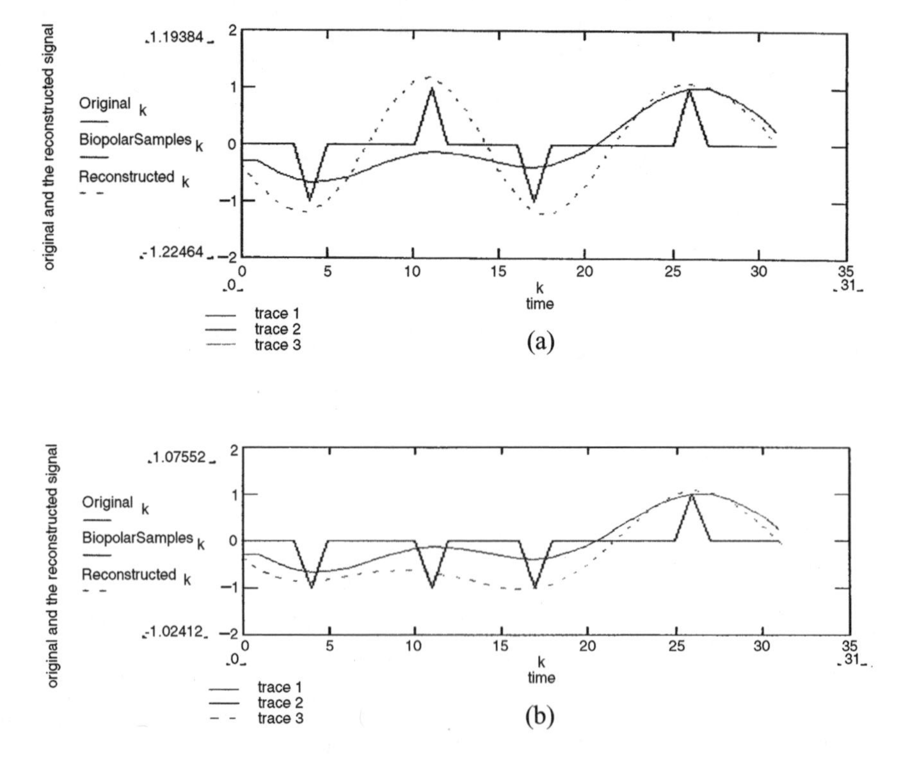

Figure 26. (a) Reconstruction from Bipolar Samples—maxima and minima points are positive or negative, respectively—correlation is 0.701. (b) Reconstruction from Bipolar Samples—extremum points are positive or negative depending on the sign of the extremum samples.

However, if the extremum points happen on the positive or negative side—Fig. 26(b), the situation would be different; the positive and negative pulses in Fig. 26(b) yield a correlation of 0.963, which is much better than the one in Fig. 26(a). For the sake of comparison, the correlation of DFT low-pass filtering of Extremum samples is 0.97, assuming the same signal as the one in Fig. 26(b). The Mathcad™ simulation for this technique is given in the accompanying CD-ROM under the file: 'Chapter4\Mathcad\Bipolar.mcd'.

4.10.12. Extremum and Zero-Crossing Positions

In this section we discuss the interpolation of a DFT band-limited signal from the position of the zero-crossings of the signal and its derivative. We shall show how the Extremum amplitudes can be derived from these zero-crossings and hence we can use Lagrange interpolation to recover the original signal.

Initially, we assume the Extremum samples and the zero-crossings of a signal are known and then we try to reconstruct the signal from the known samples. The composite nonuniform samples is the combination of the Extremum samples and the zero-crossing samples with amplitude zero. Hence the Lagrange interpolation is the same as (107) except that the samples x_i are the Extremum points but the interpolating function $\Psi_{i,k}$ consists of time positions of the zero-crossings and the Extremum points. The result is shown in Fig. 27(a) with perfect reconstruction for a time resolution of 128 points per frame, i.e., 16 times the Nyquist rate. In general, for band-limited signals such as speech, the rate of zero-crossings is slightly above half the Nyquist rate [1], and since there are more Extremum points than the zero-crossings, the total rate is always greater than the Nyquist rate.

The Mathcad™ simulation for this technique is given in the accompanying CD-ROM under the file: 'Chapter4\Mathcad\Extrema+ZC+Lagrange.mcd'.

If we only send the position of the Extremum and the zero-crossings, we should be able to reconstruct the original signal given the extra information that samples of the Extremum points have zero derivative. The Lagrange interpolation is as follows:

$$Result_k = \sum_{i=0}^{2K-2} SA_i \phi_{i,k}, \tag{111a}$$

where according to (107)

$$SA_i = x_{s_i} w_{s_i}^{N-1}, \quad \phi_{i,k} = \psi_{i,k} w_k^{-N+1}. \tag{111b}$$

Since the Extremum positions are known, from (111) we can differentiate the signal at the Extremum points to get the modified samples SA_i, i.e.,

$$\sum_{i \in S} SA_i \left(\frac{d}{dk} \phi_{i,k} \right)_{k=a_i} = 0, \tag{112}$$

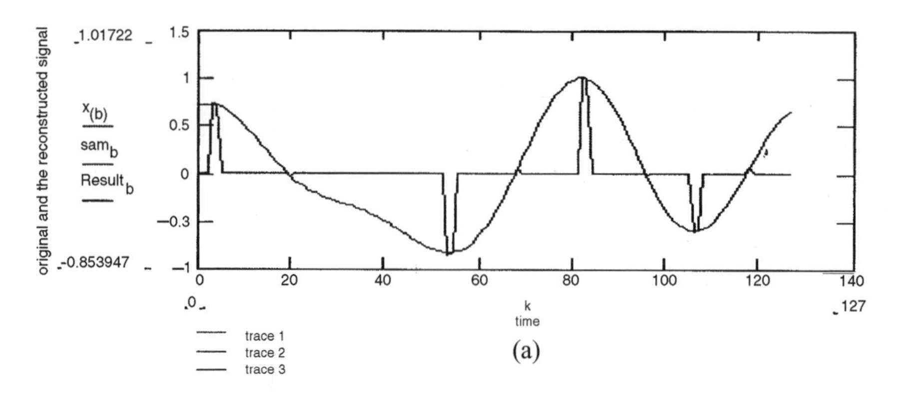

(a)

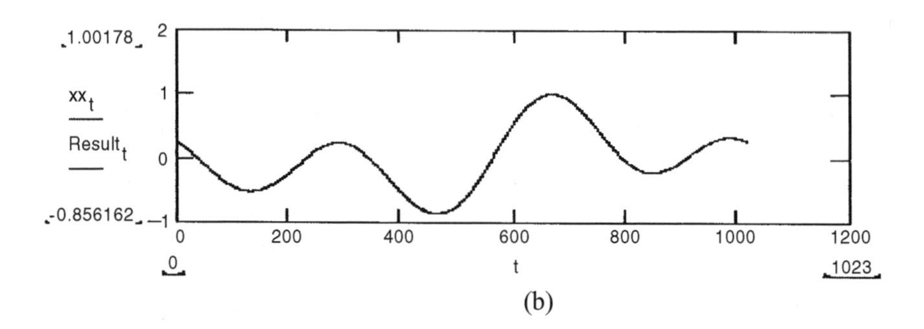

(b)

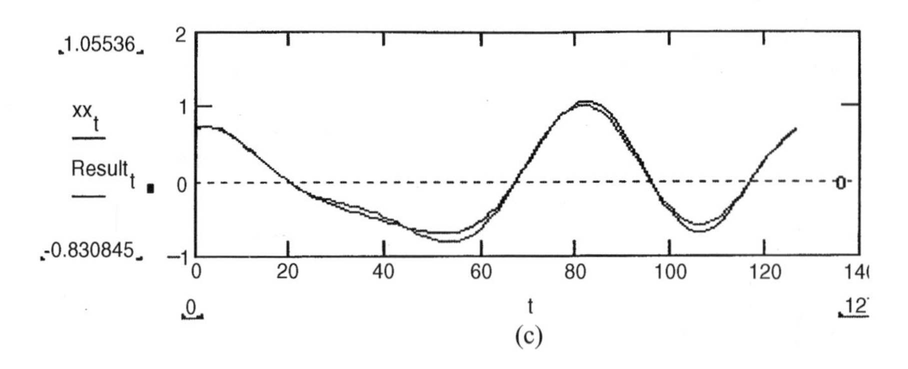

(c)

Figure 27. (a) Perfect recovery of a band-limited signal from amplitudes and positions of the extremum points and the zero-crossings of the signal using LI (107); time resolution is 128 points per frame. (b) Recovery of a band-limited signal from positions of the zero-crossings of the signal and its derivatives using Equations (112)–(113); time resolution is 1024 points per frame. (c) Recovery of a band-limited signal from positions of the zero-crossings of the signal and its derivative using (114); time resolution is 128 points per frame—the correlation is 0.975.

where $\boldsymbol{S}$ is the set of Extremum positions. Assuming any value for the first Extremum sample SA_0, we can solve for the remaining Extremum amplitudes using a set of simultaneous equations; the result is

$$SA = M^{-1}C \tag{113}$$

where

$$C_n = \frac{d}{dk}\phi_{0,k}|_{k=s_{n+1}} \cdot SA_m, \; M_{n,m} = \frac{d}{dk}\phi_{m,k}|_{k=s_{n+1}} \; for \; n = 0 \ldots E-1,$$
$$m = 0 \ldots E-1,$$

where E is the total number of Extremum points in the set $\boldsymbol{S}$. After the modified Extremum samples, $SA_{i'}$'s, are found, we can use (111) for the interpolation. The result for a time resolution of 1024 points per frame, i.e., 128 times the Nyquist rare is shown in Fig. 27(b). For lower resolutions, the zero-crossing and the Extremum positions do not accurately represent the actual positions of the signal; hence severe distortions can occur at lower sampling point resolutions.

A discrete interpolation, which depends on taking the difference of the sample points as opposed to differentiation, would be more relevant when the zero-crossings are determined while the samples have a sign change similar to DLI described in Section 4.10.7. Equations (112) and (113) can be modified by replacing the derivatives by difference equations, i.e.,

$$C_n = (\phi_{0,s_{n+1}} - \phi_{0,s_{n+1}-1})SA_0, \; M_{n,m} = \phi_{m,s_{n+1}} - \phi_{m,s_{n+1}-1}. \tag{114}$$

The simulation is shown in Fig. 27(c). In general the higher the resolution, the better the correlation of the interpolation formula. Because of taking the inverse matrix, (113)–(114) may not be stable and it is preferable to send the amplitudes of the Extremum as well and then use the Lagrange interpolation.

The Mathcad(TM) simulation for (113) and (114) are given in the accompanying CD-ROM under the files: 'Chapter4\Mathcad\Extrema+ZC-Positions1.mcd' and 'Chapter4\Mathcad\Extrema+ZC-Positions2.mcd', respectively.

4.10.13. Reed Solomon (RS) Decoding

A technique developed for error correction codes and extended to an over-sampled signal with missing samples can be used for Extremum sampling. The description of the RS decoding is given in Chapter 17 and [52]. In order to use the RS interpolation techniques, similar to the DLI method, we assume that the Extremum samples contain double information, i.e., the penultimate point prior to each Extreme sample has almost the same value as the Extremum samples. Using this extra information and using the RS interpolation (the code is in the CD-

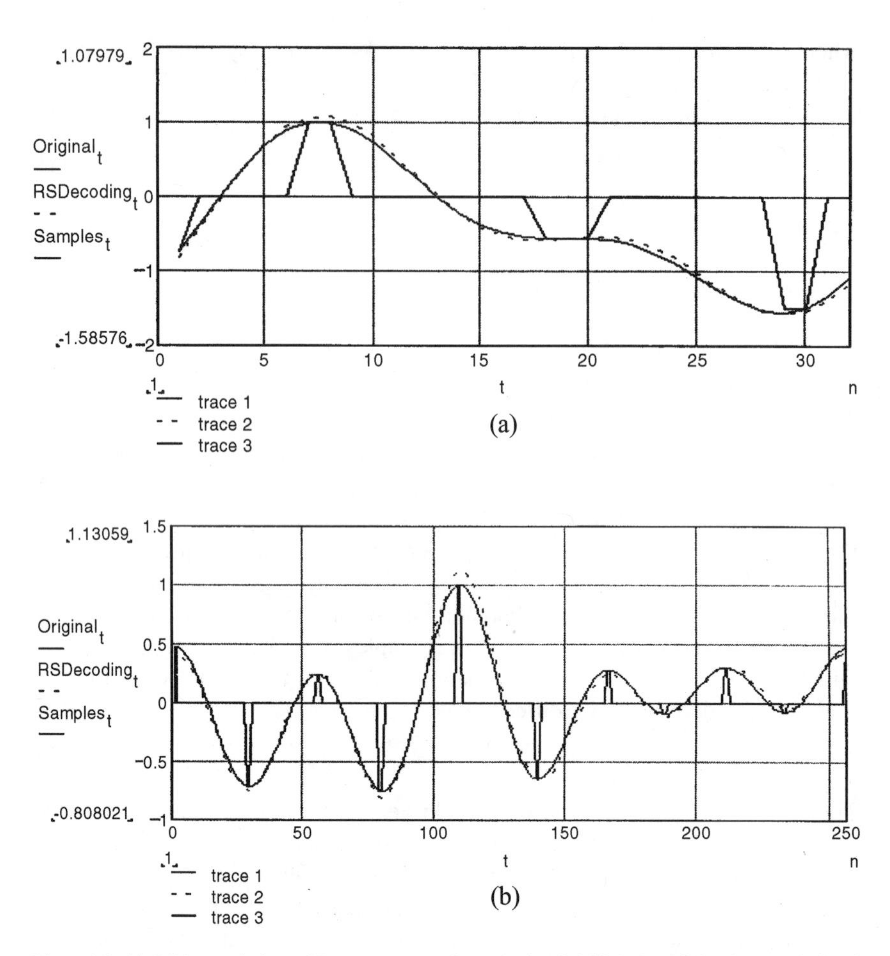

Figure 28. (a) RS interpolation of Extremum samples—the bandwidth is 3, and the time resolution is 32 points per frame—the correlation is 0.998. (b) RS interpolation of Extremum samples—the bandwidth is 6 and the time resolution is 256 points per frame—the correlation is 0.984.

ROM-file: 'Chapter4\Mathcad\RS-Extremum.mcd'), we get Fig. 28(a)–(b). As can be seen in this Fig. 28(a), the correlation for a particular random signal with a given bandwidth ($K = 3$) and a given frame resolution (32 points per frame, equivalent to about 5 times the Nyquist rate) is 0.998; which is comparable to DLI. Also we have shown in Fig. 28(b) another random signal with a higher bandwidth ($K = 6$) and time resolution (256 points per frame, equivalent to 31 times the Nyquist rate). The advantage of this method to DLI is the computational time. For a comparison of computational complexity and comparisons to Lagrange interpolation and other methods see [52].

4.10.14. Time Resolution

Resolution is defined as the number of computational points per frame of unit time. Alternatively, the number of points per Nyquist interval is a representation of its resolution. Hence, we can define resolution as

$$\text{Resolution} = \frac{N}{2KT},$$

where N is the number of points per frame size of T seconds, and K is the bandwidth. The resolution determines the accuracy of the actual Extremum samples, i.e., it is equivalent to the time quantization of the analog Extremum samples. In general, interpolation becomes more accurate at higher resolution at the expense of more computational time.

At high resolutions (more than 32 points per Nyquist interval), as expected, DLI and MDLS will approach the same performance with a correlation of close to 1. At low resolution (about 8 points per Nyquist interval), LI performs sometimes better and sometimes worse than MLDS but always worse than DLI. For example, the correlation for LI is 0.926, for MLDS is 0.883, and for DLI is 0.954. But the correlations in Fig. 22 imply that for higher resolution (64 points per Nyquist interval) MLDS is better than LI and slightly worse than DLI. In another experiment (resolution of 8 points per Nyquist interval), MLDS is better than LI but DLI is always better; correlations are, respectively, 0.795, 0.953 and 0.965. In general, the ranking of the three techniques in terms of correlation are DLI, MLDS and LI, respectively. A comparison of all the techniques are given in the Mathcad(TM) file: 'Chapter4\Mathcad\Extremum.mcd'.

4.10.15. L_2 Signals

If the sampling density of the extremum points of a band-limited L_2 signal is more than half the density—which is the case for Gaussian and speech signals—then a direct Lagrange interpolation can be derived from Rawn's formula [51] (see also Chapter 3—Theorem 11) for nonuniform samples and their derivatives; this interpolation is

$$x(t) = \sum_{n=-\infty}^{\infty} \left[x(t_n) + \left(1 - \frac{t}{t_n}\right) x_1(t_n) + \cdots + \left(1 - \frac{t}{t_n}\right)^{R-1} x_{R-1}(t_n) \right] [\psi_n(t)]^R, \tag{115}$$

where

$$x_k(t_n) = \frac{(-1)^k t_n^k}{k!} \frac{d^k}{dt^k} \left[\frac{x(t)}{[\psi_n(t)]^R} \right] \Bigg|_{t=t_n}, \quad 0 \le k \le R-1. \tag{116}$$

For Extremum sampling, $R = 2$; hence the above equation reduces to

$$x(t) = \sum_{n=-\infty}^{\infty} \left[x(t_n) + \left(1 - \frac{t}{t_n}\right) x_1(t_n) \right] [\psi_n(t)]^2, \tag{117}$$

where

$$x_1(t_n) = (-1) t_n \frac{d}{dt} \left[\frac{x(t)}{[\psi_n(t)]^2} \right] \Bigg|_{t=t_n}. \tag{118}$$

After simplification, (118) becomes

$$x_1(t_k) = 2 t_n x(t_n) \psi(t_n). \tag{119}$$

Therefore, (117) is simplified to

$$x(t) = \sum_{n=-\infty}^{\infty} x(t_n)[1 + 2(t_n - t)\psi(t_n)][\psi_n(t)]^2, \tag{120}$$

where it can be shown that

$$\psi_n(t_n) = \sum_{i \neq n} \frac{1}{t_n - t_i}.$$

Equation (120) is the L_2 Lagrange interpolation for Extremum sampling, the analog version of the $\mathbb{C}^N$ case given in (109).

4.10.16. Comparison of Extremum Sampling to the Uniform Sampling

Due to circular aliasing effect, low-pass filtering the uniform samples in the DFT domain will not yield the actual signal but rather an approximation due to aliasing distortion. see the code in the CD-ROM, file: 'Chapter4\Mathcad\Uniform.mcd'. On the other hand, if we use Lagrange interpolation or RS decoding method on the uniform samples, independent of the resolution per frame, the reconstruction is perfect as long as the sampling rate is equal to or above the Nyquist rate $(2K - 1)$.

For above example, for a specific realizations of a random signal with a bandwidth of $K = 4$, the correlation of the reconstruction using Lagrange and DFT low-pass filtering is given in Table 6. Table 6 shows that Lagrange interpolation (LI) always yields perfect recovery at or above the Nyquist rate; even at the rates lower than the Nyquist rate, the LI is better except at 4 samples, i.e., almost half the Nyquist rate. Note that the correlations for the DFT low-pass filtering also depends on the position of uniform samples; Table 6 shows the worst case.

Table 6. The Correlation of Lagrange and DFT Low-pass Filtering of Uniform Samples

No. of Samples	Lagrange Interpolation	DFT Low-Pass Filtering
4	0.521	0.73
5	0.826	0.698
6	0.917	0.802
7* Nyquist Rate	1	0.86
8	1	0.953
9	1	0.984
10	1	0.986
13	1	0.999
17	1	1

In comparison to Extremum sampling, Lagrange interpolation works for both uniform and Extremum sampling as long as the Nyquist rate is satisfied for each case. The sampling density of Extremum samples is half as much as the uniform samples if the position of nonuniform samples are not included. We shall discuss this point when we consider modulation schemes such as PPM in Chapter 16. In some applications such as speech, coding the time positions of the Extremum samples may be advantageous in terms of data compression.

4.11. Random Sampling

A theoretical study of random sampling of stochastic processes is given in Chapter 8. Here we shall discuss some practical methods of recovery using iterative or other methods for the recovery based on [1] and [32].

Assume a wide-sense stationary band-limited random signal $\boldsymbol{x}(t)$ (or a deterministic signal) is sampled randomly with a point process $\boldsymbol{s}(t)$ according to a given probability distribution function such as Poisson or uniform distribution. We want to find out whether the stochastic process can be recovered from the random samples in the mean square sense. The set of random samples is represented by

$$x_s(t) = x(t)s(t). \tag{121}$$

Since the stochastic process $\boldsymbol{x}(t)$ is independent of the random samples $\boldsymbol{s}(t)$, the power spectral density of the random samples $\boldsymbol{x}_s(t)$ for Poisson or uniform distributed $\boldsymbol{s}(t)$ is given by [1]–[2]

$$P_{x_s}(f) = \lambda^2 P_x(f) + \lambda R_x(0), \tag{122}$$

where λ is the density of the random samples $s(t)$ per unit time, $P(f)$'s are the corresponding power spectral densities, and $R_x(0)$ is the auto-correlation function evaluated at zero, i.e., the total signal power.

The above equation implies that the power spectrum of random samples consists of a term proportional to the power spectrum of the original signal embedded in a white noise with a power spectral density of $\lambda R_x(0)$. If the random samples are filtered with an ideal low-pass filter of bandwidth equal to the bandwidth of $x(t)$, the Signal-to-Noise ration SNR is given by

$$\mathrm{SNR} = \lambda / 2W, \tag{123}$$

where W is the bandwidth of the original signal. If the average density of the sampling process λ is greater than the Nyquist rate ($2W$), the SNR is greater than 1 and iterative techniques that was discussed in Section 4.10.8 can recover the stochastic process. For more details, the reader is referred to [1]–[2], [26] and [32].

For Mathcad simulation of the iterative technique for random samples, see the code in the CD-ROM, file: 'Chapter4\Mathcad\Iteration.mcd'.

References

[1] F. Marvasti. *A Unified Approach to Zero-Crossings and Nonuniform Sampling of Single and Multidimensional Signals and Systems*. Nonuniform Publication, 5 Bayberry Drive, Princeton, NJ 08540, USA.

[2] F. Marvasti. Nonuniform Sampling. In R. J. Marks II, Ed., *Advanced Topics in Shannon Sampling and Interpolation Theory*, Springer Verlag, New York, 1993.

[3] A. Requicha. The Zeros of Entire Functions: Theory and Engineering Applications. *Proc. IEEE*, 68(3):308–328, 1980.

[4] N. Levinson. *Gap and Density Theorems*. Colloquium Publication 26, American Mathematical Society, New York, 1940.

[5] F. J. Beutler. Sampling Theorems and Bases in Hilbert Space. *Information and Control*, 4:97–117 1961.

[6] E. C. Titchmarsh. *The Theory of Functions* (2nd ed.). University Press, London, England, 1939.

[7] S. O. Rice. Mathematical Analysis of Random Noise. *Bell System Technical J.*, 23:282–332, 1944 and 24:46–156, 1945.

[8] F. J. Beutler. Error Free Recovery of Signals From Irregular Samples. *SIAM Rev.*, 8:322–335, July 1966.

[9] F. Marvasti. Reconstruction of Two-Dimensional Signals From Nonuniform Samples or Partial Information. *J. Opt. Soc. Am. (JOSA)*, Sec. A, 6:52–55, January 1989.

[10] F. Marvasti. Rebuttal on the Comment on the Properties of Two-Dimensional Bandlimited Signals. *J. Opt. Soc. Am. (JOSA)*, Sec. A, 6(9):1310, September 1989.

[11] J. L. C. Sanz. On the Reconstruction of Band-Limited Multidimensional Signals From Algebraic Sampling Contours. *Proc. IEEE*, 73(8):1334–1336, August 1985.

[12] D. H. Mugler and W. Splettstosser. Reconstruction of Two-Dimensional Signals from Irregularly Spaced Samples. In *Conf. Proc. 6th Aachen Symposium on Signal Theory*, Technische Hochschule Aachen, Lehrstuhl a fur Mathematik, September 1987.

[13] J. L. Yen. On Nonuniform Sampling of Bandwidth-Limited Signals. *IRE Trans. on Circuit Theory*, **CT-3**:251–257, December 1956.

[14] J. L. Yen. On the Synthesis of Line Sources and Infinite Strip Sources. *IRE Trans. Antennas and Propagation*, 40–46, January 1957.

[15] F. Marvasti. Extension of Lagrange Interpolation to 2-D Nonuniform Samples in Polar Coordinates. *IEEE Trans. on Circuits and Systems*, 37:(4), 567–568, April 1989.

[16] H. Stark. Polar, Spiral and Generalized Sampling and Interpolation. In R. J. Marks II, Ed., *Advanced Topics in Shannon Sampling and Interpolation Theory*, Springer Verlag, New York, 1993.

[17] F. Marvasti and A. K. Jain. Zero Crossings, Bandwidth Compression and Restoration of Nonlinearly Distorted Bandlimited Signals. *J. Opt. Soc. Am. (JOSA)*, Sec. A, 3(5):651–654, May 1986.

[18] A. Papoulis. Error Analysis in Sampling Theory. *Proc. IEEE*, 54:(7), 947–955, July 1966.

[19] J. J. Clark, M. R. Palmer, and P. D. Lawrence. A Transformation Method for the Reconstruction of Functions from Nonuniformly Spaced Samples. *IEEE Trans. Acoustic, Speech, and Signal Processing*, **ASSP-33**(4), October 1985.

[20] E. Yudilevich and H. Stark. Spiral Sampling: Theory and Application to Magnetic Resonance Imaging. *J. Opt. Soc. of Am. A*, 5:542–553, April 1988.

[21] F. Marvasti. Comments on a Note on the Predictability of Band-Limited Processes. *Proc. IEEE*, 74(11):1596, November 1986.

[22] H. L. Landau. Sampling Data Transmission and the Nyquist Rate. *Proc. IEEE*, 55:1701–1706, 1967.

[23] K. Yao and J. O. Thomas. On some Stability and Interpolatory Properties of Nonuniform Sampling Expansions. *IEEE Trans. Circuit Theory*, **CT-14**(4), December 1967.

[24] R. Duffin and A. C. Schaeffer. Some Properties of Functions of Exponential Type. *Bull. Amer. Math. Soc.*, 44:236–240, 1938.

[25] H. G. Feichtinger. Wiener Amalgam Spaces and some of their Applications. In *Conf. Proc. Function Spaces*, Edwardsville, IL, Marcel Dekker, April 1990.

[26] F. Marvasti. Spectral Analysis of Irregular Samples of Multidimensional Signals. *Sixth Workshop of Multidimensional Signal Processing*. Pacific Grove, CA, September 1989.

[27] F. Marvasti, C. Liu, and G. Adams. Analysis and Recovery of Multidimensional Signals from Irregular Samples using Nonlinear and Iterative Techniques. *Signal Processing, North Holland*, 36:(1), 13–30, March 1994.

[28] F. Marvasti, P. Clarkson, M. Dokic, and L. Chuande. Reconstruction of Speech Signals from Lost Samples. *IEEE Trans. on Acoustic, Speech, and Signal Processing*, 40(12):2897–2903, December 1992.

[29] R. G. Wiley. Recovery of Bandlimited Signals from Unequally Spaced Samples. *IEEE Trans. Commun.*, **COM-26**(1):135–137, January 1978.

[30] I. W. Sandberg. On the Properties of some Systems that Distort Signals—I. *Bell Syst. Tech. J.*, 2003–2046, September 1963.

[31] F. Marvasti, M. Analoui, and M. Gamshadzahi. Recovery of Signals from Nonuniform Samples using Iterative Methods. *IEEE Trans. ASSP*, 39:872–878, April 1991.

[32] F. Marvasti. Spectral Analysis of Random Sampling and Error Free Recovery by an Iterative Method. *Trans. of IECE of Japan*, **E69**(2):79–82, February 1986.

[33] H. Feichtinger and C. Grochenig. Nonuniform Sampling. In J. J. Benedetto *et al.*, Eds., *Wavelets—Mathematics and Applications*, CRC Publications, 1994.

[34] F. Marvasti. Fast Packet Network: Data Image, and Voice Signal Recovery. In F. Froehlich and A. Kent, Eds., *Encyclopaedia of Telecommunications*, Vol. 7, Marcel Dekker Inc., 1994, pp. 453–479.

[35] M. Nafie and F. Marvasti. Implementation of Recovery of Speech with Missing Samples on a DSP Chip. *Electronic Letters*, 30(1), January 1994.
[36] R. Crochiere, S. Weber, and J. Flanagan. Digital Coding of Speech in Sub-Bands. *Bell Tech. System J.*, 55:1069–1985, October 1976.
[37] D. Esteban and C. Galand. Application of QMFs to Split Band Voice Coding Schemes. *Int. Conf. Acoust. Speech Signal Process*, Hartford, CT, May 1997, pp. 191–195.
[38] P. Vaidyanathan. Multirate Digital Filters, Filter Banks, Polyphase Networks, and Applications: A Tutorial. *Proc. IEEE*, 78:56–93, 1990.
[39] J. J. Benedetto and M. Frazier, Eds., *Wavelets: Mathematics and Applications*. CRC Press, Inc., Boca Raton, 1994.
[40] A. Papoulis. *Signal Analysis*, McGraw-Hill, New York, 1977.
[41] Robert J. Marks II. *Shannon Sampling and Interpolation Theory I: Foundations*. Springer Verlag, 1990.
[42] Kwan Cheung. A Multidimensional Extension of Papoulis's Generalized Sampling. In R. Marks II, Ed., *Advanced Topics in Shannon Sampling and Interpolation Theory*, Springer-Verlag, New York, 1993.
[43] P. P. Vaidyanathan. *Multirate Systems and Filter Banks*. Prentice-Hall, New Jersey, 1993.
[44] F. Marvasti. Interpolation of Low Pass Signals at Half the Nyquist Rate. *IEEE Journal on Signal Processing Letters*, 3(2):42–43, February 1996.
[45] D. W. Rice and K. H. Wu. Quadrature Sampling with High Dynamic Range. *IEEE Trans. Aerosp. Electron. Syst.*, **AES-18**(4):736–739, November 1982.
[46] A. Kohlenberg. Exact Interpolation of Band-Limited Functions. *J. Appl. Phys.*, 24:1432–1436, 1953.
[47] D. A. Linden. A Discussion of Sampling Theorems. *Proc. IRE*. 47:1219–1226, 1959.
[48] S. C. Scoular and W. J. Fitzgerald. Periodic Nonuniform Sampling of Multiband Signals. *Signal Processing*, 28:195–200, 1992.
[49] F. Marvasti. Nonuniform Sampling Theorems for Bandpass Signals at or Below the Nyquist Rate Density. *IEEE Trans. on Signal Processing*, 572–576, March 1996.
[50] M. D. Rawn. A Stable Nonuniform Sampling Expansion Involving Derivatives. *IEEE Trans. Information Theory*, 35(6):1223–1227, November 1989.
[51] M. D. Rawn. On Nonuniform Sampling Expansions using entire Interpolating Functions, and on the Stability of Bessel-Type Sampling Expansions. *IEEE Trans. on Information Theory*, 35(3):549–557, May 1989.
[52] F. Marvasti, M. Hasan, M. Eckhart, and S. Talebi. Efficient Algorithms for Burst Error Recovery using FFT and other Transform Kernels. *IEEE Trans. on Signal Processing*, 47(4):1065–1075, April 1999.
[53] F. Marvasti and Liu Chuande. Parseval Relationship of Nonuniform Samples of One and Two-Dimensional Signals. *IEEE Trans. on Acoustic, Speech, and Signal Processing*, 38(36):1061–1063, June 1990.
[54] H. G. Feichtinger. Parseval Relationship for Nonuniform Samples of Signals with Several Variables. *IEEE Trans. ASSP*, 40:1262–1263, 1992.
[55] F. Marvasti. Reply to Comment on Parseval Relationship of Nonuniform Samples of One- and Two-Dimensional Signals. *IEEE Trans. on Signal Processing*, 43(12): December 1995.
[56] M. Nakhai and F. Marvasti. Application of Extremum Sampling to Speech Coding. ICASSP'2000, Istanbul, Turkey, June 2000.
[57] P. L. Butzer and A. Gessinger. The Approximate Sampling Theorem, Poisson's Sum Formula, a Decomposition Theorem for Parseval's Equation and their Interconnections. In *Proc. of SampTA'95*, F. Marvasti, Ed., Jurnala, Latvia, September 1995.

5

Iterative and Noniterative Recovery of Missing Samples for 1-D Band-Limited Signals

P. J. S. G. Ferreira

5.1. Introduction

This chapter discusses a class of iterative and noniterative methods for the reconstruction of partially known band-limited signals. Band-limiting is equivalent to the vanishing of a known subset of the samples of the Discrete Fourier Transform (DFT) of the signals. The signals are partially known in the sense that only a subset of its samples is supposed to be available for measurement. The problem is to determine the signal from the available samples. Although this is in essence a nonuniform sampling problem, the methods employed differ considerably from the "infinite series" approaches. This is due to the nature of the finite-dimensional spaces in which we work, for which the matrix formulations are the most natural tool.

The chapter is written as a tutorial. Although it is not a comprehensive review of the many methods that have been proposed for the solution of the missing data and interpolation problems, it focuses on a variety of possible methods, their interrelations, possible implementations, stability, and relative performance. Understanding the connections between the methods is particularly useful since it helps the reader in navigating through the literature, and it contributes to a better grasp of the subject and of the main options and algorithms.

P. J. S. G. Ferreira • Dep. de Electrónica e Telecomunicações Universidade de Aveiro, 3810–193 Aveiro Portugal. E-mail: pjf@ieeta.pt URL: http://www.ieeta.pt/~pjf

Nonuniform Sampling: Theory and Practice, edited by Marvasti
Kluwer Academic/Plenum Publishers, New York, 2001.

The methods described can be readily extended to signal and image interpolation and extrapolation, superresolution, and other multivariate problems, but for simplicity and brevity we refrain from discussing the theory behind the multivariate problems. The next chapter of this book will cover some of the multidimensional issues.

There are four possible models for the band-limited missing data problem, which correspond to all possible combinations of discrete and continuous time and frequency. We will not insist on the L_2 or continuous time, continuous frequency problem, nor on its infinite-dimensional sampled versions (continuous time and discrete frequency, or discrete time and continuous frequency). The focus will be on finite-dimensional signals, characterized by finite (or periodic) time domain and frequency domain vectors. In the terminology of [1], this corresponds to the discrete-discrete model. Mathematically, we deal with functions $f : G \mapsto \mathbb{C}$ or $f : G \mapsto \mathbb{R}$, where G is the group of integers with addition modulo N, equipped with the counting measure. The Fourier transform in this group is the Discrete Fourier Transform (DFT).

5.1.1. Outline of the Chapter

The structure of the chapter is as follows. Band-limited signals are introduced in Section 5.2, together with the terminology and notation that will be used throughout the chapter.

The iterative methods discussed in Section 5.3 operate on the entire signal (known data and estimated data) on a step-by-step basis. The discussion centers around the finite-dimensional Papoulis–Gerchberg algorithm and its connections with other similar or equivalent methods (alternating projections and the frame algorithm, for example).

It is shown that the iterations converge if the density of the sampling set exceeds a certain minimum value which naturally increases with the bandwidth of the data. Naturally, the minimum density is related to the Nyquist–Landau density. Upper and lower error bounds are given as a function of the number of iterations. These bounds are best-possible, that is, there exist signals for which the bounds are exact. These signals are characterized, and turn out to be related to certain eigenvalue problems.

Several other issues are examined, relaxation being among them. Briefly, relaxation is a technique whereby one goes from the iteration

$$x^{(i+1)} = Tx^{(i)} + v, \tag{1}$$

to the iteration

$$x^{(i+1)} = \mu Tx^{(i)} + (1 - \mu)x^{(i)},$$

where $\mu \in \mathbb{R}$ is the relaxation parameter. We study the effect of μ on the convergence rate of the Papoulis–Gerchberg method, and the value that leads to the best convergence rate (assuming a given, fixed, sampling set).

The number of distinct sampling sets with the same density is usually large. In fact, if N is the total number of samples, and if K denotes the number of known samples, then the number of sampling sets with density K/N is given by the binomial coefficient $\binom{N}{K}$. One of the questions that is discussed in the chapter concerns the sampling sets with a given density K/N that lead to maximum or minimum convergence rates (or equivalently, to maximum or minimum numerical stability).

We will see that for low-pass signals the best convergence rates result when the distances among the missing samples are a multiple of a certain integer. The worst convergence rates generally occur when the missing samples are contiguous, as in the case of extrapolation problems. Scattered missing data usually leads to well-posed and stable reconstruction problems, whereas extrapolation leads to ill-posed, numerically difficult problems. This will be confirmed, not just qualitatively, but also quantitatively, in this section and also in Section 5.5.

In contrast to the methods studied in Section 5.3, the formulation described in Section 5.4 leads to (iterative or noniterative) algorithms of "minimum dimension". The dimensions of the vectors and matrices in the algorithm is given by the number of unknown data, whereas the iterative methods studied in Section 5.3 involve matrices and vectors of dimension N (the total number of data).

The relation between the minimum dimension methods and the iterative methods from Section 5.3 is explained. It is shown how and why minimum dimension formulations lead to more efficient iterative and noniterative solutions to the missing data problem. Both time domain and frequency domain minimum dimension formulations are described.

Any given recovery algorithm, no matter how efficient, is of limited practical use if it is not numerically stable. Stability issues are the subject of Section 5.5. They allow a proper understanding of the limitations of both iterative and noniterative methods. On one hand, the convergence rate of the linear stationary iterative method of first order (1), with iteration matrix T and iteration vector v, depends on the magnitude and distribution of the eigenvalues of T. On the other hand, the noniterative reconstruction methods that we will study require solving linear equations of the form $Ax = b$, whose numerical stability depends on the condition number of the matrix A, and hence on the magnitude of its extreme eigenvalues. The estimation of the eigenvalues of the iteration matrices and its dependence upon the sampling set is the core of Section 5.5.

The chapter closes with a brief comparison of the methods, and some hints concerning the choice of the algorithm for a specific problem.

5.2. Band-Limited Signals

The recovery of missing samples from an arbitrary signal is of course impossible. This does not represent a serious difficulty, since the signals that occur in the real world, far from being "arbitrary", often satisfy known constraints. For example, the values attained by the signal, in practice, are always bounded above and below by constants. Many other constraints convey useful information that can be used by the recovery algorithms, such as constraints on the number and magnitude of the derivatives of the signal, convexity, analyticity, duration, and so on. The constraint extensively used in this chapter is called band-limiting. Before defining it, it is necessary to introduce some concepts and notation.

The complex N-dimensional space is denoted by $\mathbb{C}^N$, with the usual inner product

$$\langle a, b \rangle := \sum_{i=0}^{N-1} a_i b_i^*$$

and norm

$$\|x\|^2 := \langle x, x \rangle.$$

A signal with N samples can be described by a N-dimensional complex vector x. The elements of the vector are denoted by $x(0), x(1), \ldots, x(N-1)$, and correspond to the samples of the signal.

The Discrete Fourier Transform (DFT) of a signal $x \in \mathbb{C}^N$ is the vector $X \in \mathbb{C}^N$ defined by

$$X(k) := \sum_{i=0}^{N-1} x(i) e^{-\mathrm{j}(2\pi/N)ik}$$

(j denotes the imaginary unit, not to be confused with i or j, which usually denote integer indexes). Because the DFT is a linear map in $\mathbb{C}^N$, it can be expressed succinctly in matrix form,

$$X = Fx,$$

where F is the $N \times N$ matrix with elements

$$F_{ab} = e^{-\mathrm{j}(2\pi/N)ab}.$$

Let F_i denote the ith column of F. Obviously, $\langle F_i, F_i \rangle = \|F_i\|^2 = N$, regardless of the value of i. On the other hand, any two distinct columns F_i and F_j with $i \neq j$ are orthogonal, that is, $\langle F_i, F_j \rangle = 0$. The Fourier matrix F has orthogonal columns and rows, and the inverse transformation (IDFT) is

$$x(k) = \frac{1}{N} \sum_{i=0}^{N-1} X(i) e^{\mathrm{j}(2\pi/N)ik}.$$

This can also be written in matrix form

$$x = F^{-1} X,$$

where

$$F_{ij}^{-1} = \frac{1}{N} e^{\mathrm{j}(2\pi/N)ij}.$$

Note that $x(k)$ and $X(k)$ should be interpreted as periodic functions of k, with period N. This fact is often convenient in computations, and we will use it tacitly from now on.

A signal x is band-limited if its discrete Fourier transform X vanishes on some fixed set, that is, if

$$X(i) = 0, \qquad i \in S,$$

where S is a fixed nonempty proper subset of $\{0, 1, \ldots, N-1\}$. The set of signals band-limited to a specific set S is a linear subspace of $\mathbb{C}^N$, of dimension equal to the cardinal of the complement of S. This dimension is often called the bandwidth of the signals. When the complement of S is the set of $2M+1$ elements

$$\{0\} \cup \underbrace{\{1, 2, \ldots, M\}}_{M \text{ elements}} \cup \underbrace{\{N-M, \ldots, N-1\}}_{M, \text{elements}},$$

or equivalently (by periodicity)

$$\{0, \pm 1, \pm 2, \ldots, \pm M\},$$

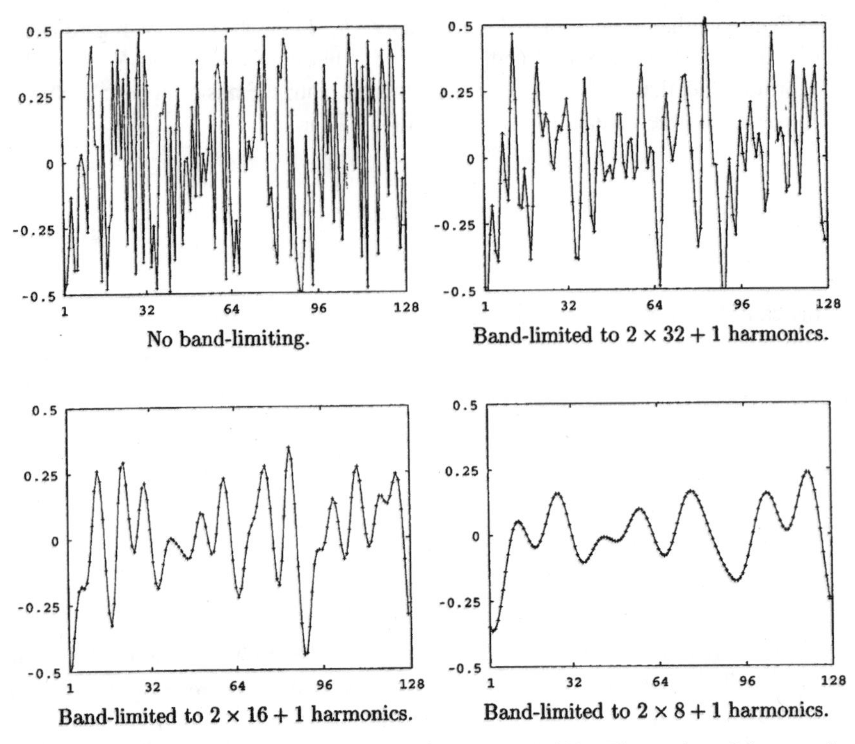

Figure 1. Band-limiting a signal to successively smaller bandwidths. The number of data samples (hence the maximum number of nonzero harmonics) is $N = 128$. The dimension of the subspace of signals band-limited to $2M + 1$ harmonics is $2M + 1$, and hence $N - (2M + 1)$ data samples are redundant.

the band-limited signal is called a low-pass signal, band-limited to $2M + 1$ harmonics. Such signals can be synthesized using only the $2M + 1$ harmonics of lowest (positive and negative) frequency,

$$\begin{aligned} x(j) &= \frac{1}{N} \sum_{i=-M}^{+M} X(i) e^{\mathrm{j}(2\pi/N)ij} \\ &= \frac{1}{N} \sum_{i=-M}^{+M} \left[\sum_{k=0}^{N-1} x(k) e^{-\mathrm{j}(2\pi/N)ki} \right] e^{\mathrm{j}(2\pi/N)ij} \\ &= \frac{1}{N} \sum_{k=0}^{N-1} x(k) \sum_{i=-M}^{+M} e^{\mathrm{j}(2\pi/N)i(j-k)} \\ &= \sum_{k=0}^{N-1} x(k) \frac{\sin[\pi(2M+1)(j-k)/N]}{N \sin[\pi(j-k)/N]}. \end{aligned} \tag{2}$$

Band-limited signals are smooth, and the smoothness increases as the bandwidth decreases. Figure 1 shows the practical meaning of this concept, and the effect of varying M. Note how the redundancy of the signal increases as the bandwidth decreases.

Equation (2) can also be written in matrix form. But before doing that, we need to introduce two linear operations defined on $\mathbb{C}^N$ which are of great importance in the context of missing data recovery.

The first operation is *sampling*. It maps a signal into another by setting to zero a certain subset of its samples. In matrix form, this corresponds to multiplication by a diagonal matrix D containing only zeros or ones. The diagonal of D will be called the sampling set associated with the sampling operation D, and D itself will be called a *sampling matrix*.

The second operation is *band-limiting*. Band-limiting is performed by a linear operator characterized by a matrix B of the form $B = F^{-1}\Gamma F$, where Γ is a sampling matrix other than the identity I, and F and F^{-1} denote the forward and inverse discrete Fourier transform matrices. Thus, according to this definition, the band-limiting operation in $\mathbb{C}^N$ is equivalent to sampling in the frequency domain.

We now observe that equation (2) can be expressed, in matrix form, as $x = Bx$, where $B = F^{-1}\Gamma F$, with the main diagonal Γ being

$$[1, \underbrace{1, \ldots, 1}_{M \text{ times}}, 0, \ldots, 0, \underbrace{1, \ldots, 1}_{M \text{ times}}].$$

The relation $B = F^{-1}\Gamma F$ translates into

$$\begin{aligned} B_{ij} &= \frac{1}{N} \sum_{k=-M}^{M} e^{-\mathrm{j}(2\pi/N)(i-j)k} \\ &= \frac{\sin[\pi(2M+1)(i-j)/N]}{N \sin[\pi(i-j)/N]}, \end{aligned} \tag{3}$$

which shows that $x = Bx$ is indeed the matrix form of (2). The signals that satisfy $x = Bx$ are low-pass signals and B is a low-pass filter matrix. Note the similarity between the previous equation and

$$f(t) = \int_{-\infty}^{+\infty} f(x) \frac{\sin w(t-x)}{\pi(t-x)} dx$$

which is satisfied by any $f \in L_2(\mathbb{R})$ if and only if its Fourier transform vanishes almost everywhere outside $[-w, w]$. A signal belonging to $L_2(\mathbb{R})$ is band-limited if its Fourier transform vanishes outside a compact (closed and bounded) set. When this set is an interval of the form $[-w, w]$ the signal is called a low-pass signal. Bandpass signals are characterized by a Fourier transform that vanishes outside $[-b, -a] \cup [a, b]$.

We are now ready to formulate the missing data problem, around which this chapter turns.

> **Problem 1** *Let $x \in \mathbb{C}^N$ be a low-pass signal with $2M + 1$ nonzero harmonics. Given the observation $y = Dx$, where D is a sampling matrix, how can one recover x from y? Is this problem well-posed?*

There is a trivial necessary condition that D has to satisfy: the number of samples of x that D preserves should be at least equal to $2M + 1$, the dimension of the subspace of low-pass, band-limited signals with $2M + 1$ nonzero harmonics. The problem becomes under-determined if this condition does not hold, and exact recovery ceases to be possible. Although this is also an interesting situation and there are several ways of dealing with it, we will restrict ourselves to the exactly or over-determined case.

5.3. Iterative Methods

5.3.1. The Papoulis–Gerchberg Iteration

One of the methods for the solution of the missing data problem is the Papoulis–Gerchberg algorithm. This is an iterative method, illustrated in Fig. 2. Each iteration of the algorithm consists of two steps: filtering, and resampling.

The filtering step imposes the frequency domain or DFT-domain constraints about the data, that is, the fact that they are band-limited to $2M + 1$ nonzero harmonics. This low-pass filtering produces an improved approximation $b^{(i)}$ from the previous approximation $x^{(i)}$, according to

$$b^{(i)} = Bx^{(i)}. \tag{4}$$

The next step restores the time domain knowledge about x, that is, the known samples, which were preserved by D but which were changed by the preceding filtering step. Resampling, or reinserting, the known data maps the output of the previous step, $b^{(i)}$, into $x^{(i+1)}$, which will be the input to the next iteration, that is

$$x^{(i+1)} = Rb^{(i)}. \tag{5}$$

We have denoted by R the resampling operator, defined by

$$Ra := (I - D)a + y,$$

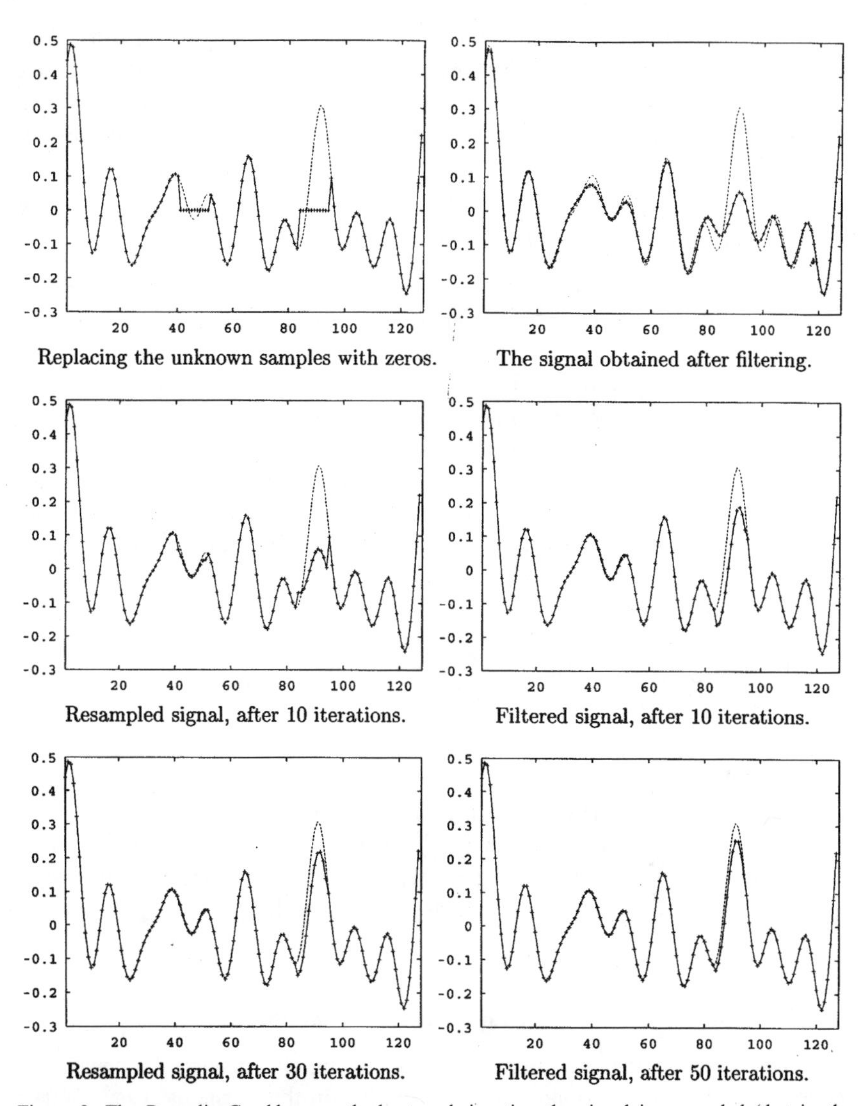

Figure 2. The Papoulis–Gerchberg method: at each iteration the signal is resampled (that is, the known samples are reinserted) and then filtered. For better comparison, the original signal has been superimposed on the reconstructed signal.

for any $a \in \mathbb{C}^N$. Recall that $y = Dx$, and that y is known. We can now write down the rule through which one proceeds from iteration i to iteration $i + 1$, by combining (4) and (5). This leads to

$$x^{(i+1)} = (I - D)Bx^{(i)} + y, \tag{6}$$

This algorithm is the finite-dimensional, or discrete-discrete version of the Papoulis–Gerchberg extrapolation algorithm, adapted to the missing data problem, originally studied by Papoulis [2] and Gerchberg [3] in a different context, for the continuous-continuous $L_2(\mathbb{R})$ extrapolation problem. See also [4].

5.3.2. Alternating Projections, POCS, and the Frame Algorithm

The Papoulis–Gerchberg iteration may also be described as an alternating projection algorithm [5], a class of algorithms based on successive projections onto subspaces.

The alternating projection algorithms, in turn, are special cases of the method known as Projections Onto Convex Sets (POCS). They can be obtained from POCS by taking all the convex sets to be subspaces. For more details on POCS see [6] and [7].

A countable set of elements f_n $(n \in J)$ belonging to a separable Hilbert space with inner product $\langle \cdot, \cdot \rangle$ and norm $\| \cdot \|$ is a *frame* if there exist two constants α and β such that

$$\alpha \|g\|^2 \le \sum_{n \in J} |\langle g, f_n \rangle|^2 \le \beta \|g\|^2.$$

for every element g of that Hilbert space. The Papoulis–Gerchberg algorithm is also closely related to the frame algorithm, a procedure through which the signal g can be recovered from the data $\langle g, f_n \rangle$. A detailed tutorial exposition of the connections between iterative signal reconstruction and frames, eigenvalue problems, and energy concentration, can be found in [8].

5.3.3. Constrained Iterative Restoration, Relaxation

The iteration (6) can also be derived using the general framework for constrained iterative restoration proposed in [9]. In this particular case, we take the sampling and band-limiting operations to be, respectively, the distortion and constraint mentioned in [9]. The original signal satisfies $x = Bx$ and the observed signal is $y = Dx$, for appropriate matrices B and D. This suggests the equation

$$\begin{aligned} x &= Bx + \mu(y - Dx) \\ &= \mu y + (I - \mu D)Bx, \end{aligned}$$

where μ is any fixed constant, and I denotes the identity matrix. To solve this equation we consider the sequence of successive approximations

$$x^{(i+1)} = T_1 x^{(i)}, \tag{7}$$

where the operator T_1 is defined by

$$T_1(\cdot) := \mu y + (I - \mu D)B(\cdot). \tag{8}$$

Note that when $\mu = 1$ the iteration (7) reduces to (6). Conversely, it is easy to introduce relaxation on the basic iteration (6), leading to (7).

It is possible to obtain a sequence of band-limited successive approximations by taking the distortion to be sampling followed by a band-limiting operation. This gives

$$x^{(i+1)} = \mu By + (I - \mu BD)x^{(i)} =: T_2 x^{(i)} \tag{9}$$

Yet another possibility is to apply the band-limiting operator to (7). This leads to

$$x^{(i+1)} = \mu By + B(I - \mu D)x^{(i)} =: T_3 x^{(i)}. \tag{10}$$

It is well-known that such iterative methods converge if and only if the spectral radius (the largest eigenvalue in absolute value) of the respective iteration matrices[1] is strictly less than unity. See, for example, [10] and [11].

Since the iteration matrices of the methods (7) and (10) are the transpose of each other, and (9) is equivalent to (10) if the first approximation is band-limited ($x^{(0)} = Bx^{(0)}$), it is sufficient to consider (7).

It will be shown that this method leads to a sequence of approximations that converges to the required solution, provided that suitable conditions are imposed upon D, B and μ. Under the same conditions the other two iterations will also converge to the same solution, provided that $x^{(0)} = Bx^{(0)}$ in the case of (9).

The major drawback of the Papoulis–Gerchberg iteration and similar methods is related to its convergence rate. The precise rate critically depends on the distribution of the missing data, and can be very slow. We will see that the minimum dimension formulations, discussed in Section 5.4, allow much more efficient algorithms.

[1]By definition, the iteration matrix of the iteration $x^{(i+1)} = Ax^{(i)} + b$ is the matrix A.

5.3.4. Convergence Analysis

In this section we will examine the convergence of (7). It turns out that if the density of the known samples is above the minimum density the iteration converges regardless of the distribution of the missing samples. The convergence was established in [12] for extrapolation problems, in which the known samples are contiguous, using a technique based on matrix partitioning. The general case requires different tools, and was treated in [13].

There are several possibilities for the analysis of (7). As in [13], we use a few well-known results in fixed-point theory. This technique is often used in signal restoration [9] and [14] and does not require the linearity of the operators involved. In fact, it is easy to see that other possibly non-linear nonexpansive constraints can be incorporated in the algorithm without disturbing the convergence.

Denote the set of subscripts of the known samples of x by

$$S_t = \{i_1, i_2, \ldots, i_s\},$$

and its complement by $\bar{S}_t$. The observed signal y satisfies $y = Dx$, where D is the sampling matrix that corresponds to S_t, that is,

$$D_{ii} = \begin{cases} 1, & i \in S_t \\ 0, & i \in \bar{S}_t. \end{cases}$$

By definition, the *density* of the sampling matrix D is s/N, s being the cardinal of S_t.

Let the set of subscripts of the known vanishing DFT coefficients of x be denoted by S_f, and its complement in $\{0, 1, \ldots, N-1\}$ by $\bar{S}_f = \{k_1, k_2, \ldots, k_q\}$.

The process of band-limiting a signal x with the band-limiting matrix B is equivalent to setting to zero a subset of its DFT coefficients X. More precisely,

$$X(i) = 0, \qquad i \in S_f.$$

By definition, the *bandwidth* of B is q/N, q being the cardinal of $\bar{S}_f$. For low-pass signals, both S_f and $\bar{S}_f$ are contiguous sets (modulo N).

Lemma 1 *Let D be a sampling operator with density d, and let B be a band-limiting operator with bandwidth w. If $d \geq w$, $x = (I - D)x$ and $x = Bx$, then x is the zero vector.*

Proof: The hypothesis $x = (I - D)x$ means that $x(i) = 0$ for all $i \in S_t$, that is, s samples of x vanish. The N equations $X = Fx$ reduce to

$$X = \sum_{i \notin S_t} F_i \, x(i),$$

where F_i denotes the ith column of F.
On the other hand, $x = Bx$ means that $X(i) = 0$ for all $i \in S_f$, and so $N - q$ samples of X vanish. Thus,

$$\sum_{i \notin S_t} F_{ki} \, x(i) = 0, \qquad k \in S_f. \tag{11}$$

The submatrix of F that appears in (11) is $(N - q) \times (N - s)$ with $(N - q) \geq (N - s)$ since $d \geq w$ (and so $s \geq q$).
Since S_f is contiguous, there is an integer r such that $k_i = (i + r) \bmod N$, and the submatrix of F that appears in (11) can be reduced to a Vandermonde matrix with linear independent columns. Thus, the unique solution to (11) is given by $x(i) = 0$, and x must be the zero vector. □

Let X be a metric space, with metric $d(\cdot, \cdot)$. Recall that a possibly nonlinear operator $A : X \mapsto X$ is *nonexpansive* when

$$d(Ax, Ay) \leq d(x, y),$$

for all $x, y \in X$. In a normed space with norm $\| \cdot \|$, this can be converted to

$$\|Ax - Ay\| \leq \|x - y\|.$$

An operator $A : X \mapsto X$ is *strictly nonexpansive* if it is nonexpansive and

$$d(Ax, Ay) = d(x, y) \Rightarrow x = y.$$

If X is a normed space and A is strictly nonexpansive then

$$\|Ax - Ay\| < \|x - y\|$$

for all distinct pairs x, y of elements of X.

If the operator A is linear and nonexpansive then

$$\|Ax\| \leq \|x\|$$

for all $x \in X$. Therefore, the norm of a linear nonexpansive operator cannot exceed unity. In $\mathbb{C}^N$, a linear nonexpansive operator is a matrix with spectral norm bounded by one. Therefore, its spectral radius is also bounded by one, and consequently none of the eigenvalues can exceed unity in absolute value.

The point $x \in X$ is a *fixed point* of the operator $A : X \mapsto X$ if $Ax = x$. For example, the fixed points of ideal low-pass filters are low-pass signals. A nonexpansive operator may have any number of fixed points, but a strictly nonexpansive operator clearly has at most one.

Lemma 2 *Let $0 \leq \mu \leq 2$. Then, the operator T_1 defined by (8) is nonexpansive.*

Proof: Consider

$$A_\mu := (I - \mu D)B. \tag{12}$$

Checking (8), we see that T_1 is nonexpansive if and only if A_μ is. The nonexpansiveness of the projection operator B in the Euclidean norm is obvious. Thus, T_1 is nonexpansive if $I - \mu D$ is, which happens if and only if $0 \leq \mu \leq 2$. Otherwise there would be an eigenvalue of the diagonal matrix $I - \mu D$ greater than one in absolute value. □

Lemma 3 *If $0 \leq \mu \leq 2$, the equation $x = T_1 x$, with T_1 defined by* (8), *has at least one solution.*

Proof: T_1 is nonexpansive, hence continuous. It maps a non-empty, compact and convex subset of $\mathbb{C}^N$ into itself (consider the set of all signals with norm less than or equal to a given bound). Any continuous operator that maps a non-empty compact and convex set into itself has the fixed-point property [9] and [15], and the result follows. □

Theorem 1 *Let D be a sampling operator with density d, and let B be a band-limiting operator with bandwidth w. If*

$$0 < \mu < 2, \tag{13}$$

$$d \geq w, \tag{14}$$

the operator T_1, defined by (8), *is strictly nonexpansive, and the equation $x = T_1 x$ has one and only one solution. This solution is given by*

$$x = \lim_{i \to \infty} T_1^i x^{(0)},$$

where the initial approximation $x^{(0)} \in \mathbb{C}^N$ is arbitrary.

Proof: Assume, to the contrary, that (13) and (14) hold but T_1 is *not* strictly nonexpansive. This means that

$$\|A_\mu v\| := \|(I - \mu D)Bv\| = \|Bv\| = \|v\|$$

for some $v \neq 0$. We will show that this assumption leads to a contradiction. The last equality implies the vanishing of $V(i)$ for $i \in S_f$, and thus $Bv = v$. This reduces the second equality to

$$\|(I - \mu D)v\| = \|v\|.$$

But

$$\begin{aligned}\|(I - \mu D)v\|^2 &= (1 - \mu)^2 \sum_{i \in S_t} v(i)^2 + \sum_{i \in \bar{S}_t} v(i)^2 \\ &= \|v\|^2 - \mu(2 - \mu) \sum_{i \in S_t} v(i)^2\end{aligned}$$

which implies the vanishing of the $v(i)$ for $i \in S_t$, since, by hypothesis, $0 < \mu < 2$ and so $\mu(2 - \mu) > 0$. By Lemma 1, v must be the zero vector, a contradiction. Thus, T_1 must be strictly nonexpansive, and a strictly nonexpansive mapping cannot have more than one fixed point. □

5.3.5. Upper and Lower Error Bounds

The iterations (7) or (6) are examples of iterative methods in $\mathbb{C}^N$ of the general form

$$x^{(i+1)} = Tx^{(i)} + v,$$

where T and v are the iteration matrix and vector. Assume that this iteration converges to a vector x such that $x = Tx + v$. The error at the ith iteration is

$$e^{(i)} = x^{(i)} - x,$$

and the errors at iterations i and $i + 1$ are related by

$$e^{(i+1)} = Te^{(i)}.$$

Clearly, the behavior of $e^{(i)}$ is crucial when the iterative method is used to approximate x.

We will now consider this issue for the iteration (7). Upper and lower bounds for the reconstruction error at any given iteration will be given, assuming that the relaxation parameter is $\mu = 1$. The bounds are best-possible, in the sense that there are initial error vectors $e^{(0)}$ for which they are exact. The initial error vectors for which the error bounds are attained will be identified.

Recall that, in agreement with (12), the matrix A_1 is given by

$$A_1 := (I - D)B.$$

Let v denote an eigenvector of $A_1^H A_1 = B(I - D)B$, pertaining to the eigenvalue λ, that is,

$$A_1^H A_1 v = B(I - D)Bv = \lambda v. \tag{15}$$

The subscript ^{H} denotes conjugate transpose. We claim that $A_1 v$ is an eigenvector of A_1 pertaining to the same eigenvalue, and thus

$$A_1^2 v = \lambda A_1 v.$$

To check that this is true, left-multiply (15) by $(I - D)B$ and use the idempotency of B.

It is now easy to derive bounds for the error $e^{(k)}$ at iteration k of the algorithm. It follows from $e^{(k)} = A_1 e^{(k-1)}$ that

$$\max_{\|e^{(0)}\|=1} \|e^{(1)}\|^2 = \max_{\|e^{(0)}\|=1} \|A_1 e^{(0)}\|^2 = \lambda_{\max}$$

and

$$\min_{\|e^{(0)}\|=1} \|e^{(1)}\|^2 = \min_{\|e^{(0)}\|=1} \|A_1 e^{(0)}\|^2 = \lambda_{\min},$$

where $\lambda_{\min}$ and $\lambda_{\max}$ denote the smallest and largest of the eigenvalues of $A_1^H A_1$. The maximum and the minimum are attained when $e^{(0)}$ equals $v_{\max}$ or $v_{\min}$, the eigenvectors of $A_1^H A_1$ that correspond to the eigenvalues $\lambda_{\max}$ and $\lambda_{\min}$, respectively.

Now, we have already seen that $A_1 v_{\max}$ and $A_1 v_{\min}$ will also be eigenvectors of A_1, and therefore

$$e^{(2)} = A_1^2 v_{\max} = \lambda_{\max} A_1 v_{\max}$$

if $e^{(0)} = v_{\max}$, and

$$e^{(2)} = A_1^2 v_{\min} = \lambda_{\min} A_1 v_{\min}$$

if $e^{(0)} = v_{\min}$. By induction it is now clear that

$$\lambda_{\min}^k \|e^{(1)}\| \leq \|e^{(k+1)}\| \leq \lambda_{\max}^k \|e^{(1)}\|. \tag{16}$$

The upper bound is attained if $e^{(1)} = v_{\max}$, the lower bound if $e^{(1)} = v_{\min}$.

In many cases one takes $x^{(0)}$, the initial approximation, to be the zero vector. This leads to $x^{(1)} = y$, and $e^{(k)} = A_1^k x$. Taking $x = v_{\max}$, we get the lowest possible asymptotic convergence rate. In this sense, the $v_{\max}$ are the signals which result in the worst possible algorithm performance. A similar argument, regarding best possible performance, holds for $v_{\min}$. Also note that $v_{\max}$ and $v_{\min}$, being eigenvectors of $A_1^H A_1$, depend on B and D, and hence on the bandwidth and sampling set. In particular, changing the sampling set may lead to dramatic changes in the convergence rate.

For examples of $v_{\max}$ and $v_{\min}$ for particular problems see Figs. 3 and 4. It is instructive to compare the effect of the sampling set and the signal on the performance of the algorithm. Roughly speaking, well distributed sampling sets lead to values of $\lambda_{\max}$ reasonably below unity, and thus to favorable convergence rates. In those cases $\lambda_{\min}$ is also not much smaller than $\lambda_{\max}$, and hence the worst and best convergence rates do not differ appreciably.

The opposite situation occurs when the sampling set has large gaps. In that case, there exist band-limited signals whose energy is well concentrated in the sampling set, but also other band-limited signals well concentrated on its complement. The signal $v_{\min}$ is an (extreme) example of the former class, and $v_{\max}$ of the latter. In practice, as the gaps increase $\lambda_{\max}$ tends to unity and the convergence rate becomes increasingly slower (the iterations do not reduce the magnitude of the components of the error signal proportional to $v_{\max}$).

When B is a low-pass matrix and the sampling set corresponds to an extrapolation problem (that is, the known data are contiguous), the analysis that we have made reduces to the singular value analysis of the operator DB and to the periodic discrete prolate spheroidal sequences (P-DPSS) discussed in [16]. In fact, setting $M = (I - D)B$ in $M^H M x = \lambda x$ leads to

$$B(I - D)Bv = \lambda v, \tag{17}$$

which is equivalent to

$$BDBv = (1 - \lambda)v \tag{18}$$

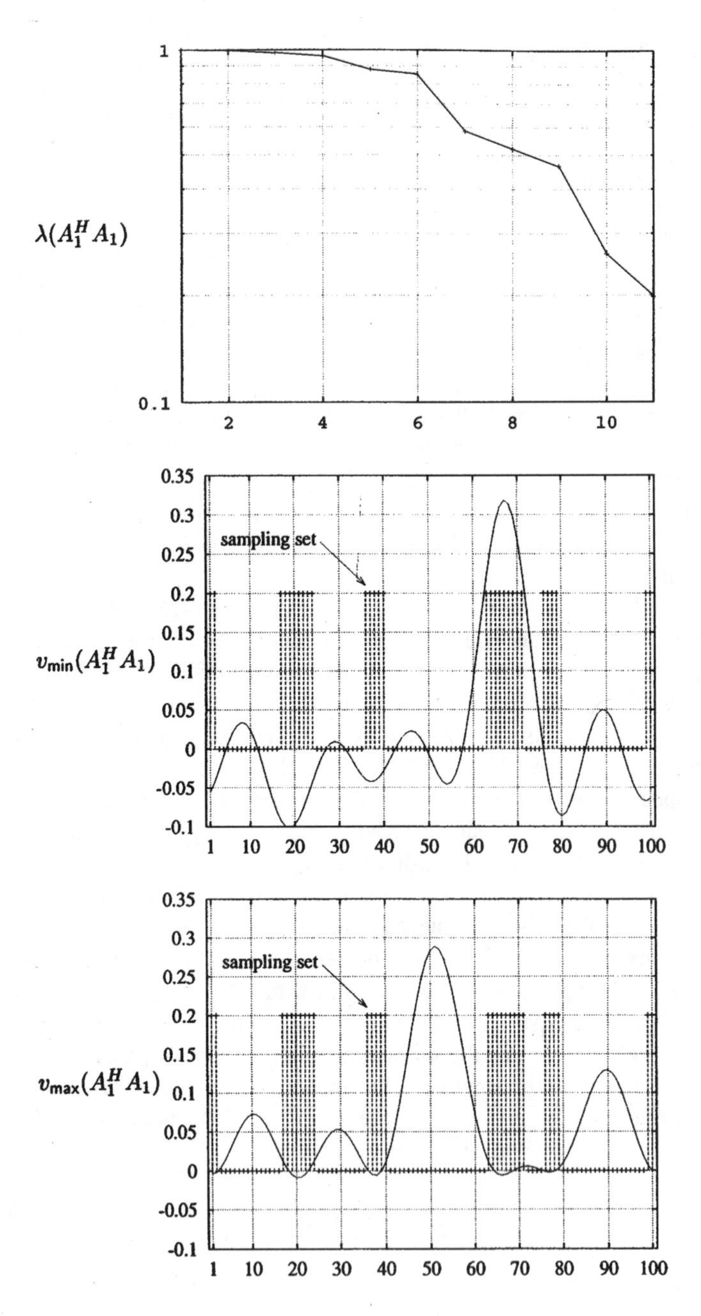

Figure 3. The eigenvalues and the eigenvectors $v_{\min}$ and $v_{\max}$ of the matrix $A_1^H A_1 = B(I - D)B$. The extreme eigenvalues dictate the best and worst (geometric) convergence rates in (16). The easiest signal to handle is $v_{\min}$, the most difficult is $v_{\max}$ (for this bandwidth and sampling set).

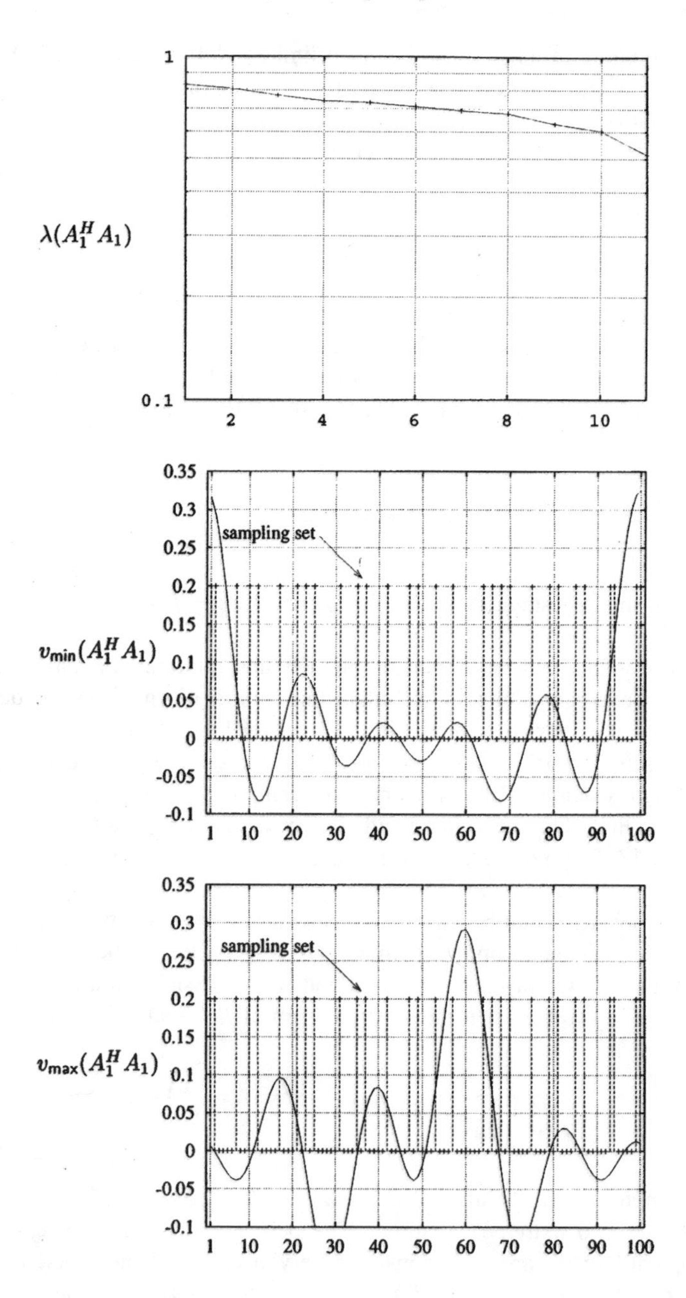

Figure 4. The eigenvalues and the eigenvectors $v_{\min}$ and $v_{\max}$ of the matrix $A_1^H A_1 = B(I - D)B$. The extreme eigenvalues dictate the best and worst (geometric) convergence rates in (16). The easiest signal to handle is $v_{\min}$, the most difficult is $v_{\max}$ (for this bandwidth and sampling set). Compare with Fig. 3.

since any solution of (17) must be band-limited (left multiply (17) by B). Equation (18) is equivalent to the equation used to define the P-DPSS in [16]. The eigenvectors of (17) $v_{\min}$ and $v_{\max}$, which correspond to the smallest and largest eigenvalues $\lambda_{\min}$ and $\lambda_{\max}$ in (17), reduce to P-DPSS in the case of extrapolation problems.

5.3.6. Other Results

There are several other questions that can be asked concerning the iterative method (7). For example, one may be interested in the norm and spectral radius of the matrix $A_1 = (I - D)B$, or in the effect of the relaxation parameter μ upon the convergence rate of the algorithm. It is not difficult to see that the optimal value of μ is

$$\mu_{\text{opt}} = \frac{2}{2 - \lambda_{\max}},$$

where $\lambda_{\max}$ is the largest eigenvalue of the matrix $A_1^H A_1 = B(I - D)B$. For details, see [13].

We have mentioned that the sampling sets that minimize or maximize the asymptotic convergence rate of the algorithm, for a given sampling density, depend on the B matrix. It is shown in [13] that for low-pass data, for example, the missing sample positions that give best-possible sampling sets define a grid with a certain spacing. Thus, the optimum sampling strategy is, in a sense, uniform sampling. On the other hand, the worst possible sampling sets are contiguous. This may not be the case for data that are neither low-pass nor high-pass.

Let us consider a number of missing samples greater than two, and let s/N be the density of the sampling matrix, and $N - s$ the number of unknown samples. What can we say about the best and worst possible sampling sets?

It can be shown [13] that the squared norm of $B(I - D)B$ is given by

$$\sup_{\|x\|=1} \left(\sum_{i \notin S_t} \langle b_i, x \rangle^2 + (1 - \mu)^2 \sum_{i \in S_t} \langle b_i, x \rangle^2 \right), \tag{19}$$

where b_i denotes the ith column of B.

There are two extreme situations. The best situation arises when all the vectors b_i with $i \notin S_t$ are orthogonal, or very nearly so. The worst possible situations occur when the b_i with $i \notin S_t$ are close to linearly dependent.

Assume for a moment that the sampling set is such that the vectors b_i, with $i \notin S_t$, are orthogonal. The squared norm of the b_i equals q/N, where q is the number of nonzero harmonics, or the cardinal of the set $\bar{S}_f$, in the notation introduced in Section 5.3.4. Therefore, one has

$$x = \frac{N}{q} \sum_{i \notin S_t} \langle b_i, x \rangle b_i$$

for any x belonging to the subspace spanned by the b_i, $i \notin S_t$. Thus

$$\|x\|^2 = \frac{N}{q} \sum_{i \notin S_t} \langle b_i, x \rangle^2,$$

or, equivalently,

$$\sum_{i \notin S_t} \langle b_i, x \rangle^2 = \|x\|^2 \frac{q}{N}.$$

It follows from (19) with $\mu = 1$ that

$$\|(I - D)B\|^2 = \sup_{\|x\|=1} \frac{q}{N} \|x\|^2 = \frac{q}{N},$$

that is, $(I - D)B$ will be a convergent matrix with squared norm q/N. In this case, the norm is independent of the number of missing samples, and equals the value that would be obtained for a single missing sample. For these "ideal" sampling sets, the recovery of one, two, or more missing samples can be achieved with a matrix whose norm is determined only by the bandwidth of the data. Also, the asymptotic convergence rate of the reconstruction algorithm would be independent of the number of missing samples. We are assuming, of course, that the conditions of Theorem 1 are satisfied, that is, that the sampling set has sufficiently high density.

To find out more about the "ideal" sampling sets, we need to see under what conditions, if any, the columns b_i of the matrix B can be orthogonal to each other. For signals with a contiguous set of nonzero harmonics, and in particular for low-pass signals, the elements of B are given by (3). Since $B^2 = B$, it follows that $\langle b_i, b_j \rangle = B_{ij}$, and so

$$\langle b_i, b_j \rangle = \frac{\sin[\pi(2M + 1)(i - j)/N]}{N \sin[\pi(i - j)/N]}.$$

One may assume, without loss of generality, that one of the missing samples is sample number zero. But because

$$\langle b_i, b_0 \rangle = \frac{\sin[\pi(2M+1)i/N]}{N \sin[\pi i/N]},$$

orthogonality is only possible if the remaining missing samples belong to the set

$$\left\{ \frac{N}{2M+1}, \frac{2N}{2M+1}, \frac{3N}{2M+1}, \ldots \right\}$$

assuming of course that $2M+1$ divides N. These quantities determine the possible positions of the missing samples which correspond to optimum sampling sets. If $2M+1$ does not divide N, they still originate sampling sets with near-optimum convergence properties.

5.4. Minimum Dimension Formulations

The iterative reconstruction algorithms mentioned in Section 5.3 are useful in connection with many reconstruction problems. They have been used, for example, in the estimation of sinusoids from incomplete time series [17], or in the removal of lipid artifacts in ^{1}H spectroscopic imaging [18].

There are, however, classes of problems for which these algorithms do not seem well suited. The extrapolation of low-pass signals is an example. In that case, and more generally whenever there are long contiguous gaps of missing samples, the convergence rate falls down to very low values.

The minimum dimension formulations presented in this section lend themselves to better and faster algorithms. The ideas behind these methods were put forward in several works, including [19] and [20], which address the finite-dimensional case, and [21] and [22], which consider finite-dimensional images. The study of the continuous-time case was initiated with [23].

The computational requirements necessary to complete one iteration of the Papoulis–Gerchberg algorithm do not decrease as the number of unknown samples decreases. The algorithm operates on N-dimensional vectors, and the computational burden per iteration is dominated by the filtering operation, which requires $O(N \log_2 N)$ arithmetic operations using fast Fourier transform (FFT) algorithms. It seems redundant to operate on sets of N samples when only a smaller number is unknown.

The minimum dimension methods lead to sets of linear equations for the unknown quantities, with as many equations as unknown samples (for the time domain formulation), or as many equations as unknown DFT coefficients (for the frequency domain formulation). Hence, the system matrix, in both cases, has minimum dimension. In the time domain case the equations yield the unknown samples directly, whereas in the frequency domain one obtains coefficients of the DFT of the signal.

The basic idea behind the minimum dimension method can also be applied to image interpolation problems, and to continuous-time missing data problems. A signal $f : \mathbb{R} \to \mathbb{R}$ (for example of finite-energy) can be recovered from its samples $f(k)$, $k \in \mathbb{Z}$ (for simplicity we are normalizing the sampling period to unity) if its Fourier transform is zero outside the interval $[-r/2, r/2]$, with $0 \leq r \leq 1$. The case $r = 1$ corresponds to sampling at the critical Nyquist–Landau density. When $r < 1$ there is oversampling, and the relevant reconstruction formula is

$$f(t) = r \sum_{k=-\infty}^{+\infty} f(k) \operatorname{sinc} r(t-k).$$

Theoretically (barring stability issues), it is possible to reconstruct f even if any finite number n of samples $f(i)$ are unknown [23], [24], [25] and [26]. Let the unknown samples be

$$f(j_1), f(j_2), \ldots, f(j_n).$$

Setting $t = j_i$ for $i = 1, 2, \ldots, n$, we get n equations

$$f(j_i) = r \sum_{k=-\infty}^{+\infty} f(k) \operatorname{sinc} \; r(j_i - k),$$

and by splitting the summation in two, one over the unknown samples, another over the known samples, we get

$$f(j_i) = r \sum_{k=1}^{n} f(j_k) \operatorname{sinc} r(j_i - j_k) + h_i.$$

In matrix form, this can be written as

$$u = Su + h,$$

with the obvious interpretations for u, S, and h. The elements h_i of the vector h can be computed because they depend only on known data. By solving these n equations we get the n unknown samples (but see [24] and [27] for some crucial stability and implementation issues). Note that the essential steps of the method are:

- Obtaining an equation of the form $x = Tx$, where x is the vector of all samples and T is a specific linear or affine operator.
- Obtaining n linear equations for the n unknown samples, by splitting x into known and unknown samples.

We are already aware of an equation of the required form, namely, equation (2). This solves the first part of the problem for the time domain case. For the frequency domain formulation, it will be necessary to determine the operator T such that $X = TX$, X being the DFT of x.

We start with the time domain formulation, expressing the vector x of the time domain samples of the signal as $x = u + v$, where

$$v_i = \begin{cases} x_i, & i \in S_t, \\ 0, & i \notin S_t. \end{cases}$$

Recall the notation introduced in Section 5.3.4: S_t is the set of known samples, and therefore the vector v is known, whereas u is not. Defining the diagonal matrix

$$D^u_{ii} = \begin{cases} 0, & i \in S_t, \\ 1, & i \notin S_t, \end{cases}$$

it is seen that $u = D^u u$ and $D^u v = 0$.

We now deal with the frequency domain case. The vector X of the samples of the DFT of the signal x is written as $X = G + H$, where

$$H_i = \begin{cases} X_i, & i \in S_f, \\ 0, & i \notin S_f. \end{cases}$$

In accordance with the notation of Section 5.3.4, S_f denotes the set of known DFT coefficients. The vector H is known, but G is not. Defining the diagonal matrix

$$D^g_{ii} = \begin{cases} 0, & i \in S_f, \\ 1, & i \notin S_f, \end{cases}$$

it is seen that $G = D^g G$, and $D^g H = 0$.

Theorem 2 *The signal $x \in \mathbb{C}^N$ that solves problem 1 of Section 5.2 satisfies $x = Bx + h$, and its DFT satisfies $X = TX + V$. The matrices B and T are* *circulant projections, with* $B = F^{-1}D^g F$ *and* $T = FD^u F^{-1}$. *Explicitly,*

$$B_{ab} = \frac{1}{N}\sum_{p \notin S_f} e^{j(2\pi/N)(a-b)p},$$

$$T_{ab} = \frac{1}{N}\sum_{p \notin S_t} e^{-j(2\pi/N)(a-b)p}.$$

Proof: The decompositions $x = u + v$ and $X = G + H$ introduced above lead to

$$\begin{aligned} x &= F^{-1}X = F^{-1}G + F^{-1}H \\ &= F^{-1}D^g G + h = F^{-1}D^g(X - H) + h \\ &= F^{-1}D^g X + h = F^{-1}D^g Fx + h. \end{aligned}$$

Setting $B = F^{-1}D^g F$ yields the equation $x = Bx + h$. Similarly,

$$\begin{aligned} X &= Fx = Fu + Fv \\ &= FD^u u + V = FD^u(x - v) + V \\ &= FD^u x + V = FD^u F^{-1}X + V, \end{aligned}$$

and setting $T = FD^u F^{-1}$ we get $X = TX + V$. □

We now face a slight difficulty. The equations $x = Bx + h$ and $X = TX + V$ cannot be solved for x or X, because the matrices $I - B$ and $I - T$ are singular. However, this obstacle can be circumvented.

Theorem 3 *Let the number of unknown time domain samples (the cardinal of* $\bar{S}_t$*) be denoted by n. Let P be the* $n \times n$ *principal submatrix of B, obtained by deleting from B all rows and columns whose indices belong to* S_t. *Denote by* $c \in \mathbb{C}^n$ *the vector with elements*

$$c_i = \sum_{j \in S_t} B_{ij} v_j + h_i \qquad (i \notin S_t).$$

Then, the vector $y \in \mathbb{C}^n$ *formed by the n time domain unknowns* x_i $(i \notin S_t)$ *satisfies*

$$y = Py + c. \tag{20}$$

Proof: The N equations $x = Bx + h$ are redundant in the sense that there are only n unknown time domain samples x_i. Therefore, we look for a subset of n equations. Since $x = u + v$, it follows that $u + v = Bu + Bv + h$, that is,

$$u_i + v_i = \sum_{j=0}^{N-1} B_{ij} u_j + \sum_{j=0}^{N-1} B_{ij} v_j + h_i.$$

Considering only $i \notin S_t$, and noting that $u_i = 0$ for all $i \in S_t$ whereas $v_i = 0$ for $i \in S_f$, we get

$$u_i = \sum_{j \notin S_t} B_{ij} u_j + \sum_{j \in S_t} B_{ij} v_j + h_i \quad (i \notin S_t),$$

which is (5.20). □

Theorem 4 *Let the number of unknown DFT coefficients (the cardinal of $\bar{S}_f$) be denoted by r. Let Q be the $r \times r$ principal submatrix of T, obtained by deleting from T all rows and columns whose indices belong to S_f. Denote by $d \in \mathbb{C}^r$ the vector with elements*

$$d_i = \sum_{j \in S_f} T_{ij} H_j + V_i \quad (i \notin S_f).$$

Then, the vector $z \in \mathbb{C}^r$ formed by the r unknown DFT harmonics X_i $(i \notin S_f)$ satisfies

$$z = Qz + d. \tag{21}$$

Proof: One can appeal to duality and the reversal of the roles of the time and frequency domains. It is also possible to consider the N equations $X = TX + V$, and substituting $X = G + H$ to obtain $G + H = TG + TH + V$, that is,

$$G_i + H_i = \sum_{j=0}^{N-1} T_{ij} G_j + \sum_{j=0}^{N-1} T_{ij} H_j + V_i.$$

Thus, just as in the case of Theorem 3,

$$G_i = \sum_{j \notin S_f} T_{ij} G_j + \sum_{j \in S_f} T_{ij} H_j + V_i \qquad (i \notin S_f),$$

which is (21). □

The following result [19] can be used to show that the matrices $I - P$ and $I - Q$, corresponding to the equations given in Theorem 3 and 4, are nonsingular, provided that the sampling sets fulfill the necessary density conditions. Recall that $I - P$ and $I - Q$ are submatrices of the singular circulant projection matrices $I - B$ and $I - T$ mentioned in Theorem 2.

Theorem 5 *Let the matrix C be $n \times n$, nonnegative definite, circulant, with z contiguous zero eigenvalues. Then every principal submatrix of C of order $p \leq n - z$ is positive definite.*

Proof: We argue by contradiction. Assume to the contrary that there exists a $n \times n$ nonnegative definite circulant matrix C with singular principal submatrices of order $p = n - z$.

This, in turn, would imply the existence of a nonzero vector $v \in \mathbb{C}^n$ satisfying $v_i = 0$ for all i belonging to a certain set $\mathscr{T}$ containing at least k distinct integers, and such that $v^H Cv = 0$.

But $v^H Cv$ can be rewritten as $V^H \Gamma V$, where V is the DFT of v, $V = Fv$, and the matrix $\Gamma = FCF^H$ is diagonal. Hence, $v^H Cv = 0$ would imply $V_i = 0$ for all i belonging to a certain contiguous set $\mathscr{F}$ with $n - k$ elements. Because of this, v would have to satisfy

$$\sum_{i \notin \mathscr{T}} F_{ki} v_i = 0, \qquad (k \in \mathscr{F}). \tag{22}$$

The matrix of this set of $n - z$ equations for no more than $n - z$ unknowns v_i has linear independent columns when $\mathscr{F}$ is contiguous (the matrix will be Vandermonde if it is square). The linear independence would imply $v_i = 0$ for all $i \notin \mathscr{T}$, leading to the conclusion that v is the zero vector, a contradiction.

□

The minimum dimension equations in the frequency domain appear in [28], and have been used as starting points for efficient methods for signal and image reconstruction [29] and [30]. The similarity (duality) between the time domain formulations found in [19] and the frequency domain formulation was discussed in [20]. The time domain equations were applied to the recovery of clipped electrocardiographic signals in [31]. Several other approaches and applications have been discussed. See, for example, [16], [24], [32] and [37–45].

5.4.1. Examples and Comparisons

The minimum dimension formulations, in the time or frequency domain, allow a variety of reconstruction algorithms (iterative and noniterative). In this section we discuss some of the possibilities, giving examples to illustrate and confirm the main findings.

We will start by considering the reconstruction problem illustrated in Fig. 5 (problem A). For this problem, we have $M = 200$ and $n = 200$. Therefore, the

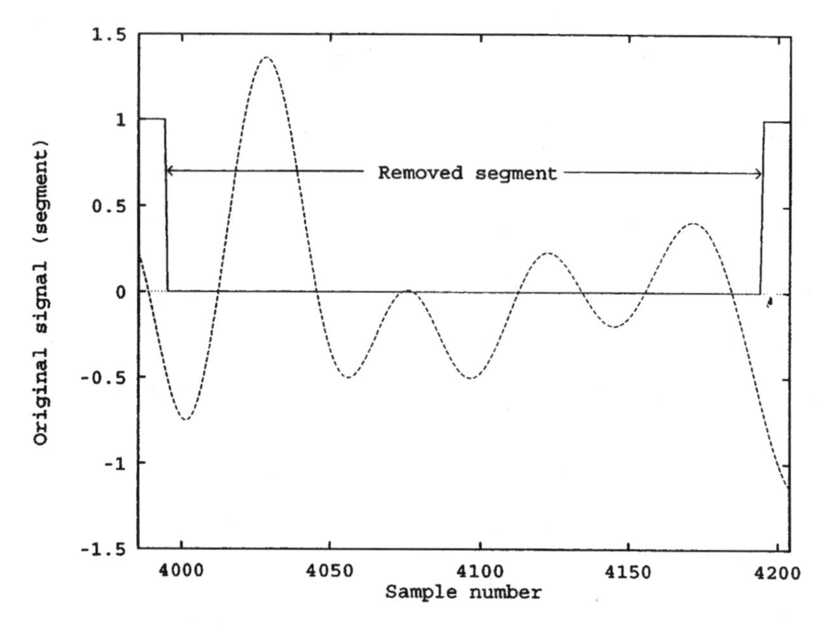

Figure 5. A detail of the data for problem A. We have $M = 200$, and the number of (contiguous) missing samples is $n = 200$.

time domain equations are of dimension $n = 200$, whereas the frequency domain equations involve vectors of dimension $2M + 1 = 401$.

The basic method is the Papoulis–Gerchberg iteration. The error evolution is illustrated in Fig. 6, which also depicts the reconstruction result. Note that the error is plotted against CPU time, not against the number of iterations. This was done in order to make the comparisons between the methods simpler. In fact, a method that achieves a certain error in L iterations is not necessarily faster than a method that needs more than L iterations to achieve the same error, because the necessary CPU time per iteration may differ between the methods.

The performance of the Papoulis–Gerchberg method can be easily improved using the minimum dimension formulations. For example, taking the minimum dimension equations in the time domain (20) and iterating one obtains the method

$$y^{(i+1)} = Py^{(i)} + c,$$

denoted by SIT (simple iteration, time domain) in Fig. 7. The error evolution and the reconstructions depicted in Fig. 7 confirm that the method is slightly more efficient. Each iteration requires one matrix-vector multiplication of size n, as

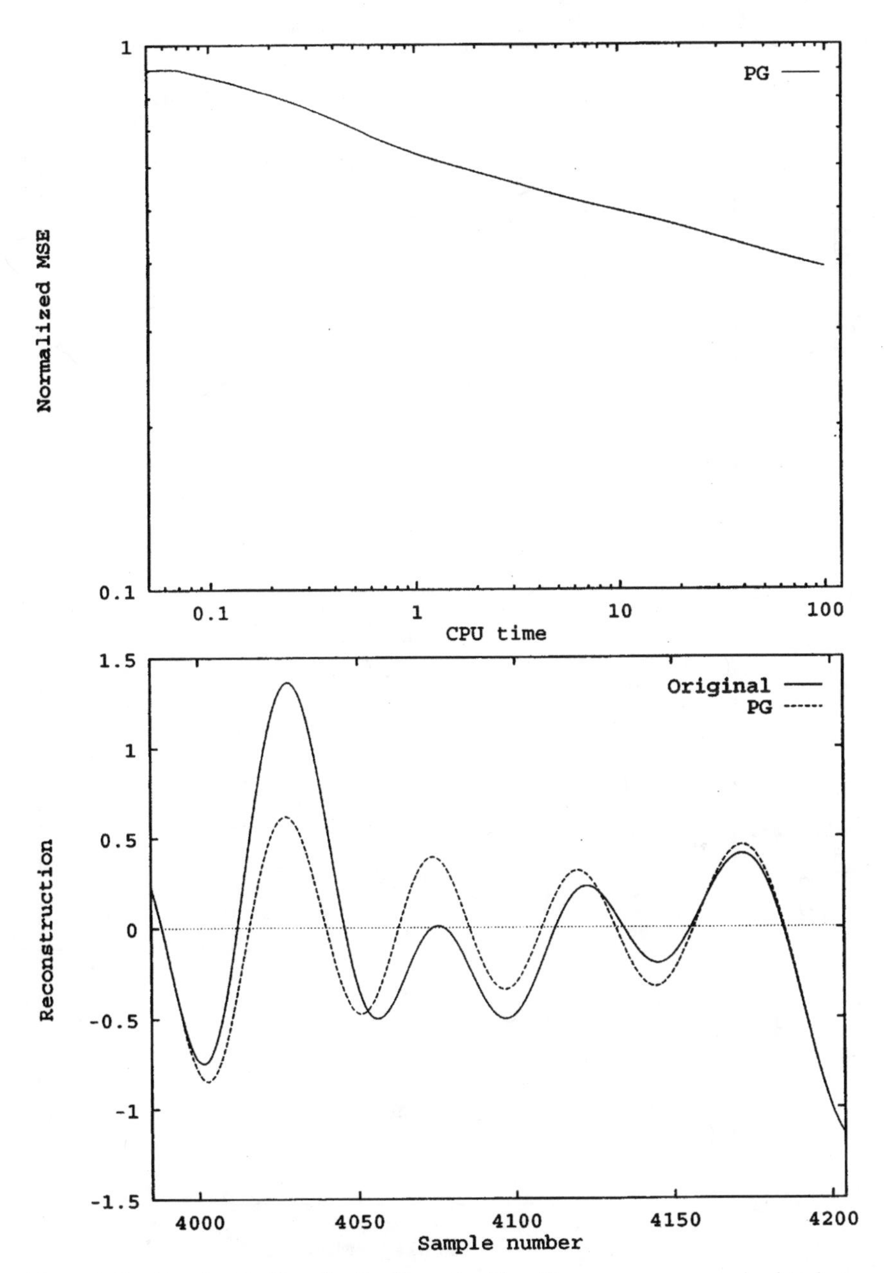

Figure 6. Results using the Papoulis–Gerchberg algorithm. The mean-square error is plotted as a function of time, not as a function of the number of iterations.

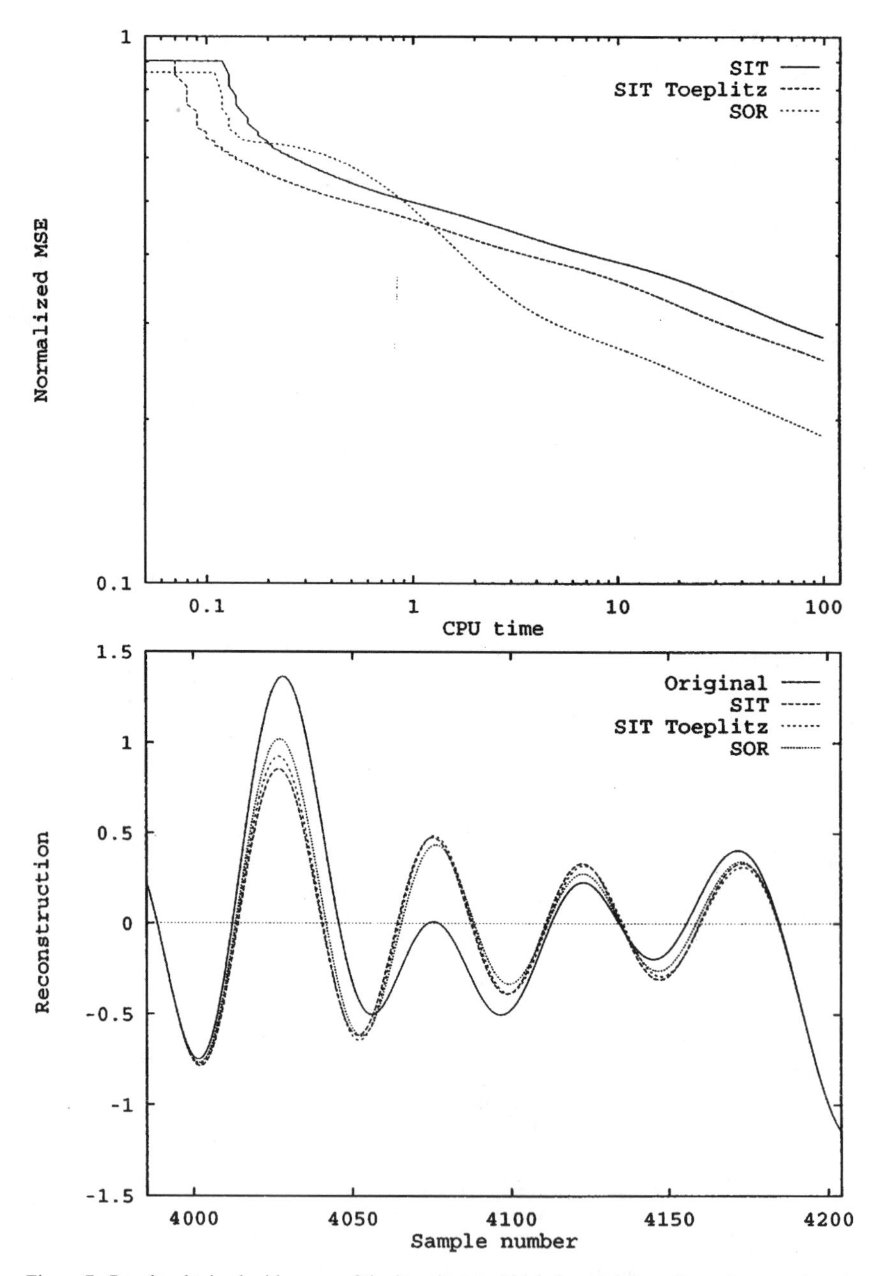

Figure 7. Results obtained with some of the iterations to which the minimum dimension equations in the time domain lead. The mean-square error is again plotted as a function of time.

opposed to $N \gg n$ in the Papoulis–Gerchberg algorithm. Thus, the iterations will in general be faster.

Any of the classical iterative methods (Jacobi, JOR, Gauss-Seidel, SOR, SSOR, among others) for the solution of linear systems can be applied, often leading to significant improvements in the convergence rate. A discussion of some of these methods can be found in [8]. Some results are given in Fig. 7.

When the iteration matrix is Toeplitz, as is the case for example when there are contiguous losses, the matrix-vector multiplications can be implemented using the FFT, leading to further speed-ups. The effect of performing the Toeplitz matrix-vector multiplications in the SIT method with the FFT is also demonstrated in Fig. 7. However, for the sampling set in the example, the convergence rate remains slow. Fortunately, there are much better alternatives.

First of all, the equations can be solved using standard *noniterative* techniques. There are two options: one is to start from the time domain equations, a particularly interesting option when the number of unknown samples n is small compared with the number of nonzero harmonics $2M + 1$. Another is to start from the frequency domain equations, which lead to more efficient methods when $2M + 1$ is sufficiently small compared with n.

Because the matrices of the minimum dimension equations are Hermitian and positive definite, Cholesky decomposition is a natural possibility (other methods, such as Gaussian elimination, LU decomposition, SVD, are of course also possible). Refer to Fig. 8 for examples using Cholesky decomposition.

Recall that when solving a set of equations $Ax = b$ using the Cholesky or LU decomposition methods, the factorization of A has to be performed only once, even if there are multiple right-hand sided vectors b that have to be processed. This is important since the time necessary to perform the decomposition dominates the total CPU time. This is clearly shown in Fig. 8. After computing the factorization of the matrix, the solution to linear problems involving the same matrix can be computed in negligible time.

Some further comments are in order. The frequency domain equations involve complex numbers, and the implementations of the frequency domain methods have to be carried out using complex arithmetic. Despite this, if $2M + 1$ is sufficiently small compared with n, the frequency domain methods can be preferable. The opposite is of course true when n is sufficiently small. There is no "faster" method in general.

The structure of the matrices should also be taken in consideration. The known frequency domain samples are contiguous (the data are low-pass), and hence the frequency domain equations are always Toeplitz. This is not necessarily the case with the time domain equations: the matrix can be Toeplitz, but for sparse or irregular distributions of unknown samples the Toeplitz structure will in general be absent.

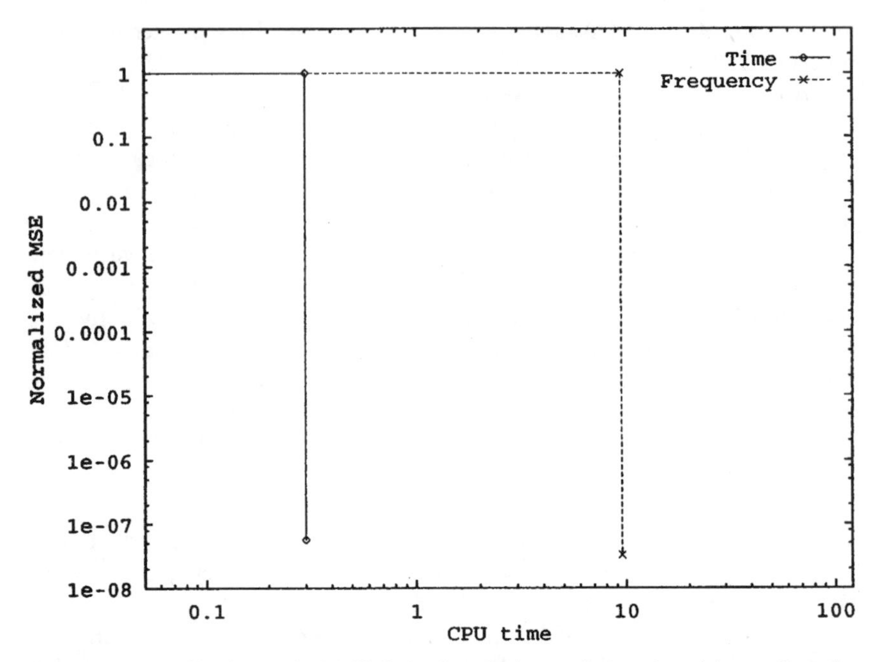

Figure 8. Two noniterative methods: Cholesky factorization applied to the minimum dimension equations, in both the time and frequency domains. The initial and final mean-square errors are plotted *versus* time. The horizontal lines correspond to the time spent in the Cholesky factorization itself, the remaining (and by comparison negligible) time corresponds to the solution of the equations. The reconstruction result is not shown (it is indistinguishable from the original).

For problem A, $n < 2M + 1$, and the two matrices P and Q that appear in the minimum dimension equations (20) (time) and (21) (frequency) are Toeplitz. Furthermore, P is real and Q complex. Therefore, the approaches based on the time domain equations should be preferable. The results, as we have seen, confirm this.

The conjugate gradient (CG) method also leads to efficient ways of solving the equations. The basic iteration can be easily modified to explore the structure of the matrix, whenever it exists, or to use preconditioning (see [46] for details).

There are classes of problems for which CG leads to very interesting performance. The combination of the conjugate gradient technique with the equations given in [28], as discussed in [29] and [30], leads to a very attractive algorithm. The equivalence between the minimum dimension formulation in the frequency domain and the equations given in [28] and [47] was established in [20].

It is of course possible to apply the CG method to either the time or frequency domain equations. This is demonstrated in Fig. 9. Is it better to use the

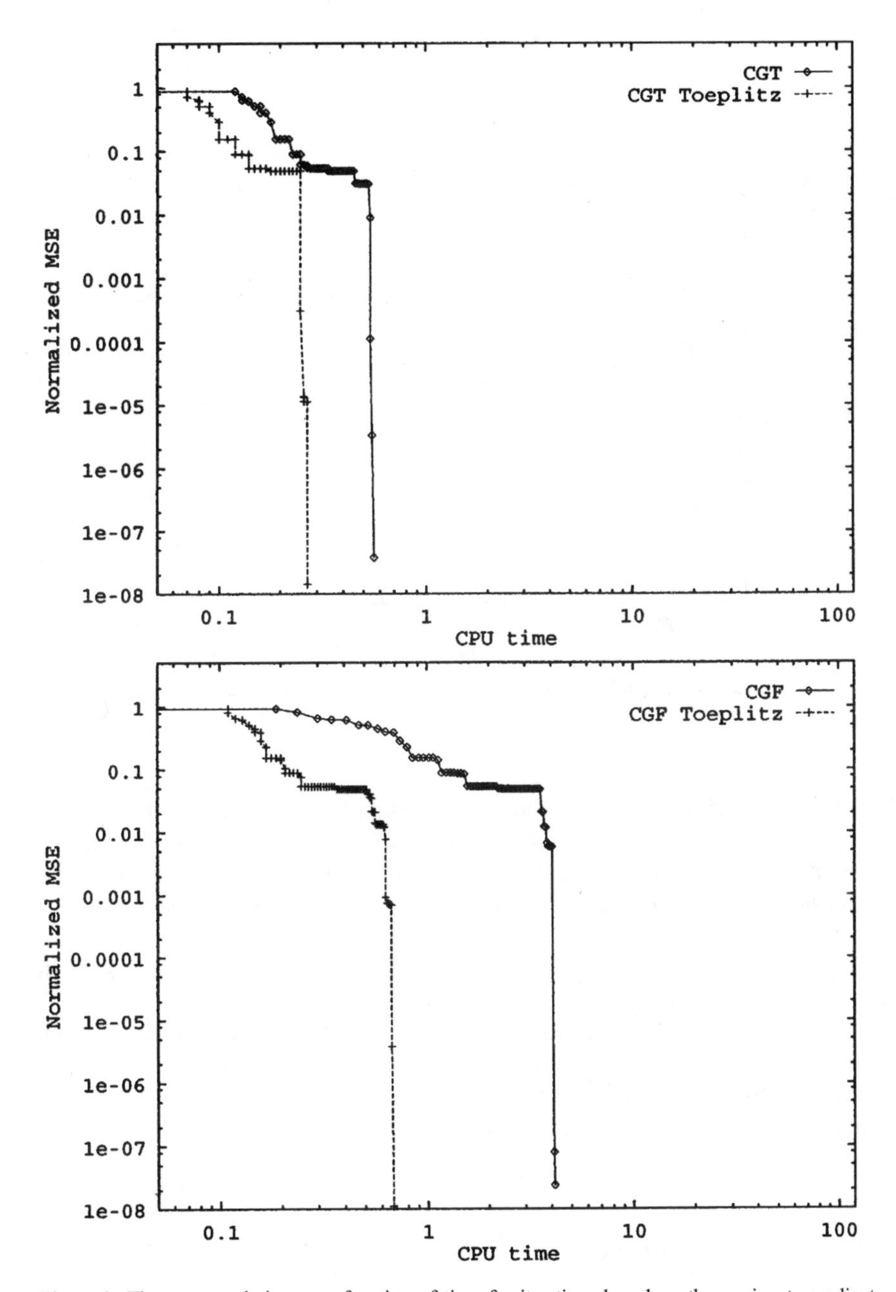

Figure 9. The error evolution as a function of time for iterations based on the conjugate gradient method, for the time domain (CGT) and frequency domain (CGF) equations, with and without taking advantage of the Toeplitz character of the matrix.

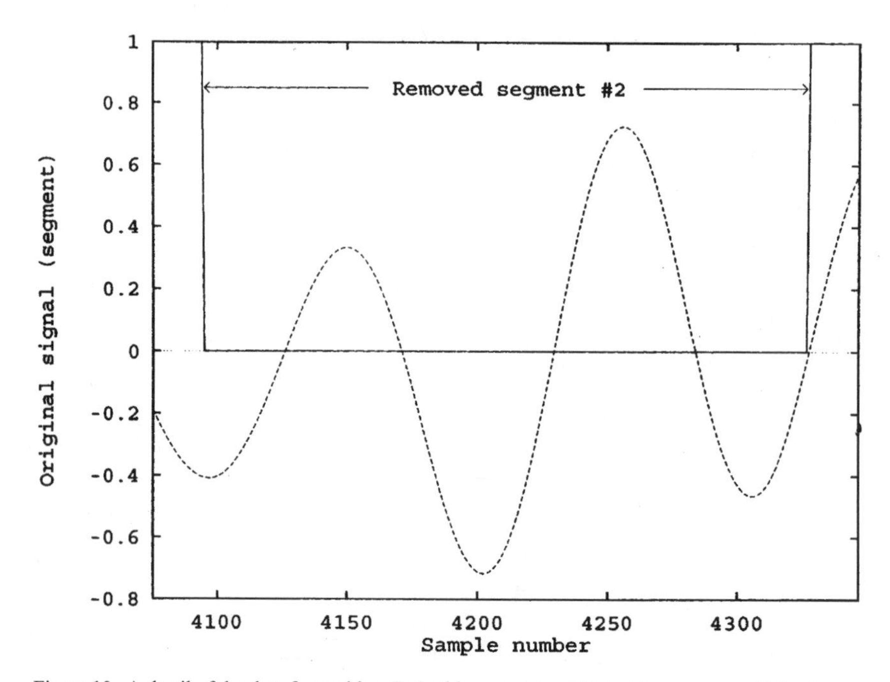

Figure 10. A detail of the data for problem B. In this case, $M = 100$, but there are $n = 700$ unknown samples, distributed in three segments of approximately equal length. The figure depicts the central segment.

time domain or the frequency domain equations? Again, there is not a unique answer, nor a "better" choice. One should at least consider the relative sizes of n and $2M + 1$, taking into account the need for complex arithmetic in the frequency domain case, and the structure of the matrices (Toeplitz or non-Toeplitz).

To demonstrate this, consider a second problem (problem B). In this case, $M = 100$ and $n = 700$, and the unknown samples are distributed in three separate segments of approximately equal length, one of which is depicted in Fig. 10. Therefore, the frequency domain equations have lower dimensionality, and the matrix of the time domain equations is not Toeplitz (although it has block structure, we did not take advantage of it). In this case, it turns out that the methods based on the frequency domain equations are preferable. This can be confirmed in Fig. 11 and Fig. 12.

5.5. Stability Analysis

The stability of the interpolation problem critically depends on the sampling set. Therefore, for both iterative and noniterative methods, assessment of the

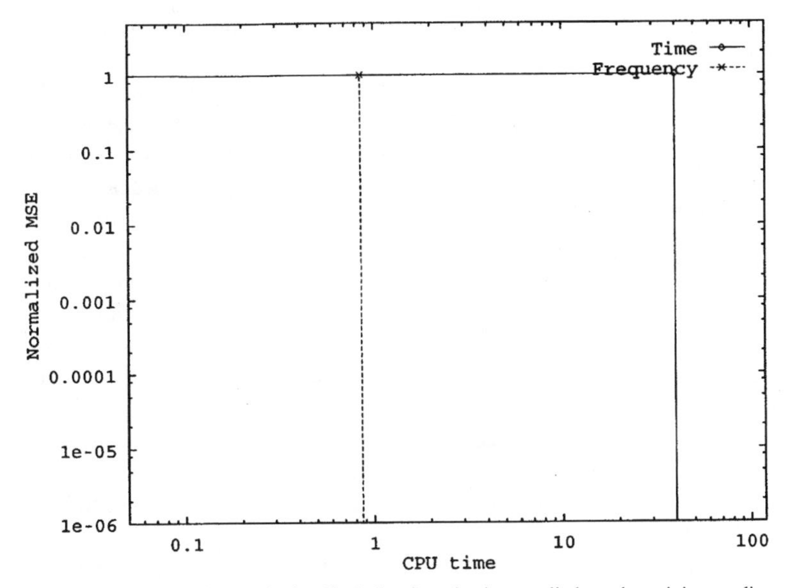

Figure 11. Two noniterative methods: Cholesky factorization applied to the minimum dimension equations, in both the time and frequency domains. The initial and final mean-square errors are plotted *versus* time. Compare with Fig. 8.

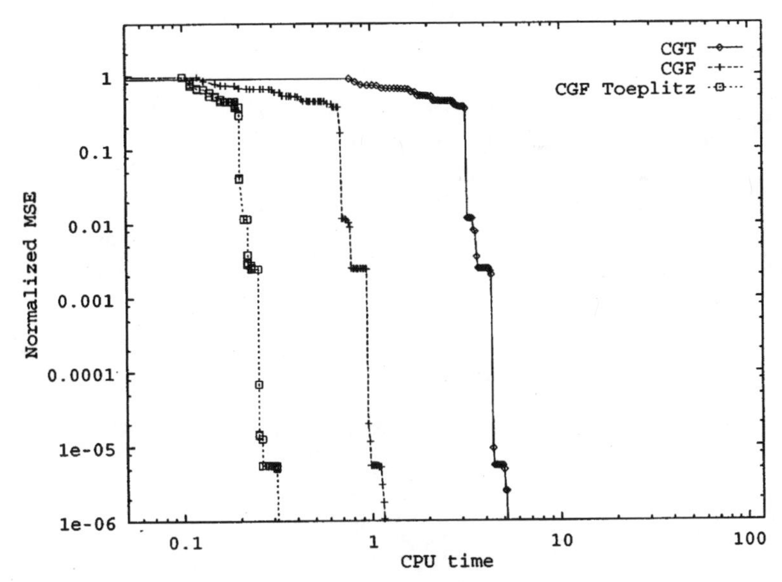

Figure 12. The error evolution as a function of time for iterations based on the conjugate gradient method, for the time and frequency domain equations. In the frequemcy domain case, the matrix is Toeplitz and it is possible to accelerate the algorithm using the FFT. Compare with Fig. 9.

sampling sets is a crucial task. Several results concerning this topic have been published.

References [23] and [27] deal with stability issues related to the interpolation of continuous-time band-limited signals from incomplete sets of samples. The approach in [48] is based on the minimization of the maximum of the mean-squared error with respect to the sampling set, whereas [24] studies, among other issues, the effect of noise and out-of-band components on the reconstruction.

Some stability aspects in the context of finite-dimensional signals were already discussed in Section 5.3, largely based on the methods employed in [13]. Other issues related to the stability of the band-limited reconstruction problem, also as a function of the sampling set, were studied in [49]. A review of several stability and eigenvalue problems in reference to the minimum dimension methods, including the 2-D case, is available in [50]. The interplay between eigenvalue problems, energy concentration, and the stability of the signal reconstruction problem, is discussed (using the language of frames) in [8].

As already seen in Section 5.3, for contiguous pass-bands (including low-pass and high-pass signals), equidistant losses lead to interpolation processes with optimum stability, and contiguous losses lead to ill posed problems. These conclusions do not necessarily hold for, say, bandpass signals.

In this section we address the stability problem using a minimum dimension formulation. Our aim is to supply accurate bounds for the eigenvalues of the interpolation matrices, in the spirit of [49]. The simplest bounds can be found in negligible time (they require only four arithmetic operations!). This dramatically simplifies the assessment of the sampling sets for this class of interpolation problems, yielding precise, quantitative numerical stability measures. It also explains the striking staircase-like behavior of the eigenvalues, illustrated in Fig. 13, which also demonstrates the effectiveness and accuracy of the bounds.

As seen in the previous sections, a band-limited signal $x \in \mathbb{C}^N$ can be reconstructed from a subset of its samples. As in Section 5.4, the unknown samples are denoted by

$$x(j_1), x(j_2), \ldots, x(j_n). \tag{23}$$

If the density of the known samples is above the critical value, the minimum dimension formulation in the time domain (see Theorem 3) leads to

$$u = Su + h, \tag{24}$$

which can then be solved for u, the vector of the n unknown samples (23). Recall that the $n \times 1$ vector h depends only on known data, and that S is a $n \times n$

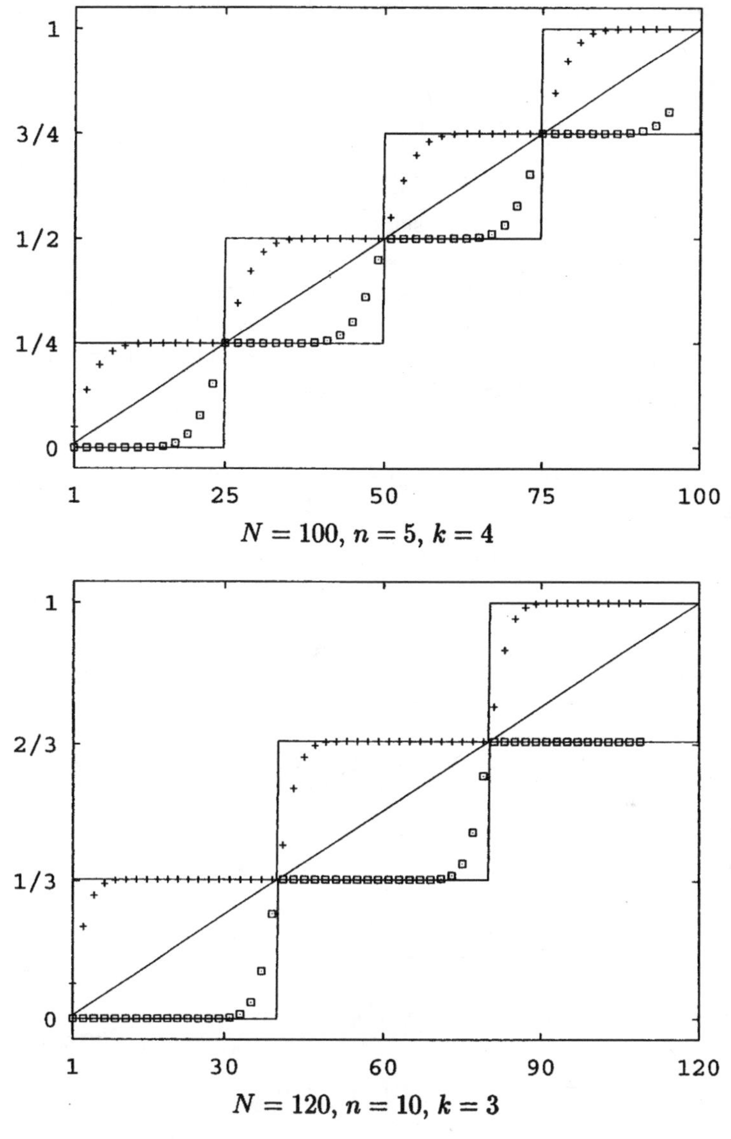

Figure 13. The extreme eigenvalues of S and the bounds predicted by Theorem 7, plotted as functions of the bandwidth $2M + 1$. The separation line $(2M + 1)/N$ predicted by Theorem 6 is also shown. The total number of samples is N, the number of unknown data is n, the gap between unknown samples is k. The bounds become increasingly tighter as N and n increase.

Hermitian matrix. If the signal is low-pass with $2M+1$ harmonics, the elements of S are

$$S_{pq} = \frac{1}{N}\sum_{\ell=-M}^{M} e^{\mathrm{j}(2\pi/N)(j_p-j_q)\ell} = \frac{\sin[\pi(2M+1)(j_p-j_q)/N]}{N\sin[\pi(j_p-j_q)/N]}. \tag{25}$$

when $p \neq q$. The diagonal elements S_{ii} equal $(2M+1)/N$, which is the normalized bandwidth of the data, denoted by w from now on.

In the following, v denotes an arbitrary nth dimensional column vector, with elements $v_1, v_2, \ldots, v_n$. The quadratic form associated with the matrix S is

$$v^H S v = \sum_{p=1}^{n}\sum_{q=1}^{n} v_p^* v_q \left[\frac{1}{N}\sum_{\ell=-M}^{M} e^{\mathrm{j}(2\pi/N)(j_p-j_q)\ell}\right] = \frac{1}{N}\sum_{\ell=-M}^{M}\left|\sum_{p=1}^{n} v_p e^{-\mathrm{j}(2\pi/N)j_p\ell}\right|^2.$$

We use the notation

$$v^H S v = \frac{1}{N}\sum_{\ell=-M}^{M} |\phi(\ell)|^2, \tag{26}$$

where

$$\phi(\ell) = \sum_{p=1}^{n} v_p e^{-\mathrm{j}(2\pi/N)j_p\ell}.$$

The set of locations of the unknown samples is denoted by U,

$$U = \{j_1, j_2, \ldots, j_n\},$$

and to emphasize the dependence of S and its eigenvalues on U we write $S(U)$ whenever convenient, or, even more explicitly, $S(\{j_1, j_2, \ldots, j_n\})$. Thus, for example, $S(\{0, 2, 7\})$ is the 3×3 matrix which arises in the recovery of samples $x(0)$, $x(2)$ and $x(7)$ of a band-limited signal. Note that the matrix S is a principal $n \times n$ submatrix of the low-pass $N \times N$ matrix B defined by (3). Thus, in this notation, $S(\{0, 1, \ldots, N-1\})$ reduces to B.

It follows from (26) that the eigenvalues of S are nonnegative. In fact, the spectrum (set of eigenvalues) of the $N \times N$ low-pass matrix B is, as we have seen, $\{0, 1\}$. Therefore, because S is a $n \times n$ principal submatrix of B, all of its eigenvalues belong to the interval $[0, 1]$. This follows from the interlacing inequalities [51].

We know that $I - S$ will be nonsingular if the number of known samples, $N - n$, exceeds the dimension of the subspace of low-pass signals with $2M + 1$ nonzero harmonics. Or, in other words, if the density of the known samples exceeds the normalized bandwidth $w = (2M + 1)/N$ of the data. This is also true for other classes of band-limited signals, provided that the associated matrix B has contiguous eigenvalues (this is the case for high-pass signals, for example). Under these conditions, the largest eigenvalue of S cannot be equal to one. However, a lot more can be said concerning the dependence of the eigenvalues of S on the sampling set U.

Theorem 6 *The minimum eigenvalue $\lambda_{\min}$ of S belongs to $[0, w]$, and the maximum eigenvalue $\lambda_{\max}$ to $[w, 1]$, with $w = (2M + 1)/N$.*

Proof: We have seen that the eigenvalues belong to $[0, 1]$. On the other hand,

$$n\lambda_{\min} \leq \sum_{i=1}^{n} \lambda_i = \sum_{i=1}^{n} S_{ii} = nw \leq n\lambda_{\max},$$

that is, $\lambda_{\min} \leq w \leq \lambda_{\max}$. □

The following result holds when U is a subset of an arithmetic progression, and gives more precise information concerning the eigenvalues. We use the following notations: $\lfloor x \rfloor$, for the largest integer smaller than or equal to x, and $\lceil x \rceil$, for the smallest integer greater than or equal to x.

Theorem 7 *If $U = \{j_1 k, j_2 k, \ldots, j_n k\}$, and N/k is an integer, the smallest and largest eigenvalues $\lambda_{\min}$ and $\lambda_{\max}$ of the matrix $S(U)$ defined by* (25) *satisfy*

$$\frac{\lfloor kw \rfloor}{k} \leq \lambda_{\min} \leq \lambda_{\max} \leq \frac{\lceil kw \rceil}{k},$$

where $w = (2M + 1)/N$.

Proof: This is true when $k = 1$ because the eigenvalues lie in $[0, 1]$. Therefore, assume that $k > 1$ and consider the matrix S as a function of M. Clearly, S

will be diagonal if and only if $(2M+1)k/N$ is an integer. The smallest value of $2M+1$ for which this can happen is

$$2M+1 = \frac{N}{k}. \tag{27}$$

Under the conditions stated, there is an integer M that solves this equation, and $S_{ii} = (2M+1)/N = 1/k$. But, when (27) holds, one has

$$v^H S v = \frac{1}{k}\sum_{i=1}^{n} |v_i|^2,$$

for any nth dimensional column vector v. This means that

$$\sum_{\ell=-M}^{M} |\phi(\ell)|^2 = \frac{1}{k}\sum_{i=1}^{n} |v_i|^2. \tag{28}$$

Now, let $0 < 2M+1 < N$. Under the hypothesis placed on U, ϕ is periodic with period N/k. The interval $[-M, M]$ contains at least

$$\left\lfloor \frac{2M+1}{N/k} \right\rfloor$$

and at most

$$\left\lceil \frac{2M+1}{N/k} \right\rceil$$

periods of ϕ. This and (28) implies

$$v^H S v = \sum_{\ell=-M}^{M} |\phi(\ell)|^2 \geq \frac{\lfloor kw \rfloor}{k}\sum_{i=1}^{n} |v_i|^2,$$
$$v^H S v = \sum_{\ell=-M}^{M} |\phi(\ell)|^2 \leq \frac{\lceil kw \rceil}{k}\sum_{i=1}^{n} |v_i|^2.$$

The conclusion now follows because the smallest and largest eigenvalues of S are respectively equal to the minimum and maximum values assumed by $v^H S v$, subject to $\|v\| = 1$. □

The bounds can be found in negligible time (they require only a total of four arithmetic operations, including the rounding operation and the computation of

w). Nevertheless, they are quite accurate, and can be combined with those of Theorem 6, as seen in Fig. 13.

We now turn to less regular sets U. Assume that there is a simple way of estimating the eigenvalues of $S(U_1)$ and $S(U_2)$, for some sets U_1 and U_2. Is it possible to combine those estimates to obtain information concerning the eigenvalues of the larger matrix $S(U_1 \cup U_2)$? The following result is based on this idea.

Theorem 8 *Let $U_{n-1} = \{j_1, j_2, \ldots, j_{n-1}\}$ and $U_n = U_{n-1} \cup \{j_n\}$. Denote by S_n and S_{n-1} the $n \times n$ and $(n-1) \times (n-1)$ reconstruction matrices associated with the sets U_n and U_{n-1}, respectively. Then,*

$$\lambda'(S_{n-1}) - \|v\| \leq \lambda(S_n) \leq \lambda'(S_{n-1}) + \|v\|, \tag{29}$$

where $v = [v_1, v_2, \ldots, v_{n-1}]$ and

$$v_k = \frac{\sin[\pi(2M+1)(j_k - j_n)/N]}{N \sin[\pi(j_k - j_n)/N]}, \qquad 1 \leq k \leq n-1. \tag{30}$$

The $\lambda'(S_{n-1})$ denote the sequence of the eigenvalues of S_{n-1}, together with w, sorted by increasing order, and $\lambda(S_n)$ denotes the eigenvalues of S_n, similarly ordered.

The proof appears in [49]. The result can also be useful when $U = U_1 \cup U_2$, U_1 being of the type addressed by Theorem 7, and U_2 arbitrary. The simplest possible case arises when U_2 has a single element (consider, for example, $U = \{0, 4, 12, 17\}$).

Corollary 1 *The smallest and largest eigenvalues $\lambda_{\min}(S_n)$ and $\lambda_{\max}(S_n)$ of S_n satisfy*

$$\lambda_{\min}(S_n) \geq \lambda_{\min}(S_{n-1}) - \|v\|,$$
$$\lambda_{\max}(S_n) \leq \lambda_{\max}(S_{n-1}) + \|v\|$$

Computing $\|v\|$ requires only $O(n)$ operations, but it can be seen that $\|v\|^2 \leq w(1-w)$, where, as always, $w = (2M+1)/N$ for low-pass signals. Several other variants and related results are discussed in [49].

The problem of obtaining bounds when the elements of the set U are subject to much weaker restrictions is less simple. Results in terms of the minimum separation distance can be found in [52], [53] and [54], in the discrete-discrete case and also in the context of L_2 signals.

5.5.1. Examples

The bounds for the eigenvalues can be applied in different contexts.

The estimation of the optimum relaxation parameter of the Papoulis–Gerchberg and similar algorithms is an example. As mentioned in Section 5.3.6, the optimum value of μ depends on the eigenvalues of certain matrices. For ill-posed or large problems, determination of the eigenvalues may be difficult or time-consuming. On the other hand, iteration with a sub-optimum relaxation parameter leads to comparatively low convergence rates.

The bounds lead to approximate values for the eigenvalues of S in (24). They also lead to estimates for the condition number of $I - S$, which give a quantitative measure of the feasibility of the reconstruction problem itself. See, for example, [49].

A simple example is given in Fig. 14 and Fig. 15. Note that in the situation depicted in Fig. 14, only eight data samples were unknown in a block of $N = 1024$, but the reconstruction error was still significant.

The situation in Fig. 15 appears to be hopeless, since the bandwidth and block length are the same, yet the number of samples to interpolated has increased by a factor of 25. However, the set U is more sparse, and the eigenvalue bounds predict a condition number of at most

$$\kappa(I - S) = \frac{1 - \lambda_{\min}}{1 - \lambda_{\max}} \leq \frac{0.5}{0.25} = 2.$$

Figure 14. Interpolating eight data samples in a record of $N = 1024$ samples with normalized bandwidth $(2M + 1)/N = 0.7$, using noisy data. The figure depicts the original data and the interpolated samples. Only 17 samples out of $N = 1024$ are shown. The indexes of the interpolated samples were multiples of two, and the bounds predicted by Theorem 7 were 0 and 1.

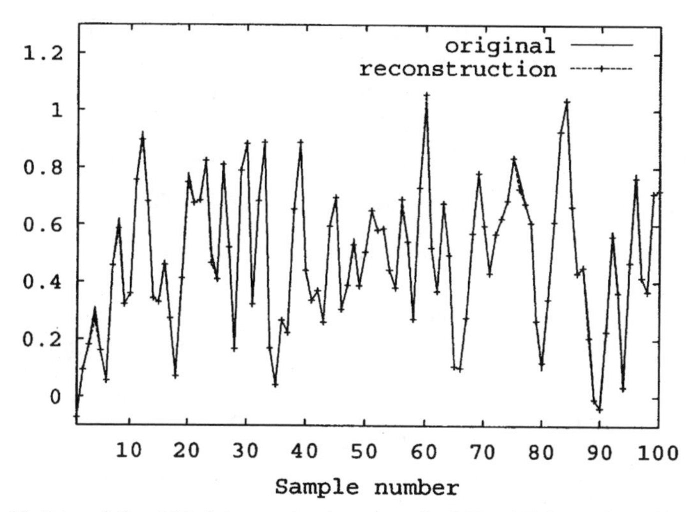

Figure 15. Interpolating 200 data samples in a record of $N = 1024$ samples with normalized bandwidth $(2M + 1)/N = 0.7$, using the same data used in Fig. 14. The figure depicts the original and reconstructed data. Only 100 samples out of $N = 1024$ are shown. The indexes of the interpolated samples were multiples of four, and the bounds predicted by Theorem 7 were 0.5 and 0.75.

The small condition number explains the unnoticeable interpolation error seen in Fig. 15.

5.6. Conclusions

In this chapter we have examined several iterative and noniterative methods for the reconstruction of partially known band-limited signals. The minimum dimension equations, in the time or frequency domains, were the starting point from which several possible implementations were derived and compared.

The minimum dimension matrices are Hermitian and nonnegative definite, and nonsingular if the density of the known data is above the Nyquist–Landau minimum density. Several possible strategies for dealing with the equations exist, and we have examined some of the several factors that the practitioner should consider when designing an algorithm for a specific problem. For the minimum dimension formulation in the time domain, the main issues and possibilities are:

- The matrices are real-valued, symmetric, $n \times n$. If the unknown data are consecutive, the matrices are also Toeplitz.
- The equations can be solved by simple iteration, or using iterative methods based on other additive splittings of the matrix (Jacobi, JOR, Gauss-Seidel, SOR, SSOR, block versions of these, etc.).
- If the matrix is Toeplitz, the Toeplitz-vector multiplications in the preceding methods can be computed using the FFT, leading to important speedups.

- The conjugate gradient method should also be considered. In this case, the matrix-vector products should of course be computed using the FFT whenever possible. Preconditioning can also have a favorable impact.
- Noniterative factorizations of the matrix (such as Cholesky decomposition, for example), or other noniterative techniques, can also be used.

The crucial points, in the frequency domain case, are as follows:

- In the low-pass case the matrices are always Toeplitz, of size $2M + 1 \times 2M + 1$. They are Hermitian and complex-valued, meaning that complex arithmetic and algorithms will have to be used.
- The equations can be solved using iterative techniques, such as Jacobi, JOR, Gauss-Seidel, SOR, SSOR, and so on. The necessary Toeplitz-vector multiplications can be carried out using the FFT.
- The preconditioned conjugate gradient method should be considered. Again, the matrix-vector products can be implemented using the FFT.
- Noniterative factorizations of the complex-valued matrix (say, Cholesky decomposition) or other standard noniterative techniques can also be used.

The algorithm for a given problem should be planned with these factors in mind. The time domain equations are of interest when the number of unknown samples n is small compared with the number of nonzero harmonics $2M + 1$. The frequency domain equations may lead to more efficient methods when $2M + 1$ is sufficiently small compared with n. The break-even points depend on the implementation details (factors such as cache size often lead to substantial performance differences).

When solving a set of equations using the Cholesky or LU decomposition methods, the factorization of the matrix has to be performed only once, even if there are multiple right-hand sided vectors that have to be processed. This is important since, as we have seen, the time necessary to perform the decomposition dominates the total CPU time. After computing the factorization of the matrix, the solution to linear problems involving the same matrix can be computed in negligible time. This can be successfully explored in image extrapolation or superresolution problems.

References

[1] J. L. C. Sanz and T. S. Huang. Discrete and Continuous Band-Limited Signal Extrapolation. *IEEE Trans. Acoust., Speech, Signal Processing*, 31(5):1276–1285, October 1983.

[2] A. Papoulis. A New Algorithm in Spectral Analysis and Band-Limited Extrapolation. *IEEE Trans. Circuits Syst.*, 22(9):735–742, September 1975.

[3] R. W. Gerchberg. Super-Resolution Through Error energy Reduction. *Optica Acta*, 21(9):709–720, 1974.

[4] H. G. Feichtinger and K. Gröchenig. Theory and Practice of Irregular Sampling. In J. J. Benedetto and M. W. Frazier, Eds, *Wavelets: Mathematics and Applications*, CRC Press, Boca Raton, 1994, pp. 305–363.

[5] D. C. Youla. Generalized Image Restoration by the Method of Alternating Orthogonal Projections. *IEEE Trans. Circuits Syst.*, 25(9):694–702, September 1978.

[6] M. I. Sezan and H. Stark. Image Restoration by the Method of Convex Projections: Part 2—Applications and Numerical Results. *IEEE Trans. Med. Imag.*, 1(2):95–101, October 1982.

[7] D. C. Youla and H. Webb. Image Restoration by the Method of Convex Projections: Part 1—Theory. *IEEE Trans. Med. Imag.*, 1(2):81–94, October 1982.

[8] P. J. S. G. Ferreira. Mathematics for Multimedia Signal Processing II—Discrete Finite Frames and Signal Reconstruction. In J. S. Byrnes, Ed., *Signal Processing for Multimedia*, IOS Press, 1999, pp. 35–54. Presented at the NATO Advanced Study Institute "Signal Processing for Multimedia", Il Ciocco, Italy, July 1998.

[9] R. W. Schafer, R. M. Mersereau, and M. A. Richards. Constrained Iterative Restoration Algorithms. *Proc. IEEE*, 69(4):432–450, April 1981.

[10] R. S. Varga. *Matrix Iterative Analysis*. Prentice-Hall, Englewood Cliffs, New Jersey, 1962.

[11] D. M. Young. *Iterative Solution of Large Linear Systems*, (3rd edn.) Academic Press, Orlando, 1971.

[12] M. C. Jones. The Discrete Gerchberg Algorithm. *IEEE Trans. Acoust., Speech, Signal Processing*, 34(3):624–626, June 1986.

[13] P. J. S. G. Ferreira. Interpolation and the Discrete Papoulis–Gerchberg Algorithm. *IEEE Trans. Signal Processing*, 42(10):2596–2606, October 1994.

[14] V. T. Tom, T. F. Quatieri, M. H. Hayes, and J. H. McClellan. Convergence of Iterative Nonexpansive Signal Reconstruction Algorithms. *IEEE Trans. Acoust., Speech, Signal Processing*, 29(5):1052–1058, October 1981.

[15] D. R. Smart. *Fixed Point Theorems*. Cambridge University Press, Cambridge, 1980.

[16] A. K. Jain and S. Ranganath. Extrapolation Algorithms for Discrete Signals with Application in Spectral Estimation. *IEEE Trans. Acoust., Speech, Signal Processing*, 29(4):830–845, August 1981.

[17] P. S. Naidu and B. Paramasivaiah. Estimation of Sinusoids from Incomplete Time Series. *IEEE Trans. Acoust., Speech, Signal Processing*, 32(3):559–562, June 1984.

[18] C. I. Haupt, N. Schuff, M. W. Weiner, and A. A. Maudsley. Removal of Lipid Artifacts in ^{1}H Spectroscopic Imaging by Data Extrapolation. *Magn. Reson. Medic.*, 35(5):678–687, May 1996.

[19] P. J. S. G. Ferreira. Noniterative and Faster Iterative Methods for Interpolation and Extrapolation. *IEEE Trans. Signal Processing*, 42(11):3278–3282, November 1994.

[20] P. J. S. G. Ferreira. Interpolation in the Time and Frequency Domains. *IEEE Sig. Proc. Letters*, 3(6):176–178, June 1996.

[21] P. J. S. G. Ferreira. The Stability of Certain Image Restoration Problems: Quantitative Results. In *Proceedings of the Second IEEE International Conference on Image Processing, ICIP-95*, volume II, Washington, D.C., U.S.A., October 1995, pp. 29–32.

[22] P. J. S. G. Ferreira and A. J. Pinho. Errorless Restoration Algorithms for Band-Limited Images. In *Proceedings of the First IEEE International Conference on Image Processing, ICIP-94*, volume III, Austin, TX, U.S.A., November 1994, pp. 157–161.

[23] P. J. S. G. Ferreira. Incomplete Sampling Series and the Recovery of Missing Samples from Oversampled Band-Limited Signals. *IEEE Trans. Signal Processing*, 40(1):225–227, January 1992.

[24] P. Delsarte, A. J. E. M. Janssen, and L. B. Vries. Discrete Prolate Spheroidal Wave Functions and Interpolation. *SIAM J. Appl. Math.*, 45(4):641–650, August 1985.

[25] R. J. Marks II. Restoring Lost Samples from an Oversampled Band-Limited Signal. *IEEE Trans. Acoust., Speech, Signal Processing*, 31(3):752–755, June 1983.

[26] R. J. Marks II. *Introduction to Shannon Sampling and Interpolation Theory.* Springer, Berlin, 1991.

[27] P. J. S. G. Ferreira. The Stability of a Procedure for the Recovery of Lost Samples in Band-Limited Signals. *Sig. Proc.*, 40(3):195–205, December 1994.

[28] K. Gröchenig. A Discrete Theory of Irregular Sampling. *Linear Algebra Appl.*, 193:129–150, 1993.

[29] T. Strohmer. On Discrete Band-Limited Signal Extrapolation. In M. E. H. Ismail, M. Z. Nashed, A. I. Zayed, and A. F. Ghaleb, Eds., *Mathematical Analysis, Wavelets, and Signal Processing*, volume 190 of *Contemporary Mathematics*, American Mathematical Society, Providence, Rhode Island, 1995, pp. 323–337.

[30] T. Strohmer. Computationally Attractive Reconstruction of Bandlimited Images from Irregular Samples. *IEEE Trans. Image Processing*, 6(4):540–548, April 1997.

[31] P. J. S. G. Ferreira. Fast Iterative Reconstruction of Distorted ECG Signals. In *Proceedings of the 14th Annual International Conference of the IEEE Engineering in Medicine and Biology Society*, volume II, Paris, France, October 1992, pp. 777–778.

[32] S. D. Cabrera and T. W. Parks. Extrapolation and Spectral Estimation with Iterative Weighted Norm Modification. *IEEE Trans. Signal Processing*, 39(4):842–851, April 1991.

[33] F. Marvasti, M. Hasan, M. Echhart, and S. Talebi. Efficient Algorithms for Burst Error Recovery Using FFT and Other Transform Kernels. *IEEE Trans. Signal Processing*, 47(4):1065–1075, April 1999.

[34] F. Marvasti and T. J. Lee. Analysis and Recovery of Sample-and-Hold and Linearly Interpolated Signals with Irregular Samples. *IEEE Trans. Signal Processing*, 40(8):1884–1891, August 1992.

[35] F. A. Marvasti, C. Liu, and G. Adams. Analysis and Recovery of Multidimensional Signals from Irregular Samples Using Nonlinear and Iterative Techniques. *Sig. Proc.*, 36:13–30, 1993.

[36] F. M. Marvasti, P. M. Clarkson, M. V. Dokic, U. Goenchanart, and C. Liu. Reconstruction of Speech Signals with Lost Samples. *IEEE Trans. Signal Processing*, 40(12):2897–2903, December 1992.

[37] M. Nafie and F. M. Marvasti. Implementation of Recovery of Speech with Missing Samples on a DSP Chip. *Electron. Letters*, 30(1):12–13, January 1994.

[38] A. Papoulis and C. Chamzas. Detection of Hidden Periodicities by Adaptive Extrapolation. *IEEE Trans. Acoust., Speech, Signal Processing*, 27(5):492–500, October 1979.

[39] L. C. Potter and K. S. Arun. Energy Concentration in Band-Limited Extrapolation. *IEEE Trans. Acoust., Speech, Signal Processing*, 37(7):1027–1041, July 1989.

[40] J. L. C. Sanz and T. S. Huang. Some Aspects of Band-Limited Signal Extrapolation: Models, Discrete Approximations, and Noise. *IEEE Trans. Acoust., Speech, Signal Processing*, 31(6):1492–1501, December 1983.

[41] J. L. C. Sanz and T. S. Huang. Unified Hilbert Space Approach to Iterative Least-Squares Linear Signal Restoration. *J. Opt. Soc. Am.*, 73(11):1455–1465, November 1983.

[42] J. L. C. Sanz and T. S. Huang. A Unified Approach to Noniterative Linear Signal Restoration. *IEEE Trans. Acoust., Speech, Signal Processing*, 32(2):403–409, April 1984.

[43] B. J. Sullivan and B. Liu. On the use of Singular Value Decomposition and Decimation in Discrete-Time Band-Limited Signal Extrapolation. *IEEE Trans. Acoust., Speech, Signal Processing*, 32(6):1201–1212, December 1984.

[44] C. K. W. Wong, F. Marvasti, and W. G. Chambers. Implementation of Recovery of Speech with Impulsive Noise on a DSP chip. *Electron. Letters*, 31(17):1412–1413, August 1995.

[45] W. Y. Xu and C. Chamzas. On the Extrapolation of Band-Limited Functions with Energy Constraints. *IEEE Trans. Acoust., Speech, Signal Processing*, 31(5):1222–1234, October 1983.

[46] G. H. Golub and C. F. Van Loan. *Matrix Computations.* The Johns Hopkins University Press, Baltimore, 1989.

[47] T. Strohmer. *Efficient Methods for Digital Signal and Image Reconstruction from Nonuniform Samples.* PhD thesis, Institut für Mathematik der Universität Wien, November 1993.

[48] D. S. Chen and J. P. Allebach. Analysis of Error in Reconstruction of Two-Dimensional Signals from Irregularly Spaced Samples. *IEEE Trans. Acoust., Speech, Signal Processing*, 35(2):173–180, February 1987.

[49] P. J. S. G. Ferreira. The Eigenvalues of Matrices which Occur in Certain Interpolation Problems. *IEEE Trans. Signal Processing*, 45(8):2115–2120, August 1997.

[50] P. J. S. G. Ferreira. A Class of Eigenvalue Problems in Interpolation, Extrapolation and Sampling. In *SampTA'95, 1995 Workshop on Sampling Theory and Applications*, Jurmala, Latvia, September 1995, pp. 125–136.

[51] R. A. Horn and C. R. Johnson. *Matrix Analysis.* Cambridge University Press, Cambridge, 1990.

[52] P. J. S. G. Ferreira. The Condition Number of Certain Matrices and Applications. In *Proceedings of the IEEE International Conference on Acoustics, Speech, and Signal Processing, ICASSP 99*, Phoenix, Arizona, U.S.A., March 1999, pp. 2043–2046.

[53] P. J. S. G. Ferreira. Superresolution, the Recovery of Missing Samples, and Vandermonde Matrices on the Unit Circle. In *Proceedings of the 1999 Workshop on Sampling Theory and Applications, SampTA'99*, Loen, Norway, August 1999, pp. 216–220.

[54] P. J. S. G. Ferreira. Stability Issues in Error Control Coding in the Complex Field, Interpolation, and Frame Bounds. *IEEE Sig. Proc. Letters*, 7(3), March 2000.

6

Numerical and Theoretical Aspects of Nonuniform Sampling of Band-Limited Images

K. Gröchenig and T. Strohmer

Abstract

This chapter presents recent developments and progress in nonuniform sampling of band-limited functions. Since the theory and practice of nonuniform sampling of 1-D signals are well understood and already treated in many articles and surveys, the emphasis will be on the nonuniform sampling of images. Both numerical, theoretical, and applied aspects of the sampling problem will be considered. For Matlab© simulations, see the CD-ROM: 'Chapter6\matlab'.

6.1. Introduction

The numerical processing of data, which are often given in the form of signals or images, requires always a discretization of the data by means of sampling. This means that instead of the signal or image f only its sampled values $y_i = f(x_i)$, where i runs through some index set, are stored. At the end, possibly after processing these samples, one wants to recover the original signal f from its samples. However, this is an ill-posed problem, since the subspace of

K. Gröchenig • Department of Mathematics, The University of Connecticut, Storrs, CT 06269-3009; E-mail: groch@math.uconn.edu
T. Strohmer • Department of Mathematics, University of California at Davis, Davis, CA 95616-8633; E-mail: strohmer@math.ucdavis.edu. T. Strohmer was supported by NFS grant 9973373.

Nonuniform Sampling: Theory and Practice, edited by Marvasti
Kluwer Academic/Plenum Publishers, New York, 2001.

functions f with $f(x_i) = y_i$ has always infinite dimension. In practical applications, signals and images are not arbitrary functions, but usually they possess some smoothness and decay properties. Usually, it is assumed that the given functions are band-limited. For band-limited functions there is an extensive theory to show how a signal or an image is well-defined by its samples and how it can be reconstructed. Moreover, it is often assumed that the signal or image is sampled uniformly on a set of the form $\alpha\mathbb{Z}$ (i.e., the set $\{\alpha k\}_{k\in\mathbb{Z}}$) or on a lattice $\alpha\mathbb{Z}^2$. This situation is well-understood and can be treated with a diligent use of the Poisson summation formula. In recent years nonuniform sampling sets appear in more and more applications. The processing of nonuniformly spaced data is required in astronomy, geophysics, medical imaging, radar, and telecommunication systems.

In this contribution we treat some recent developments in nonuniform sampling of band-limited images. Nonuniform sampling of signals in 1-D is well-understood and well documented. The reader should consult the standard references [1], [2], Chapters 3 and 4 in this book, or the SAMPTA proceedings [3] and [4] for information. Our own point of view and an efficient numerical approach are contained in [5] and in [6]. See also [7, 8], and [9].

In higher dimensions our understanding of nonuniform sampling is less complete. In this chapter we survey recent results and algorithms for nonuniform sampling in higher dimensions. The emphasis is on the numerical aspects, but we also provide theoretical results to underline the efficiency and robustness of these algorithms. Other results and methods can be found for instance in [10–13] and [14].

We have adapted most of our notation to the conventions in engineering signal processing. We avoid vector notation and spell out the coordinates (x, y) of an image pixel explicitly. Therefore we discuss everything in the context of functions in 2-D. However, all results can be extended to arbitrary dimensions by a simple adjustment of the notation. We use j for the imaginary unit. A function or signal is denoted by a small letter, its Fourier transform is denoted by the corresponding capitalized letter (e.g., F is the Fourier transform of the signal f). Sometimes it is convenient to interpret matrices as vectors. This is done by simply stacking the columns of a matrix.

6.2. Band-Limited Images and Finite-Dimensional Models

In data processing several models of band-limited functions are in use. We discuss the traditional class of band-limited functions of two variables in $\mathbb{R}^2$, discrete band-limited images, and two-dimensional trigonometric polynomials.

6.2.1. Band-limited Images

A band-limited image with spectrum in the square $[-\Omega, \Omega] \times [-\Omega, \Omega]$ is a function in $\boldsymbol{L}^2(\mathbb{R}^2)$ that can be written as

$$f(x, y) = \int_{-\Omega}^{\Omega} \int_{-\Omega}^{\Omega} F(\xi, \eta)\, e^{2\pi j(x\xi + y\eta)} d\xi d\eta. \tag{1}$$

Here F is the Fourier transform of f, defined as

$$F(\xi, \eta) = \int_{\mathbb{R}} \int_{\mathbb{R}} f(x, y) e^{-2\pi j(x\xi + y\eta)}\, dxdy$$

We write $\boldsymbol{B}_\Omega$ for the space of all band-limited functions of the form (1) with finite energy $\|f\|_2^2 = \int_{\mathbb{R}} \int_{\mathbb{R}} |f(x, y)|^2\, dxdy$.

Given a uniform or nonuniform sampling set $X = \{(x_i, y_i), i \in I\}$, the problem in many applications is to recover f from its samples $f(x_i, y_i), i \in I$.

In the case of uniform sampling on the lattice $\frac{1}{2\Omega}\mathbb{Z}^2 = \left\{\left(\frac{k}{2\Omega}, \frac{l}{2\Omega}\right), k, l \in \mathbb{Z}\right\}$, the higher dimensional versions of the Shannon-Whittaker-Kotel'nikov sampling theorem provide a reconstruction. Specifically,

$$f(x, y) = \sum_{k \in \mathbb{Z}} \sum_{l \in \mathbb{Z}} f\left(\frac{k}{2\Omega}, \frac{l}{2\Omega}\right) \frac{\sin 2\pi\Omega\left(x - \frac{k}{2\Omega}\right) \sin 2\pi\Omega\left(y - \frac{l}{2\Omega}\right)}{\pi^2\left(x - \frac{k}{2\Omega}\right)\left(y - \frac{l}{2\Omega}\right)}. \tag{2}$$

The so-called cardinal series is easily derived with the Poisson summation formula [15].

Sometimes other spectra are considered. This means that we consider images f for which F vanishes outside a compact set S. If f has its spectrum S in the rectangle $[-A, A] \times [-B, B]$, then the function $f(Ax, By)$ has its spectrum in $[-1, 1] \times [-1, 1]$. Similarly if the spectrum is a parallelogram, then a simple coordinate change transforms f into an image with spectrum in the square $[-1, 1] \times [-1, 1]$. Thus for most purposes it suffices to consider the sampling problem in the space $\boldsymbol{B}_\Omega$. In certain applications, in particular in problems with a radial symmetry, the spectrum is a circle (or a ball in higher dimensions).

6.2.2. Discrete Band-limited Images

For numerical implementations in real applications two new considerations have to be added: (a) One is forced to choose a finite-dimensional model for images and for band-limitedness, because the numerical treatment of a linear problem must be set in a finite-dimensional vector space. (b) The number of data (samples of f) is always finite.

For this very reason reconstruction formulas like the cardinal series (2) and the large number of sampling theorems in the mathematical literature, e.g. [10, 16–18], are of limited use in practical questions.

The reduction to a finite-dimensional model is usually done by representing an image by a matrix u with entries $u(k, l)$, $k, l = 0, \dots, N-1$. An image has bandwidth $M < N/2$, if it can be written as

$$u(k, l) = \sum_{m=-M}^{M} \sum_{n=-M}^{M} U(m, n) e^{2\pi j(km+ln)/N}. \tag{3}$$

In this context the sampling problem asks to find a reconstruction of u from its samples $u(k_i, l_i), i = 1, \dots, r$, where the points (k_i, l_i) are distributed nonuniformly in the square $[0, N-1] \times [0, N-1] \cap \mathbb{Z}^2$. If the sampling set is uniform, i.e., if it is a grid of the form (ak, al), then a discrete version of the Poisson summation formula yields the reconstruction formula from the samples $u(ak, al)$. See the next section, formula (7). In this case a fast and efficient numerical reconstruction can be obtained with FFT methods. In the nonuniform case the problem is much harder, since the system matrix lacks structure.

6.2.3. Trigonometric Polynomials

For an analytical approach it is useful to reinterpret the representation (3) as a discretization of a trigonometric polynomial. Let

$$p(x, y) = \sum_{m=-M}^{M} \sum_{n=-M}^{M} U(m, n) e^{2\pi j(mx+ny)}$$

be the trigonometric polynomial of degree M in both the x and y-direction. Then

$$u(k, l) = p(k/N, l/N) \tag{4}$$

is just a uniformly sampled version of p. Note that p has period one in each direction. For images on a rectangle of dimension $[-A, A] \times [-B, B]$, the scaling $p(x, y) \to p\left(\frac{x}{2A}, \frac{y}{2B}\right)$ produces trigonometric polynomials with periods $2A$ and $2B$ respectively.

We will write $\boldsymbol{P}_M$ for the vector space of all trigonometric polynomials

$$p(x, y) = \sum_{m=-M}^{M} \sum_{n=-M}^{M} a(m, n) e^{2\pi j(mx+ny)} \tag{5}$$

of order M in x and in y with coefficients $a(m, n) \in \mathbb{C}$.

For the numerical analysis of the sampling problem for images, it is most convenient to regard a band-limited image on a rectangle as trigonometric polynomial in two variables, rather than as a discrete band-limited image as in (3). There are several advantages to this point of view. The discrete model is a special case of trigonometric polynomials. However, as trigonometric polynomials are functions of the continuous variables x, y, tools from mathematical analysis become available, for instance derivatives. Moreover, whereas the sampling points in the discrete model (k_i, l_i) correspond to the points $(k_i/N, l_i/N)$ in the unit square, the (nonuniform) sampling points in the continuous and periodic model are no longer restricted to be selected from a grid.

The case where the sampling points are taken from a grid (illustrated in Fig. 1(a)) is also called *missing data problem*, since one has to reconstruct missing samples of a discrete signal. The case where the sampling values of a continuous signal can be arbitrarily located (shown in Fig. 1(b)) is usually referred to as *scattered data problem*. By using trigonometric polynomials as a finite-dimensional model we do not have to distinguish between these two types.

For completeness we state the regular sampling theorem for trigonometric polynomials. The following formula corresponds to the cardinal series in (2) and elucidates the Nyquist rate for $\boldsymbol{P}_M$. Let

$$D_M(x) = \frac{1}{2M+1} \sum_{k=-M}^{M} e^{2\pi jkx} = \frac{\sin(2M+1)\pi x}{(2M+1)\sin \pi x} \tag{6}$$

be the Dirichlet kernel, then for every $p \in \boldsymbol{P}_M$ we have the reconstruction formula from the samples on the Nyquist grid $\left(\frac{k}{2M+1}, \frac{l}{2M+1}\right)$

$$p(x,y) = \sum_{k=0}^{2M} \sum_{l=0}^{2M} p\left(\frac{k}{2M+1}, \frac{l}{2M+1}\right) \times D_M\left(x - \frac{k}{2M+1}\right) D_M\left(y - \frac{l}{2M+1}\right). \tag{7}$$

Proof: We substitute (5) and the formula (6) for the Dirichlet kernel into the right-hand side of (7). To simplify the resulting expression, we use the following formula for exponential sums:

$$\frac{1}{2M+1} \sum_{k=0}^{2M} e^{2\pi jk(m-r)/(2M+1)} = \delta_{mr},$$

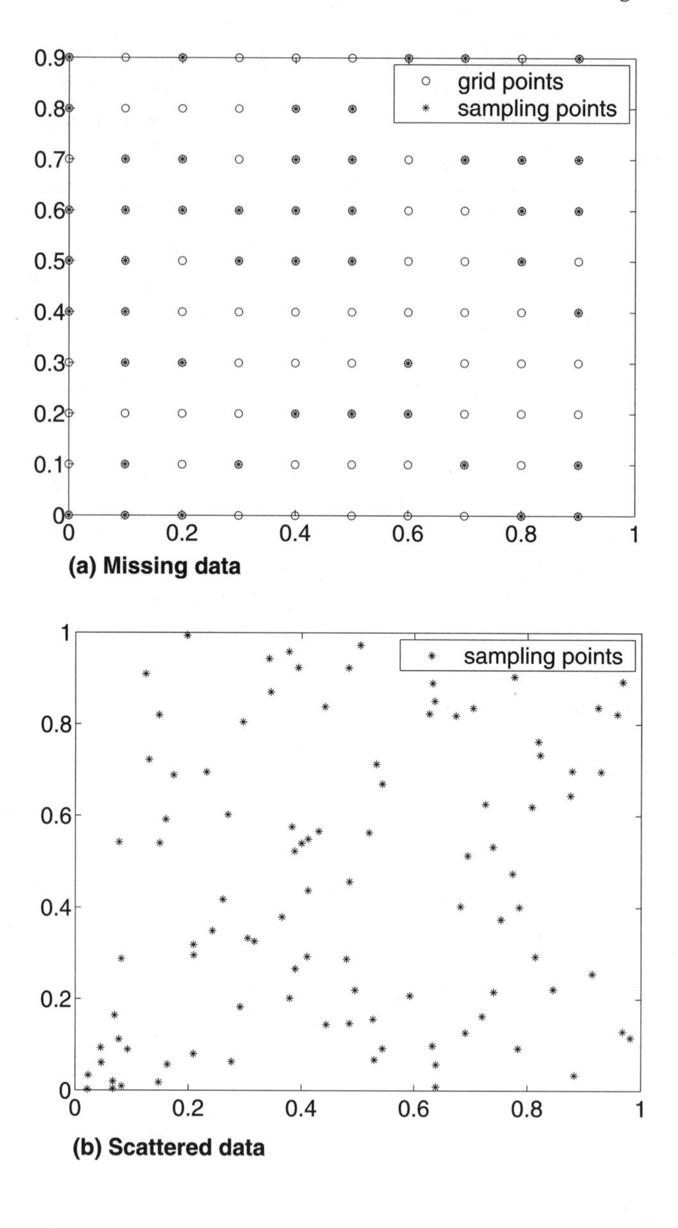

Figure 1. By using trigonometric polynomials as finite model we do not have to distinguish between the problems of missing data and scattered data.

where $\delta_{mr} = 1$ if $m = r$ and $\delta_{mr} = 0$ if $m \neq r$. We obtain

$$\frac{1}{(2M+1)^2} \sum_{k,l=0}^{2M} \sum_{m,n=-M}^{M} a(m,n) e^{2\pi j \left(\frac{mk+nl}{2M+1}\right)} \sum_{r,s=-M}^{M} e^{2\pi jr(x-k/(2M+1))} e^{2\pi js(x-l/(2M+1))}$$

$$= \sum_{m,n,r,s=-M}^{M} a(m,n) e^{2\pi j(rx+sy)} \left(\frac{1}{2M+1} \sum_{k=0}^{2M} e^{2\pi jk(m-r)/(2M+1)} \right)$$

$$\times \left(\frac{1}{2M+1} \sum_{l=0}^{2M} e^{2\pi jl(n-s)/(2M+1)} \right)$$

$$= \sum_{m,n,r,s=-M}^{M} a(m,n) e^{2\pi j(rx+sy)} \delta_{mr} \delta_{ns}$$

$$= \sum_{m,n=-M}^{M} a(m,n) e^{2\pi j(mx+ny)} = p(x,y),$$

as desired. □

6.3. Numerical Methods

Suppose that the samples of a trigonometric polynomial p of degree M are given at the r sampling points (x_i, y_i), $j = 1, \ldots, r$, in the unit square $[0, 1] \times [0, 1]$. For simplicity of notation we write $s_i := p(x_i, y_i)$ and $s = \{s_i\}_{i=1}^{r}$. Slightly more restrictive, the samples of a discrete signal u are given at r integer points (k_i, l_i), $i = 1, \ldots, r$, in $[0, N-1] \times [0, N-1]$.

6.3.1. Vandermonde Systems

Then a reconstruction of p from these samples amounts to the solution of the following system of linear equations for the unknown Fourier coefficients $a(m, n)$:

$$\sum_{m=-M}^{M} \sum_{n=-M}^{M} a(m,n) e^{2\pi j(mx_i+ny_i)} = p(x_i, y_i) = s_i \tag{8}$$

for $i = 1, \ldots, r$. These are r linear equations in the $(2M+1)^2$ unknown coefficients $a(m,n)$. From a dimension count we can deduce the following observation.

Lemma 1 *If every trigonometric polynomial $p \in \boldsymbol{P}_M$ of degree M is uniquely determined by its samples $p(x_i, y_i)$, $i = 1, \ldots, r$, then the system matrix has full rank, and thus $r \geq (2M+1)^2$.*

In (8) the system matrix V has the entries

$$V_{i,(mn)} = e^{2\pi j(mx_i + ny_i)} \tag{9}$$

and resembles a double Vandermonde matrix. Although numerical algorithms for the solution of (ordinary) Vandermonde systems are known [19, 20], experience has shown that it is rather inefficient to solve a double Vandermonde system $Va = s$ directly.

If r is much larger than the dimension $(2M+1)^2$ of $\boldsymbol{P}_M$, then the system is over-determined. This usually causes numerical problems. On the other hand, if r is close to $(2M+1)^2$, then for general sampling sets the condition number is usually extremely large for nonuniform sampling sets, which implies unstable recovery. Condition numbers of the order 10^{10}–10^{15}, say, can already be observed in the case of 1-D signals [6].

At this point there is a significant difference between 1-D and higher dimensional signals. In dimension 1, a trigonometric polynomial of degree M is uniquely determined by its samples on any r distinct points, provided that $r \geq 2M+1$, consequently the system matrix has always full rank, independent of the sampling geometry. This is no longer true in higher dimensions. We will return to this problem in Section 6.4.

6.3.2. Frames in Nonuniform Sampling

A more efficient and far-reaching approach is suggested by frame theory. Following Duffin and Schaeffer's ground breaking approach to sampling problems [16], we consider the sampled energy and compare it to the signal's true energy. The energy of $p \in \boldsymbol{P}_M$ is

$$\|p\|_2^2 = \int_0^1 \int_0^1 |p(x,y)|^2 dxdy = \sum_{k=-M}^{M} \sum_{l=-M}^{M} |a(k,l)|^2, \tag{10}$$

whereas the sampled energy equals

$$\sum_{i=1}^{r} |p(x_i, y_i)|^2.$$

For any reasonable sampling theory we need as a minimal requirement that the sampled energy is comparable to the actual energy of p. This means that there should exist positive constants $A, B > 0$, such that

$$A\|p\|_2^2 \leq \sum_{i=1}^{r} |p(x_i, y_i)|^2 \leq B\|p\|_2^2. \tag{11}$$

for all trigonometric polynomials $p \in \boldsymbol{P}_M$. If the inequalities (11) are satisfied, then the set of samples $\{(x_i, y_i)\}$ is referred to as a *set of (stable) sampling* [21]. The inequality (11) is often called the *frame condition* and A and B are called the *frame bounds*.

To find an estimate for the upper bound B is usually easy and not a problem. The real challenge is to establish a positive lower bound $A > 0$. Clearly, when A is positive, then $p(x_i, y_i) = 0$ for all $i = 1, \ldots, r$, implies that $p \equiv 0$. But (11) says more: a small change of a sampled value $p(x_i, y_i)$ induces only a small change of p itself. Consequently the reconstruction of p from its samples is stable. From a numerical view point the existence of the constants A, B is not sufficient for stability, it is also required that the ratio B/A is small, i.e., the problem is well-conditioned. These observations are of fundamental importance when working with noisy data.

6.3.3. Toeplitz Systems

In order to exploit the frame condition (11) for a numerical algorithm, we write it in terms of the coefficients $a(m, n)$ of p, rather than p itself. Substituting (5) into (11), we obtain

$$\sum_{i=1}^{r} |p(x_i, y_i)|^2 = \sum_{k,m=-M}^{M} \sum_{l,n-=-M}^{M} a(k, l)\overline{a(m, n)} \sum_{i=1}^{r} e^{2\pi j((k-m)x_i+(l-n)y_i)}. \qquad (12)$$

Now collect the coefficients $a(k, l)$ into the vector a and write

$$T_{kl,mn} = \sum_{i=1}^{r} e^{-2\pi j((k-m)x_i+(l-n)y_i)}. \qquad (13)$$

Then the matrix T with entries $T_{kl,mn}$, where $k, l, m, n = -M, \ldots, M$, is the system matrix of the sampling problem. Note that T acts on $\mathbb{C}^{(2M+1)^2}$. Thus by combining (12) and (13), the sampled energy can be written as

$$\sum_{i=1}^{r} |p(x_i, y_i)|^2 = \langle a, Ta \rangle \qquad (14)$$

for $p \in \boldsymbol{P}_M$. By using (10) and (11), the frame inequalities turn into

$$A\|a\|_2^2 \leq \langle a, Ta \rangle \leq B\|a\|_2^2. \qquad (15)$$

Structure of T: The system matrix T has a very peculiar structure: $T_{kl,mn}$ depends only on the differences $k - m$ and $l - n$. We can rearrange T into $(2M + 1) \times (2M + 1)$ blocks B_{km} of size $(2M + 1) \times (2M + 1)$ with entries

$$(B_{km})_{ln} = T_{kl,mn} = \sum_{i=1}^{r} e^{2\pi j((k-m)x_i+(l-n)y_i)}.$$

Then each block is a Toeplitz matrix and since B_{km} depends only on $k - m$, T itself is a block Toeplitz matrix.

Using $V_{i,(mn)} = e^{2\pi j(mx_i + ny_i)}$ for the entries of the Vandermonde system (8), we calculate the entries of V^*V as

$$(V^*V)_{kl,mn} = \sum_{i=1}^{r} e^{2\pi j((m-k)x_i + (n-l)y_i)} = T_{kl,mn},$$

in other words,

$$T = V^*V. \tag{16}$$

This leads to an alternative approach to the system of equations (8). Instead of solving $Va = s$, we shall solve the *system of normal equations*

$$Ta = V^*Va = V^*s \tag{17}$$

for a.

In contrast to the Vandermonde type matrix V the matrix T is always a square matrix with the correct dimension $(2M+1)^2 \times (2M+1)^2$, independent of the number of samples. Furthermore, numerical algorithms for Toeplitz matrices and block Toeplitz matrices are available in abundance and well-understood. This is one of the rare cases where it is advantageous to explicitly establish the system of normal equations (17) instead of staying with the naive approach in (8), since we gain structure and consequently numerical efficiency.

Spectrum of T: From (14) we see that T is always positive semi-definite. This means that the spectrum $\sigma(T)$, i.e., the set of all eigenvalues of T, is contained in $[0, \infty)$. If (x_i, y_i), $i = 1, \ldots, r$, is a set of stable sampling, then T is strictly positive definite and thus invertible. Furthermore, if λ_- and λ_+ denote the smallest and largest eigenvalues of T, then (14) and (15) imply that $A \leq \lambda_-$ and $\lambda_+ \leq B$, consequently $\sigma(T) \subseteq [A, B]$ and the condition number of T, defined as $\kappa(T) = \lambda_+/\lambda_-$ satisfies the estimate

$$\kappa(T) \leq B/A. \tag{18}$$

This means that reasonable estimates for the frame bounds A, B are of immediate relevance for the numerical behavior of the sampling problem. Estimates for condition numbers are usually a hard mathematical problem. Very few estimates for the block Toeplitz matrices in (13) are known in higher dimensions, a few of those will be discussed in Section 6.4.

6.3.4. An Efficient Reconstruction Algorithm

In almost any application the sampled values of an image come with noise. This means that we cannot be sure that the measured data s_i are actually the samples of a band-limited image, i.e., s_i is not necessarily of the form $p(x_i, y_i)$ for some trigonometric polynomial $p \in \boldsymbol{P}_M$. In this case the reconstruction of the image amounts to finding the trigonometric polynomial $p \in \boldsymbol{P}_M$ which yields the best approximation for the data s_i.

Formally, we have to solve the following *least square problem*.
Find $p \in \boldsymbol{P}_M$, such that

$$\sum_{i=1}^{r} |p(x_i, y_i) - s_i|^2 w_i = \text{minimum in } \boldsymbol{P}_M. \tag{19}$$

Note that, compared to the previous section, we have formulated the least squares problem in a more general fashion by introducing the weights $w_i > 0$. Such weights are useful to compensate for varying density in the sampling geometry or to fine-tune the condition number of the problem.

By increasing the bandwidth M so that $r \leq (2M+1)^2$, one obtains an underdetermined system, and then it is usually possible to find a trigonometric polynomial $p \in \boldsymbol{P}_M$ that interpolates the data exactly as $p(x_i, y_i) = s_i$. However, in the presence of noise, such a solution is usually rough and jittery and does not seem to reproduce the image accurately. Moreover the increase of the bandwidth usually reduces the stability of the reconstruction problem. In general, it is better to stay with the original bandwidth and work with oversampling, i.e., the number of samples r is larger than the estimated bandwidth M. Then the solution to the least square problem (19) is much smoother and will resemble the original image more accurately. For an illustration of this effect see the numerical examples in Section 6.6.1. The choice of the appropriate bandwidth amounts to a *regularization* procedure for the sampling problem. In this sense, the solution of the least square problem can be considered as a denoising procedure. In Section 6.5.1 we will discuss this aspect in more detail.

We have seen that the relation (15) between the sampled energy and the block Toeplitz matrix T provides the key to the numerical solution of the nonuniform sampling problem for band-limited images [6], [22], [23]. After interpreting the formulas obtained so far in a new light, we obtain the following algorithm for the solution of the least square problem (19).

Proposition 2 *Assume that for a given sampling set* $\{(x_i, y_i), i = 1, \ldots, r\} \subseteq [0,1] \times [0,1]$ *the associated block Toeplitz matrix T is invertible on* $\mathbb{C}^{(2M+1)^2}$. *Furthermore, choose a set of weights $w_i > 0$. Then the solution to the least square problem* (19) *can be computed as follows.*

Step 1: Compute the entries of T by

$$T_{kl,mn} = \sum_{i=1}^{r} w_i e^{-2\pi j((k-m)x_i+(l-n)y_i)} \tag{20}$$

and the vector $b \in \mathbb{C}^{(2M+1)^2}$ with coordinates

$$b(k,l) = \sum_{i=1}^{r} w_i s_i e^{-2\pi j(kx_i+ly_i)} \tag{21}$$

for $k,l = -M, \ldots, M$.

Step 2: Compute $a_{opt} = T^{-1}b \in C^{(2M+1)^2}$

Step 3: Compute the trigonometric polynomial $p_{opt} \in P_M$ with coefficient vector a_{opt} as follows:

$$p_{opt}(x,y) = \sum_{k=-M}^{M} \sum_{l=-M}^{M} a_{opt}(k,l) e^{2\pi j(kx+ly)} \tag{22}$$

Then for all $p \in \boldsymbol{P}_M$, $p \neq p_{opt}$,

$$\sum_{i=1}^{r} |p_{opt}(x_i,y_i) - s_i|^2 w_i < \sum_{i=1}^{r} |p(x_i,y_i) - s_i|^2 w_i \tag{23}$$

Thus p_{opt} is the unique solution to the least square problem (19).

Proof: We have to verify inequality (19). For arbitrary $p(x,y) = \sum_{k=-M}^{M} \sum_{l=-M}^{M} a(k,l) e^{2\pi i(kx+ly)} \in \boldsymbol{P}_M$ we have

$$\sum_{i=1}^{r} |p(x_i,y_i) - s_i|^2 w_i \tag{24}$$

$$= \sum_{i=1}^{r} |p(x_i,y_i)|^2 w_i - \sum_{i=1}^{r} (s_i \overline{p(x_i,y_i)} + \overline{s_i} p(x_i,y_i)) w_i + \sum_{i=1}^{r} |s_i|^2 w_i \tag{25}$$

Here the sampled energy is $\langle a, Ta \rangle$ by (14), and

$$\sum_{i=1}^{r} \overline{p(x_i,y_i)} s_i w_i = \sum_{k=-M}^{M} \sum_{l=-M}^{M} \overline{a(k,l)} \sum_{i=1}^{r} s_i e^{-2\pi j(kx_i+ly_i)} w_i = \langle b, a \rangle.$$

Consequently

$$\sum_{i=1}^{r} |p(x_i,y_i) - s_i|^2 = \langle a, Ta \rangle - \langle a, b \rangle - \langle b, a \rangle + \sum_{i=1}^{r} |s_i|^2. \tag{26}$$

Using $Ta_{opt} = b$ from Step 2 we obtain

$$\sum_{i=1}^{r}(|p(x_i, y_i) - s_i|^2 - |p_{opt}(x_i, y_i) - s_i|^2) = \tag{27}$$

$$= \langle a, Ta\rangle - \langle a, b\rangle - \langle b, a\rangle - \langle a_{opt}, Ta_{opt}\rangle + \langle a_{opt}, b\rangle + \langle b, a_{opt}\rangle \tag{28}$$

$$= \langle (a - a_{opt}), T(a - a_{opt})\rangle > 0 \tag{29}$$

If $a \neq a_{opt}$, the last expression is strictly positive, since by hypothesis T is invertible. □

Proposition 2 yields an efficient reconstruction algorithm.

Algorithm 1 *Under the assumptions of Proposition 2 the solution p_{opt} can be computed in $\mathcal{O}(kM \log M + r \log(1/\varepsilon))$, where ε is a predefined tolerance, by following algorithm:*

Step 1: Compute the entries of T and y in (20) *and* (21) *using Beylkin's unequally spaced FFT algorithm* [24]. *This takes $\mathcal{O}(M \log M + r \log(1/\varepsilon))$ operations.*

Step 2: Solve $Ta = b$ iteratively by the conjugate gradient (CG) algorithm [25]. *Choose an initial guess a_0 and compute iteratively:*

$k = 0$
$r_0 = b - Ta_0$
while $\|r_k\| > \varepsilon\|b\|$
$k = k + 1$
  if $k = 1$
    $q_1 = r_0$
  else
    $\beta = \|r_{k-1}\|^2 / \|r_{k-2}\|^2$
    $q_k = r_{k-1} + \beta_k p_{k-1}$
  end
  $\alpha_k = \|r_{k-1}\|^2 / \langle q_k, Tq_k\rangle$
  $a_k = a_{k-1} + \alpha_k q_k$
  $r_k = r_{k-1} - \alpha_k Tq_k$
end
$a = a_k$ (30)

The matrix-vector multiplication Tq_k above is carried out in $\mathcal{O}(M^2 \log M)$ operations via a 2-D FFT by embedding T in a block-circulant matrix [26, 27]. *The solution of $Ta = b$ takes $\mathcal{O}(kM^2 \log M)$ operations, where k is the number of performed iterations.*

Step 3: If the signal is reconstructed on regularly spaced nodes $\{u_i, v_i\}_{i=1}^N$, the solution $p_{opt}(u_i, v_i)$ in (22) *can be computed by a 2-D FFT. For nonuniformly spaced nodes we can again resort to Beylkin's USFFT algorithm.*

Alternatively to Beylkin's nonuniform FFT algorithm we can use Rokhlin's method [28]. For the extension of Rokhlin's algorithm to 2-D see [29].

Since the number of nodes at which the reconstructed signal is evaluated is usually independent of the number of given data and the bandwidth, the computational effort for the last step is not included in the operations count for the signal reconstruction.

The above algorithm is called the ACT-method in the literature. It incorporates adaptive weights (the choice of which will be discussed in the next section), the conjugate gradient method and the Toeplitz structure of the sampling problem, and seems to be one of the most efficient methods to treat the numerical aspects of nonuniform sampling.

Remarks

1. The main problem is now to establish conditions on the sampling set under which the system matrix T is invertible. In practice, this is "almost always" the case (although the condition number may be large), but in theory this is an extremely difficult problem and far from being completely understood.
2. In dimension $d = 1$, T is invertible, if and only if $r \geq 2M + 1$, i.e., if the number of samples is large enough. This follows from the fundamental theorem of algebra, according to which the polynomial $q(z) = z^M \sum_{k=-M}^{M} a_k z^k$ of degree $2M$ either has exactly $2M$ zeros counting multiplicity or is identical to 0.
3. In higher dimensions, the zero set of a trigonometric polynomial is an algebraic curve or an algebraic surface. For T to be invertible, the samples must not be contained in any algebraic surface. It is an open problem to efficiently characterize the invertibility of T. It follows from Lemma 1 that at least $(2M + 1)^2$ samples are required, but these need to be distributed appropriately. For instance, if the sampling set is contained in the straight line $\ell := \{(1/2, x''), x'' \in \mathbb{R}\}$, then the trigonometric polynomial $p(x', x'') = \sin 2\pi x'$ vanishes on ℓ and there can be no sampling theorem.
4. The number of iterations of CG depends essentially on the clustering of the eigenvalues of the system matrix [30]. If the sampling points stem from a perturbed regular sampling set, then the eigenvalues of T will be clustered around β, where β is the oversampling rate, cf. [31] and convergence will be extremely fast. Note that the Landweber-Richardson iteration (also known as

frame iteration [2] or Marvasti method [1] in the sampling community) cannot take advantage of this clustering property.
5. If T is not invertible then CG converges to the minimal norm least squares solution of $Ta = b$, see [32].

6.3.5. Adaptive Weights

Optimal performance of Algorithm 1 can be achieved by properly selecting weights w_i. A good choice for the weights originates from abstract sampling theorems in [18, 33].

To gain some motivation for the use of weights we first discuss the main factors which determine the condition number of the system matrix T.

(a) The lower bound of T is mainly determined by the large gaps in the sampling set. Suppose there is a large hole in the sampling set and choose a trigonometric polynomial $p \in \boldsymbol{P}_M$ which, like the prolate spheroidal functions, concentrates most of its energy in this hole. Then the sampling of p will not pick up any information about the main concentration of the polynomial energy. On the other hand, the samples outside that hole will be small. Consequently, the sampled energy will be small compared to the actual energy of p. For such a sampling set the lower frame bound A in the frame inequality (11) must be small. It is clear that this problem cannot be fixed by heavy oversampling outside the large gap.

Generically, large gaps and the ensuing lack of information always results in bad condition numbers. Without additional *a priori* information about the solution this problem cannot be fixed by any numerical method.

(b) On the other hand, clusters of sampling points within a small distance from each other will push up the upper bound B in the frame inequality, because the same local information is counted and added several times. Yet, as mentioned in (a) a cluster will not contribute much to the lower bound and to the uniqueness of the problem. In this case the condition number is large, because too much local information is given in certain areas of the image.

Problem (b) can be addressed by the *method of adaptive weights*. The idea is to compensate for the local variation of the density by using weights in the frame inequality.

We explain the idea for 1-D signals first.

Suppose that $0 \leq x_1 \leq x_2 \leq \cdots \leq x_r < 1$ is a sampling set in [0, 1]. Then the natural weights are defined as $w_i = (x_{i+1} - x_{i-1})/2$, here w_i is the length of the interval around the point x_i between the midpoints $(x_{i-1} + x_i)/2$ and $(x_{i+1} + x_i)/2$. Thus if many samples are clustered near a point x_i, then the weight w_i is small. If x_i is the only sampling point in a large neighborhood, then the corresponding weight is large.

Extensive numerical simulations have demonstrated that this method can improve the condition number of nonuniform sampling problems significantly [6, 27]. In our experience it should always be used with highly nonuniform sampling sets.

In higher dimensions the calculation of appropriate weights is a bit more delicate. We discuss two easy and practical choices for the weight function, namely Voronoi regions and cell counting.

(a) Given a sampling set, the *Voronoi region* V_i of a point (x_i, y_i) consists of all points which are closer to (x_i, y_i) than to any other sample. Formally

$$V_i = \{(x,y) : (x - x_i)^2 + (y - y_i)^2 \le (x - x_k)^2 + (y - y_k)^2 \text{ for all } k \ne i\} \quad (31)$$

Now let w_i be the area (volume) of the Voronoi region V_i. Clearly, the Voronoi regions for points in a cluster are small, therefore these points receive a small weight. For numerical recipes to calculate Voronoi regions consult [34–36].

(b) For the sampling problem the precise area of the Voronoi region is not crucial and often a cruder estimate for the local density is sufficient. According to the regular sampling theorem for $\boldsymbol{P}_M$ in (7) any trigonometric polynomial is uniquely determined by its samples on the grid $\left(\frac{k}{2M+1}, \frac{l}{2M+1}\right)$, k, $l = 0, \ldots, 2M$. Roughly speaking, we need one sampling point in each "Nyquist cell" $C_{kl} = \left(\frac{k}{2M+1}\right), \left(\frac{k+1}{2M+1}\right) \times \left(\frac{l}{2M+1}\right), \left(\frac{l+1}{2M+1}\right)$.

In the cell counting method we proceed as follows: let n_{kl} be the number of sampling points in the cell C_{kl} and set

$$w_i = \frac{1}{n_{kl}}, \qquad \text{whenever} \quad (x_i, y_i) \in C_{kl}$$

This choice of the weights guarantees that all cells contribute equally to the sampled energy.

With these choices of weights one can improve the stability and rate of convergence of the algorithm of the previous section.

Remarks

1. It is not hard to see that the invertibility of T does not depend on the choice of the weights. Let for the moment T be the matrix without using weights and T_w the matrix with weights w_i. Then $\sum_{i=1}^{r} |p(x_i)|^2 w_i = \langle a, T_w a \rangle = 0$ holds, if and only if $\sum_{i=1}^{r} |p(x_i)|^2 = \langle a, Ta \rangle = 0$. Therefore both T_w and T have the same kernel. Since a matrix is invertible, if and only if its kernel is $\{0\}$, it follows that either both T_w and T are invertible or both are singular.
2. As mentioned before, the condition number of T will be large in case of large gaps in the sampling set. Since T is of Toeplitz-type one may be tempted to use one of the many circulant preconditioners to improve upon the stability of the

system. However, in the case of large gaps, preconditioning does not help, since the preconditioned problem will still have a spectral cluster at the origin, see [31].

6.4. Theoretical Back-Up

Even without a deeper theory, the discussed model of band-limited functions and the corresponding algorithm work remarkably well and very efficiently in numerical simulations. In this section we address some of the theoretical issues.

(a) We state a number of conditions which guarantee the invertibility of the system matrix T. These come in the form of estimates for the condition numbers of the Toeplitz matrices involved. For such estimates the use of adaptive weights is indispensable.

(b) We establish a connection between the finite-dimensional model of nonuniform sampling and the classical theory for the sampling of band-limited functions. In this way we obtain sampling theorems for band-limited images in their classical form.

For details and proofs the reader should consult the cited mathematical literature.

6.4.1. Condition Numbers for Toeplitz Systems

We start with a result for Toeplitz matrices that arose from the theory of nonuniform sampling for 1-D signals.

Assume that $0 \le x_1 < x_2 < \cdots < x_r < 1$ are samples in [0, 1]. To preserve periodicity, we set $x_0 = x_r - 1$ and $x_{r+1} = x_1 + 1$. Then we consider the $(2M+1) \times (2M+1)$-Toeplitz matrix T (in the strict sense) with entries

$$T_{kl} = \sum_{i=1}^{r} w_i e^{-2\pi j(k-l)x_i} \quad \text{for } k, l = -M, \ldots, M\,.$$

For the choice of weights w_i as explained in Section 6.3.5, i.e., $w_i = (x_{i+1} - x_{i-1})/2$ for $i = 1, \ldots, r$, it is possible to derive explicit estimates for the condition number of T.

Theorem 3 ([8], Thm. 5) *Assume that* $\max_{i=1,\ldots,r}(x_{i+1} - x_i) = \delta < 1/2M$. *Then the spectrum of T is contained in*

$$\sigma(T) \subseteq [(1 - 2M\delta)^2, (1 + 2M\delta)^2], \tag{32}$$

and $\kappa(T) \le (1 + 2M\delta)^2/(1 - 2M\delta)^2$ *is a worst case estimate for the condition number of T.*

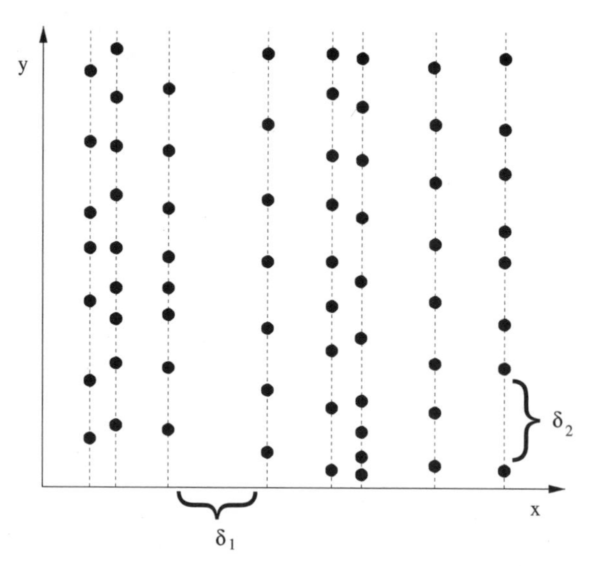

Figure 2. Line-type nonuniform sampling set, δ_1 and δ_2 denote the maximum gap in the sampling set in the y- and x-direction, respectively.

This 1-D result can be extended easily to special sampling sets and their corresponding block Toeplitz matrices in higher dimensions. In a number of applications in medical imaging and in geophysics, images are sampled line by line, but the sampling on each line is nonuniform, and the spacing between the lines is nonuniform as well. In dimension $d = 2$ such a sampling set is of the form $(x_i, y_{ik}) \in [0, 1] \times [0, 1]$, $i = 1, \ldots, r$, $k = 1, \ldots, s_i$, such that $x_i < x_{i+1}$ and $y_{ik} < y_{i,k+1}$. (Again we set $x_{r+1} = x_1 + 1, x_0 = x_r - 1$ and $y_{i,0} = x_{i,s_i} - 1$, $y_{i,s_i+1} = y_{i,1} + 1$ to reflect the periodicity of 2-D trigonometric polynomials.) A typical example for such a sampling set is illustrated in Fig. 2.

With an appropriate choice of the adaptive weights we have the following estimates.

Theorem 4 ([22, 23]) *Suppose that* $\max_{i=1,\ldots,r}(x_{i+1} - x_i) = \delta_2 < 1/2M$ *and* $\max_{i,k\in\mathbb{Z}}(y_{i,k+1} - y_{ik}) = \delta_1 < 1/2M$ *and choose* $w_{ik} = \frac{1}{4}(x_{i+1} - x_{i-1}) \times (y_{i,k+1} - y_{i,k-1})$. *Then the spectrum of the block Toeplitz matrices* T *in* (20) *satisfies*

$$\sigma(T) \subseteq [(1 - 2M\delta_1)^2(1 - 2M\delta_2)^2, (1 + 2M\delta_1)^2(1 + 2M\delta_2)^2].$$

In particular, T *is invertible for such sampling sets with a worst case estimate for the condition number*

$$\kappa(T) \leq \frac{(1 + 2M\delta_1)^2(1 + 2M\delta_2)^2}{(1 - 2M\delta_1)^2(1 - 2M\delta_2)^2}. \tag{33}$$

Note that (33) shows explicitly how, for a given sampling set, the condition number deteriorates with increasing bandwidth.

Next we state a result for arbitrary sampling sets in the unit square. Let $B_\delta(a, b) = \{(x, y) \in \mathbb{R}^2 : (x-a)^2 + (y-b)^2 < \delta^2\}$ be the disc of radius δ centered at (a, b). We say that a set $\{(x_i, y_i), i = 1, \ldots, r\}$ is *δ-dense* in $[0, 1] \times [0, 1]$, if $\bigcup_{i=1}^{r} B_\delta(x_i, y_i) \supseteq [0, 1] \times [0, 1]$. In other words, the distance of a given sample (x_i, y_i) to its nearest neighbor (x_k, y_k), $k \neq i$, is at most 2δ.

As in Theorem 3 we choose the size of the Voronoi regions V_i as the weights and use $w_i = \text{area}\,(V_i)$ in the block Toeplitz matrix T in (20). (As in the 1-D case, the weights have to be slightly modified near the boundary of the unit square $[0, 1] \times [0, 1]$ to take into account the periodicity of trigonometric polynomials. To avoid additional notation, we omit these details.) Now we have the following analogue of Theorem 3 for arbitrary sampling sets in 2-D [7].

Proposition 5 *Suppose that sampling set* $\{(x_i, y_i), i = 1, \ldots, r\} \subseteq [0, 1] \times [0, 1]$ *is δ-dense and*

$$\delta < \frac{\log 2}{4\pi M} . \tag{34}$$

Then

$$\sigma(T) \subseteq [(2 - e^{4\pi M\delta})^2, 4] . \tag{35}$$

In particular, for arbitrary δ-dense sampling sets, the block Toeplitz matrix T is invertible and Proposition 2 and Algorithm 1 are applicable.

6.4.2. Towards Nonuniform Sampling of Band-limited Images

So far our main concern was an approach to the nonuniform sampling problem which emphasized the numerical solution and its practical and algorithmic aspects. In the previous section we have shown that mathematical results confirm the efficiency and feasibility of this approach. However, nonuniform sampling of trigonometric polynomials is certainly not the problem that is usually called the "irregular sampling problem of band-limited functions". This raises the question of the nature of the solutions of (8) with respect to sampling of band-limited functions in $\boldsymbol{B}_\Omega$.

For this we assume that the samples $f(x_i, y_i)$ of a band-limited function $f \in \boldsymbol{B}_\Omega$ are given on an infinite discrete (nonuniform) set $\{(x_i, y_i), i \in I\}$. (Since there is no canonical way to enumerate a nonuniform set in $\mathbb{R}^2$, we write I for the index set rather than $\mathbb{Z}$ which would suggest some order.) For simplicity we assume that $\Omega = 1/2$, i.e. $F(\xi, \eta) = 0$ for $(\xi, \eta) \notin [-1/2, 1/2] \times [-1/2, 1/2]$.

For the solution or approximate solution of this infinite-dimensional problem we argue in the style of Galerkin or finite element methods.

1. Take just the samples in the square $[-M - 1/2, M + 1/2]^2$, approximate them by the samples of a trigonometric polynomial p_M of degree M and period $2M + 1$.

Except for a different normalization we have solved this problem in Section 6.3. Set

$$(x_i', y_i') = \left(\frac{x_i + M + 1/2}{2M + 1}, \frac{y_i + M + 1/2}{2M + 1}\right).$$

Since these normalized samples are in the square $[0, 1] \times [0, 1]$, Proposition 2 and Algorithm 1 furnish a trigonometric polynomial p of order M which minimizes the least square deviation

$$\sum_{i \in I : (x_i, y_i) \in [-M-1/2, M+1/2]^2} |p(x_i', y_i') - f(x_i, y_i)|^2 w_i = \text{ minimum in } \boldsymbol{P}_M. \tag{36}$$

Now set

$$p_M(x, y) = p\left(\frac{x + M + 1/2}{2M + 1}, \frac{y + M + 1/2}{2M + 1}\right).$$

Then $p_M(x_i, y_i) = p(x_i', y_i')$ and by (36) p_M is the best trigonometric polynomial of order M to approximate the data $f(x_i, y_i)$.

If p is defined as in (5), then we have

$$p_M(x, y) = \sum_{k=-M}^{M} \sum_{l=-M}^{M} a(k, l)(-1)^{k+l} e^{2\pi j(kx+ly)/(2M+1)}.$$

Consequently its (distributional) Fourier transform is

$$P_M(\xi, \eta) = \sum_{k=-M}^{M} \sum_{l=-M}^{M} a(k, l)(-1)^{k+l} \delta_{k/2M+1}(\xi)\delta_{l/2M+1}(\eta), \tag{37}$$

where δ_x denotes the point measure at x. Like the original band-limited image f, P_M it has its support in $[-1/2, 1/2] \times [-1/2, 1/2]$.

This observation validates our intuition that trigonometric polynomials are the correct discretization of band-limited functions.

2. By virtue of (36), p_M is a good approximation of f on the big square $[-M - 1/2, M + 1/2]^2$. On the other hand, since p_M is periodic, whereas $f \in \boldsymbol{B}_{1/2}$ decays, p_M cannot approximate f outside that cube. This is not surprising, because we have only used information from within $[-M - 1/2, M + 1/2]^2$.

In order to understand the sampling problem for band-limited functions on $\mathbb{R}^2$, we need to investigate what happens when $M \to \infty$. Both (36) and (37) suggest that the local approximations p_M should converge to the original band-limited function f. Clearly any reasonable discretization of the sampling problem should enjoy this property.

The next result shows that this property holds indeed for the model of trigonometric polynomials. For other discretizations see [31, 37].

In order to avoid certain pathologies, we impose a mild assumption on the sampling set and the associated weights.

(1) There exists $R > 0$, such that

$$\bigcup_{i \in I} B_R(x_i, y_i) = \mathbb{R}^2, \tag{38}$$

in other words, there are no arbitrarily large "holes" in the sampling set.

(2) Let V_i be the Voronoi region around (x_i, y_i) as defined in (31) and let the weight w_i be the area (volume) of $V_i \cap B_R(x_i, y_i)$.

Sampling sets and weights satisfying these properties are called permissible.

Theorem 6 ([9, 22]) *Let $\{(x_i, y_i), i \in I\}$ be a permissible sampling set in $\mathbb{R}^2$ and $f \in \boldsymbol{B}_{1/2}$. Let T_M be the block Toeplitz matrix that intervenes in the solution of the least squares problem* (36). *If, for some $\alpha > 0$, $\sigma(T_M) \subseteq [\alpha, \beta]$ for all M, then*

$$\lim_{M \to \infty} \int_{[-M,M]^d} |f(x) - p_M(x)|^2 dx = 0$$

and also $\lim_{M \to \infty} p_M(x) = f(x)$ uniformly on compact sets.

This theorem possesses several distinct facets:

1. From a practical point of view it validates the use of the algorithm based on the interpolation of data by trigonometric polynomials. The approximation by trigonometric polynomials is indeed an appropriate finite-dimensional model for nonuniform sampling of multivariate band-limited functions.
2. Note that the main assumption "$\sigma(T_M) \subseteq [\alpha, \beta]$ for all M" is automatically satisfied for the following sampling sets as a consequence of Theorems 4 and Proposition 5.
   (a) The sampling set has the product form (x_i, y_{ik}) with $\sup_i(x_{i+1} - x_i) < 1$ and $\sup_{i,k}(y_{i,k+1} - y_{ik}) < 1$, or

(b) If the sampling set is δ-dense, i.e., $\bigcup_{i \in I} B_\delta(x_i, y_i) = \mathbb{R}^2$ for a density δ satisfying $\delta < \log 2/2\pi$, then the assumptions of Theorem 6 are satisfied.

3. The block Toeplitz matrices T_M play the same role for the nonuniform sampling problem as the so-called stiffness matrices do in the theory of Galerkin methods and finite elements for partial differential equations. In this completely different field, the uniform boundedness of the stiffness matrix is a well-known necessary assumption for numerical stability.

6.4.3. Sampling Theorems for Band-limited Images

The numerical approach of Section 6.3 and the limit Theorem 6 can be turned into the more classical form of a frame inequality for $\boldsymbol{B}_{1/2}$.

Corollary 7 ([9, 38]) *Under the assumptions of Theorem 6 the frame inequality*

$$\alpha \|f\|^2 \leq \sum_{i \in I} |f(x_i, y_i)|^2 w_i \leq \beta \|f\|^2 \qquad (39)$$

holds for all $f \in \boldsymbol{B}_{1/2}$.

Since the frame bounds α and β are related to the extremal eigenvalues of the Toeplitz matrices T_M, Theorem 6 allows to approximate the frame bounds α, β by the eigenvalues of T_M.

We conclude this section with a review of the most profound and spectacular nonuniform sampling theorem in higher dimensions. It was proved around 1963, but is essentially unknown in engineering circles. We hope that its brief discussion will lead to useful applications and further investigation of related numerical issues. Beurling's theorem is a sampling theorem for band-limited functions with frequency spectrum contained in a ball.

Let $\mathcal{E}_R = \{f \in L^2(\mathbb{R}^2 : F(\xi, \eta) = 0 \quad \text{for} \quad \xi^2 + \eta^2 \geq R^2\}$

Theorem 8 ([10]) *Assume that the sampling set $\{(x_i, y_i), i \in I\}$ satisfies*

$$\inf_{i \neq k}[(x_i - x_k)^2 + (y_i - y_k)^2] > 0$$

and

$$\bigcup_{i \in I} B_r(x_i, y_i) = \mathbb{R}^2 .$$

If

$$rR < \tfrac{1}{4}$$

then there exist frame bounds $A, B > 0$, such that the frame inequalities

$$A \int_{\mathbb{R}} \int_{\mathbb{R}} |f(x,y)|^2 \, dxdy \leq \sum_{i \in I} |f(x_i, y_i)|^2 \leq B \int_{\mathbb{R}} \int_{\mathbb{R}} |f(x,y)|^2 \, dxdy$$

hold for all $f \in \mathcal{E}_R$.

6.5. Deeper Aspects

The literature offers a variety of reconstruction algorithms for the costumer of sampling and reconstruction technology. However, in the applications of these algorithms to real world problems, one often is confronted with a number of additional difficulties.

- An algorithm may ask for certain parameters, e.g., relaxation parameters in iterative algorithms, which are not readily available or are not known *a priori*.
- An algorithm gives satisfactory results only for selected examples, usually of small dimension, but it may be too expensive for large-scale real world problems.
- The theoretical assumptions, such as required sampling density, under which the algorithm is guaranteed to be efficient are not fulfilled.

In this section we address some of these problems for the reconstruction algorithms of Section 6.3.4. In this form we answer some questions that we are frequently asked by users in signal processing and industry about these algorithms.

6.5.1. Bandwidth Estimation – A Multi-level Approach

Most theoretical results and numerical algorithms for reconstructing a band-limited signal from nonuniform samples require that the bandwidth is known *a priori*. This information, however, is often not available in practice, and thus the bandwidth enters the algorithm as a hidden parameter. Yet, a good choice of the bandwidth is crucial for any reconstruction algorithm in case of noisy data. It is intuitively clear that choosing too large a bandwidth leads to over-fit of the noise in the data, while too small a bandwidth yields a smooth solution but also to under-fit of the data. We will illustrate the importance of choosing the correct bandwidth by a reconstruction problem in medical imaging in Section 6.6.1. Of course, we do not want to search for the optimal bandwidth by trial-and-error methods. Hence the problem is to design a method that can reconstruct a signal from nonuniformly spaced and noisy samples without requiring *a priori* information about the bandwidth of the signal.

6.5.1.1. Idea of Multi-Level Method

Let s_i denote an unperturbed sample and s_i^δ a perturbed sample. Here δ is the noise level given by

$$\left(\sum_{i=1}^{r} |s_i^\delta - s_i|^2\right)^{1/2} \le \delta \sum_{i=1}^{r} |s_i|^2 .$$

If the data are distorted by noise it does not make sense to ask for a solution p that satisfies $\sum_{i=1}^{r} |p(x_i, y_i) - s_i|^2 = 0$. All we should ask for is a solution p that fulfills

$$\sum_{i=1}^{r} |p(x_i, y_i) - s_i|^2 \le \delta^2 \sum_{i=1}^{r} |s_i|^2 . \tag{40}$$

The method we propose is based on the following idea. For simplicity of notation we represent the following approach without using weights, the resulting method can be easily modified to incorporate weights. Furthermore, Q_M denotes the orthogonal projection from $\boldsymbol{B}$ onto $\mathcal{P}_M$, and $\tau \ge 1$ is a fixed regularization parameter.

Let the noisy samples $s^\delta = \{s_i^\delta\}_{i=1}^r = \{f^\delta(x_i, y_i)\}_{i=1}^r$ of $f \in \boldsymbol{B}$ be given. We start with initial degree $M = 1$ and run Algorithm 1 until the iterates $p_{1,k}$ satisfy for the first time the *inner stopping criterion*

$$\sum_{i=1}^{r} |p_{1,k}(x_i, y_i) - s_i^\delta|^2 \le 2\tau(\delta \|s^\delta\| + \|Q_1 f - f\|)\|s^\delta\| . \tag{41}$$

Denote this approximation (at iteration k_*) by p_{1,k_*}. If p_{1,k_*} satisfies the *outer stopping criterion*

$$\sum_{i=1}^{r} |p_{1,k_*}(x_i, y_i) - s_i^\delta|^2 \le 2\tau\delta \|s^\delta\|^2 , \tag{42}$$

we take p_{1,k_*} as the final approximation. Otherwise we proceed to the next bandwidth $M = 2$ and run Algorithm 1 again, using p_{1,k_*} as the initial approximation by setting $p_{2,0} = p_{1,k_*}$.

In general, at level $M = N$, the inner level-dependent stopping criterion becomes

$$\sum_{i=1}^{r} |p_{N,k}(x_i, y_i) - s_i^\delta|^2 \le 2\tau(\delta \|s^\delta\| + \|Q_N f - f\|)\|s^\delta\| \tag{43}$$

while the outer stopping criterion does not change since it is level-independent.

For simplicity of representation we have chosen $M = 1, 2, \ldots$ but any increasing sequence of positive integers may be chosen instead (e.g., $M = 5, 10, 15, \ldots$).

6.5.1.2. *Stopping Criteria*

The reasons for the particular choice of the stopping criteria are the following. The error analysis in [39] shows that the stopping rule (43) guarantees that the iterations of the conjugate gradient algorithm are terminated before divergence of the iterates sets in. This stopping rule also ensures that CG does not iterate too long at a certain level, since if M is too small further iterations at this level will not lead to a significant improvement. Therefore we switch to the next level. The outer stopping criterion (42) controls over-fit and under-fit of the data.

As an additional problem, we note that $\|Q_N f - f\|$ cannot be computed exactly, since the solution f is not known. In [40] following recursive procedure to estimate $\|f - Q_N f\|$ has been proposed. Let us start at level $M = 1$ where $\|f - Q_1 f\|$ must be estimated. For a well-conditioned frame we estimate by using relation (7)

$$\|f - Q_1 f\|^2 \leq \|f - Q_0 f\|^2 = \|f\|^2 \approx \sum_{i=1}^{r} |f(x_i, y_i)|^2 w_i \,,$$

where the w_j are the weights introduced in Section 6.3.5. We run Algorithm 1 until the stopping criterion (41) is satisfied, at iteration k_*, say. By doing so an approximation p_{1,k_*} is obtained and we switch to the next level $M = 2$. To apply the multilevel algorithm at this level we first have to estimate $\|f - Q_2 f\|^2$. Similar to above we have that

$$\|f - Q_2 f\|^2 = \|f\|^2 - \|Q_2 f\|^2 \,.$$

We estimate as before the L^2-norm of f by the sampled norm $\|f\|^2 \approx \sum_{i=1}^{r} |f(x_i, y_i)|^2 w_i$ and we use the approximation $\|Q_2 f\|^2 \approx \|p_{1,k_*}\|^2$ to get $\|f - Q_2 f\|^2 \approx \sum_{i=1}^{r} |f(x_i, y_i)|^2 w_i - \|p_{k_*}\|^2$. Recall that $\|p_{1,k_*}\|^2$ is just the ℓ^2-norm of the sequence of coefficients which is calculated explicitly in Step 2 of Algorithm 1.

In general we estimate

$$\|f - Q_N f\|^2 \approx \sum_{i=1}^{r} |f(x_i, y_i)|^2 w_i - \|p_{N-1,k_*}\|^2 \,. \tag{44}$$

6.5.1.3. *Computational Efficiency*

From the viewpoint of computational efficiency it is worth mentioning that there is a strong relationship between the block Toeplitz systems for different levels M. Let T_M and T_{M+1} denote the Toeplitz matrices at level M and level $M+1$ respectively. We adopt a similar notation for the associated right hand sides b_M and b_{M+1}. Then it is easy to see that the $(2M+1) \times (2M+1)$ matrix T_M with first column $(t_0, \ldots, t_{2M})^T$ is embedded in the $(2M+3) \times (2M+3)$ matrix T_{M+1} in the following way [23]:

$$T_{M+1} = \begin{pmatrix} t_0 & \cdots & \bar{t}_{2M+2} \\ \vdots & T_M & \vdots \\ t_{2M+2} & \cdots & t_0 \end{pmatrix}.$$

A similar relation holds for b_M and b_{M+1} (but of course this is not true for the solutions a_M and a_{M+1} of these systems). Thus when we switch from one level to the next level, we only have to compute a few new entries to establish the new linear system of equations.

The stopping criteria (43) and (42) are derived based on worst-case error estimates. Numerical experiments have shown that in practice both stopping criteria are somewhat pessimistic. It is usually possible to iterate longer before stagnation or divergence sets in. Extensive numerical experiments and the heuristic arguments in [32] indicate that a good choice for the inner stopping criterion is

$$\left(\sum_{i=1}^{r} |p_{N,k}(x_i, y_i) - s_i^\delta|^2\right)^{1/2} \leq 2\tau(\delta + \|Q_N f - f\|)\|s^\delta\|, \tag{45}$$

and as outer stopping criterion we use

$$\left(\sum_{i=1}^{r} |p_{N,k}(x_i, y_i) - s_i^\delta|^2\right)^{1/2} \leq \tau\delta\|s^\delta\|, \tag{46}$$

where the choice $\tau = 1$ is usually sufficient.

These are also the stopping criteria we will use in the Algorithm 2 below which in turn has been used for several numerical experiments presented in Section 6.6.

Note that the evaluation of $\sum_{i=1}^{r} |p_{N,k}(x_i, y_i) - s_i^\delta|^2$ can be carried out efficiently in $\mathcal{O}(M^2 \log M)$ operations by making use of equation (26), which reads as

$$\sum_{i=1}^{r} |p(x_i, y_i) - s_i|^2 = \langle a, Ta \rangle - \langle a, b \rangle - \langle b, a \rangle + \sum_{i=1}^{r} |s_i|^2.$$

It is useful to set

$$\Psi(a) := \langle a, Ta \rangle - \langle a, b \rangle - \langle b, a \rangle + \sum_{i=1}^{r} |s_i|^2. \tag{47}$$

Altogether the multi-level algorithm can be described by following pseudo-code. For better readability we state the pseudo-code without utilizing estimate (44) and use $M = 0, 1, 2, \ldots$. Of course we make use of estimate (44) in the actual numerical implementation. Also it is straightforward to modify the algorithm for any increasing sequence of levels, such as $M = 2, 4, 6, \ldots$.

Algorithm 2 (Multi-level reconstruction algorithm) *Input: Sampling points* $\{(x_i, y_i)\}_{i=1}^r$ *and samples* $\{s_i\}_{i=1}^r$ *and the noise level* δ. *Choose* τ (*e.g.,* $\tau = 1$), *set* $M = 0$ *and initialize* $stop\_level = 0$, $stop = 0$.

while $stop = 0$
  while $stop\_level = 0$
    run Algorithm 1 to solve $T_M a_M = b_M$ with initial guess $a_{M,0} =: a_{M-1}$;
    (this produces the iterates $a_{M,k}$)
    if $\Psi(a_{M,k}) \leq 2\tau(\delta + \|Q_M f - f\|)^2)\|s^\delta\|$
      $stop\_level = 1$
      $a_M := a_{M,k}$
    end
  end
  if $\Psi(a_M) \leq \tau\delta\|s^\delta\|$
    $stop = 1$
    Compute the solution p_M from a_M via formula (22)
  else
    $M = M + 1$
  end
end

In order to use a_{M-1} as initial guess at the next level M we proceed as follows. a_{M-1} is a vector of length $(2M-1)^2$ whereas at level M we are computing with vectors of length $(2M+1)^2$. Hence we augment a_{M-1} to a vector of length $(2M+1)^2$ by adding zeros at those indices that correspond to frequencies of order M (a polynomial of degree M can be interpreted as polynomial of degree $M+N$ by adding N zero-coefficients). The definition $a_{M,0} := a_{M-1}$ in the algorithm above should be understood in this sense.

6.5.2. Ill-Conditioned Sampling Problems

In Section 6.4 we have seen several conditions on the sampling points (x_i, y_i) under which the block Toeplitz matrix T is invertible. However, from a numerical point of view, the invertibility alone does not guarantee a stable reconstruction, since the condition number of T can still be extremely large. Therefore the solution of $Ta = b$ may be numerically unstable, and the sampling problem may still be ill-conditioned. In general this occurs when the sampling set has large gaps, which is a common situation in astronomy and geophysics.

In case of under-sampling the reconstruction problem is even ill-posed and does not have a unique solution. From a numerical point of view, an ill-posed problem is very similar to an ill-conditioned problem. Note that these instabilities of the reconstruction problem do *not* result from an inadequate discretization of the problem as it arises when using the truncated sinc matrix approach [31, 37, 41–43].

Numerical analysis offers a large arsenal of regularization techniques to deal with ill-posed or ill-conditioned problems. In the following, we discuss a few techniques which pertain to the sampling problem.

(a) For block Toeplitz matrices there exists a large number of preconditioners that could be applied to the ill-conditioned system $Ta = b$. However, in general they do not improve upon the stability of the problem because the preconditioned problem still has a spectral cluster at the origin and preconditioning will not be efficient. In [31] this behavior has been analyzed in detail for the 1-D case, the analysis for the higher-dimensional case is similar.

(b) Fortunately there are other possibilities to obtain a stabilized solution of $T_M a_M = b_M$. As explained in Section 6.3.5, the condition number of T_M essentially depends on the ratio of the maximal gap in the sampling set to the Nyquist rate, which in turn depends on the bandwidth of the signal. We can improve the stability of the system by adapting the bandwidth M of the approximation accordingly. Thus the parameter M serves as regularization parameter that balances stability and accuracy of the solution. This technique can be seen as a specific realization of *regularization by projection*, see Chapter 3 in [44].

(c) In addition, we can use the conjugate gradient method (CG) as a regularization method for the solution of the block Toeplitz system in order to balance the approximation error and the propagated error. When solving a linear system $Ax = y$ by CG for noisy data y^δ with $\|y - y^\delta\| \leq \delta$, the following happens. Although the iterates x_k of the CG-iteration may diverge for $k \to \infty$, the error propagation remains limited in the beginning of the iteration. The quality of the approximation therefore depends on how many iterative steps can be performed before the iterates start to diverge. The goal is now to stop the iteration exactly at the point where the divergence sets in. In other words, the iteration count serves as a regularization parameter which needs to be controlled by appropriate stopping rules [32, 45]. Such a stopping rule is given by the *discrepancy principle*, which says that the iterations should be terminated when the iterates x_k satisfy for the first time

$$\|y^\delta - Ax_k\| \leq \tau\delta \tag{48}$$

for fixed $\tau > 1$ (in practice $\tau = 1$ is usually sufficient). In our context this stopping rule reads as

$$\left(\sum_{i=1}^{r} |s_i^\delta - p(x_i, y_i)|^2\right)^{1/2} \leq \tau\delta\,.$$

The multilevel method introduced in Section 6.5.1 combines both regularization techniques. By optimizing the level (bandwidth) and the number of iterations in each level it provides an efficient and robust regularization technique for ill-conditioned sampling problems. See Section 6.6 for numerical examples.

6.5.3. Additional Knowledge About the Signal

In many applications the physical process that generates the signal provides further information about its nature. Thus in general a signal is not only essentially band-limited, but physics predicts a certain rate of decay for the spectrum of the signal. For instance, geophysical potential fields have exponentially decaying Fourier transform. In case of ill-conditioned sampling problems this *a priori* knowledge can be used to improve the accuracy of the approximation. Instead of the usual regularization methods, such as Tikhonov regularization, we propose a heuristic approach that is computationally much cheaper.

Assume that the decay of the Fourier transform of f can be bounded by $|F(\omega_1, \omega_2)| \leq C\phi(\omega_1, \omega_2)$. Typical choices of ϕ for biomedical or geophysical signals are

$$\phi(\omega_1, \omega_2) = \begin{cases} e^{-\gamma\sqrt{\omega_1^2+\omega_2^2}} & \text{for some} \quad \gamma > 0; \\ 1/(1+\omega_1^2+\omega_2^2)^\alpha, & \alpha \geq 0. \end{cases} \tag{49}$$

In order to take advantage of this information and to obtain an improved reconstruction, we will solve the following modified system of equations.

For a given degree (bandwidth) M and system $Ta = b$ we construct an array ϕ of dimension $(2M+1) \times (2M+1)$ by choosing, e.g., one of the decay conditions in (49). For instance set $\phi(k, l) = 1/(1 + k^2 + l^2)$ for $k, l = -M, \ldots, M$. We interpret ϕ as vector, as usual by stacking its columns and define $D := \text{diag}(\phi)$. Then the *a priori* information about the frequency decay can be incorporated by using one of the following two algorithms.

(i) Instead of solving $Ta = b$ we solve the problem

$$DTa = Db \tag{50}$$

which has the weighted pseudo-inverse solution

$$a^{(1)} = (DT)^{+} Db$$

where $(DT)^{+}$ denotes the pseudo-inverse of DT [25].

(ii) Alternatively, we consider following problem:

$$TDc = b, \quad a = Dc \tag{51}$$

which has the solution

$$a^{(2)} = D(TD)^{+} b\,.$$

Of course, for invertible T we have $a = a^{(1)} = a^{(2)}$. However, if T is not invertible or ill-conditioned, then equations (50) and (51) lead to weighted minimal norm least squares solutions.

Algorithm 1 can be easily adapted to solve systems (50) and (51). To solve (50), for instance, we replace the conjugate gradient method applied to $Ta = b$ in Step 2 of Algorithm 1 by the preconditioned conjugate gradient method applied to $DTa = Db$ [25]. The computational effort per iteration is only slightly larger than that of Algorithm 1, since D is a diagonal matrix. However, in contrast to the intention of PCG, the purpose of introducing the matrix D is not to precondition the system, but to obtain an improved solution via incorporating *a priori* information about the frequency decay. The ordinary pseudo-inverse solution assigns each frequency (i.e., each coefficient in the solution a) the same weight, whereas by introducing the weight matrix D we "punish" errors corresponding to high frequencies in a stronger than errors corresponding to low frequencies.

Numerical experiments (see e.g., Section 6.6.2) confirm this heuristic approach. A thorough numerical analysis of the rate of convergence as well as the choice of appropriate stopping criteria for the associated CG iterations has still to be carried out.

6.6. Applications

In order to illustrate the performance of the proposed methods we will apply them to reconstruction problems arising in medical imaging, exploration geophysics, and digital image reconstruction.

6.6.1. Object Boundary Recovery in Echocardiography

Band-limited functions and trigonometric polynomials are certainly not suitable to model the shape of arbitrary objects. However, they are often useful in cases where an underlying stationary physical process imposes smoothness conditions of the object. Typical examples arise in medical imaging, for instance, in clinical cardiac studies. There the evaluation of cardiac function with parameters of left ventricular contractibility is an important part of an echocardiographic examination [46]. These parameters are derived by means of boundary tracing of endocardial borders of the Left Ventricle (LV).

The extraction of the boundary of the LV involves two steps, once the ultrasound image of a cross section of the LV is given, see Figs. 3(a)–(d). First an edge detection is applied to the ultrasound image to detect the boundary of the LV, see Fig. 3(c). However this procedure may be hampered by the presence of interfering biological structures (such as papillar muscles), by the unevenness of boundary contrast, and by various kinds of noise [47]. Thus the edge detection often provides only a set of nonuniformly spaced and perturbed boundary points rather than a connected boundary. Therefore in a second step the original boundary has to be recovered from the detected edge points, see Fig. 3(d). Since the shape of the Left Ventricle is definitely smooth, trigonometric polynomials are particularly well suited to model its boundary.

We denote the boundary of the left ventricle (LV) by f and parameterize it by $f(u) = (x_u, y_u)$, where x_u and y_u are the coordinates of f at "time" u in the x- and y-direction, respectively. Obviously, we can interpret f as a 1-D continuous, complex, and periodic function, where x_u represents the real part and y_u represents the imaginary part of $f(u)$. It follows from the approximation theorem of Stone-Weierstrass [48] that a continuous periodic function can be approximated uniformly by trigonometric polynomials. If f is smooth, we may safely assume that trigonometric polynomials of low degree provide a sufficiently accurate approximation.

Assume that we know only some arbitrary, perturbed points

$$s_k = (x_{u_k}, y_{u_k}) = f(u_k) + \delta_k, \quad k = 0, \ldots, N-1,$$

of f, and we want to recover f from these points. By a slight abuse of notation we interpret s_k as complex number and write

$$s_k = x_k + jy_k \,. \tag{52}$$

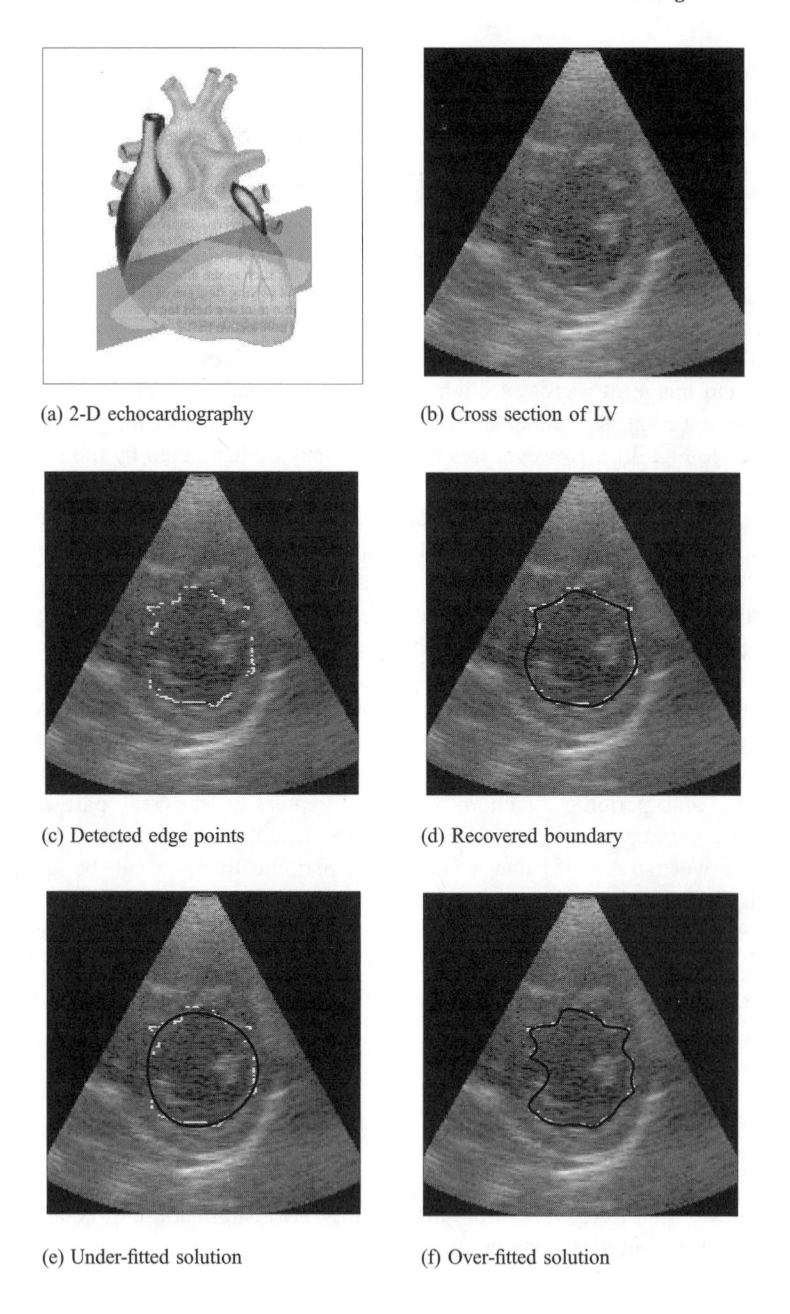

(a) 2-D echocardiography
(b) Cross section of LV
(c) Detected edge points
(d) Recovered boundary
(e) Under-fitted solution
(f) Over-fitted solution

Figure 3. The recovery of the boundary of the Left Ventricle from 2-D ultrasound images is a basic step in echocardiography to extract relevant parameters of cardiac function.

We relate the curve parameter u to the boundary points s_k by computing the distance between two successive points s_{k-1}, s_k via

$$u_0 = 0 = u_N \tag{53}$$

$$u_k = u_{k-1} + d_k \tag{54}$$

$$d_k = \sqrt{(x_k - x_{k-1})^2 + (y_k - y_{k-1})^2} \tag{55}$$

for $k = 1, \ldots, N-1$. Using the normalization $t_k = t_k/L$ with $L = u_{N-1} + d_N$, we force all sampling points to be in [0,1). Other choices for d_k in (55) can be found in [49] in conjunction with curve approximation using splines.

After the transformation (52)–(55) of the detected boundary points, we can use the 1-D version of Algorithm 1 (see [6]) to recover the boundary.

As already mentioned, the detected contour points are distorted by noise. (The noise level δ depends on the quality of the technical equipment and can be determined experimentally.) Therefore least squares approximation of the data points is preferable over exact interpolation. This raises the question of choosing the optimal degree of the approximating trigonometric polynomial. For the reconstruction of Fig. 3(d) we have used the 1-D version of the multi-level method proposed in Section 6.5.1. An alternative method was proposed in [23]. In general the multi-level method it is more stable and computationally only slightly more expensive than the method in [23], therefore it is preferable in practice.

Figures 3(e)–(f) demonstrate the importance of determining an appropriate bandwidth (= degree) M for the approximating polynomial. The approximation displayed in Fig. 3(e) has been computed by applying the 1-D version of Algorithm 1, where the bandwidth M has been chosen too small. Obviously we have under-fitted the data. The over-fitted approximation was obtained by applying the same algorithm using a too large M. The result is displayed in Fig. 3(f). The approximation shown in Fig. 3(d) has been computed by the multi-level algorithm. It provides the optimal balance between fitting the data and smoothness of the solution.

6.6.2. Image Reconstruction in Exploration Geophysics

Exploration geophysics relies on measurements of the earth's magnetic field with the goal of detecting anomalies which reveal underlying geological features. In geophysical practice, it is essentially impossible to gather data in a form that allows direct interpretation. Geoscientists, used to look at their measurements on maps or profiles and aim at further processing, need a representation of the originally irregularly spaced (scattered) data points on a regular grid.

Gridding is thus one of the first and crucial steps in the analysis of geophysical data. However, a number of practical constraints, such as measurement errors and the huge amount of data, make the development of reliable gridding methods a difficult task. It is agreed in the geophysics community that "a good image of the earth hides our data acquisition footprint" [50]. Unfortunately many gridding methods proposed in the literature fail in this aspect, and show how the data were sampled and processed, in particular when the data are

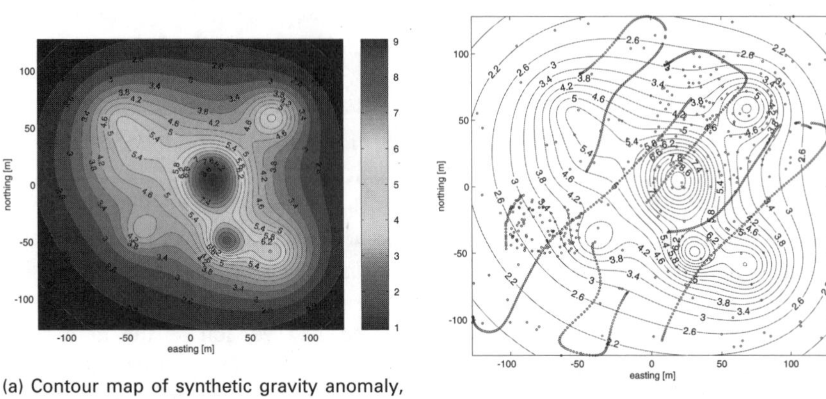

(a) Contour map of synthetic gravity anomaly, gravity is in mGal.

(b) Sampling set and synthetic gravity anomaly.

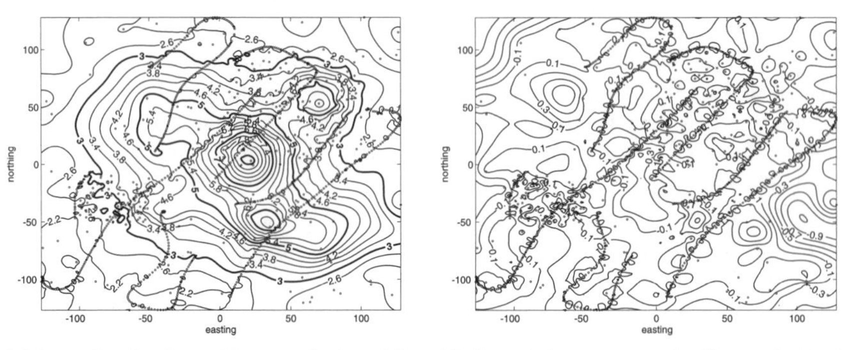

(c) Approximation from noisy samples by minimum curvature method.

(d) Error between approximation and actual anomaly.

Figure 4. Many existing reconstruction methods used in exploration geophysics produce approximations that are smooth, but which do not conform to a relevant physical model. These methods are often very sensitive to the sampling geometry, cf. Fig. 4(d).

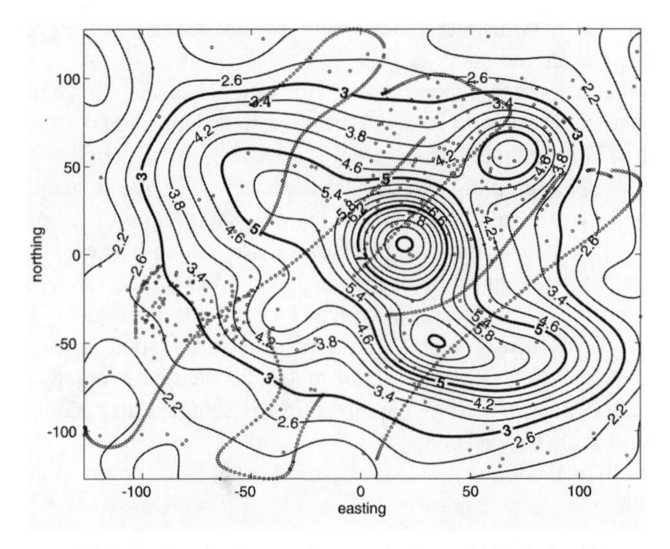

(a) Approximation from noisy samples by multi-level algorithm.

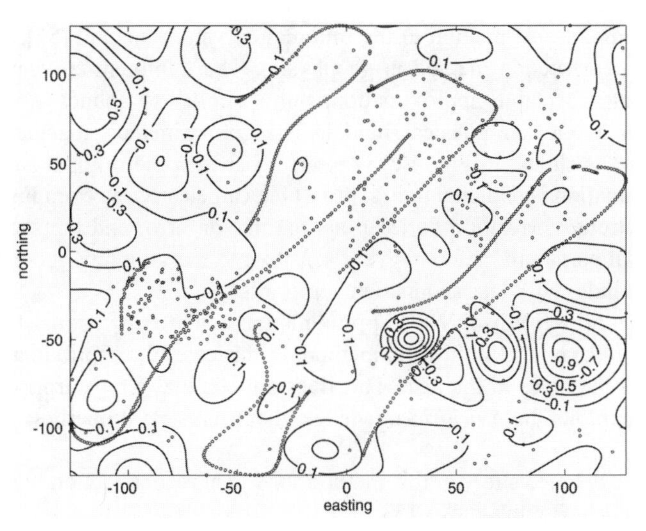

(b) Error between approximation and actual anomaly.

Figure 5. The incorporation of *a priori* knowledge about physical properties of potential fields greatly reduced the influence of the sampling geometry, as can be seen by comparing Figs. 5(a) and (b) to Fig. 4(c) and (d).

noisy. Figures 4(a)–(d) illustrate this problem for the recovery of the gravitational field from noisy scattered data.

Our synthetic anomaly represents the gravitational acceleration caused by an ensemble of buried rectangular boxes of different size, depth, and density contrast, see Fig. 4(b). This anomaly was sampled at a highly scattered set of 1000 sampling points. The sampling geometry is depicted in Fig. 4(a) together with the contour plot of the original anomaly. Note that this sampling set is composed of a line-type sampling pattern (which is characteristic for airborne or other fast-moving measurement devices), of some clusters, and of a pure random part. We chose this sampling set because it is typical for real-world settings.

Since physical measurements contains necessarily some error, we have added white noise of 5% of the ℓ^2-norm of the sampling energy. We then reconstructed the anomaly from the nonuniform samples and plotted it on a 256×256 grid with 1-m spacing.

In the numerical computation, we exploit the fact that a potential field can be modeled as an essentially band-limited function [51]. This observation allows us to use the heuristic method of Section 6.5.3. Since the essential bandwidth of the anomaly is not known we use the multi-level approach proposed in Section 6.5.1 and apply Algorithm 2 to recover the original signal.

We compare this method to the minimal curvature method [52], which is a standard reconstruction method in geophysics. The minimum curvature method produces an acceptable approximation, but – similar to splines – it does not conform to a relevant physical model. The approximation together with the sampling set is shown in Fig. 4(c). The error between the original anomaly and the approximation is depicted in Fig. 4(d). One can clearly see from Fig. 4(d) that there is a strong correlation between approximation error and the sampling set. The minimal curvature method is very sensitive to the sampling geometry and is not able to hide the "data acquisition footprints".

Figures 5(a)–(b) display the approximation of the gravitational field shown in Figs. 4(a)–(b) by the multi-level method. In this case, the approximation is less distorted by the noise in the data. This does not come as a big surprise since our algorithm exploits the *a priori* knowledge about physical properties of potential fields.

A more detailed study of this method for signal reconstruction in geophysics can be found in Rauth's thesis [53].

6.6.3. Reconstruction of Missing Pixels in Images

In practice images are rarely band-limited in the strict sense of definition (1). But images are often *essentially M-band-limited*, i.e. $F(n,m)$ is negligible for $|n|, |m| > M$.

Our test-image is the well-known "Lena"-image of size 512×512, see Fig. 7(a). The image is sampled at 150320 randomly spaced pixels, i.e., about 39% of pixels are lost (displayed as black pixels in Fig. 7(b)). The signal-to-noise ratio is 4.04 dB[1]. Since "Lena" is not band-limited, we cannot expect perfect reconstruction of the missing data.

We compare the output of Algorithm 1 (with and without using weights) to other iterative reconstruction methods proposed in the literature. These methods are the Marvasti or frame method [2, 14] and the Adaptive weights method [54]. For each method we use the same bandwidth (we set $2M + 1 = 255$) and run each method until it returns a reconstruction with a signal-to-noise ratio (SNR) of at least 30 dB. The results are illustrated in Fig. 7.

Marvasti's method (with optimal relaxation parameter) requires 2.88×10^9 floating point operations (*flops*) to achieve an SNR of 30 dB. The Adaptive weights method needs 1.5×10^9 flops. The number of flops for Algorithm 1 (referred to as ACT in Figs. 6 and 7) is 7.99×10^8 without weights and 1.08×10^9 with Voronoi-type weights. Algorithm 1 clearly outperforms the other methods in terms of computational efficiency. Note that in case of missing data the entries of the block Toeplitz matrix used in Algorithm 1 can be easily computed by one 2-D FFT, see [27].

The bottom figure in Fig. 6 shows the rate of convergence of the reconstruction methods. In each iteration we measure the normalized least square error e_n between the original signal f and the approximation f_n via $e_n = \|f - f_n\| / \|f\|$. Also in terms of convergence rates Algorithm 1 is preferable over the other two methods. Algorithm 1 with weights gives the fastest rate of convergence, it needs only 12 iterations to achieve the required SNR of 30 dB, the same algorithm without weights needs only one iteration more. The computational effort to calculate the weights and the slightly higher costs in the CG iterations due to the additional multiplication by the weights lead to a somewhat larger number of flops for the "weighted version" of Algorithm 1 in this example. The adaptive weights method needs 27 iterations and the Marvasti/frame algorithm terminates after 43 iterations.

The approximations after 15 iterations are shown in Fig. 7(c)–(f). The Marvasti or frame method [2, 14] provides an acceptable approximation of most of the lost samples, but there are still many black dots visible. The signal-to-noise ratio (SNR) after 15 iterations is 27.3 dB. The adaptive weights method [54] returns an approximation with an SNR of 29.4 dB. Both versions of Algorithm 1 provide a reconstruction which shows no visible difference to the original image. Other examples and more comparisons of methods can be found in [27].

The representation of edges in images requires a large bandwidth. This in turn implies that the area of missing samples has to be small in order to achieve

[1]The SNR of two images f_1, f_2 is defined as usual as $\mathrm{SNR} = 10 \log_{10}(\|f_1\|^2 / \|f_1 - f_2\|^2)$.

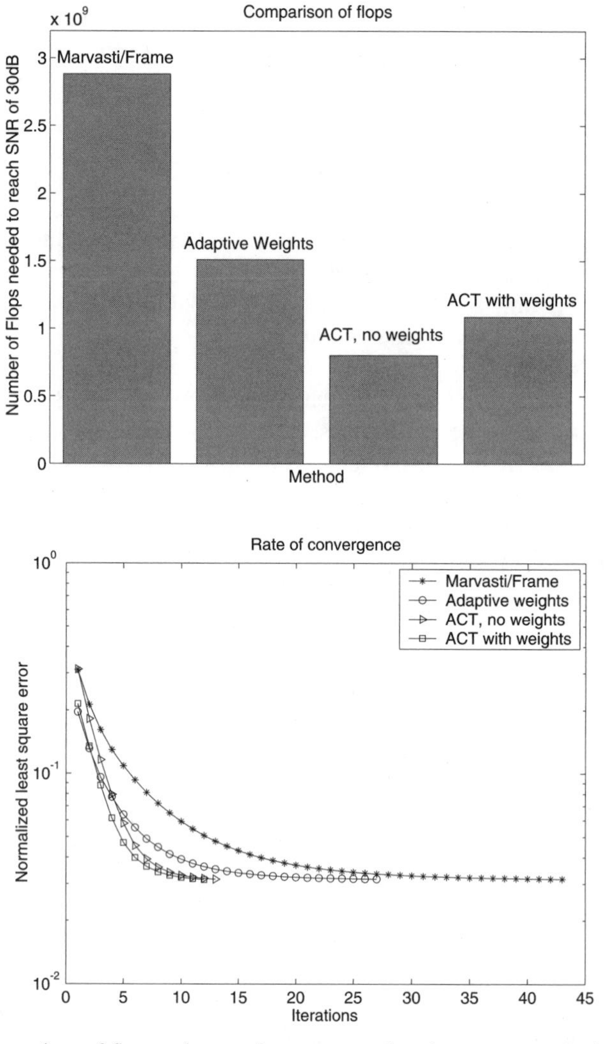

Figure 6. Comparison of flops and rates of convergence. Iterations are stopped when an SNR of 30 dB is reached (which corresponds approximately to a normalized least square error of 3%). The two versions of Algorithm 1 are the most efficient methods.

stable reconstruction. Therefore the use of band-limited functions for the reconstruction of missing pixels is somewhat restricted. Other reconstruction methods, for instance, methods based on local Fourier bases or wavelets, have to be developed to handle the restoration of large areas of missing pixels.

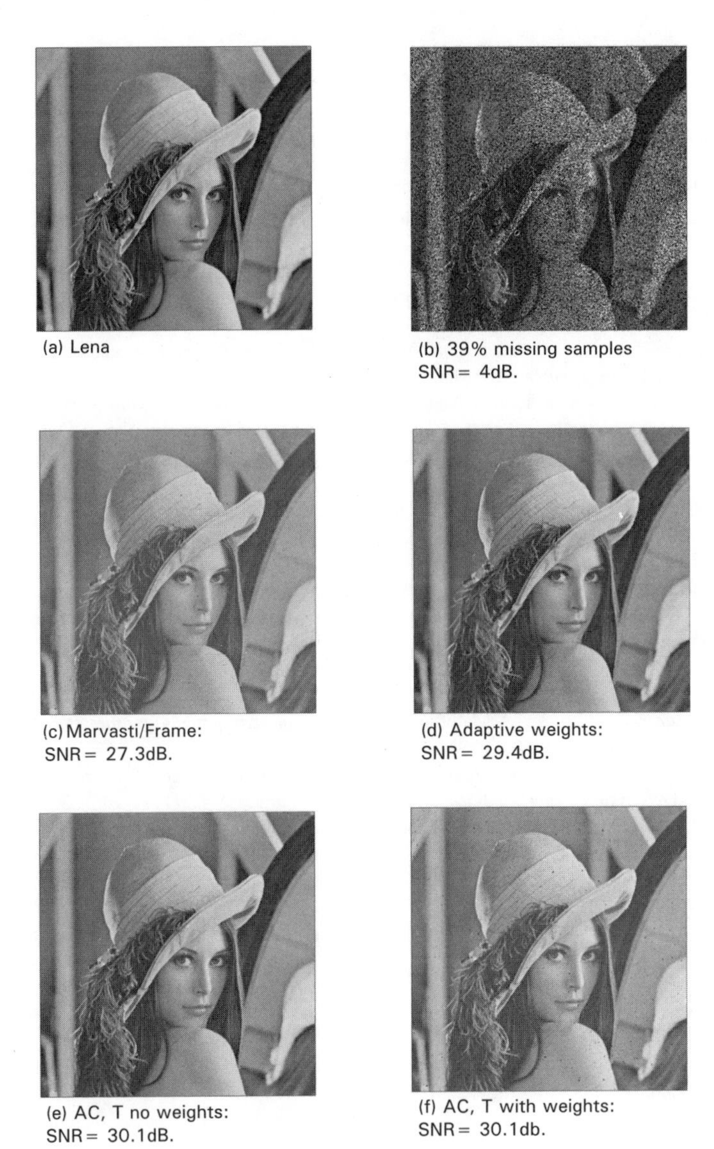

(a) Lena

(b) 39% missing samples
SNR= 4dB.

(c) Marvasti/Frame:
SNR= 27.3dB.

(d) Adaptive weights:
SNR= 29.4dB.

(e) AC, T no weights:
SNR= 30.1dB.

(f) AC, T with weights:
SNR= 30.1db.

Figure 7. Nonuniformly sampled Lena and reconstructions after 15 iterations.

References

[1] F. Marvasti. Nonuniform sampling. In R. J. Marks II, Ed., *Advanced in Shannon Sampling and Interpolation Theory*, Springer Verlag, 1993, pp. 121–156.

[2] J. Benedetto. Irregular Sampling and Frames. In C. K. Chui, Ed., *Wavelets: A Tutorial in Theory and Applications*, Academic Press, 1992, pp. 445–507.

[3] *Proc. Sampta'95–Workshop on Sampling Theory & Applications*, (Jurmala, Latvia), September 1995.

[4] *Proc. Sampta'97–Workshop on Sampling Theory & Applications*, (Aveiro, Portugal), June 1997.

[5] H. Feichtinger and K. Gröchenig. Theory and Practice of Irregular Sampling. In J. Benedetto and M. Frazier, Eds., *Wavelets: Mathematics and Applications*, CRC Press, 1994, pp. 305–363.

[6] H. G. Feichtinger, K. Gröchenig, and T. Strohmer. Efficient Numerical Methods is Non-Uniform Sampling Theory. *Numerische Mathematik* 69:423–440, 1995.

[7] K. Gröchenig. Reconstruction Algorithms in Irregular Sampling. *Math. Comp.*, 59:181–194, 1992.

[8] K. Gröchenig. A Discrete Theory of Irregular Sampling. *Linear Algebra Appl.*, 193:129–150, 1993.

[9] K. Gröchenig. Irregular Sampling, Toeplitz Matrices, and the Approximation of Entire Functions of Exponential Type. *Math. Comp.*, 68:749–765, 1999.

[10] A. Beurling. The Collected Works of Arne Beurling. Vol. 2. In L. Carleson, P. Malliavin, J. Neuberger and J. Wermer, Eds., *Harmonic analysis*. Birkhäuser Boston Inc., Boston, MA, 1989.

[11] K. D. Sauer and J. P. Allebach. Iterative Reconstruction of Band-Limited Images from Nonuniformly Spaced Samples. *IEEE Trans. CAS*, 34(12):1497–1506, 1987.

[12] P. L. Butzer and G. Hinsen. Two-Dimensional Nonuniform Sampling Expansions – An Iterative Approach. *I, II. Appl. Anal.*, 32:53–88, 69–85, 1989.

[13] J. Benedetto and H. Wu. A Multidimensional Irregular Sampling Algorithm and Applications. *IEEE Proc. ICASSP*, 1999.

[14] F. A. Marvasti, C. Liu, and G. Adams. Analysis and Recovery of Multi-Dimensional Signals from Irregular Samples Using Nonlinear and Iterative Techniques. *Signal Processing*, 36:13–30, 1994.

[15] D. Peterson and A. Middleton. Sampling and Reconstruction of Wave-Number Limited Functions in n-Dimensional Euclidean Spaces. *Inform. and Contr.*, 5:279–323, 1962.

[16] R. Duffin and A. Schaeffer. A Class of Nonharmonic Fourier Series. *Trans. Amer. Math. Soc.*, 72:341–366, 1952.

[17] K. Seip. Density Theorems for Sampling and Interpolation in the Bargmann-Fock Space. *Bull. A.M.S.*, 26:322–328, 1992.

[18] H. Feichtinger and K. Gröchenig. Non-Orthogonal Wavelet and Gabor Expansions, and Group Representations. In G. Beylkin, R. Coifman, I. Daubechies, *et al.*, Eds., *Wavelets and their applications*, Jones and Bartlett, 1992, pp. 353–376.

[19] L. Reichel, G. Ammar, and W. Gragg. Discrete Least Squares Approximation by Trigonometric Polynomials. *Math. Comp.*, 57:273–289, 1991.

[20] C. Demeure. Fast QR Factorization of Vandermonde Matrices. *Linear Algebra Appl.*, 122–124:165–194, 1989.

[21] H. Landau. Necessary Density Conditions for Sampling and Interpolation of Certain Entire Functions. *Acta Math.*, 117:37–52, 1967.

[22] K. Gröchenig. Non-Uniform Sampling in Higher Dimensions: From Trigonometric Polynomials to Band-Limited Functions. In J. Benedetto and P. Ferreira, Eds., *Modern Sampling Theory: Mathematics and Applications*, Birkhäuser, Boston, 2001, pp. 161–178.

[23] T. Strohmer. A Levinson-Galerkin Algorithm for Regularized Trigonometric Approximation. *SIAM J. Sci. Comp.* 22(4):1160–1183, 2000.

[24] G. Beylkin. On the Fast Fourier Transform of Functions with Singularities. *Appl. Comp. Harm. Anal.*, 2(4):363–381, 1995.

[25] G. Golub and C. van Loan. *Matrix Computations*, (3rd edn). Johns Hopkins, London, Baltimore, 1996.

[26] G. Strang. A Proposal for Toeplitz Matrix Calculations. *Stud. Appl. Math.*, 74:171–176, 1986.

[27] T. Strohmer. Computationally Attractive Reconstruction of Band-Limited Images from Irregular Samples. *IEEE Trans. Image. Proc.*, 6(4):540–548, 1997.

[28] A. Dutt and V. Rokhlin. Fast Fourier Transforms for Nonequispaced Data. *SIAM J. Sci. Comp.*, 14(6):1368–1394, 1993.

[29] A. Duijndam and M. Schonewille. Nonuniform Fast Fourier Transform. *Geophysics*, 64(2):539–551, 1999.

[30] A. van der Sluis and H. van der Vorst. The Rate of Convergence of Conjugate Gradients. *Numer. Math.*, 48:543–560, 1986.

[31] T. Strohmer. Numerical Analysis of the Non-Uniform Sampling Problem. *J. Comp. Appl. Math.*, 122(1–2):297–316, 2000.

[32] M. Hanke. *Conjugate Gradient Type Methods for Ill-Posed Problems*. Longman Scientific & Technical, Harlow, 1995.

[33] H. Feichtinger and K. Gröchenig. Gabor Wavelets and the Heisenberg Group: Gabor Expansions and Short Time Fourier Transform from the Group Theoretical Point of View. In C. Chui, Ed., *Wavelets–A Tutorial in Theory and Applications*, Academic Press, 1992, pp. 359–397.

[34] B. Chazelle and H. Edelsbrunner. An Improved Algorithm for Constructing *k*th-Order Voronoi Diagrams. *IEEE Trans. Comput.*, 36(11):1349–1354, 1987.

[35] A. Okabe, B. Boots, and K. Sugihara. *Spatial Essellations: Concepts and Applications of Voronoĭ Diagrams*. John Wiley & Sons Ltd., Chichester, 1992. With a foreword by D. G. Kendall.

[36] H. G. Feichtinger and T. Strohmer. Fast Iterative Reconstruction of Band-Limited Images from Irregular Sampling Values. In D. Chetverikov and W. Kropatsch, Eds., *Proc. on Computer Analysis of Images and Patterns*, Conf. CAIP, 1993, pp. 82–91.

[37] M. Harrison. Frames and Irregular Sampling from a Computational Perspective. PhD thesis. University of Maryland–College Park, 1998.

[38] K. Gröchenig. Finite and Infinite-Dimensional Models for Non-Uniform Sampling. In *SampTA – Sampling Theory and Applications*, Aveiro, Portugal, 1997, pp. 285–290.

[39] M. Hanke. Regularizing Properties of a Truncated Newton-CG Algorithm for Nonlinear Inverse Problems. *Numer. Funct. Anal. Optim.*, 18(9–10):971–993, 1997.

[40] O. Scherzer and T. Strohmer. A Multi-Level Algorithm for the Solution of Moment Problems. *Num. Funct. Anal. Opt.*, 19(3–4):353–375, 1998.

[41] D. Wingham. The Reconstruction of a Band-Limited Function and its Fourier Transform from a Finite Number of Samples at Arbitrary Locations by Singular Value Decomposition. *IEEE Trans. Circuit Theory*, 40:559–570, 1992.

[42] J. Yen. On Nonuniform Sampling of Bandwidth-Limited Signals. *IRE Trans. Circuit Theory*, CT-3:251–257, 1956.

[43] J. Benedetto and W. Heller. Irregular Sampling and the Theory of Frames, I. *Mat. Note*, 10:103–125, 1990.

[44] H. Engl, M. Hanke, and A. Neubauer. *Regularization of Inverse Problems*. Kluwer Academic Publishers Group, Dordrecht, 1996.

[45] A. S. Nemirovskiĭ. Regularizing Properties of the Conjugate Gradient Method in Ill-Posed Problems. *Zh. Vychisl. Mat. i Mat. Fiz.*, 26(3):332–347, 477, 1986.

[46] D. Wilson, E. Geiser, and J. Li. Feature Extraction in 2-Dimensional Short-Axis Echocardiographic Images. *J. Math. Imag. Vision*, 3:285–298, 1993.

[47] M. Suessner, M. Budil, T. Strohmer, M. Greher, G. Porenta, and T. Binder. Contour Detection Using Artifical Neuronal Network Presegmention. *Proc. Computers in Cardiology*, Vienna, 1995.

[48] W. Rudin. *Fourier Analysis on Groups*. Wiley Interscience, New York, 1976.

[49] P. Dierckx. *Curve and Surface Fitting with Splines*. Monographs on Numerical Analysis, Oxford University Press, 1993.

[50] J. Claerbout. *Geophysical Estimation by Example: Enviromental Soundings Image Construction: Multi-dimensional Autoregression*, 1998. This book is on the Web and can be downloaded at http://sepwww.stanford.deu.sep/prof/gee/tochtml/index.html.

[51] M. Rauth and T. Strohmer. Smooth Approximation of Potential Fields from Noisy Scattered Data. *Geophysics*, 63(1):85–94, 1998.

[52] I. Briggs. Machine Contouring Using Minimum Curvature. *Geophysics*, 39(1):39–48, 1974.

[53] M. Rauth. Gridding of Geophysical Potentials from Noisy Scattered Data. PhD thesis. University of Vienna, 1998.

[54] H. G. Feichtinger and K. Gröchenig. Error Analysis in Regular and Irregular Sampling Theory. *Applicable Analysis*, 50:167–189, 1993.

7

The Nonuniform Discrete Fourier Transform

S. Bagchi and S. K. Mitra

Abstract

In many applications, when the representation of a discrete-time signal or a system in the frequency domain is of interest, the Discrete-Time Fourier Transform (DTFT) and the z-transform are often used. In the case of a discrete-time signal of finite length, the most widely used frequency-domain representation is the Discrete Fourier Transform (DFT), which is simply composed of samples of the DTFT of the sequence at equally spaced frequency points, or equivalently, samples of its z-transform at equally spaced points on the unit circle. A generalization of the DFT, introduced in this chapter, is the Nonuniform Discrete Fourier Transform (NDFT), which can be used to obtain frequency domain information of a finite-length signal at arbitrarily chosen frequency points. We provide an introduction to the NDFT and discuss its applications in the design of 1-D and 2-D FIR digital filters. We begin by introducing the problem of computing frequency samples of the z-transform of a finite-length sequence. We develop the basics of the NDFT, including its definition, properties and computational aspects. The NDFT is also extended to two dimensions. We propose NDFT-based nonuniform frequency sampling techniques for designing 1-D and 2-D FIR digital filters, and present design examples. The resulting filters are compared with those designed by other existing methods.

S. Bagchi • Mobilian Corporation, Hillsboro, OR 97124, USA
E-mail: sonali@mobilian.com

S. K. Mitra • Department of Electrical and Computer Engineering, University of California at Santa Barbara, Santa Barbara, CA 93106, USA
E-mail: mitra@ece.ucsb.edu

Nonuniform Sampling: Theory and Practice, edited by Marvasti
Kluwer Academic/Plenum Publishers, New York, 2001.

7.1. Introduction

A widely used frequency-domain representation of a finite-length sequence is its Discrete Fourier Transform (DFT), which corresponds to equally spaced samples of its Discrete-Time Fourier Transform (DTFT), or equivalently, the samples of its z-transform evaluated on the unit circle in the z-plane at equally spaced points. In most signals, the energy is distributed nonuniformly in the frequency domain. Therefore, a nonuniform sampling scheme, tailored to the frequency-domain attributes of the signal, can be more useful and convenient in some applications. This observation has been the motivation for developing the concept of the Nonuniform Discrete Fourier Transform (NDFT), the main subject of this chapter. We thus define the NDFT of a finite-length sequence as samples of its z-transform evaluated at arbitrarily chosen distinct points in the z-plane. Basically, this concept is a generalization of the conventional Discrete Fourier Transform. In this chapter, we develop the basic framework of the NDFT representation, and demonstrate its potential in signal processing by considering two specific applications in filter design [1, 2]. Other applications of the NDFT include spectral analysis, antenna array design, and decoding of dual-tone multi-frequency (DTMF) signals [3, 4].[1]

7.2. The 1-D NDFT

7.2.1. Definition

The one-dimensional Nonuniform Discrete Fourier Transform (NDFT) of a sequence $x[n]$ of length N is defined as [1, 5]

$$X(z_k) = X(z)|_{z=z_k} = \sum_{n=0}^{N-1} x[n] z_k^{-n}, \qquad k = 0, 1, \ldots, N-1, \tag{1}$$

where $X(z)$ is the z-transform of $x[n]$, and $z_0, z_1, \ldots, z_{N-1}$ are distinct points located arbitrarily in the z-plane. We can express (1) in a matrix form as

$$\mathbf{X} = \mathbf{D}\mathbf{x} \tag{2}$$

where

$$\mathbf{X} = \begin{bmatrix} X(z_0) \\ X(z_1) \\ \vdots \\ X(z_{N-1}) \end{bmatrix}, \qquad \mathbf{x} = \begin{bmatrix} x[0] \\ x[1] \\ \vdots \\ x[N-1] \end{bmatrix}, \tag{3}$$

[1]Adapted from The Nonuniform Discrete Fourier Transform and its Applications in Signal Processing ©1999 Kluwer Academic Publishers, with the permission of the publisher.

and

$$\mathbf{D} = \begin{bmatrix} 1 & z_0^{-1} & z_0^{-2} & \cdots & z_0^{-(N-1)} \\ 1 & z_1^{-1} & z_1^{-2} & \cdots & z_1^{-(N-1)} \\ \vdots & \vdots & \vdots & \ddots & \vdots \\ 1 & z_{N-1}^{-1} & z_{N-1}^{-2} & \cdots & z_{N-1}^{-(N-1)} \end{bmatrix}. \tag{4}$$

Note that the NDFT matrix **D** is fully specified by the choice of the N points, z_k. A matrix of this form is known as a Vandermonde matrix [6]. It can be shown that the determinant of **D** can be expressed in a factored form [7] as

$$\det(\mathbf{D}) = \prod_{i \neq j, i > j} (z_i^{-1} - z_j^{-1}). \tag{5}$$

Consequently, **D** is nonsingular provided the N sampling points, $z_0, z_1, \ldots, z_{N-1}$, are distinct. Thus, the inverse NDFT exists and is unique, and is given by

$$\mathbf{x} = \mathbf{D}^{-1}\mathbf{X}. \tag{6}$$

As a special case, consider the situation when the points z_k are located at equally spaced angles on the unit circle in the z-plane. This corresponds to the conventional DFT. The matrix **D** then reduces to the conventional DFT matrix.

7.2.2. The Inverse NDFT

The problem of computing the inverse NDFT, i.e., determining **x** from a given NDFT vector **X**, is equivalent to solving the Vandermonde system in (6).

Given the NDFT **X** and the NDFT matrix **D**, the inverse NDFT **x** is found directly by solving the linear system in (2), using Gaussian elimination. This involves $O(N^3)$ operations.

The same problem can be solved more efficiently by using polynomial interpolation. In this approach, the z-transform $X(z)$ is directly determined in terms of the NDFT coefficients,

$$\hat{X}[k] = X(z_k), \qquad k = 0, 1, \ldots, N-1, \tag{7}$$

by using polynomial interpolation methods. The inverse NDFT $x[n]$ can then be identified as the coefficients of this interpolating polynomial, and can be solved using popular polynomial interpolation methods such as Lagrange and Newton interpolation.

In the Lagrange interpolation method (see Chapter 3), $X(z)$ is expressed as the Lagrange polynomial of order $N-1$,

$$X(z) = \sum_{k=0}^{N-1} \frac{L_k(z)}{L_k(z_k)} \hat{X}[k], \tag{8}$$

where $L_0(z), L_1(z), \ldots, L_{N-1}(z)$ are the fundamental polynomials, defined as

$$L_k(z) = \prod_{i \neq k} (1 - z_i z^{-1}), \qquad k = 0, 1, \ldots, N-1. \tag{9}$$

Note that

$$L_{k+1}(z) = \frac{(1 - z_{k+1} z^{-1})}{(1 - z_k z^{-1})} L_k(z), \qquad k = 0, 1, \ldots, N-1. \tag{10}$$

In the Newton interpolation method, $X(z)$ is expressed in the form

$$\begin{aligned} X(z) = c_0 &+ c_1(1 - z_0 z^{-1}) + c_2(1 - z_0 z^{-1})(1 - z_1 z^{-1}) \\ &+ \cdots + c_{N-1} \prod_{k=0}^{N-2} (1 - z_k z^{-1}), \end{aligned} \tag{11}$$

where the coefficient c_j is called the divided difference of the jth order of $\hat{X}[0], \hat{X}[1], \ldots, \hat{X}[j]$ with respect to $z_0, z_1, \ldots, z_j$. The divided differences are computed recursively as follows:

$$\begin{aligned} c_0 &= \hat{X}[0], \\ c_1 &= \frac{\hat{X}[1] - c_0}{1 - z_0 z_1^{-1}}, \\ c_2 &= \frac{\hat{X}[2] - c_0 - c_1(1 - z_0 z_2^{-1})}{(1 - z_0 z_2^{-1})(1 - z_1 z_2^{-1})}, \\ &\vdots \end{aligned} \tag{12}$$

Note that each c_j is a linear combination of the $\hat{X}[k]$, and moreover, c_j depends only on $\hat{X}[0], \hat{X}[1], \ldots, \hat{X}[j]$, and $z_0, z_1, \ldots, z_j$. In the Lagrange representation, if we include an additional point and increase the order of the interpolating polynomial, all the fundamental polynomials change and, consequently, have to be recomputed. In the Newton representation, this can be accomplished by simply adding one more term. Thus, it has a permanence property which is a characteristic of the Fourier series and other orthogonal and biorthogonal expansions.

Since the coefficients c_j are computed by solving the lower triangular system of equations,

$$\mathbf{Lc} = \mathbf{X}, \tag{13}$$

where

$$\mathbf{L} = \begin{bmatrix} 1 & 0 & 0 & \cdots & 0 \\ 1 & (1 - z_0 z_1^{-1}) & 0 & \cdots & 0 \\ 1 & (1 - z_0 z_2^{-1}) & 0 & \cdots & 0 \\ \vdots & \vdots & \vdots & \ddots & \vdots \\ 1 & (1 - z_0 z_{N-1}^{-1}) & \cdots & \prod_{k=0}^{N-2} (1 - z_k z_{N-1}^{-1}) & \end{bmatrix}, \tag{14}$$

and

$$\mathbf{c} = \begin{bmatrix} c_0 \\ c_1 \\ c_2 \\ \vdots \\ c_{N-1} \end{bmatrix}, \qquad \mathbf{X} = \begin{bmatrix} \hat{X}[0] \\ \hat{X}[1] \\ \hat{X}[2] \\ \vdots \\ \hat{X}[N-1] \end{bmatrix}, \tag{15}$$

this involves $O(N^2)$ operations. The sequence $x[n]$ can now be easily computed from the c_j. As compared to straightforward Lagrange interpolation, Newton interpolation provides a more efficient method of solving for the inverse NDFT. However, if Lagrange interpolation is modified by using (10), then it is more efficient.

7.2.3. Computation of the NDFT

We address the problem of computing the NDFT in this section. The factors to be considered are—the amount of computation needed in terms of multiplications and additions, and the number of coefficients used for computation. To establish a reference for comparing these factors, we begin by examining the direct method. This is followed by Horner's method, which requires the same amount of computation as the direct method, but a lower number of coefficients. Finally, we use the Goertzel algorithm to compute the NDFT [8]. The Goertzel algorithm achieves a reduction in computation as well as the number of coefficients used. The computational saving is obtained when the NDFT is evaluated at points on the unit circle in the z-plane.

Due to the generalized sampling inherent in the NDFT, there is no periodicity or symmetry in the complex numbers, z_k^{-n}. Consequently, algorithms as efficient as the FFT cannot be derived in the general case. In the next section, we will present the subband NDFT, which leads to a fast, approximate computation of the NDFT, when the signal has its energy concentrated in a few bands of the spectrum.

(1) Direct Method

We can compute the NDFT directly by evaluating the expression given in (1). In general, the signal $x[n]$ is a complex sequence of length N, and the sampling points $z_0, z_1, \ldots, z_{N-1}$ are also complex numbers. To compute each sample of the NDFT, we need N complex multiplications and $(N-1)$ complex additions, i.e., $4N$ real multiplications and $(4N-2)$ real additions. Therefore, the amount of computation needed for evaluating N samples of the NDFT is approximately proportional to N^2. Note that this is the same as that needed for directly computing the DFT. In the direct method, we need to compute or store N complex coefficients, $\{z_k^{-n}, \quad n = 0, 1, \ldots, N-1\}$ for evaluating each NDFT sample $X(z_k)$.

(2) Horner's Method

Let us rewrite the NDFT in (1) as

$$X(z_k) = z_k^{-(N-1)} A_k, \tag{16}$$

where

$$\begin{aligned} A_k &= \sum_{n=0}^{N-1} x[n] z_k^{N-1-n} \\ &= x[0] z_k^{N-1} + x[1] z_k^{N-2} \\ &\quad + \cdots + x[N-2] z_k + x[N-1]. \end{aligned} \tag{17}$$

To avoid using the N coefficients z_k^n, we can express (17) as a nested multiplication (also known as Horner's method [9]):

$$A_k = \{\cdots (x[0] z_k + x[1]) z_k + \cdots\} z_k + x[N-1]. \tag{18}$$

We start by evaluating the expression in the innermost parentheses of (18), and proceed to solve for A_k. $X(z_k)$ can then be found from (16). This requires a total of $4N$ real multiplications and $(4N-2)$ real additions, which is the same as in the direct method. However, in Horner's method, we need only two coefficients, z_k and $z_k^{-(N-1)}$, to evaluate the NDFT sample $X(z_k)$.

Note that the NDFT has been expressed as shown in (16), so that we can start evaluating the nested multiplication in (18) with the first sample $x[0]$, rather than the last sample $x[N-1]$. This eliminates the need for buffering the input signal.

(3) Goertzel Algorithm: Digital Filter Interpretation

Consider the NDFT sample at $z = z_k$ given by

$$X(z_k) = \sum_{r=0}^{N-1} x[r] z_k^{-r}. \tag{19}$$

Multiplying both sides of (19) by z_k^N, we have

$$z_k^N X(z_k) = \sum_{r=0}^{N-1} x[r] z_k^{N-r}. \tag{20}$$

Let us now define the sequence

$$y_k[n] = \sum_{r=-\infty}^{\infty} x[r] z_k^{n-r} u[n-r], \tag{21}$$

where $u[n]$ denotes the unit step sequence. Equation (21) is equivalent to the discrete convolution

$$y_k[n] = x[n] * z_k^n u[n]. \tag{22}$$

Since $x[n]$ is zero outside the range $0 \leq n \leq N-1$, we infer from (20) and (21) that

$$X(z_k) = z_k^{-N} \, y_k[n]|_{n=N}. \tag{23}$$

Thus, the NDFT sample $X(z_k)$ is obtained by multiplying z_k^{-N} with the Nth sample at the output of a system, whose impulse response is $z_k^n u[n]$. This is a first-order recursive system, whose system function is

$$H_k(z) = \frac{1}{1 - z_k z^{-1}}. \tag{24}$$

To compute $X(z_k)$ using this system, we need $(4N+4)$ real multiplications and $(4N+2)$ real additions. These are nearly the same as in the direct method. However, only two coefficients, z_k and z_k^{-N}, are needed.

Next, we consider a special case for which the number of multiplications can be reduced by a factor of two. Multiplying the numerator and denominator of (24) by $(1 - z_k^* z^{-1})$, where z_k^* is the complex conjugate of z_k, we obtain

$$H_k(z) = \frac{1 - z_k^* z^{-1}}{1 - 2Re\{z_k\}z^{-1} + |z_k|^2 z^{-2}}. \tag{25}$$

Let $|z_k|^2 = 1$. Thus, the NDFT is being evaluated at a point $z_k = e^{j\omega_k}$ on the unit circle in the z-plane. In this case, we can simplify (25) to obtain the system function

$$H_k(z) = \frac{1 - e^{-j\omega_k} z^{-1}}{1 - 2\cos\omega_k z^{-1} + z^{-2}}, \tag{26}$$

which is shown in Fig. 1. Thus, we can interpret the Goertzel algorithm as computing the output of this second-order recursive digital filter.

Since we want to compute only $y_k[N]$, the multiplication by $e^{-j\omega_k}$ in the feedforward section of Fig. 1 (to the right of the delay elements) need not be performed until the Nth iteration. The intermediate signals, $q_k[n]$ and $q_k[n-1]$, are computed recursively using the difference equation

$$q_k[n] = a_k\, q_k[n-1] - q_k[n-2] + x[n], \qquad n = 0, 1, \ldots, N, \tag{27}$$

where the coefficient a_k is given by

$$a_k = 2\cos\omega_k,$$

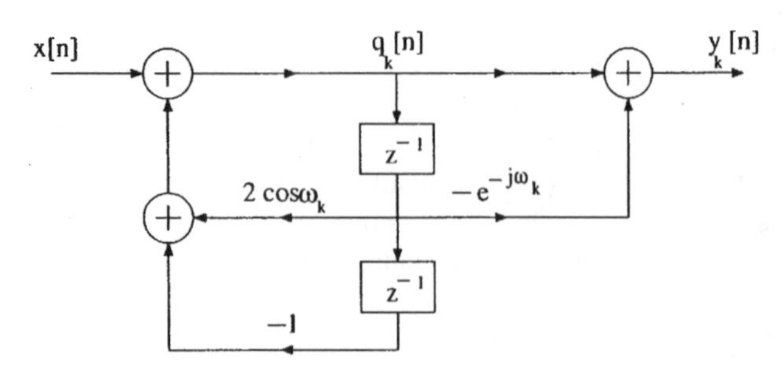

Figure 1. Goertzel algorithm as a second-order recursive computation.

and the initial conditions are

$$q_k[-1] = q_k[-2] = 0.$$

Finally, we evaluate $y_k[N]$ as follows:

$$y_k[N] = q_k[N] - e^{-j\omega_k}\, q_k[N-1]. \tag{28}$$

The total computation required for $X(z_k)$ is $(2N+8)$ real multiplications and $(4N+6)$ real additions. Thus, the number of multiplications is nearly half of that in the direct method. In addition, we need only two coefficients, $e^{-j\omega_k}$ and $e^{-j\omega_k N}$.

As expected, the Goertzel algorithm described here reduces to the one used for computing the DFT, when $\omega_k = 2\pi k/N$. The amount of computation required for the DFT and NDFT is nearly the same. Only one extra complex multiplication is needed in the case of the NDFT.

In some applications, we are interested in finding only the squared magnitude of the NDFT, i.e., $|X(z_k)|^2$. In this case, the Goertzel algorithm can be modified as follows. Since $z_k = e^{j\omega_k}$, we infer from (23) that

$$|X(z_k)|^2 = |y_k[N]|^2.$$

Substituting for $y_k[N]$ from (28), we obtain

$$|y_k[N]|^2 = q_k^2[N] + q_k^2[N-1] + a_k\, q_k[N]\, q_k[N-1]. \tag{30}$$

This modified scheme uses only one real coefficient a_k. If we have a real input signal, then complex arithmetic is avoided, and we need only $(N+4)$ real multiplications and $(2N+2)$ real additions. Note that this is equal to the computation needed for finding the squared magnitude of the DFT. This scheme has been used to detect dual-tone multi-frequency (DTMF) tones [10, 11].

7.3. The 2-D NDFT

7.3.1. Definition

As in the case of a 1-D sequence, the Nonuniform Discrete Fourier Transform of a 2-D sequence corresponds to sampling its 2-D z-transform. The 2-D NDFT of a sequence $x[n_1, n_2]$ of size $N_1 \times N_2$ is defined as

$$\hat{X}(z_{1k}, z_{2k}) = \sum_{n_1=0}^{N_1-1} \sum_{n_2=0}^{N_2-1} x[n_1, n_2] z_{1k}^{-n_1} z_{2k}^{-n_2}, \qquad k = 0, 1, \ldots, N_1N_2 - 1, \tag{31}$$

where (z_{1k}, z_{2k}) represent N_1N_2 distinct points in the 4-D (z_1, z_2) space. These points can be chosen arbitrarily, but in such a way that the inverse transform exists.

We illustrate this by a simple example. Consider the case, $N_1 = N_2 = 2$. We can express (31) in a matrix form as

$$\hat{\mathbf{X}} = \mathbf{DX}, \tag{32}$$

where

$$\hat{\mathbf{X}} = \begin{bmatrix} \hat{X}(z_{10}, z_{20}) \\ \hat{X}(z_{11}, z_{21}) \\ \hat{X}(z_{12}, z_{22}) \\ \hat{X}(z_{13}, z_{23}) \end{bmatrix}, \qquad \mathbf{X} = \begin{bmatrix} x[0, 0] \\ x[0, 1] \\ x[1, 0] \\ x[1, 1] \end{bmatrix}, \tag{33}$$

and

$$\mathbf{D} = \begin{bmatrix} 1 & z_{20}^{-1} & z_{10}^{-1} & z_{10}^{-1} z_{20}^{-1} \\ 1 & z_{21}^{-1} & z_{11}^{-1} & z_{11}^{-1} z_{21}^{-1} \\ 1 & z_{22}^{-1} & z_{12}^{-1} & z_{12}^{-1} z_{22}^{-1} \\ 1 & z_{23}^{-1} & z_{13}^{-1} & z_{13}^{-1} z_{23}^{-1} \end{bmatrix}. \tag{34}$$

In general, the 2-D NDFT matrix $\mathbf{D}$ is of size $N_1N_2 \times N_1N_2$. It is fully specified by the choice of the N_1N_2 sampling points. For the 2-D NDFT matrix to exist uniquely, these points should be chosen so that $\mathbf{D}$ is nonsingular. In the case of the 1-D NDFT, if the points z_k are distinct, the inverse NDFT is guaranteed to exist uniquely. This happens because the 1-D NDFT matrix has a determinant that can always be factored as in (5). However, there is no simple extension of this to the 2-D case. Even if the 2-D sampling points are distinct, this does not guarantee the nonsingularity of **D**. This calls for the need to make a judicious choice of sampling points. Some results have been derived on sufficient conditions under which the equivalent 2-D polynomial interpolation problem has a unique or nonunique solution when the samples are located on irreducible curves [12]. However, no set of necessary and sufficient conditions has been found. This does not pose a serious problem from the point of view of applications. For all practical purposes, we can just perform a check on the determinant of **D** to ascertain that it is nonzero for that particular choice of points. In the general case, the inverse 2-D NDFT is computed by solving a linear system of size N_1N_2, which requires $O(N_1^3N_2^3)$ operations.

Note that the definition of the NDFT can be readily extended to higher dimensions.

7.3.2. Special Cases

In general, the determinant of the 2-D NDFT matrix is not factorizable. However, we now consider special cases in which the determinant can be factored. In these cases, the choice of the sampling points is restricted in some way so that the 2-D NDFT matrix is guaranteed to be nonsingular.

(1) Nonuniformly spaced rectangular grid

In this case, the sampling points lie at the vertices of a rectangular grid in the (z_1, z_2) space. For a sequence of size $N_1 \times N_2$, the z_1-coordinates of the N_1 grid lines running parallel to the z_2 axis can be chosen arbitrarily, so long as they are distinct. Let these coordinates be denoted by $z_{10}, z_{11}, \ldots, z_{1,N_1-1}$. Similarly, the z_2-coordinates of the N_2 grid lines running parallel to the z_1 axis can be chosen arbitrarily, so long as they are distinct. Let these coordinates be denoted by $z_{20}, z_{21}, \ldots, z_{2,N_2-1}$. This distribution of points is illustrated in Fig. 2, for an example where $N_1 = 3$, $N_2 = 4$. Note that the representation of the z_1 and z_2 axes in this figure is for convenience only, since the complex variables z_1 and z_2 actually form a 4-D space.

Equation (32) can then be expressed in a simpler matrix form as

$$\hat{\mathbf{X}} = \mathbf{D}_1\mathbf{X}\mathbf{D}_2^t, \tag{35}$$

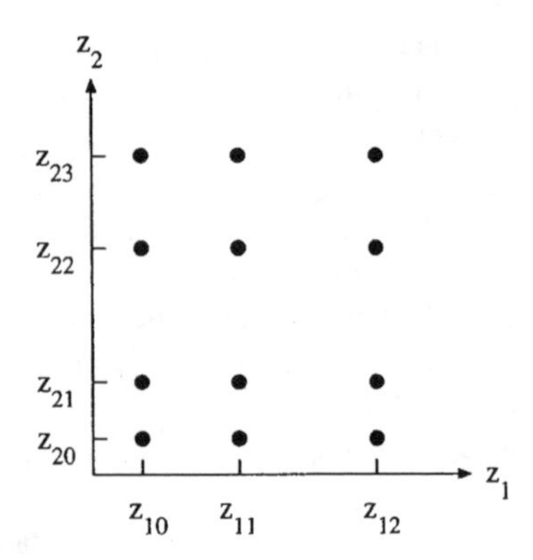

Figure 2. 2-D NDFT with a nonuniformly spaced rectangular grid for $N_1 = 3$, $N_2 = 4$.

where

$$\hat{\mathbf{X}} = \begin{bmatrix} \hat{X}(z_{10}, z_{20}) & \hat{X}(z_{10}, z_{21}) & \cdots & \hat{X}(z_{10}, z_{2,N_2-1}) \\ \hat{X}(z_{11}, z_{20}) & \hat{X}(z_{11}, z_{21}) & \cdots & \hat{X}(z_{11}, z_{2,N_2-1}) \\ \vdots & \vdots & \ddots & \vdots \\ \hat{X}(z_{1,N_1-1}, z_{20}) & \hat{X}(z_{1,N_1-1}, z_{21}) & \cdots & \hat{X}(z_{1,N_1-1}, z_{2,N_2-1}) \end{bmatrix}, \tag{36}$$

$$\mathbf{X} = \begin{bmatrix} x[0,0] & x[0,1] & \cdots & x[0, N_2-1] \\ x[1,0] & x[1,1] & \cdots & x[1, N_2-1] \\ \vdots & \vdots & \ddots & \vdots \\ x[N_1-1, 0] & x[N_1-1, 1] & \cdots & x[N_1-1, N_2-1] \end{bmatrix}, \tag{37}$$

$$\mathbf{D_1} = \begin{bmatrix} 1 & z_{10}^{-1} & z_{10}^{-2} & \cdots & z_{10}^{-(N_1-1)} \\ 1 & z_{11}^{-1} & z_{11}^{-2} & \cdots & z_{11}^{-(N_1-1)} \\ \vdots & \vdots & \vdots & \ddots & \vdots \\ 1 & z_{1,N_1-1}^{-1} & z_{1,N_1-1}^{-2} & \cdots & z_{1,N_1-1}^{-(N_1-1)} \end{bmatrix}, \tag{38}$$

$$\mathbf{D_2} = \begin{bmatrix} 1 & z_{20}^{-1} & z_{20}^{-2} & \cdots & z_{20}^{-(N_2-1)} \\ 1 & z_{21}^{-1} & z_{21}^{-2} & \cdots & z_{21}^{-(N_2-1)} \\ \vdots & \vdots & \vdots & \ddots & \vdots \\ 1 & z_{2,N_1-1}^{-1} & z_{2,N_1-1}^{-2} & \cdots & z_{2,N_2-1}^{-(N_2-1)} \end{bmatrix}. \tag{39}$$

Here, $\mathbf{X}$ and $\hat{\mathbf{X}}$ are matrices of size $N_1 \times N_2$. $\mathbf{D_1}$ and $\mathbf{D_2}$ are Vandermonde matrices of sizes $N_1 \times N_1$ and $N_2 \times N_2$, respectively. The equivalent 2-D NDFT matrix $\mathbf{D}$ can be expressed as a Kronecker (also called tensor, or direct) product

$$\mathbf{D} = \mathbf{D_1} \otimes \mathbf{D_2}, \tag{40}$$

where $\otimes$ denotes the Kronecker product [13]. Note that the Kronecker product $\mathbf{C}$ of two matrices $\mathbf{A}$ and $\mathbf{B}$ is defined as follows. Let

$$\mathbf{A} = (a_{ik}), \qquad i = 1, \ldots, I, k = 1, \ldots, K, \tag{41}$$

and

$$\mathbf{B} = (b_{jl}), \qquad j = 1, \ldots, J, l = 1, \ldots, L. \tag{42}$$

Then, the Kronecker product $\mathbf{C} = (c_{ij;kl})$ is a matrix with $I \times J$ rows and $K \times L$ columns, given by

$$c_{ij;kl} = a_{ik} b_{jl}. \tag{43}$$

Applying a property of the Kronecker product, the determinant of $\mathbf{D}$ can be written as

$$\begin{aligned} \det(\mathbf{D}) &= \{\det(\mathbf{D_1})\}^{N_2} \otimes \{\det(\mathbf{D_2})\}^{N_1} \\ &= \prod_{i \neq j, i > j} ({z_{1i}}^{-1} - {z_{1j}}^{-1})^{N_2} \prod_{p \neq q, p > q} ({z_{2p}}^{-1} - {z_{2q}}^{-1})^{N_1}. \end{aligned} \tag{44}$$

Therefore, $\mathbf{D}$ is nonsingular provided $\mathbf{D_1}$ and $\mathbf{D_2}$ are nonsingular, i.e., if the points $z_{10}, z_{11}, \ldots, z_{1,N_1-1}$ are distinct, and $z_{20}, z_{21}, \ldots, z_{2,N_2-1}$ are distinct.

For this choice of sampling points, only $N_1 + N_2$ degrees of freedom are used among the $N_1 N_2$ degrees available in the 2-D NDFT. Consequently, the inverse 2-D NDFT $\mathbf{X}$ in (35) can be computed by solving two separate linear systems of sizes N_1 and N_2, respectively. This involves $O(N_1^3 + N_2^3)$ operations, instead of $O(N_1^3 N_2^3)$ operations in the general case.

Angelides [14] has used a specific case of this sampling structure, in which the samples are placed on a nonuniform rectangular grid in the 2-D (ω_1, ω_2) plane, where $z_1 = e^{j\omega_1}$, $z_2 = e^{j\omega_2}$.

The 2-D DFT is a special case under this category, obtained when the points are chosen on a uniform grid in the (ω_1, ω_2) plane:

$$\begin{aligned} z_{1k_1} &= e^{j\frac{2\pi}{N_1}k_1}, \qquad k_1 = 0, 1, \ldots, N_1 - 1, \\ z_{2k_2} &= e^{j\frac{2\pi}{N_2}k_2}, \qquad k_2 = 0, 1, \ldots, N_2 - 1. \end{aligned}$$

(2) Nonuniform sampling on parallel lines

This is a generalization of the sampling structure used in Case 1 above. For an $N_1 \times N_2$ sequence, the samples are placed on N_1 lines parallel to the z_2 axis, with N_2 points on each line. The z_1-coordinates corresponding to the N_1 lines can be chosen arbitrarily, but distinct from each other. Let these co-ordinates be denoted by $z_{10}, z_{11}, \ldots, z_{1,N_1-1}$. Similarly, the z_2-coordinates of the N_2 points on each line can be chosen arbitrarily, as long as they are distinct. Let the z_2-coordinates of the points on the ith line be $z_{20i}, z_{21i}, \ldots, z_{2,N_2-1,i}$. Figure 3 shows an example where $N_1 = 3$, $N_2 = 4$.

In this case, the 2-D NDFT matrix can be expressed as a generalized Kronecker product [13]

$$\mathbf{D} = \{\mathbf{D_2}\} \otimes \mathbf{D_1}. \tag{45}$$

Here, $\mathbf{D_1}$ is an $N_1 \times N_1$ Vandermonde matrix as shown in (38). $\{\mathbf{D_2}\}$ denotes a set of N_1 $N_2 \times N_2$ Vandermonde matrices $\mathbf{D_{2i}}$, $i = 0, 1, \ldots, N_1 - 1$. This is represented as

$$\{\mathbf{D_2}\} = \left\{ \begin{array}{c} D_{20} \\ D_{21} \\ \vdots \\ D_{2,N_1-1} \end{array} \right\}, \tag{46}$$

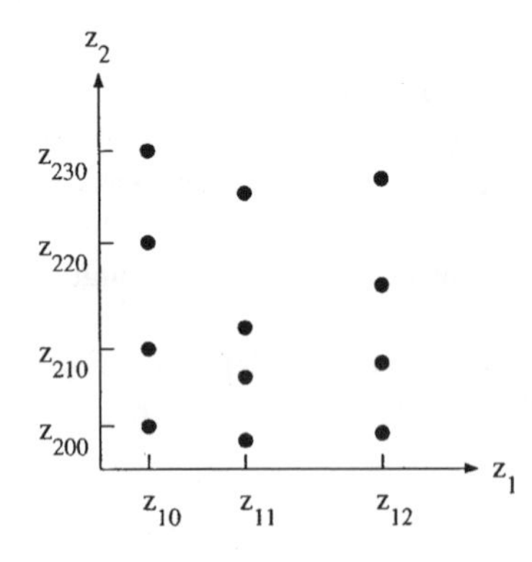

Figure 3. 2-D NDFT with nonuniform sampling on parallel lines (parallel to the z_2 axis) for $N_1 = 3$, $N_2 = 4$.

where

$$\mathbf{D_{2i}} = \begin{bmatrix} 1 & z_{20i}^{-1} & z_{20i}^{-2} & \cdots & z_{20i}^{-(N_2-1)} \\ 1 & z_{21i}^{-1} & z_{21i}^{-2} & \cdots & z_{21i}^{-(N_2-1)} \\ \vdots & \vdots & \vdots & \ddots & \vdots \\ 1 & z_{2,N_1-1,i}^{-1} & z_{2,N_1-1,i}^{-2} & \cdots & z_{2,N_2-1,i}^{-(N_2-1)} \end{bmatrix}, \quad i = 0, 1, \ldots, N_1 - 1. \tag{47}$$

Equation (45) means that

$$\mathbf{D} = \begin{bmatrix} \mathbf{D_{20}} \otimes \mathbf{d_0} \\ \mathbf{D_{21}} \otimes \mathbf{d_1} \\ \vdots \\ \mathbf{D}_{2,N_1-1} \otimes \mathbf{d}_{N_1-1} \end{bmatrix}, \tag{48}$$

where $\mathbf{d_i}$ denotes the ith row vector of matrix $\mathbf{D_1}$.

The determinant of $\mathbf{D}$ can then be written as

$$\det(\mathbf{D}) = \{\det(\mathbf{D}_1)\}^{N_2} \prod_{i=0}^{N_1-1} \det(\mathbf{D_{2i}}). \tag{49}$$

Therefore, $\mathbf{D}$ is nonsingular if the matrices $\mathbf{D_1}$ and $\mathbf{D_{2i}}$ are nonsingular. Alternatively, we can also place the samples on N_2 lines parallel to the z_1 axis with N_1 points on each line.

To compute the inverse 2-D NDFT, we have to solve for $x[n_1, n_2]$ from

$$\begin{aligned} \hat{X}(z_{1i}, z_{2ji}) &= \sum_{n_2=0}^{N_2-1} \sum_{n_1=0}^{N_1-1} x[n_1, n_2] z_{1i}^{-n_1} z_{2ji}^{-n_2} \\ &= \sum_{n_2=0}^{N_2-1} y[z_{1i}, n_2] z_{2ji}^{-n_2}, \\ & i = 0, 1, \ldots, N_1 - 1, \qquad j = 0, 1, \ldots, N_2 - 1, \end{aligned} \tag{50}$$

where

$$y[z_{1i}, n_2] = \sum_{n_1=0}^{N_1-1} x[n_1, n_2] z_{1i}^{-n_1}. \tag{51}$$

For each z_{1i}, $i = 0, 1, \ldots, N_1 - 1$, (50) represents a Vandermonde system of N_2 equations which can be solved to find $y[z_{1i}, n_2]$, $n_2 = 0, 1, \ldots, N_2 - 1$. For each

$n_2 = 0, 1, \ldots, N_2 - 1$, (51) represents a Vandermonde system of N_1 equations which can be solved to obtain $x[n_1, n_2]$, $n_1 = 0, 1, \ldots, N_1 - 1$. Therefore, the inverse 2-D NDFT can be computed by solving N_1 linear systems of size N_2 each, and N_2 linear systems of size N_1 each. This requires $O(N_1 N_2^3 + N_2 N_1^3)$ operations.

Rozwood et al. [15] have used a specific case of this sampling structure, in which the samples were placed on parallel lines in the 2-D (ω_1, ω_2) plane.

7.4. 1-D FIR Filter Design Using the NDFT

Filters are used to select or suppress components of signals at certain frequencies. In this section, we develop an application of the NDFT for designing 1-D finite-impulse-response (FIR) digital filters. Commonly used techniques for FIR filter design are the windowed Fourier series approach, frequency sampling approach, and optimal minimax designs. By using a design method based on nonuniform frequency sampling, we can design filters that are nearly equal to optimal filters designed with the same design specifications. Additionally, the filter design times are much lower than those taken by the Parks–McClellan algorithm for optimal minimax filter design [16].

7.4.1. Nonuniform Frequency Sampling Design Method

In the NDFT-based FIR filter design method, the desired frequency response is sampled at N nonuniformly spaced points on the unit circle in the z-plane [1]. An N-point inverse NDFT of these frequency samples gives the filter coefficients. We address the problem of nonuniform frequency sampling design by considering the two major issues involved: (a) the generation of the desired frequency response for the given filter specifications, and (b) the choice of the frequency-sample locations. The aim is to obtain a filter whose interpolated frequency response is nearly equiripple in each band. The approaches used for these two issues are outlined here:

(a) **Generation of the desired frequency response:** Given the filter specifications, we construct the desired response by using a separate analytic function for each frequency band. The equiripple nature of the response is obtained by using Chebyshev polynomials and a suitable transformation to map the domain of the polynomial to the frequency axis. Each function involves several parameters which are found by imposing appropriate constraints so as to obtain the desired mapping.

(b) **Choice of the frequency-sample locations:** In addressing this issue, we must note that there is no good analytic approximation for the frequency response in the transition band. We have found that the best locations for the frequency samples are the extrema of the desired equiripple response, as shown later in this chapter. Since the desired response is approximated by analytic functions, the extremal frequencies can be expressed in closed form. In addition, since the total number of extrema is related to the filter length, we do not need to place any extra samples in the transition band. However, in special cases (e.g., for designing filters with very wide transition bands), we can place a few transition samples to constrain any possible overshoot within this band.

Once we have addressed these issues, we can refer to the NDFT formulation in (1). Samples of the analytic functions generated in Step (a) are used to construct the NDFT vector **X**. The NDFT matrix **D** is constructed from the frequency-sample locations found in Step (b). The filter impulse response **x** is then obtained by solving this linear system of equations. Note that all known symmetries in the filter impulse response can be utilized to reduce the number of independent filter coefficients. This leads to a smaller linear system, and, therefore, to reduced design time.

We can use the NDFT-based method to design various filters (of length N), such as lowpass filters (Type I if N is odd, Type II if N is even), highpass filters (Type I if N is odd, Type IV if N is even), bandpass filters (Types I or III if N is odd, Types II or IV if N is even), and Mth-band filters such as third-band filters (Type I) [4]. We now describe the design method in detail by considering a Type I lowpass filter.

7.4.2. Lowpass Filter Design

Let us consider the design of a linear-phase lowpass filter of Type I. The filter has a real and symmetric impulse response

$$h[n] = h[N-1-n], \qquad n = 0, 1, \ldots, N-1, \tag{52}$$

where the filter length N is an odd integer. The frequency response of the filter is

$$H(e^{j\omega}) = \sum_{n=0}^{N-1} h[n]e^{-j\omega n}. \tag{53}$$

By applying the symmetry condition in (52), this can be expressed in the form [17]

$$H(e^{j\omega}) = A(\omega)e^{-j\omega(N-1)/2}, \tag{54}$$

where the amplitude function $A(\omega)$ is real, even, and is a periodic function of ω, given by

$$A(\omega) = \sum_{k=0}^{(N-1)/2} a[k] \cos \omega k, \tag{55}$$

and

$$\begin{aligned} a[0] &= h[(N-1)/2], \\ a[k] &= 2h[(N-1)/2 - k], \qquad k = 1, 2, \ldots, (N-1)/2. \end{aligned} \tag{56}$$

Let the filter have its passband edge at ω_p, stopband edge at ω_s, and peak ripples δ_p and δ_s in the passband and stopband, respectively. We proceed with the design by considering the two issues involved:

(a) Generation of the Desired Frequency Response

The real-valued amplitude response $A(\omega)$ is represented by analytic functions as follows [1]:

$$A(\omega) = \begin{cases} H_p(\omega) = 1 - \delta_p T_P(X_p(\omega)), & 0 \le \omega \le \omega_p, \\ H_s(\omega) = \delta_s T_S(X_s(\omega)). & \omega_s \le \omega \le \pi. \end{cases} \tag{57}$$

with $T_M(\cdot)$ denoting a Chebyshev polynomial of order M, defined as

$$T_M(x) = \begin{cases} \cos(M \ \cos^{-1}(x)), & -1 \le x \le 1, \\ \cosh(M \ \cosh^{-1}(x)), & \text{otherwise.} \end{cases} \tag{58}$$

Note that $T_M(x)$ is equiripple in the range $-1 \le x \le 1$ and monotone outside this range. Depending on whether n is even or odd, $T_M(x)$ is an even or odd function of x, respectively. The integers P and S are given by

$$P = N_p \tag{59}$$

and

$$S = N_s, \tag{60}$$

where N_p equals the number of extrema in the passband $0 \le \omega \le \omega_p$, and N_s equals the number of extrema in the stopband $\omega_s \le \omega \le \pi$. The functions, $X_p(\omega)$ and $X_s(\omega)$, are needed to map the equiripple interval $-1 \le x \le 1$ of $T_P(x)$ and

$T_S(x)$ to the passband and stopband, respectively. This mapping is obtained by using the transformations,

$$X_p(\omega) = A\cos(a\omega + b) + B, \tag{61}$$

$$X_s(\omega) = C\cos(c\omega + d) + D. \tag{62}$$

The values for the eight parameters, A, B, C, D, a, b, c, d, are obtained by imposing appropriate constraints on the functions, $H_p(\omega)$ and $H_s(\omega)$ [4, 18]:

$$H_p(\omega) = H_p(-\omega), \tag{63}$$

$$H_s(\pi + \omega) = H_s(\pi - \omega), \tag{64}$$

$$H_p(\omega_p) = 1 - \delta_p, \tag{65}$$

$$H_s(\omega_s) = \delta_s, \tag{66}$$

$$H_p(0) = \begin{cases} 1 + \delta_p, & P = \text{odd}, \\ 1 - \delta_p, & P = \text{even}, \end{cases} \tag{67}$$

$$H_s(\pi) = \begin{cases} -\delta_s, & S = \text{odd}, \\ \delta_s, & S = \text{even}, \end{cases} \tag{68}$$

$$\min(H_p(\omega)) = -\delta_s, \tag{69}$$

$$\max(H_s(\omega)) = 1 + \delta_p. \tag{70}$$

These constraints lead to the following expressions for the parameters:

$$A = \frac{1}{2}\left\{T_P^{-1}\left(\frac{1+\delta_s}{\delta_p}\right) + 1\right\}, \tag{71}$$

$$B = A - 1, \tag{72}$$

$$C = \frac{1}{2}\left\{T_S^{-1}\left(\frac{1+\delta_p}{\delta_s}\right) + 1\right\}, \tag{73}$$

$$D = C - 1, \tag{74}$$

$$a = \frac{1}{\omega_p}\cos^{-1}\left(\frac{B-1}{A}\right), \tag{75}$$

$$b = \pi, \tag{76}$$

$$c = \frac{1}{(\omega_s - \pi)}\cos^{-1}\left(\frac{D-1}{C}\right), \tag{77}$$

$$d = \pi(1 - c). \tag{78}$$

Given the filter specifications, N, ω_p, ω_s and $k = \delta_p/\delta_s$, we estimate the ripple sizes δ_p and δ_s from [19]

$$N = \frac{-10 \log_{10}(\delta_p \delta_s) - 13}{2.324(\omega_s - \omega_p)} + 1, \tag{79}$$

which gives

$$\delta_p = \sqrt{k}\, 10^{-0.1162(\omega_s - \omega_p)(N-1) - 0.65} \tag{80}$$

and

$$\delta_s = k/\delta_p. \tag{81}$$

Alternatively, if we are given ω_p, ω_s, δ_p and δ_s, we can estimate the required filter length N from (79). The Chebyshev polynomial orders P and S are then determined. Since they correspond to the number of extrema in the passband and stopband, they are found by weighting them proportionately to the sizes of the passband and stopband such that $P + S = (N - 1)/2$. Now, $P + S$ equals the total number of alternations in the filter response over the range $0 \leq \omega \leq \pi$, excluding the ones at the band edges, ω_p and ω_s. The alternation theorem states that an optimum Type I filter must have a minimum of $(L + 2)$ alternations, where $L = (N - 1)/2$ [20]. Thus, we obtain

$$P + S = L = \frac{N - 1}{2}. \tag{82}$$

Note that a Type I filter can have a maximum of $(L + 3)$ alternations [20]. For this extraripple filter, we have $P + S = (N + 1)/2$.

(b) Choice of the Frequency-Sample Locations

Since the impulse response of the filter is symmetric, as shown in (52), the number of independent filter coefficients is given by

$$N_i = \frac{N + 1}{2}. \tag{83}$$

Thus, we need only N_i samples located in the range $0 \leq \omega \leq \pi$. These samples are placed at the extrema of the desired response which has been approximated by analytic functions in Step (a).

The P extrema of $H_p(\omega)$ occur when

$$H_p(\omega) = 1 \pm \delta_p, \tag{84}$$

of

$$X_p(\omega) = \cos\left(\frac{k\pi}{P}\right), \qquad k = 1, 2, \ldots, N_p. \tag{85}$$

Using the definition for $X_p(\omega)$ in (61), we obtain the following expression for the extrema in the passband:

$$\omega_k^{(p)} = \frac{1}{a}\left\{\cos^{-1}\left[\frac{\cos\left(\frac{k\pi}{P}\right) - B}{A}\right] - b\right\}, \qquad k = 1, 2, \ldots, N_p. \tag{86}$$

Similarly, the S extrema of $H_s(\omega)$ occur when

$$H_s(\omega) = \pm\delta_s. \tag{87}$$

This leads to the following expression for the extrema in the stopband:

$$\omega_k^{(s)} = \frac{1}{c}\left\{\cos^{1}\left[\frac{\cos\left(\frac{k\pi}{S}\right) - D}{C}\right] - d\right\}, \qquad k = 1, 2, \ldots, N_s. \tag{88}$$

From (82) and (83), we observe that $P + S = N_i - 1$. Therefore, we need one more sample besides those at the $P + S$ extrema. This sample is placed either at the passband edge ω_p, or at the stopband edge ω_s.

Thus, we sample the functions generated in Step (a) at the locations obtained in Step (b), and then solve for the N_i independent filter coefficients. We can also design highpass filters using the NDFT-based method with minor modifications, by designing the function $H_s(\omega)$ for the low-frequency band and $H_p(\omega)$ for the high-frequency band.

7.4.3. A Design Example

The NDFT-based filter design method is compared with other methods by considering the design of a Type I lowpass filter with specifications: $N = 37$, $\omega_p = 0.3\pi$, $\omega_s = 0.4\pi$, $\delta_p/\delta_s = 1$. The methods used for comparison include three uniform frequency sampling methods and the Parks–McClellan algorithm:

(1) Original uniform frequency sampling (OUFS), with sample values of unity in the passband and zeros in the stopband,

(2) Uniform frequency sampling (UFS), using linear programming to optimize the values of the samples in the transition band [21],
(3) Modified uniform frequency sampling (MUFS), using analytic functions to approximate the desired frequency response [18],
(4) Parks–McClellan algorithm (PM).

If Method 2 is used with the given specifications, there is only one transition sample—its value is chosen to be 0.39 so as to minimize the peak stopband ripple. In Method 3, the analytic functions generated in the first step of the NDFT method are sampled uniformly. Since some of these samples are located in the transition band, an additional function is required to approximate the frequency response in this band. This function is chosen to be a weighted linear combination of the passband and stopband functions as described in [18].

Figure 4 shows the filter designed by the NDFT method. Table 1 provides a comparison of the band-edges, attenuation and ratio of peak ripples actually attained by these filters. A_p is the passband attenuation, defined as

$$A_p = -20 \log_{10}(1 - \delta_p), \tag{89}$$

where δ_p is the maximum ripple in the passband. A_s is the stopband attenuation, defined as

$$A_s = -20 \log_{10} \delta_s, \tag{90}$$

where δ_s is the maximum ripple in the stopband. The passband and stopband edges of the filters designed are measured as

$$\omega_p = \max\{\omega \mid H(\omega) = 1 - \delta_p \text{ and } 0 < \omega < \pi\}, \tag{91}$$

and

$$\omega_s = \min\{\omega \mid |H(\omega)| = \delta_s \text{ and } 0 < \omega < \pi\}. \tag{92}$$

The filter designed by Method 1 has large ripples near the band edges. They are decreased by introducing the transition sample in Method 2. Method 3 provides better control over the band edges. Table 1 clearly shows that among the frequency sampling methods, the NDFT method provides the closest overall match to the desired specifications. For example, although Method 2 provides the highest A_s, this is obtained at the expense of a δ_p/δ_s which far exceeds the desired value of unity. Also, the characteristics of the filters designed by the NDFT method compare very closely to those of the optimal minimax filter designed by the Parks–McClellan algorithm.

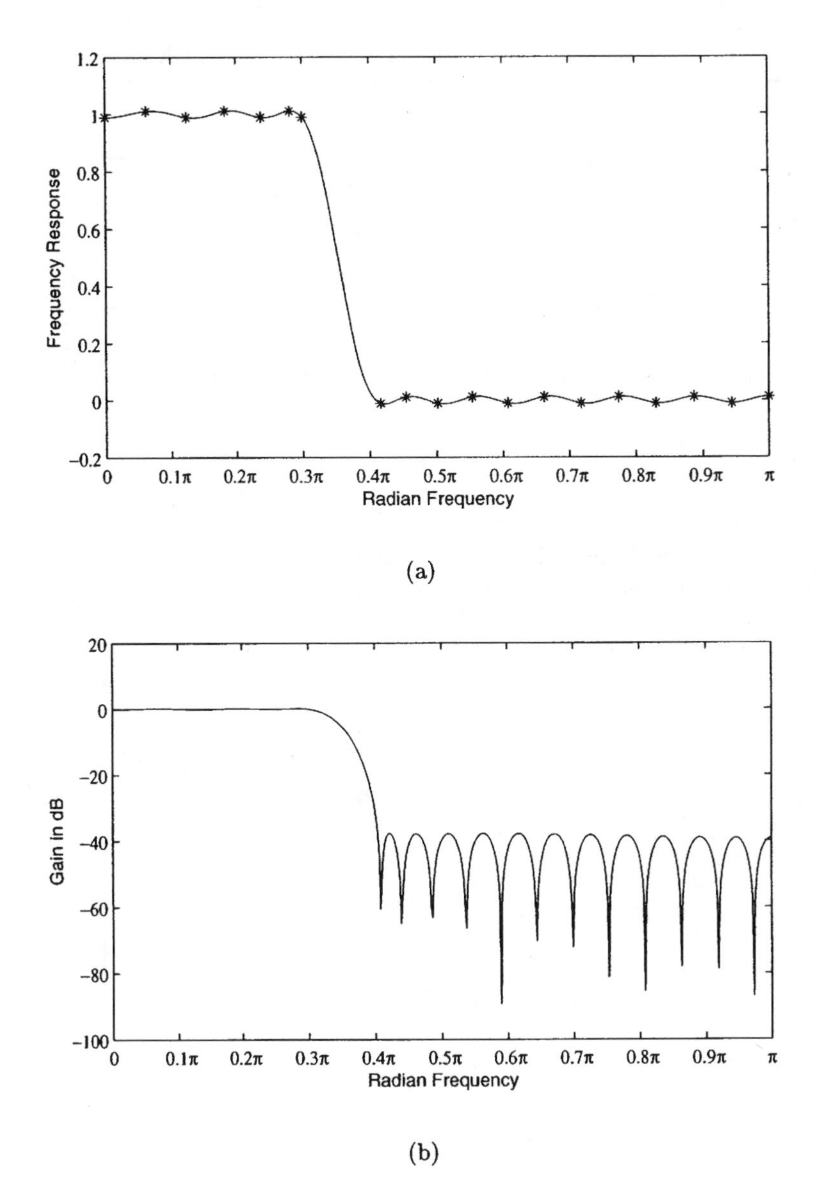

Figure 4. Type I lowpass filter of length 37, designed by the NDFT method. (a) Frequency response, with samples denoted by '*'. (b) Gain response.

Table 1. Performance Comparison for Type I Lowpass Filter Design

Method	ω_p (π)	ω_s (π)	A_p (dB)	A_s (dB)	δ_p/δ_s
OUFS	0.2790	0.3374	1.1512	16.0977	0.7921
UFS	0.2782	0.3740	0.4671	42.6863	7.1332
MUFS	0.3103	0.3963	0.2404	33.6428	1.3130
NDFT	0.3006	0.4039	0.1121	37.7473	0.9894
PM	0.3	0.4	0.1225	37.0715	1.0

For more examples with other types of filters, refer to [4].

7.4.4. Remarks

In many cases, the filters designed by the NDFT-based nonuniform frequency sampling method are *very close to optimal* equiripple filters. This was illustrated by the preceding design example. The design time required by the NDFT method is *much lower*, when compared to iterative optimization methods such as the Parks–McClellan algorithm. Although extensive testing is required for a thorough performance comparison between these methods, we have obtained some preliminary results by recording the times taken by both methods to design lowpass filters with fixed band-edges and increasing lengths. Figure 5 shows the

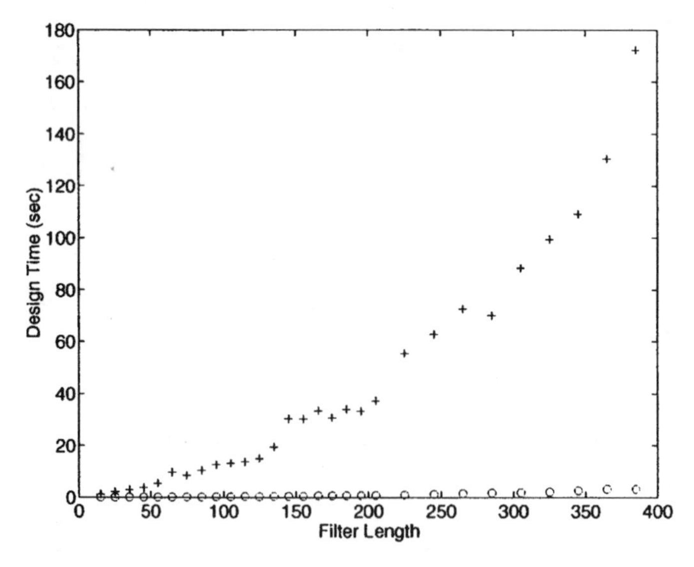

Figure 5. Plot of filter design time with filter length, for the NDFT method (o) and Parks–McClellan algorithm (+).

variation of design time with filter length for the NDFT method and Parks–McClellan algorithm. From the comparison presented in Fig. 5, it is clear that the NDFT method is particularly useful for designing long filters, since iterative routines need excessively large amounts of time in such cases. The NDFT-based method can be used to design a nearly optimal filter in a much shorter time.

All symmetries in the filter impulse response are utilized in the NDFT-based method, so that we need to solve for only the independent filter coefficients. For example, the presence of alternating zeros in the impulse response of half-band filters has been utilized, reducing the design time by a factor of two. In third-band filter design, we chose not to place any samples in the don't-care frequency band. These illustrate the flexibility of nonuniform frequency sampling.

We have proposed a choice of extremal frequencies based on a nonuniform sampling of analytic functions derived from Chebyshev polynomials. This set of extrema can also be used as a good starting point for the Parks–McClellan algorithm, if exactly optimal filters are to be designed. We performed simulations to compare the performances of this algorithm with two different starting points: (1) the proposed nonuniform choice of extrema, and (2) the standard uniform choice of extrema. Our results show that the nonuniform choice of extrema *decreases* the number of iterations required, as well as the design time [4].

7.5. 2-D FIR Filter Design Using the NDFT

Two-dimensional (2-D) digital filters find applications in diverse areas such as image processing and coding, robotics and computer vision, seismology, sonar, radar and astronomy. The 2-D NDFT defined in Section 7.3 can be used to design 2-D FIR filters by nonuniform frequency sampling of the specified frequency response.

7.5.1. Design Methods

There exist four standard approaches for designing 2-D FIR filters—windowing, frequency sampling, frequency transformation, and optimal filter design methods. In the window method, a 2-D window is typically obtained from a 1-D window by separable or nonseparable techniques. The infinite-extent impulse response of an ideal 2-D filter is then multiplied by this window to obtain a 2-D FIR filter. Although straightforward, this method suffers from a lack of control over the frequency domain specifications. Besides, it can only design filters with separable or circularly symmetric frequency responses. Frequency sampling is a conceptually simple approach to 2-D FIR filter design. It includes a variety of existing methods that use uniform as well as nonuniform frequency sampling. Filter design by frequency transformation does not have a 1-D counterpart. This is an attractive, practical method of designing a 2-D filter by applying a frequency transformation function to a 1-D filter. The function controls the shape of the contours of the 2-D frequency response. The amplitude characteristics of the 1-D filter are preserved in the 2-D filter. These three methods are not optimal. However, in contrast to the case of 1-D filters, development of a practical, reliable algorithm to design optimal 2-D filters still remains an area for active research. The problem of designing an optimal 2-D filter is much more complex than in the 1-D case. The powerful alternation theorem does not apply in 2-D, and the minimax solution is not unique. Consequently, existing iterative algorithms of the Remez exchange type are very intensive computationally, and do not always converge to a correct solution.

Earlier efforts in nonuniform 2-D frequency sampling design have involved either constrained sampling structures which reduce computational complexity [12, 14, 15, 22], or a linear least squares approach that guarantees unique interpolation [12]. Our approach involves generalized frequency sampling, where the samples are placed on contour lines that match the desired shape of the passband or stopband of the 2-D filter. The NDFT-based method produces nonseparable 2-D filters with good passband shapes and low peak ripples. Filters of good quality are obtained, even for small support sizes. This is important since such filters are most likely to be used in practical filtering applications.

7.5.2. Nonuniform Frequency Sampling Design Method

In the NDFT-based 2-D filter design method, the desired frequency response is sampled at N_i points located nonuniformly in the 2-D frequency plane, where N_i is the number of independent filter coefficients [23, 2]. All symmetries present in the filter impulse response are utilized so that N_i is typically much lower than the total number of filter coefficients, N^2. This reduces the design time, besides guaranteeing symmetry. The set of N_i linear equations, given by the 2-D NDFT formulation in (31), is then solved to obtain the filter coefficients.

As in the case of 1-D filter design, the choice of the sample values and locations depends on the particular type of filter being designed. In general, the problem of locating the 2-D frequency samples is much more complex than in the 1-D case. Our experience in designing 2-D filters with various shapes indicates that the best results are obtained when the samples are placed on *contour lines* that match the desired passband shape. For example, to design a square-shaped filter, we place the samples along a set of square contour lines in the 2-D frequency plane. Note that these results agree with the filter design results reported earlier [12, 15], where better control over shape was obtained by placing samples at the edges of the passband and the stopband. The total number of contours and number of samples on each contour have to be chosen carefully so as to avoid singularities. A necessary condition for nonsingularity is known [12, Theorem 2, p. 171], and helps to serve as a rough check. However, this condition is not sufficient to guarantee nonsingularity. This theorem asserts that if the sum of the degrees of the irreducible curves, on which the samples are placed, is small compared to the degree of the filter polynomial, then the interpolation problem becomes singular. Going back to our example of designing a square filter, it is clear that the number of square contours must be chosen appropriately with respect to the filter size.

As we locate the frequency samples along contour lines of the desired shape, the *parameters* to be chosen are: (a) the number of contours and the spacing between them, (b) the number of samples on each contour and their relative spacing, and (c) the sample values. In the NDFT-based 2-D filter design method, a particular cross-section of the desired 2-D frequency response is approximated by 1-D analytic functions based on Chebyshev polynomials, similar to those used for 1-D filter design. The samples are then placed on contours that pass through the extrema of the function approximating this cross-section. In the following section, we illustrate the NDFT-based method by considering the design of a diamond-shaped filter.

The NDFT-based method has been used to design 2-D filters with square, circular, diamond, fan, and elliptically-shaped passbands [4].

7.5.3. Diamond Filter Design

Diamond filters find important practical applications as prefilters for quincunx sampled data, and in interlaced-to-noninterlaced scanning converters for television signals [24]. Consider a diamond filter of size $N \times N$, whose amplitude-response specification has passband edge ω_p and stopband edge ω_s, as defined in Fig. 6. In other words, the diagonal line $\omega_1 = \omega_2$ in the frequency plane intersects the passband edge at (ω_p, ω_p) and the stopband edge at (ω_s, ω_s). A diamond filter exhibits an eightfold symmetry in the frequency domain [25]. Considering the frequency response $H(e^{j\omega_1}, e^{j\omega_2})$ to be zero-phase, the amplitude response $A(\omega_1, \omega_2)$ satisfies the condition

$$A(\omega_1, \omega_2) = A(-\omega_1, \omega_2) = A(\omega_1, -\omega_2) = A(\omega_2, \omega_1). \tag{93}$$

Equivalently, the impulse response $h[n_1, n_2]$ satisfies the relation

$$h[n_1, n_2] = h[-n_1, n_2] = h[n_1, -n_2] = h[n_2, n_1]. \tag{94}$$

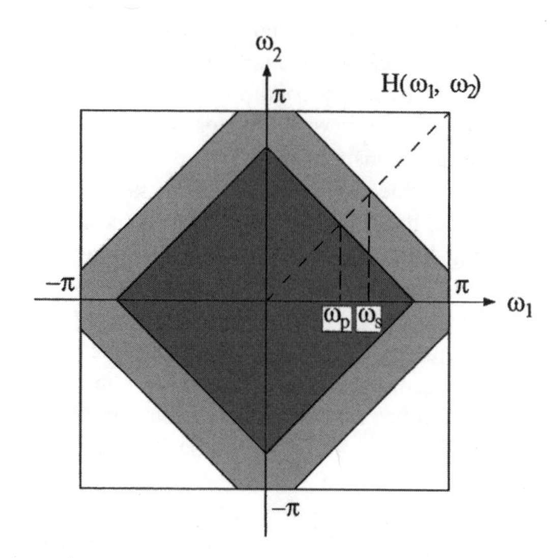

Figure 6. Amplitude-response specification of a diamond filter. Darkly shaded region: passband, lightly shaded region: transition band, unshaded region: stopband.

Besides, a diamond filter is a 2-D half-band filter. Its amplitude response $A(\omega_1, \omega_2)$ is symmetric about the point $(\omega_1, \omega_2, A(\omega_1, \omega_2)) = (\pi/2, \pi/2, 0.5)$ in frequency space:

$$A(\omega_1, \omega_2) + A(\pi - \omega_1, \pi - \omega_2) = 1. \tag{95}$$

This implies that the impulse response has alternating zeros so that the non-zero coefficients form a quincunx-like lattice. Thus, we get

$$h[n_1, n_2] = \begin{cases} 0, & n_1 + n_2 = \text{even}, \\ 0.5, & n_1 = n_2 = 0. \end{cases} \tag{96}$$

For example, the arrangement of points in the impulse response of a 9×9 half-band filter is illustrated in Fig. 7. On account of the properties in (94) and (96), the number of independent coefficients [26] in a filter of size $N \times N$ is reduced to

$$N_i = \left\lfloor \frac{P+1}{2} \right\rfloor \left\lfloor \frac{P+2}{2} \right\rfloor, \tag{97}$$

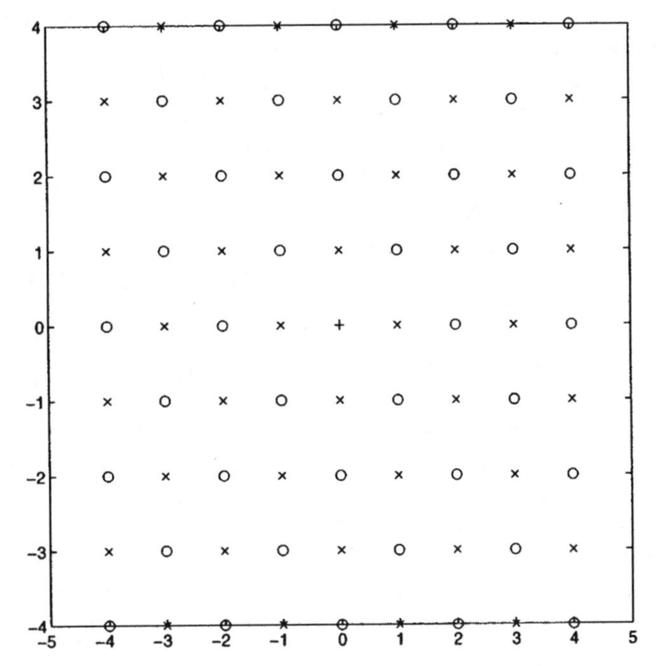

Figure 7. Arrangement of points in the impulse response of a 9×9 half-band FIR filter. Zero-valued samples are denoted by 'o', and nonzero samples are denoted by '$\times$'. The value of the sample at '+' is 0.5.

where

$$P = \frac{N-1}{2}. \tag{98}$$

The N_i independent points in the impulse response lie in a wedge-shaped region below the diagonal line $n_1 = n_2$ in the first quadrant of the (n_1, n_2) spatial plane. By using (94) and (96), the amplitude response of a diamond filter [26] is expressed as

$$\begin{aligned} A(\omega_1, \omega_2) &= 0.5 \\ &+ \sum_{n_1=1}^{\lfloor (P+1)/2 \rfloor} 2h[2n_1 - 1, 0]\{\cos(2n_1 - 1)\omega_1 + \cos(2n_1 - 1)\omega_2\} \\ &+ \sum_{n_1=1}^{\lfloor (P+1)/2 \rfloor} \sum_{n_2=1}^{\lfloor P/2 \rfloor} 4h[2n_1 - 1, 2n_2]\{\cos(2n_1 - 1)\omega_1 \cos(2n_2)\omega_2 \\ &+ \cos(2n_2)\omega_1 \cos(2n_1 - 1)\omega_2\}. \end{aligned} \tag{99}$$

Due to the eightfold symmetry in (93), and the half-band nature of the filter in (96), the only independent part of the amplitude response is a triangular area within the passband, as shown in Fig. 8. In our design method, N_i samples are placed within this region of the frequency plane. If we take a cross-section of $A(\omega_1, \omega_2)$ along the diagonal line $\omega_1 = \omega_2$, the plot looks like a 1-D half-band lowpass response. We approximate the passband of this response by a 1-D function $H_p(\omega)$, as used for 1-D half-band lowpass filter design [4]. The order P of the corresponding Chebyshev polynomial $T_p(x)$ is $(N-1)/2$. The samples are then placed on $(N-1)/2$ lines of slope -1, that pass through the extrema of $H_p(\omega)$. All samples on a particular line have the same value and are evenly spaced. The number of samples on successive lines, as we move away from the origin, is given in Table 2, for filter sizes from 7×7 to 31×31. This range of filter sizes is large enough to cover the needs of most practical applications. The given distribution of samples has been found to work well for various choices of the band-edges. Note that if there is only one sample to be placed on a particular contour, it is placed on the ω_1 axis. Finally, the N_i samples are used to solve for the impulse response coefficients in (99).

The NDFT-based design method produces diamond filters of high quality, with low peak ripple and better passband shape as compared with filters produced by other existing design methods. A comparison between these methods is presented in the next section.

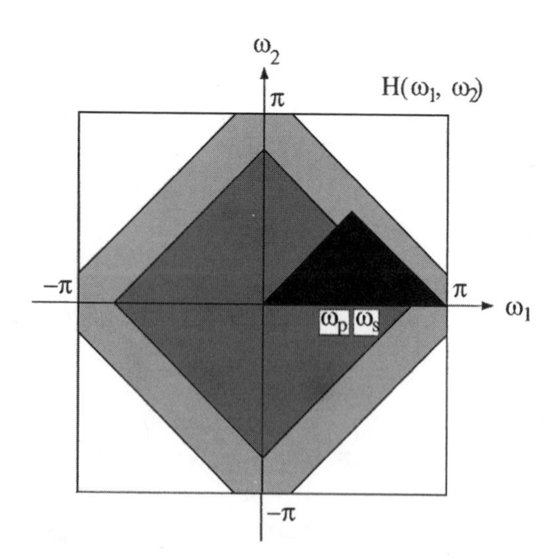

Figure 8. Amplitude response of a diamond filter. The darkly shaded region is the only independent part, because of eightfold symmetry and the half-band nature of the filter.

Table 2. Distribution of samples for diamond filter design. The last column shows the number of samples placed on $(N-1)/2$ successive contours for designing a filter of size $N \times N$, which has N_i independent coefficients.

N	N_i	Number of samples on successive contours
7	4	1, 2, 1
9	6	1, 1, 2, 2
11	9	1, 1, 2, 3, 2
13	12	1, 1, 2, 3, 3, 2
15	16	1, 1, 2, 3, 3, 4, 2
17	20	1, 1, 2, 3, 3, 4, 4, 2
19	25	1, 1, 2, 3, 3, 4, 4, 4, 3
21	30	1, 1, 2, 3, 3, 3, 4, 4, 5, 4
23	36	1, 1, 2, 3, 3, 3, 4, 4, 5, 6, 4
25	42	1, 1, 2, 3, 3, 4, 4, 4, 5, 6, 6, 5
27	49	1, 1, 2, 3, 3, 4, 4, 4, 5, 6, 6, 6, 5
29	56	1, 1, 2, 3, 3, 4, 4, 4, 5, 6, 6, 6, 7, 5
31	64	1, 1, 2, 3, 3, 4, 4, 4, 5, 6, 6, 6, 7, 7, 6

7.5.4. A Design Example

Consider the design of a diamond filter with the following specifications:

Support size $= 9 \times 9$,
Passband edge $\omega_p = 0.36\pi$, Stopband edge $\omega_s = 0.64\pi$.

Only 6 of the 81 filter coefficients are independent. Thus, 6 samples are placed as shown in Fig. 9(b), on lines that follow the diamond shape. The diagonal cross-section of the 2-D amplitude response is represented by a function $H_p(\omega)$, that has 4 extrema. Samples are placed on 4 lines passing through these extrema. The number of samples on these lines are 1, 1, 2, 2, respectively, as given in Table 2. Figs. 9(a) and (b) show the amplitude response and contour plot of the resulting diamond filter.

For comparison, we design the same filter using three existing methods for designing diamond filters:

(1) Frequency transformation [25],
(2) Method proposed by Bamberger and Smith [27],
(3) Method proposed by Chen and Vaidyanathan [28].

A performance comparison in Table 3 shows that the NDFT-based method gives the lowest peak ripples among these methods as well as good contour shapes.

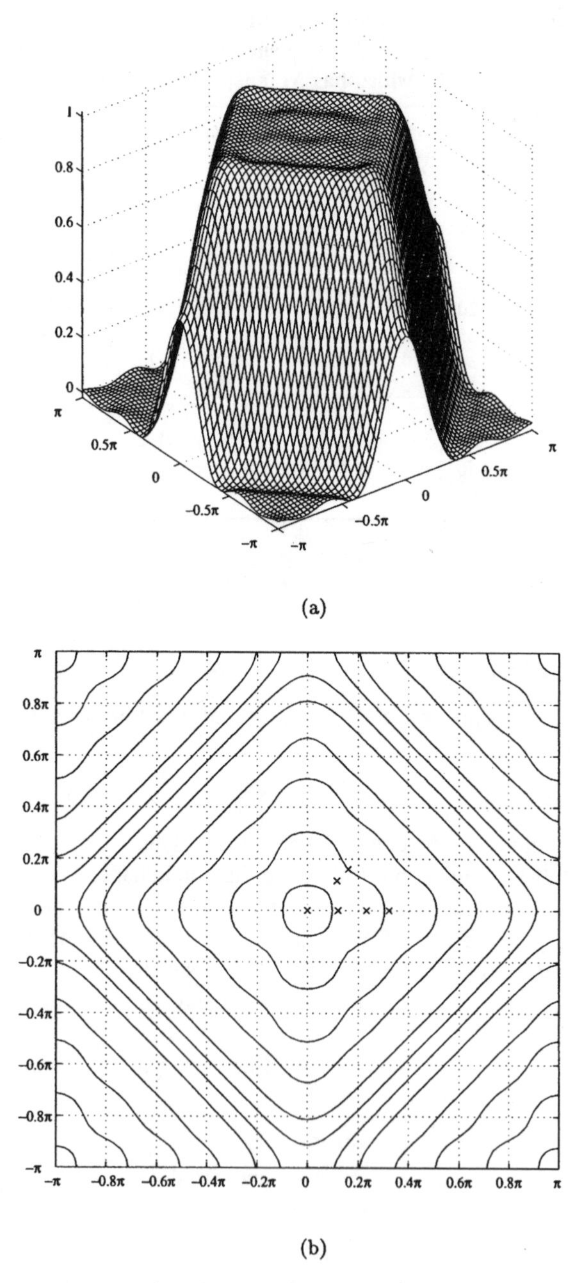

Figure 9. Diamond filter of size 9×9 designed by the NDFT method. (a) Amplitude response. (b) Contour plot with the sample locations denoted by '$\times$'.

Table 3. Performance comparison for diamond filter design.

Method	δ_p	δ_s
NDFT	0.0189	0.0184
Frequency Transformation	0.0636	0.0636
Bamberger–Smith	0.1085	0.1084
Chen–Vaidyanathan	0.0292	0.0281

In the frequency transformation method, a 1-D half-band lowpass filter is transformed to a 2-D diamond filter. The simplest transformation used to obtain the desired diamond-shaped contours is the following one of size 3×3.

$$\cos \omega = T(\omega_1, \omega_2) = \frac{1}{2} \cos \omega_1 + \frac{1}{2} \cos \omega_2. \tag{100}$$

The contours of the resulting 2-D filter are more circular rather than diamond-shaped. The shape of the contours can be improved by using a higher order transformation, but this also increases the filter size considerably. If the transformation is of size $(2P + 1) \times (2Q + 1)$, and the 1-D filter is of length $(2M + 1)$, then the resulting 2-D filter has size $(2PM + 1) \times (2QM + 1)$. Thus, in our example, if we use a 5×5 transformation and the same 1-D filter of length 9, then the 2-D filter size becomes 17×17, which is uneconomical.

In Method 2, a diamond filter is designed by rotating a checkerboard-shaped filter through an angle of 45 degrees [27]. This checkerboard filter is the sum of two square-shaped filters,

$$C(z_1, z_2) = H_0(z_1)H_0(z_2) + H_1(z_1)H_1(z_2), \tag{101}$$

where $H_0(z)$ is a 1-D half-band lowpass filter, and $H_1(z) = H_0(-z)$ is the corresponding highpass filter. Although the shape of the contours is better than with frequency transformation, the ripple is too large.

Finally, we consider the method proposed by Chen and Vaidyanathan [28]. This method can be used to design a general class of M-dimensional filters with arbitrary parallelepiped-shaped passbands. However, we shall only mention how it can be used to design a 2-D diamond filter of size $N \times N$. A 1-D half-band lowpass filter $h[n]$ of length $2N - 1$ is designed, and used to form a separable 2-D filter,

$$h_s[n_1, n_2] = h[n_1]h[n_2]. \tag{102}$$

Then, this separable filter is downsampled by a quincunx matrix,

$$\hat{\mathbf{M}} = \begin{bmatrix} 1 & 1 \\ -1 & 1 \end{bmatrix}. \tag{103}$$

to generate the diamond filter $h[n_1, n_2]$ as follows:

$$h[n_1, n_2] = 2h_s[n_1 + n_2, n_2 - n_1]. \tag{104}$$

The resulting diamond filter has a good passband shape as well as low peak ripples, comparable to the results given by our method.

7.6. Summary

In this chapter, the concept of the Nonuniform Discrete Fourier Transform (NDFT) was introduced to provide a generalized approach for nonuniform sampling in the frequency domain. The NDFT of a finite-length sequence corresponds to sampling its z-transform at arbitrarily chosen points in the z-plane. The NDFT reduces to the Discrete Fourier Transform (DFT) when these sampling points are located on the unit circle at equally spaced angles. The flexibility in sampling offered by the NDFT leads to a variable spectral resolution which can be controlled by the user. This is important in many signal processing applications, since most signals and systems tend to have their energies distributed unevenly in different regions of the spectrum. We outlined applications of the NDFT in 1-D and 2-D FIR filter design. More NDFT applications including other types of filter design, antenna pattern synthesis with prescribed nulls, and dual-tone multi-frequency tone detection, can be found in [4].

References

[1] S. Bagchi and S. K. Mitra. The Nonuniform Discrete Fourier Transform and its Applications in Filter Design: Part I–1-D, *IEEE Trans. Circuits Syst. II: Analog and Digital Signal Processing*, 43:422–433, June 1996.

[2] S. Bagchi and S. K. Mitra. The nonuniform discrete Fourier transform and its applications in filter design: Part II–2-D, *IEEE Trans. Circuits Syst. II: Analog and Digital Signal Processing*, 43:434–444, June 1996.

[3] S. Bagchi. The Nonuniform Discrete Fourier Transform and its Applications in Signal Processing. Ph.D. thesis, University of California at Santa Barbara, 1994.

[4] S. Bagchi and S. K. Mitra. *The Nonuniform Discrete Fourier Transform and Its Applications in Signal Processing*. Kluwer Academic Publishers, Norwell, MA, 1999.

[5] S. K. Mitra, S. Chakrabarti and E. Abreu. Nonuniform Discrete Fourier Transform and its Applications in Signal Processing. *Proc. EUSIPCO '92, Sixth European Signal Processing Conf.*, Brussels, Belgium, August 1992, 2:909–912.

[6] G. H. Golub and C. F. Van Loan. *Matrix Computations*. The John Hopkins University Press, Baltimore, 1983.

[7] P. J. Davis. *Interpolation and Approximation*, Dover Publications, New York, 1975.

[8] G. Goertzel. An Algorithm for the Evaluation of Finite Trigonometric Series, *American Math. Monthly*, 65:34–35, January 1958.

[9] K. E. Atkinson. *An Introduction to Numerical Analysis*. John Wiley & Sons, New York, 1978.

[10] S. Bagchi and S. K. Mitra. An Efficient Algorithm for DTMF Decoding using the Subband NDFT. *Proc. IEEE Int. Symp. on Circuits and Syst.*, Seattle, WA, May 1995, 3:1936–1939.

[11] S. Bagchi and S. K. Mitra. Efficient Robust DTMF Decoding using the Subband NDFT, *Signal Processing*, 56:255–267, February 1997.

[12] A. Zakhor and G. Alvstad. Two-Dimensional Polynomial Interpolation from Nonuniform Samples, *IEEE Trans. Acoust., Speech, Signal Processing*, ASSP-40:169–180, January 1992.

[13] P. A. Regalia and S. K. Mitra. Kronecker Products, Unitary Matrices, and Signal Processing Applications, *SIAM Review*, 31:586–613, December 1989.

[14] E. Angelides. A Novel Method for Modeling 2-D FIR Digital Filters in Frequency Domain with Nonuniform Samples. *IEEE Trans. Circuits Syst. II: Analog and Digital Signal Processing*, 41:482–486, July 1994.

[15] W. J. Rozwood, C. W. Therrien and J. S. Lim. Design of 2-D FIR Filters by Nonuniform Frequency Sampling, *IEEE Trans. Acoust., Speech, Signal Processing*, ASSP-39:2508–2514, November 1991.

[16] J. H. McClellan, T. W. Parks and L. R. Rabiner. A Computer Program for Designing Optimum FIR Linear Phase Digital Filters, *IEEE Trans. Audio Electroacoust.*, AU-21:506–526, December 1973.

[17] S. K. Mitra. *Digital Signal Processing: A Computer-Based Approach*, McGraw-Hill, New York, 1998.

[18] M. Lightstone, S. K. Mitra, I.-S. Lin, S. Bagchi, P. Jarske and Y. Neuvo. Efficient Frequency-Sampling Design of One- and Two-Dimensional FIR Filters using Structural Subband Decomposition. *IEEE Trans. Circuits Syst. II: Analog and Digital Signal Processing*, 41:189–201, March 1994.

[19] J. F. Kaiser. Nonrecursive Digital Filter Design using the I_0-sinh Window Function. *Proc. IEEE Int. Symp. on Circuits and Syst.*, San Francisco, CA, April 1974, pp. 20–23.

[20] A. V. Oppenheim and R. W. Schafer. *Discrete-Time Signal Processing*. Prentice-Hall, Englewood Cliffs, N.J., 1989.

[21] L. R. Rabiner, B. Gold and C. A. McGonegal. An Approach to the Approximation Problem for Nonrecursive Digital Filters, *IEEE Trans. Audio Electrocoust.*, AU-18:83–106, June 1970.

[22] J. E. Diamessis, C. W. Therrien and W. J. Rozwood. Design of 2-D FIR Filters with Nonuniform Frequency Samples. *Proc. IEEE Int. Conf. On Acoust., Speech, Signal Processing*, Dallas, TX, April 1987, vol. 3, pp. 1665–1668.

[23] S. Bagchi and S. K. Mitra. Nonseparable 2-D FIR Filter Design using Nonuniform Frequency Sampling. *Proc. IS&T/SPIE Symposium on Electronic Imaging: Image and Video Processing III*, San Jose, CA, February 1995, pp. 104–115.

[24] G. J. Tonge. The Sampling of Television Images. Experimental and Development Rep. 112/81, Independent Broadcasting Authority, May 1981.

[25] J. S. Lim. *Two-Dimensional Signal and Image Processing*. Prentice-Hall, Englewood Cliffs, N.J., 1990.

[26] T. Yoshida, A. Nishihara and N. Fujii. A Design Method of 2-D Maximally Flat Diamond-Shaped Half-Band FIR Filters. *Trans. IE-ICE (The Institute of Electronics, Information and Communication Engineers)*, E 73:901–907, June 1990.

[27] R. H. Bamberger and M. J. T. Smith. A Filter Bank for the Directional Decomposition of Images: Theory and Design. *IEEE Trans. Signal Processing*, 40:882–893, April 1992.

[28] T. Chen and P. P. Vaidyanathan. Multidimensional Multirate Filters and Filter Banks Derived from One-Dimensional Filters. *IEEE Trans. Signal Processing*, 41:1749–1765, May 1993.

8

Reconstruction of Stationary Processes Sampled at Random Times

B. Lacaze

Abstract

The reconstruction of a deterministic function from a finite or infinite number of its values is an old problem initially studied by I. Newton, E. Waring and J. L. Lagrange. This chapter addresses the problem of reconstructing a stationary process (rather than deterministic function) from observations at known or unknown instants. In the first case, the reconstruction depends explicitly on the known instants. In the second case, the unknown instants are modelled by random variables whose joint distributions allow to define interpolation formulas.

8.1. Introduction

8.1.1. Problem Formulation

Consider a stationary process $\mathbf{Z} = \{Z(t), t \in \mathcal{R}\}$, zero-mean, mean-square continuous, with an autocorrelation function $K_Z(\tau)$ and a power spectrum $S_Z(\omega)$ defined by [30–31]

$$K_Z(\tau) = E[Z(t)Z^*(t-\tau)] = \int_{-\infty}^{\infty} e^{j\omega\tau} dS_Z(\omega), \tag{1}$$

B. Lacaze • TéSA, 2 rue Camichel, 31071 Toulouse Cedex, France. Fax +33(0)5.61.58.82.37; E-mail: bernard.lacaze@tesa.prd.fr

Nonuniform Sampling: Theory and Practice, edited by Marvasti
Kluwer Academic/Plenum Publishers, New York, 2001.

where $E[..]$ is the mathematical expectation and $z^* = x - jy$ is the conjugate of the complex number $z = x + jy$. The right term in (1) is a Riemann-Stieltjes integral since $S_Z(\omega)$ is a real, bounded and non-decreasing function [30]. When a regular derivative $s_Z(\omega) = dS_Z/d\omega$ exists, the Riemann-Stieltjes integral reduces to the familiar Riemann integral

$$K_Z(\tau) = E[Z(t)Z^*(t-\tau)] = \int_{-\infty}^{\infty} e^{j\omega\tau} s_Z(\omega) d\omega. \tag{2}$$

Define a sequence of distinct real numbers $\mathbf{t} = \{t_k, k \in \mathcal{A}\}$, where $\mathcal{A}$ is a finite or infinite subset of $\mathcal{Z}$, the set of integers. These numbers may be random or deterministic, and constitute the "instants" where the random process $\mathbf{Z}$ is observed, usually denoted "sampling times". In the random case, the sequence $\mathbf{t} = \{t_k, k \in \mathcal{A}\}$ will be characterized by its joint distribution. The random process $\mathbf{Z}$ and the random sequence $\mathbf{t}$ are always assumed to be independent.

This paper addresses the problem of reconstructing the process $\mathbf{Z}$, i.e., to estimate the random variables $Z(t)$, $\forall t \in \mathcal{R}$, from the observation of the sequence $\mathbf{U}' = \{U_k, k \in \mathcal{A}\}$, where $U_k = Z(t_k)$. The power spectrum S_Z is assumed to be known or well measured.

Section 1.2 defines the estimators of $Z(t)$ denoted $\widetilde{Z}(t)$. The "distance" between the estimate and the process is studied in Section 1.3.

8.1.2. Estimator Classes

8.1.2.1. The Deterministic Case

Consider first the completely deterministic case. We have to approximate an ordinary function $f(t)$ from samples $f(t_k)$, where the t_k are known. Generally, we consider a class of functions $g_k(t, \mathbf{t})$, and an approximation $\widetilde{f}(t)$ is constructed as follows

$$\widetilde{f}(t) = \sum_k g_k(t, \mathbf{t}) f_k, \tag{3}$$

where $f_k = f(t_k)$, and $g_k(t, \mathbf{t})$ is a function which depends on both the time t and the sequence $\mathbf{t} = \{t_k, k \in \mathcal{A}\}$. For example, the Lagrange interpolation case is characterized by

$$g_k(t, \mathbf{t}) = \prod_{l \neq k} \frac{t - t_l}{t_k - t_l}. \tag{4}$$

For a finite set of sampling times, $g_k(t, \mathbf{t})$ is a polynomial function of t. In the infinite case, and if the infinite product (4) is convergent, (4) defines no longer a polynomial. For example, if $t_k = k, \mathcal{A} = \mathcal{Z}$, we obtain $g_k(t, \mathbf{t}) = \sin \pi(t-k)/\pi(t-k)$. Other classes of functions can be used, such as Splines. Note that (3) seems to be a linear operator with respect to f. A condition for having this linearity is that the function g does not depend on the function to be approximated.

Note that whatever the interpolating functions, we want to have the sampling property $\widetilde{f}(t_k) = f(t_k)$. Of course, $f(t_k)$ is known *a priori* but it may not be the case for other values of t. Consequently, the interpolation formula should satisfy $g_k(t_k, \mathbf{t}) = 1$ and $g_k(t_l, \mathbf{t}) = 0, k \neq l$. This trivial remark will be emphasized in the next sections.

Furthermore, in the completely deterministic case, (3) and (4) take explicitly into account the true values of sampling times t_k.

8.1.2.2. The Random Case

When a deterministic signal is sampled, two situations can occur: the sampling times can be either observed or not. In the first case, equation (3), which depends explicitely on t_k can be used to build estimates. In the second case, this method is no longer valid. The following reconstruction can then be considered

$$\widetilde{f}(t) = \sum g_k(t) f_k, \tag{5}$$

with, $f_k = f(t_k)$. The family of functions $g_k(t)$ will be chosen from the statistical characteristics of the random sequence $\mathbf{t}$. Equation (5) emphasizes that t_k cannot appear in the formula apart implicitly in f_k and g_k. For example, if $t_k = k + A_k$, where A_k is a zero-mean random variable, $g_k(t)$ can be built as follows

$$g_k(t) = \prod_{l \neq k} \frac{t-l}{k-l}.$$

However, (4) cannot be used since it depends explicitly on $\mathbf{t} = \{t_k, k \in \mathcal{A}\}$.

Both situations can occur when sampling a random process $\mathbf{Z}$. When the sampling times are observed, an estimator can be built as follows

$$\widetilde{Z}(t) = \sum g_k(t, \mathbf{t}) U_k, \tag{6}$$

where $U_k = Z(t_k)$. Of course, this way to write $\widetilde{Z}(t)$ is the right one, because the known values of t_k have to be brought in g and not elsewhere: the process Z is observed in the entity U_k. When the sampling times are unknown, the following estimator can be considered

$$\widetilde{Z}(t) = \sum g_k(t) U_k, \tag{7}$$

where $U_k = Z(t_k)$. The functions g_k have to be chosen from the distributions of $\mathbf{Z}$ and $\mathbf{t} = \{t_k, k \in \mathcal{A}\}$, and not from the (unknown) true values of t_k.

Finally, as noted before, the previous equations may show a kind of linearity, since the right hand side terms of (5), (6) and (7) are $f_k = f(t_k)$ or U_k. This linearity is verified when the interpolation functions only depend on the sequence $\mathbf{t}$ and consequently are independent of the processes to approximate. Of course, we might opt for other more complicated forms of interpolations including those with $h[f_k]$ or $h(U_k)$, for some h, instead of f_k or U_k. However, the only way to fulfill the sampling conditions $\widetilde{f}(t_k) = f_k$ or $\widetilde{Z}(t_k) = U_k$ for any $\mathbf{t}$ is to choose $h = 1$. Another possible extension consists of building estimations as follows

$$\widetilde{f}(t) = \sum g_{1k}(t, \mathbf{t}) f_k + \sum g_{2k}(t, \mathbf{t}) h[f_k], \tag{8}$$

where $g_{1k}(t_k, \mathbf{t}) = 1, g_{2k}(t_k, \mathbf{t}) = 0$ (see Section 5). A larger number of components in the second member of (8) could also be considered.

8.1.3. Error

8.1.3.1. The Deterministic Case

Consider again the completely deterministic case. Before defining the approximation error σ, we must first choose between a local or a global distance measure. A local distance measure can be defined as

$$\sigma_t = |\widetilde{f}(t) - f(t)|, \tag{9}$$

which, except for a perfect reconstruction, depends on t. In the second category, a commonly used global distance is the mean-square error

$$\sigma = c\sqrt{\int_a^b |\widetilde{f}(t) - f(t)|^2 dt}, \tag{10}$$

where a and b are the (finite) bounds of the approximation interval and c is a positive constant (for example $c = 1/\sqrt{b-a}$). In this case, σ does not depend on t. The principal advantage of this metric is that it is often possible to compute the function $\widetilde{f}(t)$ which minimizes σ. In the case where a (or b) is not finite, the weighting will change depending on the membership of f in $L^2(\mathcal{R})$, the set of square integrable functions on the real line ($L^2(\mathcal{R}) = \{f;\ \int_{-\infty}^{\infty} |f(t)|^2 dt < \infty\}$).

When the sampling times are unknown, it is necessary to take into account their statistical properties, and to make use of mathematical expectations. But we know that it is most often easier to calculate a variance than an absolute moment (if the latter exists). In this case, (9) can be replaced by

$$\sigma_t = \sqrt{E\left[\left|\tilde{f}(t) - f(t)\right|^2\right]}, \tag{11}$$

and (10) by

$$\sigma = c\sqrt{\int_a^b E\left[\left|\tilde{f}(t) - f(t)\right|^2\right] dt}. \tag{12}$$

Many other errors could be defined but (11) and (12) are adequate in most applications.

8.1.3.2. The Random Observed Case

The situation is more difficult when a random process **Z** is sampled even if the sampling instants are observed. What is the meaning of "**t** is random and observed"? If we consider two different experiments about the sequence **t**, we will know the elements of both sampling sequences (they are observed), but these elements will be different (they are random). The problem consists of choosing between a conditional error, such as

$$\sigma_{1t} = \sqrt{E_{\mathbf{t}}\left[\left|\tilde{Z}(t) - Z(t)\right|^2\right]}, \tag{13}$$

or a non conditional error such as

$$\sigma_{2t} = \sqrt{E\left[\left|\tilde{Z}(t) - Z(t)\right|^2\right]}, \tag{14}$$

where $E_{\mathbf{t}}[..]$ in (13) means that the expectation is computed at fixed **t** as in the deterministic case (it is a conditional mathematical expectation).

Note that the computation of σ_{2t} in (14) requires the knowledge of some statistical properties of the sampling sequence **t** and the random process **Z**. In the first case defined by (13), it is not necessary to know the statistical properties of **t**.

The second one demands much more about the statistics of $\mathbf{t}$ than about that of $\mathbf{Z}$. For the first case, it suffices to know the power spectrum of $\mathbf{Z}$.

The only advantage of (14) seems to be that the result σ_{2t} may be independent of t when stationarity is assumed about $\mathbf{t}$ and for a particular $\mathbf{Z}$-spectrum. σ_{2t} is the measure of the mean error when studying several experiments with several sampling sequences. The computation of this quantity is generally very difficult and can represent a very high cost for a property which is not obviously very interesting.

Finally, suppose that we use (14). Using conditional expectations, both errors are linked by

$$\sigma_{2t}^2 = E[\sigma_{1t}^2].$$

In other words, the computation of σ_{2t} often supposes the computation of σ_{1t}, which may lead to intractable expressions. The additional information (the distribution of $\mathbf{t}$) and the difficulty of this computation are, most of time, prohibitive. When we will have no doubt about the definition of the error, we will take the notation σ_t.

8.1.3.3. The Random Non-observed Case

If the sampling times are not observed, it is necessary to have some information about the distribution of $\mathbf{t}$. In usual cases, the computation of (14) requires the joint distribution of $\mathbf{t}$. The calculations will be based on an accurate study of the sequence $\mathbf{U}' = \{U_n, n \in \mathcal{Z}\}$, whose statistical characteristics depend on the process $\mathbf{Z}$ and the sampling sequence $\mathbf{t}$. The error will be defined as in (14) and generally, (13) will be not an intermediary calculation stage. There are two reasons for this. The first one is that it will be (most often) impossible to determine this preliminary result in the case of infinite sequences. The second reason is that the conditional error is not a useful quantity when the sequence $\mathbf{t}$ is not observed.

8.1.4. Error Minimization

In all cases, the problem is to find the estimator which minimizes the chosen error, inside a class of admissible functions or random variables. Consider the case of a random process $\mathbf{Z}$. Experience shows that the mean-square criterion is the most suitable here. Consequently, errors will be defined as in (13) or (14).

In the observed case, the estimator minimizes the error $\sigma_t = \sigma_{1t}$ defined in (13), for any sequence $\mathbf{t}$, the class of possibilities being given by (6) where the $g_j(t, \mathbf{t})$ depend explicitely on $\mathbf{t}$. Obviously, it is also the good estimator for the criterion σ_{2t}. This last quantity will be computed in some circumstances, knowing distributions of **t**, and taking the mathematical expectation of σ_{1t}^2.

In the unobserved case, the class of estimators is defined by (7), and the error is $\sigma_t = \sigma_{2t}$, because the sequence **t** does not appear in the approximation formula (apart from its distribution).

In both situations, the space of estimators defines a Hilbert structure, such that the search of the best estimator can be reduced to an orthogonal projection. The calculations will be greatly facilitated using fundamental isometry (Appendix 1).

The previous study deals with a mean-square error and an orthogonal projection. This study is well adapted when the set of values of **Z** is not denumerable, but not in other cases. For example, when **Z** is a binary process, $\widetilde{Z}(t)$ takes only two values. In this situation, formula (6) and (7) are not suitable. Furthermore, the probability of error and not the mean-square error will define the error [26].

8.2. Previous Work on Random Sampling

8.2.1. Interpolation Formula for Random Processes

A. L. Cauchy studied in the middle of the nineteenth century many interpolation schemes. For instance, he used the roots of $\tan x = ax$, as sampling instants. Interpolation formulas based on samples of a function and its derivative can be found in Cauchy's papers [25]. Consequently, the problem of interpolating (deterministic) functions from infinite numbers of nonuniform samples is not a new problem.

The classical sampling theorem for stationary processes was proved by Balakrishnan [6] in 1957, and for the general case, by Lloyd two years later [7]. In 1962, Balakrishnan solved the interpolation problem for unobserved jitter [5]. A generalization of this study is developed in Section 4.

In the observed random case, several estimators were studied by Leneman and Lewis in 1966 [8], the zero-order hold $\widetilde{Z}(t) = Z(t_n)$, the exponential hold $\widetilde{Z}(t) = Z(t_n)e^{-a(t-t_n)}$, $t \in [t_n, t_{n+1}]$, and the poloygonal hold, where $\widetilde{Z}(t)$ varies linearly between $(t_n, Z(t_n))$ and $(t_{n+1}, Z(t_{n+1}))$. They considered sampling sets that are uniform (periodic), uniform with skips (see Appendix 4), or random with Poisson distribution. Furthermore, Leneman and Lewis considered a deterministic function $h(t)$ so that $\overline{Z}(t)$ defined by

$$\overline{Z}(t) = \sum Z(t_n)h(t - t_n), \tag{15}$$

minimizes $\sigma_t^2 = E[(Z(t) - \overline{Z}(t))^2]$ ($Z(t)$ is real). This scheme is problematic because the error is a constant $\sigma_t^2 = \sigma_0^2$ whereas it should be zero at sampling instants when the instants are known. Even if $h(0) = 1$, (15) generally yields $\overline{Z}(t_k) \neq Z(t_k)$ unless $h(t) = 0, \forall t \geq \inf_k t_k$, since h only depends on the statistical properties of t_k and not on its observed values. The model (15) is slightly more complicated in [9] with the same drawbacks. Note that, from the viewpoint of circuit theory, (15) represents a filtering of an impulse process $U(t)$ through a linear time invariant filter of impulse response h and can be rewritten

$$\begin{cases} \overline{Z}(t) = \int_{-\infty}^{\infty} U(u)h(t-u)du \\ U(t) = \sum_k Z(t)\delta(t-t_k). \end{cases} \tag{16}$$

In (16), $U(t)$ can be decomposed into a noise component and a part linked to $\mathbf{Z}$ [11–12].

In the works described above, the random observed sampling is a Stationary Point Process (SPP). A rigorous definition of these classes of random sequences, which is not trivial, can be found in [10, 14].

8.2.2. Power Spectra Measurements

The effect of random sampling on power spectra was particularly studied in 1960 in a paper by Shapiro and Silverman [15]. This paper addressed the problem of aliasing for (real) processes with quadratically integrable spectral density. Let $\phi(n, \omega) = E[e^{j\omega(t_m - t_{m-n})}]$ be a family of characteristic functions linked to the stationary point process $\mathbf{t}$. The autocorrelation function of the observed sequence $\mathbf{U}' = \{U_n, n \in \mathcal{Z}\}$ is

$$K_U(n) = E[U_m U_{m-n}^*] = \int_{-\infty}^{\infty} \phi(n, \omega) e^{j\omega n} s_Z(\omega) d\omega, \quad n \in \mathcal{Z}. \tag{17}$$

A random sampling $\mathbf{t}$ is alias-free for a class $\mathcal{S}$ of spectra, when the sequence of $K_U(n)$, $n \in \mathcal{Z}$ given by (17) defines only one s_Z. For example, a uniform sampling of unit period is alias-free in the class of continuous spectra concentrated in $[-\pi, \pi]$, but it is not for the class of spectra with densities which are never zero on $[-2\pi, 2\pi]$. Surprisingly, the situation can be improved when sampling times have a stronger random character. Shapiro and Silverman [15] give sufficient conditions for aliasing or alias-free sampling. For example, when $t_{n+1} = t_n + \xi_n$ (with ξ_n iid), the sampling is alias-free (whatever $\mathcal{S}$), for a Poisson process ($\phi(1, \omega) = a/(a - j\omega)$), alias-free if suppressing samples so that $\phi(1, \omega) = a^2/(a - j\omega)^2$. But this property fails for $\phi(1, \omega) = a^3/(a - j\omega)^3$. Results of [15] are developed by Beutler and Masry in [16–17]. The last one derives consistent estimates for spectral densities [18–19]. Note that this problem was studied at the same period by E. Parzen in the case of missing observations [20].

8.3. The Observed Random Case

8.3.1. The Finite Case

8.3.1.1. Known Sampling Instants

Assume a finite set of N known sampling instants $\mathbf{t} = \{t_1, t_2, \ldots, t_N\}$. As noted in Section 1.3.2, the sampling instants can first be considered as deterministic. The reconstruction problem consists of determining a linear combination of $U_1 = Z(t_1), \ldots, U_N = Z(t_N)$, denoted $\widetilde{Z}(t)$ (as defined in equation (6)) which minimizes the quantity $\sigma_t^2 = (\sigma_{1t}^2) = E[|\widetilde{Z}(t) - Z(t)|^2]$.

Using the Hilbert structure of $H(\mathbf{U}')$ defined in Appendix 1, we have

$$\widetilde{Z}(t) = pr_{H(\mathbf{U}')} Z(t), \tag{18}$$

so that $\widetilde{Z}(t)$ is defined for any t by

$$\begin{cases} \widetilde{Z}(t) = \sum_{k=1}^{N} g_k(t) U_k \\ E\left[\left(Z(t) - \widetilde{Z}(t)\right) U_l^*\right] = 0, \quad l = 1, 2, \ldots, N. \end{cases} \tag{19}$$

Note that $g_k(t)$ is used instead of $g_k(t, \mathbf{t})$ to simplify the notations. The second line of (19) yields the linear system of N equations

$$\sum_{k=1}^{N} g_k(t) K_Z(t_k - t_l) = K_Z(t - t_l), \quad l = 1, 2, \ldots, N, \tag{20}$$

where $K_Z(\tau) = E[Z(t) Z^*(t - \tau)]$. These equations can be written in matrix form

$$\mathbf{B_t a_t}(t) = \mathbf{b_t}(t), \tag{21}$$

where

$$\mathbf{B_t} = \left[K_Z(t_k - t_l)\right]_{k,l}, \ \mathbf{a_t}(t) = \left[g_k(t)\right]_k, \ \mathbf{b_t}(t) = \left[K_Z(t - t_l)\right]_l.$$

$\mathbf{B_t}$ is a square matrix, $\mathbf{a_t}(t)$ and $\mathbf{b_t}(t)$ are column vectors. Of course, the system (21) has a unique solution by the uniqueness of the orthogonal projection onto a Hilbert space. To simplify, suppose that, for all t, $\mathbf{B_t}^{-1}$ exists, so that the solution $\mathbf{a_t}(t)$ is perfectly identified by (21). The error, using the Pythagorian theorem, becomes

$$\sigma_{1t}^2 = E\left[|Z(t)|^2\right] - E\left[\left|\widetilde{Z}(t)\right|^2\right],$$

or, using (1)

$$\begin{cases} \sigma_{1t}^2 = \int_{-\infty}^{\infty} \left(1 - |\alpha_t(\omega)|^2\right) dS_Z(\omega), \\ \alpha_t(\omega) = \sum_{k=1}^{N} g_k(t) e^{j\omega t_k} = \left(\mathbf{B}_\mathbf{t}^{-1}\mathbf{b}_\mathbf{t}(t)\right)' \mathbf{e}_\mathbf{t}(\omega), \end{cases} \tag{22}$$

where $\mathbf{e}_\mathbf{t}(\omega) = \left[e^{j\omega t_k}\right]$, and $\mathbf{A}'$ is the transpose of $\mathbf{A}$.

Equations (20)–(22) define the solution of the reconstruction problem, for a finite and deterministic sampling sequence $\mathbf{t} = \{t_1, \ldots, t_N\}$. Note that σ_{1t}^2 depends explicitly on t and $\mathbf{t}$.

8.3.1.2. Random Sampling Instants

Now, suppose that the sequence $\mathbf{t}$ is random and finite. Because the sampling times are observed, the best estimator (conditional to, and using the experimental values of t_j) is still given by (21), but the error has to be replaced, as explained in Section 1.3, by

$$\begin{cases} \sigma_{2t}^2 = \int_{-\infty}^{\infty} \left(1 - E\left[|\alpha_t(\omega)|^2\right]\right) dS_Z(\omega), \\ \alpha_t(\omega) = \left(\mathbf{B}_\mathbf{t}^{-1}\mathbf{b}_\mathbf{t}(t)\right)' \mathbf{e}_\mathbf{t}(\omega). \end{cases} \tag{23}$$

The last expression is a linear combination of exponentials $e^{j\omega t_k}$ and $K_Z(t - t_k)$, and a rational function of $K_Z(t_k - t_l)$. Except for some cases, the mathematical expectation of this term cannot be computed, even when the $\mathbf{t}$-distribution is known.

The case of infinite sampling times can be seen as a generalization of the previous situation. We can assume that a solution can be found when $N \to \infty$ in (21) and (22). Of course, the convergence of the created matrix series is not obvious. To develop this point, other hypotheses have to be introduced (see Section 3.2).

8.3.1.3. Example

1) Assume first that $N = 2$, and $t_1 < t < t_2$ for a fixed t. $U_1 = Z(t_1)$, $U_2 = Z(t_2)$, t_1, t_2 are observed. In this case (for convenience, $\mathbf{Z}$ is real)

$$\begin{cases} g_1(t) = \dfrac{K_Z(0)K_Z(t - t_1) - K_Z(t - t_2)K_Z(t_1 - t_2)}{K_Z^2(0) - K_Z^2(t_2 - t_1)}, \\ g_2(t) = \dfrac{K_Z(0)K_Z(t - t_2) - K_Z(t - t_1)K_Z(t_1 - t_2)}{K_Z^2(0) - K_Z^2(t_2 - t_1)}, \\ \widetilde{Z}(t) = g_1(t)U_1 + g_2(t)U_2, \\ \sigma_{1t}^2 = (1 - g_1^2 - g_2^2)K_Z(0) - 2g_1g_2K_Z(t_1 - t_2). \end{cases} \tag{24}$$

$\widetilde{Z}(t_1) = U_1$, $\widetilde{Z}(t_2) = U_2$ as expected and σ^2_{1t} depends on t_1, t_2 and t through g_1, g_2. In the particular case where $K_Z(0) = 2$, $K_Z(1) = 1, t_2 - t_1 = 1$, we obtain

$$\begin{cases} \widetilde{Z}(t) = \frac{1}{3}(2K_Z(t-t_1) - K_Z(t-t_2))U_1 + \frac{1}{3}(2K_Z(t-t_2) - K_Z(t-t_1))U_2, \\ \sigma^2_{1t} = 2 - \frac{2}{3}\left[K_Z^2(t-t_1) + K_Z^2(t-t_2) - K_Z(t-t_1)K_Z(t-t_2)\right]. \end{cases}$$

2) Now, suppose that t_1 is an (observed) random variable, uniformly distributed on $(t-1, t)$. $\widetilde{Z}(t)$ does not change, but the previous error $\sigma^2_{1t}(t_1)$ is random, like t_1. The new (global or average) error will be

$$\begin{cases} \sigma^2_{2t} = E\left[\sigma^2_{1t}\right] = \int_{t-1}^{t} \sigma^2_{1t}(t_1)dt_1, \\ \sigma^2_{2t} = 2 - \frac{2}{3}\int_0^1 \left[K_Z^2(u) + K_Z^2(-u+1) - K_Z(u)K_Z^2(-u+1)\right]du. \end{cases}$$

The new result will, as expected, be independent of t, t_1, t_2, but no longer equal to zero for $t = t_1$ and $t = t_2$. It is a measure of the error when dealing with a large number of experiments, that is to say a large number of observations of (U_1, U_2, t_1, t_2) and calculations of $\widetilde{Z}(t)$. A numerical value can be obtained when giving a particular shape to K_Z (or S_Z). For example, $K_Z(\tau) = 2e^{-|\tau|\ln 2}$ leads to a (mean) error $\sigma_{2t} = \sqrt{\frac{10}{3} - \frac{2}{\ln 2}}$.

3) Suppose that $\mathbf{Z}$ is a telegraph signal [32], or a process having the same Markov property. In this situation, the only samples which give information about $Z(t)$ are the nearest. Then, even in the infinite sampling case, the above solutions can be used. The calculation of the average error σ^2_{2t} depends on the probability laws of the random variables (t_-, t_+), the nearest to t sampling times.

8.3.2. An Example in the Infinite Observed Case

Consider a stationary process $\mathbf{Z}$ with the power spectral density

$$s_Z(\omega) = 1 - \frac{|\omega|}{2\pi}, \quad \omega \in [-2\pi, 2\pi], \text{ and } 0 \text{ elsewhere.}$$

Assume that this process is observed at instants $t_n = n + \phi$ where $n \in \mathcal{Z}$ and ϕ is a known constant in [0, 1]. We are in a classical case of uniform (infinite) sampling with aliasing, because the support of s_Z is larger than 2π. The best linear m.s estimator $\tilde{Z}(t)$ of $Z(t)$ and the related error σ_{1t}^2 are obtained as follows

$$\begin{cases} \tilde{Z}(t) = \sum_{n=-\infty}^{\infty} \left(\frac{\sin \pi(t-\phi-n)}{\pi(t-\phi-n)} \right)^2 U_n, \\ \sigma_{1t}^2 = \frac{4\pi}{3} \sin^2 \pi(t-\phi). \end{cases}$$

If ϕ is the observed value of a continuous random variable Φ, σ_{1t} is also random. The error is then defined by

$$\sigma_{2t}^2 = E[\sigma_{1t}^2] = \frac{4\pi}{3} \int_0^1 \sin^2 \pi(t-\phi) f(\phi) d\phi,$$

where $f(\phi)$ is the probability density function of Φ. For example, for a uniform density on [0, 1]

$$\sigma_{2t}^2 = 2\pi/3,$$

which is independent of t.

8.3.2.1. Remark

As explained in Section 3.1.3, the Markov property for $\mathbf{Z}$ allows us to take into account only the two nearest sampling times of t, even in the infinite case. A similar situation occurs when $\mathbf{Z}$ has a finite correlation radius. Then, there exists $a > 0$, such that $K_Z(\tau) = 0, \; |\tau| \geq a$. Given a minimum distance between the samples, say $b > 0$, the system (20) reduces to a finite number of equations, whatever N. This property reduces the size of the numerical tools for solving the system.

8.3.3. The Band-limited Infinite Case

8.3.3.1. The Levinson Theorem

Consider that $Z(t)$ is observed at an infinite set of instants $\mathbf{t} = \{t_j, \; j \in \Delta\}$, where Δ is countable. From a theoretical point of view, we could try to find the best approximation when $|\Delta| = N$ and set $N \to \infty$. Unfortunately, the limit (in the m.s sense) cannot be found easily in most applications. An interesting generalization of the classical sampling formula can however be derived for band-limited processes.

Suppose that the so-called Shannon condition is verified

$$s_Z(\omega) = 0, \quad \omega \notin (-\pi, \pi),$$

so that all the increasing points of $S_Z(\omega)$ are in $[-\pi, \pi]$. Do there exist sufficient conditions about the (observed) sampling sequence $\mathbf{t}$ to obtain an exact reconstruction of $\mathbf{Z}$?

The isometry I (Appendix 1) shows that the problem is equivalent to knowing if the sequence $\{e^{j\omega t_j},\ t_j \in \mathbf{t}\}$ of functions of ω, is complete in $\mathcal{H}(S_Z)$ or not. In the case of regular spectral densities $s_Z(\omega) = dS_Z/d\omega$, the problem is the same in $\mathcal{H}(S_Z)$ as in $L^2_{[-\pi,\pi]}$, that is to say when $s_Z(\omega) = 1,\ \omega \in [-\pi, \pi]$, and 0 outside this interval. A Levinson theorem [33, 34] asserts that the answer is affirmative when the following condition is fulfilled

$$\begin{cases} \overline{\lim}_{n\to\infty}\left(N(R) - 2R + \frac{1}{2}\ln R\right) > -\infty, \\ N(R) = \displaystyle\int_0^R \frac{\nu(t)}{t}\,dt, \end{cases} \tag{25}$$

where $\nu(R)$ is the number of t_k in $[-R, R]$. In this case, $N(R)$ characterizes the "density" of the sequence $\mathbf{t}$. Some examples satisfying condition (25) are developed below.

a) The condition (25) is satisfied when

$$\left|t_k - k\right| \leq \mu < \tfrac{1}{4}, \quad k \in \mathcal{Z}. \tag{26}$$

Such sequences refer to the "Kadec's one-quarter theorem" [3, 33]. In fact, (26) is a strong condition because it has to be verified for any j, whereas (25) is an asymptotic condition.

b) Another example is given by a random sequence in the form $t_j = j + A_j$. The random variables A_j are i.i.d (independent and identically distributed). Suppose that $\Pr[|A_j| > 1/4] \neq 0$. Then, the probability that (26) is not satisfied is one. When $t_j = (1-\varepsilon)j + A_j$, for some $\varepsilon > 0$ (the sampling rate is above the Nyquist rate), (25) is almost surely satisfied, but it is possible that (26) will be wrong with a positive probability.

c) Consider the "additive model" $t_0 = 0,\ t_j = \sum_{l=1}^{j} A_l,\ j > 0$ or $t_j = \sum_{l=j}^{-1} A_l,\ j < 0$, with i.i.d positive random variables A_l. Now, suppose that $E\left[A_l\right] < 1$ and $Var\,A_l \neq 0$. Then (25) will be almost surely satisfied, so that the knowledge of both sequences $\mathbf{U}',\ \mathbf{t}$ is theoretically sufficient to recover the process $\mathbf{Z}$.

8.3.3.2. The Reconstruction

Let $f(z)$ be the complex function defined by the infinite product

$$f(z) = \lim_{n \to \infty} \prod_{k=-n}^{n} \left(1 - \frac{z}{t_k}\right),$$

with $t_k \neq t_l$ for $k \neq l$. The existence of $f(z)$ is insured if $\lim_{n\to\infty} t_n/n$ exists. A slightly modified formula is taken when one of t_k is zero. Consider next the particular Lagrange interpolation function $F(z)$

$$F(z) = \sum_k \frac{e^{j\omega t_k} f(z)}{f'(t_k)(z - t_k)}.$$

If $F(z) = e^{j\omega z}$, $\omega \in [-\pi, \pi]$, under appropriate hypotheses (see [33]), the fundamental isometry (see Appendix 1) yields

$$Z(t) = \sum_k \frac{f(t)}{f'(t_k)(t - t_k)} Z(t_k).$$

For $t_k = k$, we find the standard sampling formula

$$Z(t) = \sum_k \frac{\sin \pi(t-k)}{\pi(t-k)} Z(k).$$

Conditions for a good convergence of Lagrange interpolation functions can be found in [33]. In these cases, we have reconstructions without error. But, the formulae (through f) are built on the complete knowledge of the infinite sequence $\mathbf{t}$, which define f.

8.4. The Unobserved Random Case

8.4.1. The Jitter Model

The problem is to reconstruct the random process $\mathbf{Z}$ as best as possible, observing the values of U_n, where $U_n = Z(t_n)$, without knowing the (random) values t_n. We note as usual $\mathbf{U}' = \{U_n,\ n \in \mathcal{Z}\}$ the observed sequence.

The jitter model studied here is defined by sampling times in the form

$$t_n = nT + A_n, \quad n \in \mathcal{Z},$$

where A_n represents an error when taking the sample of rank n in a periodic scheme. It is a realistic model when a random transit time occurs in a wave transmission [35–36]. It is also a good model at the analog-to-digital conversion, and for a lack of synchronization. The problem was first studied by A. V. Balakhrishnan [5] for special cases and then generalized by the author [21–22]. In what follows, to simplify notations, we assume $T = 1$. To take another value of T would be obviously equivalent to a time dilatation. To solve the problem of jitter, we need statistical properties for the random variables A_n. More precisely, the knowledge of the following characteristic functions

$$\begin{cases} \psi(\omega) = E\left[e^{j\omega A_n}\right], \\ \phi(q, \omega) = E\left[e^{j\omega(A_n - A_{n-q})}\right], \end{cases} \tag{27}$$

is required (which are supposed to be independent of n). This hypothesis is a little weaker for the random sequence $\mathbf{A}' = \{A_n,\ n \in \mathcal{Z}\}$ than the stationarity at the order two (for the distributions). Note that no hypothesis is made on the existence of moments nor for some bounds of the A_n. In this case, sampling times could not be ordered. The problem has the following general solution.

8.4.2. The Best Estimator

1) In the fundamental isometry linked to the observed series $\mathbf{U}'$, the best m.s estimator $\widetilde{Z}(t)$ of $Z(t)$ is defined by (see Appendix 2)

$$U_n \longleftrightarrow e^{jn\omega}, \quad \widetilde{Z}(t) \longleftrightarrow \mu_t(\omega), \tag{28}$$

where

$$\begin{cases} \mu_t(\omega) = e^{j\omega t} dM_t / dS'_U(\omega), \quad \omega \in [-\pi, \pi], \\ dM_t(\omega) = \displaystyle\sum_{k=-\infty}^{\infty} e^{2j\pi kt} \psi^*(\omega + 2k\pi) dS_Z(\omega + 2k\pi), \\ \displaystyle\int_{-\infty}^{\infty} e^{jn\omega} \phi(n, \omega) dS_Z(\omega) = \int_{-\pi}^{\pi} e^{jn\omega} dS'_U(\omega), \quad n \in \mathcal{Z}. \end{cases} \tag{29}$$

$\widetilde{Z}(t)$ is obtained explicitly by expanding $\mu_t(\omega)$ as a linear combination of complex exponentials $e^{j\omega n}$, $n \in \mathcal{Z}$, and replacing them by the observed U_n.

2) The reconstruction error is

$$\sigma_t^2 = E\left[\left|\widetilde{Z}(t) - Z(t)\right|^2\right] = K_Z(0) - \int_{-\pi}^{\pi} \frac{dM_t}{dS'_U}(\omega) dM_t^*(\omega). \tag{30}$$

In the previous expressions, $S'_U(\omega)$ is the power spectrum (on $[-\pi, \pi]$) of the stationary series $\mathbf{U}'$. It depends on the joint distribution of (A_n, A_m) and is defined by the Herglotz theorem [28]

$$E[U_l U^*_{l-n}] = \int_{-\pi}^{\pi} e^{jn\omega} dS'_U(\omega).$$

The measure $M_t(\omega)$ is the sum of weighted shifted versions of $S_Z(\omega)$. We know that this kind of expression appears also in the theory of uniform sampling (but without the term depending on ψ) [7]. Furthermore, dM_t/dS'_U is the Radon-Nicodym derivative. For regular spectral densities, $s_Z(\omega) = dS_Z/d\omega$, dM_t/dS'_U will be an ordinary function, and (29)–(30) yield

$$\begin{cases} \mu_t(\omega) = e^{j\omega t}\dfrac{m_t}{s'_U}(\omega), \quad \omega \in [-\pi, \pi], \\ m_t(\omega) = \sum\limits_{k=-\infty}^{\infty} e^{2j\pi kt}\psi^*(\omega + 2k\pi)s_Z(\omega + 2k\pi), \\ \sigma_t^2 = K_Z(0) - \displaystyle\int_{-\pi}^{\pi} \frac{|m_t|^2}{s'_U}(\omega)d\omega. \end{cases} \tag{31}$$

3) As explained above, after $\mu_t(\omega)$ has been obtained, it can be expanded into a series of exponentials and using the isometry (Appendix 1)

$$\mu_t(\omega) = \lim_{n\to\infty} \sum_{k=-n}^{n} a_{kn}(t)e^{j\omega k} \longleftrightarrow \widetilde{Z}(t) = \lim_{n\to\infty} \sum_{k=-n}^{n} a_{kn}(t)U_k. \tag{32}$$

In the most regular cases, the expansion will be a Fourier series development, so that (32) can be written

$$\widetilde{Z}(t) = \sum_{k=-\infty}^{\infty} a_k(t)U_k, \quad a_k(t) = \frac{1}{2\pi}\int_{-\pi}^{\pi} \mu_t(\omega)e^{-jk\omega}d\omega. \tag{33}$$

In fact, it is not difficult to verify that $a_k(t) = a_0(t-k)$, so that it suffices to determine only the element $a_0(t)$. This last equality is equivalent to the cyclostationarity of $\widetilde{Z}(t)$.

The proof of the previous properties can be found in Appendix 2. It is based on the following decomposition of U_n

$$U_n = G_n + V_n \quad \text{with} \quad G_n = pr_{H(\mathbf{Z})}U_n,$$

so that both Hilbert spaces $H(\mathbf{G}')$ and $H(\mathbf{V}')$ generated by the new sequences $\mathbf{G}' = \{G_n,\ n \in \mathcal{Z}\}$, $\mathbf{V}' = \{V_n,\ n \in \mathcal{Z}\}$ are orthogonal. Furthermore, $H(\mathbf{G}') \subset H(\mathbf{Z})$, $H(\mathbf{V}') \perp H(\mathbf{Z})$. The sequence $\mathbf{G}'$ is the part of $\mathbf{U}'$ which provides information about $\mathbf{Z}$ and $\mathbf{V}'$ can be seen as an additive noise. Consequently, the reconstruction of $Z(t)$ can be seen as the extraction of information contamined by additive noise.

8.4.3. Two Particular Cases

8.4.3.1. Case 1

For all m and n $(m \neq n)$, A_m and A_n are independent. The characteristic function of $A_m - A_{m-n}$ is defined by

$$\phi(n,\omega) = E[e^{j\omega(A_m - A_{m-n})}] = \begin{cases} 1, & n = 0, \\ |\psi(\omega)|^2, & n \neq 0. \end{cases}$$

Then, the sequence $\mathbf{V}'$ is a white noise, that is to say its spectral density is constant. More precisely, both spectra satisfy

$$\begin{cases} dS'_G(\omega) = \sum_{k=-\infty}^{\infty} |\psi(\omega + 2k\pi)|^2 dS_Z(\omega + 2k\pi), \\ s'_V(\omega) = dS'_V/d\omega = \int_{-\infty}^{\infty} (1 - |\psi(u)|^2) dS_Z(u). \end{cases}$$

8.4.3.2. Case 2

When the Nyquist rate is fulfilled, such that $S_Z(\pi) - S_Z(-\pi) = S_Z(\infty)$, the infinite sums of (29) or (31) are brought down to a single term, which greatly simplifies the previous formula. As an example, in the case of spectral densities

$$\begin{cases} \mu_t(\omega) = e^{j\omega t}\left[\psi^* s_Z / \left(|\psi|^2 s_Z + s_V\right)\right](\omega), \\ \sigma_t^2 = \int_{-\pi}^{\pi} \left[s_V s_Z / \left(|\psi|^2 s_Z + s_V\right)\right](\omega) d\omega. \end{cases}$$

Note that the error is independent of t, but is, most of the time, different from zero, even if the sampling rate is above the Nyquist on the average. Recall that, in the observed case (Section 3.2), $Z(t)$ can be often exactly recovered by a Lagrange type series.

8.4.4. Examples

The reconstruction strategy required to determine $\widetilde{Z}(t)$ can be summarized as follows:

a) determination of the characteristic functions $\psi(\omega)$ and $\phi(n,\omega)$ from the statistical properties of A_n.

b) determination of the measures $M_t(\omega)$ and $S'_U(\omega)$ which yield $\mu_t(\omega)$.
c) depending on the regularity of $\mu_t(\omega)$, with respect to the measure $S'_U(\omega)$, expansion of $\mu_t(\omega)$ as a Fourier series

$$\mu_t(\omega) = \sum_{k \in Z} a_k(t) e^{j\omega k}, \quad a_k(t) = a_0(t-k),$$

or use adequate computations (as explained in Example 4).

Next sections illustrate the reconstruction of $Z(t)$ through different examples.

8.4.4.1. *Example 1*

Let the sampling sequence be defined by $t_n = n + \Phi$, $\psi(\omega)$ be the characteristic function of Φ and $\phi(n, \omega) = 1$, for any n, ω. We are in the case of a delayed uniform sampling (by the unknown quantity Φ), which models a lack of synchronization in the sampling. Furthermore, suppose the band-limited case $[-\pi, \pi]$ with a spectral density

$$s_Z(\omega) = 0, \quad \omega \notin [-\pi, \pi].$$

For sufficiently regular functions $\psi(\omega)$, we obtain

$$\begin{cases} s_Z(\omega) = s'_U(\omega), \ \mu_t(\omega) = \psi^*(\omega) e^{j\omega t}, \\ a_0(t) = \dfrac{1}{2\pi} \displaystyle\int_{-\pi}^{\pi} \psi^*(\omega) e^{j\omega t} d\omega, \\ \sigma_t^2 = \displaystyle\int_{-\pi}^{\pi} \left(1 - |\psi(\omega)|^2\right) s_Z(\omega) d\omega. \end{cases}$$

We consider two particular cases:

1) Φ is degenerate so that $\Pr[\Phi = \phi] = 1, \psi(\omega) = e^{j\omega\phi}$. We have

$$\tilde{Z}(t) = \sum_{n=-\infty}^{\infty} \frac{\sin \pi(t - \phi - n)}{\pi(t - \phi - n)} U_n, \quad \sigma_t = 0,$$

since we are studying the uniform sampling with known delay (given by $\psi(\omega)$). In this case, the Shannon interpolation formula can be used.

2) Φ is defined by $\Pr[\Phi = \varepsilon] = \Pr[\Phi = -\varepsilon] = 1/2, \varepsilon \neq 0$. Then

$$\begin{cases} \mu_t(\omega) = e^{j\omega t} \cos \varepsilon\omega, \\ \tilde{Z}(t) = \displaystyle\sum_{n=-\infty}^{\infty} \left(\frac{\sin \pi(t + \varepsilon - n)}{2\pi(t + \varepsilon - n)} + \frac{\sin \pi(t - \varepsilon - n)}{2\pi(t - \varepsilon - n)} \right) U_n, \\ \sigma_t^2 = \displaystyle\int_{-\pi}^{\pi} (\sin \omega\varepsilon)^2 s_Z(\omega) d\omega. \end{cases}$$

The uniform sampling case is when $\varepsilon \to 0$. As remarked before, the error is different from zero in spite of the fact that the Nyquist condition is satisfied in the mean (and even if the support width of s_Z is really smaller than 2π). But, because of this last property, the error is independent of t.

8.4.4.2. Example 2

We take again the model $t_n = n + \Phi$, with

$$s_Z(\omega) = 1 - \frac{|\omega|}{2\pi}, \quad \omega \in [-2\pi, 2\pi] \text{ and zero elsewhere,}$$

so that we are in the situation of Section 3.1.2 (Example 2), but with unobserved sampling times (if Φ is not degenerate). We have

$$\begin{cases} s'_U(\omega) = 1, \ \omega \in [-\pi, \pi], \\ m_t(\omega) = \begin{cases} \psi^*(\omega)\left(1 - \dfrac{\omega}{2\pi}\right) + \psi^*(\omega - 2\pi)\dfrac{\omega}{2\pi}e^{-2j\pi t}, & \omega \in [0, \pi], \\ \psi^*(\omega)\left(1 + \dfrac{\omega}{2\pi}\right) - \psi^*(\omega + 2\pi)\dfrac{\omega}{2\pi}e^{2j\pi t}, & \omega \in [-\pi, 0]. \end{cases} \end{cases}$$

We obtain $\widetilde{Z}(t)$ expanding $\mu_t(\omega) = m_t(\omega)e^{j\omega t}$ in a Fourier series and replacing $e^{in\omega}$ by U_n. Of course, the result depends on t and it is the same for σ_t, because $s_Z(\omega)$ does not have the property of non-aliasing. For example, when Φ is defined by $\Pr[\Phi = \varepsilon] = \Pr[\Phi = -\varepsilon] = 1/2$, we find

$$\begin{cases} \widetilde{Z}(t) = \dfrac{1}{2}\sum\limits_{n=-\infty}^{\infty}\left(\left(\dfrac{\sin \pi(t+\varepsilon-n)}{\pi(t+\varepsilon-n)}\right)^2 + \left(\dfrac{\sin \pi(t-\varepsilon-n)}{\pi(t-\varepsilon-n)}\right)^2\right)U_n, \\ \sigma_t^2 = \dfrac{4\pi}{3} - \dfrac{1}{4\pi\varepsilon^2}\left(1 - \dfrac{\sin 4\varepsilon\pi}{4\varepsilon\pi}\right) - \dfrac{\pi}{3}\cos 2\pi t \cos 2\varepsilon\pi \\ \qquad + \dfrac{\cos 2\pi t}{4\pi\varepsilon^2}\left(\cos 2\varepsilon\pi - \dfrac{\sin 2\varepsilon\pi}{2\varepsilon\pi}\right). \end{cases}$$

We will compare these results with those of Section 3.2, where the sampling times are known, and with the classical uniform sampling ($\varepsilon = 0$) where the interpolation formula and the error are

$$\begin{cases} \widetilde{Z}(t) = \sum\limits_{n=-\infty}^{\infty}\left(\dfrac{\sin \pi(t-n)}{\pi(t-n)}\right)^2 U_n, \\ \sigma_t^2 = \dfrac{4\pi}{3}\sin^2 \pi t. \end{cases}$$

8.4.4.3. Example 3

Let $\mathbf{A} = \{A(t), t \in \mathcal{R}\}$ be a telegraph signal [32] having the values $\pm\alpha, 0 < \alpha < 1/2$, and based on a Poisson process of parameter $\lambda(\lambda > 0)$. Denote $A_n = A(n)$ and assume that $s_Z(\omega)$ is a constant on $[-\pi, \pi]$. The sampling times are linked, so that the "noise" $\mathbf{V}' = \{V_n, n \in \mathcal{Z}\}$ is not white. More precisely

$$\begin{cases} \psi(\omega) = \cos\alpha\omega, \\ \phi(n,\omega) = \cos^2\alpha\omega + e^{-2\lambda|n|}\sin^2\alpha\omega. \end{cases}$$

Then, with $s_Z(\omega) = 1/2\pi,\ \omega \in [-\pi, \pi]$, and 0 elsewhere

$$\begin{cases} \mu_t(\omega) = e^{i\omega t}\dfrac{\cos\alpha\omega}{2\pi s'_U(\omega)}, \\ 2\pi s'_U(\omega) = \cos^2\alpha\omega + \dfrac{1+e^{-4\lambda}}{2\pi}\displaystyle\int_{-\pi}^{\pi}\frac{\sin^2\alpha x dx}{1+e^{-4\lambda}+2e^{-2\lambda}\cos(\omega+x)}, \end{cases}$$

which lead to

$$\begin{cases} \widetilde{Z}(t) = \displaystyle\sum_{n=-\infty}^{\infty} a_0(t-n)U_n, \\ a_0(t) = \frac{1}{4\pi^2}\displaystyle\int_{-\pi}^{\pi}\frac{\cos\alpha\omega}{s'_U(\omega)}e^{i\omega t}d\omega, \\ \sigma_t^2 = 1 - \dfrac{1}{4\pi^2}\displaystyle\int_{-\pi}^{\pi}\frac{\cos^2\alpha\omega}{s'_U(\omega)}d\omega. \end{cases}$$

As noted above, σ_t^2 is independent of t (because of the Nyquist condition), and $a_0(t)$ determines $a_n(t) = a_0(t-n)$. Of course, for all t, σ_t^2 has to approach 0 when α approaches 0.

8.4.4.4. Example 4

In this example, $S_Z(\omega)$ is a line spectrum and the random variables A_n are independent continuous random variables

$$\begin{cases} S_Z(\omega) = \frac{1}{2}\Gamma(\omega+b) + \frac{1}{2}\Gamma(\omega-b), b > 0, \quad b \notin \pi\mathcal{Z}, \\ \psi(\pm b) \neq 0, \\ \phi(n,\omega) = |\psi(\omega)|^2, n \neq 0. \end{cases}$$

Then, with l such that $-\pi < b - 2l\pi < \pi$, we have

$$\begin{cases} \mu_t(\omega) = 0, \omega \neq \pm(b - 2l\pi), \\ \mu_t(b - 2l\pi) = e^{jbt}/\psi(b), \\ \mu_t(-b + 2l\pi) = e^{-jbt}/\psi(-b), \\ \widetilde{Z}(t) = \lim_{n\to\infty} \frac{1}{n}\sum_{k=1}^{n}\left(\frac{e^{jb(t-k)}}{\psi(b)} + \frac{e^{-jb(t-k)}}{\psi(-b)}\right)U_k, \\ \sigma_t = 0. \end{cases}$$

Here, we are outside the usual frame of spectral densities where the error is different from zero. The solution is not obtained from a Fourier series development, but by averaging samples. In fact, $\mu_t(\omega)$ is a discontinuous function and its development has to converge towards the true values at the increasing points of $S_Z(\omega)$, and towards 0 elsewhere (in the mean).

The reconstruction formula can be easily explained when we take into account the three consequences of the hypotheses:

a) $Z(t)$ has the form $Z(t) = Ae^{jbt} + Be^{-jbt}$, where A, B are uncorrelated random variables.
b) In the orthogonal decomposition of $U_n = G_n + V_n$, we have (see Appendix 2)

$$G_n = \psi(b)Ae^{jbn} + \psi(-b)Be^{-jbn}.$$

c) V_n has a continuous spectrum, so that its arithmetic mean converges towards 0.

Replacing U_n by $G_n + V_n$ in $\widetilde{Z}(t)$, one yields $\widetilde{Z}(t) = Z(t)$ and $\sigma_t = 0$.

Other cases can be studied, mixing different kinds of spectra, and taking into account the zeros of $\psi(\omega)$ (see Appendix 3 for studying the general case where $\sigma_t = 0$).

8.5. The Complex Process Case

8.5.1. A Remark about Uniform Sampling

Let $\mathbf{X} = \{X(t), t \in \mathcal{R}\}$ be a real stationary process with spectral density

$$s_X(\omega) \neq 0, \quad \omega \in [-2\pi, 2\pi], \text{ and 0 elsewhere.}$$

Now, we define $\mathbf{Y} = \{Y(t), t \in \mathcal{R}\}$ by (in some sense)

$$Y(t) = \int_{-\infty}^{\infty} \frac{X(u)}{t-u} du.$$

$Y(t)$ is the Hilbert transform of $X(t)$, characterized by the complex gain $-jsgn(\omega)$ (see the part of Appendix 1 on time invariant filters). The process $\overline{\mathbf{Z}}$ defined by $\overline{Z}(t) = X(t) + jY(t)$ is the "analytic signal" associated with $\mathbf{X}$. A fundamental property of $\overline{\mathbf{Z}}$ is

$$s_{\overline{Z}}(\omega) = 4s_X(\omega), \omega \in [0, 2\pi], \text{ and 0 elsewhere.}$$

Then, the spectral width of $\mathbf{Z}$ is 2π (without aliasing in respect to a unit sampling). Consequently, $\overline{Z}(t)$ can be recovered, observing the sequence $\overline{\mathbf{Z}}' = \{\overline{Z}(n), n \in \mathcal{Z}\}$. After that, we obtain $X(t)$, $Y(t)$ taking the real and imaginary parts of $\overline{Z}(t)$.

Now, define the process $\mathbf{Z}$ by $Z(t) = X(t) + jaY(t), a \neq 0, \pm 1$. In this case, $s_Z(\omega)$ is different from zero on $[-2\pi, 2\pi]$ in the same time that $s_X(\omega)$. It is generally impossible to retrieve $Z(t)$ linearly, using solely of the samples $Z(n), n \in \mathcal{Z}$. Nevertheless, observing the samples $X(n), Y(n), n \in \mathcal{Z}$, allows us to know the $\overline{Z}(n), n \in \mathcal{Z}$, $\overline{Z}(t)$, $X(t)$, $Y(t)$, and finally $Z(t)$ for all t.

Consequently, in certain circumstances, it is possible to reconstruct exactly a process of 4π-spectral width, by uniform sampling, using the real and imaginary parts of the samples. The real reason is that, in classical theory, the allowed interpolation formulae are in the form

$$\widetilde{Z}(t) = \lim_{n\to\infty} \sum_{k=-n}^{n} a_{kn}(t) Z(k).$$

Although we can make use of the larger class of estimators in the form

$$\widetilde{Z}(t) = \lim_{n\to\infty} \sum_{k=-n}^{n} \big(a_{kn}(t)X(k) + b_{kn}(t)Y(k)\big).$$

This remark holds for all complex processes [23] and for nonuniform sampling. It is this last case which is treated in the following section.

8.5.2. Example of Random Sampling

We are in the same situation as in the previous section, with a random sampling of $Z(t) = X(t) + jaY(t), a \in \mathcal{R} - \{0, 1, -1\}$, so that we observe the sequences $\mathbf{U}_1', \mathbf{U}_2', \mathbf{U}'$ defined by $U_{1n} = X(n + A_n)$, $U_{2n} = Y(n + A_n)$, $U_n =$

$U_{1n} + jaU_{2n}$. Then, we observe also the sequences $\mathbf{C}', \mathbf{C}'^*$ defined by $C_n = U_{1n} + jU_{2n}$, $C_n^* = U_{1n} - jU_{2n}$. Using conditional expectations, we obtain

$$E\left[C_n C_m\right] = 0, \quad m, n \in \mathcal{Z}, \quad m \neq n,$$

so that the Hilbert spaces $H(\mathbf{C}')$ and $H(\mathbf{C}'^*)$ generated by the sequences $\mathbf{C}' = \{C_n, n \in \mathcal{Z}\}$, $\mathbf{C}'^* = \{C_n^*, n \in \mathcal{Z}\}$, are orthogonal. Moreover

$$H(\mathbf{C}') \oplus H(\mathbf{C}'^*) = H(\mathbf{U}_1') + H(\mathbf{U}_2'),$$

where the sign $\oplus$ means the orthogonality of components. It is a particular property of the analytic signal. In this case of orthogonality, the best (m.s) estimators are given by

$$\begin{cases} \widetilde{X}(t) = pr_{H(\mathbf{C}')}X(t) + pr_{H(\mathbf{C}'^*)}X(t), \\ \widetilde{Y}(t) = pr_{H(\mathbf{C}')}Y(t) + pr_{H(\mathbf{C}'^*)}Y(t), \\ \widetilde{Z}(t) = \widetilde{X}(t) + ja\widetilde{Y}(t). \end{cases}$$

$\widetilde{Z}(t)$ is obtained taking into account the observation of the sequences $\mathbf{U}_1', \mathbf{U}_2'$, and not solely $\mathbf{U}$. When $s_X(\omega) = 0, \omega \neq [-2\pi, 2\pi], A_n = \Phi$, $\Pr[\Phi = \varepsilon] = \Pr[\Phi = -\varepsilon] = 1/2$, simple calculations lead to

$$\begin{cases} \widetilde{Z}(t) = \sum_{n \in \mathcal{Z}}[b_1(t-n)X(n+\Phi) + jb_2(t-n)Y(n+\Phi)], \\ b_1(t) = c(t+\varepsilon) + c(t-\varepsilon), \\ b_2(t) = d(t+\varepsilon) + d(t-\varepsilon), \\ c(t) = \dfrac{\sin \pi t}{2\pi t}(\cos \pi t + ja \sin \pi t), d(t) = \dfrac{\sin \pi t}{2\pi t}(j \sin \pi t + a \cos \pi t), \\ \sigma_t^2 = 2(1+a^2)\displaystyle\int_0^{2\pi} \sin^2(\varepsilon\omega)s_X(\omega)d\omega. \end{cases} \tag{34}$$

Note that σ_t does not depend on t. Furthermore, the coefficients $b_1(t), b_2(t)$ are independent of $s_X(\omega)$. These properties are linked to the spectral occupancy of the **X**-spectrum.

We may compare the interpolation formula (34) with the results obtained using the coarser method described in Section 4. In this case, the interpolation depends on $s_X(\omega)$, and the error on t. For example, suppose that $s_X(\omega) = 1/4\pi$, $\omega \in [-2\pi, 2\pi]$, $s_X(\omega) = 0$ elsewhere. For a small a, and $\varepsilon = 1/4$, the relative gain of this new method is about 8% at the point $t = 0$, and 22% at $t = 1/2$.

8.6. Conclusion

The classical sampling formula with $\sin x/x$ (the cardinal function) as its kernel corresponds to a uniform sampling with a sufficiently high rate. Its origin is a disputed question. J. R. Higgins [3] found it in a paper by E. Borel on analytic functions, but A. L. Cauchy results can be seen as the starting point for these formula for functions with bounded spectrum [25]. In a book by H. S. Black on modulation theory appears the idea that recovering a function is possible for a nonuniform sampling set if the mean rate is sufficient. However, such idea can be related to the papers on completion of exponential sequences written by the mathematician N. Levinson [34].

The problem of developing (or approximating) a random process as a function of its samples is not very different from the deterministic functions. In the case where sampling times are known, the difference is in the definition of the distance which is not the same for deterministic functions and random processes. In the finite case, the best linear mean-square solution is found using elementary algebra. In the infinite case, Levinson theorems, and work on analytic functions are the heart of the solutions.

In the case of unknown sampling times, we have seen that the problem is equivalent to projections onto Hilbert spaces. Advances have to be made in the definition of classes of estimates. Section 5 about complex random processes shows that it is possible to improve the usual results, using more elaborate solution forms.

8.7. Appendix

8.7.1. Appendix 1—The Fundamental Isometry

8.7.1.1. Definition of $H(\mathbf{Z})$

$\mathbf{Z}$ is the stationary process defined in Section 1, and with power spectrum $S_Z(\omega)$. Let $H(\mathbf{Z})$ be the set of random variables A in the form

$$A = \lim_{n\to\infty} \sum_{k=-n}^{n} a_{kn} Z(t_{kn}), \quad a_{kn} \in \mathcal{C}, \quad t_{kn} \in \mathcal{R},$$

where the limits are taken with respect to the mean-square convergence. $H(\mathbf{Z})$ is a Hilbert space [30] when the following scalar product is defined

$$\langle A, B \rangle_{H(\mathbf{Z})} = E[AB^*].$$

The associated distance is defined by

$$d(A, B) = \sqrt{E\big[|A - B|^2\big]},$$

which leads to the mean-square convergence. Recall that a Hilbert space is complete so that it is possible to prove the existence of a limit without knowing it.

8.7.1.2. *Definition of $\mathcal{H}(S_Z)$*

Consider the set of functions $\mathcal{H}(S_Z)$ in the form

$$a(\omega) = \lim_{n\to\infty} \sum_{k=-n}^{n} a_{kn} e^{j\omega t_{kn}}, \quad a_{kn} \in \mathcal{C}, \quad t_{kn} \in \mathcal{R},$$

where the limits are taken in $L^2(S_Z)$, the set of measurable functions $a(\omega)$, such that

$$\int_{-\infty}^{\infty} |a(\omega)|^2 \, dS_Z(\omega) < \infty.$$

In fact, the Hilbert spaces $\mathcal{H}(S_Z)$ and $L^2(S_Z)$ are equal, and the scalar product is defined by

$$\langle a, b \rangle_{\mathcal{H}(S_Z)} = \int_{-\infty}^{\infty} a(\omega) b^*(\omega) \, dS_Z(\omega).$$

8.7.1.3. *The Fundamental Isometry*

Now, we define a mapping I which links the two spaces $H(\mathbf{Z})$ and $\mathcal{H}(S_Z)$ by

$$I[Z(t)] = e^{j\omega t}, \quad t \in \mathcal{R},$$

denoted

$$Z(t) \longleftrightarrow e^{j\omega t}.$$

I is first extended by linearity to finite sums of random variables and functions, and subsequently to entire spaces. I is an isometry, that is to say

$$A \longleftrightarrow_I a, B \longleftrightarrow_I b \Rightarrow \langle A, B \rangle_{H(\mathbf{Z})} = \langle a, b \rangle_{\mathcal{H}(S_Z)},$$

and, consequently

$$E\left[|A-B|^2\right] = \int_{-\infty}^{\infty} |a-b|^2(\omega)\, dS_Z(\omega),$$

I associates to any random variable A, a function $a(\omega)$ which is limit of a sum of complex exponentials. Then, a problem of projection or of distance about the random process $\mathbf{Z}$ can be replaced by an equivalent problem concerning the Fourier analysis.

For example, assume that U and $a(\omega)$ are related by the isometry I. If $a(\omega)$ can be expanded in $\mathcal{H}(S_Z)$ as follows

$$a(\omega) = \lim_{n\to\infty} \sum_{k=-n}^{n} a_{kn} e^{j\omega t_{kn}},$$

then, the following expansion of U in $H(\mathbf{Z})$ can be written

$$U = \lim_{n\to\infty} \sum_{k=-n}^{n} a_{kn} Z(t_{kn}).$$

Consequently, when $t_{kn} = k$, a Fourier series development of $a(\omega)$ allows to interpolate U with respect to the sequence $\mathbf{Z}'$.

8.7.1.4. *Linear Filtering*

A linear invariant filter $\mathcal{F}$ is generally defined by an impulse response (see Chapter 2) $f(t)$ and a complex gain $F(\omega)$ (or frequency response), so that, in some sense

$$U(t) = \mathcal{F}[Z](t) = \int_{-\infty}^{\infty} Z(u)\, f(t-u)du, F(\omega) = \int_{-\infty}^{\infty} e^{-i\omega u} f(u)du.$$

By linearity, if the above integral is a limit of a Riemann sum, the corresponding random variable of $U(t)$ in the isometry linked to $\mathbf{Z}$ is calculated replacing $Z(t)$ by $e^{i\omega t}$ so that

$$Z(t) \longleftrightarrow e^{i\omega t}, \quad U(t) \longleftrightarrow \int_{-\infty}^{\infty} e^{i\omega u} f(t-u)du = e^{i\omega t} F(\omega).$$

In fact, because the impulse response does not always exist in the usual sense of functions, the definition of $\mathcal{F}$ by a complex gain $F(\omega)$ and the isometry allows easy computations. For example, the results given in Section 5 can be obtained, using $F(\omega) = -sign(\omega)$ $(sign(\omega) = 1, \omega > 0$, and $sign(\omega) = -1, \omega < 0)$ as the complex gain for the Hilbert filter.

8.7.1.5. *Subspaces and Projections*

We have to consider the restrictions of the previous Hilbert spaces to finite or countable infinite sets of values of parameter t. These spaces will be recognizable by a $'$ added to the notation. For example, we denote by $H(\mathbf{Z}')$ the set of random variables in the form

$$A = \lim_{n\to\infty} \sum_{k=-n}^{n} a_{kn} Z(t_{kn}), \quad a_{kn} \in \mathcal{C}, \quad t_{kn} \in \mathbf{t},$$

$\mathbf{t}$ finite or countable and the limits being taken in $H(\mathbf{Z})$.

Furthermore, we will use projections of elements of Hilbert spaces on subspaces. We know the uniqueness of this operation. The projection B of $A \in H(\mathbf{Z})$ on the subspace $H(\mathbf{Z}')$ is denoted

$$B = pr_{H(\mathbf{Z}')} A.$$

8.7.2. Appendix 2—Jitter Formulae

8.7.2.1. *An Orthogonal Decomposition*

$\mathbf{U}'$ is the sampled process at times $t_n = n + A_n$, constituted by the random variables $U_n = Z(n + A_n)$. The sequence of delays $\mathbf{A}' = \{A_n, n \in \mathcal{Z}\}$ is defined by the characteristic functions $\psi(\omega)$, $\phi(n, \omega)$ defined by (27). Now, look at the process $\mathbf{G} = \{G(t), t \in \mathcal{R}\}$ defined in the isometry I by (see the end of Appendix 1)

$$Z(t) \longleftrightarrow e^{j\omega t}, \quad G(t) \longleftrightarrow e^{j\omega t}\psi(\omega).$$

Process $\mathbf{G}$ is the result of filtering $\mathbf{Z}$ with a complex gain $\psi(\omega)$. We define now the sequence $\mathbf{G}' = \{G_n, n \in \mathcal{Z}\}$, $\mathbf{V}' = \{V_n, n \in \mathcal{Z}\}$, by

$$G_n = G(n), U_n = G_n + V_n. \tag{35}$$

Using isometry and conditional expectations (and the independence between $\mathbf{Z}$ and $\mathbf{A}'$), one gets

$$\begin{cases} E[Z(t)V_n^*] = E[G(t)V_n^*] = 0, n \in \mathcal{Z}, t \in \mathcal{R}, \\ E[Z(t)G_n^*] = \displaystyle\int_{-\infty}^{\infty} e^{j\omega(t-n)}\psi^*(\omega)dS_Z(\omega), \\ E\left[G_n G_{n-q}^*\right] = \displaystyle\int_{-\infty}^{\infty} e^{j\omega q}|\psi(\omega)|^2 dS_Z(\omega), \end{cases} \tag{36}$$

so that (35) is an orthogonal decomposition with $H(\mathbf{G}') \subset H(\mathbf{Z})$, $H(\mathbf{V}') \perp H(\mathbf{Z})$. Recall that $H(\mathbf{G}')$, $H(\mathbf{V}')$ are the Hilbert spaces generated by sequences $\mathbf{G}'$ and $\mathbf{V}'$.

8.7.2.2. The Best Estimator

The best linear mean square estimator $\widetilde{Z}(t)$ of $Z(t)$ is the element of $H(\mathbf{U}')$ which verifies

$$E\left[\left(Z(t) - \widetilde{Z}(t)\right)U_n^*\right] = 0, \quad n \in \mathcal{Z},$$

or, using (36) and (28)

$$\begin{cases} U_n \longleftrightarrow e^{jn\omega}, \widetilde{Z}(t) \longleftrightarrow \mu_t(\omega), \\ \displaystyle\int_{-\infty}^{\infty} e^{j\omega(t-n)}\psi^*(\omega)dS_Z(\omega) - \int_{-\pi}^{\pi} \mu_t(\omega)dS'_U(\omega) = 0, \quad n \in \mathcal{Z}, \end{cases} \tag{37}$$

where $S'_U(\omega)$ is the spectrum of the sequence $\mathbf{U}'$. Making use of the uniqueness of the Fourier transform, one gets (29). The error (30) is due to the application of the Pythagorian theorem.

8.7.3. Appendix 3—An Exact Reconstruction

It is possible to prove the following property [22]. A necessary and sufficient condition for an exact recovery of $Z(t)$ in the jitter case (Section 4.1) is that the three following conditions are fulfilled

1) If $\Delta = \{\omega; \psi(\omega) \neq 0\}$ then $\int_\Delta dS_Z(\omega) = K_Z(0)$.
2) The translated measures $S_Z^n(\omega) = S_Z(\omega + 2n\pi), n \in \mathcal{Z}$, are mutually singular.
3) The measures S'_G and S'_V are mutually singular.

The first condition implies the fact that $\psi(\omega)$ is (almost everywhere) non-zero on the support of S_Z. The second one is the classical non-aliasing condition which is found in uniform sampling. The latter one is generally true when dealing with continuous random variables A_n and line spectra for $Z(t)$. For an example, see Section 4.4.4.

8.7.4. Appendix 4—Skip Sampling

a) Let $\mathbf{B}' = \{B_n, n \in \mathcal{Z}\}$ be a stationary sequence independent of $\mathbf{Z}$ taking the values 0 and 1, and characterized by

$$\begin{cases} \Pr[B_n = 0] = p, \Pr[B_n = 1] = q = 1 - p, \\ E[B_n B_{n-m}] = \Pr[B_n = B_{n-m} = 1] = \displaystyle\int_{-\pi}^{\pi} e^{jm\omega} s'_B(\omega)d\omega. \end{cases}$$

Suppose that we observe the sequence $\mathbf{U}' = \{U_n, n \in \mathcal{Z}\}$ defined by $U_n = B_n Z(n)$. Then, we skip, in the uniform sampling, the samples of index n when $B_n = 0$. More generally, deleting samples in uniform or nonuniform sampling, one gets skip sampling. This situation was earlier studied in Section 2 and corresponds practically to defaults of sampler and "fading" situations when the transmitted waves are drastically weakened (bursts).

b) We look for an estimator in the form (7) but with $U_n = B_n Z(n)$, so that $\widetilde{Z}(t)$ is the orthogonal projection on $H(\mathbf{U}')$. In the isometry built on $\mathbf{U}'$, and in the case of spectral densities, the results are (the calculations are straightforward)

$$\begin{cases} U_n \longleftrightarrow e^{j\omega n}, \widetilde{Z}(t) \longleftrightarrow \mu_t(\omega), \\ \mu_t(\omega) = q e^{j\omega t} \dfrac{s'^t_{ZU}}{s'_U}(\omega), \\ \sigma_t^2 = \displaystyle\int_{-\pi}^{\pi} \left(s'_Z - q^2 \frac{|s'^t|^2}{s'_U} \right)(\omega) d\omega, \\ s'^t_Z(\omega) = \displaystyle\sum_k e^{2j\pi k t} s_Z(\omega + 2k\pi), \\ s'_U(\omega) = s'_U(\omega) \displaystyle\int_{-\pi}^{\pi} s'_B(\omega - u) s'^0_Z(u) du. \end{cases}$$

When the Nyquist condition is fulfilled ($s_Z(\omega) = 0, \omega \notin [-\pi, \pi]$), the error becomes independent of t. Moreover, for independent B_n

$$\sigma_t^2 = pK_Z(0) \int_{-\pi}^{\pi} \frac{s_Z(\omega)}{pK_Z(0) + 2\pi q s_Z(\omega)} d\omega. \tag{38}$$

c) Now, let $s_Z(\omega) = 0,\ \omega \notin [-\pi + a, \pi - a], 0 < a < \pi, (\pi - a)q > \pi$, and $Pr[Z(t) = 0] = 0$. We are in a situation where the Levinson theorem (Section 3.2.1) can be used because by construction, the sampling times are observed (they are the integers n so that $B_n = 1$), and because a condition similar to (25) is satisfied. Then, $Z(t)$ can be perfectly retrieved. Nevertheless, σ_t given by (38) is never zero. This is not a paradox since the Levinson theorem leads to a reconstruction for example in the form of a Lagrange series with supporting function (denoted f in Section 3.3.2) defined by the (random) sampling times linked to observed samples. Iterative procedures are also available in this case [11].

References

[1] A. J. Jerri. The Shannon Sampling Theorem: Its Various Extensions and its Applications. *IEEE Trans. on Inf. Th.* 65:1565–1595, 1977.

[2] R. J. Marks II. *Introduction to Shannon Sampling Theory and Interpolation Theory.* Springer-Verlag, 1991.

[3] J. R. Higgins. *Sampling Theory in Fourier and Signal Analysis. Foundations.* Oxford Sc. Pub., 1996.

[4] H. S. Black. *Modulation Theory.* Van Nostrand, 1953, pp. 41–58.

[5] A. V. Balakhrishnan. On the Problem of Time Jitter in Sampling. *IEEE Trans. on Inf. Th.*, 226–236, 1962.

[6] A. V. Balakhrishnan. A Note on the Sampling Principle for Continuous Signals. *IRE Trans. on IT,* IT-3:143–146, 1957.

[7] S. P. Lloyd. A Sampling Theorem for (Wide Sense) Stochastic Processes. *Trans. Am. Math. Soc.* 92:1–12, 1959.

[8] O. A. Z. Leneman. J. B. Lewis. Random Sampling of Random Processes, Mean-Square Comparison of Various Interpolators. *IEEE Trans. on Aut. Cont.* AC 11:396–403, 1966.

[9] O. A. Z. Leneman. Random Sampling of Random Processes. *J. of Franklin Inst.*, 281(4): 302–314, 1966.

[10] O. A. Z. Leneman. Random Sampling of Random Processes: Impulse Processes. *Inf. and Control,* 9:347–363, 1966.

[11] F. A. Marvasti. *Signal Recovery from Nonuniform Samples and Spectral Analysis on Random Nonuniform Samples. ICASSP,* Tokyo, 1986, pp. 1649–1652.

[12] F. A. Marvasti. *A Unified Approach to Zero-Crossings and Nonuniform Sampling.* Nonuniform, Oak Park IL, 1987, pp. 57–61.

[13] F. A. Marvasti, P. M. Clarkson, M. V. Dokic, U. Goenchanart, C. Lui. Reconstruction of Speech Signals with Lost Samples. *IEEE Trans. on SP.*, 2897–2903, 1992.

[14] F. J. Beutler and O. A. Z. Leneman. Random Sampling of Random Processes: Stationary Point Processes. *Inf. and Control*, 9:325–346, 1966.

[15] H. S. Shapiro and R. A. Silverman. Alias-Free Sampling of Random Noise. *J. Soc. Ind. Appl. Math.*, 8(2):225–248, 1960.

[16] F. J. Beutler. Alias-Free Randomly Timed Sampling of Stochastic Processes. *IEEE Trans. on IT,* IT-16(2):147–152, 1970.

[17] E. Masry. Random Sampling and Reconstruction of Spectra. *Inf. and Control*, 19:275–288, 1971.

[18] E. Masry and M-C. C. Lui. Discrete-Time Estimation of Continuous-Parameter Processes – A New Consistent Estimate. *IEEE Trans. on IT,* IT-22(3):298–312, 1978.

[19] E. Masry. Poisson Sampling and Spectral Estimation of Continuous-Time Processes. *IEEE Trans. on IT,* 24(2):173–183, 1978.

[20] E. Parzen. *On Spectral Analysis with Missing Observations and Amplitude Modulation. Time Series Analysis Papers*, Holden-Day, 1967, pp. 180–189.

[21] B. Lacaze. Stationary Clock Changes on Stationary Processes. *Signal Processing*, 55:191–205, 1996.

[22] B. Lacaze. A Note about Stationary Process Random Sampling. *Stat. Prob. Lett.*, 31:133–137, 1996.

[23] B. Lacaze. Periodic Bi-sampling of Stationary Processes. *Signal Processing*, 68:283–293, 1998.

[24] B. Lacaze. Modeling the HF Channel with Gaussian Random Delays. *Signal Processing*, 64(2): 215–220, 1998.

[25] B. Lacaze. La Formule d'Échantillonnage et A. L. Cauchy. *Traitement du Signal*, 15(4):289–295, 1998.

[26] R. Veldhuis. *Restoration of Lost Samples in Digital Signals*. Prentice Hall, 1990.
[27] W. Feller. *An Introduction to Probability Theory and Its Applications*. Wiley, 1950.
[28] J. L. Doob. *Stochastic Processes*. Wiley, 1953.
[29] E. Lukacs. *Characteristic Functions*. Griffin, London, 1960.
[30] H. Cramer and M. R. Leadbetter. *Stationary and Related Stochastic Processes*. Wiley, 1967.
[31] A. M. Yaglom. *Theory of Stationary Random Functions*. Prentice-Hall, 1962.
[32] A. Papoulis. *Probability, Random Variables and Stochastic Processes*. McGraw-Hill, 1991.
[33] B. Y. Levin. Zeros of Entire Functions. *Am. Math. Soc.*, 1964.
[34] N. Levinson. Gap and Density Theorems. *Am. Math. Soc. Coll. Pub.*, 1940.
[35] J. M. H. Elmirghani, R. A. Cryan and F. M. Clayton. Spectral Characterization and Frame Synchronisation of Optical Fiber Digital PPM. *El. Let.*, 28(16):1482–1483, 1992.
[36] J. M. H. Elmirghani, R. A. Cryan and F. M. Clayton. Spectral Analysis of Time Jitter Effects on the Cyclostationary Format. *Signal Processing*, 43:269–277, 1995.
[37] J. L. Yen. On Nonuniform Sampling of Bandwith-Limited Signals. *IRE on Circ. Th.*, CT-3, (12): 251–257, 1956.

9

Zero-Crossings of Random Processes with Application to Estimation and Detection

J. T. Barnett

9.1. Introduction

The origin of Rice's formula for the average level-crossing rate of a general class of random processes can be traced back to his 1936 notes on "Singing Transmission Lines," [35]. For a stationary, ergodic, random process $\{X(t)\}$, $-\infty < t < \infty$, with sufficiently smooth sample paths, the average number of crossings about the level u, per unit time, is given by Rice's formula [36]

$$E[D_u] = \int_{-\infty}^{\infty} |x'| p(u, x') dx' \tag{1}$$

where $p(x, x')$ is the joint probability density of the $X(t)$ and its mean-square derivative $X'(t)$, and D_u is the number of u level-crossings of $\{X(t)\}$ for t in the unit interval [0,1]. Rice's formula is quite general in nature, and as we shall see, has a simplified form when the process is Gaussian.

Rice's celebrated formula and the basic mathematical techniques for level-crossing analysis developed in his seminal works, *Mathematical Analysis of Random Noise* (Bell System Technical Journals Vol's 23, 24, 1944, 1945), and *Distribution of the Duration of Fades in Radio Transmission* (Bell System Technical Journal Vol. 37, 1958) [37], have been continually referenced and used for over 50 years. As noted in these early works one of the main theoretical,

J. T. Barnett • Space and Naval Warfare Systems Center, Intelligence, Surveillance, Reconnaissance Department, San Diego, California 92152

Nonuniform Sampling: Theory and Practice, edited by Marvasti
Kluwer Academic/Plenum Publishers, New York, 2001.

as well as, practical problems concerns the probability distribution of the interval length between the level-crossings, (or the number of crossings) of a given random process. It turns out it is extremely difficult, in general, to determine the exact probability distribution function of the length of such intervals, and to this day, analytic results are available for only special cases. See [29], and [25] for early pioneering work regarding the distribution of interval lengths. This remains today essentially an unsolved problem.

Since obtaining the exact distribution function is problematic, statistical information on the level-crossing counts is obtained via the moments of the level-crossing counts, usually the mean number of crossings and the variance of the number of crossings. The mean value is, of course, given by Rice's formula, the variance has been given by many authors, see [41], and [7].

Applications of Rice's type of analysis as well as the general theory of level-crossings of random processes are found in many areas outside of electrical engineering. For example, ocean engineering, problems of wave height and wave slamming, structural engineering, loading and fatigue analysis, seismology, acoustics, measurement of power in ultrasonic transducers [6], and Nonlinear Dynamics [16]. For a comprehensive historical account of the level-crossing problem for random see [9], [1]. [1] updates [9] and gives a survey of new results to that date. Both papers have extensive bibliographies which bring together a diverse set of results scattered in the mathematics and engineering literature.

For the rest of the chapter we focus on two applications pertinent to problems in electrical engineering, namely: discrete frequency estimation, and narrow band signal detection. In particular, we first review Rice's formula for a Gaussian process and some non-Gaussian processes, then present a technique for frequency estimation using zero-crossing counts of repeatedly filtered time-series. Lastly, we consider a radar detection problem using the mean level-crossing count of the envelope of a narrow band signal in narrow band noise.

9.1.1. Expected Zero-Crossing Rate of a Gaussian Process

We present formulas for the expected zero-crossing rate of a Gaussian process. Both the continuous time and discrete time cases are given.

If a zero-mean, stationary Gaussian process $\{Z(t)\}$, for $-\infty < t < \infty$, with normalized autocorrelation function $\rho(t)$ has sufficiently smooth sample paths, the average number of zero-crossings per unit time is given by

$$E[D] = \frac{1}{\pi}\sqrt{-\rho''(0)} \tag{2}$$

where D is the number of zero-crossings of $\{Z(t)\}$ for t in the unit interval $[0, 1]$, and $\rho''(0)$ is the second derivative of the normalized autocorrelation of $\{Z(t)\}$ at 0.

We start with Rice's formula (1) and note that since $Z(t)$ and $Z'(t)$ are uncorrelated, hence independent, the joint density factors and the integration is easily performed and gives (2).

In [47] Rice's formula (2) is rigorously proved under mild conditions and it is also shown that the expected number of zero-crossings is finite if and only if the autocorrelation function is twice differentiable at the origin.

The analogous formula for a discrete-time, zero-mean, stationary Gaussian sequence $\{Z_k\}$, $k = 0, \pm 1, \pm 2 \ldots$ has been obtained by many authors (see [21], [21], [47]) and is given by

$$\rho_1 = \cos \frac{\pi E[D_1]}{N-1} \tag{3}$$

or, equivalently, by the inverse form

$$\frac{E[D_1]}{N-1} = \frac{1}{\pi} \cos^{-1} \rho_1$$

where D_1 is the number of sign-changes or zero-crossings in $Z_1, \ldots, Z_N$, $\rho_k = E[Z_{k+j} Z_j]/E[Z_j^2]$ is the correlation sequence of $\{Z_k\}$, and $E[D_1]/(N-1)$ is the expected zero-crossing rate in discrete time. We refer to (3) as the "cosine formula". Observe that, because of stationarity, the expected zero-crossing rate $E[D_1]/(N-1)$ is independent of N. In general $\cos(\pi E[D_1]/N-1)$ need not be a correlation, see [20].

It is worth noting that the expected zero-crossing rate of a non-Gaussian process can in fact be quite different than that of Gaussian process even when they have identical spectral densities. For example, let $\varphi(x)$ be a strictly monotone real-valued-function defined over the real line. Let $\{Z(t)\}$ for $-\infty < t < \infty$ be a zero-mean stationary process with unit variance and autocorrelation $\rho_z(t)$. Define a new process $\{Y(t)\}$ for $-\infty < t < \infty$, with autocorrelation $\rho_y(t)$, as

$$Y(t) = \varphi(Z(t)) - \varphi(0). \tag{4}$$

$\{Y(t)\}$ is not necessarily Gaussian, and its mean need not be 0. Rice's formula yields (see [3]),

$$E[D] = \frac{1}{\pi} \sqrt{\frac{\mathrm{Var}[\varphi(Z(t))]}{E[\varphi'(Z(t))^2]}} \sqrt{-\rho_y''(0)}. \tag{5}$$

To illustrate the use of general formula (5), let $\varphi(x) = x^n$, where n is a fixed positive odd integer, and consider the process

$$Y(t) = \varphi(Z(t)).$$

Observe that, for all t, $Y(t)$ has a symmetric distribution about zero and

$$\mathrm{Var}[Y(t)] = 1 \cdot 3 \cdot 5 \cdots (2n-3)(2n-1).$$

Since

$$\varphi'(Z(t)) = nZ(t)^{n-1}$$

$$E[\varphi'(Z(t))^2] = n^2 \cdot 1 \cdot 3 \cdot 5 \cdots (2n-3),$$

thus, using (5), we obtain the expected zero-crossing rate for $Y(t)$,

$$E[D] = \frac{1}{\pi}\sqrt{\frac{2n-1}{n^2}}\sqrt{-\rho_y''(0)}. \tag{6}$$

From (6) we see that n can be chosen so that in going from $Z(t)$ to $Y(t)$ the increase in the second spectral moment can be made arbitrarily large.

It turns out there are non-Gaussian processes for which the expected zero-crossing rate per unit time is

$$\frac{\kappa}{\pi}\sqrt{-\rho''(0)}$$

with κ1 or $\kappa \geq 1$.

For monotone transformations and mixtures of a Gaussian process κ1, for products $\kappa \geq 1$. Moreover, these examples show that non-Gaussian processes exist which can have quite different zero-crossing rates—arbitrarily larger or smaller—than a Gaussian process with the same spectral density as that of the non-Gaussian process (see [4]).

9.2. Discrete Frequency Estimation via Zero-Crossings

Oscillation as observed in time-series is ubiquitous. Simply by considering a centered pure sinusoid, we see that there are two zero-crossings per cycle. This intimate connection between zero-crossings and frequency content will be the starting point. In this section we will show how zero-crossing counts and higher-order-crossing counts can be an efficient tool for performing discrete frequency estimation (i.e. frequency only, not the power)—competitive in accuracy and speed with the renowned Cooley–Tukey FFT algorithm.

Several techniques have been investigated which use parametric families of linear filters for discrete frequency estimation. The proposed methods are similar in that they use iterative filtering procedures for estimating the frequencies of underlying periodic components embedded in noise. Here we present a technique that combines parametric filtering with a contraction mapping principle to recursively estimate the frequencies of discrete spectral components. By incorporating the contraction mapping idea with parametric filtering a *fundamental property* is determined which when satisfied, guarantees the convergence of the iterative procedure. Several examples are provided which illustrate the method.

Frequency estimation is a classic problem in time series analysis. Aside from the purely mathematical interest of the problem, there are a number of engineering systems that require precise discrete frequency estimation. Communications systems, sonar receivers, and nuclear magnetic resonance spectroscopy devices are such examples.

For almost a hundred years the periodogram has been widely used for spectral estimation and analysis. The fast Fourier transform (FFT), which is an efficient algorithm for evaluating the periodogram at the Fourier frequencies, has helped to sustain the popularity of this important tool. However, over the last decade a number of authors have suggested iterative filtering techniques for discrete frequency estimation (see [14], [17], [19], [24], [28], [43] and [46]). Although there are similarities in the various methods, an important and notable aspect of the He-Kedem work is a so-called *fundamental property* required of the parametric filter family which guarantees convergence of the frequency estimates. As we will show, a number of parametric filter families can be defined which satisfy this property.

A useful mathematical model, as well as the one we use for this example, is the following mixed spectrum stationary process,

$$Z_t = \sum_{j=1}^{p}(A_j \cos(\omega_j t) + B_j \sin(\omega_j t)) + \zeta_t \tag{7}$$

where, $t = 0, \pm 1, \pm 2, \ldots,$ the A's and B's are all uncorrelated, $E(A_j) = E(B_j) = 0$, and $\mathrm{Var}(A_j) = \mathrm{Var}(B_j) = \sigma_j^2$. In general, one assumes $\{\zeta_t\}$ is colored stationary noise with mean 0 and variance σ_ζ^2, independent of the A's and B's. The noise is assumed to possess an absolutely continuous spectral distribution function $F_\zeta(\omega)$ with spectral density $f_\zeta(\omega), \omega \in [-\pi, \pi]$. For our purposes we will assume $\{Z_t\}$ to be Gaussian. However, the Gaussianity assumption is not necessary for the parametric filtering method [46].

Without loss of generality assume that the frequencies are ordered fixed constants,

$$0 < \omega_1 < \omega_2 < \cdots < \omega_p < \pi.$$

The general problem is to estimate the frequencies, $\omega_1, \omega_2, \ldots, \omega_p$, using a finite length observation from the time series, $Z_1, Z_2, \ldots, Z_N$.

In words, our basic strategy is to filter the observations $Z_1, Z_2, \ldots, Z_N$ with a filter from a given parametric family of linear filters, observe a zero-crossing statistic of the filtered output, then select another filter from the family based on this observed statistic. We show, under some conditions, that this iterative procedure converges and accurate frequency estimates may be obtained.

The basic iterative scheme for the case of a single sinusoid in Gaussian white noise is based on the so-called He-Kedem or HK algorithm [17] and uses an autoregressive order 1, $AR(1)$ filter family. Later we provide two other examples of the parametric filters applicable to this method. They are, a moving average order 1, $MA(1)$ family and an autoregressive order 2, $AR(2)$ family. The $AR(2)$ family is a particularly important example which illustrates the idea of contracting the bandwidth of the filter during the iterative procedure. The idea of shifting the center frequency of the filter and simultaneously contracting the bandwidth was first presented in [46].

9.2.1. Higher Order Crossings

Since a linearly filtered Gaussian process results in a Gaussian process, the cosine formula holds for the filtered process where the correlation coefficient and zero-crossing count of the filtered process are used in the cosine formula (3). To be precise, let $\mathcal{L}_\alpha(Z)_t$ be the output at time t of a linear time invariant filter $\mathcal{L}_\alpha$ applied to $\{Z_t\}$. Using the cosine formula (3) and the spectral representation for stationary processes, the first-order correlation coefficient, $\rho_1(\alpha)$, of the filtered process $\{\mathcal{L}_\alpha(Z)_t\}$ is given by,

$$\rho_1(\alpha) = \cos\frac{\pi E[D_\alpha]}{N-1} = \frac{\int_{-\pi}^{\pi} \cos(\omega)\,|\,H(\omega;\alpha)\,|^2 dF_Z(\omega)}{\int_{-\pi}^{\pi} |\,H(\omega;\alpha)\,|^2 dF_Z(\omega)} \tag{8}$$

where, D_α is the zero-crossing count in $\{\mathcal{L}_\alpha(Z)_1, \ldots, \mathcal{L}_\alpha(Z)_N\}$, $F_Z(\omega)$ the spectral distribution function of the process $\{Z_t\}$, and $|\,H(\omega;\alpha)\,|^2$ the squared gain of the filter $\mathcal{L}_\alpha$. The zero-crossings, D_α, of filtered time series are referred to as Higher-Order-Crossings or HOC (see [22]).

For a given zero-mean time series $\{Z_k\}$ and parametric filter family with parameter space Θ, $\{\mathcal{L}_\alpha(\cdot), \alpha \in \Theta\}$, the corresponding HOC family is denoted by $\{D_\alpha, \alpha \in \Theta\}$. The spectral representation above is one of the underpinnings of an alternate theory, developed in [21], for analyzing time-series.

9.3. The He-Kedem Algorithm

The iterative scheme described below illustrates a method for detecting a single frequency in Gaussian noise. Our model is equation (7) with $p = 1$ and $\{\zeta_t\}$

white Gaussian noise. As we will see the algorithm presented next guarantees convergence of a HOC sequence to the frequency ω_1 in our model. The filter family used is the exponential smoothing filter or autoregressive order 1, $AR(1)$ filter which is called an IIR filter in the signal processing literature.

The $AR(1)$ filter known as the (α-filter) is defined by the operation,

$$Z_t(\alpha) = \mathcal{L}_\alpha(Z)_t = Z_t + \alpha Z_{t-1} + \alpha^2 Z_{t-2} + \cdots \quad (9)$$

or equivalently in its recursive form by,

$$Z_t(\alpha) = \alpha Z_{t-1}(\alpha) + Z_t$$

where the squared gain of the filter $|H(\omega; \alpha)|^2$ is given by

$$|H(\omega; \alpha)|^2 = \frac{1}{1 - 2\alpha\cos(\omega) + \alpha^2}, \quad \alpha \in (-1, 1), \quad \omega \in [0, \pi]. \quad (10)$$

Similarly define the output noise at time t by,

$$\zeta_t(\alpha) = \mathcal{L}_\alpha(\zeta)_t$$

and the contraction factor $C(\alpha)$ by,

$$C(\alpha) = \frac{\text{Var}(\zeta_t(\alpha))}{\text{Var}(Z_t(\alpha))}. \quad (11)$$

Then for $\alpha \in (-1, 1)$,

$$0 < C(\alpha) < 1.$$

Clearly $C(\alpha)$ also depends on ω_1, but this is not included to keep the notation simple.

The following theorem from [17] provides the theoretical basis for the parametric filtering and contraction mapping method and reveals the *fundamental property* in its proof.

Theorem 1 [17]

Suppose

$$Z_t = A_1 \cos(\omega_1 t) + B_1 \sin(\omega_1 t) + \zeta_t, \quad t = 0, \pm 1, \ldots$$

where $\omega_1 \in (0, \pi), A_1, B_1$ *are uncorrelated, normal, zero-mean, variance* σ_1^2, *(i.e.* $N(0, \sigma_1^2)$*) random variables, and* $\{\zeta_t\}$ *is Gaussian white noise with mean 0 and variance* σ_ζ^2, *independent of* A_1, B_1. *Let* $\{D_\alpha\}$ *be the HOC from the AR*(1) *filter* (9). *Fix* $\alpha_1 \in (-1, 1)$, *and define*

$$\alpha_{k+1} = \cos\left(\frac{\pi E[D_{\alpha_k}]}{N-1}\right), \quad k = 1, 2, \ldots \tag{12}$$

Then, as $k \to \infty$,

$$\alpha_k \to \cos(\omega_1)$$

and

$$\frac{\pi E[D_{\alpha_k}]}{N-1} \to \omega_1 \tag{13}$$

Proof: Note that the special form (10) gives

$$\int_0^\pi |H(\omega; \alpha)|^2 d\omega = \frac{\pi}{1-\alpha^2}$$

and

$$\int_0^\pi \cos(\omega)\,|H(\omega; \alpha)|^2 d\omega = \frac{\pi}{1-\alpha^2} \times \alpha$$

Therefore, by symmetry, we obtain the *factorization*,

$$\int_{-\pi}^{\pi} \cos(\omega)\,|H(\omega; \alpha)|^2 d\omega = \alpha \times \int_{-\pi}^{\pi} |H(\omega; \alpha)|^2 d\omega \tag{14}$$

and so, from the zero-crossing spectral representation (8), and the cosine formula (3),

$$\rho_1(\alpha) = \cos\left(\frac{\pi E[D_\alpha]}{N-1}\right)$$

we have, with $dF_\zeta(\omega) = (1/2\pi)\sigma_\zeta^2 d\omega$,

$$\rho_1(\alpha) = \frac{\sigma_1^2\,|H(\omega_1; \alpha)|^2 \times \cos(\omega_1) + \int_{-\pi}^{\pi} |H(\omega; \alpha)|^2 dF_\zeta(\omega) \times \alpha}{\sigma_1^2\,|H(\omega_1; \alpha)|^2 + \int_{-\pi}^{\pi} |H(\omega; \alpha)|^2 dF_\zeta(\omega)} \tag{15}$$

or, from the (definition of $C(\alpha)$ in (11),

$$\rho_1(\alpha) = [1 - C(\alpha)] \times \cos(\omega_1) + C(\alpha) \times \alpha \tag{16}$$

We can see that $\rho_1(\alpha)$ is a convex combination of $\cos(\omega_1)$ and α and that it also can be rewritten as a contraction mapping of the form,

$$\rho_1(\alpha) = \alpha^* + C(\alpha)(\alpha - \alpha^*) \tag{17}$$

where $\alpha^* = \cos(\omega_1)$. Invoke the cosine formula, and write the recursion (12) as,

$$\alpha_{k+1} = \rho_1(\alpha_k) \tag{18}$$

Starting with $k = 1$, substitute this in (17), iteratively, to obtain,

$$\rho_1(\alpha_k) = \alpha^* + \left[\prod_{j=1}^{k} C(\alpha_j)\right](\alpha_1 - \alpha^*).$$

As $k \to \infty$, we have that $\prod_{j=1}^{k} C(\alpha_j) \to 0$, and this implies $\alpha_k \to \alpha^*$, and that α^* is a fixed point of $\rho_1(\cdot)$,

$$\alpha^* = \rho_1(\alpha^*)$$

or

$$\cos(\omega_1) = \cos\left(\frac{\pi E[D_{\alpha^*}]}{N-1}\right)$$

By the monotonicity of $\cos(x), x \in [0, \pi]$,

$$\omega_1 = \frac{\pi E[D_{\alpha^*}]}{N-1}.$$

The most important single fact in the preceeding proof is the factorization equation (14) in which the parameter α is factorized outside the integral. This factorization is the basis for extending Theorem 1.1 as was done in [24] and [46]. The fact that the parameter is "kicked out" in (14) is somewhat more apparent if we rewrite (14) as

$$\alpha = \rho_{1,\zeta}(\alpha) = \frac{\int_{-\pi}^{\pi} \cos(\omega)\,|\,H(\omega;\alpha)\,|^2 d\omega}{\int_{-\pi}^{\pi} |\,H(\omega;\alpha)\,|^2 d\omega} \tag{19}$$

where $\rho_{1,\zeta}(\alpha)$ is the first-order autocorrelation of the filtered noise. The property (19) is what we call the *fundamental property* relative to a given family of filters. Thus, the $AR(1)$ parametric filter possesses the fundamental property relative to white noise. This together with the correlation representation (16) lead to the contraction mapping (17), and eventually to the convergent HOC sequences α_k, and $\pi E[D_{\alpha_k}]/N - 1$. Fortunately, as we shall see in the next section, factorizations of the form (14) are readily available. In actual practice, the observed or empirical zero-crossing rate is used in place of $E[D_{\alpha_k}]$ at each stage in the iteration and the noise process need not be white—it simply needs to possess a sufficiently continuous spectrum. Computer simulation results using the α-filter for a single sinusoid in white Gaussian noise are given in Table 1.

Table 1 Illustration of the HK algorithm using the $AR(1)$ filter family, convergence based on the observed zero-crossing count, $(\alpha_{k+1}) = \pi\hat{D}_{\alpha_k}/(N-1)$, $k \to \infty$, towards $\omega_1 = 0.8$ as a function of SNR. $N = 10{,}000$.

k	1 dB $\alpha_1 = -0.1$	0 dB $\alpha_1 = 0.9$	-1.94 dB $\alpha_1 = 0.2$	-6.02 dB $\alpha_1 = 0.5$
1	0.8848	0.5194	0.9127	0.9291
2	0.8006	0.5904	0.8222	0.8713
3	0.7987	0.6563	0.8015	0.8411
4	0.7987	0.7142	0.7965	0.8191
5	9.7987	0.7600	0.7952	0.8053
6	0.7987	0.7864	0.7952	0.8015
7	0.7987	0.8002	0.7952	0.7990
8	0.7987	0.8065	0.7952	0.7984
9	0.7987	0.8065	0.7952	0.7977
10	0.7987	0.8065	0.7952	0.7971
11	0.7987	0.8065	0.7952	0.7971
12	0.7987	0.8065	0.7952	0.7971
.	.	.	.	.
.	.	.	.	.
.	.	.	.	.

9.3.1. Examples of Parametric Families and Contraction Mappings

Next we give two examples of parametric filter families which satisfy the *fundamental property* (19). They are, the moving average order one, $MA(1)$ (also known as a one tap FIR filter) family, which is similar to the α-filter family, and an $AR(2)$ filter family. With the $AR(2)$ filters it is possible to simultaneously shift the center frequency of the filter and contract the bandwidth. This allows for a faster rate of convergence of the HOC sequence and greater accuracy in comparison with the α-filter family. General conditions relating filter bandwidth contraction rate and the convergence rate of the HOC sequences may be found in [24] and [43].

9.3.1.1. An Example Using a MA(1) Filter

Again let our model be as in Theorem 1 with $\{Z_t\}$, a zero-mean stationary Gaussian time series defined by,

$$Z_t = A_1 \cos(\omega_1 t) + B_1 \sin(\omega_1 t) + \zeta_t, \quad t = 0, \pm 1, \ldots \tag{20}$$

where we restrict $\omega_1 \in (\pi/3, 2\pi/3)$ for convenience.

Consider the family $\{\mathcal{L}_r\}$ of moving average order one, $MA(1)$, filters indexed by parameter r, $r \in (-1, 1)$ and defined by,

$$Z_t(r) = \mathcal{L}_r(Z)_t = Z_t + rZ_{t-1}, \tag{21}$$

and whose squared gain $|H(\omega; r)|^2$ is,

$$|H(\omega; r)|^2 = 1 + 2r\cos(\omega) + r^2, \quad r \in (-1, 1), \quad \omega \in [0, \pi]. \tag{22}$$

This family consists of simple finite impulse response filters which exhibit lowpass characteristics for values of the parameter r which are positive and exhibits high-pass characteristics for values which are negative.

The *fundamental property* would require,

$$\int_{-\pi}^{\pi} \cos(\omega)\,|H(\omega; r)|^2 d\omega = r \times \int_{-\pi}^{\pi} |H(\omega; r)|^2 d\omega, \tag{23}$$

since we assume the noise to be white. However, evaluating the particular integrals yields,

$$\frac{\int_{-\pi}^{\pi} \cos(\omega)\,|H(\omega; r)|^2 d\omega}{\int_{-\pi}^{\pi} |H(\omega; r)|^2 d\omega} = \frac{r}{1 + r^2}. \tag{24}$$

Thus, we need to reparameterize. To obtain a reparameterization which will satisfy the *fundamental property*, set

$$\beta = \frac{r}{1 + r^2},$$

and solve for r in terms of β. This gives,

$$r = \frac{1 - \sqrt{1 - 4\beta^2}}{2\beta}.$$

Thus, the fundamental property is satisfied by the family reparameterized by β. Note that $\beta \in (-\frac{1}{2}, \frac{1}{2})$, hence the reason for restricting $\omega_1 \in (\pi/3, 2\pi/3)$.

9.3.1.2. An AR(2) Filter Family

The next example illustrates how to enhance the HK algorithm by selecting a parametric family that allows for adjustable narrow bandwidth filters.

Our model again will be $\{Z_t\}$ as in (20) with $\omega_1 \in (0, \pi)$. Consider the family $\{\mathcal{L}_{(\beta,\gamma)}\}$ of autoregressive order 2, $AR(2)$ filters indexed by the 2-vector parameter (β, γ) and defined by,

$$Z_t(\beta, \gamma) = \mathcal{L}_{(\beta,\gamma)}(Z)_t = \beta Z_{t-1}(\beta, \gamma) + \gamma Z_{t-2}(\beta, \gamma) + Z_t. \tag{25}$$

The squared gain of the $AR(2)$ filter, $|H(\omega; (\beta, \gamma))|^2$, is given by

$$|H(\omega; (\beta, \gamma))|^2 = \frac{1}{1 + \beta^2 + \gamma^2 + 2\beta(\gamma - 1)\cos(\omega) - 2\gamma\cos(2\omega)}, \quad \omega \in [0, \pi]. \tag{26}$$

Evaluating time integrals yields,

$$\frac{\int_{-\pi}^{\pi} \cos(\omega)\,|H(\omega; (,\gamma))|^2 d\omega}{\int_{-\pi}^{\pi} |H(\omega; (\beta, \gamma))|^2 d\omega} = \frac{\beta}{1 - \gamma} \tag{27}$$

It is seen from (27) that we need to reparameterize the filter family in order to satisfy (15).

Before reparameterizing the filter family, note that these filters have poles at

$$\frac{\beta \pm \sqrt{\beta^2 + 4\gamma}}{2},$$

which are inside the unit circle for values of the parameters given by,

$$-1 < \gamma < 0 \quad \text{and} \quad |\beta| < \frac{4\gamma}{\gamma - 1} \tag{28}$$

and approach the unit circle as $\gamma \to -1$. That is the poles go to $\exp(\pm i\theta)$ as $\gamma \to -1$ and $\theta \to \cos^{-1}(\beta/2)$. Thus, we will restrict our parameter space according to (28) to guarantee stable filters.

If γ is fixed to some $\gamma_0 \in (-1, 0)$, then $\delta = \beta/(1 - \gamma_0)$ gives a parameterization (in terms of $\beta/(1 - \gamma_0)$, with only β free) satisfying the *fundamental property*. Furthermore, by allowing γ to vary in a prescribed way with each iteration (i.e. let γ approach -1 during the iteration process for narrower bandwidths), the filters can be made to simultaneously shift there center frequency and contract the bandwidth. Thus, for suitably chosen sequences $\{\gamma_k\}$, with $\gamma_k \to -1$, we can also satisfy *fundamental property* with the added bonus of accelerated convergence of the HOC sequence and greater accuracy of the estimates (see [23], [43] and [46]). Simulation results for a single sinusoid in white Gaussian noise using the $AR(2)$ filter with $\gamma = -0.9$ are given in Table 2.

Table 2 Illustration of the HK algorithm using the $AR(2)$ filter family with $\gamma = -0.9$, convergence based on the observed zero-crossing count, $(\alpha_{k+1}) = \pi\hat{D}_{\alpha_k}/(N-1)$, $k \to \infty$, towards $\omega_1 = 0.8$ as a function of SNR. $N = 2{,}000$.

k	0 dB $\alpha + 1 = .09$	0 dB $\alpha_1 = -0.5$	$-$ 6.02 dB $\alpha_1 = 0.9$	$-$ 6.02 dB $\alpha_1 = 0.2$
1	0.5613	2.0377	0.4685	1.3349
2	0.6839	2.0000	0.4890	1.3050
3	0.7987	1.9717	0.5141	1.2830
4	0.8003	1.9198	0.5424	1.2547
5	0.8003	1.8679	0.5833	1.2390
6	0.8003	1.8050	0.6336	1.1918
7	0.8003	1.7689	0.6965	1.1516
8	0.8003	1.7233	0.7704	1.1242
9	0.8003	1.6651	0.7987	1.0676
10	0.8003	1.6274	0.8034	0.9119
11	0.8003	1.5645	0.8034	0.8302
12	0.8003	1.5016	0.8034	0.8019
.	.	.	.	.
.	.	.	.	.
.	.	.	.	.
20	0.8003	0.8003	0.8019	0.8019

9.4. Radar Detection via Level-Crossings of the Envelope Process

A radar system generally transmits a waveform which is both amplitude and phase modulated in a deterministic fashion. The transmitted signal, $S_T(t)$, is given by $S_T(t) = A(t)\cos[\omega_c t + \theta(t)]$, where the amplitude, $A(t)$, and the phase, $\theta(t)$, are known deterministic functions. The carrier frequency of the radar transmitter, ω_c, is a known constant. For a simple radar transmitter the amplitude and phase functions, $A(t)$ and $\theta(t)$, are slowly varying relative to the carrier frequency ω_c. This condition will indeed be met if ω_c is much greater than the largest frequency components in the spectra of $A(t)$ and $\theta(t)$. For this case, as we shall see, it is reasonable to identify $A(t)$ as the "envelope" of the signal $S_T(t)$.

When the transmitted signal backscatters or reflects off a source (i.e., target), the received signal, $S_R(t)$, is a randomly attenuated and phase distorted version of $S_T(t)$. The phase and amplitude modulation of $S_R(t)$ is, in part, due to radiation propagation effects and source kinematics which can modulate the radar cross-section of the target. Now if the radar receivers' noise characteristics are modeled by a sufficiently regular, ergodic, stationary process, then filtering within an ideal narrow bandpass filter, centered on the carrier frequency, ω_c, essentially converts the receiver noise to a narrow-band Gaussian process [39], [12]. The intuition here is as follows:

A wide-band process will necessarily decorrelate "fast" (i.e. have a short decorrelation time). A narrow-band filter has long memory and allows for averaging samples of the input over a long period. Thus, for a wide-band input to a narrow-band filter, the output will contain a component which is a long period averaging of essentially uncorrelated samples. With constraints on the filter weights it should not be unexpected that a central limit theorem holds. Consequently, a reasonable mathematical model for $S_R(t)$ is a narrow-band Gaussian process. That is, $S_R(t) = R(t)\cos[\omega_c t + \phi(t)]$ where $R(t)$ and $\phi(t)$ are jointly stationary random processes with Rayleigh and Uniform marginal densities respectively.

In short, by first conditioning the radar receivers output by pre-filtering with an ideal narrow-band filter, centered on ω_c, we convert the receiver noise to a narrow-band Gaussian noise process, say $N(t)$. As long as the spectrum of the received signal, $S_R(t)$, is contained in the passband of the prefilter, we preserve $S_R(t)$ as well.

The detection problem can now be stated: Determine if a narrow-band Gaussian signal, $S_R(t)$, is present or not in the receiver output $Y(t)$. This detector may be handled as a decision problem, that is, as a hypothesis testing procedure:

H_0: no signal present, noise only $[Y(t) = N(t)]$
H_1: signal plus noise $[Y(t) = S_R(t) + N(t)]$.

If we assume the received signal and noise are statistically independent and jointly Gaussian, the Neyman-Pearson likelihood ratio test is optimal. Here the optimality criterion is maximum power for a fixed size test. Stated in signal processing vernacular, maximum probability of detection for a fixed false alarm rate.

This detection problem and variations of it are known collectively as "incoherent detection" or "partially coherent detection" processing. Optimum detector structures have been derived and investigated by many authors. For a comprehensive and thorough discussion see [45], pp. 333–366.

As an alternative to the optimal procedure for detecting a narrow-band Gaussian signal in narrow-band Gaussian noise, we consider a detector based on level-crossing counts. We detail an approach first proposed by [32]. His procedure for detecting weak narrow-band signals in narrow-band Gaussian noise uses the sample mean level-crossing counts of the "envelope" of the receiver output as a test statistic for detection processing. This approach, though not optimum, can be less computationally complex than the optimum detector, with apparently little penalty paid in terms of probability of detection performance. In subsequent sections we provide the details of Rainal's detector and formally verify the performance of his detector, he observed, via computer experiments. One assumption made by Rainal, but not rigorously proved, is that the mean level-crossings of the envelope of a Gaussian process are asymptotically normal. For Gaussian processes the level-crossing counts are asymptotically normal (see [11], [26], [40]). Later, we prove asymptotic normality of the level-crossing count of the envelope of a bandpass Gaussian process and provide an integral expression for the variance of the envelope level-crossing counts, which can be numerically evaluated.

In the next section we define the envelope of a stationary random process and detail some of its properties used in subsequent sections. We shall make frequent use of Hilbert transforms of both deterministic and random functions in the sequel and next provide the definition for completeness.

The Hilbert transform of a real-valued function (non-random), $g(t)$, is defined as (see [42], pp. 119–151),

$$\hat{g}(t) = -\frac{1}{\pi}\int_{0^+}^{\infty} \frac{g(t+s) - g(t-s)}{s}\, ds \tag{29}$$

or, equivalently, as a Cauchy Principal Value integral (PV) at $s = t$.

$$\hat{g}(t) = \frac{1}{\pi}\mathrm{PV}\int_{-\infty}^{\infty} \frac{g(s)}{t-s}\, ds. \tag{30}$$

We next define the envelope of a real-valued function.

The "envelope" of the function, $g(t)$, which we denote by, $A_g(t)$, is defined as ([8], pg. 487),

$$A_g(t) = [g^2(t) + \hat{g}^2(t)]^{1/2}. \tag{31}$$

If $g(t)$ is a narrow-band function, the above definition conforms to our intuitive understanding or notion of what the envelope of the function $g(t)$ should be. For example, let $g(t) = A(t)\cos\omega_0 t$. Using the *Modulation* property of the Hilbert transform, $\hat{g}(t) = A(t)\sin\omega_0 t$, provided $\omega_0 > 0$ is outside the support of the spectrum of $A(t)$. Thus, our intuition is verified in this case by the fact that the envelope of $g(t)$ is $A(t)$. The following table of transform pairs and envelope functions further illustrates our intuitive notion of the envelope:

$g(t)$	$\hat{g}(t)$	$[g^2(t) + \hat{g}^2(t)]^{1/2}$
$A\cos(\omega t)$	$A\sin(\omega t)$	$\lvert A \rvert$
$A\sin(\omega t)$	$-A\cos(\omega t)$	$\lvert A \rvert$
$\dfrac{\sin(t)}{t}$	$\dfrac{1-\cos(t)}{t}$	$\left\lvert \dfrac{\sin(t/2)}{t/2} \right\rvert$
$\dfrac{1}{1+t^2}$	$\dfrac{t}{1+t^2}$	$\dfrac{1}{\sqrt{1+t^2}}$
$\sin(\omega t)J_n(\omega_1 t)$	$\cos(\omega t)J_n(\omega_1 t)$	$\lvert J_n(\omega_1 t) \rvert$

where $J_n(\cdot)$ is the Bessel function of order $n = 0, 1, 2, \ldots$ and $0 < \omega < \omega_1$.

The Hilbert transform of a stationary random process is defined as follows. Let $\{X(t)\}$ be a zero-mean weakly stationary process. Using the spectral representation for real-valued processes ([10], pg. 137) one can write $X(t)$,

$$X(t) = \int_{0^+}^{\infty} \cos(\omega t)\xi_1(d\omega) \int_{0^+}^{\infty} \sin(\omega t)\xi_2(d\omega) \tag{32}$$

where $\xi_1(d\omega)$ and $\xi_2(d\omega)$ are orthogonal random measures, $\xi_1(d\omega)$ is even and $\xi_2(d\omega)$ is odd. We assume the spectral distribution function, $F_X(\omega)$, of the process $\{X(t)\}$, is continuous and normalized so that

$$\int_0^{\infty} dG_X(\omega) = 1$$

where $G_X(\omega) = 2F_X(\omega)$. It then follows from (32) that the autocorrelation function, $\rho_X(\tau)$, of $\{X(t)\}$ is,

$$\rho_X(\tau) = \int_0^{\infty} \cos(\omega\tau) dG_X(\omega).$$

The "Hilbert Transform", $\hat{X}(t)$, of $X(t)$ can be defined as ([10], pg. 142),

$$\hat{X}(t) = \int_{0^+}^{\infty} \sin(\omega t)\xi_1(d\omega) - \int_{0^+}^{\infty} \cos(\omega t)\xi_2(d\omega). \tag{33}$$

From definition (33) it follows that the Hilbert transform (defined for stationary processes) is a linear operator or filter which maps stationary processes to stationary processes. The Hilbert transform can also be defined, equivalently, in the frequency domain, by its transfer function, $H(\omega)$ ([10], pp. 141–142),

$$H(\omega) = \begin{cases} -j & \omega > 0 \\ 0 & \omega = 0 \\ j & \omega < 0 \end{cases} \tag{34}$$

The pre-envelope of the process $\{X(t)\}$, also known as the analytic signal, is *defined* as the complex random process, $\{W(t)\}$, with real part $X(t)$ and imaginary part $\hat{X}(t)$,

$$W(t) = X(t) + j\hat{X}(t).$$

The envelope of the stationary process $\{X(t)\}$, which we denote by $R(t)$, is *defined* analogously as in the non-random case. The envelope process is given by,

$$R(t) = [X^2(t) \pm \hat{X}^2(t)]^{1/2} = |\, W(t) \,|.$$

It is worth noting that the above definition for the envelope process, which appears to be different than that given by Rice ([36], pp. 81–82) is in fact the same. Rice shows that the underlying Gaussian process, which he writes as $I(t)$, in our notation $X(t)$, can in fact be written as

$$I(t) = I_c(t) \cos \omega_c t - I_s(t) \sin \omega_c t = X(t) \tag{35}$$

where $I_c(t)$ and $I_s(t)$ are the so-called in-phase and quadrature components.

To see this, make the following change of variables for any $Y(t)$

$$I_c(t) = X(t) \cos \omega_c t + Y(t) \sin \omega_c t \tag{36}$$

$$I_s(t) = Y(t) \cos \omega_c t - X(t) \sin \omega_c t \tag{37}$$

Then (35) holds. If $Y = \hat{X}$, then $I_c(t)$ and $I_s(t)$ are uncorrelated. Rice then defines time envelope by,

$$R(t) = [I_c^2(t) + I_s^2(t)]^{1/2}.$$

However, upon using the Hilbert transform we obtain

$$R(t) = [I_c^2(t) + I_s^2(t)]^{1/2} = [X^2(t) + \hat{X}^2(t)]^{1/2},$$

and this is yet another justification for the use of $\hat{X}$ in defining the envelope. For an interesting survey paper on different definitions one may use for defining an envelope of narrow-band signals, see [38].

The Hilbert transform, $\{\hat{X}(t)\}$, is obtained via a linear operation on $\{X(t)\}$, so we have immediately, that, if $\{X(t)\}$ is a stationary Gaussian process, so is $\{\hat{X}(t)\}$. Generally, if $\{X(t)\}$ is zero-mean and has a continuous spectral distribution function, then $\{\hat{X}(t)\}$ is zero-mean as well, and moreover, since ξ_1 and ξ_2 are uncorrelated, has exactly the same spectrum and autocorrelation function as $\{X(t)\}$ (Cramér and Leadbetter (1967), pg. 142). So, in particular,

$$\rho_{\hat{X}}(\tau) = \rho_X(\tau).$$

We assume in the sequel that all spectral distribution functions considered are continuous unless noted otherwise.

The cross-correlation function of $\{X(t)\}$ and $\{\hat{X}(t)\}$, $\rho^*(\tau)$, is also of interest and is given by ([10], pg. 142)

$$\rho^*(\tau) = E[X(t)\hat{X}(t+\tau)] = \int_0^\infty \sin(\omega\tau)dG_X(\omega).$$

The above integral expression for the cross-correlation is in fact just the Hilbert transform of the autocorrelation function, $\rho_X(\tau)$, ([48], pg. 556, eq. 3) so that,

$$\rho^*(\tau) = \hat{\rho}_X(\tau) = \mathcal{H}[\rho_X(\tau)]. \tag{38}$$

Since the Hilbert transform of an even-function is an odd-function, $\rho^*(\tau)$ is odd and in particular $\rho^*(0) = 0$. Thus, if $\{X(t)\}$ is Gaussian, the pair of random variables, $(X(t), \hat{X}(t))$, are independent for all t.

We next consider the envelope and the squared envelope of a Gaussian process.

9.4.1. The Envelope of a Gaussian Process

We first detail the derivation originally given by [10], pp. 248–255, which is based on the work of [36], pp. 81–84, for the joint density of the envelope, $\{R(t)\}$, and its mean-square derivative, $\{R'(t)\}$. We will assume the underlying process $\{X(t)\}$ is Gaussian and for convenience mean-zero with variance one.

The main results used later are: the marginal distributions of the envelope, $R(t)$, and its mean-square derivative, $R'(t)$, are respectively, Rayleigh and Gaussian. For each t, $R(t)$ and $R'(t)$ are independent, and hence by Rice's formula

$$\int_{-\infty}^{\infty} |\dot{r}| p_{R,R'}(u, \dot{r})d\dot{r}$$

the expected u ($u > 0$) level-crossing rate per unit time of $R(t)$ is,

$$ED_u = \left(\frac{2\Delta}{\pi}\right)^{1/2} ue^{-(u^2/2)}$$

where Δ is the variance of $R'(t)$ and $D_u = N_u[0, 1]$ is the number of u level-crossings in the unit interval. Lastly, the variance of the number of level-crossings per unit time is given by the

$$\mathrm{Var}(D_u) = E[D_u] - (E[D_u])^2 + 2\int_{0^+}^{1} (1-\tau)\psi(\tau)d\tau$$

where

$$\psi(\tau) = \int_{-\infty}^{\infty} \int_{-\infty}^{\infty} |\dot{r}_1 \dot{r}_2| p_{R_1,R_1',R_2,R_2'}(u, \dot{r}_1, u, \dot{r}_2) d\dot{r}_1 d\dot{r}_2$$

and $p_{R_1,R_1',R_2,R_2'}$ is the joint density of $(R(0), R'(0), R(\tau), R'(\tau))$. The formula above is sometimes called the Bendat-Rice formula ([7], pg. 396, eq. 10—121) but has been proved by many authors.

9.4.2. The Joint Density of $R(t)$ and $R'(t)$

The joint density of $R(t)$ and its mean-square derivative, $R'(t)$, can be obtained as the limit as $\tau \to 0$ of the joint density of $R(t)$ and $1/\tau[R(t+\tau) - R(t)]$. We start with the derivation of the joint density of $R(t)$ and $R(t+\tau)$. As noted above, this derivation is essentially that given by [36], pp. 81–84.

Consider the jointly normal random variables $X(t), \hat{X}(t), X(t+\tau), \hat{X}(t+\tau)$. Again for convenience we assume mean-zero and variance one. The covariance matrix is obtained using (38) and is

$$\begin{bmatrix} 1 & 0 & \rho & \rho^* \\ 0 & 1 & -\rho^* & \rho \\ \rho & -\rho^* & 1 & 0 \\ \rho^* & \rho & 0 & 1 \end{bmatrix}$$

where $\rho = \rho_X(\tau)$ and $\rho^* = \rho^*(\tau)$. The inverse of the above covariance matrix is easily obtained as

$$A^{-1} \begin{bmatrix} 1 & 0 & -\rho & -\rho^* \\ 0 & 1 & \rho^* & -\rho \\ -\rho & \rho^* & 1 & 0 \\ -\rho^* & -\rho & 0 & 1 \end{bmatrix}$$

where $A = 1 - \rho^2 - \rho^{*2}$. Hence, the joint density of $X(t)$, $\hat{X}(t)$, $X(t+\tau)$, $\hat{X}(t+\tau)$, which we denote by, $f_{X,\tau}(x_1, x_2, x_3, x_4)$, is

$$\frac{1}{4\pi^2 A} \exp\left\{ -\frac{1}{2A} [(x_1^2 + x_2^2 + x_3^2 + x_4^2) - 2\rho(x_1 x_3 + x_2 x_4) - 2\rho^*(x_1 x_4 - x_2 x_3)] \right\}. \tag{39}$$

By changing variables,

$$x_1 = R_1 \cos\theta_1 \quad x_2 = R_1 \sin\theta_1$$

$$x_3 = R_2 \cos\theta_2 \quad x_4 = R_2 \sin\theta_2$$

and integrating over θ_1 and θ_2, the joint density of $R(t)$ and $R(t+\tau)$ is obtained (with some further coordinate transformations) as

$$\frac{R_1 R_2}{\pi A} e^{-(R_1^2+R_2^2)/2A} \int_0^{\pi} \cosh\left\{ \left[\frac{R_1 R_2 (\rho^2 + \rho^{*2})^{1/2}}{A} \right] \cos\phi \right\} d\phi. \tag{40}$$

The integral in (40) can be evaluated in terms of the zero-order modified Bessel function of the first kind, $I_0(z)$, ([2], 1972, pp. 374–378) and so finally,

$$p_{R(0),R(\tau)}(R_1, R_2) = \frac{R_1 R_2}{\pi A} e^{-(R_1^2+R_2^2)/2A} I_0 \left\{ \frac{R_1 R_2}{A} (\rho^2 + \rho^{*2})^{1/2} \right\}. \tag{41}$$

With (41) the joint distribution of $R(t)$ and $R'(t)$ is obtained and given by

$$p_{R(t),R'(t)}(R, R') = (2\pi\Delta)^{-1/2} \exp(-R'^{2/2\Delta}) R \exp^{-R^2/2} \tag{42}$$

Since the univariate density of $\{R(t)\}$ is Rayleigh and $\{R'(t)\}$ is Gaussian, mean zero, variance Δ, we see that $R(t)$ and $R'(t)$ are independent for each t.

Using (41) the covariance function of $\{R(t)\}$ can be calculated, with some work (the two-fold integration is a bit involved), and is

$$E[R(t)R(t+\tau)] = \frac{\pi}{2} {}_2F_1\left(-\frac{1}{2}, -\frac{1}{2}; 1; k_0^2\right), \tag{43}$$

where, $k_0^2 = \rho^2 + \rho^{*2}1$ and ${}_2F_1(\alpha, \beta; \gamma; x)$ is the Gaussian hypergeometric function (see [30], pp. 1076–1077) which is represented by the series

$${}_2F_1(\alpha, \beta; \gamma; x) = 1 + \frac{\alpha\beta}{\gamma}\frac{x}{1!} + \frac{\alpha(\alpha+1)\beta(\beta+1)}{\gamma(\gamma+1)2!} x^2 + \cdots |x| < 1. \tag{44}$$

Equation (43), giving the covariance in terms of the hypergeometric function, ${}_2F_1$, was originally given by [44] but is hinted at by [36], pg. 84, eq. 3.7-13 as well. For our special case, where $\alpha = \beta = -\frac{1}{2}$ and $\gamma = 1$, the covariance is given by a power series in k_0^2 (which depends on τ),

$$E[R(t)R(t+\tau)] = \frac{\pi}{2}\left(1 + \frac{k_0^2}{4} + \frac{k_0^4}{64} + \cdots\right). \tag{45}$$

The power series clearly converges for all $k_0^2 1$.

It is interesting to note by (45), that if we take the Fourier transform of the autocovariance of the envelope, $E[R(t)R(t+\tau)] - E[R(t)]^2$, noting $E[R(t)]^2 = \pi/2$, to obtain the power spectrum we see that, in general, the envelope process may not be bandlimited. To see this assume that the spectrum of $\{X(t)\}$ has a continuous component. The power spectrum of the envelope, $P_R(\omega)$, is given by the termwise Fourier transform of

$$\frac{\pi}{2}\left(\frac{k_0^2}{4} + \frac{k_0^4}{64} + \cdots\right)$$

or

$$P_R(\omega) = \frac{\pi}{2}\left(\frac{\tilde{k}_0 * \tilde{k}_0}{4} + \frac{\tilde{k}_0 * \tilde{k}_0 * \tilde{k}_0 * \tilde{k}_0}{64} + \cdots\right). \tag{46}$$

Assume the spectrum of $\{X(t)\}$ is bandlimited. Then, by considering each successive term in time series for $P_R(\omega)$, we see that each higher-order convolution of $\tilde{\rho}(\omega) = f_X(\omega)$ with itself effectively doubles the bandwidth, thus guaranteeing that the support of the spectrum is unbounded.

An example, and one we will use in the sequel, is provided by the envelope of an ideal bandpass process. Let $\{X(t)\}$ be a Gaussian process with a spectral density function which is constant over the frequency intervals $(-\omega_c - \delta, -\omega_c + \delta)$ and $(\omega_c - \delta, \omega_c + \delta)$ and zero elsewhere. The center frequency is said to be ω_c and the bandwidth is 2δ. Assume the total power of this ideal bandpass process is unity. Then, the antocorrelation function is given by

$$\rho(\tau) = \frac{\sin \delta\tau}{\delta\tau} \cos \omega_c \tau.$$

Using the Hilbert transform of the above equation, we see

$$\hat{\rho}(\tau) = \frac{\sin \delta\tau}{\delta\tau} \sin \omega_c \tau$$

and so k_0^2 is given by

$$k_0^2 \equiv k_0^2(\tau) = \rho^2(\tau) + \hat{\rho}^2(\tau) = \left(\frac{\sin \delta\tau}{\delta\tau}\right)^2.$$

Note that in this example the spectral support of $k_0^2(\tau)$ is $[-2\delta, 2\delta]$. This is in fact the spectral support of the squared envelope of the ideal bandpass process.

9.4.3. The Squared Envelope Process

For the purpose of counting mean level-crossings of the envelope process we can use the squared envelope process instead. That is, the u level-crossings of $R(t)$ are, of course, the u^2 level-crossings of $R^2(t)$.

As we saw in the last section the antocorrelation function and spectral density of the envelope process $\{R(t)\}$ are given by infinite series expansions. This is in contrast to time squared envelope process, $\{R^2(t)\}$, whose autocorrelation and spectral density are given by simpler looking expressions which are easily obtained in terms of $\rho_X(\tau)$ and $f_X(\omega)$. We will see that the autocorrelation is,

$$\rho_{R^2}(\tau) = \rho_X^2(\tau) + \hat{\rho}_X^2(\tau) \tag{47}$$

and thus, by the convolution theorem, the spectral density is

$$f_{R^2}(\omega) = f_X(\omega) * f_X(\omega) + \tilde{f}_X(\omega) * \tilde{f}_X(\omega) \tag{48}$$

where $\tilde{f}_X(\omega)$ is the Fourier transform of $\hat{\rho}_X(\tau)$.

The autocorrelation of the squared envelope, $\{R^2(t)\}$, can be computed directly and is given by

$$\rho_{R^2}(\tau) = \frac{E[R^2(t+\tau)R^2(t)] - E[R^2(t+\tau)]E[R^2(t)]}{\sigma_{R^2}^2} \tag{49}$$

$$= \rho_X^2(\tau) + \rho_{\hat{X}}^2(\tau) \tag{50}$$

If the underlying Gaussian process is an ideal bandpass process, we see from (50) that

$$\rho_{R^2}(\tau) = \rho_X^2(\tau) + \rho_{\hat{X}}^2(\tau) = \left(\frac{\sin \delta\tau}{\delta\tau}\right)^2 \tag{51}$$

and hence the spectral density is

$$f_{R^2}(\omega) = \begin{cases} \dfrac{1}{4\delta^2}\,\omega + \dfrac{1}{2\delta} & -2\delta < \omega < 0 \\ \dfrac{1}{2\delta} - \dfrac{1}{4\delta^2}\,\omega & 0 < \omega < 2\delta \\ 0 & \text{else} \end{cases} \tag{52}$$

Using (46) it is interesting to observe that for this particular example, when $\{X(t)\}$ is an ideal bandpass process, the power spectrum of $\{R^2(t)\}$ is bandlimited while the spectral support of $\{R(t)\}$ is the whole real line.

9.5. Level-Crossing Based Detector

As an alternative to the optimal procedure for detecting a narrow-band Gaussian signal in narrow-band Gaussian noise we consider a detector based on level-crossing counts of the envelope of the observed process.

Following Rainal's procedure described in [32], the envelope of the received signal is obtained and the mean level-crossing counts of the envelope are used as a test statistic for detection processing, that is, to determine whether a narrow-band Gaussian signal is present or not in the narrow-band Gaussian noise.

Rainal assumes, without proof, that the level-crossing counts are asymptotically normal (for observations over a large time interval), and consequently, a test of a given significance level is then determined by time variance (asymptotic) of the mean level-crossing count. Hence, the asymptotic variance of the crossing counts is needed to set an appropriate threshold for a fixed probability of false alarm. This is the usual Neyman-Pearson criterion. Rainal shows experimentally that his less complex level-crossing based detector is competitive with the near-optimal quadratic detector, also known as the square-law detector. Optimal linear-quadratic detectors for Gaussian systems are well known. For a general discussion, including both Gaussian and non-Gaussian systems, see [31].

In the next two sections we derive an expression for the variance of the u-level-crossings ($u > 0$) of the envelope process and then prove asymptotic normality of the crossing counts. We start with the general formula for the variance of the crossings.

9.5.1. Variance of the Level-Crossing Count

The formula for the variance of the number of zero-crossings in an interval, $[0, T]$, has its roots in the work of Rice [36]. This formula has been investigated by many authors over the last 40 years with emphasis on necessary and sufficient conditions for a finite second moment.

One of the earliest papers, if not the first, that deals with the variance of the number of zero-crossings is [41]. In their paper an explicit formula (equation 40) is given for the mean-square number of zeros in the interval $[0, T]$ of a Gaussian process. The formula for the variance includes expressions which depend on the autocorrelation function and its first two derivatives.

Let $D = N[0, T]$ be the number of zero-crossings of $\{X(t)\}$ in the interval $[0, T]$. Assume $\{X(t)\}$ is a zero-mean sufficiently smooth stationary process. Then,

$$\mathrm{Var}(D) = E[D] - (E[D])^2 + 2\int_{0+}^{T} (T - \tau)\psi(\tau)d\tau \tag{53}$$

where,

$$\psi(\tau) = \int_{-\infty}^{\infty}\int_{-\infty}^{\infty} |\dot{x}_1\dot{x}_2| p_{X_1X_1',X_2,X_2'}(0, \dot{x}_1, 0, \dot{x}_2)d\dot{x}_1 d\dot{x}_2 \tag{54}$$

and $p_{X_1,X_1',X_2,X_2'}$ is the joint density of $(X(0), X'(0), X(\tau), X'(\tau))$. To obtain the variance of the u-level-crossings of $\{X(t)\}$, for any u, simply replace $p_{X_1,X_1',X_2X_2'}(0, \dot{x}_1, 0, \dot{x}_2)$ by $p_{X_1,X_1',X_2,X_2'}(u, \dot{x}_1, u, \dot{x}_2)$.

The above formula, (53), is in fact a general formula, which is applicable to a wide class of random processes, and may be adapted to non-stationary processes as well. A thorough discussion for the general case including rigorous mathematical formulation and proof is in [10], pp. 202–212. A general treatment of higher order product moments of level-crossing counts for processes with absolutely continuous sample paths can be found in [27].

When $\{X(t)\}$ is a mean zero, unit variance, Gaussian process with autocorrelation function $\rho(\tau)$, (53) becomes ([7], pp. 398–401)

$$\mathrm{Var}(D) = E[D] - (E[D])^2 + \frac{2}{\pi^2}\int_{0+}^{T} (T - \tau)\frac{1 + g(\tau)\arctan g(\tau)}{1 - \rho^2(\tau)}\sqrt{h(\tau)}d\tau, \tag{55}$$

where

$$h(\tau) = [1 - \rho^2(\tau)][\rho''(0) - \rho''(\tau)] + 2[\rho''(0) - \rho''(\tau)\rho(\tau)]\rho'^2(\tau) + \rho'^4(\tau)$$

$$g(\tau) = \frac{[1 - \rho^2(\tau)]\rho''(\tau) + \rho'^2(\tau)\rho(\tau)}{\sqrt{(1 - \rho^2(\tau))h(\tau)}}$$

Necessary and sufficient conditions for the variance of the number of zero-crossings of $\{X(t)\}$ to be finite can be found in [15]. The conditions are given in terms of the second derivative of the autocorrelation function, $\rho(\tau)$, of $\{X(t)\}$ and are: (1) $\rho''(0)$ finite and (2)

$$\int_0^{\delta} \frac{\rho''(\tau) - \rho''(0)}{\tau} d\tau < \infty \text{ for some } \delta > 0. \tag{56}$$

9.5.2. Variance for the Envelope Process

In this section we obtain an expression for the variance of the u-level-crossing count ($u > 0$) of the envelope of a symmetric bandpass Gaussian process. We assume the underlying process is Gaussian, zero-mean, unit variance, with a one-sided power spectral (density, $g_X(\omega)$), which is symmetric about the positive midband frequency, $\omega_c > 0$. That is, for any $\delta \in [0, \omega_X]$ we have $g_X(\omega_X - \delta) = g_X(\omega_X + \delta)$.

Unless otherwise stated, we understand $\{R(t)\}$ to be the envelope of the symmetric bandpass Gaussian process, whose autocorrelation function, $\rho(\tau)$, is given by,

$$\rho(\tau) = \int_0^{\infty} \cos(\omega\tau) dG_X(\omega) = \int_0^{\infty} \cos(\omega\tau) g_X(\omega) d\omega. \tag{57}$$

By using (35) we can write

$$X(t) = I_c(t) \cos \omega_c t - I_s(t) \sin \omega_c t$$

and

$$R^2(t) = I_c^2(t) + I_s^2(t),$$

where $\{I_c(t)\}$ and $\{I_s(t)\}$ are the so-called in-phase and quadrature components respectively. $\{I_c(t)\}$ and $\{I_s(t)\}$ are, independent, identical, Gaussian processes, zero-mean, unit variance with power spectral density, $g_I(\omega)$, given by

$$g_I(\omega) = \tfrac{1}{2}\{g_X(\omega - \omega_c) + g_X(\omega + \omega_c)\}.$$

The results obtain in this section on the variance rely heavily on the work of [37]. In his paper entitled, “Duration of Fades in Radio Transmission”, Rice is concerned with obtaining the probability density function of the interval length between zero-crossings for a particular class of Gaussian processes. He also considers the probability density function for the interval length between u-level-crossings of the envelope for this same class of Gaussian processes.

Rice approximates the density function for the interval between crossings by considering related conditional probability density functions. Let $p(\tau, u)$ denote the probability density function for the length of the interval when $R(t) < u$ (i.e., $p(\tau, u)d\tau$ is the probability that the interval length is between τ and $\tau + d\tau$). Rice argued that as a "first approximation" to the density function $p(\tau, u)$, one could use the conditional probability that an upcrossing occurs at time, τ, given a downcrossing occurred at time 0, we will denote this conditional probability by $p_1(\tau, u)$. Rice maintained that $p_1(\tau, u)$ should be close to the actual density function, $p(\tau, u)$, especially for small τ. (Finding an expression for $p(\tau, u)$ was known to be difficult, and in fact, it is still an open problem today.)

Since conditional probabilities must be approached with care and depend on the limiting process itself, to be precise, all conditional probabilities are to be understood in the horizontal-window averaging sense ([18], pg. 1216, eq. 2.1 and [10], pp. 219–223). That is, we will use the following definition: The probability of the event $\{R(t) \in S\}$ conditioned on $R(0) = u$, denoted by, $\Pr[R(t) \in S \mid R(0) = u]$ is defined by the following limit,

$$\lim_{\delta \to 0} \Pr[R(t) \in S \mid R(t) = u \text{ for some } t \in [-\delta, 0]], \tag{58}$$

provided the limit exists.

The conditioning event A we need for determining $p_1(\tau, u)$ is $A = \{R(t) = u$ for some $t \in [-\delta, 0]$ and $R'(t) > 0\}$. By Korolyook's theorem the limiting behavior of this probability is given by,

$$\lim_{\delta \to 0} \Pr[R(t) = u \text{ for some } t \in [-\delta, 0] \text{ and } R'(t) > 0] \\ = \left(\frac{\Delta}{2\pi}\right)^{1/2} u e^{-(u^2/2)} \cdot \delta + o(\delta),$$

which is just the average u-level-upcrossing rate per unit time of $R(t)$ times δ plus a $o(\delta)$. Thus,

$$p_1(\tau, u) \cdot \left(\frac{\Delta}{2\pi}\right)^{1/2} u e^{-(u^2/2)} = \int_{-\infty}^{0} d\dot{r}_1 \int_{0}^{\infty} d\dot{r}_2 \, |\dot{r}_1 \dot{r}_2| \, p_{R_1, R_1', R_2, R_2'}(u, \dot{r}_1, u, \dot{r}_2). \tag{59}$$

[37], pp. 611–613, eq. 97 completes the required integration and shows that,

$$p_1(\tau, u) = \frac{u M_{22} e^{u^2/2}}{(2\pi\Delta)^{1/2}(1 - \eta^2(\tau))^2} \int_0^{2\pi} J(\gamma, k) \exp\left[-\frac{u^2(1 - \eta(\tau)\cos\phi)}{1 - \eta^2(\tau)}\right] d\phi \tag{60}$$

where

$$\eta(\tau) = \int_0^\infty g_X(\omega)\cos((\omega - \omega_c)\tau)d\omega \tag{61}$$

$$J(\gamma, k) = \frac{1}{2\pi\sqrt{1-\gamma^2}}\int_k^\infty dx \int_k^\infty (x-k)(y-k)e^z dy \tag{62}$$

$$\gamma = \frac{M_{23}}{M_{22}}\cos\phi \tag{63}$$

$$k = \frac{u\eta'(\tau)[\eta(\tau) - \cos\phi]}{1-\eta^2(\tau)}\sqrt{\frac{1-\eta^2(\tau)}{M_{22}}} \tag{64}$$

$$M_{22} = -\eta''(0)[1 - \eta^2(\tau)] - \eta'^2(\tau) \tag{65}$$

$$M_{23} = \eta''(\tau)[1 - \eta^2(\tau)] + \eta(\tau)\eta'^2(\tau) \tag{66}$$

$$z = -\frac{x^2 + y^2 - 2\gamma xy}{2(1-\gamma^2)}. \tag{67}$$

$$\Delta = -\eta''(0) \tag{68}$$

Now the first step in obtaining an expression for the variance of the u-level-crossing count of the envelope process, $\{R(t)\}$, is the computation of $\psi(\tau)$,

$$\psi(\tau) = \int_{-\infty}^\infty \int_{-\infty}^\infty |\dot{r}_1\dot{r}_2| p_{R_1,R_1',R_2,R_2'}(u, \dot{r}_1, u, \dot{r}_2)d\dot{r}_1 d\dot{r}_2. \tag{69}$$

Here again $p_{R_1,R_1',R_2,R_2'}$ is the joint density of $(R(0), R'(0), R(\tau), R'(\tau))$. From (59) and (60) we see that

$$\int_{-\infty}^0 d\dot{r}_1 \int_0^\infty d\dot{r}_2 \, |r_1\dot{r}_2| \, p_{R_1,R_1',R_2,R_2'}(u, \dot{r}_1, u, \dot{r}_2) = \tag{70}$$

$$\frac{u^2 M_{22}}{2\pi(1-\eta^2(\tau))^2}\int_0^{2\pi} J(\gamma, k)\exp\left[-\frac{u^2(1-\eta(\tau)\cos\phi)}{1-\eta^2(\tau)}\right]d\phi \tag{71}$$

And thus, we have a start on evaluating $\psi(\tau)$.

Let I_R denote the integrand $|\dot{r}_1\dot{r}_2| p_{R_1,R_1',R_2,R_2'}(u, \dot{r}_1, u, \dot{r}_2)$. (70) gives the integral of I_R over the second quadrant in the $(\dot{r}_1, \dot{r}_2)$ plane. The integral over the fourth quadrant is obtained using Rice's result for time conditional probability of a u-level-downcrossing, at time τ, given a u-level-upcrossing at time 0. Denote this conditional probability by $p_2(\tau, u)$ (taken in horizontal window sense). Then,

$$p_2(\tau, u) \cdot \left(\frac{\Delta}{2\pi}\right)^{1/2} ue^{-(u^2/2)} = \int_0^\infty d\dot{r}_1 \int_{-\infty}^0 d\dot{r}_2 \,|\, \dot{r}_1\dot{r}_2 \,|\, p_{R_1,R_1',R_2,R_2'}(u, \dot{r}_1, u, \dot{r}_2). \tag{72}$$

[37], pg. 615, eq. 107 shows

$$\begin{aligned} p_2(\tau, u) &= p_1(\tau, u) \\ &\quad + \frac{uM_{22}e^{u^2/2}}{(2\pi\Delta)^{1/2}(1-\eta^2(\tau))^2} \int_0^{2\pi} \left[(\gamma + k^2)\mathrm{erf}\left(\frac{k}{\sqrt{2}}\right) + \frac{2k}{\sqrt{2\pi}} e^{-k^2/2}\right] \\ &\quad \cdot \exp\left[-\frac{u^2(1-\eta(\tau)\cos\phi)}{1-\eta^2(\tau)}\right] d\phi \end{aligned} \tag{73}$$

where $\mathrm{erf}(x)$ is the error function,

$$\mathrm{erf}(x) = \frac{2}{\sqrt{\pi}} \int_0^x e^{-t^2} dt. \tag{74}$$

Using (70) and (73) the integral over the second and fourth quadrants (denoted $\int_{II \cup IV}$) is,

$$\begin{aligned} &\int_{II \cup IV} |\, \dot{r}_1\dot{r}_2 \,|\, p_{R_1,R_1',R_2,R_2'}(u, \dot{r}_1, u, \dot{r}_2) d\dot{r}_1 d\dot{r}_2 \\ &\quad = \frac{u^2 M_{22}}{2\pi(1-\eta^2(\tau))^2} \int_0^{2\pi} \left[2J(\gamma, k) + \left[(\gamma + k^2)\mathrm{erf}\left(\frac{k}{\sqrt{2}}\right) + \frac{2k}{\sqrt{2\pi}} e^{-k^2/2}\right]\right] \\ &\quad \times \exp\left[-\frac{u^2(1-\eta(\tau)\cos\phi)}{1-\eta^2(\tau)}\right] d\phi. \end{aligned} \tag{75}$$

Following the same type of analysis as Rice, we obtain the two remaining integrals necessary for determining $\psi(\tau)$. These are, the integral of $|\, \dot{r}_1\dot{r}_2 \,|\, p_{R_1,R_1',R_2,R_2'}(u, \dot{r}_1, u, \dot{r}_2)$, over the first and third quadrants of the $(\dot{r}_1, \dot{r}_2)$ plane. In fact, due to symmetry of the integrand, we will see that

$$\int_I |\, \dot{r}_1\dot{r}_2 \,|\, p_{R_1,R_1',R_2,R_2'}(u, \dot{r}_1, u, \dot{r}_2) d\dot{r}_1 d\dot{r}_2 = \int_{III} |\, \dot{r}_1\dot{r}_2 \,|\, p_{R_1,R_1',R_2,R_2'}(u, \dot{r}_1, u, \dot{r}_2) d\dot{r}_1 d\dot{r}_2.$$

From Rice (1958, pg. 613, eq. 92) the joint density

$$p_{R_1,R'_1,R_2,R'_2}(u,\dot{r}_1,u,\dot{r}_2) = \left(\frac{u}{2\pi}\right)^2 \int_0^{2\pi} (M_{22} - \cos^2 \phi M_{23}^2)^{-1/2} e^A d\phi \tag{76}$$

where A is a quadratic form in the variables $\{\dot{r}_1, \dot{r}_2\}$. All linear terms in the quadratic expression enter in the form, $\dot{r}_1 - \dot{r}_2$, and Rice uses the following change of variables to simplify the exponent A ([37], pg. 613, eq. 96),

$$x = -(\dot{r}_1 - a_1)[(1-\eta^2(\tau))/M_{22}]^{1/2} \quad y = (\dot{r}_2 - a_1)[(1-\eta^2(\tau))/M_{22}]^{1/2} \tag{77}$$

where,

$$a_1 = \frac{u\eta'(\tau)(\cos\phi - \eta(\tau))}{1-\eta^2(\tau)}. \tag{78}$$

With this final change of variables Rice then arrives at (70). Key to the right hand side of (70) is the function $J(\gamma, k)$, which is obtained by integrating the right side of (76) with respect to x and y using (77).

Following the same analysis that led to (76), the integral of I_R over the first quadrant can be evaluated and is given by,

$$\int_0^\infty d\dot{r}_1 \int_0^\infty d\dot{r}_2 \,|\,\dot{r}_1\dot{r}_2\,|\, p_{R_1,R'_1,R_2,R'_2}(u,\dot{r}_1,u,\dot{r}_2) \tag{79}$$

$$= \frac{u^2 M_{22}}{2\pi(1-\eta^2(\tau))^2} \int_0^{2\pi} J^*(\gamma,k) \exp\left[-\frac{u^2(1-\eta(\tau)\cos\phi)}{1-\eta^2(\tau)}\right] d\phi \tag{80}$$

where

$$J^*(\gamma,k) = \frac{1}{2\pi\sqrt{1-\gamma^2}} \int_k^{-\infty} dx \int_k^\infty (x-k)(y-k)e^z dy. \tag{81}$$

(80) is the same form as (70) except that $J(\gamma, k)$ is replaced by $J^*(\gamma, k)$.

Finally, the integral of I_R over the third quadrant yields,

$$\int_{-\infty}^0 d\dot{r}_1 \int_{-\infty}^0 d\dot{r}_2 \,|\,\dot{r}_1\dot{r}_2\,|\, p_{R_1,R'_1,R_2,R'_2}(u,\dot{r}_1,u,\dot{r}_2) \tag{82}$$

$$= \frac{u^2 M_{22}}{2\pi(1-\eta^2(\tau))^2} \int_0^{2\pi} J^{**}(\gamma,k) \exp\left[-\frac{u^2(1-\eta(\tau)\cos\phi)}{1-\eta^2(\tau)}\right] d\phi \tag{83}$$

where,

$$J^{**}(\gamma, k) = \frac{1}{2\pi\sqrt{1-\gamma^2}} \int_k^{\infty} dx \int_k^{-\infty} (x-k)(y-k)e^z dy. \tag{84}$$

However, recall that

$$z = -\frac{x^2 + y^2 - 2\gamma xy}{2(1-\gamma^2)},$$

so the integrand in (84) is symmetric in (x, y), and therefore $J^{**}(\gamma, k) = J^{*}(\gamma, k)$. Collecting terms from (75), (80), and (83),

$$\begin{aligned}\psi(\tau) = {} & \frac{u^2 M_{22}}{2\pi(1-\eta^2(\tau))^2} \int_0^{2\pi} \\ & \times \left[2J(\gamma, k) + 2J^{*}(\gamma, k) + \left[(\gamma + k^2)\operatorname{erf}\left(\frac{k}{\sqrt{2}}\right) + \frac{2k}{\sqrt{2\pi}} e^{-k^2/2}\right]\right] \\ & \cdot \exp\left[-\frac{u^2(1-\eta(\tau)\cos\phi)}{1-\eta^2(\tau)}\right] d\phi. \end{aligned} \tag{85}$$

Further simplification of (85) occurs using the result of [34, Eqn. (5)], which relates $J^{*}(\cdot)$ to $J(\cdot)$ by

$$J^{*}(\gamma, k) = J(\gamma, k) + \frac{k}{\sqrt{2\pi}} \exp^{-k^2/2} - \frac{k^2 + \gamma}{2}\left[1 - \operatorname{erf}\left(\frac{k}{\sqrt{2}}\right)\right], \tag{86}$$

and allows us to write (85)

$$\begin{aligned}\psi(\tau) = {} & \frac{2u^2 M_{22}}{\pi(1-\eta^2(\tau))^2} \int_0^{2\pi} \\ & \times \left[J(\gamma, k) + \frac{k}{\sqrt{2\pi}} e^{-k^2/2} + \frac{(\gamma + k^2)}{4}\left[2\operatorname{erf}\left(\frac{k}{\sqrt{2}}\right) - 1\right]\right] \\ & \cdot \exp\left[-\frac{u^2(1-\eta(\tau)\cos\phi)}{1-\eta^2(\tau)}\right] d\phi. \end{aligned} \tag{87}$$

To see that $\psi(\tau)$ is well behaved, assume baseband autocorrelation function, $\eta(\tau)$, is at least four times continuously differentiable (which is certainly true for a symmetric bandpass process). Expanding $\eta(\tau)$ about zero we have

$$\eta(\tau) = 1 - \frac{\Delta\tau^2}{2!} + \frac{\kappa\tau^4}{4!} + o(\tau^4), \tag{88}$$

where Δ and κ are the second and fourth spectral moments, respectively, for the baseband spectrum $g_I(\omega)$.

We need to examine the behavior of $\psi(\tau)$ as τ approaches zero from above. Note that as $\tau \to 0^+$, using (88), we have the following limits

$$\frac{M_{22}}{(1 - \eta^2(\tau))^2} \to \frac{B}{4\Delta} \tag{89}$$

$$\gamma \to \cos\phi \tag{90}$$

$$k \to \frac{-2u}{\sqrt{B}} \frac{(1 - \Delta\tau^2/2\cos\phi)}{\tau^2} \tag{91}$$

where $B = \kappa - \Delta^2$.

Thus, for small τ, the contribution to $\psi(\tau)$ from the first term in the integrand of (87),

$$\int_0^{2\pi} J(\gamma, k) \exp\left[-\frac{u^2(1 - \eta(\tau)\cos\phi)}{1 - \eta^2(\tau)}\right] d\phi, \tag{92}$$

as $\tau \to 0$ is given by (see [37], pg. 614, and [33], using the change of variable $\phi = \sqrt{\Delta}\tau x$,)

$$\int_0^{2\pi} J(\gamma, k) \exp\left[-\frac{u^2(1 - \eta(\tau)\cos\phi)}{1 - \eta^2(\tau)}\right] d\phi$$
$$\to \tau \cdot \int_0^{\infty} J\left[1, \frac{u\Delta}{\sqrt{B}}(1 - x^2)\right] \exp[-u^2x^2/2] dx$$

so that

$$\int_0^{2\pi} J(\gamma, k) \exp\left[-\frac{u^2(1 - \eta(\tau)\cos\phi)}{1 - \eta^2(\eta)}\right] d\phi \to O(\tau). \tag{93}$$

Also, for small τ, we see

$$\int_0^{2\pi}\left[\frac{k}{\sqrt{2\pi}}e^{-k^2/2}+\frac{(\gamma+k^2)}{4}\left[2\,\mathrm{erf}\left(\frac{k}{\sqrt{2}}\right)-1\right]\right]$$

$$\cdot\exp\left[-\frac{u^2(1-\eta(\tau)\cos\phi)}{1-\eta^2(\tau)}\right]d\phi$$

$$\to\tau\int_0^{\infty}\left[\frac{k_0}{\sqrt{2\pi}}e^{-k_0^2/2}+\frac{(1+k_0^2)}{4}\left[2\mathrm{erf}\left(\frac{k_0}{\sqrt{2}}\right)-1\right]\right]$$

$$\cdot\exp[-u^2x^2/2]dx\to O(\tau^4) \tag{94}$$

where $k_0=(u\Delta/\sqrt{B})(1-x^2)$.

It is interesting to note ψ is still well-behaved about the origin even when the baseband autocorrelation function is singular. Suppose that $\eta(\tau)$ is twice differentiable but has a discontinuous third derivative at the origin. So that

$$\eta(\tau)=1-\frac{\Delta\tau^2}{2!}+\frac{\vartheta\,|\,\tau^3\,|}{3!}+o(\tau^3) \tag{95}$$

Table 3. Sampled crossing rates for an ideal bandpass Gaussian process.

Theor. value	Sample Mean $N_X(1000)$ 577.4	Sample Variance N_X (1000) 284
Run #1	573.4	285.8
2	573.8	277.7
3	572.9	278.3
4	573.1	278.5
5	572.7	272.3

Table 4. Sampled crossing rates for the envelope of an ideal bandpass Gaussian process.

Theor. value	Sample Mean $N_R(1000)$ 513.5	Sample Variance $N_R(1000)$ 265
Run #1	411.5	255.5
2	411.8	263.9
3	411.4	256.1
4	411.4	268.6
5	411.7	264.1

Then using (95) it can be shown that ([33])

$$\frac{M_{22} \cdot \tau}{(1 - \eta^2(\tau))^2} \to \frac{2\vartheta}{3\Delta} \tag{96}$$

$$\gamma \to \frac{\cos\phi}{2} \tag{97}$$

$$k \to 0 \tag{98}$$

So that using (87) as before we see that as $\tau \to 0$

$$\psi(\tau) \to \psi_0 > 0. \tag{99}$$

So $\psi(\tau)$ does not go to zero but some finite value and thus still well-behaved, i.e., continuous (see [25], pgs. 557, 572 on how this relates to the distribution of the interval between crossings).

We evaluated $\psi(\tau)$ for the particular process of interest, namely, the envelope of an ideal narrow-band Gaussian process. For this special case we performed numerical quadrature to evaluate $\psi(\tau)$, and then found the the variance using (53) and (87). These numerical results based on the analytic expression (87) were then compared within computer simulations for the envelope of a Gaussian process.

Computer simulations of the envelope of bandlimited white Gaussian noise were obtained by two different methods. The first method used an approximation of continuous time bandlimited white Gaussian noise (BLWGN), which was synthesized by first generating a discrete time sequence of independent, identically distributed, pseudo-Gaussian numbers, and then using the sampling theorem to approximate the continuous time BLWGN. The envelope was then obtained via the Hilbert transform of the continuous time process (33). The second method used simply synthesized the in-phase, $I_c(t)$, and quadrature, $I_s(t)$, components of BLWGN and then used Rice's equivalent definition for the envelope, $R(t)$,

$$R(t) = [I_c^2(t) + I_s^2(t)]^{1/2}.$$

Results of computer experiments for the mean and variance of the crossing counts are presented in Tables 3 and 4 and in Figs. 1, 2, 3, and 4. We compare theoretical values with the sample statistics obtained from the simulations for both BLWGN and the envelope of BLWGN.

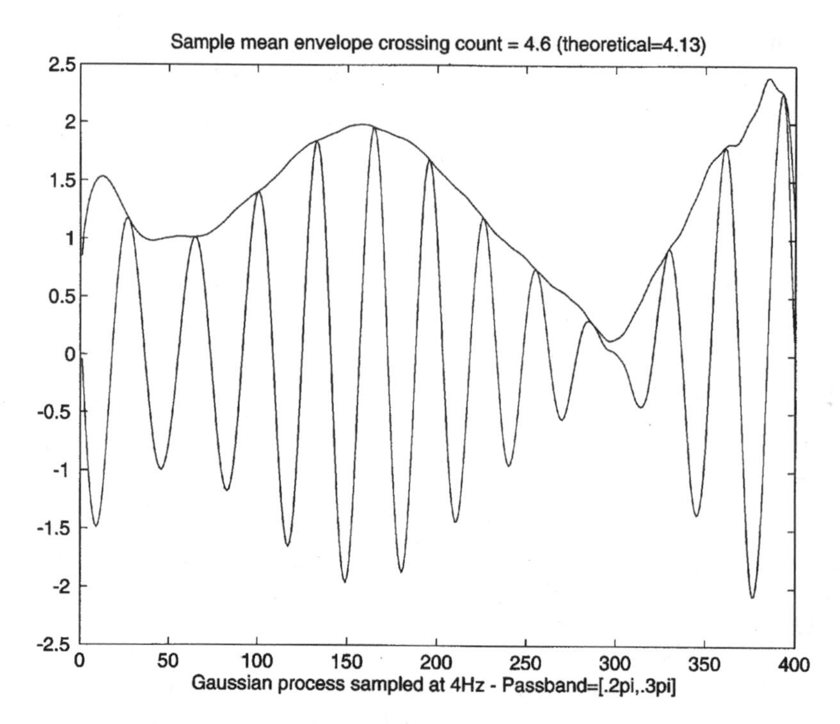

Figure 1. Lowpass Gaussian process and envelope sampled at 4 Hz.

9.6. Asymptotic Normality for the Level-Crossings of the Envelope of a Gaussian Process

In this last section we prove asymptotic normality for the level-crossing counts of the envelope of a narrow-band Gaussian process. Throughout we will assume the underlying Gaussian process, $\{X(t)\}$, is separable, and whose one-sided spectral density, $g_X(\omega)$, is symmetric about the center frequency of the passband ω_c.

The approach taken here parallels the proof given by [11] of the asymptotic normality for the zeros of a differentiable Gaussian process. Cuzick's proof is based on the paper of [26], whereby Malevich proves asymptotic normality of the zero-crossings with restrictive assumptions on the spectrum.

Both Cuzick's and Malevich's proofs use a sequence of M-dependent processes which converge in mean-square to the underlying Gaussian process $\{X(t)\}$. By using a CLT for M-dependent processes (see [13] 1953, 1955) the sequence of zero-crossings of the approximating sequence are shown to be asymptotically normal, so it is enough to show uniform convergence of the

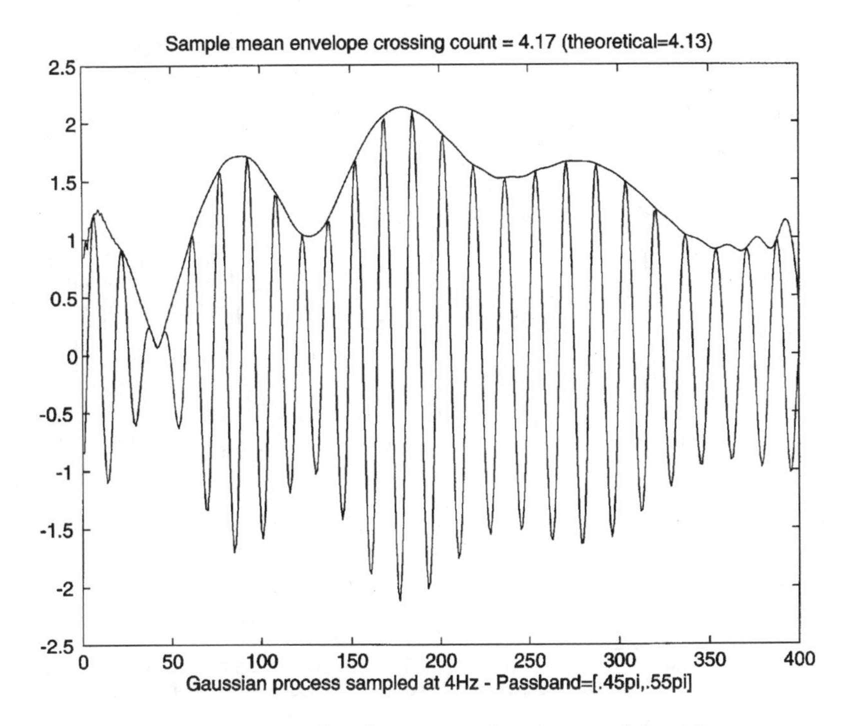

Figure 2. Bandpass Gaussian process and envelope sampled at 4 Hz.

M-dependent processes to $\{X(t)\}$, in mean-square, and show that the zero-crossing counts, as well, converge uniformly in mean-square.

Cuzick shows that some associated correlation functions and cross-correlation functions between the M-dependent processes and the underlying Gaussian process converge in mean-square and this enables him to prove a CLT under less restrictive conditions than those used by Malevich. However, still even less restrictive assumptions were needed by [40] (Theorem 3, pg. 353) to prove asymptotic normality of the crossing counts. Using the powerful stochastic calculus of multiple Wiener–Ito expansions and under the least restrictive assumptions to date, ($\rho(\tau) \in L^2(-\infty, \infty)$ and $\rho''(\tau) \in L^2(-\infty, \infty)$ along with the indispensable [56], Slud proves asymptotic normality of the level-crossings and guarantees nondegeneracy of asymptotic variance with a useful positive lower bound.

In the next section we prove a CLT for the level-crossings of the envelope process. The method used is an adaptation of Cuzick's and Malevich's proofs whereby we approximate the in-phase and quadrature components of the underlying Gaussian process by M-dependent Gaussian processes. Since each compo-

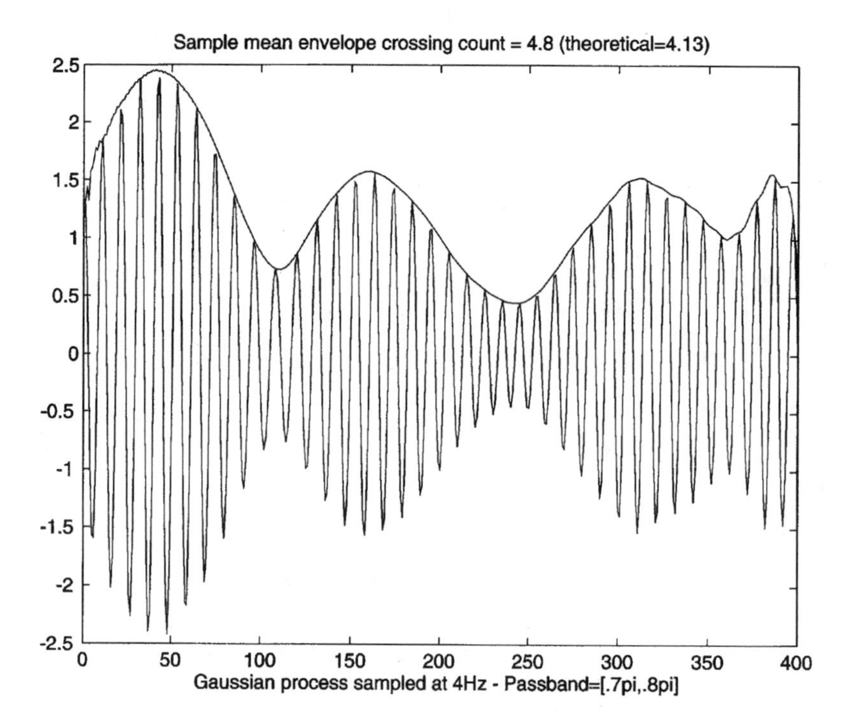

Figure 3. Bandpass Gaussian process and envelope sampled at 4 Hz.

nent is Gaussian we can readily apply a number of results of Cuzick and Malevich to aid in our proof.

9.6.1. Preliminaries

Let $\{X(t)\}$ be our standard Gaussian bandpass process, zero-mean, and unit variance, whose one-sided spectral density is symmetric about the midband frequency ω_c. Using Rice's representation, we can write

$$X(t) = I_c(t) \cos \omega_c t - I_s(t) \sin \omega_c t,$$

and as before, time envelope $R(t)$ is

$$R(t) = [I_c^2(t) + I_s^2(t)]^{1/2}.$$

We define M-dependent (i.e. autocorrelation function vanishes for $|\tau| > 4M$) Gaussian processes which are approximations to $\{I_c(t)\}$ and $\{I_s(t)\}$. For the

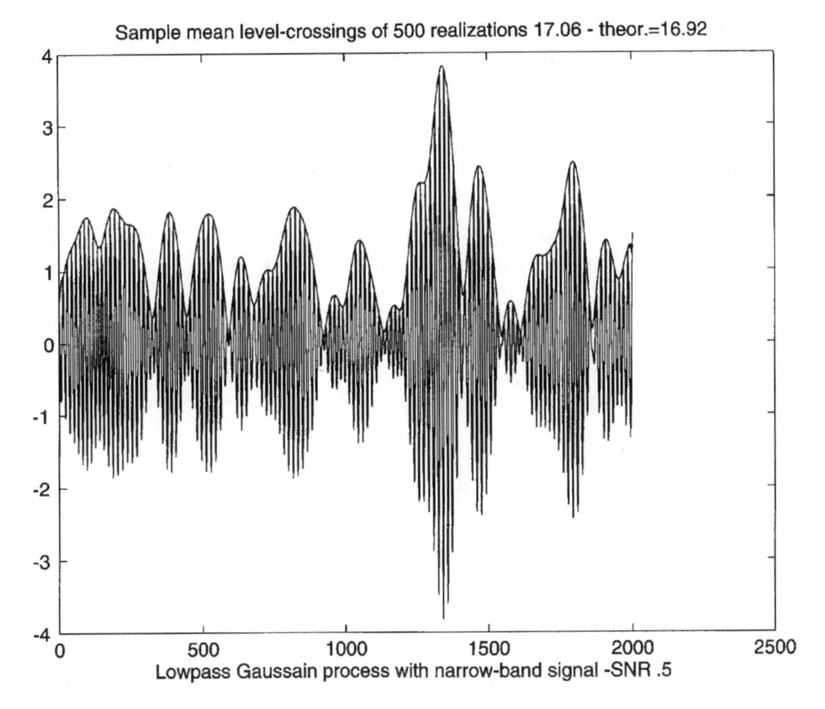

Figure 4. Superposition of two Gaussian process and the envelope sampled at 4 Hz.

in-phase component $\{I_c(t)\}$ define the M-dependent approximation, $\{I_{c,M}(t)\}$ by

$$I_{c,M}(t) = \int_{-\infty}^{\infty} \cos \omega t[(g_I * P_M)(\omega)]^{1/2} dB_c(\omega) \tag{100}$$

where $dB_c(\omega)$ is a Gaussian white noise process, $g_I(\omega)$ the spectral density of $\{I_c(t)\}$ (and $\{I_s(t)\}$ as well), and

$$P_M(\omega) = K \cdot M \left[\frac{\sin M\omega}{M\omega} \right]^4. \tag{101}$$

Above the $*$ denotes convolution and K is a normalization constant such that

$$\int_{-\infty}^{\infty} P_M(\omega) d\omega = 1.$$

By the convolution theorem the autocorrelation, $\rho_{I,M}(\tau)$, of $\{I_{c,M}(t)\}$ is given by pointwise product

$$\rho_{I,M}(\tau) = \rho_I(\tau) \cdot \tilde{P}_M(\tau), \tag{102}$$

where $\rho_I(\tau)$ is the autocorrelation function of $\{I_c(t)\}$ (and $\{I_s(t)\}$), and $\tilde{P}_M(\tau) = \mathcal{F}\{P_M(\omega)\}$ is the Fourier transform of $P_M(\omega)$. It follows that (see pg. 549, Cuzick 1976):

(1) $\tilde{P}_M(\tau)$ is piecewise cubic,
(2) $\tilde{P}_M(\tau) = 1 - (K_0/M^2)\tau^2 + O(|\tau|^3)$ as $\tau \to 0$, with $K_0 > 0$,
(3) $\tilde{P}_M(\tau) = 0$ for $|\tau| > 4M$.

So by (3) above we see that $\{I_{c,M}(t)\}$ is an M-dependent Gaussian process. Now take $\{I_{s,M}(t)\}$ to be defined in an analogous fashion

$$I_{s,M}(t) = \int_{-\infty}^{\infty} \cos \omega t[(g_I * P_M)(\omega)]^{1/2} dB_s(\omega),$$

where $dB_s(\omega)$ is again Gaussian white noise, but is independent of $dB_c(\omega)$. This implies that $\{I_{c,M}(t)\}$ and $\{I_{s,M}(t)\}$ are independent random processes.

Define the M-dependent approximation to the envelope by

$$R_M(t) = [I_{c,M}^2(t) + I_{s,M}^2(t)]^{1/2}. \tag{103}$$

Observe that since $\{I_{c,M}(t)\}$ and $\{I_{s,M}(t)\}$ are independent processes, and each M-dependent, it follows that $\{R_M(t)\}$ is an M-dependent process as well. This is easily seen to be the case using the expressions (43) and (45) for the covariance of the envelope

$$E[R_M(t)R_M(t+\tau)] = \frac{\pi}{2}\left(1 + \frac{k_0^2}{4} + \frac{k_0^4}{64} + \cdots\right), \tag{104}$$

and noting $k_0 = k_0(\tau) = \rho_{I,M}(\tau)$. The squared envelope process, $\{R_M^2(t)\}$ is M-dependent as well by observing,

$$\rho_{R_M^2}(\tau) = \rho_I^2(\tau) \cdot \tilde{P}_M^2(\tau)) = \rho_{I,M}^2(\tau).$$

Next with the help of the following two lemmas from ([11], pg. 549) we prove mean-square convergence of $\{R_M^2(t)\}$ and its derivative $\{2R_M(t)R'_M(t)\}$. We show

$$R_M^2(t) \xrightarrow{L^2} R^2(t) \tag{105}$$

$$2R_M(t)R'_M(t) \xrightarrow{L^2} 2R(t)R'(t). \tag{106}$$

Lemma 1 *If $f \geq 0, f_n \geq 0$, and $f_n^2 \to f^2$ in $L^1(-\infty, \infty)$ then $f_n \to f$ in $L^2(-\infty, \infty)$.*

Lemma 2 *If f in $L^2(-\infty, \infty)$, then $(f * P_M)^{1/2} \to \sqrt{f}$ in $L^2(-\infty, \infty)$ and $\omega \cdot (f * P_M)^{1/2} \to \omega\sqrt{f}$ in $L^2(-\infty, \infty)$.*

Since $g_I \in L^1$, and $P_M \in L^1$, then $g_I * P_M \in L^1$ so by virtue of Lemma 1 we have

$$(g_I * P_M)^{1/2} \xrightarrow{L^2} \sqrt{g_I}.$$

Consequently,

$$\int_{-\infty}^{\infty} \cos\omega t[(g_I * P_M)(\omega)]^{1/2} dB_c(\omega) \xrightarrow{L^2} \cos\omega\sqrt{g_I(\omega)}dB_c(\omega) \tag{107}$$

$$\tag{108}$$

so

$$I_{c,M}(t) \xrightarrow{L^2} I_c(t) \tag{109}$$

and similarly for $I_{s,M}(t)$, so that

$$I_{s,M}(t) \xrightarrow{L^2} I_s(t). \tag{110}$$

The convergence in both cases is uniform in t. From the last two equations we easily obtain (since $E[(I_{\{\cdot\}}(T))^2] < \infty$)

$$I_{c,M}^2(t) \xrightarrow{L^2} I_c^2(t) \tag{111}$$

$$I_{s,M}^2(t) \xrightarrow{L^2} I_s^2(t), \tag{112}$$

again uniformly in t. Thus, we have mean-square convergence, uniformly in t, of the approximating squared envelope process,

$$R_M^2(t) \xrightarrow{L^2} R^2(t). \tag{113}$$

We also have uniform convergence of the approximating sequence of derivatives also. Since the mean-square derivative of $\{R^2(t)\}$ is $\{2R(t)R'(t)\}$ and likewise the derivative of $\{R_M^2(t)\}$ is $\{2R_M(t)R_M'(t)\}$. Using Lemma 4.2 we get

$$\omega[g_I * P_M]^{1/2} \xrightarrow{L^2} \omega\sqrt{g_I}$$

so that

$$-\int_{-\infty}^{\infty} \omega \sin \omega t[(g_I * P_M)(\omega)]^{1/2} dB_c(\omega) \xrightarrow{L^2} -\int_{-\infty}^{\infty} \omega \sin \omega \sqrt{g_I(\omega)} dB_c(\omega)$$

and we have convergence uniformly in t

$$I'_{c,M}(t) \xrightarrow{L^2} I'_c(t).$$

Similarly for the quadrature component, $I_s(t)$, we get

$$I'_{s,M}(t) \xrightarrow{L^2} I'_s(t), \tag{114}$$

uniformly in t, and since all second-moments are finite,

$$I_{c,M}(t)I'_{c,M}(t) \xrightarrow{L^2} I_c(t)I'_c(t) \tag{115}$$

$$I_{s,M}(t)I'_{s,M}(t) \xrightarrow{L^2} I_s(t)I'_s(t). \tag{116}$$

Therefore, the sequence of M-dependent derivatives converge uniformly in t

$$2R_M(t)R'_M(t) \xrightarrow{L^2} 2R(t)R'(t). \tag{117}$$

Definition 1 *Denote the number of u^2-level-crossings of $R^2(t)$ for $t \in [0, T]$ by $N_{R^2}(T)$. We define the centered normalized u^2-level-crossings, $Z(T)$, by*

$$Z(T) \doteq T^{-1/2}[N_{R^2}(T) - E[N_{R^2}(T)]].$$

Similarly, we define the u^2-level-crossings of $R_M^2(t)$ for $t \in [0, T]$ by $N_{R_M^2}(T)$ and

$$Z_M(T) \doteq T^{-(1/2)}[N_{R_M^2}(T) - E[N_{R_M^2}(T)]].$$

With these preliminary results and definitions we are ready to prove a CLT for the u level-crossings of the envelope of a sufficiently smooth Gaussian process.

Theorem 1 *Let $I_c(t)$ and $I_s(t)$ be independent, identical, Gaussian processes, mean-zero, variance 1. Suppose their autocorrelation function, $\rho_I(\tau)$, is four times continuously differentiable at the origin, both $\rho_I(\tau)$ and $\rho_I''(\tau)$ are $\in L^2(-\infty, \infty)$.*

Then for $u > 0$ the u^2-level-crossings of $R^2(t) = I_c^2(t) + I_s^2(t)$ are asymptotically normal. That is,

$$T^{-1/2}[N_{R^2}(T) - E[N_{R^2}(T)]] \xrightarrow{Law} Normal(0, \sigma^2),$$

where

$$\sigma^2 = E[Z_{R^2}(1)] + 2\int_0^\infty [\psi(\tau) - (E[Z_{R^2}(1)])^2]d\tau. \tag{118}$$

Steps in the Proof: First note that since $\rho_I(t)$ is four times continuously differentiable at the origin, we have the following expansion about the origin for small τ,

$$\rho_I(\tau) = 1 - \frac{\Delta\tau^2}{2!} + \frac{\kappa\tau^4}{4!} + o(\tau^4)$$

and the indispensable (56),

$$\int_0^\delta \frac{\rho_I''(\tau) - \rho_I''(0)}{\tau} d\tau < \infty \text{ for some } \delta > 0, \tag{119}$$

is satisfied.

Now following Cuzick's argument, to prove asymptotic normality of $Z(T)$ as $T \to \infty$ it is enough to show that

(A) $Z_M(t) \xrightarrow{L^2} Z(T)$ uniformly in T as $M \to \infty$,

(B) $Z_M(T) \xrightarrow{Law} Normal(0, \sigma_M^2)$ for each M as $T \to \infty$,

and

(C) $\lim_{T\to\infty} T^{-1}\,\mathrm{Var}[N_{R^2}(t)] \to V_0 > 0$.

If we assume our one-sided spectrum is symmetric about a midband frequency ω_c then our expression for the variance of the envelope crossings guarantees that

$$\lim_{T\to\infty} T^{-1}\,\mathrm{Var}[N_{R^2}(t)] \to V_0 > 0.$$

For the more general case we will assume we have non-degeneracy.

From (A) and (B) we that

$$\sigma_M^2 \to \sigma^2$$

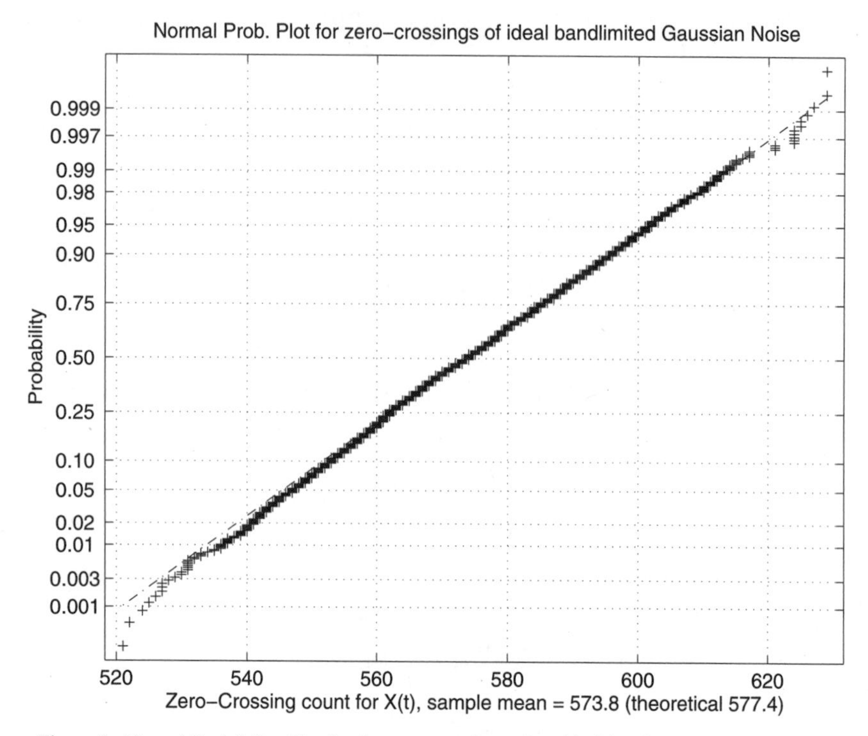

Figure 5. Normal Probability Plot for the zero-crossings of an ideal bandpass Gaussian process.

and so by (C)

$$\sigma_M^2 > 0. \tag{120}$$

Once we have (120) we can use [13], which gives a central limit theorem for M-dependent sequences and obtain the asymptotic normality of $\{Z_M(t)\}$ given as (B).

For details of the steps in the proof, see [5]

Normal probability plots from computer simulations are given in Figs. 5 and 6.

9.7. Summary

In this chapter a parametric filtering technique was presented for application to the problem of discrete frequency estimation. By incorporating a contraction mapping principle with parametric filtering a theoretical basis for this new

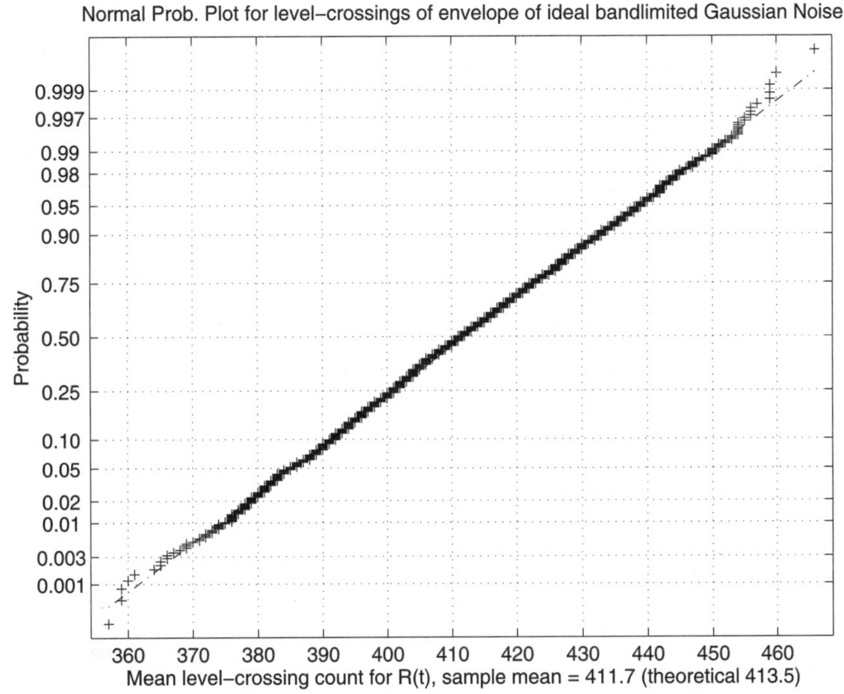

Figure 6. Normal Probability Plot for the mean-level-crossings of the envelope of an ideal bandpass Gaussian process.

method was established (Theorem 1). The theorem also provides a *fundamental property* (19), which places a condition on the parametric filter family guaranteeing convergence of the iterative filtering method for frequency estimation. Two contrasting examples were given which illustrate the utility of the method.

We also derived an expression for the variance of the level-crossings of the envelope of a Gaussian process possessing a symmetric one-sided spectral density. The integral expression obtained for the variance was evaluated numerically for the case of an ideal bandpass process and the results were compared with computer simulations. The theoretical values were found to be in good agreement with the Monte Carlo computer experiments.

Lastly, we presented an outline of the proof that time level-crossings of the envelope are asymptotically normal. This fact was assumed in Rainal (1966) and then used to devise a radar detector based on the mean-level-crossing counts of the envelope. The advantage of the level-crossing detector is that it is within 1 dB of the square law detector but less computationally complex and immune to gain fluctuations. The central limit theorem for the envelope level-crossings formalizes the computer analysis by Rainal and verifies his intuition.

References

[1] J. Abrahams. A Survey of Recent Progress on Level-Crossing Problems for Random Processes. In *Communications and Networks*, I. F. Blake and H. V. Poor, Eds., Springer-Verlag, New York, 1986.

[2] M. Abramowitz and I. A. Stegun. *Handbook of Mathematical Functions*, Dover, New York, 1972.

[3] J. T. Barnett and B. Kedem. Zero-Crossing Rates of Functions of Gaussian Processes. *IEEE Transactions on Information Theory*, 37(4):1188–1194, July 1991.

[4] J. T. Barnett and B. Kedem. Zero-Crossing Rates of Mixtures and Products of Gaussian Processes, *IEEE Transactions on Information Theory*, 44(4):1672–1677, July 1998.

[5] J. T. Barnett and B. Kedem. Asymptotic Normality of the Level-Crossing Counts for Time Envelope of a Gaussian Process, *Journal of Fourier Analysis and Applications*, submitted 1999.

[6] J. T. Barnett, R. B. Clough, and B. Kedem. Power Considerations in Acoustic Emission, *J. Acoust. Soc. Am.*, 98(4):2070–2081, October 1995.

[7] J. S. Bendat. *Principles and Applications of Random Noise Theory.* John Wiley & Sons, New York, 1958.

[8] J. S. Bendat and A. G. Piersol. *Random Data Analysis and Measurement Procedures.* John Wiley & Sons, New York, 1986.

[9] I. F. Blake and W. C. Lindsey. Level-Crossing Problems for Random Processes. *IEEE Transactions on Information Theory*, 19:295–315, 1973.

[10] H. Cramér and M. R. Leadbetter. *Stationary and Related Stochastic Processes.* John Wiley & Sons, New York, 1967.

[11] J. Cuzick. A Central Limit Theorem for the Number of Zeros of a Stationary Gaussian Process. *Ann. Prob.*, 4(4):547–556, 1976.

[12] H. Davis. A Central Limit Theorem for the Response of Filters with Vanishing Bandwidths. *Journal of Mathematical Analysis and Applications*, 3:605–618, 1961.

[13] P. H. Diananda. Some Probability Limit Theorems with Statistical Applications. *Proc. Cambridge Philos.*, 49:239–346, 1953.

[14] M. V. Dragoević and S. S. Stanković. A Generalized Least Squares Method for Frequency Estimation. *IEEE Trans. Acoust., Speech, Signal Process.*, 37(6):805–819, 1989.

[15] D. Geman. On Time Variance of the Number of Zeros of a Stationary Gaussian Process. *Ann. Math. Stat.* 43:977–982, 1972.

[16] S. M. Hammel, J. T. Barnett, and N. Platt. *Diagnosing Intermittency, Applied Nonlinear Dynamics and Stochastic Systemns Near the Millenium.* James B. Kadtke, Adi Bulsara, Eds., AIP Press, July 1997, pp. 51–56.

[17] S. He and B. Kedem. Higher Order Crossings of an Almost Periodic Random Sequence in Noise. *IEEE Trans. Infor. Theory*, 35:360–370, March 1989.

[18] M. Kac and D. Slepian. Large Excursions of Gaussian Processes. *Annals of Mathematical Statistics*, 30:1215–1228, 1959.

[19] SM. Kay. Accurate Frequency Estimation at Low Signal-to-Noise Ratio. *IEEE Trans. Acoust., Speech, Signal Process.*, 32(3):540–547, June 1984.

[20] B. Kedem. Elliptically Symmetric Orthant Probabilities. *The American Statistician*, 45:256, 1991.

[21] B. Kedem. *Time Series Analysis by Higher Order Crossings.* IEEE Press, New Jersey, 1994.

[22] B. Kedem. Spectral Analysis and Discrimination by Zero-Crossings. *Proc. of IEEE*, 74:1477–1493, November 1986.

[23] T. Li. Multiple Frequency Estimation in Mixed Spectrum Time Series by Parametric Filtering, Doctoral Dissertation. Dept. of Mathematics, University of Maryland, College Park, 1992.

[24] T. Li. and B. Kedem. Improving Prony's Estimator for Multiple Frequency Estimation by a General Method of Parametric Filtering. *ICASSP-93*, IV April 1993, pp. 256–259.

[25] M. S. Longuet-Higgins. The Distribution of Intervals Between Zeros of a Stationary Random Function. *Phil Trans Royal Society (A)*, 254:557–599, May 1962.

[26] T. L. Malevich. Asymptotic Normality of the Number of Crossings of Level Zero by a Gaussian Process. *Theory of Probability Appl.*, 14:287–295, 1981.

[27] M. B. Marcus. Level Crossings of a Stochastic Process with Absolutely Continuous Sample Paths. *Annals of Probability*, 5(1):52–71, 1977.

[28] M. R. Matauek, S. S. Stanković, and D. V. Radović. Iterative Universe Filtering Approach to the Estimation of Frequencies of Noisy Sinusoids. *IEEE Tr. on Acoust. Speech Sig. Proc., ASSP-31*, 6:1456–1463, 1983.

[29] J. A. McFadden. The Axis-Crossing Intervals of Random Functions. *IRE Trans. on Information Theory*, 146–150, December 1956.

[30] D. Middleton. *Introduction to Statistical Communication Theory*. McGraw-Hill, New York, 1960.

[31] B. Picinbono and P. Duvant. Optimal Linear-Quadratic Systems for Detection and Estimation. *IEEE Trans. on Information Theory, IT-34*, 2:304–311, March 1988. 5(1):38–43, 1966.

[32] A. J. Rainal. Zero-crossing Principle for Detecting Narrow-Band Signals. *IEEE Trans. on Instrum. and Measurement*, 15(1):38–43, March 1966.

[33] A. J. Rainal. Zero-Crossing Intervals of Envelopes of Gaussian Processes. John Hopkins University Carlyle Barton Laboratory, Baltimore Maryland, Technical Report No. AF-110, June 1964.

[34] A. J. Rainal. Axis-Crossing Intervals of Rayleigh Process. *Bell System Technical Journal*, 44:1219–1224, July–August 1965.

[35] A. J. Rainal. Origin of Rice's Formula. *IEEE Trans. Info. Theory*, 134(6):1383–1387, 1988.

[36] S. O. Rice. Mathematical Analysis of Random Noise, reprinted in *Noise and Stochastic Processes*. Nelson Wax, Ed., Dover, 1954.

[37] S. O. Rice. Distribution of the Duration of Fades in Radio Transmission: Gaussian Noise Model. *Bell Systems Technical Journal*, 37(3):581–635, May 1958.

[38] S. O. Rice. Envelopes of Narrow-Band Signals. *Proceedings of the IEEE*, 70(7):692–699, July 1982.

[39] M. Rosenblatt. Some Comments on Narrow Band-Pass Filters. *Quarterly of Applied Mathematics*, 18(4):387–393, 1961.

[40] E. V. Slud. Multiple Wiener-Ito Integral Expansions for Level-Crossing-Count Functionals. *Probability Theory and Related Fields*, 87:349–364, 1991.

[41] H. Steinberg, P. M. Schultheiss, C. A. Wagrin, and F. Zweigh. Short-Time Frequency Measurements of Narrow-Band Ramdon Signals by Means of a Zero-Counting Process. *J. Appl. Phys.*, 26:195–201, February 1955.

[42] F. C. Titchmarsh. *Introduction to the Theory of Fourier Integrals*. Oxford University Press, 1948.

[43] J. F. Troendle. An Iterative Filtering Method of Frequency Detection in a Mixed Spectrum Model. Doctoral Dissertation. Department of Mathematics, University of Maryland, College Park, 1991.

[44] G. E. Uhlembeck. Theory of Random Process. *MIT Radiation Laboratory Report 454*, Oct. 15, 1943.

[45] H. L. Van Trees. *Detection, Estimation and Modulation Theory*. Wiley, 1968.

[46] S. Yakowitz. Some Contributions to a Frequency Location Method Due to He and Kedem. *IEEE Trans. Infor. Theory*, 37(4):1177–1182, 1991.

[47] D. N. Ylvisaker. The Expected Number of Zeros of a Stationary Gaussian Process. *Annals of Stat.*, 36:1043–1046.

[48] M. Zakai. Second-Order Properties of Time Pre-Envelope and Envelope Processes. *IRE Trans. on Inform. Theory*, IT-6:556–557, 1960.

10

Magnetic Resonance Image Reconstruction from Nonuniformly Sampled *k*-space Data

F. T. A. W. Wajer, G. H. L. A. Stijnman, M. Bourgeois, D. Graveron-Demilly, D. van Ormondt

10.1. Introduction

Medical doctors exert perennial pressure on scanner manufacturers to reduce the measurement time of Magnetic Resonance Imaging (MRI). Applications requiring measurement time reduction are, among others, real-time imaging of the heart, bolus tracking, contrast-agent uptake, and functional imaging. To achieve this goal, one can either devise faster measurement techniques or skip substantial numbers of sample points, or both. Often, the adopted tactics amount to nonuniform undersampling. Unfortunately, this complicates reconstruction of the MR image.

MRI signals are sampled in the multi-dimensional *k*-space in the presence of a spatially-dependent magnetic field. To help appreciate MRI sampling strategies, Section 2 provides an elementary treatment of some basic MRI principles. It should become clear that sample positions are often nonuniform and do not satisfy the Nyquist criterion. Section 3 deals with image reconstruction from raw sparse *k*-space samples. Section 3.2 treats the case of radially sampled data, Section 3.3 the recalculation of sample values from a nonuniform grid to a

F. T. A. W. Wajer, G. H. L. A. Stijnman, D. van Ormondt • Applied Physics Department, Delft University of Technology, The Netherlands, **M. Bourgeois, D. Graveron-Demilly** • Lab. RMN, CNRS D2057, Université LYON I - CPE, 69622 Villeurbanne, France.
E-mail: ormo@si.tn.tudelft.nl

Nonuniform Sampling: Theory and Practice, edited by Marvasti
Kluwer Academic/Plenum Publishers, New York, 2001.

uniform rectangular grid, and Section 3.4 Bayesian reconstruction from arbitrary sparse sample distributions in k-space. Section 4 provides examples of scan time reduction for Cartesian, radial, and spiral sampling; also pseudo-random sampling is covered. Finally, conclusions are drawn in Section 5.

Computer X-ray tomography (CT) [1, 2] is beyond the scope of this chapter. Here, we merely point at some fundamental differences between MRI and CT: 1—MRI is best at imaging light nuclei, notably hydrogen nuclei which are abundantly present in soft tissue [3]. In addition, when fine-tuned, MRI can distinguish between molecules of different chemical composition. Sampling necessarily takes place in the Fourier space (k-space). 2—CT is best at imaging heavier nuclei, such as calcium which is abundant in bones. Projections of the object are sampled directly in the image space. The image is then to be reconstructed from its projections.

10.2. Sampling Strategies in MRI

10.2.1. The Basic MRI Experiment

An MRI scanner uses three magnetic fields [3–7]. These are

1. A strong, homogeneous, and static magnetic field $\vec{B}_0 = B_0\vec{e}_B$ which aligns magnetic nuclei in a patient; $\vec{e}_B$ is a unit vector.
2. A computer-controlled gradient field $\vec{B}_{x,y,z}//\vec{B}_0$, depending linearly on the spatial coordinates x, y, z. The gradient field distinguishes between nuclei at different locations.
3. A computer-controlled radiofrequency field $\vec{B}_{\text{rf}}$, to manipulate the nuclear alignment. For instance, a $\vec{B}_{\text{rf}}$ pulse of proper duration and frequency can rotate the nuclear alignment over a desired angle.

At thermal equilibrium all magnetic nuclei are aligned, resulting in a net magnetization $\vec{M}$ along $\vec{B}_0$. With the aid of a $\vec{B}_{\text{rf}}$ pulse, called 'excitation', one can orient $\vec{M}$ perpendicular to $\vec{B}_0$. Once this has been achieved, $\vec{M}$ starts to precess about $\vec{B}_0$, at the so-called Larmor frequency $\nu_{\text{Larmor}} = \gamma B_0/(2\pi)$, γ being the gyromagnetic ratio of the isotopic species under study. The precessing magnetization, in turn, induces a radio frequency signal $s(t)$,

$$s(t) \propto \exp(-2\pi j \nu_{\text{Larmor}} t), \tag{1}$$

in an antenna surrounding the patient. $j = \sqrt{(-1)}$ and t is time. Equation (1) represents the basic MR signal. Note that the argument of the exponent equals j times the accumulated phase angle of the precessing magnetization. (In actual fact, the antenna feeds two channels, one for the cosine component and one for the sine component. The two components can be combined to produce (1)).

In (1) we have ignored relaxation effects, which destroys the signal according to e^{-t/T_2^*} [3].

10.2.1.1. 1-D Object

Suppose one has a 1-D object (*e.g.*, water-filled capillary), oriented parallel to the x-axis and containing $\rho_n(x)dx$ magnetic nuclei in a segment dx at position x, the concentration $\rho_n(x)$ of the nuclei being zero for $|x| > x_0$ ('band-limit'). $\vec{B}_0$ is applied along the z-axis. Once thermal equilibrium of the magnetic nuclei has been attained, one rotates $\vec{M}$ perpendicular to $\vec{B}_0$ with the aid of a $\vec{B}_{rf}$ pulse. Subsequently, one applies a gradient field $\vec{B}_{x,y,z} = xG\vec{e}_B$, where G is a constant. Under the influence of xG, the Larmor precession frequency in the capillary becomes linearly dependent on the position x, according to

$$2\pi\,\nu_{\mathrm{Larmor}}(x) = \gamma(B_0 + xG). \tag{2}$$

The signal emanating from the capillary follows from integrating over the contributions from all its segments dx, *i.e.*,

$$s(k) \propto \exp(-2\pi j \nu_{\mathrm{Larmor}} t) \int_{-\infty}^{\infty} \rho_n(x) \exp(-2\pi jkx) dx, \tag{3}$$

with

$$k = \gamma\, Gt/(2\pi). \tag{4}$$

Equation (3) represents the basic MRI signal $s(k)$ for a 1-D object. After removal of the 'carrier' signal $\exp(-2\pi j\, \nu_{\mathrm{Larmor}} t)$, which is a simple technical matter, the remainder appears proportional to the Fourier transform of the 1-D object.

The measurement space is called k-space. Note that the variable k evolves proportionally to the real-time t; its physical meaning is seen to be the 'accumulated phase angle offset due to the gradient field per metre'. Sampling (acquisition) of $s(k)$ takes place at N appropriate points of time, subject to the Nyquist criterion. Omission of samples cannot yield reduction of measurement time in 1-D experiments.

10.2.1.2. n-D Object

A 2-D object can be thought of as an array of parallel 1-D objects. For example, a flat object in the x, y-plane can be decomposed into an array of capillaries parallel to the x-axis. After applying the same procedure as in Section 2.1.1, the resulting image is the projection of all capillaries on the x-axis. Clearly, an additional gradient field is required to distinguish between the individual capillaries. In other words, a 2-D measurement is called for.

Setting up an appropriate 2-D measurement [3, 6, 7] is beyond the scope of this chapter. Suffice it to say that one applies the procedure of Section 2.1.1 N

times, but each new signal acquisition is preceded by application of a y-gradient field (i.e., linear dependence of the field on y). The duration of these additional gradient fields is kept constant but at each new turn the strength is incremented by a fixed amount, subject to the Nyquist criterion. This procedure results in an $N \times N$ matrix of samples. Subsequent 2-D IFFT of the matrix yields the desired 2-D image [8]. By adding a third gradient, along the z axis, one can extend this method to 3-D. In the same vein, 3-D IFFT of the corresponding 3-D data matrix provides a 3-D image. In the sequel, the n-D measurements just discussed will be referred to as (conventional) Cartesian scanning. In Cartesian scanning, the samples reside on a uniform rectangular grid.

It is of crucial importance to realize that after each 1-D signal acquisition one has to *wait for the restoration of thermal equilibrium*. This is time-consuming. Thus, omitting a number of increments of the y- and z-gradient fields yields reduction of the measurement time. However, such omission entails local undersampling.

10.2.2. Alternative k-Space Trajectories

Now we turn to various alternative evolutions (trajectories [4]) of k. The choice of a particular evolution depends on the physics of the measurement. Some insight in these matters is useful for appreciating the constraints on sampling in MRI.

The simplest example illustrating the notion of a 'k-space trajectory' is (4) which shows the evolution of k for a 1-D experiment along the x-axis. A k-space trajectory along an arbitrary unit vector $\vec{e}_G$ in 3-D k-space is given by

$$\vec{k}(t) = \gamma \vec{G} t/(2\pi), \tag{5}$$

where $\vec{G} \stackrel{\text{def}}{=} G\vec{e}_G$. In both examples of (4) and (5), the gradient field $\vec{B}_{x,y,z} = (\vec{r}.\vec{G})\vec{e}_B$ is constant over time while switched on, resulting in a straight trajectory starting at $\vec{k} = \vec{0}$. However, it is often advantageous to vary the direction of the trajectory during its course. This can be brought about by varying the direction (and strength) of $\vec{G}$. In that case (5) is to be modified into an integral over time [4, 5], according to

$$\vec{k}(t) = \int_0^t \gamma \vec{G}(t')/(2\pi) dt'. \tag{6}$$

Proper control of $\vec{G}$ over time [9–11] enables one to create many alternative k-space trajectory shapes, such as Cartesian [4, 5] radial [12–17], spiral [18–21], circular [22–24], rosette [25–27], and stochastic [28–30]. Each trajectory shape has its merits in MRI practice. Methods to check the intended shape of a trajectory are treated in, e.g., [31, 32]. Fig. 1 shows examples of Cartesian, radial,

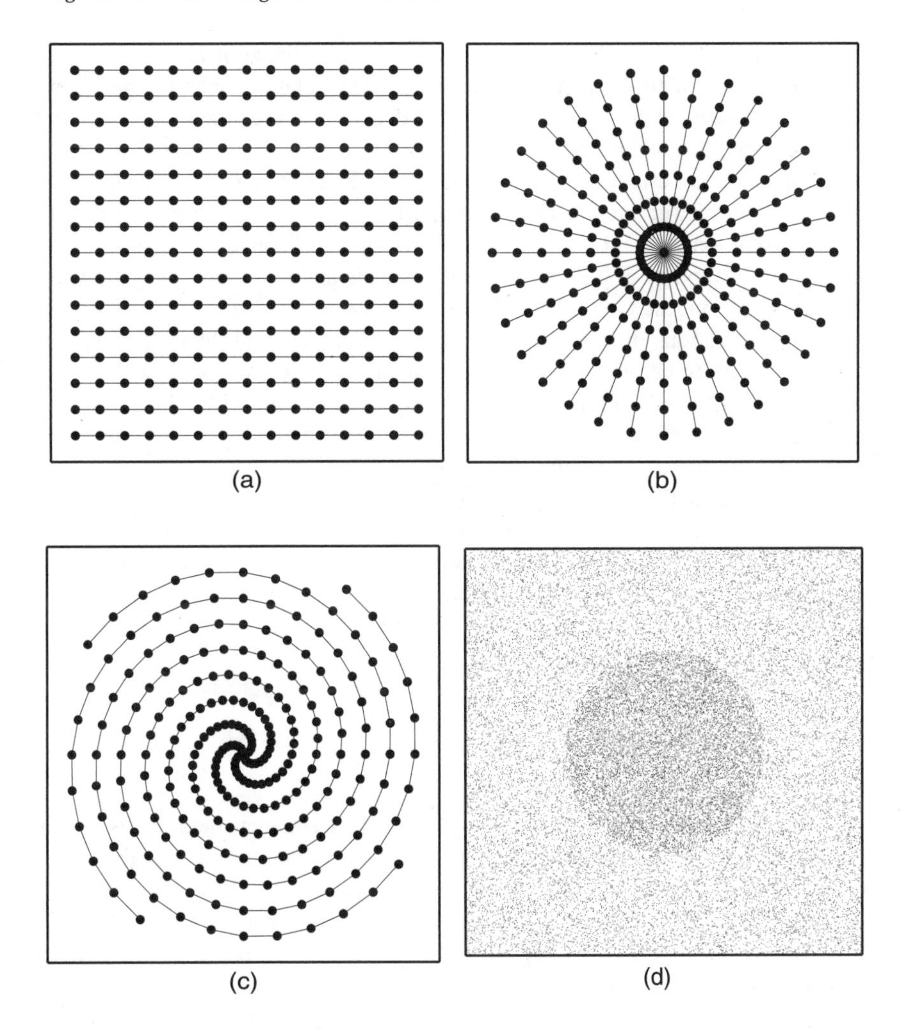

Figure 1. 2-D k-space trajectories of four alternative scans. Horizontal axis k_x, vertical axis k_y, each in the range $-0.5, \ldots, 0.5$. (a) Cartesian: straight horizontal lines starting at the edge. (b) Radial: straight lines starting at $\vec{k} = \vec{0}$ if the signal decays fast. Otherwise, they can start at the edge and pass over to the other side via $\vec{k} = \vec{0}$. 'PR' variant (see also Fig. 2). (c) Spiral: starting at $\vec{k} = \vec{0}$, if the signal decays fast. For clarity of presentation, (a,b,c) show fewer trajectories and sample positions than actually used. (d) Random: beyond $|k| = 0.25$ undersampled by 60%.

spiral, and stochastic sampling. Some considerations in favour of one or another are given in the next two paragraphs.

As for the measurement time, recall that it is the number of trajectories that counts most. This is because thermal equilibrium must be restored between subsequent trajectories. Once a trajectory has been initiated, it is pursued to the edge of k-space. Without spending extra time, one can take as many samples on a trajectory as deemed necessary. If the signal decays fast, the time needed to finish the trajectory must be limited, leading to radial trajectories. However, if the decay of the signal is not critical, spiral trajectories can be chosen too. An advantage of spirals is that fewer are needed to properly cover k-space. Hence, the total measurement time of all required spirals is smaller than that of all required radials. An advantage of both radials and spirals is that they start at $\vec{k} = \vec{0}$, where the signal is strongest. On the other hand, a notable disadvantage of the two is that the sample positions do not coincide with a rectangular grid. In fact, radial and spiral data have to be recalculated from the actual sample positions to rectangular positions prior to n-D IFFT. This process is called gridding [33, 34]. Note that gridding works well in the case of stochastic oversampling [29], but not in the case of stochastic undersampling [30, 35, 36].

'Conventional' Cartesian sampling (as in Section 2.1.2) is advantageous in that all sample positions reside on a rectangular grid, thus requiring only n-D IFFT for image reconstruction. However, for physical reasons Cartesian sampling, as depicted in Fig. 1, is not the method of choice for imaging dynamic objects. We point out that a Cartesian trajectory starts at the edge of k-space, where the signal is weakest.

10.2.3. Encoding

The above discussion assumes that the radiofrequency field $\vec{B}_{\mathrm{rf}}$ is homogeneously applied, manipulating the magnetization at each location in the region of interest in equal manner. Since image reconstruction can be brought about with IFFT, one speaks of 'Fourier Encoding' the signal. Alternatively, one may make $\vec{B}_{\mathrm{rf}}$ spatially dependent according to certain patterns. Depending on the pattern, this adapts the signal to SVD (Singular Value Decomposition) or Wavelet reconstruction of the image. One then speaks of 'SVD Encoding' or 'Wavelet Encoding'. The latter development [37–39] is beyond the scope of this chapter.

10.2.4. Sampling Summary

To summarize the previous Sections, the following facts and considerations are important for sampling in MRI.

- Sampling is done in k-space, trajectory-wise.
- Often, sample positions do not coincide with a rectangular (Cartesian) grid.

- The number of trajectories required to satisfy the Nyquist criterion everywhere in k-space, depends on the shape of the trajectories. For instance, one needs fewer spiral trajectories than radial trajectories.
- Each trajectory shape has its advantages and disadvantages. Technical realization of particular shapes is not a trivial matter.
- The measurement time is proportional to the number of trajectories. Hence, omission of trajectories results in reduction of measurement time.
- Omission of trajectories entails local undersampling, which in turn causes artefacts in the image. Strategic distribution of omissions is subject to study.

10.3. Image Reconstruction from Raw k-Space Data

10.3.1. Introduction

In Section 2, we pointed out that certain physical conditions may force one to sample along non-Cartesian trajectories. Once such trajectories are chosen, the possibility of fast image reconstruction by n-D IFFT has been forfeited. Here we treat three alternatives: 1—The reconstruction exploits the properties of the sample-position distribution in question [12, 25, 40, 41]. 2—New samples are estimated on a uniform rectangular grid through convolution of the measured samples with a suitable window. These new samples are subsequently transformed by n-D IFFT [33, 34, 42]. 3—The reconstruction is based on Bayesian estimation which can accommodate general prior knowledge of the image [43, 44]. In practice, all three methods are used, each having advantages and disadvantages. Common features of methods 1 and 2 are that the density of sample positions needs to be estimated and taken into account (density correction) at each sample position and that the Nyquist criterion must be satisfied. Properties of method 3 are: i—it can accommodate local undersampling and arbitrary sample positions, ii—it obviates the sampling density correction, iii—it is iterative and therefore needs more computing time.

10.3.2. Radial Samples on a Square or Cube

This Section treats image reconstruction from two recently introduced radial sampling distributions. The reconstruction exploits special properties of the sample positions.

Traditionally, the radial trajectories of a 'Projection Reconstruction' (PR) scan are evenly distributed in k-space and hence the sample positions reside on radial trajectories as well as on concentric circles; see Fig. 1(b). In this case, the image is often obtained by so-called Filtered Back-Projection [12]. Since Filtered Back-Projection entails interpolation and is relatively computationally intensive, the so-called *linogram* method [16], which was developed in X-ray computer tomography, has been adapted to MRI [17]. With linograms, the trajectories are again radial, but their angular distribution and density of samples are chosen such that the samples reside on concentric squares (or cubes) rather than circles (or spheres); see Fig. 2(b). This choice obviates interpolation and enables reconstruction by the chirp-z transform, thus leading to increased accuracy and reduced computation. Recently, a method called Sparse Radial Scan and Fourier Transform (SRS-FT) has been proposed [14–15]. With SRS-FT, the number of positions that coincide with the rectangular grid is maximized; see Fig. 2(a). The latter property enables fast reconstruction of a preliminary image by 2-D IFFT.

In the sequel, we treat a reconstruction method applicable to radial square (or cubic) sample positions as occurring in linograms and SRS-FT [41]. The method is based on FFT and at the same time avoids interpolation which is inherent to Filtered Back-Projection.

With both SRS-FT and linogram, the relation between k_x and k_y on a radial trajectory is governed by $k_y/k_x = \tan\phi = m/l$ and l and m are integers satisfying $|m/l| \leq 1$ so that $|\phi| \leq \pi/4$. The k-space with $|\phi| > \pi/4$ is covered by an appropriate symmetry operation. The number of ϕ-values and their distributions differ in the two methods. SRS-FT uses *all* values for l and m such that $|l|$, $|m| \leq N_p \ll N$, where N_p is the number of directions and N is the Cartesian gridsize. This implies that no Cartesian samples are omitted in the centre of k-space, i.e., for $|k_x|$ and $|k_y| \leq N_p/N$. With the linogram, the trajectories are more evenly distributed over k-space: $|m| \leq N/2$ and $|l| = N/2$, but coincidence with Cartesian sampling points is reduced with respect to SRS-FT.

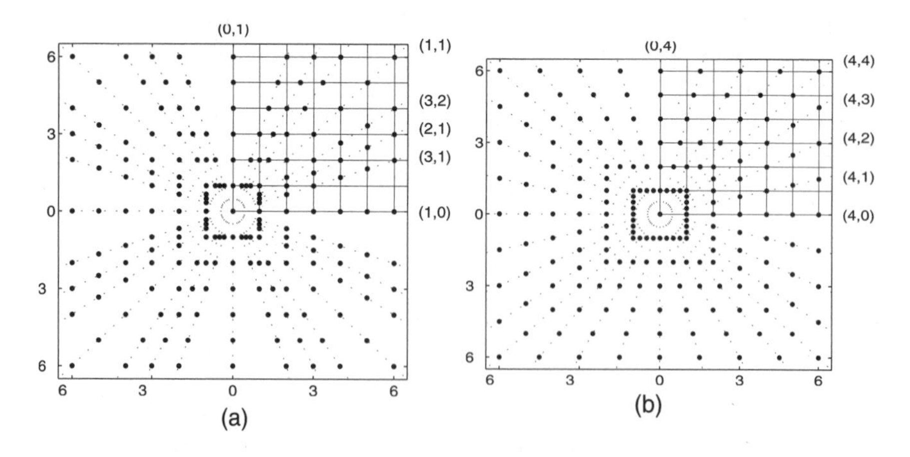

Figure 2. Sample positions for (a) SRS-FT and (b) linogram.

Now we transform from Cartesian to radial coordinates using the expressions in Table 1 and discretize the inverse Fourier transformation

$$I(x, y) = \iint s(k_x, k_y) \exp(2\pi j(xk_x + yk_y)) dk_x dk_y, \tag{7}$$

in which $I(x, y)$ and $s(k_x, k_y)$ stand for image and k-space signal, respectively. This results in

$$I_{x,y} = \sum_{l,m} I_{x,y}^{l,m}, \tag{8}$$

where

$$I_{x,y}^{l,m} = \Delta\phi^{l,m} \sum_k s_k^{l,m} \exp\left(2\pi j\left(x + \frac{m}{l}y\right)k\right) J_k^{l,m}, \tag{9}$$

$s_k^{l,m} \overset{\text{def}}{=} s(k, km/l)$ stands for the data along the radial trajectory; the Jacobian $J_k^{l,m} \overset{\text{def}}{=} J(k, \text{atan}(m/l))$ and the angular increment $\Delta\phi^{l,m}$ account for the change of variables (see Table 1). Furthermore, the quantities kN, x, and y each vary over the range $-N/2, -N/2 + 1, \ldots, N/2 - 1$, and N is a power of 2.

Equations (8–9) constitute an extension of the Filtered Back-Projection reconstruction. Equation (9) is to be calculated for all $N \times N$ values of (x, y). Note that one can resort to IFFT for the calculation in the x-dimension, at any y. This is because the argument of the exponent in (9) contains the appropriate factor $2\pi jxk/N$. This condition is not satisfied for PR. As for the y-dimension, use of the relation

$$I_{x,y+l}^{l,m} = I_{x+m,l}^{l,m}, \tag{10}$$

which can easily be verified by replacing y in (9) by $y + l$, obviates the need to repeat the calculation for all N y-values. Rather, it needs to be done for only l y-values. Interpolation is not necessary. These properties lead to reduced computation and increased accuracy compared to Filtered Back-Projection. For further details we refer the reader to [13, 45].

Table 1. Values of k_x, k_y, and the Jacobian for alternative 2-D radial sampling strategies. k and ϕ are the independent variables.

Method	k_x	k_y	Jacobian
PR	$k \cos \phi$	$k \sin \phi$	k
SRS-FT, linogram	k	$k \tan \phi$	$k/\cos^2 \phi$

Finally, we point out that the product $\Delta\phi^{l,m} J_k^{l,m}$ represents the k-space area to be attributed to sample $s_k^{l,m}$. The inverse of this area equals the sampling density alluded to in Section 3.1.

10.3.3. Gridding of Nonuniform Sampling Distributions

10.3.3.1. Introduction

This Section treats conversion of samples measured on a non-rectangular grid to new samples on a uniform rectangular (Cartesian) grid, called gridding [13, 33–34, 42]. Gridding starts off by convolving the measured samples with a suitable window, yielding a continuous signal. Next, one takes new samples on a Cartesian grid followed by n-D IFFT. In order to reduce ensuing aliasing, the step size of the Cartesian grid is often chosen well below the Nyquist value that corresponds to the desired field of view (FOV). This procedure results in an increased FOV and is called overgridding. Only the desired FOV is retained and corrected for the shape of the convolution window. An important aspect of gridding is the trade-off between computation time and accuracy, given the signal-to-noise ratio (SNR). Below, we discuss a graphical and mathematical description of 1-D gridding, the convolution window, and overgridding.

10.3.3.2. Graphical Description of 1-D Gridding

Figure 3 provides a graphical description of 1-D gridding [46]. Extension to n-D gridding does not entail new principles. The top row of Fig. 3 shows the continuous signal $s(k)$ and corresponding image $I(x)$, related to each other by the continuous (inverse) Fourier transform. The nonuniform sampling grid is represented by a series of δ-functions, $\mathcal{S}(k)$, and the set of samples is the multiplication of $s(k)$ and $\mathcal{S}(k)$. The samples are to be divided by the sampling density ρ, which equals the inverse of the area spanned by a sample position. In the following, the latter is referred to as density correction; see second row of Fig. 3. ρ is estimated in Section 10.3.2 for square radial sampling, in Section 10.4.4.1 for spiral sampling, and in Section 10.4.5.2 for pseudorandom sampling. Note that multiplication ($\times$) in k-space corresponds to convolution ($*$) in image space and vice versa. $C(k)$ and $c(x)$ are the convolution/multiplication window in k-space and image space respectively; see row 4 of Fig. 3. In order to limit the computation time, $C(k)$ is strictly zero beyond a certain region $|k| = L/2$. The latter implies that $c(x)$ extends to infinity, but this is barely discernible in Fig. 3 at the applied scale. III is the comb function which consists of δ-functions on a uniform grid. The (inverse) Fourier transform of a comb function is again a comb function; see row 6 of Fig. 3. Convolution of the comb function with the continuous image $I''(x)$ lowers the accuracy in the FOV due to aliasing. Choosing a finer comb III(k)

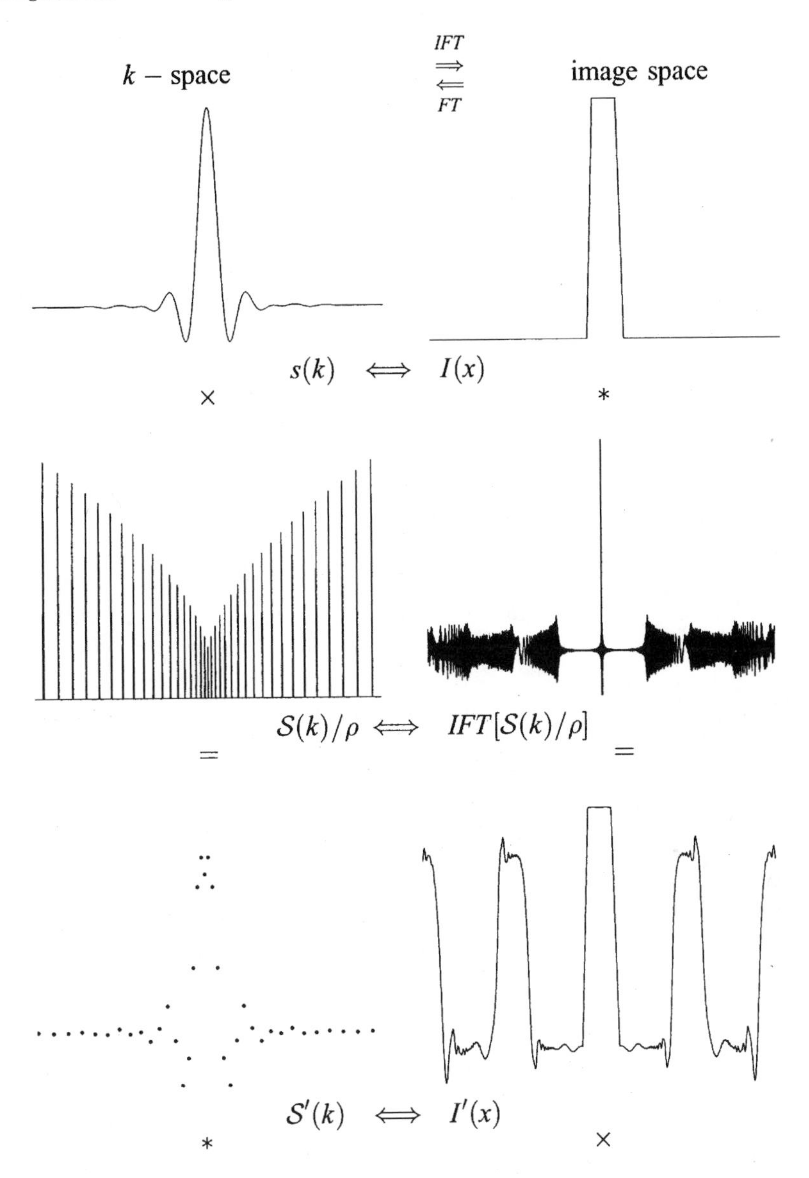

Figure 3. Graphical description of 1-D convolution gridding with a Kaiser-Bessel window and image reconstruction. See Section 10.3.3.2.

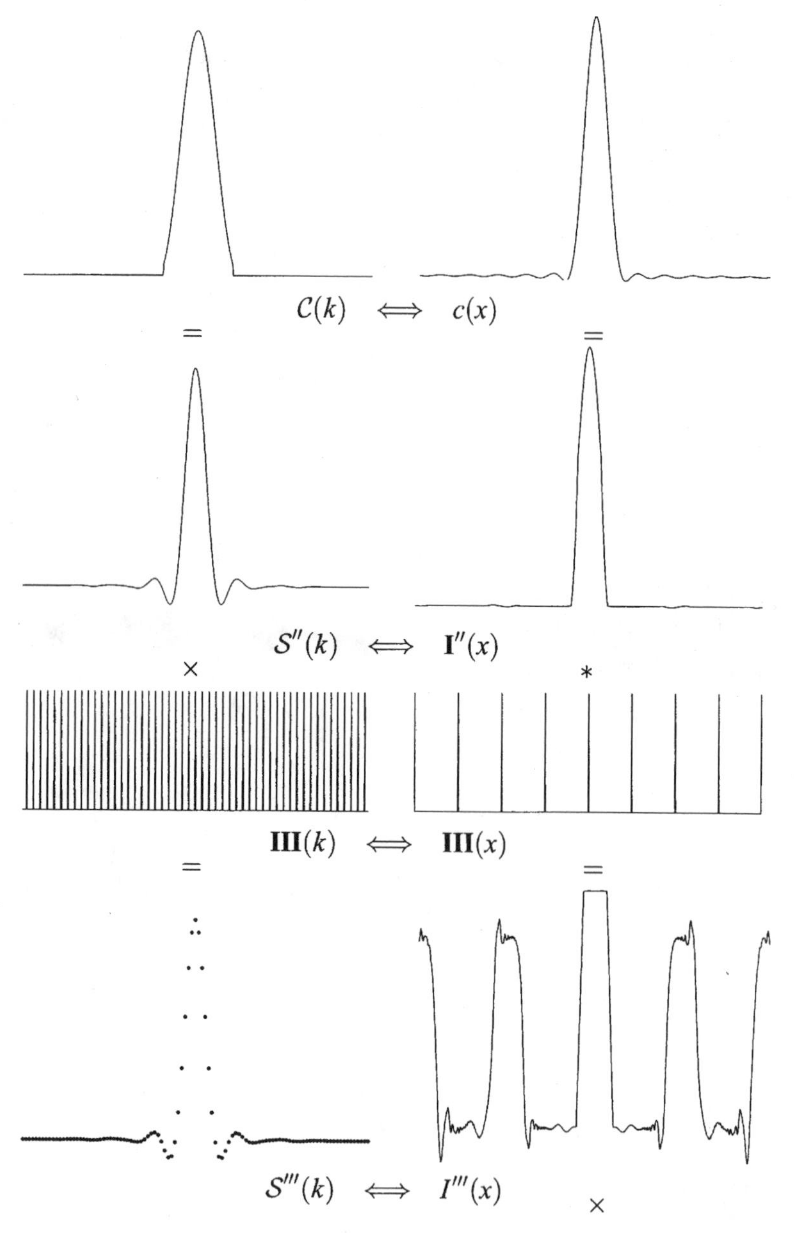

Figure 3. (*continued*)

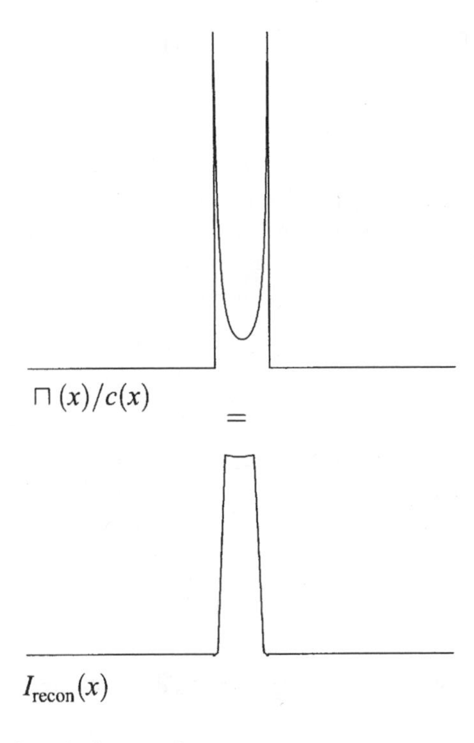

Figure 3. *(continued)*

(overgridding) results in a wider comb III(x), and thus reduces aliasing. However, overgridding increases the computing time. Finally, multiplication by the rectangular function ⊓ zeroes the image outside the desired FOV, and division by the Kaiser-Bessel window $c(x)$ undoes the earlier multiplication by that window; see rows 8 and 9 of Fig. 3. The next Section describes gridding algebraically.

10.3.3.3. Algebraic Description of Gridding

The above graphical description is complemented here by an algebraic one. We start off with the inverse Fourier integral,

$$I(x) = \int s(k) e^{2\pi jkx} dk. \tag{11}$$

A discrete approximation of (11) is

$$I(x) \simeq \sum_n s(k_n) e^{2\pi j k_n x} \Delta k_n. \tag{12}$$

Errors are incurred due to aliasing and to estimating appropriate values of Δk_n (density correction) in the case of irregular sampling. The latter is not a trivial matter, especially for n-D data with $n > 1$. Next, we write the convolution window in k and image space.

$$c(x) = \int C(k) e^{2\pi j k x} dk. \tag{13}$$

As will be explained in the next Section, the window $C(k)$ is strictly zero beyond a certain width $|k| = L/2$. This implies that $c(x)$ has infinite extent. Multiplying both sides of (13) with $e^{2\pi j k_n x}$, one gets

$$e^{2\pi j k_n x} c(x) = \int C(k) e^{2\pi j (k+k_n) x} dk = \int C(k - k_n) e^{2\pi j k x} dk. \tag{14}$$

Now we replace the right-hand side of (14) by a discrete approximation, yielding

$$e^{2\pi j k_n x} c(x) \simeq \sum_m C(k_m - k_n) e^{2\pi j k_m x} \Delta k_m. \tag{15}$$

Here, the estimation of Δk_m poses no problem because the k_m are chosen on a uniform rectangular grid. In addition, division by $c(x)$ in the FOV is permitted because this function is chosen greater than zero in that region.

$$e^{2\pi j k_n x} \simeq \frac{1}{c(x)} \sum_m C(k_m - k_n) e^{2\pi j k_m x} \Delta k_m. \tag{16}$$

From (16) and (12) one gets

$$I(x) \simeq \sum_n s(k_n) e^{2\pi j k_n x} \Delta k_n \simeq \sum_n s(k_n) \left(\frac{1}{c(x)} \sum_m C(k_m - k_n) e^{2\pi j k_m x} \Delta k_m \right) \Delta k_n. \tag{17}$$

Exchanging then the summations, one arrives at

$$I(x) \simeq \frac{1}{c(x)} \sum_m \left(\sum_n s(k_n) \Delta k_n C(k_m - k_n) \right) e^{2\pi j k_m x} \Delta k_m. \tag{18}$$

Equation (18) reveals the gridding process. The inner summation is the discrete convolution of the window with the nonuniformly sampled signal. The quantities Δk_n are to be estimated by a separate procedure, referred to as density correction. The outer summation is the subsequent IFFT. Finally, division by $c(x)$ corrects for the shape of this window in the FOV.

10.3.3.4. The Kaiser-Bessel Window

A well-known candidate window for convolution in k-space is the sinc function. Its Fourier transform is a rectangular function which effectively avoids aliasing. However, convolution with a sinc function is computationally expensive, disqualifying it for use in MRI. Another window considered in the past is the prolate spheroidal wavefunction [47]. It too was rejected for computational reasons. Following [33–34, 42], we recommend the so-called Kaiser-Bessel window, a truncated bell-shaped function given by

$$C(k) = \begin{cases} \frac{1}{L} I_0(B(1 - (2k/L)^2)^{\frac{1}{2}}) & -\frac{L}{2} \leq k < \frac{L}{2}, \\ 0 & \text{otherwise,} \end{cases} \tag{19}$$

where I_0 is the zero-order modified Bessel function of the first kind, L the width of the window, and B a shape parameter. The larger L, the longer the computation time. Usually, we choose $L = 3$ or 5. The shape parameter B is used to optimize the accuracy of the reconstructed image. Using an 'overgridding factor' $f_{og} = 2$, one finds $B_{opt} = 1.5\pi L$ for 2-D applications [34]; see also Section 10.3.3.5.

The inverse Fourier transform of $C(k)$ is given by

$$c(x) = \frac{\sin(\pi^2 L^2 x^2 - B^2)^{1/2}}{(\pi^2 L^2 x^2 - B^2)^{1/2}}, \tag{20}$$

which is an oscillating function that extends to infinity. As already noted above, convolution with the comb function results in aliasing which in turn affects the accuracy of the image reconstruction in the region of interest. The next Section discusses how overgridding reduces the effect of aliasing and can serve to make the final step of the reconstruction suitable for FFT.

10.3.3.5. Overgridding

The stepsize $\Delta k_{\text{Nyquist}}$ required to uniformly sample the desired FOV at the Nyquist rate, is the standard with respect to which we measure overgridding. If the stepsize of $\text{III}(k)$ equals $\Delta k_{\text{Nyquist}}$, the overgridding factor f_{og} equals unity.

Raising f_{og} above unity narrows the comb III(k) and correspondingly widens III(x). Since $|c(x)|$ decreases with increasing $|x|$, widening III(x) reduces aliasing effects. At the same time, however, the computation time which is proportional to $f_{og}L$, increases. Therefore, a trade-off is called for. It has been found that at the usual SNR, choosing $f_{og} = 2$ provides acceptable accuracy [34].

The fact that the size of f_{og} is not critical around 2 is useful for limiting the total reconstruction time. Note that the number of new, uniformly distributed sampling points is proportional to the area (volume) covered in k-space by the scanner and to f_{og}. To limit the time of the subsequent n-D IFFT, this number should be expressible as a power of 2. It follows that f_{og} is not only useful for increasing the accuracy but also for making the number of new sample points suitable for FFT. This property in turn calls for a simple procedure to adapt the shape parameter B for optimal accuracy. Using numerical analysis, we arrive at

$$B_{opt} = (f_{og} - 0.5)\pi L. \tag{21}$$

As an example, we quote a computation time of 2.8 seconds on a Sun Ultra 5 for gridding of spiral data comprising 60 trajectories with 1736 samples each, to 256×256 points on a uniform rectangular grid. The gridding parameters were $L = 3$ and $f_{og} = 2.0$.

Finally, we point out that gridding cannot work in undersampled regions of k-space. In addition, if sampling is random, estimation of the sampling density ρ requires computation of a Voronoi cell [48] around each sample position which may be cumbersome (see Section 10.4.5.1).

10.3.4. Bayesian Image Reconstruction

10.3.4.1. Introduction

The previous Sections assumed that sampling satisfied the Nyquist criterion. If this condition is not satisfied, additional information is required to arrive at the image. We choose to exploit the fact that the histogram of differences of intensity of nearest neighbour pixels in the image domain has Lorentzian shape (Cauchy shape, in mathematics) [43, 49–51]. This property is not limited to MRI. A suitable vehicle for imposing prior information such as mentioned here is Bayesian estimation [52]. Another favourable property of the Bayesian algorithm is that correction for sampling density fluctuations, treated in Section 10.3.3, is avoided. This has to do with the fact that the algorithm transforms from a uniform rectangular grid to a nonuniform grid, rather than in the usual reverse direction.

10.3.4.2. The Most Probable Image

A Bayesian reconstruction algorithm aims at finding the image I_{MAP}, which maximizes the posterior probability density function $\text{P}(I|S)$, where I and S are the image matrix and measured (i.e., ignoring omitted sample positions) k-space data matrix, respectively. This implies that I_{MAP} is the most probable image given S. Using Bayes-theorem,

$$\text{P}(I|S) = \frac{\text{P}(S|I)\text{P}(I)}{\text{P}(S)}, \tag{22}$$

and maximizing $\log \text{P}(I|S)$ instead of $\text{P}(I|S)$, one arrives at the following problem

$$\max_I \{\log \text{P}(S|I) + \log \text{P}(I)\}. \tag{23}$$

In the above, the evidence $\text{P}(S)$ was treated as a constant because it does not depend on I. The prior, $\text{P}(I)$, which describes any *a priori* knowledge we have about the image, will be elaborated on below. $\text{P}(S|I)$ is the likelihood. This term describes the influence of the measurement noise. Specification of this term requires modelling of the k-space data S.

10.3.4.3. The Likelihood

For ease of notation, we shall rewrite the data and image matrices as vectors, $\underline{\mathbf{s}}$ and $\underline{\mathbf{i}}$, by stacking their respective columns on top of each other. As shown in Section 2, the k-space signal and image are related to each other through the 2-D Fourier transform. In addition, the samples in k-space are contaminated by additive, white, Gaussian noise with zero mean and standard deviation σ. One can therefore model the data as

$$\underline{\mathbf{s}} = T\underline{\mathbf{i}} + \underline{\mathbf{n}}, \tag{24}$$

in which $T_{m,n} = \exp(-2\pi j(x_n k_{x_m} + y_n k_{y_m}))$ and $\underline{\mathbf{n}}$ is the measurement noise. (x_n, y_n) are the locations of the pixels, which are assumed to be distributed uniformly over a rectangular grid. As mentioned in the introduction, arbitrary, irregular sample positions (k_{x_m}, k_{y_m}) are allowed. When the (k_{x_m}, k_{y_m}) are indeed irregular, straight-forward calculation of $T\underline{\mathbf{i}}$ is very expensive. A solution to this problem will be given below. Should the (k_{x_m}, k_{y_m}) coincide with a rectangular grid, T reduces to a 2-D FFT. We point out that in any case T transforms *from a uniform rectangular grid* in image space to the measurement grid in k-space. Hence, correction for the nonuniform sampling density needed in Section 10.3.3, where the reverse direction was followed, is not required here.

The Gaussian assumption of the measurement noise results in the following expression for the likelihood-term

$$\log \mathrm{P}(S|I) = -\frac{|\underline{\mathbf{s}} - T\underline{\mathbf{i}}|^2}{2\sigma^2}, \tag{25}$$

where we have ignored terms independent of the image.

If sufficient data are available, a unique image $\underline{\mathbf{i}}$ can be found that minimizes $|\underline{\mathbf{s}} - T\underline{\mathbf{i}}|^2$ in (25). This is equivalent to the maximum likelihood method. Such a solution is impossible if the number of k-space data is smaller than the number of pixels in the image, for instance when deliberately omitting k-space samples which in turn results in ill-conditioning of T. However, maximizing the posterior probability (23), allows incorporation of *a priori* information (the $\mathrm{P}(I)$ term) about the image. This will regularize the reconstruction and compensate any ill-conditioning of T; see [53] for an alternative approach based on SVD. The next subsection describes our choice of the prior.

10.3.4.4. The Prior

$\mathrm{P}(I)$ is the prior which enables incorporation of prior knowledge about the image into the reconstruction algorithm. We start off (see also [43]) by expanding $\mathrm{P}(I)$ as follows

$$\mathrm{P}(I) = \mathrm{P}(I_{0,0})\mathrm{P}(I_{1,0}|I_{0,0})\mathrm{P}(I_{2,0}|I_{0,0}, I_{1,0}) \ldots, \tag{26}$$

where $I_{0,0}$ is the intensity of a pixel in a corner of the image. We now assume that $I_{x,y}$ depends (directly) only on its left-hand neighbour $I_{x-1,y}$, i.e., a 1st order Markov process. Hence,

$$\mathrm{P}(I_{x,y}|I_{x-1,y}, I_{x-2,y}, \ldots) = \mathrm{P}(I_{x,y}|I_{x-1,y}). \tag{27}$$

Substituting (27) into (26), and ignoring the boundaries, one gets

$$\mathrm{P}(I) = \prod_{x,y} \mathrm{P}(I_{x,y}|I_{x-1,y}). \tag{28}$$

Next, we assume that $P(I_{x,y}|I_{x-1,y})$ is a function of *only* the intensity difference between the corresponding pixels. This difference is written as ${}^{\mathrm{x}}\Delta_{x,y} \stackrel{\mathrm{def}}{=} I_{x,y} - I_{x-1,y}$.

In [49], Fuderer shows that the probability density function of ${}^{x}\Delta_{x,y}$ is Lorentz-shaped; see also [43, 50, 51, 54]. This leads to the following expression for the prior

$$\mathrm{P}(I) = \prod_{x,y} \frac{a}{\pi(a^2 + ({}^{x}\Delta_{x,y})^2)}. \tag{29}$$

The prior just described ignores the dependence of a pixel on its neighbour in the vertical direction. Therefore, from now on we call (29) the 1-D prior. To illustrate both the horizontal and vertical dependences, Fig. 4 shows a contour plot of the 2-D histogram of the $(N-1) \times (N-1)$ pairs ${}^{x}\Delta_{x,y}, {}^{y}\Delta_{x,y}$ for an $N \times N$ image of a human brain; ${}^{y}\Delta_{x,y} \stackrel{\text{def}}{=} I_{x,y} - I_{x,y-1}$. The contours are approximately circular, which indicates that the intensity difference distribution does not significantly depend on the angle $\text{atan}({}^{y}\Delta_{x,y}/{}^{x}\Delta_{x,y})$ (isotropic). The histogram shown appears representative of a whole class of MRI images.

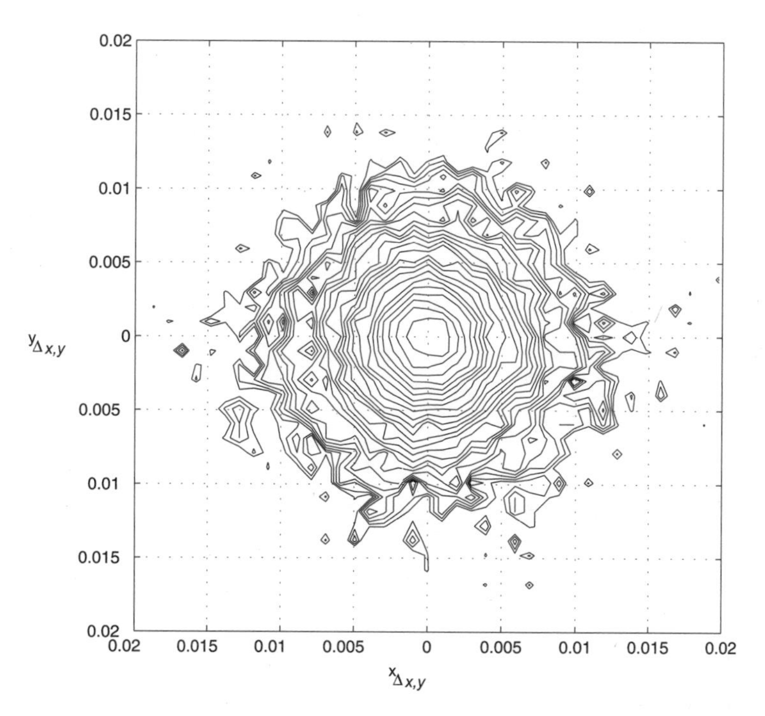

Figure 4. Contour plot of the 2-D histogram of the horizontal and vertical intensity differences, ${}^{x}\Delta_{x,y}$ and ${}^{y}\Delta_{x,y}$, of an image of a human brain. The contours are circular, indicating that the histogram is an isotropic function of ${}^{x}\Delta_{x,y}$ and ${}^{y}\Delta_{x,y}$. This histogram appears representative of a whole class of MRI images.

To model this histogram, one could multiply two 1-D priors, one for the x-direction and one for the y-direction, as in

$$\frac{a^2}{\pi^2(a^2 + ({}^{x}\Delta_{x,y})^2)(a^2 + ({}^{y}\Delta_{x,y})^2)}. \tag{30}$$

However this function is not isotropic. Next, we try

$$\frac{a}{2\pi(a^2 + ({}^{x}\Delta_{x,y})^2 + ({}^{y}\Delta_{x,y})^2)^{1.5}}, \tag{31}$$

which is isotropic. Furthermore, integration of (31) over ${}^{y}\Delta_{x,y}$ appropriately yields an expression that is consistent with the 1-D distribution function (29).

Figure 5 shows the central slice through the histogram in Fig. 4 and the corresponding least-squares fit of the distribution function of (31). The quality of

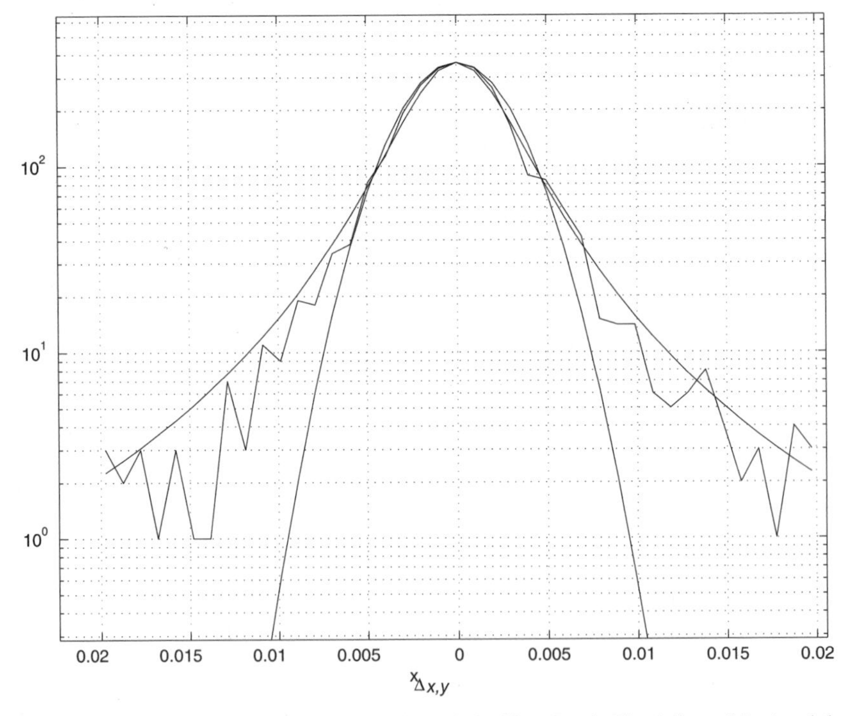

Figure 5. Plot of the central slice (${}^{y}\Delta_{x,y} = 0$) through the histogram in Fig. 4 (jagged line) and the corresponding least-squares fit of the distribution function (31) (upper smooth line). Also shown is a least-squares fit of a Gaussian distribution function (lower smooth line) which is too narrow for larger values of ${}^{x}\Delta_{x,y}$.

the fit is good. The highest values of the histogram are in the centre. This means that the most probable values are ${}^{x}\Delta_{x,y} = 0$ and ${}^{y}\Delta_{x,y} = 0$. These values correspond to the homogeneous regions in the image. The tails of the histogram correspond to intensity edges in the image. The power of 1.5 in the denominator of (31) appeared essential for obtaining a good fit. In (30), the corresponding power is only 1, which is too small to follow the descent of the histogram at large ${}^{x}\Delta_{x,y}$ (not shown). This constitutes another reason to reject (30) as a model for the histogram in Fig. 4. Also shown in Fig. 5 is a least-squares fit of a Gaussian distribution function. The latter fit is only good at small values of ${}^{x}\Delta_{x,y}$ but is rather too narrow for larger values of ${}^{x}\Delta_{x,y}$.

Using (31), our 2-D prior becomes

$$\mathrm{P}(I) = \prod_{x,y} \frac{a}{2\pi\left(a^2 + ({}^{x}\Delta_{x,y})^2 + ({}^{y}\Delta_{x,y})^2\right)^{1.5}}. \tag{32}$$

Substituting then (32) in (23), while ignoring any constant terms, we arrive at the following optimization problem

$$\min_{I} \left\{ \frac{|\underline{\mathbf{s}} - T\underline{\mathbf{i}}|^2}{2\sigma^2} + \frac{3}{2} \sum_{x,y} \log\{a^2 + ({}^{x}\Delta_{x,y})^2 + ({}^{y}\Delta_{x,y})^2\} \right\}. \tag{33}$$

The image I_{MAP} resulting from (33) is our reconstructed image. The minimisation is done with the conjugate gradient method.

10.3.4.5. Computational Details

The computation time required for the Bayesian reconstruction algorithm described above is 3.5 minutes on a Sun Sparc Ultra 1 for the random sample positions in Fig. 1(d), producing a 128×128 image in 32 iterations. The reconstruction of a 256×256 image from sparse spiral data in 38 iterations requires 28 minutes on the same machine. These times are rather too long for routine use in a clinic.

The computation time of (33) is dominated by the likelihood term. The size of the transformation matrix T equals $N_s \times N^2$, where N_s is the number of samples and N^2 the number of pixels. Therefore, for reasonably sized images, straight-forward calculation of $T\underline{\mathbf{i}}$ is very expensive. T represents a discrete Fourier transform from the image to the (nonuniform) sample positions in k-space. Hence T amounts to the same operation as a convolution of a 2-D sinc-function with the 2-D FFT of the image. For the same reasons as in Section 10.3.3 we approximate the 2-D sinc-convolution by replacing the sinc with a 2-D version of the Kaiser-Bessel window.

This convolution step needs to be carried out only once, prior to the minimization. To this end we expand the likelihood term as follows

$$\frac{|\underline{\mathbf{s}} - T\underline{\mathbf{i}}|^2}{2\sigma^2} = \frac{\underline{\mathbf{s}}^\dagger \underline{\mathbf{s}} - 2\mathrm{Re}(\underline{\mathbf{i}}^\dagger T^\dagger \underline{\mathbf{s}}) + \underline{\mathbf{i}}^\dagger T^\dagger T\underline{\mathbf{i}}}{2\sigma^2} \tag{34}$$

where $\underline{\mathbf{s}}^\dagger \underline{\mathbf{s}}$ is a constant that can be eliminated from the minimization process. The second term of (34) contains

$$(T^\dagger \underline{\mathbf{s}})(x, y) = \sum_{(k_x, k_y)} s(k_x, k_y) e^{2\pi j(xk_x + yky)}, \tag{35}$$

which equals a nonuniform 2-D IDFT. Its straightforward computation being too expensive, it is approximated as follows. First, convolve $s(k_x, k_y)$ to a uniform grid, then use 2-D IFFT to get to the image domain, and finally correct for the convolution step.

The right-hand part of the third term of (34) can be rewritten as

$$\begin{aligned}(T^\dagger T\underline{\mathbf{i}})(x, y) &= \sum_{(x', y')} \left(\sum_{(k_x, k_y)} e^{2\pi j((x-x')k_x + (y-y')ky)} \right) I(x', y') \\ &= \sum_{(x', y')} Q(x - x', y - y') I(x', y')\end{aligned} \tag{36}$$

This equation represents a 2-D convolution of $Q(x, y)$ with the image $i(x, y)$; it can be calculated efficiently using the 2-D FFT. $Q(x, y) = \sum_{(k_x, k_y)} e^{2\pi j(xk_x + yky)}$ has exactly the same form as (35), so we calculate it using the same approximation technique. Note, however, that the size of $Q(x, y)$ should equal two times the size of the image since $(x - x')$ and $(y - y')$ have twice the range of x and y, respectively.

To speed up the convolution, the 2-D FFT of $Q(x, y)$ is stored rather than $Q(x, y)$ itself. Moreover, using Parsevals relation, the inner product $\underline{\mathbf{i}}^\dagger (T^\dagger T\underline{\mathbf{i}})$ is calculated in k-space. Once $T^\dagger \underline{\mathbf{s}}$ and the 2-D FFT of $Q(x, y)$ are known, the likelihood term can be calculated using only one 2-D FFT.

10.4. Applications

10.4.1. Introduction

As mentioned in Section 10.2, measurement time reduction usually entails both nonuniform sampling and local undersampling. Nonuniform sampling arises when applying curved or nonparallel sampling trajectories or when nonuniformly omitting Cartesian trajectories. Undersampling occurs when omitting trajectories of any shape. This Section concerns optimal locations of omitted trajectories for Cartesian and spiral sampling. Furthermore, square-radial undersampling and gridding of full spiral sampling are treated. Finally, we turn to pseudorandom sampling due to patient motion. A general result is that Bayesian reconstruction appears capable of handling all undersampling situations presented.

10.4.2. Sparse Cartesian Sampling

10.4.2.1. Uniform Omission of Trajectories

Presently, Cartesian sampling is the most commonly used scheme for routine investigations in a clinic. In a 2-D measurement of a slice one uses typically $N = 256$ equidistant trajectories, i.e., $Nk_y = -128, -127, \ldots, 127$, each with N samples, at $Nk_x = -128, -127, \ldots, 127$. With a state of the art scanner, this takes about 1 minute. The simplest and cheapest way to reduce the measurement time is to simply omit almost all trajectories on one side of the k_x-axis, i.e.., $Nk_y = -128, -127, \ldots, -17$ [8]. This is illustrated in the upper halve of Fig. 6(a); the black rectangle represents such a one-sided block of omissions. 1-D IFFT with respect to k_x of the remaining trajectories is still possible of course. In this Section, we tacitly assume that this 1-D IFFT with respect to k_x has been carried out.

The data omitted on one side can, in principle, be retrieved at minimal computational cost by exploiting the Hermitian symmetry operation [8].

$$s(x, k_y) = s^*(x, -k_y), \tag{37}$$

which holds under ideal conditions. In actual practice, inhomogeneity of the net magnetic field in the slice of interest perturbs the symmetry of (37). However, the data in the limited region $|Nk_y| \leq 16$ permit a Hermiticity test and estimation of a low-resolution field inhomogeneity map [8]. This in turn enables an approximate field inhomogeneity correction of the data such that the above symmetry relation can still be used to obtain the missing data. This method results in 44% measurement time reduction. Note that this so-called half-k-space or one-sided scan avoids nonuniform sampling. Consequently, the extra computation time required to retrieve missing data is limited and does not prohibit application in a clinical environment.

10.4.2.2. Nonuniform Omission of Trajectories

Uniform, one-sided omission entails minimal loss of information and allows fast image reconstruction. However, it cannot supersede a maximum measurement time reduction of some 44%. (Experimental limitations prevent reaching the ideal 50% reduction.) If yet more reduction is desired, the first issue is how to limit concomitant loss of information. Traditionally, one tended

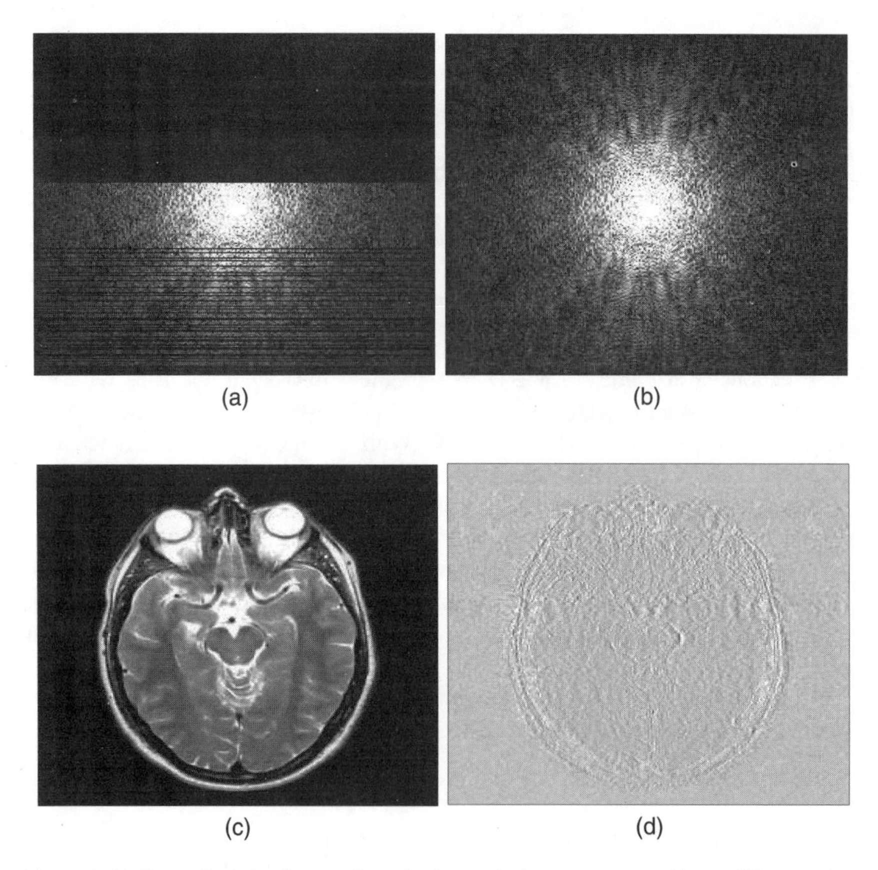

Figure 6. (a) Sparse Cartesian k-space data of a human brain scan corresponding to 57% scan time reduction [43]. The black rectangle and lines represent omissions. Note that omissions avoid the region where the signal is strongest, *i.e.*, around $\vec{k} = \vec{0}$. (b) k-space data after Bayesian reconstruction (2-D FFT of reconstructed image). (c) Bayesian reconstruction of the image. (d) Difference between the original image (not shown), *i.e.*, from full data and that from sparse data in (c).

to adhere to uniform sampling, simply truncating the measurement of the otherwise fully sampled half of k-space, and retrieving the missing data by, e.g., extrapolation [8]. Later, it was found that distributing omitted trajectories nonuniformly limits loss of information [55]. The latter result emerged from investigating thousand randomly chosen distributions of $0.7 \times 256 = 179$ trajectories on a regular grid of $Nk_y = -128, -127, \ldots, 127$. The grid points $|Nk_y| \leq 32$ were not available for omission, to assure adequate estimation of field inhomogeneity; however, each distribution was chosen symmetric with respect to $k_y = 0$, to avoid possibly unwarranted use of (37). Next, real-world

MRI 256 × 256 data matrices of three different human brains, a liver, a lumbar, and a knee were made sparse according to these thousand trajectory distributions, and the corresponding images were estimated with the Bayesian method described in Section 10.3.4 [56]. Perusal of the resulting image qualities revealed that: 1– sampling distributions yielding the highest image quality are nonuniform, 2– the ranking of sampling distributions in terms of image reconstruction quality holds for a variety of image shapes. Figure 7 shows the five best and five worst sampling distributions of the set of thousand.

Figure 6(a) shows a successful combination of one-sided and irregular omission of trajectories, amounting to 57% scan time reduction. Bayesian reconstruction yields the image Fig. 6(c), and the FFT of this image, in turn, retrieves the omitted samples as shown in Fig. 6(b). The quality of the reconstruction can be judged in Fig. 6(d) which shows the difference between the original image (no omissions) and Fig. 6(c).

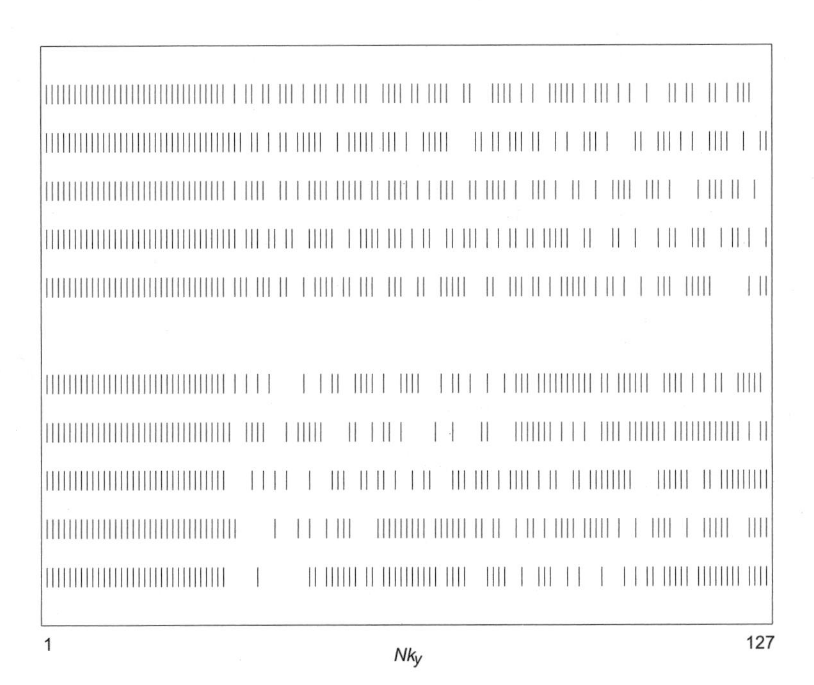

Figure 7. Right-hand sides of ten distributions in descending order of the image reconstruction quality for a human brain. The upper five are the best distributions, the lower five are the worst distributions. Note that the k_y-direction is here plotted horizontally.

10.4.3. Sparse Radial Sampling

10.4.3.1. Introduction

Radial scanning is useful whenever the decay of the signal is fast. This condition occurs, e.g., when imaging lungs with hyperpolarized ^{3}He gas [57]. Recall from Section 10.3.2 that at least three variants exist, namely, Projection Reconstruction (PR) [12], Linogram [40], and 'Radial Scanning adapted to FFT' (RS-FT, or SRS-FT if scanning is sparse) [14]. With PR, the angles between successive trajectories are equal and so are the sample intervals on all trajectories. Consequently, the sample positions reside on intersections of straight lines and circles. See Fig. 1(b). If $N \times N$ is the size of the uniform rectangular grid, PR requires about πN trajectories to satisfy the Nyquist criterion in a circle of radius $N/2$. With Linogram and SRS-FT, the angles between successive trajectories and the sample intervals on trajectories are adapted such that the sample positions reside on intersections of straight lines and squares (or cubes), making reconstruction amenable to fast Fourier techniques; see Fig. 2. For linogram, $4N$ trajectories are required to satisfy the Nyquist criterion within an $N \times N$ square. For SRS-FT, this number is even higher because the angles between successive trajectories are less well distributed. However, the latter point is not relevant since, in practice, SRS-FT measurements are sparse.

It follows from the above that although radial sampling can cope with fast decay of a signal along a single trajectory, the required number of trajectories is relatively high. Consequently, reconstruction techniques that can cope with omission of trajectories are welcome.

10.4.3.2. Omission of Radial Trajectories

All trajectories start at $\vec{k} = \vec{0}$ where the signal is strongest. Hence, contrary to the Cartesian case, there are no individual radial trajectories yielding much more or much less information than others. The consequences of this for optimal omission of radial trajectories are yet to be examined. Here we present the case of SRS-FT for which omission is done such that the number of sample points coinciding with the uniform rectangular grid be maximal at the centre of k-space [14]. This can be achieved in only one way. Figure 8(a) shows a measurement of a water-filled flower-shaped object (phantom) with typical omissions amounting to a scan time reduction of as much as 80%. In the outer regions of k-space the sparseness of sampling is rather unevenly spread. The sampling density correction for such sampling requires further study; see also [13, 45].

Figure 8(c) shows the reconstruction of the image with the method described in Section 10.3.2. The image is sharp but omission artefacts are clearly visible.

Figure 8. (a) Sparse radial k-space data of a water-filled object (phantom); only 96 trajectories were used, in SRS-FT pattern [14] and corresponding to about 80% measurement time reduction. (b) k-space data after Bayesian reconstruction (2-D FFT of (d)). (c) Image reconstruction using the method described in Section 10.3.2. (d) Bayesian reconstruction of the image.

Subsequently, this image is used as a starting value for the Bayesian reconstruction resulting in the image of Fig. 8(d). Omission-related artefacts have now disappeared. Application of 2-D FFT to the image retrieves the omitted data, as shown in Fig. 8(b).

Note that the size of the original FOV is preserved in the above experiment. However, if only a small part of the object is of interest, a substantial reduction of the FOV is acceptable. The latter enables symmetric omission of trajectories of a PR or linogram scan. This in turn opens the way to substantional scan time reduction and fast artefact-free reconstruction [58].

10.4.4. Spiral Sampling

10.4.4.1. Gridding of Fully Sampled Data

Since a spiral trajectory covers more area of k-space than does a straight line, fewer of them are needed to satisfy the Nyquist criterion. If the criterion is indeed satisfied everywhere—full sampling—then gridding followed by n-D IFFT suffices to reconstruct an image. The attendant density correction, alluded to in Section 10.3.3.2, can be derived from the formula of the spiral trajectories [18, 19]

$$\vec{k}(t) = K_{\max}\, t^{\gamma} \exp(2\pi j(N_{\text{turn}} t^{\beta} + n_s/N_s)) \quad 0 \leq t \leq 1, \tag{38}$$

where $K_{\max}$ is the maximum value of $|\vec{k}(t)|$, N_{turn} is the number of turns covered by a trajectory, N_s is the number of spiral trajectories, n_s is the index of a spiral trajectory, and β and γ are shape parameters. Here, we use $\beta = 1$ and $\gamma = 1$ while acquiring $N_{ss} + 1$ samples at uniform time intervals along each trajectory [46]. Since the sample positions reside on circles as well as spirals, and $t = 0, 1/N_{ss}, 2/N_{ss}, \ldots, 1$, the area per sample equals $2\pi|\vec{k}(t)|\Delta|\vec{k}(t)|/N_s$ which can be rewritten as [46]

$$\rho^{-1} \approx \text{area} \approx \pi K_{\max}^2 n^{\gamma}((n+1)^{\gamma} - (n-1)^{\gamma})/(N_{ss}^{2\gamma} N_s), \tag{39}$$

for sample No n, $0 < n \leq N_{ss}$, on any trajectory. For $t=0$ and $t=1$, one needs separate expressions. For $t=0$ we use $\rho^{-1} \approx 2\pi|\vec{k}(1/(2N_{ss}))|^2$ and for $t=1$, $\rho^{-1} \approx$ the area between the last and penultimate circles divided by the number of trajectories N_s [46].

In the case of full sampling, i.e., the Nyquist criterion is satisfied everywhere, the above procedure yields good images. However, when trajectories are omitted to reduce the measurement time, i.e., sampling sparsely, one has to resort to Bayesian estimation; this will be discussed in more detail in the next Section.

10.4.4.2. Sparse Spiral Sampling

The number of spiral trajectories required to satisfy the Nyquist criterion depends on the curvature of the trajectories. Going from $\vec{k} = \vec{0}$ to the edge of k-space, the direction of the trajectories used here changes by 4π. At this rate of change, one needs about 60 trajectories to avoid undersampling throughout k-space. This already constitutes a substantial reduction of measurement time with respect to Cartesian and radial scanning. Here, the aim is to reduce the measurement time further, to 2/3, by omission of 20 trajectories.

The more the symmetry of the object under investigation is circular, the fewer spiral or radial trajectories are needed. A difference between sparse spiral and sparse radial scanning becomes apparent from the shapes of the regions representing omitted trajectories: As $|\vec{k}|$ increases, spiral omissions tend to have circular symmetry (see Fig. 9(d)), whereas radial omissions retain their radial appearance (see Fig. 8(a)). This consideration leads to the assumption that the effect of an arbitrary rotation of the omission pattern is small; in fact, it amounts

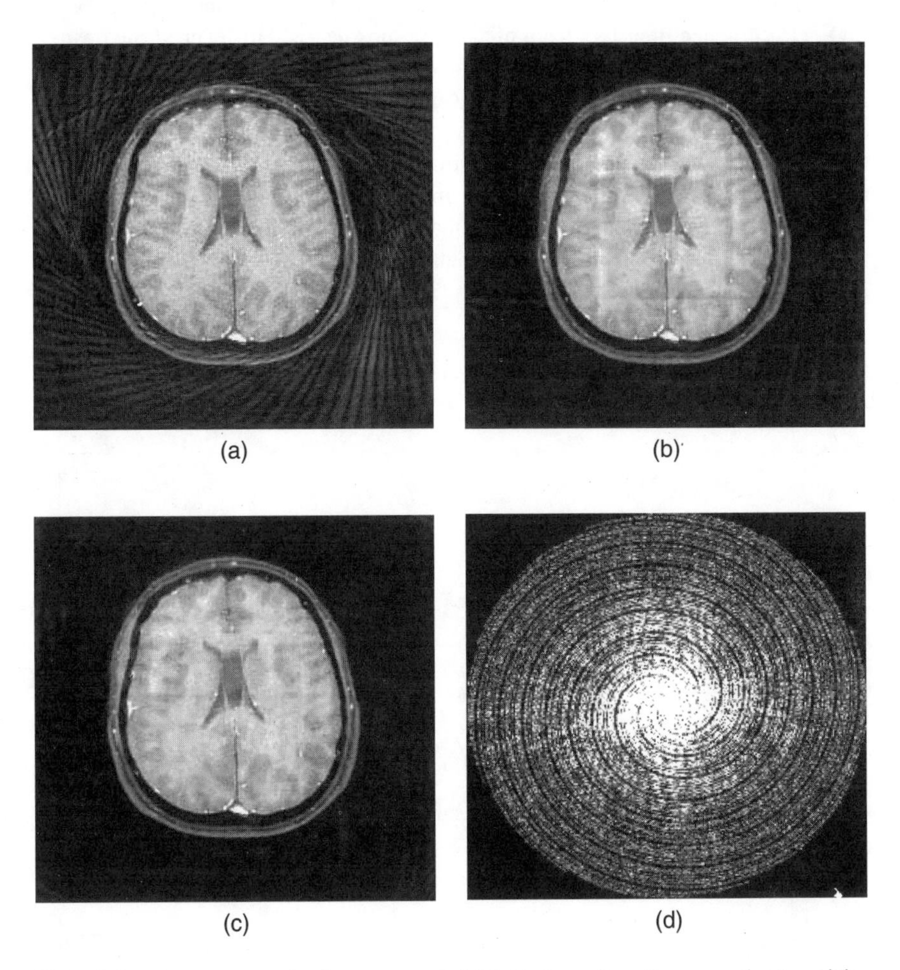

Figure 9. Bayesian reconstruction from sparse spiral data for three alternative approaches to omitting trajectories each corresponding to 33% reduction of measurement time: (a) The remaining trajectories are evenly spread over k-space. (b) Every third trajectory of the full set is omitted. (c) Trajectories are omitted at random from the full set; the best-performing sparse set is chosen. (d) Best-performing sparse data set, pertaining to (c).

to a few tenths of a dB, and allows limitation of the search for optimal omission patterns.

We present results for three alternative approaches [59]: (a) The remaining 40 trajectories are evenly spread over *k*-space, (b) every third trajectory of the full set of 60 is omitted, (c) trajectories are omitted at random from the full set of 60; this is repeated for 1000 different random omissions.

The three approaches were tested by omitting trajectories from a full spiral scan of a human brain. Approach (a) causes a reduction of the Field of View (FOV) [58]; hence it suffers from aliasing in most cases. However, if the desired part of the object is smaller than the reduced FOV, it can be made successful [58]. Although the image investigated here (Fig. 9) is relatively small, its edges do cross the boundaries of the reduced FOV, as is apparent in Fig. 9(a). As expected, the quality of the unaffected part of the image is optimal. On the whole, we do not recommend approach (a) for full-sized images.

Approach (b) is found to be less than optimal for Cartesian scanning; see Section 10.4.2. It remains to be seen whether the same is true for spiral scanning.

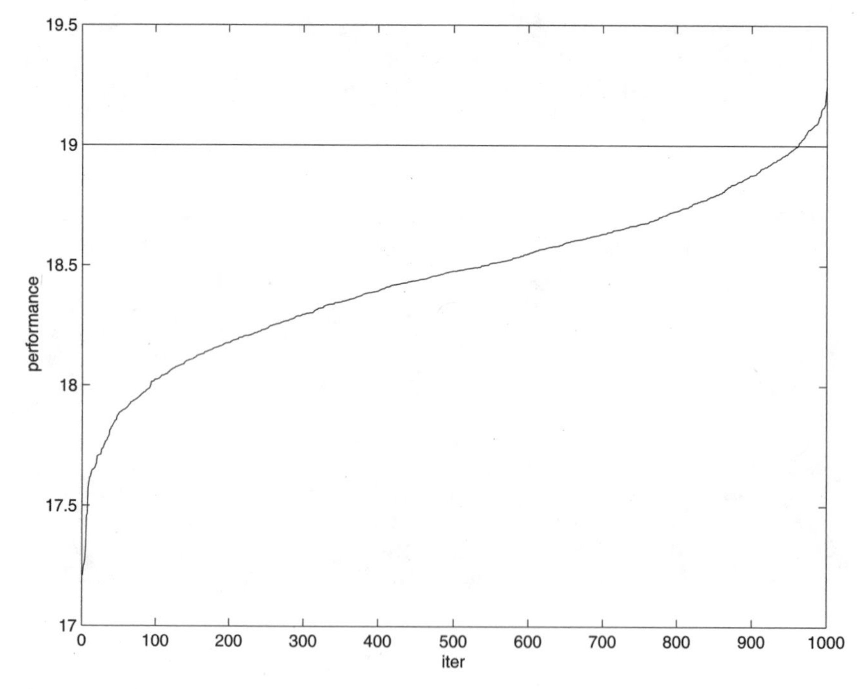

Figure 10. Image quality obtained with 1000 alternative random omissions of 20 spiral trajectories from a full set of 60 trajectories. The results have been sorted according to increasing performance. Horizontal axis: index of sorted random distribution. Horizontal line: performance for regular omission of 20 trajectories (*i.e.*, every 3rd trajectory in a set of 60). Vertical axis: Image quality in dB.

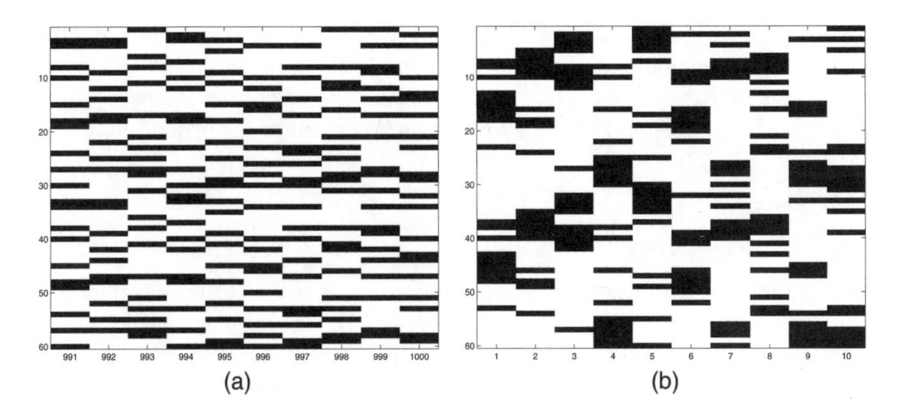

Figure 11. Random omission (black) of 20 spiral trajectories from a full set of 60 trajectories. Vertical axis: angular starting position of spiral trajectory. Horizontal axis: index of random distribution (sorted according to image quality): (a) Ten best random omissions. (b) Ten worst random omissions.

Surprisingly, the performance appears rather good. However, a clearly visible square artefact perturbs the image. The cause of this artefact can be traced to imperfections of the measurements at $\vec{k} = \vec{0}$. With simulations, no such artefact appears. It follows that approach (b) is applicable if experimental conditions could be improved.

Approach (c) is optimal for Cartesian scanning; see Section 10.4.2. We have investigated the case of spiral scanning by perusing the performances of a thousand random distributions. The performances are plotted in Fig. 10, sorted according to increasing value. Also indicated is the performance of approach (b). It follows that a number of random omissions perform better than regular omission. Figure 11 shows the ten best and ten worst random distributions. Apparently, some clustering of omissions is favourable. Figure 9(c) shows the image obtained from the best random omission. Strong reduction of the square-type artefact is immediately apparent. This implies that the effect of experimental imperfections at $\vec{k} = \vec{0}$ is averaged out when spreading the omissions randomly.

10.4.5. Random Sampling

10.4.5.1. Introduction

In Section 10.2, we pointed out that the shape of MRI scan trajectory is determined by time-dependent gradient fields. Physical considerations show that smooth trajectories are less demanding on the hardware. Hence the irregular, random sample positions to be treated in this section are not designed intentionally but arise due to experimental conditions beyond the control of the scanner operator. Such conditions can have serious consequences in *functional* MRI (fMRI). However, Bayesian image reconstruction can cope with regionally 60% undersampled random scanning as depicted in Fig. 1 [30]. In the following, we show that Bayesian image reconstruction can accommodate samples whose positions have become pseudorandom through correction for patient motion in fMRI [35, 36, 60].

10.4.5.2. Motion Correction in fMRI

In an fMRI study, one obtains a time-series of MR images during a stimulation or task paradigm of interest [61, 62]. Those image pixels whose intensity variations correlate with the temporal behaviour of the paradigm correspond to the functional stimulation. Hence, one speaks of functional MRI, abbreviated as fMRI. In the last few years, one of the most significant developments is to noninvasively map human cortical function [63, 64], giving rise to a steep increase of brain fMRI applications.

fMRI relies on the detection of small localized variations in signal intensity, only slightly higher than the noise. Ideally, only those variations related to neuronal activation should be detected. In practice, many other sources contribute to intensity variation, which leads to artefacts in the resultant functional maps [65]. These sources include system instability, subject motion, and normal physiological motions. In particular, artefacts due to *gross* subject motion have been recognized as possible source of false activation [66]. But brain motion and other effects caused by respiration and cardiac pulsation contribute also. In summary, motions will influence fMRI maps through their amplitudes and their correlation to the stimulation paradigm.

Although emerging ultrafast imaging methods [3] are increasingly applied, conventional Cartesian scanning is still current in fMRI. We now present a method to estimate gross motion during individual Cartesian scans in a time-series, called intrascan motion [36]. The method pertains to the 2-D case and estimates rotation motions using the correlation between a (motion-free) reference k-space scan and the corrupted k-space scan. The reference k-space scan can be for instance the first or any other of the fMRI time-series. See [67] for motion correction in anatomic MRI.

An in-plane gross motion can be described in terms of a rotation α and a translation T. We assume that random motions do not occur during sampling along a trajectory but only during the time slot between subsequent trajectories. For a grid size N, N translations and N rotations need to be estimated. One can express the samples $S_{\alpha_i,\vec{T}_i}(\vec{k}_i)$ of the corrupted i^{th} trajectory as a function of the reference samples $S(\vec{k}_i)$

$$S_{\alpha_i,\vec{T}_i}(\vec{k}_i) = R_{\alpha_i} S(\vec{k}_i) e^{-2\pi j R_{\alpha_i}^{-1} \vec{k}_i . \vec{T}_i}, \tag{40}$$

where $\vec{T}_i$ is the translation vector and R_{α_i} is the rotation operator of angle α_i about the centre of the FOV describing the gross motion of the object.

It can be seen from (40) that 1—a translation of the object induces a phase $\phi_{\vec{T}_i}$ linear in k and depending also on the rotation operator, 2—a rotation R_{α_i} in image space is equivalent to a rotation R_{α_i} in k-space. This means that when intrascan rotations occur, trajectories are acquired on nonparallel directions leading to *pseudorandom* sampling. Thus, correction of rotation motions requires resampling of the k-space data to a uniform rectangular grid. Figure 12 shows an example of motion-corrected sample positions for a simulated case. Also shown are the areas associated with each corrected sampling position, as estimated with the Voronoi algorithm. These areas are the inverse of the sampling density ρ, mentioned in Section 10.3.3.2. Unfortunately, the inverse Fourier transform of $\mathcal{S}(\vec{k})/\rho$ similar to that in row 2 in Fig. 3, is perturbed inside the FOV for the given pseudorandom sampling. In such a case, one can resort to Bayesian Image reconstruction because it obviates the sampling density correction and can handle local undersampling.

For a simulated case, Fig. 13 shows that Bayesian image reconstruction can adequately handle pseudorandom sample positions caused by gross motion [36]. Figure 13(a) displays the reference image, reconstructed with 2-D IFFT from the

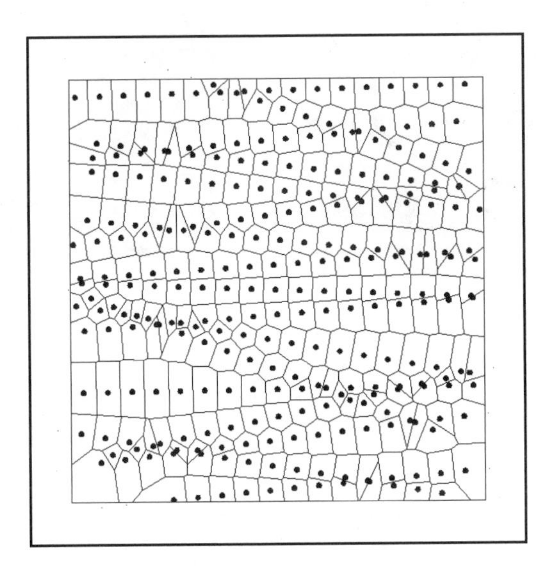

Figure 12. Correction for random rotations that occurred in the time slots between successive k-space trajectories, leading to pseudorandom sampling distribution. Also shown are the areas associated with each sample, i.e., the inverse of the sampling density, computed with the Voronoi algorithm [48].

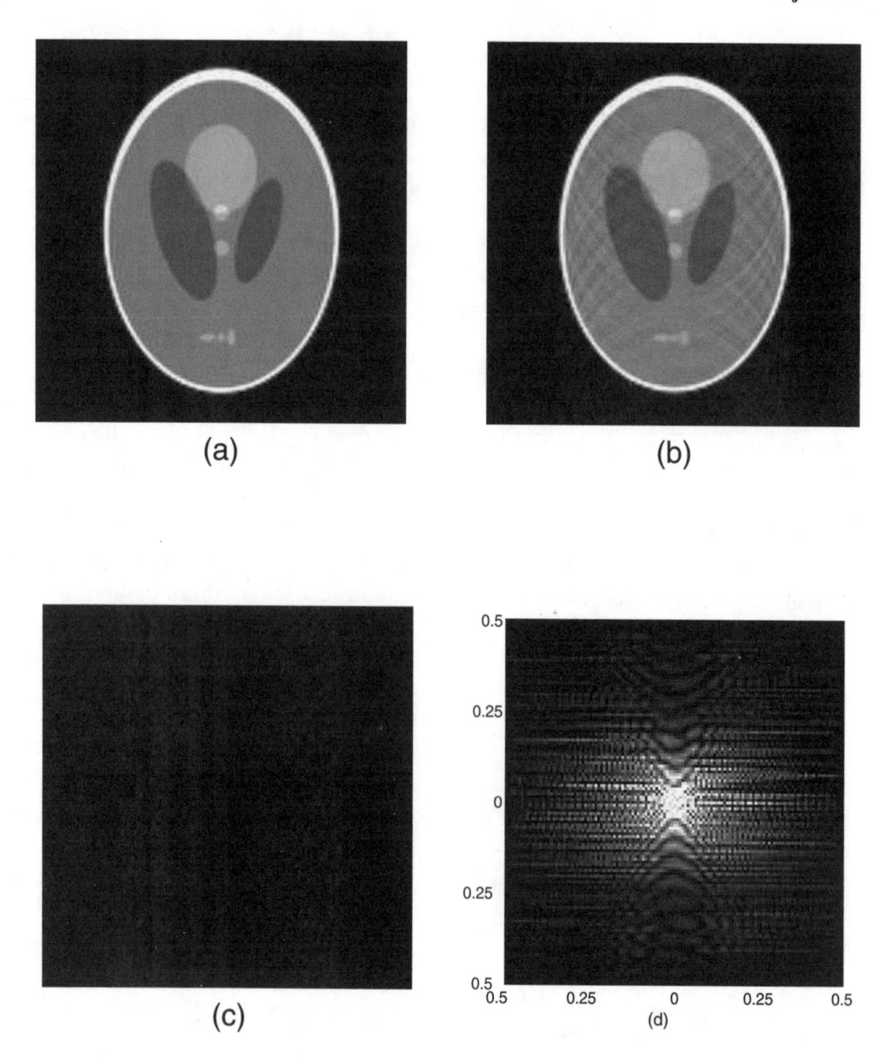

Figure 13. Correction for intrascan rotations using a simulation according to Shepp-Logan [1]. Trajectories were rotated over random angles having a Gaussian distribution ($\mu = 0.0$, $\sigma = 1.0$ degree): (a) Reference image. (b) Distorted image reconstructed using 2-D IFFT and showing blurring and ghosting. (c) Difference between (a) and (b) showing motion distortions. (d) The *k*-space data with rotation artefacts. (e) Bayesian reconstruction of the image. (f) The *k*-space data after Bayesian reconstruction.

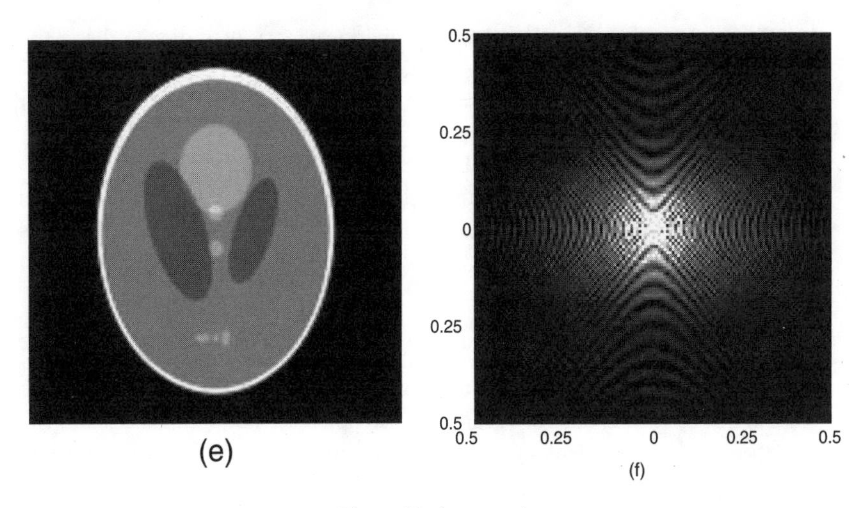

Figure 13. *(continued)*

original, unperturbed data. Subsequently, we perturb the data by random intrascan motion. Ignoring the effect of motion, the reconstruction was done again with 2-D IFFT. Figure 13(b) shows that the motion results in ghosting in the vertical direction and in blurring. The latter artefacts are clearly visible in the difference between Fig. 13(a) and Fig. 13(b), shown in Fig. 13(c). The next step is to estimate the intrascan motion using the method described above; see Fig. 13(d). Reconstruction by gridding and 2-D IFFT of the data in Fig. 13(d) induces artefacts (not included in the Figure) worse than those visible in Fig. 13(b). However, as shown in Fig. 13(e), Bayesian reconstruction is capable of producing a good image. Finally, the FFT of the latter image, displayed in Fig. 13(f), shows the result in the k-space.

10.5. Summary and conclusions

This chapter concerns the topic of nonuniform sampling and image estimation (reconstruction) from the raw data, in MRI. MRI data are sampled in the Fourier space, also called k-space. As for sampling proper, the following facts and conclusions were established.

- Sampling is done along 'trajectories' in k-space, e.g., along straight lines or spirals.

- Physical conditions rather than information theory may dictate the shape of the trajectories. Often, this results in nonuniform sampling. Conventional MRI in clinics uses sample positions on a uniform rectangular (Cartesian) grid.
- The scan time is determined by the number of trajectories.
- Preferably, the number of trajectories is chosen such that all k-space is covered subject to the Nyquist criterion. If the available scan time is insufficient, trajectories must be omitted, in spite of concomitant local undersampling.

In image reconstruction from the raw data in k-space, the most important points are

- If the samples reside on a Cartesian grid, image reconstruction is done simply by n-D IFFT. However, if the samples are not on a Cartesian grid, more elaborate means are required. In most methods for non-Cartesian data, estimation of the density of sample positions plays an important role.
- Various nonuniform sampling distributions, governed by simple equations, are amenable to image reconstruction by special methods. An example of this is the linogram sampling distribution and attendant chirp-z image reconstruction.
- General nonuniform sampling distributions that satisfy the Nyquist criterion can be handled by gridding. This method resamples the data to a Cartesian grid, after which n-D IFFT is applied.
- Bayesian image reconstruction can handle arbitrary sample positions that locally violate the Nyquist criterion. Advantages of this method are

  - Estimation of the sampling density is obviated.
  - Reduction of the scan time by judicious omission of trajectories is possible.
  - Motion effects incurred during scanning can be corrected.

  A disadvantage of the method is the amount of computing involved.

Acknowledgments

The authors are supported by the Dutch Technology Foundation (STW) and the EU Programme TMR/Networks, project No ERB-FMRX-CT97-0160. Y. Crémillieux (UCB Lyon I) performed the radial MRI measurement. J. Groen and M. Fuderer (Philips Medical Systems) performed the spiral MRI measurement. G. J. Marseille (presently with KNMI) initiated our research on Bayesian image reconstruction and performed the Cartesian applications. The authors acknowledge stimulating discussions with R. de Beer, A. Briguet,

M. Décorps, R. A. J. de Jong, R. Lethmate, L. T. Martinez, A. F. Mehlkopf, J. van Osch, M. Roth, C. Segebarth, E. Woudenberg, and J. H. Zwaga.

References

[1] A. C. Kak and M. Slaney. *Principles of Computerized Tomographic Imaging*. IEEE Press, 1987.

[2] P. -E. Danielsson. Algorithms for Reconstructions from Projections. In E. J. H. Kerckhofs, P. M. A. Sloot and J. F. M. Tonino, Eds., *Proc., ASCI*, Lommel, Belgium, June 1996, pp. 1–14.

[3] M. T. Vlaardingerbroek and J. A. den Boer. *Magnetic Resonance Imaging: Theory and Practice*. Springer, Berlin, (2nd edn), 1999.

[4] D. B. Twieg. The *k*-Trajectory Formulation of the NMR Imaging Process with Application and Synthesis of Imaging Methods. *Medical Physics*, 10:610–621, September 1983.

[5] K. F. King and P. R. Moran. A Unified Description of NMR Imaging, Data-Collection Strategies and Reconstruction. *Medical Physics*, 11:1–14, January 1984.

[6] *Basic Principles of MR Imaging*. Philips Medical Systems, Best, The Netherlands, 1995.

[7] G. A. Wright. Magnetic Resonance Imaging. *IEEE Signal Processing Magazine*, 14:56–66, January 1997.

[8] Z. P. Liang, F. E. Boada, R. T. Constable, E. M. Haacke, P. C. Lauterbur, and M. R. Smith. Constrained Reconstruction Methods in MR Imaging. *Reviews of Magnetic Resonance in Medicine*, 4(2):67–185, 1992.

[9] K. F. King, T. K. F. Foo, and C. R. Crawford. Optimized Gradient Waveforms for Spiral Scanning. *Magnetic Resonance in Medicine*, 34:56–160, August 1995.

[10] J. H. Duyn and Y. H. Yang. Fast Spiral Magnetic Resonance Imaging with Trapezoidal Gradients. *Journal of Magnetic Resonance*, 128:130–134, October 1997.

[11] B. Aldefeld and P. Bornert. Effects of Gradient Anisotropy in MRI. *Magnetic Resonance in Medicine*, 39:606–614, April 1998.

[12] G. H. Glover and D. C. Noll. Consistent Projection Reconstruction (CPR) Techniques for MRI. *Magnetic Resonance in Medicine*, 28:345–351, December 1993.

[13] M. L. Lauzon and B. K. Rutt. Polar Sampling in *k*-Space – Reconstruction effects. *Magnetic Resonance in Medicine*, 40:769–782, November 1998.

[14] D. Graveron-Demilly, Y. Crémillieux, S. Cavassila, and C. Mauger. A 3-D Fourier Imaging Method Based on Radial Scan. In *Proc. Int. Society of Magnetic Resonance in Medicine, Third Scientific Meeting*, Nice, France, August 1995, p. 671.

[15] D. Graveron-Demilly, G. J. Marseille, Y. Crémillieux, S. Cavassila, and D. van Ormondt. SRS-FT: A Fourier Imaging Method Based on Sparse Radial Scanning and Bayesian Estimation. *Journal of Magnetic Resonance B*, 112:119–123, August 1996.

[16] P. Edholm and G. T. Herman. Linograms in Image Reconstruction from Projections. *IEEE Trans. Medical Imaging*, MI-6(4):301–307, December 1987.

[17] N. Gai and L. Axel. A Dual Approach to Linogram Imaging for MRI. *Magnetic Resonance in Medicine*, 38:337–341, August 1997.

[18] E. Yudilevich and H. Stark. Spiral Sampling: Theory and an Application to Magnetic Resonance Imaging. *J. Optical Society of America*, 5:542–553, April 1988.

[19] C. H. Meyer, B. S. Hu, D. G. Nishimura, and A. Macovski. Fast Spiral Coronary Artery Imaging. *Magnetic Resonance in Medicine*, 28:202–213, December 1992.

[20] Y. H. Yang, G. H. Glover, P. van Gelderen, V. S. Mattay, A. K. S. Santha, R. H. Sexton, N. F. Ramsey, C. T. W. Moonen, D. R. Weinberger, J. A. Frank, and J. H. Duyn. Fast 3D Functional Magnetic Resonance Imaging at 1.5 T with Spiral Acquisition. *Magnetic Resonance in Medicine*, 36:620–626, October 1996.

[21] S. Lai and G. H. Glover. Three-Dimensional Spiral fMRI Technique – A Comparison with 2D Spiral Acquisition. *Magnetic Resonance in Medicine*, 39:68–78, January 1998.

[22] H. Azhari, O. E. Denisova, A. Montag, and E. P. Shapiro. Circular Sampling: Perspective of a Time-Saving Scanning Procedure. *Magnetic Resonance Imaging*, 14(6):625–631, 1996.

[23] X. H. Zhou, Z. P. Liang, S. L. Gewalt, G. P. Cofer, P. C. Lauterbur, and G. A. Johnson. Fast Spin-Echo Technique with Circular Sampling. *Magnetic Resonance in Medicine*, 39:23–27, January 1998.

[24] N. E. Myridis and C. Chamzas. Sampling on Concentric Circles. *IEEE Transactions on Medical Imaging*, 17:294–299, April 1998.

[25] N. E. Myridis and C. Chamzas. Rhombus Hankel Transform Reconstruction (RHTR) and Rosette Trajectories. In *Proc. Int. Society of Magnetic Resonance in Medicine, Sixth Scientific Meeting*, Sydney, Australia, April 1998, p. 677.

[26] D. C. Noll. Multishot Rosette Trajectories for Spectrally Selective MR Imaging. *IEEE Transactions on Medical Imaging*, 16:372–377, August 1997.

[27] D. C. Noll, S. J. Peltier, and F. E. Boada. Simultaneous Multislice Acquisition Using Rosette Trajectories (SMART) – A New Imaging Method for Functional MRI. *Magnetic Resonance in Medicine*, 39:709–716, May 1998.

[28] K. Scheffler and J. Hennig. Frequency Resolved Single-Shot MR Imaging using Stochastic *k*-Space Trajectories. *Magnetic Resonance in Medicine*, 35:569–576, April 1996.

[29] R. Proksa and V. Rasche. Reconstruction of MR Images from Data Sampled Along Arbitrary *k*-Space Trajectories. In *Proc. Int. Society of Magnetic Resonance in Medicine, Sixth Scientific Meeting*, Sydney, Australia, April 1998, p. 668.

[30] F. T. A. W. Wajer, R. de Beer, M. Fuderer, A. F. Mehlkopf, and D. van Ormondt. Bayesian Image Reconstruction from Arbritrarily Sampled *k*-Space without Density Correction. In *Proc. Int. Society of Magnetic Resonance in Medicine, Sixth Scientific Meeting*, Sydney, Australia, April 1998, p. 667.

[31] G. F. Mason, T. Harshbarger, H. P. Hetherington, Y. Zhang, G. M. Pohost, and D. B. Twieg. A Method to Measure Arbitrary *k*-Space Trajectories for Rapid MR Imaging. *Magnetic Resonance in Medicine*, 38:492–496, September 1997.

[32] X. P. Ding, J. Tkach, P. Ruggieri, J. Perl, and T. Masaryk. Improvement of Spiral MRI with the Measured *k*-Space Trajectory. *Journal of Magnetic Resonance Imaging*, 7:938–940, September–October 1997.

[33] J. D. O'Sullivan. A Fast Sinc Function Gridding Algorithm for Fourier Inversion in Computer Tomography. *IEEE Trans. Medical Imaging*, 4:200–207, December 1985.

[34] J. I. Jackson, C. H. Meyer, D. G. Nishimura, and A. Macovski. Selection of a Convolution Function for Fourier Inversion Using Gridding. *IEEE Trans. Medical Imaging*, 10:473–478, September 1991.

[35] M. Bourgeois, F. Wajer, G. Stijnman, Y. Crémillieux, D. van Ormondt, A. Briguet, and D. Graveron-Demilly. Reduction of Random in Plane Intrascan Rotation Artefacts. In *Proc. Int. Soc. Magn. Reson. Med, Sixth Sci. Meeting*, Sydney, Australia, April 1998, p. 1963.

[36] M. Bourgeois, F. T. A. W. Wajer, D. van Ormondt, and D. Graveron-Demilly. Reconstruction of MRI Images from Non-Uniform Sampling: Application to Intrascan Motion Correction in Functional MRI, Birkhauser. In P. J. Ferreira and J. J. Benedetto, Eds., *Sampling Theory and Applications*, Birkhauser, Boston, 2001, pp. 343–363.

[37] G. P. Zientara, L. P. Panych, and F. A. Jolesz. Dynamically Adaptive MRI with Encoding by Singular Value Decomposition. *Magnetic Resonance in Medicine*, 32:268–274, August 1994.

[38] L. P. Panych. Theoretical Comparison of Fourier and Wavelet Encoding in Magnetic Resonance Imaging. *IEEE Trans. Medical Imaging*, 15:141–153, April 1996.

[39] G. P. Zientara, L. P. Panych, and F. A. Jolesz. Applicability and Efficiency of Near-Optimal Spatial Encoding for Dynamically Adaptive MRI. *Magnetic Resonance in Medicine*, 39:204–213, February 1998.

[40] N. Gai and L. Axel. Characterization of and Correction for Artifacts in Linogram MRI. *Magnetic Resonance in Medicine*, 37:275–284, February 1997.

[41] G. H. L. A. Stijnman, D. Graveron-Demilly, F. T. A. W. Wajer, and D. van Ormondt. MR Image Estimation from Sparsely Sampled Radial Scans. In *ProRISC IEEE Workshop on Circuits, Systems and Signal Processing*, Mierlo, The Netherlands, November 1997, pp. 603–610.

[42] H. Schomberg and J. Timmer. A Gridding Method for Image Reconstruction by Fourier Transformation. *IEEE Trans. Medical Imaging*, 14(3):596–607, September 1995.

[43] G. J. Marseille. MRI Scan Time Reduction through Nonuniform Sampling. Ph.D. Thesis. Delft University of Technology, NL, 1997.

[44] F. T. A. W. Wajer, G. H. L. A. Stijnman, R. de Beer, A. F. Mehlkopf, D. Graveron-Demilly, and D. van Ormondt. MRI Measurement Time Reduction by Undersampling and Bayesian Reconstruction Using a 2D Lorentzian Prior. In B. M. ter Haar Romeny, D. H. J. Epema, J. F. M. Tonino, A. A Wolters, Eds., *Proc., ASCI*, Lommel, Belgium, June 1998, pp. 147–153.

[45] M. Magnusson. Linogram and Other Direct Fourier Methods for Tomographic Reconstruction. Ph.D. Thesis. Linköping University, SE, 1993.

[46] J. H. Zwaga. 'MR Image Reconstruction from Nonuniform Samples using Convolution Gridding.' M.S. Thesis. Delft University of Technology, NL, 1997.

[47] D. Slepian. Some Comments on Fourier Analysis, Uncertainty, and Modeling. *SIAM Rev.*, 25:379–393, July 1983.

[48] A. Okabe, B. Boots and K. Sugihara. *Spatial Tessellations: Concepts and Applications of Voronoi Diagrams*. Wiley, New York, 1992.

[49] M. Fuderer. Ringing Artefact Reduction by an Efficient Likelihood Improvement Method. In *Proc. SPIE*, April 1989, pp. 84–90.

[50] A. H. Lettington and Q. H. Hong. Image Restoration Using a Lorentzian Probability Model. *J. Modern Optics*, 42:1367–1376, July 1995.

[51] B. McNally. Lorentzian Probability Model. Internal Report, pp. 16–17, King's College, London, U.K., 1996.

[52] D. S. Sivia. *Data Analysis. A Bayesian Tutorial*, Clarendon Press, Oxford, 1996.

[53] D. Rosenfeld. An Optimal and Efficient New Gridding Algorithm Using Singular Value Decomposition. *Magnetic Resonance in Medicine*, 40:14–23, July 1998.

[54] T. Hebert and R. Leahy. A Generalized EM Algorithm for 3D Bayesian Reconstruction from Poisson Data Using Gibbs Priors. *IEEE Trans. Medical Imaging*, 8:194–202, June 1989.

[55] G. J. Marseille, M. Fuderer, R. de Beer, A. F. Melkopf, and D. van Ormondt. Reduction of MRI Scan Time Through Nonuniform Sampling and Edge-Distribution Modeling. *J. Magnetic Resonance B*, 103:292–295, March 1994.

[56] G. J. Marseille, M. Fuderer, R. de Beer, A. F. Melkopf, and D. van Ormondt. Non-Uniform Phase-encode Distributions for MRI Scan Time Reduction. *J. Magnetic Resonance B*, 111:70–75, April 1996.

[57] M. S. Albert and D. Balamore. Development of Hyperpolarized Noble Gas MRI. *Nuclear Instruments & Methods in Physics Research Section A-Accelerators Spectrometers Detectors & Associated Equipment*, 402:441–453, January 1998.

[58] K. Scheffler and J. Hennig. Reduced Circular Field of View Imaging. *Magnetic Resonance in Medicine*, 40:474–480, September 1998.

[59] F. T. A. W. Wajer, G. H. L. A. Stijnman, D. Graveron-Demilly, M. Fuderer, A. F. Mehlkopf, and D. van Ormondt. Sparse Sampling Distributions for MRI. In *ProRISC, IEEE Benelux*, Mierlo, The Netherlands, November 1998, pp. 603–608.

[60] M. Bourgeois, F. T. A. W. Wajer, G. H. L. A. Stijnman, Y. Crémillieux, D. van Ormondt, A. Briguet, and D. Graveron-Demilly. Réduction des Artéfacts de Rotation Aléatoire Intra-Image. In *GRAMM*, Toulouse, France, February 1998, p. 28.

[61] S. Ogawa, R. S. Menon, S. G. Kim, and K. Ugurbil. On the Characteristics of Functional Magnetic Resonance Imaging of the Brain. *Annual Review of Biophysics and Biomolecular Structure*, 27:447–474, 1998.

[62] L. R. Frank, R. B. Buxton, and E. C. Wong. Probabilistic Analysis of Functional Magnetic Resonance Imaging Data. *Magnetic Resonance in Medicine*, 39:132–148, January 1998.

[63] S. Ogawa, T. M. Lee, A. R. Kay, and D. W. Tank. Brain Magnetic Resonance Imaging with Contrast Dependent on Blood Oxygenation. *Proc. Natl. Acad. Sci. USA*, 87:9868–9872, December 1990.

[64] J. W. Belliveau, B. R. Rosen, H. L. Kantor, R. R. Rzedzian, D. N. Kennedy, R. C. McKinstry, J. M. Vevea, M. S. Cohen, I. L. Pykett, and T. J. Brady. Functional Cerebral Imaging by Susceptibility-Contrast NMR. *Magnetic Resonance in Medicine*, 14:538–546, June 1990.

[65] X. Hu, T. H. Le, T. Parrish, and P. Erhard. Retrospective Estimation and Correction of Physiological Fluctuation in Functional MRI. *Magnetic Resonance in Medicine*, 34:201–212, August 1995.

[66] J. V. Hajnal, R. Myers, A. Oatridge, J. E. Schwieso, I. R. Young, and G. M. Bydder. Artifact due to Stimulus Correlated Motion in Functional Imaging of the Brain. *Magnetic Resonance in Medicine*, 31:283–291, March 1994.

[67] C. Weerasinghe and H. Yan. Correction of Motion Artefacts in MRI Caused by Rotations at Constant Angular Velocity. *Signal Processsing*, 70:103–114, October 1998.

11

Irregular and Sparse Sampling in Exploration Seismology

A. J. W. Duijndam, M. A. Schonewille and C. O. H. Hindriks

11.1. Introduction

The aim of exploration seismology is to obtain an image of the subsurface by probing it with seismic waves. These waves are generated at the surface, by using a suitably powerful source, at various locations. The waves propagate downwards through the subsurface, and are reflected at interfaces between geological layers. They subsequently propagate upwards to the surface, where they are detected and recorded. The seismic response depends of course on the source and the receiver location. In practice, the response for one source location is detected by hundreds of receivers, typically arranged along several receiver lines. In common use today is three-dimensional (3-D) surface seismology, where a volume image is obtained, rather than a cross-section image as in 2-D seismology. The sources and receivers can, for the sake of simplicity, be assumed to be located in a horizontal plane, at the surface. The recorded seismic wavefield then is a function of five variables, $p(x_s, y_s, x_r, y_r, t)$, with two horizontal coordinates, x_s and y_s, for the source position, two horizontal positions, x_r and y_r, for the receiver positions and the variable time, denoted by t. In 2-D seismology the signal $p(x_s, x_r, t)$ is a function of two spatial coordinates and the time coordinate. For readers, not familiar with exploration seismology, we will give a brief overview of some notions, required to understand this chapter, in the next few subsections.

A. J. W. Duijndam • Philips Medical Systems, P.O. Box 10000, 5680 DA Best, The Netherlands,
M. A. Schonewille • PGS Onshore, 738 Highway 6 South, Houston, TX 77079, USA,
C. O. H. Hindriks • Delft University of Technology, Lorentzweg 1, 2628 CJ Delft, The Netherlands.

Nonuniform Sampling: Theory and Practice, edited by Marvasti
Kluwer Academic/Plenum Publishers, New York, 2001.

11.1.1. Spatial Sampling in Exploration Seismology

Whereas the time coordinate is regularly sampled (after anti-alias filtering), all four spatial coordinates are typically irregularly sampled. This is, amongst others, due to the presence of obstacles in land seismology and limited navigation capabilities and the presence of currents in marine seismology. Moreover, seismic data is sparsely sampled along the spatial coordinates. Consider, for example, a seismic survey of 10×10 km and a complete, regular sampling of this area by sources and receivers with a sampling interval of 10 m along each spatial coordinate. This would yield an impractical number of 10^{12} combinations of sources and receivers, i.e., spatial samples. Limiting the maximum distance between source and receivers to a more practical 3 km, one still obtains approximately 10^{11} spatial samples. In practice however, a survey of this size typically consists of 10^7 spatial samples, a factor 10^4 less than ideally required. (Note that, with 2000 samples along the time coordinate, and four bytes per sample, the total data volume still is in the order of 100 Gbyte.) Despite the spatial undersampling, quite satisfactory images can nevertheless be obtained, due to a particular design of the sampling geometries, in combination with seismic data processing.

However, conventional seismic processing algorithms do not handle irregular and sparse spatial sampling in an optimal way. Simple approaches used in practice for handling missing traces in otherwise regularly sampled subsets of seismic data are: (1) copying or linearly interpolating neighbouring traces, and (2) ignoring them altogether (by inserting dead traces). For certain processing steps in 3-D seismic processing the data is regularized by collecting seismic traces within squares and subsequently stacking them, thereby disregarding their actual positions. This is clearly not optimal. In this chapter, reconstruction algorithms are discussed that address the irregular and sparse sampling of seismic data. It is clear that, due to the sheer size of the data volumes, an integral approach in which the four spatial coordinates are treated simultaneously is, practically speaking, impossible. Therefore, algorithms for one and two spatial dimensions are discussed which can be applied to various subdomains of seismic data and in various stages of seismic data processing.

11.1.2. Subdomains

A 1-D subdomain, in the context of this chapter, is a data set with one varying spatial coordinate, and the time coordinate. For example, in 2-D seismology, the data set for one fixed source position (a so-called Common Shot (CS) gather) is a function of the receiver coordinate and the time coordinate: $p(x_r, t)$. A 2-D subdomain is a data set with two varying spatial coordinates. An example is a CS-gather in 3-D seismology: $p(x_r, y_r, t)$. 1-D and 2-D Common Receiver (CR) gathers can be defined in a similar way.

The source and receiver coordinates are often transformed into midpoint (x_m and y_m) and (half) offset (x_h and y_h) coordinates according to:

$$x_m = \frac{x_s + x_r}{2}, \tag{1}$$

$$y_m = \frac{y_s + y_r}{2}, \tag{2}$$

$$x_h = \frac{x_s - x_r}{2}, \tag{3}$$

and

$$y_h = \frac{y_s - y_r}{2}. \tag{4}$$

The latter are, in turn often transformed into absolute offset h and azimuth θ:

$$h = \sqrt{(2x_h)^2 + (2y_h)^2}, \tag{5}$$

$$\theta = \tan^{-1} x_h/y_h. \tag{6}$$

In this way we can also construct Common Midpoint (CMP) amd Common Offset (CO) gathers, useful in seismic data processing.

11.1.3. General Strategy for Reconstruction

In handling irregularly sampled data, one may wish to reconstruct the data in a transform domain, corresponding to some desired transformation of the data or to regularise the data, i.e., reconstruct the data on a regular grid in the spatial domain. The underlying principles for solving these two problems are equivalent. For the first problem, a straightforward discretization of the forward transformation corresponding to the irregular grid at hand will usually lead to unacceptably large errors. Instead, one can take the exact inverse transform from the regularly sampled transform domain to the spatial domain (for any spatial location) and use this as a forward model in an inverse problem that is subsequently solved. In exploration seismology, this approach has been taken for example for a hyperbolic transform [1], for the linear Radon transform [1], and for the parabolic Radon transform [2]. If desired, the estimated data in the transform domain can be transformed to the regularly sampled spatial domain, thus solving the regularisation problem. For efficiency reasons the transformation back to the spatial domain may also be incorporated in the forward model, which then describes the transformation from the regularly sampled spatial domain to the irregularly sampled spatial domain. This forward model is implicitly constrained by the limited region in the transform domain. The transform with the most compact description of the data yields the best reconstruction. These general principles will be illustrated for the Fourier transform.

The Fourier domain reconstruction methods do not require any geological or geophysical assumptions concerning the data other than being spatially band limited. They can handle arbitrarily irregular sampling geometries. The basic idea behind the method of estimating the Fourier spectrum from irregularly spaced samples has been published in the applied mathematics literature (see, amongst others, Marvasti [3], Marvasti [4], Feichtinger et al. [5] and many references given there). In contrast and in addition to these publications we will in this chapter: (1) develop the basic method from discrete inverse theory, (2) give a more sound foundation for the weighting schemes used, (3) introduce stabilization schemes based on the physics of the problem, (4) adapt the algorithms to seismic data through optimal parameterizations, (5) discuss edge effects and aliasing, (6) derive practical uncertainty and quality control measures, and (7) give very fast implementations of the algorithms.

In Section 11.2, the reconstruction of bandlimited seismic signals, irregularly sampled along one spatial coordinate, is discussed. The algorithm is based on least squares estimation of Fourier coefficients. In Section 11.3, the reconstruction is extended to two spatial coordinates. The issue of sparse sampling is discussed in Section 11.4. The approach here is based on a 2-D Fourier or mixed Fourier-Radon transform, in combination with a judicious choice of the Region of Support (ROS) in the transform domain. In Section 11.5 efficiency issues are discussed.

11.2. 1-D Reconstruction Using the Fourier Transform

11.2.1. Introduction

Application of the approach described above for the Fourier transform has been proposed, e.g., by Kar *et al.* [6], Marvasti [3], [4], and in a series of related papers containing many useful results by Feichtinger, Grochenig and coworkers (see [7]), who base their work on functional analysis and in particular on the theory of frames. A comparison with our more compact fomulation is beyond the scope of this chapter.

We define the continuous forward temporal Fourier transform of a wave-field $p(x, t)$ as[1]

$$P(x,f) = \int_{-\infty}^{\infty} p(x, t)e^{-j2\pi ft}\, dt, \tag{7}$$

where x and t are the space and time coordinate respectively, and f is the temporal frequency.[2] The continuous forward *spatial* Fourier transform is defined as

$$\tilde{P}(k_x,f) = \int_{-\infty}^{\infty} P(x,f)e^{-j2\pi k_x x}\, dx, \tag{8}$$

where k_x is the wavenumber or spatial frequency.[3] The inverse spatial Fourier transform is given by

$$P(x,f) = \int_{-\infty}^{\infty} \tilde{P}(k_x,f)e^{j2\pi k_x x}\, dk_x. \tag{9}$$

For signals, regularly sampled along x, the integral in equation (8) is replaced by a summation:

$$\hat{P}(k_x,f) = \sum_{n=0}^{N-1} P(n\Delta x,f)e^{-j2\pi k_x n\Delta x}\Delta x, \tag{10}$$

where Δx should be chosen small enough to avoid aliasing in the spatial Fourier domain. In the presence of irregular sampling a straightforward approach to obtain the data in the Fourier domain is using the sum corresponding to the actual sample locations $(x_0, \ldots, x_{N-1})$, to wit

$$\hat{P}(k_x,f) = \sum_{n=0}^{N-1} P(x_n,f)e^{-j2\pi k_x x_n}\Delta x_n, \tag{11}$$

[1] In this chapter, we will use lower case letters for functions of space and time, upper case letters for functions of space and temporal frequency, and upper case letters, combined with a tilde for functions of spatial and temporal frequency.
[2] In the text and figures we will also use the circular frequency $\omega_c = 2\pi f$.
[3] The convention in geophysical literature is to use the *positive* sign in the exponent. For consistency within this book we will use the negative sign in the forward transform.

where Δx_n is defined as

$$\Delta x_n = l_{n+1} - l_n = \frac{x_{n+1} - x_{n-1}}{2}, \tag{12}$$

in which $l_n = (x_n + x_{n-1})/2$ is the midpoint between two samples. We will refer to the transform (11) as the Nonuniform Discrete Fourier Transform (NDFT). The NDFT does not reproduce the Fourier spectrum exactly as will be demonstrated later.

11.2.2. Forward Model

The following approach is proposed in order to improve upon the NDFT. Let us consider, for practical purposes, the Fourier domain with sampling interval Δk_x and the Fourier domain is restricted to the interval $[-M\Delta k_x, M\Delta k_x]$, i.e., bandlimited data. Note that shifted versions or even nonadjacent intervals of the same total size can be chosen as well. The inverse discrete Fourier transform for any spatial location x is given by

$$P(x,f) = \Delta k_x \sum_{m=-M}^{M} \tilde{P}(m\Delta k_x, f)e^{j2\pi m\Delta k_x x} \tag{13}$$

and is exact, apart from spatial aliasing, which can be controlled by the choice of Δk_x. The combination of N versions of equation (13) for the irregularly spaced sample locations $(x_0, \ldots, x_{N-1})$ can be written in vector notation as[4]

$$\mathbf{y} = \mathbf{Ap}, \tag{14}$$

with

$$y_n = P(x_n, f), \qquad n = 0, \ldots, N-1, \tag{15}$$

$$p_m = \tilde{P}(m\Delta k_x, f), \qquad m = -M, \ldots, M, \tag{16}$$

and

$$A_{nm} = \Delta k_x e^{j2\pi m\Delta k_x x_n}. \tag{17}$$

[4] In this chapter we use lower case boldface characters for vectors and upper case boldface characters for matrices.

Within the context of inverse theory, or, parameter estimation, (14) is regarded to be a "forward model". Given the spatial frequency components, the irregularly spaced data in the spatial domain can be computed. In practice the data will not be exactly band limited and some spatial frequency components exist beyond the bandwidth that we specify the data to have. These components constitute the errors or "noise" in the forward model and should be incorporated in a noise term $\mathbf{n}$:

$$\mathbf{y} = \mathbf{A}\mathbf{p} + \mathbf{n}. \tag{18}$$

11.2.3. MAP Estimator

We now recognize a standard linear inverse problem where in this case the unknown vector $\mathbf{p}$, containing the data in the regularly sampled Fourier domain is to be estimated from the data vector $\mathbf{y}$ containing the irregularly sampled data in the spatial domain. Any desired parameter estimation technique can be used. Within the Bayesian formalism ([8], [9] and [10]) and under the assumptions of Gaussian distributions for the noise $\mathbf{n} = N(\mathbf{0}, \mathbf{C}_n)$ (expected value $E\mathbf{n} = \mathbf{0}$ and covariance matrix $\mathbf{C}_n$) and a priori information $\mathbf{p} = N(\mathbf{p}^i, \mathbf{C}_p)$ (prior model $\mathbf{p}^i$ and associated covariance matrix $\mathbf{C}_p$) the well known MAP (Maximum of the A Posteriori probability density function) estimator is the least squares solution

$$\hat{\mathbf{p}} = (\mathbf{A}^H\mathbf{C}_n^{-1}\mathbf{A} + \mathbf{C}_p^{-1})^{-1}(\mathbf{A}^H\mathbf{C}_n^{-1}\mathbf{y} + \mathbf{C}_p^{-1}\mathbf{p}^i), \tag{19}$$

where $\mathbf{A}^H$ denotes the complex conjugate transpose of $\mathbf{A}$.

Simplifications of this estimator are obtained as follows. For seismic data we do not usually have an a priori model, therefore $\mathbf{p}^i = \mathbf{0}$. When, according to standard discrete Fourier theory, Δk_x is chosen as $\Delta k_x = 1/X$ (or somewhat smaller, see below), with $X = \sum_n \Delta x_n$, the registration interval, then we have no reason to assume correlations between the spatial frequencies and $\mathbf{C}_p$ will be diagonal. By varying the diagonal, we could specify a preference for certain spatial frequencies to others, for example when flat, events are most likely to occur. Usually, however, $\mathbf{C}_p$ will take the form $\mathbf{C}_p = \sigma_p^2\mathbf{I}$, with σ_p to be specified below.

11.2.4. Weighting

An accurate representation of the noise in $\mathbf{C}_n$ is expensive and typically not realistic in the sense that detailed information on the noise is not available. On the other hand, the specification of uncorrelated noise through $\mathbf{C}_n = \sigma_n^2\mathbf{I}$ would be an oversimplification for closely spaced samples since the noise is approximately band limited. Duijndam *et al.* [11] have shown that a very practical approximation taking negligible computation time is obtained by replacing $\mathbf{C}_n$ by the diagonal matrix $c^2\mathbf{W}^{-1}$, with c^2 a positive constant to be specified below. The diagonal elements of $\mathbf{W}$ correspond to the distances between the samples:

$$W_{ii} = \Delta x_i. \tag{20}$$

The estimator now becomes

$$\hat{\mathbf{p}} = (\mathbf{A}^H\mathbf{W}\mathbf{A} + k^2\mathbf{I})^{-1}\mathbf{A}^H\mathbf{W}\mathbf{y}, \tag{21}$$

with $k^2 = c^2/\sigma_p^2$ to be specified below. Closely spaced samples have less weight than widely spaced samples. Two intuitive arguments for this choice can be given. The first is that for band-limited noise, closely spaced samples contain almost the same signal and noise values hence they do not represent independent information and should instead be regarded as one piece of information or sample. The weight matrix $\mathbf{W}$ achieves this. Another argument arises when we evaluate the term $\mathbf{A}^H\mathbf{W}\mathbf{y}$ on the right hand side of (21). We obtain

$$[\mathbf{A}^H\mathbf{W}\mathbf{y}]_m = \Delta k_x \sum_{n=0}^{N-1} P(x_n, f)e^{-j2\pi m\Delta k_x x_n}\Delta x_n, \qquad m = -M, \ldots, M. \tag{22}$$

and observe that, apart from the constant factor Δk_x, this term represents the sum (NDFT) of equation (11) and is a first approximation to the answer. The operator $(\mathbf{A}^H\mathbf{W}\mathbf{A} + k^2\mathbf{I})^{-1}$ in equation (21) "deconvolves" or improves the output of the NDFT. Without the weight matrix $\mathbf{W}$ the term in (22) would have been the poorer approximation $\mathbf{A}^H\mathbf{y}$ and the corresponding correction process described by the operator $(\mathbf{A}^H\mathbf{A} + k^2\mathbf{I})^{-1}$ would have been less stable. Note that the usage of $\mathbf{W}$ has been proposed by Feichtinger and Grochenig [5] who disregard the noise aspects but base its use solely on the convergence and stability of iterative schemes for the solution of the inverse problem. It is shown by Duijndam *et al.* [11] that the usage of $\mathbf{W}$ and a more correct statistical approach are equivalent for very dense clusters of samples.

11.2.4.1 Stabilization

If one is prepared to specify the ratio F of the expected energies of the noise and the signal in the input data, then it can be shown [11] that the stabilization constant can be approximated by

$$k^2 = \frac{c^2}{\sigma_p^2} = F\Delta k_x \frac{M_p}{N}, \tag{23}$$

where $M_p = 2M + 1$ is the total number of Fourier coefficients. Otherwise, a rule of thumb such as taking 1% of the value of the diagonal for k^2 or trial and error procedures can be used.

The estimator (21) for the general situation of irregular sampling reduces to a standard result for the special case of regular sampling. It can easily be shown that for the choice $\Delta k_x = \frac{1}{N\Delta x}$, it becomes (see also [12])

$$\hat{\mathbf{p}} = (\Delta k_x + k^2)^{-1}\mathbf{A}^H\mathbf{W}\mathbf{y}. \tag{24}$$

Setting the stabilization term to zero and substituting the definitions of **p**, **A** and **y** in equation (24) indeed yields the standard DFT

$$\tilde{P}(k_x, f) = \sum_{n=0}^{N-1} P(n\Delta x, f)e^{-j2\pi k_x n\Delta x}\Delta x. \tag{25}$$

11.2.5. Uniqueness and Stability

Important related questions are: (1) To what extent is the estimate uniquely determined by the data? (2) How stable is the inversion procedure without stabilization constant? (3) How many samples are needed? (4) How large can the gaps in the sampling be set?

A straightforward answer to these questions is that we simply have to have at least as many distinct samples as Fourier coefficients to uniquely solve the inverse problem. This can be seen by rewriting the basic equation that we solve, equation (13), in the form

$$P(x, f) = \Delta k_x \sum_{m=-M}^{M} \tilde{P}(m\Delta k_x, f)w^m \tag{26}$$

with $w = e^{j2\pi\Delta k_x x}$. The $M_p = 2M + 1$ unknown Fourier coefficients $\tilde{P}(m\Delta k_x, f)$ to be determined are the coefficients of a polynomial function of order $M_p - 1$, [13]. They can uniquely be determined when we have at least M_p distinct samples. Correspondingly, the square matrix **A** for $N = M_p$ is the product of a diagonal matrix and the so-called Vandermonde matrix and can be shown to have a determinant unequal to zero when the samples are distinct [14]. Generalizing this to a larger number of samples and including the weighting scheme, it can be shown that the weighted normal equations $\mathbf{A}^H\mathbf{W}\mathbf{A}\mathbf{p} = \mathbf{A}^H\mathbf{W}\mathbf{y}$ can be solved uniquely for $N \geq M_p$, with N the number of distinct samples.

This implies that the only condition for unique reconstruction is that we have to have enough samples and that there is no restriction to their location, allowing for large gaps in data acquisition! Exciting as this may seem, it is of academic

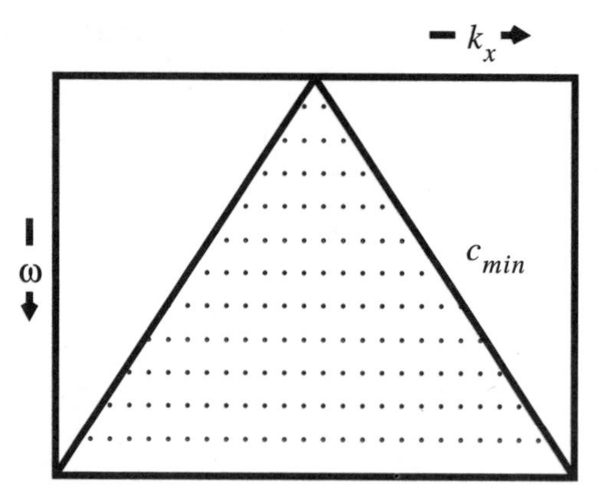

Figure 1. Parameterization: temporal frequency dependent spatial bandwidth determined by the minimum apparent velocity.

interest only since the system to be solved becomes highly unstable when gaps in the data become too large. As will be demonstrated below the signal can be reconstructed in gaps up to three times the Nyquist interval given enough input samples. For larger gaps stabilization is required which forces the reconstructed signal to stay close to zero in the gaps.

11.2.6. Reconstruction Per Time Slice or Per Temporal Frequency Component

The reconstruction method we have described so far is carried out for each temporal frequency f separately. This implies that a temporal Fourier transform, according to a discretized version of (7) is carried out before the actual reconstruction. It is fairly obvious that the problem can be solved per time slice (seismic data for a constant value of the time variable) as well. Seismic data however, usually is reasonably characterized by a minimum apparent velocity, giving rise to f-dependent spatial bandwidth, see Fig. 1.[5] For low f the spatial bandwidth is small, leading both to better efficiency and stability and to a better capability to reconstruct the signal in gaps. For higher f the problem becomes more difficult to solve.

[5] Seismic data can be regarded as a superposition of linear events $p(x, t) = s(x - ct)$, with $s(t)$ the wavelet. The Fourier transform of such a linear event is $\tilde{P}(k_x, f) = S(f + ck_x)$, and is a straight line, defined by $k_x = -f/c$, through the origin of the Fourier domain. The lower the absolute value of the velocity of the event, the further the line deviates from the axis $k_x = 0$.

With reconstruction per time slice the parameterization is identical to the parameterization for the highest f and we loose both the advantages of efficiency and stability. Only for strongly time dependent spatial sampling, as may arise after muting of parts of the seismic data, may reconstruction in the time domain be warranted. An alternative method may be reconstruction using slightly overlapping time windows with a temporal frequency scheme per time window.

11.2.7. Reconstruction in Space Versus Reconstruction in the Fourier Domain

Given the estimator (21) in the Fourier domain, the regularly sampled signal in the spatial domain $\mathbf{y}_r$ is easily obtained by

$$\hat{\mathbf{y}}_r = \mathbf{A}_r\hat{\mathbf{p}} = \mathbf{A}_r(\mathbf{A}^H\mathbf{W}\mathbf{A} + k^2\mathbf{I})^{-1}\mathbf{A}^H\mathbf{W}\mathbf{y}, \tag{27}$$

with $\mathbf{A}_r$ the matrix performing the inverse Fourier transformation similar to the definition in equation (17) but now for regularly spaced samples $x_n = n\Delta x$, $n = -N_x, \ldots, N_x$. In practice of course the FFT can be used.

A more direct way leading to the same result can be obtained for the special case of the Nyquist sampling interval $\Delta x = \dfrac{1}{(2M+1)\Delta k_x}$. First the discrete Fourier transformation is written in the matrix form:

$$\mathbf{p} = \frac{\Delta x}{\Delta k_x}\mathbf{A}_r^H\mathbf{y}_r. \tag{28}$$

Substitution in the forward model (equation (18)) results in

$$\mathbf{y} = \frac{\Delta x}{\Delta k_x}\mathbf{A}\mathbf{A}_r^H\mathbf{y}_r + \mathbf{n} = \mathbf{S}\mathbf{y}_r + \mathbf{n}. \tag{29}$$

The elements of $\mathbf{S} = \dfrac{\Delta x}{\Delta k_x}\mathbf{A}\mathbf{A}_r^H$ are given by

$$S_{ij} = \frac{\sin \pi \dfrac{x_i - j\Delta x}{\Delta x}}{(2M+1)\sin \pi \dfrac{x_i - j\Delta x}{(2M+1)\Delta x}}. \tag{30}$$

Equation (29), without the noise term, for $N_x \to \infty$, and written as:

$$P(x_i, f) = \sum_{j=-\infty}^{\infty} P(j\Delta x, f) \frac{\sin \pi \frac{x_i - j\Delta x}{\Delta x}}{(2M+1) \sin \pi \frac{x_i - j\Delta x}{(2M+1)\Delta x}} \tag{31}$$

can be recognized as the discrete version of the Whittaker-Kotelnikov-Shannon (WKS) reconstruction theorem. The continuous form follows from letting $2M + 1 \to \infty$ while keeping the bandwidth $(2M + 1)\Delta k_x$ fixed:

$$P(x_i, f) = \sum_{j=-\infty}^{\infty} P(j\Delta x, f) \frac{\sin \pi \frac{x_i - j\Delta x}{\Delta x}}{\pi \frac{x_i - j\Delta x}{\Delta x}}. \tag{32}$$

These models relate the band-limited regularly sampled signal to the irregularly spaced samples. The inverse problem cast by (29) can be solved in the same way as the "Fourier domain inverse problem" by

$$\hat{\mathbf{p}}_r = (\mathbf{S}^H \mathbf{W} \mathbf{S} + k^2 \mathbf{I})^{-1} \mathbf{S}^H \mathbf{W} \mathbf{y}. \tag{33}$$

By substitution of $\mathbf{S} = \frac{\Delta x}{\Delta k_x} \mathbf{A} \mathbf{A}_r^H$, it can be demonstrated that for $N_x = M$ equation (33) is identical to equation (27). Note that when a f-dependent spatial bandwidth as proposed in Section 11.2.6 is used, Δx becomes f-dependent as well. After estimation of $\mathbf{y}_r$ an interpolator as described in (31) can be used to go to a finer grid. The direct utilization of a small Δx in the spatial domain estimator implies a larger bandwidth and consequently a less well determined system.

For the regularisation problem, reconstruction in the spatial domain may seem attractive since it saves an inverse FFT but for seismic data the route via the Fourier domain is more efficient as will be demonstrated in the section on efficiency.

11.2.8. Edge Effects

Like Fourier analysis for regular sampling the reconstruction of signals for irregular sampling suffers from signal truncation. A truncation corresponds to convolution with a sinc function in the Fourier domain and broadens the spatial bandwidth. In order to avoid significant edge effects, the signal can be tapered towards the edges in the spatial domain before reconstruction is applied.

The discretization in k_x, corresponds to periodicity in the spatial domain. Aliasing effects may occur when the periodicity interval as defined by $X_p = 1/\Delta k_x$ is not chosen large enough (somewhat larger than the spatial ROS). The difference between the discrete and continuous versions of the WKS reconstruction formula manifests itself in the edge effects. The former corresponds to the sampled Fourier domain and may show wrap around effects in x, while the latter may show edge effects corresponding to zero signal outside the registration interval.

11.2.9. Signal Outside the Bandwidth

The spatial frequency components outside the bandwidth that we specify for the signal, are regarded as noise. In the familiar case of regular sampling, this noise shows up in an organized way due to the periodicity of the Fourier spectrum (aliasing). For random sampling, the NDFT $\mathbf{A}^H\mathbf{W}\mathbf{y}$ in (21), can be regarded as a Monte Carlo approach to the evaluation of the Fourier integral. In this case the noise is distributed evenly over the spectrum and has an RMS value of [15]

$$\sqrt{E|\tilde{N}(k_x)|^2} \approx \frac{X}{N}\sqrt{\sum_{i=1}^{N} |n(x_i)|^2}, \tag{34}$$

where $n(x_i)$ are the samples of the noise. The amplitude is proportional to $1/\sqrt{N}$ for uncorrelated (wide band) noise and thus decreases when the number of samples increases. After the NDFT the input samples have been mapped on the Fourier components within the bandwidth and the correction factor $(\mathbf{A}^H\mathbf{W}\mathbf{A} + k^2\mathbf{I})^{-1}$ cannot distinguish between signal and noise. The precise error propagation depends on the specific sampling, but in general and especially for the case of many input samples without extreme gaps, the noise components will stay more or less evenly distributed over the spectrum. The noise suppression property of the NDFT allows a suboptimal but more efficient way of signal reconstruction for limited bandwidth in case of many samples.

11.2.10. Uncertainty Analysis and Quality Control

A proper usage of the stabilization constant ensures a robust way of processing and the proposed algorithm can be made part of a standard processing flow without extra need for quality control (QC). If desired however, a number of quantities can be computed that can be used for QC as well as for quality assurance (QA) during data acquisition.

11.2.10.1 Fourier Domain

Within Bayesian parameter estimation theory, the most natural quantity to use for uncertainty analysis for the linear Gaussian case is the a posteriori covariance matrix [8]

$$\mathbf{C}_{\langle p\rangle} = (\mathbf{A}^H \mathbf{C}_n^{-1} \mathbf{A} + \mathbf{C}_p^{-1})^{-1}, \tag{35}$$

where $\mathbf{C}_n$ and $\mathbf{C}_p$ are as defined in (19). With our particular choice of $\mathbf{C}_p = \sigma_p^2 \mathbf{I}$ and $\mathbf{C}_n = F\Delta k_x \frac{M_p}{N} \sigma_p^2 \mathbf{W}^{-1}$, this becomes

$$C_{\langle p\rangle} = \sigma_p^2 \left(\frac{N}{FM_p \Delta k_x} \mathbf{A}^H \mathbf{W} \mathbf{A} + \mathbf{I} \right)^{-1} \tag{36}$$

Note that for the computation of $\mathbf{C}_{\langle p\rangle}$, the a priori variance σ_p^2 or its counterpart for the noise c^2 is needed whereas for the estimator (21) the ratio between signal and noise energy F is sufficient. Usually σ_p^2 and c^2 are not available independent of the data but as practical procedures either σ_p^2 can be approximately determined using an NDFT (assuming that the aliased energy is significantly smaller), or c^2 can be estimated from the data mismatch $\mathbf{y} - \mathbf{A}\hat{\mathbf{p}}$.

When assessment of the full covariance matrix is regarded to be too cumbersome, one may restrict oneself to the variances on the diagonal of this matrix. The square root of the variances yields the standard deviations and they can directly be compared to: (a) one another, (b) the estimated Fourier components in $\hat{\mathbf{p}}$, and (c) the a priori standard deviation σ_p. Regarding the last option one may also study the ratio

$$\frac{[\mathbf{C}_{\langle p\rangle}]_{ii}}{\sigma_p^2} = \left[\left(\frac{N}{FM_p \Delta k_x} \mathbf{A}^H \mathbf{W} \mathbf{A} + \mathbf{I} \right)^{-1} \right]_{ii} \tag{37}$$

for which only F is required. This quantity should be much smaller than 1 when the data determines the estimated value and the stabilization plays no significant role. Otherwise, the value will be close to 1.

11.2.10.2 Spatial Domain

From the study of uncertainty and resolution from the data in the Fourier domain, it is hard to recognize that the data typically cannot be reconstructed in large gaps (this fact is expressed in the off-diagonal elements in the a posteriori covariance matrix). For this reason it will be usually more convenient to compute the corresponding matrices in the spatial domain using transformation rules for Gaussian distributions:

$$\mathbf{C}_{p_r} = \mathbf{A}_r \mathbf{C}_p \mathbf{A}_r^H \tag{38}$$

and

$$\mathbf{C}_{\langle p_r \rangle} = \mathbf{A}_r \mathbf{C}_{\langle p \rangle} \mathbf{A}_r^H. \tag{39}$$

Again, if desired, only the standard deviations derived from the diagonals can be studied. As the most simple and practical measure of all, we propose the ratio of the standard deviations of the estimated and the a priori model in the spatial domain, denoted as $\sigma_{\langle p_r \rangle}(i)$ and $\sigma_{p_r}(i)$, respectively, i.e.:

$$\frac{\sigma_{\langle p_r \rangle}(i)}{\sigma_{p_r}(i)} = \sqrt{\frac{[\mathbf{C}_{\langle p_r \rangle}]_{ii}}{[\mathbf{C}_{p_r}]_{ii}}}. \tag{40}$$

It will be close to zero when there is sufficient data to reconstruct the regularly spaced samples and close to one otherwise.

The computation of these quantities is relatively expensive but can of course be carried out for a few selected temporal frequencies only. During data acquisition especially the indicator in equation (40) can be used to determine whether enough spatial samples have been acquired or whether more samples are required.

11.2.11. Applications and Results

11.2.11.1 Synthetic Example

Figure 2(a) shows a synthetic data set, with a regular sampling interval but 50 traces (a trace is a registration at a particular spatial location) are missing out of the original 128 traces. A gap of 4 traces is present around position $x = 940$ m. This represents a severe test. A set of linear dipping events is chosen since this allows a clear illustration in the Fourier domain. Note that the reconstruction algorithm is not at all restricted to such data sets, nor to regularly sampled data sets with missing traces. Figure 2(b) shows the Fourier-spectrum computed with the NDFT. Distortion is clearly visible. Figure 2(d) shows the Fourier-spectrum computed with the reconstruction algorithm where it was assumed that the energy of the aliased part of the signal is 1% of the desired part of the signal (stabilization). The number of spatial Fourier coefficients estimated is proportional to the temporal frequency and determined by the minimum apparent velocity of 2600 m/s. The maximum number is 119 at the temporal frequency of

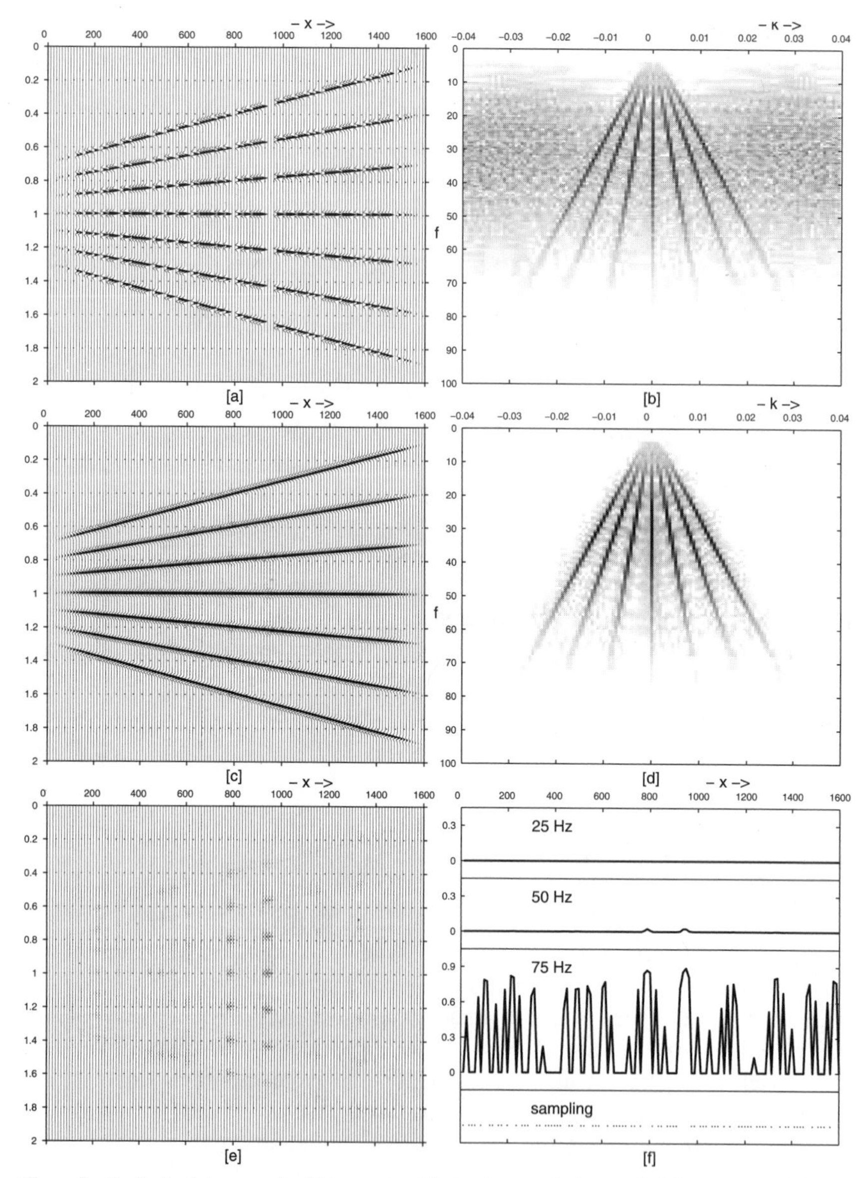

Figure 2. Synthetic data example; 50 traces are missing in a regularly sampled data set consisting of a total of 128 traces; (a) Input data; (b) *fk*-spectrum obtained by NDFT; (c) reconstructed signal; (d) Amplitude spectrum of reconstructed signal in the *fk*-domain; (e) difference of original data set and reconstructed data set; (f) Uncertainty analysis with the quantity as defined in equation (40) for $f = 25\,\text{Hz}, f = 50\,\text{Hz}$ and $f = 75\,\text{Hz}$. The sampling operator is given in the bottom figure.

90 Hz. For the lower temporal frequencies, the reconstruction is perfect. For the highest frequencies there were not enough samples present to allow a complete reconstruction of the signal. The reconstructed signal in the $x - t$ domain (Fig. 2(c)) looks close to perfect. The difference between the original and reconstructed data set is given in Fig. 2(e) on the same scale as Figs. 2(a) and (c). The largest difference occurs in the largest gaps. The maximum amplitude is $1/3$ of the amplitude in the original data set. The QC indicator as defined in equation (40) is displayed in Fig. 2(f) for frequencies of 25, 50 and 75 Hz and shows the increasing uncertainty in the estimate in the largest gaps for increasing frequency. For 75 Hz there are more parameters than samples. The indicator clearly shows that for this frequency the signal can not be reconstructed in the gaps. Overall, the algorithm is quite capable of handling this data set despite the severe irregularities in the sampling. Numerical experiments showed good stability in the presence of noise (guaranteed by the use of the stabilization constant).

11.2.12. Field Data Example

In Fig. 3 the reconstruction is shown on a field data set, a shot-record (the combination of traces for one shot position) in a marine environment. One gap of three traces, two gaps of two traces and two other single traces are missing. An asymmetric fk-spectrum was specified. In the reconstruction result in Fig. 3(b) the locations of the gaps are very difficult to find, except for the rightmost gap of two traces. The difference between input and output traces is very small and is due to the signal component outside the specified bandwidth. If desired, one may also choose to put the reconstructed traces in the gaps in the original data set and continue with this data set. Although not consistent with the theory of recon-

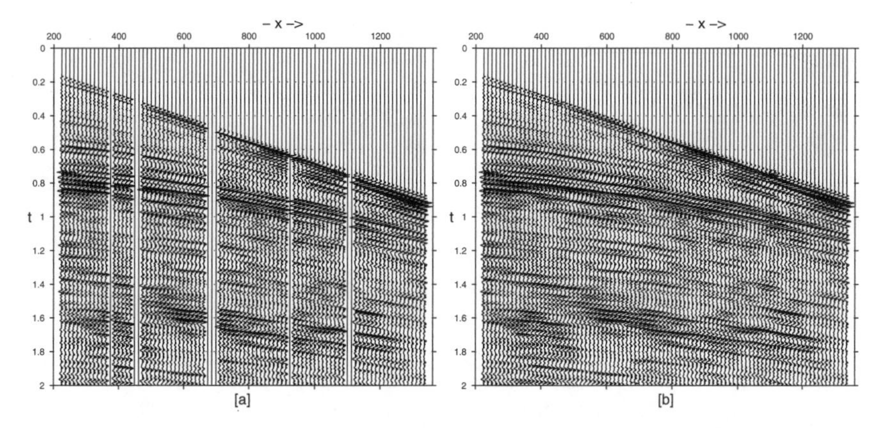

Figure 3. Reconstruction of a shot-gather. (a) Input data; (b) reconstructed signal.

struction this is sometimes preferred in practice out of fear of losing features in the original input data.

11.3. Two-Dimensional Reconstruction of Irregularly Sampled Signals

11.3.1. Introduction

In this section the reconstruction method described above will be extended to two coordinates (see also [16]). Like the method for one coordinate, it does not require any geological or geophysical assumptions concerning the data other than being spatially band limited. It is based on Fourier theory and parametric inversion only and can handle arbitrarily irregular sampling geometries. To some extent, the extension to more dimensions is straightforward. However, as discussed below, some non-trivial issues arise in the definition of the "sample area" and the parameterization, or, the sampling of the transform domain.

11.3.2. Forward Model

We define the continuous two-dimensional forward spatial Fourier transform as

$$\tilde{P}(k_x, k_y, f) = \int_{-\infty}^{\infty} \int_{-\infty}^{\infty} P(x, y, f) e^{-j2\pi(k_x x + k_y y)}\, dx\, dy, \tag{41}$$

where x and y are the spatial variables and k_x and k_y are the corresponding wavenumbers, or, spatial frequency components. The inverse transform is given by

$$P(x, y, f) = \int_{-\infty}^{\infty} \int_{-\infty}^{\infty} \tilde{P}(k_x, k_y, f) e^{j2\pi(k_x x + k_y y)}\, dk_x\, dk_y, \tag{42}$$

The (direct) Nonuniform Discrete Fourier Transform (NDFT) for irregularly sampled data is given by

$$\tilde{P}(k_x, k_y, f) = \sum_{n=0}^{N-1} P(x_n, y_n, f) e^{-j2\pi(k_x x_n + k_y y_n)} \Delta S_n, \tag{43}$$

where ΔS_n is the area corresponding to the sample (x_n, y_n). In contrast to the one-dimensional problem, the assignment of this area is not trivial. Triangulation algorithms can be used to determine them.

The following approach is proposed in order to improve upon the NDFT. Let us consider, for practical purposes, a regular sampling of the Fourier domain where the data in the Fourier domain is restricted to a certain limited area (bandlimited in the two dimensions). The discrete inverse Fourier transform for any spatial location is given by

$$P(x, y, f) = \Delta S_F \sum_{m=0}^{M-1} \tilde{P}(k_{x,m}, k_{y,m}, f) e^{j2\pi(k_{x,m}x + k_{y,m}y)}, \tag{44}$$

where the precise specification of $(k_{x,m}, k_{y,m})$ and the elementary area ΔS_F in the Fourier domain depend on the specific parameterization chosen, see below. With the exception of spatial aliasing, the discrete inverse Fourier transform is exact, provided the parameterization is such that the so-called Region of Support (hereafter referred to as ROS), i.e., the region in the Fourier domain where the signal is located, is fully covered. The combination of N versions of equation (44) for the irregularly spaced sample locations $(x_0, y_0), \ldots, (x_{N-1}, y_{N-1})$ can be written in vector notation as

$$\mathbf{y} = \mathbf{A}\mathbf{p}, \tag{45}$$

with

$$y_n = P(x_n, y_n, f), \tag{46}$$

$$p_m = \tilde{P}(k_{x,m}, k_{y,m}, f), \tag{47}$$

and

$$A_{nm} = \Delta S_F e^{j2\pi(k_{x,m}x_n + k_{y,m}y_n)}. \tag{48}$$

The errors due to spatial aliaising and possible violations of the assumptions concerning the ROS are incorporated in a noise term $\mathbf{n}$:

$$\mathbf{y} = \mathbf{A}\mathbf{p} + \mathbf{n}. \tag{49}$$

11.3.3. Least Squares Inversion

The MAP estimator under Gaussian assumptions is given by the least squares solution

$$\hat{\mathbf{p}} = (\mathbf{A}^H \mathbf{C}_n^{-1} \mathbf{A} + \mathbf{C}_p^{-1})^{-1} (\mathbf{A}^H \mathbf{C}_n^{-1} \mathbf{y} + \mathbf{C}_p^{-1} \mathbf{p}^i). \tag{50}$$

A practical approximation for the covariance matrix $\mathbf{C}_n$ is

$$\mathbf{C}_n = c^2 \mathbf{W}^{-1}, \tag{51}$$

where $\mathbf{W}$ is a diagonal matrix containing the areas associated with the spatial samples:

$$W_{ii} = \Delta S_i \tag{52}$$

For $\mathbf{C}_p$ we will take

$$\mathbf{C}_p = \sigma_p^2 \mathbf{I}, \tag{53}$$

and we assume that no specific prior model is available ($\mathbf{p}^i = \mathbf{0}$). The least squares estimator then becomes

$$\hat{\mathbf{p}} = (\mathbf{A}^H \mathbf{W} \mathbf{A} + k^2 \mathbf{I})^{-1} \mathbf{A}^H \mathbf{W} \mathbf{y}, \tag{54}$$

with $k^2 = c^2/\sigma_p^2$. If one is prepared to specify the ratio F between the expected energies of the noise and the signal in the input data, then it can be shown [16] that the stabilization constant can be approximated by

$$k^2 = \frac{c^2}{\sigma_p^2} = F \Delta S_F \frac{M}{N}, \tag{55}$$

where M is the total number of Fourier coefficients. Otherwise, a rule of thumb such as taking k^2 equal to 1% of the value of the diagonal of $\mathbf{A}^H \mathbf{W} \mathbf{A}$, or trial and error procedures, can be used.

Many aspects of the two-dimensional reconstruction are straightforward extensions of the one-dimensional case. Therefore, we will emphasize the aspects that are non-trivial.

11.3.3.1 Parameterization

The parameterization is again crucial. Even for regular sampling of the Fourier domain, as considered here, it is less straightforward to specify than in the 1-D problem. There, a sample interval and the specification of the one-dimensional interval in the Fourier domain (through one or two minimum apparent velocities) suffices. Regular sampling in two dimensions, whether in the spatial or the Fourier domain, can be specified through two base vectors, each with two components. These base vectors determine the repetition of the region of support in the other domain. For a discussion of regular sampling in multiple dimensions, see [17], from which we repeat some results for two-dimensional sampling here. The position of a sample in the Fourier domain is determined by two base vectors $\mathbf{v}_1$ and $\mathbf{v}_2$, given by

$$\begin{pmatrix} k_x \\ k_y \end{pmatrix} = m_1 \mathbf{v}_1 + m_2 \mathbf{v}_2, \qquad (m_1, m_2) \in Z^2. \tag{56}$$

The vectors $\mathbf{v}_1$ and $\mathbf{v}_2$ determine how the inverse Fourier transformation of (44) is discretized. The area surrounding the samples in the Fourier domain ΔS_F is determined by the crossproduct of $\mathbf{v}_1$ and $\mathbf{v}_2$:

$$\Delta S_F = |\mathbf{v}_1 \times \mathbf{v}_2|, \tag{57}$$

and the forward model of equation (44) can be rewritten as

$$P(x, y, f) = |\mathbf{v}_1 \times \mathbf{v}_2| \sum_{m_1} \sum_{m_2} \tilde{P}(m_1\mathbf{v}_1 + m_2\mathbf{v}_2) e^{j2\pi\mathbf{x}\cdot(m_1\mathbf{v}_1 + m_2\mathbf{v}_2)}, \tag{58}$$

with

$$\mathbf{x} = \begin{pmatrix} x \\ y \end{pmatrix}. \tag{59}$$

By discretizing the inverse Fourier transform, the signal will become periodic in the spatial domain. The spatial periodicity is described by two base vectors $\mathbf{u}_1$ and $\mathbf{u}_2$, which are given by:

$$\mathbf{u}_i\mathbf{v}_j = \delta_{ij} \qquad i, j = 1, 2, \tag{60}$$

where δ_{ij} is the Kronecker delta. The orientation and length of the vectors $\mathbf{v}_1$ and $\mathbf{v}_2$ should be chosen such that overlap, or, aliasing, will not occur in the spatial domain. For a rectangular ROS in the spatial domain with sides X and Y along the x- and y-axis, respectively, we may allow a periodicity described by (Fig. 4(a))

$$\mathbf{u}_1 = \begin{pmatrix} X \\ 0 \end{pmatrix}, \qquad \mathbf{u}_2 = \begin{pmatrix} 0 \\ Y \end{pmatrix}, \tag{61}$$

which corresponds with

$$\mathbf{v}_1 = \begin{pmatrix} 1/X \\ 0 \end{pmatrix}, \qquad \mathbf{v}_2 = \begin{pmatrix} 0 \\ 1/Y \end{pmatrix}. \tag{62}$$

In case the area has an elliptic ROS with axes parallel to the the x-axis and y-axis, the most optimal periodicity is determined by a closest packing of ellipses (Fig. 4(b)). The base vectors describing the repetition in the spatial domain are then defined as

$$\mathbf{u}_1 = \begin{pmatrix} \frac{1}{2}\sqrt{3}X \\ -\frac{1}{2}Y \end{pmatrix}, \qquad \mathbf{u}_2 = \begin{pmatrix} 0 \\ Y \end{pmatrix}. \tag{63}$$

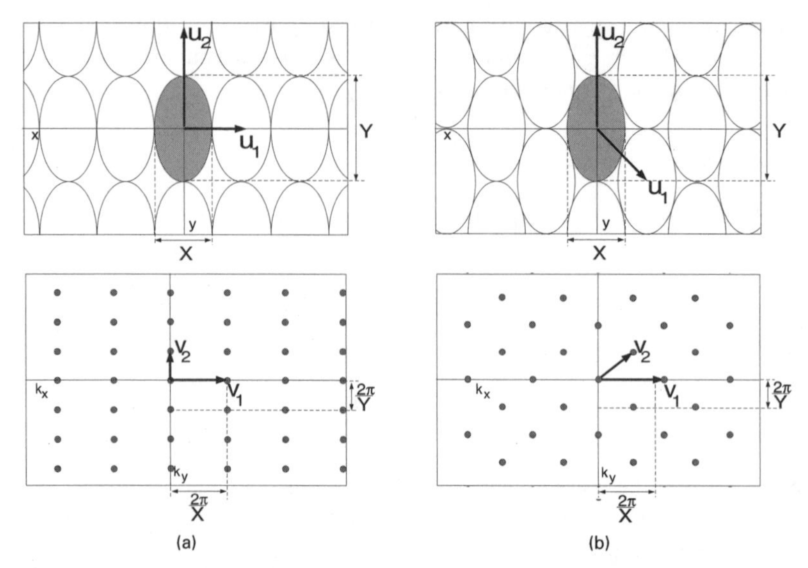

Figure 4. The relation between the Region of Support in the spatial domain and the sampling grid in the Fourier domain. (a) A rectangular Region of Support leads to orthogonal sampling. (b) An elliptical Region of Support leads to hexagonal sampling.

which leads to the base vectors describing the sampling in the Fourier domain:

$$\mathbf{v}_1 = \begin{pmatrix} \frac{2}{3}\sqrt{3}\frac{1}{X} \\ 0 \end{pmatrix}, \qquad \mathbf{v}_2 = \begin{pmatrix} \frac{1}{3}\sqrt{3}\frac{1}{X} \\ \frac{1}{Y} \end{pmatrix}. \tag{64}$$

This is a less dense sampling, compared to the orthogonal sampling and results in a better determination of the estimator in (54).

Apart from the grid, the size and shape of the ROS has to be specified as well and again we have more degrees of freedom than in the 1-D case. For many cases in seismology, however, it will be reasonable to assume and specify a minimum absolute apparent velocity that does not vary with azimuth.[6] When the reconstruction problem is solved per temporal frequency component, this gives rise to the cone-shaped ROS as depicted in Fig. 5.

[6] Seismic data with two spatial variables can be regarded as a superposition of plane events $p(x, t) = s(t - x/c_x - y/c_y)$, with $s(t)$ the wavelet. The Fourier transform of such a linear event is $\tilde{P}(k_x, f) = S(f + c_x k_x + c_y k_y)$, and is a straight line, defined by $c_x k_x + c_y k_y = -f$, through the origin of the Fourier domain. The lower the absolute value $\sqrt{c_x^2 + c_y^2}$ of the velocity of the event, the further the line deviates from the f-axis.

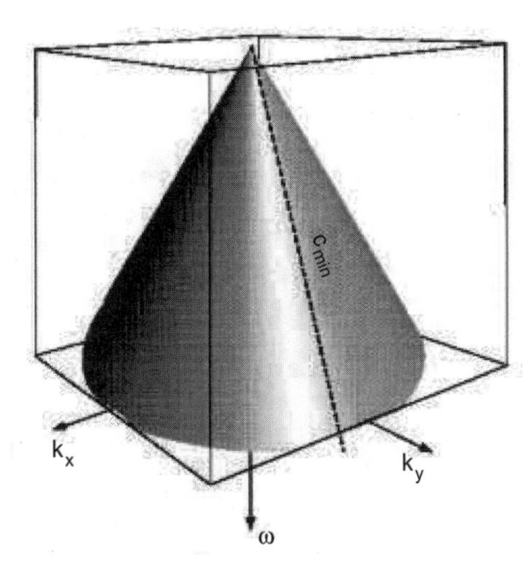

Figure 5. The cone shaped Region of Support in the Fourier domain, determined by the minimum absolute apparent velocity.

As in the 1-D case, it is important to realize that for low f less parameters are required and the inversion problem is easier to solve than for high f. This, besides efficiency reasons, is a strong argument to solve the problem in the temporal frequency domain instead of the time domain.

11.3.4. Examples of 2-D Reconstruction

A synthetic example is shown in Fig. 6. The data set consists of 6 plane dipping events in the spatial-temporal domain Fig. 6(b). The Fourier transform of one plane event is a line through the origin in the $k-f$ domain. The sixth event (with half the amplitude of the others) has an absolute apparent velocity that is lower than the minimum apparent velocity specified for the reconstruction and thus falls outside the spatial bandwidth and constitutes noise in the context of this example. The total number of spatial samples is sufficient for the reconstruction of the five desired events. In Fig. 7 the $k_x - k_y$-plots for a single temporal frequency (58.5 Hz) are shown for the signal model (a), for the conventional processing method of binning and stacking[7] (b), for the NDFT (c) and for the least squares reconstruction (d). The ideal result would be five points in these

[7] The conventional procedure of "binning and stacking" consists of dividing the spatial ROS in a set of adjacent rectangles, with sides Δx and Δy, and adding the seismic traces that fall within a rectangle. The position of the result is defined as the middle of the rectangle.

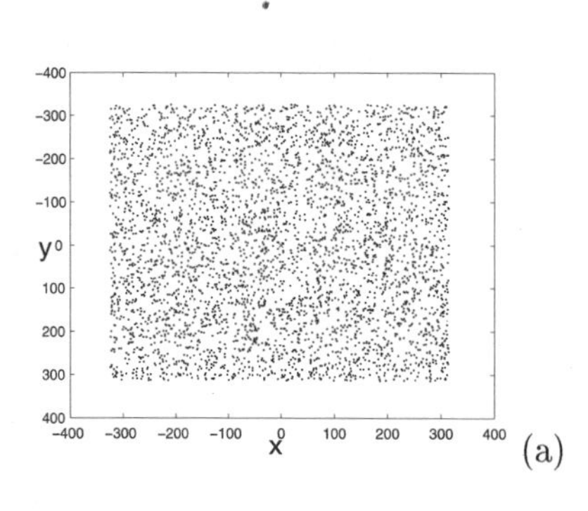

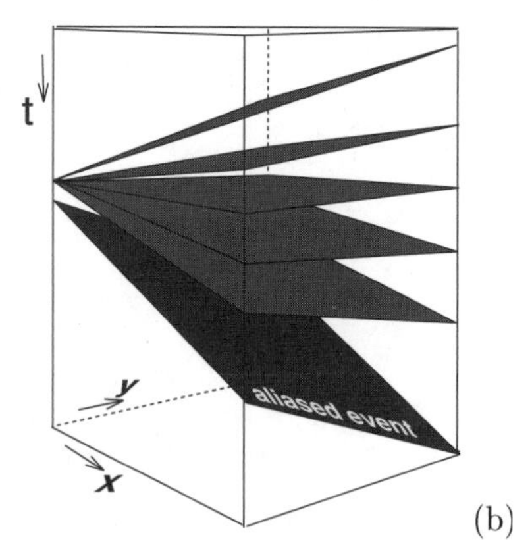

Figure 6. Synthetic experiment: (a) The sampling geometry. (b) The model used, consisting of six plane dipping events.

pictures but the sixth noise event deteriorates the results. It is clear that the binning and stacking procedure gives significant distortion. The NDFT already improves upon this but the least squares reconstruction gives the best results. Fig. 8 shows $x - t$-plots of the results for the line $y = 0$. A field data example is given in [16].

11.4. Sparse Sampling

11.4.1. Introduction

Apart from the desired primary reflections, seismic data also consists of a variety of undesired events like multiples, ground roll and refracted waves. In order to suppress these noise events, there are several filtering techniques like Fourier domain filtering, linear, and parabolic Radon domain filtering.

Standard processing algorithms based on these transforms are typically applied to Common Midpoint (CMP) gathers (see section 11.1.2), after "Normal MoveOut" (NMO) correction.[8] However, in 3-D seismology the offset direction

[8] For plane, dipping, reflectors, the arrival times of seismic reflections in a CMP-gather approximately follow a hyperbolic trajectory. NMO-correction removes this "hyperbolic move-out" and transforms the reflections into horizontally aligned events in the CMP gathers. Other events, such as "multiples" (events that are reflected at multiple boundaries, before arriving at the surface) are typically not horizontally aligned. By the subsequent, conventional, procedure of "stacking" (adding the traces within the CMP-gather), the primary reflections are thus enhanced and the other events are suppressed.

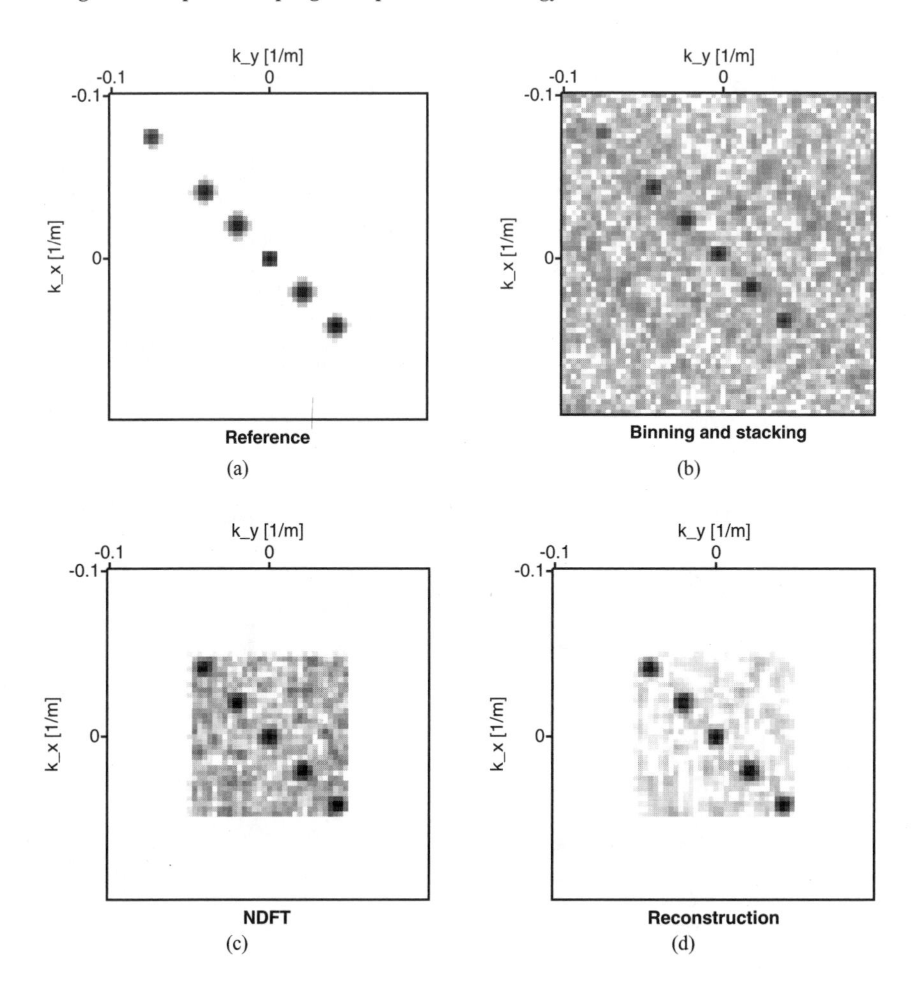

Figure 7. The amplitude spectra for $f = 58.5$ Hz, plotted on a logarithmic scale with a 43 dB dynamic range: (a) Reference data; (b) the result after binning and stacking; (c) NDFT; (d) least squares reconstruction.

is often sparsely (and irregularly) sampled per CMP-gather. This leads to aliasing in the transformation and consequently, CMP-based decomposition into signal and noise often gives unsatisfactory results. It is possible to achieve sufficient sampling of the offset coordinate by using a number of CMP-gathers concurrently. This however necessitates the use of a transform for two spatial dimensions.

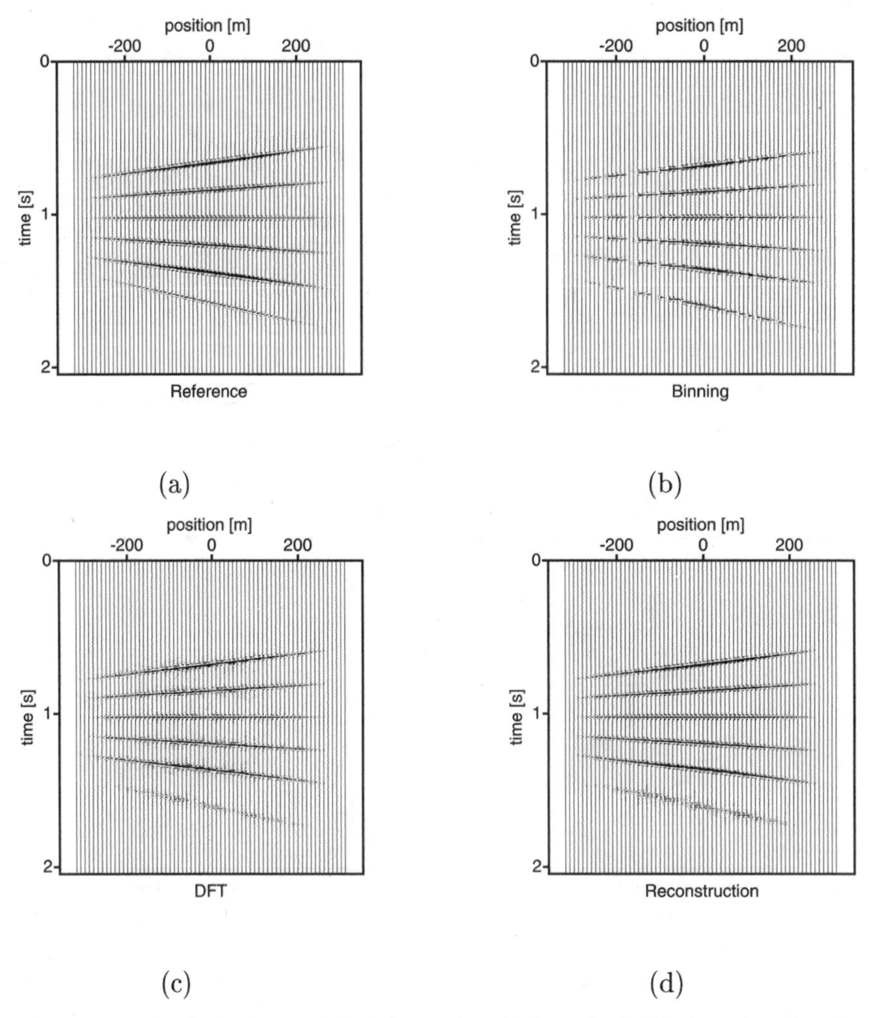

Figure 8. x-A-plots for the line y = 0: (a) Reference data; (b) the result after binning and stacking; (c) NDFT; (d) least squares reconstruction.

In this section a method is presented to estimate the 2-D transform domain for a mixed Fourier–Radon transform. By a judicious choice of the Region of Support in this domain, the aliasing problem is avoided. The technique nicely illustrates the general principle that several transforms can be used and that the one with the most compact description of the data yields the best result. It also illustrates that different transformations can be used for different spatial directions.

First Radon transforms for irregular sampling are briefly reviewed and the aliasing conditions are discussed. The mixed Fourier/Radon transform is subsequently introduced and illustrated with a synthetic data example.

11.4.2. Radon Transform

The continuous (linear) Radon transform of the signal $p(x, t)$ is defined as

$$\tilde{p}(\alpha, \tau) = \int_{-\infty}^{\infty} p(x, \tau + \alpha x)\, dx, \tag{65}$$

and is basically a decomposition into plane "events" with slope α and intercept time τ. Similarly, the parabolic Radon transform is defined as

$$\tilde{p}(q, \tau) = \int_{-\infty}^{\infty} p(x, \tau + qx^2)\, dx, \tag{66}$$

and is basically a decomposition into parabolae, with curvature q. After a Fourier transformation over the intercept time, we obtain for the linear Radon transform

$$\tilde{P}(p, f) = \int_{-\infty}^{\infty} p(x, f) e^{j2\pi f \alpha x}\, dx, \tag{67}$$

which can be recognized to be equivalent to a spatial Fourier transform. For the parabolic Radon transform in the temporal Fourier domain we have

$$\tilde{P}(q, f) = \int_{-\infty}^{\infty} p(x, f) e^{j2\pi f q x^2}\, dx. \tag{68}$$

For sampled data, a straightforward discretization of the integrals in (67) and (68) could be used. However, for improved resolution and in order to handle irregularly spaced data, the Radon transforms are generally implemented using a least squares formulation, with the *inverse* Radon transform as a *forward* model (see, e.g., [2]). The discrete inverse Radon transform (per temporal frequency f, from the data in the Radon domain to the data in the spatial domain can be written as:

$$P(x_n, f) = \sum_m \tilde{P}(m\Delta q, f) e^{-j2\pi f m \Delta q x_n^r}, \tag{69}$$

where $r = 1$ for the linear, and $r = 2$ for the parabolic Radon transform, and $m\Delta q$ is the slowness (for the linear Radon transform) or the curvature (for the Parabolic Radon Transform (PRT)).

By writing equation (69) in matrix notation:

$$\mathbf{y} = \mathbf{L}\mathbf{p}, \tag{70}$$

with

$$y_n = P(x_n, f), \qquad n = 0, \ldots, N-1, \tag{71}$$

$$p_m = \tilde{P}(m\Delta q, f), \qquad m = -M, \ldots, M-1, \tag{72}$$

and

$$L_{nm} = e^{j2\pi m\Delta q x_n^r}. \tag{73}$$

The forward transformation can be found by least squares inversion:

$$\hat{\mathbf{p}} = (\mathbf{L}^H\mathbf{L})^{-1}\mathbf{L}^H\mathbf{y}, \tag{74}$$

where $\mathbf{L}^H$ denotes the complex conjugate transpose of $\mathbf{L}$. Note that x_n need not be regularly spaced.

11.4.3. Aliasing Conditions

The aliasing conditions for the linear Radon transform are well known. For the conventional parabolic Radon transform, where Δq is taken independent of f, they are:

$$\Delta q \leq \frac{1}{x_{\max}^2 f_{\max}}, \tag{75}$$

where $x_{\max}$ is the maximum offset, and $f_{\max}$ is the maximum frequency in the data. This condition should hold in order to avoid aliasing in the spatial domain.

To avoid the aliasing in the forward transform, the maximum allowable curvature ($q_{\max}$), in the data has to satisfy:

$$q_{\max} \leq \frac{1}{(\Delta(x^2))_{\max} f_{\max}}. \tag{76}$$

The problem of insufficient sampling is illustrated with a synthetic example using an acquisition geometry closely resembling that of 3-D marine survey with dual source acquisition. Figure 9(a) shows the geometry. The source interval is 50 m

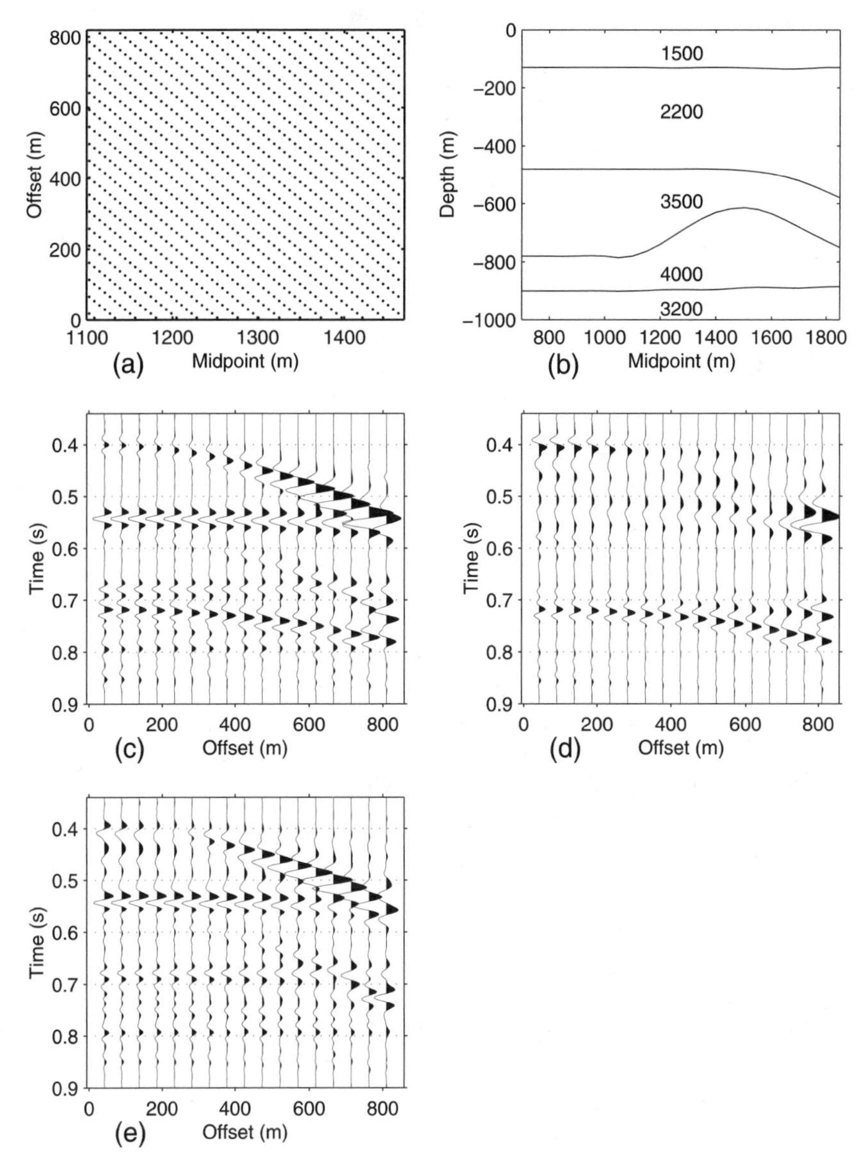

Figure 9. A synthetic example: (a) Geometry; (b) Subsurface; (c) Original CMP-gather; (d) Estimated noise conventional, 1-D, PRT; (e) Original CMP-gather minus estimated noise 1-D PRT.

and the receiver interval is 25 m. Per CMP-gather only a quarter of all offsets is available. Synthetic data was generated with finite difference software using the velocity model shown in Fig. 9(b). Figure 9(c) shows a CMP-gather after NMO correction. The primary events corresponding to the second, third and fourth interfaces are visible as well as several multiple events. Conventional parabolic Radon filtering is applied to this gather. Due to the limited number of offsets the maximum curvature $q_{\max}$ has to be limited, according to (76). Figure 9(d) shows the estimated multiples and Fig. 9(e) the CMP-gather after subtraction of these multiples. Due to the limited q-range, only a part of the multiple energy has been removed. The multiples with strong curvature are still present.

11.4.4. The Mixed Fourier-Radon Transform

In order to circumvent the problems caused by the sparse offset sampling per CMP-gather, the 1-D Radon, and Fourier transforms can be combined in the following 2-D transform:

$$P(x_k, h_k, f) = \Delta k_x \sum_m \sum_n \tilde{P}(m\Delta k_x, n\Delta q_h, f) \exp(j2\pi(m\Delta k_x x_k - fn\Delta q_h h_k^r)), \quad (77)$$

where $P(x_k, h_k, f)$ is a data point in the spatial domain (with any position), and $\tilde{P}(m\Delta k_x, n\Delta q_h, f)$ is a data point in the Fourier–Radon domain. Typically, x is a midpoint and h the offset coordinate. In this formulation orthogonal sampling with a rectangular region of support (ROS) in the transform domain is used. In order to allow flexible sampling, (77) is modified to

$$P(x_k, h_k, f) = \Delta k_x \sum_l \tilde{P}(k_{x,l}, q_{h,l}, f) \exp(j2\pi(k_{x,l} x_k - f q_{h,l} h_k^r)), \quad (78)$$

where $k_{x,l}$ and $q_{h,l}$ can be located in an arbitrary part of the transform domain, or Region of Support.

Equation (78) can be written in matrix notation as:

$$\mathbf{p} = \mathbf{Q}\tilde{\mathbf{p}}, \quad (79)$$

with

$$\mathbf{p}_k = P(x_k, h_k, f), \quad (80)$$

$$\tilde{\mathbf{p}}_l = \tilde{P}(k_{x,l}, q_{h,l}, f), \quad (81)$$

and

$$\mathbf{Q}_{k,l} = \Delta k_x \exp(j2\pi(k_{x,l}x_k - fq_{h,l}h_k^p)). \tag{82}$$

In order to find the regularly sampled transform domain, a stabilized least squares solution can be calculated

$$\hat{\mathbf{p}} = (\mathbf{Q}^H\mathbf{Q} + k^2\mathbf{I})^{-1}\mathbf{Q}^H\mathbf{p}, \tag{83}$$

where $\hat{\mathbf{p}}$ is the estimate in the mixed Fourier–Radon domain.

11.4.5. Region of Support

The sparse offset sampling in a midpoint gather results in a limited q-range. Often, a part of the multiple energy is located beyond that range, and conventional CMP-based algorithms cannot correctly estimate these noise events. An important advantage of the method presented here is related to the ROS of the method. The ROS is the specific area in the transform domain that is estimated.

With CMP-based algorithms, in general for each midpoint the same q-range and also the same demarcation between signal and noise is used. Therefore the Region of Support can be given in the $q/x/f$-domain as shown in Fig. 10 (and Fig. 11(a) for one frequency). With the method presented in this chapter, the shape of the ROS can arbitrarily be chosen, with the only limitation that the least squares inversion does not depend too much on the stabilization. The optimal choice will be that ROS on which most of the signal and noise energy is mapped. Of course, this part of the transform domain will be dependent on the data.

In many cases, however, high-energy *noise* events (e.g. groundroll, water bottom multiples) are generated by the shallow part of the subsurface. Often, the

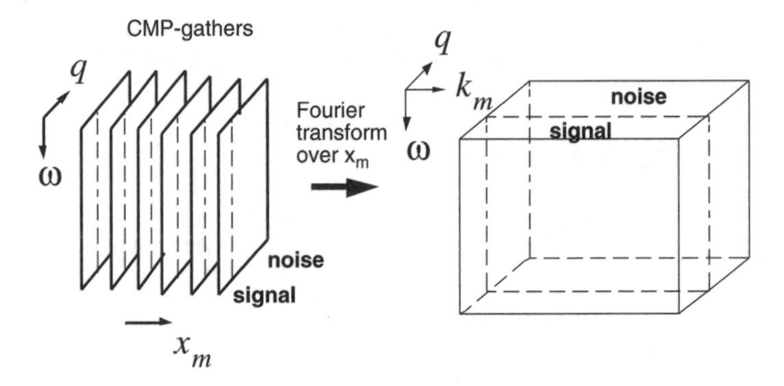

Figure 10. ROS for CMP-based algorithms.

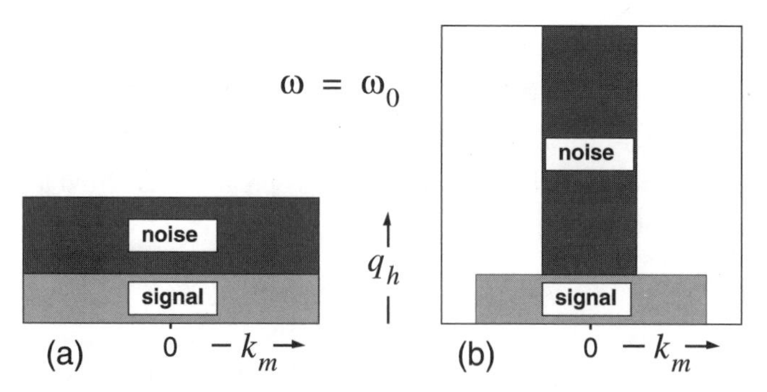

Figure 11. ROS for a single temporal frequency f.

variations along the midpoint coordinate are not strong for the shallow subsurface and therefore these noise events are mapped onto a region in the transform domain which is limited by small k_x-values. The q-range should be chosen sufficiently large so as to cover the actual curvature of the noise in the data.

The *signal* will ideally map to $q = 0$. Due to imperfect NMO and AVO effects, there will be some spreading of energy, but mostly to small q-values. The k-range of the signal is larger than the k-range of the noise, but limited by:

$$k_{\max} = \frac{f}{c_{\min}}, \tag{84}$$

where $c_{\min}$ is the minimum apparent velocity (often the velocity in the top layer).

A good choice for the ROS will therefore be as shown (for a single frequency) in Fig. 11(b). The frequency dependent $k_{\max}$ for the signal, and a similar effect for the noise (i.e. the k-range is more limited for lower frequencies) yields a frequency dependent ROS as shown in Fig. 12. It is clear, that with the method presented here, noise events with high q-values can properly be estimated in the transform domain while that is not possible using CMP-based algorithms.

11.4.6. Sampling in the q Domain

Instead of using the anti-aliasing condition in (75), a frequency dependent sampling is possible which gives us a larger Δq for frequencies smaller than $\omega_{c,\max}$:

$$\Delta q \leq \frac{1}{x_{\max}^2 f}. \tag{85}$$

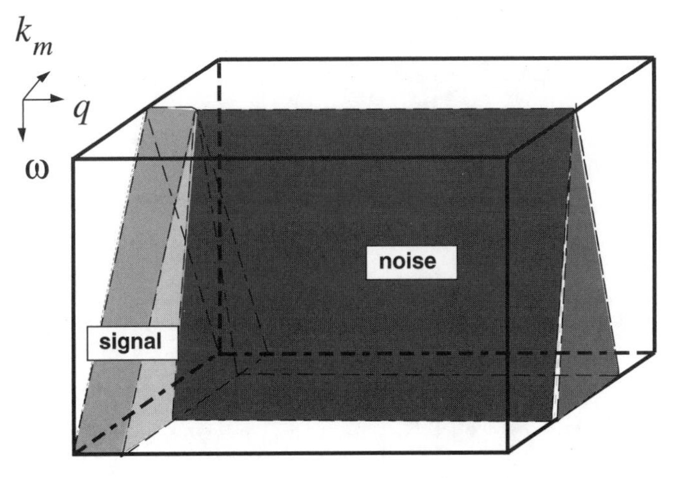

Figure 12. Frequency dependency of ROS.

Because of the smaller ROS and the larger Δq for lower frequencies, the number of q/k-values to be estimated is smaller for low temporal frequencies, and therefore we will typically have a well-posed problem, while for high temporal frequencies the problem may become ill-posed and stabilization may be required.

11.4.7. Results

The mixed Fourier/Radon technique is first illustrated using the synthetic example shown above. Figure 13(a) shows the same input CMP-gather as in Fig. 9(c). In Fig. 13(b) the estimated noise with the mixed Fourier/Radon technique is shown for this CMP-gather. Note that a whole range of CMP-gathers is used for the transform! Figure 13(c) shows the CMP-gather after subtraction of the estimated noise. The multiple energy is almost completely removed. For comparison Fig. 13(d) repeats the result of the conventional 1-D parabolic Radon filtering (Fig. 9(e)). The improvement of the mixed Fourier/Radon noise removal (hereafter referred to as FRNR) is clear.

11.4.8. Conclusions

Conventional CMP-based signal and noise separation techniques can fail due to sparse sampling. By treating several CMP gathers concurrently using a 2-D transform, noise events with high curvature (or, using linear Radon, strong dips) can properly be estimated, by making use of an optimal ROS and frequency dependent sampling in the transform domain. The method can be applied to

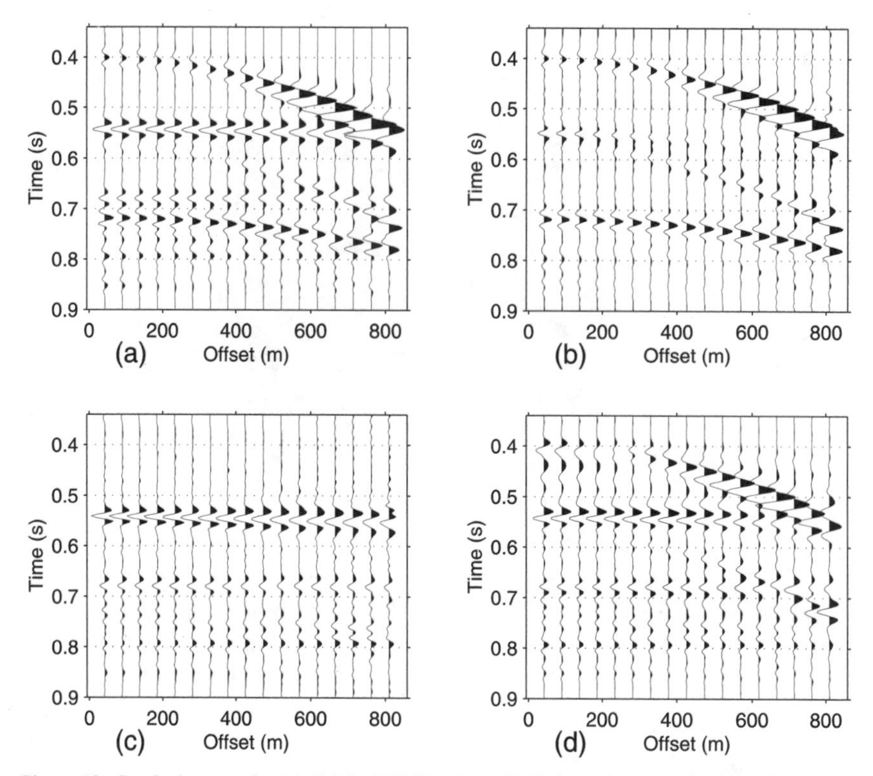

Figure 13. Synthetic example: (a) Original CMP-gather; (b) Estimated noise, mixed Fourier–Radon transform; (c) Original CMP-gather minus estimated noise mixed Fourier–Radon transform; (d) Original CMP-gather minus estimated noise 1-D PRT (from Fig. 9(e)).

irregularly sampled data (in 2-D). In this case linear- and parabolic Radon transforms and the Fourier transform can be chosen for offset and midpoint coordinates, respectively. The technique is an illustration of the general principle that several transforms can be used for reconstruction and that the one with the most compact description of the data yields the best result. It also illustrates that different transformations can be used for different spatial directions. The method has been successfully tested on synthetic and field data sets.

11.5. Efficiency of Reconstruction Algorithms

For practical application an efficient implementation of the reconstruction algorithms described in the previous sections is mandatory since we are dealing with a number of parameters and samples that may be in the order of hundreds or thousands. The formulation of these algorithms can be generalized to multiple (L) dimensions. All Fourier, linear/parabolic-Radon transforms (or mixtures of those) solve the signal in the transform domain $\tilde{P}$, from the forward model, now in L dimensions,

$$P(\mathbf{x}_n, f) = \sum_{m=0}^{M-1} \tilde{P}(\mathbf{k}_m, f) e^{-j2\pi \mathbf{k}_m \cdot \mathbf{s}_n} \Delta S_T, \tag{86}$$

in a least squares sense. ΔS_T is an elementary area or volume surrounding the samples in the transform domain. Although not strictly necessary we will from this point onwards take it to be constant. Equation (86) can be written in matrix notation as

$$\mathbf{y} = \mathbf{A}\mathbf{p}, \tag{87}$$

with

$$y_n = P(\mathbf{x}_n, f), \tag{88}$$

$$A_{nm} = \Delta S_T e^{-j2\pi \mathbf{k}_m \cdot \mathbf{s}_n}, \tag{89}$$

and

$$p_m = \tilde{P}(\mathbf{k}_m, f). \tag{90}$$

The unknown sampled signal in the transform domain $\mathbf{p}$ is estimated from the irregularly sampled signal in the spatial domain using a weighted stabilized least squares estimator through the solution of the system

$$(\mathbf{A}^H \mathbf{W} \mathbf{A} + k^2 \mathbf{I})\mathbf{p} = \mathbf{A}^H \mathbf{W} \mathbf{y}, \tag{91}$$

where $\mathbf{W}$ is a diagonal matrix and k^2 the stabilization constant, which may depend on the temporal frequency f. We will write (91) in shorthand notation as:

$$\mathbf{H}\mathbf{p} = \mathbf{b}, \tag{92}$$

with $\mathbf{H} = \mathbf{A}^H \mathbf{W} \mathbf{A} + k^2 \mathbf{I}$ and $\mathbf{b} = \mathbf{A}^H \mathbf{W} \mathbf{y}$. The right hand side of equation (92) equals the NDFT times a constant:

$$b_m = [\mathbf{A}^H \mathbf{W} \mathbf{y}]_m = \Delta S_T \sum_{n=0}^{N-1} P(\mathbf{x}_n, f) e^{j2\pi \mathbf{k}_m \cdot \mathbf{s}_n} \Delta S_n, \tag{93}$$

and is a first approximation to the solution. The inversion for $\mathbf{H}$ improves upon $\mathbf{b}$.

For most practical problems the transform domain will be regularly sampled with the sample positions determined by multiples of L base vectors. The sampling should be fine enough to avoid aliasing in the spatial domain. Orthogonal sampling, where the base vectors are orthogonal, is an often used special case. In L dimensions the sample positions are then given by $(m_1 \Delta k_1, \ldots, m_L \Delta k_L)$, where Δk_l is the sample interval along dimension l. In the rest of this section we will limit ourselves to orthogonal sampling. Note that this does not necessarily imply a rectangular Region of Support (ROS).

11.5.1. Computational Complexity

The straightforward computation of $\mathbf{H}$ and $\mathbf{b}$ and the solution of $\mathbf{Hp} = \mathbf{b}$ usually would be very expensive, but fortunately dramatic improvements can be obtained because of the specific mathematical structure of the problem. First of all the computation of $\mathbf{b}$ can be solved by the nonuniform fast Fourier Transform (NFFT), [18]. This algorithm requires

$$O\left(N(-0.72 \ln \epsilon)^L + \left(\prod_{l=1}^{L} f_0 M_l\right) \sum_{l=1}^{L} \log f_0 M_l\right) \tag{94}$$

operations, where M_l is the number of Fourier components along dimension L, ϵ is the required accuracy, and f_0 is an oversampling factor, which typically has the value 2. This is a dramatic improvement over the $O(N \prod_{l=1}^{L} M_l)$ operations required for the direct evaluation of the NDFT. Note that $\mathbf{b}$ also has the mathematical form of the NDFT for the parabolic Radon transform and therefore the NFFT can be used for this problem as well. Secondly, the elements of the matrix $\mathbf{H}$ are given by

$$H_{mn} = \Delta S_t^2 \sum_{l=0}^{N-1} \Delta S_l e^{j2\pi(\mathbf{k}_m - \mathbf{k}_n)\cdot \mathbf{s}_l} + k^2(f)\mathbf{I}. \tag{95}$$

A number of conclusions follow directly:

1. The first part of $\mathbf{H}$, corresponding to $\mathbf{A}^H\mathbf{WA}$ is f-independent, provided the sample intervals Δk are chosen independent of f. Since Δk is related to the periodicity and therefore aliasing in the spatial domain, this can indeed be taken independent of f. Consequently $\mathbf{H}$ needs to be computed only once for the largest ROS required (for the highest temporal frequency f). For a smaller ROS the required part can be selected and the possibly f-dependent term can be added to the diagonal.
2. For orthogonal sampling $\mathbf{H}$ has a Toeplitz structure. For the 1-D case $\mathbf{H}$ is strictly Toeplitz. For the 2-D case $\mathbf{H}$ has a block Toeplitz Toeplitz block (BTTB) structure. This extends to higher dimensions. The consequence is that the multiplication can be carried out very efficiently by an L-dimensional FFT, which is advantageous for iterative methods for solving $\mathbf{Hp} = \mathbf{b}$, like the conjugate gradient (CG) method, see the next section.

3. Due to the special structure, only a very limited amount of elements of **H** is distinct and needs to be stored.
4. The computation of **H** has the mathematical structure of an NDFT and can be efficiently computed with the NFFT.

For the 1-D case the well known Levinson scheme can be used to solve $\mathbf{Hp} = \mathbf{b}$, which requires $2M^2$ operations. However, in many cases the conjugate gradient scheme, discussed below, is even faster.

11.5.2. Conjugate Gradient Schemes for General Irregular Geometries

A particularly useful method for solving $\mathbf{Hp} = \mathbf{b}$ for $\mathbf{p}$ is the conjugate gradient (CG) method. This scheme is initialized with a suitable starting model, in our case

$$\mathbf{p}_0 = 1/\Delta S_T \mathbf{b} \tag{96}$$

with corresponding residual $\mathbf{r}_0 = \mathbf{b} - \mathbf{Hp}_0$ and using $\mathbf{d}_0 = \mathbf{0}$, the CG iteration, starting at $n = 1$ is:

$$\begin{aligned} \mathbf{d}_n &= \mathbf{d}_{n-1} + \frac{\mathbf{r}_{n-1}}{\|\mathbf{r}_{n-1}\|^2} \\ \mathbf{p}_n &= \mathbf{p}_{n-1} + \frac{\mathbf{d}_n}{\mathbf{d}_n^T \mathbf{H} \mathbf{d}_n} \\ \mathbf{r}_n &= \mathbf{r}_{n-1} + \frac{\mathbf{H}\mathbf{d}_n}{\mathbf{d}_n^T \mathbf{H} \mathbf{d}_n}. \end{aligned} \tag{97}$$

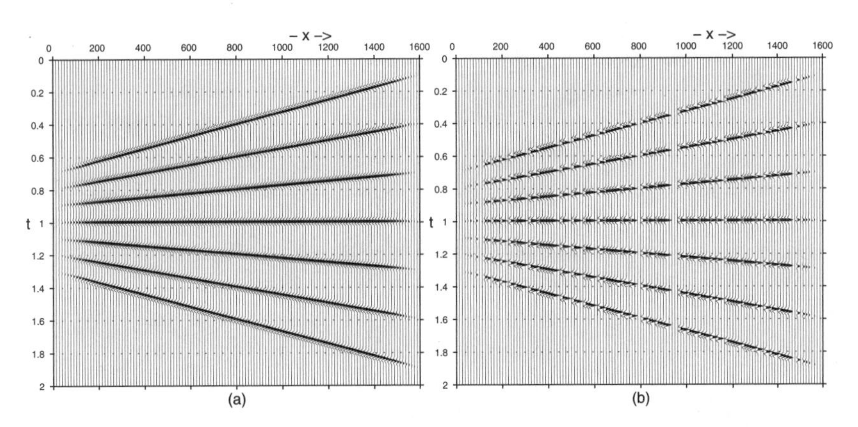

Figure 14. (a) Original full data set; (b) irregular data set with missing traces.

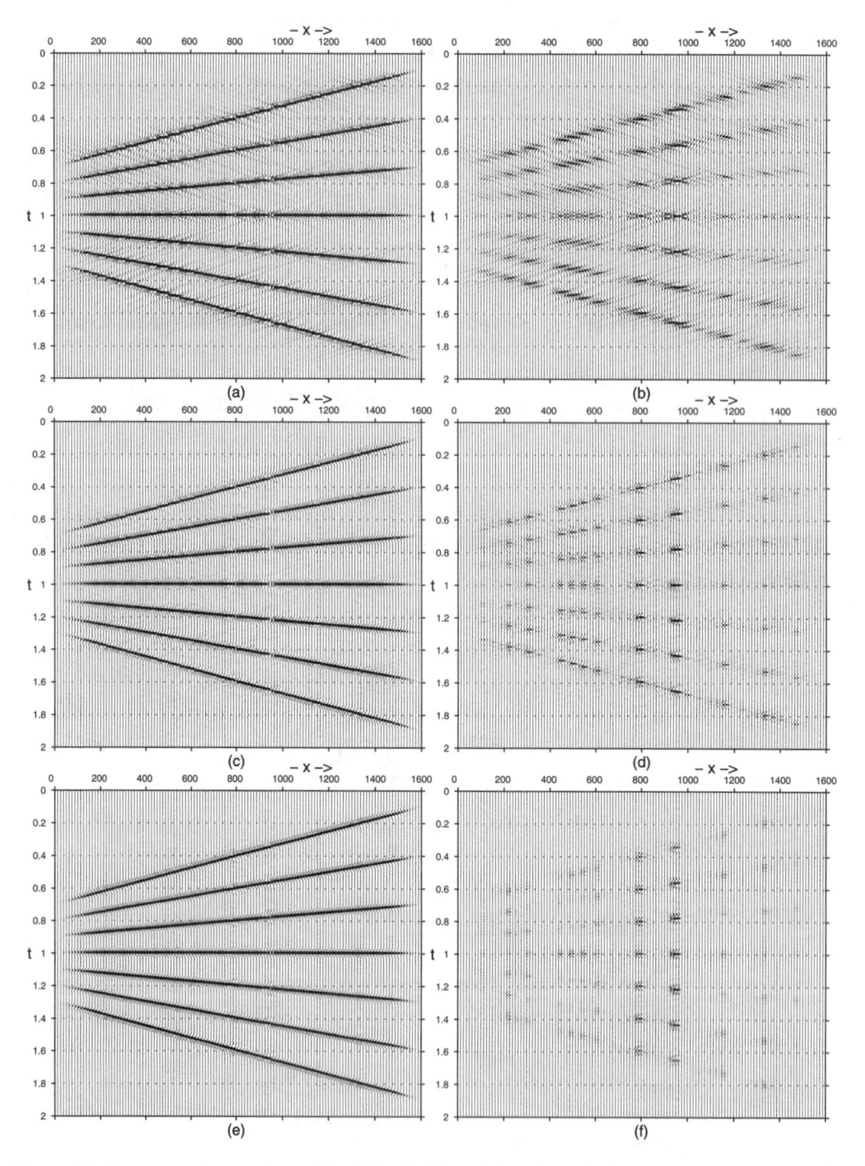

Figure 15. Reconstruction results using CG: (a) After 0 iterations; (b) the error of (a); (c) after 2 iterations; (d) the error of (c); (e) after 5 iterations; (f) the error of (e).

The only computationally intensive part in the iteration is the evaluation of $\mathbf{Hd}_n$, which can be carried out using FFTs. An important point to realize is that the CG scheme, like most iterative schemes produces a better result after each iteration and the scheme can be stopped when desired. Stopping criteria are the relative norm of the residual $\|\mathbf{r}_n\|^2/\|\mathbf{b}\|^2$ and a maximum number of iterations. (For infinite precision computing, the CG scheme stops after M iterations, where M is the dimension of $\mathbf{H}$.)

11.5.3. Illustration of the CG Scheme for 1-D Fourier Reconstruction

The iterative nature of the CG scheme and the possibility to stop after a limited number of iterations is illustrated for the 1-D Fourier reconstruction with a synthetic data set. It involves a regularly sampled gather (Fig. 14(a)) with a relatively large number of traces missing (Fig. 14(b)), plotted as empty traces. The limited number of traces available makes this a difficult reconstruction problem and the condition number will be high, which in turn implies that the CG scheme cannot be expected to converge very rapidly. However even in this case a relatively small number of iterations gives satisfactory results. In Fig. 15 the left columns (a), (c) and (e) show the results after 0 (the NDFT result), 2 and 5 iterations. The right columns (b), (d) and (f) depict the differences between the original full data set and the reconstruction results on the same scale as the reconstruction results. After two iterations the result is already significantly better than that of NDFT. After 5 iterations, the result is very reasonable. After 10 iterations (not shown), the error does not decrease significantly anymore.

References

[1] J. R. Thorson and J. F. Claerbout. Velocity-Stack and Slant-Stack Stochastic Inversion. *Geophysics*, 50(12):2727–2741, 1985.

[2] D. Hampson. Inverse Velocity Stacking for Multiple Elimination. *J. Can. Soc. Expl. Geophys.*, 22(1):44–55, 1986.

[3] F. Marvasti. *A Unified Approach to Zero Crossings and Nonuniform Sampling of Single and Multidimensional Signals and Systems*. Nonuniform Publication, Oak Park, IL, 1987.

[4] F. Marvasti, Nonuniform Sampling. In R. Marks II, Ed., *Advance Sharman Sampling Theory*, Springer Verlag, New York, 1993.

[5] H. G. Feichtinger and K. Grochenig. *Theory and Practice of Irregular Sampling*. In J. Benedetto, and M. Easien, Eds., *Wavelets: Mathematics and Applications*. CRC Press, 1993, pp. 305–363.

[6] M. L. Kar, J. O. Hornkohl, and W. Farmer. A New Approach to Fourier Analysis of Randomly Sampled Data Using Linear Regression. *IEEE, Conf. on Acoustics, Speed and Signal Processing*, 1981, pp. 89–93.

[7] H. G. Feichtinger, K. Grochenig, and T. Strohmer. Efficient Numerical Methods in Non-Uniform Sampling Theory. *Numerische Mathematik*, 69(4):423, 1995.
[8] Y. Bard. *Nonlinear Parameter Estimation*. Academic Press, 1974.
[9] A. Tarantola. *Inverse Problem Theory*. Elsevier, 1987.
[10] A. J. W. Duijndam. Bayesian Estimation in Seismic Inversion. Part I—Principles. *Geophys. Prosp.*, 36(8):878–898, 1988.
[11] A. Duijndam, M. Schonewille, and C. Hindriks. Reconstruction of Band-Limited Signals, Irregularly Sampled along one Spatial Direction. *Geophysics*, 64(2):524–538, 1999.
[12] M. D. Sacchi and T. J. Ulrych. Estimation of the Discrete Fourier Transform, a Linear Inversion Approach. *Geophysics*, 61(4):1128–1136, 1996.
[13] S. Bagchi and S. K. Mitra. *The Nonuniform Discrete Fourier Transform and its Applications in Signal Processing*. Kluwer Academic Publishers, 1999.
[14] A. Basilevsky. *Applied Matrix Algebra in the Statistical Sciences*. North-Holland, 1983.
[15] W. H. Press, B. P. Flannery, S. A. Teukolsky, and W. T. Vetterling. *Numerical Recipes*. Cambridge University Press, 1986.
[16] C. Hindriks and A. Duijndam. Reconstruction of 3d Seismic Signals, Irregularly Sampled Along Two Spatial Coordinates. *Geophysics*, 65(1):253–263, 2000.
[17] P. D. Peterson and D. Middleton. Sampling and Reconstruction of Wavenumber Limited Functions in n-Dimensional Euclidean Spaces. *Information and Control*, 5:279–323, 1962.
[18] A. Duijndam and M. Schonewille. Nonuniform Fast Fourier Transform. *Geophysics*, 64(2):539–551, 1999.

12

Randomized Digital Optimal Control

W. L. De Koning and L. G. van Willigenburg

12.1. Introduction

In sampled data systems the sampling periods are almost always assumed to be deterministic, i.e., known in advance. However, in practice, the sampling periods should often be conceived as stochastic variables. This so called stochastic sampling phenomena occurs in many areas of sampled data systems such as in biological control systems or when a human being selects the sampling instants. Data may be absent at sampling instants when the sampling is governed by a stochastic process like a radar or sonar echo, or when the sampling mechanism fails. Also technical imperfections in the instrumentation may cause stochastic sampling.

Stochastic sampling may also be applied intentionally for instance when, for economy, a digital computer is time-shared in a stochastic manner, or to prevent jamming in communications. Other applications of intentional stochastic sampling are elimination of hidden oscillations between sampling instants, decreasing the influence of intelligent disturbances, and increasing stabilizability [1].

In this chapter we consider the important class of sampled-data systems where a continuous-time system is controlled by a digital computer, called digital control systems. We are interested especially in the influence of the presence of

W. L. De Koning • Department of Information Technology and Systems, Delft University of Technology, P.O. Box 5031, 2600 GA Delft, The Netherlands.
E-mail: w.l.deKoning@its.tudelft.nl

L. G. van Willigenburg • Wageningen University, Agrotechnion, Bomenweg 4, 6703 HD Wageningen, The Netherlands.
E-mail: Gerard.vanWilligenburg@user.aenf.wau.nl

Nonuniform Sampling: Theory and Practice, edited by Marvasti
Kluwer Academic/Plenum Publishers, New York, 2001.

stochastic sampling on the existence of a stationary optimal controller, on the stability of the optimal controlled system and on the control cost. Two important cases are distinguished. Firstly the case where the presence of stochastic sampling is taken into account and secondly where it is not taken into account in the determination of the optimal controller. Among other things it will be shown that in the first case stochastic sampling may increase or restore stabilizability and that in the second case stabilizability may decrease or be lost.

Bergen [2], and Leneman [3–4] studied deterministic scalar digital control systems with unity feedback, constant gain and independent identically distributed (IID) sampling periods. Results were found concerning spectral densities, stability, and response using convolution integrals or a direct approach. Kusher and Tobias [5] and Agniel and Jury [6–7] investigated deterministic single-input/single-output (SISO) digital control systems with unity feedback, constant gain, different types of nonlinearities and IID sampling periods. They obtained stability and boundedness results using stochastic Lyapunov functions. Darkhovsky and Leybovich [8] investigated the stability and response of deterministic digital control systems with complete state information, unity feedback and IID sampling periods using Kronecker products. An analysis of the stability in the mean sense of a deterministic SISO digital control system was conducted by Dannenberg and Melsa [9] using expected transforms in analogy with z-transforms. Optimal control in digital control systems with IID sampling periods was studied by Kalman [10], Gunckel and Franklin [11] and Davidson [12]. They all assumed complete state information and a quadratic cost criterion. Kalman [10] succeeded in finding an implicit condition for the existence of a stationary optimal controller in the case of a cost criterion without a control term and a continuous-time system which contains at least one integration. Davidson [12] considered the influence of stochastic sampling on the stationary control cost without bothering about the existence and stability of a stationary control system. Gunckel and Franklin [11] extended the result of Kalman [10] to a general continuous-time system and criterion in the finite horizon case. Chang [13] studied also the finite-horizon case but assumed incomplete state information. He considered in essence only the suboptimal filtering of continuous-time systems in the case of stochastic sampling.

In this chapter we consider digital stationary optimal control in the general case of linear stochastic continuous-time systems, quadratic integral criteria, incomplete state information and where the sampling periods are IID stochastic variables. The digital stationary optimal control problem is transformed to a discrete-time stationary optimal control problem for linear discrete-time systems and quadratic sum criteria, both with stochastic parameters, and with incomplete state information. This latter problem is then solved using the notions of mean-square stabilizability and mean-square detectability [14–15]. Also the criterion value is determined when the feedback is not optimal but arbitrary. Conditions are stated for the existence and stability of the stationary optimal linear estimator and convergence of the sample mean of the estimator error covariance in the case of a particular stochastic sampling scheme using the notions of uniform stabilizability and uniform detectability [16]. Given our interest in randomized digital optimal control the particular sampling scheme is an intentional stochastic sampling scheme. Finally the influence of stochastic sampling on the criterion value is investigated by means of some illustrative examples. It will appear that stochastic sampling may increase or even restore stabilizability. This makes intentional stochastic sampling a useful tool in the design of digital control systems. However, if the presence of stochastic sampling is not taken into account in the determination of the optimal controller, then stability may decrease or even be lost. Thus unintentional stochastic sampling because of, e.g., imperfections of instrumentation may have dramatic consequences in digital control systems if not considered in the design procedure.

12.2. Optimal Control Problem

In this section the digital optimal control problem for linear continuous-time systems and quadratic integral criteria, in the case of stochastic sampling and incomplete state information, is stated. This problem is transformed to a discrete-time stationary optimal control problem for linear discrete-time systems and quadratic sum criteria, both with stochastic parameters and with incomplete state information.

Consider a digital control system, consisting of a continuous-time system connected with a digital computer by means of a sample-and-hold at the input and a sampler at the output, see Fig. 1.

The sampling instants of both samplers are $t_0, t_1, \ldots$. The continuous-time system, the sample-and-hold operation and the observations at the sampling instants are described by respectively,

$$\dot{x}(t) = Ax(t) + Bu(t) + v(t),\, t \geq 0, \tag{1a}$$

$$u(t) = u(t_i),\, t_i \leq t < t_{i+1},\, i = 0, 1, \ldots, \tag{1b}$$

$$y(t_i) = Cx(t_i) + w(t_i),\, i = 0, 1, \ldots, \tag{1c}$$

where $x(t) \in R^n$ is the state, $u(t) \in R^m$ is the control, $y(t_i) \in R^l$ is the observation, $v(t) \in R^n$ is the system noise, $w(t_i) \in R^l$ is the observation noise and A, B and C are the known real system matrices of appropriate dimensions. The initial

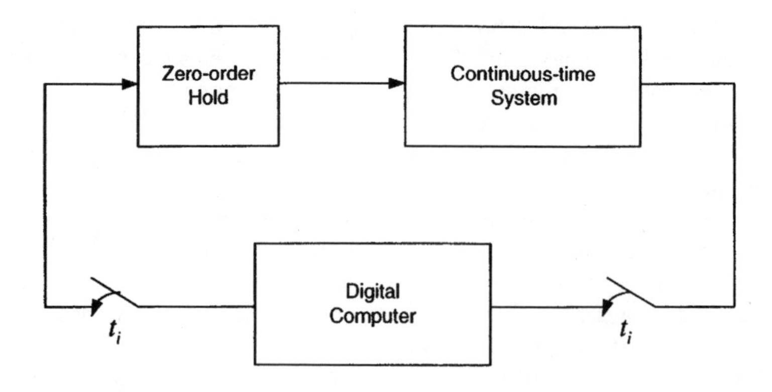

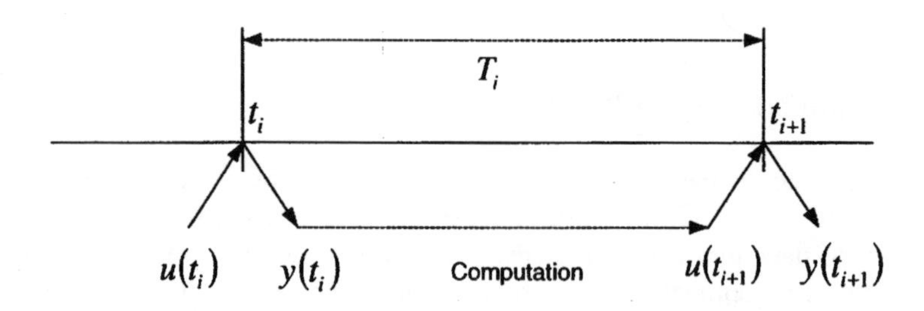

Figure 1. Digital control system and the task sequence of the computer.

condition $x(t_0) = x_0$ and the processes $\{v(t)\}$ and $\{w(t_i)\}$ are independent with known means and covariances

$$E\{x_0\} = \bar{x}_0, E\{(x_0 - \bar{x}_0)(x_0 - \bar{x}_0)^T\} = G, \tag{2a}$$

$$E\{v(t)\} = 0, E\{v(t)v^T(t)\} = V\delta(t - s), \tag{2b}$$

$$E\{w(t_i)\} = 0, E\{w(t_i)w^T(t_i)\} = W\delta_{ij}, \tag{2c}$$

where $\delta(.)$ and δ denote, respectively, the Dirac and Kronecker delta function. G, V and W are real symmetric matrices with $G \geq 0, V \geq 0$ and $W \geq 0$. Equation (1a) can be defined in terms of a stochastic integral equation and $v(t)$ as the formal derivative of an independent increments process $\beta(t)$ [17].

Consider also the long-term average integral criterion

$$\sigma_\infty(U_\infty) = \lim_{t_f \to \infty} \frac{1}{t_f - t_0} \int_{t_0}^{t_f} E\{[x^T(t)Qx(t) + u^T(t)Ru(t)]dt\}, \tag{3}$$

where Q and R are known real symmetric matrices of appropriate dimensions with $Q \geq 0$ and $R \geq 0$ and U_i denotes the control sequence $\{u(t_0), \ldots, u(t_i)\}$.

Define the sampling periods T_i by

$$T_i = t_{i+1} - t_i, \quad i = 0, 1, \ldots. \tag{4}$$

Then we assume that the process $\{T_i\}$ is a sequence of independent stochastic variables with known constant statistics independent of x_0, $v(t)$ and $w(t_i)$. Moreover

$$0 < \alpha \leq T_i \leq \beta < \infty, \quad i = 0, 1, \ldots, \tag{5}$$

where α and β are real scalars.

Let Y_i denote the observation sequence $\{y(t_0), \ldots, y(t_i)\}$. Then the digital stationary optimal control problem is defined as follows.

Definition 1 *Assume that $u(t_i)$ is a deterministic function of Y_{i-1}, U_{i-1}, $i = 0, 1, \ldots,$ and that the system* (1), *the criterion* (3) *and the sampling period process $\{T_i\}$ is given. Then the problem of finding the control sequence $U_\infty^* = \{u^*(t_0), u^*(t_1), \ldots\}$ which minimizes $\sigma_\infty(U_\infty)$ and of finding the minimal value σ_∞^* is called the digital stationary optimal control problem.*

From definition 1, we see that at time t_i the computer is supposed to send the control $u(t_i)$ and to receive the observation $y(t_i)$. Within the interval $[t_i, t_{i+1})$ the next control $u(t_{i+1})$ has to be calculated on the basis of the observations Y_i and the controls U_i.

We wish to transform the digital stationary optimal control problem into a discrete-time one, i.e., where only the behavior at the sampling instants is involved. First we need the following lemma concerning the sampling process.

Lemma 1 $t_i \to \infty \Leftrightarrow i \to \infty$

Proof: From (4) we have $t_i = t_0 + \sum_{k=0}^{i-1} T_i$, thus with (5) $t_0 + i\alpha \leq t_i \leq t_0 + i\beta$. From the first inequality it follows that $i \to \infty \Rightarrow t_i \to \infty$ and from the second one $t_i \to \infty \Rightarrow i \to \infty$.

Let x_i, u_i, y_i and w_i denote, respectively, $x(t_i)$, $u(t_i)$, $y(t_i)$ and $w(t_i)$. Let an overbar denote expectation. If the statistics of a stochastic variable s_i is independent of i, then $\bar{s}_i$ is denoted by $\bar{s}$. The following theorem concerns the transformation of the system (1).

Theorem 1 *The behavior of system* (1) *at the sampling instants is identical to the behavior of the discrete-time system*

$$x_{i+1} = \Phi_i x_i + \Gamma_i u_i + v_i, \quad i = 0, 1, \ldots, \tag{6a}$$

$$y_i = Cx_i + w_i, \quad i = 0, 1, \ldots, \tag{6b}$$

where

$$\Phi_i = \Phi(T_i) = e^{AT_i}, \tag{7a}$$

$$\Gamma_i = \Gamma(T_i) = \int_0^{T_i} \Phi(s)B\, ds, \tag{7b}$$

$$v_i = v(t_i + T_i, t_i) = \int_{t_i}^{t_i+T_i} \Phi(t_i + T_i - s)d\beta(s). \tag{7c}$$

The initial condition x_0, $\{v_i\}$ *and* $\{w_i\}$ *are independent with*

$$E\{x_0\} = \bar{x}_0, \quad E\{(x_0 - \bar{x}_0)(x_0 - \bar{x}_0)^T\} = G, \tag{8a}$$

$$E\{v_i\} = 0, \quad E\{v_i v_j^T\} = \bar{V}\delta_{ij}, \tag{8b}$$

$$E\{w_i\} = 0, \quad E\{w_i w_j^T\} = W\delta_{ij}, \tag{8c}$$

where $\bar{V}$ *denotes the known constant matrix* $E\{V_i\}$ *with*

$$V_i = V(T_i) = \int_0^{T_i} \Phi(s)V\Phi^T(s)ds \tag{9}$$

and $V_i \geq 0$. *Also* $\{v_i\}$ *is a sequence of independent stochastic vectors. The processes* $\{\Phi_i\}$ *and* $\{\Gamma_i\}$ *are sequences of independent random matrices with known constant statistics, independent of* $\{w_i\}$ *and* x_0. *Moreover,* Φ_i *and* Γ_i *are independent of* v_j, $i \neq j$ *and uncorrelated with* v_i.

Proof: See De Koning [17].

The asymptotic behavior of system (1) and system (6) expressed in theorem 1 goes still further as the following theorem shows.

Theorem 2 *The asymptotic behavior of system* (1) *at the sampling instants is identical to the asymptotic behavior of the system* (6).

Proof: Follows immediately from Theorem 1 and Lemma 1.

Now we turn to the transformation of the criterion (3).

Theorem 3 *Assume that u_i of system* (1) *is a deterministic function of Y_{i-1} and U_{i-1}. Then the value of criterion* (3)*, if it exists, is identical to the value of the long-term average sum criterion*

$$\sigma_\infty(U_\infty) = \frac{1}{\bar{T}} \lim_{N\to\infty} \frac{1}{N} E\left\{ \sum_{i=0}^{N-1} x_i^T \bar{Q} x_i + 2x_i^T \bar{M} u_i + u_i^T \bar{R} u_i \right\} + \frac{1}{\bar{T}} \bar{\gamma}, \tag{10}$$

where $\bar{Q}$, $\bar{M}$, $\bar{R}$, $\bar{\gamma}$, $\bar{T}$ denote, respectively, the known constant matrices $E\{Q_i\}$, $E\{M_i\}$, $E\{R_i\}$ and the scalars $E\{\gamma_i\}$, $E\{T_i\}$ with

$$Q_i = Q(T_i) = \int_0^{T_i} \Phi^T(s) Q \Phi(s) ds, \tag{11a}$$

$$M_i = M(T_i) = \int_0^{T_i} \Phi^T(s) Q \Gamma(s) ds, \tag{11b}$$

$$R_i = R(T_i) = \int_0^{T_i} [R + \Gamma^T(s) Q \Gamma(s)] ds, \tag{11c}$$

$$\gamma_i = \gamma(T_i) = \int_0^{T_i} tr[V(s) Q] ds, \tag{11d}$$

and $\bar{Q} \geq 0$, $\bar{R} \geq 0$ and $\bar{\gamma} \geq 0$.

Proof: Consider the criterion (3). Put the factor $1/(t_f - t_0)$ inside the expectation. Then, using Lemma 1, we may replace t_f by t_N and $t_f \to \infty$ by $N \to \infty$. Hence (3) may be written as

$$\sigma_\infty(U_\infty) = \lim_{N\to\infty} E\left\{ \frac{1}{t_N - t_0} \int_{t_0}^{t_N} [x^T(t) Q x(t) + u^T(t) R u(t)] dt \right\}.$$

By the Lebesque dominated convergence theorem we may write this, if it exists, as

$$\sigma_\infty(U_\infty) = E\left\{ \lim_{N\to\infty} \frac{1}{t_N - t_0} \int_{t_0}^{t_N} [x^T(t) Q x(t) + u^T(t) R u(t)] dt \right\}. \tag{12}$$

Let a.s. denote almost surely (with probability one). By (4) and the strong law of large numbers [18] $(1/N)(t_N - t_0) = (1/N)\sum_{i=0}^{N-1} T_i \to \bar{T}$ a.s. as $N \to \infty$. Hence (12) may be written as

$$\sigma_\infty(U_\infty) = \frac{1}{\bar{T}} E\left\{ \lim_{N\to\infty} \frac{1}{N} \int_{t_0}^{t_N} [x^T(t) Q x(t) + u^T(t) R u(t)] dt \right\},$$

and applying the Lebesque dominated convergence theorem again we have

$$\sigma_\infty(U_\infty) = \frac{1}{\bar{T}} \lim_{N\to\infty} \frac{1}{N} E\left\{ \int_{t_0}^{t_N} [x^T(t)Qx(t) + u^T(t)Ru(t)]dt \right\}.$$

From [17] we may write the expectation part as

$$E\left\{ \sum_{i=0}^{N-1} [x_i^T Q_i x_i + 2x_i^T M_i u_i + u_i^T R_i u_i] \right\} + N\bar{\gamma},$$

where Q_i, M_i and R_i are independent of x_i and u_i, and might therefore be replaced by

$$E\left\{ \sum_{i=0}^{N-1} [x_i^T \bar{Q} x_i + 2x_i^T \bar{M} u_i + u_i^T \bar{R} u_i] \right\} + N\bar{\gamma}.$$

The fact that $\bar{Q} \geq 0$, $\bar{R} \geq 0$ and $\bar{\gamma} \geq 0$ follows directly from (11).

Now we define a discrete-time control problem.

Definition 2 *Assume that* u_i *is a deterministic function of* Y_{i-1}, U_{i-1}, $i = 0, 1, \ldots$ *and given the system* (6) *and the criterion* (10), *then the problem of finding the control sequence* $U_\infty^* = \{u_0^*, u_1^*, \ldots\}$ *which minimizes* $\sigma_\infty(U_\infty)$ *and of finding the minimal value* σ_∞^* *is called the equivalent discrete-time stationary optimal control problem.*

The relation of this problem with the one in Definition 1 is as follows.

Theorem 4 *The solution of the digital stationary optimal control problem is identical to the solution of the equivalent discrete-time stationary optimal control problem.*

Proof: Follows from theorem 1, 2 and 3.

12.3. Optimal Control

In this section the stationary optimal control problem is solved via the solution of the equivalent discrete-time stationary optimal control problem. Furthermore, the criterion value is determined when the feedback is not optimal but arbitrary.

First we repeat some results concerning stabilizability and detectability from De Koning [14–15] for easy reference.

Consider the open loop system

$$x_{i+1} = \Phi_i x_i + \Gamma_i u_i, \quad i = 0, 1, \ldots, \tag{13}$$

where $x_i \in R^n$ is the state, $u_i \in R^m$ the control and Φ_i, Γ_i are real random matrices of appropriate dimensions. The processes $\{\Phi_i\}$ and $\{\Gamma_i\}$ are sequences of independent random matrices with constant statistics. The initial value x_0 is deterministic. System (13) is characterized by the pair (Φ_i, Γ_i). Suppose

$$u_i = -Lx_i, \tag{14}$$

where L is a real matrix with appropriate dimensions, then we have from (13) the closed loop system

$$x_{i+1} = (\Phi_i - \Gamma_i L)x_i. \tag{15}$$

Let ms denote mean square.

Definition 3 *System* (15) *is called ms-stable if* $\overline{\|x_i\|^2} \to 0$ *as* $i \to \infty$, $\forall x_0$.

Definition 4 (Φ_i, Γ_i) *is called ms-stabilizable if* $\exists L$ *such that system* (15) *is ms-stable.*

Let S^n denote the linear space of real symmetric $n \times n$ matrices and define the transformation A_L: $S^n \to S^n$ by

$$A_L X = \overline{(\Phi - \Gamma L)^T X(\Phi - \Gamma L)}, \quad X \in S^n. \tag{16}$$

Note that the statistics of $(\Phi_i - \Gamma_i L)^T X(\Phi_i - \Gamma_i L)$ are independent of i, so index i may be deleted in (16) as agreed in Section 2.

Lemma 2 A_L *is linear.* $X \geq 0 \Rightarrow A_L^i X \geq 0$, $i = 0, 1, \ldots$.

Proof: See [14].

Let ρ denote spectral radius.

Theorem 5 *System* (15) *is ms-stable* $\Leftrightarrow \rho(A_L) < 1$.

Proof: See [14].

Let $\otimes$ denote the Kronecker product [19]. Then it is easy to show that $\rho(A_L) = \rho\big((\Phi - \Gamma L) \otimes (\Phi - \Gamma L)\big)$ which is not difficult to calculate. Note that if $\Phi_i = \Phi$ and $\Gamma_i = \Gamma$ where Φ and Γ are deterministic and constant, then $\rho(A_L) = \rho^2(\Phi - \Gamma L)$.

In view of detectability, consider the system

$$x_{i+1} = \Phi_i x_i, \quad i = 0, 1, \ldots, \tag{17a}$$

$$y_i = C_i x_i, \quad i = 0, 1, \ldots, \tag{17b}$$

where $x_i \in R^n$ is the state, $y_i \in R^l$ the observation, and Φ_i, C_i are real random matrices of appropriate dimensions. The processes $\{\Phi_i\}$ and $\{C_i\}$ are sequences of independent random matrices with constant statistics. The initial value x_0 is deterministic. System (17) is characterized by the pair (Φ_i, C_i).

Definition 5 *(Φ_i, C_i) is called ms-detectable if $\overline{\|y_i\|^2} = 0,\ i = 0, 1, \ldots \Rightarrow \overline{\|x_i\|^2} \to 0$ as $i \to \infty$.*

There exist explicit conditions for ms-stabilizability [14] and for ms-detectability [15] which are easy to calculate. Furthermore we have the following two results. Let θ denote the zero matrix.

Theorem 6 $\rho(A_\theta) < 1 \Rightarrow (\Phi_i, \Gamma_i)$ *ms-stabilizable,* (Φ_i, C_i) *ms-detectable.*

Proof: Follows directly from Theorem 5 and Definitions 3, 4 and 5.

Theorem 7 *If Φ_i, Γ_i and C_i are deterministic and constant then ms-stability, ms-stabilizability and ms-detectability is identical to respectively stability, stabilizability and detectability in the usual sense.*

Proof: See De Koning [14-15].

Now we turn our attention to the equivalent discrete-time stationary optimal control problem. Consider system (6) and criterion (10) which are repeated here for convenience.

$$x_{i+1} = \Phi_i x_i + \Gamma_i u_i + v_i, \quad i = 0, 1, \ldots, \tag{18a}$$

$$y_i = C x_i + w_i, \quad i = 0, 1, \ldots, \tag{18b}$$

$$\sigma_\infty(U_\infty) = \frac{1}{\bar{T}} \lim_{N \to \infty} \frac{1}{N} E\left\{ \sum_{i=0}^{N-1} x_i^T \bar{Q} x_i + 2 x_i^T \bar{M} u_i + u_i^T \bar{R} u_i \right\} + \frac{1}{\bar{T}} \bar{\gamma}, \tag{19}$$

Assume that $\bar{R} > 0$. Define the transformation B_L: $S^n \to S^n$ by

$$B_L X = \overline{\Phi^T X \Phi} + \bar{Q} - (\overline{\Phi^T X \Gamma} + \bar{M})L - L^T(\overline{\Gamma^T X \Phi} + \bar{M}^T) + L^T(\overline{\Gamma^T X \Gamma} + \bar{R})L, \quad X \in S^n, \tag{20}$$

and the transformation $B_* : S^n \to S^n$ by

$$B_* X = B_{L_x} X, \quad X \in S^n, \tag{21}$$

where B_{L_x} is defined by (20) and L_X by

$$L_X = (\overline{\Gamma^T X \Gamma} + \bar{R})^{-1}(\overline{\Gamma^T X \Phi} + \bar{M}^T), \quad X \in S^n. \tag{22}$$

The transformation B_* is a generalized discrete-time Riccati transformation. The usual Riccati transformation is obtained after deletion of all the overbars in equation (20) and (21) [1]. Equation (21) may also be written as

$$B_* X = \overline{\Phi^T X \Phi} + \bar{Q} - L_X^T(\overline{\Gamma^T X \Gamma} + \bar{R})L_X, \quad X \in S^n. \tag{23}$$

We will also need the transformations B_L and B_* in a different form. Define the matrices Φ_i', Q_i' and L' by

$$\Phi_i' = \Phi_i - \Gamma_i \bar{R}\bar{M}^T, \tag{24a}$$
$$\bar{Q}' = \bar{Q} - \bar{M}\bar{R}^{-1}\bar{M}^T, \tag{24b}$$
$$L' = L - \bar{R}^{-1}\bar{M}^T, \tag{24c}$$

and the transformation $A_{L'}'$: $S^n \to S^n$

$$A_{L'}' X = \overline{(\Phi' - \Gamma L')^T X (\Phi' - \Gamma L')}, \quad X \in S^n. \tag{25}$$

It can be proven that $\bar{Q}' \geq 0$ [20]. One may easily verify that (20) may be written as

$$B_L X = A_{L'}' X + \bar{Q}' + L'^T \bar{R} L', \quad X \in S^n, \tag{26}$$

and $L_X' = L_X - \bar{R}^{-1}\bar{M}^T$ as

$$L_X' = (\overline{\Gamma^T X \Gamma} + \bar{R})^{-1}\overline{\Gamma^T X \Phi'}, \quad X \in S^n. \tag{27}$$

The relation between $A_{L'}'$ and A_L is as follows.

Lemma 3 *Assume that $\bar{R} > 0$. Then $A'_{L'}X = A_L X$, $X \in S^n$.*

Proof: From (24) we have $\Phi'_i - \Gamma_i L' = \Phi_i - \Gamma_i L$. Then the lemma follows from the definitions of A_L and $A'_{L'}$.

Finally, let $\hat{x}_i$ denote the minimum variance optimal estimator of x_i given Y_{i-1} and U_{i-1}, and let the estimator error covariance P_i be defined by

$$P_i = E\{(x_i - \hat{x}_i)(x_i - \hat{x}_i)^T\}. \tag{28}$$

Note that $\hat{x}_i$ is a deterministic function of Y_{i-1} and U_{i-1}. An arbitrary, not necessarily optimal, estimator $\hat{x}'_i$ is called variance neutral [21] if the associated estimator error covariance P'_i is not a function of U_{i-1}. The subject of estimation will be considered in Section 12.4.

Now we are in a position to state the solution of the equivalent discrete-time stationary optimal control problem.

Theorem 8 *Assume that $\bar{R} > 0$, that $\hat{x}_i$ is variance neutral for $i = 0, 1, \ldots$ and that $P = \lim_{i\to\infty}(1/i)\sum_{k=0}^{i-1} P_k$ exists. Then (Φ'_i, Γ_i) ms-stabilizable $\Rightarrow S = \lim_{i\to\infty} B_*^i \theta$ exists, S is the minimum nonnegative definite solution of the equation $S = B_* S$, $U_\infty^* = \{-L_S\hat{x}_0, -L_S\hat{x}_1, \ldots\}$ and*

$$\sigma_\infty^* = \frac{1}{\bar{T}} tr[\bar{V}S + L_S^T(\overline{\Gamma^T S \Gamma} + \bar{R})L_S P] + \frac{1}{\bar{T}}\bar{\gamma} \tag{29a}$$

$$= \frac{1}{\bar{T}} tr[\bar{V}S + (\overline{\Phi^T S \Phi} + \bar{Q} - S)P] + \frac{1}{\bar{T}}\bar{\gamma}, \tag{29b}$$

where the minimization is with respect to all control sequences for which (19) *exists.*

Proof: The result for S follows directly from (21), (26), (27) and [14]. In fact variance neutrality of $\hat{x}_i$ for $i = 0, 1, \ldots$ is not needed here. Consider the criterion

$$\sigma_N(U_{N-1}, H) = \frac{1}{\bar{T}N} E\left\{\sum_{i=0}^{N-1}(x_i^T\bar{Q}x_i + 2x_i^T\bar{M}u_i + u_i^T\bar{R}u_i) + x_N^T H x_N\right\} + \frac{1}{\bar{T}}\bar{\gamma}, \tag{30}$$

where H is a real symmetric matrix of appropriate dimensions with $H \geq 0$. This criterion is the same as criterion (19) except that N has a finite value and that the term $x_N^T H x_N$ has been inserted. The solution of the equivalent discrete-time stationary optimal control problem with criterion (10) replaced by criterion

(30), assuming variance neutrality of $\hat{x}_i$ for $i = 0, 1, \ldots$ gives for the minimum $\sigma_N^*(H)$ [22]

$$\sigma_N^*(H) = \frac{1}{\bar{T}N}\left\{\bar{x}_0^T S_0 \bar{x}_0 + tr(S_0 G) + \sum_{i=0}^{N-1} tr[\bar{V}S_{i+1} + L_i^T(\overline{\Gamma_i^T S_{i+1}\Gamma_i} + \bar{R})L_i P_i]\right\} + \frac{1}{\bar{T}}\bar{\gamma},$$

where $S_{i+1} = B_*^{N-i-1}H$, $S_0 = B_*^N H$ and $L_i = L_{S_{i+1}}$. The minimizing control sequence is $U_{N-1}^* = \{-L_0\hat{x}_0, \ldots, -L_{N-1}\hat{x}_{N-1}\}$. Let $H = \theta$ then, due to the existence of S and P, we have $S_{i+1} \to S$, $S_0 \to S$, $L_i \to L_S$ as $N \to \infty$ and $\lim_{N\to\infty} \sigma_N^*(\theta)$ equals the right part of equation (29a), say ϕ. For the proof of $\sigma_\infty^* = \sigma(U_\infty') = \phi$ we refer to Kushner [23]. Finally using (23) and $S = B_*S$, equation (29a) may be written as (29b).

The equation $S = B_*S$ is a generalized discrete-time Riccati equation. If complete state information is available then variance neutrality of $\hat{x}_i$ for $i = 0, 1, \ldots$ and the existence of P is not needed, in fact $\hat{x}_i = x_i$ and $P_i = P = \theta$ for $i = 0, 1, \ldots$. We see that u_i^* is a deterministic function of $\hat{x}_i$. A control problem with this property is called separable [21]. It has been proven by De Koning [22] that variance neutrality of $\hat{x}_i$, for $i = 0, 1, \ldots$ is a sufficient condition for separability of the equivalent discrete-time stationary optimal control problem. The existence of P will be considered in Section 12.4.

The closed loop system may not be ms-stable. The next theorem states the conditions for which it is in the case of complete state information.

Theorem 9 *Assume* $\bar{R} > 0$ *and that* (Φ_i', Γ_i) *is ms-stabilizable. Then* $(\Phi_i', \bar{Q}'^{1/2})$ *ms-detectable* $\Rightarrow \rho(A_{L_S}) < 1$, S *is the unique nonnegative definite solution of the equation* $S = B_*S$.

Proof: From (21), (26), (27) and [15] the result for S and $\rho(A'_{L'_S}) < 1$ is obtained. Then by Lemma 3 $\rho(A_{L_S}) < 1$.

The statement $\rho(A_{L_S}) < 1$ is by Theorem 5 equivalent to the statement that the closed loop system is ms-stable in the case of complete state information with $u_i = -L_S x_i$, $i = 0, 1, \ldots$. Of course, the stability of the control system with incomplete state information depends also on the stability of the estimator. The estimator will be considered in Section 12.4.

Assume $\bar{R} > 0$, that $\hat{x}_i$ is variance neutral and that P exists. Then summarizing loosely we may state that if (Φ_i', Γ_i) is ms-stabilizable, then the equivalent discrete-time stationary optimal control problem has a solution. If in addition $(\Phi_i', \bar{Q}'^{1/2})$ is ms-detectable, then the solution is unique and the closed loop system is ms-stable in the case of complete state information.

A useful relation between stability and convergence of $B_*^i\theta$ is as follows.

Corollary 1 *Assume $\bar{R} > 0$ and that $(\Phi_i', \bar{Q}'^{1/2})$ ms-detectable. Then (Φ_i', Γ_i) ms-stabilizable $\Leftrightarrow$ $B_*^i\theta$ converges as $i \to \infty$.*

Proof: Follows from (21), (26), (27) and [15].

Thus under very weak conditions, ms-stabilizability of (Φ_i', Γ_i) is not only sufficient but also necessary for the convergence of $B_*^i\theta$.

Furthermore in this section we are interested in the value σ_∞^L of $\sigma_\infty(U_\infty)$ when the feedback matrix is not optimal but has the arbitrary value L.

Theorem 10 *Assume $\bar{R} > 0$, that $P = \lim_{i\to\infty} (1/i) \sum_{k=0}^{i-1} P_k$ exists and that $U_\infty = \{-L\hat{x}_0, -L\hat{x}_1, \ldots\}$. Then $\rho(A_L) < 1 \Rightarrow S_L = \lim_{i\to\infty} B_L^i\theta$ exists, $S_L \geq 0$, S_L is the unique solution of the equation $S_L = B_L S_L$ and*

$$\sigma_\infty^L = \frac{1}{\bar{T}} tr[\bar{V} S_L + ((\overline{\Phi^T S_L \Gamma} + \bar{M})L + L^T(\overline{\Gamma^T S_L \Phi} + \bar{M}^T) - L^T(\overline{\Gamma^T S_L \Gamma} + \bar{R})L)P] + \frac{1}{\bar{T}}\bar{\gamma}, \tag{31a}$$

$$= \frac{1}{\bar{T}} tr[\bar{V} S_L + (\overline{\Phi^T S_L \Phi} + \bar{Q} - S_L)P] + \frac{1}{\bar{T}}\bar{\gamma}. \tag{31b}$$

Proof: Applying Lemma 2 and Lemma 3 we have from (26) by induction $B_L^i\theta = \sum_{k=0}^{i-1} A_L^k(\bar{Q}' + L'^T\bar{R}L')$. Thus $\rho(A_L) < 1 \Rightarrow S_L = \lim_{i\to\infty} B_L^i X = \sum_{k=0}^{\infty} A_L^k(\bar{Q}' + L'^T\bar{R}L')$ exists.

Because $A_L^k(\bar{Q}' + L'^T\bar{R}L') \geq 0$, $k = 0, 1, \ldots,$ also $S_L \geq 0$. Moreover $B_L^{i+1} = B_L B_L^i\theta$, thus taking the limits we have $S_L = B_L S_L$. Suppose $\tilde{S}_L \in S^n$ is any other solution of the equation $S_L = B_L S_L$ then $S_L - \tilde{S}_L = A_L(S_L - \tilde{S}_L)$ and by induction $S_L - \tilde{S}_L = A_L^i(S_L - \tilde{S}_L)$. Let $i \to \infty$ then $S_L = \tilde{S}_L$, thus S_L is the unique solution. Note that the existence of P is not needed so far. Consider criterion (30). Choosing $u_i = -L\hat{x}_i, i = 0, 1, \ldots$ and deleting the minimizations in the derivation of $\sigma_N^*(H)$, [22] gives for the value $\sigma_N^L(H)$:

$$\sigma_N^L(H) = \frac{1}{\bar{T}N}\left\{\bar{x}_0^T S_0 \bar{x}_0 + tr(S_0 G) + \sum_{i=0}^{N-1} tr[\bar{V} S_{i+1} + ((\overline{\Phi^T S_{i+1} \Gamma} + \bar{M})L + L^T(\overline{\Gamma^T S_{i+1} \Phi} + \bar{M}^T) - L^T(\overline{\Gamma^T S_{i+1} \Gamma} + \bar{R})L)P_i]\right\} + \frac{1}{\bar{T}}\bar{\gamma},$$

where $S_{i+1} = B_L^{N-i-1}H$. Let $H = \theta$ then due to the existence of S_L and P, we have $S_{i+1} \to S_L$, $S_0 \to S_L$ as $N \to \infty$ and $\lim_{N\to\infty} \sigma_N^L(\theta) = \sigma_\infty^L$ where σ_∞^L is defined by (31a). Finally using (20) and $S_L = B_L S_L$, (31a) may be written as (31b).

Note that variance neutrality of $\hat{x}_i$ for $i = 0, 1, \ldots$ is not needed in this theorem because there is no minimization involved as in Theorem 8. Observe also the resemblance between (29b) and (31b). If $S_L \to S$ in (31b) then $\sigma_\infty^L = \sigma_\infty^*$.

Finally we remark that the derivation of $\sigma_N^*(H)$ in Theorem 8, and therefore all the results in this section are still valid if $\hat{x}_i$ for $i = 0, 1, \ldots$ is restricted to be linear. Also observe that knowledge of T_i after t_i for $i = 0, 1, \ldots$ does not change the result of this section. It does affect the optimal estimator as will be seen in Section 12.4.

12.4. Optimal Estimation

In this section, conditions are stated for the existence and stability of the stationary optimal linear estimator in the case of a particular sampling scheme. Moreover, convergence of the sample mean of the estimator error covariance is considered.

Consider the discrete-time system (6). Let $\hat{x}'_i$ denote an arbitrary, not necessarily optimal, estimator of x_i. Then we have the following fact concerning $\{T_i\}$.

Lemma 3 *Assume $B \neq \theta$. Then $\hat{x}'_0, \ldots, \hat{x}'_{i+1}$ variance neutral $\Rightarrow T_0, \ldots, T_i$ deterministic.*

Proof: From De Groot and De Koning [24], we have $\hat{x}'_0, \ldots, \hat{x}'_{i+1}$ variance neutral $\Rightarrow \Gamma_0, \ldots, \Gamma_i$ deterministic. Assuming $\beta \neq \theta$ implies $T_0, \ldots, T_i$ deterministic.

Thus whatever we do with the data, there is no hope in finding any variance neutral estimators $\hat{x}'_0, \ldots, \hat{x}'_{i+1}$ if $T_0, \ldots, T_i$ are not deterministic. To apply the solution of the digital stationary optimal control problem, we have to determine successively the variance neutral estimators $\hat{x}'_0, \hat{x}'_1, \ldots$. Then from Lemma 3 it is necessary that for $i = 0, 1, \ldots$ the realization of T_i is known before the determination of $\hat{x}'_{i+1}$. Due to this, and also due to our interest in randomized digital control, we assume in the case of incomplete state information that the stochastic sampling occurs intentionally where the computer generates T_i first in the interval $[t_i, t_{i+1})$ and then determines $\hat{x}'_{i+1}$ on the basis of Y_i, U_i and $T_0, \ldots, T_i$.

Starting from the assumed intentional stochastic sampling scheme, we conclude that, as far as estimation is concerned, T_i may be assumed to be known in system (18). Thus we have the system

$$x_{i+1} = \Phi_i x_i + \Gamma_i u_i + v_i, \quad i = 0, 1, \ldots, \tag{32a}$$

$$y_i = C x_i + w_i, \quad i = 0, 1, \ldots, \tag{32b}$$

which is the same as system (18) except that $\Phi_i = \Phi(T_i)$ and $\Gamma_i = \Gamma(T_i)$ are known and given by (7a) and (7b), and that

$$E\{v_i\} = 0, \quad E\{v_i v_j^T\} = V_i \delta_{ij},$$

where $V_i = V(T_i)$ is known and given by (8). The system (32) is discrete-time with time-dependent parameters.

From system (32) and its properties it follows that the standard linear estimation theory [1, 25] applies immediately. Let $\hat{x}_i$ denote the linear minimum variance or optimal linear estimator of x_i given the observations Y_{i-1} and the controls U_{i-1}, and let P_i denote the estimator error covariance defined by (28). Then the optimal linear estimator is given by

$$\begin{aligned}\hat{x}_{i+1} &= \Phi_i \hat{x}_i + \Gamma_i u_i + K_i(y_i - C\hat{x}_i) \\ &= (\Phi_i - K_i C)\hat{x}_i + \Gamma_i u_i + K_i y_i, \quad \hat{x}_0 = \bar{x}_0, \end{aligned} \tag{34a}$$

$$K_i = \Phi_i P_i C^T (C P_i C^T + W)^+, \tag{34b}$$

$$\begin{aligned}P_{i+1} &= \Phi_i P_i \Phi_i^T + V_i - K_i(C P_i C^T + W) K_i^T \\ &= (\Phi_i - K_i C) P_i (\Phi_i - K_i C)^T + V_i + K_i W K_i^T, \quad P_0 = G,\end{aligned} \tag{34c}$$

where the superscript + denotes the Moore-Penrose pseudo inverse [26]. It is apparent that the estimator $\hat{x}_i$ is variance neutral for $i = 0, 1, \ldots$. Due to the assumed intentional stochastic sampling scheme, the Riccati equation (34c) must be solved on-line, one step at each sampling interval. Suppose that x_0, $\{v(t)\}$ and $\{w(t_i)\}$ are jointly normal, then x_0, $\{v_i\}$, and $\{w_i\}$ of system (32) are also jointly normal. In that case $\hat{x}_i$ given by (34) is also the optimal estimator, i.e. not restricted to be linear.

In order to find *a priori* conditions for the existence and stability of the stationary optimal linear estimator, we may of course not assume that T_i is deterministic for $i = 0, 1, \ldots$. However, we may conceive system (32) as a realization of system (18) dependent on a particular realization of the sequence $\{T_i\}$, and $\hat{x}_i$ given by (34) as the associated optimal linear estimator. Then, using the fact that Φ_i, Γ_i and V_i are bounded due to (5), (7) and (9), we have from [16] the following result concerning stationary behavior of the optimal linear estimator (34).

Theorem 11 *(Φ_i, C) a.s. uniformly detectable $\Rightarrow$ P_i a.s. bounded.*

Theorem 12 *Assume $W > 0$ and that (Φ_i, C) a.s. uniformly detectable. Then $(\Phi_i, V_i^{1/2})$ a.s. uniformly stabilizable $\Rightarrow$ estimator (34) a.s. exponentially stable.*

Corollary 2 *Assume $W > 0$ and $(\Phi_i, V_i^{1/2})$ a.s. uniformly stabilizable. Then (Φ_i, C) a.s. uniformly detectable $\Leftrightarrow$ P_i a.s. bounded.*

Note that the a.s. exponential stability of system (34) means that $x_{i+1} = (\Phi_i - K_i C)x_i$ is a.s. exponentially stable. If Φ_i and C_i are deterministic and constant, then a.s. exponential stability, a.s. uniformly stabilizability and a.s. uniformly detectability is identical to, respectively, stability, stabilizability and detectability in the usual sense [1, 25].

Furthermore in this section, we consider the convergence of the sample mean of P_i which was needed in Theorems 8 and 10.

Theorem 13 *(Φ_i, C) a.s. uniformly detectable $\Rightarrow \lim_{i\to\infty}(1/i)\sum_{k=0}^{i-1} P_k$ a.s. exists where the limit is in the mean square sense.*

Proof: Let $\bar{P}_i$ denote the expectation of P_i w.r.t. $T_0, \ldots, T_{i-1}$. From Theorem 11, P_i is a.s. bounded, thus by the weak law of large numbers [18] a.s. $(1/i)\sum_{k=0}^{i-1} P_k \to \lim_{i\to\infty}(1/i)\sum_{k=0}^{i-1}\bar{P}_k = P$ in the mean square sense as $i \to \infty$. Because P_i a.s. bounded, $\bar{P}_i$ is bounded and the right limit exists.

Note that P, if it exists, has the same value for any realization of $\{T_i\}$. So we might calculate P off-line by evaluating P_i for one single realization of $\{T_i\}$.

Lastly, we remark that the assumed intentional stochastic sampling scheme in this section does not affect the control result in Section 12.3. If complete state information is available then this particular intentional sample scheme is not needed, in fact $\hat{x}_i = x_i$ and $P_i = P = \theta$ for $i = 0, 1, \ldots$.

12.5. Influence of Stochastic Sampling

In this section the influence of stochastic sampling on the criterion value is investigated by means of some illustrative examples.

Consider the continuous-time system (1) and the corresponding equivalent discrete-time system (5). Assume that the sampling periods T_i are uniformly distributed with mean $\bar{T}$ and variance $\mathrm{var}(T) = E\{(T_i - \bar{T})^2\}$. Then because $T_i > 0$ we have

$$\mathrm{var}(T) < \frac{1}{3}\bar{T}^2. \tag{35}$$

Suppose $\mathrm{var}(T) = 0$ then (a.s.) $T_i = T$, $\Phi_i = \bar{\Phi}(T)$, $\Gamma_i = \bar{\Gamma}(T)$, thus T_i, Φ_i and Γ_i are deterministic and constant. Then due to Theorem 11 and the remark after Corollary 2, stabilizability (in the usual sense) and ms-stabilizability of (Φ_i, Γ_i) are identical, and detectability (in the usual sense) and a.s. uniform detectability of (Φ_i, C) are identical.

Assume that system (1) represents a second order single-input/single-output system and that (A, B) is reachable (in the usual sense) and (A, C) is observable (in the usual sense) [1]. Then in the case $\mathrm{var}(T) = 0$, (Φ_i, Γ_i) is reachable and

(Φ_i, C) is observable if and only if the eigenvalues of A are real, or the eigenvalues λ and λ^* of A are conjugate complex and [27]

$$\bar{T} \neq \frac{k\pi}{\mathrm{Im}(\lambda)}, \quad k = \pm 1, \pm 2, \ldots \tag{36}$$

Thus if the eigenvalues of A are complex, then reachability of (Φ_i, Γ_i) and observability of (Φ_i, C) are lost for values of $\bar{T}$ given by (36). If in addition these eigenvalues have positive real parts (unstable), then stabilizability of (Φ_i, Γ_i) and detectability of (Φ_i, C) are lost also for the values of $\bar{T}$ given by (36). In the case $\mathrm{var}(T) > 0$ there is in general no loss of ms-stabilizability of (Φ_i, Γ_i) and a.s. uniform detectability of (Φ_i, C) as will be illustrated by an example in this section.

In order to get more insight in the influence of stochastic sampling on the criterion value, we would like to compare this value for a particular system in three different cases. Firstly, from (29b), we may calculate σ^*_∞ for $\mathrm{var}(T) = 0$, i.e. if the sampling is deterministic, and secondly for $\mathrm{var}(T) > 0$, i.e. if the sampling is stochastic. Thirdly, from (31b), we may calculate $\sigma^{L^D}_\infty$ where L_D is the control law L in the case $\mathrm{var}(T) = 0$. The value $\sigma^{L^D}_\infty$ is the criterion value if the presence of stochastic sampling is not taken into account as far as the design of the control part is concerned.

Assume that system (1) is given by

$$A = \begin{bmatrix} 0.01 & 1 \\ 0 & 0.01 \end{bmatrix}, \quad B = \begin{bmatrix} 0 \\ 1 \end{bmatrix}, \quad C = [1 \quad 0], \quad V = \begin{bmatrix} 1 & 0 \\ 0 & 1 \end{bmatrix}, \quad W = [0.01],$$

and criterion (2) by

$$Q = \begin{bmatrix} 1 & 0 \\ 0 & 1 \end{bmatrix}, \quad R = [0].$$

The pair (A, B) is reachable and (A, C) is observable. The eigenvalues 0.01 and 0.01 of A are unstable but real. Thus in the case $\mathrm{var}(T) = 0$, (Φ_i, Γ_i) is reachable (and thus stabilizable) and (Φ_i, C) observable (and thus detectable). In the case $\mathrm{var}(T) > 0$ we may calculate that $\bar{R} > 0$, (Φ_i, Γ_i) is ms-stabilizable and (Φ_i, C) is a.s. uniformly detectable, which are the conditions to apply Theorem 8 and 13 in order to determine σ^*_∞. Furthermore we may calculate that $\rho(A_L) < 1$, so Theorem 10 may be applied to determine $\sigma^{L^D}_\infty$. In Fig. 2, σ^*_∞ is plotted for $\mathrm{var}(T)/\bar{T} = 0$ and for $\mathrm{var}(T)/\bar{T} = 0.1$, and $\sigma^{L^D}_\infty$ for $\mathrm{var}(T)/\bar{T} = 0.1$ as a function of $\bar{T}$ in the case of complete state information, i.e. $P = \theta$. Fig. 3 is the same except that there is incomplete state information. Comparing Figs. 2 and 3 we see that incomplete state information makes σ^*_∞ about ten times higher w.r.t. state

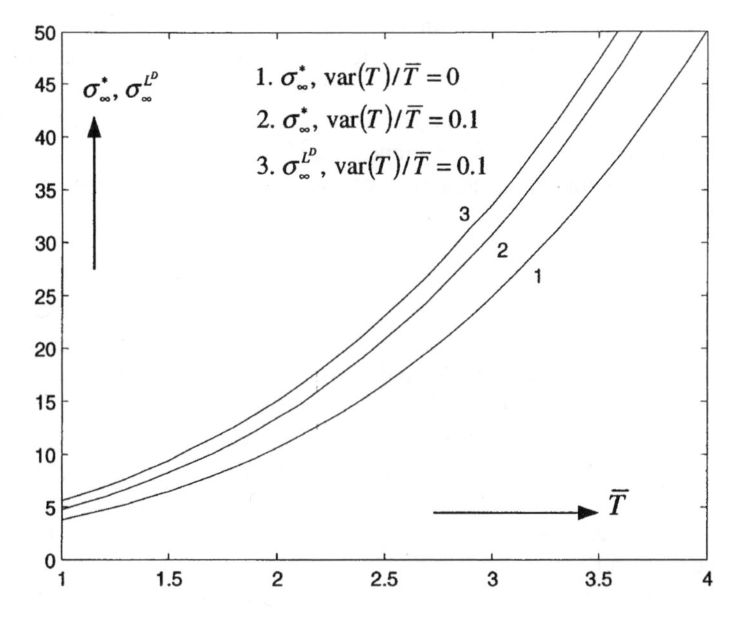

Figure 2. Complete state information.

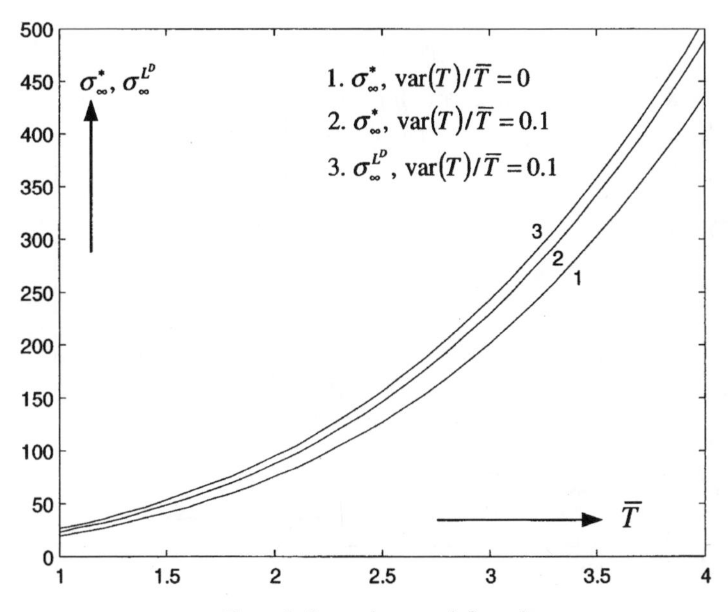

Figure 3. Incomplete state information.

information. Compared to deterministic sampling, i.e. $\mathrm{var}(T)/\bar{T} = 0$, the stochastic sampling, i.e., $\mathrm{var}(T)/\bar{T} = 0.1$, increases σ^*_∞ with about 15%. From the values of $\sigma^{L^D}_\infty$ in Figs. 2 and 3 observe that not taking into account the presence of stochastic sampling increases the criterion value with another 7%. Furthermore all the functions in Figs. 2 and 3 are monotonic increasing w.r.t. $\bar{T}$.

Now we assume the same system and criterion as above except that

$$A = \begin{bmatrix} 0.01 & -1 \\ 1 & 0.01 \end{bmatrix}.$$

The pair (A, B) is reachable and (A, C) is observable. The eigenvalues $0.01 + j$ and $0.01 - j$ of A are unstable and conjugate complex. Thus in the case $\mathrm{var}(T) = 0$, (Φ_i, Γ_i) is not stabilizable and (Φ_i, C) is not detectable for $\bar{T} = k\pi, k = 1, 2, \ldots$. In spite of this, in the case $\mathrm{var}(T) > 0$ we may calculate that $\bar{R} > 0$, (Φ_i, Γ_i) is ms-stabilizable and (Φ_i, C) is a.s. uniformly detectable thus σ^*_∞ exists for any $\bar{T} > 0$. However, $\rho(A_{L^D}) \geq 1$ for certain values of $\bar{T}$, so Theorem 10 may not be applied for these values to determine $\sigma^{L^D}_\infty$. In Fig. 4 σ^*_∞ is plotted for $\mathrm{var}(T)/\bar{T} = 0$ and $\mathrm{var}(T)/\bar{T} = 0.1$ and $\sigma^{L^D}_\infty$ for $\mathrm{var}(T)/\bar{T} = 0.1$ as a function of $\bar{T}$ in the case of complete state information and in Fig. 5 in the case of incomplete state information. Comparing Figs. 4 and 5, we see that incomplete state information increases σ^*_∞ with about 50 % w.r.t. complete state information.

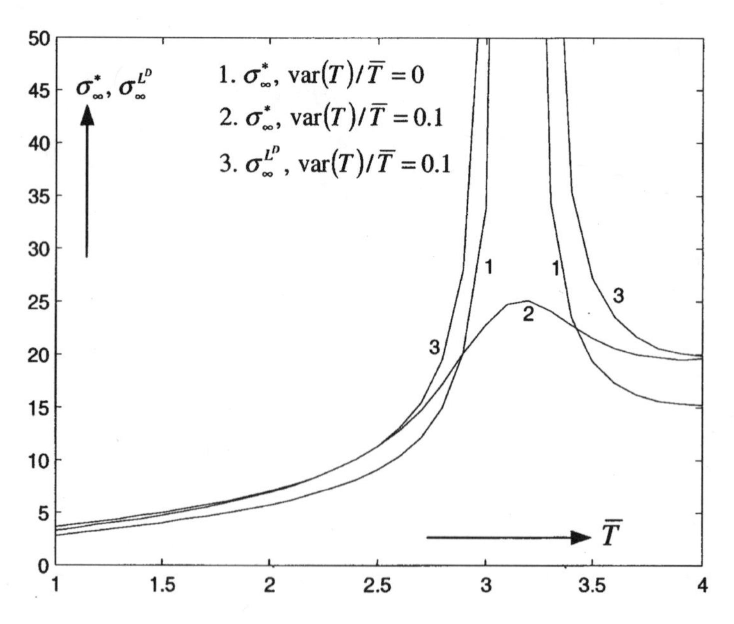

Figure 4. Complete state information.

In the case of deterministic sampling, i.e. $\mathrm{var}(T)/\bar{T} = 0$, the criterion value σ^*_∞ has an asymptote for $\bar{T} = \pi$ corresponding with the loss of stabilizability of (Φ_i, Γ_i) and detectability of (Φ_i, C). The stochastic sampling, i.e. $\mathrm{var}(T)/\bar{T} = 0.1$, increases σ^*_∞ with about 10% for values of $\bar{T}$ which are not in the neighborhood of π. However, now σ^*_∞ does not have an asymptote for $\bar{T} = \pi$, which corresponds to the fact that in the case $\mathrm{var}(T) > 0$ there is no loss of ms-stabilizability of (Φ_i, Γ_i) and a.s. uniform detectability of (Φ_i, C). Thus in the neighborhood of $\bar{T} = \pi$, the value σ^*_∞ for $\mathrm{var}(T)/\bar{T} = 0.1$ is much lower than for $\mathrm{var}(T)/\bar{T} = 0$. From the values of σ^{LD}_∞ in Figs. 4 and 5 observe that not taking into account the presence of stochastic sampling has almost no effect on σ^*_∞ for values of $\bar{T}$ which are not in the neighborhood of $\bar{T} = \pi$. However, in the neighborhood of $\bar{T} = \pi$ the value σ^{LD}_∞ goes to infinity which corresponds to the fact that $\rho(A_{LD}) \geq 1$.

In connection with the foregoing two examples we make some remarks. It appears from the computer calculations that the effects of stochastic sampling illustrated by the two examples in Figs. 2–5 arise for general continuous-time systems with (A, B) stabilizable and (A, C) detectable and general distributions of T_i. Especially, it appears that (Φ_i, Γ_i) ms-stabilizable and (Φ_i, C) a.s. uniformly detectable for any $\bar{T} > 0$. Thus there are no values of $\bar{T}$ for which σ^*_∞ has an asymptote. However, if the stochastic sampling is not taken into account, then in

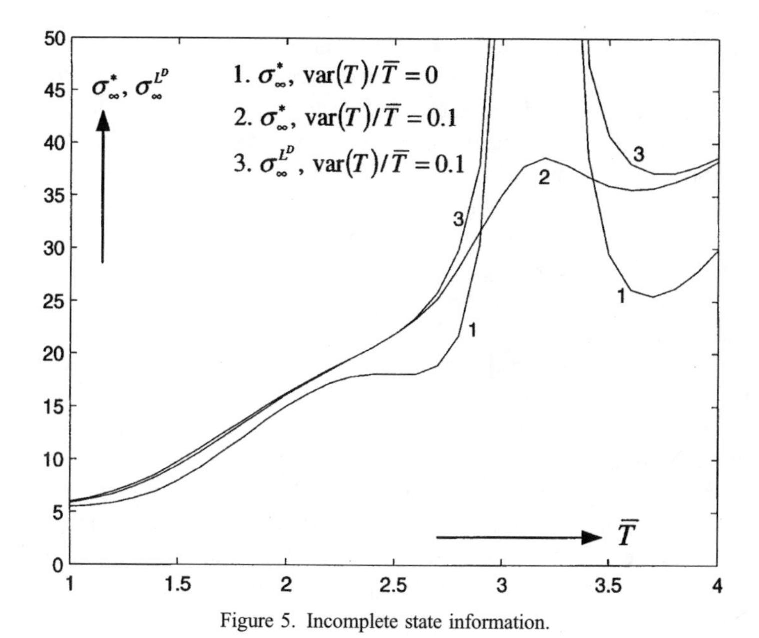

Figure 5. Incomplete state information.

general the area of $\bar{T}$ for which σ_∞^{LD} exists is smaller than the area for which σ_∞^* exists.

Finally in this section we make a practical remark concerning the calculation of σ_∞^* and σ_∞^{LD} from respectively (29) and (31). In these equations there arise terms like $\overline{\Phi^T X \Gamma}$ for an arbitrary matrix X, which may equally be written as $st^{-1}[\overline{(\Gamma \otimes \Phi)}^T st(X)]$, where st^{-1} means the inverse of the stack operator st [19]. Hence $\overline{(\Gamma \otimes \Phi)}$ need only be calculated once, while the st and st^{-1} operations involve only the renumbering of computer memory locations.

12.6. Conclusions

The digital stationary optimal control problem in the general case of linear stochastic continuous-time systems, quadratic integral criteria, incomplete state information and stochastic IID sampling periods has been solved, using the notions of mean-square stabilizability and mean-square detectability. Also the criterion value has been determined when the feedback is not optimal but arbitrary.

Existence and stability of the stationary optimal linear estimator and convergence of the sample mean of the estimator error covariance in the case of a particular stochastic sampling scheme have been considered, using the notions of uniform stabilizability and uniform detectability. The particular sampling scheme is an intentional stochastic sampling scheme.

It has been shown by means of two illustrative examples that stochastic sampling may increase or even restore stabilizability and may decrease or even destroy stability if the stochastic sampling is not taken into account in the determination of the optimal controller. In view of these effects intentional stochastic sampling is a useful tool in the design of digital control systems. On the other hand, if stochastic sampling is present but neglected in the design procedure, then the performance of the control system may become unacceptable.

Given our interest in randomized digital optimal control, in the case of incomplete state information, a particular intentional stochastic sampling scheme was considered. As this paper demonstrated, if the state information is incomplete and if the stochastic sampling is not intentional, the associated equivalent discrete-time stationary optimal control problem is not separable. This more difficult problem has been considered and solved by De Koning [28]. The same problem has also been considered and solved by De Koning and Van Willigenburg [29], when in addition, the dimension of the controller state is constrained to be *a-priori* fixed, and less than that of the system state. The same authors have presented a similar result which applies to non-stationary digital optimal control problems considered over a finite horizon [30].

References

[1] H. Kwakernaak and R. Sivan. *Linear Optimal Control Systems*. Wiley, New York, 1972.
[2] A. R. Bergen. On the Statistical Design of Linear Random Sampling Systems. *Proc. 1st IFAC Congr.*, Moscow, 1960, pp. 430–436.
[3] O. A. Z. Leneman. Random Sampling of Random Processes: Mean Square Behavior of a First Order Closed-Loop System. *IEEE Trans. Autom. Control*, AC-14:429–432, 1968.
[4] O. A. Z. Leneman. A Note on Mean-Square Behavior of a First Order Random Sampling System. *IEEE Trans. Autom. Control*, AC-13:450–451, 1968.
[5] H. J. Kusher and L. Tobias. On the Stability of Randomly Sampled Systems. *IEEE Trans. Autom. Control*, AC-14:319–324, 1969.
[6] R. G. Agniel and E. I. Jury. Stability of Nonlinear Randomly Sampled Systems. *Proc. Allerton Conf. on Circ. and Syst. Th.*, Urbana, 1969, pp. 710–720.
[7] R. G. Agniel and E. I. Jury. Almost sure Boundedness of Randomly Sampled Systems. *Siam J. Control*, 9:372–384, 1971.
[8] B. S. Darkhovskiy and V.S. Leybovich. Statistical Analysis of Sampled data Systems with Random Sampling. *Eng. Cybernetics*, 8:767–772, 1971.
[9] K. D. Dannenberg and J. L. Melsa. Stability Analysis of Randomly Sampled-Data Digital Control Systems. *Automatica*, 11:99–103, 1975.
[10] R. E. Kalman. Analysis and Synthesis of Linear Systems Operating on Randomly Sampled Data, Ph.D. Thesis. Columbia Univ., New York, 1957.
[11] T. L. Gunckel and G. F. Franklin. A General Solution for Linear Sampled Data Control. *ASME J. Bas. Eng.*, 85-D:197–201, 1963.
[12] C. Davidson. Random Sampling in Optimal Control Systems. *Ericsson Technics*, 186–219, 1974.
[13] S. S. L. Chang. Optimum Filtering and Control of Randomly Sampled Systems. *IEEE Trans. Autom. Control*, AC-12:537–546, 1967.
[14] W. L. De Koning. Infinite Horizon Optimal Control of Linear Discrete–Time Systems with Stochastic Parameters. *Automatica*, 18:443–453, 1982.
[15] W. L. De Koning. Detectability of Linear Discrete-Time Systems with Stochastic Parameters. *Int. J. Control*, 38:1035–1046, 1983.
[16] B. D. O. Anderson and J. B. Moore. Detectability and Stabilizability of Time-Varying Discrete-time Linear Systems. *SIAM J. Control*, 19:20–32, 1981.
[17] W. L. De Koning. Equivalent Discrete-Time Optimal Control Problem for Randomly Sampled Digital Control Systems. *Int. J. Systems Sci.*, 11:841–850, 1980.
[18] C. W. Burrill. *Measure, Integration and Probability*. McGraw-Hill, New York, 1972.
[19] R. Bellman. *Introduction to Matrix Analysis*. McGraw-Hill, New York, 1970.
[20] A. H. Levis, R. A. Schlueter and M. Athans. On the Behaviour of Optimal Linear Sampled Data Regulators. *Int. J. Control*, 13:343–361, 1971.
[21] Y. Bar-Shalom and E. Tse. Dual Effect, Certainty Equivalence, and Separation in Stochastic Control. *IEEE Trans. Autom. Control*, AC-19:494–500, 1974.
[22] W. L. De Koning. On the Optimal Control of Randomly Sampled Linear Stochastic Systems. *Proc. 1st IFAC Symp. on CAD of Contr. Syst.*, Zurich, 1979, pp. 107–112.
[23] H. J. Kushner. *Introduction to Stochastic Control*. Holt Rhinehart and Winston, New York, 1971.
[24] G. P. M. De Groot and W. L. De Koning. On Variance Neutrality within General Random Parameters. *IEEE Trans. Autom. Control*, AC-28:101–103, 1983.
[25] B. D. O. Anderson and J. B. Moore. *Optimal Filtering*. Prentice Hall, Englewood Cliffs, 1979.
[26] S. Barnett. *Matrices in Control Theory*. Van Nostrand and Reinold, London, 1971.
[27] R. E. Kalman. Mathematical Description of Linear Dynamical Systems. *SIAM J. Control*, 1:152–192, 1963.

[28] W. L. De Koning. Compensatibility and Optimal Compensation of Systems with White Parameters. *IEEE Trans. Autom. Control*, AC-37(5):579–588, 1992.

[29] W. L. De Koning and L. G. van Willigenburg. Numerical Algorithms and Issues Concerning the Discrete-Time Optimal Projection Equations for Systems with White Parameters. *Proc. UKACC Int. Conf. Control* '98, Univ. of Swansea, UK, 2:1605–1610, 1–4 September 1998.

[30] L. G. van Willigenburg and W. L. De Koning. Optimal Reduced–Order Compensators for Time-Varying Discrete-Time Systems with Deterministic and White Parameters. *Automatica*, 35:129–138, 1999.

[31] I. Bilinskis and A. Mikelsons. *Randomized Signal Processing*. Prentice-Hall International, Hemel Hempstead, Herts, 1992.

13

Prediction of Band-Limited Signals from Past Samples and Applications to Speech Coding

D. H. Mugler and Y. Wu

13.1. Introduction

This chapter is concerned with the prediction of the *next* value of a band-limited signal from its past, uniformly-sampled or periodic nonuniformly-sampled discrete values. The idea of a next value assumes a pattern of sampling that is not completely arbitrary, but has a periodic pattern. In addition to past samples that are uniformly-spaced, nonuniform but periodic sampling will be considered here, and each of these patterns will be related to the Nyquist criterion.

Both linear and nonlinear methods will be discussed in this article, although the focus is on linear prediction methods. The basic form of linear prediction, based on samples which are uniformly-spaced, gives a prediction of a band-limited signal $x(t)$ based on the samples $x(t-nT)_{n=1,\dots,N}$ in the form

$$x(t) = \sum_{n=1}^{N} a_n x(t-nT), \tag{1}$$

where the constants $\{a_n\}_{n=1,\dots,N}$ are the prediction coefficients. Determining the prediction coefficients which best apply to the class of band-limited signals is the subject of this paper.

D. H. Mugler • Divison of Applied Mathematics, University of Akron, Akron, Ohio, USA 44325-4002. E-mail: dmugler@uakron.edu.
Y. Wu • Department of Mathematics and Computer Science, Georgia Southern University, Statesboro, GA 30760-8093, USA.

Nonuniform Sampling: Theory and Practice, edited by Marvasti
Kluwer Academic/Plenum Publishers, New York, 2001.

An overview of linear prediction that was encompassing at its time was done by Makhoul [6]. A more recent review of mathematical methods in linear prediction, particularly for band-limited signals, was given by Butzer and Stens in [1]. The original work of this chapter is based on previous work of the authors in [10–15]. Some of the initial development of this area was done by Splettstößer in [23].

If the samples are from a finite energy band-limited signal, then we assume that the Fourier representation of the signal is that given by the Paley-Wiener theorem, i.e.,

$$x(t) = \int_{-W}^{W} X(f)e^{j2\pi ft}\,df, \tag{2}$$

for $X \in L^2[-W, W]$. Note that the highest frequency, W (in hertz), in the signal is known.

The class of "band-limited" signals is traditionally thought of as containing only finite energy signals, but it may be enlarged through distribution theory to include those which have unlimited energy and are of polynomial growth [19]. In particular, we suppose that there is a frequency limit W and constants N and γ so that

$$|x(z)| \leq \gamma(1 + |z|)^N e^{2\pi W|Im(z)|} \tag{3}$$

when x is extended as a function of a complex variable z. Functions which satisfy (3) have a Fourier transform representation that is similar to the above Paley-Wiener representation for finite energy signals (2). In this chapter, we follow the notation of [19] for signals of polynomial growth. In particular, if a function satisfies (3), then there is a distribution u of order N with support in $[-W, W]$ such that

$$x(t) = u(e_{-t}), \tag{4}$$

where e_{-t} is a notation for an exponential function (of f) which is defined to mean $e_{-t}(f) := e^{-j2\pi ft}$. In this paper, we begin the derivations of Section 13.2 with methods for finite energy signals and then extend those methods to the more general type of functions considered in (4).

In Section 13.2, we derive prediction coefficients which minimize the error over the class of band-limited signals. In addition to showing that the error-minimizing prediction coefficients arise from the solution of a system of equations, it is shown that they can be computed from the eigenvectors of a matrix related to that system of equations. Then a new method involving a linear combination of eigenvectors is introduced and noted to have some superior properties for the prediction problem. The new linear combination of eigenvectors method is then used in subsequent sections to determine the prediction coefficients.

The general case of nonuniform sampling is considered in Section 13.3 as demonstrated through the periodic nonuniform sampling pattern of interlaced sampling. The Nyquist rate is shown for this case to apply to prediction and to be a fraction of the usual rate. Similar to the computations of Section 13.2, the prediction coefficients may be found from eigenvectors of a certain matrix, and connections to the discrete prolate spheroidal sequences (DPSS) of Slepian [22] are discussed. Attention is given to a numerical technique for computing the eigenvectors through the use of a commuting tridiagonal matrix that is involved in a recurrence relation for the DPSS. An example is presented to show that the prediction method for interlaced sampling is quite accurate.

Section 13.4 provides a brief overview of some nonlinear techniques as applied to the problem of prediction, and Section 13.5 presents the application of the method of Section 13.2 to speech signals. Such signals are commonly low-pass filtered for communication purposes, so that they then qualify for an application of the methods developed in the earlier sections for band-limited signals. Examples of the new prediction method applied to speech signals are presented and the results are compared to a similar application of Linear Prediction Coding (LPC) methods. Some initial development of the method of Section 13.2 towards compression is given. One advantage of the method developed in the earlier sections and applied in this section to speech signals is that the prediction coefficients apply to the entire class of band-limited signals and do not need to be recomputed. This is similar to the standard Nyquist criterion in that once the sampling rate, sampling interval, and highest frequency are known, the method applies to the prediction of the next value of any band-limited signal with those same characteristics.

13.2. Prediction from Past Samples Based on Nyquist-type Criterion

13.2.1. Derivation of the Prediction Coefficients

The basic form of linear prediction based on past uniform samples was noted earlier (1) as $x(t) = \sum_{n=1}^{N} a_n x(t - nT)$, where $\{a_n\}_{n=1,\ldots,N}$ are the prediction coefficients. We lay the groundwork for the extension to nonuniform samples by listing the location of the sample points as t_n, for $n = 1, \ldots, N$, where t_n is a general form for the nth time sample. We also list Δ_n as the difference of that time sample from the prediction point. In particular, $t_n = t - nT$ and $\Delta_n = t - t_n = nT$ for $n \geq 1$ for the uniformly-spaced case. Thus, we write

$$x(t) = \sum_{n=1}^{N} a_n x(t - \Delta_n) \tag{5}$$

for the general form of the linear prediction, where the prediction point t is assumed to be outside an interval containing the sample values.

The (absolute) error in the prediction is expressed by $\epsilon = |x(t) - \sum_{n=1}^{N} a_n x(t - \Delta_n)|$. Assuming at the beginning that the signal is a finite-energy signal and using the Fourier representation (2) and the Schwarz formula, we find a bound on the squared error ϵ^2 given by

$$\epsilon^2 \leq \left\{ \int_{-W}^{W} |X(f)|^2 df \right\} \left\{ \int_{-W}^{W} |ds(f)|^2 df \right\}, \tag{6}$$

where $ds(f) = e^{j2\pi ft} - \sum_{n=1}^{N} a_n e^{j2\pi f(t-\Delta_n)}$. Note that (6) has the form

$$\epsilon^2 \leq \|x\|^2 \cdot \epsilon_I, \tag{7}$$

where ϵ_I is the second integral in (6) and $\|x\|$ is the energy, or the L^2-norm of the signal x. In particular

$$\epsilon_I = \int_{-W}^{W} \left| 1 - \sum_{n=1}^{N} a_n e^{-j2\pi f \Delta_n} \right|^2 df. \tag{8}$$

Note that the error integral (8) may be considered as an N-dimensional function $E(a_1, a_2, \ldots, a_N)$ of the prediction coefficients.

13.2.1.1. The Prediction Coefficients as the Solution of a System of Equations

The prediction coefficients are determined by minimizing the error over the whole class of band-limited signals in the prediction. In particular, by minimizing the part of the error in the error bound (6) as expressed by the function E. Using a standard method of minimization for $E(a_1, a_2, \ldots, a_N)$, see [14], gives a system of equations for the prediction coefficients as $\sum_{n=1}^{N} a_n \mathrm{sinc}(2W(\Delta_n - \Delta_j)) = \mathrm{sinc}(2W\Delta_j)$ for $j = 1, \ldots, N$, where the standard definition of sinc(t) is $\mathrm{sinc}(t) := \sin(\pi t)/(\pi t)$. If t_0 is the prediction point, then $\Delta_n - \Delta_j = (t_0 - t_n) - (t_0 - t_j) = t_j - t_n$ when Δ_j is the difference from the prediction point defined above.

It is important to realize that the prediction coefficients are independent of the particular time of prediction. The prediction coefficients depend on the relative spacing between the prediction point and the data time values on which the prediction is based, as well as the knowledge of the underlying type of signal. Since the focus of this article is on the prediction of a *next* value from past samples, this means that there is no need to update the prediction coefficients when marching ahead to predict a set of values from updated data points. This differs from the techniques used in linear predictive coding and these methods will be contrasted in a later section.

That the prediction coefficients are independent of the particular time of prediction can be seen from the error expression (8) above, or the equivalent form given for the function $E(a_1, a_2, \ldots, a_N)$. In each of those forms, the expression depends on differences Δ_n for $n = 1, \ldots, N$ instead of on a specific dependence on t.

For the case of uniform samples, the system of equations to solve for the prediction coefficients is a Toeplitz system. For this case, $\Delta_n - \Delta_j = (n-j)T$, where T is the sampling interval, and $\Delta_j = jT$ for $j = 1, \ldots, N$, as noted above. The system for the prediction coefficients becomes $\sum_{n=1}^{N} a_n \mathrm{sinc}(2TW(n-j)) = \mathrm{sinc}(2TWj)$ for $j = 1, \ldots, N$, using the property of the sinc function as an even function. With the Nyquist parameter defined as $\tau = 2TW$, this gives the system as

$$\sum_{n=1}^{N} a_n \mathrm{sinc}((n-j)\tau) = \mathrm{sinc}(j\tau), \tag{9}$$

for $j = 1, \ldots, N$, which is a Toeplitz system. If the system matrix is labeled H and the right side vector $\mathbf{b}$, then we may find the prediction coefficients by solving the system

$$H\mathbf{a} = \mathbf{b},$$

where $\mathbf{a}$ is the $N \times 1$ vector of prediction coefficients and $\mathbf{b}$ is of $N \times 1$ size with $\mathbf{b} = [\mathrm{sinc}(\tau), \mathrm{sinc}(2\tau), \ldots, \mathrm{sinc}(N\tau)]^T$, where the superscript T stands for matrix transpose.

For the general case of arbitrarily-located samples, the system of equations to solve in order to find the prediction coefficients is

$$\sum_{n=1}^{N} a_n \mathrm{sinc}(2W(t_j - t_n)) = \mathrm{sinc}(2W(t_0 - t_j)), \tag{10}$$

for $j = 1, \ldots, N$. We again label this system as $H\mathbf{a} = \mathbf{b}$ and note that the system is symmetric but not necessarily Toeplitz.

13.2.1.2. The Prediction Coefficients From Eigenvectors

The integral error expression ϵ_I of (8) can be written as an inner product, $\langle Av, v\rangle$ where v is a vector with $N+1$ entries, [10], and where A and v will be described further below. That is,

$$\epsilon_I = \langle Av, v\rangle = v^T Av \tag{11}$$

using either form of the inner product. One can show, [10], that the matrix A is related to the system of equations (10) by

$$A = \begin{bmatrix} 1 & \mathbf{b}^T \\ \mathbf{b} & H \end{bmatrix} \tag{12}$$

It is important for the vector v in (11) to have its first component equal to 1, i.e., $v_1 = 1$ in order to obtain $\epsilon_I = \langle Av, v \rangle$.

The form of the matrix A in the inner product is given by (12) for either case of uniform or nonuniform samples, with the convention that t_0 is the prediction point, as noted earlier. In particular, the matrix A in (12) is of size $(N+1) \times (N+1)$ and has entries given by the formula

$$A(i,j) = \text{sinc}(2W(t_{j-1} - t_{i-1})) \tag{13}$$

for $1 \le i, j \le N+1$. Written in matrix form, this gives

$$A = \begin{bmatrix} s(t_0 - t_0) & s(t_1 - t_0) & s(t_2 - t_0) & \ldots & s(t_N - t_0) \\ s(t_0 - t_1) & s(t_1 - t_1) & s(t_2 - t_1) & \ldots & s(t_N - t_1) \\ s(t_0 - t_2) & s(t_1 - t_2) & s(t_2 - t_2) & \ldots & s(t_N - t_2) \\ \ldots & \ldots & \ldots & \ldots & \ldots \\ s(t_0 - t_N) & s(t_1 - t_N) & s(t_2 - t_N) & \ldots & s(t_N - t_N) \end{bmatrix} \tag{14}$$

where $s(t) := \text{sinc}(2Wt)$. Note that this matrix is symmetric and Toeplitz for uniform samples, and is symmetric for the general case, since sinc(t) is an even function. Also, the matrix is diagonally dominant since the entries on the main diagonal are all equal to 1 since sinc(0) = 1. More important, matrix (14) is positive definite, as can be seen from the relation to the error integral through (11).

For the case of uniform samples, the general entry in the matrix A reduces to

$$A(i,j) = \text{sinc}(2TW(j-i)) = \text{sinc}((j-i)\tau) \tag{15}$$

for $1 \le i, j \le N+1$. In the next section, we will discuss the relation of these entries to the discrete prolate spheroidal sequences.

In order to find the best prediction coefficients using the relation of the prediction error term ϵ_I to the inner product above, it is thus necessary to find the *smallest* value of $v^T Av$ over all vectors with first component equal to one. Over all normalized vectors with norm 1, the minimum of $v^T Av$ is given by the smallest *eigenvalue* of matrix A, [17], p. 216. Since A is positive definite, all the eigenvalues are positive. The eigenvector v_N associated with the smallest eigenvalue, λ_N is only unique up to a constant multiple. What is needed here is the vector v that gives the minimum of $v^T Av$, and this must be normalized so that $v_1 = 1$, as noted above. There are different approaches that may be used at this point.

13.2.1.2.1. Smallest Eigenvalue Approach. The first approach we discuss for the minimization of the error in the form of the error integral (8) using eigenvectors involves the single eigenvector corresponding to the smallest eigenvalue only. Let v_N be the eigenvector associated with the smallest eigenvalue and suppose that $||v_N|| = 1$, since a unit norm vector is the usual default when computing eigenvectors from commercial software. In this case, $\langle Av_N, v_N \rangle = \lambda_N$, where λ_N is the smallest eigenvalue. If v_{norm} has first component $v_{norm}(1) = 1$ and is obtained by $v_{norm} = v_N / v_{N1}$, where v_{N1} is the first component of the unit norm eigenvector, it follows that

$$\langle Av_{norm}, v_{norm} \rangle = \lambda_N / v_{N1}^2. \tag{16}$$

Note that, as described above, it is important to have the vector in (11) have its first component equal to 1 in order to obtain $\epsilon_I = \langle Av, v \rangle$. Since that is satisfied for v_{norm}, the resulting prediction coefficients $\{a_i\}_{i=1,\ldots,N}$ are the second through $N+1$st entries of this normalized eigenvector, v_{norm}.

The prediction error depends on ϵ_I and on the signal energy, as in (7). When using the prediction coefficients calculated from v_{norm}, as described above, then $\epsilon_I = \langle Av_{norm}, v_{norm} \rangle = \lambda_N / v_{N1}^2$, see (11) and (16). The eigenvalue λ_N is typically very small, and we have found in many numerical trials that the value of λ_N / v_{N1}^2 is also quite small. This means that the entries of the eigenvector v_{norm} can be used to determine prediction coefficients that give predictions with small error, independent of the point of prediction.

13.2.1.2.2. Minimum Error Using a Linear Combination of the Eigenvectors. Next, we present an approach to minimize the error integral ϵ_I for the prediction problem that uses a linear combination of all the eigenvectors. The values of prediction coefficients in this case do yield the minimum value of ϵ_I, so that the results give coefficients which are equivalent to those found by solving the system as noted in Section 13.2.1.1. That is, using the complete set of eigenvectors makes the linear combination of eigenvectors method equivalent to the system solution method of Section 13.2.1.1. However, we will demonstrate that using a linear combination of a subset of the complete set of eigenvectors has some advantages, and we will contrast this approach to the other methods in the next section.

Again, the goal is to find the smallest value of the quadratic form $v^T Av$ over the set of vectors whose first component equals 1. The complete set of eigenvectors of A span the $N+1$-dimensional space of vectors, and we assume we begin with the set of eigenvectors that is normalized to have unit norm. Since A is symmetric, these vectors are orthonormal. As noted above, since A is positive definite, all of the corresponding eigenvalues are distinct and positive. Conforming to Slepian's notation in [22], we let λ_0 be the largest eigenvalue, and suppose the ordering of the eigenvalues is

$$0 < \lambda_N < \cdots < \lambda_1 < \lambda_0 < 1,$$

and that the index of the normalized eigenvector v_i corresponds to its associated eigenvalue λ_i for each i.

To find the minimum of ϵ_I is actually a constrained minimization problem. Suppose $v = \sum_{i=0}^{N} \alpha_i v_i$ is a linear combination of the set of orthonormal eigenvectors. In this case, $v^T A v = \sum_{i=0}^{N} \alpha_i v^T A v_i = \sum_{i=0}^{N} \alpha_i \lambda_i \left(\sum_{j=0}^{N} \alpha_j v_j^T v_i\right) = \sum_{i=0}^{N} \alpha_i^2 \lambda_i$, where the λ_i are the eigenvalues of A. The minimization problem is: *Find coefficients α_i for $i = 0, \ldots, N$ such that*

$$f(\alpha) = \sum_{i=0}^{N} \alpha_i^2 \lambda_i$$

is minimized, subject to the single constraint equation $\sum_{i=0}^{N} \alpha_i v_{i1} = 1$, where v_{i1} is the first component of vector v_i for each i.

If just the smallest eigenvector v_N is used, then it follows that the solution is $\alpha_N = 1/v_{N1}$ and the minimum value of the quadratic form is given by $\min(\epsilon_I) = \lambda_N / v_{N1}^2$, where this is the value we obtained when we use just the smallest eigenvector. This eigenvector has some nice properties, such as a symmetry (actually odd symmetry) about the center point, and we will come back to these properties later.

Next suppose that just the first two smallest eigenvectors are used, so that $v = \alpha_{N-1} v_{N-1} + \alpha_N v_N$. One can show that the α values which minimize the above problem in such a case have a nice symmetry to them as can be seen by the formulas

$$\alpha_N = \frac{v_{N1}}{v_{N1}^2 + \left(\dfrac{\lambda_N}{\lambda_{N-1}}\right) v_{(N-1)1}^2}$$

$$\alpha_{N-1} = \frac{v_{(N-1)1}}{\left(\dfrac{\lambda_{N-1}}{\lambda_N}\right) v_{N1}^2 + v_{(N-1)1}^2}$$

and, further, that the minimum value of the quadratic form is given by

$$\min(\epsilon_I) = \frac{\lambda_N}{v_{N1}^2 + \left(\dfrac{\lambda_N}{\lambda_{N-1}}\right) v_{(N-1)1}^2},$$

where v_{i1} is the first component of vector v_i for each i. Note that this minimum value reduces to the one when just using the single smallest eigenvector as described above when the ratio λ_N/λ_{N-1} is near zero, since then the minimum value is near λ_N/v_{N1}^2.

For the general case that includes all the eigenvectors, the solution of the minimization problem still has an elegant symmetric form. In this case, one can show that the formula for the coefficients is

$$\alpha_i = \frac{v_{i1}}{\sum_{n=0}^{N}\left(\frac{\lambda_i}{\lambda_n}\right)v_{n1}^2} \tag{17}$$

for each i and that the minimum of the quadratic form is given by

$$\min(\epsilon_I) = \frac{\lambda_N}{\sum_{n=0}^{N}\left(\frac{\lambda_N}{\lambda_n}\right)v_{n1}^2}. \tag{18}$$

Note again that if the smallest eigenvalue λ_N is relatively much smaller than the other eigenvalues, then for each n other than N the ratio λ_N/λ_n is near zero, so that the minimum value attained is essentially the value λ_N/v_{N1}^2 as when just using the smallest eigenvector as discussed above.

Since the error integral ϵ_I is minimized when using the complete set of eigenvectors in a linear combination with coefficients given by (17), the resulting vector sum will give the same prediction coefficients as by solving the system of equations in Section 13.2.1.1. In addition, this method has resulted in the formula (18) for the actual minimum value of the error integral, which can be used in combination with signal energy in (7) to determine an error bound on the prediction, independent of the prediction point. As in Section 13.2.1.2.1, the first component of the linear combination has been set to 1 from the constraint equation, and it is the second through $N+1$st entries of the linear combination that gives the prediction coefficients.

13.2.1.2.3. Contrasting the Smallest Eigenvector Approach to a Combination of Eigenvectors Method. The prediction coefficients that are derived in the previous sections using two different methods can give different results, especially for small quantities N of prediction coefficients. For the case of uniformly-spaced samples, Table 1 provides a comparison of prediction coefficient values for $N=4$ coefficients. For the three different values of the Nyquist parameter τ listed, the table presents the coefficients on the left side for

Table 1. A comparison ($N=4$) of prediction coefficients using the complete set of eigenvectors method (left) and the smallest eigenvector approach (right).

τ	0.05	0.10	0.15	0.05	0.10	0.15
a_1	3.9758	3.9037	3.7855	3.9789	3.9162	3.8137
a_2	−5.9485	−5.7964	−5.5506	−5.9579	−5.8332	−5.6314
a_3	3.9696	3.8794	3.7330	3.9789	3.9162	3.8137
a_4	−0.9969	−0.9875	−0.9719	−1.0000	−1.0000	−1.0000

the complete set of eigenvectors. On the right side, the values listed in the table are the coefficients obtained by using the smallest eigenvector approach.

As can be seen from the entries in Table 1, the prediction coefficients are not quite the same for the two different methods. Perhaps easiest to recognize is that the last coefficient for the smallest eigenvector approach always equals -1.0 for $N=4$ and that this approach yields coefficients that have a symmetry about their center point. We will comment more on these patterns in the next section. Although the values of the coefficients are different, corresponding coefficients are similar to each other.

There is something of a compromise to be made in determining which method to use for determining the prediction coefficients. Methods that are available as discussed above include either the smallest eigenvalue approach, the complete set of eigenvectors approach, or an approach that uses some combination in between. We have found in extensive numerical trials that the smallest eigenvalue method has a better capability of predicting the "peak" values of a signal than does the complete set of eigenvectors approach. The complete set of eigenvectors provides the same prediction coefficients as are obtained by solving the system (10) of Section 13.2.1.1 by Gaussian elimination, and the set of predictions obtained from this approach just does not match the highs and lows of the signal as well. On the other hand, the prediction coefficients that come from the smallest eigenvalue method are larger in magnitude, as can be seen from Table 1, than those from the complete eigenvector approach. This leads to more sensitivity to noise for the smallest eigenvector method and favors the complete eigenvector approach.

The method we choose for prediction is a combination of the two different approaches. This is a balance between accurately predicting peak values of the signal while employing prediction coefficients that are not too large in magnitude. This method will be presented for the band-limited signals illustrated in the following, particularly for speech signals in Section 13.5. Based primarily on numerical trials, our compromise uses the three smallest eigenvectors and combines them using the minimizing procedure identified by the linear combination of eigenvectors with coefficient α_i, for $i = 1, \ldots, 3$ given by (17).

13.2.1.3. Prediction Coefficients for Band-limited Signals of Polynomial Growth

The same prediction coefficients are valid for band-limited signals of polynomial growth as are valid for band-limited signals with finite energy. It is the very similar form of the extended Paley-Wiener representation (4) for these functions that results in the derivation of the prediction coefficients in this case yielding the same values as for band-limited functions with finite energy. Speech signals typically fall into this extended category of signals, which for example includes $\sin(2\pi Wt)$ and $\cos(2\pi Wt)$. That the prediction coefficients as derived above apply not just to finite energy signals is quite important for the applications to prediction of speech signals in Section 13.5.

The concept of a band-limited signal was discussed at the beginning of this chapter. The condition on growth uses the extension of the signal to the complex plane and was listed earlier in (3). For ease of reference we repeat it here as

$$|x(z)| \leq \gamma(1 + |z|)^N e^{2\pi W|Im(z)|}. \tag{19}$$

For a signal satisfying this growth condition, the extension of the standard Paley-Wiener theorem, see [19], says that there is a distribution u of order N with "band-limited" suppport in $[-W, W]$, such that the representation

$$x(t) = u(e_{-t}),$$

given previously in (4), is valid, where $e_{-t}(f) := e^{-j2\pi tf}$. For example, a function such as $\sin(2\pi Wt)$ satisfies the growth condition since $|\sin(2\pi Wz)| = |Im[e^{2\pi jWz}]| \leq e^{2\pi W|Im(z)|}$. In addition, $x(t) = t\sin(2\pi Wt)$ satisfies the condition, since $|x(z)| = |z\sin(2\pi Wz)| = |z||\sin(2\pi Wz)| \leq (1 + |z|)e^{2\pi|Im(z)|}$.

The derivation of the prediction coefficients for the extension of the band-limited signals to include those of polynomial growth involves distributions and substitutes the representation $x(t) = u(e_{-t})$ for the Paley-Wiener representation (2). More details of this derivation may be found in our previous work in [10], but we provide in this chapter some examples, particularly in Section 13.5, of predictions for band-limited signals of polynomial growth which use exactly the same set of prediction coefficients as described above. That the same set of prediction coefficients apply to band-limited signals of polynomial growth is quite important for these applications.

One of the simplest cases for which the growth condition (19) applies is when the signal is actually a polynomial. In this case, the appropriate value of the Nyquist parameter τ is 0 since τ depends on the highest frequency W in the signal. Note that $W = 0$ for a polynomial, where the Fourier transform is a distribution. We conclude this section by considering this special case and show that standard polynomial prediction is covered in the limiting case as τ tends to zero.

For the special case of polynomials, the Nyquist parameter τ is 0 but this value cannot be substituted directly into matrix A (14) as that value of τ would make the matrix completely singular. Thus we consider the limiting case as $\tau \to 0$.

It is not difficult to show that the prediction coefficients in a formula of form $x(t) = \sum_{n=1}^{N} a_n x(t - nT)$, which *exactly* reproduce polynomials up to degree $N-1$ are the coefficients which are alternating binomial coefficients,

$$a_n = (-1)^{n+1}\binom{N}{n}, \tag{20}$$

for $n = 1, \ldots, N$. The proof of this result can be shown by proving the result for polynomials $p(t)$ with $p(t) = t^k$ for each k with $k \leq N-1$, and then using the linearity of the prediction formula to conclude the result for arbitrary polynomials up to degree $N-1$. In particular, for polynomials $p(t) = t^k$ of this form,

$$\begin{aligned}\sum_{n=1}^{N} a_n p(t - nT) &= \sum_{n=1}^{N} a_n (t - nT)^k \\ &= \sum_{n=1}^{N} a_n \sum_{j=0}^{k} \binom{k}{j} t^{k-j}(-T)^j n^j \\ &= \sum_{j=0}^{k} \binom{k}{j} t^{k-j}(-T)^j \left[\sum_{n=1}^{N} (-1)^{n+1}\binom{N}{n} n^j\right].\end{aligned}$$

The bracketed expression at the end of the sum, $\sum_{n=1}^{N}(-1)^{n+1}\binom{N}{n}n^j$, can be shown to be equal to 1 if $j = 0$ and equal to 0 if $1 \leq j \leq k$, $k \leq N-1$. This can be proved by considering the binomial expansion of the function $1 - (1-x)^N$ and successively evaluating derivatives of this function at $x = 1$. This shows that the final sum reduces to t^k, as desired, proving that the alternating binomial coefficients are the prediction coefficients which exactly reproduce polynomials.

Numerical calculations confirm the result that the prediction coefficients converge to the alternating binomial coefficients (for $n \geq 1$) as $\tau \to 0$. The computations presented in Table 2 show values of τ decreasing towards 0 and the corresponding set of prediction coefficients. This set of coefficients comes from the smallest eigenvector method, although the limits are the same for this method as for the combination of eigenvectors method. These limits are the same because, as described in Section 13.2.1.2 and specifically by (17), as τ tends to 0, the smallest eigenvalue λ_N of the matrix becomes relatively small in comparison to the other eigenvalues and the smallest eigenvector method and the combination of eigenvectors method produce the same results.

Note that the actual values of the appropriate alternating binomial coefficients are $4, -6, 4, -1$ for $N = 4$ and $5, -10, 10, -5, 1$ for $N = 5$, and that the prediction coefficients as listed in Table 2 are converging to these values as $\tau \to 0$.

Table 2. Prediction coefficients as $\tau \to 0$ for $N=4, 5$.

	$N=4$				
τ	a_1	a_2	a_3	a_4	
0.050	3.9789	− 5.9579	3.9789	− 1.0000	
0.030	3.9924	− 5.9848	3.9924	− 1.0000	
0.010	3.9992	− 5.9983	3.9992	− 1.0000	
0.005	3.9998	− 5.9996	3.9998	− 1.0000	
	$N=5$				
τ	a_1	a_2	a_3	a_4	a_5
0.050	4.9726	− 9.9181	9.9181	− 4.9726	1.0000
0.030	4.9901	− 9.9704	9.9704	− 4.9901	1.0000
0.010	4.9989	− 9.9967	9.9967	− 4.9989	1.0000
0.005	4.9997	− 9.9992	9.9992	− 4.9997	1.0000

The size of the prediction coefficients fortunately decreases as τ increases, and this can be seen by looking at Table 2 from bottom to top. The size of the binomial coefficients is large for large N and thus they magnify any noise in signal samples. However, the magnitudes of the coefficients decrease as τ increases, so the resulting coefficients do not magnify the noise of the prediction of truly oscillatory functions as much as when $\tau = 0$. The well-known result that prediction of pure polynomial signals of large degree is a procedure that is sensitive to noise can be seen from the size of the prediction coefficients for such functions. That this sensitivity to noise decreases as τ increases is helpful to the prediction of speech signals, as will be discussed in Section 13.5.

13.2.2. The Nyquist Condition and the Prediction Problem

Among the many results of sampling theory for signal reconstruction, the relation between the sampling rate and the highest frequency in a signal may be the most widely known. Known as the Nyquist condition and applied primarily to interpolation, this is the statement that the sampling rate $1/T$ must be greater than twice the highest frequency W present in the signal. As an equation, the condition is $1/T > 2W$, or equivalently $2TW < 1$. In the discussion above, we defined a Nyquist parameter as $\tau = 2TW$, and we write the Nyquist condition simply as the requirement that $\tau < 1$.

It is not nearly as well-known that the Nyquist condition also applies to the *prediction* of a signal from its past samples. This was shown for uniform samples in [23], and the authors have shown that a generalization of it applies to the prediction problem when samples include derivative samples [11]. Since the Nyquist condition is usually applied to interpolation problems, the extension of this condition to the one-sided sampling of prediction may be surprising. In Section 13.3, the question of a Nyquist rate for interlaced samples will be considered.

13.2.3. The Discrete Prolate Spheroidal Sequences and Prediction

There is a connection between the problem of the prediction of values of a band-limited signal and some sequences introduced by Slepian [22] as "discrete prolate spheroidal sequences," hereafter referred to as DPSS. In this section, we provide some background on the DPSS and in Section 13.3.2 show that the prediction coefficients for interlaced sampling often retain many of the properties of the DPSS. Whereas the primary interest in filtering has been connected with the DPSS associated with the largest eigenvalue of a certain system, we will show that the prediction problem is connected with the completely opposite case of the DPSS associated with the *smallest* eigenvalue of that system.

The DPSS for the largest eigenvalue have been called the best possible digital filters used for removing frequencies outside a particular frequency band [9]. They were developed by Slepian as the discrete case of prolate spheroidal wave functions and follow a sequence of papers that includes [21] and [22]. As defined in these papers, the DPSS are bi-infinite sequences, and the sequence index-limited to integers in $[0, N-1]$ is just considered a restriction. In this paper, we label the finite-length vectors as DPSS also, and will make clear if we are considering the infinite extension of these vectors. This is slightly different than what was introduced in [21] but will hopefully not cause confusion and will be useful in the application to filtering. For the application to filters, the DPSS can be computed accurately [4], even for very long sequences, using a connection to tridiagonal matrices and special numerical methods, and this property will be considered further in Section 13.3.2.2.

The DPSS for filtering applications result [9] from trying to maximize the energy concentration of the power transfer function in some fixed interval relative to the Nyquist interval. To describe this connection, we adapt the work in [9] to the notation relating to matrix (15) in Section 13.2 by defining a filter with output $y(t) = \sum_{n=0}^{N} a_n x(t - nT)$ related to input $x(t)$. Then for corresponding Fourier transforms, $Y(f) = H(f) \cdot X(f)$, with $H(f) = \sum_{n=0}^{N} a_n e^{-j2\pi nTf}$. The power transfer function is $S(f) = H(f) \cdot H^*(f)$, where * represents the complex conjugate. Let W_N be the Nyquist frequency chosen so that $2TW_N = 1$, and let W be the highest frequency present in the signal, as used in Section 13.1. The maximization problem is to maximize the ratio

$$\alpha = \int_{-W}^{W} S(f)df \Big/ \int_{-W_N}^{W_N} S(f)df,$$

for appropriate real numbers $\{a_n\}_{n=0,\ldots,N}$. As discussed in [9], it can be shown that this maximization problem has solution depending on eigenvectors of a matrix with entries $\tau A(i,j) = \sin((i-j)\pi\tau)/(\pi(i-j))$, where τ is used throughout this paper as $\tau = 2TW$, see Section 13.1. Dividing the entries in the above matrix by τ gives the matrix A in (15), with entries as $\mathrm{sinc}(i-j)\tau$ and this division does not affect normalized eigenvectors. For the maximum energy problem, the coefficients are chosen [9] as the normalized eigenvector associated with λ_0, the *largest* eigenvalue of the matrix. The eigenvector of this matrix, i.e., of A of Section 13.1, are the DPSS, and as noted above, the normalized eigenvector associated with the largest eigenvalue is appropriate for this calculation.

The general definition of DPSS as given by [22] p. 1376, involves a system of equations which we adapt slightly to match the size of matrix A in (15). In that case, the DPSS are defined for each $i = 0, 1, \ldots, N$ as the real solution to the system of equations $\sum_{n=0}^{N} \sin((i-j)2\pi W)/(\pi(i-j))v_n^{(i)} = \lambda_i v_n^{(i)}$, for $n = 0, \pm 1, \pm 2, \ldots$, normalized so that $||v^{(i)}||^2 = \sum_{n=0}^{N}[v_n^{(i)}]^2 = 1$, and $\sum_{n=0}^{N} v_n^{(i)} \geq 0$, where the λ_i are eigenvalues for the discrete prolate spheroidal wave functions. Note that Slepian [22] defines the sequences for all n and the vector obtained by index-limiting the $v_n^{(i)}$ to the values $n = 0, 1, \ldots, N$ are eigenvectors for the matrix.

The matrix A (15) for the prediction coefficients and the matrix as described above for the maximization of energy are basically the same matrices, with only a constant multiple separating the forms of the system of equations. As noted above, the prediction problem is involved with the eigenvector for the smallest eigenvalue, while most research has concentrated on the energy question and the eigenvector associated with the largest eigenvalue. The DPSS were defined for the uniform case, and the matrix (15) essentially extends these ideas to nonuniform samples. For the case of nonuniform samples, we call the eigenvectors of matrix A the "nonuniform DPSS" or NUDPSS. The sequences of prediction coefficients involved in nonuniform "interlaced" sampling extend the standard DPSS in that, for certain cases or parameters, the new sequences reduce to the standard DPSS. The interlaced sampling case will be discussed in Section 13.3. In this next section, we will review some of the other properties of the standard DPSS and investigate analogous properties of some of the NUDPSS.

13.3. Prediction from Past Periodic Nonuniform Samples

13.3.1. Introduction

As stated in the main introduction, the object of this paper is to consider methods to predict the *next* sample from a set of past signal samples. In particular, some of the results of Sections 13.2 and 13.3 will be explored in Section 13.5 for prediction of speech signal values. The idea of a "next" sample assumes a pattern for sampling, and we will broaden the pattern of sampling in this section from the uniform samples considered in the previous section to include a particular case of nonuniform sampling.

Authors have considered very arbitrary patterns of sampling and signal reconstruction. For example, in [24], Wingham considered completely random patterns of sampling. See also [25], where the problem is to reconstruct missing samples in terms of some fairly arbitrary sampling patterns. However, because we are considering prediction from a periodic pattern of sampling, we concentrate on the basic case of prediction at the next point when the sampling pattern is "interlaced" sampling, as in Fig. 1. This term is used for a nonuniform sampling set where two subsets of uniform samples which begin at slightly different places are combined as one set of samples.

Interlaced sampling is a periodic pattern, and we consider prediction of the next value for this case of nonuniform sampling. Thus, we will concentrate on the case when two close samples are followed by a uniform delay, then two more close samples are taken, and so on. From this representative case, we expect that the analysis will follow similarly for the extended case of periodic nonuniform sampling with $m \geq 2$ samples closely spaced together. For example, Papoulis [17], p. 194, shows that interpolation from an infinite set of samples is possible for the general case where a band-limited function is "sampled at $1/m$ times the Nyquist rate, but in each interval not one but m samples are used." We will show that something very similar occurs for the one-sided sampling of prediction.

The interlaced sampling pattern will involve prediction at two different types of points. We will refer to the point that is the second of the close samples in the periodic pattern as the "far" point, as it is the farthest away from the main set of previous samples. The first of the close samples will be called the "near" point, as

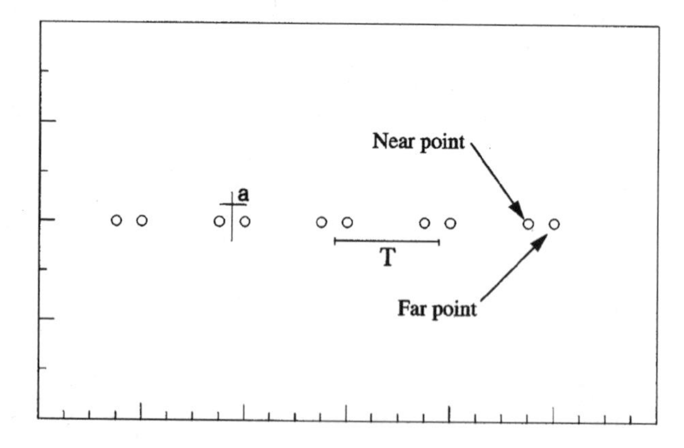

Figure 1. Positions of interlaced sampling points. Circles represent sample points. Spacing T and α (as a) are pictured.

it is nearer to the other sets of two close samples. In each case, we will refer to t_0 as the time location for the prediction. We suppose that samples are at the points $t_0 - kT \pm \alpha$ for integers k, so that the general spacing between each set of two nonuniform samples is T while α is the distance of each of the close samples from the point halfway between them; see Fig. 1.

13.3.1.1. Location of Sample Points

First, we present formulas for the location of sample points for a prediction at the near point. Also, we need the differences between these points for use in matrix (14), whose eigenvalues and eigenvectors will lead to the prediction coefficients. The sample times for the near point are at t_n given by (21) where t_0 is the point of prediction and

$$t_n = \begin{cases} t_0 - \left(\left\lceil \frac{n}{2} \right\rceil T - 2\alpha\right) & \text{if } n \text{ is odd} \\ t_0 - \left(\frac{n}{2}\right) T & \text{if } n \text{ is even} \end{cases} \tag{21}$$

where $\lceil x \rceil$ is the ceiling function.

Similarly, the location of sample points for a prediction at the far point are

$$t_n = \begin{cases} t_0 - \left(\left\lfloor \frac{n}{2} \right\rfloor T + 2\alpha\right) & \text{if } n \text{ is odd} \\ t_0 - \left(\frac{n}{2}\right) T & \text{if } n \text{ is even} \end{cases} \tag{22}$$

where $\lfloor x \rfloor$ is the floor function.

13.3.1.2. The Nyquist Rate for Interlaced Sampling

The mathematical interpretation of the Nyquist rate of sampling for uniformly-sampled signals was given in Section 13.2.2 as $\tau < 1$. This is equivalent to $1/T > 2W$, where T is the sampling interval and W is the highest frequency (in Hertz) of the signal, since $\tau = 2TW$. The inequality $1/T > 2W$ provides the standard interpretation that the rate of sampling must be twice the highest frequency of the signal to guarantee reconstruction. As noted above, similar to what Papoulis has shown for interpolation from periodic nonuniform samples, we show in this section that interlaced sampling allows for increasing T by a factor of 2. In particular, we show that for interlaced sampling and prediction, we need $1/T > W$, or $\tau < 2$, where the time interval T for interlaced sampling is defined in the previous section such that the T for interlaced sampling is basically the main period between the sets of close nonuniform periodic samples.

The proof of this result for prediction from interlaced samples is based on nonharmonic Fourier series [26]. In particular a result from that book (see p. 116), adapted to the notation here, is the following: If $\{\delta_n\}$ is a sequence of positive real numbers for which

$$\limsup_{n\to\infty} \frac{n}{\delta_n} > 2B, \tag{23}$$

then $\{e^{j2\pi\delta_n f}\}$ is complete in $L^2[-B, B]$. We present the proof for prediction at the near point, and note that the proof for the far point is similar. We let

$$\delta_n = \begin{cases} \left\lceil \frac{n}{2} \right\rceil T - 2\alpha & \text{if } n \text{ is odd} \\ \left(\frac{n}{2}\right) T & \text{if } n \text{ is even} \end{cases} \tag{24}$$

which is just the set of differences used for prediction at the near point based on the positions of the samples as defined in Section 13.3.1.1. A list of the set of values of the ratio n/δ_n in order gives the first few entries of that sequence as follows:

$$\left\{ \frac{1}{T-2\alpha}, \frac{2}{T}, \frac{3}{2T-2\alpha}, \frac{4}{2T}, \frac{5}{3T-2\alpha}, \frac{6}{3T}, \ldots \right\}$$

from which it can be seen that the *lim sup* of this sequence is $2/T$. Thus the condition from (23) becomes

$$\lim sup_{n\to\infty} \frac{n}{\delta_n} = \frac{2}{T} > 2W.$$

The value of the constant B in (23) is chosen as $B = W$ because we are concerned with the prediction error integral as given by (8). In particular, the above inequality is satisfied as long as $(2/T) > 2W$, or $(1/T) > W$, or $\tau = 2TW < 2$. When this is satisfied, the above result says that the constant function 1 will be included in the space spanned by $\{e^{i\delta_n f}\}$ which means that the error integral (8) goes to zero as $N \to \infty$. Thus the prediction will be valid if $\tau = 2TW < 2$, as claimed. This completes the proof that shows that the sampling rate is halved for interlaced sampling.

The special case of uniform sampling is included in interlaced sampling for the value of the parameter α as $\alpha = T/4$. In that case, the interlaced samples have expanded so far that they are actually equally-spaced, with a sampling interval T_1

such that $T_1 = T/2$. But then the Nyquist rate as given by $2TW < 2$ becomes the old rate since we then have $2TW = 4T_1W < 2$ which is equivalent to the usual $2T_1W < 1$. We will refer to this special case of interlaced sampling when comparing results for interlaced sampling to the standard case of uniform sampling.

13.3.2. DPSS for Interlaced Sampling

As mentioned in Section 13.2, we will investigate the properties of some of the Nonuniform Discrete Prolate Spheroidal Sequences (NUDPSS) after first reviewing some of the main properties of the standard DPSS. This is important to prediction for the nonuniform case of interlaced sampling because of the eigenvector method mentioned in Section 13.2.1.2 for determining the prediction coefficients.

13.3.2.1. Main Properties of the DPSS

There are many properties of the DPSS. The work of Slepian, as described earlier in Section 13.2.3, uses this term for the bi-infinite sequences, but we concentrate on the part of those sequences for the finite case with $0 \le n \le N$.

First, we present a list of four properties which were summarized nicely in [9] but presented earlier in [22], of the DPSS: (1) orthogonality, (2) symmetry, (3) complementary relations, and (4) recurrence relations. We will discuss the properties for the new sequences for the interlaced case in light of these known properties of the DPSS.

Suppose that $a_n^{(i)}$ represents the ith DPSS ($0 \le i \le N$) eigenvector with each DPSS eigenvector having $N+1$ elements ($0 \le n \le N$). The ordering of the DPSS eigenvector is as described in Section 13.2.1.2. with $a_n^{(0)}$ corresponding to the largest eigenvalue and $a_n^{(N)}$ to the smallest eigenvalue. Note that we use the parameter N here as the number of prediction coefficients, as described in Section 13.1. The properties are as follows:

1. Orthogonality. This means that $\sum_{n=0}^{N} a_n^{(i)} a_n^{(j)} = \delta_{ij}$, where δ_{ij} is the Kronecker delta. Orthogonality indicates that the DPSS can be used in a linear combination for the complete set of sequences on $N+1$ terms.
2. Symmetry. The graph of a DPSS eigenvector has a similar look on each side of its center point. Written as an equation, the symmetry is

$$a_n^{(i)} = (-1)^i a_{N-n}^{(i)}, \tag{25}$$

$0 \le n \le N/2$, which is technically either even symmetry or odd symmetry, depending on the index i.

3. Complementary relations. The DPSS depend on the Nyquist parameter τ and there is a complementary parameter defined, see [9], to be $\tau' = 1 - \tau$. If $b_n^{(i)}$ is the DPSS for the complementary parameter, then it is known that

$$a_n^{(i)} = (-1)^n b_n^{(N-i)} \tag{26}$$

for each index i, $0 \le i \le N$. For example, $a_n^{(0)} = (-1)^n b_n^{(N)}$.
4. Recurrence relations. The elements of a DPSS eigenvector satisfy a three-term recurrence relation:

$$\begin{aligned}\tfrac{1}{2}(n)(N+1-n)a_{n-1}^{(i)} + ((N/2-n)^2 \cos \pi\tau - \mu_i)a_n^{(i)} \\ + \tfrac{1}{2}(n+1)(N-n)a_{n+1}^{(i)} = 0\end{aligned} \tag{27}$$

where τ is the Nyquist parameter as described earlier and μ_i is a constant described in the following.

13.3.2.2. Analogous Properties of the NUDPSS

The set of vectors we call the NUDPSS were introduced in Section 13.2.3 as the eigenvectors of the symmetric matrix A (14). The acronym NUDPSS stands for nonuniform discrete prolate spheroidal sequences, and we cover here the nonuniform case of interlaced sampling with sample locations as defined in Section 13.3.1.1. We shall use the smallest eigenvector of the matrix as the vector needed to determine the prediction coefficients, as discussed in Section 13.2.1.2.

Many of the properties of the DPSS as summarized in Section 13.3.2.1 are shared by the NUDPSS. Some of the similarities between the NUDPSS and the DPSS can be seen from the examples in Table 3, which shows the complete set of NUDPSS eigenvectors for predictions at the "far" point for two different nonuniformly spaced sets of samples corresponding to values of $\alpha = 0.05, 0.10$ with $\tau = 1, T = 0.2$, as well as the complete set of DPSS eigenvectors which come from equally-spaced samples. In the table, the time t points are listed for convenience, and the vectors are listed left-to-right from "smallest" eigenvectors, v_N, to "largest", v_0. The most striking similarities that one can see immediately from Table 3 are the symmetries of each vector about its middle and the common pattern of the sets of plus/minus signs among the different sets of vectors.

To discuss properties of these NUDPSS relative to the list of properties presented in Section 13.3.2.1 for the DPSS, we consider those properties presented in the same order as in the Section 13.3.2.1.

1. The property of orthogonality carries over directly to the NUDPSS. That is, each set of NUDPSS vectors is an orthonormal set, just as the

Table 3. Lists of NUDPSS eigenvectors for $\alpha=0.05$ and $\alpha=0.10$ spacing values (top two tables) and of DPSS eigenvectors for the equally-spaced case ($\alpha=0.25$, bottom table). Each case uses $\tau=1$ and $T=0.2$.

		$\alpha=0.05$			
t=	−0.0200	−0.2000	−0.2200	−0.4000	−0.4200
v_5	v_4	v_3	v_2	v_1	v_0
0.2337	−0.4859	−0.4574	−0.4574	0.4859	0.2337
−0.2815	0.5113	0.3992	−0.3992	0.5113	0.2815
0.6051	−0.0501	0.3624	0.3624	0.0501	0.6051
−0.6051	−0.0501	−0.3624	0.3624	−0.0501	0.6051
0.2815	0.5113	−0.3992	−0.3992	−0.5113	0.2815
−0.2337	−0.4859	0.4574	−0.4574	−0.4859	0.2337

		$\alpha=0.10$			
t=	−0.0400	−0.2000	−0.2400	−0.4000	−0.4400
v_5	v_4	v_3	v_2	v_1	v_0
0.2094	−0.4677	−0.4872	−0.4872	0.4677	0.2094
−0.3051	0.5206	0.3686	−0.3686	0.5206	0.3051
0.6025	−0.1011	0.3560	0.3560	0.1011	0.6025
−0.6025	−0.1011	−0.3560	0.3560	−0.1011	0.6025
0.3051	0.5206	−0.3686	−0.3686	−0.5206	0.3051
−0.2094	−0.4677	0.4872	−0.4872	−0.4677	0.2094

		$\alpha=0.25$			
Equally spaced case. t=	−0.1000	−0.2000	−0.3000	−0.4000	−0.5000
v_5	v_4	v_3	v_2	v_1	v_0
0.1329	−0.3694	−0.5881	−0.5881	0.3694	0.1329
−0.3766	0.5415	0.2549	−0.2549	0.5415	0.3766
0.5835	−0.2653	0.2985	0.2985	0.2653	0.5835
−0.5835	−0.2653	−0.2985	0.2985	−0.2653	0.5835
0.3766	0.5415	−0.2549	−0.2549	−0.5415	0.3766
−0.1329	−0.3694	0.5881	−0.5881	−0.3694	0.1329

set of DPSS vectors is orthonormal. This follows easily from the derivation of the NUDPSS as the normalized eigenvectors of a symmetric matrix.

2. The symmetry property (25) is exactly the same for the NUDPSS when there are an odd number of sample points N, such as $N = 5$ for Table 3.

This is valid for predictions at either the far or near point, although Table 3 only lists these sets of vectors for the far point case. The symmetry does not hold when N is even. The equation expressing the even symmetry (or odd symmetry) for the interlaced case is exactly as in (25). Examples of this can be seen in Table 3 where corresponding vectors have entries that share the same sign.

3. Complementary relations (26) between different sets of vectors for complementary τ values hold only for a particular value of τ for the NUDPSS. The relations (26) depend on the parameter τ and the complementary value τ', but this τ parameter has a different upper bound for the nonuniform case, as discussed in Section 13.3.1.2, where it was shown that this rate has been halved for interlaced sampling.

   The NUDPSS satisfy the complementary relation only for the special "midpoint" τ value of $\tau = 1$, for which τ' is also equal to 1. Consider the complementary relation for the DPSS as given by (26) for the "midpoint" τ value of $\tau = 1/2$. For this special τ value, the complementary parameter τ' is also equal to $1/2$ so that the two sets of DPSS vectors come from the *same* matrix of eigenvectors. For example, $a_n^{(0)}$ is the DPSS for the largest eigenvalue associated with the τ parameter, and it equals $(-1)^n b_n^{(N)}$ for the DPSS for the smallest eigenvalue associated with the τ' parameter. This example is illustrated in Table 3 as the example at the bottom of that table, where $a_n^{(0)}$ is v_0 and $b_n^{(N)}$ is v_5. One can see from that table that $v_0 = (-1)^n v_5$, as the complementary relation requires.

   For the NUDPSS, $\tau' = 2 - \tau$ so that $\tau' = 1$ for the value of $\tau = 1$. As can be seen from Table 3 for the top two examples, the complementary relation holds for the interlaced case in precisely the same way, $a_n^{(i)} = (-1)^n b_n^{(N-i)}$, as for the DPSS. Property 3 of Section 13.3.2.1 notes that the complementary relation holds for the DPSS for all τ in (0, 1), but this analogous property is not valid for the NUDPSS for values of τ other than 1.

4. The three-term recurrence relation listed as (27) is the relation as discovered by Slepian [22] for the DPSS, but it is not unique. Such a relation may be derived from any tridiagonal matrix which commutes with the matrix A from (14) whose columns are the vectors for a complete set of DPSS vectors. The values listed as μ_i in the recurrence relation (27) are the eigenvalues of this tridiagonal commutator.

Similar to the relation (27) for the DPSS, this section includes a form for interlaced sampling for a tridiagonal matrix T which commutes with matrix A (14) in the sense that $TA = AT$. The T matrix leads to a three-term recurrence relation for the NUDPSS. The tridiagonal commutator is assumed to be symmetric, as is matrix A, but as noted earlier, such a matrix is not unique. Given one such matrix, an arbitrary multiple of it is still a tridiagonal commutator. Also, the addition of any multiple of the identity matrix to it would also be a tridiagonal commutator matrix. The form we will derive for a tridiagonal commutator appears quite different than the one used by Slepian [22] for the uniformly-sampled case.

For the NUDPSS, we present a commuting tridiagonal matrix for small values of N ($N = 2, 3$) but not for large values. The advantage of having a commuting tridiagonal matrix is that the set of eigenvectors is the same for the original matrix as for the tridiagonal case, and the computation of the eigenvectors can be done much more stably for the tridiagonal matrix. See [9] where the authors mention this as a crucial property of the DPSS. Considerable research has been done on classifying the Toeplitz matrices which have commuting tridiagonal matrices, see [18], but those results do not directly apply to the matrix A (14) for the nonuniform case, since the matrix is not Toeplitz in that case. Since we are using the eigenvectors to compute the prediction coefficients, more accurate ways of calculating the eigenvectors give better prediction coefficients.

We present these results for the interlaced sampling case for prediction at the "far" point and assume that the case of the near point is similar. The matrix A (14) from Section 13.2.1.2 is based on differences of time sample locations, and for interlaced sampling this set of differences $t(i) - t(j)$ for the matrix are

$$\{0, aT, T\},$$

for $N = 2$ and

$$\{0, aT, T, (1+a)T\}$$

for $N = 3$, where $a = 2\alpha$, $a \leq 1/2T$, and α is as described in Section 13.3.1.1 on interlaced sampling. Here, the use of N is consistent with the use of this variable throughout this paper so that N stands for the number of sample points. The matrix A has size $(N+1) \times (N+1)$, as described in an earlier section.

The commutator matrix has the same eigenvectors as the matrix A, and since we are employing the eigenvectors of the matrix for the prediction coefficients, the commutator matrix gives a new and more stable way of calculating those coefficients.

The 3×3 matrix and a tridiagonal commutator

In this case, the symmetric tridiagonal matrix has the form

$$H = \begin{bmatrix} x_1 & y_1 & 0 \\ y_1 & x_2 & y_2 \\ 0 & y_2 & x_3 \end{bmatrix}. \tag{28}$$

Let $s(x) = \mathrm{sinc}(x\tau)$, where $\tau = 2TW$ and W is the highest frequency. Note that this is an even function. It can be shown that entries in one form of a commuting tridiagonal matrix may be given by

$$\begin{aligned} y_1 &= \frac{s(1-a)}{s(1)^2 - s(1-a)^2} \\ y_2 &= \frac{s(a)}{s(1)^2 - s(a)^2} \\ x_1 &= x_2 + \frac{s(1)}{s(1)^2 - s(a)^2} \\ x_3 &= x_2 + \frac{s(1)}{s(1)^2 - s(1-a)^2} \end{aligned} \tag{29}$$

and x_2 is arbitrary. In addition, note that $y_1 = (s(a)/s(1))(x_1 - x_2)$ and $y_2 = (s(1-a)/s(1))(x_3 - x_2)$.

If $\tau = 1$ then $s(1) = 0$, and the above equations force $x_1 = x_2 = x_3$ and thus y_1 and y_2 are the only non-zero elements.

As will be discussed for the 4×4 case, the above formulas for the entries in the commutator are valid even for the uniform case. That case corresponds to the spacing $a = 2\alpha$ between close samples for $a = 1/2$. In this case the values for the entries in the commutator given by the above formulas give $y_1 = y_2$ and $x_1 = x_3$ since $a = 1 - a = 1/2$. For this case, it also follows from the above formulas that $x_2 = 0$ and $x_1 = (s(1)/s(0.5))y_1$. But this form then matches the one given by Slepian in (27) as can be seen through the following steps.

For the equally-spaced case, the matrix has the form

$$\begin{bmatrix} \frac{s(1)}{s(1/2)} y_1 & y_1 & 0 \\ y_1 & 0 & y_1 \\ 0 & y_1 & \frac{s(1)}{s(1/2)} y_1 \end{bmatrix}. \tag{30}$$

Next, multiplying the matrix by the constant $1/y_1$ (which preserves the commutator property) gives

$$T = \begin{bmatrix} \frac{s(1)}{s(1/2)} & 1 & 0 \\ 1 & 0 & 1 \\ 0 & 1 & \frac{s(1)}{s(1/2)} \end{bmatrix}. \tag{31}$$

Finally, trigonometric identities give $s(1)/s(1/2) = \sin(2 * 0.5\pi\tau)/(2\sin(0.5 * \pi\tau)) = \cos(0.5 * \pi\tau)$ as given by the center term of the three-term recurrence equation in (27). This shows that the above formulas give a form equivalent to that three-term recurrence relation from (27), since the τ value used here is halved in the case that the samples are actually equally-spaced.

We present two examples for the general case.

For example,

Example 1 When $\tau = 0.5$ and $a = 0.05$, the matrix A is

$$\mathrm{A1} = \begin{bmatrix} 1.0000 & 0.9990 & 0.6366 \\ 0.9990 & 1.0000 & 0.6681 \\ 0.6366 & 0.6681 & 1.0000 \end{bmatrix}. \tag{32}$$

for prediction point and sample differences {0, .025, .5} and the corresponding tridiagonal commutator matrix given by the set of equations above is

$$\mathrm{T1} = \begin{bmatrix} 1.0742 & 16.2863 & 0 \\ 16.2863 & 0 & 1.6856 \\ 0 & 1.6856 & 15.5198 \end{bmatrix}. \tag{33}$$

Matrix A1 has condition number 1.53e + 04 while matrix T1 has much smaller condition number 1.18. As noted above, matrix A1 and T1 share the same eigenvectors.

As another example,

Example 2 When $\tau = 0.2$ and with a, T and sample differences {0, .025, .5} as in Example 1 above, then the matrix A is

$$\mathrm{A2} = \begin{bmatrix} 1.0000 & 0.9998 & 0.9355 \\ 0.9998 & 1.0000 & 0.9417 \\ 0.9355 & 0.9417 & 1.0000 \end{bmatrix}. \tag{34}$$

and the corresponding tridiagonal commutator matrix given by the above set of formulas is

$$\mathrm{T2} = \begin{bmatrix} 7.5121 & 81.2048 & 0 \\ 81.2048 & 0 & 8.0288 \\ 0 & 8.0288 & 80.6720 \end{bmatrix}. \tag{35}$$

In this case the condition number of T2 is 1.1536 and the condition number of A2 is much larger at 7.06e + 05.

The 4 × 4 matrix and a tridiagonal commutator
In this case, the symmetric tridiagonal matrix has the form

$$T = \begin{bmatrix} x_1 & y_1 & 0 & 0 \\ y_1 & x_2 & y_2 & 0 \\ 0 & y_2 & x_3 & y_3 \\ 0 & 0 & y_3 & x_4 \end{bmatrix}.$$

The prediction point difference (at $t_0 - t_0 = 0$) and sample time differences are $\{0, aT, T, (1+a)T\}$, as needed to correspond to matrix A (14) for this case. Again, where $s(x) = \text{sinc}(x\tau)$, it can be shown that the entries of a tridiagonal commutator matrix for this case are given by

$$\begin{aligned} y_1 &= \frac{s(1+a)+s(1-a)}{s(1+a)^2 - s(1-a)^2} = \frac{1}{s(1+a)-s(1-a)} \\ y_2 &= \frac{s(a)}{s(1)^2 - s(a)^2} \\ y_3 &= y_1 \\ x_1 &= x_2 + \frac{s(1)}{s(1)^2 - s(a)^2} \\ x_3 &= x_2 \\ x_4 &= x_1 \end{aligned} \tag{36}$$

and x_2 is arbitrary. Note that the formulas for x_1 and y_2 are exactly as for the 3 × 3 case presented above.

For example,

Example 1 When $\tau = 0.1$ and $a = 0.05$, the matrix A, see (14), *is*

$$\text{A} = \begin{bmatrix} 1.0000 & 1.0000 & 0.9836 & 0.9820 \\ 1.0000 & 1.0000 & 0.9852 & 0.9836 \\ 0.9836 & 0.9852 & 1.0000 & 1.0000 \\ 0.9820 & 0.9836 & 1.0000 & 1.0000 \end{bmatrix}. \tag{37}$$

where values are rounded to four decimal places. This matrix A has the rather large condition number of 1.05e + 10. Differences $(t_0 - t_j)_{j=0,\ldots,3}$ of prediction point and sample locations are $\{0, 0.02, 0.40, 0.42\}$ using $T = 0.40$.

The form obtained by using the formulas for the commutator entries as given above for a symmetric tridiagonal commutator is

$$\mathrm{T1} = \begin{bmatrix} 30.3716 & 306.9902 & 0 & 0 \\ 306.9902 & 0 & 30.8757 & 0 \\ 0 & 30.8757 & 0 & 306.9902 \\ 0 & 0 & 306.9902 & 30.3716 \end{bmatrix} \tag{38}$$

which has the simple condition number of 1.22 compared to the much larger condition number of A listed above as 1.05e + 10.

Example 2 The next example will show how the form of the equations for the entries in a tridiagonal commutator for the 4×4 case reduces to the matrix of Slepian for the case of equally-spaced samples. The case of equally-spaced samples results from choosing the parameter a to have the value $a = 1/2$. This can be seen since the general sample differences $\{0, aT, T, (1+a)T\}$ become $\{0, \frac{1}{2}T, T, \frac{3}{2}T\}$, which are equally-spaced.

From the form of the tridiagonal entries given above when $a = \frac{1}{2}$, we find the following:

Choose $x_2 = 0$, and then $x_4 = x_1 = s(1)/[s(1)^2 - s(1/2)^2)]$, $y_2 = s(1/2)/[s(1)^2 - s(1/2)^2)]$, and $y_3 = y_1 = 1/[s(3/2) - s(1/2)]$. This uses the function defined earlier as $s(x) = \mathrm{sinc}(x\tau)$. The ratio of y_2 to y_1 looks as if it depends on τ, since

$$y_2/y_1 = \frac{s(1/2)s(3/2) - s(1/2)^2}{s(1)^2 - s(1/2)^2}$$

but with the help of a few trigonometric identities this reduces to

$$\begin{aligned} y_2/y_1 &= \frac{\frac{1}{3}[\cos(\pi\tau) - \cos(2\pi\tau)] - (1 - \cos(\pi\tau))}{\frac{1}{4}(1 - \cos(2\pi\tau) - (1 - \cos(\pi\tau))} \\ &= \frac{4}{3} \cdot \frac{\cos(\pi\tau) - \frac{1}{4}\cos(2\pi\tau) - 3/4}{\cos(\pi\tau) - \frac{1}{4}\cos(2\pi\tau) - 3/4} = \frac{4}{3}, \end{aligned}$$

so that $y_2/y_1 = 4/3$ is actually a constant. Further, $x_1 = s(1)/s(1/2) \cdot y_2$, so that the matrix has the form

$$\begin{bmatrix} \frac{s(1)}{s(1/2)}y_2 & y_1 & 0 & 0 \\ y_1 & 0 & y_2 & 0 \\ 0 & y_2 & 0 & y_1 \\ 0 & 0 & y_1 & \frac{s(1)}{s(1/2)}y_2 \end{bmatrix}. \tag{39}$$

Finally, multiplying the matrix by the constant $2/y_2$ (which preserves the commutator property) gives

$$T = \begin{bmatrix} 2\dfrac{s(1)}{s(1/2)} & 1.5 & 0 & 0 \\ 1.5 & 0 & 2 & 0 \\ 0 & 2 & 0 & 1.5 \\ 0 & 0 & 1.5 & 2\dfrac{s(1)}{s(1/2)} \end{bmatrix}. \tag{40}$$

which is equivalent to the Slepian form of a tridiagonal commutator as given through the three-term recurrence relation of (27), after possibly subtracting a multiple of the identity, for the 4×4 case.

In all of these examples, the important point is that each of these commuting tridiagonal matrices leads to a three-term recurrence relation for the interlaced sampling case that is similar to the three-term recurrence relation (27) for the DPSS.

Prediction for the particular interlaced case of nonuniform sampling may be done using the smallest eigenvector method of calculating the prediction coefficients, similar to what was discussed for the case of uniform sampling in Section 13.2.1.2. The comparison of the NUDPSS to the DPSS given in this section provides analogies between these different sets of eigenvectors for the nonuniform case relative to the uniform case. At least for small N, the eigenvectors can be computed very accurately through the use of a tridiagonal commutator, as presented in this section. Further work casts doubt on the existence of such a commutator for larger values of N, so that other methods for accurate computation may be needed in those cases. Next, we present an example of prediction applied to this particular nonuniform sampling case.

13.3.3. Example of Prediction Using Interlaced Sampling

The methods described above for prediction based on interlaced samples are exhibited in this section for a specific example. It is not possible to present several examples here, but we can report on some of the other findings we can observe from many numerical trials.

The theoretical limit of Section 13.3.1.2 for the parameter τ was derived there to be $\tau = 2$ and we have found reasonably accurate results for values of τ up to about $\tau = 1.5$. The difference between these numbers is probably due to numerical calculations.

The accuracy of the prediction at the far point is generally better than prediction at the near point, essentially because we are using the correct value for the near point when computing the predicted value for the far point. In a juxtaposition of the use of the words, the far point turns out to be closer to (at least one) data point than does the near point, which can be relatively far from the data points.

The example whose graph is given as Fig. 2 represents the prediction methods of the previous sections applied to interlaced sampling for the function $f(t) = 30t^2 \text{sinc}(4t)$. This function is not band-limited where the sense of that word implies finite energy, but it is band-limited (to $W = 2$) as in Section 13.2.1.3 where band-limited functions of polynomial growth were considered.

Each prediction represented by a starred point on the plot in Fig. 2 represents the predicted value of this signal based on the previous correct values of the signal. The order of the predictive filter used in this example was $N = 61$, and that order was used for predictions both at the near and at the far points. Two sets of prediction coefficients must be stored for use here, where one set is for the prediction at the near point and one set is for the prediction at the far point. However, each set simply depends on the parameter τ and can be stored after its computation. The coefficients do not need to be recalculated at any time.

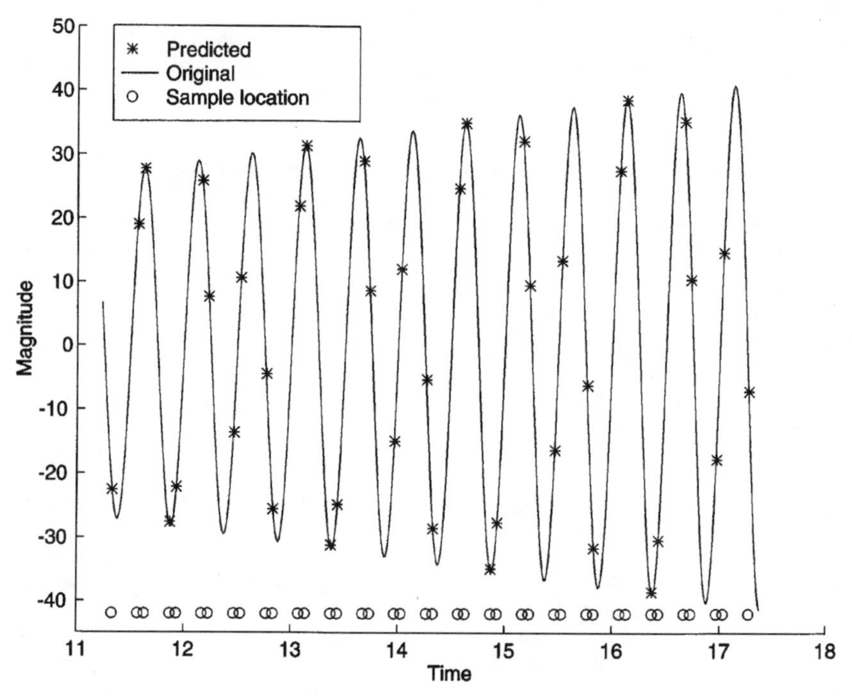

Figure 2. The prediction of a band-limited signal from interlaced sampling. $T = 0.3$, $\tau = 1.2$, $\alpha = 0.03$. The signal has polynomial growth since it is $f(t) = 30t^2\ \text{sinc}(4t)$.

For the particular example given in Fig. 2, the spacing between pairs of samples was $T = 0.3$, and the spacing between centers of the pairs of samples was $\alpha = T/10 = 0.03$. With $W = 2$, the parameter was $\tau = 1.2$. The results listed in Fig. 2 are presented in numbered form in Table 4.

13.4. Prediction from Past Samples Using Nonlinear or Other Methods

Nonlinear methods have not been employed to a great extent towards the main problem considered in this paper, where that problem is to accurately predict the "next" value of a band-limited signal in a periodic sampling scheme from a finite set of samples. The methods discussed in Sections 13.2 and 13.3 are likewise *linear* prediction methods, and there are certainly advantages in being able to work with a sum of two functions by working with the functions in an individual manner through a linear formula. Few authors have generalized either the linear prediction formula (1) or the one (5) for nonuniform data to include a nonlinear function g_{nl} of N variables. That is, to a nonlinear prediction formula of form $f(t) = g_{nl}(f(t_1), f(t_2), \ldots, f(t_N))$, where $f(t)$ is the next value and $\{f(t_n)\}_{n=1,\ldots,N}$ are the past samples. Besides the inherent problems in then dealing with even the simple case of a constant multiple of the signal $f(t)$, there are difficulties involved in choosing an appropriate nonlinear function g_{nl}.

Table 4. Predictions using interlaced sampling at far f_k and near n_k points. This data is for the first seventeen points as displayed in Fig. 2.

point	true	predicted	Rel. error
f_0	−2.2838e + 001	−2.2578e + 001	1.1e − 002
n_1	2.1283e + 001	1.8978e + 001	1.1e − 001
f_1	2.7710e + 001	2.7685e + 001	9.1e − 004
n_2	−2.8282e + 001	−2.7618e + 001	2.3e − 002
f_2	−2.1945e + 001	−2.2165e + 001	1.0e − 002
n_3	2.4531e + 001	2.5771e + 001	5.1e − 002
f_3	7.2610e + 000	7.6413e + 000	5.2e − 002
n_4	−1.0959e + 001	−1.3625e + 001	2.4e − 001
f_4	1.1012e + 001	1.0619e + 001	3.6e − 002
n_5	−7.5816e + 000	−4.5237e + 000	4.0e − 001
f_5	−2.5861e + 001	−2.5610e + 001	9.7e − 003
n_6	2.4042e + 001	2.1781e + 001	9.4e − 002
f_6	3.1284e + 001	3.1272e + 001	3.8e − 004
n_7	−3.1856e + 001	−3.1274e + 001	1.8e − 002
f_7	−2.4704e + 001	−2.4937e + 001	9.4e − 003
n_8	2.7554e + 001	2.8882e + 001	4.8e − 002
f_8	8.1515e + 000	8.5394e + 000	4.8e − 002

In this section, we review some nonlinear methods that have been used to reconstruct or extrapolate band-limited signals and suggest some ways that those methods may be applied to the prediction problem considered here. The methods discussed here are time varying if not strictly nonlinear.

One of the most well-known methods for extrapolating portions of a band-limited signal is the Papoulis-Gerchberg algorithm, described in [7], pp. 260 ff, for a continuous interval of given data values. Reference to the discrete Papoulis-Gerchberg algorithm is given in [3], which also includes a wealth of information about previous research. This algorithm, as well as the other techniques discussed in this section, addresses a significantly larger problem related to the extrapolation of a band-limited signal than is discussed in this chapter, namely that those methods are trying to determine all values of the signal outside of the known range as opposed to just predicting the next value of the signal as is done in this chapter.

The Papoulis-Gerchberg algorithm is also different from the focus of this chapter in that the algorithm is iterative. It proceeds approximately as follows: the known interval of values is Fourier-transformed, then band-limited, then inverse transformed, with the result then combined in a clever way with the original part of the function so as to extend the known signal segment to the entire axis (i.e., to extrapolate). This process is iterated and can be shown to converge.

An iterative, time-varying method due to Marvasti [8] recovers a signal from nonuniform samples. In that technique, suppose $\{t_k\}$ are the signal samples with $-\infty < k < \infty$, and suppose $\theta(t)$ is defined to be a function such that $\theta(t_k) = t_k - kT$. In this way, $\theta(t_k)$ is the deviation of the kth sample from the uniformly-spaced position kT. The complete set of samples of a signal $x(t)$ can be written as $x_s(t) = x(t) \cdot \sum_k \delta(t - t_k)$.

This method involves the two steps of (1) low-pass filtering the nonuniform samples $x_s(t)$, and (2) dividing the resulting function $x_{slp}(t)$ by a function derived from the $\theta(t)$ function described above. In particular, the second step gives the signal as the quotient of $x_{slp}(t)$ and $(1 - \theta'(t))/T$. Although the method was initially derived for an infinite number of samples in both negative and positive t directions, in [8], it is noted that this method has produced impressive results in simulations to recover missing samples in speech data. This procedure can be iterated for further improvements. When appropriately adjusted for a finite set of samples, this particular nonlinear method may apply to the problem considered in this paper of the prediction of the next value of a band-limited signal from periodically-sampled data.

Other methods that are focused on a finite set of discrete signal values have been described recently by Dharanipragada and Arun in [2]. They review four algorithms for extrapolation, again in the more general problem of reconstructing the complete set of band-limited function values rather than the problem of this paper. Those four algorithms are (i) a minimum energy extrapolation, (ii) an algorithm of Kolba and Parks, (iii) an "energy concentrated extrapolation", and (iv) an algorithm based on spectral discretization. Each of those algorithms is linear in the data. In addition, a new algorithm is also presented in [2] where that algorithm is called a "reduced dimension reconstruction", since it works with a low-dimensional linear subspace to approximate the set of band-limited functions which are also essentially time-limited. This algorithm involves use of the singular value decomposition, but it again is different from the focus of this paper, since the purpose of that algorithm is to extrapolate the complete signal rather than to predict the next value of the signal from the periodically-sampled data. For simulations and papers, see CD-ROM: 'Chapter13\'.

13.5. Speech Signals – Prediction and Compression

The analysis of the previous sections applies to signals which are band-limited, and speech signals for telecommunications are especially appropriate for this assumption. Due to some common practices of filtering for telecommunications, the speech signals considered in this section fall into the class of band-limited signals to which the prediction methods presented above apply. In particular, it is common [5] to sample a speech signal at 8 kHz and filter it so that it is band-limited to 3.40 kHz. For the Nyquist parameter of the previous sections, a sampling inteval of $T = 1/8000$ and highest frequency of $W = 3.4$ make the value of τ to be $\tau = 0.85$, and this is the value of the Nyquist parameter which will be used for applications to speech signals in the following.

The particular speech sample that will be used for examples in the following appears as in Fig. 3.

Linear prediction is an important and well-developed area in relation to speech signals, and this section will include only basic results related to the prediction formulas of sections 13.2 and 13.3. The standard prediction method for *speech* signals includes an excitation term [5], and that type of term is not yet included in the analysis of the previous sections. In particular, instead of the simple prediction scheme of (1), a prediction formula for speech may include a factor such as $Gx_1(t)$ for excitation [5], in the form $x(t) = G \cdot x_1(t) + \sum_{n=1}^{N} a_n x(t - nT)$. Developments for speech signals also include backward prediction, adaptive prediction methods, and more, but this paper will only address basic standard prediction results as related to the methods of Sections 13.2 and 13.3.

In this section, we first show that the methods of this chapter provide relatively highly accurate predictions for speech signals of the type described above. One advantage of the approach presented here is that the set of prediction coefficients applies to the entire signal, even a portion of it whose bandwidth may

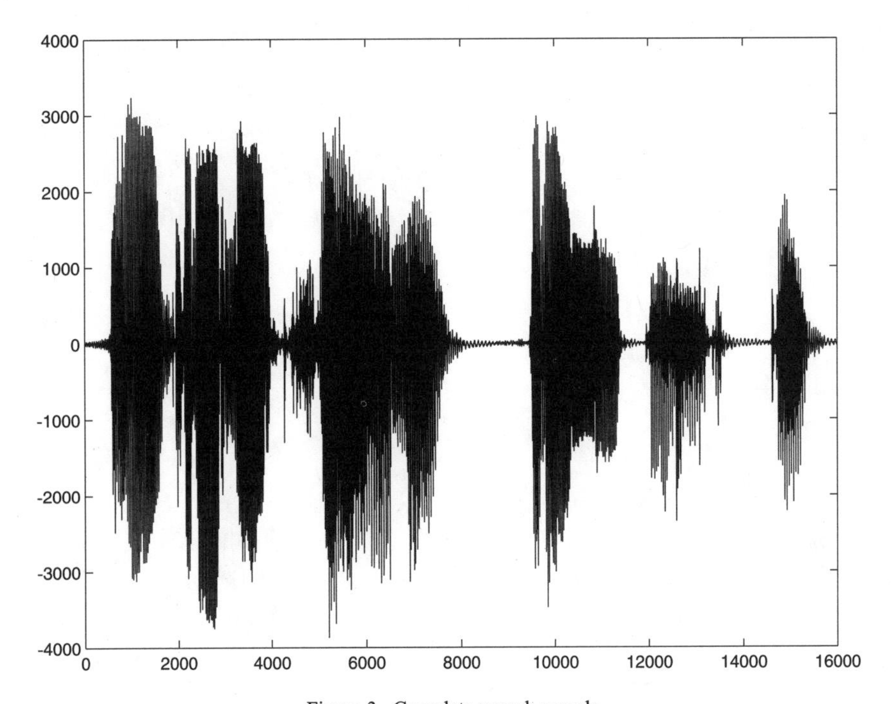

Figure 3. Complete speech sample.

be smaller, since the methods of this chapter apply as long as the signal bandwidth falls below the value $W = 3.4$ used for the calculation of the coefficients. Then we consider a very simple approach in applying the methods of this chapter to data compression of a speech signal. Many deep and complex methods provide high compression for speech signals, and this section will not provide a comparison to those, but will provide an example of a straightforward way of using the prediction methods of this chapter for data compression.

13.5.1. An Application of the Linear Prediction Methods to Speech Signals

This section shows some examples of applying the combination of eigenvectors prediction method of section 13.2.1.2 to the speech signal presented in Fig. 3. The figure presents the low-pass filtered version of the signal sampled initially at about 8 kHz. The signal was filtered as described above so that the highest frequency in the signal is 3.40 kHz.

The prediction coefficients used for the predictions for this speech signal were as described in Section 13.2. They were computed only once and stored,

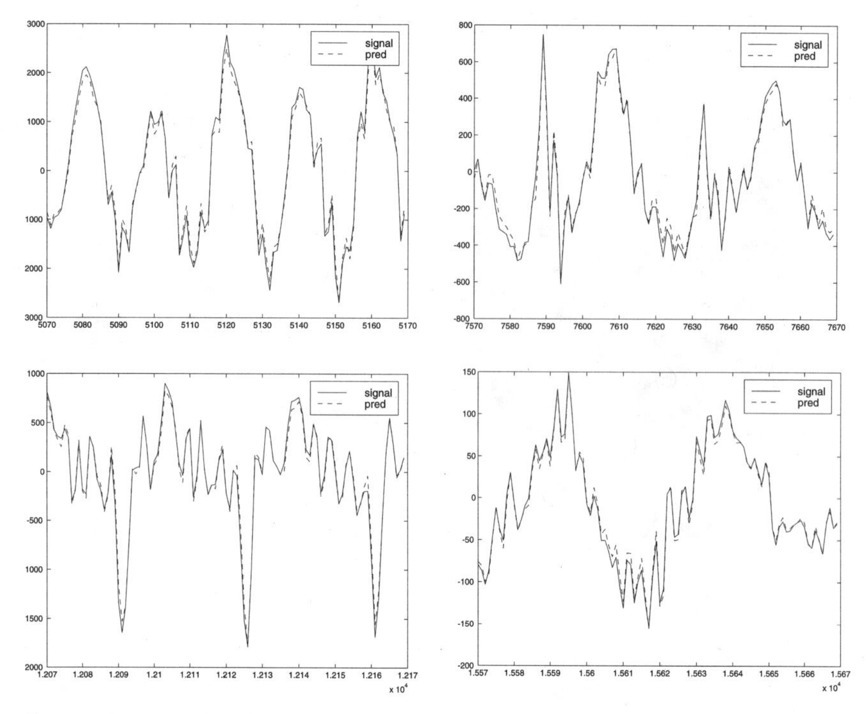

Figure 4. Four examples using the methods of this chapter for the prediction of 100 consecutive points each. Starting point within speech sample is 5 k (upper left), 7.5 k (upper right), 12 k (lower left), and 15.5 k (lower right). Order of predictor $N = 70$. Note vertical scale on each plot.

based on a value of $\tau = 0.85$, as noted above. That is, the same prediction coefficients were used for each of the results listed in the subsequent figure.

The four subplots in Fig. 4 show the result of applying the linear prediction approach of Section 13.2 to a consecutive set of 100 points. Each of these graphs was done using the order of the predictor as $N = 70$, and the graphs show the set of predictions as the dotted line and the actual signal as the solid line. The predictions used the actual signal samples in every case, and the different plots show the results of the predictions starting at various points in the signal. For example, the plot in the upper left corner of this figure starts with data at the 5,000 point in the approximately 16,000 sample set. Note the vertical scale on each subplot, since the different graphs differ widely in vertical scale, even though the graphs are normalized to show the result on the scale listed.

The graphs show the relatively high accuracy present in the signal predictions for this speech signal. The predictions become more accurate, up to a point, with an increase in the order N of the predictor, so that the graphs would not look as accurate for $N = 40$, say, as for these plots with $N = 70$. In the following, examples will be given for smaller values of N, and the results compared to predictions obtained by the Linear Predictive Coding (LPC) method. For the graphs as given, the dotted line of the prediction in the graphs often coincides with the solid line of the signal, since the prediction is so accurate.

There is a balance to be chosen between the accuracy of the prediction and sensitivity of the prediction to noise. The larger the order N of the predictor, the more accurate the prediction, but also the larger are the magnitudes of the prediction coefficients. For example, for $N = 70$, a few of the prediction coefficients have magnitudes on the order of 10^5, while for $N = 40$ the largest coefficients have magnitude of approximately 285. Large coefficients amplify any noise in the signal, including quantization noise, so that large values of N give coefficients with a dynamic range that would amplify noise present in the signal. Low-pass filtering the signal, even a quantized signal, would remove this difficulty and allow the prediction to be made with relatively large N.

13.5.2. Comparison to Well-known LPC Methods

The method of linear prediction presented in Section 13.2 differs from the well-known statistically-based method of "LPC" in that the prediction coefficients of the present method depend primarily on the band-limited nature of the signal and thus on the Nyquist parameter τ and not on the particular set of signal samples. Once the Nyquist parameter is known, the prediction coefficients can be computed and stored and never recalculated. The standard LPC method, however, uses prediction coefficients that are not based on an assumption that the signal is band-limited. The prediction coefficients for LPC are recalculated for speech signals based on an updated set of samples, so that they are often updated every 10–30 ms [16]. In extensive numerical trials, the methods of this chapter have given consistently accurate predictions for any segment of the speech signal, once the complete signal is filtered to 3.4 kHz and the Nyquist parameter τ is fixed.

The graphs presented in Figs. 5, 6, and 7 show a comparison between some predictions using the standard LPC method and predictions with the combination of eigenvectors method of Section 13.2. The LPC method was used with an order of predictor of the same size, $N = 70$, as the combination of eigenvectors method of this paper, see Section 13.2.1.2, for these two sets of plots, each using the signal filtered to 3.4 kHz. For the final plots, Fig. 7, each of the plots used the much lower value of $N = 12$. The combination of eigenvectors method appears superior to the LPC method from our results using $N = 70$, while errors appear comparable for smaller values of N such as $N = 8$ or $N = 12$, as in Fig. 7.

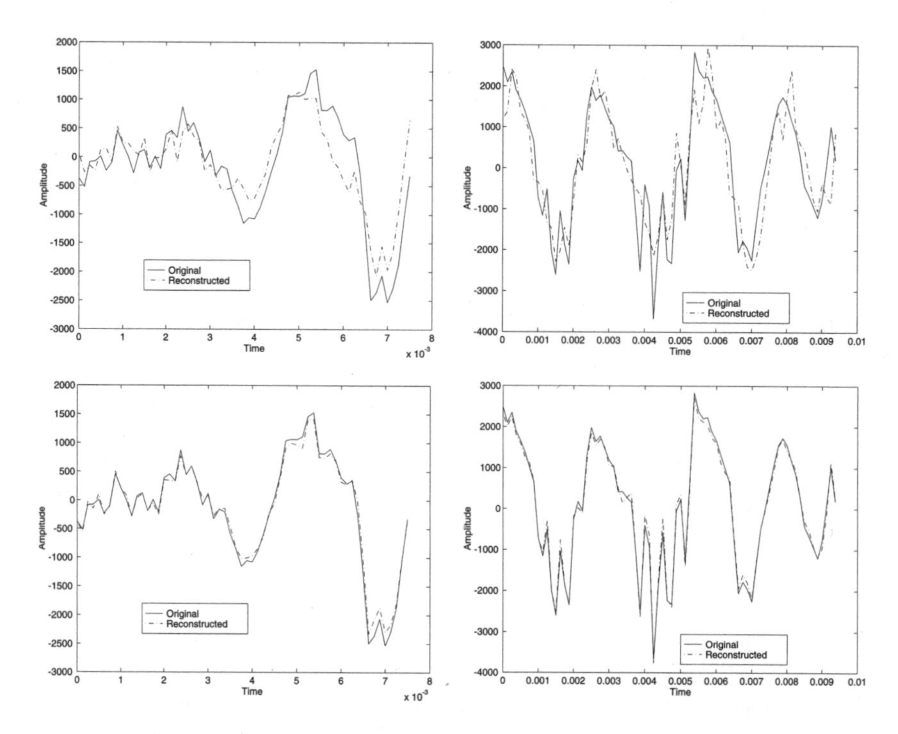

Figure 5. Two examples comparing LPC-method (top two plots) to corresponding plots made using methods of this paper (bottom two plots) for prediction of 60 points each. Starting point within speech sample is 2.11 k (left plots) and 5.325 k (right plots). Order of predictor $N = 70$ is the same for each method. Note vertical scale on each plot.

13.5.3. Linear Prediction and Data Compression

Data compression is a well-developed subject, particularly for speech signals, and the methods of the previous sections for prediction are quite different from the well-known prediction methods where the prediction coefficients come from the LPC technique. That technique is part of the process in the well-known DPCM (differential pulse code modulation) system for data compression, which has even further generalizations, see [20]. This section will give only a preliminary overview of how the methods of this chapter may be applied to data compression.

The fundamental idea of data compression using prediction is that the differences between the actual values and the predicted values may be stored instead of the actual signal samples. Because of the relative accuracy of the predictions obtained by methods of this paper, it is natural to wonder if these methods could be used for data compression. As noted in the graphs presented

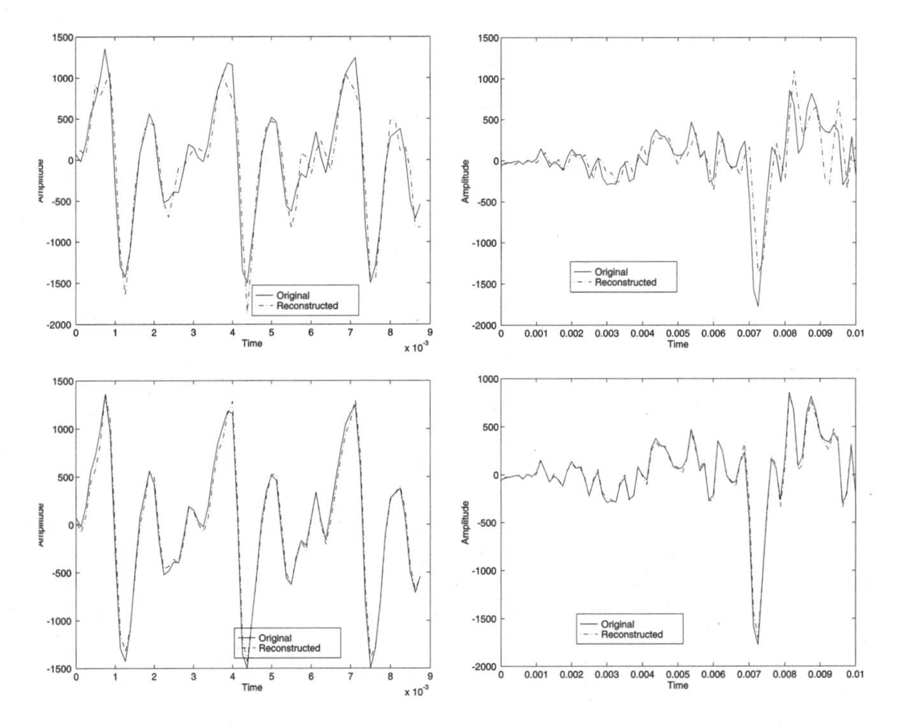

Figure 6. Two additional examples comparing LPC-method (top two plots) to corresponding plots made using methods of this paper (bottom two plots) for prediction of 60 points each. Starting point within speech sample is 10.99 k (left plots) and 12 k (right plots). Order of predictor $N = 70$ is the same for each method. Note vertical scale on each plot.

earlier in this section, the methods presented in this chapter often provide more accurate predictions than the LPC method. The advantage of accuracy allows for a smaller error, and a smaller error requires fewer quantization levels for storage of the differences between the predictions and the signal samples.

If the actual signal values are used, then the methods of this chapter provide more accurate predictions as the order N of the predictor is increased. As noted earlier, we assume the signal samples to be equally-spaced and to have been low-pass filtered to 3.40 kHz. With a sampling rate of 8 kHz, this results in a Nyquist parameter τ value of 0.85, which is near the Nyquist limit of $\tau = 1$ and for which N may need to be rather large to obtain highly accurate predictions. Using the combination of eigenvectors method of Section 13.2, there is the disadvantage that the dynamic range of the prediction coefficients grows as N increases, which could make the predictions sensitive to noise in the data for large N.

Although the signal samples have been low-pass filtered, for data compression there is the additional step that the data must be quantized. For the speech

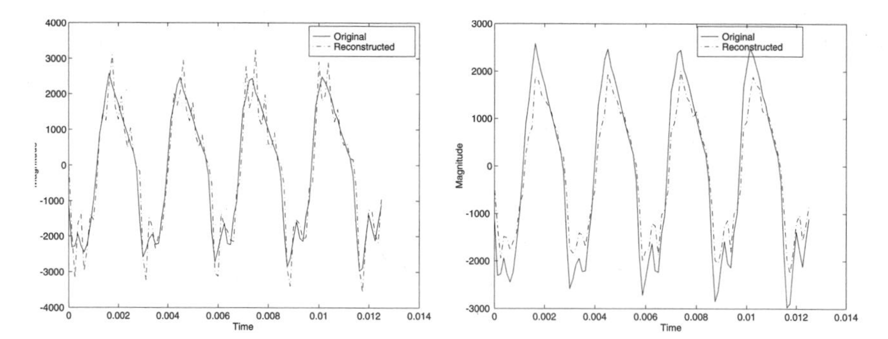

Figure 7. Comparison of LPC method (left) to the combination of eigenvectors method of this paper (right), both with $N = 12$.

signal, see Fig. 3, and the examples in this section we assume a 13-bit quantizer. In general, the methods of this chapter work better with a fine quantizer than with a coarse one. More quantization levels introduce less error into the signal samples. Suppose $\{\hat{f}(t_n)\}$ respresents the quantized data, $n = 1, \ldots$. At the transmitter, instead of a prediction $\sum_{n=1}^{N} a_n f(t_n)$ that uses the non-quantized data, we would use

$$p_{N+k} = \sum_{n=1}^{N} a_n \hat{f}(t_{N+k} - nT), \tag{41}$$

for $k = 1, \ldots$, i.e., the quantized data would be used to calculate the prediction.

At the transmitter, for data compression with the method of this chapter, the steps in data compression could go as follows:

1. The first N quantized samples of the signal $\{\hat{f}_j\}_{j=1,\ldots,N}$ are stored.
2. The first N quantized samples and the actual prediction coefficients $\{a_n\}$ are used to generate the predicted values for the remainder of the sampled signal as shown in (41). The prediction coefficients $\{a_k\}$ would be computed prior to the data compression sequence, based on the value of τ, and then recalled from storage to be used in the prediction formula.
3. The next step is to quantize the prediction value, i.e., to form $\hat{p}_{N+k}$ for each prediction, and then calculate the error at kth prediction point by

$$e_k = \hat{f}_{N+k} - \hat{p}_{N+k} \tag{42}$$

   for $k = 1, \ldots$ Note that the error is then automatically quantized.
4. As one approach, the transmitter could then send or store the first N quantized signal values and all the quantized error values.

At the receiver, the decoder collects the first N quantized signal values, $\hat{f}_1, \ldots, \hat{f}_N$ and the quantized error values (42). The receiver has the correct set of prediction coefficients, not quantized, since the receiver can generate those coefficients internally. The receiver's first step is to calculate the prediction

$$p_{N+1} = \sum_{n=1}^{N} a_n \hat{f}(t_{N+1} - nT),$$

based on the first set of quantized data. Then the receiver uses the same quantizer as the transmitter to quantize this prediction, i.e., to calculate $\hat{p}_{N+1}$. Then the receiver calculates the correct next quantized signal value through the formula

$$\hat{f}_{N+1} = e_{N+1} + \hat{p}_{N+1}.$$

Note that the receiver obtains the exact value of the quantized signal, with no accumulation of error.

The receiver can continue this process to generate $\hat{f}_{N+2}, \hat{f}_{N+3}, \ldots,$ i.e., to generate the remainder of the signal. Note that the remainder of the signal may be very long, since the original set of error values (42) can be for a lengthy signal.

Advantages of this method over the LPC method occur in two places. First, the same set of prediction coefficients is used for step 2 for the entire set of signal samples. The prediction coefficients never need to be recomputed. In addition, the prediction coefficients need not be quantized or transmitted to a receiver, since the same set of prediction coefficients could be generated at the receiver simply based on the knowledge of the Nyquist parameter τ. Some advantages of the LPC method over this method are that the LPC method is less sensitive to noise in the data, because of the smaller dynamic range of the LPC coefficients. An example of the method of this chapter applied to quantized data is given in Fig. 8.

Error in this process is entirely contained in the accuracy of the original set of predictions. As an example of the approximation error, Fig. 8 presents plots illustrating application of the prediction method of this chapter to quantized data. The upper plot in the figure is a plot of the speech signal and the prediction and the lower plot is just that of the error. These plots work with data from the original speech sample of Fig. 3 beginning at the point 9401. The plots work with 100 points just to make the dotted line of the prediction curve more visible, as the number of points at which one can calculate the predictions is not limited. As can be seen from Fig. 8, the error is significantly smaller in magnitude than the original signal samples. As noted above, the quantization of the data makes it so that one cannot directly use the methods of this chapter to increase the accuracy by increasing the order N of the predictor, and the accuracy of the prediction in Fig. 8 is not as good as one obtains with non-quantized data. It may be that there are further methods that could modify the methods of this chapter and retain its

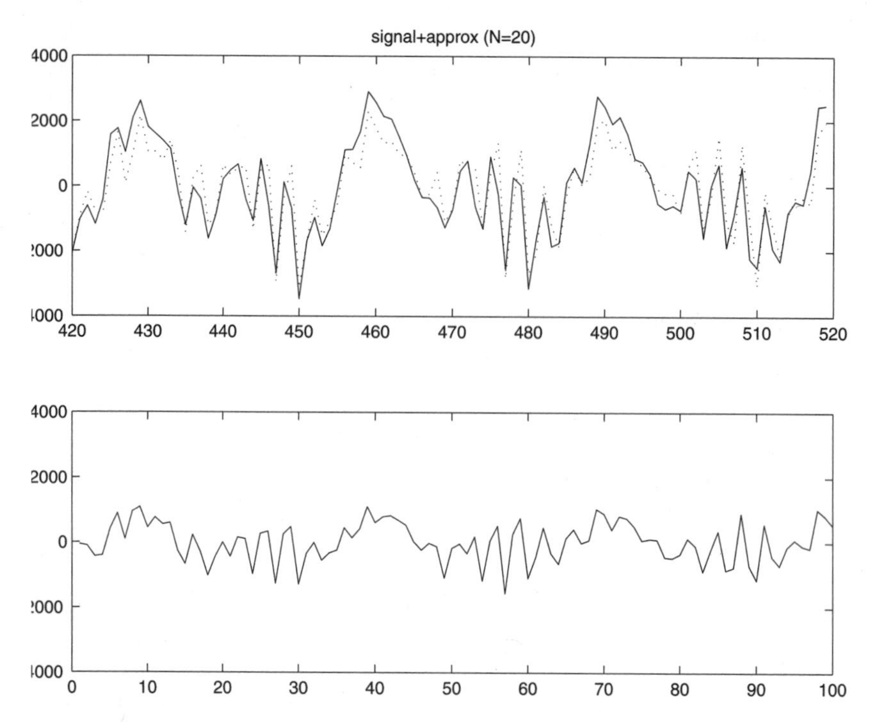

Figure 8. Prediction using quantized data. Signal is a set of 100 points from the original speech sample at location 9401. Top plot shows the signal and prediction with order of predictor $N = 20$. Bottom plot shows the error.

advantages while still reducing the size of the error. The authors have preliminary methods that may help in this regard, but these need further testing. This work is at this point quite introductory and does not have the sophistication of the well-developed methods for speech compression, such as LPC.

13.6. Summary

This chapter has presented both computational methods and applications for the prediction of a band-limited signal from past, periodically-sampled values. The focus of this chapter is on linear prediction methods, although an overview of nonlinear prediction and "extrapolation" was given in Section 13.4. The error-minimizing prediction coefficients for linear prediction from uniform samples were determined in Section 13.2, where alternative computational methods involving either the solution of the linear system (9) or a linear combination of eigenvectors of matrix (12) were discussed. As noted in Section 13.2, the new linear combination of eigenvectors method was the method chosen for the prediction examples that followed.

The definition of band-limited signals was not limited to finite energy signals in this paper, even though the initial derivations in Section 13.2 were carried out under this assumption. In Section 13.2.1.3, signals of even polynomial growth which satisfy the extended band-limited condition were shown to be appropriate for the prediction methods. The prediction coefficients depending on the Nyquist parameter $\tau = 2TW$ were even shown in Section 13.2.1.3 to converge to those which exactly reproduce polynomials in the limit as $\tau \to 0$.

For the new linear combination of eigenvectors method of determining the prediction coefficients, accurate computation of the eigenvectors is related to properties of the discrete prolate spheroidal sequences (DPSS), and many of the connections between the standard DPSS and the new DPSS for nonuniformly spaced sampling were explored in Section 13.3. In particular, interlaced sampling was the focus of that section. Numerical examples for prediction of a band-limited signal of polynomial growth under interlaced sampling were also presented.

An application of the methods of Section 13.2 to speech signals was presented in the final section. Such signals are particularly appropriate for the methods developed in earlier sections since it is a common practice to low-pass filter speech signals for telecommunications. The methods developed in this chapter were compared to standard LPC methods and shown to be competitive in accuracy, with the new method having the advantage that the prediction coefficients never need to be recomputed since they apply to predictions for the entire band-limited signal. Some initial ideas on how the new method might be applied to data compression were also presented in Section 13.5.3.

Further work in this area could generate additional advances in prediction, such as in the area of prediction methods for nonuniform but periodically-sampled data. The authors expect and look forward to those advances.

References

[1] P. L. Butzer and R. L. Stens. Linear Prediction by Samples from the Past. In R. J. Marks, II, Ed., *Advanced Topics in Shannon Sampling and Interpolation Theory*, Springer-Verlag, 1993, pp. 157–184.

[2] S. Dharanipragada and K. S. Arun. Bandlimited Extrapolation using Time-Bandwidth Dimension. *IEEE Trans. Sig. Proc.*, 45:2951–2966, 1997.

[3] P. J. S. G. Ferreira. Interpolation and the Discrete Papoulis-Gerchberg Algorithm. *IEEE Trans. Sig. Proc.*, 42:2596–2606, 1994.

[4] D. M. Gruenbacher and D. R. Hummels. A Simple Algorithm for Generating Discrete Prolate Spheroidal Sequences. *IEEE Trans. Sig. Proc.*, 42:3276–8, 1994.

[5] A. M. Kondoz. *Digital Speech*. John Wiley & Sons, 1994.

[6] J. Makhoul. Linear Prediction: A Tutorial Review. *Proc. IEEE*, 63:561–580, 1975.

[7] R. J. Marks, II. *Introduction to Shannon Sampling and Interpolation Theory*, Springer-Verlag, 1991.

[8] F. Marvasti. Nonuniform Sampling. In R. J. Marks, II, Eds., *Advanced Topics in Shannon Sampling and Interpolation Theory*, Springer-Verlag, 1993, pp. 121–156.

[9] J. Mathews, J. Breakall, and G. Karawas. The Discrete Prolate Spheroidal Filter as a Digital Signal Processing Tool. *IEEE Trans, Acoust. Speech Signal Processing* ASSP-33:1471–78, 1985.

[10] D. H. Mugler and Y. Wu. An Integrator for Time-Dependent Systems with Oscillatory Behavior. *Comp. Meth. Appl. Mech. & Eng*, 171:25–41, 1999.

[11] D. H. Mugler, Y. Wu, and W. Splettstößer. Nyquist Rates and the Linear Prediction of Band-Limited Signals Based on Signal and Derivative Samples. *Proc. International Workshop on Sampling Theory and Applications*. Portugal, 321–326, 1997.

[12] D. H. Mugler and Y. Wu. Linear Prediction of Band-limited Signals from Signal and Derivative Samples. *Proc. IASTED Signal and Image Proc.* SIP-96:18–22, 1996.

[13] D. H. Mugler. Linear Prediction of a Band-Limited Signal from Past Samples at Arbitary Points: An SVD-based Approach. In *Proc. International Workshop on Sampling Theory and Applications*, Latvia, 1995, pp. 113–118.

[14] D. H. Mugler. Computational Aspects of an Optimal Linear Prediction Formula for Band-Limited Signals. *Computational and Applied Mathematics, I*, Elsevier Science Publishers, 1992, pp. 351–356.

[15] D. H. Mugler. Computationally-efficient Linear Prediction of a Band-Limited Signal and its Derivative. *IEEE Trans. on Information Theory*, IT-36, 1990, pp. 589–596.

[16] F. J. Owens. *Signal Processing of Speech*. McGraw-Hill, New York, 1993.

[17] A. Papoulis. *Signal Analysis*. McGraw-Hill, London, 1981.

[18] R. Perline. Toeplitz Matrices and Commuting Tridiagonal Matrices. *SIAM J. Matrix Anal. Appli.*, 12:321–326, 1991.

[19] W. Rudin. *Functional Analysis*. McGraw-Hill, New York, 1973.

[20] K. Sayood. *Data Compression*. Morgan Kaufman, San Francisco, 1996.

[21] D. Slepian. Prolate Spheroidal Wave Function, Fourier Analysis, and Uncertainty-I. *Bell Syst. Tech. J.*, 40:43–64, 1961.

[22] D. Slepian. Prolate Spheroidal Wave Functions, Fourier Analysis, and Uncertainty-V: The Discrete Case. *Bell Syst. Tech. J.*, 57:1371–1430, 1978.

[23] W. Splettstößer. On the Prediction of Band-limited Signals from Past Samples. *Inform. Sci.*, 28:115–130, 1982.

[24] D. J. Wingham. The Reconstruction of a Band-Limited Function and its Fourier Transform From a Finite Number of Samples at Arbitrary Locations by Singular Value Decomposition. *IEEE Trans. Sig. Proc.*, 40:559–570, 1992.

[25] C. Xiao. Reconstruction of Bandlimited Signal with Lost Samples at its Nyquist Rate – The Solution to a Nonuniform Sampling Problem. *IEEE Trans. Sig. Proc.*, 43:1008–9, 1995.

[26] R. M. Young. *An Introduction to Nonharmonic Fourier Series*. Academic Press, New York, 1980.

14

Frames, Irregular Sampling, and a Wavelet Auditory Model

J. J. Benedetto and S. Scott

Abstract

A Wavelet Auditory Model (WAM) is constructed in terms of wavelet frames and an irregular sampling algorithm for Fourier frames. Its theoretical effectiveness is demonstrated in the context of speech coding, and its original formulation is found in [8–9]. The presentation of WAM in this chapter emphasizes its underlying mathematical ideas, and, in particular, develops the notions from the theory of frames and irregular sampling that arise naturally in constructing WAM.

14.1. Introduction

We shall develop some of the basic theory of frames, as well as consequences of that theory from the area of irregular sampling. Moreover, we shall see how these ideas play a natural role in modelling parts of the mammalian auditory process with the purpose of using this process to devise signal reconstruction algorithms in the field of speech coding.

The theory of frames is due to Duffin and Schaeffer [22], and it was developed to address problems in non-harmonic Fourier series. Prior to [22], these problems were concerned with finding criteria on real sequences $\{t_n\}$ so that the closed linear span, $\overline{span}\{e_{t_n}\}$, of exponentials $e_{t_n}(f) = e^{2\pi j t_n f}$ would be equal to the space $L^2[-\Omega, \Omega]$ of finite energy signals defined on $[-\Omega, \Omega]$. The origins of

J. J. Benedetto and S. Scott • Department of Mathematics, University of Maryland, College Park, MD 20742; E-mail: jjb@math.umd.edu and sscott@math.umd.edu

Nonuniform Sampling: Theory and Practice, edited by Marvasti
Kluwer Academic/Plenum Publishers, New York, 2001.

these problems go back at least to G. D. Birkhoff (1917) and J. L. Walsh (1921), and the basic and highly non-trivial theory associated with such problems was established by Paley and Wiener in [35]. The theory was further developed in remarkable ways by Levinson [30], Pollard [36], Beurling and Malliavin [14–15], and others, see [38, 48]. The impact of the ideas in [22] to address issues in signal reconstruction was not fully appreciated until the work of Daubechies, Grossmann, and Meyer [21], as well as subsequent work by Daubechies in her book [20]. The 1990s have seen a plethora of contributions to the theory of frames, e.g., see the Duffin memorial issue of the Journal of Fourier Analysis and Applications (Volume 3, 1997).

The formulation of irregular sampling algorithms in terms of so-called Fourier frames (Definition 10) is a natural topic in the theory of frames, see [7] from 1990; and there are several expositions on the subject including [2] and [23], cf., Marvasti's book [33] for many classical irregular sampling results and a comprehensive bibliography. The relationship between sampling theory and other types of signal decompositions, including wavelet and Gabor frames, has also become a highly developed area, see [5].

The mammalian auditory system possesses excellent abilities to detect, separate, and recognize speech and environmental sounds. In recent decades, these capabilities have been the subject of theoretical and experimental research, particularly with a view towards applying auditory functional principles in the design of man-machine communication links, e.g., [16, 18, 24, 41, 45]. As indicated above, we shall use frames and irregular sampling methods to construct a wavelet and Fourier frame based mathematical model for the mammalian auditory system. It is called the Wavelet Auditory Model (WAM) [9]. Our purpose is to present the mathematical underpinnings of WAM more fully than in [8] with a goal of gaining a deeper understanding of the role of irregular sampling in the reconstruction of speech signals.

After introducing the wavelet transform and some notation in Section 14.2, we shall describe WAM in Section 14.3. In fact, we shall trace the processing of a speech signal in a mammalian auditory system, and construct a corresponding mathematical model, viz., WAM. This model will exhibit some of the mathematical structure associated with wavelet frames, and in the process of exploiting this structure we shall see the role of irregular sampling in reconstructing a signal y.

Because of the point of view developed in Section 14.3, we shall present the elements of the theory of frames in Section 14.4 and related results about irregular sampling and Fourier frames in Section 14.5.

We begin Section 14.4 with the definition of frames (Definition 1). The Frame Decomposition Theorem (Theorem 2) which follows is reformulated in Proposition 4 as a reconstruction formula for a signal y in terms of its "sampled values" Ly. This reformulation gives rise to a reconstruction algorithm (Algorithm 8), which implements a Gram operator whose entries can theoretically be stored off-line.

Algorithm 8 is a perfect reconstruction theorem, and its implementation on WAM data guarantees excellent speech signal reconstruction under ideal conditions. Because of the inherent wavelet frame structure in WAM which emerged in Section 14.3, we close Section 14.4 with a wavelet frame calculation (Example 11) that will play a role in Section 14.7. The calculation also shows how Fourier frames can arise in this context, thereby giving explicit motivation for the material in Section 14.5.

In Theorem 12 of Section 14.5, we give basic criteria for irregularly spaced modulations and translations of a signal to be a frame for $L^2(\mathbb{R})$, the space of finite energy signals on $\mathbb{R}$. Then we use a corollary of this result to prove the Yao-Thomas irregular sampling formula in Theorem 14. The Yao-Thomas Theorem is really a result about so-called exact frames (Definition 1).

A critical feature in the success of mammalian auditory systems is the fact that the cochlea has the equivalent of a sophisticated and effective filter bank on its basilar membrane. This filter bank is discussed in Section 14.3, and we shall use the Paley-Wiener Logarithmic Integral Theorem in Section 14.6 to design the corresponding filters in WAM. We shall then see that the wavelet and Fourier frame approach to mammalian auditory systems corroborates the fact that white noise is often naturally reduced in such systems using such filters during speech processing. We shall also include related calculations for some other noises.

In our final section, Section 14.7, we shall integrate the modelling of Section 14.3, the theoretical results of Sections 14.4 and 14.5, and the construction of Section 14.6 to apply WAM to a typical problem in speech coding. The mathematical success of this application is illustrated, and its potential practical value is the subject of [9].

14.2. Mathematical Background

Let $L^2(\mathbb{R})$ be the space of complex-valued finite energy signals defined on the real line $\mathbb{R}$. The *Fourier transform* Y or $\hat{y}$ of $y \in L^2(\mathbb{R})$ is

$$Y(f) = \hat{y}(f) = \int y(t)e^{-2\pi jtf}\,dt$$

for $f \in \hat{\mathbb{R}}(=\mathbb{R})$, where integration is over $\mathbb{R}$. The Fourier pairing between y and Y is designated by $y \leftrightarrow Y$. If Y is defined on $\hat{\mathbb{R}}$, then formally one has

$$y(t) = Y^{\vee}(t) = \int Y(f)e^{2\pi jtf}\,df,$$

where integration is over $\hat{\mathbb{R}}$, see [3] for conditions for the validity of this formula. $Y^{\vee}$ is the *inverse Fourier transform* of Y.

For $s > 0$, the L^2-*dilation* operator D_s is defined by $D_s y(t) = s^{1/2} y(st)$ for $y \in L^2(\mathbb{R})$, and the Fourier transform of $D_s y$ is

$$(D_s y)^\wedge(f) = s^{-1/2} Y(s^{-1} f) = D_{1/s} Y(f).$$

For $u \in \mathbb{R}$, the *translation* operator τ_u is defined by $\tau_u y(t) = y(t-u)$ for $y \in L^2(\mathbb{R})$. As such, $(\tau_u y)^\wedge(f) = e^{-2\pi j u f} Y(f) = e_{-u}(f) Y(f)$, where

$$e_{-t}(f) = e^{-2\pi j t f}.$$

The *convolution* of $x, \ y \in L^2(\mathbb{R})$ is

$$x * y(t) = \int x(t-u) y(u) du = \int x(u) y(t-u) du;$$

and the *inner product* of x and y is $\langle x, y \rangle = \int x(t) \overline{y(t)}\, dt$.

For a fixed $g \in L^2(\mathbb{R})$, the *wavelet transform* of $y \in L^2(\mathbb{R})$ is the function

$$W_g y(t, s) = (y * D_s g)(t) \tag{1}$$

defined on the *time-scale plane* $t \in \mathbb{R}, s > 0$. By a straightforward calculation, we obtain

$$W_g y(t, s) = \langle y, \tau_t D_s \tilde{g} \rangle,$$

where $\tilde{g}$ is the *involution* of g defined as $\tilde{g}(u) = \bar{g}(-u)$. If the derivative $\partial_t g$ is an element of $L^2(\mathbb{R})$, we define $W_{\partial_t g} y$ analogously to the definition of $W_g y$ in (1). In this case, if $W_{\partial_t g} y$ converges uniformly on time intervals for each fixed scale $s > 0$, and if a mild smoothness condition is satisfied, then

$$\partial_t W_g y(t, s) = s W_{\partial_t g} y(t, s). \tag{2}$$

(2) is true for the causal filters G and signals y under consideration in this chapter.

Notationally, we follow standard notation in mathematical analysis, see, e.g., [42]. In particular, $L^2[-\Omega, \Omega]$ is the space of finite energy signals defined on the interval $[-\Omega, \Omega]$, and PW_Ω is the *Paley-Wiener space* (bandlimited finite energy signals), defined as

$$PW_\Omega = \{y \in L^2(\mathbb{R}) : \text{supp } Y \subseteq [-\Omega, \Omega]\},$$

where *supp* Y is the support of Y. Let

$$2\Omega \,\text{sinc}\, 2\Omega t = d_{2\pi\Omega}(t) = \frac{\sin 2\pi\Omega t}{\pi t}, \quad t \in \mathbb{R},$$

where "d" is for Dirichlet. The characteristic function of the interval $[-\Omega, \Omega] \subseteq \hat{\mathbb{R}}$ is denoted by $\mathbf{1}_{(\Omega)}$, i.e.,

$$\Pi\left(\frac{f}{2\Omega}\right) = \mathbf{1}_{(\Omega)}(f).$$

Thus, we have the Fourier pairing $d_{2\pi\Omega} \leftrightarrow \mathbf{1}_{(\Omega)}$.

Finally, $\mathbb{Z}$ denotes the ring of integers, and $l^2(\mathbb{Z}^d)$ is the space of finite energy sequences indexed by $\mathbb{Z}^d$, the direct product of d-tuples of integers.

14.3. Wavelet Auditory Model

14.3.1. Setting

In a mammalian auditory system, an acoustic signal or sound wave y induces vibrations in the ear drum, which travel through the middle ear to the cochlear fluid of the inner ear. These vibrations then cause traveling waves on the basilar membrane of the cochlea. As the waves propogate into the spiral shaped cochlea, they produce a pattern of displacements W of the basilar membrane at different locations for different frequencies. Displacements for high frequencies occur at the basal end; for low frequencies they occur at the wider apical end inside the spiral, see e.g., [24]. The basilar membrane records frequency responses between 200 and 20,000 Hz. For comparison, telephone speech bandwidth deals with the range 300–4,000 Hz. The cochlea analyzes sound in terms of these traveling waves much like a parallel bank of filters – in this case a band with 30,000 channels. The impulse responses of these filters along most of the interior length of the cochlea are related by dilation. Consequently, their transfer functions are invariant except for frequency translation along the approximately logarithmic axis of the cochlea, see, e.g., [39–40].

Mathematically, this dilational relationship between impulse responses can be expressed by the assertion that there is a function $g : \mathbb{R} \mapsto \mathbb{C}$ such that the set of impulse responses is of the form $\{D_s g : s \in [s_1, s_2] \subseteq (0, \infty)\}$. Thus, we identify the displacements W, due to the stimulus y, with the output of the cochlear filter bank having the impulse responses $\{D_s g\}$, i.e., we set $W = W_g y(t, s)$, where g is a fixed causal impulse response. Specifically, because of the structure of the cochlear filter bank, we fix $a > 1$ and set $s_m = a^m, m \in \mathbb{Z}$. As such, in WAM the signal y is first transformed into a pattern of displacements,

$$W_g y(t, s_m), \quad m \in \mathbb{Z},$$

for a discrete set of points (t, s_m) in the time-scale plane, where t is written more explicitly in (3) below, see the box labelled Wavelet Transform in Fig. 1.

The shape of $|G|$ is critical for the effectiveness of the auditory model. Generally, G should be a causal filter that has a "shark-fin" shaped amplitude. The design problems for such filters are dealt with in Section 14.6. In the case of properly designed filters, the high frequency edges of the cochlear filters $D_{1/s}G$ act as abrupt "scale" delimiters. Thus, a sinusoidal stimulus will propogate up to the appropriate scale and die out beyond it.

The auditory system does not receive the wavelet transform directly, but rather a substantially modified version of it. In fact, the output of each cochlear filter is effectively highpassed by the velocity coupling between the cochlear membrane and the cilia of the hair cell transducers that initiate the electrical nervous activity by a shearing action on the tectorial membrane. Hence, the mechanical motion of the basilar membrane is converted to a receptor potential in the inner hair cells. It is reasonable to approximate this stage by a time derivative, obtaining the ouput $\partial_t W_g y(t, s_m)$, see the box labelled ∂_t in Fig. 1. The extrema of the wavelet transform $W_g y(t, s_m)$ become the zero-crossings of the new function $\partial_t W_g$; and so one output of the auditory process is

$$\forall m \in \mathbb{Z}, \quad Z_m = \{t(n; s_m) : \partial_t W_g y(t(n; s_m), s_m) = 0\}, \tag{3}$$

where n is used to index the domain of the zero-crossings for a given m, i.e., the box labelled Zeros in Fig. 1.

14.3.2. Sigmoidal Operation

At the next step in the auditory process, an instantaneous sigmoidal non-linearity R is applied, followed by a low pass filter with impulse response h. These operations model the threshold and saturation that occur in the hair cell channels and the leakage of electrical current throughout the membranes of these cells [34, 39]. The resulting cochlear output,

$$C_{h,R}(t, s) = (R \circ \partial_t W_g y(\cdot, s)) * h(t),$$

where "$\circ$" denotes composition of functions and where convolution $*$ is with respect to time, is a planar auditory nerve pattern sent to the brain. Typically, the composition by R can be represented by functions

$$R_T(x) = \frac{e^{Tx}}{1 + e^{Tx}},$$

parameterized by T. Obviously, $\lim_{T\to\infty} R_T = U$, the Heaviside step function. Approximations to the Heaviside step function are reasonable since the nerve fibers from the inner hair cells to the auditory nervous system fire at positive rates, and since this action cannot process above a certain limit, i.e., the

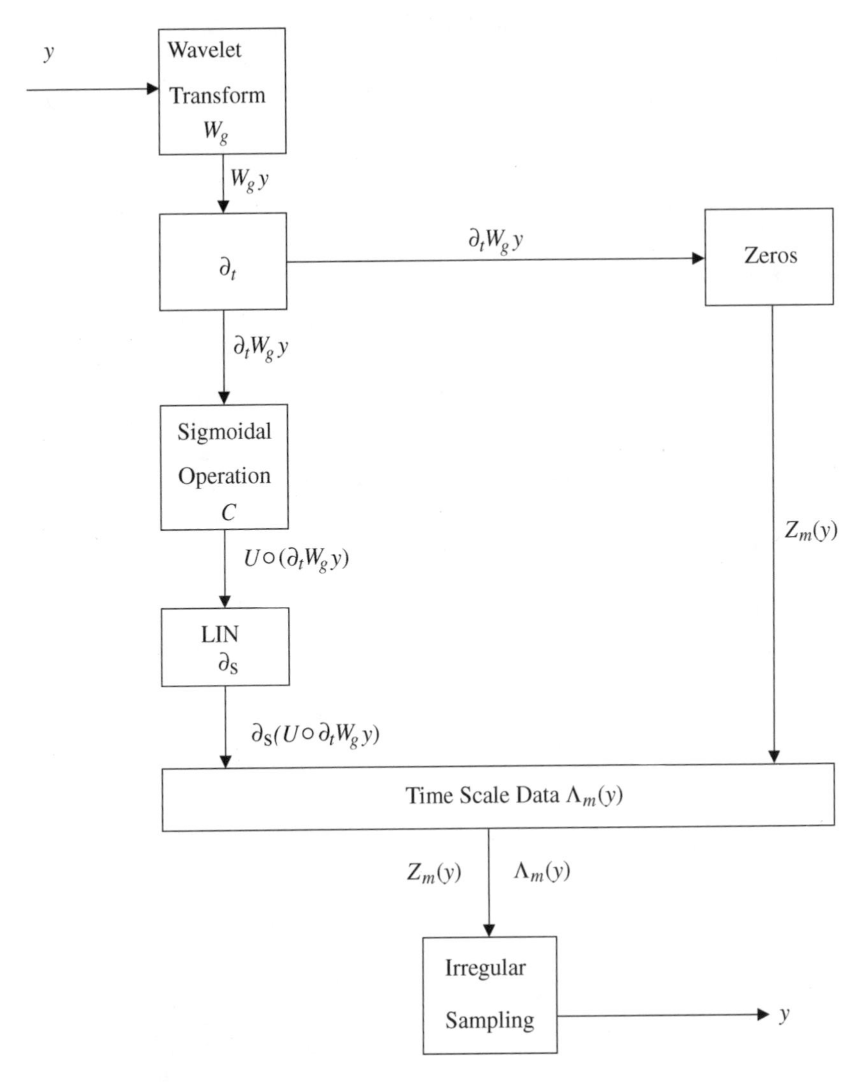

$$Z_m(y) = \{(t(n; s_m) : \partial_t W_g y(t(n; s_m), s_m) = 0\}$$

$$\Lambda_m(y) = \{\partial_s \partial_t W_g y(t(n; s_m), s_m): t(n; s_m) \in Z_m(y)\}$$

Figure 1. Wavelet auditory model.

aforementioned saturation. For computational convenience in WAM, we take R to be U and set $h = \delta$, the Dirac δ-measure (impulse), even though δ does not give rise to a low pass filter. Thus, $C_{h,R}(t, s)$ above is replaced by the *cochlear output*,

$$C(t, s) = U \circ \partial_t W_g y(t, s),$$

i.e., the output of the box labelled Sigmoidal Operation C in Fig. 1.

14.3.3. Lateral Inhibitory Network

The mammalian auditory nerve patterns determined by W, ∂W, and Z_m are now processed by the brain in ways that are not completely understood. One processing model, the lateral inhibitory network (LIN), has been closely studied with a view to extracting the spectral pattern of the acoustic stimulus [34, 40, 46]; and we shall implement it in our alogorithm. Scientifically, it reasonably reflects proximate frequency channel behavior, and mathematically it is relatively simple.

For a given acoustic signal y, constant $a > 1$, and properly designed causal filter G, we generate Z_m and the set

$$\Lambda_m = \{\partial_s \partial_t W_g y(t(n; s_m), s_m) : t(n; s_m) \in Z_m\} \tag{4}$$

for each $m \in \mathbb{Z}$, i.e., the box labelled Time-Scale Data $\Lambda_m(y)$ in Fig. 1. The elements of Λ_m are the coefficients of a natural linear expansion of $\partial_s(U \circ \partial_t W_g y)$, see page 5 of [8] for details. The scaling partial ∂_s reflects the action of LIN, indicated by the box in Fig. 1 called LIN ∂_s.

14.3.4. The WAM Problem and Solution

Let G be a properly designed finite energy causal filter and let $a > 1$. Suppose an unknown acoustic signal y has generated the set

$$\Lambda_m = \{\partial_s \partial_t W_g y(t(n; s_m), s_m) : t(n; s_m) \in Z_m\}$$

and that the receiver has knowledge of this set or of some subset. This irregularly spaced array in the $t - s$ plane is called *WAM data*. It is natural for the receiver to attempt to reconstruct the signal y in terms of this data. The *WAM problem* is to effect this reconstruction by means of irregular sampling formulas. We shall outine the solution given in [8], but stress the mathematical development more than appeared there. In particular, we shall present irregular sampling formulas developed by Benedetto and Heller, [2, 7], and give perspective in terms of the theory of frames and other irregular sampling criteria. The fact that the WAM problem can be solved theoretically by means of such formulas leaves open the problem of effective implementation, see Section 14.7.

14.3.5. WAM Wavelet Frame

Observe that for $t(n; s_m) \in Z_m$,

$$\partial_s \partial_t W_g y(t(n; s_m), s_m)$$
$$\approx \frac{\partial_t W_g y(t(n; s_m), s_{m+1}) - \partial_t W_g y(t(n; s_m), s_m)}{s_{m+1} - s_m}.$$

The second term in the numerator of the right side vanishes, since $t(n; s_m) \in Z_m$. Hence,

$$\frac{\partial_t W_g y(t(n; s_m), s_{m+1}) - \partial_t W_g y(t(n; s_m), s_m)}{s_{m+1} - s_m} = \frac{\partial_t W_g y(t(n; s_m), s_{m+1})}{s_{m+1} - s_m}$$
$$= \frac{s_{m+1} W_{\partial_t g} y(t(n; s_m), s_{m+1})}{s_{m+1} - s_m},$$

where the last equality follows from (2). Writing this approximation as an equality, we have

$$\partial_s \partial_t W_g y(t(n; s_m), s_m) = \frac{s_{m+1}}{s_{m+1} - s_m} W_{\partial_t g} y(t(n; s_m), s_{m+1})$$
$$= -\frac{1}{a-1} \langle y, \tau_{t(n;s_m)} D_{s_{m+1}} (\partial_t \tilde{g})(u) \rangle,$$

where $a = s_m / s_{m+1}$. Because of this equation and the frame-theoretic point of view of the next section (Section 14.4), we define

$$\psi_{m,n} = -\frac{1}{a-1} \tau_{t(n;s_m)} D_{s_{m+1}} (\partial_t \tilde{g}) \tag{5}$$

and the mapping

$$L : \mathbb{H} \to l^2(\mathbb{Z} \times \mathbb{Z})$$
$$y \mapsto \{\langle y, \psi_{m,n} \rangle\},$$

where $\mathbb{H}$ is a Hilbert subspace of $L^2(\mathbb{R})$ containing the class of acoustic signals to be analyzed. Each function $\psi_{m,n}$ corresponds to an element $t(n; s_m) \in Z_m$. Note that (5) can be rewritten as

$$\psi_{m,n} = -\frac{1}{a-1} D_{s_{m+1}} \tau_{s_{m+1} t(n;s_m)} (\partial_t \tilde{g}).$$

In particular, $\{\psi_{m,n}\}$ depends on a given acoustic signal y and the known filter G.

The discussion in the previous paragraph leads naturally into the theory of wavelet frames, that we shall develop in Section 14.4. The dependence of $\{\psi_{m,n}\}$ on y is not amenable to a global theory of frames but such a theory is not essential for our purpose. The degree to which the sequence $\{\psi_{m,n}\}$ can be considered as a wavelet frame will be analyzed in Example 11.

14.4. Theory of Frames

In this section we review the theory of frames that was introduced by Duffin and Schaeffer [22], see also [20, 48], and Chapters 3 and 7 of [6]. Let $\mathbb{H}$ be a separable Hilbert space with inner product $\langle x, y\rangle$ and norm $\|x\| = \langle x, x\rangle^{\frac{1}{2}}$.

Definition 1 Frames

a. *A sequence $\{x_n : n \in \mathbb{Z}^d\} \subseteq \mathbb{H}$ is a frame for $\mathbb{H}$ if there exist $A, B > 0$ such that*

$$\forall y \in \mathbb{H}, \quad A\|y\|^2 \leq \sum |\langle y, x_n\rangle|^2 \leq B\|y\|^2.$$

A and B are frame bounds, and a frame is tight if $A = B$. A frame is exact if it is no longer a frame whenever any one of its elements is removed.

b. *The frame operator of the frame $\{x_n\}$ is the function $S : \mathbb{H} \to \mathbb{H}$ defined as $Sy = \sum \langle y, x_n\rangle x_n$ for all $y \in \mathbb{H}$.*

c. *It can be shown that if $\{x_n\} \subseteq \mathbb{H}$ is a frame, then $S = L^*L$, where*

$$\begin{aligned} L : \mathbb{H} &\to l^2(\mathbb{Z}^d) \\ x &\mapsto \{\langle x, x_n\rangle\} \end{aligned}$$

is called the Bessel map associated with $\{x_n\}$ and where the adjoint map

$$L^* : l^2(\mathbb{Z}) \to \mathbb{H}$$

of L is the reconstruction map.

Theorem 2 Frame Decomposition Theorem *Let $\{x_n : n \in \mathbb{Z}^d\} \subseteq \mathbb{H}$ be a frame for $\mathbb{H}$ with frame bounds A and B.*

a. *The frame operator S is a one-to-one, onto, and continuous map with inverse $S^{-1} : \mathbb{H} \to \mathbb{H}$. $\{S^{-1}x_n\} \subseteq \mathbb{H}$ is a frame with frame bounds B^{-1} and A^{-1}, and*

$$\forall y \in \mathbb{H}, \quad y = \sum \langle y, S^{-1}x_n\rangle x_n = \sum \langle y, x_n\rangle S^{-1}x_n \quad \text{in } \mathbb{H}.$$

$\{S^{-1}x_n\}$ is called the dual frame of $\{x_n\}$, and it is easy to see that S^{-1} is the frame operator of $\{S^{-1}x_n\}$.

b. *If $\{x_n\}$ is a tight frame for $\mathbb{H}$, if $\|x_n\| = 1$ for all n, and if $A = B = 1$, then $\{x_n\}$ is an orthonormal basis for $\mathbb{H}$, see [2] for the original formulation of this result by Vitali.*

c. *If $\{x_n\}$ is an exact frame for $\mathbb{H}$, then $\{x_n\}$ and $\{S^{-1}x_n\}$ are biorthonormal, i.e.,*

$$\forall m, n, \quad \langle x_m, S^{-1}x_n \rangle = \begin{cases} 1 & \text{if } m = n, \\ 0 & \text{otherwise,} \end{cases}$$

and $\{S^{-1}x_n\}$ is the unique sequence in $\mathbb{H}$ which is biorthonormal to $\{x_n\}$

d. *If $\{x_n\}$ is an exact frame for $\mathbb{H}$, then the sequence resulting from the removal of any one element is not complete in $\mathbb{H}$, i.e., the linear span of the resulting sequence is not dense in $\mathbb{H}$.*

Theorem 3 Characterization of Frames

a. *A sequence $\{x_n : n \in \mathbb{Z}^d\} \subseteq \mathbb{H}$ is a frame for $\mathbb{H}$ with frame bounds A and B if and only if the map*

$$L : \mathbb{H} \to l^2(\mathbb{Z}^d)$$
$$y \mapsto \{\langle y, x_n \rangle\}$$

is a topological isomorphism of $\mathbb{H}$ onto a closed subspace of $l^2(\mathbb{Z}^d)$. In this case,

$$\|L\| \leq B^{1/2} \quad \text{and} \quad \|L^{-1}\| \leq A^{-1/2},$$

where L^{-1} is defined on the range $L(\mathbb{H})$. Thus, in the case of a frame, L is the associated Bessel map.

b. *A sequence $\{x_n : n \in \mathbb{Z}^d\} \subseteq \mathbb{H}$ is a frame for $\mathbb{H}$ if and only if there is $C > 0$ such that for all $y \in \mathbb{H}$*

$$\sum |\langle y, x_n \rangle|^2 < \infty,$$

$$\exists\, c_y = \{c_n\} \in l^2(\mathbb{Z}^d), \quad \text{such that} \quad y = \sum c_n x_n \in \mathbb{H},$$

and

$$\|c_y\|_{l^2(\mathbb{Z}^d)} \leq C\|y\|.$$

Part a is proved in Theorem 7.15 of [6]*; and part b is proved in Remark 3.9 of* [6]*, cf., the treatment of part b in* [10].

If $\{x_n\}$ is a frame with Bessel map L and given data $c \in l^2(\mathbb{Z}^d)$ is of the form $Ly = \{\langle y, x_n \rangle\} = c$, where y is not explicitly known, then c can be thought of as "sampled data" of some signal y, which must then be reconstructed in terms of c. The following result is the first step in developing this point of view.

Proposition 4 Frame Reconstruction Formula *Let $\{x_n : n \in \mathbb{Z}^d\} \subseteq \mathbb{H}$ be a frame for $\mathbb{H}$ with frame operator S, frame bounds A and B, and Bessel map $L : \mathbb{H} \to l^2(\mathbb{Z}^d)$. Then*

$$\forall y \in \mathbb{H}, \quad y = (S^{-1}L^*)Ly, \tag{6}$$

cf., (11).

Equation 6 can be viewed as a reconstruction formula for signals y in which discrete "sampled data" Ly is given, as indicated in the remark introducing Proposition 4. In fact, using the hypotheses of Proposition 4, Equation (6) and the Neumann expansion

$$S^{-1} = \frac{2}{A+B}\sum_{k=0}^{\infty}\left(I - \frac{2}{A+B}S\right)^k$$

can be combined to provide an iterative reconstruction algorithm for signal reconstruction, see Algorithm 8.

Definition 5 Gram Operator *Let $\{x_n : n \in \mathbb{Z}^d\} \subseteq \mathbb{H}$ be a frame for $\mathbb{H}$ with frame operator S, frame bounds A and B, and Bessel map $L : \mathbb{H} \to l^2(\mathbb{Z}^d)$.*

a. *The Gram operator associated with $\{x_n\}$ is the map $R = LL^* : l^2(\mathbb{Z}^d) \to l^2(\mathbb{Z}^d)$.*
b. *Let L' and R' denote the Bessel map and Gram operator, respectively, associated with the dual frame $\{S^{-1}x_n\}$.*
c. *It is easy to check that R restricted to $L(\mathbb{H})$ is a one-to-one map onto $L(\mathbb{H})$. If R^{-1} denotes the inverse defined on $L(\mathbb{H})$, then we can extend R^{-1} to $l^2(\mathbb{Z}^d)$ by defining the pseudo-inverse $R^\dagger$* of *R* as

$$R^\dagger = R^{-1}P_{L(\mathbb{H})} : l^2(\mathbb{Z}^d) \to L(\mathbb{H}) \subseteq l^2(\mathbb{Z}^d),$$

where $P_{L(\mathbb{H})}$ is the orthogonal projection operator onto the image of L.

Lemma 6 *Let $\{x_n : n \in \mathbb{Z}^d\} \subseteq \mathbb{H}$ be a frame for $\mathbb{H}$, with Gram operator R, frame bounds A and B, and Bessel map $L : \mathbb{H} \to l^2(\mathbb{Z}^d)$. If $0 < \lambda < 2/B$, then $\|I - \lambda R\|_{L(\mathbb{H})} < 1$. We may take $\lambda = 2/(A+B)$.*

Proof: Since $(L')^*$ is surjective (onto), for any $y \in \mathbb{H}$ there is a $c \in L'(\mathbb{H})$ so that $y = (L')^* c$. This together with the fact that $\{S^{-1}x_n\}$ is a frame for $\mathbb{H}$ yields

$$B^{-1}\langle c, R'c\rangle \leq \langle c, (R')^2 c\rangle \leq A^{-1}\langle c, R'c\rangle.$$

Letting $c = (R')^{\dagger} d$ for some $d \in l^2(\mathbb{Z}^d)$, we have $B^{-1}\langle Rd, d\rangle \leq \langle d, d\rangle \leq A^{-1}\langle Rd, d\rangle$. For all nonzero $d \in L'(\mathbb{H})$ this means

$$A \leq \frac{\langle Rd, d\rangle}{\langle d, d\rangle} \leq B.$$

Thus, we have for $\lambda > 0$ that

$$1 - \lambda B \leq \frac{\langle (I - \lambda R)d, d\rangle}{\langle d, d\rangle} \leq 1 - \lambda A,$$

and, since $I - \lambda R$ is self-adjoint,

$$\|I - \lambda R\|_{L(\mathbb{H})} = \sup_{c \in L(\mathbb{H})} \frac{|\langle (I - \lambda R)c, c\rangle|}{\langle c, c\rangle} \leq \max\{|1 - \lambda A|, |1 - \lambda B|\}. \quad (7)$$

We would like to find λ such that $\|I - \lambda R\|_{(L(\mathbb{H})} < 1$. This condition is satisfied for all $\lambda \in (0, 2/B)$. In particular, if $\lambda = 2/(A + B)$ then $|1 - \lambda A| = |1 - \lambda B| = (B - A)/(A + B) < 1$. For this choice of λ we have proved that $\|I - \lambda R\|_{L(\mathbb{H})} < 1$. □

Proposition 7 Frame Reconstruction Formula *Let $\{x_n : n \in \mathbb{Z}^d\} \subseteq \mathbb{H}$ be a frame for $\mathbb{H}$, with frame operator S, Gram operator R, frame bounds A and B, and Bessel map $L : \mathbb{H} \to l^2(\mathbb{Z}^d)$. If $\lambda \in (0, 2/B)$, e.g., if $\lambda = 2/(A + B)$, then*

$$\forall y \in \mathbb{H}, \quad y = \lambda \sum_{i=0}^{\infty} L^*(I - \lambda R)^i L y, \quad (8)$$

where $L^* c = \sum c_n x_n$ for $c = \{c_n\} \in l^2(\mathbb{Z}^d)$.

Proof: Since $\langle Lx, c\rangle = \langle x, L^* c\rangle$ and $\langle Lx, c\rangle = \sum \overline{c_n}\langle x, x_n\rangle$, we obtain the formula for $L^* c$.

Because of the Neumann expansion

$$S^{-1} = \frac{2}{A + B} \sum_{k=0}^{\infty} \left(I - \frac{2}{A + B} S\right)^k$$

and the fact that $S = L^*L$, it is sufficient to prove

$$\lambda \sum_{i=0}^{\infty} L^*(I - \lambda R)^i Ly = \sum_{i=0}^{\infty} (I - \lambda L^*L)^i (\lambda L^*L)y, \tag{9}$$

where the sums are well-defined by Lemma 6. The $i = 0$ terms are clearly the same in (9). Assume

$$\lambda L^*(I - \lambda R)^i Ly = (I - \lambda L^*L)^i (\lambda L^*L)y. \tag{10}$$

Then, using (4.5), compute

$$\begin{aligned}\lambda L^*(I - \lambda R)^{i+1} Ly &= \lambda L^*(I - \lambda R)^i Ly - \lambda L^*(I - \lambda R)^i \lambda R(Ly) \\ &= \lambda(I - \lambda L^*L)^i L^*Ly - \lambda(I - \lambda L^*L)^i L^*L(\lambda L^*Ly) \\ &= \lambda(I - \lambda L^*L)^i (I - \lambda L^*L)L^*Ly = \lambda(I - \lambda L^*L)^{i+1} L^*Ly\,,\end{aligned}$$

and the result follows by induction. □

Proposition 7 leads directly to the following theorem (Algorithm 8), which provides an iterative reconstruction procedure for the recovery of a signal y from its "sampled values" Ly. This iterative procedure converges at an exponential rate.

Algorithm 8 Frame Reconstruction Algorithm *Let* $\{x_n : n \in \mathbb{Z}^d\} \subseteq \mathbb{H}$ *be a frame for* $\mathbb{H}$*, with Gram operator* R*, frame bounds* A *and* B*, and Bessel map* L*. Let* $y \in \mathbb{H}$ *and set* $c_{(0)} = Ly \in l^2(\mathbb{Z}^d)$, $y_0 = 0$, $\lambda = 2/(A+B)$, *and* $\alpha = \|I - \lambda R\|_{L(\mathbb{H})} < 1$. *Define* $y_m, u_m \in \mathbb{H}$, *and* $c_{(m)} \in L(\mathbb{H})$, $m = 0, 1 \ldots,$ *recursively, as*

$$u_m = \lambda L^* c_{(m)}, \quad c_{(m+1)} = c_{(m)} - Lu_m,$$

and

$$y_{m+1} = y_m + u_m.$$

Then

$$\forall m \in N, \quad \|y - y_m\| < \alpha^m \frac{B}{A} \|y\|,$$

and, in particular, $\lim_{m\to\infty} y_m = y$ *in* $\mathbb{H}$.

Proof:

(i) An elementary induction argument shows that

$$\forall\ m = 0, 1, \ldots, \quad y_{m+1} = \lambda L^* \left(\sum_{i=0}^{m} (I - \lambda R)^i \right) c_{(0)}.$$

Consequently, by Proposition 7, we have $\lim_{m\to\infty} y_m = y$ in $\mathbb{H}$.

(ii) For any fixed $m \geq 0$ and for any $k \in N$,

$$\|y - y_m\| = \|y + (y_{m+1} - y_m) + \cdots + (y_{m+k+1} - y_{m+k}) - y_{m+k+1}\|$$
$$\leq \|y - y_{m+k+1}\| + \sum_{i=0}^{k} \|y_{m+i+1} - y_{m+i}\|.$$

Using part i and taking the lim sup as $k \to \infty$, see e.g., page 278 of [3], we obtain

$$\|y - y_m\| \leq \sum_{k \geq m} \|y_{k+1} - y_k\|,$$

from which we compute, using part i for the first step and (7) for the last, that

$$\|y - y_m\| \leq \sum_{k \geq m} \left\| \lambda L^*(I - \lambda R)^k L y \right\|$$
$$\leq \sum_{k \geq m} \lambda \|L^*\| \left\| (I - \lambda R)^k \right\|_{L(\mathbb{H})} \|L\| \|y\|$$
$$\leq \lambda B \left(\sum_{k \geq m} \alpha^k \right) \|y\| = \left(\frac{\alpha^m}{1 - \alpha} \right) \lambda B \|y\| \leq \alpha^m \frac{B}{A} \|y\|. \qquad \square$$

Algorithm 8 underscores the importance of the discrete nature of the Gram operator R in the reconstruction process. Also, formally, we may rewrite (8) as

$$y = (L^* R^{-1}) L y, \tag{11}$$

cf., (6). It should be pointed out that the implementation of Algorithm 8, in terms of finite matrices approximating R, is sometimes difficult, see [17, 26, 43].

A crucial element in the proof of Algorithm 8 is the fact that the sampled data $c_{(0)}$ has the form $c_{(0)} = Ly$. If $c_{(0)}$ is not entirely in $L(\mathbb{H})$, then the algorithm will not converge. An analysis of this latter situation is found in Section 14.6 of [44], and it is related to noise reduction.

Definition 9 Wavelet Systems and Frames *Let $\psi \in L^2(\mathbb{R})$. The affine system* or *wavelet system for ψ is the sequence $\{\psi_{m,n} : (m, n) \in \mathbb{Z} \times \mathbb{Z}\}$, where, for $a > 1$,*

$$\psi_{m,n}(t) = a^{m/2} \psi(a^m t - n).$$

The Fourier transform of $\psi_{m,n}$ is computed to be

$$\Psi_{m,n}(f) = a^{-m/2} e^{-2\pi j n (f/a^m)} \Psi(f/a^m) = a^{-m/2} (e_{-n} \Psi)(f/a^m),$$

where

$$a^{-m/2}(e_{-n}\Psi)(f/a^m) = a^{-m/2}e_{-n}(f/a^m)\Psi(f/a^m).$$

If the wavelet system $\{\psi_{m,n}\}$ *for* ψ *is a frame, then it is a wavelet frame.*

Definition 10 Fourier Frames

a. *Let* $\{b_m\} \subseteq \mathbb{R}$. *If* $\{e_{b_m}\}$ *is a frame for* $L^2[-T, T]$, *it is called a* ***Fourier frame*** for $L^2[-T, T]$. In this situation we also say that $\{b_m\}$ is a Fourier frame for $PW_T \subseteq L^2(\hat{\mathbb{R}})$.
b. *Given* $g \in L^2(\mathbb{R})$ *and sequences* $\{a_n\}, \{b_m\} \subseteq \mathbb{R}$. *If* $\{e_{b_m}\tau_{a_n}g\}$ *is a frame for* $L^2(\mathbb{R})$, *it is called a weighted Fourier frame* for $L^2(\mathbb{R})$ with weight g or a *Gabor frame* for $L^2(\mathbb{R})$ depending on the window function g.

Example 11 WAM Wavelet Frame Analysis
The following calculation illustrates to what extent $\{\psi_{m,n}\}$, *defined in (3.3), can be considered a wavelet frame for some sufficiently robust Hilbert space* $\mathbb{H} \subseteq L^2(\mathbb{R})$, *cf., the critical observation in Section 14.3 that* $\{\psi_{m,n}\}$ *depends on y. We first compute*

$$\sum_n |\langle y, \psi_{m,n}\rangle|^2 = \frac{1}{(a-1)^2}\sum_n \left|\langle \overline{YD_{s_{m+1}^{-1}}(\partial\tilde{g})^\wedge}, e_{-t(n;s_m)}\rangle\right|^2.$$

Then, we assume that for each $m \in \mathbb{Z}$, $\{-e_{-t(n;s_m)} : n \in \mathbb{Z}\}$ *is a Fourier frame with frame bounds* A_m, B_m. *Thus,*

$$\sum_m A_m \left\| YD_{s_{m+1}^{-1}}(\partial\tilde{g})^\wedge \right\|^2 \leq (a-1)^2 \sum_m \sum_n |\langle y, \psi_{m,n}\rangle|^2$$
$$\leq \sum_m B_m \left\| YD_{s_{m+1}^{-1}}(\partial\tilde{g})^\wedge \right\|^2.$$

Consequently, if we suppose that

$$0 < A \leq \frac{1}{(a-1)^2}A_m \leq \frac{1}{(a-1)^2}B_m \leq B < \infty,$$

0 for some A, B, *then by a simple calculation and the Parseval-Plancherel Theorem (Rayleigh Theorem), we have*

$$A\left(\inf_f \sum_m \left|D_{s_{m+1}^{-1}}(\partial\tilde{g})^\wedge(f)\right|^2\right)\|y\|^2$$
$$\leq |\langle y, \psi_{m,n}\rangle|^2 \leq B\left\|_m \left|D_{s_{m+1}^{-1}}(\partial\tilde{g})^\wedge(f)\right|^2\right\|_\infty \|y\|^2. \tag{12}$$

The inequalities in (4.7) lead to frame properties of $\{\psi_{m,n}\}$ *if*

$$H(f) = \sum_n \left| D_{s_{m+1}^{-1}} (\partial \tilde{g})^\wedge (f) \right|^2 \tag{13}$$

is bounded above and bounded below away from 0 independently of f. In any case, the function in (13) must be quantified to obtain effective frame decompositions by means of Theorem 2; and it should be noted that the scaling constant a plays a role in (13). The mammalian cochlear filter G satisfies (13), see the examples in [8].

14.5. Irregular Sampling and Fourier Frames

In this section we shall state and prove an irregular sampling expansion by frame methods. We begin with the following result, see e.g., [2, 7].

Theorem 12 Fourier and Gabor Frames *Let* $g \in PW_\Omega$ *for a given* $\Omega > 0$. *Assume that* $\{a_n\} \subseteq \mathbb{R}$, $\{b_m\} \subseteq \hat{\mathbb{R}}$ *are real sequences for which*

$$\{e_{a_n}\} \text{ is a Fourier frame for } L^2[-\Omega, \Omega],$$

and that there exist $A, B > 0$ *such that*

$$0 < A \leq H(f) \leq B < \infty \text{ almost everywhere on } \hat{\mathbb{R}},$$

where

$$H(f) = |G(f - b_m)|^2.$$

Then $\{e_{a_n} \tau_{b_m} G\}$ *is a frame for* $L^2(\hat{\mathbb{R}})$ *with frame operator S; and* $\{e_{a_n} \tau_{b_m} G\}$ *is a tight frame for* $L^2(\hat{\mathbb{R}})$ *if and only if* $\{e_{a_n}\}$ *is a tight frame for* $L^2[-\Omega, \Omega]$ *and H is constant almost everywhere on* $\hat{\mathbb{R}}$.

Corollary 13 *Let us assume the hypotheses and notation of Theorem 12, and set* $I_m = [-\Omega, \Omega] + b_m$. *Then, for each fixed m,* $\{\tau_{b_m} e_{a_n}\}$*is a frame for* $L^2(I_m)$ *with frame operator* S_m, *and*

$$\forall y \in L^2(\mathbb{R}), \quad SY = \sum (\tau_{b_m} G) S_m (Y \tau_{b_m} \overline{G}) \quad \text{in } L^2(\hat{\mathbb{R}}).$$

Proof: We compute

$$\begin{aligned}
SY &= \sum_m \sum_n \langle Y, e_{a_n}\tau_{b_m}G\rangle e_{a_n}\tau_{b_m}G \\
&= \sum_m (\tau_{b_m}G)\Big(\sum_n \langle Y, e_{a_n}\tau_{b_m}G\rangle e_{a_n}\Big)\mathbf{1}_{I_m} \\
&= \sum_m (\tau_{b_m}G)\Bigg(\sum_n \langle Y\tau_{b_m}\overline{G}, e_{a_n}\rangle_{I_m} e_{a_n}\mathbf{1}_{I_m}\Bigg) \\
&= \sum_m (\tau_{b_m}G)S_m(Y\tau_{b_m}\overline{G}). \qquad \square
\end{aligned}$$

We can now prove the Yao-Thomas irregular sampling theorem in terms of exact frames, see [47].

Theorem 14 Yao-Thomas Formula *Let $\{e_{a_n}\}$ be an exact frame for $L^2[-\Omega, \Omega]$ for a given $\Omega > 0$ and sequence $\{a_n\} \subseteq \mathbb{R}$. Define the sampling function s_n in terms of its involution $\tilde{s}_n(t) = \overline{s_n}(-t)$, where*

$$\forall t \in \mathbb{R}, \quad \tilde{s}_n(t) = \int_{-\Omega}^{\Omega} \overline{h_n}(f)e^{2\pi jtf}\,df,$$

and where $\{h_n\} \subseteq L^2[-\Omega, \Omega]$ is the unique sequence for which $\{e_{a_n}\}$ and $\{h_n\}$ are biorthonormal. (In particular, $\tilde{s}_n \in PW_\Omega$). If $t_n = -a_n$, then

$$\forall y \in PW_\Omega, \quad y = \sum y(t_n)s_n \quad \text{in } L^2(\mathbb{R}).$$

Proof: Let $g = (2\Omega)^{-1/2}d_{2\pi\Omega}$, and set $b_m = 2m\Omega$. Since $\{e_{a_n}\}$ is a frame we can apply Theorem 12 and, hence, $\{e_{a_n}\tau_{b_m}G\}$ is a frame for $L^2(\hat{\mathbb{R}})$ with frame operator S. In particular,

$$\forall h \in L^2(\mathbb{R}), \quad \hat{h} = \sum \langle h, e_{a_n}\tau_{b_m}G\rangle S^{-1}(e_{a_n}\tau_{b_m}G) \quad \text{in } L^2(\hat{\mathbb{R}}). \tag{14}$$

We obtain

$$\langle \hat{y}, e_{a_n}\tau_{b_m}G\rangle = \begin{cases} (2\Omega)^{-1/2}y(-a_n) & \text{if } m = 0 \\ 0 & \text{if } m \neq 0 \end{cases} \tag{15}$$

for $y \in PW_\Omega$. By means of Corollary 13 we can then verify that

$$S^{-1} = 2\Omega S_0^{-1} \quad \text{on } L^2[-\Omega, \Omega],$$

where S_0 is defined in Corollary 13. Thus, since $g \in PW_\Omega$, we compute

$$S^{-1}(e_{a_n}\tau_{b_0}G) = (2\Omega)^{1/2}S_0^{-1}(e_{a_n}\mathbf{1}_{(\Omega)}),$$

so that, by the exactness hypothesis and part *a* of Theorem 2, the right side is

$$(2\Omega)^{1/2}\sum_m \langle e_{a_n}, h_m\rangle_{[-\Omega,\Omega]} h_m = (2\Omega)^{1/2} h_n.$$

Combining these two equalities with (5.1) and (5.2) gives the reconstruction,

$$\forall y \in PW_\Omega, \quad \hat{y} = \sum_n (2\Omega)^{-1/2} y(-a_n)(2\Omega)^{1/2} h_n \quad \text{in } L^2(\mathbb{R}), \qquad \square$$

and the result follows.

Remark 15 Perspective on Fourier Frames and Irregular Sampling

a. *The Yao-Thomas result, formulated in terms of exact frames, is an irregular sampling theorem in that the coefficients of the decomposition are really sampled values. The assertion about biorthonormality in Theorem* 14 *shows the relation betwen Fourier frames and the Yao-Thomas decomposition. The in-depth study of the Fourier frame case is due to Beurling, e.g.,* [13], *and Landau* [29], *and is treated in terms of multidimensional irregular sampling in* [11, 12].

b. *The three assertions in this remark deal with a density criterion measuring the uniform distance between* $n/(2\Omega)$, $n \in \mathbb{Z}$, *and elements* a_n *of the sampling set. This particular idea is due to Duffin and Schaeffer* [22] *for frames, but goes back to Wiener* (1927) *for determining the closure of linear spans. Although seemingly weaker than obtaining frame decompositions, some of the most formidable analysis of the 20th century is associated with such closure issues, e.g.,* [14–15, 30, 35]. *A beautiful exposition of the following assertions is due to Young* [48].

(i) *Let* $\{a_n\} \subseteq \mathbb{R}$. *Levinson* [30] *proved that if*

$$\sup\left|a_n - \frac{n}{2\Omega}\right| \leq \frac{1}{4}\left(\frac{1}{2\Omega}\right),$$

then $\overline{span}\{e_{a_n}\} = L^2[-\Omega, \Omega]$.

(ii) *Further, Kadec's* $\frac{1}{4}$*-Theorem (1964) asserts that if*

$$\sup\left|a_n - \frac{n}{2\Omega}\right| \leq L < \frac{1}{4}\left(\frac{1}{2\Omega}\right),$$

then $\{e_{a_n}\}$ *is an exact frame for* $L^2[-\Omega, \Omega]$

(iii) *Kadec's* $\frac{1}{4}$*-Theorem is sharp in the sense that there exists a sequence* $\{a_n\} \subseteq \mathbb{R}$ *such that*

$$\sup\left|a_n - \frac{n}{2\Omega}\right| = \frac{1}{4}\left(\frac{1}{2\Omega}\right),$$

and therefore $\{e_{a_n}\}$ *is complete in* $L^2[-\Omega, \Omega]$, *but* $\{e_{a_n}\}$ *is not an exact frame for* $L^2[-\Omega, \Omega]$

c. *If* $\{t_n\} \subseteq \mathbb{R}$ *be a strictly increasing sequence for which* $\lim_{n\to\pm\infty} t_n = \pm\infty$, *and for which*

$$\exists d > 0 \quad \text{such that } \forall m \neq n, \quad |t_m - t_n| \geq d,$$

then $\{t_n\} \subseteq \mathbb{R}$ *is said to be uniformly discrete. A uniformly discrete sequence* $\{t_n\}$ *is uniformly dense with uniform density* $\Delta > 0$ *if*

$$\exists L > 0 \quad \text{such that } \forall n \in \mathbb{Z}, \left| t_n - \frac{n}{\Delta} \right| \leq L.$$

d. *Using a theorem due to Duffin-Schaeffer* [22] *for one direction, Jaffard* [28] *has provided the following characterization of frames* $\{e_{-t_n}\}$ *for* $L^2[-\Omega, \Omega]$. *Let* $\{t_n\} \subseteq \mathbb{R}$ *be a strictly increasing sequence for which* $\lim_{n\to\pm\infty} t_n = \pm\infty$, *and let* $I \subseteq \mathbb{R}$ *denote an interval.*

   1. *The following two assertions are equivalent:*

      (a) *There is* $I \subseteq \mathbb{R}$ *for which* $\{e_{-t_n}\}$ *is a frame for* $L^2(I)$.

      (b) *The sequence* $\{t_n\}$ *is a disjoint union of a uniformly dense sequence with uniform density* Δ *and a finite number of uniformly discrete sequences.*

   2. *In the case assertion b of part 1 holds, then* $\{e_{-t_n}\}$ *is a frame for* $L^2(I)$ *for each* $I \subseteq \mathbb{R}$ *for which* $|I| < \Delta$.

The Classical Sampling Theorem, often associated with the names Whittaker, Kotel'nikov, Shannon et al., goes back to Cauchy (1841), *e.g.,* [4], *and it provides a sampling formula of the form*

$$y = \sum Ty(nT)\tau_{nT}\theta$$

when $y \in PW_\Omega$, $2T\Omega \leq 1$, *and the sampling function* θ *satisfies some natural conditions, see Theorem 3.10.10 of* [3]. *Theorem* 17 *below was proved with Heller in* [7], *and gives an analogue of the Classical Sampling Theorem for irregularly spaced sampling sets* $\{t_n\}$.

Lemma 16 *Given* $y, y_n \in L^2(\mathbb{R})$, *and assume* $y = y_n$ *in* $L^2(\mathbb{R})$. *If* $x \in L^\infty(\mathbb{R})$ *then* $xy = \sum xy_n$ *in* $L^2(\mathbb{R})$.

Theorem 17 An Irregular Sampling Theorem *Suppose* $\Omega > 0$ *and* $\Omega_1 > \Omega$, *and let* $\{t_n\} \subseteq \mathbb{R}$ *have the property that* $\{e_{-t_n}\}$ *is a Fourier frame for* $L^2[-\Omega_1, \Omega_1]$ *with frame operator S. Further, let* $\theta \in L^2(\mathbb{R})$ *have the properties that* $\hat{\theta} \in L^\infty(\mathbb{R})$,

$$\text{supp } \hat{\theta} \subseteq [-\Omega_1, \Omega_1],$$

and $\hat{\theta} = 1$ *on* $[-\Omega, \Omega]$. *Then*

$$\forall y \in PW_\Omega, \quad y = \sum c_n(y)\tau_{t_n}\theta \quad \text{in } L^2(\mathbb{R}), \tag{16}$$

where

$$c_n(y) = \langle S^{-1}(Y\mathbf{1}_{(\Omega_1)})e_{-t_n}\rangle. \tag{17}$$

Proof: Since $\{e_{-t_n}\}$ is a Fourier frame for $L^2[-\Omega_1, \Omega_1]$ and *supp* $Y \subseteq [-\Omega, \Omega]$, we have

$$Y = Y\mathbf{1}_{(\Omega_1)} = \sum \langle S^{-1}(Y\mathbf{1}_{(\Omega_1)}), e_{-t_n}\rangle_{[-\Omega_1,\Omega_1]} e_{-t_n}\mathbf{1}_{(\Omega_1)} \quad \text{in } L^2(\mathbb{R}). \tag{18}$$

In this expression, we note that S^{-1}, being positive, is self-adjoint so that the frame expansion in Theorem 2 gives rise to (18). Also, the convergence in $L^2[-\Omega_1, \Omega_1]$ from our frame hypothesis can be taken in $L^2(\hat{\mathbb{R}})$ by extending all functions to be zero outside $[-\Omega_1, \Omega_1]$.

We have $Y = Y\Theta$ on $\hat{\mathbb{R}}$ since $\Theta = 1$ on $[-\Omega, \Omega]$ and $Y = 0$ off of $[-\Omega, \Omega]$. Further,

$$\begin{aligned} &\Theta \sum \langle S^{-1}(Y\mathbf{1}_{(\Omega_1)}), e_{-t_n}\rangle_{[-\Omega_1,\Omega_1]} e_{-t_n}\mathbf{1}_{(\Omega_1)} \\ &\quad = \sum \langle S^{-1}(Y\mathbf{1}_{(\Omega_1)}, e_{-t_n}\rangle_{[-\Omega_1,\Omega_1]} e_{-t_n}\mathbf{1}_{(\Omega_1)}\Theta \quad \text{in } L^2(\hat{\mathbb{R}}) \end{aligned}$$

by Lemma 16. Thus, since *supp* $\Theta \subseteq [-\Omega_1, \Omega_1]$, we obtain

$$\begin{aligned} Y &= Y\Theta \\ &= \sum_n \langle S^{-1}(Y\mathbf{1}_{(\Omega_1)}), e_{-t_n}\rangle_{[-\Omega_1,\Omega_1]} e_{-t_n}\Theta \quad \text{in } L^2(\hat{\mathbb{R}}). \end{aligned}$$

Taking the inverse Fourier transform gives (16). □

Remark 18 *Let $\{e_{-t_n}\}$ be a Fourier frame for $L^2[-\Omega_1, \Omega_1]$ with frame bounds A and B and frame operator S. In general we cannot write $c_n(y) = y(t_n)$ in (16). However, Theorem 17 is an irregular sampling theorem in the sense that the coefficients $c_n(y)$ can be described in terms of values of y on the irregularly spaced sampling set $\{t_n\}$. From the Neumann expansion,*

$$S^{-1} = \frac{2}{(A+B)} \sum_{k=0}^{\infty} \left(I - \frac{2}{(A+B)} S\right)^k,$$

we have

$$c_n(y) = \frac{2}{(A+B)} \sum_{k=0}^{\infty} \left\langle \left(I - \frac{2}{(A+B)}\right)^k (Y\mathbf{1}_{(\Omega_1)}), e_{-t_n} \right\rangle. \tag{19}$$

If we truncate this expression after the $k = 0$ term, we obtain the sampled value

$$\frac{2}{(A+B)} y(t_n) \tag{20}$$

as an approximation of (19).

Remark 19 In the case of regular sampling we can use a frame analysis similar to the proof of Theorem 17 to prove the formula

$$\forall y \in L^2(\mathbb{R}), \quad y = T\sum_{m,n} \langle Y, e_{nT}\tau_{mb}\Theta\rangle \tau_{-nT}(e_{mb}s) \quad \text{in } L^2(\mathbb{R}), \qquad (21)$$

where $T, \Omega > 0$ are constants for which $0 < 2T\Omega \leq 1$, $\theta \in PW_{1/(2T)}$ has the properties that $\Theta \in L^\infty(\hat{\mathbb{R}})$ and $\Theta = 1$ on $[-\Omega, \Omega]$, and, in case $2T\Omega < 1$, Θ is continuous and

$$|\Theta| > 0 \quad \text{on} \left(-\frac{1}{2T}, -\Omega\right] \cup \left[\Omega, \frac{1}{2T}\right).$$

In dealing with high frequency information, with close fluctuations, it is necessary to sample closely in order to capture all of the fluctuations. By definition, then, in the case with very high frequencies, thought of as "infinite frequencies", and hence nonbandlimited, we cannot reconstruct the function with a discrete set of samples. However, the frame reconstruction formula (21) gives the Classical Sampling Theorem for bandlimited functions, as well as giving signal representation for nonbandlimited functions. In this latter case, there is added complexity in the coefficients necessary to deal with "infinite frequencies". Equation (21) also allows us to interpret aliasing in a quantitative way, see [2, 7].

14.6. Filter Design

14.6.1. Cochlear Filters

As mentioned in Section 14.3, the shape of $|G|$, where g is the impulse response of the cochlear system, is critical for the effectiveness of the auditory process, and generally G has an asymmetrical "shark-fin" shaped amplitude with faster rate of decay on the high frequency side than on the low frequency side. All realizable systems, such as our filter bank with "shark-fin" shaped amplitudes, are necessarily causal. In particular, the cochlear filter bank cannot characterize (reconstruct) future utterances in terms of known (present) speech signals. As such, we design *causal filters* $G \in L^2(\hat{\mathbb{R}})$, i.e., $\operatorname{supp} g \subseteq [0, \infty)$, for which G has the required "shark-fin" shaped amplitude consistent with mammalian auditory models. Our point of view is that such filters provide a realistic mathematical model for the cochlear filters described in Section 14.3, and are therefore the proper filters for optimizing the reconstruction process inherent in WAM.

The starting point for the design of such causal filters is the Paley-Wiener Logarithmic Integral Theorem, i.e., Theorem XII of [35].

Theorem 20 Paley-Wiener Logarithmic Integral Theorem *Let* $A \in L^2(\hat{\mathbb{R}})\backslash\{0\}$ *be non-negative on* $\hat{\mathbb{R}}$. $A(f) = |G(f)|$ *almost everywhere for some causal filter* $G \in L^2(\hat{\mathbb{R}})$ *if and only if*

$$\int \frac{|\log A(f)|}{1+f^2}\,df < \infty. \tag{22}$$

Let $A \in L^2(\hat{\mathbb{R}})$ *satisfy* (22), *and define*

$$\phi(x,f) = \frac{1}{\pi}\int \frac{x \log A(\lambda)}{x^2 + (f-\lambda)^2}\,d\lambda.$$

Clearly, ϕ *is harmonic in the half-plane* $x > 0$. *If* θ *is a conjugate harmonic function of* ϕ, *then it is unique up to an additive constant; and we shall construct below a particular* θ *in* (25). *The functions* ϕ *and* θ *satisfy the Cauchy-Riemann equations, and* $K(z) = \phi(x,f) + j\theta(x,f)$, where $z = x + jf$, *is an analytic function in the half-plane* $x > 0$. *We let*

$$p(f) = \frac{1}{\pi}\frac{1}{(1+f^2)}$$

and consider the L^1*-dilation (by* $1/x$*),*

$$p_{1/x}(f) = \rho(x,f) = \frac{1}{\pi}\frac{x}{x^2+f^2}, \quad x > 0.$$

Thus, $\lim_{x\to 0} p_{1/x} = \delta$ *distributionally, in fact, in the* $\sigma(M_b, C_0)$ *topology, where* C_0 *is the space of continuous functions vanishing at* $\pm\infty$ *and* M_b *is the space of bounded Radon measures on* $\hat{\mathbb{R}}$, *e.g.,* [1]. *By the definition of* ϕ *we have*

$$\phi(x,f) = p_{1/x} * (\log A)(f), \quad x > 0, \tag{23}$$

and, because of the approximate identity $p_{1/x}$, *a classical calculation yields*

$$\lim_{x\to 0^+} \phi(x,f) = \log A(f) \quad \textit{almost everywhere}; \tag{24}$$

see e.g., [27, 37, 42].

The harmonic function

$$K(x,f) = \frac{-1}{\pi}\frac{f}{x^2+f^2}, \quad x > 0$$

is a conjugate harmonic function of ρ *and so the Cauchy-Riemann equations,* $\partial_x\rho = \partial_f K$ *and* $\partial_f \rho = -\partial_x K$, *are valid in the half-plane* $x > 0$. *Using* (23), *the equations,*

$$\partial_x\phi = (\partial_x\rho)*_\lambda \log A$$

and

$$\partial_f\phi = (\partial_f\rho)*_\lambda \log A, \quad x > 0,$$

follow from (24), *where "$*_\lambda$" designates convolution in the second variable of* ρ. *Thus, we define*

$$\theta = K *_\lambda \log A, \quad x > 0. \tag{25}$$

The function

$$J(z) = e^{K(z)}, \quad z = x + jf,$$

is analytic in the half-plane $x > 0$, *and provides the solution asserted in Theorem 20 in the following sense. By* (24), *we formally compute*

$$J(jf) = A(f)e^{j\theta(0,f)} \quad \textit{almost everywhere,} \tag{26}$$

and note, by (25), *that*

$$\theta(0,f) = \frac{-1}{\pi} \int \frac{\log A(\lambda)}{f - \lambda} d\lambda \tag{27}$$

is formally the Hilbert transform of $-\log A$. *It turns out that condition* (22) *allows us to assert the existence of a causal filter* $G \in L^2(\hat{\mathbb{R}})$ *for which* $G(f) = J(jf)$ *almost everywhere. The actual filter design is a consequence of* (26) *and* (27), *and is formulated in the following result.*

Theorem 21 Construction of Causal Filter *Let* $A \in L^2(\hat{\mathbb{R}})\backslash\{0\}$ *be nonnegative on* $\hat{\mathbb{R}}$, *and assume condition* (22). *Then the function*

$$G = Ae^{-j\mathcal{H}(\log A)} \tag{28}$$

is a causal filter in $L^2(\hat{\mathbb{R}})$, *i.e.,* $g \in L^2(\mathbb{R})$ *and* $\operatorname{supp} g \subseteq [0, \infty)$, *where* $\mathcal{H}(\log A)$ *is the Hilbert transform of* $\log A$.

Example 22 WAM Filter *Take* $F(f) = mf\mathbf{1}_{[0,\Gamma)}(f)$. *Let* $\rho \geq 0$ *be compactly supported with the property that* $\int \rho(f)df = 1$. *Then consider the nonnegative function* $A_\rho = F * \rho$. *The cochlear filters for WAM use Theorem 21 and* A_ρ *in the following way. Note that* $\operatorname{supp} F = [0, \Gamma]$, *and choose* $N \gg \Gamma$. *Let* $\epsilon(f)$ *be an even function on* $(-\infty, -N] \cup [N, \infty)$ *defined by*

$$\forall f \geq N, \quad \epsilon(f) = e^{f/(\log^2 f)}.$$

Then, for the function A in Theorem 21, we set $A = A_\rho$ *on* $[0, \Gamma)$, *and let* $A(f) = A_\rho(0)$ *on* $[-N, 0] \cup [\Gamma + \epsilon, N]$, *where* $\epsilon > 0$ *is small. Finally, let* $A(f) = e^{-f/\log^2 |f|}$ *if* $|f| > N$. *Clearly,* $A \in L^2(\hat{\mathbb{R}})$ *and* (22) *is valid. Thus, the causal cochlear filter G can be defined by* (22) *in Theorem 21.*

14.6.2. Other Filters

The cochlear filter suppresses white noise in a way we shall comment on in Section 14.7, but other filter designs can be implemented to reduce other types of noises. Given a signal y and a set $\{\psi_{m,n}\}$ of functions, we say that y is coherent or "non-noisy" with respect to $\{\psi_{m,n}\}$ if y may be effectively approximated by a linear combination of a relatively small number of elements of $\{\psi_{m,n}\}$, cf., [32]. With this point of view, noise in a signal is that part which lacks coherence with respect to $\{\psi_{m,n}\}$. If $\{\psi_{m,n}\}$ is a frame for a large enough space of finite energy signals, then this view of coherent signal versus noise admits a nonlinear thresholding algorithm, inspired by mammalian auditory systems, which allows for the recovery of a coherent signal embedded in noise.

For simplicity, we shall assume additive noise, i.e., if y_0 is coherent with respect to $\{\psi_{m,n}\}$ and η is noise, then we express the signal y as $y_\eta = y_0 + \eta$. We interpret the norm equivalence property of frames as an approximate energy preservation between the signal y_η and its digitized version or sampled values $Ly_\eta = Ly_0 + L\eta$, where L is the Bessel map associated with $\{\psi_{m,n}\}$, i.e., $Ly = \{\langle y, \psi_{m,n}\rangle\}$ and

$$\langle y, \psi_{m,n}\rangle = -\frac{1}{a-1} W_{\partial_t g} y(t(n; s_m), s_{m+1}).$$

With this interpretation and the fact that $\{\psi_{m,n}\}$ depends on g, we have the following problem. For a given noise η construct a filter G so that the coefficients $\{L\eta(m, n)\}$ are small.

For a given time t, scale s, and parameter $a > 1$, we have

$$\begin{aligned} -\frac{1}{a-1} W_{\partial_t g} y(t, s) &= \frac{\sqrt{s}}{1-a} \int y()(\partial g)(s(t-))d \\ &= C_s(y)(t). \end{aligned} \tag{29}$$

Thus, for a noise η, we can use the Parseval-Plancherel Theorem to compute

$$C_s(\eta)(t) = \frac{2\pi j}{1-a}\frac{1}{s} \int f\hat{\eta}(f)(D_{s^{-1}}G)(f)e^{2\pi jtf}\,df. \tag{30}$$

We shall consider noises η_α which are $1/f$-*processes*, e.g., [19]. These are noises whose generalized power spectra $S_\alpha(f)$ satisfy

$$\frac{b}{|f|}\alpha \le S_\alpha(f) \le \frac{c}{|f|}\alpha, \tag{31}$$

where $\alpha \in [0, 2]$ is fixed. In a very simple model, if $b = c = 1$, then η_0 is white noise and η_2 is Brownian motion.

14.6.2.1. 1/f-noises

Let $y_{\eta_\alpha} = y_0 + \eta_\alpha$, where η_α is a $1/f$*-process* defined by (31). We assume that η_α is uncorrelated with respect to the signal y_0, and, hence, $y_0 * \tilde{\eta}_\alpha$ is small, where $\tilde{\eta}_\alpha$ is the involution of η_α. Thus, recalling that $\{\psi_{m,n}\}$ is dependent on g, our goal is to construct G so that the coefficients

$$C_{s_{m+1}}(\eta_\alpha)(t(n; s_m)), \quad m, n \in \mathbb{Z}, \tag{32}$$

defined by (30), are small. Such a construction presents theoretical difficulties, and so, because of the above assumption on the noncorrelation of y_0 and η_α, we shall solve the design of G problem in the case that the coefficients

$$C_{s_{m+1}}(\eta_\alpha * \tilde{\eta}_\alpha)(t(n; s_m)) \tag{33}$$

are small. A minimization of (32) yields noise suppression for η_α-noise contaminating signals y_0. Our minimization of (33) provides an approximation to the desired minimization of (32). Note that

$$\begin{aligned} C_s(\eta_\alpha * \tilde{\eta}_\alpha)(t) &= \frac{2\pi j}{1-a}\frac{1}{s}\int f \frac{1}{|f|^\alpha}(D_{s^{-1}}G)(f)e^{2\pi jtf}\,df \\ &= \frac{2\pi j}{1-a}\frac{1}{\sqrt{s}}[[(\operatorname{sgn} f)|f|^{1-\alpha}]^\vee(u) * g(us)](t). \end{aligned} \tag{34}$$

For the case of white noise $\alpha = 0$, the term $(\operatorname{sgn} f)|f|^{1-\alpha}$, on the right side of (34), becomes f and we have the Fourier pairing,

$$\frac{1}{2\pi j}\delta'(t) \leftrightarrow f,$$

where δ' is the derivative of the impulse, i.e., the dipole, at the origin. Thus,

$$\begin{aligned} C_{s_{m+1}}(\eta_0 * \tilde{\eta}_0)(t) &= \frac{1}{(1-a)\sqrt{s_{m+1}}}(\delta'(u) * g(us_{m+1}))(t) \\ &= \frac{1}{1-a}D_{s_{m+1}}g'(t). \end{aligned} \tag{35}$$

We want to construct g so that

$$C_{s_{m+1}}(\eta_0 * \tilde{\eta}_0)(t(n; s_m)) \tag{36}$$

is small whenever

$$W_{\partial_t g}(\eta_0 * \tilde{\eta}_0)(t(n; s_m), s_m) = 0. \tag{37}$$

The points $t(n; s_m)$ are defined by (37). By (29), this means we want the quantity defined in (36) to be small whenever

$$C_{s_m}(\eta_0 * \tilde{\eta}_0)(t(n; s_m)) = 0. \tag{38}$$

The left side of (38) is

$$\frac{1}{1-a} D_{s_m} g'(t(n; s_m))$$

by (35). Thus, if $D_{s_m} g'(t) = a^{m/2} g'(a^m t) = 0$, we want to conclude that

$$\frac{1}{1-a} D_{s_{m+1}} g'(t) = \frac{a^{(m+1)/2}}{1-a} g'(a^{m+1} t)$$

is small, and this is the criterion used to define g.

14.6.2.2. Noise reduction for $1/f$*-noise*

For the case of $1/f$*-noise* $\alpha = 1$, the term $(sgn\, f)|f|^{1-\alpha}$, on the right side of (34), becomes $sgn f$ and we have the Fourier pairing,

$$-\frac{1}{\pi j} pv\left(\frac{1}{t}\right) \leftrightarrow sgn\, f,$$

where $pv(1/t)$ is the first order distributional principal value. Thus,

$$C_{s_{m+1}}(\eta_1 * \tilde{\eta}_1)(t)$$

is the Hilbert transform $\mathcal{H}$ of $D_{s_{m+1}} g(u)$:

$$C_{s_{m+1}}(\eta_1 * \tilde{\eta}_1)(t) = \frac{2}{a-1} p \int \frac{(D_{s_{m+1}} g)(u)}{t-u} du. \tag{39}$$

As before, we want to construct g so that the right side of (39) is small whenever

$$pv \int \frac{(D_s g)(u)}{t-u} du = 0.$$

To attack this problem, we can use the bounds of the form

$$|\mathcal{H}F(t)| \leq \frac{4}{\pi}(\log 2)\sqrt{Mm},$$

where $m = \|F'\|_{L^\infty(\mathbb{R})}$ and $|\int_b^c F(t)dt| \leq M$ for all b, c, see [31].

14.7. WAM Implementation and an Application to Speech Coding

14.7.1. WAM Implementation

Now that we have established some results from the theory of frames, as well as proving a Fourier frame based irregular sampling formula, we can address the WAM problem posed in Section 14.3.4. Given a cochlear system with impulse response g, let y be a speech signal to be processed. From (3) and (4) the corresponding WAM data is

$$\Lambda_m = \{\partial_s \partial_t W_g y(t(n; s_m), s_m) : t(n; s_m) \in Z_m\},$$

where

$$Z_m = \{t(n; s_m) : \partial_t W_g y(t(n; s_m), s_m) = 0\}.$$

Also, from Section 14.3.5 we have

$$\begin{aligned}\partial_s \partial_t W_g y(t(n; s_m), s_m) &= \frac{-1}{a-1} \langle y, \tau_{t(n;s_m)} D_{s_{m+1}}(\partial_t \tilde{g})\rangle \\ &= \langle y, \psi_{m,n}\rangle,\end{aligned}$$

where

$$\psi_{m,n} = -\frac{1}{a-1} \tau_{t(n;s_m)} D_{s_{m+1}}(\partial_t \tilde{g}).$$

Thus, if we assume $\{\psi_{m,n}\}$ to be a frame for $\mathbb{H}$ with Bessel map L, then $Ly = \langle y, \psi_{m,n}\rangle = \partial_s \partial_t W_g y(t(n, s_m), s_m)$.

By Proposition 4,

$$y = (S^{-1}L^*)Ly,$$

i.e., the signal y can be reconstructed from the discrete WAM data Ly. As mentioned in the remark following Proposition 4,

$$S^{-1} = \frac{2}{A+B}\sum_{k=0}^{\infty}\left(I - \frac{2}{A+B}S\right)^k,$$

and, hence,

$$y = (S^{-1}L^*)Ly = \sum_{i=0}^{\infty}(I - \lambda S)^i(\lambda S)y,$$

where $\lambda = 2/(A+B)$ and A, B are frame bounds. Moreover, Proposition 7 asserts that

$$y = \lambda\sum_{i=0}^{\infty}L^*(I - \lambda R)^i(Ly), \tag{40}$$

where R is the Gram operator, i.e., $R = LL^*$.

We now apply Algorithm 8 to compute the signal y when we are given the WAM data Ly. Thus, we let $c_{(0)} = Ly$ and $y_0 = 0$; and then define

$$u_m = \lambda L^* c_{(m)},$$

$$c_{(m+1)} = c_{(m)} - Lu_m,$$

and

$$y_{m+1} = y_m + u_m.$$

The algorithm gives

$$\lim_{m\to\infty} y_m = y,$$

and so we can reconstruct y from Ly.

Since our data is irregularly spaced, (40) is really an irregular sampling formula which we can think of in the following "one-dimensional" way. First, we can consider $\theta = D_{s_{m+1}}(\partial_t\tilde{g})$ as a sampling function and suppose that $\{e_{-t(n;s_m)}\}$ is a Fourier frame with frame operator S. Then we can apply Theorem 17 to obtain, as in (16),

$$\forall y \in PW_\Omega, \quad y = \sum c_{m,n}\tau_{t(n;s_m)}(D_{s_{m+1}}\partial_t\tilde{g}),$$

where $c_{m,n} = \langle S^{-1}Y, e_{-t(n;s_m)}\rangle_{L^2[-\Omega_1,\Omega_1]}$ and $0 < \Omega < \Omega_1$. The coefficients can be computed by means of (17) and several algorithms have been developed for this purpose, e.g., [26, 43]. The coefficients $c_{m,n}$ are sampled values in the sense of the approximation (20).

It is more analytically precise to think of (40) as a "two-dimensional" wavelet frame expansion using the calculations in Example 11. By "two-dimensional", we mean that we are dealing with $\{Z_m : m \in \mathbb{Z}\}$ defined in (3) as a sampling set in two-dimensional $t - s$ space. In this case, the wavelet frame expansion, which is a consequence of Example 11, is a two-dimensional irregular sampling formula whose coefficients are sampled values of y in the same way that (19) and (20) are related. In any case, the actual implementation is based on Proposition 7, i.e., (40), and Algorithm 8.

14.7.2. An Application to Speech Coding

In this subsection, we shall show how to use WAM processing in speech coding. Let y be an acoustic signal on a time interval I of duration $|I|$, and let $Ly = \{\langle y, \psi_{m,n}\rangle\}$ be the corresponding WAM data defined in Sections 14.3.4 and 14.3.5. For this discussion we shall also refer to the WAM data as the set of WAM coefficients. A basic problem in speech coding is to designate a bit rate b_r and a bit allocation b_c for transmitting sets of coefficients corresponding to a speech signal y, and to reconstruct y at this given bit rate and bit allocation. There are different goals for different versions of the problem, but one criterion of success is to obtain good reconstruction using low bit rates and, even more, to obtain good reconstruction of a given signal y embedded in certain types of noises. Naturally, since we are dealing with WAM, the sets of coefficients to be transmitted will be WAM coefficients.

Because of the auditory modelling of Section 14.3, the reconstruction theory developed in Sections 14.4 and 14.5, and the properties of the mammalian auditory filter bank given in Section 14.6, it turns out that WAM coefficients solve the aforementioned coding problem at a mathematical level, cf., [9] for its practical implementation. In particular, signal reconstruction is theoretically perfect because of the frame decompositon theory and irregular sampling formulas we have given. Further, the structure of the mammalian auditory filter bank, coupled with the thresholding technique defined below, allow for signal reconstruction of signals embedded in certain levels of white noise, since the WAM coefficients of white noise tend to spread out over the $t - s$ plane and to have low amplitude.

In order to justify these claims, suppose we are given a bit rate of b_r bits per second, and that we are also specified an allocation of b_c bits per WAM coefficient $\langle y, \psi_{m,n}\rangle$. The signal y is defined on I, but is only known to the receiver to the extent that it will receive the coefficients $\{\langle y, \psi_{m,n}\rangle\}$ at the bit rate b_r with bit allocation b_c. Since b_r and b_c are fixed, we can define a fixed transmittal coefficient rate $c_r = b_r/b_c$, i.e., given b_r and b_c, we send c_r WAM coefficients per second to the receiver. Consequently, with the coefficient rate fixed and specified, the maximum number of coefficients n_c that we are able to transmit for the function y of duration $|I|$ is

$$n_c = c_r |I|,$$

i.e., n_c is the maximum number of coefficients with which the signal y may be represented. With respect to WAM data, n_c can be related to a threshold value δ in the following way. We define the distribution function

$$\lambda(\delta) = \#\{\langle y, \psi_{m,n} \rangle \geq \delta\}$$

for $\delta \in [0, M]$, where $M = \sup\{|y, \psi_{m,n}|\}$ and # denotes cardinality. Note that we have neglected negative coefficients. The distribution function λ is monotonically decreasing and continuous from the left. As such, we may define an inverse λ^{-1} as

$$\forall n \in \mathbb{N}, \quad \lambda^{-1}(n) = \inf\{ \in [0, M] : \lambda() < n\}.$$

Hence, if we choose a threshold value δ as

$$\delta = \lambda^{-1}(n_c),$$

then we are within our bit rate and bit allocation constraint for encoding the signal y.

We now reconstuct y by the WAM implementation method of Section 14.7.1 using the set

$$\{\langle y, \psi_{m,n} \rangle \geq \delta\}$$

of thresholded WAM coefficients as the initial sequence $c_{(0)}$ (in Section 14.7.1). Many examples demonstrating the effectiveness to reconstruct y appear in [8].

Acknowledgments

The authors gratefuly acknowledge support from AFOSR Contract F49G20-96-1-0193. The first named author would also like to thank the Maryland Industrial Partnerships Program (MIPS) for its support.

References

[1] J. J. Benedetto. *Real Variable and Integration*. Stuttgart, Teubner, 1976.

[2] J. J. Benedetto. Irregular Sampling and Frames. In C. Chui, Ed., *Wavelets: A Tutorial in Theory and Applications*, Academic Press, Boston, 1992, pp. 445–507.

[3] J. J. Benedetto. *Harmonic Analysis and Applications*. CRC Press Inc., Boca Raton, FL, 1997.

[4] J. J. Benedetto. Frames, Sampling, and Seizure Prediction. In Ka-Sing Lau, Ed., *Advances in Wavelets*, Springer Verlag, New York, 1999.

[5] J. J. Benedetto and P. J. S. G. Ferreira, Eds., *Modern Sampling Theory: Mathematics and Applications*. Birkhäuser, Boston, 2001.

[6] J. J. Benedetto and M.W. Frazier, Eds., *Wavelets: Mathematics and Applications*. CRC Press Inc., Boca Raton, FL, 1994.

[7] J. J. Benedetto and W. Heller. Irregular Sampling and Frames. *Mat. Note*, 10(Supp. 1):103–125, 1990.

[8] J. J. Benedetto and A. Teolis. A Wavelet Auditory Model and Data Compression. *Applied and Computational Harmonic Analysis*, 1:3–28, 1993.

[9] J. J. Benedetto and A. Teolis. Nonlinear Methods and Apparatus for Coding and Decoding Acoustic Signals with Data Compression and Noised Suppression using Cochlear Filters, Wavelet Analysis, and Irregular Sampling Reconstructions. U. S. Patent 5,388,182, 1995.

[10] J. J. Benedetto and O. M. Treiber. Wavelet Frames: Multiresolution Analysis and Extension Principles. In L. Debnath, Ed., *Wavelet Transforms and Time-Frequency Signal Analysis*, Birkhäuser, Boston, 1999.

[11] J. J. Benedetto and H. C. Wu. A Multidimensional Irregular Sampling Algorithm and Applications. *ICASSP*, Phoenix, Arizona, 1999.

[12] J. J. Benedetto and H. C. Wu. Fourier Frame Theoretic Multidimensional Irregular Sampling, to appear.

[13] A. Beurling. *Collected Works Volume II*. Birkhäuser, Boston, 1989.

[14] A. Beurling and P. Malliavin. On Fourier Transforms of Measures with Compact Support. *Acta Math.*, 107:291–309, 1962.

[15] A. Beurling and P. Malliavin. On the Closure of Characters and the Zeros of Entire Functions. *Acta Math.*, 118:79–93, 1967.

[16] D. Childers *et al.* The Past, Present, and Future of Speech Processing. *IEEE Signal Processing Magazine*, 24–48, May 1998.

[17] O. Christensen, Finite-Dimensional Approximation of the Inverse Frame Operator. *Journal of Fourier Analysis and Applications*, 6:79–91, 2000.

[18] J. Cohen. Application of an Auditory Model to Speech Recognition. *J. Acoust. Soc. Amer.*, 85:2623–2629, 1989.

[19] D. Colella. Detection of Signals in $1/f$-*Noise* Using Wavelets, MITRE TR, 1993.

[20] I. Daubechies. *Ten Lectures on Wavelets*. Philadelphia, Pennsylvania: CBMS/NSF Series on Applied Math SIAM Publ., no. 61, 1992.

[21] I. Daubechies, A. Grossmann, and Y. Meyer. Painless Nonorthogonal Expansions. *J. Math. Phsyics*, 27:1271–1283, 1986.

[22] R. J. Duffin and A. C. Schaeffer. A Class of Nonharmonic Fourier Series. *Trans. Amer. Math. Soc.*, 72:341–366, 1952.

[23] H. G. Feichtinger and K.-H. Gröchenig. Theory and Practice of Irregular Sampling. In J. J. Benedetto and M. W. Frazier, Eds., *Wavelets: Mathematics and Applications*, CRC Press Inc, Boca Raton, FL, 1994.

[24] S. Greenberg. Acoustic Transduction in the Auditory Periphery. *Journal of Phonet.*, 16:3–18, 1988.

[25] K.-H. Gröchenig. A Discrete Theory of Irregular Sampling. *Linear Algebra and its Applications*, 193:129–150, 1993.

[26] M. L. Harrison. Frames and Irregular Sampling from a Computational Perspective. Ph.D. thesis. Dept. of Math., Univ. of Maryland, College Park, Maryland, May 1998.

[27] K. Hoffman. *Banach Spaces of Analytic Functions.* Prentice Hall, Inc., Englewood Cliffs, NJ, 1962.

[28] S. Jaffard. A Density Criterion for Frames of Complex Exponentials. *Michigan Math. J.*, 38:339–348, 1991.

[29] H. Landau. Necessary Density Condition for Sampling and Interpolation of Certain entire Functions. *Acta Math.*, 117:37–52, 1967.

[30] N. Levinson. *Gap and Density Theorems.* Amer. Math. Soc., Colloquium Publications, Providence, RI, 1940.

[31] B. Logan. Hilbert Transform of a Function having a Bounded Integral and a Bounded Derivative. *SIAM J. Math. Analysis*, 14:247–248, 1983.

[32] S. Mallat and Z. Zhang. Matching Pursuits with Time-Frequency Dictionaries. *IEEE Trans. Signal Processing*, 41:3397–3415, 1993.

[33] F. A. Marvasti. *A Unified Approach to Zero-Crossing and Nonuniform Sampling of Single and Multidimensional Systems.* Nonuniform, Oak Park, IL 60304, 1987.

[34] I. Morishita and A. Yajima. Analysis and Simulation of Networks of Mutually Inhibiting Neurons. *Kybernetik*, 11:154–165, 1972.

[35] R. E. A. C. Paley and N. Wiener. *Fourier Transforms in the Complex Domain. Amer. Math. Soc.*, Colloquium Publications, Providence, RI, 1934.

[36] H. Pollard. Completeness Theorems of Paley-Wiener Type. *Annals of Math.*, 45:738–739, 1944.

[37] P. Porcelli. *Linear Spaces of Analytic Functions.* Rand McNally, Chicago, 1966.

[38] R. Redheffer. Completeness of Sets of Complex Exponentials. *Advances in Math.*, 24: 1–63, 1977.

[39] S. Shamma. Speech Processing in the Auditory System. I. The Representation of Speech Sounds in the Responses of the Auditory Nerve. *J. Acoust. Soc. Amer.*, 78:1612–1621, 1985.

[40] S. Shamma. Speech Processing in the Aauditory System. II. Lateral Inhibition and the Central Processing of Speech Evoked Activity in the Auditory Nerve. *J. Acoust. Soc. Amer.*, 78:1622–1632, 1985.

[41] S. Shamma. Auditory Cortical Representation of Complex Acoustic Spectra as Inferred from the Ripple Analysis Method. *Network: Computation in Neural Systems*, 7:439–476, 1996.

[42] E. Stein and G. Weiss. *Fourier Analysis on Euclidean Spaces.* Princeton, Princeton University Press, NJ, 1971.

[43] T. Strohmer. Computationally Attractive Reconstruction of Band-Limited Images from Irregular Samples. *IEEE Trans. on Image Processing*, 6:540–548, 1997.

[44] A. Teolis and J. J. Benedetto. Local Frames and Noise Reduction. *Signal Processing*, 45:369–387, 1995.

[45] A. Waibel and K. F. Lee, Eds., *Readings in Speech Recognition.* San Mateo, Morgan Kaufmann Publishers, Inc., CA, 1990.

[46] X. Yang, K. Wang, and S. A. Shamma. Auditory Representations of Acoustic Signals. *IEEE Trans. Inform. Theory*, 38:2, 824–839, March 1992.

[47] K. Yao and J. Thomas. On Some Stability and Interpolatory Properties of Nonuniform Sampling Expansions. *IEEE Trans. Circuit Theory*, 14:404–408, 1967.

[48] R. M. Young. *An Introduction to Nonharmonic Fourier Series.* Academic Press, New York, 1980.

15

Application of the Nonuniform Sampling to Motion Compensated Prediction for Video Compression

A. Sharaf, F. Marvasti and M. Hasan

15.1. Introduction[1]

Motion Compensation (MC) is an essential component in most of today's video compression algorithms. It is a predictive process where a current frame is predicted from one or more previously coded frames in a sequence (also known as interframe prediction). In general, it helps to reduce the overall amount of data required to code a video sequence compared to using intraframe coding alone (where frames are coded with no reference to each other) [1].

The majority of techniques used for MC belong to the well known class of algorithms called the Block Matching Algorithms (BMA). It is well known that the BMA suffers from blocking and mosquito effects, as well as the inability to cope with uncovered background situations and non-translational motion, such as rotation and uneven stretching.

Recently, there has been an increasing interest in a more generalized class of MC algorithms based on Spatial Transformations (ST), which includes BMA as its simplest subclass. The concept behind these algorithms is known in other areas of applications such as computer graphics, image registration, optical flow

[1]This chapter is reprinted, with permission, from the paper titled: A. Sharaf and F. Marvasti, "Motion compensation using spatial transformations with forward mapping", Signal Processing Image Communication 14, pp. 209–227, 1999; © Signal Processing: Image Communications.

A. Sharaf, F. Marvasti and M. Hasan • Multimedia Laboratory, King's College London, Strand, London WC2R 2LS, UK. E-mail: farokh.marvasti@kcl.ac.uk

Nonuniform Sampling: Theory and Practice, edited by Marvasti
Kluwer Academic/Plenum Publishers, New York, 2001.

analysis, and image warping. The idea is to model motion by transformation functions which map the co-ordinates of the pixels of one image to form another image in a new co-ordinate system. The advantages of using ST based MC over the simpler BMA include the reduction of bitrate, the increased prediction quality, and the elimination of blocking and mosquito artifacts.

There are two different, but mathematically equivalent, techniques for implementing an ST to digital images. The first technique is known as the *backward mapping* [2]. It is the most widely used technique due to its implementational simplicity and the existence of fast algorithms for its computation. The second technique, known as the *forward mapping*, has received much less attention than backward mapping. This is despite the fact that it can provide adaptive ways of MC which are not possible with backward mapping [3]. The problem with its use is partially due to the fact that the spatially transformed image consists of *irregularly spaced samples*. In this chapter, we describe some techniques for recovering a regularly sampled image given a set of irregularly spaced samples.

A more general problem with the use of ST for MC is the computational load involved in finding the optimal motion parameters. In this chapter, we describe a novel fast search algorithm based on a logarithmic search tactic to calculate the ST motion parameters, which can be implemented in parallel (see the accompanying thesis of the first author on the CD-ROM, file: /Chapter15/PhDThesis.zip).

15.2. A Review of the Previous Work on ST Based MC

15.2.1. The Different Models of ST

The idea of ST is a well known mathematical concept. A function in a two-dimensional co-ordinate system can be geometrically transformed into another function such that the co-ordinates of the transformed function are functions of the co-ordinates of the original function. Thus, a function $\hat{I}(x, y)$ in a two-dimensional co-ordinate system (x, y) is transformed into another function $I(u, v)$ in a new two-dimensional co-ordinate system (u, v) such that

$$\hat{I}(x, y) = I(u, v) = I(f_1(x, y), f_2(x, y)), \tag{1}$$

where $u = f_1(x, y)$ and $v = f_2(x, y)$ are the mapping functions. The mapping functions can be linear or nonlinear functions. An example of simple linear mapping functions is

$$u = x + a_{00}, \tag{2}$$

$$v = y + b_{00}, \tag{3}$$

where a_{00} and b_{00} describe horizontal and vertical translational mappings of x and y, respectively. Hence, this model can provide two degrees of freedom, vertical and horizontal translations. In fact, the BMA is a translational ST being applied to the individual square blocks of a current frame with respect to a previously coded frame. Furthermore, in the BMA, the mapping parameters are usually restricted to integer translational steps, i.e. $(a_{00}, b_{00}) \in \mathbf{Z}^2$. Clearly, these restrictions make this model insufficient to accurately describe real three-dimensional motion in video signals. A more accurate description of motion can be obtained through the use of more powerful ST functions having more degrees of freedom.

The Affine Spatial Transformation (AST) is next to the translational model in terms of complexity and the number of degrees of freedom. The mapping functions of the AST can be written as

$$u = a_{10}x + a_{01}y + a_{00}, \tag{4}$$

$$v = b_{10}x + b_{01}y + b_{00}, \tag{5}$$

where a_{ij}, b_{ij} are the mapping parameters, with $i, j \in \{0, 1\}$. Thus, the AST can provide six degrees of freedom; three degrees in each direction. Apart from translational motion; the AST can effectively describe shear, rotation, and scaling in a two-dimensional plane. Moreover, lines and parallel lines in all orientations are preserved through an AST. It can be easily shown that the translational ST is a special case of the AST where $a_{10} = b_{01} = 1$ and $a_{01} = b_{10} = 0$.

Moving up on the number of degrees of freedom, we get the Bilinear Spatial Transformation (BST) that has eight degrees of freedom and its mapping functions can be written as:

$$u = a_{10}x + a_{01}y + a_{11}xy + a_{00}, \tag{6}$$

$$v = b_{10}x + b_{01}y + b_{11}xy + b_{00}, \tag{7}$$

where a_{ij}, b_{ij} are the mapping parameters with $i, j \in \{0, 1\}$. The BST handles the four-corner mapping for non-planar quadrilaterals, and like the AST, it preserves the lines that are horizontal or vertical. Unlike the AST, however, lines which are not oriented along these two directions are not preserved as lines, but rather as quadratic curves. The AST is a special case of the BST where $a_{11} = b_{11} = 0$.

A more complex nonlinear ST model is the Second-Order ST (SOST) which has twelve degrees of freedom. The mapping functions of the SOST can be written as

$$u = a_{20}x^2 + a_{10}x + a_{11}xy + a_{02}y^2 + a_{01}y + a_{00}, \tag{8}$$

$$v = b_{20}x^2 + b_{10}x + b_{11}xy + b_{02}y^2 + b_{01}y + b_{00}, \tag{9}$$

where a_{ij}, b_{ij} are the mapping parameters with $i, j \in \{0, 1, 2\}$. The SOST model is more powerful and general than the previous ST models. It permits the mapping of straight lines of any orientation into quadratic curves and can handle cases of uneven stretching. The BST can be considered as a special case of the SOST where $a_{20} = b_{20} = a_{02} = b_{02} = 0$. In fact, the AST and the SOST are the first two members of a whole family of mapping functions known as the *bivariate polynomial transformations*, which can be written as

$$u = \sum_{i=0}^{N} \sum_{j=0}^{N-i} a_{ij} x^i y^j, \tag{10}$$

$$v = \sum_{i=0}^{N} \sum_{j=0}^{N-i} b_{ij} x^i y^j, \tag{11}$$

where N is the order of the transformation.

In ST based MC, one ST model is usually adopted to represent the motion which takes place between a previously coded frame and a current frame to be coded. The result of the ST operation is a "warped" version of the previous frame known as the *prediction frame* which is desired to be as close as possible to the current frame. For the purpose of this chapter, we choose the BST to model motion since, as we explain later, it can be easily integrated with the existing video coding standards such as H.261 and MPEG.

15.2.2. Backward Versus Forward Mapping

A digital image is a two-dimensional function sampled at discrete and regularly spaced points of a two-dimensional plane. The use of ST mapping as described earlier will most likely result in images which are not necessarily defined at the discrete and regularly spaced sampling points. There are two possible methods in which an ST can be applied to construct a prediction of a current frame using the pixels of a previous frame. In the first method, known as *backward mapping*, the mapping is assumed from a *discrete* two-dimensional plane in the prediction frame (a destination image), to a *continuous* two-dimensional plane in the previous frame (a source image), see Fig. 1(a). Thus, the pixels of a prediction frame $I_d(u, v)$ are constructed from the pixels of a previous frame $I_s(x, y)$ such that

$$I_d(u, v) = I_s(f_1(u, v), f_2(u, v)) = I_s(x, y), \tag{12}$$

where $x = f_1(u, v)$ and $y = f_2(u, v)$ are the transformation functions, $(u, v) \in \mathbf{Z}^2$, and $(x, y) \in \mathbf{R}^2$. Therefore, in order to reconstruct the pixel amplitudes in the

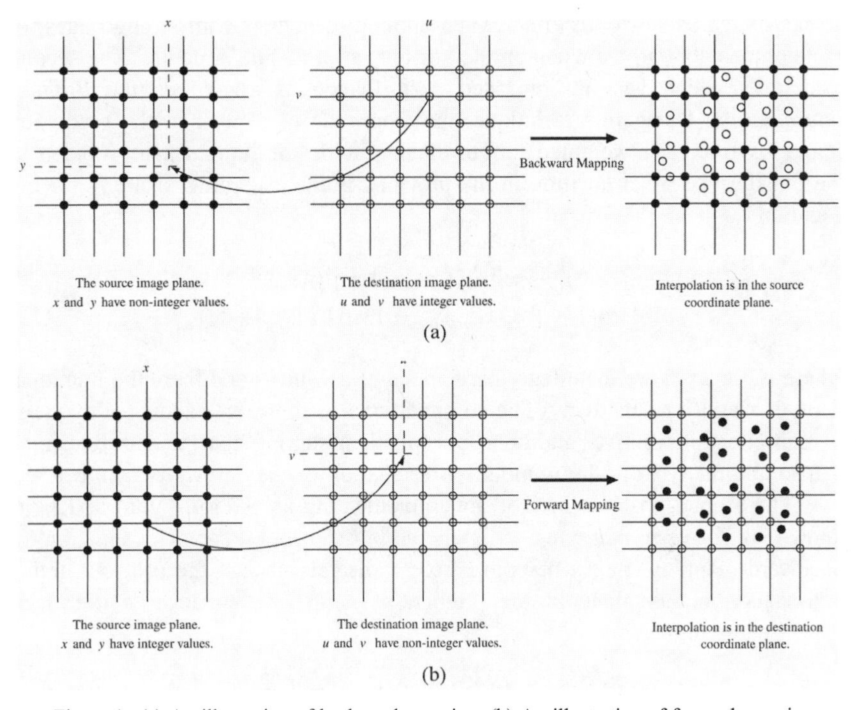

Figure 1. (a) An illustration of backward mapping; (b) An illustration of forward mapping.

prediction frame, interpolation is carried out in the previous frame plane. In other words, a pixel in the prediction frame is defined by looking back at the previous frame to see where it came from.

Backward mapping is generally the preferred method of implementing ST, since it is possible to interpolate the pixels of the desired prediction frame in a scan-line order and there are fast and efficient algorithms for this purpose [2]. Two techniques are typically used for frame reconstruction when using backward mapping ST. The first technique is known as the *Nearest Neighbor Substitution*. Using this technique, the amplitude of a pixel in the prediction frame is assigned the amplitude of the pixel nearest to its transformed pixel position in the previously coded frame. This is achieved by simply rounding the values of the transformed co-ordinates to their nearest integer values. Thus the pixels of a prediction frame are interpolated from the pixels of a previously coded frame as follows

$$I_d(u, v) = I_s(f_1(u, v), f_2(u, v)) = I_s(x_{rnd}, y_{rnd}), \tag{13}$$

where x_{rnd} and y_{rnd} are x and y rounded to their nearest integer value. This technique is fast and computationally very simple. However, it can introduce artifacts such as sawtooth effect when applied to images with strong edges, or broken lines and curves when applied to images with fine structure. The second technique, which has an improved performance, is known as the *Bilinear Interpolation*. Using this technique, the amplitude of a pixel in the prediction frame is written as a weighted sum of the amplitudes of the four nearest pixels to the transformed pixel position in the previous frame according to the following equation

$$I_d(u, v) = (1 - \alpha)[(1 - \beta)I_s(x_{int}, y_{int}) + \beta I_s(x_{int} + 1, y_{int})] \\ + \alpha[(1 - \beta)I_s(x_{int}, y_{int} + 1) + \beta I_s(x_{int} + 1, y_{int} + 1)], \tag{14}$$

where x_{int} and y_{int} are the integer parts of x and y, while α and β are the fractional parts of x and y respectively. The computational complexity of this technique is 6 multiplications and 5 additions per pixel, assuming that α and β can be extracted directly from the number representation of the hardware architecture.

The second possible method for implementing an ST for digital images is known as *forward mapping*. This method assumes the reverse situation of backward mapping, i.e., a mapping from a discrete grid in the previous frame onto a continuous plane in the prediction frame. Mathematically, it can be expressed as follows

$$I_s(x, y) = I_d\,(f_3(x, y), f_4(x, y)) = I_d(u, v), \tag{15}$$

where $u = f_3(x, y)$ and $v = f_4(x, y)$ are the forward transformation functions, $(x, y) \in \mathbf{Z}^2$, and $(u, v) \in \mathbf{R}^2$. Thus, a prediction frame is formed by transforming the positions of pixels of a previous frame (source image) to lie anywhere in a continuous two-dimensional plane (destination image), see Fig. 1(b). Clearly, this will most likely result in a prediction frame where the intensity values are scattered over a continuous two-dimensional plane, i.e., an *irregularly sampled* image. What is needed here, is to recover the intensity values at the discrete two-dimensional lattice points of the prediction frame, i.e., only where $(u, v) \in \mathbf{Z}^2$. As opposed to backward mapping, interpolation here is done in the prediction plane itself where we wish to recover the pixel amplitudes at regularly spaced positions given irregularly spaced sample amplitudes. In the next section, we present various efficient techniques which can be used to reconstruct a regularly sampled prediction frame from a given set of irregularly spaced samples.

15.2.3. Estimation of the ST Motion Parameters

Another problem about the use of ST in MC is how to calculate the ST motion parameters which describe the motion information that takes place between two frames. In BMA (translational ST), the motion vector is two-dimensional and the search space can be limited to a finite window in the neighborhood of the block in the previous frame and full search can be carried out for a finite set of integer translational steps in the horizontal and vertical directions. However, for other ST models with more degrees of freedom, the number of full search possibilities that exist can be enormous to be performed in real-time. There are two main methods for estimating the ST motion parameters. The first method is a gradient based technique known as the Gauss-Newton Iterative (GNI) minimization which was suggested in [3–4]. This technique, which provides a framework for minimizing functions which are sums of squares, is used iteratively to improve an initial estimate for the required motion vector using the Mean Squared Error (MSE) as an error criterion. The number of iterations required to produce satisfactory results vary according to the kind of motion present, but an average of 6 iterations is enough to obtain reasonable results [4]. However, the GNI is a very demanding operation to be performed in real-time.

The second method is a fast, but a less accurate, technique known as the Grid Point Tracking (GPT) algorithm [5]. It relies on tracking the *translational* motion of certain pixels at known locations in the frame, known as the *grid points* or control points. These grid points usually lie on the vertices of the patches that divide the frame. The motion of a grid point is tracked from one frame to another by conducting a BMA-type search of the block which includes the grid point as its center. The motion parameters of a patch can then be found by substituting the co-ordinates of its vertices (grid points) in the current frame and their respective positions in the previous frame in the relevant ST model and solving two resulting systems of simultaneous equations. In order to ensure a sufficient number of simultaneous equations, the number of grid points tracked per patch must be equal to *at least* half the number of degrees of freedom of the chosen ST model. For example, suppose that we wish to compute the BST motion parameters of a quadrilateral patch. Since the BST has eight degrees of freedom, then the translational motion of at least 4 different grid points (usually, the vertices of the quadrilateral) must be tracked from one frame to the next. Suppose that the motion parameters of the quadrilateral patch are written as

$$\mathbf{d}_a = [a_{10}\ a_{01}\ a_{11}\ a_{00}] \quad \text{and} \quad \mathbf{d}_b = [b_{10}\ b_{01}\ b_{11}\ b_{00}].$$

Then, by substituting the values of the positions of the vertices before tracking (x_g, y_g), where $g = \{1, 2, 3, 4\}$, and their respective positions after tracking (u_g, v_g) in (6) and (7), we obtain two sets of four simultaneous equations. The first set of simultaneous equations corresponding to the substitution in (6) can be rewritten in matrix multiplication form as

$$\mathbf{d}_a \mathbf{P}_g = \mathbf{u}_g \tag{16}$$

where $\mathbf{u}_g = [u_1\ u_2\ u_3\ u_4]$ and

$$\mathbf{P}_g = \begin{bmatrix} x_1 & x_2 & x_3 & x_4 \\ y_1 & y_2 & y_3 & y_4 \\ x_1y_1 & x_2y_2 & x_3y_3 & x_4y_4 \\ 1 & 1 & 1 & 1 \end{bmatrix}.$$

Thus, we can solve for $\mathbf{d}_a$ by evaluating and post multiplying $\mathbf{P}_g^{-1}$ with both sides of (16)

$$\mathbf{d}_a = \mathbf{u}_g\mathbf{P}_g^{-1}. \tag{17}$$

Similarly, the second set of simultaneous equations obtained by substituting in (7) can be rewritten as

$$\mathbf{d}_b\mathbf{P}_g = \mathbf{v}_g, \tag{18}$$

where, $\mathbf{v}_g = [v_1\ v_2\ v_3\ v_4]$. Therefore, $\mathbf{d}_b$ can be solved for using

$$\mathbf{d}_b = \mathbf{v}_g\mathbf{P}_g^{-1}. \tag{19}$$

One advantage of the GPT algorithm is that the motion parameters can be unambiguously defined using only translational motion vectors. Therefore, in the case of using the BST, the GPT algorithm has the same computational complexity as traditional BMA, with the added complexities of computing the BST motion parameters from the translational motion vectors (by finding the inverses of 4×4 matrices for each patch and multiplying the result with two 4-dimensional vectors), spatially transforming each pixel position in the frame, and finally, reconstructing the prediction frame using a chosen interpolation technique.

The GPT algorithm is a fast and simple technique but results in a sub-optimal performance. This is because the pixel tracking procedure does not explicitly minimize the ST prediction error. Also the translational motion of a few grid points does not take into account the motion of the other pixels in the patch. Hence, in order to *refine* the estimates of motion parameters obtained using the GPT algorithm, some search techniques have been proposed. An exhaustive full search method, known as the Hexagonal Matching Algorithm (HMA), was suggested in [3] which is a computationally demanding process. To speed up the computation process, it was also suggested in [3] to use the gradient based technique in order to *locally* refine the estimates of the motion vectors obtained by the initial tracking algorithm. This reduces the computational burden but results in a slightly inferior results. A fast search refinement technique to reduce the computational burden using a cross-search algorithm was also suggested in [7]. In the following, we propose a new alternative fast search refinement algorithm based on a logarithmic search tactic. The algorithm also lends itself easily to parallel implementation and which can further reduce the coding delay.

15.3. Prediction Frame Reconstruction from Irregularly Spaced Samples

In this section, we consider various techniques for reconstructing regularly sampled images from irregularly spaced samples. We would like to emphasize that the techniques presented here could be applied not only to reconstruct forward mapped spatially transformed images, but also to *any* other form of irregularly sampled images (e.g., images with jitter). Four techniques are discussed in this section. The first two of these techniques are novel, while the other two have been previously proposed in [8–9]. However, the performance of these techniques have not been previously evaluated for a *general case of irregular sampling.*

Since the focus of this chapter is in the context of motion compensated prediction, we assume that the motion parameters describing the mappings of the different patches of a previous frame are already found (e.g., using the GPT) and that the forward mapping transformation is applied to the previous image to form an irregularly sampled prediction image. Thus, what remains of the MC process is to reconstruct the regularly sampled prediction frame from its irregularly spaced samples.

15.3.1. The Nearest Voronoi Substitution Technique

In contrast to backward mapping where interpolation takes place in the previous frame plane, the interpolation plane in forward mapping is the prediction plane itself where we wish to interpolate unknown amplitudes of a regularly spaced set of pixels from the known amplitudes of an irregularly set of pixels. We rewrite the forward mapped frame as

$$I_s(x, y) = I_d(f_3(x, y), f_4(x, y)) = I_d(u, v), \tag{20}$$

where I_d is the forward mapped frame and $(u, v) \in \mathbf{R}^2$ are the irregularly spaced positions of the transformed samples. Our aim now is to reconstruct the pixel amplitudes at the regularly spaced points $(m, n) \in \mathbf{Z}^2$ from pixel amplitudes at the irregularly spaced points $(u, v) \in \mathbf{R}^2$. As mentioned previously, the simplest rule for interpolation is to substitute the amplitude of the nearest known neighbor. In the forward mapping case, this becomes a Nearest Voronoi Substitution. Thus, the unknown sample amplitude at a uniform sampling position takes the amplitude of the sample having the minimum Euclidean distance to it. The Euclidean distance between two pixel positions $\mathbf{a} = [m \;\; n]^T$ and $\mathbf{b} = [u \;\; v]^T$ is defined as

$$r(\mathbf{a}, \mathbf{b}) = ||\mathbf{a} - \mathbf{b}|| = \sqrt{(m - u)^2 + (n - v)^2}. \tag{21}$$

Hence to construct a regularly sampled frame $I_d(m, n)$, where $(m, n) \in \mathbf{Z}^2$, from a collection of irregularly spaced pixels $I_d(u, v)$, where $(u, v) \in \mathbf{R}^2$, using a nearest Euclidean distance rule, we set

$$I_d(m, n) = I_d(u_p, v_p), \tag{22}$$

such that

$$r(\mathbf{a}, \mathbf{b}_p) \; \leq \; r(\mathbf{a}, \mathbf{b}) \quad \forall \; \mathbf{b} \in \mathcal{S}_d^2, \tag{23}$$

where $\mathcal{S}_d^2$ is the finite set of all irregularly spaced samples. This technique can be slightly more computationally demanding than the nearest neighbor interpolation in backward mapping. A square domain of ± 0.5 pixels vertically and horizontally of a regularly sampled position is used to detect the nearest irregularly sampled pixels. The size of this search space is reasonable based on the statistical results presented in a later section. In the squre domain of ± 0.5 pixels around the pixel position to be interpolated, there are three possibilities (see Fig. 2):

- there is only one irregularly sampled pixel falling into that domain,
- there are more than one pixel falling into that domain, or,
- there are no pixels falling into that domain.

For the first possibility, the computational complexity of this technique is equivalent to the nearest neighbor substitution in backward mapping. The other

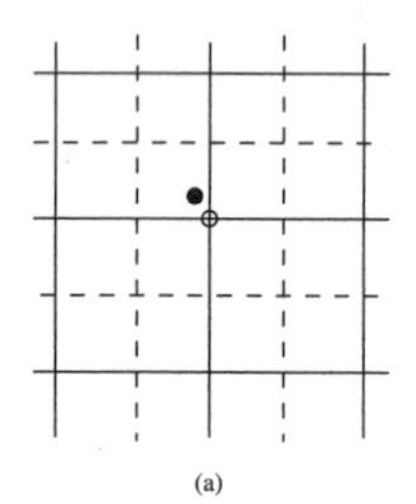

(a)

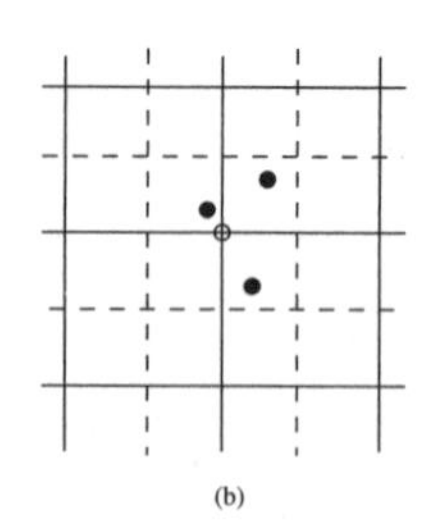

(b)

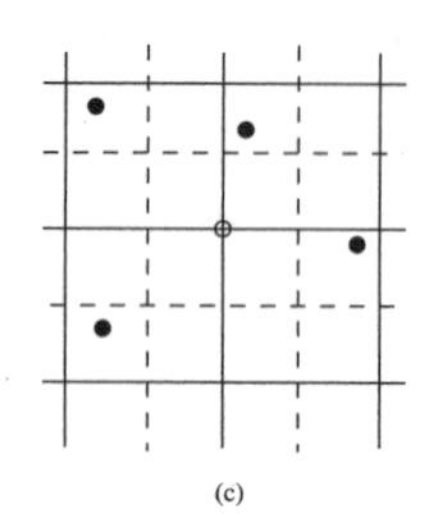

(c)

● Known pixel value.

o Unknown pixel value.

Figure 2. The three different possibilities for the position of a transformed pixel relative to a pixel position we wish to interpolate for.

two possibilities require the computation of the distances of their nearest samples and the comparison of those distances. The overall computational complexity of this technique will, therefore, depend on the nature of the resulting irregular sampling process. For our application, the results of the statistics of the irregular sampling show that the first possibility is the most likely possibility.

As with the nearest neighbor substitution in backward mapping, this technique can introduce artifacts such as the sawtooth effect along the edges or broken lines and curves when applied to images with very fine structure.

15.3.2. The Inverse-Distance Weighting Interpolation Technique

As with the Bilinear Interpolation in backward mapping, a better estimate of the value of an unknown sample is obtained by incorporating knowledge of more than one sample in its neighborhood. One way to achieve this is to write the unknown sample amplitude as a weighted sum of its nearest neighbors. A simple rule is to weight each neighbor by its *inverse distance* to the unknown sample. Thus, the closer the known sample is, the more weight it has towards the final value of the unknown amplitude. An inverse-distance-weighting formula is given below where each of neighboring samples is weighted by its inverse Euclidean distance to the unknown sample position and the total sum is then normalized by the total sum of the inverse distances.

$$I_d(m, n) = \frac{\sum_{p=1}^{l} r^{-1}(\mathbf{a}, \mathbf{b}_p)\, I_d(u_p, v_p)}{\sum_{p=1}^{l} r^{-1}(\mathbf{a}, \mathbf{b}_p)}, \tag{24}$$

where $r(\mathbf{a}, \mathbf{b}_p) \neq 0$ and l is the total number of the nearest samples to be used in the interpolation. This technique has an improved performance over the nearest neighbor interpolation. In general, it works as a low-pass filter and the greater the value of l, the more the low-pass filtering effect becomes. Experimentally, it was found that setting $l = 4$ gives a good quality at a reasonable computational complexity.

15.3.3. The Time Varying Technique

This technique was developed by one of the authors [10] for the recovery of continuous signals from a set of irregularly spaced samples. The technique was described in Section 6.1 of Chapter 4 of this book.

The block diagram shown in Fig. 3 shows a physical realization of the Time Varying technique. The irregularly sampled stream passes through two branches. In the lower branch, the irregular samples are passed through an absolute value operator and a hard limiter to give a stream of unit impulses at the irregular

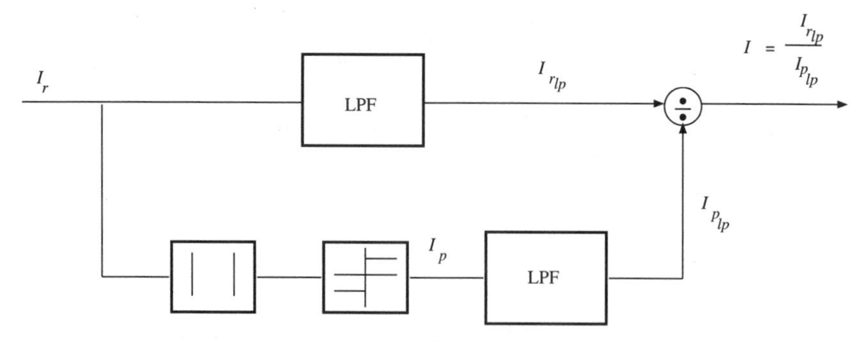

Figure 3. A block diagram of the Time Varying (Nonlinear (NL) Division) technique.

sampling positions, which in turn is fed to a low-pass filter. In the upper branch, the irregularly spaced signal is low-pass filtered and the result is then divided by the output of the lower branch to give the original continuous signal.

From Fig. 3, the output can be written as

$$I(m, n) = \frac{I_{r_{lp}}(u, v)}{I_{p_{lp}}(u, v)}, \tag{25}$$

where $I_{r_{lp}}(u, v)$ and $I_{p_{lp}}(u, v)$ represent the low-pass filtered version of the irregular sampling set and the low-pass filtered version of the set of unit impulses at the irregular sampling positions, respectively. Interpolation of the pixel values at a discrete set of pixels (m, n) is easily carried out by applying the filtering operations only at the points of interest. This technique has proved quite effective in experiments for image recovery from random pixel losses, which is a special case of irregular sampling [8].

The computational complexity of this technique depends on the duration of the filter used. A cosine-windowed *Sinc* filter was used in our simulations with a duration of 3 pixels in each direction. Due to the irregular nature of the sampling process, it is not guaranteed that 9 pixels $(= 3^2)$ will be found in the domain of the filter.

15.3.4. The Iterative Technique

Another technique of recovering signals from irregular samples is a successive approximation technique proposed by one of the authors and others [8, 11]. This technique was also described in more detail in Section 6.2 of Chapter 4 of this book (see also Fig. 14 in Chapter 4). An adaptation of Fig. 14 in Chapter 4 is shown in Fig. 4, described below. The "Module" could be either a low-pass

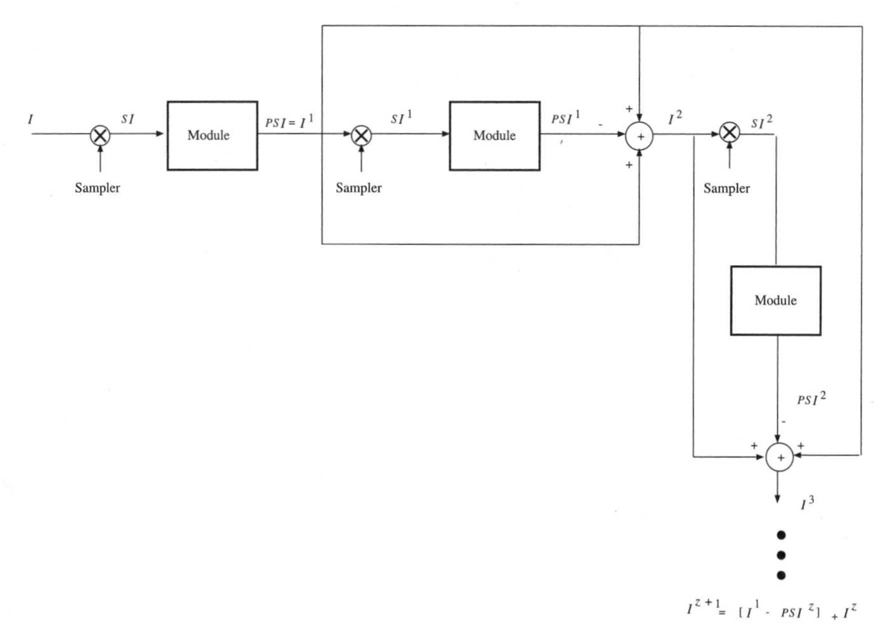

Figure 4. A block diagram of the iterative technique.

filter or the Time Varying technique. Previous experiments with images with sample losses (a special case of irregular sampling) [8] show that with the use of a low-pass filter, the speed of convergence is linear with respect to the number of iterations and is exponential when the Time Varying technique is used. We, therefore, choose to use the Time Varying (Nonlinear Division) technique as the band-limiting operator in the modules shown in Fig. 4. The value of the relaxation parameter λ can also affect the speed of convergence and is usually found experimentally. In our experiments, we found that setting $\lambda = 1$ gives reasonable results. It was experimentally found that, for the purpose of this application, almost no significant improvement is observed beyond 3 time varying iterations.

15.4. A Fast BST MC Algorithm

In this section we describe the overall MC algorithm used in our experiments as well as the proposed fast search refinement algorithm for finding the BST motion parameters.

15.4.1. The MC Algorithm

An overall block diagram of the MC algorithm is shown in Fig. 5. The inputs to the algorithm are two frames; a current frame to be predicted I_n and its previous

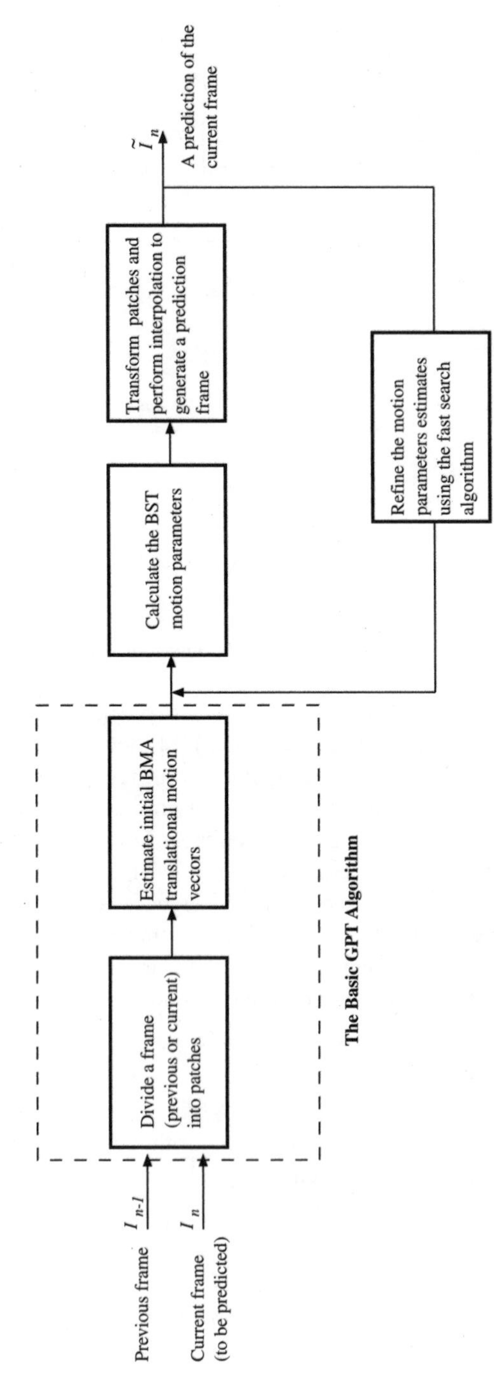

Figure 5. A block diagram of the overall MC process using BST with the fast search refinement algorithm.

frame I_{n-1}. In general, the MC process starts by dividing one of the two frames (either the current frame or the previous frame) into different patches. This is mainly because different parts of the frames are likely to have different motion parameters. In backward mapping, it is the current frame which is divided into patches and different motion parameters are estimated for each patch. Since the decoder does not have any information about the current frame, the shape of those patches should be regular, uniform, and known to the decoder. Otherwise, if any nonuniform segmentation procedure applied to the current frame, the encoder must transmit overhead information to decoder in order to describe it. In fact, the conventional BMA can be considered as a simple backward mapping ST operation. This is because the current frame is divided into uniform square blocks and each block is predicted by searching backwardly in the previous frame. Also, when the BMA is applied with half-pixel accuracy, the interpolation is carried out in the previous frame plane (the source image plane). In forward mapping, on the other hand, it is the previous frame which is divided into patches and different motion parameters are estimated for each patch. Since both the encoder and the decoder are assumed to have the previous frame, it is then possible to apply any segmentation technique which can be equally performed in the encoder as well as the decoder. Thus, there will be no need to transmit any overhead information describing the segmentation applied to the previous frame. However, in order to have a fair comparison at the same data rate and to compare the different interpolation techniques in forward and backward mappings, we adopt the same uniform division into square patches of both the current frame (in the case of backward mapping) and the previous frame (in the case of forward mapping). Also, in order to have a fair comparison at the same data rate with the BMA-based algorithms, such as the video coding standard H.261 for ISDN, the layout of the grid points (which define the vertices of the adjacent square patches) used in our experiments for both forward and backward mappings, is similar to that of [6]. The spacing of the grid points using this layout is 16 pixels in each direction which gives a uniform overlay of grid points on the previous frame in the case of forward mapping, or the current frame in case of backward mapping. Thus, resulting in the same amount of motion overhead as conventional BMA. Also, the search space was limited to ± 15 pixels in each direction for all experiments. Next, the GPT algorithm is used where the translational motion vector of each grid point is found by conducting a BMA-type search of the block which includes the grid point as its center. If only the basic GPT algorithm is used to find the motion parameters, the MC algorithm stops at this point and the translational motion vectors are directly transmitted to the decoder. The decoder then uses the positions of the grid points before and after tracking to find the BST motion parameters of the different patches according to (17) and (19). The prediction frame is then constructed by applying the BST to the different patches of the frame (current if backward mapping is used or previous if forward mapping is used).

However, as previously stated, the basic GPT algorithm does not necessarily generate the motion parameters which minimize the errors with respect to the BST. In the next section, we describe the fast search algorithm which refines the initial estimates of motion parameters obtained using the basic GPT algorithm.

15.4.2. The Fast-search Refinement Algorithm of Estimated ST Motion Vectors

The fast search algorithm is depicted as a three-block loop as shown in the block diagram in Fig. 5. The first two blocks in the loop are identical to what happens in the decoder; the BST motion parameters are calculated from the translational motion vectors, the patches are transformed, and the prediction frame is reconstructed. The next step in the loop is to apply the fast search refinement algorithm. The algorithm is a combination of the conjugate direction search technique [12], and a logarithmic search tactic. It is executed twice for each translational motion vector; once over its horizontal component and once over its vertical component; and it is identical in both cases. It can also be executed over many stages, where in each stage, a *search step* is halved relative to the previous one. The procedure of the algorithm is identical in all stages except for the first stage, which has two parts. In part one of the first stage, the value of one component of a translational motion vector (either vertical, or horizontal) is changed twice; once by adding a search step to it and once by subtracting a search step from it. The new values correspond to two directional changes in the value of the motion vector component (left and right, if horizontal; top and down, if vertical). These translational changes correspond to new BST motion parameters. The new BST motion parameters are computed, the patches are transformed twice, and the prediction errors are computed for both cases. Any measure of error can be used to measure the prediction errors, but for the purpose of our experiments, we adopt the MSE as our error criterion. Different actions are then taken depending on the direction that the prediction error is minimized. The assumption here is that the prediction error will be monotonically decreasing along one direction. If the prediction error is not minimized along any direction, the first stage terminates and the value of the motion vector is fixed until the next stage. Part two of the first stage is executed if the prediction error is decreased along one particular direction. The motion vector is changed to the value giving the minimum prediction error and a pursue procedure along the direction in which the prediction error was minimized is carried out. The value of the motion vector is changed twice again. This time, both changes take place along the particular direction in which the error is minimized. The value is changed once by adding (or subtracting; depending on the direction) one search step, and once by adding (or subtracting) two search steps. The new BST motion parameters are then calculated, the patch is transformed, and the prediction errors are calculated. If the prediction error is not minimized by changing the current position of the motion vector, the second part (and the first stage) terminate and the position of the grid is fixed. Also, if the prediction error is minimized one search step away but not two search steps away, then the component is changed by one search step and part two (and the first stage) terminates. However, if the prediction error is still minimized two search steps away from the current position, then the component is changed by two search steps and part two of the first stage is repeated again.

The second stage of the algorithm is identical to part one of the first stage, except that the search step is reduced by half. Thus the new value of the component, obtained from the previous stage, is varied in both directions (with half the value of the previous search step), the new BST motion parameters are calculated, the patch is transformed and the errors are evaluated and compared. The value of the component is then changed to that giving the minimum prediction error and the second stage terminates. Third and fourth stages are possible where each time the search step is reduced to half that of the previous stage. Each stage in the algorithm either reduces the prediction error or leaves it unchanged. An example of optimizing the position of a single grid point is also explained graphically in Fig. 6. The example is explained as follows:

Stage 1:

Part 1: The value of the search step is set to 2 pixels and the current horizontal position of the grid point, obtained the GPT algorithm, is denoted by a_0. The position of the grid point is changed to a_1 and a_2 which are two pixels away to the left and the right from a_0, respectively. The errors at both positions $Er(a_1)$ and $Er(a_2)$ are evaluated after reconstruction. Since $Er(a_2) < Er(a_0)$ and $Er(a_2) < Er(a_1)$, the position of the grid point is changed to a_2.

Part 2: A pursuit of the direction in which the error is minimized (the right in this case) is carried out. Errors are evaluated at a_3 and a_4 which are two and four pixels (one search step and two search steps) away from a_2, respectively. Since $Er(a_4) < Er(a_3)$ and $Er(a_4) < Er(a_2)$, the decision is to pursue the same direction again. The position of the grid point is changed to a_4 and errors are evaluated again at a_5 and a_6 since $Er(a_5) < Er(a_6)$ and $Er(a_5) < Er(a_4)$. This stage terminates with the position of the grid point at a_5.

Stage 2:

The search step is reduced to 1 pixel and b_0 is set to the value of a_5, which was obtained from the previous stage. The position of the grid point is varied to the left b_2 and to the right b_1; the patch is transformed and the errors are evaluated. Since $Er(b_2) < Er(b_0)$ and $Er(b_2) < Er(b_1)$, this stage terminates with the horizontal component of the grid point having the value of b_2.

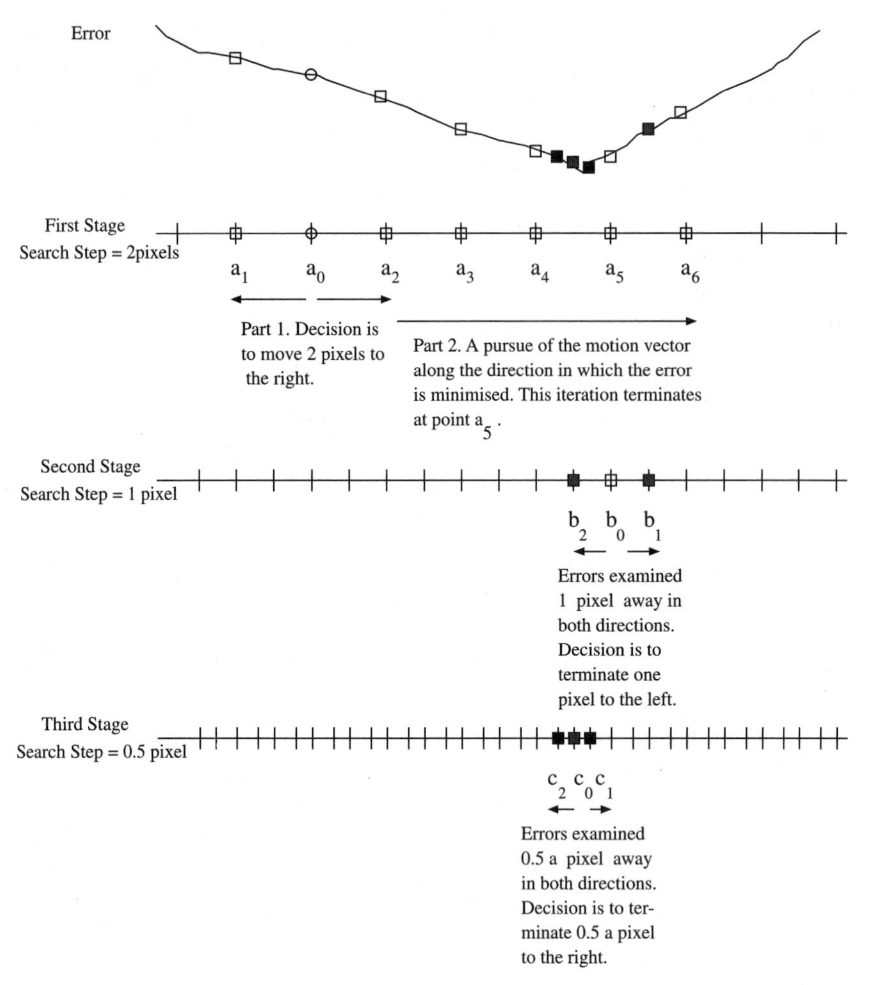

Figure 6. Three stages of the fast search technique being applied in the horizontal direction.

Stage 3:

The search step is reduced to 0.5 pixel and c_0 is set to the value of b_2. The position of the grid point is varied to the left c_2 and to the right c_1, respectively, which are half a pixel away from c_0. Errors are evaluated again at both new positions. Since $Er(c_1) < Er(c_0)$ and $Er(c_1) < Er(c_2)$, the algorithm finally halts with the horizontal component of the grid point at c_1.

Obviously, this search strategy will result in an enormous increase in the computational load if the frame has to be reconstructed each time a grid point

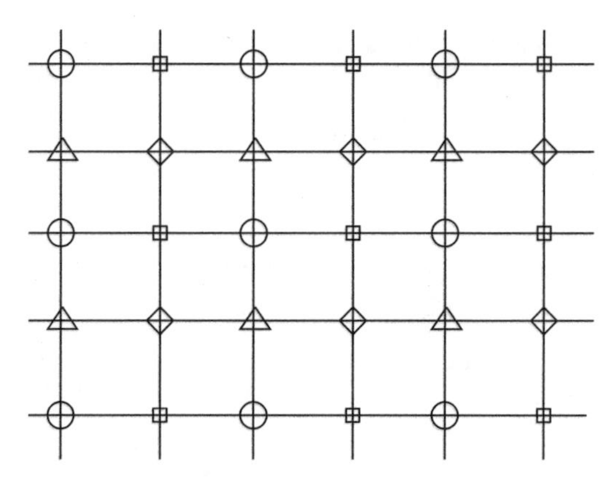

Figure 7. Parallelizing the fast search algorithm. The algorithm is applied to the grid points sharing the same symbols simultaneously.

changes its position. However, it is possible to reduce this computational load using parallel processing. Since each change in a grid point's position affects only four adjacent patches, it is possible to optimize the positions of "every other" grid point, horizontally and vertically, simultaneously such that no overlap occurs. This is demonstrated in Fig. 7. The positions of the grid points sharing the same symbol are optimized simultaneously while the positions of the other grid points are kept constant, and thus, the search algorithm can be applied over four successive batches of grid points sharing the same symbols as shown in Fig. 7. This parallel processing reduces the number of frame reconstructions and error evaluations by a factor of 4.

We would like to point out that the use of this algorithm is not limited to forward mapping BST and can be equally used with the backward mapping BST and other ST models provided that enough points are tracked per patch.

15.5. Simulation Results

15.5.1. The Test Sequences

Simulations were carried out using the luminance components of the QCIF resolution (176 × 144) of "Miss America" and "Claire" sequences. The frame rate was set to 10 Hz in each sequence by skipping two out of every three frames of the original 30 Hz sequences. Both sequences were band-limited using a two-dimensional low-pass filter with a cut-off frequency of 0.8 of the original sampling frequency. This is to ensure an over-sampled sequence to guarantee convergence of the Time Varying and Iterative techniques. We note that the search resolution in all algorithms is kept at 1 pixel in order to realize the same rate of motion overhead information. Also, the initial resolution of the GPT algorithm was set to 4 pixels, with two more stages of refinement corresponding to 2 and 1 pixels.

15.5.2. The Statistics of the Irregular Sampling Process Resulting from the Use of Forward Mapping BST for MC

We start by presenting some statistical measurements of the irregular sampling process caused by the use of the forward BST for constructing the prediction frames. These statistics are fairly important in order to assure the suitability of the Time Varying and Iterative techniques which are sensitive to large "gaps" in the irregular sampling process. The graphs in Figs. 8 and 9 show the maximum, mean and standard deviation of the distances measured between each regularly sampling point and its closest irregularly sampling point for each prediction frame in both sequence. These statistics are derived for the case where the BST motion parameters are calculated using the basic GPT algorithm without using the fast refinement algorithm. The statistics show that the maximum distance of an irregular sample from a regular sampling point is always below 1 pixel in any frame for both sequences. The mean and the standard deviations

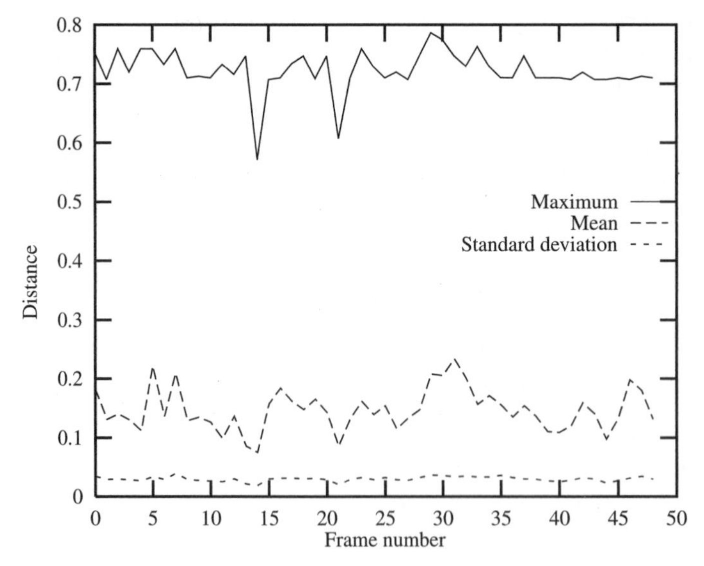

Figure 8. The above curves show the maximum, mean and standard deviation of the distances measured between each regularly sampling point and its closest irregularly sampling point for each prediction frame in the sequence "Claire".

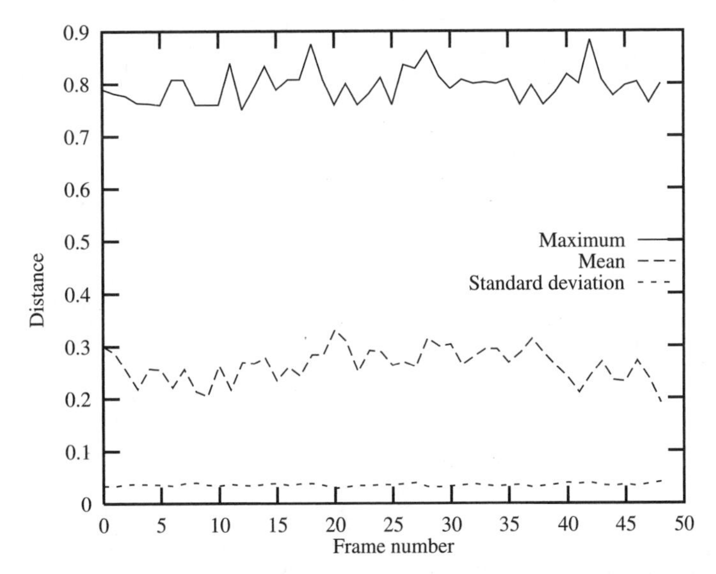

Figure 9. The above curves show the maximum, mean and standard deviation of the distances measured between each regularly sampling point and its closest irregularly sampling point for each prediction frame in the sequence "Miss America".

confirm that, on the average, an irregularly sampled point is not deviating from its closest regularly sampled position by more than 0.5 a pixel. It is interesting to note that the average distances calculated for "Miss America" is more than that of "Claire". This is due to the faster motion action in "Miss America" relative to "Claire".

15.5.3. Comparison of Prediction PSNRs

In order to evaluate the efficiency of the proposed techniques, we first compare the relative performance of the interpolation algorithms in forward mapping when the motion parameters are calculated using the GPT algorithm. The interpolation technique having the best performance is then compared to both the standard BMA and backward mapping BST (using Bilinear Interpolation) when the fast search algorithm is applied.

The graphs in Figs. 10 and 11 show the PSNR (Peak-to-Peak Signal to Noise Ratio) for the prediction frames of forward mapping for each sequence against the frame number using the GPT algorithm for finding the motion parameters. In Table 1 we give the corresponding average prediction PSNRs taken over both

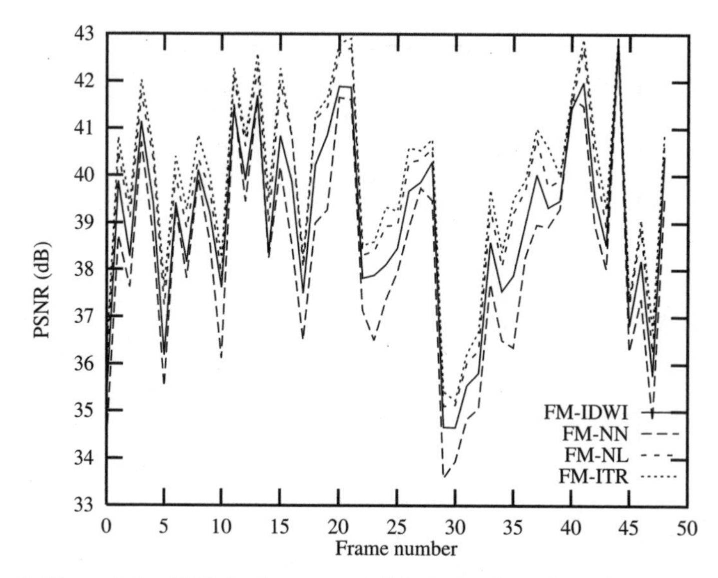

Figure 10. The prediction SNR's for the sequence "Claire" when forward mapping bilinear ST is used together with the Nearest Neighbor interpolation (NN), Inverse Distance Weighted Interpolation (IDWI), Time Varying (Nonlinear (NL) Division) Interpolation and Iterative Interpolation (ITR). The BST motion parameters here are found using the GPT algorithm.

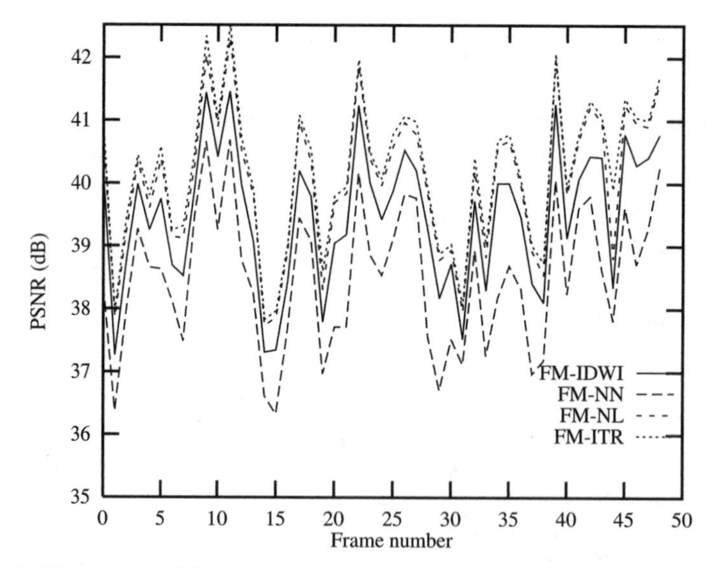

Figure 11. The prediction SNR's for the sequence "Miss America" when forward mapping bilinear ST is used together with the Nearest Neighbor interpolation (NN), Inverse Distance Weighted Interpolation (IDWI), Time Varying (Nonlinear (NL) Division) Interpolation and Iterative Interpolation (ITR). The BST motion parameters here are found using the GPT algorithm.

Table 1. The average PSNRs of the predictions of "Claire" and "Miss America" sequences taken over 50 frames using the various techniques.

Technique	PSNR for "Claire" (dB)	PSNR for "Miss America" (dB)
BMA	38.923	39.008
FM-NN (GPT)	38.192	38.495
FM-IDWI (GPT)	38.963	39.483
FM-NL (GPT)	39.563	40.041
FM-ITR (GPT)	39.845	40.198
FM-NL (FSA)	41.187	41.201
BM-BLI (FSA)	41.325	41.269

sequences using the various techniques. The PSNR is calculated according to Chapter 18.

The graphs show the superiority of the IDWI over the simple Nearest Neighbor Substitution method by about 1 dB on the average for both sequences. Moreover, both the Time Varying (Nonlinear Division) and the Iterative-Time Varying (Nonlinear Division) techniques show a superiority of about 0.5 dB over the IDWI for both sequences. While the Time Varying technique shows a gain of about 0.4 dB on the average, the Iterative-Time Varying (Nonlinear Division) technique shows very slight improvements over the Time Varying technique. This is explained by the fact that all further errors are due to motion compensation and very little gain can be achieved by using a more sophisticated interpolation technique. Since the Iterative technique shows very slight improvement over the Time Varying technique at almost double the computational cost (see Table 2), we therefore choose the Time Varying technique as the interpolation method for our subsequent comparisons with the backward mapping when the fast search algorithm is applied and with the conventional BMA.

The curves shown in Figs. 12 and 13 show the PSNR performance of the forward mapping prediction (with the Time Varying interpolation) and the backward mapping prediction (with bilinear interpolation) when the fast search algorithm is applied. The performance of conventional BMA prediction is also plotted for comparison. If we compare the forward mapping curves with the fast search refinement with their GPT based counterparts, we find an improvement of

Table 2. A comparison of the computational complexities of the various algorithms.

Technique	BMA	FM-NN (GPT)	FM-IDWI (GPT)	FM-NL (GPT)	FM-ITR (GPT)	FM-NL (FSA)	BM-BLI (FSA)
Complexity Estimate (relative to BMA)	1	1.2	1.3	1.7	3.1	85.0	16.0

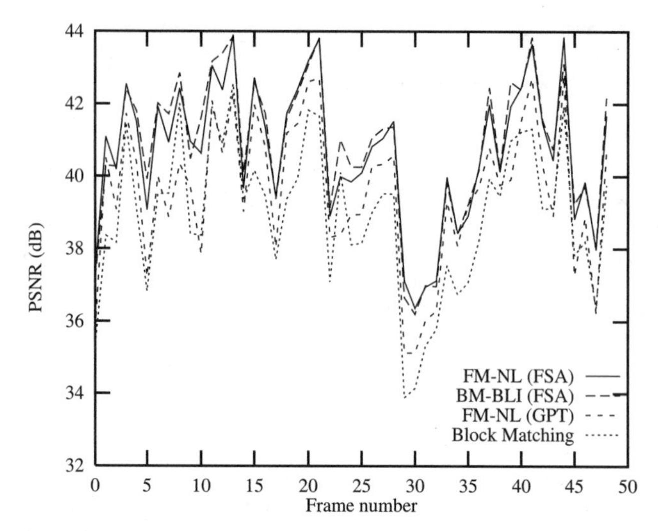

Figure 12. The prediction PSNR's for 50 frames of the sequence "Claire" when forward mapping BST with Time Varying (Nonlinear (NL) Division) interpolation (FM-NL) and backward mapping BST with the bilinear interpolation (BM-BLI) are used with the fast search algorithm (FSA). Both show a very close performance which is better than that of forward mapping BST using GPT and BMA at the same motion overhead.

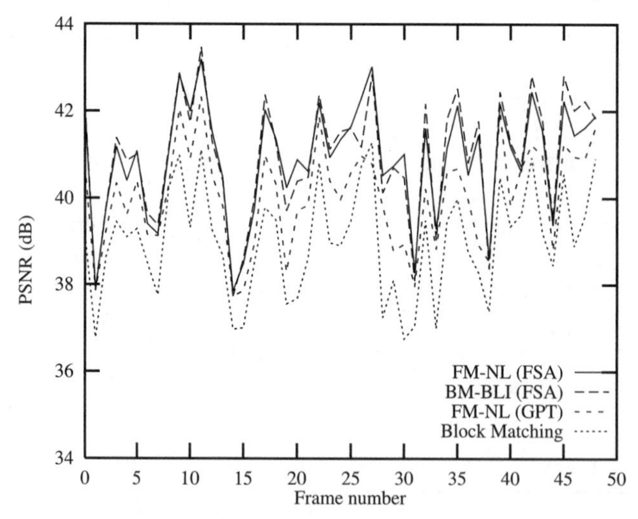

Figure 13. The prediction PSNR's for 50 frames of the sequence "Miss America" when forward mapping BST with Time Varying (Nonlinear (NL) Division) interpolation (FM-NL) and backward mapping BST with the bilinear interpolation (BM-BLI) are used with the fast search algorithm (FSA). Both show a very close performance which is better than that of forward mapping BST using GPT and BMA at the same motion overhead.

around 1.3 dB on the average. This ascertains the gain obtained using the fast search technique for calculating the BST motion parameters as compared to using the simple GPT algorithm alone. The performance of both forward and backward mappings are very close in both sequences with very little difference between them. Both techniques are superior to conventional BMA with 2.2 dB on the average for both sequences. Figures 14 and 15 show a previous and a current frame of the sequence "Claire". Figures 16–18 show prediction frames using the BMA, the BST with forward mapping and Time Varying (Nonlinear (NL) Division) interpolation, and the BST with backward mapping and bilinear interpolation, respectively. The absence of blocking artifacts in the prediction frames based on the BST is clear and a superiority in subjective quality is apparent.

15.5.4. Comparison of Computational Complexities of the Various Algorithms

In this section, we compare the overall computational complexities of the MC schemes using the BMA, BST with backward mapping using bilinear inerpolation, as well as the BST with forward mapping using the various interpolation techniques. Complexity estimates of the various algorithms relative to BMA (MSE based) are given in Table 2. These estimates take into account all parts of each algorithm, including the conversion from translational motion vectors to the 8-dimensional BST motion parameters, the transformation and the reconstruction of each pixel, and the calculation of prediction errors during the fast search refinement algorithm. The estimates give an approximate overall "order of magnitude" by which each technique differ in complexity relative to BMA, and were calculated based on the average times taken for the computer simulations of both sequences. As we can see from Table 2, the increased

Figure 14. A previous frame of the sequence "Claire".

Figure 15. A current frame of the sequence "Claire".

Figure 16. Prediction of the current frame using BMA.

Figure 17. Prediction of the current frame using forward mapping BST with Time Varying interpolation.

Figure 18. Prediction of the current frame using backward mapping BST with bilinear interpolation.

objective performance of the increased computational complexities of the first three interpolation methods (the Nearest Neighbor, the IDWI and the Time Varying technique) over the conventional BMA may justify their increased performance. The relatively modest gain of the Iterative technique over the Time Varying technique, however, may not be justified by an almost twofold increase in computational complexity. This is why the Time Varying technique is chosen as the means of further comparison using the fast search refinement algorithm. The final two results confirm that the gain obtained using forward mapping, although objectively comparable with backward mapping, may be unjustified given the huge increase in computational complexity.

15.5.5. Conclusions

In this chapter, we have evaluated the use of various techniques to reconstruct regularly sampled frames from irregularly sampled frames resulting from the use of forward mapping BST for MC. The performance of the techniques is close to that obtained using backward mapping BST. The techniques were applied in a motion compensation scheme which is intended for video coding. However, the use of the new reconstruction techniques may not be limited to video coding. The full potential of forward mapping may not be realized yet for video coding. A new algorithm which combine segmentation of previously coded frames with forward mapping should be considered. We also proposed a new fast search algorithm for estimating the BST motion parameters. The new algorithm proved quite effective in increasing the subjective and objective qualities of the prediction frames.

References

[1] B. Girod. The Efficiency of Motion-Compensating Prediction for Hybrid Coding of Video Sequences. *IEEE J. Select. Areas in Commun.*, SAC-5(7):1140–1154, August 1987.
[2] G. Wolberg. Digital Image Warping. *IEEE Computer Society Press*, Los Alamos, CA, 1990.
[3] Y. Nakaya and H. Harashima. Motion Compensation Based on Spatial Transformations. *IEEE Transactions on Circuits and Systems for Video Technology*, 4(3):339–356, June 1994.
[4] C. Papadopolous and T. Clarkson. Motion Compensation Using 2nd Order Geometric Transformations. *IEEE Transactions on Circuits and Systems for Video Technology*, 5(4):319–331, 1995.
[5] H. Bruesewitz. Motion Compensation with Triangles. *Proceedings of the 3rd International Workshop on 64 kbits/s Coding of Moving Video*, free session, September 1990.
[6] J. Nieweglowski, T. G. Campbell, and P. Haavisto. A Novel Video Coding Scheme Based on Temporal Prediction Using Digital Image Warping. *IEEE Transactions on Consumer Electronics*, 39(3):141–150, August 1993.
[7] V. Seferidis and M. Ghanbari. A General Approach to Block Matching Motion Estimation. *Optical Engineering*, 32(7):1464–1474, July 1993.
[8] F. Marvasti, L. Chuande, and G. Adams. Analysis and Recovery of Multidimensional Signals from Irregular Samples Using Nonlinear and Iterative Techniques. *Signal Processing*, 36:13–30 1994.
[9] F. Marvasti. A Chapter Contribution. In F. Froehlich and A. Kent, Eds., *The Froehlich/Kent Encyclopedia of Telecommunications*, 7:453–479, Marcel Dekker Inc., 1994.
[10] F. Marvasti. Spectrum of Nonuniform Samples. *Electronic Letters*, 20:896–897, 1984.
[11] F. Marvasti. Spectral Analysis of Random Sampling and Error Free Recovery by an Iterative Method. *IECE Trans. Inst. Electron. Commun. Engs.*, Japan (section E), E 69(2):79–82, February 1986.
[12] R. Srinivasan and K. Rao. Predictive Coding Based on Efficient Motion Estimation. *IEEE International Communications Conference (ICC'84)*, 1984.

16

Applications of Nonuniform Sampling to Nonlinear Modulation, A/D and D/A Techniques

F. Marvasti and M. Sandler

16.1. Introduction

One of the important applications of nonuniform sampling is in the area of telecommunications, specifically in the field of modulation and coding. The modulation aspect is treated in this chapter, and the coding part will be discussed in Chapter 17.

Most modulation schemes are related to the crossings of the intersection of the modulating (original or baseband) signal with a periodic waveform [1]. Indeed Amplitude Modulation (AM) Suppressed Carrier that is sometimes called Double Side Band (DSB) modulation can be generated from the crossings of a cosine wave and the modulating signal. The zero-crossings of Phase Modulation (PM) is nothing but the crossings of the intersection of the modulating signal and a sawtooth wave; these crossings also generate Pulse Position Modulation (PPM). Frequency Modulation (FM) zero-crossings are also equivalent to the intersection of a sawtooth-wave with the integral of the modulating signal. The crossings of the modulating signal with an adaptive signal yields Delta Modulation (DM). There are other related modulation methods that can be indirectly described by the intersection of two signals, namely, Pulse Width Modulation (PWM), Phase Shift Keying (PSK), and Frequency Shift Keying (FSK). The last two are the

F. Marvasti and M. Sandler • Department of Electronics Engineering, King's College London, Strand WC2R 2LS. E-mail: farokh.marvasti@kcl.ac.uk

Nonuniform Sampling: Theory and Practice, edited by Marvasti
Kluwer Academic/Plenum Publishers, New York, 2001.

digital version of PM and FM, respectively. The techniques developed in Chapter 4 on Extremum sampling will then be used for demodulation of PPM, Sine Wave Crossings (SWC), FM, PWM, and DM.

PWM and a modification of DM called Sigma Delta Modulation (SDM) are used in Digital-to-Analog (D/A) and Analog-to-Digital conversion (A/D). In the classical approach in D/A conversion, simple low-pass filtering is used; the distortion of recovery implies that higher than the Nyquist rate has to be used for acceptable distortion. However, if nonuniform sampling recovery techniques, that will be discussed in this chapter, are used, the sampling rate for PWM and SDM can approach the Nyquist rate. Naturally, the cost will be the increased complexity of the circuitry.

16.2. Pulse Position Modulation (PPM)

PPM is widely used in fibre optic communications [2]. The main advantage of PPM is that it has constant height and width, the information lies in the position of the pulses. PPM can be regarded as the pulses at the position of the intersection of any band-limited signal with a sawtooth-wave—(Fig. 1). The band-limited signal could be regarded as a signal with finite support in L_2 or in $\mathbb{C}^N$—the class of finite dimensional discrete signals. PPM in classical communication systems is always considered to be in L_2. We first analyse the PPM from the classical point of view and then suggest a superior method of analysis and recovery based on nonuniform sampling theory.

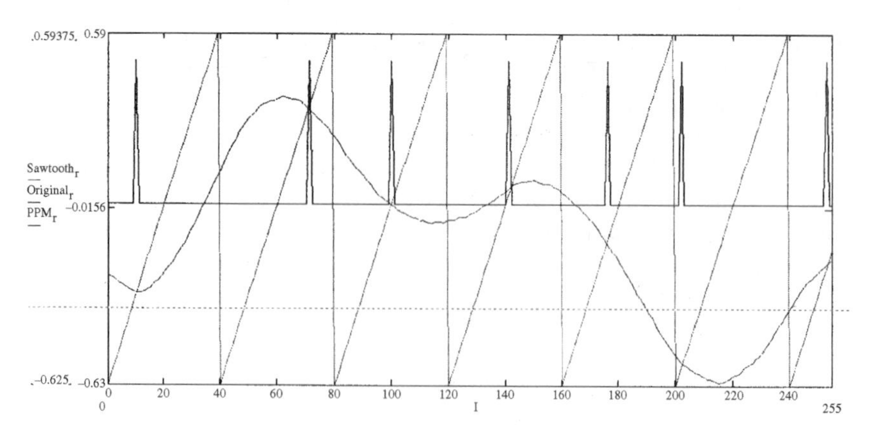

Figure 1. PPM pulses generated for an L_2 signal.

16.2.1. Analysis and Recovery in L_2

The position of the pulses from Fig. 1 can be written in terms of the amplitude of the band-limited signal as

$$t_k = kT + cx(t_k), \tag{1}$$

where t_k is the nonuniform position, T is the period of the sawtooth-wave, and c is the inverse slope of the sawtooth-wave. The sum of delta functions can be expanded in the Fourier series as shown in [1]. The summary of analysis is as follows:

$$\sum_k \delta(t - t_k) = \frac{|1 - c\dot{x}|}{T} \sum_n \exp\left\{\frac{2\pi}{T} jn[t - cx(t)]\right\}, \tag{2}$$

where $\dot{x}$ is the derivative of the modulating signal $x(t)$.

Due to conjugate symmetry of (2), the right side of (2) turns into sum of cosines. The above equation implies that the sum of PPM impulses consists of infinite number of Phase Modulated (PM) signals at carrier frequencies of $2\pi n f_0$, where f_0 is the inverse of T—the fundamental frequency of the sawtooth-wave. The baseband and the fundamental frequency of (2) can be written as

$$Baseband + Fundamental = \frac{|1 - c\dot{x}|}{T} + \frac{2|1 - c\dot{x}|}{T} \cos\left\{\frac{2\pi}{T}[t - cx(t)]\right\}. \tag{3}$$

If the sawtooth wave fundamental frequency $1/T$ is large enough compared to the bandwidth of $x(t)$, the baseband term can be separated from the first PM term. Actually, the approximate bandwidth of the PM term is proportional to the bandwidth of $x(t)$ and c/T. Since the carrier frequency of PM ($1/T$) also changes with $1/T$, the most important factor for lack of frequency interference between the baseband and the PM term is c—the inverse of the slope. This implies that the larger the slope of the sawtooth-wave, the smaller the bandwidth of the PM component in (3) and hence the less the aliasing in the baseband.

This analysis is the basis of the traditional low-pass filtering of the PPM signal to demodulate the original signal. By removing the DC term of the baseband component, one can integrate and recover the band-limited signal $x(t)$. Indeed this low-pass filtering method is the predominant way of demodulating the PPM signal in communication systems. We shall show that by using the nonuniform sampling theory developed in [3], one can get almost perfect results. In either case, large transmission bandwidth is assumed for the PPM so that the time positions are accurately transmitted; we shall discuss this point again in Section 16.4.2.

16.2.2. Nyquist Rate in PPM

From the nonuniform sampling theory [2], we know that a band-limited signal can be recovered from its nonuniform samples if the average Nyquist density is satisfied. But according to the first author's publications [4–5] that are summarized in Sections 4.8 and 4.9 of this book, if the sample positions are related to the amplitude of the samples, it is possible to recover the original signal from the nonuniform samples at half the Nyquist density. Now, the PPM impulses have to be at the Nyquist rate. This is equivalent to half the Nyquist density since the amplitudes of PPM do not convey any information, i.e., we can combine two PPM pulses to form a nonuniform sample where the position of the nonuniform sample represents the position of the first pulse and the amplitude of the nonuniform sample represents the position of the second pulse.

Since the product of the PPM impulses and the sawtooth-wave yields a set of nonuniform samples at the Nyquist density, this implies that if nonuniform sampling techniques are employed, perfect recovery is possible. The simple low-pass filtering discussed in the previous section does not yield acceptable results at the Nyquist rate. Below are some other sophisticated techniques for the recovery of a band-limited signal from PPM pulses at or close to the Nyquist density.

16.2.3. Recovery Techniques Based on Sampling Theory

Time Varying or iterative techniques as discussed in [3] or (Chapter 4 of this book—Section 4.6) are some of the methods used in recovery of a band-limited signal from a set of nonuniform samples. The use of these techniques to PPM pulses is straightforward. The demodulator consists of converting the PPM pulses back to a set of nonuniform samples by multiplying the pulses by the sawtooth-wave; the demodulator then becomes recovery of the original band-limited signal from this set of nonuniform samples. Simulation results show significant improvement of the quality of the demodulated signal [6]; a reprint file is in the CD-ROM, file: 'Chapter16\Conference-paper.wpd'. In these simulations, a signal such as speech is band-limited using *sinc* filters. Iterations are achieved using low-pass filters with *sinc* impulse responses. Simulation results for $\mathbb{C}^N$ signals using FFT low-pass filters will be discussed in the following sections.

16.2.4. Analysis and Recovery in Finite Dimensional $\mathbb{C}^N$

A novel approach to using PPM is to use DFT band-limited discrete $\mathbb{C}^N$ signals as opposed to ideal band-limited continuous L_2 signals; for the definition of $\mathbb{C}^N$, see Chapter 4, Section 4.3 and Chapter 5. The N samples in $\mathbb{C}^N$ space could be regarded as a block of sampling points of a continuous signal (over-sampled at a much higher rate than the Nyquist rate) that is band-limited in the

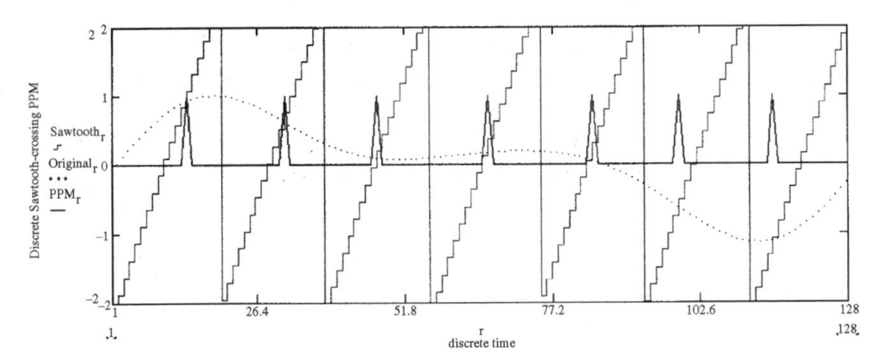

Figure 2. Discrete PPM as represented by discrete sawtooth-wave crossings for a band-limited DFT signal.

DFT domain, i.e., the high frequency DFT coefficients are made to be zero. PPM in this case means the position of samples that are the closest to discrete sawtooth-wave as shown in Fig. 2. Since the original samples are discrete, the time position of the PPM pulses represent a quantized version of the nonuniform sample amplitudes at the position of PPM pulses. Alternatively, the Discrete PPM pulses could be regarded as a time quantized (jittered) version of the actual nonuniform samples of the original analog signal.

Based on the analysis of the Extremum sampling as given in Chapter 4—(106), if the DFT bandwidth is $K,\ 2K-1$ samples in the time domain are needed to determine all the DFT coefficients. Hence, $2K-1$ PPM pulses are needed to fully determine the block of N samples. The relation is

$$T = r\frac{N}{2K-1},$$

where r is the time interval of two points, and T as before is the period of the sawtooth-wave. The advantage of PPM to that of $2K-1$ random nonuniform samples is that since the amplitude of the PPM pulses is a constant, only the position needs to be encoded.

For demodulating PPM pulses, there are different methods such as the Matrix Approach (MA), the Lagrange Interpolation (LI), and other methods such as the Reed Solomon (RS) and Iterative Methods (IM). These methods are the subjects of the next sections.

16.2.4.1. The Matrix Approach (MA)

The problem of interpolating a signal from the PPM pulses can be regarded as recovering the signal from a set of nonuniform samples for $\mathbb{C}^N$ signals. Since the time intervals for $\mathbb{C}^N$ signals are discrete, the interpolation problem reduces to the recovery of a signal with missing samples (see Chapters 5 and 17). Thus the matrix approaches discussed in Chapter 5 as well as the one in Section 4.10.5, can be used. The main problem with this approach is the complexity of finding the inverse of a matrix. We therefore discuss other methods that are simpler to implement. These methods are the Lagrange Interpolation, the Iterative technique, and the Reed Solomon decoding to be discussed below.

16.2.4.2. The Lagrange Interpolation (LI)

The LI method discussed for the Extremum sampling in Section 4.10.4 still holds for PPM. The main difference is that the nonuniform positions t_n given by (1) are quantized to $n.r$, where

$$nr \leq t_n < (n+1)r,$$

and r is the time interval between two sampling points. For our case, the actual sample values x_{s_i} in (3) of Section 4.10.4 are equal to $(t_i - iT)/c$.

The LI and its simulation are given in (4) and Fig. 3, respectively. Fig. 3 shows that for a time resolution of 36 points per the sawtooth period (Nyquist interval), the Lagrange interpolation yields a correlation of 0.999 right at the Nyquist rate (see footnote 4 of Section 4.10.4 for the definition of correlation function—a correlation of 1 implies perfect recovery). If we decrease the time resolution down to 18, the correlation decreases down to 0.98. If we over-sample

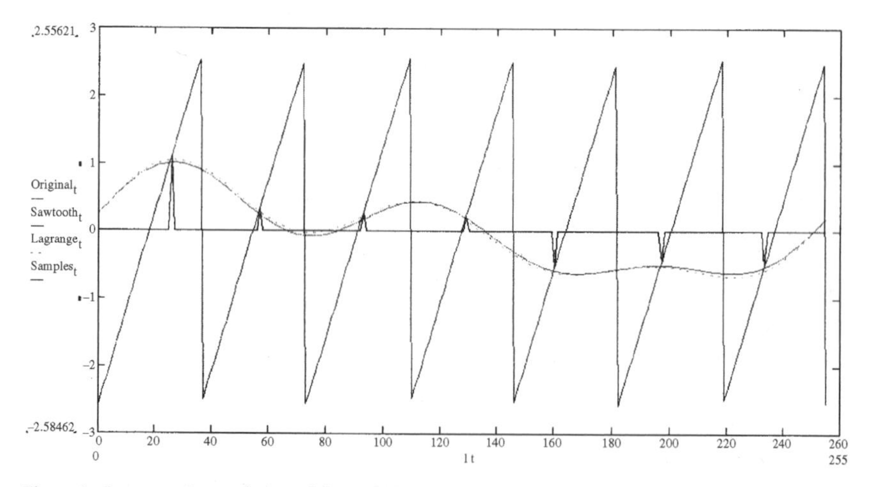

Figure 3. Lagrange interpolation of discrete PPM at the Nyquist rate with a resolution of 36 time slots in each time period T.

the signal, the Lagrange interpolation is still valid but does not seem to have any advantage to the Nyquist rate except it may be a protection against additive noise; for more discussion please see the section on RS decoding (Section 16.4.4).

$$x_k = \frac{1}{cw_k^{K-1}} \sum_{i=0}^{2K-1} (t_i - iT)(w_{t_i})^{K-1} \psi_{i,k}, \tag{4}$$

where

$$\psi_{i,k} = \frac{\prod_j (w_k - w_{t_j})}{\left[\prod_{j \neq i} (w_{t_i} - w_{t_j})\right](w_k - w_{t_i})},$$

$$T = r\frac{N}{(2K-1)} \quad \text{and} \quad w_k = \exp\left(\frac{2\pi kj}{N}\right).$$

The parameters N, K, r, and T are as defined before, namely, the number of points per frame, the bandwidth, the time interval between two points in seconds, and the period of the sawtooth-wave, respectively. Equation (4) shows that from the position of the PPM pulses, t_i, one can interpolate and get the original discrete signal to within the time quantization error—see Section 16.4.2.5.

The Mathcad™ simulation is given in the accompanying CD-ROM under the directory 'Chapter16\Ppm\Lagrange-PPM.mcd'.

16.2.4.3. The Iterative Technique (IT)

The IT was discussed in Chapters 4 and 5. Although the IT method outperforms simple low-pass filtering and other methods such as low-pass filtering of the quantized difference of the sawtooth-wave and the modulating signal, the amount of the computation is not worth the effort compared to the following technique in terms of the complexity and the quality of the reconstruction. For instance, for the example in the next section, the correlation of the reconstruction for low-pass filtering is 0.971 at the Nyquist rate while for the IT method after 20 iterations is only 0.973, and for the RS method to be discussed in the next section is 0.994. The performance of the IT improves if both the sampling rate and the pulse time resolution are increased, which is also true for the other techniques.

A generic iterative algorithm simulated by Mathcad™ is given in the accompanying CD-ROM under the directory 'Chapter16\Iteration.mcd'. This iterative algorithm is also valid for PPM, where the product of PPM pulses with the sawtooth-wave generates a set of nonuniform samples.

16.2.4.4. The Reed Solomon (RS) Decoding Approach

The RS method has been successfully implemented for the missing sampling problem [7] and was also discussed in Section 4.10.13 on Extremum sampling. We shall show in this section that this method can also be used very efficiently to demodulate a PPM signal. Based on our analysis in [7] that is also discussed in Chapter 17 on coding and error concealment, the PPM demodulation is stable and the time quantization, which is translated into sample quantization is demodulated into finite quantization error. The zero diagram (see Section 17.6.2 and the corresponding Figs. 12–14) for the error locator polynomial of a specific PPM signal shown in Fig. 2 is depicted in Fig. 4. The diagram shows that the zeros are around the circle and hence the demodulation should be stable—see Chapter 17. Indeed the h parameters calculated from the missing locator polynomial (Equation (5) in Chapter 17) are less than 1 and are well behaved as shown in Fig. 5—the simulation for the error locator polynomial and the h parameter is given in the accompanying CD-ROM under the directory 'Chapter16\RS-PPM.mcd'.

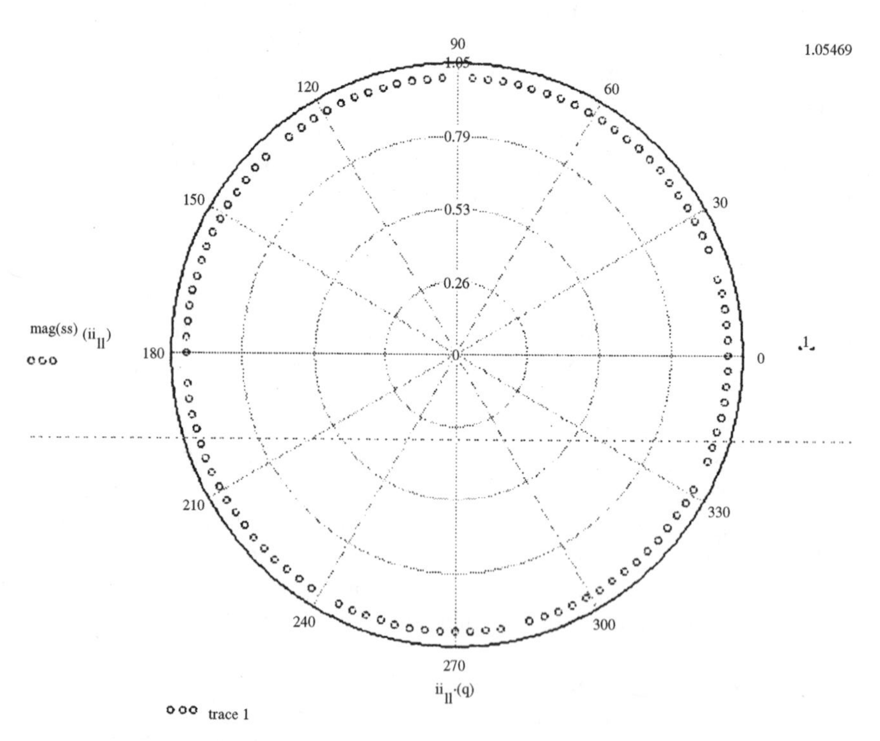

Figure 4. The zeros of the Error Locator Polynomial of the PPM signal shown in Fig. 2; see Fig. 12 in Chapter 17.

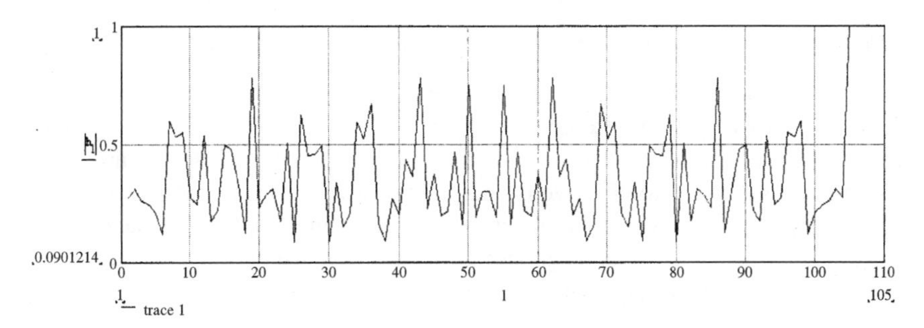

Figure 5. The h values derived from the missing locator polynomial; see Fig. 11 in Chapter 17.

The demodulated signal using this method is shown in Fig. 6. The correlation of the reconstructed signal as shown in Fig. 6 is 0.994; the error is due to the time quantization of the PPM pulses. The pulse time resolution is represented by 4 bits (16 time slot) at the Nyquist rate (7 samples per frame). If we increase the time resolution (transmission bandwidth), the correlation will approach 1.

Alternatively, we can increase the sampling rate at a given resolution between the sample intervals. Using this approach, we conclude that it is more advantageous to increase the resolution at the Nyquist rather than increasing the sampling rate. For example, if we increase the sampling rate by a factor of 2 and decrease the resolution down to 2 bits per sampling interval (the total bit rate per frame remains unchanged), the correlation of the demodulated signal decreases from 0.994 down to 0.97. Over-sampling may only become advantageous in case of additive channel noise. The redundancy of over-sampling becomes a protection against channel noise. The Mathcad(TM) simulation for RS decoding is given in the accompanying CD-ROM under the directory 'Chapter16\Ppm\RS-PPM.mcd'.

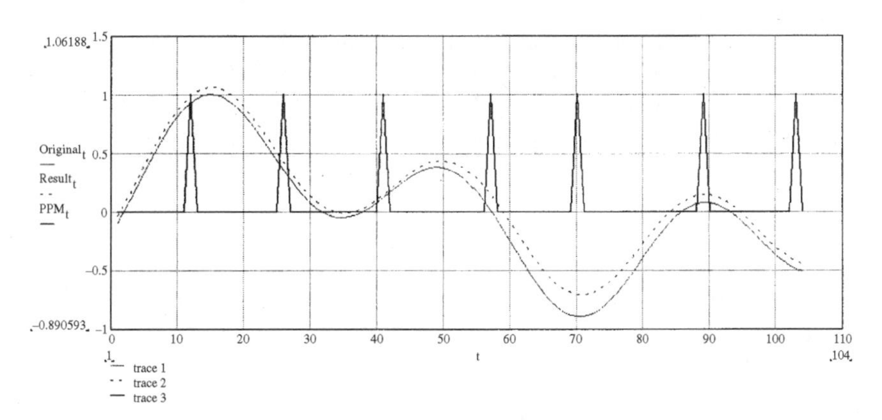

Figure 6. The original signal, the PPM pulses and the reconstructed signal using the RS method.

16.2.4.5. Quantization Error of Discrete PPM

Since the real intersection of a sawtooth-wave with a continuous signal may lie between two discrete samples $x(nr)$ and $x[(n+1)r]$ that are separated by r units in time, the position of one of the adjacent samples that has the closest distance with the sawtooth-wave is chosen for PPM. This time jitter is translated into quantization noise in the generated nonuniform sample. The quantization error magnitude $|e|$ cannot be greater than the difference between the amplitudes of samples of the two adjacent samples, i.e.,

$$|e| \leq r/c, \tag{5}$$

where $r = t_n - t_{n-1}$ and $1/c$ is the slope of the sawtooth-wave, which is assumed to be positive. To assure that we have only one intersection per interval, we need the sawtooth-wave to have an amplitude and slope higher than the highest amplitude and slope of the continuous signal (which depends on its bandwidth), i.e.,

$$1/c \geq |\max \textit{slope of } x|. \tag{6}$$

The proof of (5) is simple since if e violates (5), then we can find two points on the continuous signal $x(t)$, such that the magnitude of the slope of the line passing through the two points is larger than the slope of the continuous signal. The two points consist of one of the chosen points $x(nr)$ or $x[(n+1)r]$ and the intersection-point of the sawtooth-wave and the continuous signal $x(t)$, i.e., $x(t_n)$; t_n is somewhere between nr and $(n+1)r$. According to the Mean Value Theorem, we can find a point between t_n and nr or $(n+1)r$ such that its slope is the same as the slope of this line. But this violates our assumption of (6); hence (5) must be true.

We can show that if condition (6) is satisfied, the amplitude condition is satisfied for all practical cases. The proof is as follows: The amplitude of the sawtooth-wave A is determined by the slope $1/c$ and the period T, i.e., $A = T/c$. But the amplitude A must be greater than the maximum amplitude of the signal $x(t)$, hence we have

$$T/c \geq Max\, x(i). \tag{7}$$

It is not difficult to show that the maximum value of $x(i)$ is related to its bandwidth K by

$$Max\, x(i) \leq \frac{|X_{\max}(k)|(2K-1)}{N}. \tag{8}$$

Also, by the same analysis, the slope of $x(i)$ is upper bounded by

$$slope = \left|\frac{dx(i)}{di}\right| \leq \frac{2\pi}{N} | X_{\max}(k) | (K - 1), \tag{9a}$$

or a tighter bound

$$slope = \left|\frac{dx(i)}{di}\right| \leq \frac{2\pi}{N^2} | X_{\max}(k) | K(K + 1). \tag{9b}$$

Now if (9) is satisfied, we can show rather easily that (7) is satisfied since $K > 1$, $K < N$ and T is given by (4).

Other periodic waveforms such as triangular waves as opposed to sawtooth-wave can be used, the analysis is similar and will not be discussed here; see [1].

The noise analysis of PPM would be similar to Sine Wave Crossings (SWC) that will be discussed in Section 16.5.2.3. Also, for more details about PPM and the noise analysis, you may refer to the MSc thesis and the Matlab(TM) codes in the accompanying CD-ROM, directory: 'Chapter16\Ppm\MSc-Thesis*.*'.

16.3. Pulse Width Modulation (PWM)

PWM is a pulse time modulation where information lies in the width of each transmitted pulses. This can be achieved by passing the difference of modulating signal and the sawtooth-wave through a hard limiter. The classical demodulation is to low-pass the PWM pulses at the receiver to yield the original modulating signal. Since PPM pulses are derived from PWM pulses, PWM can be converted into PPM pulses and thus the above techniques for PPM demodulation can be used for PWM demodulation. PPM modulation performs better than PWM in case of channel noise but this is beyond the scope of this book. PWM has applications in Digital-to-Analog conversions and will be discussed again in Section 16.7.4.

16.4. FM Demodulation

Traditionally, FM demodulation has been very important for radio, television, and analog video recording. Some ad-hoc classical demodulation techniques has been the use of zero-crossing counts; this can be explained by the spectral analysis of [9] that will be discussed in following section. Wiley was the first investigator that related FM zero-crossings to nonuniform sampling reconstruction [10]. Below, we shall discuss the spectral analysis, analog reconstruction methods, and a discrete version of FM demodulation for the sake of simulations. This discrete implementation may lead to new ways of demodulation of digital modulation techniques such as Frequency Shift Keying (FSK) and Phase Shift Keying (PSK). Also, it may lead to new ways of modulation of discrete time but analog signals.

16.4.1. FM Spectral Analysis and Lagrange Interpolation (LI)

An FM signal is represented by the following equation:

$$x_c(t) = \sin\left(\omega_c t + 2\pi f_\Delta \int_{-\infty}^{t} x(\lambda)d\lambda + \theta\right). \tag{10}$$

In the above equation, $x_c(t)$ is the FM modulated signal, ω_c is the carrier frequency in radians, f_Δ is the index of modulation, and θ is a constant phase. The FM zero-crossings can be used for demodulation. Since the set of FM zero-crossings is related to a set of nonuniform samples, we can use the standard nonuniform sampling techniques for FM demodulation. The FM zero-crossings are derived by setting (10) to zero and solving for t; the result is

$$t_k = \frac{kT}{2} - Tf_\Delta \int^{t_k} x(\lambda)d\lambda - \frac{T\theta}{2\pi}, \tag{11}$$

where T is the period of the carrier frequency given by $T = 2\pi/\omega_c$ and k is any integer. If $\theta = \pi/2$, (10) becomes a cosine term and the last constant term of (11) becomes $T/4$. If the integral of $x(t)$ in (11) is called $y(t)$, we can rewrite (11) as

$$t_k = kT_1 + cy(t_k) - \varphi, \tag{12}$$

where $T_1 = T/2$, c is a constant equal to $-Tf_\Delta$, $y(t_k)$'s are the nonuniform samples of the integral of $x(t)$, which is also band-limited, and φ is a constant shift—equal to the last term of (11)–that does not affect the structure of the zero-crossings t_k. A comparison of (12) to that of (1) shows that the zero-crossings of an FM signal is equivalent to a PPM signal provided that the modulating signal $x(t)$ is integrated first; i.e., the intersection of a sawtooth-wave with the integral of $x(t)$ yields the FM zero-crossings.

Therefore, spectral analysis and all the techniques discussed in Section 16.2.1 on PPM demodulation can be used for FM demodulation. Below we will derive direct demodulation of FM from its zero-crossings based on Lagrange interpolation. Since $y(t)$ is band-limited, as long as the nonuniform sampling conditions for Lagrange interpolation are satisfied, we can recover $y(t)$ from the nonuniform samples $y(t_k)$'s using the following interpolation:

$$y(t) = \sum_{k=-\infty}^{\infty} y(t_k)\Psi_k(t), \tag{13}$$

where $\Psi_k(t)$ is as defined Chapters 2–3 and repeated in (90) of Section 4.8.3; see also (107.b) in Section 4.10.4. From (12), (13) and the fact that $x(t)$ is the derivative of $y(t)$, we get

$$x(t) = \sum_{k=-\infty}^{\infty} \frac{-t_k + kT_1 + \varphi}{c} \dot{\Psi}_k(t, t_k). \tag{14}$$

The above equation is equivalent to Equation (13) in [1].

16.4.2. Simulation Results

16.4.2.1. Band-limited L_2 Signals

It has been shown that the process of low-pass filtering the impulses at the zero-crossings yields approximately the original signal [8]–[9]. This may be obvious from the fact that the FM zero-crossings convey the same information as the intersection points of a sawtooth-wave with the integral of the modulating signal. Hence equations (2) and (3) for PPM also hold for FM zero-crossings. The modification is the replacement of $x(t)$ with its integral $y(t)$ and T with T_1 as defined in (12). Equation (3) implies that for the FM demodulation, just low-pass filtering the zero-crossing impulses yields a baseband signal which is proportional to the original signal since the derivative of $y(t)$ is equal to $x(t)$.

Wiley was the first investigator that used the iterative method and a smeared version of nonuniform samples to demodulate FM signal [10]. The author has repeated Wiley's iteration on actual impulsive nonuniform samples with good results. Below, we will describe the results for $\mathbb{C}^N$ signals.

16.4.2.2. Band-limited $\mathbb{C}^N$ Signals

Similar to the L_2 case, the $\mathbb{C}^N$ signal can be represented by a discrete version of (10), i.e.,

$$x_c(n) = \sin(\omega_c n + 2\pi f_\Delta y(n) + \theta), \tag{15}$$

where $y(n)$ is the discrete version of the integral of the modulating signal, $x(t)$, modified by FFT low-pass filtering. If $y(n) = x(n)$, then we get the discrete version of Phase Modulated (PM) signal. The modulator converts (15) into a continuous signal by spline or sinc interpolation. At the receiver, the zero-crossings of the received signal convey the same information as the discrete PPM pulses that was discussed in previous section. In fact these zero-crossings, depending on the carrier frequency ω_c, are over-sampled and can be down-sampled to the Nyquist rate.

Any of the techniques discussed in the previous section on PPM can be used for the demodulation of FM. Equations (15) and its *sinc* interpolated signal are shown in Fig. 16.7. Alternatively, the band-limited $\mathbb{C}^N$ samples can be sampled-and-held and then a conventional FM modulator could be used on the sampled-and-held modulating signal. At the receiver end, the retrieved zero-crossings of the FM signal is equivalent, in information content, to the uniform PPM as discussed in [1].

Discrete PM and FM demodulation using Lagrange interpolation (14) for $\mathbb{C}^N$ signals are shown in Figs. 8 and 9, respectively. Figure 8 shows that the discrete PM signal can be demodulated via the Lagrange Interpolation of the PPM with a correlation of 0.994 for a time resolution of 27 points per Nyquist interval. For the discrete FM, similar to the one shown in Fig. 7, the modulating signal $x(n)$ goes through a "circular" accumulator to get $y(n)$ before the PM modulation shown in (15). This accumulator is the equivalent of an integrator for the analog signals in L_2; the impulse response is equal to 1 for $n = 0, \ldots, N/2$ and is equal to 0 for $n = N/2, \ldots, N-1$. $y(n)$ is demodulated exactly like the

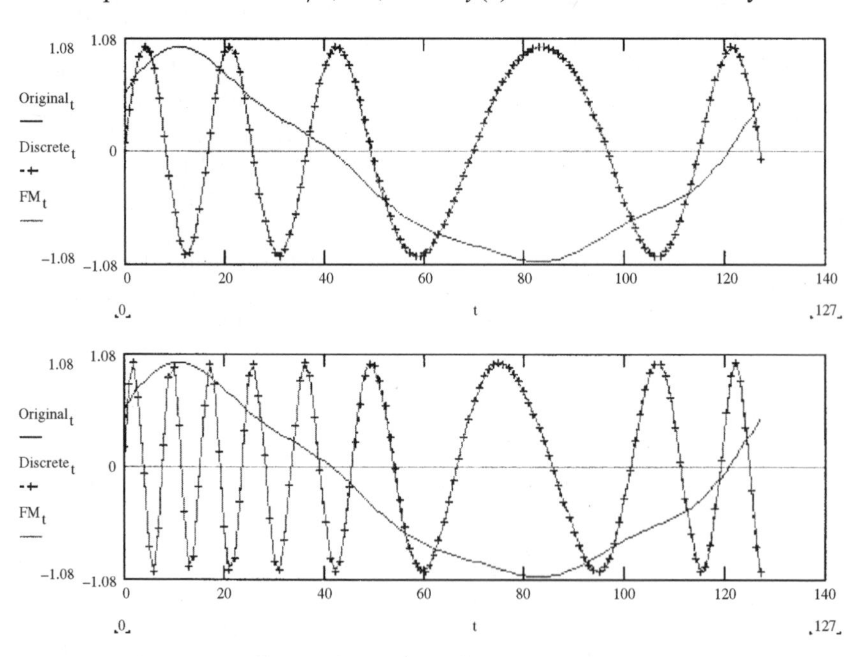

Figure 7. (a) The original DFT band-limited signal, discrete FM and the *sinc* interpolated discrete FM with zero-crossings at the Nyquist rate. (b) The original DFT band-limited signal, discrete FM and the *sinc* interpolated discrete FM with zero-crossings at twice the Nyquist rate.

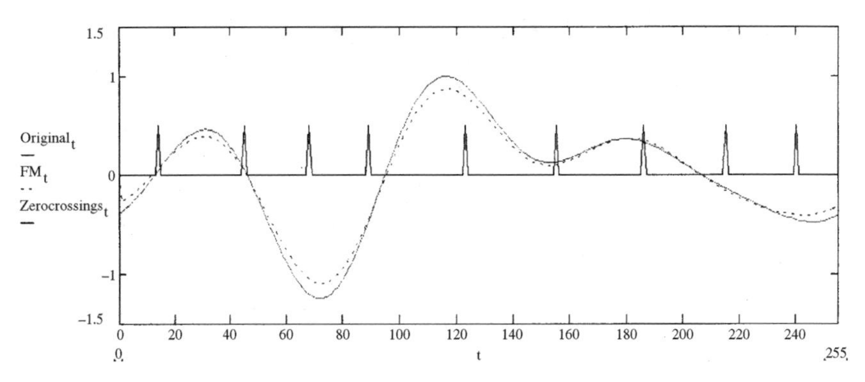

Figure 8. Discrete PM recovery using Lagrange interpolation from the zero-crossings of PM at the Nyquist rate—correlation is 0.994.

discrete PM discussed for Fig. 8 and then $x(n)$ is retrieved by either going into an inverse filter or by the approximation $x(n) \approx y(n) - y(n-1)$. The demodulated signal shown in Fig. 9 is the result of using an inverse filter to retrieve $x(n)$ from $y(n)$.

The Mathcad(TM) simulation for PM and FM demodulation using Lagrange interpolation is given in the accompanying CD-ROM under the director 'Chapter16\Fm\FM-Lagrange.mcd'.

16.5. Sine-Wave Crossings (SWC)

A sine (or cosine) wave crossing with a band-limited signal is represented by the following equations and is shown in Fig. 10.

$$A\cos(\omega_c t) = x(t). \tag{16}$$

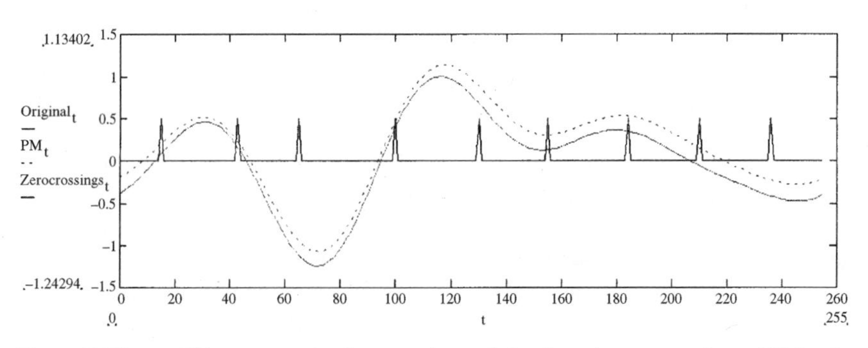

Figure 9. Discrete FM recovery using Lagrange interpolation from the zero-crossings of FM at the Nyquist rate—correlation 0.997.

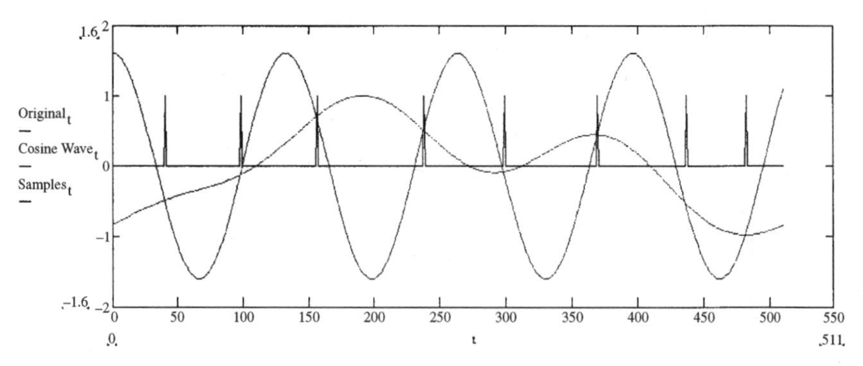

Figure 10. Sine-wave crossings (SWC) of a band-limited signal at the Nyquist rate.

The solution to the above equation leads to two sets of equation:

$$t_k = 1/\omega_c \cos^{-1}[x(t_k)/A] + k\pi, \quad k = \text{even integers} \tag{17a}$$

$$t_k = -1/\omega_c \cos^{-1}[x(t_k)/A] + k\pi, \quad k = \text{odd integers} \tag{17b}$$

where the inverse cosine is the Principal value defined in the interval $-T/2 < t < T/2$.

If the amplitude of the sine-wave is larger than the maximum amplitude of a band-limited signal and if the carrier frequency of the sine-wave is greater than the bandwidth of the modulating signal, then recovery from the sine-wave crossings is possible since the average Nyquist rate for nonuniform sampling is satisfied. An explicit mathematical derivation for representing a band-limited signal by its sine (cosine) wave crossings is given below [11]

$$x(t) = [x(0) - A] \prod_{k=-\infty}^{k=\infty} (1 - t/t_k) + A \cos \omega_c t, \tag{18}$$

where the parameters are as defined in (16). The above formula is an extension of Titchmarsh infinite expansion for a band-limited signal as given in [1, see equations (4)–(7) and (10) in that reference]. One could imagine a new mixed band-limited signal (sum of an L_2 and a periodic signal), $z(t) = x(t) - A\cos(\omega_c t)$. The zeros of $z(t)$ are the same as (18) and $z(t)$ is equal to the product term of (18). The above formula does not yield a practical method for reconstruction. Below, we will discuss practical methods for signal recovery based on the analysis similar to that of PPM for both L_2 and $\mathbb{C}^N$ band-limited signals.

16.5.1. Analysis for L_2 Band-limited Signals

Equations (2) for PPM together with (17a) become:

$$\sum_{k=even} \delta(t-t_k) = \left|1 - \frac{1}{\omega_c}\frac{d}{dt}\arccos\frac{x(t)}{A}\right| \sum_{n\in Z} \delta\left(t - \frac{1}{\omega_c}\arccos\frac{x(t)}{A} - \frac{2\pi n}{\omega_c}\right) \quad (19)$$

The Fourier series expansion of (19) is

$$Fourier - Series = \left|f_c - \frac{1}{2\pi}\frac{d}{dt}\arccos\frac{x(t)}{A}\right| \sum_{n} \exp\left(\omega_c jn\left[t - \frac{1}{\omega_c}\arccos\frac{x(t)}{A}\right]\right). \quad (20a)$$

Likewise, for the odd component of (17b), the summation of (19) and (20a) become

$$\sum_{k=odd} \delta(t-t_k) = \left|f_c + \frac{1}{2\pi}\frac{d}{dt}\arccos\frac{x(t)}{A}\right| \sum_{n} \exp\left(\omega_c jn\left[t + \frac{1}{\omega_c}\arccos\frac{x(t)}{A} - \frac{\pi}{\omega_c}\right]\right). \quad (20b)$$

The baseband and the fundamental components of (20a) similar to equation (3) for PPM become

$$= \left|f_c - \frac{1}{2\pi}\frac{d}{dt}\arccos\frac{x(t)}{A}\right| + \left|f_c - \frac{1}{2\pi}\frac{d}{dt}\arccos\frac{x(t)}{A}\right| \cos\left(\omega_c t - \arccos\frac{x(t)}{A}\right). \quad (21a)$$

Similarly, for part (20b), we have

$$= \left|f_c + \frac{1}{2\pi}\frac{d}{dt}\arccos\frac{x(t)}{A}\right| - \left|f_c + \frac{1}{2\pi}\frac{d}{dt}\arccos\frac{x(t)}{A}\right| \cos\left(\omega_c t + \arccos\frac{x(t)}{A}\right). \quad (21b)$$

If we assume the odd impulses in (17b) that are generated from negative slope of the cosine-wave are negative, i.e., (21a) is subtracted from (21b) and then integrate the impulses, we get

$$= -\frac{1}{\pi}\arccos\left(\frac{x(t)}{A}\right) + \frac{1}{\pi}\frac{x(t)}{A}\sin(\omega_c t). \quad (22)$$

Derivation of (22) is based on the assumption that ω_c and A are large enough such that the terms inside the magnitude signs in (21) are always positive. The above equation implies that a proper band-pass (even low-pass) filter of a rectangular waveform (see Fig. 10) derived from the integral of the interlaced positive and negative impulses generated from the sine-wave crossings yields a Double Side Band DSB modulated signal. Hence, a proper DSB demodulation can be used to retrieve the original band-limited signal from the sine-wave crossings.

16.5.2. Analysis for $\mathbb{C}^N$ Band-limited Signals

If we take a block of N samples and use an FIR or an IIR filter to make it a low-pass $\mathbb{C}^N$ signal, its Sine-wave crossings at or above the Nyquist rate should be enough to approximately recover the original low-pass signal. The approximation is due to the fact that the digital filters are not exact and do not have sharp roll-offs beyond their cutoff frequencies. Simple recovery techniques such as low-pass filtering the SWC samples can still be used; the low-pass filtering result is shown in Fig. 11(a) at twice the Nyquist rate (with a correlation of 0.971). Right at the Nyquist rate, the recovery of the original signal from the nonuniform samples is poor and the correlation for the random signal given in Fig. 11(a) is about 0.65. If the sampling rate is increased, this simple technique improves. However, at higher frequencies (> twice the Nyquist rate), we can get better results by just low-pass filtering the SWC samples as depicted by a rectangular wave in Fig. 11(b) derived from hard-limiting the difference between the cosine-wave and the original band-limited signal. At twice the Nyquist rate the correlation for the method shown in Fig. 11(b) is about the same as the previous simple low-pass filtering technique and it is 0.968. As the sampling rate increases beyond twice the Nyquist rate, the method of Fig. 11(b) outperforms that of Fig. 11(a).

The reason why low-pass filtering works comes from (22); the output of the low-pass filter is

$$\text{Lowpass filter of the rectangular wave} = -\frac{1}{\pi}\arccos\left(\frac{x(t)}{A}\right).$$

Assuming large A compared to the maximum value of $x(t)$, we get the following approximation from the above formula

$$\approx -\frac{1}{2} + \frac{x(t)}{A}. \tag{23}$$

Equation (23) explains the reason for DC offset when we use low-pass filtering in our simulations.

The above approach could be extended to DFT low-pass filtering. In this case, by just converting the high frequencies to zeros, we achieve low-pass

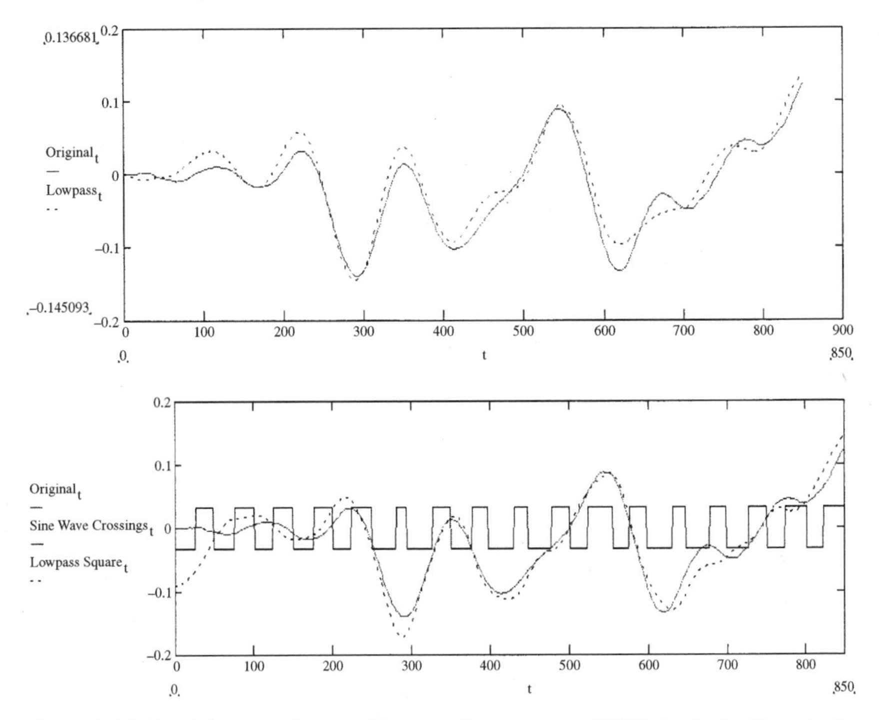

Figure 11. (a) Simple low-pass filtering of the nonuniform samples of SWC at twice the Nyquist rate–the correlation is 0.971. (b) Low-pass filtering the SWCs represented by the rectangular curve at twice the Nyquist rate—the correlation is 0.968.

filtering. In many practical applications, this way of low-pass filtering creates ringing distortion, which can be minimized by increasing the frame (block) size, increasing the effective bandwidth by reducing the number of zeros in the DFT domain, and by windowing the DFT coefficients. In any case if we low-pass filter a discrete signal in this fashion, we can then use Lagrange and RS decoding approaches to demodulate from the sine-wave crossings. The advantage is that we can achieve almost perfect recovery at the Nyquist rate. These two methods will be discussed in the next sections.

The Mathcad(TM) simulations for SWC demodulation using Low-pass filtering (as shown in Figure 16.11) are given in the accompanying CD-ROM under the directory 'Chapter16\SWC\Lowpass-SWC.mcd'.

16.5.2.1. Lagrange Interpolation (LI)

The results and the method of recovery for sine-wave crossings are the same as the PPM case. The difference is that a sawtooth-wave is replaced by a sine-

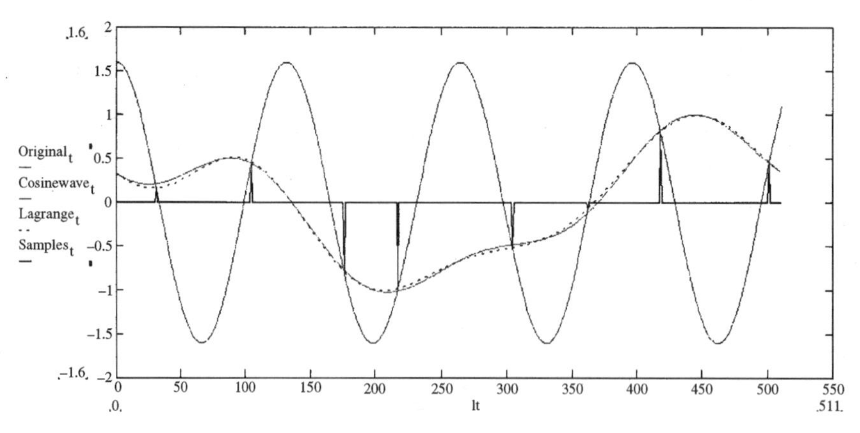

Figure 12. Lagrange interpolation (LI) for a Sine-wave Crossings (SWC)—the correlation is 0.999.

wave. The results for the LI method is shown in Fig. 12. For the Nyquist crossings of a band-limited random signal given in Fig. 12 with a time resolution of 33 time slots per Nyquist interval, the correlation of the LI method is 0.999 while low-pass filtering the nonuniform samples is 0.964 and the Nonlinear Method (see Chapter 4, Section 4.6.1) yields a correlation of 0.96. We should bear in mind that these numbers are only for the signal given in Fig. 12 but for other realizations of discrete random signals of the same bandwidth, the relative correlation numbers change. In fact the orders for some occasions may be reversed. Only statistically, we would say that, most of the time, LI is better than the Nonlinear method and the latter is better than simple low-pass filtering the nonuniform samples generated from SWC.

The Mathcad(TM) simulation for SWC demodulation using Lagrange interpolation is given in the accompanying CD-ROM under the directory 'Chapter16\SWC\SWC-Lagrange.mcd'.

16.5.2.2. RS Decoding Method

All the issues discussed in Section 16.2.4.4 on RS decoding for PPM signal is also valid for SWC. The result of RS decoding of this type of modulation is shown in Fig. 13. The correlation of reconstruction is 0.995 for a time resolution of 64 slots per Nyquist interval. The results are essentially comparable to the LI methods except that the RS decoding requires much less computation.

The Mathcad(TM) simulation for SWC demodulation using Lagrange interpolation is given in the accompanying CD-ROM under the directory 'Chapter16\SWC\Cosine-RS.mcd'.

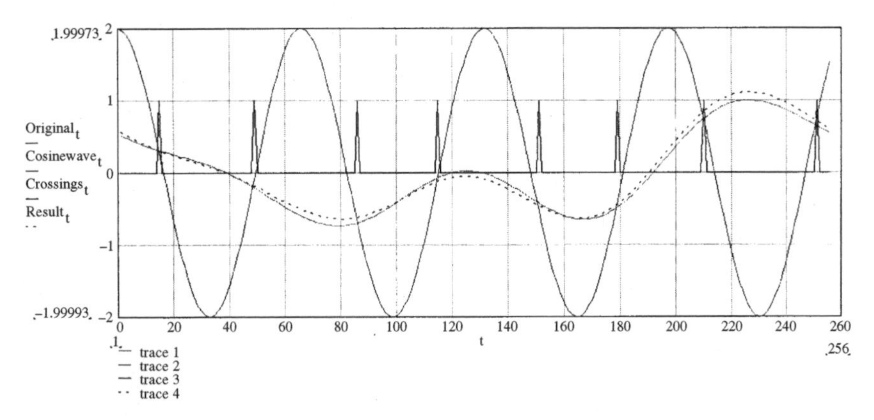

Figure 13. RS decoding based on nonuniform samples of SWC—the correlation is 0.995.

16.5.2.3. Noise Analysis

If the pulses generated from the SWC crossings are corrupted with additive Gaussian noise and then low-pass filtered, the exact positions of the received pulses are estimated by thresholding the corrupted pulses. The thresholding effect creates some uncertainty on the time positions; i.e., it generates misplaced pulses due to band-limiting and additive noise corruption. In addition, there is a possibility of generating false pulses from the additive noise component.

Figure 14(a) shows the effect of noise and band-limiting distortion on the sine-wave crossings and reconstruction from the corrupted pulses at the Nyquist rate. In this figure, the Signal-to-Noise Ratio is about 26 dB; the correlation of the reconstruction is 0.977 for the signal shown. If we increase the sampling rate, then we have the liberty of increasing the threshold; this would improve the quality since the probability of false pulses decrease and as long as we get enough samples per block (Nyquist rate), we would be able to reconstruct the original signal; Fig. 14(b) shows this phenomenon for a different realization of the band-limited signal. In other words, there is a possibility of improving the quality of recovery by over-sampling without increasing the time resolution. Notice that the assumptions made for Fig. 14(b) is not the same as part (a); therefore the two figures cannot be really compared.

For the case of over-sampling as shown in Fig. 14(b), the main problem with the reconstruction of the signal from the sine-wave crossings is the stability of reconstruction irrespective of whether one uses LI or RS decoding approaches. The problem is similar to the stability problem of burst error recovery for error correction or error concealment methods discussed in [7] and [12] and will be discussed in Chapter 17. We are currently doing research on the feasibility of using burst error recovery techniques to address demodulation of PPM and SWC

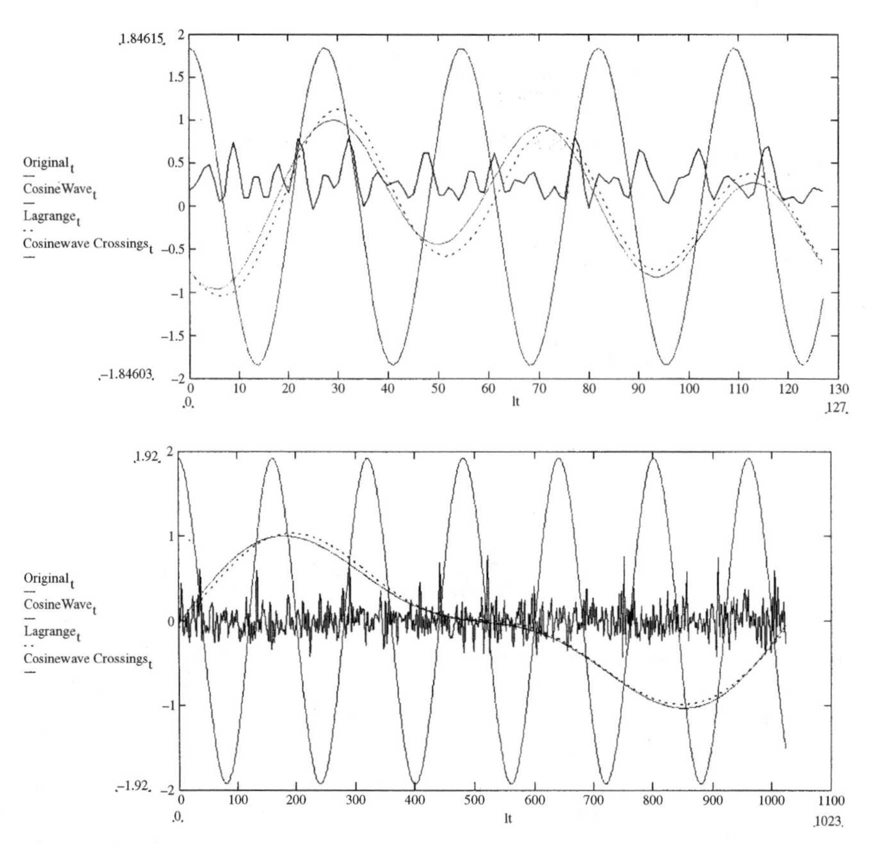

Figure 14. (a) Recovery from SWC based on corrupted and band-limited pulses at the Nyquist rate—the correlation is 0.977. (b) Recovery from SWC based on corrupted and band-limited pulses at two and a half times the Nyquist rate.

under noisy environment. This would ultimately lead into a unified and joint approach towards modulation, error control and error concealment.

The Mathcad™ simulation for Fig. 14 is given in the accompanying CD-ROM under the directory 'Chapter16\SWC\SWC-Lagrange-Noise.mcd'; the RS decoding for noisy SWC is given in 'Chapter16\SWC\SWC-RS-Noise.mcd'.

16.6. Delta Modulation (DM)

Variations of DM can be used in Analog-to-Digital (A/D) conversion [13]. DM is an adaptive procedure based on the modulating (baseband) signal and the previous analysis methods cannot be used for spectral analysis and recovery of a DM signal. A typical DM signal is shown in Fig. 15(a). This figure shows a staircase function that approximates the original signal. The information lies in the crossings of this approximate signal with the original modulating signal; these crossings are shown in Fig. 15 as a rectangular waveform. The mathematical relationship that generates the Mathcad simulations in Fig. 15 are shown below. Assuming a step size of d, we have the following Mathcad® relations[1]:

$$k \equiv 0 \ldots f \cdot L$$
$$\text{sign}(a) \equiv (a > 0) - (a < 0)$$
$$\begin{pmatrix} S_0 \\ M_0 \end{pmatrix} \equiv \begin{pmatrix} d \\ -d \end{pmatrix}$$
$$\begin{pmatrix} S_{k+1} \\ M_{k+1} \end{pmatrix} \equiv \begin{bmatrix} S_0 \cdot \text{sign}[F((k+2) \cdot T) - (M_k + S_k)] \\ M_k + S_k \end{bmatrix}, \tag{24}$$

where f and L are, respectively, the sampling frequency and the duration of the signal; k is the discrete time index varying from 0 to $f \cdot L$; M is the staircase approximation to the modulating signal F, S is the rectangular waveform that is actually transmitted, and sign($\cdot$) is the signum function. The integral (accumulator in the discrete case) of this rectangular waveform yields the staircase function.

The classical method of recovery of the original signal from the rectangular waveform is to integrate and then use a low-pass filter as shown in Fig. 15(b). This approximation is accurate only at a high resolution; i.e., high clock rates which require large bandwidth. An alternative method to improve the quality of recovery is to use an Adaptive DM (ADM) as shown in Fig. 16. This figure shows that the step size is adapted to match the slope of the modulating signal F at different times. By adapting the step size, the *slope overload* and the *granular noise* of DM are decreased; for Mathcad simulation, the Mathcad[2] relation for adapting a DM is shown below:

$$j \equiv 0 \ldots f \cdot L$$
$$\begin{pmatrix} M_0 \\ s_0 \\ e_0 \end{pmatrix} \equiv \begin{pmatrix} 0 \\ d_{min} \\ 1 \end{pmatrix}$$
$$\begin{pmatrix} M_{j+1} \\ s_{j+1} \\ e_{j+1} \end{pmatrix} \equiv \begin{bmatrix} M_j + s_j \\ |s_j| \cdot \text{sign}[F((j+2) \cdot T) - (M_j + s_j)] + s_0 \cdot e_j \\ \text{sign}[F((j+2) \cdot T) - (M_j + s_j)] \end{bmatrix}, \tag{25}$$

where d_{min} is the minimum distance and e is the adaptation parameter; the other parameters are as defined in (24).

[1] With permission from the Electrical Engineering Handbook of Mathcad®.
[2] With permission from the Electrical Engineering Handbook of Mathcad®.

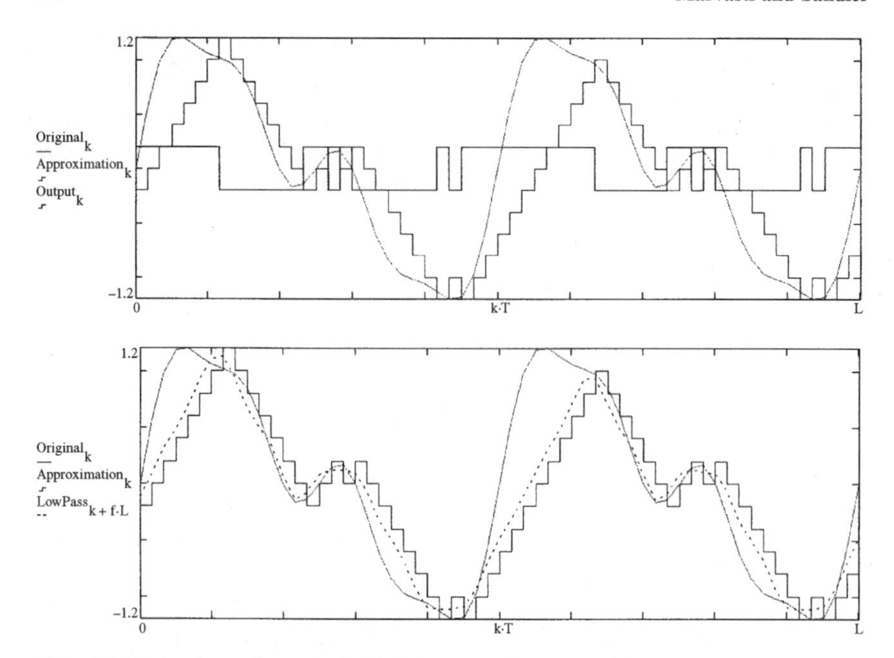

Figure 15. (a) A staircase discrete Delta Modulator DM. (b) Recovered signal by low-pass filtering the staircase approximate function—the correlation is 0.916.

If nonuniform sampling recovery techniques are used, the step size and to some degree the clock rate can be reduced; this reduction decreases the effect of quantization noise. Also, we do not have to worry about slope overload; which is a problem in DM if the step size is reduced; also we can bypass ADM is nonuniform sampling recovery is used. At the decoder, quantized nonuniform samples are generated by sampling the staircase function at the negative transitions of the rectangular waveform as shown in Fig. 17(a). As long as the set of nonuniform samples has a density above the Nyquist rate, recovery is

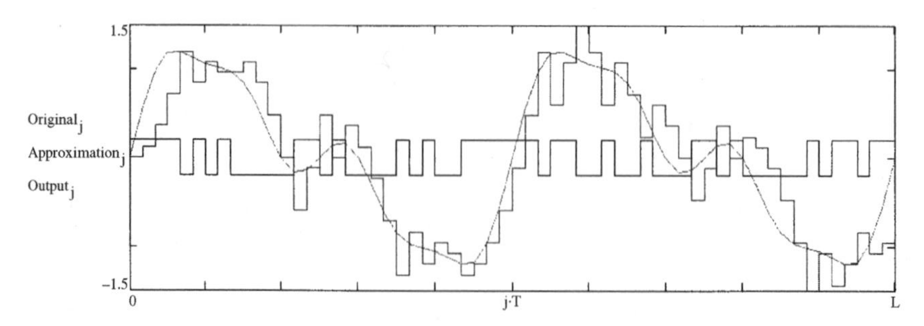

Figure 16. An adaptive DM.

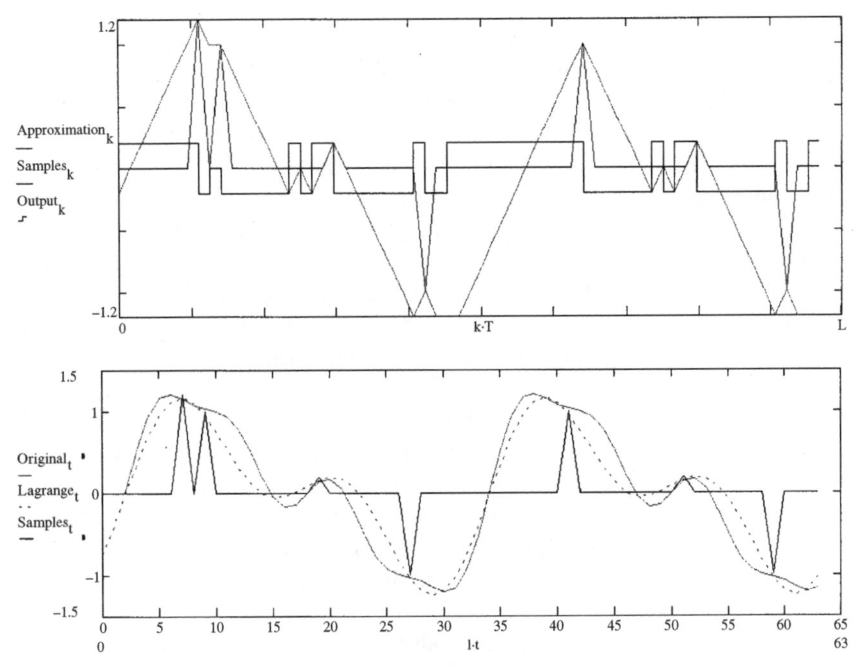

Figure 17. (a) Nonuniform samples generated from the negative going edges of the rectangular waveform. (b) Recovery of DM using LI on nonuniform samples derived from (a)—the correlation is 0.955.

possible. The error of recovery is only due to the quantization noise and not the approximation of the recovery method. Depending on the type of band-limited signal (L_2 and $\mathbb{C}^N$), we can use the methods discussed in the previous sections. Fig. 17(b) shows the recovery using Lagrange interpolation. The correlation in this case is 0.955 which is better than that of low-pass filtering shown in Fig. 15(b). For L_2 band-limited signals, the simulation results for iterative techniques are given in [14].

The Mathcad(TM) simulation for Figs. 15–17 are given in the accompanying CD-ROM under the directory 'Chapter16\DM\'; there are two files in that directory. The file 'DM.mcd' generates the DM figures and the file 'DM-Lagrange.mcd' generates the Lagrange interpolation—Fig. 17(b).

16.7. Modulation Techniques for Data Conversion

16.7.1. Introduction

We have discussed time-varying, iterative, and non-linear techniques for the recovery of various signals. Here we discuss the use of two modulation techniques that find application in Digital-to-Analog (D/A or DAC) and Analog-to-Digital (A/D or ADC) data conversion. Conventionally, both rely on linear recovery techniques (i.e., low-pass filtering). However, it is possible to make use of the previously discussed time-varying recovery techniques, with the intention of reducing reliance on over-sampling to reduce the distortion.

Traditionally, D/A conversion consists of PCM type digital-to-analog conversion followed by analog amplification. These processes are costly to implement for low noise or high dynamic range signals. To remedy these problems, Sigma Delta Modulation (SDM) and analog Pulse Width Modulation (PWM), especially the former, have found use in data conversion, typically in applications where high dynamic range is combined with moderate signal bandwidths.

Although SDM and PWM have been used in a multitude of applications, it is in their application to audio (speech and particularly music) that they have come to prominence. Both these techniques rely on over-sampling (i.e., sampling well in excess of the Nyquist limit) to enable them to overcome the inherent distortion if linear demodulation techniques (low-pass filtering) are used. Note that PWM needs additional techniques to reduce the distortion, mainly because less over-sampling can be applied.

By means of the over-sampling, both techniques are effective because they trade resolution against bandwidth. In each case the modulated signal requires a greater bandwidth than either the original or an equivalent PCM coding. For example, a 16 bit CD-quality signal (i.e., sampled at 44.1 kHz) requires typically 64 times over-sampling for a one-bit SDM to represent it at equivalent quality—i.e., a data expansion of 4 times. For a PWM representation, 8 times over-sampling with 8-bit data would be required, which also amounts to a four-fold data expansion.

It is apparent that this is the price paid for specifying a linear demodulation technique (low pass filtering) for a non-linear modulator. This raises the issue of better utilizing channel bandwidth if non-linear or time-varying demodulation techniques discussed in the previous sections are used.

In the remainder of this section, we first discuss conventional data converters that operate at or near the Nyquist sampling. This is developed into a discussion of over-sampled data converters and multi-rate filtering. Both of these underpin the subsequent discussion in more depth of the structures and performance of the modulators. Starting with SDMs, we then look at PWM.

16.7.2. Preliminaries

Conventional data converters, both ADCs (Analog-to-Digital Converters) and DACs (Digital-to-Analog Converters), use no (or mild) over-sampling and as such convert signals at or just over the Nyquist sampling limit. A general

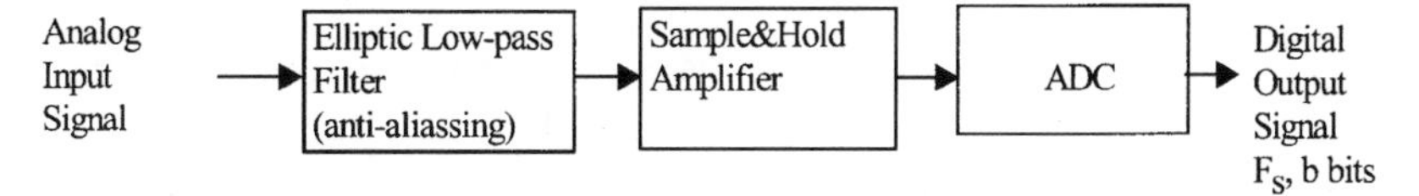

Figure 18. Analog to Digital Conversion (ADC) at the Nyquist rate.

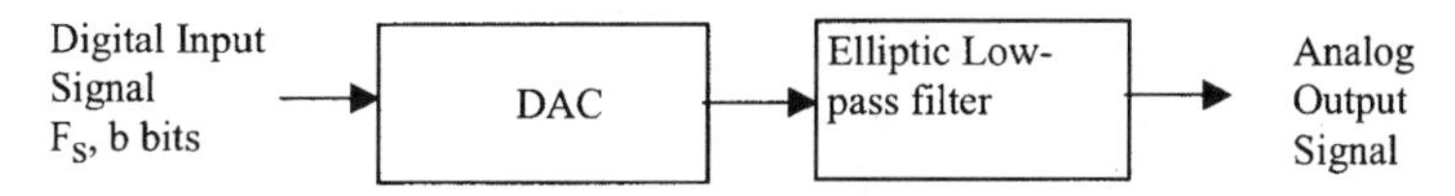

Figure 19. Digital to Analog Converter DAC at the Nyquist rate.

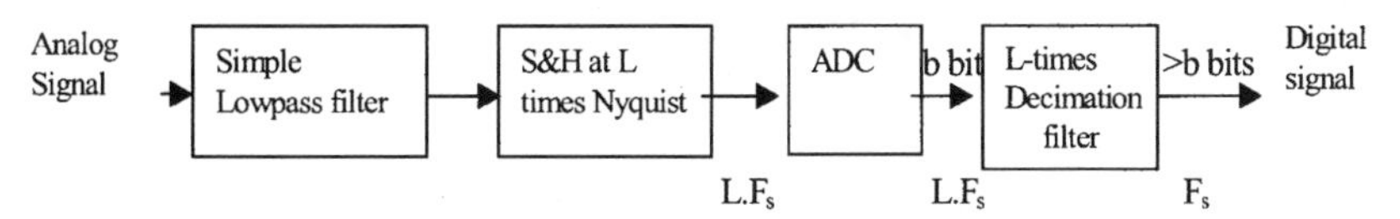

Figure 20. Over-sampling Analog to Digital Conversion structure.

structure for a Nyquist ADC is shown in Fig. 18 and that of a Nyquist DAC is shown in Fig. 19. Typically, for a 20 kHz bandwidth, the sampling and conversion will take place at a rate of 44.1 kHz for CD quality.

A generic over-sampling ADC structure is shown in Fig. 20, where the key point to note is that the core ADC will typically produce fewer bits per sample than the final digital output. The extra bits in the output are regained by the decimation filter which reduces the sampling rate back down to (or close to) the Nyquist rate.

The reason the structure of Fig. 20 works is that the quantization error power of a core ADC is determined solely by the number of bits it uses. However, with over-sampling, this power is spread more thinly through the spectrum, so that the noise power per Hz is reduced and in-band SNR is improved. Because the decimation filter is used to extract only the desired baseband signal, it is possible to recover the additional bits. The SNR improvement is 3 dB for every doubling of sampling rate (over and above the Nyquist rate), or an extra bit for a 4 times over-sampling, 2 extra bits for 16 times over-sampling, etc.

Over-sampling like this is a simple way to trade amplitude resolution for time resolution, but as shown in the preceding sections, if time varying/non-linear techniques are used, improved gains can be made.

Figure 21 represents a generic over-sampling DAC. The principal reason for choosing such a structure is to ease the specification of the reconstruction filter, which is analog and in Nyquist DACs may introduce unwanted phase distortion. The purpose of the interpolation filter is to increase the sampling rate of the incoming digital signal prior to conversion. The signal at the output of that block

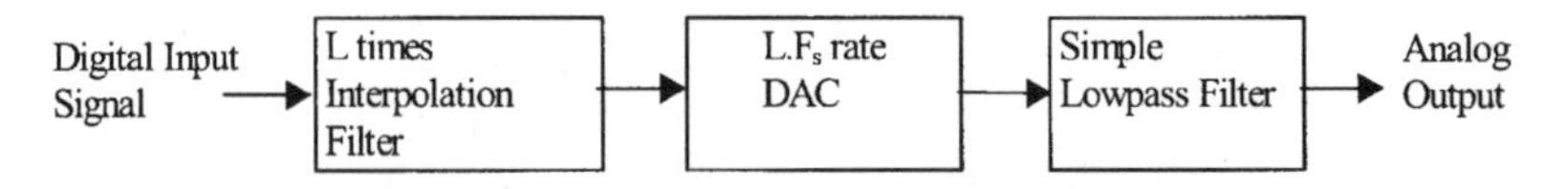

Figure 21. The over-sampling Digital to Analog structure.

should approximate closely the digital signal that would have been obtained had the analog signal been sampled at the higher rate rather than the Nyquist rate.

Together, interpolation filters and decimation filters are known as multirate filters [22]. This is because they operate at more than one sampling frequency. As we will see, multirate filtering is crucial to the operation of both PWM and SDM systems. It is by operating these modulation systems in excess of the Nyquist sampling rate that they are able to perform well.

16.7.3. Sigma Delta Modulation (SDM)

Sigma-Delta (or in many texts Delta-Sigma) Modulation is closely related to Delta Modulation (Section 16.6). However, SDM both simplifies the decoder and encoder by including an integrator in the encoder's feed-forward path. The development of SDM is generally attributed to Cutler [36], whose paper actually deals with an error diffusion approach. The structure now known as Sigma Delta Modulator is due to Inose et al [37] who first reported it in the early 1960s.

16.7.3.1. General Description

Sigma Delta Modulation works by converting a signal into a sequence of pulses, again whose short term average follows that of the modulating signal. In this case, the pulses are of fixed duration, and their number is proportional to the signal amplitude; it is thus sometimes referred to as Pulse Density Modulation (PDM) and has much in common with Error Diffusion Techniques, as used in image half-toning.

An example of the pulse stream that results from SDM is shown in Fig. 22, and a generic structure for a SDM-DAC is shown in Fig. 23. Note that the number of levels that an SDM pulse stream can attain is determined by the number of discrete levels of its internal quantizer. For a single bit quantizer, there are two levels, normally denoted as $+1$ and -1. Note that often it is clearer and/or more convenient to label the negative pulse value as zero, so that pulse values are 1 and 0.

To date most SDM-ADCs have used the single bit quantizer approach. This is because this eases the analog design requirements of the overall system. However, multibit SDMs are now becoming of interest to the community, and will be of particular relevance in this book, where a modified form of multibit SDM is used within the PWM-DAC structure.

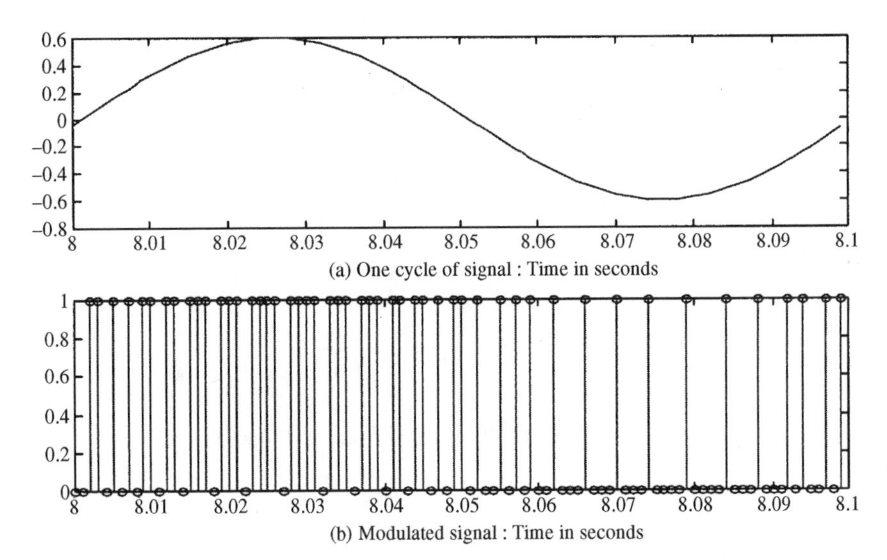

Figure 22. First order Sigma Delta Modulation using 1 Hz sine wave, 50 times over-sampled.

16.7.3.2. SDM Linear Model Analysis

The non-linearity within the SDM block diagram makes precise analysis somewhat intractable, though there have been some excellent publications on exact non-linear analysis of SDMs [17]. This approach leads to accurate but restricted results, so it is common to approximate the SDM by a linearized method in which the quantizer is replaced by an additive noise source. Fig. 24 shows the linear model.

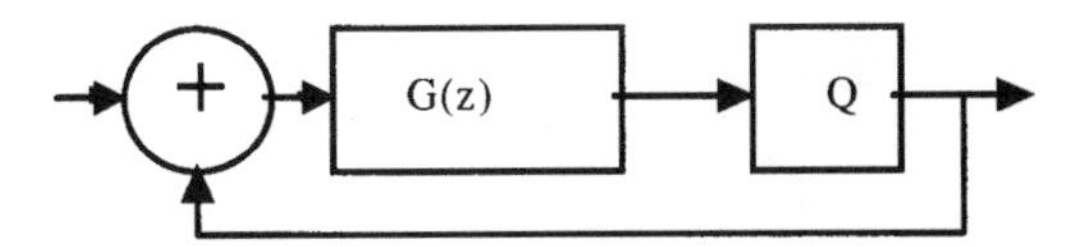

Figure 23. Sigma Delta Modulator.

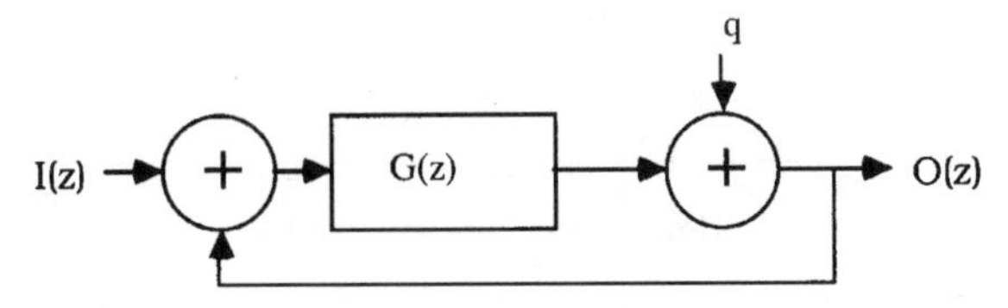

Figure 24. Linear model for Sigma Delta Modulator suitable for analysis.

Overall the output, $O(z)$, is a combination of the quantization error, $q(n)$, and the feedback error, $I(z) - O(z)$, i.e.,

$$O(z) = G(z)(I(z) - O(z)) + Q(z).$$

Setting $Q(z) = 0$, we find the output as a function of the input only, and leads to the Signal Transfer Function (STF),

$$\text{STF}(z) = \frac{O(z)}{I(z)} = \frac{G(z)}{1 + G(z)}.$$

Setting the input $I(z) = 0$, we find the Noise Transfer Function (NTF)

$$\text{NTF}(z) = \frac{O(z)}{Q(z)} = \frac{1}{1 + G(z)}.$$

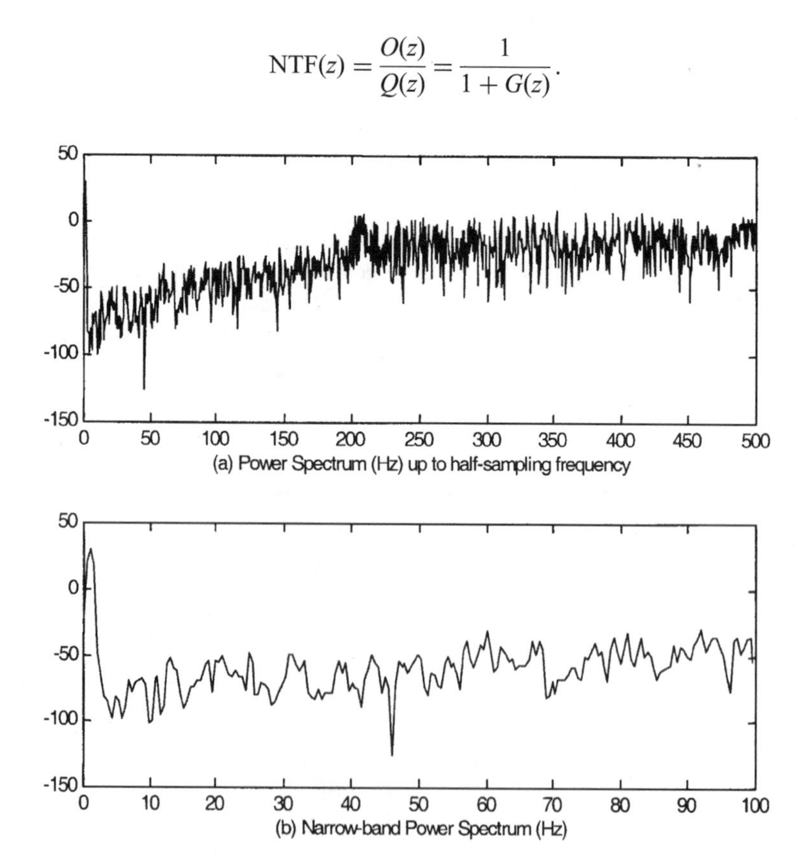

Figure 25. Power Spectral Analysis of 1st order SDM for 500 times over-sampled signal. Note the good signal resolution at low frequencies in (a) magnified in (b).

In its simplest form, $G(z)$ is an integrator, which has high gain at low frequencies so that $\text{STF}(z) \approx 1$ whereas the NTF at low frequencies will be small, so that the quantization error (manifested as noise) is greatly suppressed. This is demonstrated in Fig. 25, which presents the spectra for the signal of Fig. 22 sampled at 500 times the Nyquist rate. The noise suppression at baseband is clear and we can see that, in spite of only a single bit representation of each sample, the modulated representation provides a good Signal-to-Noise ratio over a useful bandwidth.

Higher order SDMs use cascaded integrators in the loop filter, which increase the noise suppression at low frequencies. The order of the filter is used to describe the order of the modulator, e.g., a 3rd order filter is used in a 3rd order modulator. The effect of increasing the order is demonstrated in Fig. 26 for a 2nd order modulator, with the same input signal and over-sampling rate (500) as in Fig. 25. We can see greater baseband resolution and/or a wider usable bandwidth. For all SDMs this is at the cost of amplifying noise at high frequencies.

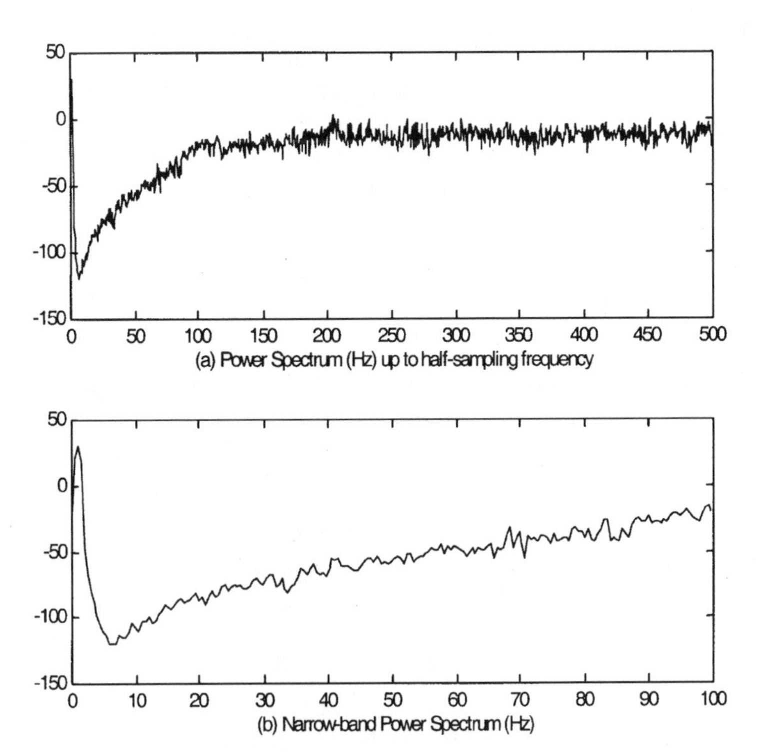

Figure 26. Power Spectral Analysis of 2nd order SDM for the same input as Fig. 25. Note the improved signal resolution at baseband compared to the 1st order modulator.

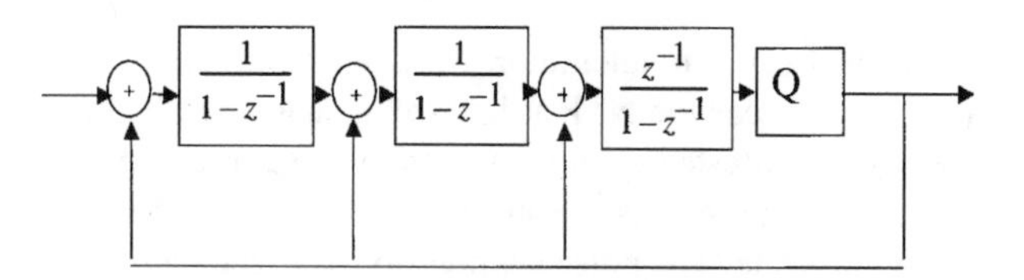

Figure 27. Alternative multiple feedback SDM structure.

The generic structure of Fig. 22 is often known as an interpolative SDM. Many alternative topologies exist, but perhaps the most popular is the so-called multiple feedback structure. This is shown in Fig. 27. The great advantage of this structure is that it is tolerant to coefficient and data path quantization because the filter is constructed from a cascade of integrators.

16.7.3.3. SDM-ADC

SDM is now well established as a means for converting between analog and digital domains. Some excellent texts have been written, including [15]–[16]. These should be consulted for more detail on this technique.

Figure 28 shows more precisely how an SDM-ADC is configured. Here the overall loop quantizer is implemented as the cascade of the single-bit ADC (a comparator or a signum function) and the single-bit DAC in the feedback loop. Also shown is the decimation filter that extracts the baseband signal and recovers extra bits to provide a multibit PCM representation.

16.7.3.4. SDM-DAC

Figure 29 shows an overall system for an SDM-DAC including the interpolation filter used to increase the sampling frequency from near-Nyquist to a multiple of Nyquist (e.g., 64 times). Also shown is the final filtering stage, which is normally implemented as a switched capacitor filter.

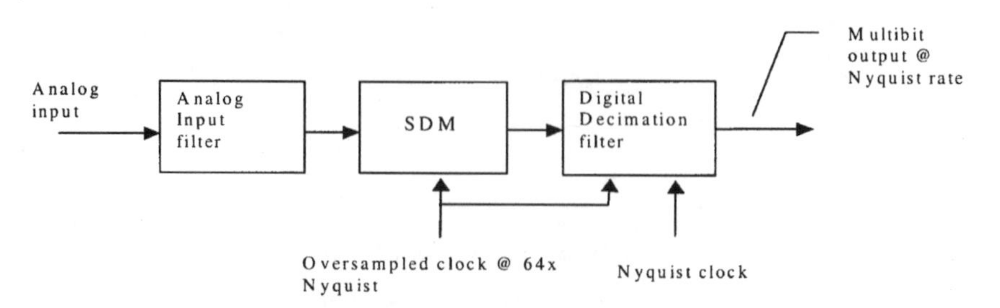

Figure 28. SDM-ADC System.

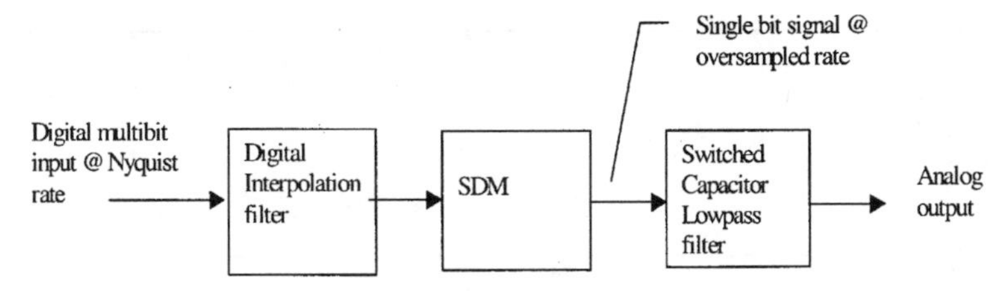

Figure 29. SDM-DAC System.

Finally, Fig. 30 presents another alternative structure for reducing the word-length of over-sampled signals. This is known by various names, including Noise Shaper, which ought to be a generic description of SDMs as well. The obvious difference is that it is not the output signal which is fed back to be compared against the input, it is now the error due to quantization—this is the structure first described by Cutler in [36]. The disadvantage of these structures for ADCs is firstly that a difference signal must be generated, and secondly that the feedback is not single-bit even when the quantizer is single-bit. However, these structures have found use in DACs and in particular have found application in PWM-DACs (Section 16.7.4.3) for reducing the word-length down to 8 bits from 16 or more.

16.7.4. Pulse Width Modulation (PWM)

One of the earliest works to provide an in-depth analysis of PWM is [18] which also links it to PPM as discussed earlier in Section 16.2. Subsequently, much of the rest of the valuable early literature is Russian in origin, as exemplified by [25] and [26] which discuss the precise form of various PWM forms under tone modulation. Crucially, these works distinguish between two sampling modes for PWM—uniform and natural nonuniform sampling. The former corresponds to 'normal' sampling where sample instants are uniformly spaced in time, whereas the latter samples the message at instants that are defined by the nature of the message.

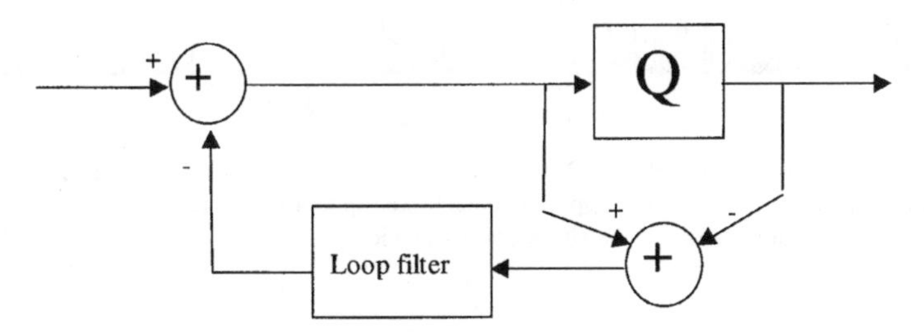

Figure 30. Error Diffusion (or Feedback) Structure—also known as Noise Shaper.

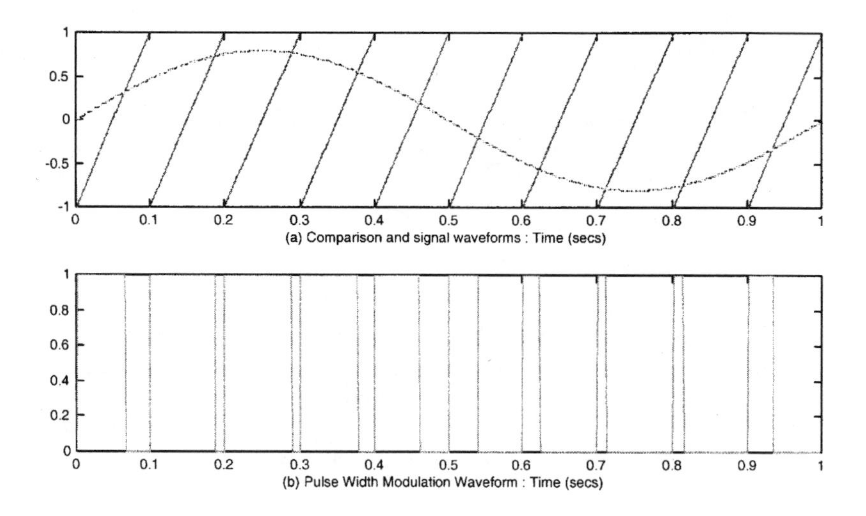

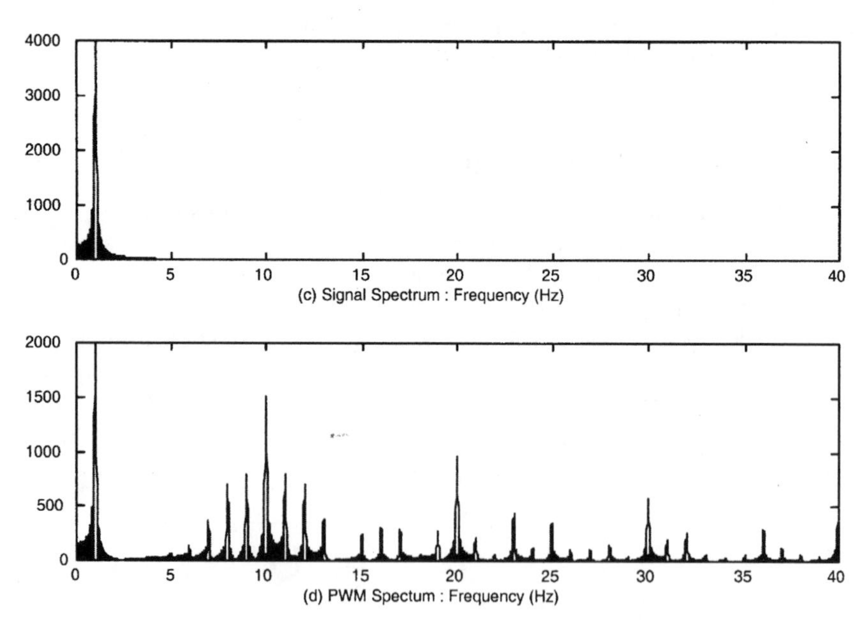

Figure 31. One-sided, trailing-edge PWM with natural nonuniform sampling. The sine is 1 Hz, the sawtooth comparison is 10 Hz and the spectra depict power up to 4 times the carrier. Note the signal-carrier inter-modulation terms and the absence of harmonics.

16.7.4.1. General Description

Pulse Width Modulation, as briefly discussed in Section 16.3, works by converting a signal sample into a pulse whose width is proportional to the sample amplitude. Strictly speaking it is the pulse area that is proportional to amplitude, but in most cases (and in the sequel) it is assumed that pulses are of equal and constant height.

An example of the pulse stream that results from modulating a single sine-wave cycle is shown in Fig. 31, along with its spectrum. Here we used natural nonuniform sampling. In the spectrum of Fig. 31(d), we note the distortion terms which are indicative of a non-linear modulation scheme. Note that the spectra were obtained using a rectangular window in the analysis; hence there are analysis artefacts in Fig. 31(c) and not distortion terms. Note more importantly that there are harmonics of the carrier at 10, 20, 30 and 40 Hz (and, not shown, beyond). There are also inter-modulation products around the carrier terms, but no harmonic distortion terms.

Conventionally, demodulation is achieved by low-pass filtering the pulse waveform, whose short term D.C. level follows that of the original signal. This simple technique as we discussed in the PPM section which is also clear from the spectrum of Fig. 31(d) creates distortion. To reduce the distortion, either we have to over-sample by increasing the fundamental frequency of the sawtooth-wave or as mentioned in the PPM section we can use the nonuniform sampling recovery methods at the Nyquist rate. The cost that one has to pay is obviously the complexity of the decoder.

Figure 32 shows, in principal, how PWM is performed. The core is a comparator. This figure depicts a sawtooth waveform $c(t)$ that will modulate the trailing edge of each pulse. An obvious alternative is to modulate the leading edge or both edges of a pulse. Leading and trailing edge (i.e., one-sided) modulation

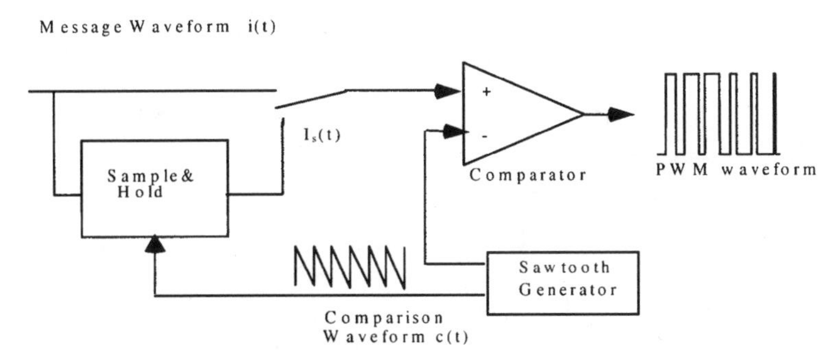

Figure 32. Analog PWM/Class D amplifier. For a two-sided modulation scheme (i.e., both edges of each pulse are modulated by the signal) the sawtooth generator is replaced by a triangular wave generator.

are equivalent in performance terms, and the choice boils down to convenience of implementation. All one-sided modulation schemes perform worse in distortion terms than two-sided schemes, all other modulation parameters being equal. The two-sided modulation class could be obtained in the figure by choosing $c(t)$ to be a triangular waveform instead of a sawtooth, so that it measures $i(t)$ twice per carrier period.

To the upper input of the comparator, the message signal, either $i(t)$ or $i_s(t)$ is fed. The signal $i_s(t)$ is the result of passing $i(t)$ through a zero-order sample-and-hold device. The sampling rate is the frequency of the sawtooth waveform. When the upper input is the message signal $i(t)$, we get natural Nonuniform sampled PWM (NPWM). When the upper input is the sampled-and-held signal $i_s(t)$, we get the Uniform sampled PWM (UPWM).

Figure 33 shows in close up the difference between uniform and natural nonuniform sampling PWM, where the original continuous time signal (designated by the solid line) and its sampled-and-held version (designated by the piecewise-horizontal, dashed line) are shown. Also shown are the comparison waveform, $c(t)$ (a sawtooth for clarity, to give a one-sided modulation) and the samples of the original continuous time waveform, $i[n]$ (depicted by small black blobs, positioned on the vertical coordinates of the sawtooth).

In the central sawtooth period, the points at which the continuous waveform, $i(t)$, and the sample-and-hold waveform, $i_s(t)$, cross the rising part of the sawtooth are highlighted and called NPWM and UPWM cross points, respectively. At these points the PWM waveform changes state, say from high to low (and we assume

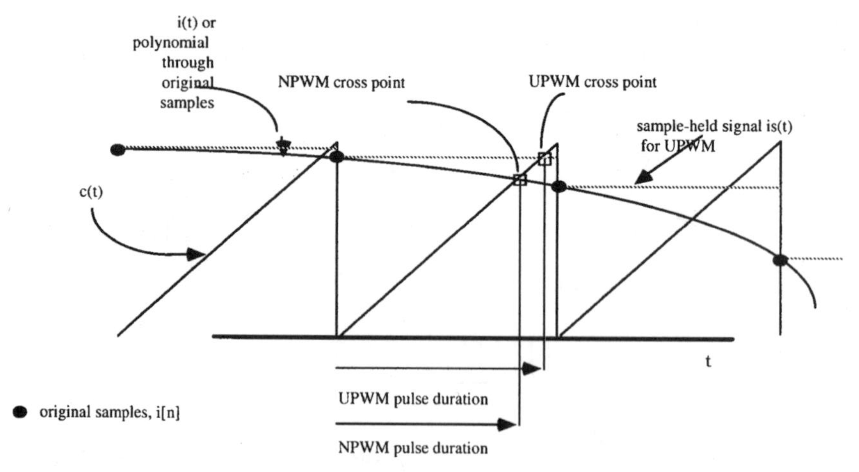

Figure 33. Uniform and natural nonuniform sampling cross points—the basis for a low distortion PWM-PDAC algorithm.

that the low to high transition occurs regularly at the times of the discontinuities of $c(t)$). It is clear, as shown, that the durations of the pulses are different.

Although the UPWM and NPWM waveforms are probably indistinguishable on an oscilloscope, UPWM suffers from harmonic distortion whereas NPWM does not. The harmonic distortion levels for various classes of UPWM can be quite high. Full details of the theoretical tone modulation spectra for both classes of modulation can be found in many texts (see [20]–[21] for example). An example of the spectrum was shown in Fig. 31.

Both types of PWM suffer from another distortion mechanism, which is inter-modulation between the carrier (or comparison) waveform and the signal. Both distortion mechanisms are dependant on the relative frequencies of message and carrier waveform and it has been shown in [21] that if the ratio of carrier frequency to the signal bandwidth, p, is made large enough, the harmonic distortion can be made negligible. Likewise, the carrier inter-modulation distortion can also be made negligible. We can increase the ratio p by means of an interpolation filter.

Although PWM-DACs and PDACs do not use analog PWM structures, as will become clear in the following subsection, this structure is important in understanding how PWM works and how to tackle the distortion that arises from the non-linear behavior.

16.7.4.2. PWM-ADC

Figure 32 shows in principle how PWM can be used as an analog-to-digital converter. Because, as we will see, the distortion performance of UPWM is worse than NPWM, a PWM-ADC would use NPWM, i.e., $i(t)$ is fed to the comparator in Fig. 32. The sample points occur at the times when $i(t) = c(t)$; clearly, these are signal dependent, and not uniformly spaced in time. To extract a digital signal, one needs to count the duration of each individual pulse. The sequence so obtained then needs resampling to provide a correct estimate of $i(n)$, i.e., the uniformly sampled sequence that would have been obtained had a conventional ADC and quantizer been used.

Although, in theory, it is possible to use PWM within an ADC, no works have been published on this, and PWM is used solely within DACs—in Power DACs in particular. Probably this is because it relies on counting to high accuracy. If the signal is to be captured at CD quality, some over-sampling is needed—at least 4 times the Nyquist rate. Thus the pulse repetition frequency is 176.4 kHz. The resolution needed in counting the duration of pulse has to be $(4 \times 44100 \times 2^{16})^{-1}$ sec. This corresponds to a digital clock of around 11 GHz–clearly not impossible, but rather a costly solution. For PWM-ADCs to be viable, our nonuniform sampling approach is needed to reduce the sampling rate and hence the clock rate. This is an area for further research.

16.7.4.3. PWM-DAC

The first known discussion of PWM as a means for effecting Digital-to-Analog conversion appears in [19] that focuses on uniformly sampled PWM. Later publications which developed this theory include those by Sandler and co-workers [20]–[21], [24] and [27], and by others [29]–[31]. The first known discussion of PWM as a means for effecting Digital-to-Analog conversion appears in [19] that focuses on uniformly sampled PWM. Later publications which developed this theory include those by Sandler and co-workers [20]–[21], [24] and [27], and by others [29]–[31].

16.7.4.3.1. Power DACs (PDAC). A Power DAC (PDAC) is a system which converts directly from a digital signal representation (typically Pulse Code Modulation—PCM) to an analog representation capable of driving a power-hungry load, such as a motor or loudspeaker. In such a system, there is no intermediate DAC, no intermediate analog amplifier, and there is normally a Class D power switching stage to provide the power to the load. Figure 32 can also be used as an analog PWM/Class D amplifier to highlight the key processing stages. Note that for clarity the power switching stage is not shown.

16.7.4.3.2. Uniformly Sampled PWM (UPWM) DACs. For a DAC or PDAC, all digital PWM systems must be implemented with a UPWM modulator, since the digital signals are assumed to be derived by a uniform sampling process (i.e., conventional PCM). As mentioned, UPWM carries a distortion penalty. However, it is possible, with some pre-processing of the digital input, to emulate the behavior of NPWM with a core UPWM converter. Acceptably low distortion can be obtained for moderate DSP.

A digital UPWM is a little more than a counter. A new digital value is loaded at regularly spaced time intervals (e.g., at a rate of 176.4 kHz) and sets the counter output high. It counts down, and on reaching zero, resets the counter output low. Again, in principle, an extra-ordinary clock frequency of a few GHz seems necessary. However, by using noise-shaping (a close relative of Sigma Delta modulation), this can be reduced to around 100 MHz, which is entirely achievable.

A good conversion system for 16 bit signals can be constructed from two-sided UPWM modulation. There will be some residual distortion, but depending on the signal frequency and amplitude, rarely worse than 100 dB below peak signal level. This would be at a minimum value of p (i.e., for the highest signal frequency) of 16, corresponding to an increase in sampling rate by 8. This has been confirmed in hardware [24]. This reference (Graph 12) shows that the harmonic distortion is present at levels commensurate with conventional DACs. To improve performance, either p has to be increased further or a version of NPWM must be used.

16.7.4.3.3. Nonuniformly sampled PWM (NPWM) DACs. NPWM gives adequate baseband performance for 16-bit conversion with one sided modulation, and over-sampling between 4 and 8 times. For a two-sided scheme the over-sampling could be lower, or the signal resolution improved. We now outline how to reduce UPWM distortion by approximating the behavior of NPWM using signal processing.

In practice, only samples $i[n]$ and not the original waveform, $i(t)$, are available. We can approximate NPWM if an approximation to $i(t)$ can be interpolated through the set of points $i[n]$, then calculate the time instant at which the continuous waveform will cross the comparison waveform, and finally feed this to the uniform modulator's counter so that an approximation to NPWM is produced.

A system to implement this is shown in Fig. 34. The box labelled Noise Shaping Filter is used to reduce the data word-lengths, so that the digital counters in the core UPWM are realizable; it does not greatly affect the performance. Noise shaping is closely related to SDM; it shapes quantization noise into the higher frequency part of the modulation spectrum. The box labelled Cross Point Detector (XPD) implements the approximation technique above. Further details can be found in [33] and [35]. The complete process has come to be known as Pseudo-Natural PWM (PNPWM).

Reference [34] (Fig. 10 in [34]) shows that if only a one-sided final modulator is used with a 3rd order interpolation filter at 8 times the Nyquist rate, there is no harmonic distortion. It is possible to use lower order interpolation filters, but the same reference shows that harmonic distortion levels will rise above the inherent noise floor for some combinations of input signal amplitude and frequency.

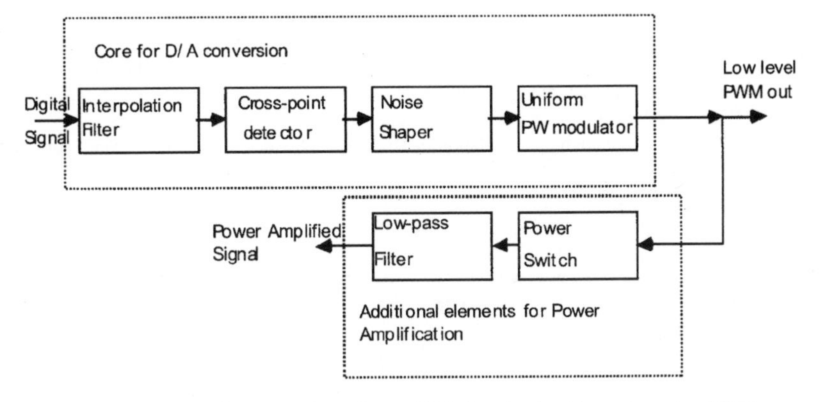

Figure 34. Overall structure for a Power Digital to Analog Converter using PWM.

References

[1] F. Marvasti. *A Unified Approach to Zero-Crossings and Nonuniform Sampling of Single and Multidimensional Signals and Systems*. Nonuniform Publication, Oak Park, IL, 1987.

[2] Z. Ghassemlooy. A Low Cost Subcarrier Multiplexed PPM for Optical Fibre Communication Systems. *IEEE Trans. Consumer Electronics*, 44(1):73–81, February 1998.

[3] F. Marvasti. Nonuniform Sampling. In Robert Marks II, Ed., *Advanced Topics in Shannon Sampling and Interpolation Theory*, New York: Springer Verlag, January 1993.

[4] F. Marvasti. Interpolation of Lowpass Signals at Half the Nyquist Rate. *IEEE Journal on Signal Processing Letters* 3(2): February 1996.

[5] F. Marvasti. Nonuniform Samples at or Below the Nyquist Rate for Bandpass Signals. *IEEE Trans on SP*, 44(3):572–576, March 1996.

[6] V. Koshi and F. Marvasti. PPM and PDM Demodulation using Novel Techniques. In the *Proc of Symposium on Communications*, Coventry, UK, December 11–13, 1994. (Correspond with the University of Lancaster, Telecom Research Centre, Lancaster, UK.)

[7] F. Marvasti, M. Hasan, M. Eckhart, and S. Talebi. Efficient Algorithms for Burst Error Recovery Using FFT and Other Transform Kernels. *IEEE Trans on Signal Processing*, 47(4): 1065–1075, April 1999.

[8] F. Marvasti. Transmission and Reconstruction of Signals Using Functionally Related Zero-Crossings. PhD dissertation. RPI, Troy, NY, USA, May 1973.

[9] F. Marvasti. Reconstruction of a Signal from the Zero-Crossings of an FM Signal. *Trans. of IECE of Japan*, p. 650, October 1985.

[10] R. G. Wiley, H. Schwarzlander and D. D. Weiner. Demodulation Procedure for Very Wideband FM. *IEEE Trans. Commun.*, COM-25, pp. 318–327, March 1977.

[11] J. Bar-David. An Explicit Sampling Theorem for Bounded Bandlimited Functions. *Information and Control*, 24:36–44, 1974.

[12] F. Marvasti, M. Hasan, and M. Eckhart. A Novel Method for Burst Error Correction. *SamPTA'97*, Aveiro, Portugal, July 1997.

[13] S. Hein and A. Zakhor. *Sigma Delta Modulators*. Kluwer Academic Publishers, 1993.

[14] M. Gamshadzahi. Bandwidth Reduction in Delta Modulation Systems using an Iterative Reconstruction Scheme. M.S. Thesis, Illinois Institute of Technology, Dept. Elec., Comput. Eng., December 1989.

[15] J. C. Candy and G. C. Temes, Eds. *Oversampling Delta-Sigma Converters*, IEEE Press, New York, 1991.

[16] S. Norsworthy, R. Schreier, and G. C. Temes, Eds. *Delta Sigma Data converters: Theory, Design and Simulation*. IEEE Press, New York, 1997.

[17] O. Feely and L. O. Chua. The Effect of Integrator Leak in Sigma-Delta Modulation. *IEEE Trans. On Circuits and Systems*. 38(11):1293–1305, November 1991.

[18] H. S. Black. *Modulation Theory*. D. Van Nostrand Co., Inc., Princeton, N.J., 1953.

[19] M. B. Sandler. Investigation by Simulation of Digitally Addressed Audio Power Amplifier. PhD thesis. Essex University, 1983.

[20] M. B. Sandler. Towards a Digital Power Amplifier. *76th Convention of the Audio Engineering Society*, New York, preprint no. 2135, October 1984.

[21] M. B. Sandler. Progress towards a Digital Power Amplifier. *80th Convention of the Audio Engineering Society*, Montreux, preprint no. 2361, March 1986.

[22] R. W. Schafer and L. R. Rabiner. A Digital Signal Processing Approach to Interpolation. *Proc. IEEE*, 61(6):692–702, June 1973.

[23] J. D. Martin. Theoretical Efficiencies of Class D Power Amplifiers. *Proc. IEE*, 117(6):1089–1090, June 1970.

[24] R. E. Hiorns and M. B. Sandler. Power Digital to Analogue Conversion using Pulse Width Modulation and Digital Signal Processing. *IEE Proc-G*, 140(5):329–338, October 1993.

[25] N. A. Baturin. Full Wave Modulation in Class D Amplifiers. *Telecommunication and Radio Engineering*, Pt 2, 29:122–123, 1974.

[26] N. A. Baturin and V. N. Plyusnin. Combination Distortion in a Two-Sided Single Ended Class D Power Amplifier. *Telecommunication and Radio Engineering*, Pt 2, 28(8):123–126, 1973.

[27] J. Goldberg and M. B. Sandler. Noise Shaping for a Digital Power Amplifier. *Convention of the Audio Engineering Society*, New York, preprint 2832, October 1989.

[28] M. Gerzon and P. Craven. Optimal Noise Shaping and Dither of Digital Signals. *87th Convention of the Audio Engineering Society*, New York, preprint no. 2822, 1989.

[29] P. H. Mellor, S. P. Leigh, and B. M. G. Cheetham. Reduction in Spectral Distortion in Class D Amplifiers by an Enhanced Sampling Process. *Proc IEE*, Part G, 138(4):441–448, August 1991.

[30] M. O. J. Hawksford. Dynamic Model-Based Linearization of Quantized PWM for Applications in Digital to Analogue Conversion and Digital Power Amplifier Systems. *Journal of the Audio Engineering Society*, 40(4):235–252, April 1992.

[31] P. Craven. Towards the 24 Bit DAC: Novel Noise Shaping Topologies Incorporating Correction for the Non-Linearity in a PWM Output Stage. *Journal of Audio Engineering*, 51(5):291–313, May 1993.

[32] R. E. Hiorns, R. G. Bowman, and M. B. Sandler. A PWM DAC for Digital Audio Conversion: from Theory to Performance. *IEE International Conference on Digital to Analogue and Analogue to Digital Conversion*, Swansea, September 1991, pp. 142–147.

[33] J. M. Goldberg and M. B. Sandler. Pseudo-Natural PWM for High Accuracy Digital to Analogue Conversion. *Electronics Letters, 1*, 27(16):1491–1492, August 1991.

[34] M. B. Sandler. Digital-to-Analogue Conversion using Pulse Width Modulation. *Electronics and Communication Engineering Journal*, 5(6):339–348, December 1993.

[35] J. M. Goldberg. Signal Processing for High Resolution PWM Based Digital to Analogue Conversion. PhD Thesis. University of London, 1993.

[36] C. Cutler. Transmission Systems Employing Quantization. U.S. Patent No. 2,927,962, March 8 1960 (filed 1954).

[37] H. Inose, Y. Yasuda, and J. Murakami. A Telemetering System by Code Modulation–Σ Modulation. *IRE Trans. Space Electron. Telemetry*, SET-8:204–209, September 1962.

17

Applications to Error Correction Codes

F. Marvasti

17.1. Introduction[1]

Sampling a band-limited signal above the Nyquist rate could be an alternative to error correction codes [1]–[2], and [33]–[34]. In fact we can show the equivalence of the sampling theorem and the fundamental theorem of information theory [1]. Essentially, over-sampling is equivalent to a convolutional code using fields of real numbers as opposed to finite Galois fields. The encoder is a simple interpolator (a low pass filter) as shown in Fig. 1.

This figure represents a convolutional encoder of rate $\frac{1}{2}$ of infinite constraint length and infinite precision per symbol. Using the nonuniform sampling theorems, see [3]–[4], we can have a new interpretation of the fundamental theorem of information theory (i.e., as long as the source information rate is less than the channel capacity, there exists a code such that errorless transmission is possible).

The nonuniform sampling theorem states that as long as the average rate of the samples is above the Nyquist rate, signal recovery is possible. In other words, if we over-sample a signal uniformly and a percentage of the samples is dropped, as long as the average rate of the remaining samples (which forms a set of nonuniform samples) is above the Nyquist rate, all the lost samples can be recovered. The main difference between the two interpretation is that in the latter

[1]Portions of Sections 17.1–17.3 and Section 17.9.1 are the modified reprint with permission of [1]; © 1993 IEICE.

F. Marvasti • Multimedia Laboratory, King's College London, Strand WC2R 2LS
E-mail: farokh.marvasti@kcl.ac.uk

Nonuniform Sampling: Theory and Practice, edited by Marvasti
Kluwer Academic/Plenum Publishers, New York, 2001.

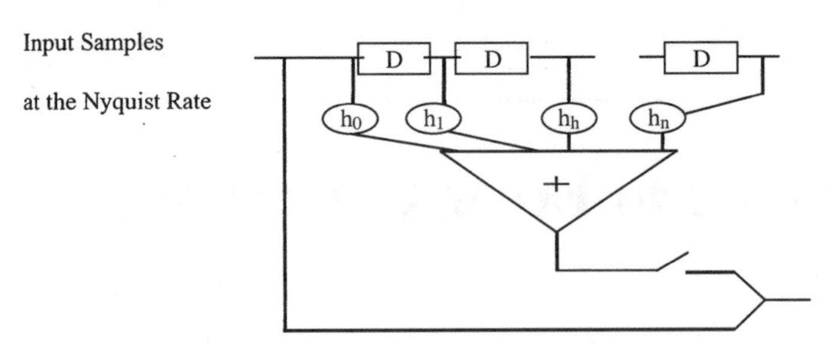

Figure 1. Over-sampling viewed as a convolutional code.

case the code is the over-sampled signal while the Shannon theorem only shows the existence of a code.

If we denote P_ε as the probability of erasure, then $1 - P_\varepsilon$ could be interpreted as either the percentage of correct (remaining) samples or channel capacity of erasure channel per symbol per bits[2]. The inverse of the channel capacity per symbol per bits is the sampling rate (normalised by the Nyquist rate) needed for errorless transmission. For example, if the percentage of sample loss (probability of erasure) is 0.5, then the sampling rate should be at least twice the Nyquist rate to have errorless transmission. If $\frac{2}{3}$ of the samples are lost, then the

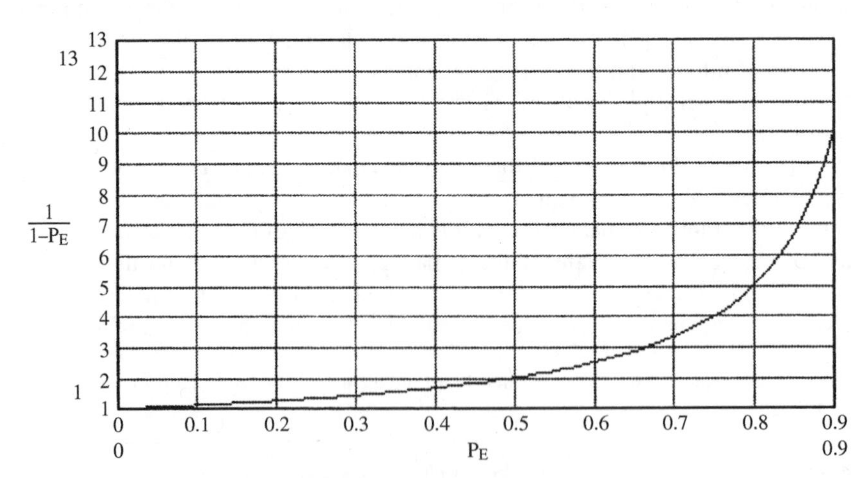

Figure 2. Sampling rate vs. the probability of erasure.

[2]The channel capacity for erasure for q bits per symbol is $c = (1 - P_\varepsilon)q)$. Thus, the capacity per symbol per bits is c/q which is finite even if q goes to infinity.

sampling rate should be at least $1/(1-\frac{2}{3}) = 3$ times the Nyquist rate for errorless transmission. This inverse is plotted in Fig. 2. This figure shows that for any sampling rate, e.g., twice the Nyquist rate, any sample loss of less than 50% is acceptable. In terms of information theory, this region is equivalent to the case where the source rate is less than the channel capacity.

In case of additive impulsive noise, errors could be detected from the side information that there are frequency gaps in the original over-sampled signal (syndrome). Alternatively, for certain sampling rates such as twice the Nyquist rate, two independent uniform sampling sets could be formed. Comparison of these two band-limited signals at the sample points show the approximate position of the errors [1] (for exact recovery, see Section 17.9 on Recovery in an Additive Impulsive Noise Environment). These erroneous samples are then discarded and the reconstruction becomes recovery of the signal with erasures. Another approach is to add parity bits to samples and discard erroneous samples when error is detected.

A discrete version of the over-sampled signal is a block of samples such that in the transform (e.g., Discrete Fourier Transform (DFT)) domain a contiguous number of high frequency components are zero. As the analog case resembled convolutional codes, the discrete version is equivalent to block codes in the Field of real numbers. In general, the zeros do not have to be the high frequency components or contiguous [5]. However, it can be shown that the DFT of BCH and Reed Solomon (RS) codes in the GF(q^m) has $\tau = 2t$ contiguous zeros, where t is the maximum number of correctable errors [6]. Therefore, the over-sampled signal in the DFT version is a special case of the RS codes in the Field of real numbers.

The relationship between DFT and error correction codes has been recognized by many authors [5]–[7], [32], [46]–[47]. Reference [5] has shown that under certain conditions, DFT is capable of detecting and correcting errors. Redinbo [32], [46]–[47] has done extensive work in the area of real-number block and convolutional codes for fault-tolerant computer systems; see also two of his papers included in the CD-ROM, directory: Chapter 17, fault-tolerant. In this chapter, we suggest iterative methods developed for nonuniform sampling problem [1], [3]–[4] and [8] for erasure and impulsive noise recovery. The advantage is the use of FFT algorithm for each iteration and simplicity of hardware implementation. Specifically, we propose a time varying iterative method for erasure recovery which significantly improves the convergence speed. We shall also show that the RS decoding of DFT block codes is quite efficient for erasure channels.

In this chapter, we shall show the erasure and impulsive noise correction capability of discrete transforms, specifically, Discrete Fourier Transform (DFT), Discrete Cosine Transform (DCT), and Discrete Walsh Transform (DWT); Discrete Wavelet and Hartley Transforms have been also simulated for erasure channels. We shall show a successive iterative approximation, a time varying technique, and RS like decoding for error recovery.

17.2. Erasure and Impulsive Channels

An erasure channel implies that the position of the sample loss (or error) is known but its amplitude is unknown. Examples of an erasure channel is the loss of packets in an ATM network or the drop of speech frames in a wireless environment at the decoder. When many consecutive samples are erased, the loss is called bursty; we shall see later that it is more difficult to recover bursty losses due to accumulation of errors. This issue is also discussed in Chapter 5 on stability of missing samples.

An impulsive channel is related to the case where some samples are corrupted but the position and the amplitude of the samples are unknown. Similar to the finite Galois Fields, this problem can also be solved in the Field of real numbers; however stability is a big issue in the Fields of real and complex numbers. Erasure channels are discussed in Chapters 5 and 6 under 1-D and 2-D signals with missing samples. Below we shall discuss coding of real/complex numbers under erasure channels, bursty erasure losses, and additive impulsive noise environments.

17.3. Erasure Recovery Using Discrete Transform Techniques

In this section, we shall discuss the erasure capability of orthonormal discrete transforms such as Discrete Fourier Transform (DFT), Discrete Cosine Transform (DCT), Discrete Walsh Transform (DWT), Discrete Hartley Transform (DHT), and Discrete Wavelet Transform (DWAT). However, the emphasis will be on DFT since more work has been done on DFT (see, e.g., Chapters 4–6) because of Fast Fourier Transform (FFT) and because of its similarity to Reed Solomon (RS) codes, which implies guaranteed error capability due to its Vandermonde matrix—to be discussed later. Other transform techniques may not have the Vandermonde property of DFT.

Let us take a finite number k of discrete (real or complex) samples and take its DFT, the so called "frequency" domain and pad m (or insert on a random basis) zeros to the DFT. Now, this code (n, k), under certain conditions, is capable of correcting m erasures where $m = n - k$. The polynomial relationship of the DFT definition is shown below

$$x(l) = \frac{1}{n}\sum_{i=0}^{n-1} X(i)e^{j2\pi li/n}, \tag{1}$$

where *x(l)* and *X(i)* are the time and frequency domain representation, respectively. Since m components are zero in the frequency domain, (1) becomes

$$x(l) = \frac{1}{n} \sum_{v=0}^{k-1} X(i_v) e^{j2\pi l i_v / n}. \tag{2}$$

The above equation is a finite polynomial in terms of $s = e^{j2\pi l/n}$, i.e.,

$$x(l) = \frac{1}{n} \sum_{v=0}^{k-1} X(i_v) s^{i_v} \tag{3}$$

Now, there are k equations and k unknowns, i.e., $X(i_v)$ for $v = 0, \ldots, k-1$ can be found from k independent linear equations derived from (2) when l is equal to the position of k different non-erased values of *x(l)*. This observation suggests that this code is capable of correcting $m = n - k$ erasures. Later, we will suggest some practical methods of implementing this error correction capability.

Similar to BCH and Reed Solomon codes [5]–[6], we can simplify (3) if we assume the m padded zeros in the frequency domain are contiguous. Specifically, for the sake of simplicity, without any loss of generality, assume that the last m components are zero. For this special case, i_v in (2) reduces to v. The order of the polynomial of (2) is now $k - 1$ or less. Thus, $x(l)$ can be found from the Lagrange interpolation as given below:

$$x(l) = \sum_{m=0}^{k-1} x(\sigma_{i_m}) \frac{H(\sigma_l)}{\dot{H}(\sigma_{i_m})(\sigma_l - \sigma_{i_m})}, \tag{4}$$

where

$$H(\sigma_l) = \prod_{m=0}^{k-1} (\sigma_l - \sigma_{i_m}) \quad \text{and} \quad \sigma_l = e^{j2\pi l/N}, \qquad \sigma_{i_m} = e^{j2\pi i_m/N}.$$

A simulation of the Lagrange interpolation as given in (4) shows perfect recovery; see [1, Table 1]. Coding under erasure channel is equivalent to the missing sampling problem discussed in Chapter 5; see also [16], which is a special case of Lagrange interpolation for random sampling discussed in Fig. 21(a) in Chapter 4. The results, as expected, are exact. If we operate at lower than the Nyquist rate, we have some graceful degradation when the erasure is not bursty as shown in Fig. 21(b) Chapter 4; see also [1, Table 2]. But when we have a burst of errors, the degradation becomes appreciable in that region. The Lagrange interpolation becomes computationally intensive as n increases.

For bursty losses in an erasure channel and for large block sizes, the above methods are sensitive to truncation errors and specially not appropriate for bursty losses. The following section discusses methods that are more robust against quantization noise and additive channel noise specially for bursty losses.

17.4. Burst Error Recovery in an Erasure Channel Using Block Codes[3]

17.4.1. Introduction

Samples of a speech signal can be lost in an erasure channel or due to cell losses in an ATM network. Since cell losses due to buffer overflow in ATM environments result into bursts of errors rather than isolated errors, we shall propose a new technique to recover these bursts of errors in an erasure channel by over-sampling the original speech signal before packetization. If the signal is over-sampled, the original signal can be recovered as long as the average sampling rate is above the Nyquist rate [4]. This implies that if a simple scheme can be devised, recovery from over-sampled signal becomes an alternative to error correction codes [1], [9] and [10].

Many schemes have been proposed to recover the missing samples using the remaining samples of the signal. Some of these schemes implement iterative and time varying techniques in the recovery process [4] and [9]. Reed-Solomon algorithm is mentioned in [5] and [6] as an error correction code for complex FFT values without any extensive simulation results or any study of its computational load or sensitivity to noise. Below, we shall describe various implementations of a robust error recovery scheme using techniques similar to Peterson's BCH decoding [5] and [11] , Forney's decoding [5] and [12], and some matrix-based techniques.

We will also simulate the algorithms for real numbers to recover bursts of missing speech samples. Also, we shall present the Mathcad and DSP simulation results of the implemented techniques when applied to speech signals. We shall compare the computational complexity of the proposed technique to other recovery schemes such as Lagrange interpolation and the Conjugate Gradient method [13]–[14]. We shall also compare the results obtained for the proposed technique with those for the iterative techniques described in [9]. We shall study the sensitivity of the simulated techniques to additive and quantization noise and propose a new class of transformation kernels to reduce the sensitivity of the proposed algorithm to noise.

17.4.2. The Peterson Recovery Technique—RS Decoding

In this section we will describe a method that is valid for both random and bursty erasures of real and complex samples. This method in [10] is called the Burst Error Recovery Technique (BERT). This robust error recovery technique is similar to Peterson's method for BCH decoding [11], which is also used for Reed Solomon (RS) decoding.

[3]Portions of Sections 17.4–17.8 are the modified reprint with permission of [10]; © 1999 IEEE.

Let us assume that a signal such as speech is sampled at the Nyquist rate yielding a discrete signal $x_{\text{org}}(i)$, where $i \in Z$. If we form a block of k samples and take its Discrete Fourier Transform (DFT), we get k complex samples in the frequency domain ($X_{\text{org}}(j), j = 1, \ldots, k$). Similar to the discussion before (4), if we insert m consecutive zeros (including the cyclic shift) to get n samples ($X(j), j = 1, \ldots, n = k + m$) and take its inverse DFT, we end up with an over-sampled version of the original signal ($x(i), i = 1, \ldots, n$) with n complex samples.

In order to get real samples in the time domain, we need to preserve the complex conjugate symmetry in the frequency domain. Therefore, $m/2$ of the zeros are added at the beginning (where x denotes the smallest integer equal to or greater than x) and the rest at the end so that the symmetry is preserved. Alternatively, $m/2$ zeros could be padded after the mid point (the $k/2 + 1$ position) and $m/2$ zeros before the $k/2 + 1$ position of $X_{\text{org}}(j)$.

As an example for the first case, if $k = 8$, and $m = 7$, the first coefficient in the frequency domain is repeated at the 9th position to preserve the symmetry; $m/2 = 4$ zeros are added at the beginning, and $7 - 4 = 3$ zeros are added at the end ($X(j) = 0$ for $j = 1, \ldots, 4$ and $j = 14, \ldots, 16$ while $X(m) = X_{\text{org}}(m - 4)$ for $m = 5, \ldots, 12$ and $X(13) = X_{\text{org}}(1)$). We now have a set of $n = 16$ coefficients that have complex conjugate symmetry in the frequency domain and hence real inverse FFT values are obtained. Taking the inverse FFT of the n coefficients spectrum we get $x(i)$, which is an over-sampled version of the original signal $x_{\text{org}}(i)$. This (16,8) code is capable of correcting 7 erasures in the over-sampled signal $x(i)$.

The following algorithm is a new method for error recovery reminiscent of RS decoding algorithm [6]. This algorithm will be compared with other iterative and time varying techniques developed for signal error recovery [15]–[25]. The missing samples are denoted by $e(i_m) = x(i_m)$, where i_m denote the positions of the lost samples; for i not equal to i_m, $e(i) = 0$. For τ lost samples, the polynomial locator for the erasure samples is

$$H(s_i) = \prod_{m=1}^{\tau}\left(s_i - \exp\left(\frac{j2\pi \cdot i_m}{n}\right)\right) = \sum_{t=0}^{\tau} h_t \cdot s_i^{\tau-t} \tag{5}$$

$$H(s_m) = 0, \qquad m = 1, 2, \ldots, \tau, \tag{6}$$

where

$$s_i = \exp\left(\frac{j2\pi \cdot i}{n}\right), \qquad s_m = \exp\left(\frac{j2\pi \cdot i_m}{n}\right), \qquad i = l, \ldots, n.$$

The polynomial coefficients $h_t = 1, h_1, \ldots, h_\tau$ (τ is the total number of erasures) can be found from the product in (5).

In the DSP implementation [9] and [26] it is easier to find h_t by obtaining the inverse FFT of $H(s)$.

By multiplying (6) by $e(i_m) \cdot (s_m)^r$ and then summing over m, we get

$$\sum_{t=0}^{\tau} h_t \cdot \sum_{m=1}^{\tau} e(i_m) \cdot (s_m)^{\tau+r-t} = 0 \tag{7}$$

Since the inner summation is the DFT of the missing samples $e(i_m)$, we get

$$\sum_{t=0}^{\tau} h_t \cdot E(\tau + r - t) = 0, \tag{8}$$

for $r = \lceil m/2 \rceil + 1 \cdots - \lceil m/2 \rceil - 1$. Note that $E(j)$ is the Fourier transform of $e(i)$. It is now obvious that the reason for defining the variable s_i to be a root of unity is to convert the inner summation in (7) to the DFT of the missing samples yielding the expression in (8). The received samples, $d(i)$, can be thought of as the original over-sampled signal, $x(i)$, minus the missing samples $e(i_m)$. The error signal, $e(i)$, is the difference between the corrupted and the original over-sampled signal and hence is equal to the values of the missing samples for $i = i_m$ and is equal to zero otherwise.

In the frequency domain we have

$$E(j) = X(j) - D(j), \quad j = 1, \ldots, n. \tag{9}$$

Since $X(j) = 0$ for $j = 1, \ldots, \lceil m/2 \rceil$ and $j = -\lceil m/2 \rceil, \ldots, n$, then

$$E(j) = -D(j), \qquad j = 1, \ldots, \lceil m/2 \rceil \quad \text{and} \quad j = -\lceil m/2 \rceil, \ldots, n. \tag{10}$$

The remaining values of $E(j)$ can be found from (8), by the following recursion

$$E(r) = -\left[\sum_{t=1}^{\tau-1} E(\mathrm{mod}_n(r+t)) \cdot h_{\tau-t} - E(\mathrm{mod}_n(r+\tau)) \cdot z \right] \cdot p, \tag{11}$$

where $r = \lceil m/2 \rceil - 1, \ldots, \lceil m/2 \rceil + 1$,

$$z = \sqrt{n} \cdot \exp\left(j2\pi \cdot \frac{\tau}{n} \right), \quad \text{and} \quad p = \frac{1}{h_\tau},$$

and where z is the complex scaling factor related to the Mathcad FFT algorithm. After finding $E(j)$ values, the spectrum of the recovered over-sampled speech signal $X(j)$ can be found by applying (9) and hence the original signal can be recovered by under-sampling $x(i)$. The above algorithm is capable of correcting any combination of erasures.

17.4.3. Forney's Decoding Algorithm

In this section, we describe a modified version of Forney's method for decoding in the field of real and complex numbers [12]. We shall see that the following polynomial approach also holds for any cyclic codes such as FFT based codes. We shall use the same notation used in [11]. For our case, *Syndromes* are the errors in the FFT domain as in (9) and the error locator polynomial in [11] is the inverse of our polynomial in (5). The values of h_t in (5) can still be used, and the error evaluator polynomial [11] is modified as

$$\xi(s) = \sum_{j=1}^{\tau} s^j \left[\sum_{i=0}^{i-1} \bar{h}_{i+1} E(j-i) + \bar{h}_{i+1} \right] + 1, \tag{12}$$

where $E(j)$ is the syndrome as defined in (9)–(11) and $\bar{h}$ is the complex conjugate of h_t obtained from (5). The error signal $e(i)$ can be found from

$$e(i_m) = \frac{\xi(s_m)}{\prod_{i=1}^{\tau}\left(1 - \frac{s_m}{s_i}\right)}, \tag{13}$$

where $i_m, m = 1, \ldots, \tau$ are the positions of the lost samples and s_i, s_m are as defined in (5) and (6). The simulation results and noise analysis of Forney's method will be discussed in Section 17.5.1.

17.4.4. Convolutional Methods for Erasure Channel

Essentially, over-sampling is equivalent to a convolutional code using fields of real numbers as opposed to Galois fields. The encoder is a simple interpolator (a low pass filter) as shown in Fig. 1. This figure represents a convolutional encoder of rate $\frac{1}{2}$ of infinite constraint length and infinite precision per symbol. The nonuniform sampling theorem states that as long as the average rate of the samples is above the Nyquist rate, signal recovery is possible. In other words, if we over-sample a signal uniformly and a percentage of the samples is dropped, as long as the average rate of the remaining samples (which forms a set of nonuniform samples) is above the Nyquist rate, all the lost samples can be recovered.

We have implemented the convolutional method for erasure channel on a DSP chip. The results are somehow similar to the block method [9].

17.5. Simulation Results

We simulated the equations for the RS decoding technique and Forney's method using Mathcad to make sure the algorithms are working for the case of bursty erasures. For random erasures, we implemented the RS decoding algorithm on a Lucent DSP32 card for a long speech signal to measure the SNR and the Mean Opinion Score (MOS) [9], [26]. We shall first discuss the Mathcad simulation and then go over the DSP implementation.

17.5.1. Mathcad Simulations

The results of the Mathcad simulations (see program file: Chapter 17 RS-Decoding.mcd in the CD-ROM) for the RS decoding implementation for $n = 32$, $m = 15$, and $t = 15$ erasures (a burst of 15 consecutive missing samples from position 1 to 15) are shown in Fig. 3.

Comparing the result of Fig. 3 to the results of other methods [10], we conclude that the performance of the proposed burst error recovery method is better for the case of signal recovery in an erasure channel for a given complexity. Since consecutive sample losses are the worst case [10], the proposed method

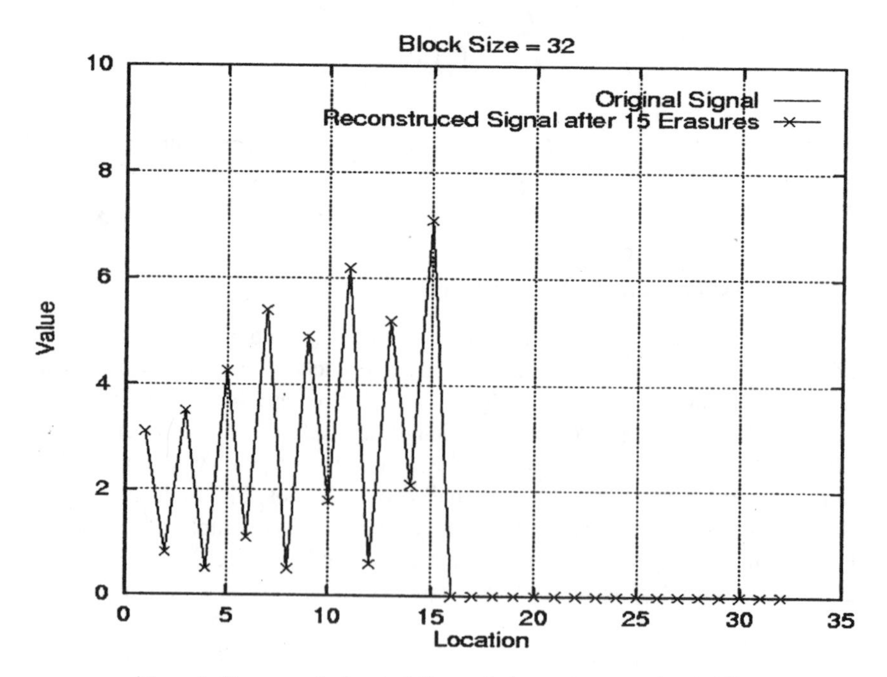

Figure 3. Recovery of a burst of 15 sample losses (Lost samples: 1–15).

seems to be very efficient. In practice, it was found that the error recovery capability of this technique degrades with the increase of the block size due to the accumulation of round-off errors.

For example, if the block length is increased to $n = 64$ with $m = 31$ and $\tau = 31$ (31 consecutive missing samples from position 1 to 31 as shown in Fig. 4), the round-off error increases. To reduce the round-off error, the total number of consecutive missing samples should be reduced as shown in Fig. 5. This figure shows the result when 23 consecutive erasures instead of 31 erasures are considered. It is clear that the round-off error for the case of 23 consecutive erasures is less than that for the case of 31 erasures. If the sample losses are random or if we have a combination of consecutive and isolated losses, the round-off error is negligible and we can perfectly recover 22 erasures as shown in Fig. 6. The round-off error—which becomes more predominant for the case of consecutive losses—is due to the large dynamic range for the values of h_t which results in loss of accuracy due to memory overflow. A study of the sensitivity of the RS decoding technique to round-off errors and additive noise is presented in Section 6. Figure 7 shows the simulation results of Forney's method for a block size of 32 and with 16 consecutive losses; see Chapter 17\Forney.mcd in the attached CD-ROM.

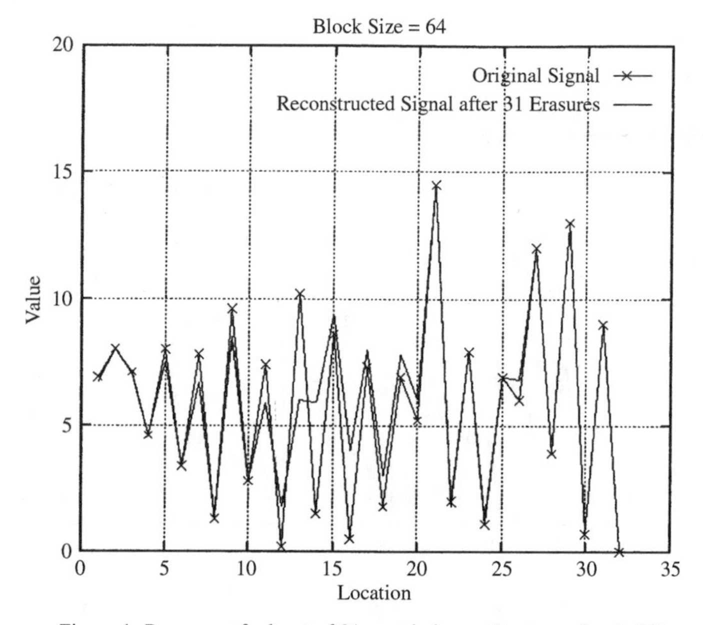

Figure 4. Recovery of a burst of 31 sample losses (Lost samples: 1–31).

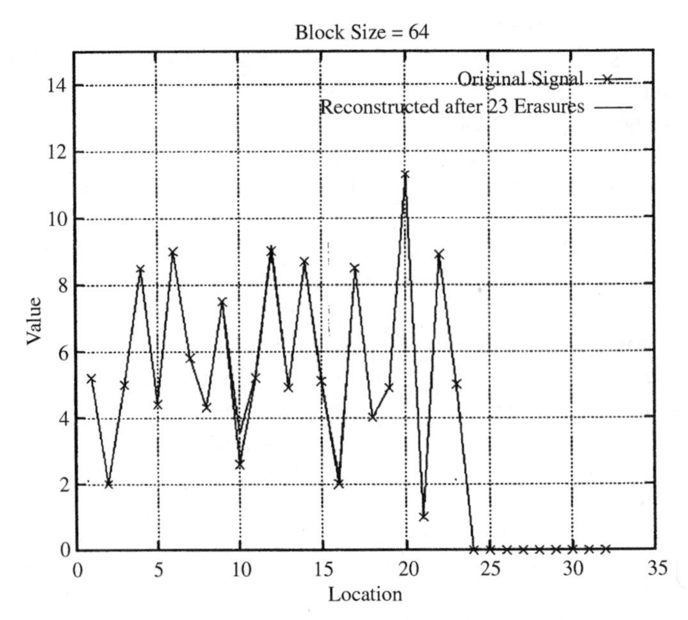

Figure 5. Recovery of a burst of 23 losses (Lost samples: 1–23).

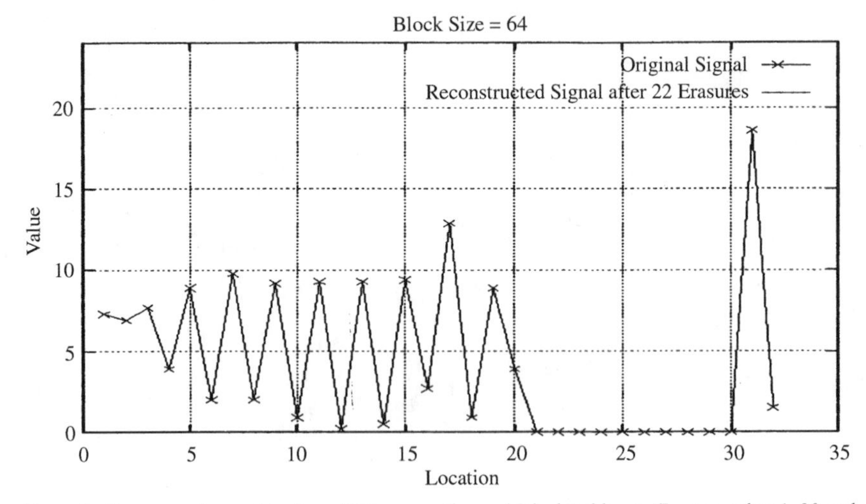

Figure 6. Recovery of a combination of 20 consecutive and 2 isolated losses (Lost samples: 1–20 and 31–33).

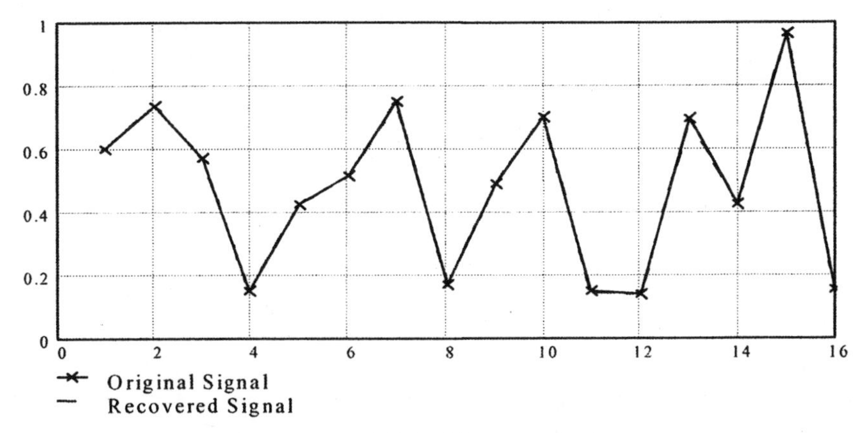

Figure 7. Forney's Method for consecutive losses.

For random losses as opposed to bursty losses, we implemented the BERT technique on a DSP card. We shall see that despite the accumulation of errors in the case represented in Fig. 4, the SNR and MOS results are still better than that for the iterative methods [25].

17.5.2. DSP Simulations

We implemented the RS decoding algorithm on a Lucent DSP32 card. Unlike the Mathcad implementation, block sizes greater than 32 cannot be implemented on this kind of DSP cards. This limitation is due to the processing delay of the DSP card, which increases as the block size increases. However, it is possible to increase the block size to 64 by padding the zeros in the time domain rather than the DFT transform domain.

We have implemented the recovery algorithm for a speech signal with block sizes of 16 and 32 samples. The erasure pattern is implemented using a pseudo-random generator that produces isolated losses. In the DSP implementation, we disable the algorithm and use a simple low pass filter whenever the total number of errors is larger than m, where m is equal to the number of padded zeros. This is due to the limited capability of the code to correct more than m erasures. This procedure allows graceful degradation in the signal quality.

For the case of a long burst of erasure, the performance of this scheme [36] is better than that of the time varying and iterative methods discussed in [22]–[25]. For the case of random erasures, the proposed burst error recovery method is, in general, better than these iterative and time varying methods. The objective and subjective evaluations for random erasures are given below.

17.5.3. Objective Evaluation of the DSP RS Decoding Implementation

The SNR for a speech signal after applying the proposed burst error recovery algorithm is

$$\mathrm{SNR} = \frac{\sum_{k=0}^{N-1} |x(k)|^2}{\sum_{k=0}^{N-1} |x(k) - r(k)|^2}, \tag{14}$$

where $x(k)$ is the input speech sample, $r(k)$ is the reconstructed speech sample after the DSP processing and N is the total length of the speech samples. Figure 8 shows a plot of SNR of the recovered speech signal using the RS decoding technique versus the percentage of the lost samples for block sizes of 16 and 32 (labelled as BERT(16) and BERT(32), respectively). In addition, Fig. 8 shows the plot of the SNR of the same speech signal recovered using a simple band pass filter versus the percentage of the lost samples for block sizes of 16 and 32 (labelled BPF (16) and BPF (32), respectively).

It can be noticed from Fig. 8 that a significant improvement in the speech signal quality can be achieved if the proposed algorithm is implemented in real-time on a DSP card. In general, the larger the block size the better is the SNR. However, when the block size increases, the accumulated round-off error increases which degrades the efficiency of the code. For percentage losses less than 7%, the proposed burst error recovery technique performs better for small block sizes (BERT (16)) than for large block sizes (BERT (32)). For 10% sample loss, the performance for large block sizes (BERT (32)) is about 70 dB better than that for small block sizes (BERT (16)).

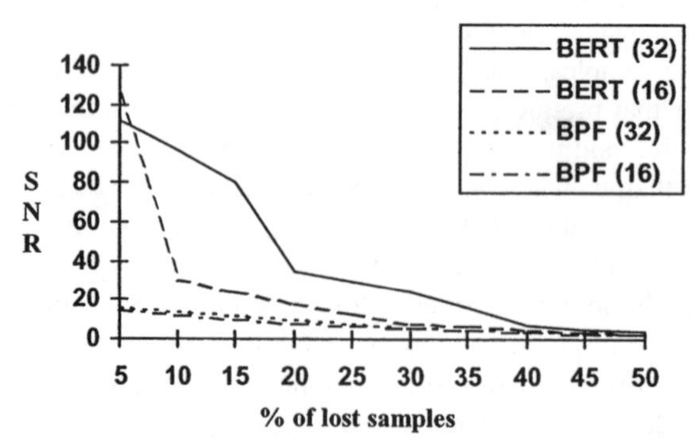

Figure 8. SNR of the recovered signal using the RS decoding (BERT) technique and the bandpass filtered signal (BPF) for block sizes of 16 and 32.

17.5.4. Subjective Evaluation of the DSP RS Decoding Implementation

The MOS scores for the four different curves in Fig. 8 are given in Fig. 9. The speech signal used in the MOS test is a conversation between a man and a woman. At 10% sample loss, the improvement for large block sizes (BERT (32)) relative to smaller ones (BERT (16)) is negligible despite the fact that Fig. 8 shows a 70 dB SNR improvement. At 30% sample loss, the performance of the BERT technique for large block sizes (BERT (32)) is significantly better than that for small ones (BERT (16)) despite the fact that Fig. 8 shows only 20-dB SNR improvement. The reason is obviously due to the sensitivity of the human ear to different SNR values.

In our previous work on erasure recovery, we had concluded that iterative techniques (linear, time varying and hybrid) are the best methods for the recovery of a signal from random erasure [25]. We now compare the new results with the results of one of the best iterative methods discussed in [25]. The comparison for a block size of 32 is shown in Fig. 10. Clearly, at 30% sample loss, the proposed

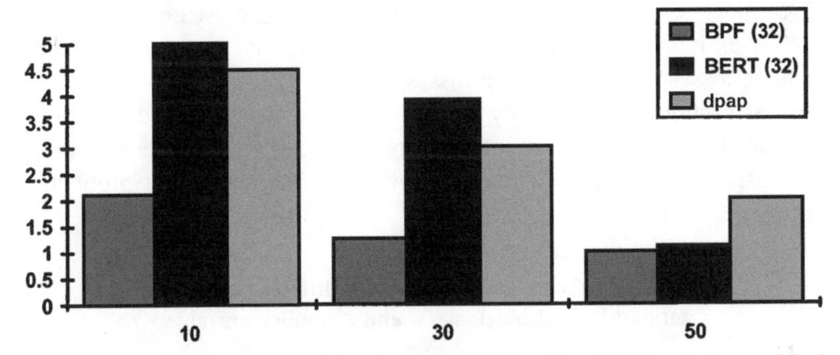

Figure 9. MOS score for the recovered signal using the RS decoding (BERT) technique and the band-pass filtered signal (BPF) for block sizes of 16 and 32.

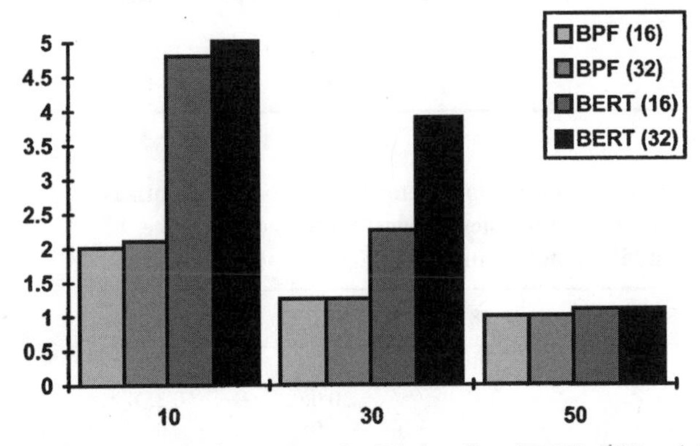

Figure 10. MOS score comparison among the RS decoding (BERT), BPF and the best technique (dpap) in [23].

method is better than the iterative technique called *dpap* [25]. The iterative technique is only better at higher sample loss rates. This is due to the fact that for more than 50% sample losses, the performance of the iterative techniques degrades gracefully while the RS decoding technique completely fails (the average sampling rate should be greater than or equal to the Nyquist rate). In the next section, the effect of noise on the recovery process is presented. More details about the noise sensitivity issue are in [10] and [20].

17.6. Noise Sensitivity of Error Correction Codes in the Galois Field of Real/Complex Numbers in an Erasure Channel

As seen from the results in the previous section, the proposed technique is sensitive to the accumulated round-off error which points to the fact that it is also sensitive to quantization and additive noise. This issue has been briefly discussed in [6], [26] and [34]. The sensitivity to quantization and additive noise increases for larger block sizes as can be noticed from the results presented in Tables 1, 2 and 4. Also, it can be noticed that the sensitivity is more accentuated for consecutive losses than for isolated losses.

In this section we shall study the sensitivity of the RS decoding algorithm and explain why some patterns of loss are more sensitive to noise than other patterns. An eigenvalue-based matrix analysis of the sensitivity problem was presented in Chapter 5.

Table 1. Correlation factors for different quantization bits and block sizes, and * implies unstable recovery.

Block Size	$Q=8$ bits	$Q=16$ bits	$Q=32$ bits
8	0.997	1.0	1.0
16	0.442	1.0	1.0
32	0.0*	0.0*	0.969

Table 2. Correlation factors for different quantization bits and block sizes with the presence of additive noise, and * implies unstable recovery.

Block Size	$Q=8$ bits	$Q=16$ bits	$Q=24$ bits
8	1.0	1.0	1.0
16	0.179	0.06	0.357
32	0.0*	0.0*	0.0*

17.6.1. Sensitivity to Quantization and Additive Noise

For the case of consecutive erasures, we simulated the proposed technique for different block sizes with half of the samples being lost. Since we padded an equal number of zeros to the original signal, the resulting code is capable of recovering a maximum of half of the transmitted samples when they are lost.

In our simulations, we assess the performance of the RS decoding technique under the maximum recovering capability and with the presence of noise. The over-sampled signal was quantized using different quantization bits before transmission. The numbers in Table 1 represent the correlation between the reconstructed signal and the original signal for different block sizes and quantization bits, Q.

The correlation factors presented in Table 1 are the *Pearson's Product Moment Correlation Factors,* ρ, which can be calculated using the following equation

$$? = \frac{N\sum_{i=0}^{N-1} x(i)y(i) - \sum_{i=0}^{N-1} x(i)\sum_{i=0}^{N-1} y(i)}{\sqrt{\left(N\sum_{i=0}^{N-1} x^2(i) - \left[\sum_{i=0}^{N-1} x(i)\right]^2\right)\left(N\sum_{i=0}^{N-1} y^2(i) - \left[\sum_{i=0}^{N-1} y(i)\right]^2\right)}}, \quad (15)$$

where $x(i)$ and $y(i)$ are the original and the recovered signals, respectively. N is the number of samples considered for both signals.

Similarly, Table 2 shows that the proposed technique is also sensitive to additive noise. For 8, 16 and 24-bit quantization of the transmitted signal, we added to the transmitted signal a white random noise of uniform distribution with an average amplitude of $1/100$ of the transmitted signal (SNR = 25 dB). The correlation factors obtained for the simulated quantization bits and block sizes with the presence of additive noise are shown in Table 2.

Tables 1 and 2 show that when losing a burst of half of the samples, the proposed RS decoding technique is robust against quantization and additive noise only for block sizes equal to 8. The RS decoding technique completely fails for block sizes greater than or equal to 16 with the presence of additive noise. For the case of isolated random losses or if the number of missing samples is less than the maximum number of erasures the technique is capable of correcting, the recovery process is relatively stable and the proposed technique is no longer as sensitive to quantization and/or additive noise as it is for the case of bursty erasures. Table 3 shows the correlation factors obtained for the case of isolated erasures with the presence of quantization and additive noise.

Table 4 presents the results obtained in the presence of quantization and additive noise where the number of consecutive missing samples is less than the

Table 3. Correlation factors for isolated erasures with added quantization and additive noise.

Block Size	$Q = 8$ bits	$Q = 16$ bits	$Q = 24$ bits
8	1.0	1.0	1.0
16	1.0	1.0	1.0
32	1.0	1.0	1.0

Table 4. Correlation factors in the presence of quantization and additive noise where the number of consecutive erasures is less than the maximum capability of the code, and * implies unstable recovery.

Block Size	$Q = 8$ bits	$Q = 16$ bits	$Q = 24$ bits
8	1.0	1.0	1.0
16	0.924	0.969	0.984
32	0.0*	0.0*	0.0*

maximum number of erasures that the technique is capable of recovering (about half the maximum capability of the code).

17.6.2. Analysis

We start with a heuristic explanation of the noise sensitivity and then we present a mathematical analysis of the problem. As seen in Fig. 11, consecutive losses produce large dynamic range for the error locator polynomial $H(s)$ as well as the polynomial coefficients h_t. This is due to the fact that the zeros of $H(s)$ are all concentrated on one side of the semi-circle. Figure 12 represents the error locator polynomial zeros diagram for the case of 16 consecutive losses in a 32-sample block. $H(s_i)$ values are zeros for $i = 1, \ldots, 16$ and very large around $i = 24$

The previous statement can be verified by calculating the value of $H(s_{24})$ which is equal to the product of all the 16 vectors emanating from the position $i = 24$ on the unit circle towards the zeros of $H(s)$ on the upper half of the unit circle as in Fig. 12. The minimum magnitude of these 16 vectors is $\sqrt{2}$ and the maximum magnitude is 2. Thus, we expect to have a magnitude between 2^8 and 2^{16}; the actual value is 1.109×10^4. This creates a large dynamic range for the values of h_t, starting from 1 for h_τ and peaking at $h_{\tau/2}$ as shown in Fig. 11.

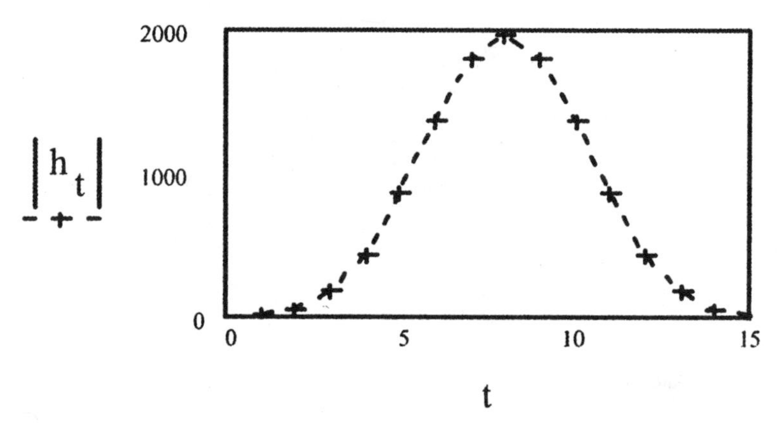

Figure 11. The behaviour of h_t for the case of consecutive losses.

Typical values of h_t for a stable reconstruction of the isolated losses are usually less than 1. This observation can be explained using the difference equation in (11). The solution for this difference equation is equivalent to the zero input response of an IIR filter with the initial conditions given in (10). The Z-transform of this IIR filter has poles identical to the zeros of the error locator polynomial and the initial conditions affect only the zeros of the filter. The solution of (11) is thus the inverse transform of this Z-transform, which has the same shape as the eigenfunctions of the impulse response of the IIR filter. For the case of consecutive losses, these complex eigenfunctions (modified in amplitude by the initial conditions) are added in an accumulative manner and hence huge errors can be produced when additive or quantization noise is present. Figure 13 shows all the possible zeros of the error locator polynomial, $H(s_i)$, for a block size of 32.

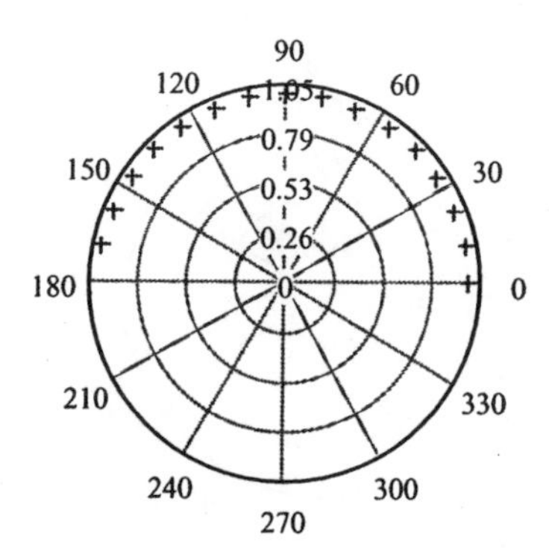

Figure 12. The zero pattern of the error locator polynomial, $H(s_i)$ for the case of consecutive losses.

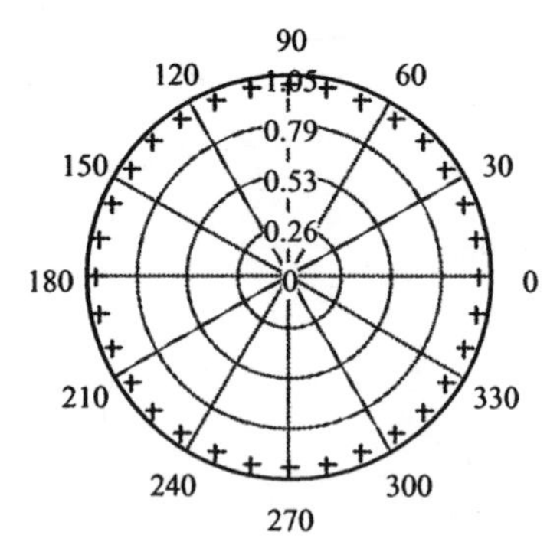

Figure 13. All the possible zeros of the error locator polynomial, $H(s_i)$.

For the case of isolated losses (every other loss), the locations of the zeros of the error locator polynomial (and hence the poles of the IIR filter representing (11)) are symmetrically distributed on the unit circle as shown in Fig. 14 and hence $H(s_i)$ and h_t values are well behaved. In fact, $H(s_i)$ values are equal to either two or zero while h_t values are all zeros except for h_t which is equal to 1. Due to the symmetry of the poles of the IIR filter on the unit circle, the eigenfunctions cancel each other in most of the cases and hence we have a stable situation. In this case, any quantization or additive noise is also cancelled most of the time due to the symmetry and hence such noise has a negligible effect on the stability of the proposed recovery technique.

From a mathematical point of view, the sensitivity to noise of an IIR filter increases dramatically whenever the poles and/or zeros of the filter are clustered [27]. The analysis in [27] shows that any small variations in the IIR filter coefficients can cause very large variations at the filter output if the filter poles

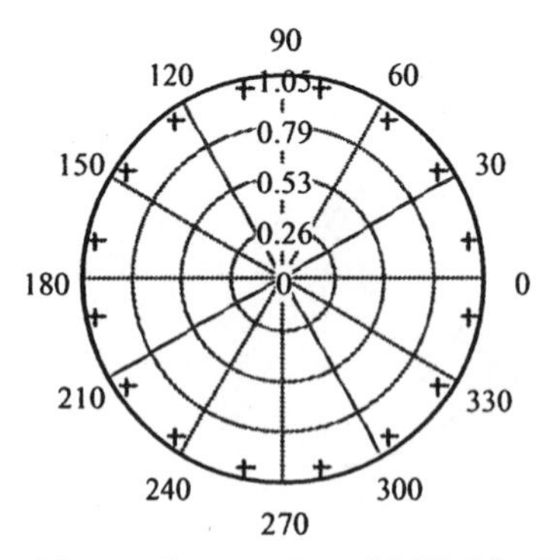

Figure 14. The zero pattern of the error locator polynomial, $H(s_i)$ for the case of isolated losses (every other loss).

and/or zeros are clustered and very close to each other. Taking the unilateral Z-transform of the difference equation (8) with the initial conditions given in (10), we can write

$$\hat{E}(z) = \frac{\sum_{k=1}^{\tau} \sum_{r=1}^{k} h_{\tau-h}/h_{\tau} \cdot E(r) z^{k-r+1}}{1 + \sum_{k=1}^{\tau} h_{\tau-h}/h_{\tau} \cdot z^{k}}, \tag{16}$$

where $\hat{E}(z)$ is the Z-transform of $E(r)$.

The pole/zero clustering case in the canonical IIR filter is equivalent to the case of bursty losses in our technique and hence the analysis in [27] offers an alternative explanation for the sensitivity to noise of the proposed technique. From [27], we can define the zeros of $\hat{E}(z)$ in (16) to be $z_i + \Delta z_i$ where i is indexed from $1, \ldots, \tau$ and Δz_i is the error in the ith zero due to the quantization and additive noise in the $E(r)$. The numerator of (16) is

$$A(z) = \sum_{k=1}^{\tau} \sum_{r=1}^{k} \frac{h_{\tau-k}}{h_{\tau}} \cdot E(r) \cdot z^{k-r+1} = \prod_{j=1}^{\tau} (1 - z_j z^{-1}). \tag{17}$$

The error in $\hat{E}(z)$ zeros, Δz_i, can be expressed in terms of the error in the $E(r)$ as

$$\Delta z_i = \sum_{r=1}^{\tau} \frac{\Delta z_i}{\Delta E(r)} \cdot \Delta E(r) \tag{18}$$

By using the two forms of $A(z)$ in (17) and the fact that

$$\left(\frac{\partial A(z)}{\partial z_i} \right)_{z=z_i} \cdot \frac{\partial z_i}{\partial E(r)} = \left(\frac{\partial A(z)}{\partial E(r)} \right)_{z=z_i}, \qquad \frac{\partial z_i}{\partial E(r)} = \frac{-\sum_{k=1}^{\tau} \frac{h_{\tau-k}}{h_{\tau}} \cdot z_i^{k-r+\tau+1}}{\prod_{\substack{j=1 \\ j \neq i}}^{\tau} (z_i - z_j)}, \tag{19}$$

it follows that

$$A(z) = \sum_{k=1}^{\tau} \sum_{r=1}^{k} \frac{h_{\tau-k}}{h_{\tau}} \cdot E(r) \cdot z^{k-r+1} = \prod_{j=1}^{\tau} (1 - z_j z^{-1}), \tag{20}$$

for $i = 1, \ldots, \tau$ and $r = 1, \ldots, \tau$. From (18) and (20), the error in $\hat{E}(z)$ zeros, Δz_i, can be written as

$$\Delta z_i = \sum_{r=1}^{\tau} \left(\frac{-\sum_{k=1}^{\tau} \frac{h_{\tau-k}}{h_\tau} \cdot z_i^{k-r+\tau+1}}{\prod_{\substack{j=1,\\ j\neq 1}}^{\tau} (z_i - z_j)} \cdot \Delta E(r) \right). \tag{21}$$

As mentioned earlier for the case of consecutive losses, the values of h_t increases to very high values around $h_{\tau/2}$ and hence the inner summation in (21) becomes very large. This means that any small changes in the values of *E(r)* due to quantization or additive noise cause drastic variations in the locations of the zeros of (16) which change the behaviour of the IIR filter representing the recursion in (11). We also observed from our simulations that the zeros of (16), which are represented as poles in (21), are clustered together for the case of consecutive losses. This observation further enhances the zero displacement in (21) in case of additive or quantization noise.

Similarly, we can show that the poles of (16) are sensitive to the round-off errors in h_t by the following equation

$$\Delta z_p = \sum_{t=1}^{\tau} \left(\frac{-z_p^{\tau-t}}{\prod_{\substack{j=1,\\ j\neq p}}^{\tau} (z_p - z_j)} \cdot \Delta h_t \right). \tag{22}$$

From the above equation, small variations in h_t cause very large variations in z_i which represent the poles of (16) for the case of consecutive losses where we have clustered poles.

The above analysis implies that there are other IIR filter structures such as the parallel and series structures that can offer less sensitivity to quantization and additive noise. But our simulation results show the improvement is not significant and other methods have to be considered [29].

17.6.3. Forney's Method

Depending on the lost pattern, Forney's approach may or may not be better than the RS decoding approach in terms of the sensitivity to quantization and additive noise. Simulation results show that both approaches present similar recovery performance in noise-free environments.

17.7. Other Transform Techniques

Based on the above analysis, the sensitivity of the proposed technique can be significantly reduced by introducing a new class of kernels instead of the FFT kernel used in equations (5)–(11) as the transformation kernel. For example, the matrix elements of one of these transformation kernels can be written as

$$t_{k,l} = \exp(-j \cdot k \cdot l), \tag{23}$$

where $M = [t_{k,l}]$ is the new transformation matrix used. Compared to the FFT kernel, the kernel in (23) reduces the sensitivity of the RS decoding algorithm to noise for the case of bursty losses. The proposed kernel causes zeros corresponding to consecutive losses to be spread evenly in the four quadrants of the unit circle as seen in Fig. 15 and Fig. 16. For this kernel, the zeros corresponding to each two consecutive losses lie in separate quadrants of the unit circle; see Fig. 16.

For this transformation kernel, the values of $H(s_i)$ behave properly and do not build up to huge values. The behaviour of the zero pattern of the modified error locator polynomial when using the proposed kernel in (23) offers a simple explanation for the reduced sensitivity of the RS decoding technique to noise. Correlation factors and simulation results obtained for the proposed kernel for different quantization bits and block sizes with the presence of additive noise are shown in Table 5.

By comparing the results of Tables 2 and 5, we conclude that the FFT kernel fails completely for block sizes greater than 8 with the presence of noise whereas the kernel in (23) provides exact recovery of the missing samples even for large blocks of size 64 samples. From the above results, we conclude that the kernel in

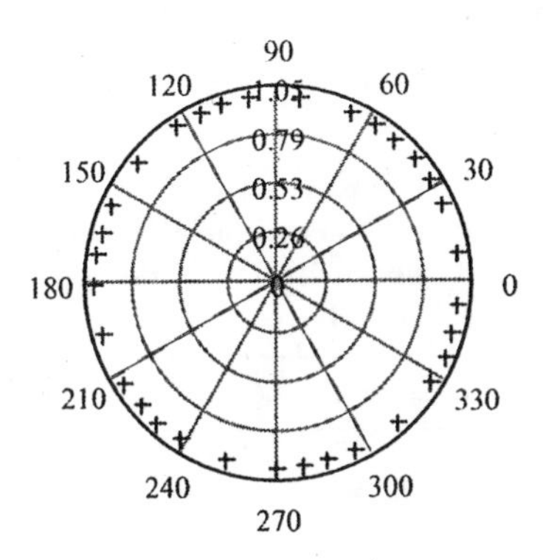

Figure 15. All the possible zeros of the error locator polynomial $H(s_i)$ for $e^{-j(k \cdot l)}$ transformation kernel.

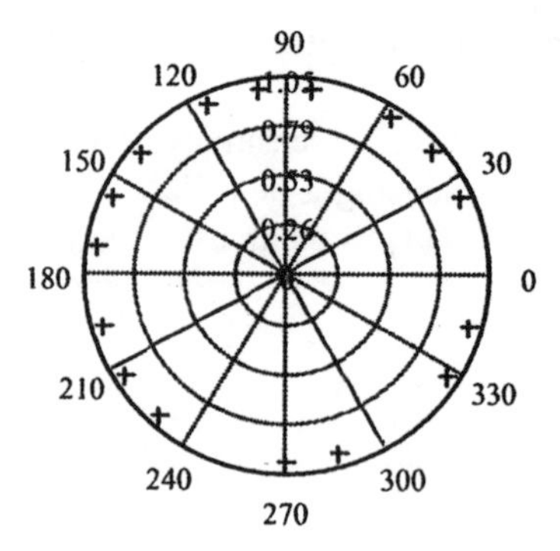

Figure 16. The zero pattern of the error locator polynomial $H(s_i)$ for the case of consecutive losses for the $e^{-j(k \cdot l)}$ transformation kernel.

(23) is much more robust to quantization and additive noise than the classical FFT kernel for the case of bursty losses.

Another approach is to choose a DFT based kernel but sorted according to

$$t_{k,l} = e^{j2\pi klq/n}, \tag{24}$$

where q is an integer relatively prime to n. If we insert zeros in the middle of the transform domain using the above kernel, we get a code that is very robust against long bursts of erasures. The above kernel is equivalent to DFT and sorting the DFT coefficients by a mod rule: mod($q \cdot i$, n), where i's are the index of the DFT coefficients. The advantage of the kernel given in (24) is that the RS decoding method can be used for the recovery of a big lost consecutive samples. For more information on this topic, see Section 17.11.

For the kernels presented in (23)–(24), there may be certain patterns of random losses which cause all the corresponding zeros to fall into one side of the

Table 5. Correlation factors of the proposed technique using the $e^{-jk.l}$ kernel for different quantization bits and block sizes with the presence of additive noise.

Block Size	$Q = 8$ bits	$Q = 16$ bits	$Q = 24$ bits
8	1.0	1.0	1.0
16	1.0	1.0	1.0
32	1.0	1.0	1.0
64	0.999	1.0	1.0

unit circle. Therefore, depending on the application, the kernels in (23)–(24) may or may not be useful. If the losses are consecutive such as speech frame losses in a wireless environment or packet losses in ATM networks, then the kernel shown in (24) is ideal.

Discrete Cosine Transform (DCT), Discrete Wavelet Transform (DWAT), Discrete Hartley Transform (DHT), and Discrete Walsh Transform (DWT) are other potential transforms that can be used instead of DFT. The problem with these transforms is that since they are not exponential like DFT or (23), the RS decoding recovery technique is not applicable. Also, since the transform matrices are not Vandermonde, recovery from random or bursty erasures is not always guaranteed[4]. Some of these simulations were done in [1] and it was shown that whenever the erasure recovery using an iterative technique worked, DFT convergence rate was faster, in general. For the simulation details please refer to [1].

For the above kernels and other non-exponential kernels, we can not use the RS decoding approach, equations (5)–(11), to recover the missing samples. Instead, the following equations can be used to find the recovered signal:

$$\mathbf{X} = \mathbf{T} \cdot \mathbf{x}, \tag{25}$$

where $\mathbf{X}$ is an n-vector containing the values of X(j) and $\mathbf{x}$ is another n-vector containing the values of the original over-sampled signal x(i). $\mathbf{X}$ is the transform-domain signal of $\mathbf{x}$ as defined before. The correctly received samples are grouped in a new vector y, then a new sub-matrix $\mathbf{T_s}$ is formed by eliminating the rows of the $\mathbf{T}$ matrix which correspond to the padded zeros (as mentioned in Section 17.3) and the columns which correspond to the erasure locations in the resultant matrix. As a result, we have:

$$\mathbf{X_s} = \mathbf{T_s} \cdot \mathbf{y}, \tag{26}$$

where $\mathbf{X_s}$ is a vector containing the transform-domain values of the original signal, $x_{\mathrm{org}}(i)$.

DWT and DCT under impulsive environment are discussed in Sections 17.9.2–17.9.3.

17.8. Computational Complexity

The RS decoding technique is very efficient in recovering bursts of errors with a performance better than the other techniques such as Lagrange interpola-

[4]DCT is an exception to be discussed in Section 17.9.3.

tion [25]. In this section, we shall compare the computational load of the proposed technique to Lagrange interpolation and the method of Conjugate Gradient (CG) using adaptively chosen weights and block Toeplitz matrices [9] and [14].

17.8.1. Lagrange Interpolation

Let $x(z_m)$ be the missing speech samples to be recovered by Lagrange interpolation, $\{m \in L$: the lost sample location indices; $m = 1, \ldots, \tau\}$ and $\{l \in S$: the remaining sample location indices; $l = \tau + 1, \ldots, n\}$. From the previous definitions, the missing sample locations can be indicated by z_m and the remaining sample locations can be indicated by z_l. The total number of the missing samples is τ and the total number of the remaining samples is k where $k = n - \tau$. The recovered samples using Lagrange interpolation [25] can be found from

$$x(z_m) = \sum_{\ell \in S} x(z\ell) \cdot \Psi_\ell(z_m), \qquad \text{where } m \in L, \text{ and } z_m = e^{\frac{j2\pi m}{n}} \tag{27}$$

and

$$\Psi_\ell(z_m) = \frac{\pi_\ell(z_m)}{\pi_\ell(z_\ell)} = \prod_{\substack{r \in S \\ r \neq \ell}} \left(\frac{z_m - z_r}{z_\ell - z_r} \right), \; \Psi_\ell(z_\ell) = 1, \Psi_\ell(z_r) \in 0, \; \ell \neq r, \tag{28}$$

where

$$\pi_\ell(z) = \prod_{\substack{r \in S \\ r \neq \ell}} (z - z_r). \tag{29}$$

Comparing (27–29) with (5–11), one can notice the following points:

- In (11), all the computations are done in the frequency domain to recover missing samples in the time domain whereas in (27), computations are carried out in the time domain directly.
- The recursive signal in (11) is the spectrum of the error signal whereas in the Lagrange interpolation, we interpolate in the time domain for the missing samples themselves.
- Although (29) resembles the error locator polynomial in (5), the main difference between the two equations is that equation (29) uses the remaining samples locations to calculate the coefficients $\Psi_l(z_m)$ used in the interpolation process while equation (5) uses the missing samples locations to compute the error locator function and hence h_t used in

the interpolation process. This implies that depending on the percentage of lost samples, (27–29) technique become more efficient than the other method (5–11); this point is discussed as minimum dimension in Chapter 5.

The number of additions and multiplications involved in Lagrange interpolation is definitely much higher than the number of additions and multiplications required for the proposed technique. For τ missing samples, the overall number of Multiplications for Lagrange (*ML*) and Additions for Lagrange (*AL*) are, respectively [10]

$$ML_{\text{total}} = 8 \cdot \tau \cdot k \cdot (k - 1) \tag{30}$$

and

$$AL = 2 \cdot \tau \cdot (4k + 1)(k - 1). \tag{31}$$

An attractive way of implementing Lagrange interpolation is to store the calculated values of $\pi_l(i_m)$ and $\pi_l(i_l)$ in a memory buffer and use them to calculate subsequent values of $\psi_l(z_m)$. Further reductions in the number of multiplications and additions involved can be achieved by calculating $\pi_l(i)$, incrementally using

$$\pi_\ell(z) = \pi_{\ell-1}(z) \cdot \frac{(z - z_{\ell-1})}{(z - z_\ell)}. \tag{32}$$

The total number of Multiplications for Optimized Lagrange (*MOL*) and Additions for Optimized Lagrange (*AOL*) are, respectively [10]

$$MOL_{\text{total}} = 4 \cdot (k - 1)(4\tau + 3) \tag{33}$$

and

$$AOL_{\text{total}} = 2 \cdot (k - 1)(6\tau + k + 3). \tag{34}$$

Although the Optimized Lagrange requires fewer number of calculations, the proposed way of implementing Lagrange interpolation requires $2(k + 1)$ memory locations and $2\tau \cdot (k + 1)$ read and write cycles to access the stored values.

17.8.2. The Conjugate Gradient (CG) Technique

This technique recovers the missing samples using the method of Conjugate Gradient (CG) to solve a linear system of trigonometric polynomials formed to represent the spectrum of the speech samples. The discussion of the algorithm is in Chapter 6 of this book and the reader may refer to [13] and [14] for more details. To obtain a fair computational complexity comparison between the proposed technique and the CG scheme, we shall consider the 1-D case of the CG scheme assuming that the pre-conditioning weights have already been determined. For $n = 2M + 1$ samples where τ samples are missing and M is the bandwidth of the speech signal, the total number of Multiplications for the CG method (*MCG*), and the Additions for the CG method (*ACG*), can be expressed by [10]

$$MCG_{\text{total}} = 3\tau + (2M+1)[4(\tau+1)\log(2M+1) + 16\tau + 1] \tag{35}$$

and

$$ACG_{\text{total}} = -3\tau + (2M+1)[6(\tau+1)\log(2M+1) + 16\tau]. \tag{36}$$

17.8.3. RS Decoding

For the proposed burst error recovery technique, the total number of Multiplications for the RS decoding (*MRS*) and Additions for the RS decoding (*ARS*) are, respectively

$$MRS = (4.\tau.(n+m) + 4.n.\log(n) + 2.(n-m).\log(n-m) - 4.(n-m)) \tag{37}$$

and

$$ARS = (4.\tau.(n+m) + 6.n.\log(n) + 3.(n-m).\log(n-m)). \tag{38}$$

17.8.4. Comparison of the Techniques

For $n = 64$, $m = 32$, $k = 32$ and $\tau = 32$, and $2M + 1 = 64$, a summary of the comparison of different techniques is shown in Table 6.

From the above calculations, we conclude that Lagrange interpolation requires about 17 times the number of multiplications and additions required by the proposed technique while the CG technique requires about 13 times the number of multiplications and additions required by the proposed burst error recovery technique. Although the optimized implementation of Lagrange interpolation algorithm requires almost similar number of arithmetic operations

Table 6. The computational complexity of the RS decoding compared to the other techniques.

Recovery Method	Multiplications	Additions
RS decoding	14,016	15,072
CG	165,632	215,104
Lagrange	253,952	255,936
Optimized Lagrange	16,244	14,074

compared to the proposed technique, it requires more memory storage and access time.

17.9. Recovery in an Additive Impulsive Noise Environment

Since error correction capability of codes in the field of real numbers depends on the number of errors per block, additive Gaussian noise would not be a suitable environment for these types of codes. Impulsive noise on the other hand corrupts a finite number of samples per block and therefore is suitable for codes in the field of real numbers. In this section, we show that it is possible to recover an over-sampled signal from impulsive noise as long as certain conditions are satisfied. We shall discuss the discrete transforms DFT, DWT and DCT.

17.9.1. The DFT Case

Assume that m zeros are padded at the end of the DFT and assume that there are t impulsive errors such that $2t = m$. At the receiver, the received signal in the frequency domain in the last m components are due to impulsive noise only. The inverse DFT of this part is the "syndrome" that can be used for error detection and correction. The analysis is as follows:

$$y(i) = x(i) + e(i), \tag{39}$$

where y, x and e are the received signal, the transmitted signal, and the impulsive noise, respectively. The syndrome is

$$s(i) = \frac{1}{n} \sum_{j=n-2t}^{n-1} Y(j)\Phi(i,j) = \frac{1}{n} \sum_{j=n-2t}^{n-1} E(j)\Phi(i,j), \tag{40}$$

where Y and E stand for the Discrete Transform of the corresponding y and e in time. Φ is the basis function which is equal to an exponential for DFT and cosine for DCT, etc. Since E is defined as

$$E(j) = \sum_{m=0}^{t-1} e(i_m)\Phi^*(i_m, j), \tag{41}$$

where * is the complex conjugate notation, from (40) and (41) we have

$$s(i) = \frac{1}{n}\sum_{m=0}^{t-1} e(i_m) \sum_{j=n-2t}^{n-1} \Phi(i, j)\Phi^*(i_m, j). \tag{42}$$

The above equation can be written as

$$s(i) = \frac{1}{n}\sum_{m=0}^{t-1} e(i_m) \sum_{l=0}^{2t-1} \Phi(i, l-2t)\Phi^*(i_m, l-2t). \tag{43}$$

For the DFT case, $\Phi(i, l) = \exp(j2\pi il/n)$, after some manipulations, we get

$$s(i) = \frac{1}{n}\sum_{m=0}^{t-1} e(i_m) \frac{\sin\left(\frac{2\pi}{n}(i-i_m)t\right)}{\sin\left(\frac{\pi}{n}(i-i_m)\right)} e^{-j\frac{\pi}{n}(i-i_m)(2t+1)}. \tag{44}$$

Equation (44) is the "syndrome" which represents $2t$ simultaneous equations in $2t$ unknowns—t error positions (i_m)and t amplitudes (e_m). We can convert the above equation into a linear set of simultaneous equations by assuming t is known and by using the following substitutions.

From (4), similar to (11) for erasure channels, we derive

$$H(s) = s^t + h_1 s^{t-1} + \cdots + h_t. \tag{45}$$

Since $H(s_m) = 0$, we get

$$\sigma_m^t + h_1\sigma_m^{t-1} + \cdots + h_t = 0 \tag{46}$$

for $m = 0, 1, \ldots, t-1$.

By multiplying by $e(i_m)\sigma_m^r$ and summing over m and invoking (9), we get

$$E(r+t) + h_1 E(r+t-1) + \cdots + h_t E(r) = 0. \tag{47}$$

for $r = 0, 1, \ldots, n-1$. For $r = n-2t$ to $n-t-1$, $E(r+t)$ through $E(r)$ in (47) are known. Therefore, the t unknowns $\{h_i, i = 1, \ldots, t\}$ can be solved from the t simultaneous equations. The unknowns $\{E(r), r = 0, \ldots, K-1\}$ can then be found recursively from (47) for $r = K - 1, \ldots, 0$. In the above the assumption has been that the number of errors t is known but since this assumption is not true in practice, one has to use the Berlekamp Massey's algorithm to find the number of errors.

It is to be noted that decoding of the received code block can also be done by Peterson's direct solution [6] but this is very computationally intensive especially for block sizes $N > 8$.

Reference [5] has shown the result of recovery from *one* impulse using BCH decoding algorithm. In [1], we proposed a sub-optimal but practical method which used a combination of nonlinear and iterative method (see Chapters 4 and 5) to recover from multiple impulsive noise. The main idea is to detect the impulses and remove the corrupted samples. The procedure is as follows: We first take the FFT of the received block of samples plus impulsive noise. We insert zeroes in the last m samples (i.e., low-pass filtering) and then take the resultant inverse FFT. The difference between this inverse and the received samples (syndrome) is a rich source for detecting errors. This is equivalent to reconstructing 2 analog signals from 2 sets of samples decimated from the redundant samples and then comparison of the two analog signals would show the position of errors; for proofs see [1]. If the absolute value of the difference samples is above a threshold, we consider the samples at those positions as unreliable. After discarding the unreliable samples, the problem becomes signal recovery with erasures. That reference also shows that if we add some random noise to the impulsive noise; the technique seems to be robust to additive noise if the SNR (14) is above 4–5 dB.

17.9.1.1. The Berlekamp Massey's Algorithm for the Codes in the Field of Real Numbers

A technique developed independently by Berlekamp and Massey [6] and [28] can also be used for real and complex numbers to detect the number and position of errors. For the implementation issues and the C code, please refer to the accompanying CD-ROM—under file name 'Chapter17\Berlekamp-Massey*.*'; also see [42], which is also given in the CD-ROM—under file name 'Chapter17\Berlekamp-Massey\MScThesis.zip'. The algorithm is described below.

The Berlekamp-Massey algorithm. The two independent algorithms: the Massey's algorithm and the Berlekamp's algorithm are so closely related that they are often treated as one[5]. Berlekamp described the algorithm first and

[5]Massey's algorithm was first published as a paper in [43]. Berlekamp's algorithm was presented at a conference [44] a year earlier.

Massey introduced the relation between the syndrome values with a Linear Feedback Shift Register (LFSR) . The equation (47) repeated below

$$E(r) + h_1 E(r-1) + \cdots + h_i E(r-t) = 0$$

describes a Linear Feedback Shift Register with the following initial states:

$$E(j), E(j+1), \ldots, E(j+2t-1),$$

where $j = n - 2t$ and having tap connections $C_i = h_i$. A diagram of the LFSR can be seen in Fig. 17. This figure is identical to the analysis of the IIR filter representation given in Section 8 of this chapter for real/complex number decoding.

A general LFSR of length t, consists of a cascade of t unit delay cells, or stages, with provision to form a linear combination of the cell contents, which then serves as the input to the first stage. The output of the LFSR is assumed to be in the last stage. The initial contents of the t stages coincide with the first t output digits and the remaining output digits are uniquely determined by a recursive equation. The output digits and the feedback coefficients C_1, $C_2, \ldots,$ C_t are assumed to be the field of real numbers. The last coefficient can be zero, i.e., it may not have a tap. Looking at the diagram, one can extract the equations that the LFSR implements:

$$E(r) = -C_1 E(r-1) - C_2 E(r-2) - \cdots - C_t E(r-t), \quad \text{for } r = n-2t \text{ to } n-1.$$

Observe the similarity between this and (47). Because of this similarity, Massey transformed one problem into a simpler one. Instead of purely trying to obtain the syndrome values, he correlated the process of synthesising the LFSR with a minimum length to the extraction of the syndromes. That is, the algorithm uses the extracted syndrome values and by calculating the tap weights and comparing them to the syndrome values, it generates a polynomial that can later

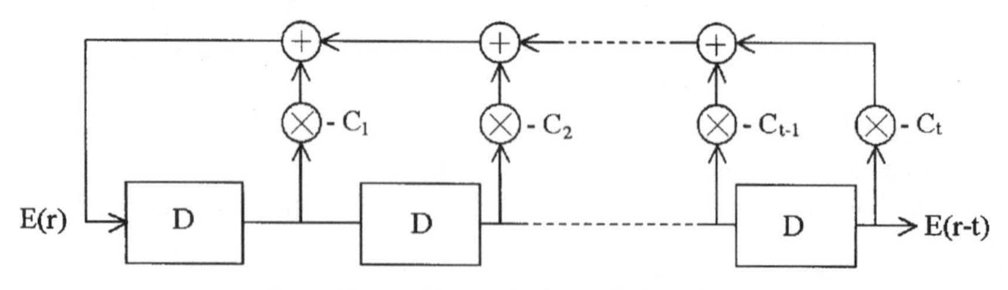

Figure 17. An IIR or a Feedback Shift Register.

be used for error correction. We thus define a feedback polynomial as the representation of the syndrome coefficients:

$$C(x) = 1 + C_1 x + C_2 x^2 + \cdots + C_t x^t$$

Thus, instead of finding $H(s)$ in (5) we find $C(x)$. The *synthesis algorithm* is given below: the parameters are defined at the end of the algorithm.

0. Initialise:

$$C(D) = 1, B(D) = 1, x = 1, b = 1, L = 0, N = 0$$

1. If N = length of the syndrome (number of padded zeros in the DFT domain), stop; otherwise go to 2.
2. Calculate:

$$d = S_N + \sum_{i=1}^{L} C_i S_{N-i}$$

3. If $d = 0, x + 1 \rightarrow x$ and go to 6.
4. If $d \neq 0$ and $2L > N$, then

$$C(D) - db^{-1} D^x B(D) \rightarrow C(D)$$

$$x + 1 \rightarrow x$$

5. If $d \neq 0$ and $2L \leq N$, then

$$T(D) = C(D)$$

$$C(D) - db^{-1} D^x B(D) \rightarrow C(D)$$

$$N + 1 - L \rightarrow L$$

$$B(D) = T(D)$$

$$b = d$$

$$x = 1$$

6. $N + 1 \rightarrow N$

   and go to 1,
   where

- d is the "discrepancy" that explains how much different is the syndrome value from the polynomial coefficient value,
- b is the value of the discrepancy in the previous iteration,
- L is the length of the LFSR,

- N is the length of Syndrome,
- $B(D)$ is the value of the C polynomial in a previous iteration,
- x is the amount of shift by which the B polynomial is displaced.

The algorithm sets out to slowly build up $C(D)$ which is the LFSR connection polynomial of the lowest degree that will be able to generate the needed syndrome sequence. It is first initialised in its simplest form as "1" and is modified as needed in every step. When a discrepancy is encountered, the polynomial is changed appropriately using the correction factor $db^{-1}D^xB(D)$.

Each iteration takes in a separate syndrome value, which is computed before the iterations begin, and is compared with the LFSR generated value of the same position using $C(D)$ of the previous iteration (i.e., $B(D)$). When step 2 is executed for the first time, the discrepancy is s_0 even though there were no previous iterations. The whole process of reconstructing $C(D)$ each time the discrepancy is non-zero, is the core of the algorithm. Not only the discrepancy is zeroed in this way but also the connection polynomial that is going to be used for the next iteration will be able to produce all the correct syndrome values. Hence, previous syndrome values have to be checked whenever $C(D)$ is modified.

An advantage of this process is that it produces a polynomial of the smallest possible order. If the algorithm terminates with an LFSR length L of less than t, then the error locator polynomial may not be correct and error detection is announced. That is, the algorithm ends with a value of L equal to the number of errors that were found.

The algorithm for real numbers is slightly modified. Unlike the finite Galois Fields, discrepancy may never get to zero but rather close to it due to round off and additive noise. In this case, we need to have a threshold. For example, when d gets smaller than 0.01, we assume it is zero. The code and some explanations derived from [42] is explained in greater detail in the accompanying CD-ROM—Chapter17\Berlakamp-Massey*.* where a thorough discussion of the C code for the simulation and the assembler code of a DSP implementation is discussed; the MSc thesis with more detail explanation is also in the same directory.

17.9.2. Application of Walsh Transform to Forward Error Correction[6]

In this section, we present a novel class of codes constructed using the Discrete Walsh Transform (DWT). They are a class of double-error correcting codes defined on the field of real numbers. An iterative decoding algorithm for DWT codes is developed and implemented. The error correcting performance of these codes over an AWGN channel is evaluated. Selected DWT code parameters are compared to those of BCH codes.

[6]Portions of Sections 17.9.2 are the modified reprint with permission of [30]; © 1999 IEEE.

The advantage of DWT is that its elements are either 1 or -1; hence its transform and inverse transform are the same, and can be implemented very efficiently. Below we shall discuss a single and a double correcting codes using DWT [30] and [38]; for more details see also the files in the accompanying CD-ROM, directory: 'Chapter17\Walsh'.

17.9.2.1. Encoding of DWT

In order to get a block code of size *n, k* information samples are chosen such that

$$k = n - \log_2(n) - 1, \tag{48}$$

where n is a power of 2. To encode these k information samples, $n - k$ zeros are inserted in the code of length n in the following positions

$$i = 2^m - 1 \qquad \text{for } m = 0, 1, \ldots, \log_2(n). \tag{49}$$

The position of zeros (49) is chosen for the possibility of detection and correction of single and almost all double errors. The remaining symbols of the block of size n consist of the k information samples. For example, for a block size *8* the 0th, 1st, 3rd, and 7th positions are set to zero and in the remaining positions, 4 information samples are inserted. The DWT of size 8 of this block is the (8,4) DWT code.

17.9.2.2. Decoding the Walsh Codes

Like any error correcting codes, the decoding consists of syndrome calculation, detecting the number and position of errors, and finding the magnitude of the errors. We shall discuss each one in the following:

Syndrome Calculation. If the channel noise is additive, the Walsh transform—which is equivalent to the inverse Walsh transform save a constant of dimensionality—of the received code will yield the syndrome at the position of zeros (49). Thus the syndrome vector has a length of $n - k = \log_2(n) + 1$. If the syndrome vector is all zero, then there is no error. In case of a single error, the absolute value of all the elements of the syndrome vector are equal to the absolute value of the error magnitude. If the elements of the syndrome vector do not show any pattern as mentioned above, there are two or more errors.

The analysis for error detection and correction is as follows: Let X_n be the transmitted code vector of block size n and Y_n be the received code vector. Since noise is assumed to be additive, we have

$$Y_n = X_n + E_n, \tag{50}$$

where E_n is the error signal. The syndrome is defined by the set of $n - k$ equations:

$$s_i = \sum_{j=0}^{n-1} Y_j WAL(i,j) = \sum_{j=0}^{n-1} E_j WAL(i,j), \tag{51}$$

where $i = 2^m - 1$ for $m = 0, 1, \ldots, \log_2(n)$, i is the position of zero as given in (49). If there are no errors, $s_i = 0$, if there is only one error at position p, then the syndrome becomes:

$$s_i = \sum_{j=0}^{n-1} E_j WAL(i,j) = E_p WAL(i,p), \tag{52}$$

for all values of i defined in (49). The above equation implies that, depending on i and p, the syndrome is equal to $\pm E_p$; s_0 is always equal to E_p, the magnitude of the error at position p. A surprisingly simple algorithm can be used to determine the position p. If we normalise the syndrome vector by s_0, and then convert 1's into 0's and -1's into 1's, a binary representation of the syndrome yields the position p. For example, if $n = 8$, the normalised syndrome matrix for a single error at positions $p = 0, 1, \ldots, 7$ is in the following form:

$$s = \begin{matrix} 0 & 0 & 0 & 0 & 0 & 0 & 0 & 0 \\ 0 & 0 & 0 & 0 & 0 & 0 & 0 & 0 \\ 0 & 0 & 0 & 0 & 1 & 1 & 1 & 1 \\ 0 & 0 & 1 & 1 & 0 & 0 & 1 & 1 \\ 0 & 1 & 0 & 1 & 0 & 1 & 0 & 1 \end{matrix}, \tag{53}$$

where the first column of S represents the normalised and converted syndrome vector when $p = 0$, and the second column represents the syndrome vector when $p = 1$, etc. As it can be seen in (53), the binary representation of each column determines the position p. Once p is known, the single error $s_0 = E_p$ is subtracted from the received code vector Y_n at position p to get the actual transmitted code vector X_n. If there are two errors, the syndrome will be unique for any pattern of loss provided that the absolute values of the two errors are not the same, i.e., $|E_p| \neq |E_q|$. The syndrome of a DWT code (4,1) is given in Table 7.

Table 7. Syndrome of the (4,1) Walsh transform code with double errors.

Position p	Position q	Syndrome e_0	Syndrome e_1	Syndrome e_2
0	1	$E_p + E_q$	$E_p + E_q$	$E_p - E_q$
0	2	$E_p + E_q$	$E_p - E_q$	$E_p + E_q$
0	3	$E_p + E_q$	$E_p - E_q$	$E_p - E_q$
1	2	$E_p + E_q$	$E_p - E_q$	$-E_p + E_q$
1	3	$E_p + E_q$	$E_p - E_q$	$-E_p - E_q$
2	3	$E_p + E_q$	$-E_p - E_q$	$E_p - E_q$

This table shows that if the magnitudes of the errors are not equal, the syndrome uniquely represents any pattern of two losses. In general, a systematic algorithm can be used to detect the position of the two errors, p and q; the algorithm is given in [30].

17.9.2.3. Erasure Channels

For erasure channels, the positions of errors are known and there are no ambiguities in case the absolute values of double errors are equal. Therefore, DWT codes can always correct for two erased (lost) samples. Many triple and more erasures may also be corrected depending on n and k of the $(n\,,\,k)$ DWT code.

17.9.2.4. Performance Evaluation

The performance of DWT codes is evaluated over AWGN channel using 32ary MFSK modulation with non-coherent detection and hard-decision decoding. The bit error rates with respect to signal-to-noise ratio of different DWT codes are shown in [30]. The code rates of DWT codes compared to those of double error correcting BCH codes are better as shown in [30].

17.9.2.5. Conclusion

A novel class of forward error correcting codes has been constructed using DWT. These codes can detect and correct all patterns of single errors and almost all patterns of double errors, and definitely all patterns of double erasures. The block size of the DWT codes is $n = 2^m$ for any integer $m \geq 2$, the number of information symbols is $k = n - m - 1$, and the code rate is $1 - (1 + m)/2^m$. The fast DWT is more efficient than FFT since the operations are all real with additions and subtractions only.

17.9.3. Discrete Cosine Transform (DCT)

Several papers have tried to discuss the use of DCT kernel as opposed to DFT [1] and [37]. The advantage of DCT kernels to those of DFT is that it is real. The disadvantage of DCT codes compared to DFT based codes is that it is not cyclic and the transformation matrix is not Vandermonde. There are various versions of DCT for image compression [39]; here we choose the DCT type II for the syndrome calculation of (43):

$$\Phi(i,j) = \begin{cases} \sqrt{\frac{2}{n}}\cos\frac{(2j+1)i*\pi}{2n} & \text{if } j \neq 0 \\ \sqrt{\frac{1}{n}} & \text{if } j = 0 \end{cases}. \tag{54}$$

Derivation of a formula for (44) is not instructive for DCT.

Although the DCT matrix is not Vandermonde, any sub-matrix generated for the evaluation of the missing samples (26) can be shown to be the product of Vandermonde and triangular matrix, and therefore it is invertible [37]. The same paper also discusses the use of Berlekamp Massey algorithm for DCT based codes.

17.9.3.1. Simulation Results for the DCT Case

In [1] we did some simulation for DCT as well as some other discrete transform such as Discrete Hartley transform and Discrete Wavelet Transform using iterative techniques (see Chapters 4 and 5) under erasure channel environment. In that paper we showed that DFT outperforms other transform techniques in a random channel erasure using iterative methods. Using the matrix approach discussed in Section 7 of this chapter (25)–(26), the results for DCT approach under consecutive losses (bursty erasure) are shown in Figs. 18 and 19. Figure 18 shows the recovered code (identical to the transmitted code, i.e., the DCT transform of the original signal with padded zeros), and the position of erased samples. Figure 19 shows the recovered original signal. The reason the recovery is so good is because the padded zeros in the original signal are inserted in a modular fashion (mod($4i$, 1), where $i = 0, 1, \ldots, n-1$, and $n = 128$ in Figs. 18 and 19). If we had padded the zeros at the end or beginning, the recovery technique would fail for bursty losses. This sensitivity to noise is more severe than the DFT case.

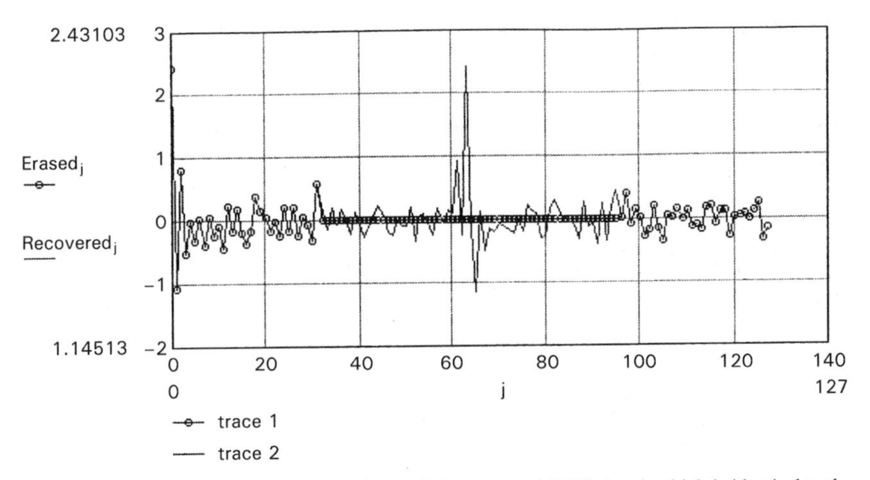

Figure 18. The erased samples at the receiver and the recovered DCT signal, which is identical to the transmitted one.

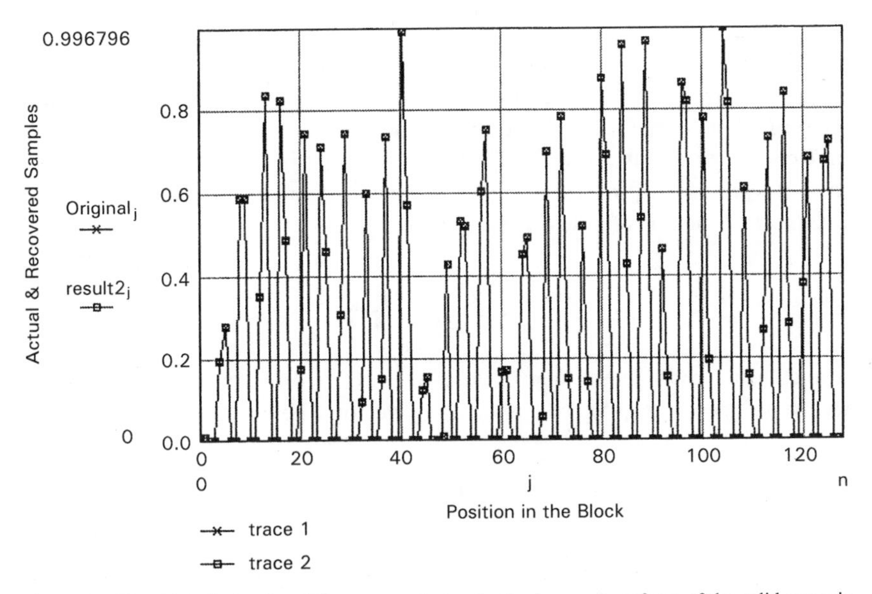

Figure 19. The original signal and the recovered signal—the inverse transform of the solid curve in Fig. 18.

Wu et al [37] have simulated the DCT codes under impulsive noise environment. Using the Berlekamp Massey's algorithm that was used for DFT in Section 9.2, they managed to find the position of errors. Similar to the DFT case, their results show that under small background noise, error recovery under impulsive channel environment is possible.

17.10. Similarity of the Generator and Parity Check Matrices with Our Approach

In coding theory it is customary to use a generator matrix of dimension $k \times n$ to generate an (n, k) code y for a block of k symbols x:

$$y = x \cdot G. \tag{55}$$

This is equivalent to inserting zeros in the information vector x and then take its transform, exactly the way we have described the generation of codes throughout this chapter. A parity check matrix is an $(n - k) \times n$ matrix such that

$$G \times H^T = 0. \tag{56}$$

Marshal [31] has shown that for real numbers, unitary transform matrices represented by T can be decomposed into a generator matrix G and a parity matrix H. If T is an $n \times n$ orthonormal transform matrix, G is a $k \times n$ sub-matrix of T consisting of any k rows of T. H is the sub-matrix consisting the remaining $(n - k) \times n$ rows of T. It can be easily shown that because the matrix T is orthonormal then, (56) holds. Also, we have

$$G \cdot \bar{G}^T = I, \tag{57}$$

where I is a $k \times k$ identity matrix. In case of errors or erasure, y in (55) will be received as $r = y + e$, where e is the error vector. The syndrome vector is

$$S = r \cdot H^T = e \cdot H^T. \tag{58}$$

The syndrome vector as defined above is identical to that represented in (42). The matrix solution in (26) is also equivalent to the solution of (58) if the position of errors are known.

17.11. 2-D Erasure Recovery

The RS recovery technique based on coding developed for 1-D signal (47) could be generalized to 2-D signals. This generalization is very powerful for slice losses of an image, which is very typical in compressed images. A summary of the analysis and results [41] is given below.

Let us assume that a 2-D signal such as an image is sampled at the Nyquist rate yielding a discrete signal ($x_{\text{org}}(i, k), i, k = 1, \ldots, U$). We use a new transform such that the kernel of the transform is equal to

$$\exp\left(-j\frac{2\pi}{N}\cdot m \cdot i \cdot q_1 - j\frac{2\pi}{N}\cdot n \cdot k \cdot q_2\right), \tag{59}$$

where q_1 and q_2 are positive prime integers with respect to N. It can be shown that this kernel is a sorted kernel of DFT. The transform of the image is called $X_{\text{org}}(m, n)$ for $m, n = 1, \ldots, U$. For the sake of clarity, we shall call this transform the Sorted Discrete Fourier Transform (SDFT). The $X_{\text{org}}(m, n)$ can be expanded by inserting τ rows and columns of zeros around it to achieve a new SDFT matrix $X_{\text{over}}(m, n)$ for $m, n = 1, \ldots, (N = U + \tau)$. An inverse SDFT will lead to an over-sampled version of the original signal $x_{\text{over}}(i, k)$ for $i, k = 1, \ldots, N$ with $(N \times N)$ complex samples. This(U, N) code is capable of correcting a block of matrix $(\tau \times \tau)$ where $\tau = N - U$.

The missing part of the over-sampled signal is denoted by $e(i_m, k_n) = x(i_m, k_n)$, where (i_m, k_n) specifies the positions of the lost pixels. The value of $e(i, k)$ for any position except (i_m, k_n) is zero. For the $(\tau \times \tau)$ lost pixels, the polynomial error locator is

$$\begin{aligned} H(s_i, p_k) &= \prod_{m=1}^{\tau}\left(s_i - \exp\frac{(j2\pi i_m q_1)}{N}\right)\prod_{n=1}^{\tau}\left(p_k - \exp\frac{(j2\pi k_n q_2)}{N}\right) \\ &= \sum_{t=0}^{\tau}\sum_{f=0}^{\tau} h_{t,f}\cdot s_i^{\tau-t} p_k^{\tau-f}, \end{aligned} \tag{60}$$

$$H(s_{i_m}, p_{k_n}) = 0 \tag{61}$$

where

$$s_i = \exp\frac{(j2\pi i q_1)}{N} \quad \text{and} \quad p_k = \exp\frac{(j2\pi k q_2)}{N}, \qquad i, k \in \{1, \ldots, N\}.$$

The polynomial coefficients ($h_{t,f}, t, f = 0, \ldots, \tau$) can be found from the product in (60). For the DSP implementation, it is easier to find $h_{t,f}$ by obtaining the inverse SDFT of $H(s, p)$. Similar to the 1-D case (6)–(7), multiplication of (61) by $e(i_m, k_n) \cdot (s_m)^r \cdot (p_n)^d$, summation over m, n, and invoking the SDFT of the missing samples $e(i_m, k_n)$ yield

$$\sum_{t=0}^{\tau} \sum_{f=0}^{\tau} h_{t,f} \cdot E(\tau + r - t, \tau + d - f) = 0, \tag{62}$$

where $(r, d) = \lceil \tau/2 \rceil + l, \ldots, N - \lceil \tau/2 \rceil - l$ and $E(r, d)$ is the SDFT of $e(i, k)$. The 2-D difference equation (62) with non-zero initial condition can be solved by a recursive method provided that the boundary conditions are given only in an L-shaped region [45]. This can be achieved by inserting zeros in the original SDFT matrix. Considering that the $X_{\text{org}}(m, n), m, n = 1, \ldots, k$ is a bi-periodic SDFT matrix, the zeros are inserted around the original matrix $X_{\text{org}}(m, n)$. From now on, $X_{\text{over}}(m, n)$ denotes this special over-sampled SDFT matrix. In the region of inserted zeros, we have

$$E(i, k) = -D(i, k),$$

where $D(., .)$ is the the SDFT of the corrupted image. The remaining values of $E(r, d)$ can be found from (62) by the following recursion

$$E(r, d) = -\frac{1}{h_{0,0}} \cdot \sum_{t=0}^{\tau} \sum_{f=0}^{\tau} h_{t,f} \cdot E(r - t, d - f), \tag{63}$$

where

$$(r, d) \in \frac{\tau}{2} + 1, \ldots, N - \frac{\tau}{2}.$$

To determine suitable values for q_1 and q_2 and their effects on the behaviours of the algorithm, at first we choose $q_1 = q_2 = 1$. For this choice the SDFT is identical to DFT. In many problems, the number of pixels are large, so that bursty losses produce large dynamic range in the error locator polynomial $H(s, p)$ as well as the polynomial coefficients $h_{t,f}$. The large dynamic ranges are due to the concentration of zeros on one side of the semi-sphere. $H(s_i, p_j)$ values are zero for $(i, j = 1, \ldots, 32)$ and very large around $(i, j = 48)$. This can be verified by calculating the value of the $H(s, p)$ which is equal to the product of all the (32×32) vectors emanating from the position $(i, j = 48)$ on the upper half of the unit sphere.

For the case of bursty losses, $h_{t,f}$ has a large dynamic range that creates a large computational error in (63). Therefore, the implementation of the algorithm for a large block size of losses is impossible. Since the coefficients $h_{t,f}$ have a very small dynamic range in the case of isolated losses; it would be beneficial to transform the bursty losses into isolated losses by proper choices of q_1 and q_2. Therefore, to determine the coefficients $q_1$and q_2, two points should be considered. Firstly, q_1 and q_2 have to be prime with respect to N. Secondly, based on the size of the block losses, the q_1 and q_2 values must be chosen such that the locations of the zeros of the error locator polynomial are approximately distributed symmetrically around the unit sphere. For example for an even number of N, when the block size of losses is equal to $(N/2, N/2)$, the best choice for q_1 and q_2 is $(N/2 - 1)$.

The SDFT is actually derived from DFT and the fast algorithm can still be used. Because the SDFT transform can be handled by DFT and sorting; the sorting of the elements is as follows:

$$a_{\mathrm{SDFT}_{m,n}} = a_{\mathrm{DFT}_{\mathrm{mod}(q1 \cdot m,N),\mathrm{mod}(q2 \cdot n,N)}}$$

for $m, n = 1, \ldots, N$, a_{SDFT} is the element of the SDFT transform matrix, and a_{DFT} is the element of the DFT transform matrix. The inverse transform of the frequency coefficients is equivalent to the inverse sorting and the inverse DFT, respectively. A small dynamic range of the coefficient $h_{t,f}$, justifies this algorithm for a large image size and a large number of losses.

The proposed technique is very efficient in recovering bursts of errors with a performance better than the other techniques such as the method of Conjugate Gradient (CG) using adaptively chosen weights and block Toeplitz matrices [14], which is also discussed in Chapter 5. But we should consider that for the recovery of losses of size $(\tau \times \tau)$, the proposed technique need more added zeros in the frequency domain in comparison to the CG method. In the best situation, the ratio between the recovery pixels and the added zeros is $\frac{1}{3}$.

17.11.1. Simulation Results

A (256×256) image is used for the simulation of the algorithm, Fig. 20(a). The image is transformed using SDFT. A number of zeros are inserted in the rows and columns of the SDFT matrix and a new matrix of size of (512×512) is derived. By taking an inverse SDFT, the over-sampled image is produced, Fig. 20(b). This image does not appear to have any similarity with the original image. According to the algorithm, the block size of the losses is limited to (256×256) in this case. Therefore, a block of this size is erased from the image as shown in Fig. 20(d). The corresponding image of Fig. 20(d) prior to the recovery method is shown in Fig. 7(c). The over-sampled signal after reconstruction and its corresponding image are shown in Figs. 20(f) and 20(e), respectively. The Mean Squared Error for this simulation is equal to 1.09×10^{-15}. None of the previous methods (see Chapter 5) are able to recover the block size of $(N/2, N/2)$. We have shown in [41] that the above SDFT is very robust against quantization and additive noise.

In conclusion, we have shown that the new algorithm has three advantages. Firstly, it is ideal to recover the missing pixels for large blocks of bursty errors.

Secondly, in terms of complexity, it is simpler than other techniques; thirdly, it is very robust in correcting bursts of errors with respect to additive and quantization noise. But the disadvantage of this method is that in the best situation the ratio between the recovery pixels and the added zeros is $\frac{1}{3}$. If we are interested in full recovery, either matrix approaches or 1-D techniques have to be used.

The Mathcad™ program for the above algorithm is in the accompanying CD-ROM, file: Chapter17\SDFT.mcd. Additional files on this topic is under the directory Chapter\Siamak.

17.12. A Comparison Between Error Correction and Error Concealment

In this chapter we have discussed the relation between nonuniform sampling theory with error correction codes as an extension of Galois Fields to real and complex numbers. In error correction codes we need to add redundancy in the form of padding zeros in the frequency domain. This is equivalent to over-sampling in the time domain before transmitting the signal.

Speech and Image signals on the other hand have some inherent correlation and therefore redundancy. This implies that even if we do not add redundancy to the signal, in case of sample losses, it may be possible to somehow partially recover the signal. If the recovery is acceptable for speech and video signals, then the recovery is called error concealment. We shall see in the next chapter that most of the techniques we have discussed in this section is also applicable to error concealment for both speech and image signals. Therefore, what happens at the transmitter is a combination of signal compression (source coding) and error correction (channel coding), but at the receiver a sequential processing of channel decoding, source decoding and error concealment may be needed. We shall see that for some types of error concealment, we need to pre-distort the signal at the encoder before going through source and channel encoding. Indeed for a specific pre-distortion of speech signals (inserting zeros for certain frequencies) and recovery from frame losses for GSM mobile telephony is the patent application the first author has applied [40]; see also the CD-ROM, directory: Chapter17\ Error-Concealment.

Error concealment for video and sampling theory will be the subject of the next chapter.

17.13. Conclusion

Over-sampling of a band-limited signal is an alternative to error detection and correction. A signal can be low-pass filtered using analog or digital filters.

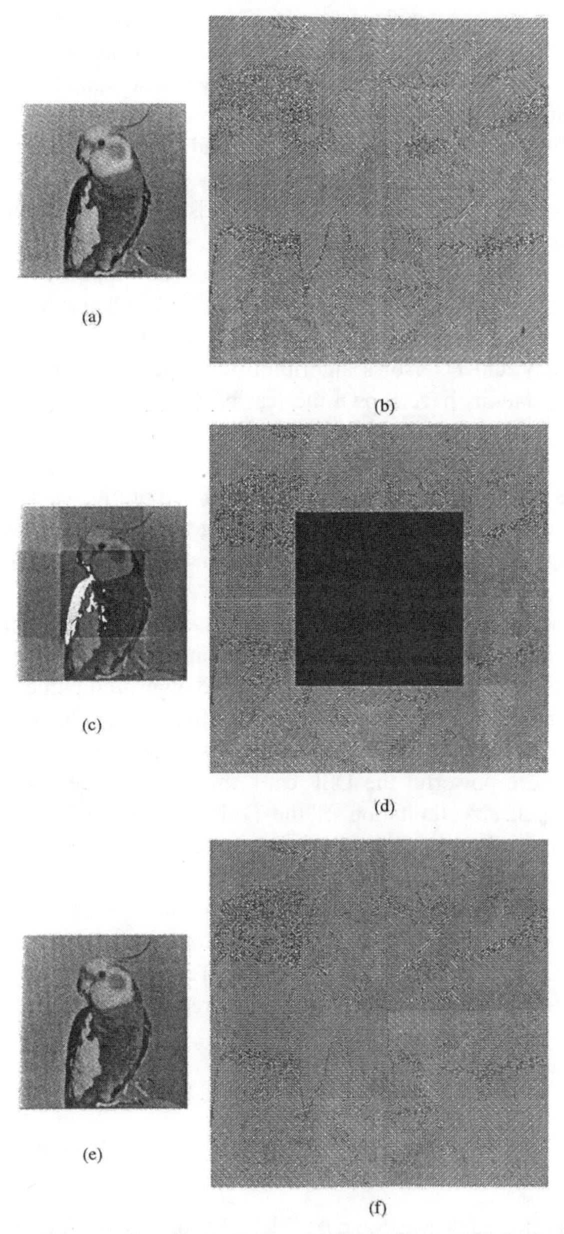

Figure 20. Image Recovery from Bursty Losses: (a) Original image, (b) the image after the padded zeros in the SDFT domain, (c) the corrupted image corresponding to the block loss in (d), (d) the block loss of the image shown in (a), (e) the recovered image, (f) the image after the padded zeros in the SDFT domain.

The over-sampled signal after the low-pass filtering is equivalent to a convolutional code in the field of real numbers. In case of erasures some nonuniform sampling recovery methods can be used to recover lost samples. If a block of samples is transformed into the frequency domain using any Discrete Transform such as DFT, DWT, DCT etc., and then if we pad zeros in different regions of frequencies, we can achieve low-pass or other types of filters. This is equivalent to over-sampling using block codes in Galois fields of real or complex numbers. DFT is the most efficient way for erasure detection and correction.

The recovery techniques could be the ones developed for nonuniform sampling such as iteration on FFT and Inverse FFT (Chapters 4 and 5). Or alternatively, we could use well developed techniques in coding such as Berlekamp Massey and Peterson's algorithm developed for Reed Solomon and BCH codes. Simulations have shown the feasibility of this method for erasures, impulsive noise, and moderate amount of random noise.

In speech and video applications, there is a possibility of frame losses, which translates into a large consecutive burst of sample losses specially when the signal is compressed and there is a danger of error propagation. An efficient technique specifically developed for bursty erasures was discussed in this chapter.

The proposed burst error recovery algorithm was simulated using Mathcad for the case of bursty losses and implemented on a DSP card for the case of isolated losses. For 30% sample loss, the proposed algorithm outperforms all the known iterative methods developed by the author for less complexity. Despite the accumulation of error, the SNR increases with increasing the block size. Obviously, the more powerful the DSP card, the better is the performance. To overcome the block size limitation of the DSP implementation, zeros can be padded in the time domain instead of the frequency domain. This approach eliminates an extra IFFT.

In terms of the sensitivity to quantization and additive noise, this technique was found to be sensitive to noise with a performance comparable to other decoding techniques such as Forney's method. The sensitivity to noise of the proposed decoding technique can be significantly reduced by the utilization of the newly proposed transform kernels.

Acknowledgments

I would like to thank Mr. Siamak Talebi for providing Fig. 20 and proofreading the chapter. Also, I would like to thank Dr. J. Vieira from the University of Aveiro for the fruitful discussions I had with him while I was in Portugal and for providing his PhD thesis in Portuguese on the accompanying CD-ROM, directory: 'Chapter17\Vieira*.*. He has also provided the Matlab files in the same directory. Finally, I would like to thank Prof. Robert Redinbo from the University of California Davis for contributing two papers on block and convolutional codes for real numbers to the CD-ROM.

References

[1] F. Marvasti and M. Nafie. Sampling Theorem: A Unified Outlook on Information Theory, Block and Convolutional Codes, Special Issue on Information Theory and its Applications. *IEICE Trans. of Japan on Fundamentals of Electronics, Commun. and Computer Sciences*, Section E, September 1993.

[2] F. Marvasti and L. Chuande. Equivalence of the Sampling Theorem and the Fundamental Theorem of Information Theory. *Proc. International Communication Conference, ICC' 92*, Chicago, IL, June 1992.

[3] F. Marvasti. Nonuniform Sampling. In Robert Marks II, Ed., *Advanced Topics in Shannon Sampling and Interpolation Theory*, Springer Verlag, New York, January 1993.

[4] F. Marvasti. *A Unified Approach to Zero-Crossings and Nonuniform Sampling of Single and Multidimensional Signals and Systems*. Nonuniform Publication, Oak Park, IL, 1987.

[5] J. K. Wolf. Redundancy, the Discrete Fourier Transform, and Impulse Noise Cancellation. *IEEE Trans. of on Commun.*, 31(3):458–461, March 1983.

[6] R. E. Blahut. Transform Techniques for Error Control Codes COM-31(3). *IBM J. of Res. Develop.*, 23:299–315, May 1979.

[7] D. Mandelbaum. On Decoding of Reed-Solomon Codes. *IEEE Trans. Info. Theory*, IT-17:707–712, 1971.

[8] F. Marvasti and M. Analoui. Recovery of Signals from Nonuniform Samples Using Iterative Methods. *IEEE Proc. of International Conference on Circuits and Systems*, Oregon (ISCAS' 1989), vol 2, July, 1989, pp. 1021–1024.

[9] M. Nafie and F. Marvasti. Implementation of Recovery of Speech with Missing Samples on a DSP Chip. *Electronic Letters*, 30(1):12–13, Jan 6, 1994.

[10] F. Marvasti, M. Hasan, M. Eckhart, and S. Talebi. Efficient Algorithms for Burst Error Recovery Using FFT and Other Transform Kernels. *IEEE Trans. on Signal Processing*, 47(4):1065–1075, April 1999.

[11] J. B. Anderson and S. Mohan. *Source and Channel Coding,* Kluwer, 1991, pp. 171–174.

[12] G. D. Forney, Jr. On Decoding BCH Codes. *IEEE Trans. on Information Theory*, IT-11:549-557, 1965.

[13] H. G. Feichtinger, K. Grochenig, and T. Strohmer. Efficient Numerical Methods in Non-Uniform Sampling Theory. *Numerische Mathematik*, 69:423–440, 1995.

[14] T. Strohmer. Computationally Attractive Reconstruction of Band-Limited Images from Irregular Samples. *IEEE Trans. on Image Processing*, 6(4):540–548, April 1997.

[15] C. Cenker, H. Feichtinger, and M. Hermann. Iterative Algorithms in Irregular Sampling: A First Comparison of Methods. *Proc. of Phoenix Conference on Computers and Communications,* Scottsdale, Arizona, March 1991.

[16] F. Marvasti, P. Clarkson, M. Dokic, and L. Chuande. Reconstruction of Speech Signals from Lost Samples. *IEEE Trans. on Acoustic, Speech, and Signal Processing*, 40(12):2897–2903, December 1992.

[17] F. Marvasti, M. Analoui, and M. Gamshadzahi. Recovery of Signals from Nonuniform Samples Using Iterative Methods. *IEEE Trans. ASSP*, 39(4):872–878, April 1991.

[18] P. J. S. G. Ferreira. Interpolation and the Discrete Papoulis-Gerchberg Algorithm. *IEEE Trans. on Signal Processing*, 42(10):2596–2606, October 1994.

[19] P. J. S. G. Ferreira. Noniterative and Faster Iterative Methods for Interpolation and Extrapolation. *IEEE Trans. on Signal Processing*, 42(11):3278–3282, November 1994.

[20] P. J. S. G. Ferreira. The Stability of a Procedure for the Recovery of Lost Samples in Band-Limited Signals. *IEEE Trans. on Signal Processing*, 40(3):195–205, December 1994.

[21] P. J. S. G. Ferreira. The Eigenvalues of Matrices Which Occur in Certain Interpolation Problems. *IEEE Trans. on Signal Processing*, 45(8):2115–2120, August 1997.

[22] H. G. Feichtinger and T. Strohmer. Fast Iterative Reconstruction of Band-Limited Images from Irregular Sampling Values. *Proc. Conf. CAIP, Computer Analysis of Images and Patterns,* D. Chetverikov and W. Kropatsch, Eds., Budapest, 1993, pp. 82–91.

[23] K. Gröchenig. Reconstruction Algorithms in Irregular Sampling. *Math. Comput.*, 59:181–194, 1992.

[24] R. Gerchberg. Super-Resolution Through Error Energy Reduction. *Optica Acta.*, 21(9):709–720, 1974.

[25] F. Marvasti. Fast Packet Network: Data Image, and Voice Signals Recovery. In F. Froehlich and A. Kent, Eds., *The Encyclopaedia of Telecommunications*, vol. 7, Marcel Dekker Inc., 1994, pp. 453–479.

[26] C. K. W. Wong, F. Marvasti, and W. G. Chambers. Implementation of Recovery of Speech with Impulsive Noise on a DSP Chip. *IEE Electronics Letters*, 31(17):1412–1413, 17 August, 1995.

[27] Alan V. Oppenheim and Ronald W. Schafer. *Discrete-Time Signal Processing*. Prentice-Hall, Inc., 1989.

[28] A. M. Michelson and A. H. Levesque. *Error Control Techniques for Digital Communication*, John Wiley & Sons, Inc., 1985, pp. 191–199.

[29] S. Talebi, F. Marvasti, and M. Hasan. Sensitivity of the Burst Error Recovery to Additive Noise. *Proc. Communication Theory and Systems, the 7th Iranian Conference on Electrical Engineering*, Tehran, Iran, May 1999, pp. 19–25.

[30] F. Marvasti, H. Ng, and M. R. Nakhai. Application of Walsh Transform to Error Correction Codes. *ICASSP' 99*. IEEE, March 1999.

[31] T. G. Marshall, Jr. Coding of Real-Number Sequences for Error Correction: A Digital Signal Processing Problem. *IEEE Trans. Select. Areas Commun.*, SAC-2, March 1984.

[32] R. Redinbo. Decoding Real Block Codes: Activity Detection, Wiener estimation. *IEEE Trans on IT*, 46(2):609–623, March 2000.

[33] F. Marvasti and C. Liu. Oversampling as an Alternative to Error Correction Codes in Digital Communication Systems. *SIAM Annual Meeting*, Chicago, IL, July 1990, p. A9.

[34] F. Marvasti. FFT as an Alternative to Error Correction Codes. *IEE Colloquium on DSP Applications in Communication Systems*, March 22, 1993.

[35] F Marvasti and M Hasan. Noise Sensitivity Analysis for Novel Error Correcting Codes. *ISCAS' 98, IEEE*, Monterey, May 1998.

[36] F. Marvasti, M. Hasan, and M. Eckhart. An Efficient Burst Error Recovery Technique. *ICT' 98*, Porto Carras, Greece, June 1998.

[37] J. L. Wu and J Shiu. Discrete Cosine Transform in Error Control Coding. *IEEE Trans. on Commun.*, 43(5):1857–1861, May 1995.

[38] Jiun Shiu and Ja-Ling Wu. Classes of Majority Decodable Real-Number Codes. *IEEE Trans. on Commun.*, 44(3):281–283, March 1996.

[39] K. R. Rao and J. J. Hwang. *Techniques & Standards for Image. Video & Audio Coding*. Prentice Hall, 1996.

[40] F. Marvasti. A Novel Technique for Error Control and Error Concealment for Speech Signals for GSM Environments. Lucent patent application, September 1998.

[41] S. Talebi and F. Marvasti. A Novel Method for Burst Error Recovery of Images. *SampTA 99*, Loven, Norway, August 1999.

[42] P. Livanos. Implementation of a Novel Error Correction Algorithm on a DSP Card. MSc Thesis. King's College London, January 1997.

[43] J. Massey. Shift Register Synthesis and BCH Decoding. *IEEE Trans. on Information Theory*, IT-15(7):122–127, January 1969.

[44] E. R. Berlekamp. The Nonbinary BCH Decoding. Abstract in the *Proc. of the International Symposium on Information Theory*, 1968.

[45] D. E. Dudgeon and R. M. Mersereau. *Multidimensional Digital Signal Processing*. Prentice-Hall, 1984.

[46] R. Redinbo. Reliability Levels for Fault-Tolerant Linear Processing Using Real Number Error Correction. *IEE Proc.-Comput. Digit. Tech*, 143(6):355–363, November 1996.

[47] R. Redinbo. Decoding Real-Number Convolutional Codes: Change Detection, Kalman Estimation. *IEEE Trans. on IT*, 43(6):1864–1876, November 1997.

18

Application of Nonuniform Sampling to Error Concealment

M. Hasan and F. Marvasti

18.1. Introduction[1]

Despite the huge advances in the communication networks industry, the new breed of multimedia applications and services is pushing the existing networks to their bandwidth limits. In ATM networks, the bandwidth is dynamically allocated according to the requirements of the transmitted services. ATM cells are usually stored in the switch buffers before being routed to the destination node. In practice, ATM buffers have limited sizes and they may overflow in the case of congested traffic. When ATM buffers overflow, low priority cells are dropped to ensure transmission continuity of the service using the rest of the cells. Also, ATM cells may be mis-routed due to non-correctable bit errors in the routing information of the cells. When ATM cells carry in their payload a compressed image, cell losses due to buffer overflow or cell mis-routing can cause major degradation in the quality of the transported image. This degradation is usually in

[1]Some sections of this chapter are a modified reprint, with permission, from the papers titled: M. Hasan, A. Sharaf and F. Marvasti, "Subimage Error Concealment Techniques," Proceedings of the IEEE International Symposium on Circuits and Systems (ISCAS 98), Monterey, CA, USA, vol. 4, pp. 245–248, June 1998 and M. Hasan, A. Sharaf and F. Marvasti, "Novel Error Concealment Techniques for Images in ATM Environments," Proceedings of the IEEE International Conference on Acoustics, Speech and Signal Processing (ICASSP 98), Seattle, USA, vol. 5, pp. 2833–2836, May 1998. ©IEEE.

M. Hasan and F. Marvasti • Multimedia Laboratory, King's College London, Strand, London WC2R 2LS, UK. E-mail: mohd_hasan@hotmail.com; farokh.marvasti@kcl.ac.uk

Nonuniform Sampling: Theory and Practice, edited by Marvasti
Kluwer Academic/Plenum Publishers, New York, 2001.

the form of corrupted blocks or macroblocks. Losing one cell may result into the loss of several blocks of the compressed image.

Block losses can also occur in the compressed images transmitted via the third generation mobile channels such as the *Universal Mobile Telecommunications System* (*UMTS*). When the non-correctable errors in the transmitted frame affect the most significant bits, the whole frame is dropped causing several blocks of the compressed image to be lost. Also, due to the fading phenomenon in wireless channels, images transmitted over such channels suffer from frame dropouts and bursts of losses that may corrupt many blocks or macroblocks in the transmitted image.

In storage channels, uncompressed images stored on different storage media may suffer from random losses of their pixels. This can be due to the occurrence of non-correctable bit errors during the read or write cycles. Furthermore, in optical recording devices, bit errors may occur in the stored images due to the noisy optical sensors. Most of the optical recording devices have a verification cycle after the write cycle. The verification cycle may be disabled to speed up the recording process. In such case, the thermal noise—produced by the electronic components of the recording device—may cause bit errors that are not detected and corrected due to the absence of the verification cycle. This leaves the stored image with several corrupted pixels. In storage and recording devices, random bit errors may change the gray level or completely damage the corrupted pixels. This type of losses is called *isolated random losses*.

In the rest of the chapter, we will describe several novel error concealment techniques that are used to recover isolated random losses as well as block losses. These techniques rely on the *Nonuniform Sampling Theory* in the recovery process.

18.2. Novel subimage error concealment techniques

In this section, we employ the nonuniform sampling theory in the development of a new approach for block loss concealment. This approach converts the corrupted image into a set of subimages that can be considered as small images corrupted by isolated losses. Losses in these subimages can be concealed by the application of the iterative and the time varying (nonlinear division) techniques-previously described in Chapter 4—that are designed to conceal isolated random losses.

As mentioned earlier, the iterative and the nonlinear division techniques are not suitable for the recovery of bursty losses. Therefore, these techniques cannot be directly used to recover the lost blocks in the corrupted image. By converting the image into the equivalent subimages, the iterative and the nonlinear division

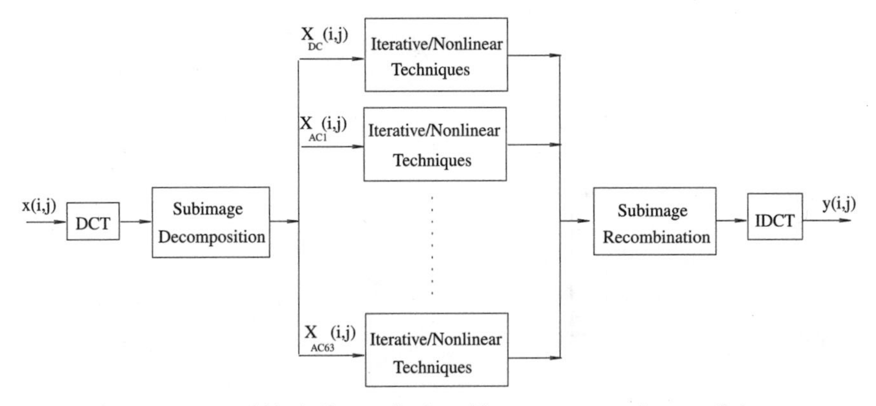

Figure 1. A general block diagram for the subimage error concealment techniques.

techniques can be used to conceal the isolated losses in the formed subimages and hence recover the lost blocks in the original image. A general block diagram of the proposed algorithm is shown in Fig. 1.

18.2.1. Subimage Decomposition

In this subsection, we present a new way of decomposing images in the DCT domain. After detecting the corrupted blocks in the decoded image, we decompose the corrupted image into DCT subimages with equal resolutions. This is done by obtaining the 8×8 DCT transform of the original corrupted image. The corresponding DCT coefficients of all the blocks are then grouped together to form 64 DCT subimages. For example, the DC coefficients of all the blocks are grouped to form the first subimage, then the first AC coefficient (AC1) of each block is used to form the second subimage and so on. The subimage decomposition process is shown in Fig. 2.

For an $N \times N$ image decomposed using 2-D 8×8 DCT transform, the final output of the decomposition process is a set of 64 subimages of $N/8 \times N/8$ dimensions. The decomposition process is aimed to convert the erroneous blocks in the corrupted image into isolated erroneous coefficients in the generated subimages. Therefore, the generated subimages should contain isolated losses instead of bursty ones. These isolated random losses can be recovered by applying any error concealment technique suitable for isolated random losses to the generated subimages.

In the following, we employ the iterative and the time varying (nonlinear division) techniques—described in Chapter 4—to recover the isolated losses in the generated subimages.

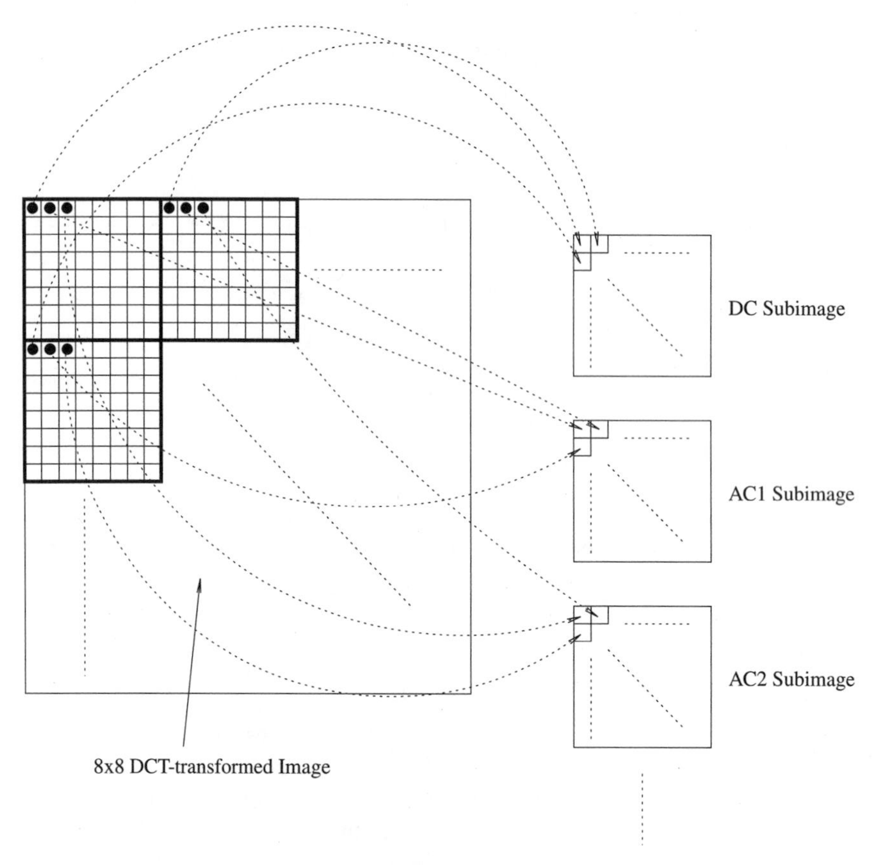

Figure 2. Subimage decomposition.

Figure 3 shows the corrupted 256 × 256 "Claire" image with 30% block losses and the corresponding first 6 DCT subimages.

18.2.2. The Iterative Technique

After generating the subimages, we employ the iterative technique to recover the corrupted subimages [1]. The corrupted image is first transferred to the DCT domain using an 8 × 8 DCT transform. Then, the transformed image is decomposed to the equivalent 64 subimages as described in the pervious subsection. Every subimage is considered as an $N/8 \times N/8$ image corrupted by isolated losses. In order to recover these isolated losses, the iterative technique is applied to every subimage, separately. In the iterative technique, the band-limiting operator—which is used to filter the input subimages in every iteration—is

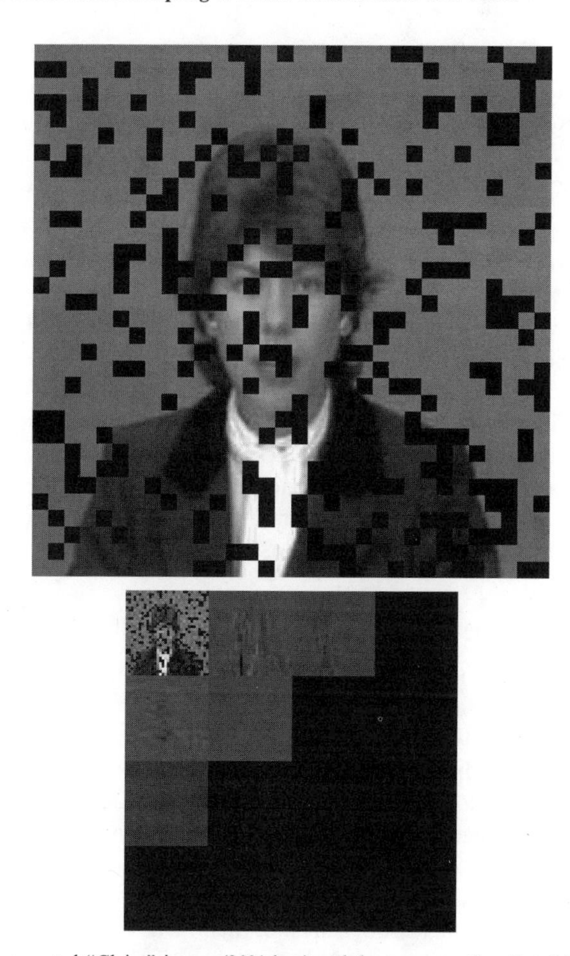

Figure 3. The corrupted "Claire" image (30% loss) and the corresponding first 6 DCT subimages.

designed to adapt to the spectral shape of the filtered subimage. The process of applying the band-limiting operator to any subimage involves the following four steps:

1. The DCT subimage is transformed to the FFT domain.
2. A 2-D mask is created using the FFT-transformed subimage. This is done by normalizing and thresholding a copy of the FFT-transformed subimage in order to eliminate any noise or undesirable frequency components. The thresholded image is then hard-limited to yield the mask to be used later in the filtering process.

3. The FFT-transformed subimage is multiplied by the 2-D mask in order to generate the FFT spectrum of the filtered subimage. This step reduces any erroneous frequency components and enhances the frequency spectrum of the filtered subimage.
4. Finally, the filtered subimage is transformed back to the DCT-domain.

The above process generates a variable filtering mask which is dependent on the subimage to be filtered. Figure 4 shows the process of generating this variable mask.

After every iteration, all the subimages are recombined to form the DCT-transformed image. The filtered DCT image is then inverse-DCT transformed to the spatial domain where a median filter is applied to the resulting image for further quality improvement. In our simulations, we conceal the losses in the first 6 subimages (the low frequency subimages, i.e. the DC subimage and the first five AC subimages) since most of the energy of the image is localized in these subimages. Also, the relaxation parameter of the iterative technique is determined and optimized experimentally for every loss percentage.

18.2.3. The Time Varying (Non-linear Division) Technique

In this technique, the corrupted image is transformed to the DCT domain and then decomposed to the equivalent subimages as previously discussed. The block diagram in Fig. 1 can be modified by replacing the "*Iterative/Non-linear Techniques*" block with the "*Time Varying (Non-linear Division)*" block. The subimages are then recombined to yield the recovered DCT-transformed image. Finally, the output image is median filtered for further image enhancement. This technique is supposed to achieve an acceptable image quality in one step and without any iterations. This technique assumes that the DCT subimages are band-limited which is almost the case for most of them. Due to the negligible effect of considering the high frequency subimages, the first 6 subimages are only considered in the recovery process.

18.2.4. The Iteration with Overhead Technique

In this variation, instead of generating the masks in the iterative technique, the exact masks used to filter the first 6 subimages are sent to the destination node

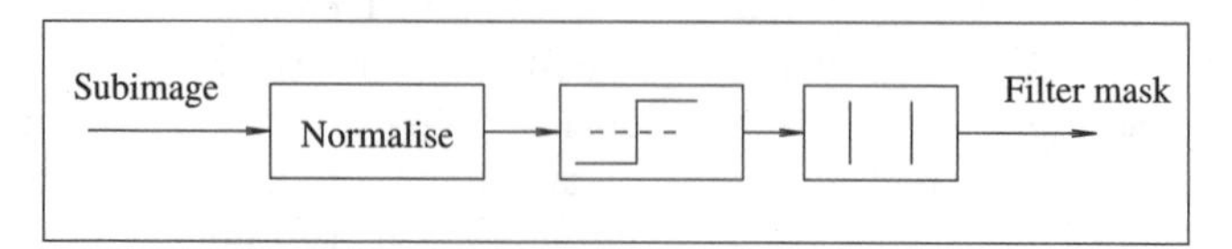

Figure 4. The filter mask generation process.

in the form of protected side information [1]. The structure of this technique is exactly the same as the structure of the iterative technique with only one exception: the filtering masks are sent with the encoded image data instead of being generated at the receiver. For 256×256 images and with 8-bit quantization levels, the total number of bits required to represent the quantized image pixels is equal to 524,288 bits. If we only consider the first 6 subimages in the concealment process, the number of bits required to represent the six generated 32×32 masks—using 4-bit quantization levels—is equal to 24,576 bits. Although overhead information represents about 4.7% of the total bit stream, it yields an average improvement of at least 1.0 dB in the PSNR of the recovered image. Therefore, sending the overhead information can be justified by the improvement obtained in the PSNR of the recovered image, and the elimination of the processing time required to generate the filtering masks.

18.2.5. The Iteration with DCT-Based Filtering Technique

Another way of implementing the iterative technique is to use DCT-based filtering rather than the previously described DFT-based filtering. Again, the structure of this technique is exactly the same as that of the original iterative technique. The only difference between the two techniques lies in the low-pass filtering process applied to the subimages. In this variation, the corrupted subimages are recovered using simple averaging techniques that take into consideration the direction of the maximum correlation among the DCT coefficients of every subimage. Figure 5 shows the basis images associated with the 8×8 DCT coefficients [2].

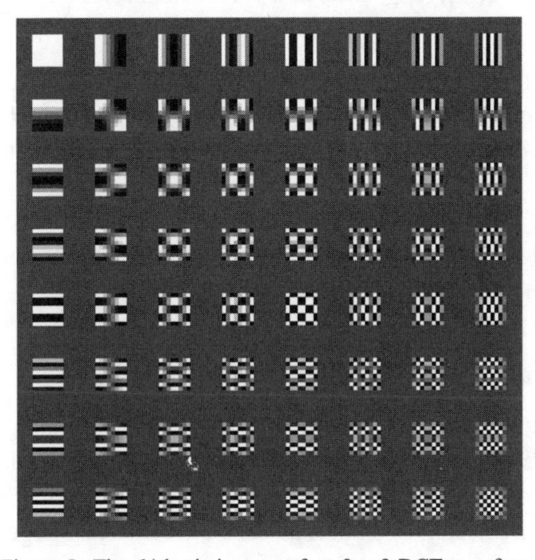

Figure 5. The 64 basis images of an 8×8 DCT transform.

By zigzag scanning the basis images in Fig. 5, it can be noticed that while the correlation among all the DC coefficients is the same in the horizontal and the vertical directions, the first AC coefficients have more correlation in the vertical direction than in the horizontal direction. On the other hand, the second and the third AC coefficients have more correlation in the horizontal direction than in the vertical one. Based on these properties of the DCT basis images, we can recover the losses in every subimage by averaging the proper neighbor coefficients to obtain the corrupted coefficient. For example, since the correlation among the DC coefficients is the same in the horizontal and the vertical directions, the corrupted DC coefficients are recovered using the following formula

$$DC(i,j) = \frac{DC(i-1,j) + DC(i+1,j) + DC(i,j-1) + DC(i,j+1)}{4}. \tag{1}$$

Similarly, we can recover the isolated losses in the first AC subimage by exploiting the vertical correlation among the AC_1 coefficients. The corrupted AC_1 coefficients can be recovered using

$$AC_1(i,j) = \frac{AC_1(i,j-1) + AC_1(i,j+1)}{2}. \tag{2}$$

The second AC subimage can be recovered in the same manner but by exploiting the horizontal correlation among the AC_2 coefficients. The corrupted AC_2 coefficients can be recovered using

$$AC_2(i,j) = \frac{AC_2(i-1,j) + AC_2(i+1,j)}{2}. \tag{3}$$

The rest of the subimages are recovered using the same concept. Figure 6 shows the DCT coefficients used to recover any corrupted DC, AC_1, AC_2 and AC_4 coefficients. The dark blocks represent the lost DCT coefficients, while the light blocks represent the neighboring DCT coefficients used in the recovery process. After recovering every subimage, all the subimages are recombined and the resulting image is then median filtered. In our simulations, we use 20 iterations of the technique to recover the first 6 subimages only.

18.2.6. The Iterative/Time Varying Hybrid Technique

To enhance the performance of the iterative technique, the contraction operator $(I - \lambda PS)$ in Chapter 4 is modified by replacing the low-pass operator P with another robust operator, F. The operator F is composed of a low-pass filter followed by a division operation in which the filtered image is divided by the nonuniform sampling image. After the time varying operation, the resulting

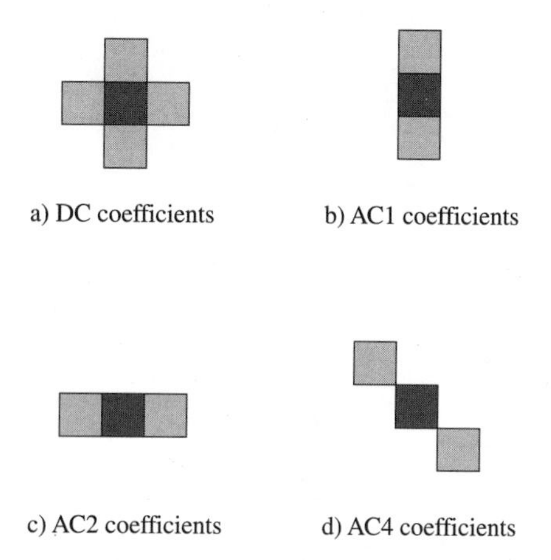

Figure 6. The neighboring coefficients used to recover the corrupted: (a) DC coefficients; (b) AC_1 coefficients; (c) AC_2 coefficients; (d) AC_4 coefficients.

image is low-pass filtered to guarantee the convergence of the modified iterative technique. Thus, the modified iterative technique is a hybrid of the iterative and the time varying techniques. The new operator introduced in this hybrid technique can be expressed mathematically by the following

$$Fx(i,j) = P\frac{x(i,j)}{D}, \tag{4}$$

where D is the result of low-pass filtering the nonuniformly positioned samples with unit amplitudes. The new contraction operator $(I - \lambda FS)$ ensures faster convergence rate than the classical $(I - \lambda PS)$ operator. This is due to the enhanced recovering capability of the F operator compared to that of the P operator. Therefore, if we define the error to be the difference between the original image and the corresponding filtered image, then the error produced using the F operator must be smaller than the error produced when the P operator is used.

$$\|I - \lambda FS\| < \|I - \lambda PS\|. \tag{5}$$

This is evident from our simulations which proved that the time varying technique has a better performance than simple low-pass filtering.

Figure 7 shows a block diagram of the proposed hybrid technique. The technique [3] has the same structure as the iterative technique but with few modifications in the filtering process. The output subimage of every iteration can be described by the following equation

$$z_{k+1}(i,j) = \lambda FSx(i,j) + (I - \lambda FS)z_k(i,j), \tag{6}$$

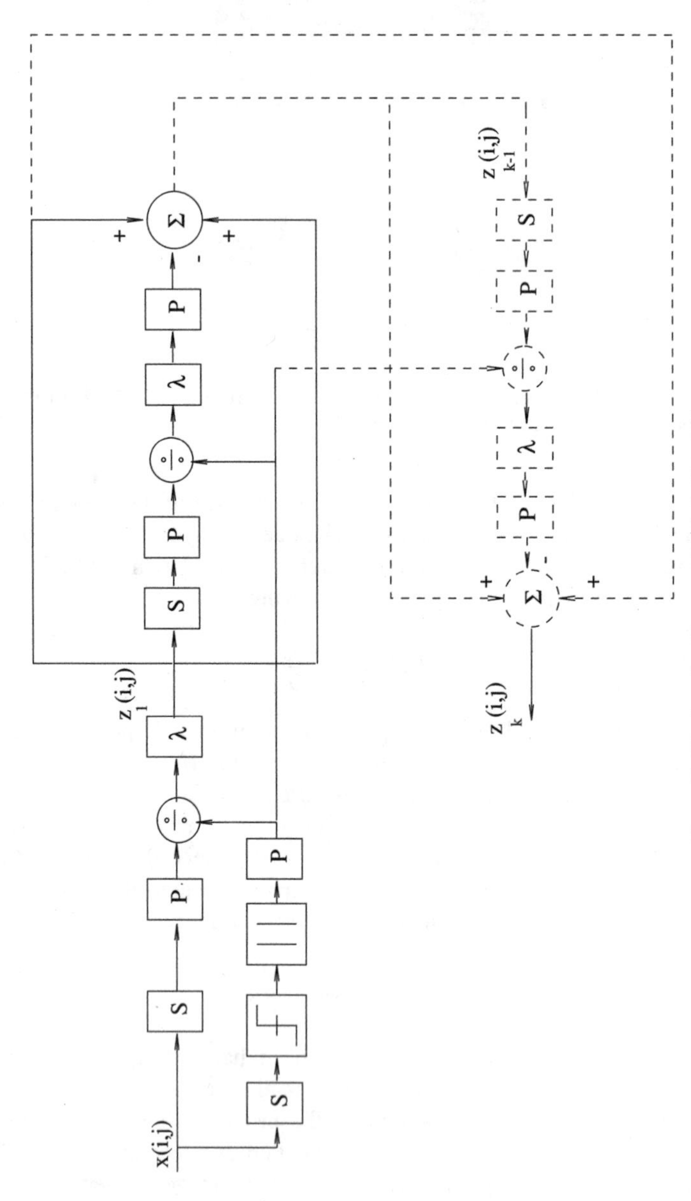

Figure 7. The iterative/time varying hybrid technique.

where $z_{k+1}(i, j)$ is the output subimage of the current iteration, F is the modified 2-D band-limiting operator, S is an ideal nonuniform sampling operator, and λ is the relaxation parameter. The relaxation parameter λ is determined experimentally for every loss percentage in order to maximize the convergence rate of the new hybrid technique. In the rest of this chapter, we shall propose two implementations of a novel error concealment technique that we have developed to conceal random isolated losses as well as bursty losses. This powerful technique is derived from the RS decoding technique previously presented in Chapter 17, [4–6]. In the RS decoding technique, if the original signal is oversampled, the corrupted signal can be recovered as long as the average sampling rate is above the Nyquist rate [7]. The error concealment technique proposed in the following sections is derived from the RS decoding technique and can be used to conceal any type of losses corrupting low-pass filtered images. However, the original image should be pre-processed before transmission or storage in order to be able to use this novel error concealment technique.

18.3. The recursive error concealment (REC) TECHNIQUE

The error concealment technique derived from the BERT technique can be implemented in many different ways. In this section, we describe a recursive way of implementing the technique similar to Peterson's BCH decoding [8, 9]. We shall call this method the *recursive error concealment* (*REC*) method. Later in this chapter, we will also describe a matrix-based implementation of the technique in order to achieve better SNR quality. The REC technique has the capability to conceal random and bursty losses in low-pass filtered images stored on any media or transmitted via any error-prone environment. In our simulations, we consider block losses in the image as an example of bursty losses. The images handled by this technique has to be low-pass filtered in a special way. Therefore, this technique requires pre-processing of the images before storage or transmission. Let us assume that the original image is sampled at the Nyquist rate and $x(i)$ is the 1-D signal formed by scanning the original image line by line. It is essential to divide the signal $x(i)$ into blocks of length n. This technique uses block sizes of 8, 16, 32 and 64. Larger block sizes are not used due to the sensitivity of the REC technique to noise. Every block of length n in the signal $x(i)$ is low-pass filtered before storage or transmission. This is done by transforming the block to the FFT domain to yield the signal $X(j)$ of which the high frequency components are set to zero.

$$X_{LPF}(j) = \begin{cases} X(j) & \text{if } j = 1 \ldots B \text{ and } j = n - B + 2 \ldots n, \\ 0 & \text{if } j = B + 1 \ldots n - B + 1, \end{cases} \tag{7}$$

where B is the bandwidth of the low-pass filtered image. The maximum number of losses per a block of size n that the REC technique is capable of correcting can be expressed by $Z = n - 2B + 1$ which is also equal to the number of the 1-D FFT coefficients set to zero during the low-pass filtering process. For images transmitted via error-prone environments, the block size n and the bandwidth B of the low-pass filtered image are protected and transmitted to the receiver so that they can be used by the error concealment algorithm. Figure 8 shows the 1-D FFT spectrum of the original block ($n = 8$) and the low-pass filtered version of the same block for $B = 3$.

The above filtering process creates blocking artifacts in the low-pass filtered image. This can be explained by the implicit assumption of periodicity of the input block signal by the FFT algorithm. Since the input block signal is not periodic and some of the high frequency coefficients are set to zero during the low-pass filtering process, discontinuities at the block boundaries occur which causes the blocking effect to appear. In order to reduce the effect of setting the high frequency coefficients to zero, the low-pass filtered image can be median filtered before being displayed. Median filtering reduces the blocking artifacts in the low-pass filtered image but it does not eliminate it. Therefore, the number of coefficients set to zero should be controlled in order to obtain a low-pass filtered image with an acceptable visual quality [10].

Due to channel noise, network losses or storage media losses, the low-pass filtered signal $x_{LPF}(i)$ may become corrupted with random isolated losses or block losses. The corrupted signal can be recovered by applying the following steps of the REC technique:

1. After detecting the locations of the erroneous pixels or blocks, the block size n and the bandwidth B of the decoded image are extracted in order to

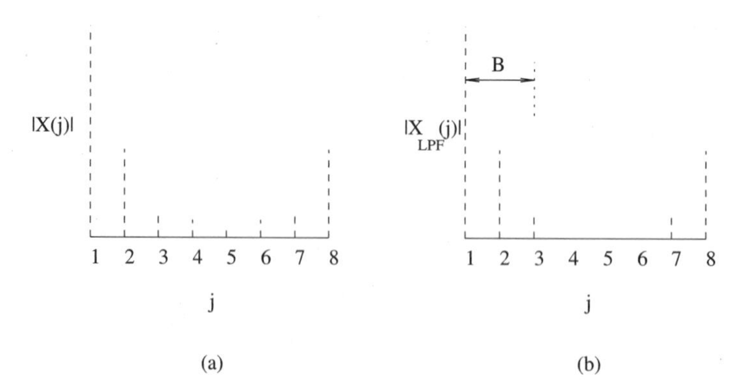

Figure 8. (a) The 1-D FFT spectrum $|X(j)|$ of the original block ($n = 8$); (b) The FFT spectrum $|X_{LPF}(j)|$ of the low-pass filtered block ($B = 3$).

be used in the concealment process. The corrupted image is scanned line by line forming a 1-D corrupted signal which is divided into blocks of length n. We denote the corrupted block by $d(i)$ where $i = 1 \ldots n$.

2. If the number of losses in the corrupted block is greater than the maximum number of correctable losses per block ($Z = n - 2B + 1$), the block is not passed to the rest of the REC technique for further processing. Instead, the corrupted pixels are recovered using the MRROM technique [10] for the case of isolated random losses. In the MRROM technique, any corrupted pixel is replaced by the *Rank Ordered Mean* (*ROM*) value of the good neighboring pixels and the estimated values of the corrupted neighboring pixels. For the case of lost blocks of 8×8 pixels, the CMAP-LP method described in [10] (or any other method) can be used to recover the corrupted blocks. The CMAP-LP technique is a cascade of the classical *Maximum A Posteriori* (*MAP*)-based technique [11] and the linear prediction technique [12]. In the MAP-based technique, two stages of median filtering are applied to the pixels of the corrupted blocks in order obtain an estimate which is then fed to the linear prediction technique. The linear prediction technique predicts the values of corrupted pixels from the previously recovered neighbors using some optimal coefficients. We chose the MRROM and the CMAP-LP methods due to their simplicity and excellent performances.
3. For every corrupted block $d(i)$ with number of losses equal to or less than Z, the error locator polynomial $H(s_i)$ is calculated using (5) in Chapter 17 and the polynomial coefficients h_t are determined by obtaining the inverse FFT transform of the discrete signal $H(s_i)$. The h_t coefficients are totally dependent on the loss pattern of the corresponding corrupted block.
4. After determining the h_t coefficients, the corrupted signal $d(i)$ is transformed to the FFT-domain yielding the signal $D(j)$ where $j = 1 \ldots n$.
5. Since the values of the original transformed signal $X(j)$ for $j = B + 1 \ldots n - B + 1$ are set to zeros during the low-pass filtering process, the corresponding values of the transformed error signal $E_k(j)$ are known since they can be determined from equation (10) in Chapter 17 to be

$$E_k(j) = -D(j), \quad j = B + 1 \ldots n - B + 1 \tag{8}$$

In our notation, we use $E_k(j)$ to represent the *known* values of the error spectrum and $E_u(j)$ to represent the *unknown* values of the spectrum.

6. The unknown values of the transformed error signal $E_u(j)$ can be determined using the recursion in (11) in Chapter 17 but for a different

range for the variable r.

$$E_u(r) = -\left[\frac{\sum_{t=1}^{\tau-1} h_{\tau-t} \cdot \hat{E}_k(\text{mod}_n(r+t)) - \hat{E}_k(\text{mod}_n(r+\tau))}{h_\tau}\right], \tag{9}$$

where $r = B \ldots 1$ and $r = n \ldots n - B + 2$.

7. After finding the unknown values of the error spectrum $E_u(j)$, the total spectrum of the error signal is formed by combining $E_u(j)$ and $E_k(j)$ as follows

$$E(j) = \begin{cases} E_u(j) & \text{if } j = 1 \ldots B \text{ and } j = n - B + 2 \ldots n, \\ E_k(j) & \text{if } j = B + 1 \ldots n - B + 1. \end{cases} \tag{18.10}$$

8. Finally, the spectrum of the recovered signal is obtained by adding $D(j)$ to the obtained $E(j)$ values for $j = 1 \ldots n$. The recovered signal spectrum is then transformed to the spatial domain to yield the recovered block.

The REC technique proved to be very effective in concealing any type of losses but it suffers from the same problems of the RS decoding technique. For bursty losses, the performance of the REC technique degrades due to round-off errors that affect the accuracy of calculating the spectrum of the error signal. As evident from the analysis in [4], any small deviations in the $\hat{E}_k(j)$ values yield large deviations in the corresponding $E_u(j)$ values. Moreover, the sensitivity of the REC technique to round-off errors and noise is accentuated when large block sizes are used. It was found from our simulations that for 256×256 images corrupted by isolated pixel losses, the performance of the REC technique degrades gracefully for block sizes larger than 64. This can be explained using the error analysis presented in Section 17.6.2.

18.4. The matrix-based error concealment (MEC) technique

In this section, we present another way of implementing the REC technique. The new matrix-based method improves the objective quality of the recovered image over the quality obtained using the REC method. One factor that increases the effect of round-off errors in the REC technique is the dependency of the $E_u(j)$ values on the recursion in (9). This causes the round-off error to accumulate causing the $E_u(j)$ values to deviate more from their correct values. As a result, the accuracy of the REC technique decreases and hence the PSNR performance of the technique degrades. In the matrix approach, the dependence of the current $E_u(j)$ values on the previously recovered $E_u(j)$ values is eliminated. The unknown $E_u(j)$ values are determined using only the known $E_k(j)$ values. Similar to the REC method, the MEC method can be used to conceal random and burst losses corrupting low-pass filtered images. The original image $x(i)$ is low-pass filtered before being stored or transported via the error-prone channel. Once the 1-D corrupted signal is formed by scanning the corrupted image line-by-line, it is divided into blocks of size n. For every block $d(i)$ in the corrupted image, the polynomial locator coefficients for the MEC technique are determined from (11) in Chapter 17 as in the REC technique. Then, we find the FFT transform of the signal $\{d(i);\ i = 1 \ldots n\}$ using the matrix approach. In this approach, a 1-D vector $\mathbf{d}$ of length n is formed using the values of the corrupted signal $d(i)$. Then, the vector $\mathbf{d}$ is multiplied by the FFT coefficients matrix $\mathbf{T}$ to yield the vector $\mathbf{D}$ which represents the FFT-transformed $d(i)$ signal.

$$\mathbf{D} = \mathbf{T} \cdot \mathbf{d} = \begin{bmatrix} t_{1,1} & t_{1,2} & \cdots & t_{1,n} \\ t_{2,1} & t_{2,2} & & \\ \vdots & & \ddots & \\ t_{n,1} & & & t_{n,n} \end{bmatrix} \cdot \begin{bmatrix} d(1) \\ d(2) \\ \vdots \\ d(n) \end{bmatrix} \tag{11}$$

where $t_{k,l} = e^{j(2\pi\, k \cdot l/n)}$ and $\mathbf{T} = [t_{k,l}]$ for $k, l = 1 \ldots n$.

As in the REC technique, if the number of losses in the corrupted block being processed is larger than the maximum number of losses the technique can correct per block ($Z = n - 2B + 1$), the block is not passed to the MEC technique for further processing. Instead, the corrupted pixels are recovered using MRROM for the case of isolated losses. For the case of lost blocks of 8×8 pixels, the CMAP-LP method is used to recover the corrupted blocks if the number of losses is greater than Z.

If the number of losses is less than or equal to Z, the values of h_t are determined in the same fashion as in the REC method but using the following matrix approach instead of inverse FFT.

$$\mathbf{h} = \mathbf{T}^{-1} \cdot \mathbf{H} = \begin{bmatrix} h_\tau \\ h_{\tau-1} \\ \vdots \\ h_1 \\ h_0 \\ 0 \\ 0 \\ \vdots \\ 0 \\ 0 \end{bmatrix} = \mathbf{T}^{-1} \cdot \begin{bmatrix} H(1) \\ H(2) \\ H(3) \\ \vdots \\ \\ \vdots \\ \\ \vdots \\ H(n-1) \\ H(n) \end{bmatrix}. \tag{12}$$

After determining the h_t coefficients and the transformed values $D(j)$, the known values of the error spectrum $E_k(j)$ are determined using (8). Instead of using the recursion in (9), the unknown values of the error spectrum $E_u(j)$ can be found using the following equation

$$\mathbf{E_u} = \mathbf{L}^{-1}(\mathbf{M} \cdot \mathbf{E_k}), \tag{13}$$

where

$$\mathbf{E_u} = \begin{bmatrix} E_u(B) \\ E_u(B-1) \\ \vdots \\ E_u(1) \\ E_u(n) \\ E_u(n-1) \\ \vdots \\ E_u(n-B+2) \end{bmatrix}, \quad \mathbf{E_k} = \begin{bmatrix} E_k(B+1) \\ E_k(B+2) \\ \\ \vdots \\ \\ \vdots \\ \\ E_k(n-B+1) \end{bmatrix}. \tag{14}$$

The $(2B-1) \times (2B-1)$ matrix $\mathbf{L}$ and the $(n-2B+1) \times (2B-1)$ matrix $\mathbf{M}$ can be expressed as

$$\mathbf{L} = \begin{bmatrix} h_\tau & 0 & 0 & 0 & \cdots & \cdots & \cdots & 0 \\ h_{\tau-1} & h_\tau & 0 & 0 & \cdots & \cdots & \cdots & 0 \\ \vdots & \ddots & \ddots & 0 & 0 & \cdots & \cdots & 0 \\ h_0 & h_1 & \cdots & h_\tau & 0 & 0 & \cdots & 0 \\ 0 & h_0 & h_1 & \cdots & h_\tau & 0 & \cdots & 0 \\ 0 & 0 & \ddots & \ddots & \cdots & \ddots & \cdots & 0 \\ \vdots & \vdots & & \ddots & \ddots & \cdots & \ddots & 0 \\ 0 & 0 & \cdots & 0 & h_0 & h_1 & \cdots & h_\tau \end{bmatrix} \tag{15}$$

and

$$\mathbf{M} = -\begin{bmatrix} h_{\tau-1} & h_{\tau-2} & \cdots & \cdots & h_0 & 0 & \cdots & 0 \\ h_{\tau-2} & h_{\tau-3} & \cdots & h_0 & 0 & \cdots & \cdots & 0 \\ \vdots & \ddots & & 0 & 0 & \cdots & \cdots & 0 \\ \vdots & & \ddots & 0 & \cdots & \cdots & \cdots & 0 \\ h_0 & 0 & 0 & 0 & \cdots & \cdots & \cdots & 0 \\ 0 & 0 & 0 & 0 & \cdots & \cdots & \cdots & 0 \\ \vdots & \vdots & \vdots & \vdots & \cdots & \cdots & \cdots & \vdots \\ 0 & 0 & 0 & 0 & \cdots & \cdots & \cdots & 0 \end{bmatrix}. \tag{16}$$

After obtaining the $\mathbf{E_u}$ vector using (13), the error spectrum $E(j)$ can be formed using (10). The spectrum of the recovered signal is obtained by adding $D(j)$ to the obtained $E(j)$ values for $j = 1 \ldots n$. The recovered signal spectrum is then transformed to the spatial domain to yield the recovered block.

It can be noted from equations (13)–(16) that the unknown $E_u(j)$ values are obtained independently which results into a better quality of the recovered image. The quality improvement obtained by the MEC technique is at the expense of the increased computational load due to the matrix inversion and multiplication operations.

One of the advantages of the matrix-based method is that it gives us the freedom to choose the transform kernel by controlling the elements of the matrix $\mathbf{T}$. This allows us to experiment with different transform kernels in order to improve the performance of the MEC technique and to reduce its sensitivity to round-off errors and noise.

Before we report any results for the REC and the MEC techniques, the quality of the low-pass filtered image is studied and assessed. In our simulations, we use 256 × 256 "Claire" image with 8, 16, 32 and 64 block sizes. Block sizes larger than 64 are not used because both the REC and the MEC techniques fail for very large block sizes.

For 256 × 256 "Claire" image, it was found that for an acceptable quality of the low-pass filtered image, the minimum bandwidth that can be used for a block size of length 8 is $B = 3$. For block size 16, the minimum bandwidth that can be used is $B = 5$ while for block sizes 32 and 64, the minimum bandwidths that can be used are $B = 10$ and $B = 19$, respectively. Table 1 summarizes the previous results:

The third and fourth columns in Table 1 give vital information about the maximum recovering capabilities of the REC and the MEC techniques. The recovering capability, Z, specifies the maximum number of pixels the REC or the MEC techniques can recover in every block given that the subjective quality of the low-pass filtered image is acceptable. It is also equal to the number of the 1-D FFT coefficients set to zero during the low-pass filtering process.

Table 1. The minimum bandwidths used when filtering "Claire" image using different block sizes.

Block size (n)	Minimum Bandwidth (B)	Recovering Capability ($Z = n - 2B + 1$)	$Z/n \times 100\%$
8	3	3	37.50%
16	5	7	43.75%
32	10	13	40.63%
64	19	27	42.19%

For example, for block size 64 and with an acceptable visual quality of the low-pass filtered image, both the REC and the MEC techniques cannot correct more than 42.2% of the losses in any corrupted block. This means that any block with 64 pixels of which more than 42.2% are corrupted should not be processed by the REC or the MEC techniques. Instead, such block should be passed to the MRROM or the CMAP-LP techniques as explained before.

Since the values of (Z/n) in Table 1 are almost the same for all the block sizes used, we can compare the performances of the REC and the MEC techniques for different block sizes. From the discussion in [4], we conclude that if (Z/n) is maintained constant, the quality of the recovered image using the REC technique improves as the block size increases.

18.5. Experimental results

In this section, we will extensively study the subjective and the objective performances of the error concealment techniques described in this chapter.

For block losses, we simulate the transmission of JPEG-coded images over ATM networks. When an ATM cell is lost, the decoder can identify the location and the size of the damaged area in the JPEG-coded image since the first block in every cell is absolutely addressed. Usually, this type of losses results into the loss of consecutive blocks in the image. In the following sections, we will describe the error concealment techniques that we use for such block losses. The compressed JPEG image is packed into ATM cells before transmission. When an ATM cell is lost, the DC and AC coefficients packed into this cell are lost. Therefore, the pixels of the corresponding lost blocks are replaced with zeros in order to allow the decoding of the next cell payload to continue. Since the DC coefficients of JPEG-coded blocks are differentially encoded, the loss of an ATM cell does not only affect the corresponding blocks packed into that cell, but it also affects all the DC coefficients of the blocks packed in the succeeding cells. Figure 9 shows the effect of cell losses on the DC levels of the blocks succeeding the lost ones. The mean cell loss rate P used to obtain Fig. 9 is 5×10^{-3} with a mean burst length $B_L = 2$.

As can be seen in Fig. 9, the DC levels of all the blocks succeeding the corrupted blocks are affected. In other words, the error in the DC levels propagates through all the blocks succeeding the lost blocks. Such error propagation makes error concealment very difficult. In order to avoid this problem, the DC coefficient of the first block in any ATM cell is fully coded without any dependence on the previous block DC coefficient. Moreover, the first block in every cell is absolutely addressed. This eliminates the dependency of any cell payload on the previous cell payload. For the rest of the blocks in any cell, the

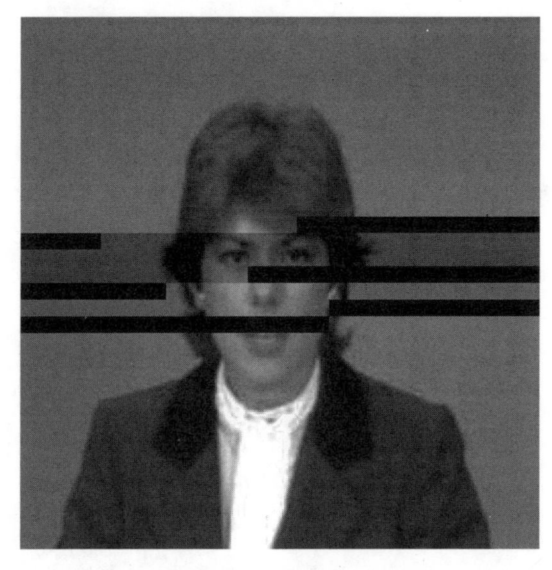

Figure 9. The DC level error propagation in JPEG-coded "Claire" image transmitted via ATM networks.

DC coefficients are still differentially encoded. Also, the payload of every cell should contain the data of complete blocks only. Therefore, if the encoded data of a certain block could not be fitted in the current cell payload, the whole block data should be moved to the next cell and the current cell payload will be filled with stuffing bits. This guarantees the complete independence of the cells payloads on each other.

Figure 10 shows the packetizing scheme described earlier. According to our simulations using "Claire" image, this packetizing scheme reduces the compression efficiency of the baseline JPEG codec by 15%. Despite the degradation in the compression efficiency of the modified JPEG codec, the proposed modifications are essential since they enable the error concealment techniques to produce high quality recovered images. For the same parameters (P and B_L) used to produce Fig. 9, we simulated the transmission of the encoded "Claire" image using the modified JPEG codec and the packetization scheme described in Fig. 10. Figure 11 shows the corrupted image after using the proposed changes to the JPEG codec and the packetization scheme.

As Fig. 11 clearly shows, the error propagation in the DC level has completely been eliminated after using the packetization scheme and the modifications described earlier. Since many of the error concealment techniques described in the literature deal with isolated block losses, we implemented a *pseudo-random* interleaving technique to convert the consecutive block losses

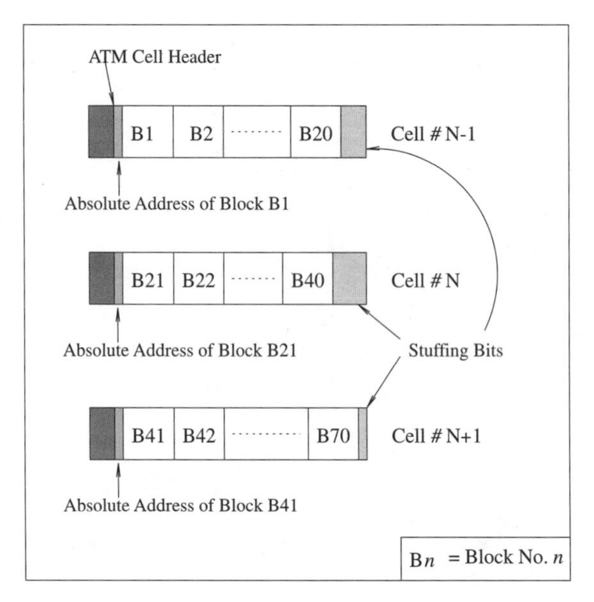

Figure 10. The ATM cell packetization scheme used to reduce error propagation.

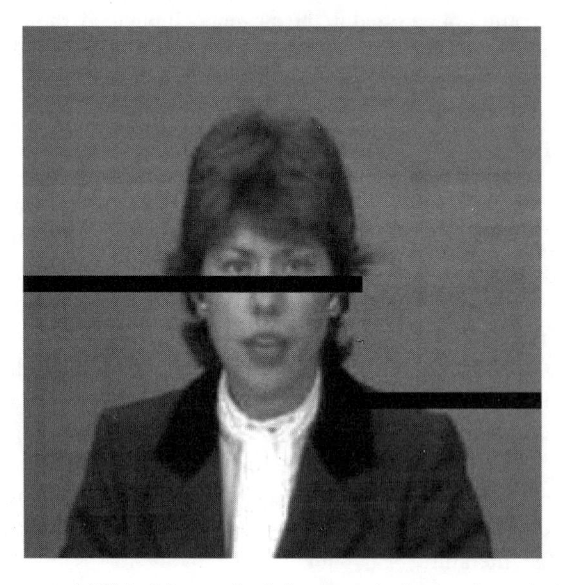

Figure 11. The corrupted "Claire" image after being transmitted using the packetization scheme.

into isolated block losses. This interleaving technique enhances the performance of the error concealment techniques that assume the case of isolated block losses.

The pseudo-random interleaving technique relies on the idea of shuffling the places of the encoded blocks before transmission. Therefore, the technique requires a buffer to store the shuffled blocks data. For every encoded block, the technique generates a random number from 1 to 20 using a *pseudo-random* number generator with a constant seed. Since the seed for the random numbers is known, the same numbers are exactly re-generated at the receiver. The technique starts by grouping all the codewords that represent the DC coefficients of the blocks and packing these codewords into cells with guaranteed delivery. This guarantees that the DC levels of all the blocks are error free. Then, the interleaving technique populates the rest of the cells with the codewords that represent the shuffled blocks of AC coefficients. For every encoded block, the generated random number indicates the cell where the codewords corresponding to the AC coefficients of that block should be packed. The reference number for every cell (1–20) as well as the absolute address of the first block packed in the cell are sent with the cell payload. This information enables the decoder to re-shuffle the blocks and identify the correct locations of the decoded blocks. The interleaving technique produces cells with randomly distributed blocks and therefore, every cell is likely to contain geometrically displaced blocks. If a cell is lost, the decoder will continue to decode the payloads of the other cells and place the decoded blocks in their correct locations using the absolute addresses of the first blocks in the cells as well as the re-generated random numbers for every block location. Therefore, the decoded image will end up with isolated block losses as Fig. 12 shows.

For the case of isolated random errors, the MMEM error detection technique [13] is used to detect the erroneous pixels in the corrupted images before applying the REC or MEC error concealment techniques. In the MMEM error detection technique, the pixels with quantized values equal to the maximum or the minimum quantized values in the $N_w \times N_w$ filtering window are first discarded. This step is supposed to eliminate any suspicious neighboring pixels from the $N_w \times N_w$ window. The average of the non-discarded neighboring pixels is then compared to the value of the inspected pixel in order to determine whether the pixel is corrupted or not.

For the case of isolated block losses, we corrupt the JPEG-coded images by multiplying the decoded image with the appropriate loss pattern. The result is an image with isolated block losses. On the other hand, for the case of consecutive block losses, we implement the previously described packetization and the pseudo-random interleaving schemes before transmitting the image encoded using the modified-JPEG algorithm. After decoding the transmitted bitstream, the resulting image should have consecutive blocks missing.

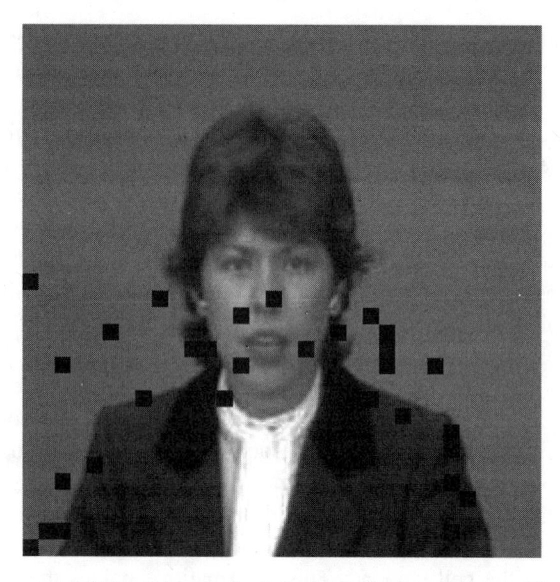

Figure 12. The corrupted "Claire" image after being transmitted using the *pseudo-random* interleaving scheme.

In order to obtain an accurate assessment of the compared error concealment techniques, we conduct 10 experiments for every loss percentage. In every experiment, the test images are corrupted by the same loss pattern for all the compared techniques. However, the 10 experiments use 10 different loss patterns of the same loss percentage. For every loss percentage, the objective results of the corresponding 10 experiments are averaged.

To assess the objective quality of the recovered images, we will use the Peak Signal-to-Noise Ratio (PSNR) as our measure. In our simulations, we use 256×256 "Claire" image ($N = 256$) to assess the performances of the compared error concealment techniques.

18.5.1. The Subimage Error Concealment Techniques

In this subsection, we will study the behavior of the proposed iterative and time varying techniques for both cases of isolated and consecutive block losses. All the iterative techniques are run for 20 iterations and only 6 subimages are used in the recovery process. By analyzing the behavior of the iterative techniques, Fig. 13 and Fig. 14 confirm that the iteration with overhead technique has the best PSNR performance among all the iterative techniques for both isolated and consecutive block losses. This is expected since in the iteration with overhead technique, the filtering masks—which convey more information about the corrupted image—are sent as overhead information with the coded image. It

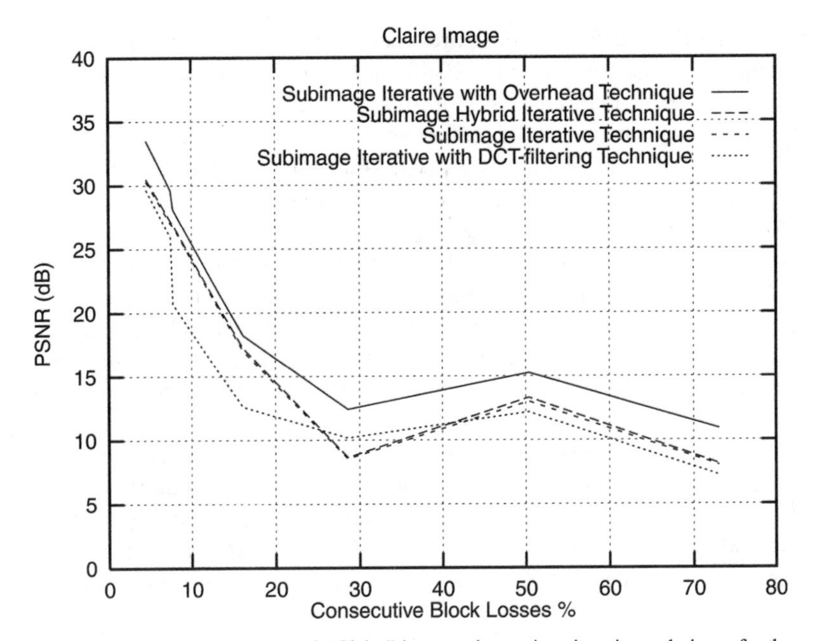

Figure 13. The PSNR of the recovered "Claire" image using various iterative techniques for the case of consecutive block losses (20 iterations).

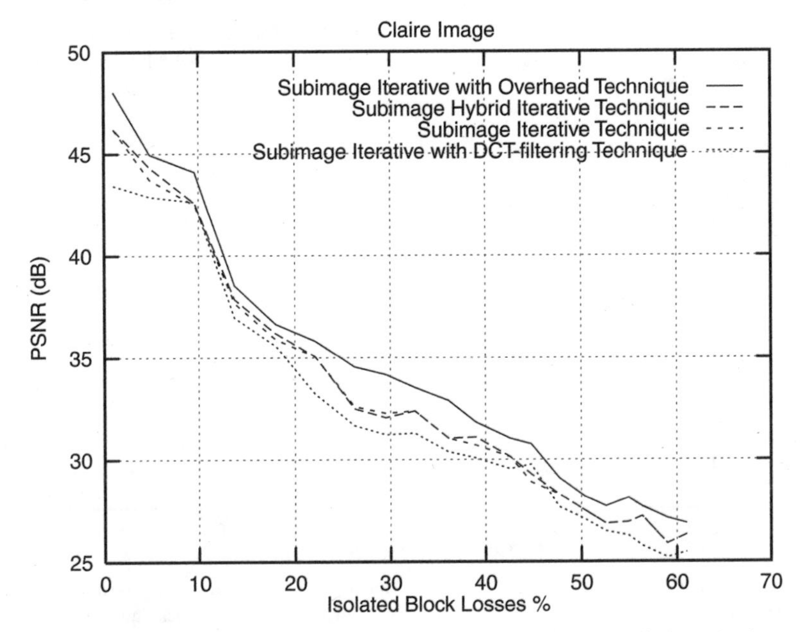

Figure 14. The PSNR of the recovered "Claire" image using the iterative techniques for the case of isolated block losses (20 iterations).

can also be noticed from both figures that the performances of all the iterative techniques degrade by more than 10 dB for the case of consecutive losses when compared to their performances with isolated block losses.

This is expected since the performance of the iterative technique—which is the backbone of all the iterative techniques—degrades for the case of consecutive losses [14]. Furthermore, the hybrid iterative technique has a comparable performance to the iterative technique for both consecutive and isolated block losses as Figs. 13 and 14 suggest. This is because the hybrid iterative technique is developed to enhance the convergence rate of the iterative technique.

The iteration with DCT filtering technique has the worst PSNR performance among all the iterative techniques. The previous statement is more evident with consecutive block losses since the DCT filter has a very poor performance if the surrounding blocks are also corrupted.

For consecutive block losses, the time varying technique is worse than the iterative technique. This is evident from Fig. 15 which compares the PSNR performances of the time varying and the iterative techniques for consecutive block losses. Moreover, for isolated block losses, Fig. 16 proves that the iterative technique has a better PSNR performance than the time varying technique.

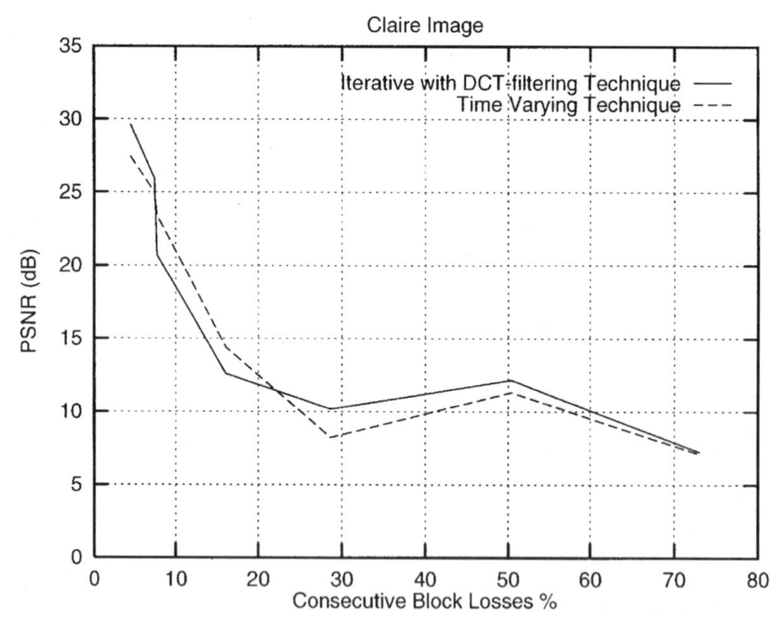

Figure 15. The PSNR of the recovered "Claire" image using the iterative with DCT-filtering and the time varying techniques for the case of consecutive block losses. All iterative techniques are run for 20 iterations.

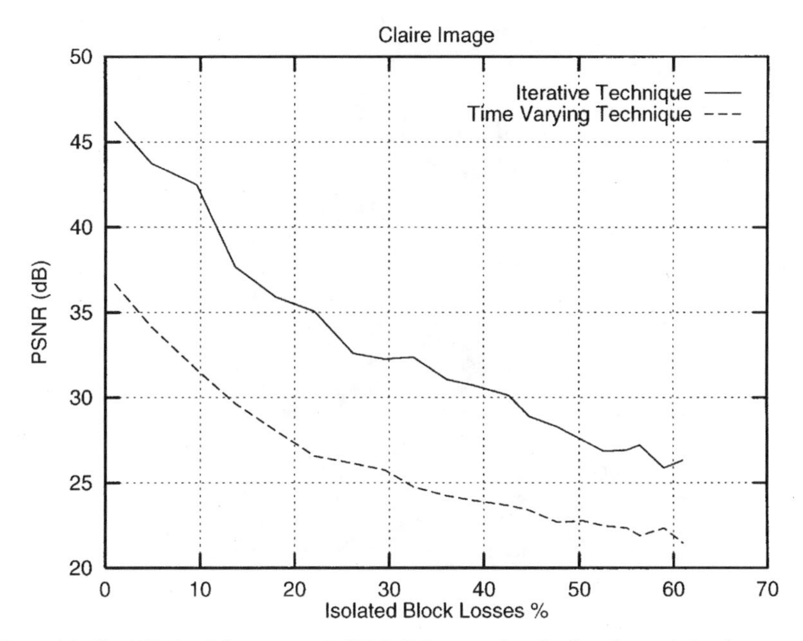

Figure 16. The PSNR of the recovered "Claire" image using the iterative and the time varying techniques for the case of isolated block losses. All iterative techniques are run for 20 iterations.

In the following, we will assess the subjective qualities of "Claire" images recovered by the subimage error concealment techniques for the cases of isolated and consecutive block losses. We will start with the subjective qualities for isolated block losses. Figures 17 and 18 show the original "Claire" image and the corrupted image with 20% of isolated block losses, respectively.

In Figs. 19 and 20, the subjective quality obtained using the iteration with overhead technique is compared to the quality obtained using the iterative technique. Clearly, the quality obtained using the iteration with overhead technique is better than that obtained using the iterative technique. The quality improvement can be noticed by observing certain regions in the recovered images such as (from the viewer's prospective): the left side of the head, the right side of the neck, the right shoulder and different regions in the right side of the jacket. The hybrid iterative technique is expected to have a similar subjective quality to the quality obtained using the iterative technique.

Although the iterative with DCT-filtering technique has an inferior PSNR performance compared to the iterative technique as the results in Fig. 14 show, it produces better subjective quality than the iterative technique. By comparing Figs. 20 and 21, the viewer can notice an improvement in the quality of the recovered image in Fig. 21 in the following regions (from the viewer's prospec-

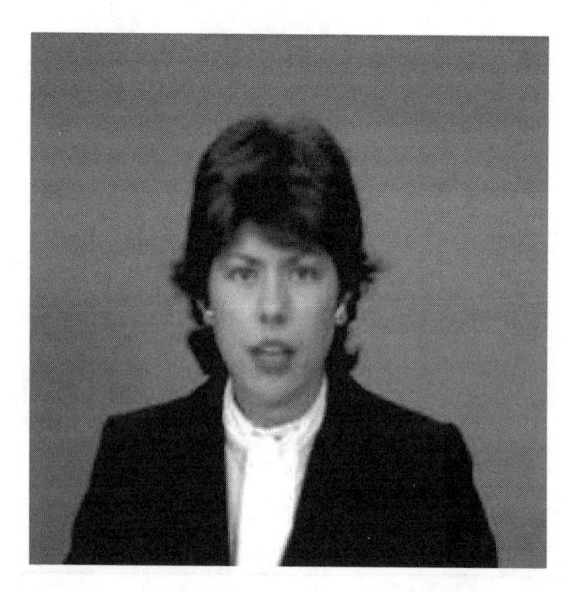

Figure 17. The original "Claire" image.

Figure 18. The corrupted "Claire" image (20% block losses).

Figure 19. The recovered "Claire" image using the iteration with overhead technique.

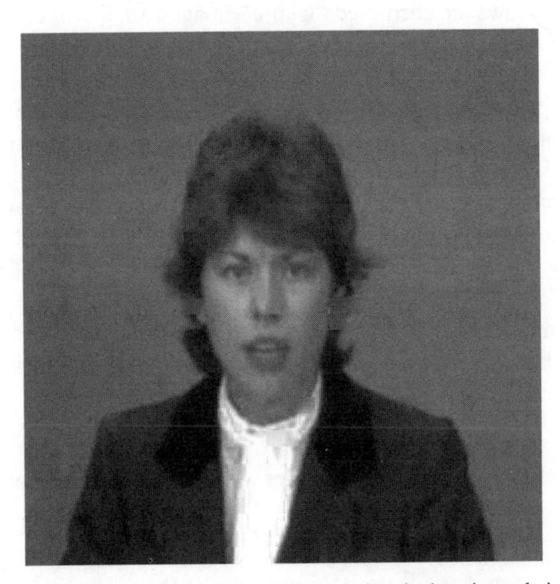

Figure 20. The recovered "Claire" image using the iterative technique.

Figure 21. The recovered "Claire" image using the iterative with DCT-based filtered technique.

tive): the left ear, the left side of the head, the right side of the neck and the right shoulder. The quality obtained using the iterative with DCT-filtering technique is comparable to the quality obtained using the iteration with overhead technique.

Figure 22 shows the corrupted "Claire" image with 8% of consecutive block losses. As can be seen from Fig. 22, complete lines of blocks are lost due to the loss of one or two ATM cells.

The comparison between Figs. 23 and 24 reveals that the iteration with overhead technique has a better subjective performance than the iterative technique. This confirms the objective results presented earlier. Furthermore, the comparison between Figs. 24 and 25 shows that the iterative technique produces better visual quality than the iterative with DCT-based filtering technique which is consistent with the objective results obtained for consecutive block losses. The above discussions drive us to conclude that for consecutive block losses, the iterative with overhead technique produces the best subjective results for "Claire" image among the other compared techniques. The iterative technique comes in the second place.

18.5.2. The REC and MEC Techniques Applied to Isolated Random Losses

For isolated random losses, we employ the REC and the MEC techniques—with different block sizes—to conceal the losses corrupting the low-pass filtered 256×256 "Claire" image. We use the optimal bandwidth corresponding to

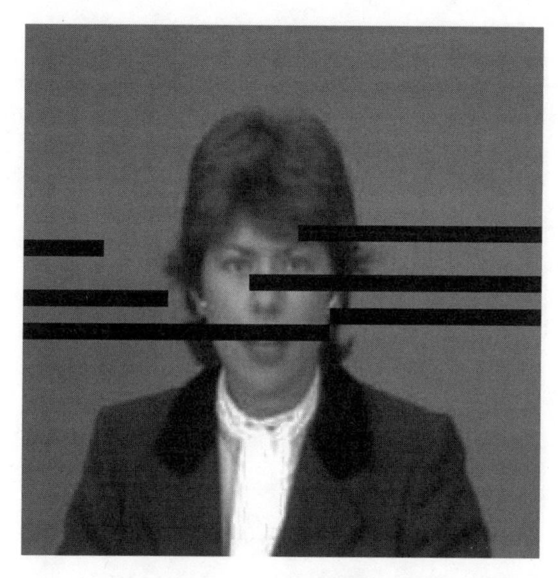

Figure 22. The corrupted "Claire" image (8% consecutive block losses).

Figure 23. The recovered "Claire" image using the iteration with overhead technique (consecutive block losses).

Figure 24. The recovered "Claire" image using the iterative technique (consecutive block losses).

Figure 25. The recovered "Claire" image using the iterative with DCT-based filtering technique (consecutive block losses).

every block size in the low-pass filtering process. These optimal bandwidths were previously presented in Table (18.1) in Section 18.4.

Figure 26 presents the results of employing the REC technique—with different block sizes—to recover the random losses corrupting the band-limited "Claire" image. As can be noticed from the figure, the objective quality of the recovered image increases as the block size increases. This confirms the discussion in the previous subsection. As the loss percentage increases, the quality of the recovered image degrades and all the PSNR curves in Fig. 26 start approaching each other till they merge together. This is expected since, as the loss percentage increases, the average number of losses in the defined blocks increases till it exceeds Z. When such case occurs for a defined block, the REC technique is disabled and the recovery process is handled by the MRROM technique.

Similar to the REC technique, the MEC technique can be employed to recover the isolated random losses corrupting the low-pass filtered "Claire" image. Figure 27 presents the PSNR results obtained by the MEC technique for different block sizes. For any loss percentage below 50%, the PSNR of the image

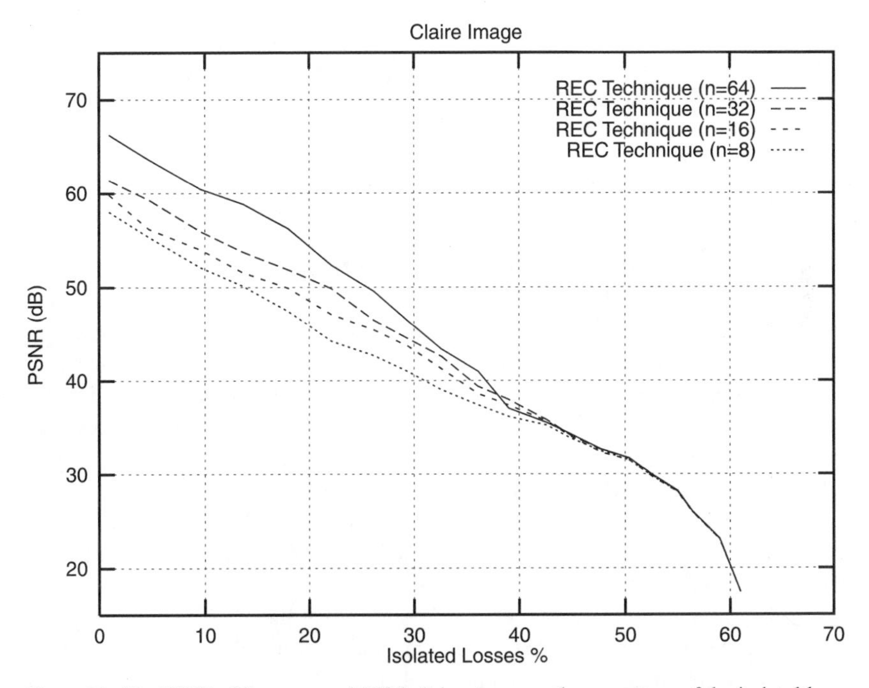

Figure 26. The PSNR of the recovered "Claire" image versus the percentage of the isolated losses. The REC technique is employed in this figure to recover the isolated losses in the image using different block sizes, n.

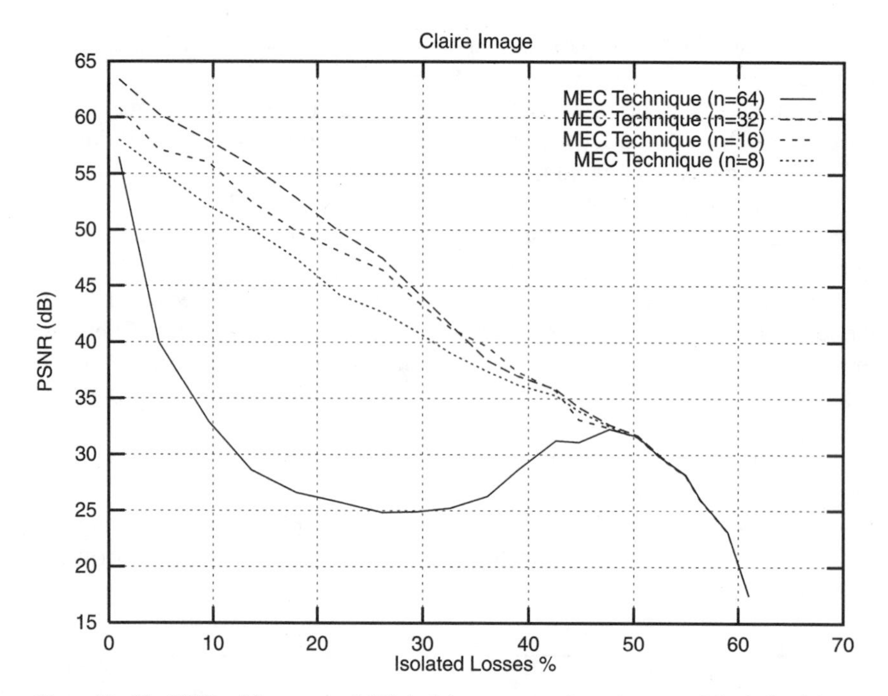

Figure 27. The PSNR of the recovered "Claire" image versus the percentage of the isolated losses. The MEC technique is employed in this figure to recover the isolated losses in the image using different block sizes n.

recovered using the MEC technique increases as the block size increases. The previous statement is valid for all block sizes, n, except for $n = 64$ where the MEC technique fails and its performance gracefully degrades. This phenomenon is due to the inaccuracy in obtaining the $\mathbf{L}^{-1}$ matrix elements when block sizes equal to or greater than 64 are used. Although the accumulated round-off errors, caused by the dependency of the $E_u(j)$ values on the previously found values, are reduced in the MEC technique, another source of round-off errors is introduced for large block sizes due to the inversion of the $\mathbf{L}$ matrix.

At this point, we compare the PSNR performance of the REC technique to that of the MEC technique. Since the MEC technique fails for block sizes larger than or equal to 64, we will limit the comparison between the two techniques by using block sizes 16 and 32 only. Figure 28 shows that the MEC technique gives better PSNR quality than the REC technique for both $n = 16$ and $n = 32$. The quality improvement is more pronounced when blocks of size 32 are used.

As can be depicted from Fig. 29, the performances of both the REC and the MEC techniques approach that of the MRROM technique as the loss percentage

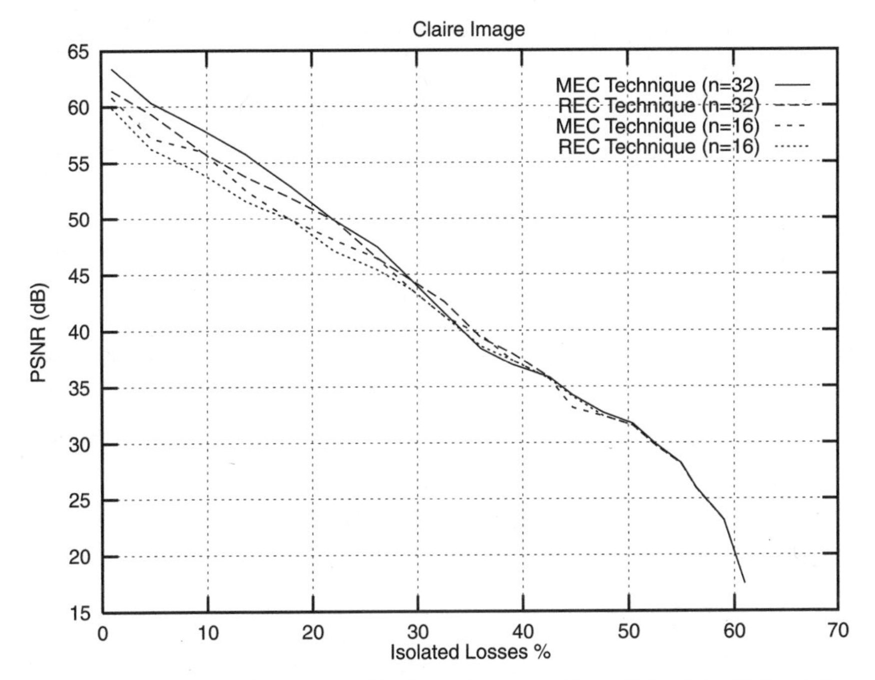

Figure 28. A comparison between the objective performances of the REC and the MEC techniques with block sizes 16 and 32 (isolated random losses case).

increases. In this figure, the objective qualities of the recovered images using the REC, the MEC and the MRROM techniques are compared. The block sizes used for the REC technique are 32 and 64 while only block size 32 is used for the MEC technique since it fails for larger block sizes. For the sake of comparison, the MRROM technique is also applied to the low-pass filtered "Claire" image using block size 64. It can be concluded from Fig. 29 that the REC technique with block size 64 gives the best PSNR results among the other techniques compared. For loss percentages less than 40%, both the REC and the MEC techniques achieve better PSNR performances compared to the MRROM technique. The improvement in the PSNR performance reaches 8 dB for 1% of random losses and decreases till it vanishes when the maximum recovering capabilities of the REC and the MEC techniques are exceeded.

Figure 30 shows the low-pass filtered "Claire" image with 30% of isolated random losses. For all the figures presented in this subsection, blocks of size 32 are used during the low-pass filtering process.

The subjective qualities of "Claire" images recovered using the REC and the MEC techniques can be depicted in Figs. 31 and 33. The quality of these images

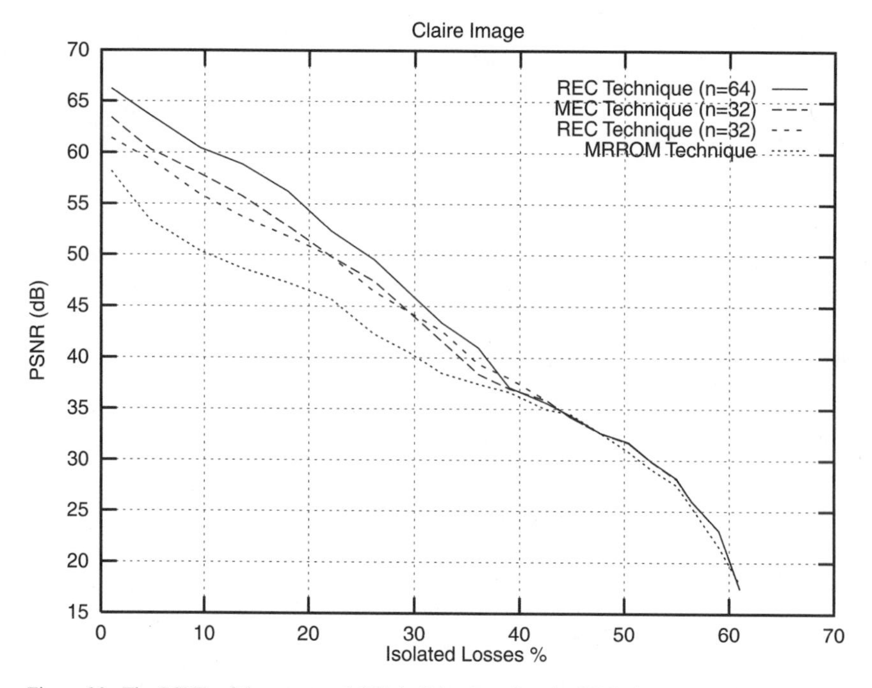

Figure 29. The PSNR of the recovered "Claire" image using the REC, the MEC and the MRROM techniques for the case of isolated random losses.

Figure 30. The corrupted "Claire" image (30% random losses).

Figure 31. "Claire" image recovered from isolated random losses using the REC technique with $n = 32$ and $B = 10$.

can be compared to the quality of the original low-pass filtered image shown in Fig. 32.

It can be noticed that both the REC and the MEC techniques produce similar visual qualities of the recovered images. The subjective qualities obtained using the REC and the MEC technique are comparable to the quality of the original low-pass filtered image. This is because if the number of losses in any block is less than or equal to Z, both techniques should be able to recover all the corrupted pixels but with different accuracies. The human eye is usually insensitive to the small differences between the recovered pixels and the corresponding original ones. Therefore, the visual qualities of the recovered images using the REC and the MEC techniques are comparable to that of the original low-pass filtered image.

In Figs. 31–34, the visual quality obtained using the MRROM technique is compared to the visual qualities obtained using the REC and the MEC techniques. As can be seen from the figures, the subjective qualities obtained using the three techniques are comparable.

18.5.3. The REC and MEC Techniques Applied to Isolated Block Losses

For bursty losses, we consider the case of 8×8 block losses which results into bursts of lost samples in the $d(i)$ signal. In this case, the minimum length of any burst of lost samples is 8. Therefore, the selected block sizes to be used in the

Figure 32. The low-pass filtered "Claire" image with $n = 32$ and $B = 10$.

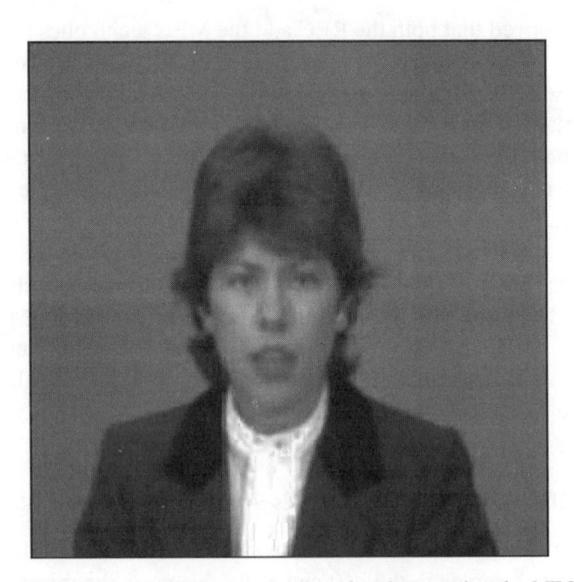

Figure 33. "Claire" image recovered from isolated random losses using the MEC technique with $n = 32$ and $B = 10$.

Figure 34. "Claire" image recovered from isolated random losses using the MRROM technique.

REC and the MEC techniques should satisfy the following condition: "For an acceptable visual quality of the low-pass filtered image, the recovering capability, Z, for the corresponding block size n should be equal to or greater than 8."

By inspecting Table 1, we conclude that we can only use block sizes 32 and 64 to recover the corrupted 8×8 blocks. Blocks of sizes 8 and 16 cannot be used for this propose since the maximum recovering capabilities corresponding to these block sizes are 3 and 7, respectively.

In our simulations, we employ the REC and the MEC techniques with block sizes 32 and 64 to recover the 8×8 lost blocks in the low-pass filtered "Claire" image. We use the optimal bandwidth corresponding to every block size in the low-pass filtering process. In Fig. 35, we notice that as the loss percentage increases, the PSNR curves of the REC and the MEC techniques for $n = 32$ start approaching each other till they merge in one curve. As explained before, this phenomenon is due to switching the recovery process to the CMAP-LP technique whenever the number of losses in the defined blocks exceeds Z.

For bursty losses, the sensitivity of the REC and the MEC techniques to round-off errors is more accentuated. This causes both techniques to fail for block sizes larger than or equal to 64 as Fig. 35 shows. Furthermore, the performance of the MEC technique is slightly better than that of the REC technique for $n = 32$.

By comparing the performances of the REC and the MEC techniques to the performance of the CMAP-LP technique, we notice from Fig. 35 that for the case of isolated block losses and with $n = 32$, both the REC and the MEC techniques

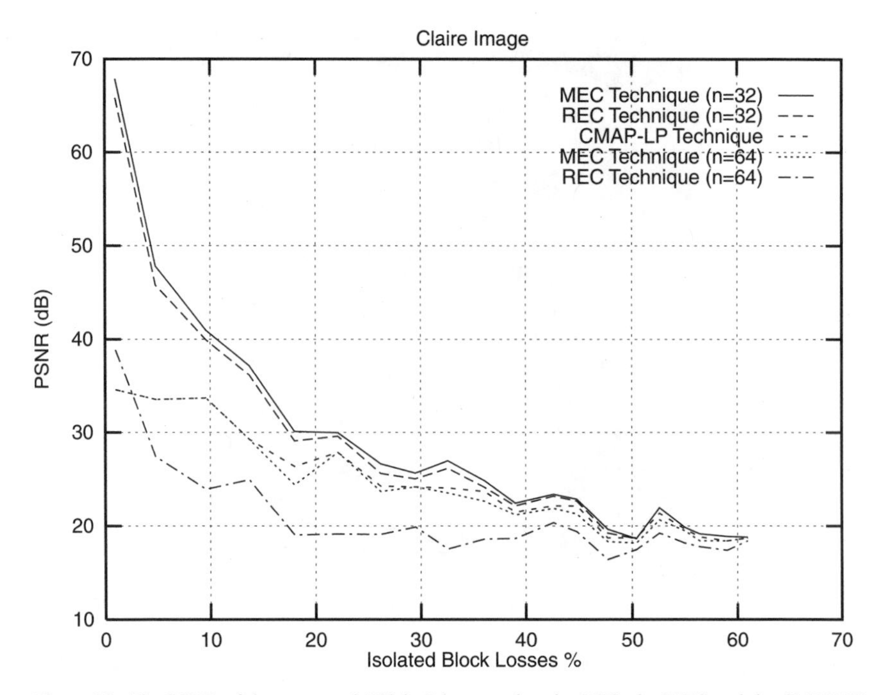

Figure 35. The PSNR of the recovered "Claire" image using the REC, the MEC and the CMAP-LP techniques for the case of corrupted 8×8 blocks.

outperform the CMAP-LP technique for all loss percentages less than 47%. The performance improvement ranges from 34 dB to less than 1 dB as can be depicted from the same figure.

After studying the subjective performances of the REC and the MEC techniques for isolated random losses, we study—in this subsection—their subjective performances for isolated block losses. Figure 36 shows the low-pass filtered "Claire" image with 10% of 8×8 isolated block losses. Again, blocks of size 32 are used during the low-pass filtering process.

By comparing Figs. 37 and 38, we conclude that the REC and the MEC techniques produce comparable subjective results for isolated block losses. This is because if the number of losses in any block is less than or equal to Z, both techniques should be able to recover all the corrupted samples but with different accuracies. Since the human eye is insensitive to the small differences between the recovered pixels and the original ones, the viewer will not be able to notice any differences among the original image and the images recovered using the REC and the MEC techniques.

The subjective quality of the image recovered using the CMAP-LP technique is shown in Fig. 39. We can easily notice the differences among the image

Figure 36. The corrupted "Claire" image (10% isolated block losses).

Figure 37. "Claire" image recovered from isolated block losses using the REC technique with $n = 32$ and $B = 10$.

Figure 38. “Claire” image recovered from isolated block losses using the MEC technique with $n = 32$ and $B = 10$.

obtained using the CMAP-LP technique and the images obtained using the REC and the MEC techniques by referring to areas such as (from the viewer’s perspective): the right side of the head, the right ear, the right shoulder and different regions in the right side of the jacket.

18.5.4. The REC and MEC Techniques Applied to Consecutive Block Losses

For consecutive block losses, the application of the REC and the MEC techniques requires large block sizes to be used in order to recover all the lost blocks. As was previously suggested, the REC and the MEC techniques fail for large block sizes. Therefore, these techniques cannot be used to recover consecutive block losses. In order to use the REC and the MEC techniques to recover the losses in ATM and wireless environments, the nature of these block losses should be converted from consecutive block losses to isolated block losses. One way to do this is to implement a powerful block interleaving mechanism before the transmission of the decoded images.

In order to use the REC and the MEC techniques for consecutive block losses, we low-pass filter the image before transmission as was suggested before. Then, we transpose the low-pass filtered image and encode it using the modified

JPEG encoder as explained in a previous section. The resulting encoded image is transmitted via an error-prone network. When the corrupted image is received and decoded, the decoded image is a corrupted transposed version of the original low-pass filtered image. As expected, the corruption in this decoded image is in the form of horizontally-oriented consecutive block losses. Then, we transpose this image in order to restore the original orientation of the transmitted image. The result is a corrupted version of the original low-pass filtered image with the consecutive block losses oriented in the vertical direction. Figure 40 shows the resulting image after implementing this procedure.

The image in Fig. 40 contains vertically-oriented consecutive block losses with a loss percentage of 7.8%. Since the original image was low-pass filtered before the transpose operation, the REC and the MEC techniques can be used to recover the image in Fig. 40 because block sizes of 32 and 64 can be used in this case.

Figure 41 presents a comparison between the performances of the REC and the MEC techniques in recovering consecutive block losses corrupting "Claire" image. The figure clearly shows that both the REC and MEC techniques produce worse PSNR results when blocks of size 64 are used than when using blocks of size 32. This is due to the sensitivity of the REC and the MEC techniques to round-off errors. Furthermore, the MEC technique has a better PSNR perfor-

Figure 39. "Claire" image recovered from isolated block losses using the CMAP-LP technique with $n = 30$ and $B - 10$.

Figure 40. The corrupted "Claire" image (7.8% consecutive block losses).

mance than the REC technique for the case of consecutive block losses. This is expected since the process of transposing the image converts the horizontally-oriented consecutive block losses to vertically-oriented consecutive block losses. This causes the REC and the MEC techniques to deal with such losses as if they are isolated block losses. Since the results in the previous section confirm that the MEC techniques has a better PSNR performance than the REC technique for isolated block losses, the results in Fig. 41 are verified and conform with the previously reported results.

The REC and the MEC techniques produce much better PSNR results than the CMAP-LP technique even for the vertically-oriented consecutive block losses. Figure 42 supports this claim. For example, the amount of PSNR improvement of the MEC technique over that of the CMAP-LP technique is more than 15 dB for 5 block losses. For more than 50% of consecutive block losses, the REC technique has a comparable performance to that of the CMAP-LP technique.

Finally, we study the subjective performances of the REC, the MEC and the CMAP-LP techniques in combating consecutive block losses. Similar to the results obtained for isolated block losses, blocks of size 32 are used during the low-pass filtering process for the case of consecutive block losses.

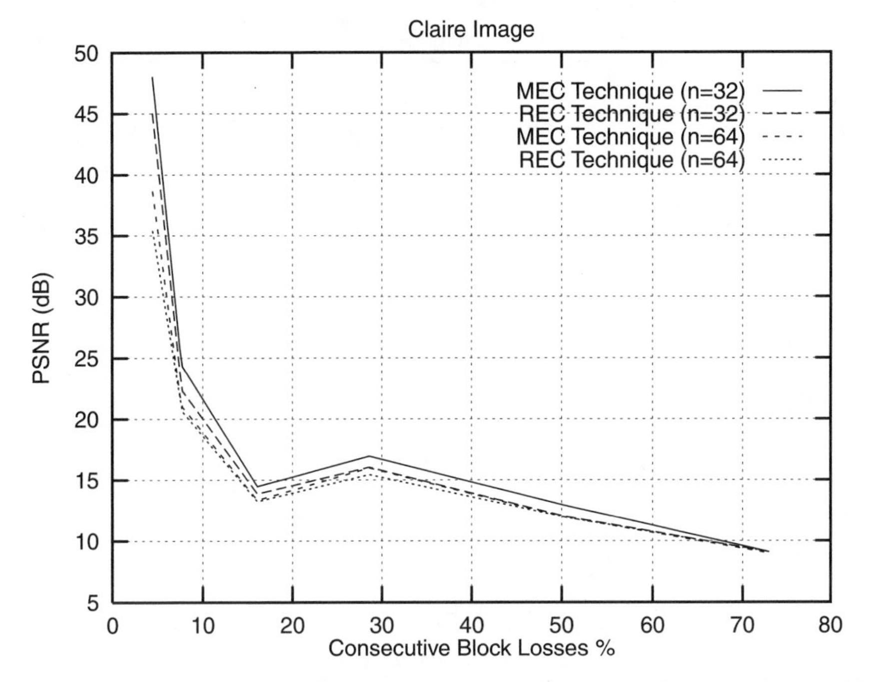

Figure 41. The PSNR of the recovered "Claire" image using the REC and the MEC techniques with block sizes of 32 and 64 for the case of consecutive block losses.

As mentioned before, the REC and the MEC techniques completely fail in recovering horizontally-oriented consecutive block losses and therefore, we convert these losses to vertically-oriented block losses as was explained before.

Figures 43–45 present the subjective qualities of the recovered "Claire" images using the REC, the MEC and the CMAP-LP techniques after being corrupted by the vertically-oriented consecutive block losses shown in Fig. 40. By comparing Figs. 43 and 44, we notice that the REC and the MEC techniques produce comparable subjective results for consecutive block losses. This can be explained using the same reasoning presented in the previous subsection for isolated block losses.

Both, the REC and the MEC techniques, produce better subjective qualities compared to the CMAP-LP technique. Figures 43–45 support this claim. We can easily notice the differences among the image obtained using the CMAP-LP technique and the images obtained using the REC and the MEC techniques by referring to areas such as (from the viewer's prospective): near the right-hand ear of "Claire" and the right-hand side of her jacket. It is obvious that the CMAP-LP technique has a very poor performance in recovering the lost blocks located near

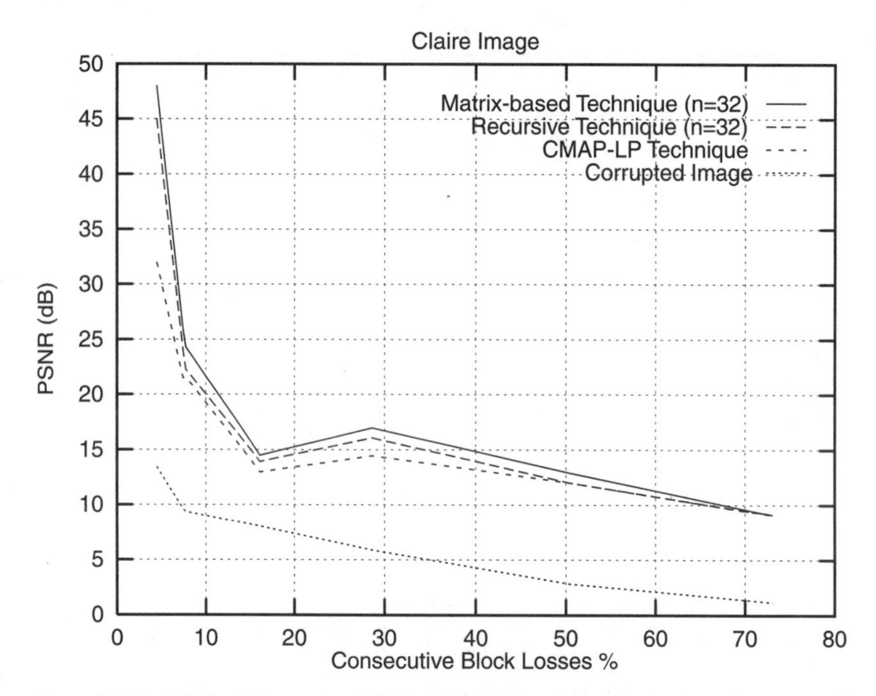

Figure 42. The PSNR of the recovered "Claire" image using the REC, the MEC and the CMAP-LP techniques with block size 32 for the case of consecutive block losses.

Figure 43. "Claire" image recovered from consecutive block losses using the REC technique with $n = 32$ and $B = 10$.

Figure 44. "Claire" image recovered from consecutive block losses using the MEC technique with $n = 32$ and $B = 10$.

Figure 45. "Claire" image recovered from consecutive block losses using the CMAP-LP technique.

any vertical edges while the REC and the MEC techniques can recover such areas with a better quality. The simulation programs are in the CD-ROM: 'Chapter18\'.

Acknowledgment

The authors wish to thank Dr. Atif Sharaf for suggesting the idea of the subimage error concealment technique while he was working towards his Ph.D. in our Multimedia Lab.

References

[1] M. Hasan, A. Sharaf, and F. Marvasti. Subimage Error Concealment Techniques. *Proceedings of the IEEE International Symposium on Circuits and Systems (ISCAS 98), Monterey, CA, USA*, 4:245–248, June 1998.
[2] A. Narula and J. Lim. Error Concealment Techniques for an All-Digital High-Definition Television System. *Proceedings of the SPIE Conference on Visual Communications and Image Processing*, 2094:304–315, 1993.
[3] M. Hasan, A. Sharaf, and F. Marvasti. Novel Error Concealment Techniques for Images in ATM Environments. *Proceedings of the IEEE International Conference on Acoustics, Speech and Signal Processing (ICASSP 98), Seattle, USA*, 5:2833–2836, May 1998.
[4] F. Marvasti, M. Hasan, M. Echhart, and S. Talebi. An Efficient Technique for Burst Error Recovery. *IEEE Transactions on Signal Processing*, 47(4):1065–1075, April 1999.
[5] M. Hasan, F. Marvasti, and M. Echhart. Speech Recovery with Bursty Losses. *Proceedings of the International Conference of Telecommunications (ICT 98), Greece*, vol. II, July 1998, pp. 264–268.
[6] F. Marvasti, M. Hasan, and M. Echhart. An Efficient Technique for Burst Error Recovery. *Proceedings of the International Workshop on Sampling Theory and Applications (SAMPTA'97), Aveiro, Portugal*, June 1997, pp. 161–168.
[7] F. Marvasti. A Unified Approach to Zero-Crossings and Nonuniform Sampling of Single and Multidimensional Signals and Systems. *Oak Park, IL: Nonuniform Publications*, (1st edn), 1987.
[8] R. Blahut. *Algebraic Methods for Signal Processing and Communications Coding*. Springer-Verlag, 1992.
[9] J. Anderson and S. Mohan. *Source and Channel Coding*. Kluwer Academic Publishers: Norwell, MA, pp. 171–174, 1991.
[10] M. Hasan. Error Concealment Techniques Suitable for Still Images in Erasure Channels. Ph.D. Thesis, King's College London, UK, December 1999.
[11] P. Salama, N. Shroff, and E. Delp. A Fast Suboptimal Approach to Error Concealment in Encoded Video Streams. *Proceedings of the International Conference on Image Processing*, Santa Barbara, California, USA, October 1997, vol. 2, pp. 101–104.
[12] M. Hasan and F. Marvasti. Error Concealment of Isolated Random Losses using Linear Prediction. *Proceedings of the International Conference of Telecommunications (ICT 99), Korea*, June 1999.

[13] W. Han and J. Lin. Minimum-Maximum Exclusive Mean (MMEM) Filter to Remove Impulse Noise from Highly Corrupted Images. *Electronic Letters*, 33(2): 124–125, January 1997.

[14] F. Marvasti. Applications of Non-Uniform Sampling. *Proceedings of the Sampling Theory and Applications Workshop (SAMPTA'95)*, Jurmala, Latvia, September 1995, pp. 19–22.

19

Sparse Sampling in Array Processing

S. Holm, A. Austeng, K. Iranpour, and J.-F. Hopperstad

Abstract

Sparsely sampled irregular arrays and random arrays have been used or proposed in several fields such as radar, sonar, ultrasound imaging, and seismics. We start with an introduction to array processing and then consider the combinatorial problem of finding the best layout of elements in sparse 1-D and 2-D arrays. The optimization criteria are then reviewed: creation of beampatterns with low mainlobe width and low sidelobes, or as uniform as possible coarray. The latter case is shown here to be nearly equivalent to finding a beampattern with minimal peak sidelobes.

We have applied several optimization methods to the layout problem, including linear programming, genetic algorithms and simulated annealing. The examples given here are both for 1-D and 2-D arrays. The largest problem considered is the selection of $K = 500$ elements in an aperture of 50 by 50 elements. Based on these examples we propose that an estimate of the achievable peak level in an algorithmically optimized array is inversely proportional to K and is close to the estimate of the average level in a random array.

Active array systems use both a transmitter and receiver aperture and they need not necessarily be the same. This gives additional freedom in design of the thinning patterns, and favorable solutions can be found by using periodic patterns with different periodicity for the two apertures, or a periodic pattern in combination with an algorithmically optimized pattern with the condition that

S. Holm, A. Austeng, K. Iranpour, and J.-F. Hopperstad • Department of Informatics, University of Oslo, P.O. Box 1080, N-0316 Oslo, Norway.
E-mail: sverre.holm@ifi.uio.no

Nonuniform Sampling: Theory and Practice, edited by Marvasti
Kluwer Academic/Plenum Publishers, New York, 2001.

there be no overlap between transmitter and receiver elements. With the methods given here one has the freedom to choose a design method for a sparse array system using either the same elements for the receiver and the transmitter, no overlap between the receiver and transmitter or partial overlap as in periodic arrays.

19.1. Introduction

Sparse arrays are antenna arrays that originally were adequately sampled, but where several elements have been removed. This is called thinning, and it results in the array being under-sampled. Such undersampling, in traditional sampling theory, creates aliasing. In the context of spatial sampling, and if the aliasing is discrete, it is usually referred to as grating lobes. In any case this is unwanted energy in the sidelobe region.

Why would one want to use sparse arrays rather than full arrays? The main reason is economy. Each of the elements needs to be connected to a transmitter and a preamplifier for reception, in addition to receive and transmit beamformers. Medical ultrasound imaging, the field where most of the work to be presented here was done, illustrates this: Conventional 2-D scans are done with 1-D arrays with between 32 and 192 elements. 3-D ultrasound imaging is now in development and this requires 2-D arrays in order to perform a volumetric scan without mechanical movement. Such arrays require thousands of elements in order to cover the desired aperture.

The purpose of the work presented here is to give a coherent presentation of sparse array properties and sparse array design. Both topics have been active areas of research for at least the last thirty years as documented in, for instance, the books [1] and [2]. We have chosen to let the terminology of this chapter be consistent with the latter reference. The main contribution of this work is the application of optimization methods such as genetic programming and simulated annealing to the problem of element placement in 1-D and 2-D arrays. These methods enable one to find solutions that are believed to be near the optimal limits in terms of sidelobe performance. They also make it possible to estimate the lower limits for peak sidelobe level for layout optimized arrays. The estimate for these limits is proportional to $1/K$, where K is the number of remaining elements in the array.

This chapter starts with an introduction to array processing based on the analogy to sampling in the time domain. Topics that do not have their parallels in time-domain sampling such as the effect of the element response, steering, grating lobes, and the coarray are covered. The important distinction between one-way and two-way responses is described and later used to give more degrees of freedom in the optimization. Theory for random arrays and a subclass of random arrays called binned arrays is then covered.

We then move on to optimization of either element weights or element layout or both. The layout problem is shown to be a combinatorial problem of such a large magnitude that an exhaustive search will never be possible. Different criteria for optimization are then reviewed and we show through an example that criteria in the coarray domain are nearly equivalent to minimizing sidelobe level. Examples of weight and layout optimization for relatively small-size 1-D arrays are then given. Some new results with a lower sidelobe level than previously reported for the problem of finding the best 25 elements in an aperture with 101 elements are then given. Large 2-D array problems are then considered and it is shown that the optimization region in the angular domain has to include some invisible regions in order for the array to be steerable. Some results obtained from simulated annealing and genetic optimization are then presented. Finally we give some results where the two-way beampattern is optimized, allowing one to use different sampling patterns for the transmitter and receiver.

In the appendix, the three optimization methods used here, i.e., linear programming, simulated annealing, and the genetic algorithm, are briefly described.

19.2. Theory

19.2.1. Introduction to Array Processing

19.2.1.1. The Array Pattern as a Spatial Frequency Response

In time-frequency signal processing, a filter is characterized by values of its impulse response, h_m, spaced regularly with a time T between samples. A linear shift-invariant system is also characterized by the frequency response

$$H(e^{j\omega T}) = \sum_{m=0}^{M-1} h_m e^{-jm\omega T} \tag{1}$$

which is given in (1) for a finite length impulse response with M samples. The relationship between the sampling interval, T and the angular frequency, ω, in order to avoid ambiguities, is that the argument in the exponent satisfies $\omega T \leq \pi$. This is a statement of the sampling theorem.

In array signal processing, the aperture smoothing function plays the same role in characterizing an array's performance. Assume that the M elements are regularly spaced with a distance d and are located at $x_m = m \cdot d$ for $m = 0, \ldots, M-1$ as in Fig. 1. This is a 1-dimensional linear array or a uniform

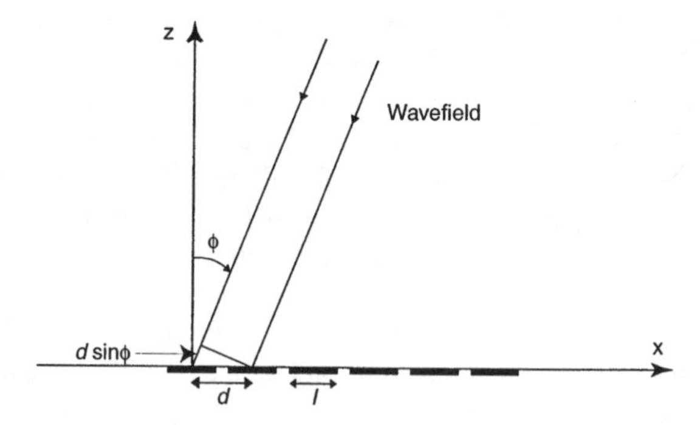

Figure 1. Uniform linear array with element distance d, element length l, and a wave arriving from direction ϕ.

linear array. Its aperture smoothing function when each element is weighted by the scalar w_m is

$$W(u) = \sum_{m=0}^{M-1} w_m e^{-jm2\pi(u/\lambda)d} \tag{2}$$

The variable u is defined by $u = \sin\phi$ where ϕ is the angle between broadside of the array and the direction from the wavefield (usually called azimuth angle), λ is the wavelength, and the weights, w_m, is a standard window function [3]. With reference to Fig. 1, (2) can be found from geometry. For a wave coming from an infinite distance, the difference in travel-distance between two neighbor elements is $d \sin\phi$. When this is converted to phase angle, where one wavelength of travel-distance corresponds to 2π, one gets the expression in the exponent of (2).

The aperture smoothing function is, therefore, the output after weighting and summing all elements in the array for a wave from infinite distance hitting the array at an angle of incidence ϕ. The aperture smoothing function determines how the wavefield Fourier transform is smoothed by observation through a finite aperture [2], just like the frequency response determines how the received signal spectrum is smoothed by the filtering operation. The condition for avoiding aliasing is that the argument in the exponent satisfies

$$2\pi \frac{|u|}{\lambda} d = |k_x| \cdot d \leq \pi \tag{3}$$

where $k_x = 2\pi u/\lambda$ is the x-component of the wavenumber.

The relationship between the array pattern for a regular 1-D array and a filter frequency-response is now

$$\begin{array}{ccc} \omega & \leftrightarrow & k_x = 2\pi(u/\lambda) \\ T & \leftrightarrow & d \\ h_n & \leftrightarrow & w_n \end{array}$$

By using these parallels, the time-frequency sampling theorem $T \leq \pi/\omega_{\max}$ translates into the spatial sampling theorem $d \leq \lambda_{\min}/2$.

19.2.1.2. Array pattern for arbitrary geometry

The spatial frequency k_x can be generalized for an array with elements located anywhere in space and with arbitrary irregular geometry. Let the wavenumber vector be $\vec{k} \in \mathbb{R}^3$ with norm $|\vec{k}| = 2\pi/\lambda$, and let it be directed from the source towards the array as in Fig. 2. This figure also defines a unit direction vector $\vec{s}_{\phi,\theta} = (\sin\phi\cos\theta,\ \sin\phi\sin\theta,\ \cos\phi) = (u, v, \cos\phi)$ in rectangular coordinates. These angles are usually called azimuth angle for ϕ and elevation angle for θ. These terms come from sidelooking radar, but are used in other applications also.

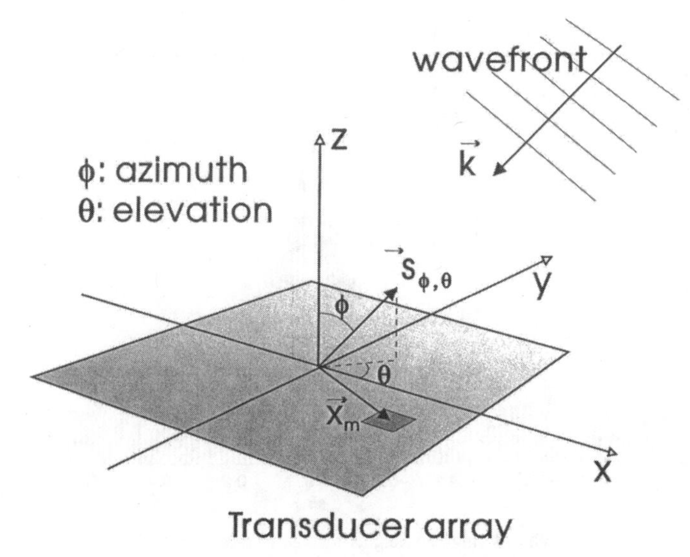

Figure 2. A 2-D planar array with coordinate system.

The wavenumber vector is now $\vec{k} = -2\pi\vec{s}_{\phi,\theta}/\lambda$ and the array pattern can be generalized to

$$W(\vec{k}) = \sum_{m=0}^{M-1} w_m e^{j\vec{k}\cdot\vec{x}_m} = \sum_{m=0}^{M-1} w_m e^{-j2\pi/\lambda(ux_m+vy_m+\cos\phi z_m)} \tag{4}$$

where the array element locations are $\vec{x}_m = (x_m, y_m, z_m) \in \mathbb{R}^3$ with the corresponding weights $w_m \in \mathbb{R}$. The weighting function is often called windowing, shading, tapering or apodization. The relationship between the general array pattern and that for the linear 1-dimensional array (2) can be found by setting the element position to be on the x-axis only: $\vec{x_m} = (m \cdot d, 0, 0)$.

In the following, the notation $W(\vec{k})$ will be used for the array pattern for a general geometry while $W(u)$ will be used for a one-dimensional geometry, with $u = \sin\phi$. When a 2-D planar array is considered, one usually uses $W(u, v)$ where $(u, v) = (\sin\phi\cos\theta, \sin\phi\sin\theta)$.

An example of the array pattern of a 1-D array with uniform weighting is shown in Fig 3. The array pattern is characterized by properties of the main lobe and the sidelobes. The main lobe width is usually measured either at the -3 dB point or the -6 dB point. In the chapter we will use the latter which for a 1-D array is given by $W(u_{-6\,\mathrm{dB}}/2) = W(\sin\phi_{-6\,\mathrm{dB}}/2) = 0.5$. For a full array with uniform weights, the beamwidths are given as $\phi_{-3\,\mathrm{dB}} \approx 0.89\lambda/D$ and $\phi_{-6\,\mathrm{dB}} \approx 1.22\lambda/D$ where D is the extent of the aperture. The sidelobe region is characterized by e.g. the peak value.

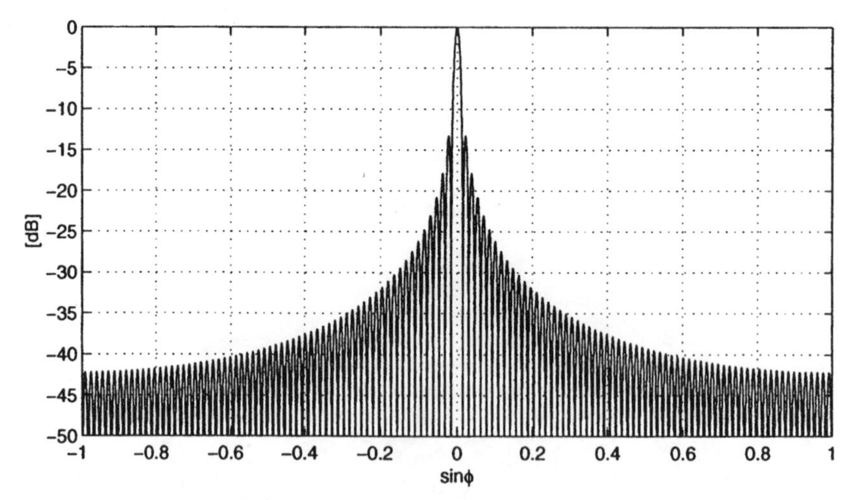

Figure 3. Array pattern with rectangular weights for 128 element array with $\lambda/2$ spacing, maximum sidelobe level -13.3 dB, beamwidth (-6 dB) 1.09°.

19.2.1.3. Periodic Arrays and Grating Lobes

For the important class of arrays that have their elements on an underlying regular grid, aliasing just like in time-frequency signal processing occurs. Equation (2) gives the array pattern. Like all regularly sampled systems, the array pattern is periodic, and the periodicity is given by the argument in the exponent repeating itself by 2π. This is equivalent to

$$u_n = u_0 + n\frac{\lambda}{d} \quad \text{for } n = \ldots, -1, 0, 1, \ldots \tag{5}$$

where u in (2) is now called u_n due to the possible repetitition in the array response. The distance between the elements relative to the wavelength is what matters. Recall now the spatial sampling theorem, $d = \lambda/2$. In this case $u_n = u_0 + 2 \cdot n$. Since u_0 is the sine of an angle, and in order for there to be aliasing, u_n must also be a valid sine of angle, the sampling theorem implies that only $n = 0$ is possible and there is no ambiguity in the array pattern.

This is changed if the system is undersampled. Let for instance $d = \lambda$. Now $u_n = u_0 + n$. A system with a response at $u_0 = 0$ will repeat the response at $u_{-1} = -1$ and $u_2 = 1$ as shown in Fig. 4. This example is actually a thinned array made from that in Fig. 3 by removing every second element. The two extra

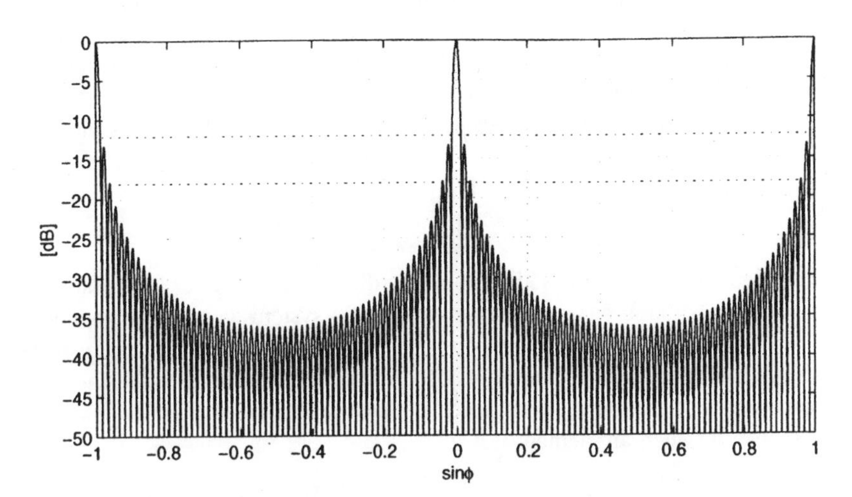

Thinning pattern (50.0% thinned):
0101010101010101010101010101010101010101010101010101010101010101
0101010101010101010101010101010101010101010101010101010101010101

Figure 4. Array pattern with grating lobes due to every other element missing ($d = \lambda$), beamwidth (-6 dB) $1.09°$.

responses are called grating lobes due to the parallel with a similar phenomenon in optical diffraction gratings.

19.2.1.4. Element response

Consider the situation in time-domain sampling where an analog signal is sampled by a non-ideal sample-and-hold circuit. Instead of impulse sampling, the sampler will average over a small time window, resulting in a low-pass filtering of the sampled data. The low-pass response will be multiplied with the spectrum of the data in order to get the final spectrum.

This has a parallel in array processing in the element response. So far it has been assumed that there are M point elements, each of them being omnidirectional. However, each element may, due to its size, have its own directivity. This is described by the element response

$$W_e(\vec{k}) = \int_{-\infty}^{\infty} w(\vec{k}) e^{j\vec{k}\cdot\vec{x}} d\vec{x} \tag{6}$$

The extent of the aperture is determined by the support for the aperture weighting function, $w(\vec{k})$. For a regular, linear array with element distance d, and non-overlapping elements, the element may be slightly smaller than the element distance, i.e., it is defined in the interval $\langle -l/2, l/2 \rangle$ where $l \leq d$ (see Fig. 1).

As in time-domain sampling, the total response for the array system is the combined effect of the element response and the array pattern. In the case that the elements are equal and one operates in the far-field of the array, it is the product of the two

$$W_{\text{total}}(u, v) = W_e(u, v) \cdot W(u, v) \tag{7}$$

These conditions are only satisfied for a uniform linear array or for a planar linear array. Arrays that are curved are examples of a system where (7) does not hold.

An example of an array response with grating lobes and element response is shown in Fig. 5. This example is based on the array of Fig. 4 and the element response of a uniformly weighted element with size $l = \lambda/2$. The element response for such an element is

$$W_e(u) = \frac{\sin(\pi(l/\lambda)u)}{\pi(l/\lambda)u} = \operatorname{sinc}\left(\frac{lv}{\lambda}\right) \tag{8}$$

Note the similarity between Fig. 5 and Fig. 3, however, this similarity vanishes when the array is steered as will be seen in the next section.

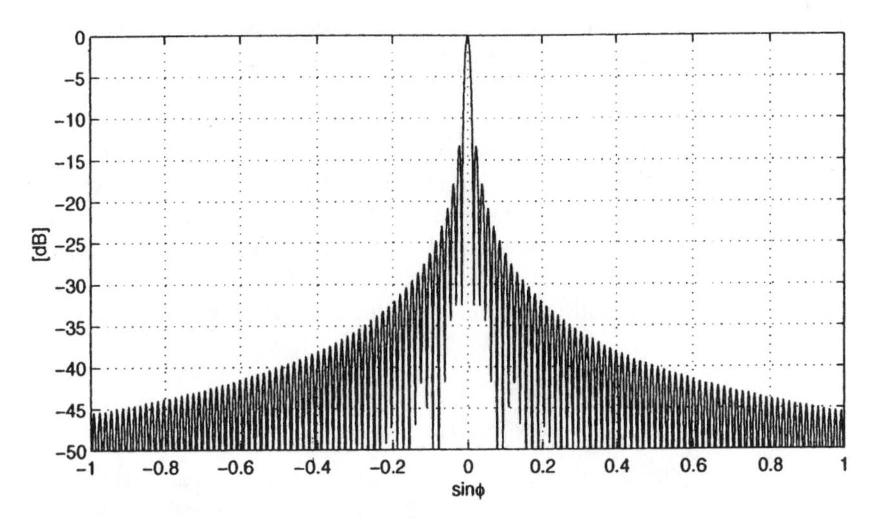

Figure 5. Response for an array with grating lobes and element response ($d = \lambda, l = \lambda/2$).

19.2.1.5. Beampattern

A beamformer sums each output from the array with appropriate delays and weights. The delays are found by considering a certain direction given by $\vec{k}^0$ and compensating for the difference in travel time between the elements. In a 1-dimensional array this corresponds to a certain direction, ϕ^0. A beamformer may also be used to focus the beam on a point in the nearfield of the array given by both the angle and the distance. This is routinely done in medical ultrasound imaging. In any case the delays are found from geometrical considerations by taking into account the velocity of propagation in the medium. In most applications such as ultrasound, radar and sonar, the medium can be assumed to be homogenous with a constant velocity of propagation.

If the array output is processed by a beamformer with delays set to match a certain direction and wavelength given by $\vec{k}^0$, the beampattern will simply be a shifted version of the array pattern

$$W(\vec{k} - \vec{k}^0) \tag{9}$$

Consider a regular linear array with element spacing d. The vector-product in (4) then simplifies to $(\vec{k} - \vec{k}^0) \cdot \vec{x}_m = (2\pi/\lambda)md(-u)$ where

$$u = \sin\phi - \sin\phi^0 \tag{10}$$

for ϕ^0 defined as the angle between the broadside direction and the steered direction. In this special case the beampattern is given by (2) with u defined by (10).

The total response for the array system is the combined effect of the element response and the beamforming. When they are separable, it is given by $W_e(u, v) \cdot W(u - u^0, v - v^0)$. Note that only the beampattern is affected by the steering, the element response is not possible to change by beamforming. This is illustrated in Fig. 6 which is the array of Figs. 4 and 5 with steering. Now the grating lobes reappear. If we had instead steered the full array of Fig. 3, the result would instead have been just a translation of the response along the axis.

19.2.2. One-Way and Two-Way Beampatterns

The array patterns and beampatterns discussed so far have all been for the case depicted in Fig. 1, that is a source that transmits a signal which is received by the array. This is a one-way scenario. Due to reciprocity in a linear medium, it could equally well have been the other way around, i.e., the array could have been configured as a transmitter and the receiver could have been located in the far-field of the array. In either case, the array patterns and beampatterns would have described the transfer function.

A two-way scenario is when a signal is transmitted by the array, it is reflected off a target, and then received by the array. In this case, the two-way beam pattern

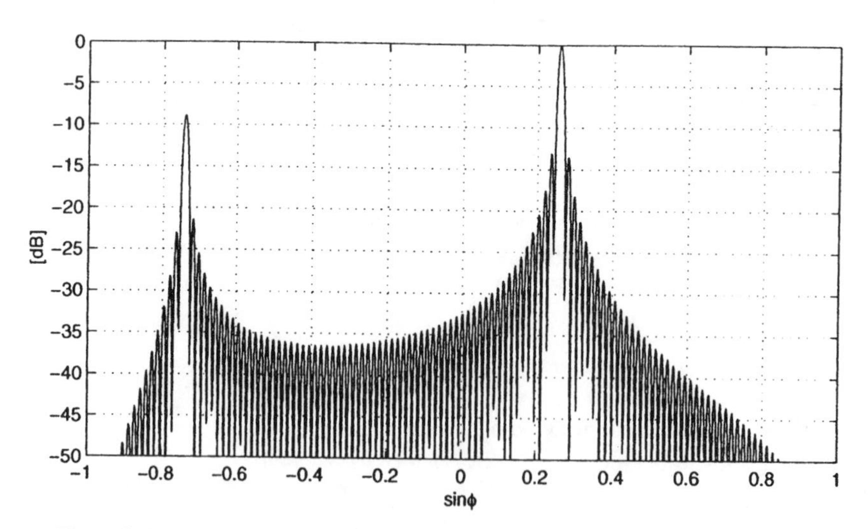

Figure 6. Beampattern with steering to $\phi = 25°$ with grating lobes and element response.

is the product of the receiver's and the transmitter's beam patterns. If the same array is used, they will be equal and one gets

$$|W_{TR}(u, v)| = |W_e(u, v) \cdot W(u - u^0, v - v^0)|^2 \tag{11}$$

where the subindex TR stands for transmit and receive. We are rarely interested in the phase of the beam patterns, and therefore only the magnitude is shown here.

The previous array patterns and beampatterns in Figs. 3–6, all show the one-way pattern. By simply squaring them, i.e., doubling the dB-axis, the two-way patterns can be found.

Equation (11) is valid under the condition that the receiver and transmitter arrays are the same, that the waveform is a continuous single-frequency wave, that the reflecting target is in the far-field of the array, and that the medium is linear. These conditions will be more or less satisfied in different kinds of imaging systems.

When optimizing the response from sparse arrays, more degrees of freedom are obtained if one lets the receiver and transmitter arrays be different. In that case the two-way beampattern will be

$$|W_{TR}(u, v)| = |W_{e,T}(u, v) \cdot W_T(u - u^0, v - v^0) \cdot W_{e,R}(u, v) \cdot W_R(u - u^0, v - v^0)| \tag{12}$$

where the subindices T and R stand for transmitter and receiver respectively.

19.2.2.1. The coarray and sparse arrays

As an alternative description of an array, the coarray may be used. Rather than describing the angular response, it describes the morphology of the array. It was first introduced in [4] and for arrays with elements on a regular grid, the coarray is defined as the autocorrelation of the element weights

$$c(l) = \sum_{m=0}^{M-|l|-1} w_m w_{m+|l|} \tag{13}$$

The coarray describes the weight with which the array samples the different lags of the incoming field's correlation function. For a linear array with element distance d, the coarray is related to the squared array pattern through the Fourier transform

$$|W(k)|^2 = \sum_{l=-(M-1)}^{M-1} c(l) e^{jl2\pi(u/\lambda)d} \tag{14}$$

A full array has a smooth-looking coarray and with unity weights it is triangular. Our interest here is to study it for sparse arrays with K remaining elements out of the original M elements in the full aperture.

In order to characterize the coarray, some definitions are required. A redundant lag, l, is when the coarray of that lag is greater than unity, $c(l) > 1$. The opposite is a hole. In that case the coarray is zero at that lag, $c(l) = 0$. In order to have an even sampling of the incoming wave field, it seems natural to require a coarray with the same weight for all lags. A perfect array is such an array. It is defined as an array with a coarray with no holes or redundancies except for lag zero. Unfortunately, perfect arrays only exist for four or fewer elements in the array. Therefore, we study arrays that approximate perfect arrays; the Minimum Redundancy (MR) and the Minimum Hole (MH) arrays. They are defined by the number of redundancies, R, and holes, H. Minimum redundancy arrays are those element configurations that have no holes and minimize the number of redundancies. Minimum hole arrays minimize the number of holes in the coarray without any redundancies. These arrays are also known as Golomb rulers [5].

The smallest minimum hole and minimum redundancy arrays are given in Table 1 where the results have been taken from [6] and [7]. Finding such arrays is a formidable task as the aperture grows. The largest proven minimum hole array that has been published is of size $K = 19$, [6] (as of this writing there are claims on the World Wide Web that the $K = 20$ and $K = 21$ minimum hole arrays also have been proven to be optimal). The largest known minimum redundancy array is of size $K = 17$ [7].

Table 1. Table of the first set of minimum hole and minimum redundancy arrays.

K	Minimum hole	Minimum redundancy
3	1101 (perfect)	1101 (perfect)
4	1100101 (perfect)	1100101 (perfect)
5	110010000101	1100100101
	100110000101	1001000111
6	11001000001010001	1100110000101
	11001000001000010 1	1100001001010 1
	11000000100101000 1	1110001000100 1
	11000000100010100 1	—
7	1100100000100000000100001 01	1100100000101001 01
	1100000100010000000001001 01	1110001000100010 01
	1011000000010000001000010 001	1111000010000100 01
	1010001001000010000000000 11	1110000001000101 001
	1001100000000100000010000 101	1100000001001010 101
8	11001000010000001000000010000000000101	110010000010000010100101
	—	111000000000100010010 0101

Examples of coarrays of minimum redundancy and minimum hole arrays are shown in Fig. 7. A generalization of the coarray that corresponds to the case when different apertures for transmit and receive are used is also possible. In this case one must build on (12) which describes the two-way beampattern. Disregarding the element patterns and the steering one gets

$$|W_{TR}(u)| = |W_T(u) \cdot W_R(u)| \tag{15}$$

The inverse Fourier transform of this expression corresponds to the sum coarray [8] which is equivalent to the convolution of the transmit and receive apertures. In other contexts this has been called the effective aperture [9].

$$c_{12}(l) = \sum_{m=0}^{M-|l|-1} w_m w_{m-l} \tag{16}$$

An example of such a coarray is shown in Fig. 8. An aperture of 128 elements is assumed for this example. Every second element is used for transmission (as in Fig. 4), and every third element for reception. Due to reciprocity, the roles of the receiver and transmitter arrays could have been exchanged without any effect on the result.

19.2.3. Random Arrays

Strictly speaking, a random array is described by a probability density function, $p_{\vec{x}}(\vec{x})$ which determines the random sensor positions. This differentiates it from a sparse array which is based on a conventional array with regular spacing between elements, and where a certain fraction of the elements are removed at random. The array patterns of the two have the same statistical properties and one often uses the term random to apply to both of them [1]. This will be done here also.

Assuming that the elements are unweighted, the one-way array pattern is

$$W(\vec{k}) = \sum_{m=0}^{K-1} e^{j\vec{k} \cdot \vec{x}_m} \tag{17}$$

The position variables are now random variables. The randomness disappears when $\vec{k} = \vec{0}$, and then $|W(\vec{0})|^2 = K^2$. This occurs as one looks broadside to a 1-D array or a 2-D planar array, i.e., for $\phi = 0$ in the direction vector. For other directions, one should sum K unit random vectors. In the case that they are uncorrelated, the power sum is K. This applies to the sidelobe region well away

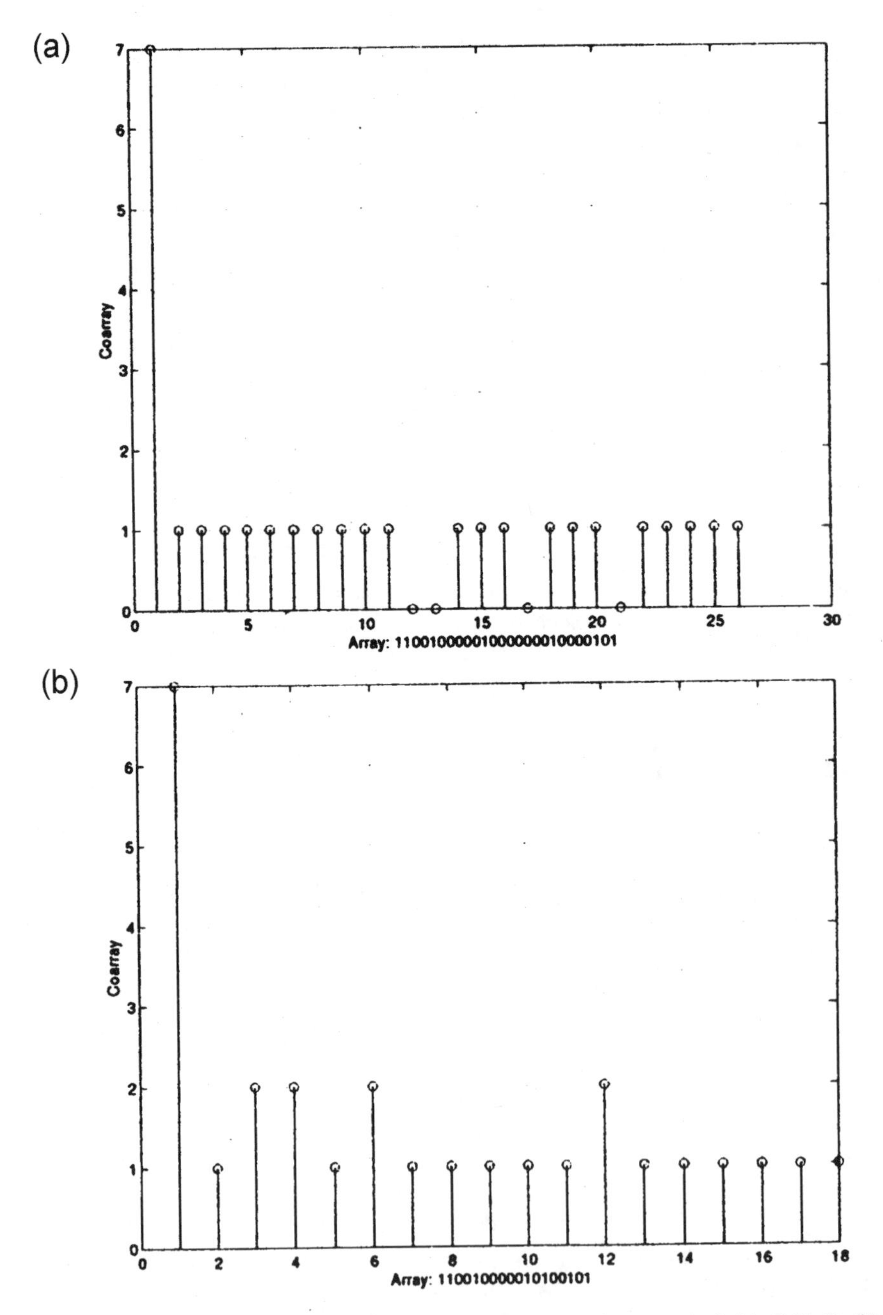

Figure 7. (a) Coarray of minimum hole and minimum hole array (first entry for k-7 in Table 1); (b) (array ofminimum redundancy array (first entry for k-7 in Table 1).

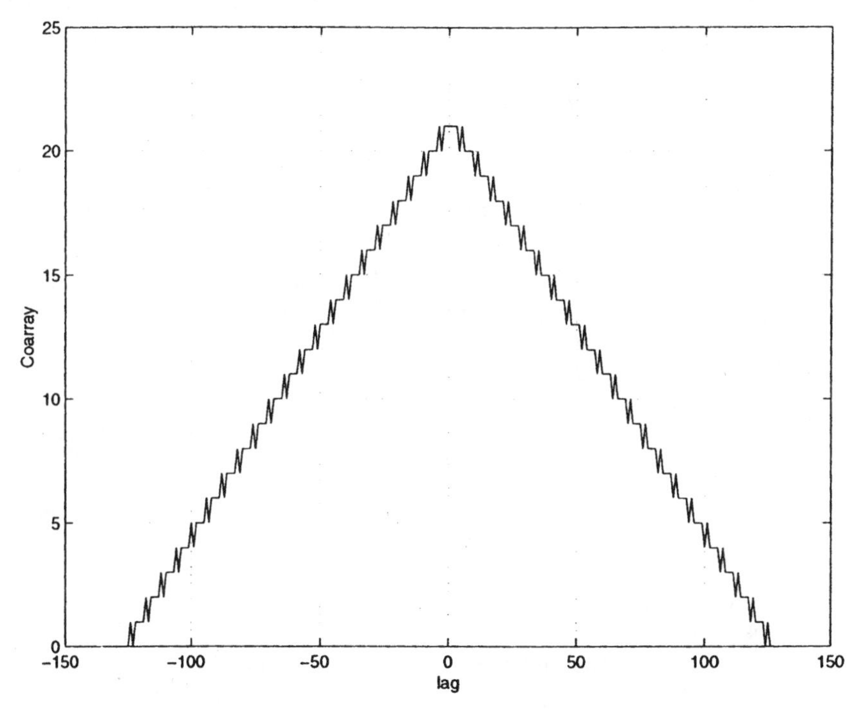

Figure 8. Coarray for a 128 element array with every other element used for transmission and every third element for reception.

from the mainlobe. Thus the ratio of average sidelobe power to main lobe power is [10], [1]

$$\text{Norm. Avg. Sidelobe} = K/K^2 = 1/K \tag{18}$$

By taking the expected value of the array pattern from (17), a more accurate statistical analysis can be performed

$$E[W(\vec{k})] = K \cdot E[e^{j\vec{k}\cdot\vec{x}_m}] = K \cdot \int_{\text{aperture}} p_{\vec{x}}(\vec{x})e^{j\vec{k}\cdot\vec{x}}d\vec{x} \tag{19}$$

The average array pattern for a random array is, therefore, equal to the array pattern of a continuous aperture of the same size with the probability density function playing the same role as the weighting function—compare to (6).

The variance is

$$\begin{aligned}\mathrm{var}[W(\vec{k})] &= K \cdot \mathrm{var}[e^{j\vec{k}\vec{x}_m}] \\ &= K \cdot E[|e^{j\vec{k}\vec{x}_m}|^2] - K \cdot |E[e^{j\vec{k}\vec{x}_m}]|^2 \\ &= K - \frac{1}{K}|E[e^{j\vec{k}\vec{x}_m}]|^2 \end{aligned} \tag{20}$$

In order to discuss these results, let us consider an example where the elements are uniformly distributed over a 1-dimensional linear aperture of length L. In this case the average array pattern is

$$E[W(u)] = K \cdot \mathrm{sinc}(Lu/\lambda) \tag{21}$$

and the variance is

$$\mathrm{var}_U[W(u)] = K \cdot (1 - \mathrm{sinc}^2(Lu/\lambda)) \tag{22}$$

Thus for small arguments the average array pattern is K and the variance is 0. For large values of u, however, the average array pattern is about 0, and the variance is close to K. This confirms the result given previously for the ratio of average sidelobe power to main lobe power to be $1/K$. For the uniform distribution, this result is valid approximately after the first null of the average array pattern, or for $|u| > \lambda/L$. Similar results can be found for other probability density distributions.

A comparison can be seen in Fig. 9 with uniform distribution of the element positions and the one in Fig. 10 with a triangular distribution. The latter has a wider mainlobe beam and lower first sidelobes. The conclusion is, therefore, that the probability distribution of the random distribution of the elements, or that of the thinning for a sparse array, determines the mainlobe shape and the first few sidelobes. Further away from the mainlobe the sidelobes can only be described in a statistical sense and the number of elements determines the average level.

An estimate of the relative peak level of a 1-D random array, derived in [11] and [12], is $\sqrt{(K \ln K)}$. This estimate gives in our experience a fairly good estimate of the peak level and is, therefore, plotted with the estimate $1/K$ for the average value on all of the beampatterns for thinned 1-D arrays.

Steinberg ([13], [1]) has developed a statistical description for the sidelobe pattern and the expected peak sidelobe level in the random array response. His theory suggests that the amplitude of the peak sidelobe is logarithmically proportional to the number of independent samples in the sidelobe region. Accordingly, the peak sidelobe amplitude is expected to be 3 dB higher in planar arrays with the same number of elements and the same dimensions compared to linear arrays.

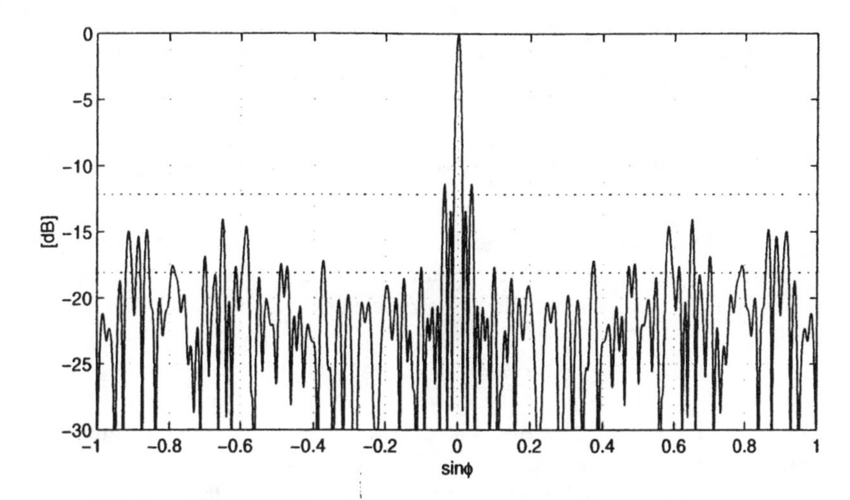

Thinning pattern (50.0% thinned):

11011101011001001111011000101110100001000110001010101101001111111
00101100100110010101101000100000100101100011010101101010101011001

Figure 9. One-way array pattern of sparse array thinned from $M = 128$ to $K = 64$ elements with uniform probability density distribution. In this and the following sparse array beampatterns an estimate of the average and the peak levels are also plotted.

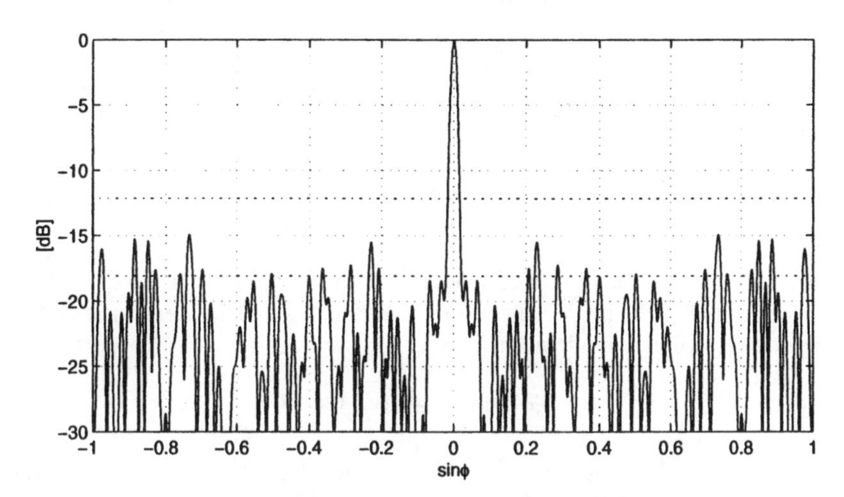

Thinning pattern (50.0% thinned):

1100010000001010000001100010011011011110110100111101011111011111
1111111110111001110000101101000001010001010010000001010100000101

Figure 10. One-way array pattern of sparse array thinned from $M = 128$ to $K = 64$ elements with triangular probability density distribution.

The hypothesis of our work here is that the achievable peak sidelobe level in an algorithmically optimized sparse array is proportional to the average value in the random case or $1/K$ for 1-D. For 2-D arrays our estimate is twice as high.

19.2.4. The Binned Random Array

An interesting variant of the random array is the binned random array. It is equivalent to jittered random time-domain sampling. Consider a one-dimensional aperture of length L. Divide the aperture into N equal-size, non-overlapping bins of length $w = L/N$. The position of each element can be found from

$$x_m = -L/2 + m \cdot w + y_m \quad \text{for } m = 0, \ldots, K-1 \tag{23}$$

The random variable y_m is distributed in the interval $(0, w)$ according to some probability density function.

The average array pattern cannot in general be found except for the important case of a uniform distribution in each bin. Statistically, this is equivalent to a uniform distribution over the full aperture, and the average array pattern is the same as for a random array, i.e., eq. (21) applies [11]. Therefore, the mainlobe and the nearest sidelobes are the same as for a random array with uniform distribution of the position of the elements.

Under the same conditions, the variance can also be found

$$\begin{aligned}
\mathrm{var}_B[W(u)] &= \sum_{m=0}^{K-1} \mathrm{var}[e^{j2\pi(-L/2+m\cdot w+y_m)u}] \\
&= \sum_{m=0}^{K-1} |e^{j2\pi(-L/2+m\cdot w)u}|^2 \cdot \mathrm{var}[e^{j2\pi y_m u}] \\
&= K\,\mathrm{var}[e^{j2\pi y_m u}] = K\mathrm{var}[e^{j2\pi K\cdot y_m u/K}]
\end{aligned} \tag{24}$$

Because $K \cdot y_m$ is uniform distributed over the interval $(0, L)$ just like the random variable in (22), the final result is that the variance is a scaled version of that for a uniform distributed random linear array

$$\mathrm{var}_B[W(u)] = \mathrm{var}_U[W(u/K)] \tag{25}$$

This is a remarkable result because the variance does not reach a full maximum until the first zero of $\sin(\pi(L/\lambda)u/K)$ or for $|u| > K\lambda/L$. This means that the binned array has a much larger region around the steered direction where the effect of the randomness is small. In Fig. 11 it means that the variance does not reach its full value until $|u| = 0.5$. Note also that the nature of the binning is such that not more than two elements can be clustered. This makes the binned array

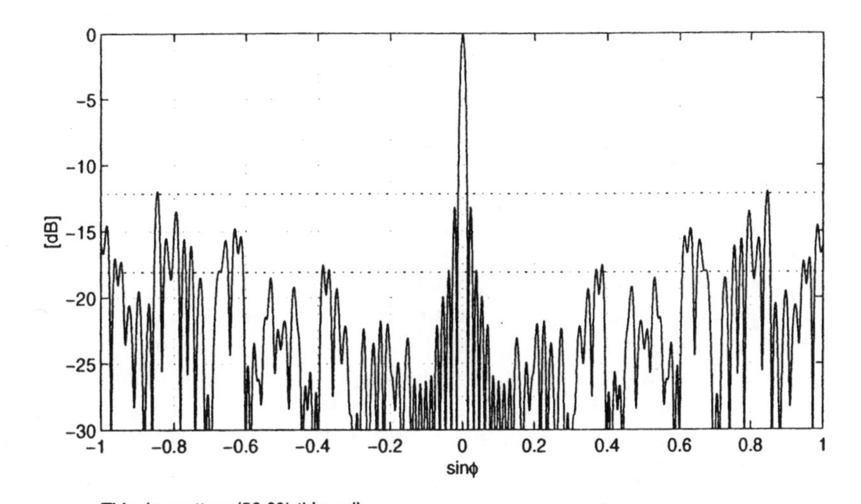

Thinning pattern (50.0% thinned):
0110101010011010101010101001011010010101101010101001101001100110
1001011001100101011010010101011010011001011001101001011010100110

Figure 11. One-way array pattern of a binned sparse array thinned from $M = 128$ to $K = 64$ elements, bin size $w = 2$. The number of elements is the same as in Figs. 9 and 10.

resemble an array with a nearest neighbor restriction, and actually the sidelobe depression just described was first reported for nearest neighbor restricted arrays in [14].

19.3. Optimization of Sparse Arrays

Given that a random sparse array has such a large variation in peak sidelobe level, it is natural to ask if it is possible to find arrays with good sidelobe behavior. Because we are dealing with sparse arrays we are concerned with arrays with elements on a regular grid. Further we restrict ourselves to one or two dimensions and uniform linear arrays or uniform planar arrays. There is no principal problem in also dealing with arrays with their elements uniformly distributed along a regular curve, such as a part of circle (curved linear arrays) or a 3-D array with elements on a spheroid. The variables to optimize can be the element weights (w_m in (4)), or the active element positions.

Let us first consider element weighting for sparse random arrays. This resembles the design of weighting functions for fully sampled arrays or for time series as for instance discussed in the overview paper by Harris [3]. Many different criteria for optimization are to be found there, but the two most relevant ones are minimization of the maximum sidelobe and minimization of the sidelobe energy. For a full array, the first criterion leads to Dolph-Chebyshev weighting, and the latter leads to the prolate-spheroidal weighting, which can be approximated by the Kaiser-Bessel window.

In spectral analysis, the first criterion minimizes the effect of spectral leakage from discrete frequency components. The second criterion is related to the estimation of a low spectral level in a background of broad-band noise at the other frequencies. This situation is not so common in spectral estimation. In imaging systems such as medical ultrasound systems, minimization of the maximum sidelobe is a criterion which is related to imaging of a strong reflecting point target in a non-reflecting background containing other point targets. A typical scenario is imaging of point targets in water. Although this is not a clinically relevant imaging scenario, it is typical for testing of imaging systems. However, in certain organs of the human body, the imaging scenario may approximate this situation. This applies for instance to imaging of valve leaflets inside the fluid-filled cardiac ventricles. The alternative criterion of minimization of the integrated sidelobe energy is directly related to image contrast when imaging a non-reflecting area like a cyst or a ventricle in a background of reflecting tissue. This is found much more often in the human body than the previous scenario. The minimum sidelobe energy criterion must be combined with a restriction on the peak sidelobe for it to be tractable, see [15]. Some results on weight optimization for 1-D arrays using this criterion and quadratic optimization have been reported in [16].

Here we will first find the properties of arrays based on minimization of the peak sidelobe, because this has been the most common criterion so far, and it is straightforward to formulate optimization algorithms for it. In [17] we showed that it is possible to find apodization functions or element weights for a given thinning pattern that give the beampattern optimal properties. An important result is that these functions have little or no resemblance with the corresponding full array's apodization function. A limitation of this work was that is was not possible to optimize the full angular extent of the sidelobe region for a sparse array. This was due to the algorithm used (Remez exchange algorithm). In [18] and [19] this approach was extended from 1-D to 2-D arrays, and improved results were reported. By using the linear programming algorithm for optimization, it was possible to optimize the whole sidelobe region. In this way it was possible to find properties of the beampattern of such arrays. Of special interest is to determine the minimum peak sidelobe level and compare it with the predictions from random theory.

It is possible to search either for real weights or for complex, unit norm weights. The latter is an optimization of phase and has been done for full arrays in [20]. The disadvantage is that it is essentially a single-frequency optimization.

The phases will be different for different frequencies, while real weights are valid for broadband signals. Therefore, real weight optimization will be the approach used here.

We will also consider optimization of the element positions of a sparse array. The array will then no longer be random, and it is more relevant to call it algorithmically optimized. This problem is considerably more difficult than weight optimization. Joint optimization of positions and weights is also possible, usually by iterating over a sequence of position optimization followed by weight optimization [19, 21]. The reason why element position optimization is so difficult can be seen by considering the number of combinations to search. For an array with M elements, the number of combinations when a subset of K elements are to be picked are

$$\binom{M}{K} = \frac{M!}{(M-K)!K!} \tag{26}$$

An array with 50×50 elements is typical of the requirement for a 2-D array for medical ultrasound imaging. If between 10% and 50% of the elements are to be kept, such an array gives between 10^{350} and 10^{750} combinations. Considering that the estimated number of electrons in the universe is about 10^{80}, it is easy to understand why an exhaustive search is out of the question.

There are several ways that this number can be reduced. First, due to the property that real functions have the same Fourier transform when they are mirrored, we can reduce the number of combinations to half. However, this reduction does not really contribute much to making the combinatorial problem more tractable. The second is to require symmetry in the array, this will result in the array pattern becoming a real function. In fact this is required for all optimization using linear programming. This will reduce the number of elements to search over to 50% (M and K will both be reduced to 50%). For the previous example the result is that between about 10^{175} and 10^{375} combinations will have to be searched.

Another way, that especially applies to 1-D arrays, is to require the end elements always to be active. This is a way to ensure that the aperture of the thinned array is maintained and that the algorithm does not just degenerate to finding an array with all the elements clustered at the center of the aperture of the original array. Such an array would have excellent sidelobe properties, but the width of the mainlobe would be inferior. The search space in this case does not diminish significantly since all it means is that M and K in (26) are both decreased by 2. For a 2-D array, it is hard to think of a similar way to fix the ends of the aperture. In any case, this is not a significant source of reduction of the size of the combinatorial problem.

A final way would be to require that the array should be a binned array. In this case M has to be divisible by K and the number of combinations to search is reduced to K independent problems, each of the size of a bin, M/K. The number of combinations is

$$\binom{M/K}{1}^K = \left(\frac{M}{K}\right)^K \tag{27}$$

If 10% or 50% of $M = 2500$ elements are to be kept, this gives 10^{250} and 10^{376} which is a considerable reduction over the full problem.

Other related work on joint optimization of thinning pattern and weights has been reported in the context of sonar arrays in [21] and [22]. Like all of the previously cited papers, our approach is based on allowing elements only on a fixed underlying grid of positions as opposed to what was done in [23]. The approach taken there is that they leave out the weights and search for the element positions that give minimum peak sidelobe levels. However, due to limitations in the fabrication process, such arrays are very difficult to manufacture in many applications, for instance as a transducer for ultrasound imaging. That is why we stick to a fixed underlying grid here.

The optimization criteria are the same for the layout problem as for the weight problem, i.e.

- Minimize maximum sidelobe in the beampattern with a condition on the maximum mainlobe width
- Minimize integrated sidelobe energy in the beampattern with a condition on the peak sidelobe and/or on the maximum mainlobe width

In addition, there are some criteria that relate to the coarray. They are

- Minimize number of holes in the coarray
- Minimize the number of redundancies in the coarray

The four criteria given here are in two different domains and very few investigators have compared them. In [24] we did that through an exhaustive search of small arrays of aperture $M = 18$ with $K = 7$ active elements and of aperture $M = 26$ with $K = 7$ active elements. There exist five different minimum redundancy arrays for the first case (Table 1). A full search of all possible arrays (4368 different ones when the end elements are fixed, Eq. (26), results in a plot of peak sidelobe versus -6 dB beamwidth as shown in Fig. 12. The interesting cases are those that have the lowest sidelobe and the smallest beamwidth, i.e., those that lie on the lower and left boundary of this figure. In Fig. 13 a line has been drawn through this optimal boundary and the positions of the five minimum redundancy arrays have been shown. Only three of the five are on the boundary and it turns

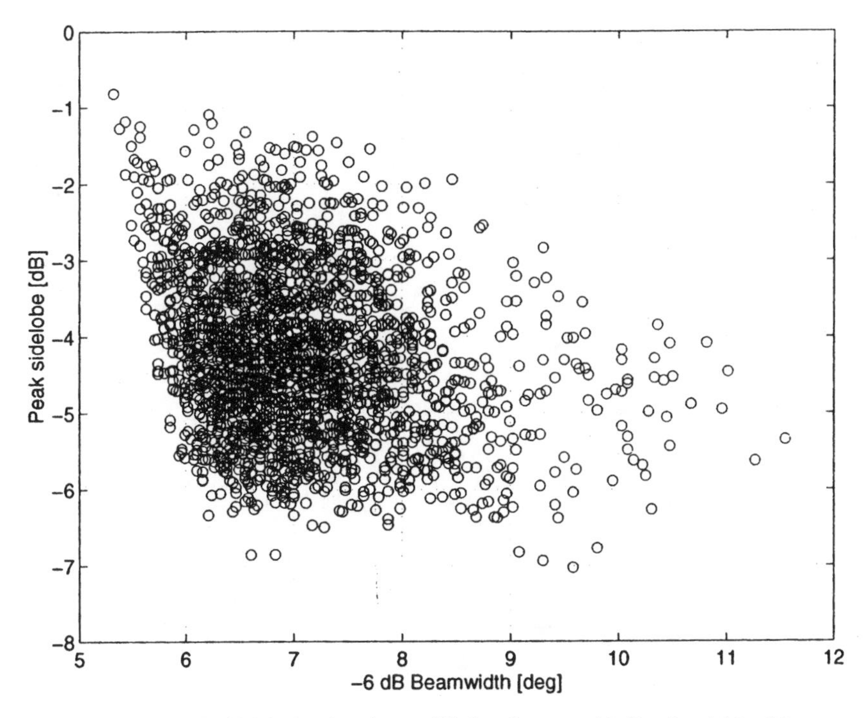

Figure 12. Peak sidelobe level vs. beamwidth for all arrays with $K = 7$ and $M = 18$.

out that these are the ones that have a redundancy of not more than 2. For larger arrays, the distribution of the redundancies over the lag domain also plays a role in determining whether a minimum redundancy array will have performance on the optimal boundary. It is in particular important to avoid periodicities in the redundancies. In Fig. 14 a similar search has been done for the five different minimum hole arrays that exist for the second case with $M = 26$ and $K = 7$ (Table 1). In this case, there were 42504 different possible thinning patterns to search. Now one can see that all five minimum hole arrays are on the optimal boundary. We have concluded from this empirical study that the minimum peak sidelobe criterion seems to be equivalent to the minimum hole criterion and also that it is close to the minimum redundancy criterion. Whether this can be proved mathematically or not, is not known.

19.3.1. 1-D Arrays

In [19] and [25] a method for optimizing the weights and/or the layout of a sparse array is described. It uses linear programming and is based on the array

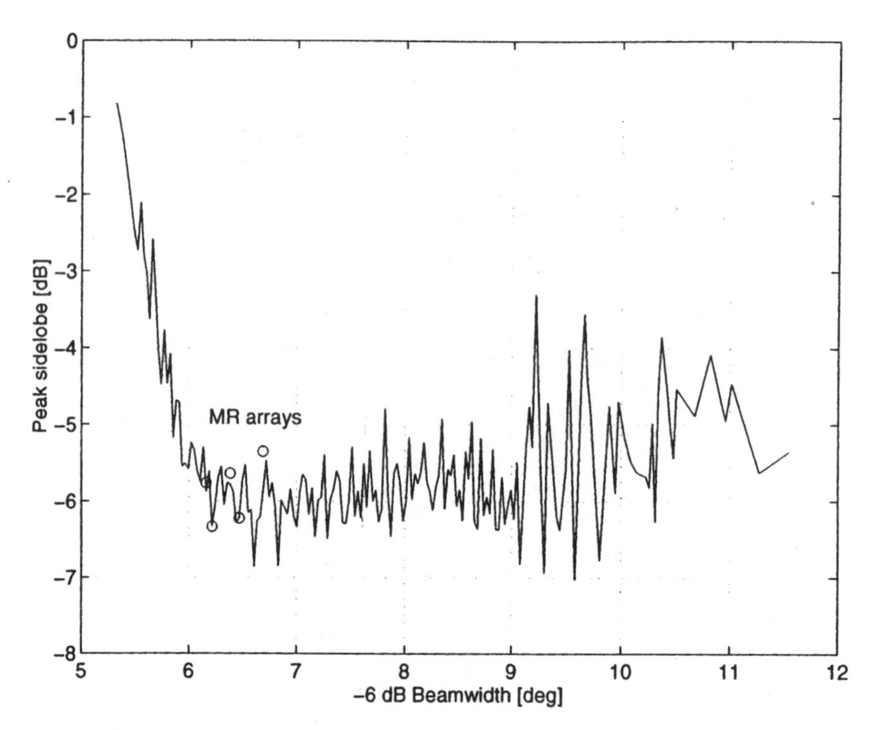

Figure 13. Optimal peak sidelobe level vs. beamwidth relative to the minimum redundancy arrays with $K = 7$ and $M = 18$.

being symmetric. Some of the 1-D examples from that paper will be described here.

An array with half wavelength spacing, 64 elements and Gaussian random thinning to 48 elements was optimized. An example of the beam patterns before and after optimization are shown in Fig. 15. The optimization of the element weights was done by minimizing the peak sidelobe in a region extending from a start angle, ϕ_1, to 90 degrees. Due to the symmetry of the beam pattern, only positive angles were required. In this way, the mainlobe is not affected by the optimization. However, the peak sidelobe level is very sensitive to the start angle. This parameter influences the trade-off between the beamwidth and the peak sidelobe level after optimization. The details of the optimization algorithm are given in Appendix A. Several optimizations were performed for various beamwidths and thinning patterns. For each thinning pattern, the start angle ϕ_1 was varied and an optimization was performed. The resulting peak sidelobe and -6 dB beamwidth is plotted in Fig. 16. Each curve is the result of between 5 and 18 such optimizations. Figure 16 shows two dash-dot lines which are the results

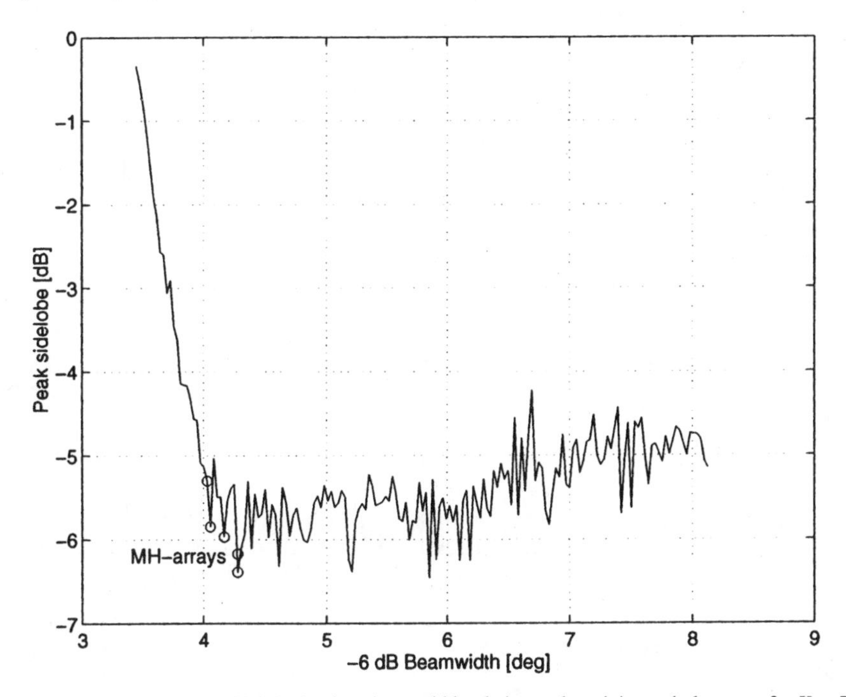

Figure 14. Optimal peak sidelobe level vs. beamwidth relative to the minimum hole arrays for $K = 7$ and $N = 26$.

of optimizing the weights to give uniform sidelobe levels for the full arrays. The left-hand one (smallest beamwidth) is the performance for a full 64-element array, and the right-hand one (largest beamwidth) for a full 48-element array. Only thinned arrays with performance better than the 48-element curve are of interest. All the remaining curves are for a 64-element array thinned to 48 elements. The upper solid line shows performance for the worst symmetric thinning that could be found, giving a minimum sidelobe level of about -13 dB. This array has almost a periodic thinning. If it had not been for the requirement for symmetry, this would have been a periodically thinned array with fully developed grating lobes.

The two dashed lines are two realizations of random Gaussian thinning. Both of them start leveling off at -17 to -18 dB sidelobe level. This is in the vicinity of the mean sidelobe level predicted for a random array (18) given as the inverse of the number of remaining elements which is $-10 \log 48 = -16.8$ dB. However, with the optimization used here this value is achieved as a peak value instead.

Finally, the two lower solid curves are the results from optimizing the weights for two near-optimal thinning patterns. They were obtained with a

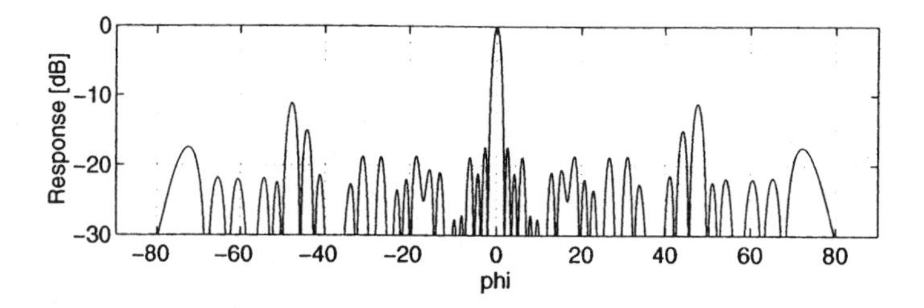

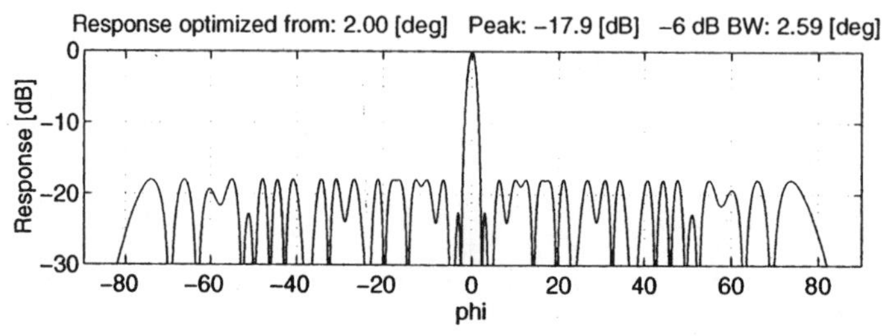

Figure 15. One-way array pattern before and after optimization for 64-element array randomly thinned to 48 elements. Thinning and weights are shown in the center panel of Fig. 17 (from [19], © 1997 IEEE).

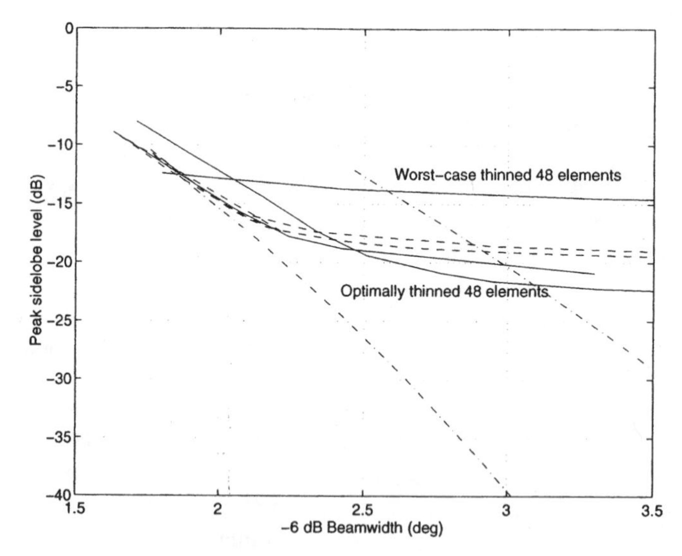

Figure 16. Result of optimizing weights given as sidelobe level as a function of beamwidth. Shown are uniform sidelobe level 64-element and 48-element full arrays (dash-dot lines), two realizations of random 25% thinning of the 64-element array (dashed lines), and worst-case and optimally 25% thinned arrays (solid lines) (from [19], © 1997 IEEE).

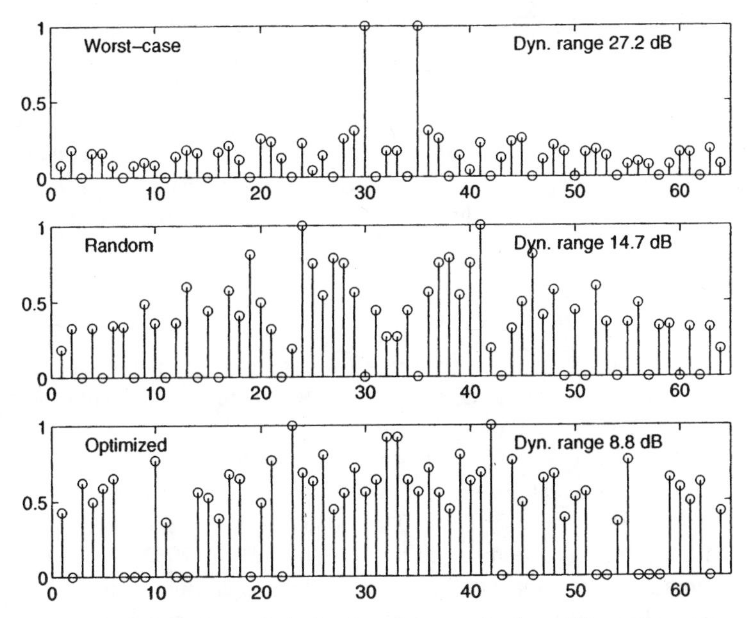

Figure 17. Weights found after optimization from 2 degrees for three different element layouts. The beampattern of the random layout is shown in the lower panel of Fig. 15 (from [19], © 1997 IEEE).

combined weight and layout optimization algorithm with sidelobe targets of -18 and -19.5 dB. The other values in their curves were obtained by keeping the layout and then optimizing the weights only for different values of start-angles in the optimization. With such thinnings the peak sidelobe level can be improved down to the range -17 to -20 dB.

All the thinning patterns are shown in Table 2. Examples of the weights required are shown in Fig. 17. They are quite different from the much smoother weight functions that are obtained for full arrays (see the Dolph-Chebyshev

Table 2. Left-hand part (32 elements) of symmetric 64-element arrays. All references to relative position are to the right-hand part of curves in Fig. 16 (from [19], © 1997 IEEE).

Elements enabled	Comment
11011101110111011101110111011101	Worst-case symmetric array
11010110110110101111011111111011	Random 1 (upper dashed curve)
11011011011111111001010111110111	Random 2 (lower dashed curve)
10111100011001111101101111111111	Optimized 1, (– 18 dB) (upper solid curve)
00101001111101111011101111111111	Optimized 2, (– 19.5 dB) (lower solid curve)

weights of Figs. 46–49 of [3]). The limitation of a symmetric array that the linear programming algorithm imposes is really unnecessary. If a heuristic method like the genetic search algorithm or simulated annealing is used instead, any array can be optimized both for weight and/or layout. However, with these algorithms, one does not have any guarantee that a global minimum is reached. On the other hand, the linear programming method is limited in that it can only solve small problems. The rest of the results here will, therefore, be found with heuristic methods.

19.3.2. 1-D Layout Optimization

In [21], simulated annealing was used to minimize the maximum sidelobe level for an array with aperture $L = 50\lambda$ ($M = 101$) with $N = 25$ active elements lying on a grid with $\lambda/2$ element distance. This problem has $1.9146 \cdot 10^{22}$ solutions ($M = 99$ and $K = 23$ inserted in (26) since the end elements are fixed). This is a problem that has been optimized since the sixties, and a table of the solutions obtained is given in Table 3 based on [21]. For a description of the simulated annealing algorithm, refer to Appendix B. In [29] this problem was solved with a simulated annealing procedure which is an improvement over that of [21, 30, 31] in several respects. First of all it is faster since an incremental procedure is used for finding the array pattern. The evaluation of (17) requires a discrete Fourier transform that can be implemented by a Fast Fourier Transform (FFT) algorithm in order to speed it up. A further speed increase can be obtained by the observation that the simulated annealing algorithm consists in perturbing just a single element at a time. Therefore, the array pattern of the perturbed array can be found by subtracting the contribution of the element that was moved and adding the contribution of that which was added. When all contributions from all the elements at all the angles are precomputed and stored in memory this results in a speed increase. In [29] it is shown that for an $N = 256$ point evaluation, this results in 6.7 times faster execution than when the FFT algorithm is used. The faster evaluation of the array pattern means that one can evaluate more configurations and therefore the simulated annealing algorithm of [29] allows perturba-

Table 3. Table of solutions to the problem of finding the best 25 unit weight element positions in an aperture of $L = 50\lambda$.

Year	Min sidelobe level	Optimization method	Reference
1964	−8.8 dB	Dynamic programming	[26]
1966	−8.9 dB	Space-tapering	[27]
1968	−10.14 dB	Dynamic programming	[28]
1996	−12.07 dB	Simulated annealing	[21]
1998	−12.36 dB	Simulated annealing	[29]

tion at an arbitrary location in the array, rather than within the interval given by the neighbors on the right-hand and left-hand sides of the element to be perturbed as in [21]. The resulting sidelobe level depends on the sampling of the array pattern. When $N = 4096$ points are used for evaluating it, the optimal solution of [21] has a sidelobe level of -12.03 dB and a -6 dB beamwidth of $2.10°$ (see Fig. 18). Ten solutions that are slightly better are given in [29]. Two of the better ones are shown in Figs. 19 and 20. The first represents a search over $2.0 \cdot 10^5$ configurations and has a sidelobe level of -12.06 dB and a -6 dB beamwidth of $1.71°$. The second solution is the result of a search over $2.9 \cdot 10^7$ configurations and it has a sidelobe level -12.36 dB and a -6 dB beamwidth of $2.10°$. Here the sidelobe level is improved by 0.32 dB over the reference. The other eight solutions are between the two given here, i.e., with some sidelobe level improvement and some beamwidth improvement over the reference.

Other solutions using the simulated annealing algorithm are shown in Figs. 21 and 22. If one accepts a widening of the mainlobe, the last figure shows that it is actually possible to find a sparse array that has a peak level which is equivalent to the mean value for a random sparse array ($1/K = -13.97$ dB). However, in this case, most of the elements are clustered near the center of the array and the mainlobe suffers. The results of Figs. 20 and 21 are probably the best that can be obtained in terms of sidelobe level. The beamwidths (-6 dB) are $2.10°$ and $2.77°$.

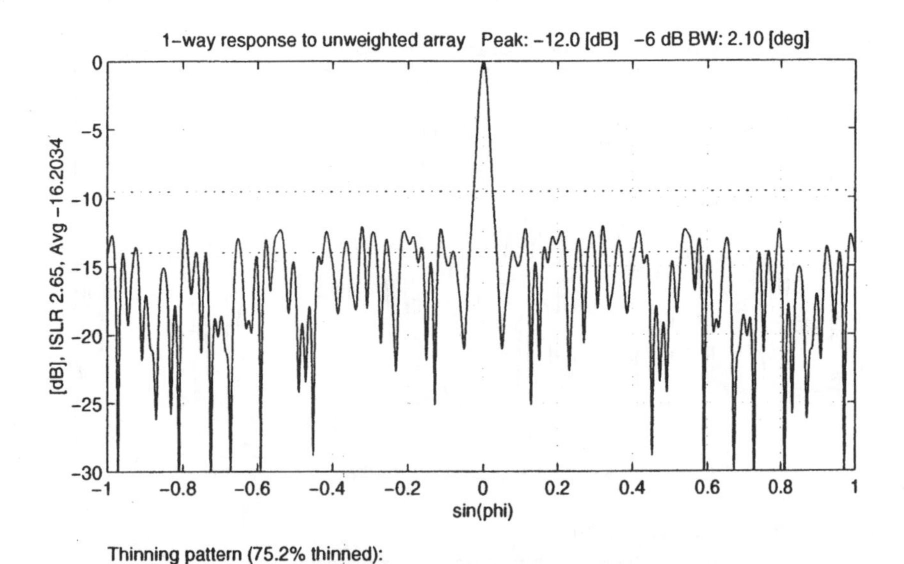

Thinning pattern (75.2% thinned):

1000000000000001000000000000001000000000001000001011111 1
0000110001100010111010001100000010010000000000000001

Figure 18. Array pattern for optimized array with 25 elements out of 101, based on [21].

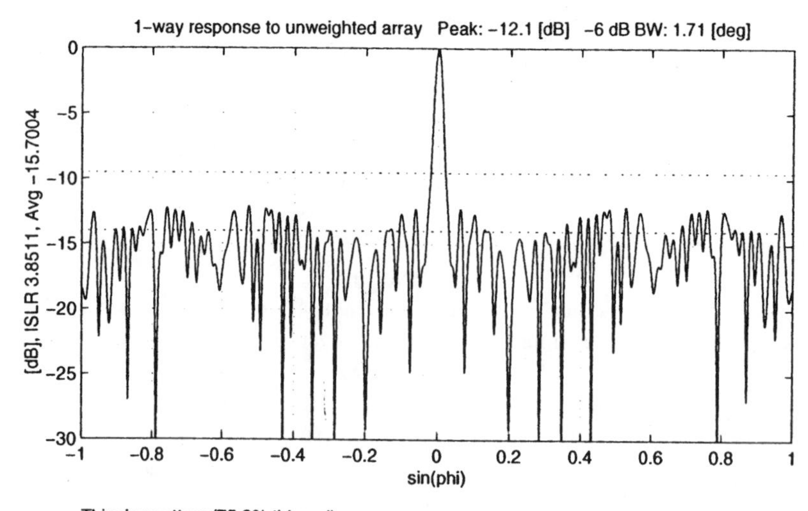

Thinning pattern (75.2% thinned):

10000100000000000000000010000100000000000001010000001 1
000100100101011111001000100000100100100000000110001

Figure 19. One-way array pattern for optimized array with 25 elements out of 101, optimized for minimum beamwidth.

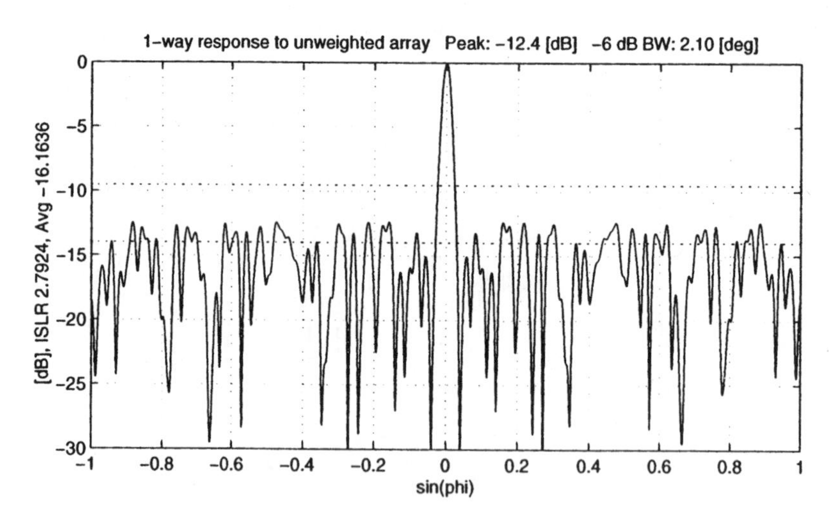

Thinning pattern (75.2% thinned):

1000000100000000000000001111010100010110100000101100 11 0
001100100011000001001000000000000000000000000000001

Figure 20. One-way array pattern for optimized array with 25 elements out of 101, optimized for minimum sidelobe level.

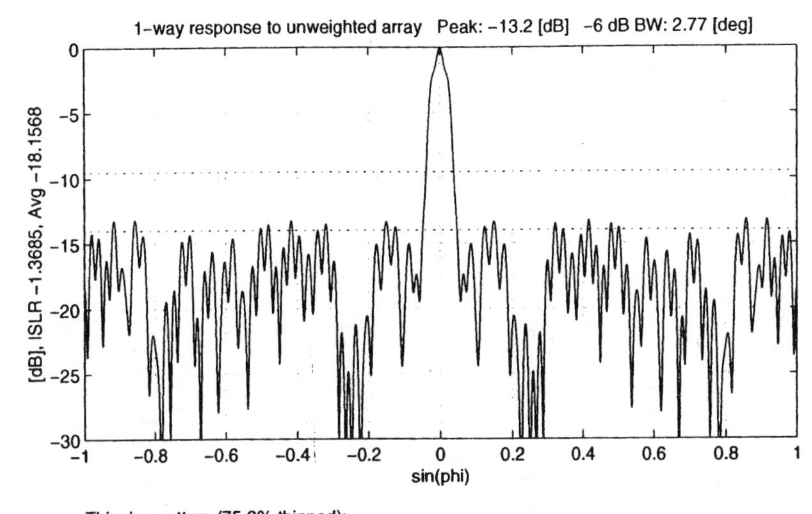

Thinning pattern (75.2% thinned):

10001010110011100111110111001011111101000000000000 0
00000000000000000000000000000000000000000000000001

Figure 21. Result of optimization. One-way array pattern for 25 elements out of 101, peak sidelobe level – 13.2 dB.

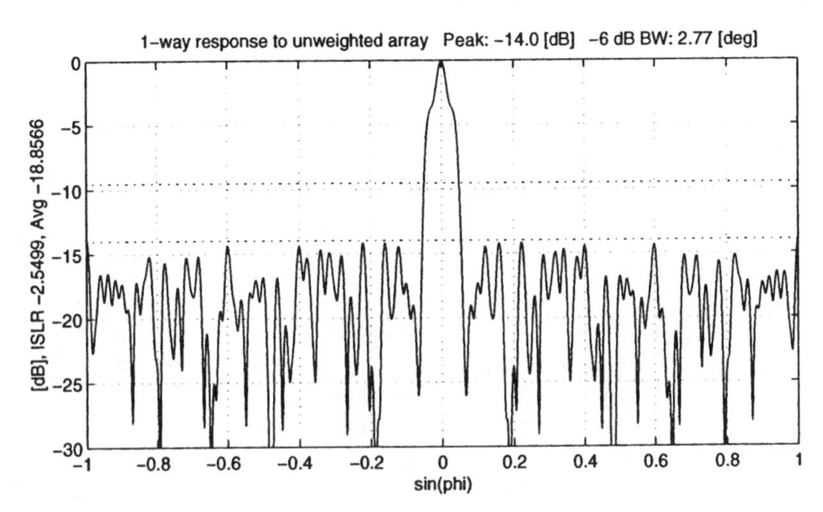

Thinning pattern (75.2% thinned):

10000000000000000000000000000000000001001000111111 1
11110101110111110000100000000000000000000000000001

Figure 22. Result of optimization. One-way array pattern with 25 elements out of 101, peak sidelobe level – 14.0 dB which is similar to the average sidelobe level for a random array.

This is 51% and 99% over the beamwidth of the full array ($1.21\lambda/D = 1.39°$). The sidelobe levels are -12.36 dB and -13.2 dB which is 0.8–1.6 dB above the average level of a random array and corresponds to a level of $1.2/K$–$1.5/K$. This is our best estimate of the peak sidelobe level for algorithmically optimized 1-D arrays.

19.4. 2-D Array Optimization

A 2-D array is considerably harder to optimize than a 1-D array due simply to its size and the vast increase in the number of combinations. In [19] we were successful in using linear programming to find weights for a thinned 2-D array. However, the much harder problem of finding layouts for arrays with thousands of elements is simply too hard a problem for linear programming at present. It is, however, a much more important problem than weight optimization. This can be illustrated by ultrasound imaging. Two-dimensional ultrasound arrays at present have a sensitivity problem that makes it unattractive to weight the individual elements. Furthermore, realization of thousands of accurate weights in a hardware implementation is very undesirable. The layout optimization problem for 2-D arrays is really the problem one would like to solve. The methods that can be used are the genetic algorithm and the simulated annealing algorithm. Before showing results, it is necessary to discuss the implications on optimization when the 2-D arrays not only look broadside, but also are required to be steered. This means that the beam pattern of Section 19.2.1.5 must be optimized, not just the array pattern of Section 19.2.1.2.

19.4.1. Optimization of Steered Arrays

For optimization of the unsteered beam pattern, the sidelobe level should be minimized over all visible angles, except those where the mainlobe is located. This is the region defined by all elevation angles and with the azimuth angle in the range $\phi \in [\phi_1, \pi/2]$, where ϕ_1 is the boundary between the mainlobe and sidelobe regions. Because of the correspondence $(k_x, k_y) = 2\pi/\lambda \cdot (u, v) = 2\pi/\lambda \cdot (\sin\phi\cos\theta, \sin\phi\sin\theta)$, this corresponds to an annular region in $\vec{k}$-space of radius $|k| = 2\pi/\lambda$ centered at the origin, except the small mainlobe region in the center, as shown in Fig. 23. The mainlobe region is defined by a circle of radius $|k| = 2\pi/\lambda \sin\phi_1$. Due to the sampling of the aperture, the beampattern will be repeated for argument of k_x and k_y larger than $2\pi/\lambda$. This means that the circles will repeat along the k_x-axis and the k_y-axis. When the element distance is $\lambda/2$, the circles will exactly touch along the direction of the axis. If the element distance is larger than $\lambda/2$, there is undersampling and the circles will partly overlap. Grating lobes may be explained in this way. When steering is applied to the array, the beampattern is $W(k_x - k_x^0, k_y - k_y^0)$ (9). The visible region will shift to have its center at the steering direction (k_x^0, k_y^0), while the optimized region from the array is still centered at the origin. There is, therefore, no longer full overlap between the optimized region and the visible region. In order to deal properly with steering, one must therefore, optimize a larger region. For an array with element distance $\lambda/2$, and for all possible steering angles, one must optimize over the area not covered by the pattern of repeating circles, i.e., the square region shown in Fig. 23.

For a 1-D array, this is greatly simplified. The only relevant variable is k_x, and when there is steering, the argument in the beampattern is

$$k_x - k_x^0 = 2\pi/\lambda \cdot (\sin\phi - \sin\phi^0) = 2\pi/\lambda \cdot u \tag{28}$$

First there is always symmetry with respect to $u = 0$. When, in addition the element locations are all on a grid with distance $\lambda/2$, there will also be symmetry with respect to $u = 1$. In this case optimization over the region $u \in [\sin\phi_1, 1]$ ensures that the array can be steered to any azimuth angle [21]. A larger element distance requires a smaller region, and a smaller element distance requires a larger region than $u \in [\sin\phi_1, 1]$.

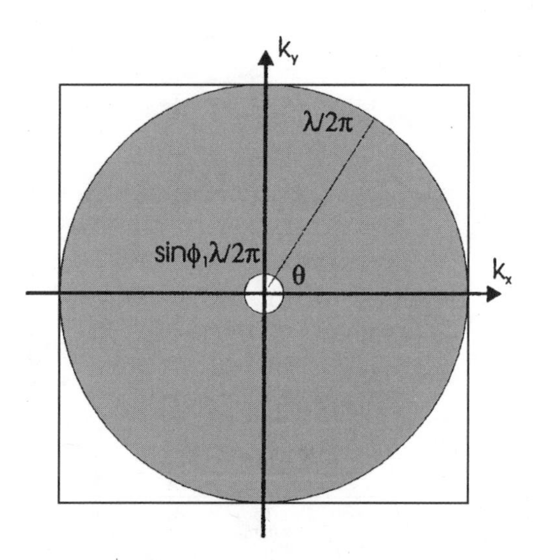

Figure 23. The optimization region in k-space, containing the visible region—everything inside the radius $|k| = 2\pi/\lambda$ except the mainlobe region which is inside a radius of $|k| = 2\pi/\lambda \sin\phi_1$ (from [19], © 1997 IEEE).

19.4.2. 2-D Array Optimized with Simulated Annealing

The simulated annealing algorithm of [29] has been used for a 2-D array of size 50×50 elements with element spacing $\lambda/2$. The algorithm with precomputed contributions from each element to the array pattern at all angles was used. For 64 points in the u and v directions, this requires several hundred Mbytes of RAM for storage. One of the best 500 element thinning patterns found is shown in Fig. 24. Finding this solution took 46 hours on a single CPU of a Silicon Graphics Power Challenge computer using a MATLAB implementation of the simulated annealing algorithm. The array pattern is shown in Fig. 25. It has a -6 dB beamwidth of $3.05°$ and a maximum sidelobe level of -21.5 dB. In comparison, the full array has a beamwidth of $2.81°$. The algorithm searched over 500 iterations and 5000 perturbations for each iteration, i.e., a total of 2.5 million configurations.

19.4.3. 2-D Array Optimized with Genetic Optimization

The genetic algorithm is also well suited for searching for solutions to large 2-D array layout problems. The general principles for the genetic algorithm are

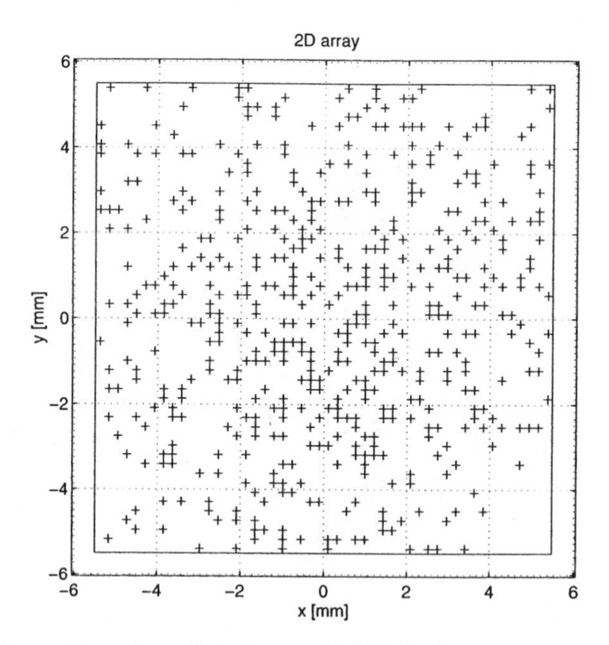

Figure 24. Element layout for optimized array with 500 elements out of 50×50, optimized for minimum beamwidth. The grid is shown for a frequency of 3.5 MHz and a velocity of sound of 1540 m/s corresponding to a wavelength of $\lambda = 0.44$ mm.

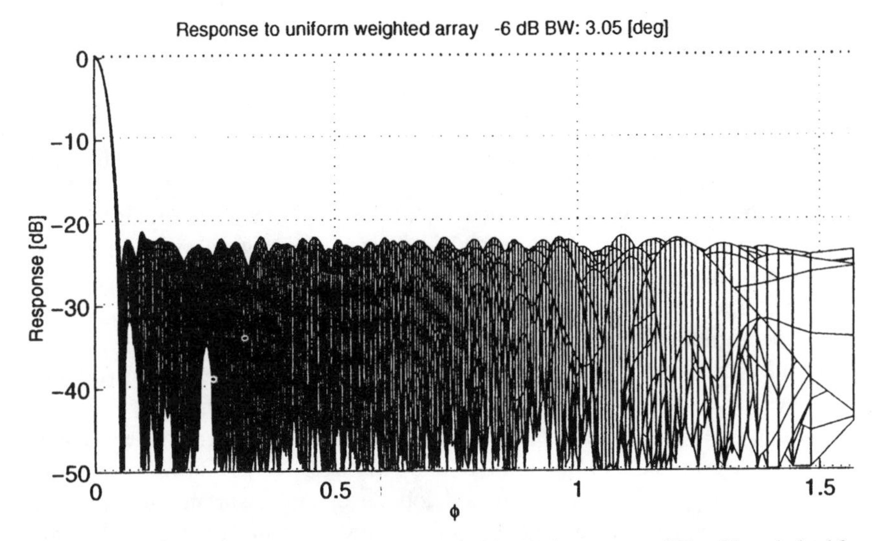

Figure 25. One-way array pattern for optimized array with 500 elements out of 50×50, optimized for minimum beamwidth. Peak sidelobe is -21.5 dB and the -6 dB beamwidth is 3.05°. The array is steered to $\theta = \phi = 0$ and the response is shown as a function of azimuth angle and seen from the side in 3-D space, i.e., the peak values over all elevation angles are seen.

given in Appendix C. Each gene is coded with 1s and 0s indicating whether an element is included or not. The gene is found by scanning the 2-D array row by row. In our implementation a parent selection strategy is used where the candidates are ranked according to their sidelobe level. Proportionate selection based on the ranking (roulette-wheel selection) is then used. A probability is assigned which is proportional to the ranking, so that the best individual has a high probability and the poorest one 0 probability. An individual is accepted as a parent with this probability, resulting in an elitist polygamous strategy. A second distinguishing feature of our implementation is that a parent dominant reproduction mechanism is used. A probability, p, is assigned to the reproduction so that with this probability the dominant parent is reproduced, and with probability $1 - p$ the rest is contributed from the other parent. The value of p is assigned so that a large number, say on average $N - 2$, where N is the number of elements in the array, are taken from the dominant parent. A low value for the mutation probability is also used, usually about $1/N$, so that mutation plays a minor role in the algorithm.

The cross-over scheme of our algorithm ensures rapid convergence, but gives an increased probability of convergence to a local minimum. This can be overcome by using an improved initialization method. Genetic algorithms are usually initialized with uniform probability distributions over the array and according to random array theory (19), one should expect first sidelobes in the -13 dB range. The main operation in the genetic algorithm is the cross-over. However, this operation does not significantly alter the probability distribution, so that the probability density distribution is still close to a uniform one. The randomness introduced by the mutation operation is often not large enough to significantly alter the probability density distribution. Therefore one should initialize the search with density functions that already have the desirable sidelobe properties in the Fourier domain. This improves convergence time and more importantly makes convergence to a good solution possible.

The previous 2-D example is now optimized using the genetic algorithm. The first example shown here (Fig. 26) is initialized with a circular symmetric probability distribution that has a Chebyshev-type Fourier transform (uniform sidelobes). It results in a sidelobe level of -22.2 dB and a beamwidth of 4.0°. A layout that gives a narrower mainlobe and higher sidelobes comparable to that of Fig. 25 is also possible to obtain (3.1° and -21.8 dB). This shows that one has freedom to trade-off beamwidth and sidelobe level. An important observation is that the sidelobe level of the first example is very close to 3 dB higher than that predicted for a 1-D array, $10 \log 1.5/K = -25.2$ dB, and the beamwidth is 42% higher than the full array's beamwidth. Thus this example seems to confirm the hypothesis that a value close to $-10 \log 1.5K + 3$ is an estimate of the achievable peak level in algorithmically optimized 2-D arrays when the beamwidth is allowed to increase by 50% over that of the full array.

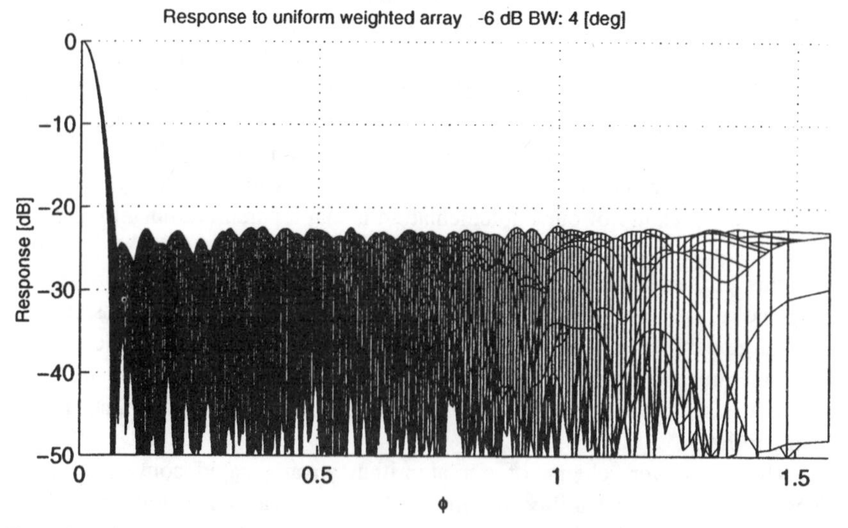

Figure 26. Optimized array response for 500 elements chosen from a 50×50 array. Peak sidelobe level is -22.2 dB and -6 dB beamwidth is 4.0°.

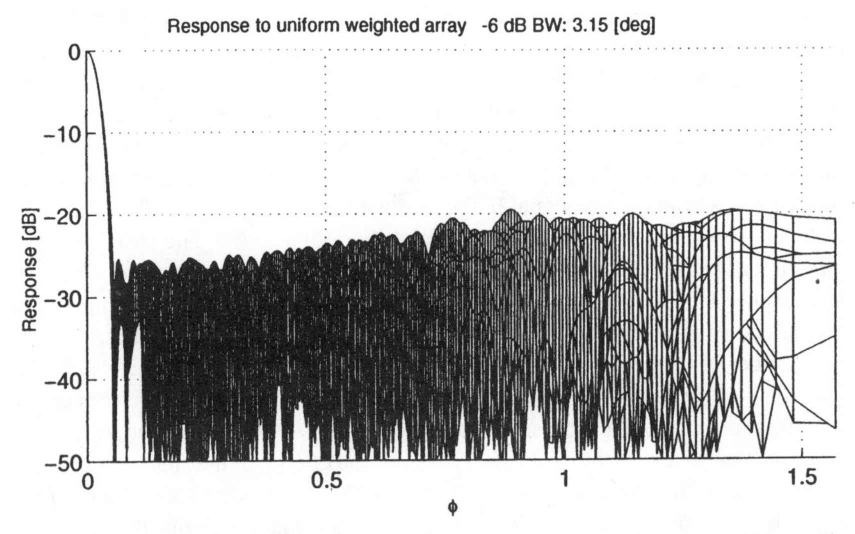

Figure 27. Designed probability density function for 500 elements chosen from a 50×50 array. The result has suppressed sidelobe response depending on the distance from the mainlobe. Beamwidth is 3.15°.

The next example (Fig. 27) shows the versatility of the method in that it allows for the sidelobe level to increase with angle away from the mainlobe. This could serve as a partial compensation for the element response (7).

19.5. Optimization of the Two-Way Beampattern

The 1-D and 2-D optimized arrays shown so far have all been one-way responses. By simply squaring them according to (11), one can find the two-way responses when the receiver and transmit array layouts are the same. More degrees of freedom can be obtained in the optimization if one allows the layouts to be different. We will still assume that the receiver and transmitter arrays are located in the same position, but allow for partial or no overlap at all between the selected elements.

The simplest way to utilize this freedom is to use the observation of [32] which was elaborated by [9] and [33], that good two-way responses can be obtained from periodic receiver and transmitter arrays if the two periodicities are different. A simple 1-D example will illustrate the idea. Assume that the transmitter periodicity is two, as in Fig. 4, and that the receiver periodicity is three. Due to the periodicity, this array will also have a fixed overlap between the receiver and transmitter elements. Every sixth element will be shared, or a total of

$128/6 = 21$ elements out of $128/2 = 64$ transmitter elements and $128/3 = 42$ receiver elements. The coarray (16) which is the convolution of the two aperture functions or effective aperture) is shown in Fig. 8. Note that it has a triangular shape which is similar to the one obtained from full, unweighted apertures. In addition it has some undesirable ripples. They may be reduced by weighting of the periodic apertures [34]. The transmitter has grating lobes that are a distance $|u| = 1$ away from the mainlobe. In the receiver the grating lobes will be located a distance $|u| = 2/3$ away from the mainlobe according to (5). The two-way array pattern will be as in Fig. 28. In the sidelobe region, the peak values are -32.7 dB at $|\sin\phi| = 1$ and -35.2 dB at $|\sin\phi| = 2/3$. This result should be compared to the randomly thinned one-way array patterns of Figs. 9 and 10. In these figures, the number of elements is 64 after random thinning from 128. The peak sidelobe value is -14 to -15 dB. If the same 64 elements are used both for the receiver and the transmitter, the peak sidelobe value of the two-way array pattern will be twice as much, i.e., -28 to -30 dB. However, the energy in the sidelobe region is much higher. The sparse binned array of Fig. 11 is somewhat comparable to the periodic array in that there is less energy near the mainlobe, but it has a peak value in the two-way pattern of about -24 dB. The downside of the periodic array approach is the existence of the discrete sidelobes at $|\sin\phi| = 1$ and $|\sin\phi| = 2/3$, due to the partly suppressed grating lobes. When this array is steered, the first grating lobe will move even closer to the broadside direction. For instance, if the array is steered to $\phi^0 = 30°$, the first grating lobe will move to $\sin\phi^0 - 2/3$ corresponding to an angle of $-9.6°$.

This approach is simple to extend to the 2-D planar case by using the same periodicity in both axes. An example of such an array based on a square 50×50 array with $\lambda/2$ element spacing where the corner elements are unused to make it a

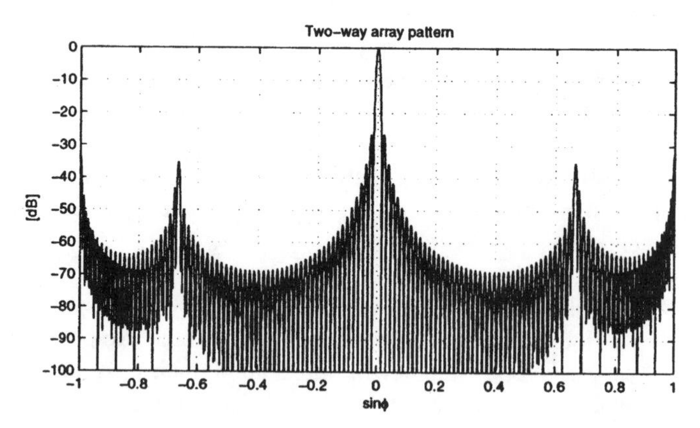

Figure 28. Two-way array pattern for a 128 element array with every other element used for transmission and every third element for reception.

circle (about $50 \cdot 50\pi/4 \approx 1963$ elements) is given here. Every second element in both directions over a reduced aperture is used for the transmitter (a total of 253 elements), and every third element in both directions is used for the receiver (a total of 241 elements). The layout of this array is shown in Fig. 29 and the two-way array pattern is shown in Fig. 30. This is a two-dimensional extension of Fig. 28. In [35] we made an attempt to get rid of the discrete sidelobes of the periodic array. This is done by combining the transmitter pattern with a periodicity of two with an algorithmically optimized receiver pattern using 256 receiver elements. Such an array is shown in Fig. 31. The receiver pattern is designed from a criterion of minimizing peak sidelobes in the two-way beampattern using the genetic algorithm. The resulting beampattern is shown in Fig. 32. Compared to Fig. 30, the new response does not have the discrete peaks along the axes at $|u| = 2/3$ and $|v| = 2/3$. This is an advantage when the array is steered because then the whole response is shifted in the (u, v)-plane. An example of this is shown in Fig. 33 where steering to $\phi = 30°$ and $\theta = 30°$ is shown. The downside of this approach is the increased average background sidelobe level.

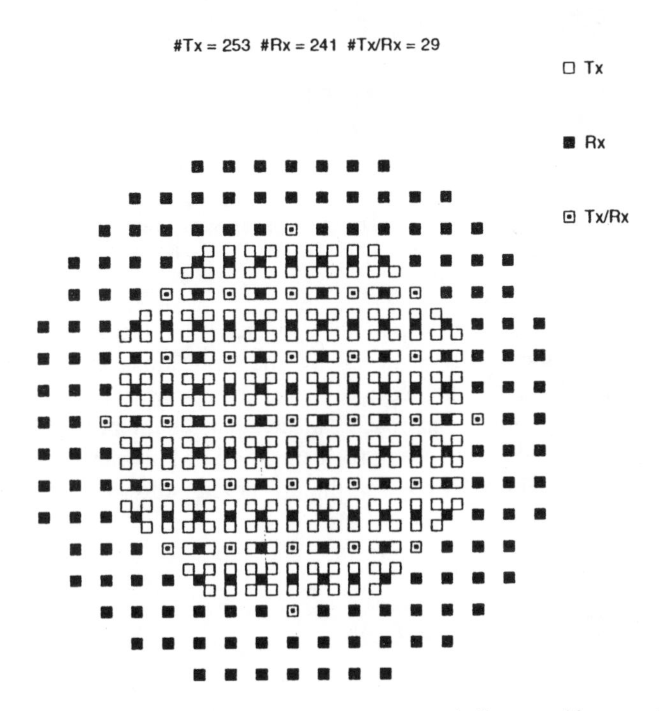

Figure 29. Layout of transmit and receive elements for a 2-D periodic array with every other element used for transmission and every third element for reception.

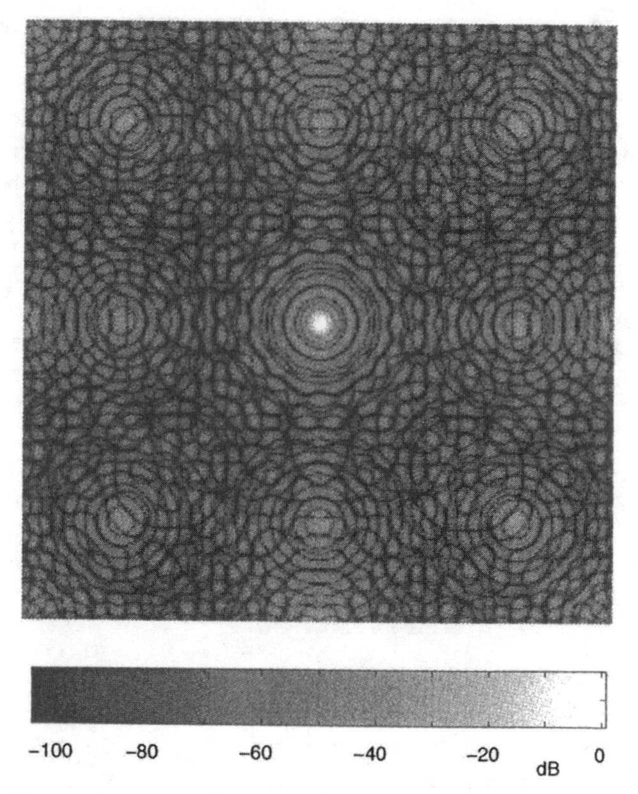

Figure 30. Two-way beampattern for a 2-D periodic array with every other element used for transmission and every third element for reception (from [35], © 1997 IEEE).

Another advantage over the periodic approach is the flexibility in the choice of the receiver and transmitter elements. In some applications it may be important to have separate transmitter and receiver elements due to restrictions in cabling or electronics. The algorithmic approach satisfies that requirement. The proposed array in Fig. 31 was designed using the genetic algorithm with a constraint that elements already occupied by the transmitter were not allowed.

19.6. Conclusion

Sparse arrays have traditionally been designed with two main objectives in mind: creation of beampatterns with low mainlobe width and small sidelobes, or best possible sampling of a random field. In the latter case the correlation

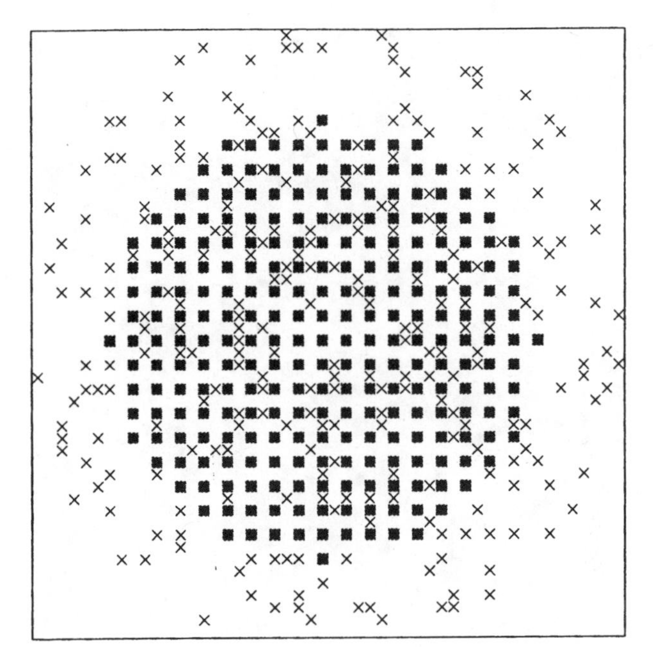

Figure 31. Element layout with periodic transmitter pattern designated by filled squares and algorithmically optimized receiver pattern designated by $\times$. There is no overlap (from [35], © 1997 IEEE).

function of the array (coarray) should be optimized and be as uniform as possible. This case is shown here to be very close to finding a beampattern with minimal peak sidelobes.

Since the search space for layout optimization for sparse arrays is so vast, heuristic search methods such as genetic optimization and simulated annealing have been applied to this problem. Both methods need to be tuned to this problem in order to speed convergence and sometimes even to make convergence possible. Both 1-D and 2-D examples have been shown here. We propose that the estimate of the average level in a random array, $1/K$, is in fact very close to an estimate of the achievable peak level in an algorithmically optimized 1-D array. A value of about $1.5/K$ is our best estimate for the peak value when the beamwidth is allowed to increase by 50% over that of the full array. This is 1.8 dB over the average value of a random array. For 2-D arrays the estimate is twice as large, or about $3/K$, based on peak sidelobe theory for sparse arrays and the examples given here.

When different array layouts for the transmitter and the receiver are allowed, one gets additional freedom in the design of the thinning patterns, as aliasing in one of the beampatterns can be cancelled by zeroes in the other one and vice

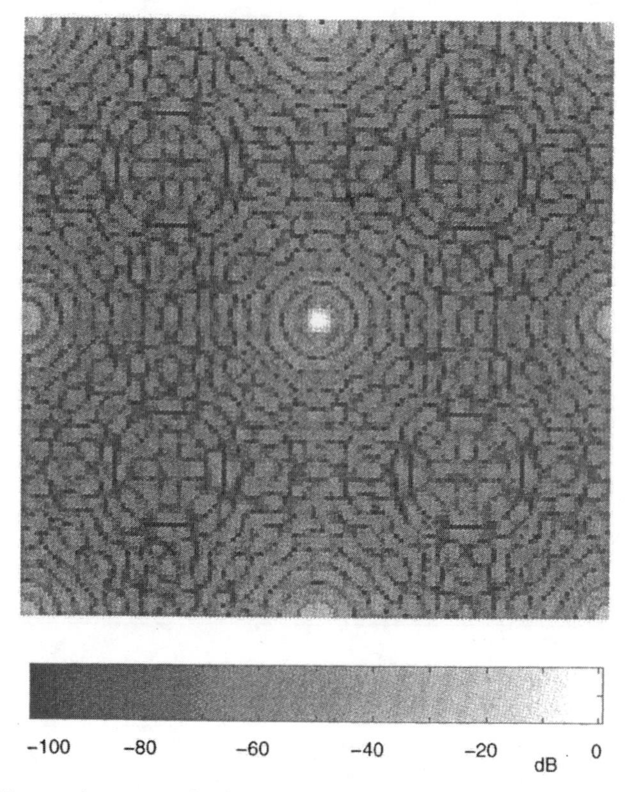

Figure 32. Two-way beampattern for the 2-D array shown in Fig. 31 (from [35], © 1997 IEEE).

versa. This has been exploited in methods based on periodic arrays and design in the coarray (effective aperture) domain. This method is compared with the previous methods.

With the methods given here, one has the freedom to choose a design method for a sparse array system using either the same elements for the receiver and the transmitter, no overlap, or partial overlap as in the periodic arrays.

Acknowledgment

This work was partly sponsored by the ESPRIT program of the European Union under contract EP 22982, and by the Norwegian Research Council. This report was completed while SH was on a sabbatical leave from the University of Oslo at the GE Corporate Research and Development Center, Schenectady, New York. I would like to thank Dr. K. Thomenius, manager of the ultrasound

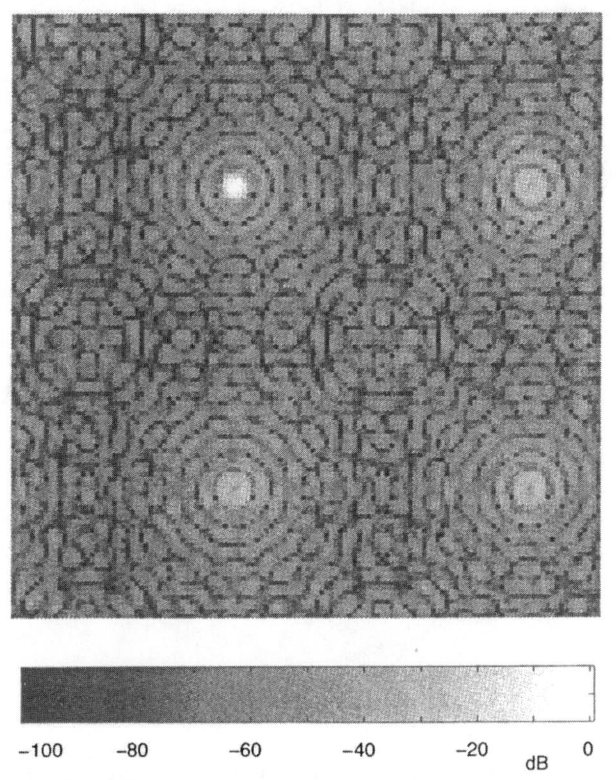

Figure 33. Two-way steered beampattern with $\phi = 30^\circ$ and $\theta = 30^\circ$. Array as in Figs. 31 and 32 (from [35], © 1997 IEEE).

program, for making that stay possible. We would also like to thank Dr. N. Aakvaag for making his genetic optimization code available to us.

Appendix: Optimization Methods

A. Linear Programming

A linear programming (LP) problem is the minimization of a linear function subject to a set of linear inequalities and linear equations [36, 37]. In matrix form an LP problem may be written as

$$\begin{array}{ll} \text{minimize} & \mathbf{c}^T\mathbf{x} \\ \text{subject to} & A\mathbf{x} \leq \mathbf{b} \end{array} \tag{29}$$

where **x** is a vector of n variables, and the data is given by the $m \times n$-matrix A and the vectors **c** and **b**. The weight optimization problem can be put in this form by letting the unknown weights and the unknown minimum sidelobe level be **x**, and the complex exponentials of (2) or (4) be contained in **c** for all the angles in the sidelobe region. The normalization that ensures that $W(0) = 1$ is handled by A and **b** [19]. One limitation is that the objective function has to be real and linear in the variables to minimize. This implies that the array pattern has to be a real function, i.e., that the array has to have symmetry.

The problem of optimizing the layout of an array is a mixed integer linear programming problem, i.e., a linear programming problem where some or all variables are required to be integers. In general, mixed integer LP problems are computationally very difficult. The layout optimization problem is hard even for moderate size problems. This is mainly due to the complex structure of the matrix V since none of its elements are zero.

Small-scale mixed-integer problems may be solved by the branch and bound method. This is a general method for solving such problems where the feasible region is gradually divided into finer subregions for which a linear programming problem is solved.

Linear programming has been applied to the sparse optimization problem in [22] and [19].

B. Simulated Annealing

Simulated annealing is a stochastic optimization method where the analogy with thermodynamics as in a metal that cools and anneals is used [38]. At high temperatures the molecules move freely. As the temperature falls, the molecules slow down and are finally lined up in a crystal, which is the state of minimum energy. In stochastic optimization this state corresponds to the optimal solution. In our problem, the energy, E, is proportional to the peak sidelobe level. A temperature function is slowly decreased for each iteration, and the solutions are randomized by changing one of the active elements at a time. The energy of the perturbed system is found and compared with last iteration's best solution. The new solution may be accepted even if it is inferior to the previous one, based on a probability function. In this way it is ensured that the algorithm does not get stuck in a local minimum. As the system cools off, the probability for accepting an inferior solution is reduced, and eventually the system converges on the final solution which, if the optimization parameters are chosen well, may be close to the optimal solution.

The temperature function used in our optimizations is $T_i = T_0/i$ where i is the iteration number. For each iteration a probability density function which is proportional to $p = e^{\Delta E/T}$ is computed. The energy difference ΔE is the change in sidelobe level due to a perturbation of the array configuration. The Metropolis algorithm used for deciding which array configuration to use next is

- If $\Delta E < 0$, the new configuration is better and it is used as a starting point for the next perturbation
- If $\Delta E > 0$, the new configuration will be used with a probability p.

Simulated annealing has been applied to optimization of sparse arrays in [21], [30], [31], and [29].

C. Genetic optimization

The genetic algorithm is an iterative process that operates on a set of individuals (population) [39, 40]. Each member of the population represents a potential solution to the problem. Initially, the population is randomly generated. The individuals are evaluated by means of a fitness function, a measure of its fitness with respect to some predefined evaluation function (environment). The presence of a sensor is indicated by one, and its absence by zero. The steps taken in the algorithm are

- Selection based on the fitness
- Reproduction
- Replacement

The reproduction stage usually consists of two separate operations, cross-over and mutation. In the cross-over operation two or more individuals (parents) are crossed according to some method to produce one or two new individuals (offspring). Some of the new individuals are then subject to a mutation process where one or more of the bits (genes) are flipped. The resulting offspring are then either ignored or are included in the new population depending on their fitness. The algorithm stops either after reaching a predefined number of generations, or by converging to a point of no improvement.

The core of the genetic algorithm resulting in production of new individuals, is the cross-over operation. Mutation operates as a secondary step where only a few elements of the array change values.

Genetic optimization has been applied to optimization of sparse arrays in [41], [42], [43], and [35].

References

[1] B. Steinberg. *Principles of Aperture and Array System Design*. Wiley, New York, 1976.

[2] D. H. Johnson and D. E. Dudgeon. *Array Signal Processing*. Englewood Cliffs, NJ, Prentice-Hall, 1993.

[3] F. J. Harris. On the Use of Windows for Harmonic Analysis with the Discrete Fourier Transform. *Proc. IEEE*, 66:51–83, January 1978.
[4] R. A. Haubrich. Array Design. *Bull. Seismological Soc. Of Am.*, 58:977–991, 1968.
[5] G. S. Bloom and S. W. Golomb. Application of Numbered Undirected Graphs. *Proc. IEEE*, 65:562–570, April 1977.
[6] A. Dollas, W. T. Rankin, and D. McCracken. A New Algorithm for Golomb Ruler Derivation and Proof of the 19 Mark Ruler. *IEEE Trans. Inf. Theory*, 44(1), 1998.
[7] D. A. Linebarger, I. H. Sudborough, and I. G. Tollis. Difference Bases and Sparse Sensor Arrays. *IEEE Trans. Inf. Theory*, 39:716–721, March 1993.
[8] R. T. Hoctor and S. A. Kassam. The Unifying Role of the Coarray in Aperture Synthesis for Coherent and Incoherent Imaging. *Proc. IEEE*, 78:735–752, April 1990.
[9] G. R. Lockwood, P.-C. Li, M. O'Donnell, and F. S. Foster. Optimizing the Radiation Pattern of Sparse Periodic Linear Arrays. *IEEE Trans. Ultrason.. Ferroelect.. Freq. Contr.*, 43:7–14, January 1996.
[10] Y. T. Lo. A Mathematical Theory of Antenna Arrays with Randomly Spaced Elements. *IEEE Trans. Antennas Propagat.*, 257–268, May 1964.
[11] W. J. Hendricks. The Totally Random Versus the Bin Approach for Random Arrays. *IEEE Trans. Antennas Propagat.*, 39:1757–1761, December 1991.
[12] G. Benke and W. J. Hendricks. Estimates for Large Deviation in Random Trigonometric Polynomials. *SIAM J. Math. Anal.*, 24:1067–1085, July 1993.
[13] B. Steinberg. The Peak Sidelobe of the Phased Array having Randomly Located Elements. *IEEE Trans. Antennas Propagat.*, AP-20:129–136, March 1972.
[14] R. L. Fante, G. A. Robertshaw, and S. Zamoscianyk. Observation and Explanation of an Unusual Feature of Random Arrays with a Nearest-Neighbor Constraint. *IEEE Trans. Antennas Propagat.*, 39:1047–1049, July 1991.
[15] J. W. Adams. A New Optimal Window. *IEEE Trans. Signal Processing*, 39:1753–1769, August 1991.
[16] S. Holm. Maximum Sidelobe Energy Versus Minimum Peak Sidelobe Level for Sparse Array Optimization. In *Proc. IEEE Nordic Signal Processing Symp., Espoo, Finland*, September 1996, pp. 227–230.
[17] J. O. Erstad and S. Holm. An Approach to the Design of Sparse Array Systems. In *Proc. IEEE Ultrason. Symp., Cannes. France*, 1994, pp. 1507–1510.
[18] S. Holm and B. Elgetun. Optimization of the Beampattern of 2D Sparse Arrays by Weighting. In *Proc. IEEE Ultrason. Symp., Seattle. WA*, 1995, pp. 1345–1348.
[19] S. Holm, B. Elgetun, and G. Dahl. Properties of the Beampattern of Weight- and Layout-Optimized Sparse Arrays. *IEEE Trans. Ultrason.. Ferroelect.. Freq. Contr.*, 44:983–991, September 1997.
[20] J. F. DeFord and O. P. Gandhi. Phase-Only Synthesis of Minimum Peak Sidelobe Patterns for Linear and Planar Arrays. *IEEE Trans. Antennas Propagat.*, AP-36:191–201,February 1988.
[21] V. Murino, A. Trucco, and C. S. Regazzoni. Synthesis of Unequally Spaced Arrays by Simulated Annealing. *IEEE Trans. Signal Processing*, 44:119–123, January 1996.
[22] R. M. Leahy and B. D. Jeffs. On the Design of Maximally Sparse Beamforming Arrays. *IEEE Trans. Antennas Propagat.*, AP-39:1178–1187, August 1991.
[23] H. Schjær-Jacobsen and K. Madsen. Synthesis of Nonuniformly Spaced Arrays Using a General Nonlinear Minimax Optimization Method. *IEEE Trans. Antennas Propagat.*, AP-24:501–506, July 1976.
[24] J.-F. Hopperstad and S. Holm. The Coarray of Sparse Arrays with Minimum Sidelobe Level. In *Proc. IEEE NORSIG-98. Vigs, Denmark*, June 1998, pp. 137–140.
[25] S. Holm, B. Elgetun, and G. Dahl. Weight- and Layout-Optimized Sparse Arrays. In *Proc. Int. Workshop on Sampling Theory and Applications, Aveiro. Portugal*, June 1997, pp. 97–102.

[26] M. I. Skolnik, G. Nemhauser, and J. W. Sherman III. Dynamic Programming Applied to Unequally Spaced Arrays. *IEEE Trans. Antennas Propagat.*, AP-12: 35–43, January 1964.

[27] Y. T. Lo and S. W. Lee. A Study of Space-Tapered Arrays. *IEEE Trans. Antennas Propagat.*, AP-14:22–30, January 1966.

[28] R. K. Arora and N.C. V. Krishnamacharyulu. Synthesis of Unequally Spaced Arrays Using Dynamic Programming. *IEEE Trans. Antennas Propagat.*, 593–595, July 1968.

[29] J.-F. Hopperstad. Optimization of Thinned Arrays (in Norwegian). Master's thesis. Department of Informatics, University of Oslo, May 1998.

[30] A. Trucco and F. Repetto. A Sochastic Approach to Optimising the Aperture and the Number of Elements of an Aperiodic Array. *Proc. OCEANS '96*, 3:1510–1515, September 1996.

[31] A. Trucco. Synthesis of Aperiodic Planar Arrays by a Stochastic Approach. *Proc. OCEANS' 97*, 1997.

[32] S. Bennett, D. Peterson, D. Corl, and G. Kino. A Real-Time Synthetic Aperture Digital Acoustic Imaging System. *Acoust. Imaging*, 10:669–692, 1980.

[33] G. R. Lockwood and F. S. Foster. Optimizing the Radiation Pattern of Sparse Periodic Two-Dimensional Arrays. *IEEE Trans. Ultrason., Ferroelect., Freq. Control.*, 43:15–19, January 1996.

[34] S. S. Brunke and G. R. Lockwood. Broad-Bandwidth Radiation Pattern of Sparse Two-Dimensional Vernier Arrays. *IEEE Trans. Ultrason. Ferroelect., Freq. Contr.*, 44:1101–1109, September 1997.

[35] A. Austeng, S. Holm, P. Weber, N. Aakvaag, and K. Iranpour. 1D and 2D Algorithmically Optimised Sparse Arrays. In *Proc. IEEE Ultrason. Symp., Toronto. Canada*, 1967, pp. 1683–1686.

[36] V. Chvátal. *Linear Programming*. Freeman. San Francisco, CA (1983).

[37] G. Strang. *Linear Algebra and its Applications.* (2nd edn) Academic, New York, 1980.

[38] S. Kirkpatrick, C. D. Gelatt Jr. and M. P. Vecchi. Optimization by Simulated Annealing. *Science*, 220:671–680, May 1983.

[39] J. H. Holland. Genetic Algorithms. *Sci. Amer.*, 66–72, July 1992.

[40] D. E. Goldberg. *Genetic Algorithms*. Addison-Wesley, 1989.

[41] R. L. Haupt. Thinned Arrays Using Genetic Algorithms. *IEEE Trans. Antennas Propagat.*, 42:993–999, July 1994.

[42] P. Weber, R. Schmitt, B. D. Tylkowski, and J. Steck. Optimization of Random Sparse 2-D Transducer Arrays for 3-D Electronic Beam Steering and Focusing. *Proc. IEEE Ultrason. Symp.*, 3:1503–1506, 1994.

[43] D. O'Neill. Element Placement in Thinned Arrays Using Genetic Algorithms. *Proc. OCEANS '94*, 2:301–306, September 1994.

20

Fractional Delay Filters—Design and Applications

V. Välimäki and T. I. Laakso

Abstract

In numerous applications, such as communications, audio and music technology, speech coding and synthesis, antenna and transducer arrays, and time delay estimation, not only the sampling frequency but the actual sampling instants are of crucial importance. Digital fractional delay (FD) filters provide a useful building block that can be used for fine-tuning the sampling instants, i.e., implement the required bandlimited interpolation. The FD filters can be designed and implemented flexibly using various established techniques that suit best for the particular application. In this review article, the generic problem of designing digital filters to approximate a fractional delay is addressed. We compare FIR and all-pass filter approaches to FD approximation. Time- and frequency-domain characteristics of various designs are shown to illustrate the nature of different approaches. Special attention is paid to time-varying FD filters and the elimination of induced transients. Also, nonuniform signal reconstruction using polynomial filtering techniques is discussed. Several applications, ranging from synchronization in digital communications to music synthesis, are described in detail. An extensive list of references is provided.

V. Välimäki • Helsinki University of Technology, Laboratory of Acoustics and Audio Signal Processing, Espoo, Finland. E-mail: vesa.valimaki@hut.fi

T. I. Laakso • Helsinki University of Technology, Signal Processing Laboratory, Espoo, Finland. E-mail: timo.laakso@hut.fi

Nonuniform Sampling: Theory and Practice, edited by Marvasti
Kluwer Academic/Plenum Publishers, New York, 2001.

20.1. Introduction

The sampling rate must satisfy the Nyquist criterion in order for a sample set to represent adequately the original continuous signal. This problem has been addressed in the sampling theory literature [55, 81, 84]. However, the appropriate sampling rate alone is not sufficient for many applications—also the *sampling instants* must be properly selected. For example, in digital communications, the decisions of the received bit or symbol values are made based on samples of the received continuous-time pulse sequence which should be taken exactly at the middle of each pulse to minimize probability of erroneous decision. This requires that both the sampling frequency and the sampling instants must be synchronized to the incoming signal.

Another class of problems is modeling of dynamical physical systems which usually involves discretization of complex sets of differential equations. Particularly in multidimensional problems, this results in massive computations where controlling the accuracy of the result may be hard. Specifically, problems are caused by exact fitting of the boundary constraints, e.g., in simulating an acoustical tube that is precisely as long as required or a vibrating membrane that has exactly circular shape of given diameter, regardless of the employed sampling grid.

Both examples are typical applications of *fractional delay* (*FD*) *filters*, i.e., situations where uniform sampling is used and interpolation between samples is required. Fractional delay means, assuming uniform sampling, a delay that is a noninteger multiple of the sample interval. Employing fractional delay filters facilitates the use of traditional well-known methods developed for uniformly sampled signals and yet the observation of signal values at arbitrary locations between the existing samples. Thus, FD filters can be viewed as a first step towards nonuniformly sampled discrete-time systems.

One may claim that this is nothing new: interpolation methods have been known for centuries. Nevertheless, the FD filters have proven useful in providing a systematic framework for solving different interpolation problems and in providing an engineering building block whose properties can be controlled in a desired way.

In this chapter, we review the theory and applications of fractional delay filters. First we introduce the FD filter approximation problem and survey the known techniques for designing nonrecursive (FIR) and recursive (IIR, especially all-pass) filters approximating a given FD value. We focus on frequency-domain design methods. This has the advantage of enabling the use of the well-developed toolbox of linear filter design algorithms which often are specified in the frequency domain. Special attention is paid to the implementation of time-varying FD filters and to the elimination of transients in time-varying recursive FD filters.

The filtering of nonuniformly sampled signals is also briefly addressed. The simultaneous reconstruction and noise suppression using polynomial filtering techniques is discussed with examples.

Finally, we proceed to applications. Several fields are discussed where FD filters have been found useful. A selection of applications is considered in more detail, e.g., synchronization of digital modems, conversion between arbitrary sampling frequencies, and simulation of time-varying acoustic propagation delay in virtual audio environments. An extensive list of references is provided.

20.2. Ideal Fractional Delay

We start by investigating the properties of continuous and discrete-time ideal fractional delay systems. We concentrate on processing of one-dimensional signals that are functions of time.

20.2.1. Continuous-Time System for Arbitrary Delay

Consider a delay element, i.e., a linear system whose purpose is to delay an incoming continuous-time signal $x_c(t)$ by τ (in seconds). Here the subscript 'c' refers to 'continuous-time'. The output signal $y_c(t)$ of this system can be expressed as

$$y_c(t) = x_c(t - \tau). \tag{1}$$

Additional insight can be gained by considering the delay in the frequency domain. The Fourier transform $X_c(F)$ of signal $x_c(t)$ is defined as

$$X_c(F) = \int_{-\infty}^{\infty} x_c(t)e^{-j2\pi Ft}dt. \tag{2}$$

where F is frequency in Hz. The Fourier transform $Y_c(F)$ of the delayed signal $y_c(t)$ can be written in terms of $X_c(F)$ as

$$Y_c(F) = \int_{-\infty}^{\infty} y_c(t)e^{-j2\pi Ft}dt = \int_{-\infty}^{\infty} x_c(t-\tau)e^{-j2\pi Ft}dt = e^{-j2\pi F\tau}X_c(F). \tag{3}$$

The frequency response $H_{id}(F)$ of the delay element can be expressed by means of Fourier transforms $X_c(F)$ and $Y_c(F)$. This yields

$$H_{id}(F) = \frac{Y_c(F)}{X_c(F)} = e^{-j2\pi F\tau}. \tag{4}$$

Hence, in the frequency domain the delay τ corresponds to a complex exponential factor $e^{-j2\pi F\tau}$. In some applications, it is desired to approximate a given delay directly in the continuous-time domain by analog filters [114]. However, in this chapter we focus on discrete-time delay systems only.

20.2.2. Discrete-time System for Arbitrary Delay

If the Fourier transform $X_c(F)$ is non-zero only on a finite interval $-W \leq F \leq W$, the continuous-time signal $x_c(t)$ is said to be *bandlimited.* According to the sampling theorem, signal $x_c(t)$ may then be expressed by its samples $x(n) = x_c(nT)$, where n is the sample index ($n = \ldots, -1, 0, 1, 2, \ldots$) and $T = 1/F_s$ is the sampling interval when the sampling frequency is $F_s > 2W$. For simplicity, we omit T and use $x(n)$ to denote the samples of the discrete-time signal.

We want to express the *discrete-time version* of the delay operation for a sampled bandlimited signal. The outcoming discrete-time signal $y(n)$ would ideally be written as

$$y(n) = x(n - D) \tag{5}$$

where $D = \tau/T$ is the desired delay normalized with respect to the sampling interval. Note that τ/T is generally irrational since τ is usually not an integral multiple of sampling interval T. The delay D may thus be written in the form $D = \mathrm{floor}(D) + d$, where $0 \leq d < 1$ is the *fractional delay* and the floor function returns the greatest integer less than or equal to D.

Unfortunately, as $x(n)$ and $y(n)$ are sequences whose values are defined for integer argument values only, equation (5) is meaningful only for integral values of D. Then the samples of the output sequence $y(n)$ are equal to the delayed samples of the input sequence $x(n)$, and the delay element is called a *digital delay line.* However, if D were real-valued, the delay operation would not be this simple, since the output value would lie somewhere between the existing samples of $x(n)$. In this case, the sample values of $y(n)$ have to be generated based on the sequence $x(n)$ by using *interpolation*. This is known as the fractional delay problem in digital signal processing.

The spectrum of a discrete-time signal can be expressed by means of the *discrete-time Fourier transform* (DTFT). In this integral transform, the time variable is discretized, but the frequency variable is continuous. The DTFT of signal $x(n)$ is defined as (see, e.g., [53], pp. 140–151)

$$X(e^{j2\pi f}) = \sum_{n=-\infty}^{\infty} x(n) e^{-j2\pi fn}, \tag{6}$$

where f is the normalized frequency, $f = F/F_s$, and thus $f = 1/2$ corresponds to the Nyquist limit. The frequency response of the ideal discrete-time delay element can be given as

$$H_{id}(e^{j2\pi f}) = \frac{Y(e^{j2\pi f})}{X(e^{j2\pi f})} = e^{-j2\pi fD}. \tag{7}$$

This result is comparable to (4)—only now the frequency response is periodic due to discretization in time. To be consistent with the z-transform notation used commonly in digital signal processing literature, we express the transfer function as

$$H_{id}(z) = \frac{Y(z)}{X(z)} = \frac{z^{-D}X(z)}{X(z)} = z^{-D}, \tag{8}$$

where D is the delay in samples. Note that the z-transform representation in (8) is used in the Fourier transform sense so that $z = e^{j2\pi f}$. In principle, the z-transform is defined only for integral powers of z and thus, if D were real-valued, the term z^{-D} should be written as an infinite series making the notation unnecessarily involved.

To understand how to produce a fractional delay using a discrete-time system, it is necessary to discuss interpolation techniques. Interpolation of a discrete-time signal is based on the fact that the amplitude of the corresponding continuous-time *bandlimited* signal changes smoothly between the sampling instants.

20.2.3. Fractional Delay and Signal Reconstruction

The fractional delay d can in principle have any value between 0 and 1. Therefore, to produce an arbitrary fractional delay for a discrete-time signal $x(n)$, one needs to know a way to compute the value of the underlying continuous-time signal $x_c(t)$ for all t. This leads us to the problems of sampling and reconstruction. According to the sampling theorem, a uniformly sampled signal can be reconstructed from its samples as follows [55]:

$$x_c(t) = \sum_{n=-\infty}^{\infty} x(n)\mathrm{sinc}[F_s(t - nT)]. \tag{9}$$

The sinc function is defined as $\mathrm{sinc}(t) = \sin(\pi t)/\pi t$.

According to (9) the ideal bandlimited interpolator has a continuous-time impulse response

$$h_c(t) = \mathrm{sinc}(F_s t). \tag{10}$$

This impulse response converts a discrete-time signal to a continuous-time one.

In delay applications, however, it is necessary to know the value of a signal at a single time instant between the samples. The desired result may be obtained by shifting the impulse response (10) by D and then sampling it at equidistant points. The output $y(n)$ of the ideal discrete-time fractional delay element is computed as

$$y(n) = x(n - D) = \sum_{k=-\infty}^{\infty} x(k)\mathrm{sinc}(n - D - k). \tag{11}$$

In conclusion, a fractional delay requires reconstruction of the discrete-time signal and shifted resampling of the resulting continuous-time signal. In (11) these two operations have been combined. The impulse response of an ideal discrete-time delay element can be expressed as

$$h_{id}(n) = \mathrm{sinc}(n - D). \tag{12}$$

The properties of this ideal filter are reviewed in the following.

20.2.4. Characteristics of the Ideal Fractional Delay Element

The *magnitude response* of the ideal fractional delay element can be obtained from the frequency response (7) as

$$|H_{id}(e^{j2\pi f})| = 1. \tag{13}$$

Thus, the magnitude response of the ideal delay element is flat. Its *phase response* is $\theta_{id}(f) = -2\pi f D$. Its *phase delay* and *group delay* are, respectively,

$$\tau_{p,id}(f) = -\frac{\theta_{id}(f)}{2\pi f} = D \tag{14}$$

and

$$\tau_{g,id}(f) = -\frac{1}{2\pi}\frac{d\theta_{id}(f)}{df} = D. \tag{15}$$

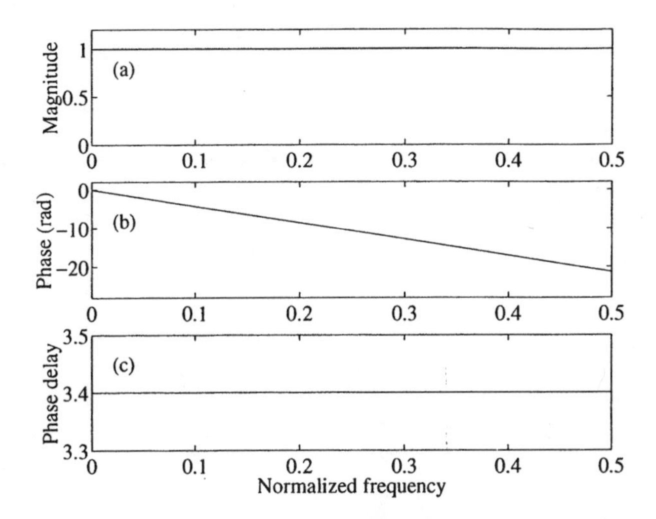

Figure 1. Frequency-domain characteristics of an ideal FD element for $D = 3.4$ samples: (a) magnitude response, (b) phase response, and (c) phase delay.

Both the phase delay and the group delay are measures for the delay of the system. Their difference is typically illustrated by considering an amplitude-modulated signal for which the phase delay and the group delay describe the delay experienced by the carrier signal and the envelope, respectively [110]. For a linear-phase system, the measures yield an identical result, which is independent of frequency—just like in (14) and (15).

Figure 1 illustrates the characteristics of an ideal FD transfer function. The responses are shown up to the Nyquist frequency, which is equivalent to $F_s/2$, or normalized frequency $f = 1/2$. The flat magnitude response, the linear phase response, and the constant phase delay (equivalent to the group delay in this case) are characteristics of a linear-phase all-pass system. However, a fractional-delay element is not a linear-phase system in the traditional sense[†] since the impulse response (12) is generally not symmetric. Oppenheim and Schafer (see [108]), Sec. 5.7) use the term "generalized linear phase" in the context of fractional delay filters.

If D is an integer ($d = 0$), the impulse response of the delay element (12) is zero at all sampling points except at $n = D$. An example is presented in Fig. 2(a) where both the continuous-time and the sampled impulse response of the ideal FD filter is shown. In this case, a chain of unit delays, as discussed earlier, can implement the delay element.

[†]Conventional linear-phase FIR filters are known to have a symmetric or anti-symmetric impulse response with respect to its midpoint (see, e.g., [108] or [53]).

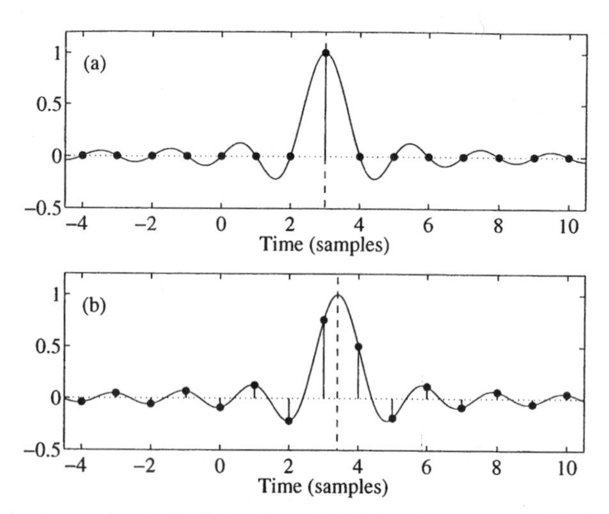

Figure 2. Continuous-time (solid line) and sampled (•) impulse response of the ideal fractional delay filter, when the delay is (a) $D = 3.0$ samples and (b) $D = 3.4$ samples. The vertical dashed line indicates the midpoint of the continuous-time impulse response in each case.

When D is a fractional number, i.e., $0 < d < 1$, the impulse response has non-zero values at all index values n. Fig. 2(b) gives an example where $D = 3.4$ samples. In this case, the impulse response $h_{\mathrm{id}}(n)$ of the discrete-time delay element, which is a shifted and sampled version of the sinc function, is infinitely long in both directions. For this reason, the impulse response corresponds to a *noncausal* filter which cannot be made causal by a finite shift in time. In addition, the filter is not stable since the impulse response (12) is not absolutely summable. The ideal filter is thus *nonrealizable*. To produce a realizable fractional delay filter, some finite-length, causal approximation for the sinc function has to be used.

Before considering the approximations, we notify a particular property of the ideal frequency response that makes the FD approximation difficult. The imaginary part of the ideal transfer function (7) at the Nyquist frequency ($f = 1/2$) is $-\sin(\pi D)$. This implies that when D has a non-zero fractional part d, the transfer function has a complex value at $f = 1/2$. However, discrete-time filters with real coefficients have the property that their frequency response is real-valued at $f = 1/2$. Thus, the approximation error cannot be smaller than $|\sin(\pi D)|$ at the Nyquist frequency [12, 161]. This strongly influences the behavior of FD filters and facilitates understanding of design obligations.

20.3. Design of FIR Fractional Delay Filters

As discussed in the previous section, the ideal FD filter has an infinitely long impulse response which is not realizable in practical systems. In this section, we focus on the approximation of the fractional delay using realizable filters, the simplest class of which are finite impulse response (FIR) filters. For a more detailed discussion and examples on FIR FD filter design, see [12] and [71].

The transfer function of an FIR filter is of the form

$$H(z) = \sum_{n=0}^{N} h(n)z^{-n}, \tag{16}$$

where N is the order of the filter, and $h(n)$ ($n = 0, 1, \ldots, N$) are the real-valued coefficients that form the impulse response of the FIR filter. Note that the length of the impulse response, i.e., the number of the filter coefficients, is $L = N + 1$. The block diagram of the FIR filter is shown in Fig. 3.

In the design procedure our aim is to minimize the error function which is defined as the difference of the ideal frequency response and the approximation, i.e., the frequency-response error

$$E(e^{j2\pi f}) = H_{\mathrm{id}}(e^{j2\pi f}) - H(e^{j2\pi f}). \tag{17}$$

The choice of error norm affects the choice of the design method. In the following, we consider the three most common approximation criteria:

1. minimization of the L_2 *norm* of the error, or the error power,
2. the *maximally-flat* criterion, and
3. minimization of the L_∞ *norm* of the error (minimax or Chebyshev criterion).

However, before going into details, let us discuss the important concept of polyphase filters and their relation to FD approximation.

20.3.1. Polyphase FIR Filters

The polyphase filter structure has been developed for multirate signal processing. It is a straightforward way to implement interpolation and decimation filters. It also offers an easy method to design FIR fractional delay filters using optimization techniques for linear-phase FIR filters [7, 18, 19, 96, 124, 131, 133, 172]. This method only allows the design of FD filters for rational fractions of a unit delay. For example, to split a unit delay into Q steps, one can design a Qth-

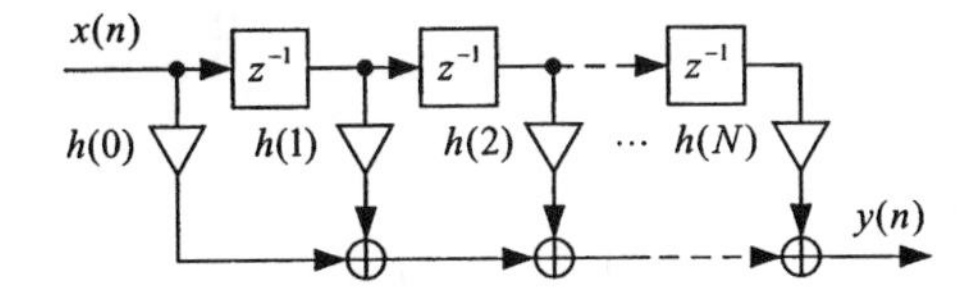

Figure 3. Block diagram of an FIR filter of order N.

band lowpass filter with the normalized bandwidth of $1/2Q$ and form the Q-branch polyphase structure by picking up every Qth sample value from the filter. Each of these branches approximates a fractional delay d_k so that

$$d_k = \frac{k}{Q}, \tag{18}$$

where $k = 0, 1, 2, \ldots, Q-1$. Thus, each polyphase branch can be used as a fractional delay filter. In order to achieve a comparable frequency response for every branch, the length of the prototype filter should be a multiple of Q plus 1.

For example, if it is desired to divide a sampling interval into 4 steps, i.e., $Q = 4$, and the length of each filter is chosen to be $L = 10$, the prototype filter to be designed should be of length $QL + 1 = 41$. An example of the impulse response of a linear-phase prototype filter of this length is shown in Fig. 4, and Fig. 5 presents the four FD filters obtained from the prototype. Note that the filter of Fig. 5(c) approximating the delay $d = 0.5$ has a symmetric impulse response, which implies that it is a linear-phase FIR filter. The impulse responses in Fig. 5(b) and (d) are time-reversed versions of each other.

The lowpass prototype filter should be linear-phase, but it can be designed with any available method. However, the optimality of the prototype filter is not shared with the branch filters. For example, equiripple magnitude characteristic will be lost. A feature that limits the usefulness of this approach is that only rational fractional delays are available. Naturally, one can obtain an arbitrarily good resolution of fractional delays by making Q large but this leads to very high-order FIR filter design. Since there are many other methods available for fractional delay filter design, the polyphase approach seems outdated for this purpose. However, rather than for approximation, the multirate approach can be used for implementing high-quality wideband FD filters, as proposed in [100]. Using a polyphase implementation with two branches, the accuracy of approximation can be increased without excessive total processing delay.

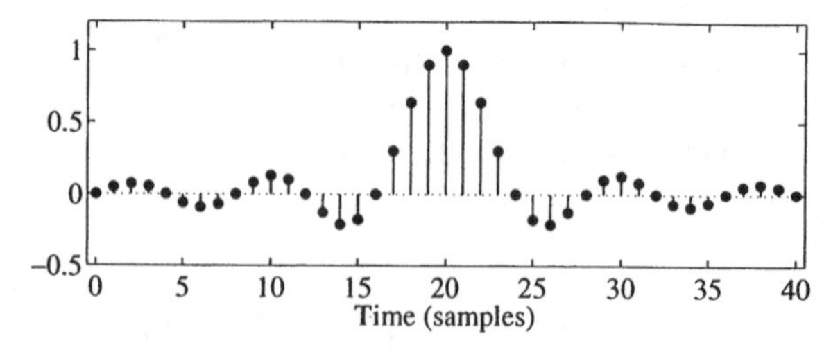

Figure 4. Prototype linear-phase Qth-band FIR filter used for creating a polyphase structure ($Q = 4$, $L = 10$).

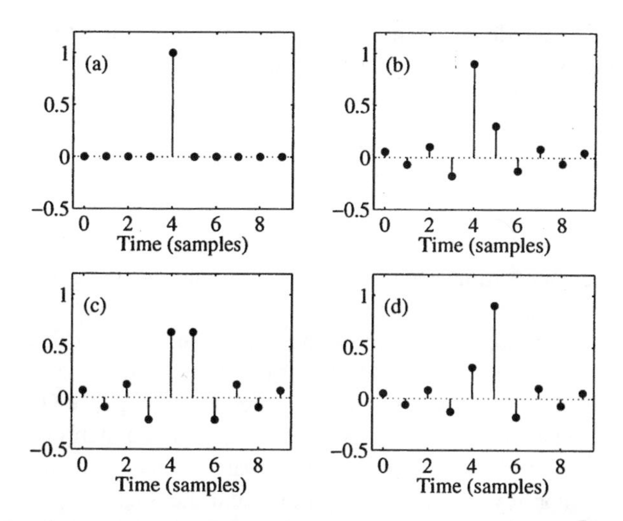

Figure 5. Impulse responses (length $L = 10$) of four fractional delay filters obtained by collecting every fourth sample value with different offsets from the prototype impulse response of Fig. 4. The corresponding fractional delay values are (a) 0.0, (b) 0.25, (c) 0.5, and (d) 0.75.

20.3.2. Least Squared Integral Error Design

The intuitively most attractive method for designing realizable FD filters for arbitrary delay values is the truncation of the ideal impulse response defined by (12). This method minimizes the least squared (LS) error function E_{LS} which is equal to the L_2 norm (integrated squared magnitude) of the frequency-response error (17), i.e.,

$$E_{\mathrm{LS}} = 2\int_0^{1/2} |E(e^{j2\pi f})|^2 df = 2\int_0^{1/2} |H_{\mathrm{id}}(e^{j2\pi f}) - H(e^{j2\pi f})|^2 df. \tag{19}$$

Using Parseval's relation, we get

$$E_{\mathrm{LS}} = \sum_{n=-\infty}^{\infty} [h_{\mathrm{id}}(n) - h(n)]^2 = \sum_{n=-\infty}^{\infty} [h_{\mathrm{id}}^2(n) + h^2(n) - 2h(n)h_{\mathrm{id}}(n)]. \tag{20}$$

From (20), it is possible to derive a closed-form solution for the squared integral error in the case of fractional delay approximation. According to Parseval's relation, the total sum of the ideal impulse response is equal to 1. The second and third term of (20) include the coefficients of the Nth-order FIR filter and thus the summation indices can be limited. The closed-form solution is

$$E_{\text{LS}} = 1 + \sum_{n=0}^{N} [h^2(n) - 2h(n)\text{sinc}(n - D)]. \tag{21}$$

The optimal solution for an Nth-order FIR FD filter in the L_2 sense is the one with $N + 1$ coefficients truncated symmetrically around the maximum value, i.e., the central point of $h_{\text{id}}(n)$. Truncating the shifted sinc function is an easy way to design FIR FD filters. This approach has been proposed, e.g., by Sivanand *et al.* [147, 148] and Cain *et al.* [12]. However, it is often not useful since truncation of the impulse response introduces ripple in the frequency response. This is called the *Gibbs phenomenon* (see, e.g., [53]). It causes the maximum deviation from the ideal frequency response to remain approximately constant irrespective of the filter order. This applies to both the magnitude and the phase response.

We present an example where the impulse response of the ideal FD filter for 3.4 samples—see Fig. 2(b)—has been used as a prototype. Eight coefficient values, $h_{\text{id}}(0)$, $h_{\text{id}}(1)$, $h_{\text{id}}(2), \ldots,$ and $h_{\text{id}}(7)$, have been truncated symmetrically about the midpoint. The impulse response of the resulting filter is displayed in Fig. 6(a). These are the coefficients of an LS FIR filter of order $N = 7$ (or, length $L = 8$), which approximates the delay $D = 3.4$ samples. The magnitude of the frequency-response error $E(e^{j2\pi f})$ computed using (17) is presented in Fig. 6(b).

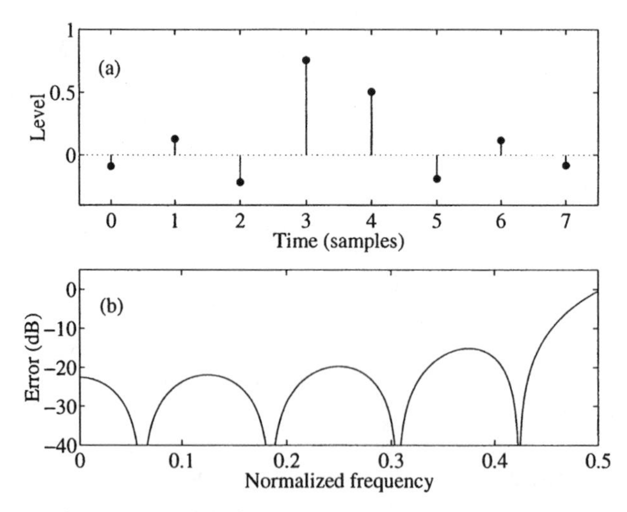

Figure 6. (a) Impulse response and (b) frequency-response error of the LS FIR FD filter obtained by truncating the impulse response of the ideal FD filter ($N = 7$, $D = 3.4$) shown in Fig. 2(b).

Note how the local maxima of the error function increase towards high frequencies.

Rather than the error function itself, engineers often find it more useful to examine the magnitude response and the phase delay or group delay of the filter. The magnitude response of the above filter is shown in Fig. 7(a). The ripple caused by the Gibbs phenomenon can be seen: The oscillation of the response grows towards higher frequencies, and the phase-delay response shown in Fig. 7(b) exhibits similar behavior. Note that the phase delay curve reaches 3.0 samples at the Nyquist frequency (i.e., normalized frequency 0.5).

A variation of the LS design technique is to use a *lowpass* interpolator as a prototype filter instead of the fullband fractional delay filter [71]. The Gibbs phenomenon will be reduced considerably, as desired, but the usable bandwidth will contract as well. Another variation of the basic LS design is to use a reduced bandwidth with a *smooth transition band* function (see, e.g., [112], pp. 63–70). This will make the impulse response of the FD element decay fast. The impulse response will still be infinitely long and must be truncated, but the Gibbs phenomenon is guaranteed to be reduced, since the discontinuity is not as sharp as originally. A good choice for the transition band is a low-order spline multiplied by $e^{-j2\pi fD}$ [71].

A well-known method to reduce the Gibbs phenomenon in FIR filter design is to multiply the ideal impulse response with a bell-shaped non-negative finite-length weighting function, i.e., a window function (see, e.g., [112], pp. 71–83).

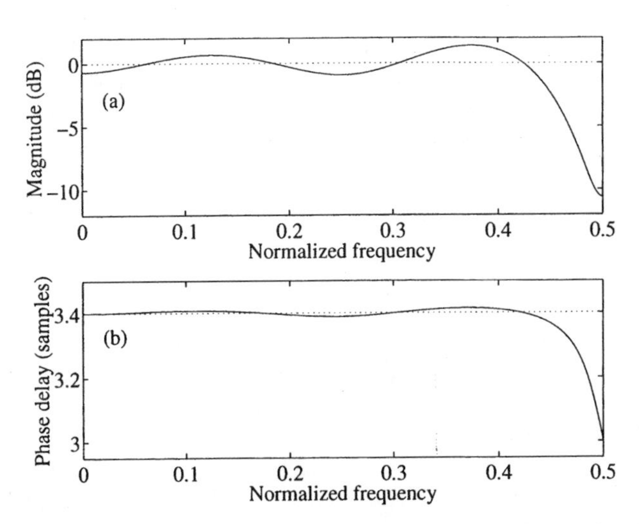

Figure 7. (a) Magnitude response (solid line) and (b) phase delay (solid line) of the LS FIR FD filter of Fig. 6(a). The dotted lines indicate the corresponding ideal characteristics.

Since the truncation yields the optimal solution in the LS sense, any windowing method is bound to be worse. Thus, the window-based design method does not minimize the LS error measure, but it is an *ad hoc* modification of the LS technique.

The impulse response of an FIR filter designed by the windowed LS method can be written in the form

$$h(n) = \begin{cases} w(n-D)\mathrm{sinc}(n-D) & \text{for } 0 \le n \le N \\ 0 & \text{otherwise.} \end{cases} \tag{22}$$

Note that the midpoint of the window function $w(n)$ of length $N+1$ has been shifted by D so that the shifted sinc function will be windowed symmetrically with respect to its center. The window function $w(n-D)$ is generally *asymmetric*, however.

Many window functions, such as the Hamming and Hanning windows, can be easily delayed by a fractional value D [12, 13, 71], whereas Dolph–Chebyshev or Saramäki windows are defined in the frequency domain and only approximative shifting techniques are known [67]. Special window functions for FD applications have been developed by Cain and Yardim *et al.* [13, 195–196].

20.3.3. Weighted Least Squared Integral Error FIR Approximation of a Complex Frequency Response

In principle, the FIR fractional delay filter with the smallest LS error in the defined approximation band is accomplished by defining the response only in that part of the frequency band and by leaving the rest out of the error measure as a "don't care" band [71]. This scheme also enables frequency-domain weighting of the LS error. This technique minimizes the following error function

$$E_{\mathrm{GLS}} = 2\int_0^{1/2} W(f)|E(e^{j2\pi f})|^2 df = 2\int_0^{1/2} W(f)|H(e^{j2\pi f}) - H_{\mathrm{id}}(e^{j2\pi f})|^2 df, \tag{23}$$

where the error is defined in the lowpass frequency band $[0, F_s/2]$ only, and $W(f)$ is the nonnegative frequency-domain weighting function (not to be confused with the time-domain window function $w(n)$).

The optimal solution can be obtained by solving a set of $N+1$ linear equations, as presented in [71]. Numerical problems may arise, particularly in narrowband approximation [71]. However, in FD filter design this is not typical. If the weighting function is $W(f) = 1$ for $f \le \alpha$ and $W(f) = 0$ for $f > \alpha$, where $0 < \alpha \le 1$, the solution can be given in closed form [174]. Also the effect of the input signal spectrum can be taken into account: Oetken *et al.* [106] proposed a *stochastic LS approach* to the design of interpolating filters. In this technique, the design criterion is the minimum expected mean squared output error. The design technique enables a great deal of flexibility and it is recommended whenever the filter properties need to be tailor made with care.

20.3.4. Maximally-flat FIR FD Design: Lagrange Interpolation

In many applications, it is important that the delay is approximated accurately at low frequencies. This can be attained by setting the error function and its N derivatives to zero at the zero frequency. This is the *maximally-flat* (MF) design at $f = 0$. It is interesting to notice that the FIR filter coefficients obtained by this method are the same as the weighting coefficients in the classical *Lagrange interpolation* for uniformly sampled data:

$$h(n) = \prod_{\substack{k=0 \\ k \neq n}}^{N} \frac{D-k}{n-k} \quad \text{for } n = 0, 1, 2, \ldots, N. \tag{24}$$

The first-order ($N = 1$) Lagrange interpolation is equivalent to *linear interpolation*. The coefficient formulas for low-order Lagrange interpolators are given in Table 1. It can be shown that the impulse response of the Lagrange interpolator (24) converges to the shifted and sampled sinc function (12), the ideal FD filter, as N approaches infinity (for a proof, see [81], pp. 100–102).

Lagrange interpolation has been studied in a large number of papers in the signal processing literature up to the present day [6, 16, 23, 24, 47, 48, 55, 63, 64, 72, 79, 80, 87, 95, 107, 130, 142, 147, 156, 164, 173, 174].

We present a design example with the same design parameters as in Section 20.3.2 ($N = 7$, $D = 3.4$). The impulse response of the Lagrange interpolator, computed using (24), is shown in Fig. 8(a). The corresponding frequency-response error magnitude, shown in Fig. 8(b), is now very small at low

Table 1. Coefficients of low-order Lagrange interpolation filters.

	$h(0)$	$h(1)$	$h(2)$	$h(3)$
$N=1$	$1-D$	D		
$N=2$	$(D-1)(D-2)/2$	$-D(D-2)$	$D(D-1)/2$	
$N=3$	$-(D-1)(D-2)(D-3)/6$	$D(D-2)(D-3)/2$	$-D(D-1)(D-3)/2$	$D(D-1)(D-2)/6$

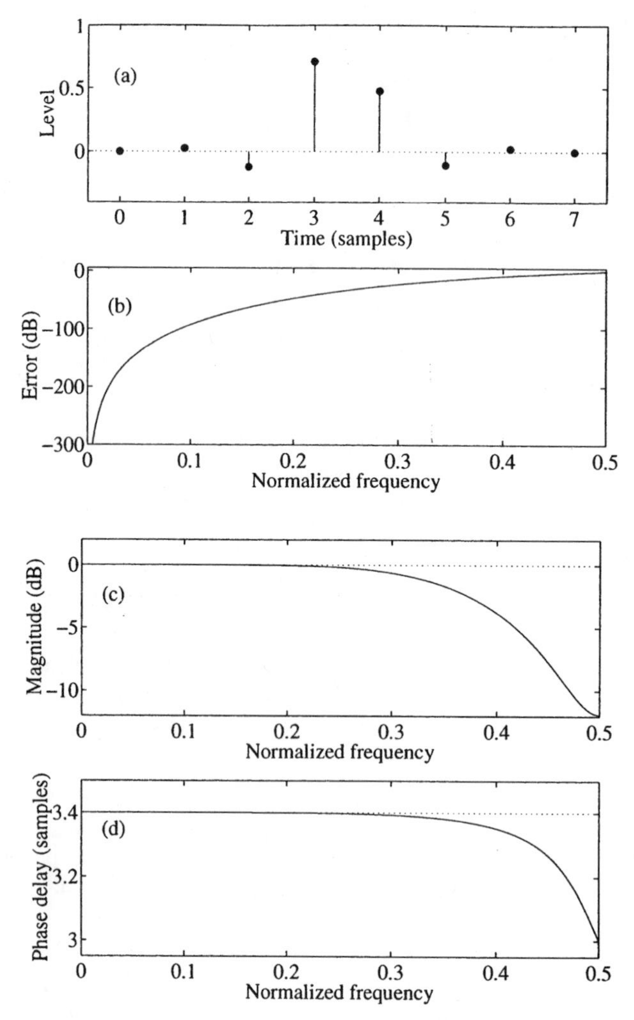

Figure 8. (a) Impulse response, (b) frequency-response error, (c) magnitude response, and (d) phase delay of a Lagrange interpolating filter ($N = 7$, $D = 3.4$). The dotted lines indicate the ideal characteristics in (c) and (d).

frequencies ($-\infty$ dB at $f = 0$), but smoothly increases in a monotonic fashion. The magnitude and phase-delay responses of the filter—see Fig. 8(c) and (d)—exhibit an increasing deviation from their ideal values, which are obtained at zero frequency.

20.3.5. Minimax Design of FIR FD Filters

If it is desired to minimize the peak approximation error, the *minimax*, or *Chebyshev*, error criterion should be used. This approximation problem can usually be solved only by iterative techniques, which are based on either an iterative weighted least squares algorithm or a modification of the Remez exchange algorithm. The minimax optimization problem is rendered particularly demanding since the target frequency response is complex-valued. (Note that the Remez algorithm originally assumes that the function to be approximated is real-valued.) Many advanced algorithms for complex approximation with minimax error characteristics have been presented [58, 77, 112, 117, 144]. The application of various minimax approximation methods to FIR FD filters have been tackled by Oetken [107], Pyfer and Ansari [121], Putnam and Smith [120], and Brandenstein and Unbehauen [11]. For a more detailed discussion on minimax FIR FD filters, see [71]. The method proposed by Oetken is particularly simple [107]. It is based on an odd-order linear-phase Chebyshev prototype filter ($d = 0.5$), whose zeros are modified with a matrix operation to obtain a fractional delay filter for a given delay parameter d.

Figure 9 gives an example of FD filter design according to Oetken's method. The delay parameter is the same as in the previous examples, $D = 3.4$. The length of the filter is 8, and the approximation bandwidth is $[0, 0.4F_s]$. It is seen that the impulse response is only slightly different from those of other FIR filters. However, the largest peaks of the frequency-response error are now maintained at the level of about -27 dB in the approximation band, i.e., from 0 up to the normalized frequency 0.4. Thus, the equiripple design is better in this sense than the fullband LS FIR filter whose error curve was displayed in Fig. 6(b)—the maximum error of the LS filter in the same frequency band is about -15 dB. The Lagrange interpolating filter is still worse in this kind of comparison: its frequency-response error (see Fig. 8(b)) at normalized frequency 0.4 is only -8.7 dB.

20.3.6. Fractional Delay Filters Based on Splines

Splines are a class of polynomial interpolation techniques that have several applications in numerical computation. The B-spline interpolator can be implemented as an all-pole or FIR filter. It has been found effective in many DSP applications, such as image processing [26, 52, 171], sampling-rate conversion [14, 20, 204, 205], and signal reconstruction [169, 170]. Aldroubi *et al.* have proved that a cardinal spline interpolator converges toward the ideal interpolator as the order of the filter approaches infinity [4]. The properties of B-spline and

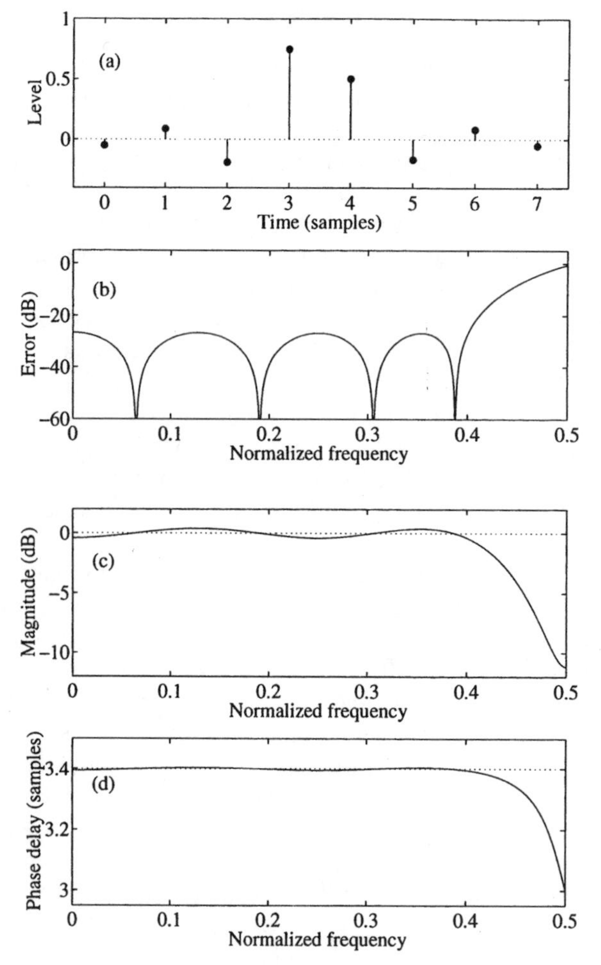

Figure 9. (a) Impulse response, (b) frequency-response error, (c) magnitude response, and (d) phase delay of an equiripple FIR FD filter designed using Oetken's method ($N = 7$, $D = 3.4$). The dotted lines indicate the ideal characteristics in (c) and (d).

Lagrange interpolation have been compared in [78]. A generalization of B-spline interpolators and an efficient implementation structure have been proposed in [30].

20.3.7. General Properties of FIR FD Filters

Some general properties of FIR FD filters can be summarized as follows:

1. The best approximation for a given design method and a given filter order N is obtained when $(N-1)/2 \leq D < (N+1)/2$ (see, e.g., [60] or

[71])[†]. This implies that the best FD filters are those whose impulse response is almost symmetric. This also means that the total delay of the filter is close to $N/2$ samples.

2. The approximation error decreases as the order N is increased (see, e.g., [142] or [71]). However, it is typical in practical applications that the smallest acceptable filter order N is used to save resources and to keep the total delay small. Thus, it becomes essential to choose the right method to design an FD filter that accomplishes the required task with the smallest possible order N.
3. Since FIR FD filters do not include any feedback, they are always stable. Furthermore, the transients due to abrupt changes in the input signal or due to changes in filter coefficient values are always finite-length. This is important in applications where the FD value needs to be controlled.

20.4. Design of IIR Fractional Delay Filters

As is well known from the general filter approximation theory, the approximation of a given frequency response with a recursive (IIR) filter achieves the same quality with a lower complexity (e.g., a smaller number of multiplications and additions) than an FIR filter [112]. However, the design of general IIR fractional delay filters having a different numerator and denominator polynomials has not been much discussed in the literature. An example is the study by Tarczynski and Cain where reduced-bandwidth IIR fractional delay filters are optimized iteratively [161]. The design of optimal IIR filters is a difficult problem. One particular difficulty with recursive filters is that they may be unstable whereas FIR filters are always stable. The design procedure has to account for this possibility and ensure the stability of the IIR filter.

Since the magnitude response of an ideal FD element is perfectly flat, we consider here only the so-called *all-pass filters*. Their magnitude response is always exactly flat irrespective of the filter coefficients. All-pass filters are typically used for phase equalization and other signal processing tasks where the phase characteristics are of greatest concern. Since the FD approximation is essentially a phase approximation problem, the all-pass filter is particularly well suited to this task.

[†]In the case of Lagrange interpolation, the consequences of interpolation outside the central interval have been discussed in [142] and [174].

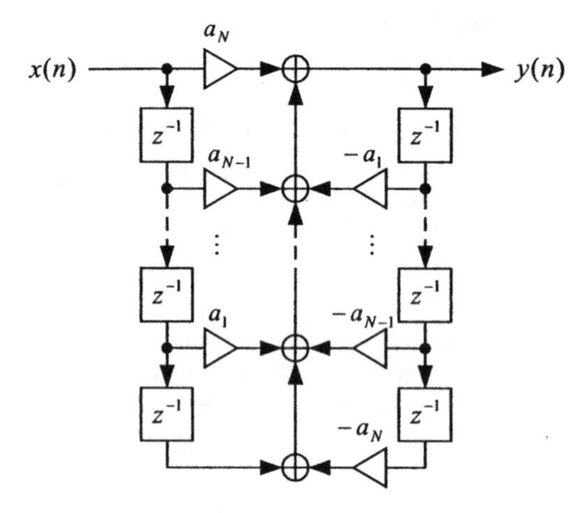

Figure 10. Direct form I implementation of an Nth-order all-pass filter.

20.4.1. Discrete-time All-pass Filter

A discrete-time all-pass filter has the transfer function

$$A(z) = \frac{z^{-N}D(z^{-1})}{D(z)} = \frac{a_N + a_{N-1}z^{-1} + \cdots + a_1 z^{-(N-1)} + z^{-N}}{1 + a_1 z^{-1} + \cdots + a_{N-1}z^{-(N-1)} + a_N z^{-N}}, \tag{25}$$

where N is the order of the filter, $D(z)$ is the denominator polynomial, and the filter coefficients a_k $(k = 1, 2, \ldots, N)$ are real-valued. The numerator of the all-pass transfer function is a mirrored version of the denominator polynomial. The direct-form I implementation structure of the all-pass filter is shown in Fig. 10. The output signal $y(n)$ of the all-pass filter is computed using the following difference equation:

$$\begin{aligned} y(n) &= a_N x(n) + a_{N-1}x(n-1) + \cdots + a_1 x(n-N-1) + x(n-N) \\ &\quad - a_1 y(n) - \cdots - a_{N-1}y(n-N+1) - a_N y(n-N) \\ &= a_N[x(n) - y(n-N)] + a_{N-1}[x(n-1) - y(n-N+1)] + \cdots \\ &\quad + a_1[x(n-N-1) - y(n)] + x(n-N). \end{aligned} \tag{26}$$

The latter form indicates that one output sample of the Nth-order all-pass filter can be computed with N multiplications and $2N$ additions (or subtractions).

The poles of a stable all-pass filter are located inside the unit circle in the complex plane. Since the same coefficients are used in the numerator, its zeros are located outside the unit circle so that their angle is the same but the radius is the inverse of the corresponding pole. For this reason the magnitude response of an all-pass filter is flat. It is expressed as

$$|A(e^{j2\pi f})| = \left| \frac{e^{-j2\pi f N} D(e^{-j2\pi f})}{D(e^{j2\pi f})} \right| = 1. \tag{27}$$

Obviously, the name 'all-pass' filter comes from the above property that this filter passes signal components of all frequencies without attenuating or boosting them. The frequency response of the all-pass filter can be written in the form $A(e^{j2\pi f}) = \exp[j\theta_A(f)]$. This form stresses the fact that the main feature of an all-pass filter is its *phase response* $\theta_A(f)$, which can be expressed as

$$\theta_A(f) = \arg\{A(e^{j2\pi f})\} = -2N\pi f + 2\theta_D(f), \tag{28}$$

where $\theta_D(f)$ is the phase response of $1/D(e^{j2\pi f})$, i.e.,

$$\theta_D(f) = \arg\left\{\frac{1}{D(e^{j2\pi f})}\right\} = \text{unwrap}\left[\arctan\left\{\frac{\sum_{k=1}^{N} a_k \sin(2\pi f k)}{1 + \sum_{k=1}^{N} a_k \cos(2\pi f k)}\right\}\right], \tag{29}$$

where the 'unwrap' function produces the desired monotonically decreasing phase function by subtracting the necessary amount of multiples of π. The *phase delay* $\tau_{p,A}(f)$ of an all-pass filter can be expressed according to (28) as

$$\tau_{p,A}(f) = -\frac{\theta_A(f)}{2\pi f} = N - 2\tau_{p,D}(f), \tag{30}$$

where $\tau_{p,D}(f)$ is the phase delay of $1/D(e^{j2\pi f})$. The corresponding *group delay* is given by

$$\tau_{g,A}(f) = -\frac{1}{2\pi}\frac{d\theta_A(f)}{df} = N - 2\tau_{g,D}(f), \tag{31}$$

where $\tau_{g,D}(f)$ is the group delay of $1/D(e^{j2\pi f})$.

20.4.2. Design of All-pass Fractional Delay Filters

There is a noteworthy difference between the design of FIR and all-pass filters: the coefficients of an FIR filter are easily obtained by the inverse discrete-time Fourier transform of the frequency-domain specifications, since the coefficients of an FIR filter are equal to the samples of its impulse response. However, the relationship between the transfer function coefficients and the impulse response of an all-pass filter or any other recursive filter is not that simple.

Hence, most of the design techniques for all-pass filters are iterative. Many design methods for FD all-pass filters are counterparts of FIR design methods discussed previously in this chapter.

A time-domain approach to FD approximation using an all-pass filter can be obtained with the impulse-response matching technique proposed by Strube [157]. According to our experiments, the technique yields poor results when the truncated impulse response of the ideal FD element (12) is used as the prototype (various integer parts floor(D) $< N + 1$ were tested). In the following our focus is on frequency-domain design methods.

The desired or ideal phase response in the fractional delay approximation problem is $-2\pi f D$, as can be seen from (7). The phase error of an all-pass filter can be defined as the deviation from the desired phase function $\theta_{\mathrm{id}}(f)$ as

$$\Delta\theta(f) = \theta_{\mathrm{id}}(f) - \theta_A(f). \tag{32}$$

20.4.2.1. LS Design of All-pass FD Filters

The weighted least-squares *phase* error that is to be minimized is defined as

$$E_{\mathrm{LS}} = 2\int_0^{1/2} W(f)|\Delta\theta(f)|^2 df, \tag{33}$$

where $W(f)$ is a nonnegative weighting function. Lang and Laakso have introduced an iterative algorithm for the design of the all-pass filter coefficients [75]. The algorithm typically converges to the desired solution, but this cannot be guaranteed. The stability is not guaranteed either.

The LS *phase delay* error can be defined as

$$E_{\mathrm{LS}} = 2\int_0^{1/2} W(f)|\Delta\tau_{\mathrm{p}}(f)|^2 df = 2\int_0^{1/2} \frac{W(f)}{(2\pi f)^2}|\Delta\theta(f)|^2 df. \tag{34}$$

In other words, the phase delay error solution is obtained by introducing an additional weighting function $1/(2\pi f)^2$ to the phase error (33). An eigenfilter formulation of this all-pass design problem was presented by Nguyen *et al.* [105].

20.4.2.2. Maximally-flat Design of All-pass FD Filters

The maximally-flat group delay design is the only known all-pass FD filter design technique that has a closed-form solution. Thiran proposed an analytic solution for the coefficients of an *all-pole* lowpass filter with a maximally-flat group delay response at the zero frequency [166]. A drawback of Thiran's design technique is that the magnitude response of the all-pole lowpass filter cannot be controlled. However, the same coefficients can be used also for the numerator polynomial in the reverse order, according to (35), to obtain an all-pass filter [34, 71, 146]. As can be seen in [31], the group delay of an all-pass filter is twice that of its all-pole counterpart. Thus, Thiran's solution may easily be applied to the design of all-pass filters by making the substitution $d' = d/2$. The solution for the coefficients of a *maximally-flat* (MF) *all-pass filter* can be written in the form [34]

$$a_k = (-1)^k \binom{N}{k} \prod_{n=0}^{N} \frac{d+n}{d+k+n} \quad \text{for } k = 0, 1, 2, \ldots, N. \tag{35}$$

Note that $a_0 = 1$, and thus the coefficient vector need not be scaled. This closed-form solution is called the *Thiran all-pass filter*. It approximates a group delay of $N + d$ samples, when the filter order is N and the fractional delay parameter is d. Since this parameterization may be rather impractical, we substitute $d = D - N$ into (35) so that the parameter D refers to the actual delay rather than the offset from N samples [174]. The coefficient formulas for the Thiran all-pass filters of orders $N = 1, 2$, and 3 are presented in Table 2. The closed-form design of the first-order all-pass filter was derived by using Taylor series expansion in papers by Jaffe and Smith [54], and Smith and Friedlander [153].

Thiran's proof of stability [166] implies that this all-pass filter will be stable when $D > N - 1$. At $D = N - 1$, one of the poles (and one zero) of the Thiran all-pass filter will be on the unit circle. When $D < N - 1$, one or more poles will be outside the unit circle and the filter will be unstable.

The suggested range of values for delay D to be approximated with the Thiran all-pass filter is $[N - 0.5, N + 0.5)$. This choice is close to the optimum that minimizes the average frequency-response error for low-order all-pass filters [174].

In the following, a design example is presented that can be compared with the FIR filter examples of Section 20.3. An MF all-pass filter of order 4 is designed, to approximate a delay of 4.4 samples. The first 11 samples of its

Table 2. Coefficients of the Thiran all-pass filter for $N = 1, 2$, and 3.

	a_1	a_2	a_3
$N = 1$	$-\frac{D-1}{D+1}$		
$N = 2$	$-2\frac{D-2}{D+1}$	$\frac{(D-1)(D-2)}{(D+1)(D+2)}$	
$N = 3$	$-3\frac{D-3}{D+1}$	$3\frac{(D-2)(D-3)}{(D+1)(D+2)}$	$-\frac{(D-1)(D-2)(D-3)}{(D+1)(D+2)(D+3)}$

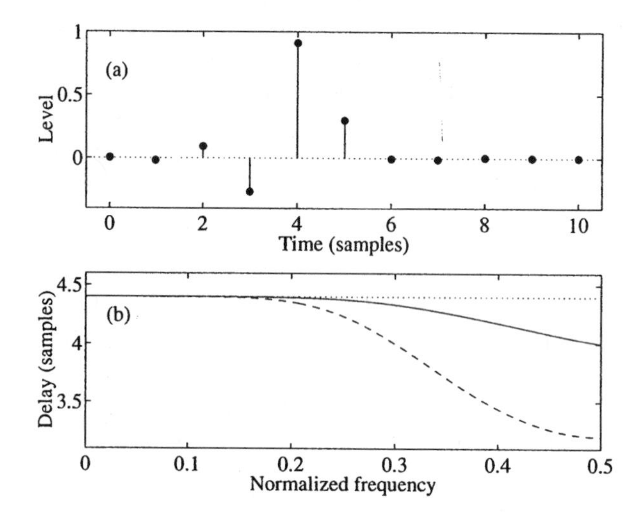

Figure 11. (a) The beginning of the impulse response, and (b) the phase delay (solid line) and group delay (dashed line) of a Thiran all-pass filter of order $N = 4$ approximating a delay of $D = 4.4$ samples.

impulse response are displayed in Fig. 11(a). Note that the impulse response—although of infinite length in theory—decays rapidly, and it does not appear much different from those of FIR FD filters.

Figure 11(b) gives both the phase and the group delay of the all-pass filter. They are practically equivalent at very low frequencies. Note that the phase delay behaves similarly to that of the Lagrange interpolator (see Fig. 8(d)).

Figure 12(a) shows the frequency-response error of the all-pass filter and that of a Lagrange interpolator of Fig. 8. Note that we have chosen the all-pass filter order so that very nearly the same error characteristics are obtained as in our previous Lagrange interpolator example. The all-pass filter appears to be slightly better at normalized frequencies below 0.1 and slightly worse at higher frequencies, as illustrated by the error difference curve of Fig. 12(b). However, the all-pass filter is more efficient in terms of computations, since only 4 multiplications and 7 additions are required per output sample; the Lagrange interpolator requires 8 multiplications and 7 additions. On the other hand, the delay caused by the all-pass filter is one sampling interval more than that of the Lagrange interpolator.

20.4.2.3. Minimax Design of All-pass FD Filters

There exists a large variety of algorithms for equiripple or minimax all-pass filter design in terms of phase, group delay, or phase delay [71, 124, 132]. New iterative techniques for equiripple phase error and phase delay error design of all-

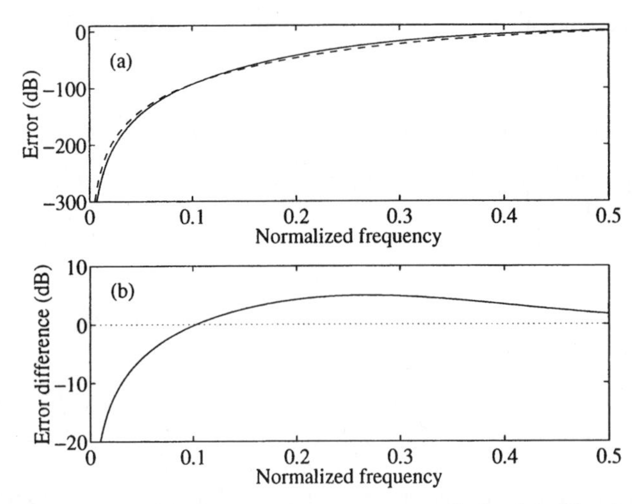

Figure 12. (a) Comparison of the frequency-response error magnitude of the 4th-order all-pass filter (solid line) of Fig. 11, and the 7th-order Lagrange interpolator (dashed line) of Fig. 8. (b) The difference of the error curves in (a).

pass fractional delay filters are described in [71]. Lang has developed a Remez-type algorithm that is guaranteed to converge to the optimal Chebyshev solution [74]. The method also alleviates the numerical problems that have been a major concern in designing high-order all-pass filters.

20.4.3. Discussion

This section has considered the all-pass filter approximation of fractional delay. A digital all-pass filter is a good choice for this task since its magnitude response is exactly flat and the design can concentrate entirely on the delay characteristics. The Thiran all-pass filter was discussed in detail since it is easy to design with closed-form formulas and is very accurate at low frequencies. One problem in the use of recursive FD filters is the *transient phenomenon* that occurs when the filter characteristics are changed during operation. This situation will be discussed in the following.

20.5. Time-Varying Fractional Delay Filters

It is important to investigate the situation where the delay parameter of an FD filter is changed during operation, since in many real-time signal processing applications the desired delay is not constant but varies with time. In this section we study the implementation of a time-varying digital delay line using FIR and IIR filters.

20.5.1. Consequences of Changing Filter Coefficients

We now discuss how the change of the filter coefficients affects the output signal of the filter. A single change of the filter coefficients at time index $n = n_c$ is considered†. There may be two kinds of consequences at the same time [174, 180]:

1. The output signal may suffer from a *transient* that starts at $n = n_c$, and
2. the output signal may experience a *discontinuity* at $n = n_c$.

The *transient* at the filter output is observed if the state variables of the filter contain intermediate results related to the former coefficient set, or if the state variables are cleared. In the case of nonrecursive (FIR) filters this problem does not exist when the filter has been implemented in direct form (as shown in Fig. 3), but in the case of recursive (IIR) filters the occurrence of transients must be accounted for somehow. Figure 13(a) gives an example: a sinusoidal signal is filtered with an all-pass filter, and at time index 30 the coefficients of the all-pass filter are changed. The transient is easily visible in the output signal of the all-pass filter: the signal waveform deviates from the sinusoidal form for a few samples after time index 30.

There is also a *discontinuity* at the same time instant caused by the change of delay. A discontinuity means a sudden change in the corresponding continuous-time signal. The discontinuity of the output signal is simply caused by the fact that after the coefficients have been changed, the output signal will be a result of a different of filtering operation than before the change. Usually this is, of course, a desired result, but the discontinuity may still be harmful in some applications. An example is given in Fig. 13(b): the transient in the signal of Fig. 13(a) has now been completely cancelled but the discontinuity remains. This case corresponds to time-varying IIR filtering with ideal transient cancellation or time-varying FIR filtering where transients do not occur. Note that there is only a small and abrupt single change in Fig. 13(b): the sample value at time index 30 is not part of the sampled sine wave displayed between indices 0 and 29; however, the waveform after time index 30 is seen to be a regular, periodically sampled sine wave. Thus, the signal is free from defects before and after the discontinuity.

Both the transient signal and the discontinuity depend on the input signal of the filter and the magnitude of coefficient change (and also on the time of change); a small change in the values of the coefficients causes a transient with smaller amplitude and a smaller discontinuity than a large change in coefficient values [180]. The severity of the transients and discontinuity can be decreased by

†Note that in this context the subscript 'c' denotes 'change'.

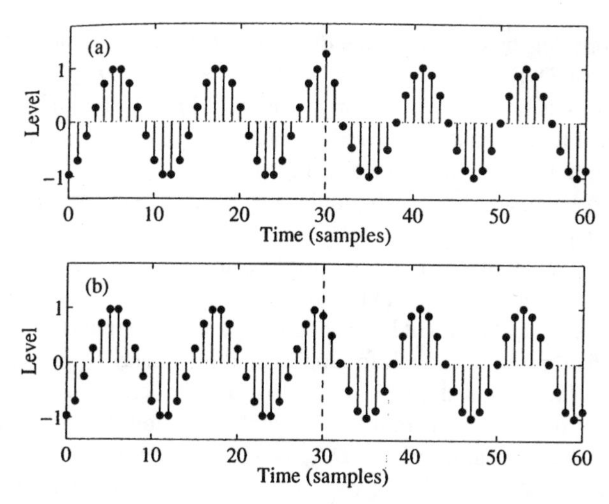

Figure 13. Examples of (a) a transient and (b) a discontinuity in a signal waveform caused by time-varying fractional delay filtering. The vertical dashed line indicates the point of change.

making the change in filter parameters smaller, e.g., by allowing the filter a reasonably long transition time when the values of the coefficients are gradually changed (by interpolation) from the initial to the target values.

20.5.2. Implementing Variable Delay Using FIR Filters

We now examine the realization of a variable-length digital delay line using an FIR interpolating filter. Thereafter, design methods for time-varying FIR FD filters are reviewed.

20.5.2.1. Time-varying Digital Delay Line

Figure 14 shows the practical implementation of a given fractional delay using an FIR filter; part of the delay is implemented with a delay line of length M and the rest with an FIR filter that approximates the delay of D samples. The overall delay experienced by the signal $x(n)$ is thus $M + D$.

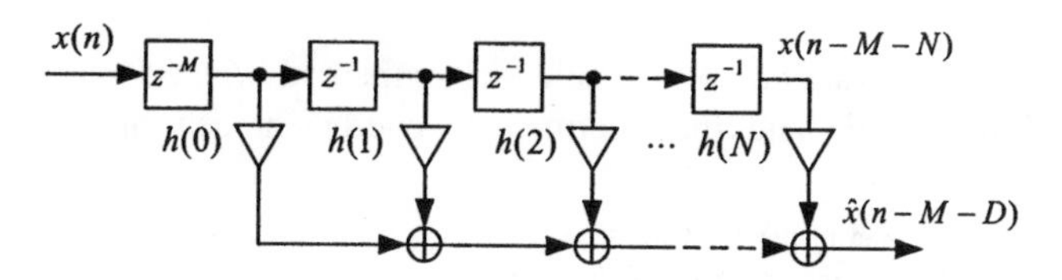

Figure 14. A digital delay line implemented using FIR interpolation.

Let us consider the case when the delay parameter is a function of time, for example, so that $D(n) = D_1$ for $n < n_c$ and $D(n) = D_2$ for $n \geq n_c$. If the next delay value D_2 does not fulfill the requirement $(N-1)/2 \leq D_2 < (N+1)/2$, the taps of the FIR FD filter must be moved to obtain the best approximation, i.e., the delay-line length M must be changed in Fig. 14.

Care must be taken to correctly implement the delay change when $D_2 > D_1$ and the filter taps are moved; the delay line must be long enough so that the delayed sample values needed for filtering are immediately available. A good practical solution is to define a maximum delay D_{max} that is the assumed largest delay value (in samples) that will be needed in a particular application. The delay-line length is then set long enough for approximating a delay D_{max} using an Nth-order FIR FD filter. This implies that the delay-line length must be at least $M_{tot} = D_{max} + \text{floor}(N/2)$, where the extra floor$(N/2)$ unit delays are needed by the FIR filter when the largest delay D_{max} is approximated.

Note that a small delay $D_2 < (N-1)/2$ cannot be approximated with good accuracy[†]. In this case, one should either use a filter of lower order N, or use the filter of original order outside its optimal range of delay parameter D. Examples with Lagrange interpolation have been presented in [174].

Thus, with careful design of the implementation, the transient signal can be completely eliminated when using an FIR interpolating filter. This is natural, since there is no feedback in the system in Fig. 14.

20.5.2.2. Implementation of Time-Varying FIR FD Filters

There are two basic approaches to implementing a time-varying FIR filter, 1) a *table lookup method* and 2) *on-line computation of coefficients* [31, 32, 43]. In the table lookup method, the coefficients of all FD filters that will be needed are computed and stored in advance. When it is needed to change the delay parameter, the coefficients corresponding to the new delay value are retrieved from the memory. The main disadvantage of this approach is the need for a large memory that can be used for fast retrieval of data. For example, if the sampling interval is needed to divide into Q parts, and each fractional delay is approximated with an FIR filter of length L, the amount of coefficients to be stored is QL. A trivial way to reduce this amount by 50% is to store only the coefficients for the interval $0 < d < 0.5$, since the coefficients of FD filters for $0.5 < d < 1$ can be obtained from them by time reversal, as seen by comparing Fig. 5(b) and Fig. 5(d).

There is also a variation of the table lookup method, where some coefficients are stored in a table and the intermediate coefficient sets are obtained by

[†] This discussion is irrelevant for the case $N = 1$, e.g., linear interpolation, since for the first-order filter the smallest delay in the optimal range is zero.

polynomial interpolation [2, 122, 154]. Even the use of simple linear interpolation may reduce the amount of coefficient memory by more than 50% [154].

The main advantage of on-line computation of filter coefficients is the saving of memory. However, the computational complexity of coefficient update is usually larger than in the table lookup method. Laakso *et al.* [71] discussed the Fourier-transform method and a general Oetken's method for changing the delay of an FD filter. During the last decade, the Farrow structure has become the most popular method for implementing time-varying FD filters [31–32, 43–44, 173–174, 186–189, 192]. This method is discussed in detail in the following sections.

20.5.2.3. Farrow Structure for Lagrange Interpolation

We now discuss a particularly efficient implementation structure for time-varying FIR FD filtering. Farrow [32] suggested that every filter coefficient of an FIR FD filter could be expressed as an Nth-order polynomial in the variable delay parameter D. This results in $N+1$ FIR filters with constant coefficients. In this section we present an efficient structure for real-time implementation of Lagrange interpolation. This derivation has been adapted from [173] and [174]. The Farrow structure for Lagrange interpolation has been proposed first by Erup *et al.* [31].

Lagrange interpolation is usually implemented using a direct-form FIR filter. An alternative structure is obtained by approximating the continuous-time function $x_c(t)$ by a polynomial in D, which is the interpolation interval or fractional delay. The interpolants, i.e., the new samples, are now represented by the following function:

$$y(n) = \hat{x}_c(n - D) = \sum_{k=0}^{N} c(k)D^k, \tag{36}$$

that takes on the value $x(n)$ when $D = n$. The above approach to Lagrange interpolation is seen to be related to Farrow's idea of having $N+1$ constant filters.

The alternative implementation for Lagrange interpolation is obtained by formulating the polynomial interpolation problem in the z-domain as $Y(z) = H(z)X(z)$ where $X(z)$ and $Y(z)$ are the z-transforms of the input and output signal, $x(n)$ and $y(n)$, respectively, and the transfer function $H(z)$ is now expressed as a polynomial in D (instead of z^{-1}).

$$H(z) = \sum_{k=0}^{N} C_k(z)D^k. \tag{37}$$

The familiar requirement that the output sample should be one of the input samples for integer D may be written in the z-domain as $Y(z) = z^{-D}X(z)$ for $D = 0, 1, 2, \ldots, N$. Together with (36) and (37) this leads to the following $N + 1$ conditions

$$\sum_{k=0}^{N} C_k(z)D^k = z^{-D} \quad \text{for } D = 0, 1, 2, \ldots, N. \tag{38}$$

This may be expressed in matrix form $\mathbf{Uc} = \mathbf{z}$ where the matrix $\mathbf{U}$ has the Vandermonde structure and thus it has an inverse matrix $\mathbf{U}^{-1}$. The solution of (38) can thus be written as $\mathbf{c} = \mathbf{U}^{-1}\mathbf{z}$.

The transfer function $C_0(z)$ is equal to 1 regardless of the order of the interpolator. The other transfer functions $C_n(z)$ are Nth-order polynomials in z^{-1}, i.e., they are Nth-order FIR filters. This implementation technique is called the *Farrow structure of Lagrange interpolation*. A remarkable feature of this form is that the transfer functions $C_n(z)$ are *fixed* for a given order N. The interpolator is directly controlled by the fractional delay D, i.e., no computationally intensive coefficient update is needed when D is changed.

The Farrow structure is most efficiently implemented using *Horner's rule* (see, e.g., [51], p. 28), i.e.,

$$\sum_{k=0}^{N} C_k(z)D^k = C_0(z) + [C_1(z) + \cdots + [C_{N-1}(z) + C_N(z)\overbrace{D]D\cdots]D}^{N}. \tag{39}$$

With this method, N multiplications by D are needed. A general Nth-order Lagrange interpolator that employs the suggested approach is shown in Fig. 15. Since there is no need for the updating of coefficients, this structure is particularly well suited to applications where the fractional delay D is changed often, even after every sample interval.

Figure 16 shows the Farrow structure for first-order Lagrange interpolation. In this case the subfilters are $C_0(z) = 1$ and $C_1(z) = z^{-1} - 1$. The overall transfer function of the filter is $H(z) = C_0(z) + DC_1(z) = 1 + D(z^{-1} - 1)$.

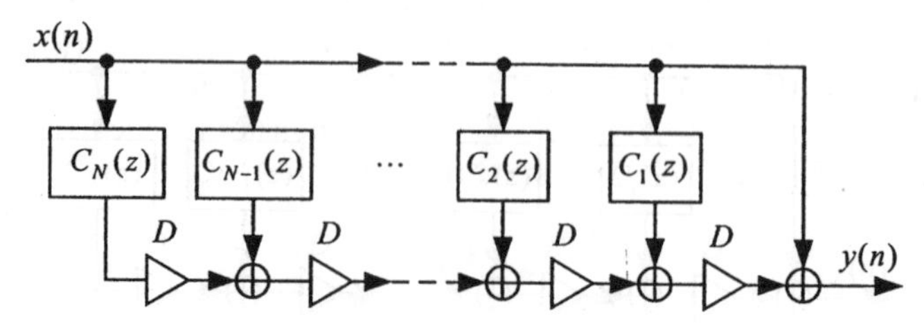

Figure 15. The Farrow structure of Lagrange interpolation implemented using Horner's rule.

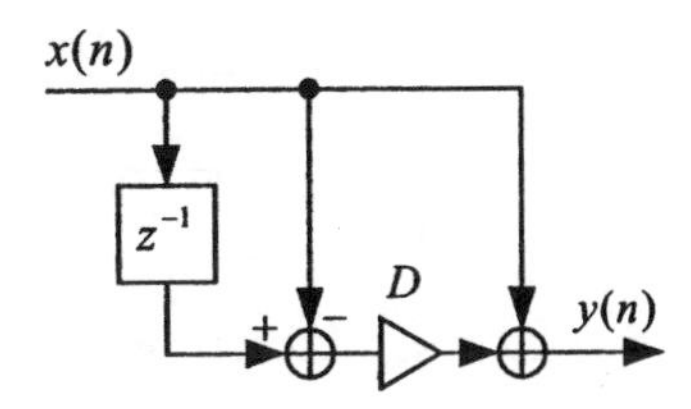

Figure 16. The Farrow structure for linear interpolation.

If the delay is constant or updated infrequently, it may be more efficient to implement Lagrange interpolation using the standard FIR filter structure. Namely, with the FIR filter structure, $N + 1$ multiplications and N additions are needed. In Farrow's structure, there are N pieces of Nth-order FIR filters which results in $N(N + 1)$ multiplications and N^2 additions. There are also N multiplications by D and N additions. Altogether this means $N^2 + 2N$ multiplications and $N^2 + N$ additions per output sample, which is more than in the case of the standard FIR filter realization.

20.5.2.4. Design Methods for the Farrow Structure

In general, the order of the polynomials $C_n(z)$ in (37) and that of the time-varying FIR filter need not be the same in the Farrow structure [32], although this is the case when Lagrange interpolation is used. It is also possible to use optimization methods to design the polynomials that represent the coefficients of a time-varying FIR filter. This results in an FD filter whose delay characteristics can be controlled with the delay parameter D. Farrow gave an example where a seventh-order (8-tap) FIR FD filter was implemented using optimized polynomials of order three [32]. Laakso *et al.* [71] also proposed a general approach, where a set of polyphase filters is first designed, and a polynomial is thereafter fitted through each coefficient in the least squares sense, for example. This method is very general, since the number and length of polyphase filters, and the order of the polynomials can be chosen independently of each other.

Tarczynski *et al.* have suggested a technique where the polynomials $C_n(z)$ are designed by minimizing the weighted least-squares error criterion [162]. In the example given by Tarczynski *et al.*, the order of polynomials was 7 while the order of the FIR filter was 67, and the weighting function was a step-like function divided into five frequency bands [162]. The approximation error of the designed filter was kept less than -100 dB for $0 < d \leq 1$ at frequencies below $0.45F_s$.

Vesma and Saramäki [186–188] have presented a modified Farrow structure where the polynomials are functions of $2D - 1$ instead of D. The advantage is that the coefficients of the polynomial $C_n(z)$ will be symmetric or anti-symmetric, which may lead to computational savings in the implementation of the time-varying structure, and makes the optimization of the filter coefficients easier. Vesma and Saramäki have developed an optimization procedure for designing the polynomials so that the time-varying FIR FD filter will have equiripple properties [186, 189, 190].

Recently, Harris stated that the subfilters of the Farrow structure are derivative filters [43]. The first subfilter $C_0(z)$ has ideally a flat magnitude response, $C_1(z)$ is a first-order differentiator, $C_2(z)$ is a second-order differentiator, and so on [43]. This knowledge may be used to facilitate the design of optimal subfilters for the Farrow structure. Interestingly, Sudhakar *et al.* have earlier tackled the design of interpolation filters based on FIR differentiators of different degrees [158].

20.5.2.5. Discussion

The use of a time-varying FIR FD filter (as opposed to a recursive one) is favorable in the sense that *no disturbing transients* will be generated when the delay D is changed. This is true, of course, only when the FIR filter has been implemented so that the input samples that are needed for computing the output of the filter after the change in the delay value are available (for example, a transpose form FIR filter causes transients if the filter coefficients are changed at the same time). Anyhow, a discontinuity is observed in the output signal of the filter, but it can be made small by making the coefficient changes small. The Farrow structure, or its modification proposed by Vesma and Saramäki [186–188], is an excellent candidate for realization of time-varying FIR filters.

20.5.3. Time-varying Recursive FD Filters

An abrupt change in the coefficient values of an IIR filter gives rise to disturbances in the future values of the internal state variables and transients in the output values of the filter. These disturbances depend on the filter's impulse response so that they are in principle infinitely long. In the case of stable filters, however, the disturbances decay exponentially and can be treated as having a finite length. In practice, these transients are often serious enough to cause trouble, such as clicks in audio applications, and they are typically the most serious problem in the implementation of tunable or time-varying recursive filters.

Despite the importance of the problem, there exists only a few research reports on strategies for elimination of transients in time-varying recursive filters (for references, see [174] and [180]). Zetterberg and Zhang presented the most general of these approaches by means of a state-space formulation of the recursive filter [200]. Their main result was that, assuming a stationary input signal, every change in the filter coefficients should be accompanied by an appropriate change in the internal state variables. This guarantees that the filter switches directly from one state to another without any transient response. The Zetterberg–Zhang (ZZ) method can *completely eliminate the transients* but it does require that all the past input samples are known. For this reason, the ZZ algorithm is impractical as such but has provided a good starting point for a more efficient approximate algorithm described in [174, 180–181].

The transient-cancellation algorithm uses two filters in parallel, as shown in Fig. 17: one for filtering the signal with current filter coefficients, and another with the next filter coefficients; the output signal of the secondary filter is not used at this time. Two filters must be executed in parallel only for a finite number of sampling intervals, then the outputs of the filters can be switched and the old filter can be stopped. This is motivated by the fact that the impulse response of a stable recursive filter decays exponentially and can thus be regarded as finite-length. The knowledge of the effective length of the impulse response from the input to the state vector helps to estimate how many past input samples need to be taken into account in updating the state vector of the filter to be used next. Thus, the advance time may then be set equal to $N_a = N_P + N$ where N_P is the effective

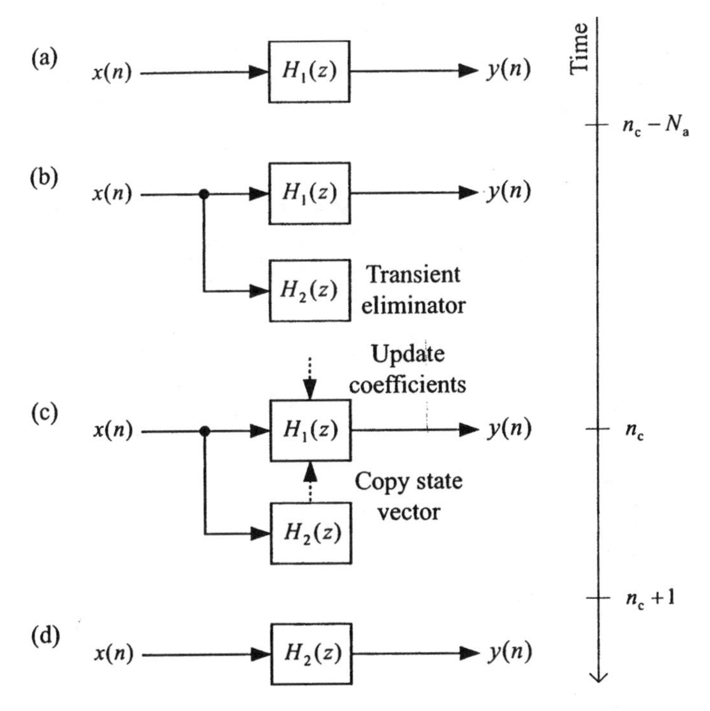

Figure 17. Different phases of the transient suppression scheme for a single change of filter coefficients (adopted from [180]).

length of the impulse response and N is the order of the filter [180]. This choice of N_a ensures that the updated state vector suffers sufficiently little from the truncation of the input signal, according to the same criterion that was used to determine N_P. In practice, it is desirable to choose N_a to be the smallest integer that yields sufficient suppression, since this minimizes the implementation costs of the transient cancellation algorithm. The effective length of an infinite-length impulse response can be determined using an energy-based method [69].

For a single coefficient change, the algorithm requires that two filters run in parallel for N_a sample intervals, as illustrated in Fig. 17. Thus, when multiple changes are required and it is fast enough to update filter coefficients at every N_ath sample interval, there is no need to run more than two filters in parallel at any time. The main advantages of this technique is that now the computation of the transient cancellation vector only takes finite time and need not be updated all the time in parallel with the filtering operation. Also the accuracy of transient cancellation can be controlled with parameter N_a: the larger the value of N_a, the more transient suppression is achieved.

20.5.3.1. Examples on Transient Suppression

We present an example that illustrates the transient suppression method. We filter a low-frequency sine wave (0.0454 times the sampling frequency F_s) with a second-order all-pass filter (direct-form II) that approximates a constant group delay. Initially, the filter coefficients are $a_1 = 0$ and $a_2 = 0$ (corresponding to a constant delay of 2 samples) and at time index 30 they are changed to values $a_1 = 0.4$ and $a_2 = -0.028571$, which gives a group delay of 1.5 samples at low frequencies. We present the output and transient signals of the filter in two cases: without transient cancellation and when the cancellation method is used with parameter value $N_a = 4$. These output signals are compared with the "ideal" output signal—see the open circles in Fig. 18(a) and Fig. 19(a)—which has been computed using the output-switching method (by running two filters in parallel and changing the output at time $n = 30$). The transient signal shown in the lower part of the figures in both cases is the difference of the output signals of the time-varying and ideal filter. Obviously, in Fig. 19 ($N_a = 4$) the maximum amplitude of the transient has been suppressed with respect to Fig. 18. More suppression can be achieved by using a larger value for N_a.

20.5.4. Conclusions

We have discussed time-varying fractional delay filters. The FIR filters do not suffer from transients when their coefficient values or locations of filter taps are changed. The output signal of an FIR filter will, however, be discontinuous at the time of the change. A strategy for efficient implementation of time-varying

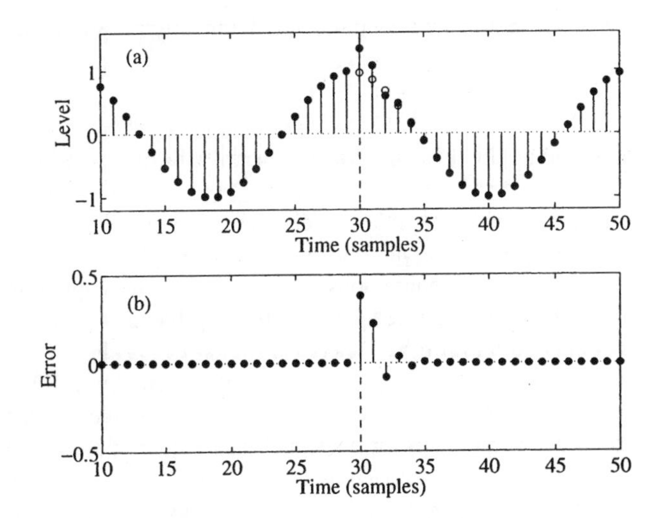

Figure 18. (a) Output signal and (b) transient of a second-order all-pass filter (DF II structure) when the coefficients are changed at time index 30 without transient elimination.

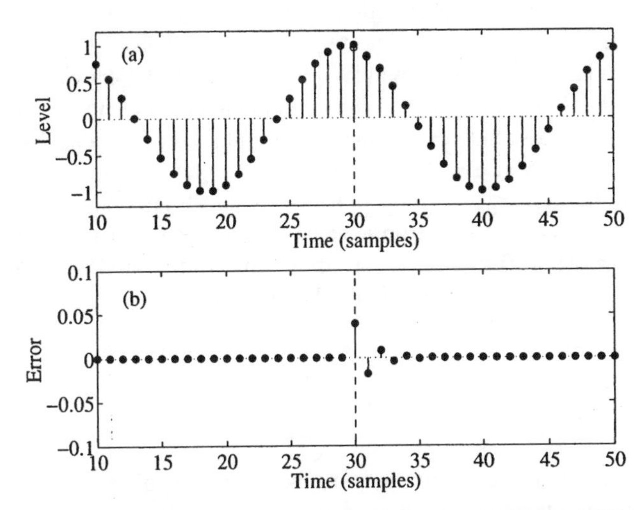

Figure 19. (a) Output signal and (b) transient of a second-order all-pass filter (DF II structure) when the coefficients are changed at time index 30 with the transient suppresser with $N_a = 4$.

FIR FD filters, known as the Farrow structure, was presented. Techniques for designing time-varying FIR FD filters were also surveyed.

Recursive filters, such as all-pass FD filters, suffer from two problems: transients and discontinuities in the output signal. There is a method that can suppress the transient as much as needed by increasing the advance time parameter. The required amount of computations is directly proportional to this parameter.

20.6. Fractional Delay Filters for Nonuniformly Sampled Signals

In this section, we consider fractional delaying of a nonuniformly sampled signal, where the sampling interval changes from sample to sample. The reconstruction of the original continuous-time signal or resampling onto a different (usually uniform) grid are relevant practical problems where interpolation techniques similar to uniform FD filters can be used. We discuss practical approaches to interpolation of nonuniformly sampled signals by a *generalized fractional delay filter* which takes as its input the signal samples with timing information, and produces estimates for the signal values at the desired sampling instants.

The uniform FD filtering problem discussed in previous sections can be viewed as an application of the reconstruction formula (9) of bandlimited uniformly sampled signals. If we retain the bandlimited signal but assume nonuniform samples, the reconstruction problem can still be solved based on the sinc series. However, the solution becomes more involved both in terms of computational complexity and numerical problems [84, 99]. Several approximate methods have been proposed which often include iterative techniques [85].

One approach to circumvent the numerical problems of the ideal bandlimited sinc reconstruction is to use *polynomial interpolation*. In the most straightforward case it reduces to Lagrange interpolation, i.e., fitting of a polynomial of order N through a set of $L = N + 1$ sample values. The coefficients of the general Lagrange interpolator filter for a set of N uniformly sampled input signal values t_n (where $n = 0, 1, 2, \ldots, N$) are obtained as

$$h(t - t_n) = \prod_{\substack{k=0 \\ k \neq n}}^{N} \frac{t - t_k}{t_n - t_k} \quad \text{for } n = 0, 1, 2, \ldots, N, \tag{40}$$

where t is the value of the desired output time. The interpolated output signal is obtained via nonuniform convolution as

$$x(t) = \sum_{n=0}^{N} h(t - t_n)x(t_n). \tag{41}$$

Note that now the filter coefficients are time-varying—unlike those of the uniform Lagrange interpolation (see Section 20.3.4)—and generally depend on all the sampling instants on the particular set of L samples. If an additional fractional delay t_D is desired, the time variable t should be replaced with $(t - t_D)$. The use of Lagrange interpolation for reconstruction of nonuniformly sampled signals has been studied by Murphy *et al.* [102].

The problem with Lagrange interpolation (as with the ideal sinc reconstruction as well) is that it makes an exact match with the signal at the sample instants. This is often unnecessary, and may even be harmful in practice when signal samples to be processed contain additive noise. In such a case, we would instead like to suppress the noise and reconstruct only the desired part of the signal.

Knowing what the'desired part of the signal' is, requires *a priori* knowledge. Often the desired part of the signal is slowly varying, which can be modeled by assuming the signal to be bandlimited up to a certain frequency. Alternatively, the signal can be assumed to be a polynomial of a certain degree P (less than $N - 1$). This is a generalization of Lagrange interpolation which assumes that the sequence of N samples can be modeled by a polynomial of order $N - 1$ and implements exact curve fitting at the desired sample values. This kind of polynomial filtering, besides being less intensive computationally than Lagrange interpolation, enables simultaneous reconstruction of the nonuniform signal and suppression of wideband noise [68].

The polynomial model for the signal can be expressed in the form

$$p(t) = \sum_{k=0}^{P} c_k (t - t_0)^k, \tag{42}$$

where P is the order of the polynomial, c_k are the polynomial coefficients, and t_0 is the time instant of the first sample. Minimizing the mean squared error (MSE) of the polynomial model and signal samples results in a set of normal equations for the polynomial coefficients the solution of which can be expressed in matrix form as

$$\mathbf{c}_{\mathrm{opt}} = [\mathbf{U}^{\mathrm{T}}\mathbf{U}]^{-1}\mathbf{U}^{\mathrm{T}}\mathbf{x}, \tag{43}$$

where $\mathbf{c}_{\mathrm{opt}}$ is the vector of the $P + 1$ polynomial coefficients, $\mathbf{U}$ is a matrix depending on the input sampling instants, and $\mathbf{x}$ is the vector of N input signal values (for details, see [68]). The new sample values are obtained by evaluating the polynomial (42) at the desired sampling instant. Similarly to uniformly sampled signals, the best results are obtained when the interpolation is carried out for samples near the center of the time window of the input samples.

Let us consider an example. We assume a nonuniformly sampled signal generated by a *jittered* sampling process, i.e. a nominally uniform sampling rate $F_s = 1/T$ which is distorted by random timing errors that have a nonuniform distribution with $\Delta = T/4$. The signal consists of a sum of low-frequency sinusoids at frequencies $f_1 = 0.070F_s$ and $f_2 = 0.123F_s$ with total power equal to unity. In addition, the signal is corrupted with additive uncorrelated Gaussian noise of variance 0.15, i.e., the signal-to-noise ratio (SNR) is 8.2 dB.

Figure 20(a) shows the original sinusoids without noise (uniformly sampled) and the nonuniformly sampled signal corrupted with noise. Only the noisy nonuniformly sampled signal is available in a practical situation, and it is used to produce the uniformly sampled reconstructions shown in Fig. 20(b). Note that, using the polynomial interpolation technique described above, the signal could have been also delayed by a fraction of the nominal sampling interval, or oversampled on a more dense grid, but here we focus on the simple case. Seven-tap ($L = 7$) polynomial filters were used. We first applied Lagrange interpolation (i.e., polynomial order $P = L - 1 = 6$) and then the polynomial filter with $P = 3$. It is seen that the Lagrange interpolation produces a signal that is quite a faithful reconstruction of the noisy one, whereas the polynomial filtering result is more similar to the noiseless original signal. The corresponding MSE values are 0.1486 and 0.0480 (corresponding to noise reduction by 0.041 dB and 4.9 dB, respectively), demonstrating that the polynomial filter is able to suppress the noise by almost 5 dB and reconstruct the desired part of the signal accurately at the same time.

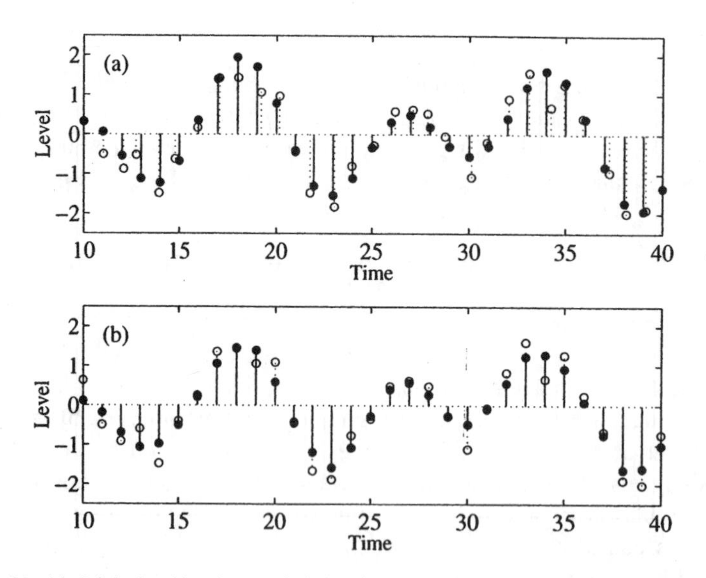

Figure 20. (a) Original uniformly sampled signal (•) and its noisy, nonuniformly sampled version obtained by jittered random sampling (○). (b) Comparison of seven-tap polynomial reconstruction using 6th-order Lagrange interpolation (○) and 3rd-order polynomial interpolation (•).

20.7. Applications of Fractional Delay Filters

Fractional delay filters can be utilized in many areas of digital signal processing. In the following, we discuss some technical fields where FD filters are useful or necessary, including sampling-rate conversion, timing adjustment of digital modems, music synthesis, virtual audio reality, array beamforming, and digital filter design. Other examples where fractional delay filters have turned out to be helpful are time delay estimation [27–29, 153], speech coding [3, 15, 65, 82, 83, 88, 89], stabilization of feedback systems [160], timing correction in multichannel data acquisition systems [6], and image resampling [26, 52, 62]. The key ideas in the usage of FD filters in several applications were also discussed in the review article by Laakso *et al.* [71].

20.7.1. Sampling-Rate Conversion

Changing the sampling rate is one of the standard problems in digital signal processing. A well-known example is the conversion between the different sampling frequencies used in digital audio, which include 44.1 kHz (used in the CD-quality audio), 48 kHz (used by the DAT recorders), 32 kHz, 22.05 kHz, and 11.025 kHz among others (see, e.g., [116]).

The sampling-rate conversion is straightforward if the ratio of the input and output sampling rates is an integer or a ratio of small integers [19, 172]. Efficient polyphase filter structures are known for the implementation. As discussed in Section 20.3.1, each branch of the polyphase filter can be interpreted as a lowpass FD filter approximating an integer number of fractions of the input sample interval.

The sampling-rate conversion for incommensurate ratios is more complicated. In practice, the situation is often made even harder by the fact that the sampling-rate ratios are not only irrational but also *time-varying*, which is caused, for instance, by variations in clock frequencies due to temperature, aging, or external disturbances.

A straightforward solution to the incommensurate sampling-rate conversion problem can be obtained via an extension of the polyphase implementation. Any ratio can, of course, be approximated to a desired precision by a ratio of large integers. Hence, one just needs sufficiently many polyphase branches to have a dense grid of preset polyphase branches from which to choose. In other words, a large bank of predesigned FD filters is needed. In addition, in order to avoid *folding*, care must be taken in the case of downsampling that the signal to be resampled is bandlimited below the final Nyquist frequency: then the polyphase filters must be fractional delay lowpass filters.

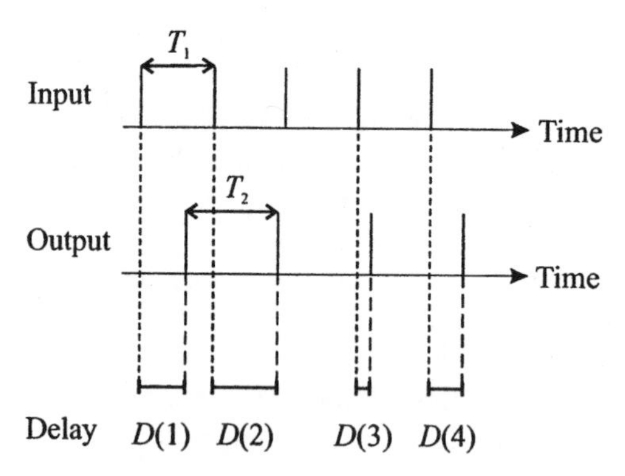

Figure 21. Illustration of the sampling grid of two signals sampled at rates f_{s1} and f_{s2}, respectively, whose sampling intervals are T_1 and T_2, and the fractional delays $D(n)$, between the contiguous samples of these signals.

From the polyphase approach there is conceptually only a short step to using time-varying FD filters; instead of storing a large number of filter coefficients one can design them on-line, e.g. by employing the above presented Farrow structure.

The incommensurate sampling-rate conversion problem can be illustrated by Fig. 21 where the sampling time instants of both the input and output sampling rates are shown. It is obvious that every output sample can be obtained by applying an FD filter approximating an appropriate delay. The essential tasks are thus to implement a *control unit* that computes the required delay values $D(n)$ and a computationally efficient FD lowpass filter whose coefficients change for every output sample. A block diagram is shown in Fig. 22.

Previous work on incommensurate sampling-rate conversion includes papers by Smith and Gossett [154], Ramstad [122–123], de Carvalho and Hanson [14], Cucchi *et al.* [20], Park *et al.* [111], Adams and Kwan [2], Tarczynski *et al.* [163], Zölzer and Boltze [204–205], Laakso *et al.* [70], Saramäki and Ritoniemi [134], and Murphy *et al.* [103].

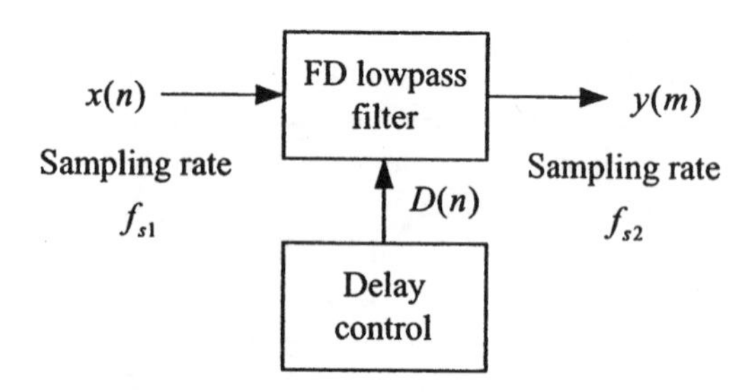

Figure 22. Block diagram of an incommensurate-ratio sampling-rate converter.

Arbitrary changes of sampling rate are also required in wavetable music synthesis, where a sampled waveform is stored in a buffer and during playback the buffer is read repeatedly to synthesize a periodic tone [10, 86]. The pitch of the note during playback is controlled by the increment of the read pointer to the buffer. For example, if a tone one octave higher than that stored in the buffer is required, the increment size is 2 samples. An increment size of 4 would result in a tone that is 2 octaves higher than the one stored, and so on. For lower tones and for other intervals than simple integer ratios, a technique called fractional addressing must be used [46]; then the increment size is not an integer. In such cases, simply selecting the nearest sample in the table causes errors in the synthetic waveform which may not be acceptable [21, 86, 98]. Fractional delay filtering techniques can be used to approximate the signal value between the stored samples during playback [21, 98, 193].

Another example of a related audio signal processing application, where the signal must be resampled but the sampling rate does not necessarily change, is restoration of old recordings, particularly the elimination of wow in gramophone disc or magnetic tape recordings [40–41].

20.7.2. Synchronization in Digital Receivers

In digital data transmission, the transmitter transmits data symbols using analog waveforms which carry one or more bits of information each. The main task of the receiver is to detect these symbols as reliably as possible. To this end, the receiver must be synchronized to the symbols of the incoming data signal. Even though the receiver usually knows the nominal symbol rate of the transmitter, the analog oscillators that are used to generate the frequency in practice have a limited precision, and the frequency tends to vary with temperature and aging. Furthermore, in mobile communications frequency shifts due to the Doppler effects must also be accounted for. Hence, in practice the synchronization requires constant monitoring and adjustment.

The symbol synchronization has traditionally been implemented by using an analog feedback or feedforward control loop to adjust the phase of a local clock at the receiver so that the sampling frequency and the sampling instants are adapted to the incoming data signal (see, e.g., [93]). An example of this kind of a receiver is illustrated in Fig. 23.

Since the postprocessing of the sampled signal (matched filtering and data detection) is usually performed digitally anyway, it is also advantageous to implement the synchronization using digital techniques. A basic digital solution is outlined in Fig. 24. The local oscillator is now independent of the received signal and often operates at a higher rate than the nominal symbol rate (e.g. two or four times oversampled). The raw samples are now postprocessed by an appropriate FD filter. Especially in the case of oversampling, simple low-order

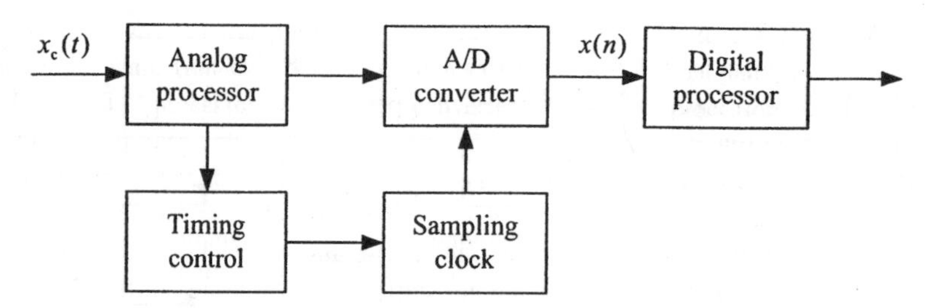

Figure 23. Analog synchronization via controlling the sampling time (adapted from [36]).

FD filters (such as a 4-tap Lagrange interpolator) are usually sufficient [70]. For on-line control of the delay value, the Farrow structure has been found practical.

For excellent tutorials to synchronization in digital receivers, see [31, 36, 90] and [94]. Related research papers include [5, 32, 39, 45, 70, 109, 168, 185, 187, 188, 192, 194, 197, 201–202] and [61]. For a more efficient implementation and lower overall delay, filter structures which integrate the FD filter and other filtering functions, e.g., matched filtering and interference suppression, have been investigated in [66, 125, 127–129, 198–199], and [191].

20.7.3. Music Synthesis Using Digital Waveguides

Fractional delay filters are essential in music synthesis based on digital waveguide modeling [54, 59, 149–152, 159, 175]. A one-dimensional acoustic resonator, such as a vibrating string, or the tube of a wind instrument, can be modeled with a bi-directional delay line. In the computational model, the two delay lines can usually be combined into one. Generally, a fractional delay filter is required for fine-tuning the delay, which determines the pitch of the synthetic sound.

Figure 25 presents a plucked string synthesis model. The loop filter is a low-order digital filter (e.g., a first-order recursive filter) that simulates the attenuation experienced by the string vibration. The propagation delays of transversal waves

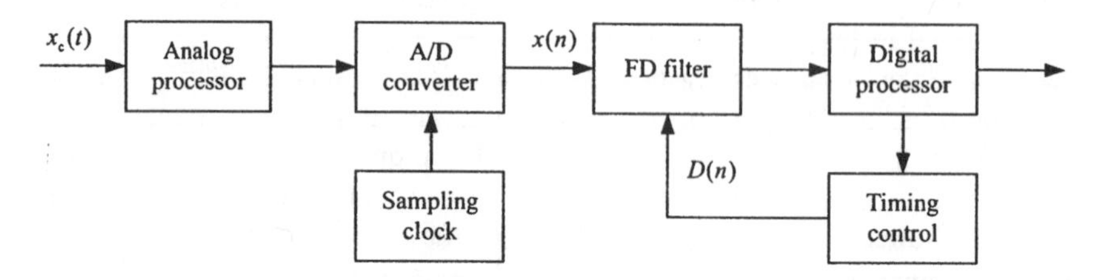

Figure 24. Digital synchronization via an FD filter (adapted from [36]).

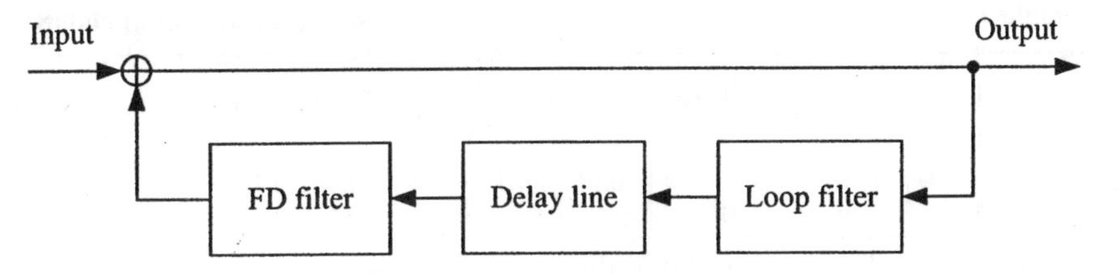

Figure 25. Block diagram of a plucked string synthesis model (after [54]).

along the string have been combined into a single delay line. The FD filter adjusts the overall loop delay so that the pitch of the synthetic tone is correct. The input signal of the synthesis model can be a short burst of noise, but for more sophisticated synthesis the excitation signal can be extracted from a recorded plucked string tone using an inverse filtering technique [175]. The fractional delay filter used in plucked string models is usually either a first-order all-pass filter [54] or a low-order Lagrange interpolation filter [59, 175].

If it is desired to vary the pitch of the synthetic tone to produce for example glissando or vibrato effects and a recursive FD filter, i.e., an all-pass filter, is used, transient problems will occur. This will appear as so-called zipper noise in the output audio signal. The transient cancellation method addressed previously in this chapter has turned out to be helpful reducing this effect [181]. Van Duyne and others have proposed a technique which combines the transient elimination method and cross-fading [184].

Fractional delay filter techniques have also been useful in adjusting the length of a vocal-tract model based on the transmission-line method, which is very similar to the digital waveguide approach [72, 156]. The locations of scattering junctions in a vocal tract model can also be fine-adjusted using interpolation techniques [174, 176–178]. This approach is called fractional delay waveguide modeling. The interpolated scattering junctions can be generalized for three waveguide branches, which is useful for implementing accurately the location of finger holes in woodwind instrument synthesis models, e.g., for the flute [17, 141, 174, 179].

Recently, a new concept in physics-based music synthesis was introduced, the passive nonlinearity [115]. It refers to a nonlinear phenomenon which enables energy dissipation such as in natural linear systems. This kind of phenomenon is very pronounced in metallic percussion instruments, such as gongs, but has also been discovered in string and wind instruments.

In synthesis models, passive nonlinearities can be realized using time-varying fractional delay filters with signal-dependent coefficients. Synthesis models based on this approach have been proposed that incorporate a longitudinally yielding (springy) end-point of strings [115], modulation of tension of a vibrating string [167, 183], and nonlinear sound propagation caused by high pressure in brass instruments [97, 165]. The time-varying FD filter may be based on a first-order all-pass filter [115] or an FIR filter [182].

20.7.4. Other Musical Applications of Fractional Delays

Fractional delays are useful in digital implementation of traditional electronic musical effects, such as chorus and flanger, which are based on summing the original musical signal and its delayed version [22, 33]. A new way of implementing a time-varying delay line for musical applications using a circular buffer has been proposed by Rocchesso [126].

Also, a surround sound system can be constructed using two loudspeakers by filtering their input signals in an appropriate way [38, 56–57]. The signal processing part of the system is composed of digital filters that approximate head-related transfer functions of the listener, and fractional delay filters that accurately model acoustic propagation delays (see [38], pp. 65–73).

Two advanced applications where FD filters are needed in the simulation of acoustic propagation delay are addressed in the following.

20.7.4.1. Doppler Effect in Virtual Audio Environments

A special feature of acoustic signals is their relatively slow propagation speed in air (ca. 340 m/s). For this reason sound is considerably delayed when it propagates from the sound source to the listener. This phenomenon needs to be modeled using delay lines and fractional delay filters in virtual reality applications that try to provide a realistic, immersive sound field [136]. When the sound source and the listener are allowed to move fast, the Doppler effect is heard. The Doppler shift refers to the rise or fall of the pitch respectively when the distance between the listener and source is increasing or decreasing fast enough. More generally, the Doppler effect means a scaling of the spectrum of a sound signal observed when the distance to the source varies over time. Figure 26 illustrates how the propagation delay of sound varies when a car passes a listener at a speed of 100 kilometers per hour. The shortest distance between them is 4 meters which

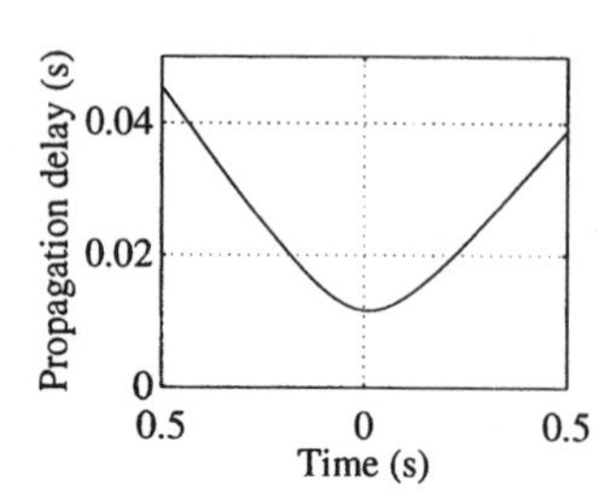

Figure 26. Variation of propagation delay of sound from a car when it passes a listener with the speed of 100 kilometers per hour (after [155]).

occurs at time $t = 0$ s. The time-varying delay given in Fig. 26 should be implemented with the help of fractional delay filters to bring about the Doppler shift. Strauss has discussed the use of interpolation filters in simulation of the Doppler effect in an auditory virtual reality application [155].

20.7.4.2. Simulation of Wave Propagation in Multiple Dimensions

A sophisticated application of fractional delay filtering is a new method for simulation of multidimensional wave propagation using a finite difference mesh [135, 137–140]. The basic model, the digital waveguide mesh, is based on a multidimensional extension of digital waveguides discussed in the previous section. It is a computationally efficient way of discretizing a membrane or an acoustic space, but the problem is that sample updates only occur in a limited number of directions, e.g., only 4 in the two-dimensional rectangular mesh. This limitation appeared to be the cause for fluctuation in the wave propagation speed in different directions and at different frequencies. In other words, the simulated mesh suffers from direction-dependent dispersion. Savioja and Välimäki extended the waveguide mesh method by applying fractional delay filters in multiple dimensions to devise a sample update procedure that renders the mesh more homogeneous. Improved performance has been obtained in both two [140] and three dimensions [135]. The method is applicable to simulation of drums and computation of the impulse response of acoustic spaces, e.g., listening rooms or concert halls.

20.7.5. Interpolated Array Beamforming

A classical method of beamforming in antenna or transducer arrays is based on delaying and summing the input signals of the array elements. Pridham and Mucci introduced *digital interpolation beamforming* in the 1970s [118–119]. Their approach was to use polyphase FIR filters to accurately implement the required time delays. A similar solution has been proposed also recently [143]. Interpolated beamforming is applicable to sonar [118–119], antenna arrays [76], and microphone arrays [42, 101].

Figure 27 shows the configuration for traditional delay and sum beamforming. For the array elements, we use the symbol of a microphone but they could also be hydrophones, antenna elements, or other detectors. Notice that we also do not show the AD converters and other electronics required, but simply assume that the input signal obtained from each element is a discrete sequence $x_k(n)$ for $k = 1, 2, \ldots, M$, where M is the number of array elements. We also assume for simplicity that the spacing of array elements is uniform. The time delays D_k should be selected so that the different channels are in the same phase for a plane-wave signal which is incident from a specific direction θ. The summing causes the

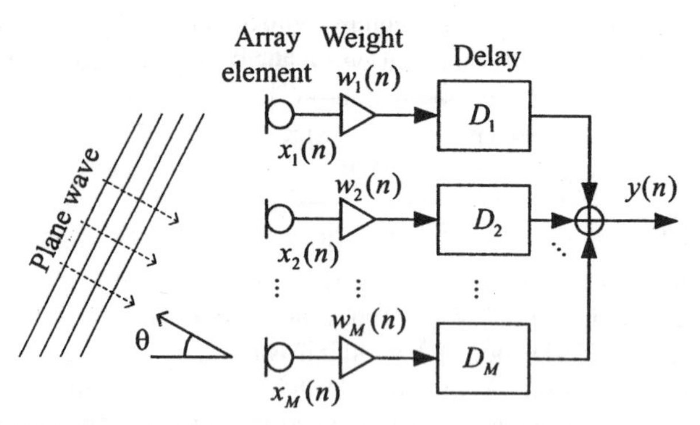

Figure 27. Delay and sum beamforming using an array of sensors, such as microphones.

signal components incident from other directions to sum destructively. Obviously, the time delays D_k used in delay and sum beamforming must be realized accurately to obtain good directivity properties. If integer-length delay lines are used, it is possible to obtain accurate directivity in M directions between 0 and π (or between $-\pi$ and 0). When fractional delay filters are used to implement delay lines, a continuum of directions between 0 and π is obtained. The quality of the beamformer then depends on the quality of the fractional delay filters used.

We present a design example where the array consists of 21 omnidirectional sensors whose distance has been set so that it corresponds to 1 sampling interval. The steering angle has been selected to be $\pi/5$, or 0.63 rad. With this choice all but one of the delays have a non-integer value. The first delay line can be neglected and replaced with unity gain, of course. The weighting coefficients for the signals of each sensor were taken from a Hamming window function. Figure 28 shows with a solid line the beam pattern of the array evaluated at the normalized frequency 0.25 as a function of incident angle. The beam pattern has been normalized so that the maximum is at 0 dB. We also incorporated Lagrange interpolation filters into the delay lines, and repeated the beam pattern evaluation with two different orders, 2 and 10, shown with dashed and dash-dot lines in Fig. 28. Note that in this case the main lobe of the beam pattern is not much affected by fractional delay filters but the attenuation of interfering directions is improved generally about 2 to 3 dB. The difference between the performances of the two choices of Lagrange FD filter orders is small, because at the normalized frequency 0.25 both filters provide a good approximation.

20.7.6. Design of Special Digital Filters

The fractional delay property can be combined with different digital filter characteristics. The control of delay of arbitrary FIR filters has been tackled by Adams [1] and Laakso *et al.* [71]. Hermanowicz and Rojewski [49] have derived

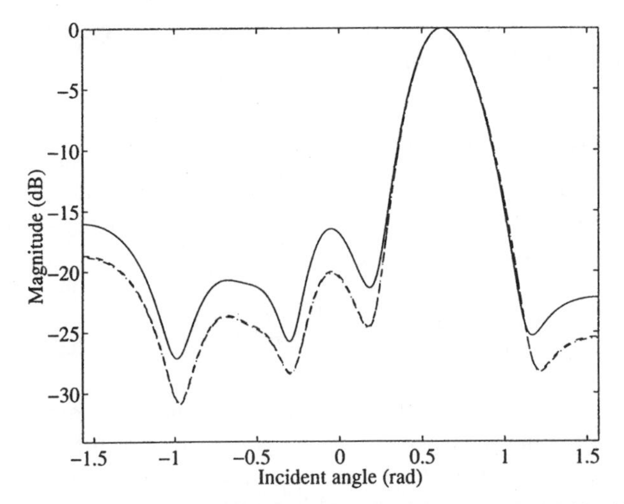

Figure 28. Beam pattern of a 21-element array at the normalized frequency 0.25 with integer-length delay lines (solid line) and with second (dashed line) and tenth-order (dash-dot line) Lagrange fractional delay filters. The steering angle is 0.63 rad.

closed-form equations for maximally-flat FIR differentiators with arbitrary fractional delay. Recently, Hermanowicz *et al.* have discussed also other digital filters with fractional delay, such as the Hilbert transformer, the half-band filter, and a complex-valued filter for computing the analytic signal [50]. Braileanu has discussed digital filters with explicit interpolated output [9].

Nazra has developed a design method for recursive fractional delay lowpass filters [104]. The method is based on using the Thiran fractional delay all-pole filter (discussed earlier in this chapter) as the recursive part of the filter, and then introducing zeros by designing an FIR compensator. The overall filter has lowpass characteristics with an equiripple stopband and adjustable fractional delay.

Selesnick generalized the maximally flat all-pass filter design method into a filter that uses only some degrees of freedom for flatness constraints at the zero frequency and the rest to flatten the response at the Nyquist frequency [145]. A parallel sum of two such all-pass filters can be used to implement lowpass (or highpass) filters with a few multiplications. The resulting structure is numerically robust against quantization effects.

Recently, the use of a polyphase FIR structure that implements fractional delays was found to be a fundamental issue in the design of a delayless sub-band adaptive filter [25, 91–92].

Fractional delay filters can also facilitate the design of special comb filters, which we discuss in the following.

20.7.6.1. Fractional Delay Comb Filter

Pei and Tsang have used fractional delay filters to control the frequencies of notches in comb filters' transfer function [113]. The transfer function of the fractional delay comb filter is

$$H_{\mathrm{fdc}}(z) = \frac{1 - H(z)}{1 - \rho^D H(z)}, \tag{44}$$

where $H(z)$ is the transfer function of a fractional delay filter approximating a delay D, and ρ is the radius of the poles used to control the flatness of the magnitude response between to the notch frequencies ($0 < \rho < 1$). Fractional delay comb filters can be used to cancel harmonic disturbances, such as power line interference in electrocardiogram (ECG) signals (see [113] for an example of this application).

We present an example on the use of the fractional delay comb filter. Let us assume that the sampling rate is 490 Hz. In Europe, the power line interference occurs at 50 Hz and its multiples. The four lowest normalized frequencies where the interference may occur are 0.1020, 0.2041, 0.3061, and 0.4082. If an integer-length delay line is used, the notches are closest to these frequencies for a comb filter order $N = 10$. The normalized frequencies of its transfer-function zeros are 0, 0.1, 0.2, 0.3, 0.4, and 0.5. The magnitude response of the comb filter with $\rho = 0.98$ is shown in Fig. 29(a) together with the dashed vertical lines that indicate the four interfering frequencies. The match between the notches and the interferences is quite poor giving an attenuation of 4.6 dB, 1.2 dB, and 0.2 dB for the three lowest frequencies, and an amplification of 0.24 dB for the fourth one. Note that the gain of the comb filter exceeds unity between the notches, the maximum gain being about 1.1 or 0.83 dB.

Instead of a delay of 10 samples—as in the comb filter of Fig. 29(a)—a delay of $1/0.1020$ or 9.8 samples is required to exactly cancel a periodic disturbance of fundamental normalized frequency 0.1020. The fractional delay can be realized using one of the techniques discussed previously. We have chosen a Lagrange interpolation filter of order 4 in this example. The fractional delay parameter of the Lagrange interpolation filter must be set to -0.2 samples, so that the effective delay of the comb filter is shortened to 9.8 samples, as desired.

Figure 29(b) shows the magnitude response of the fractional delay comb filter together with the interfering frequencies. It is seen that the match is excellent at the lowest two frequencies while at the third and fourth ones it is not perfect, but it is still better than in Fig. 29(a). Note also that the third and the

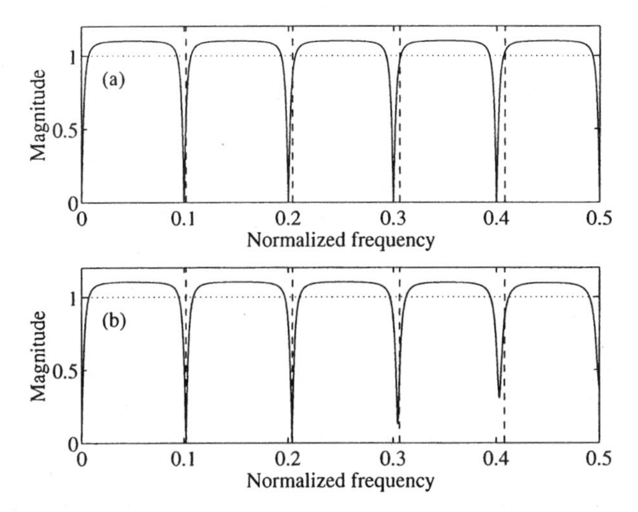

Figure 29. Comparison of the magnitude responses of (a) a conventional comb filter ($N = 10$, $\rho = 0.98$) and (b) a fractional delay comb filter ($\rho = 0.98$, $d = -0.2$), which contains a Lagrange interpolator of order 4. The vertical dashed lines indicate the disturbing frequencies, which should be cancelled. The horizontal dotted line indicates the unity gain.

fourth notch do not go to zero. This is caused by the lowpass character of the Lagrange interpolator. The attenuations at the four frequencies of interest are 49 dB, 20 dB, 6.6 dB, and 1.4 dB. The improvement over the conventional comb filter performance is dramatic at low frequencies, but at high frequencies the advantage is smaller. The result could be improved using a fractional delay design technique that optimizes the performance at the frequencies of interest, or alternatively, a general wideband FD filter design method could be used.

The behavior of the FD comb filter of this example could also be desirable if the interfering signal were also of lowpass character, i.e., having a strong fundamental component plus weaker harmonic components whose amplitude is inversely proportional to frequency; then it would be useful to have more attenuation at low frequencies where the strongest interfering harmonics appear.

The main advantage of the example design presented in Fig. 29(b) is that the remarkable improvement is achieved with a simple FD filter which can be adapted using only one parameter, the fractional delay d.

20.8. Conclusions

This chapter covered the theory and applications of fractional delay filters. Fractional delay filters are helpful in a wide range of signal processing applications where the temporal resolution of a fixed sampling interval is not enough. This is the case in problems involving detection, synthesis, or synchronization of signals, such as in time delay estimation, array beamforming, sampling-rate conversion, speech and music synthesis, time-domain simulation of acoustic systems, and synchronization of digital modems. Fractional delay filters can often be used as building blocks in various digital filters, especially adaptive or tunable filter structures.

Several FIR and all-pass filter design techniques for approximating an arbitrary delay were discussed. The target response corresponds to the well-known sampled and shifted sinc function. The maximally-flat approximations are easy to use, since the coefficients can be expressed in closed form for both FIR and all-pass filters. These techniques are called Lagrange interpolation and the Thiran all-pass filter, respectively. Least squares design methods require solving a matrix equation but offer flexibility since the approximation bandwidth can be specified and a frequency-dependent error weighting function can be used. Equiripple approximation methods yield a filter with optimal properties in the minimax sense. In the case of all-pass filters, the optimization criterion may be based on either the phase or the phase delay function.

In many cases, the fractional delay element must be time-variant in the sense that the delay parameter changes over time. Methods for implementing time-varying fractional delay filters were described. For FIR filters, the Farrow structure is a good choice. Since all-pass filters are recursive, transients may cause problems in time-varying situations. Examples of transient elimination using a recently developed technique were given.

Research on the design of FD filters and their applications is currently in a very active phase, which is demonstrated by the fact that more than 80 contributions have been published after the earlier review article [71]. The research has been active, e.g., in the fields of one- and multidimensional digital waveguides, synchronization of digital receivers, and time-delay estimation.

The bandlimited interpolation theory of one-dimensional functions is well established and well known so that the research focus is shifting from re-inventing the Lagrange interpolation to ever-diversifying fields of applications and efficient implementation structures. Another trend seems to be the interpolation in more than one dimension, such as modeling of multidimensional differential equations, where fractional delay filters can provide help to reduce the often extremely heavy computational requirements involved.

Acknowledgments

The work of the first author has been financed by the Academy of Finland. The authors are grateful to Mr. Matti Airas, Dr. Gerry Cain, Dr. Paulo Diniz,

Table 3. MATLAB files used to produce the figures of this chapter.

Figure(s)	MATLAB M file	Uses functions
1	IDFDCHAR.M	–
2	SHFTSINC.M	–
4, 5	POLYFIGS.M	–
6, 7	LSFDFIGS.M	–
8(a), (b), (c), (d)	LAGRFIGS.M	HLAGR2.M
9(a), (b), (c), (d)	OETKENFIGS.M	INIHEQ2.M, HEQRIP2.M
11, 12	THIRFIGS.M*	APFLAT2.M
13	DISCONT.M	APFLAT2.M
18	APNOSTUP3.M	APFLAT2.M
19	APNSTUP3.M	APFLAT2.M
20	NUPOLFIG.M	EXPOLF1.M, POLF1.M, POLJIT.M
26	DOPPLERD.M	–
28	ARRCOMP1.M	BPIARR.M, FINTARR2.M
29	FDCOMB.M	HLAGR2.M

*Note: You must run LAGRFIGS.M before THIRFIGS.M in order to be able to produce Figures 12(a) and 12(b).

Dr. Ewa Hermanowicz, Dr. Farokh Marvasti, Dr. Markku Renfors, Dr. Tapio Saramäki, Dr. Julius Smith, Dr. Andrzej Tarczynski, Dr. Tero Tolonen, and Dr. Jussi Vesma for helpful discussions and criticism.

MATLAB Files

The MATLAB files listed in Table 3 (see the accompanying CD-ROM, file: 'Chapter20\MATLAB-files'). were used to produce the signal figures in this chapter. MATLAB version 5.2 for Windows was used. The MATLAB Signal Processing ToolBox (version 4.1 was used) is required to run these programs.

References

The following keywords have been found useful when searching for literature on this topic: fractional delay, fractional sample delay, interpolation filter, phase shifter, polynomial interpolation, and resampling.

[1] J. W. Adams. Alternate Approach to Digital Phase Shift Filters. In *Proc. Int. Symp. Signal Process.: Theories, Implementations and Applications*, Brisbane, Australia, 160–165, August 1987, pp. 160–165.

[2] R. Adams and T. Kwan. Theory and VLSI Architectures for Asynchronous Sample-Rate Converters. *J. Audio Eng. Soc.*, 41(7/8):539–555, July–August 1993.

[3] M. Akamine, K. Miseki, and M. Oshikiri. Improvement of ADP-CELP Speech Coding at 4 kbit/s. *Proc. Global Telecommun. Conf.*, 3:1869–1873, Phoenix, AZ, December 1991.

[4] A. Aldroubi, M. Unser, and M. Eden. Cardinal Spline Filters: Stability and Convergence to the Ideal Sinc Interpolator. *Signal Process.*, 28(2):127–138, August 1992.

[5] J. Armstrong and D. Strickland. Symbol Synchronization Using Signal Samples and Interpolation. *IEEE Trans. Commun.*, 41(2):318–321, February 1993.

[6] A. Barwicz, D. Bellemare, and R. Z. Morawski. Digital Correction of A/D Conversion Error due to Multiplexing Delay. *IEEE Trans. Instr. Meas.*, 39(1):76–79, February 1990.

[7] M. G. Bellanger, G. Bonnerot, and M. Coudreuse. Digital Filtering by Polyphase Network: Application to Sample-Rate Alteration and Filter Banks. *IEEE Trans. Acoust., Speech, Signal Process*, 24(2):109–114, April 1976.

[8] C. Bertrand and P. Sehier. A Novel Approach for Full Digital Modems Implementing Asynchronous Sampling Techniques. *Proc. IEEE Global Telecommun. Conf.*, 2:1320–1324, London, UK, November 1996.

[9] G. Braileanu. Digital Filters with Implicit Interpolated Output. *IEEE Trans. Signal Process.*, 45(10):2551–2560, October 1997.

[10] R. Bristow-Johnson. Wavetable Synthesis 101, a Fundamental Perspective. Presented at the *Audio Eng. Soc. 101st Conv.*, Los Angeles, CA (November 1996), preprint no. 4400.

[11] H. Brandenstein and R. Unbehauen. Fractional Delay FIR Filter Design with the Darlington Transform. *IEEE Trans. Circ. Syst.—Part II*, 46(1):208–211, January 1999.

[12] G. D. Cain, N. P. Murphy, and A. Tarczynski. Evaluation of Several FIR Fractional-Sample Delay Filters. In *Proc. IEEE Int. Conf. Acoust., Speech, Signal Process.*, 3:621–624, Adelaide, Australia, April 1994.

[13] G. D. Cain, A. Yardim, and P. Henry. Offset Windowing for FIR Fractional-Sample Delay. In *Proc. IEEE Int. Conf. Acoustics, Speech, Signal Process.*, 2:1276–1279, Detroit, MI, May 1995.

[14] J. M. de Carvalho and J. V. Hanson. Efficient Sampling Rate Conversion with Cubic Splines. In *Proc. SBT/IEEE Int. Telecommun. Symp.*, September 1990, pp. 439–442.

[15] H. Chen, W. C. Wong, and C. C. Ko. Comparison of Pitch Prediction and Adaptation Algorithms in Forward and Backward Adaptive CELP Systems. *IEE Proc.—I: Commun., Speech and Vision*, 140(4):240–245, August 1993.

[16] A.J. R. M. Coenen, Novel Generalized Optimal Fractional Delay Filter Design for Navigational Purposes. In *Proc. 9th Int. Symp. Personal Indoor and Mobile Radio Commun.*, Boston, MA, September 1998 pp. 481–485.

[17] P. R. Cook. Integration of Physical Modeling for Synthesis and Animation. In *Proc. Int. Computer Music Conf.*, Banff, Canada, September 1995, pp. 525–528.

[18] R. E. Crochiere, L. R. Rabiner, and R. R. Shively. A Novel Implementation of Digital Phase Shifters. *Bell Syst. Tech. J.*, 54(8):1497–1502, October 1975.

[19] R. E. Crochiere and L. R. Rabiner. *Multirate Digital Signal Processing*. Prentice-Hall, Englewood Cliffs, NJ, 1983.

[20] S. Cucchi, F. Desinan, G. Parladori, and G. Sicuranza. DSP Implementation of Arbitrary Sampling Frequency Conversion for High Quality Sound Application. In *Proc. IEEE Int. Conf. Acoust., Speech, Signal Process.*, 5:3609–3612, Toronto, ON, Canada, May 1991.

[21] R. B. Dannenberg. Interpolation Error in Waveform Table Lookup. In *Proc. Int. Computer Music Conf.*, Ann Arbor, MI, October 1998, pp. 240–243.

[22] J. Dattorro. Effect Design, Part 2: Delay-Line Modulation and Chorus. *J. Audio Eng. Soc.*, 45(10):764–788, October 1997.

[23] A. G. Dempster and N. P. Murphy. Lagrange Interpolator Filters and Binomial Windows. *Signal Process.*, 76:81–91, July 1999.

[24] A. G. Dempster and N. P. Murphy. Efficient Interpolators and Filter Banks Using Multiplier Blocks. *IEEE Trans. Signal Process*, 48(1):257–261, January 2000.

[25] P. S. R. Diniz, R. Merched, and M. Petraglia. A New Delayless Subband Adaptive Filter Structure. In *Proc. IEEE Int. Symp. Circ. Syst.*, 5:162–165, Monterey, CA, May 1998.

[26] N. A. Dodgson. Quadratic Interpolation for Image Resampling. *IEEE Trans. Image Process.*, 6(9):1322–1326, September 1997.

[27] S. R. Dooley and A. K. Nandi. Fast Frequency Estimation and Tracking Using Lagrange Interpolation. *Electron. Lett.*, 34(20):1908–1910, October 1998.

[28] S. R. Dooley and A. K. Nandi. On Explicit Time Delay Estimation Using the Farrow Structure. *Signal Process.*, 72(1):53–57, January 1999.

[29] S. R. Dooley and A. K. Nandi. Adaptive Subsample Time Delay Estimation Using Lagrange Interpolators. *IEEE Signal Process. Lett.*, 6(3):65–67, March 1999.

[30] K. Egiazarian, T. Saramäki, H. Chugurian, and J. Astola. Modified B-Spline Interpolators and Filters: Synthesis and Efficient Implementation. In *Proc. IEEE Int. Conf. Acoust., Speech, Signal Process.*, 3:1743–1746, Atlanta, GA, May 1996.

[31] L. Erup, F. M. Gardner, and R. A. Harris. Interpolation in Digital Modems—Part II: Implementation and Performance. *IEEE Trans. Commun.*, 41(6):998–1008, June 1993.

[32] C. W. Farrow. A Continuously Variable Digital Delay Element. In *Proc. IEEE Int. Symp. Circ. Syst.*, 3:2641–2645, Espoo, Finland, June 1988.

[33] P. Fernández-Cid and F. J. Casajús-Quirós. Enhanced Quality and Variety for Chorus/Flange Units. In *Proc. COST-G6 Digital Audio Effects Workshop*, Barcelona, Spain, November 1998, pp. 35–39.

[34] A. Fettweis. A Simple Design of Maximally Flat Delay Digital Filters. *IEEE Trans. Audio and Electroacoust.*, 20(2):112–114 (June 1972).

[35] D. Fu and A. N. Willson, Jr., Interpolation in Timing Recovery Using a Trigonometric Polynomial and its Implementation. In *Proc. IEEE Global Telecommun. Conf.*, Sydney, Australia, November 1998, pp. 173–178.

[36] F. M. Gardner. Interpolation in Digital Modems—Part I: Fundamentals. *IEEE Trans. Commun.*, 41(3):501–507, March 1993.

[37] F. M. Gardner, Difficulties with Fractional-Delay Filters. *IEEE Signal Process. Mag.*, 13(4):16, July 1996.

[38] W. G. Gardner. 3-D audio using loudspeakers. Doctoral Dissertation. Massachusetts Institute of Technology, Cambridge, MA, September 1997.

[39] M. C. Gill and L. P. Sabel. On the Use of Interpolation in Digital Demodulators. *Australian Telecommun. Review*, 27(2):25–32, 1993.

[40] S. J. Godsill. Recursive Restoration of Pitch Variation Defects in Musical Recordings. In *Proc. IEEE Int. Conf. Acoust., Speech, Signal Process*, 2:233–236, Adelaide, Australia, April 1994.

[41] S. J. Godsill and P. J. W. Rayner. *Digital Audio Restoration*. Springer Verlag, London, UK, 1998.

[42] M. M. Goodwin and G. W. Elko. Constant Beamwidth Beamforming. In *Proc. IEEE Int. Conf. Acoust., Speech, Signal Process.*, 1:169–172, Minneapolis, MN, April 1993.

[43] F. Harris. Performance and Design of Farrow Filter Used for Arbitrary Resampling. In *Proc. 13th Int. Conf. Digital Signal Process* 2:595–599, July 1997.

[44] F. Harris. Performance and Design Considerations of the Farrow Filter when Used for Arbitrary Resampling of Sampled Time Series. In *Proc. 31st Asilomar Conf. Signals, Syst., Computers*, 2:1745–1749, Pacific Grove, CA, November 1997.

[45] F. Harris, I. Gurantz, and S. Tzukerman. Digital T/2 Nyquist Filtering Using Recursive All-Pass Two-Stage Resampling Filters for a Wide Range of Selectable Signalling Rates. In *Proc. 26th Asilomar Conf. Signals, Syst., Computers*, 2:676–680, Pacific Grove, CA, October 1992.

[46] W. M. Hartmann. Digital Waveform Generation by Fractional Addressing. *J. Acoust. Soc. Am.*, 82(6):1883–1891, December 1987.

[47] E. Hermanowicz. Explicit Formulas for Weighting Coefficients of Maximally Flat Tunable FIR Delayers. *Electron. Lett.*, 28(20):1936–1937, September 1992.

[48] E. Hermanowicz. Weighted Lagrangian Interpolating FIR Filter. In *Proc. European Signal Process. Conf.*, 2:1203–1206, Trieste, Italy, September 1996.

[49] E. Hermanowicz and M. Rojewski. Design of FIR First Order Digital Differentiators of Variable Fractional Sample Delay Using Maximally Flat Error Criterion. *Electron. Lett.*, 30(1):7–18, January 1994.

[50] E. Hermanowicz, M. Rojewski, G. D. Cain, and A. Tarczynski. Special Discrete-Time Filters Having Fractional Delay. *Signal Process.*, 67(3):279–289, June 1998.

[51] F. B. Hildebrand. *Introduction to Numerical Analysis. Second Edition*. McGraw-Hill, New York, 1974. Also: Dover, New York, 1989, p. 669.

[52] H. S. Hou and H. C. Andrews. Cubic Splines for Image Interpolation and Digital Filtering. *IEEE Trans. Acoust., Speech, Signal Process.*, 26(6):508–517, December 1978.

[53] L. B. Jackson. *Digital Filters and Signal Processing. Third Edition*. Kluwer Academic Publishers, Norwell, MA, 1996, p. 502.

[54] D. Jaffe and J. O. Smith. Extensions of the Karplus-Strong plucked string algorithm. *Computer Music J.*, 7(2):56–69, 1983.

[55] A. J. Jerri. The Shannon Sampling Theorem—Its Various Extensions and Applications: A Tutorial Review. *Proc. IEEE*, 65(11):1565–1596, November 1977.

[56] J.-M. Jot, V. Larcher, and O. Warusfel. Digital Signal Processing Issues in the Context of Binaural and Transaural Stereophony. Presented at *The Audio Eng. Soc. 98th Conv.*, Paris, France, February 1995, preprint no. 3980.

[57] J.-M. Jot, S. Wardle, and V. Larcher. Approaches to Binaural Synthesis. Presented at the *Audio Eng. Soc. 105th Conv.*, San Francisco, CA, September 1998, preprint no. 4861.

[58] L. J. Karam and J. H. McClellan. Complex Chebyshev Approximation for FIR Filter Design. *IEEE Trans. Circ. Syst.—Part II*, 42(3):207–216, March 1995.

[59] M. Karjalainen and U. K. Laine. A Model for Real-Time Sound Synthesis of Guitar on a Floating-Point Signal Processor. In *Proc. IEEE Int. Conf. Acoust., Speech, Signal Process.*, 5:3653–3656, Toronto, ON, Canada, May 1991.

[60] S. Kay. Some Results in Linear Interpolation Theory. *IEEE Trans. Acoust., Speech, Signal Process.*, 31(3):746–749, June 1983.

[61] D. Kim, M. J. Narasimha, and D. C. Cox. Design of Optimal Interpolation Filter for Symbol Timing Recovery. *IEEE Trans. Commun.*, 45(7):877–884, July 1997.

[62] P. Knutson, D. McNeely, and K. Horlander. An Optimal Approach to Digital Raster Mapper Design. *IEEE Trans. Consumer Electron.*, 37(4):746–752, September 1991.

[63] C. C. Ko and Y. C. Lim. Approximation of a Variable-Length Delay Line by Using Tapped Delay Line Processing. *Signal Process*, 14(4):363–369, June 1988.

[64] P. J. Kootsookos and R. C. Williamson. FIR Approximation of Fractional Sample Delay Systems. *IEEE Trans. Circ. Syst.—Part II*, 43(3):269–271, March 1996.

[65] P. Kroon and B. S. Atal. On the Use of Pitch Predictors With High temporal resolution. *IEEE Trans. Signal Process.*, 39(3):733–735, March 1991.

[66] T. I. Laakso, L. P. Sabel, A. Yardim, and G. D. Cain. Optimal Receive Filters for the Suppression of ISI and Adjacent Channel Interference. *Electron. Lett.*, 32(15):1346–1347, July 1996.

[67] T. I. Laakso, T. Saramäki, and G. D. Cain. Asymmetric Dolph-Chebyshev, Saramäki, and Transitional Windows for Fractional Delay FIR Filter Design. In *Proc. 38th Midwest Symp. Circ. Syst.*, 1:580–583, Rio de Janeiro, Brazil, August 1995.

[68] T. I. Laakso, A. Tarczynski, N. P. Murphy, and V. Välimäki. Polynomial Filtering Approach to Reconstruction and Noise Reduction of Nonuniformly Sampled Signals. *Signal Process.*, 80(4):567–575, April 2000.

[69] T. I. Laakso and V. Välimäki. Energy-Based Effective Length of the Impulse Response of a Recursive Filter. *IEEE Trans. Instr. Meas.*, 47(6):7–17, February 1999.

[70] T. I. Laakso, V. Välimäki, and J. Henriksson. Tunable Downsampling Using Fractional Delay Filters with Applications to Digital TV Transmission. In *Proc. IEEE Int. Conf. Acoust., Speech, Signal Process.*, 2:1304–1307, Detroit, MI, May 1995.

[71] T. I. Laakso, V. Välimäki, M. Karjalainen, and U. K. Laine. Splitting the Unit Delay-Tools for Fractional Delay Filter Design. *IEEE Signal Process. Mag.*, 13(1):30–60, January 1996.

[72] U. K. Laine, Digital Modelling of a Variable-Length Acoustic Tube. In *Proc. Nordic Acoustical Meeting*, Tampere, Finland, June 1988, pp. 165–168.

[73] U. Lambrette, K. Langhammer, and H. Meyr. Variable Sample Rate Digital Feedback NDA Timing Synchronization. In *Proc. IEEE Global Telecommun. Conf.*, 2:1348–1352, London, UK, November 1996.

[74] M. Lang. All-pass Filter Design and Applications, *IEEE Trans. Signal Process.*, 46(9):2505–2514, September 1998.

[75] M. Lang and T. I. Laakso. Simple and Robust Method for the Design of All-pass Filters Using Least-Squares Phase Error Criterion. *IEEE Trans. Circ. Syst.—Part II*, 41(1):40–48, January 1994.

[76] S.-H. Leung and C. W. Barnes. State-Space Realization of Fractional-Step Delay Digital Filters with Applications to Array Beamforming. *IEEE Trans. Acoust., Speech, Signal Process*, 32(2):317–380, April 1984.

[77] Y. C. Lim, J.-H. Lee, C. K. Chen, and R.-H. Yang. A Weighted Least Squares Algorithm for Quasi-Equiripple FIR and IIR Digital Filter Design. *IEEE Trans. Signal Process.*, 40(3):551–558, March 1992.

[78] T. J. Lim and M. D. Macleod. On-Line B-Spline and Lagrange Interpolation, Technical Report CUED/F-INFENG/TR 202, Univ. Cambridge, UK, December 1994.

[79] G.-S. Liu and C.-H. Wei. Programmable Fractional Sample Delay Filter with Lagrange Interpolation. *Electron. Lett.*, 26(19):1608–1610, September 1990.

[80] G.-S. Liu and C.-H. Wei. A New Variable Fractional Sample Delay Filter with Nonlinear Interpolation. *IEEE Trans. Circ. Syst.—Part II*, 39(2):123–126, February 1992.

[81] R. J. Marks II, *Introduction to Shannon Sampling and Interpolation Theory.* Springer Verlag, New York, NY, 1991.

[82] J. S. Marques, J. M. Tribolet, I. M. Trancoso, and L. B. Almeida. Pitch Prediction with Fractional Delays in CELP Coding. In *Proc. European Conf. Speech Commun. and Tech.*, 2:509–512, Paris, France, September 1989.

[83] J. S. Marques, I. M. Trancoso, J. M. Tribolet, and L. B. Almeida. Improved Pitch Prediction with Fractional Delays in CELP Coding. In *Proc. IEEE Int. Conf. Acoust., Speech, Signal Process.*, 2:665–668, Albuquerque, NM, April 1990.

[84] F. Marvasti. *A Unified Approach to Zero-Crossings and Nonuniform Sampling.* Nonuniform Publishing, Oak Park, IL, 1987.

[85] F. Marvasti, M. Analoui, and M. Gamshadzahi. Recovery of Signals from Nonuniform Samples Using Iterative Methods. *IEEE Trans. Signal Process.*, 39(4):872–878, April 1991.

[86] D. Massie. Wavetable Sampling Synthesis. In M. Kahrs and K. Brandenburg, Eds., *Applications of Digital Signal Processing to Audio and Acoustics.* Kluwer, Norwell, MA, 1998, pp. 311–341.

[87] N. E. Mastorakis. A New Approach for the Fractional Sample Delay Filters. In *Proc. Third IEEE Int. Conf. Electron., Circ. Syst.*, 1:502–505, October 1996.

[88] S. McClellan, J. D. Gibson, and B. K. Rutherford. Efficient Pitch Filter Encoding for Variable Rate Speech Processing. *IEEE Trans. Audio and Speech Process.*, 7(1):18–29, January 1999.

[89] Y. Medan, Using Super Resolution Pitch in Waveform Speech Coders. In *Proc. IEEE Int. Conf. Acoust., Speech, Signal Process.*, 1:633–636, Toronto, ON, Canada, May 1991.

[90] U. Mengali and A. N. D'Andrea. *Synchronization Techniques for Digital Receivers.* Plenum Press, New York, NY, 1997.

[91] R. Merched, P. S. R. Diniz, and M. Petraglia. A Delayless Alias-Free Subband Adaptive Filter Structure. In *Proc. IEEE Int. Symp. Circ. Syst.*, 4:2329–2332, Hong Kong, June 1997.

[92] R. Merched, P. S. R. Diniz, and M. Petraglia. A delayless alias-free subband adaptive filter structure. *IEEE Trans. Signal Process.*, 47(6):1580–1591, June 1999.

[93] H. Meyr and G. Ascheid. *Synchronization in Digital Communications, Vol. 1*. Wiley, New York, 1990.

[94] H. Meyr, M. Moeneclaey, and S. A. Fechtel. *Digital Communication Receivers—Synchronization, Channel Estimation, and Signal Processing*. Wiley, New York, 1998.

[95] S. Minocha, S. C. Dutta Roy, and B. Kumar. A Note on the FIR Approximation of a Fractional Sample Delay. *Int. J. Circ. Theor. Appl.*, 21(3):265–274, May-June 1993.

[96] F. Mintzer. On Half-Band, Third-Band, and Nth-Band FIR Filters and their Design. *IEEE Trans. Acoust., Speech, Signal Process.*, 30:734–738, October 1982.

[97] R. Msallam, S. Dequidt, S. Tassart, and R. Caussé. Physical Model of the Trombone Including Nonlinear Propagation Effects. In *Proc. Int. Symp. Musical Acoust.*, 2:419–424, Edinburgh, Scotland, August 1997.

[98] F. R. Moore. Table Lookup Noise for Sinusoidal Digital Oscillators. *Computer Music J.*, 1(2):26–29, 1977. Reprinted in C. Roads and J. Strawn. *Foundations of Computer Music*. MIT Press, Cambridge, MA, 1985, pp. 326–334.

[99] D. H. Mugler. Linear Prediction of a Bandlimited Signal from Past Samples at Arbitrary Points: An SVD-Based Approach. In *Proc. Workshop Sampling Theory and Appl.*, Riga, Latvia, pp. 113–118, September 1995.

[100] N. P. Murphy, A. Krukowski, and I. Kale. Implementation of Wideband Integer and Fractional Delay Element. *Electron. Lett.*, 30(20):1658–1659, September 1994.

[101] N. P. Murphy, A. Krukowski, and A. Tarczynski. An Efficient Fractional Sample Delayer for Digital Beam Steering. In *Proc. IEEE Int. Conf. Acoust., Speech, Signal Process.*, 3:2245–2248, Munich, Germany, April 1997.

[102] N. P. Murphy, T. I. Laakso, and G. Allen. Fast Resampling of Nonuniformly Sampled Signals Using Interpolative FIR Filters. In *Proc. Workshop Sampling Theory and Appl.*, Riga, Latvia, September 1995, pp. 241–246.

[103] N. P. Murphy, A. Tarczynski, and T. I. Laakso. Sampling-Rate Conversion Using a Wideband Tuneable Fractional Delay Element. In *Proc. Nordic Signal Process. Symp.*, Espoo, Finland, September 1996, pp. 423–426.

[104] S. N. Nazra. Linear phase IIR filter with equiripple stopband. *IEEE Trans. Acoust. Speech, Signal Process.*, 31(3):744–746, June 1983.

[105] T. Q. Nguyen, T. I. Laakso, and R. D. Koilpillai. Eigenfilter approach for the design of all-pass filter approximating a given phase response. *IEEE Trans. Signal Process.*, 42(9):2257–2263, September 1994.

[106] G. Oetken, T. W. Parks, and H. W. Schüßler. New results in the design of digital interpolators. *IEEE Trans. Acoust. Speech, Signal Process.*, 23(3):301–309, June 1975.

[107] G. Oetken. A new approach for the design of digital interpolating filters. *IEEE Trans. Acoust. Speech, Signal Process.*, 27(6):637–643, December 1979.

[108] A. V. Oppenheim and R. W. Schafer. *Discrete-Time Signal Processing*. Prentice-Hall, Englewood Cliffs, NJ, 1989.

[109] D. O'Shea and S. McGrath. A novel interpolation method for maximum likelihood timing recovery. *Wireless Personal Commun.*, 4:315–324, 1997.

[110] A. Papoulis. *Signal Analysis*. McGraw-Hill, New York, 1977.

[111] S. Park, G. Hillman, and R. Robles. A Novel Structure for Real-Time Digital Sample-Rate Converters With Finite Precision Error Analysis. In *Proc. IEEE Int. Conf. Acoust., Speech, Signal Process.*, 5:3613–3616, Toronto, ON, Canada, May 1991.

[112] T. W. Parks and C. S. Burrus. *Digital Filter Design*. Wiley, New York, 1987.

[113] S.-C. Pei and C.-C. Tseng. A Comb Filter Design Using Fractional-Sample Delay. *IEEE Trans. Circ. Syst.—Part II*, 45(6):649–653, June 1998.

[114] L. D. Philipp, A. Mahmood, and B. L. Philipp. An Improved Refinable Rational Approximation to the Ideal Time Delay. *IEEE Trans. Circ. Syst.—Part II*, 46(5):637–640, May 1999.

[115] J. R. Pierce and S. A. Van Duyne. A Passive Nonlinear Digital Filter Design which Facilitates Physics-Based Sound Synthesis of Highly Nonlinear Musical instruments. *J. Acoust. Soc. Am.*, 101(2):1120–1126, February 1997.

[116] K. C. Pohlman. *Principles of Digital Audio. Third Edition*. McGraw-Hill, New York, 1995.

[117] K. Preuss. On the Design of FIR Filters by Complex Approximation. *IEEE Trans. Acoust., Speech, Signal Process.*, 37(5):702–712, May 1989.

[118] R. G. Pridham and R. A. Mucci. A Novel Approach To Digital Beamforming. *J. Acoust. Soc. Am.*, 63(2):425–434, February 1978.

[119] R. G. Pridham and R. A. Mucci. Digital Interpolation Beamforming for Lowpass and Bandpass Signals. *Proc. IEEE*, 67:904–919, June 1979.

[120] W. Putnam and J. O. Smith. Design of Fractional Delay Filters Using Convex Optimization. In *Proc. IEEE Workshop Appl. Signal Process. Audio and Acoustics*, New Paltz, NY, October 1997.

[121] M. F. Pyfer and R. Ansari. The Design and Application of Optimal FIR Fractional-Slope Phase Filters. *Proc. IEEE Int. Conf. Acoust., Speech, Signal Process.*, 2:896–899, Dallas, TX, April 1987.

[122] T. Ramstad. Digital methods for conversion between arbitrary sampling frequencies. *IEEE Trans. Acoust., Speech, Signal Process.*, 32(3):577–591, June 1984.

[123] T. Ramstad. Fractional rate decimator and interpolator design. *Proc. European Signal Process. Conf.*, 4:1948–1952, Rhodes, Greece, September 1998.

[124] P. A. Regalia. Special filter designs. In S. K. Mitra and J. F. Kaiser, Eds., *Handbook of Digital Signal Processing*. Wiley, New York, pp. 907–980, 1993.

[125] S. Ries and M. Roeckerath-Ries. Combined Matched Filter/Interpolator for Digital Receivers. *Proc. European Signal Process. Conf.*, 1:627–630, Trieste, Italy, September 1996.

[126] D. Rocchesso. Fractionally-Addressed Delay Lines, *Proc. COST-G6 Digital Audio Effects Workshop*. Barcelona, Spain, pp. 40–43, November 1998.

[127] L. P. Sabel. On the BER Performance of Digital Demodulators Using Fixed Point FIR Interpolators. In *Proc. 3rd Annual Int. Conf. Universal Personal Commun.*, September/October 1994, pp. 215–219.

[128] L. P. Sabel, T. I. Laakso, A. Yardim, and G. D. Cain. Improved Delay Root-Nyquist Filters for Symbol Synchronisation in PCS Receivers. *Proc. IEEE Global Telecommun. Conf.*, 2:1302–1306, Singapore, November 1995.

[129] L. P. Sabel, A. Yardim, G. D. Cain, and T. I. Laakso. Effects of Delay-Root-Nyquist Filters on the Performance of Digital Demodulators. In *Proc. Int. Conf. Telecommun.*, Bali, Indonesia, April 1995, pp. 75–79.

[130] S. Samadi and H. Iwakura. Variable Taylor Realization and Improved Approximation of FIR Fractional Delay Systems. *Int. J. Circ. Theor. Appl.*, 26(5):513–522, September–October 1998.

[131] T. Saramäki, Finite Impulse Response Filter Design. In S. K. Mitra and J. F. Kaiser, Eds., *Handbook of Digital Signal Processing*. Wiley, New York, pp. 155–277, 1993.

[132] T. Saramäki and M. Renfors. A Remez-Type Algorithm for Designing Digital Filters Composed of All-Pass Sections Based on Phase Approximations. *Proc. 38th Midwest Symp. Circ. Syst.*, 1:571–575, Rio de Janeiro, Brazil, August 1995.

[133] T. Saramäki and M. Renfors. Nth-Band Filter Design. *Proc. European Signal Process. Conf.*, 4:1943–1947, Rhodes, Greece, September 1998.

[134] T. Saramäki and T. Ritoniemi. An Efficient Approach for Conversion Between Arbitrary Sampling Frequencies. *Proc. Int. Symp. Circ. Syst.*, 2:285–288, Atlanta, GA, May 1996.

[135] L. Savioja. Improving the Three-Dimensional Digital Waveguide Mesh by Interpolation. *Proc. Nordic Acoust. Meeting*, Stockholm, Sweden, September 1998, pp. 265–268.

[136] L. Savioja, J. Huopaniemi, T. Lokki, and R. Väänänen. Creating Interactive Virtual Acoustic Environments. *J. Audio Eng. Soc.*, 47(9):675–705, September 1999.

[137] L. Savioja and V. Välimäki, The Bilinearly Deinterpolated Waveguide Mesh. In *Proc. Nordic Signal Process. Symp.*, Espoo, Finland, September 1996, pp. 443–446.

[138] L. Savioja and V. Välimäki. Improved Discrete-Time Modeling of Multi-Dimensional Wave Propagation Using the Interpolated Digital Waveguide Mesh. *Proc. IEEE Int. Conf. Acoust., Speech, Signal Process.*, 1:459–462, Munich, Germany, April 1997.

[139] L. Savioja and V. Välimäki. Reduction of the Dispersion Error in the Interpolated Digital Waveguide Mesh Using Frequency Warping. *Proc. IEEE Int. Conf. Acoust., Speech, Signal Process.*, 2:973–976, Phoenix, AZ, March 1999.

[140] L. Savioja and V. Välimäki. Reducing the Dispersion Error in the Digital Waveguide Mesh Using Interpolation and Frequency-Warping Techniques. *IEEE Trans. Speech and Audio Process.*, 8(2):184–194, March 2000.

[141] G. Scavone and P. R. Cook. Real-Time Computer Modeling of Woodwind Instruments. In *Proc. Int. Symp. Musical Acoust.*, Leavenworth, WA, June/July 1998.

[142] R. W. Schafer and L. R. Rabiner. A Digital Signal Processing Approach to Interpolation. In *Proc. IEEE*, 61(6):692–702, June 1973.

[143] D. W. E. Schobben and P. C. W. Sommen. Increasing Beamsteering Directions Using Polyphase Decomposition. In *Proc. 8th IEEE Signal Process. Workshop on Statistical Signal and Array Process.*, June 1996, pp. 117–120.

[144] M. Schulist. Improvements of a Complex FIR Filter Design Algorithm. *Signal Process.*, 20(1):81–90, May 1990.

[145] I. W. Selesnick. Lowpass Filters Realizable as All-Pass Sums: Design Via a New Flat Delay Filter. *IEEE Trans. Circ. Syst.—Part II*, 46(1):40–50, January 1999.

[146] S. Signell. Design of Maximally Flat Group Delay Discrete-Time Recursive Filters. *Proc. IEEE Int. Symp. Circ. Syst.*, 1:192–196, Montreal, QC, Canada, May 1984.

[147] S. Sivanand, J.-F. Yang, and M. Kaveh. Focusing filters for wide-band direction finding. *IEEE Trans. Signal Process.*, 39(2):437–445, February 1991.

[148] S. Sivanand. Variable-Delay Implementation Using Digital FIR Filters. In *Proc. Int. Conf. Signal Process. Appl. Theor.*, Boston, MA, November 1992, pp. 1238–1243.

[149] J. O. Smith. Techniques for digital filter design and system identification with application to the violin. Doctoral Dissertation. Report STAN-M-14, CCRMA, Dept. of Music, Stanford University, Stanford, CA, June 1983.

[150] J. O. Smith. Physical Modeling Using Digital Waveguides. *Computer Music J.*, 16(4):74–91, 1992. Available online at http://www-ccrma.stanford.edu/~jos/.

[151] J. O. Smith. Physical Modeling Synthesis Update. *Computer Music J.*, 20(2):44–56, 1996. Available online at http://www-ccrma.stanford.edu/~jos/.

[152] J. O. Smith. Acoustic Modeling Using Digital Waveguides. In C. Roads, S. T. Pope, A. Piccialli, and G. De Poli, Eds., *Musical Signal Processing*. Swets & Zeitlinger, Lisse, The Netherlands, 1997, pp. 221–263. Available online at http://www-ccrma.stanford.edu/~jos/.

[153] J. O. Smith and B. Friedlander. Adaptive Interpolated Time-Delay estimation. *IEEE Trans. Aerospace and Electronic Syst.*, 21(3):180–199, March 1985.

[154] J. O. Smith and P. Gossett. A Flexible Sampling-Rate Conversion Method. *Proc. IEEE Int. Conf. Acoust., Speech, Signal Process.*, 2:19.4.1–4, San Diego, CA, March 1984.

[155] H. Strauss. Implementing Doppler Shifts for Virtual Auditory Environments. Presented at the *Audio Eng. Soc. 104th Conv.*, Amsterdam, the Netherlands, May 1998, preprint no. 4687.

[156] H. W. Strube. Sampled-data Representation of a Nonuniform Lossless Tube of Continuously Variable Length. *J. Acoust. Soc. Am.*, 57(1):256–257, January 1975.

[157] H. W. Strube. How to Make an All-Pass Filter with a Desired Impulse Response. *IEEE Trans. Acoust., Speech, Signal Process.*, 30(2):336–337, April 1982.

[158] R. Sudhakar, R. C. Agarwal, and S. C. Dutta Roy. Time Domain Interpolation Using Differentiators. *IEEE Trans. Acoust., Speech, Signal Process*, 30(6):992–997, December 1982.

[159] C. S. Sullivan. Extending the Karplus-Strong Algorithm to Synthesize Electric Guitar Timbres with Distortion and Feedback. *Computer Music J.*, 14(3):26–37, 1990.

[160] A. Tarczynski and G. D. Cain. Stabilisation of a Class of Feedback Systems Using Fractional-Sample Delayors. *Applied Signal Processing*, 1(1):55–58, 1994.

[161] A. Tarczynski and G. D. Cain. Design of IIR Fractional-Sample Delay Filters. In *Proc. 2nd Int. Symp. DSP for Commun. Syst.*, Adelaide, Australia, April 1994.

[162] A. Tarczynski, G. D. Cain, E. Hermanowicz, and M. Rojewski. WLS Design of Variable Frequency Response FIR Filters. *Proc. Int. Symp. Circ. Syst.*, 4:2244–2247, Hong Kong, June 1997.

[163] A. Tarczynski, W. Kozinski, and G. D. Cain. Sampling Rate Conversion Using Fractional-Sample Delay. *Proc. IEEE Int. Conf. Acoust., Speech, Signal Process.*, 3:285–288, Adelaide, Australia, April 1994.

[164] S. Tassart and Ph. Depalle. Analytical Approximation of Fractional Delays: Lagrange Interpolators and All-pass Filters. *Proc. IEEE Int. Conf. Acoust., Speech, Signal Process.*, 1:455–458, Munich, Germany, April 1997.

[165] S. Tassart, R. Msallam, Ph. Depalle, and S. Dequidt. A Fractional Delay Application: Time-Varying Propagation Speed in Waveguides. *Proc. Int. Computer Music Conf.*, Thessaloniki, Greece, 256–259, September 1997.

[166] J.-P. Thiran. Recursive Digital Filters with Maximally Flat Group Delay. *IEEE Trans. Circ. Theory*, 18(6):659–664, November 1971.

[167] T. Tolonen, V. Välimäki, and M. Karjalainen. Modeling of Tension Modulation Nonlinearity in Plucked Strings. *IEEE Trans. Speech and Audio Process.* 8(3):300–310, May 2000.

[168] V. Tuukkanen, J. Vesma, and M. Renfors. Efficient Near Optimal Maximum Likelihood Symbol Timing Recovery in Digital modems. In *Proc. Int. Symp. on Personal, Indoor and Mobile Radio Commun.*, Helsinki, Finland, September 1997, pp. 825–829.

[169] M. Unser, A. Aldroubi, and M. Eden. Polynomial Spline Signal Approximations: Filter Design and Asymptotic Equivalence with Shannon's sampling theorem. *IEEE Trans. Information Theory*, 38(1):95–103, January 1992.

[170] M. Unser, A. Aldroubi, and M. Eden. B-Spline Signal Processing: Part I—Theory. *IEEE Trans. Signal Process.*, 41(2):821–833, February 1993.

[171] M. Unser, A. Aldroubi, and M. Eden. B-spline signal processing: Part II-Efficient Design and Applications. *IEEE Trans. Signal Process.*, 41(2):834–848, February 1993.

[172] P. P. Vaidyanathan. *Multirate Systems and Filter Banks*. Prentice-Hall, Englewood Cliffs, NJ, 1993.

[173] V. Välimäki. A New Filter Implementation Strategy for Lagrange Interpolation. *Proc. IEEE Int. Symp. Circ. Syst.*, 1:361–364, Seattle, WA, April-May 1995.

[174] V. Välimäki. Discrete-time modeling of acoustic tubes using fractional delay filters. Doctoral dissertation. Report no. 37, Helsinki University of Technology, Laboratory of Acoustics and Audio Signal Processing, Espoo, Finland, December 1995. Available at http://www.acoustics.hut.-fi/~vpv/publications/vesa_phd.html.

[175] V. Välimäki, J. Huopaniemi, M. Karjalainen, and Z. Jánosy. Physical Modeling of Plucked String Instruments with Application to Real-Time Sound Synthesis. *J. Audio Eng. Soc.*, 44(5):331–353, May 1996.

[176] V. Välimäki and M. Karjalainen. Implementation of Fractional Delay Waveguide Models Using All-pass Filters. *Proc. IEEE Int. Conf. Acoust., Speech, Signal Process.*, 2:1524–1527, Detroit, MI, May 1995.

[177] V. Välimäki, M. Karjalainen, and T. Kuisma. Articulatory Speech Synthesis Based on Fractional Delay Waveguide Filters. *Proc. IEEE Int. Conf. Acoust., Speech, Signal Process.*, 1:585–588, Adelaide, Australia, April 1994.

[178] V. Välimäki, M. Karjalainen, and T. I. Laakso. Fractional Delay Digital Filters. *Proc. IEEE Int. Symp. Circ. Syst.*, 1:355–358, Chicago, IL, May 1993.

[179] V. Välimäki, M. Karjalainen, and T. I. Laakso. Modeling of woodwind bores with finger holes, *Proc. Int. Computer Music Conf.*, Tokyo, Japan, September 1993, pp. 32–39.

[180] V. Välimäki and T. I. Laakso. Suppression of Transients in Variable Recursive Digital Filters with a Novel and Efficient Cancellation Method. *IEEE Trans. Signal Process*, 46(12):3408–3414, December 1998.

[181] V. Välimäki, T. I. Laakso, and J. Mackenzie. Elimination of Transients in Time-Varying All-pass Fractional Delay Filters with Application to Digital Waveguide Modeling. In *Proc. Int. Computer Music Conf.*, Banff, Canada, September 1995, pp. 327–334.

[182] V. Välimäki, T. Tolonen, and M. Karjalainen. Signal-Dependent Nonlinearities for Physical Models Using Time-Varying Fractional Delay Filters. In *Proc. Int. Computer Music Conf.*, Ann Arbor, MI, October 1998, pp. 264–267.

[183] V. Välimäki, T. Tolonen, and M. Karjalainen. Plucked-String Synthesis Algorithms with Tension Modulation Nonlinearity. *Proc. IEEE Int. Conf. Acoust., Speech, Signal Process*, 2:977–980, Phoenix, AZ, March 1999.

[184] S. A. Van Duyne, D. A. Jaffe, G. P. Scandalis, and T. S. Stilson. A Lossless, Click-Free, Pitchbend-able Delay Line Loop Interpolation Scheme. In *Proc. Int. Computer Music Conf.*, Thessaloniki, Greece, September 1997, pp. 252–255.

[185] D. Verdin and T. C. Tozer. Interpolator Filter Structure for Asynchronous Timing Recovery Loops. *Electron. Lett.*, 29(5):490–492, March 1993.

[186] J. Vesma. Optimization and Applications of Polynomial-Based Interpolation Filters. Doctoral dissertation. Tampere University of Technology, Tampere, Finland, May 1999.

[187] J. Vesma, M. Renfors, and J. Rinne. Comparison of Efficient Interpolation Techniques for Symbol Timing recovery. *Proc. IEEE Global Telecommun. Conf.*, 2:953–957, London, UK November 1996.

[188] J. Vesma and T. Saramäki. Interpolation Filters with Arbitrary Frequency Response for All-Digital Receivers. *Proc. Int. Symp. Circ. Syst.*, 2:568–571, Atlanta, GA, May 1996.

[189] J. Vesma and T. Saramäki. Optimization and Efficient Implementation of Fractional Delay FIR Filters. In *Proc. Int. Conf. Electron., Circ. Syst.*, Rhodes, Greece, October 1996, pp. 546–549.

[190] J. Vesma and T. Saramäki. Optimization and Efficient Implementation of FIR Filters with Adjustable Fractional Delay. *Proc. Int. Symp. Circ. Syst.*, 4:2256–2259, Hong Kong, June 1997.

[191] J. Vesma, T. Saramäki, and M. Renfors. Combined Matched Filter and Polynomial-Based Interpolator for Symbol Synchronization in Digital Receivers. In *Proc. IEEE Int. Symp. Circ. Syst.*, Orlando, FL, May 1999, pp. 94–97.

[192] J. Vesma, V. Tuukkanen, and M. Renfors. Maximum Likelihood Feedforward Symbol Timing Recovery Based on Efficient Interpolation Techniques. In *Proc. Nordic Signal Process. Symp.*, Espoo, Finland, September 1996, pp. 183–186.

[193] D. Wise, J. Barish, and E. Lindemann. DSP Strategies for Faster Polynomial Interpolation. Presented at the *Audio Eng. Soc. 101st Conv.*, Los Angeles, CA, November 1996, preprint no. 4370.

[194] H.-K. Yang and M. Snelgrove. Symbol Timing Recovery Using Oversampling Techniques. *Proc. Int. Conf. Commun.*, 3:1296–1300, Dallas, TX, June 1996.

[195] A. Yardim, G. D. Cain, and P. Henry. Optimal Two-Term Offset Windowing for Fractional Delay. *Electron. Lett.*, 32(6):526–527, March 1996.

[196] A. Yardim, G. D. Cain, and A. Lavergne, Performance of Fractional-Delay Filters Using Optimal Offset Windows. *Proc. IEEE Int. Conf. Acoust., Speech, Signal Process.*, 3:2233–2236, Munich, Germany April 1997.

[197] A. Yardim, L. J. Karam, J. H. McClellan, and G. D. Cain. Performance of Complex Chebyshev Approximation in Delay-Root-Nyquist Filter Design. *Proc. IEEE Int. Symp. Circ. Syst*, 2:169–172, Atlanta, GA, May 1996.

[198] A. Yardim, T. I. Laakso, L. P. Sabel, and G. D. Cain. Design of Efficient Receive FIR Filters for Joint Minimization of Channel Noise, ISI and Adjacent Channel Interference. *Proc. IEEE Global Telecommun. Conf.*, 2:948–952, London, UK, November 1996.

[199] A. Yardim, T. I. Laakso, L. P. Sabel, and G. D. Cain. Frequency-Sampling Design of Delay-root-Nyquist filters. *Electron. Lett.*, 33(1):35–37, January 1997.

[200] L. H. Zetterberg and Q. Zhang. Elimination of Transients in Adaptive Filters with Application to Speech Coding. *Signal Process.*, 15(4):419–428, 1988.

[201] H. Zhang. Interpolator for All-Digital Receivers. *Electron. Lett.*, 33(4):261–262, February 1996.

[202] H. Zhang and M. Renfors. A Study of Performance of All-Digital Symbol Synchronizers with Polynomial Interpolators. *Proc. Int. Conf. Commun. Tech.*, Beijing, China, 2:1013–1016, May 1996.

[203] V. Živojnović and H. Meyr. Design of Optimum Interpolation Filters for Digital Demodulators. *Proc. IEEE Int. Symp. Circ. Syst.*, 1:140–143, Chicago, IL, May 1993.

[204] U. Zölzer and T. Boltze Interpolation Algorithms: Theory and Application, Presented at the *Audio Eng. Soc. 97th Conv.*, San Francisco, CA, November 1994, preprint no. 3898.

[205] U. Zölzer. *Digital Audio Signal Processing*. Wiley, Chichester, UK, 1997.

Author Index

Subject Index

Table of Contents for the Attached CD-ROM

*Note only chapters which have code or additional information have been compiled in this CD-ROM.

Chapter 15

Chapter 16

Chapter 17

FigPadeTabalg.eps
FigPadeTabalg.vsd
FigPadeTabalg.wmf
FigPadeTable.eps
FigPadeTable.vsd
FigPadeTable.wmf
gerafigcompsind.m
gerafigforney.m
GeraFigGrochenig.m
gerafigodftall.m
gerafigodftf.m
gerafigodftfbi.m
gerafigodftfbi_limiar.m
GeraFigOdftPaulo.m
gerafigodftt.m
gerafigodfttbi.m
gerafigodfttbi_limiar.m
gerafigOdftTc.m
Figuras Cap3.zip
circulo.vsd
exp552.m
exp552a.m
fig552.eps
fig552a.eps
fig552b.eps
figconds.eps
FigCondS.m
figconds2.eps
FigCondS2.m
figcondsm.eps
FigCondSm.m
figcondt.eps
FigCondT.m
Figuras Cap4.zip
DistHamming.vsd
DistHamming.wmf
FigBancoCorrector.vsd
FigBancoCorrector.wmf
FigBancoDoisCanais.vsd
FigBancoDoisCanais.wmf
Figconvul15.vsd
Figconvul15.wmf
FigConvulDef.vsd
FigConvulDef.wmf
FigDefFilterBank.vsd
FigDefFilterBank.wmf
FigInterDecim.vsd
FigInterDecim.wmf
reedmuller16_11vieira.m
reedmuller16_11vieira2.m
Theses-main-Vieira.zip
tesemain.pdf
[Walsh]
MSc-Ng.zip
dissert.ps
WTC.CPP
WTC.EXE
WTCD.C
WTCD.OUT

Chapter 18

2layer-paper-.zip
2layer.ps
PhD.zip
all.ps
[Iterative]
iterative.c
iterative.h
literative.c
[MEC]
Burst.c
Burst.h
lBurst.c
[NonLinear]
LNONLINR.C
NONLINR.C
NONLINR.H
[REC]
Burst.c
Burst.h
lBurst.c

Chapter 19

HOLM-PUB.ZIP
- Holm-chap19.html
- index.html
- [papers]
  - 00_ICASSP.pdf
  - 00_ICASSP_B.pdf
  - 94_ultra.pdf
  - 95_ICA.pdf
  - 95_NORSIG.ps
  - 95_ultra.ps
  - 96_NORSIG.pdf
  - 97_NORSIG.pdf
  - 97_SampTA.pdf
  - 97_UFFC.pdf
  - 97_ultra_B.pdf
  - 98_NORSIG.pdf
  - 99_InstSciTech.pdf
  - 99_NORSIG_B.pdf
  - 99_Rotterdam.html
  - 99_SampTA.pdf

Chapter 20

[MATLAB-files]
- APFLAT2.M
- apnostup3.m
- APNSTUP3.M
- ARRCOMP1.M
- BPIARR.M
- DISCONT.M
- DOPPLERD.M
- EXPOLF1.M
- FDCOMB.M
- FINTARR.M
- FINTARR2.M
- HEQRIP2.M
- HLAGR2.M
- IDFDCHAR.M
- INIHEQ2.M
- LAGRFIGS.M
- LSFDFIGS.M
- NUPOLFIG.M
- Oetkenfigs.m
- POLF1.M
- POLJIT.M
- POLYFIGS.M
- **README**.TXT
- SHFTSINC.M
- STRUBEFD.M
- THIRFIGS.M